La Selva

LA SELVA

*Ecology and
Natural History of a
Neotropical Rain Forest*

Edited by
Lucinda A. McDade
Kamaljit S. Bawa
Henry A. Hespenheide
and
Gary S. Hartshorn

The University of Chicago Press
Chicago and London

The University of Chicago Press, Chicago 60637
The University of Chicago Press, Ltd., London
© 1994 by The University of Chicago
All rights reserved. Published 1994
Printed in the United States of America
03 02 01 00 99   2 3 4 5
ISBN: 0-226-03950-1 (cloth)
ISBN: 0-226-03952-8 (paper)

Library of Congress Cataloging-in-Publication Data

La Selva : ecology and natural history of a neotropical rainforest / edited by Lucinda A. McDade . . . [et al.].
        p.   cm.
    Includes bibliographical references (p.   ) and index.
    1. Natural history—Costa Rica—La Selva Biological Station Region.   2. Rain forest ecology—Costa Rica—La Selva
Biological Station Region.   3. Human ecology—Costa Rica—La Selva Biological Station Region.   4. La Selva Biological Station (Costa Rica)
    I. McDade, Lucinda A.
    QH108.C6S44   1994
    574.5′2642′0972864—dc20   93-1776
                                                                                                CIP

CONTENTS

Foreword / vii
Rodrigo Gámez

Preface / ix
Lucinda A. McDade

1 Introduction / 3
Kamaljit S. Bawa, Lucinda A. McDade, and Henry Hespenheide

2 La Selva Biological Station / 6
Lucinda A. McDade and Gary S. Hartshorn

PART I   ABIOTIC ENVIRONMENT AND ECOSYSTEM PROCESSES / 15
Commentary by Lucinda A. McDade

3 Climate, Geomorphology, and Aquatic Systems / 19
Robert L. Sanford, Jr., Pia Paaby, Jeffrey C. Luvall, and Eugenie Phillips

4 Soils and Soil Process Research / 34
Phillip Sollins, Freddy Sancho M., Rafael Mata Ch., and Robert L. Sanford, Jr.

5 Soil Fertility, Nutrient Acquisition, and Nutrient Cycling / 54
Geoffrey G. Parker

PART II   THE PLANT COMMUNITY: COMPOSITION, DYNAMICS, AND LIFE-HISTORY PROCESSES / 65
Commentary by Kamaljit S. Bawa and Lucinda A. McDade

6 Vegetation Types and Floristic Patterns / 73
Gary S. Hartshorn and Barry E. Hammel

7 Plant Demography / 90
Deborah A. Clark

8 Patterns of Density and Dispersion of Forest Trees / 106
Milton Lieberman and Diana Lieberman

9 Tree-fall Gap Environments and Forest Dynamic Processes / 120
Julie Sloan Denslow and Gary S. Hartshorn

10 Physiological Ecology of Plants / 128
Ned Fetcher, Steven F. Oberbauer, and Robin L. Chazdon

11 Diversity of Long-term Flowering Patterns / 142
L. E. Newstrom, G. W. Frankie, H. G. Baker, and R. K. Colwell

12 Flowering Plant Reproductive Systems / 161
W. John Kress and James H. Beach

PART III   THE ANIMAL COMMUNITY / 183
Commentary by Henry A. Hespenheide

13 Patterns of Butterfly Diversity and Promising Topics in Natural History and Ecology / 187
Philip J. DeVries

14 Ecological Aspects of the Fish Community / 195
William A. Bussing

15 Amphibian Diversity and Natural History / 199
Maureen A. Donnelly

16 The Reptile Fauna: Diversity and Ecology / 210
Craig Guyer

17 Birds: Ecology, Behavior, and Taxonomic Affinities / 217
Douglas J. Levey and F. Gary Stiles

18 The Mammal Fauna / 229
Robert M. Timm

19 An Overview of Faunal Studies / 238
Henry A. Hespenheide

20 Population Biology: Life Histories, Abundance, Demography, and Predator-Prey Interactions / 244
H. Elizabeth Braker and Harry W. Greene

PART IV   PLANT-ANIMAL INTERACTIONS / 257
Commentary by Henry A. Hespenheide

21 Plant-Herbivore Interactions: Diversity, Specificity, and Impact / 261
Robert J. Marquis and H. Elizabeth Braker

22 Frugivory: An Overview / 282
    *Douglas J. Levey, Timothy C. Moermond, and Julie
    Sloan Denslow*

PART V   LA SELVA'S HUMAN ENVIRONMENT / 295
    Commentary by Lucinda A. McDade

23 The Regional Context: Land Colonization and
    Conservation in Sarapiquí / 299
    Rebecca P. Butterfield

24 Agricultural Systems in the La Selva Region / 307
    Florencia Montagnini

25 Forestry in Costa Rica: Status, Research Priorities, and
    the Role of La Selva Biological Station / 317
    Rebecca P. Butterfield

26 Prospects for a Comparative Tropical Ecology / 329
    Gordon H. Orians

APPENDIXES

1   Patterns of Research Productivity, 1951–1991 / 341
    Lucinda A. McDade and Kamaljit S. Bawa

2   Administration and Governance / 345
    Lucinda A. McDade

3   Vascular Plants: An Interim Checklist / 350
    Robert L. Wilbur and Collaborators

4   Fishes / 379
    William A. Bussing

5   Amphibians / 380
    Maureen A. Donnelly

6   Reptiles / 382
    Craig Guyer

7   Birds of La Selva and Vicinity / 384
    F. Gary Stiles and Douglas J. Levey

8   Mammals / 394
    Robert M. Timm

Acronyms / 399

Bibliography / 401

Contributors / 459

Part-title photograph legends / 463

Index / 465

FOREWORD

When I was in high school in the early 1950s, Puerto Viejo de Sarapiquí, the nearest town to the La Selva Biological Station, was eight hours away from my hometown of Heredia although the places are not more than 90 km apart in a straight line. The journey to Sarapiquí required several hours by Jeep on a tortuous gravel and dirt road, followed by four hours or so on horseback along muddy trails. In those years Sarapiquí was the frontier, a remote area on the plains of northern Costa Rica still covered with the luxuriant vegetation typical of tropical rain forests. It was considered the land of the future, for the luxuriance of these forests throughout this extensive area was thought to be a clear indication of the richness of its soils and, consequently, the obvious place for the future agricultural development of the province of Heredia.

In the 1990s Puerto Viejo is merely two hours away on a paved road, and my generation has witnessed how the dense forests that once covered the Sarapiquí plains have disappeared in three decades. Gone also are the dreams of the richness of its soils and the agricultural potential of these plains. Largely covered with partially degraded pasture lands, disfunctional agroecosystems, and vestiges of its once-pristine beautiful forests, Sarapiquí is, for the most part, still looking for reasonable agricultural and productive vocation. It is a good example of what Peter Raven describes as our ignorance of how to convert a tropical land into sustainable agricultural and forestry systems that would have the capability of supporting human lives at a level of dignity that we might consider reasonable.

The volume presented here tells another story that occurred in Sarapiquí at La Selva, just a few kilometers from Puerto Viejo. It is about the search for knowledge to recognize and understand the extraordinary biological diversity of tropical forests and the complex interactions that take place in these habitats, interactions that possibly have no parallel anywhere else on the planet.

The results of many years of pioneering research on the ecology and natural history of the Sarapiquí rain forests of La Selva are authoritatively documented in this volume, which represents a major contribution to the knowledge and understanding of tropical biology. It also exemplifies a dramatic contradiction of modern society with tragic consequences for tropical rain forests. On the one hand, biologists have been spending years of intellectual and physical effort and substan-

Note: This foreword was written in 1986 while the author was director of the Cellular and Molecular Biology Research Center at the University of Costa Rica.

tial material resources in trying to understand the complexity of these forests and searching for ways that would permit their sustainable use for society's benefit. On the other hand, domestic and foreign socioeconomic and political pressures have promoted the indiscriminate and rapid destruction of these forests with few parallel scientific and technical efforts to find sustainable uses of these natural resources.

The irreversible consequences for biology and humanity that destruction of tropical forests and their rich biodiversity will bring are now becoming clear. One of the obvious challenges and responsibilities faced by our generation of biologists is the need to embed tropical nature in the minds of people from both tropical and temperate societies. As a Costa Rican, I have witnessed an increased awareness of OTS scientists and students that their intellectual role is not limited to the boundaries of La Selva. They share the more universal responsibility to make the world understand that, in the words of Daniel Janzen, "tropical nature is an integral part of human life." The scientific contributions from La Selva have broadened and strengthened Costa Rica's knowledge of its natural history and are now an integral part of our national efforts to preserve the outstanding biological diversity of this country.

The history of La Zona Protectora is a good example of the comprehension of this profound responsibility. As the forests around La Selva disappeared, biologists realized that the reserve soon would become a small island in a sea of pasture. This decline would inevitably be followed by extinction of many species in the reserve. Especially vulnerable were the species that moved seasonally along the altitudinal transect that connects La Selva with Braulio Carrillo National Park. Clearly, the ecological boundaries of both La Selva and the park extended well beyond their respective legal boundaries. Connecting them with unbroken forest was necessary for the preservation of both. The successful campaign for the acquisition of La Zona Protectora, in which Costa Rican and U.S. conservation and philanthropic organizations, including OTS, participated, led to its incorporation into Braulio Carrillo. The park and La Selva were linked by a narrow protected corridor for numerous species of birds, mammals, and insects and their periodic migrations between the highlands and the lowlands. Such action has received international recognition as one of the major recent accomplishments in biological conservation.

We cannot put fences around the parks and reserves and forget about what occurs outside them. Communities around these areas must both understand fully the reasons for their existence and become direct beneficiaries of their presence as well. The information presented in this book needs to be conveyed to communities in Sarapiquí, both for their comprehension of the reason to preserve and study natural diversity and for their intellectual stimulation and material gain.

If La Selva and Braulio Carrillo are to exist for the benefit of future generations, we must find ways to develop sustainable agricultural and forestry systems that can satisfy the basic economic needs of people living in Sarapiquí and surrounding areas. This point is addressed in the final chapters of this book, but the topic demands more extensive and independent treatment in the near future. Our ability to use natural resources in a wise, rational, and sustainable manner is one of the main issues that will determine the fate of the biological diversity of the earth.

As a Costa Rican, I am proud that the natural as well as the social and political conditions of our peaceful, small country have permitted intellectual activities of the scope presented in this book. Research at La Selva also is a fine example of international collaboration. Nevertheless, Costa Rican intellectual participation in studies of tropical biology of the kind presented in this volume is, with a few notable exceptions, still meager. Research on Costa Rican biodiversity and ecology must become a priority in our own academic institutions and OTS's role in the promotion and stimulation of this effort should become a topic of increasing importance. We obviously need indigenous ecologists with close affinities to their regional problems and the capacity to formulate appropriate solutions to the environmental problems of our country. I hope that future publications on the natural history of Costa Rica will include increasing numbers of contributions from my fellow compatriots.

As an "herediano," I hope to see the present and future generations of my province recognize, thanks to books of this nature, that the real wealth of Sarapiquí, as well as most of Costa Rica, is to a large extent in its forests. We must realize that the dreams of the extraordinary potential of the Sarapiquí region could still be fulfilled, not by destroying the remaining forests but by preserving and restoring them and by learning how to use for our own benefit and the benefit of life on earth the numerous species that are contained in these extraordinary habitats.

Rodrigo Gámez

PREFACE

Because of the prominent role that La Selva has played in my professional life, the completion of this book is a source of great personal satisfaction. My first visit to La Selva was as a student attending a course offered by the Organization for Tropical Studies (OTS) in July of 1976. During our two-week stay more than 1 m of rain fell, the Río Puerto Viejo rose about 10 m, and a hastily scrawled sign on the blackboard warned that only strong swimmers should attempt a visit to the arboretum. Despite the somewhat adverse conditions, I fell in love with the site and its forest at once and promised myself that I would return someday, somehow.

Little did I then imagine that I would be back the following year as station manager! The job was not too tough, however, because Jim Beach, Tom Sherry, and I were the only researchers in residence for much of my watch. I spent a weekend or two alone in the old River Station (then the only habitable building) during the peak of the wet season, an experience I will always treasure. Among the richest aspects of the overall experience was my daily interaction with Rafael Chavarría, the foreman of Leslie Holdridge's Finca La Selva who stayed on the job when OTS acquired the property. Rafael taught me a great deal about La Selva and even more about finding common ground that acknowledges the diversity of the human experience. I left La Selva in 1978 wondering if I would ever be back. In fact, my work with OTS courses in the early 1980s and my seven-year stint as OTS' Educational Coordinator have taken me back to La Selva many times since 1978. My own research work on neotropical Acanthaceae has taken me far away from La Selva as a primary research site, but the forest there remains in many ways the source of my inspiration.

I have been fortunate that my continuing contact with La Selva has allowed me to observe at close hand the remarkable development of the station that took place in the 1980s and continues into the 1990s. These developments have totally changed La Selva's "personality" as described in some detail in chapter 2. For many "old-timers," the changes at La Selva evoke nostalgia. "Old" La Selva, however, could never have functioned as a major center for research on tropical forests. And no one can doubt the importance of establishing such centers, given the rapid pace at which humans are altering all of the earth's ecosystems and especially the tropics.

The idea of preparing a book synthesizing research productivity at La Selva Biological Station arose in parallel a number of times. In the late 1970s when the present North American Office of OTS at Duke University first began developing proposals for support and development of the facility,

the need for documentation of research productivity was apparent. In support of this effort Gary Hartshorn assembled a core of factual material for the 1977 proposal to the National Science Foundation. Over the years Gary kept alive the idea of developing this core into a full-fledged book, especially as knowledge of La Selva expanded rapidly during the 1980s. In the mid-1980s long-time La Selva researcher Kamal Bawa formally proposed to OTS that the project be undertaken. The idea was enthusiastically endorsed by OTS's governing bodies, and the real work began. The editorial team was assembled quickly, with Henry Hespenheide and me added to round out the teams' expertise and to provide a link to OTS (I was the staff member in charge of education at OTS from 1985 to 1992).

It proved remarkably easy to agree upon the subject matter to be included. The book's contents reflect the dual goal of presenting a synthesis that is comprehensive and emphasizes La Selva's research strengths. As might be expected, while the book was in preparation, rapid progress on several research fronts necessitated some adjustment of chapter contents. For example, La Selva's soils were originally to be treated as part of chapter 3 by Sanford et al. The wealth of information yielded by the complete soil survey by Freddy Sancho and Rafael Mata in 1985, however, called for a separate chapter devoted to the results of the survey and to research on soil processes. Similarly, the single planned chapter on applied ecology developed into three chapters in a two-stage process. Initially, land use was split from agriculture and forestry. Later, the proliferation of forestry research at La Selva during the late 1980s demanded that these latter subjects be treated separately.

These adjustments, in turn, required corresponding adjustments in other chapters with the result that a number of chapters went through several versions. Time passed as these changes were made so that still other chapters required updating. At some point, this process must stop if a book is ever to be completed. The book's contents congealed, but research progress had gone on and, in fact, has accelerated as more and more researchers use La Selva each year. One can already faintly hear calls for a future volume (and I have some advice for its editors!).

It will be impossible to acknowledge everyone who contributed to *La Selva*, but I will attempt it with apologies in advance to those omitted. First, I thank all of the chapter contributors for their work, their patience, and their nearly always rapid and genial responses to requests for clarifications and revisions.

Each chapter was substantially improved by the comments and suggestions of outside reviewers. The constructive and collegial manner in which these reviewers did the job requested of them gives one confidence in the often maligned peer-review system. We thank Steve Anderson, Peter Ashton, Nicholas Brokaw, Anne Brooke, Steven Bullock, David B. Clark, Martha L. Crump, Julie S. Denslow, Diane De Steven, Thomas C. Emmel, Peter Feinsinger, Christopher Field, Mercedes S. Foster, Alwyn Gentry, Michael Grayum, Russell Greenberg, Harry W. Greene, Craig Guyer, Carol Horvitz, Egbert Leigh, Jr., John T. Longino, Svata M. Louda, John G. Lundberg, Jeffrey C. Luvall, Roy W. McDiarmid, Craig Martin, Eugene S. Morton, Barry Osmond, Cheryl A. Palm, Geoffrey G. Parker, Steward T. A. Pickett, Mary E. Power, Richard B. Primack, Francis (Jack) Putz, William A. Reiners, Daniel Richter, Jr., Robert K. Robbins, Michael J. Ryan, William Schlesinger, Charles E. Schnell, Jack C. Schultz, Norman J. Scott, Jr., Thomas W. Sherry, Beryl B. Simpson, Phillip Sollins, Donald E. Stone, R. Jean Stout, Boyd Strain, Donald R. Strong, M. D. Swaine, Joseph Travis, Laurie Vitt, Henry M. Wilbur, Andrea Worthington, and several anonymous reviewers for their careful work. Egbert Leigh contributed detailed and extremely useful comments on most of the chapters of the vegetation section with very positive results. Joseph Wright and John Terborgh reviewed the chapters in the fauna section as a set and made many helpful suggestions. Gordon Orians read the entire manuscript on the way to writing chapter 26 and helped greatly to pull the entire manuscript together. Finally, two anonymous reviewers read the entire manuscript, and we are especially grateful to the second for his or her careful and constructive suggestions.

Many OTS staff have contributed to this effort in one way or another, often in many ways. We thank especially Amy Barbee, Ana Lorena Bolaños, Ana Carter, Deborah Clark, David Clark, Charles Schnell, Beverly Stone and Donald Stone. I am grateful to my husband, John Lundberg, for bearing with me through the years that this book has been in preparation and for helping to haul the "monster in a box" between home and office.

Finally, I am grateful to those at the University of Chicago Press who have helped turn the monster into a real book. I have newfound respect and appreciation for the people who make ideas and words on countless sheets of paper into books that one can hold, read, and treasure. My special thanks to Susan Abrams, whose confidence in the project was unshakable.

All of us who have worked hard on this project are motivated by a love of the tropics and concern for the fate of tropical forests and of the people who live in tropical countries. We share a desire to see our science advance and to see the ecosystems that we study protected from exploitation that is not sustainable. May this book help to inspire future tropical biologists, to convince them that it is not only possible to be a tropical scientist but critically important.

Lucinda A. McDade

LA SELVA

INTRODUCTION

Introduction

Kamaljit S. Bawa, Lucinda A. McDade, and Henry Hespenheide

The diversity and complexity of life in the tropics have fascinated Western biologists since the first European explorers returned with unfamiliar plants and animals. Nineteenth-century naturalists—including Darwin, Bates, Von Humboldt, Wallace, and Belt—contributed books that described tropical diversity and documented such quintessential tropical phenomena as mimicry and the protection of plants by ants. It is clear that experiencing the tropics radically changed these naturalists' perspectives of the living world. The exuberance of tropical diversity perhaps functioned as a shock treatment that forced those accustomed to the temperate zone to think in new ways. There can be little doubt, for example, that tropical experiences were the key to the development by Wallace and Darwin of the theory of evolution. Despite the importance of the tropics in advancing the theoretical framework of biology it is only recently that biologists have begun to explore systematically the composition, structure, and function of tropical forest biotas. Ironically, just as the talent and resources necessary for progress are being focused on describing and understanding these forests, their very survival is threatened. In the 1970s and 1980s accelerating deforestation has imperiled tropical forests in many parts of the world. The deforestation crisis is also responsible for a crisis of conscience within tropical biology. Some biologists have responded by working to quicken the pace of efforts to increase basic knowledge of tropical forests. Others have responded by shifting their research goals toward the practical problems that cause deforestation and that result from it. Biologists of all research persuasions increasingly find themselves deeply involved in political and financial campaigns to protect the forests that they study.

It seems clear that real progress will require a synthesis of these approaches. For example, effective conservation of tropical species depends upon basic knowledge of the spatial distribution of organisms, their population dynamics, and the complex interactions among species within communities. Similarly, a purely empirical approach to the development of sustainable land-use practices for areas that are already deforested would require an unrealistic investment of time and resources. Knowledge derived from intact forests, including community structure, energy flow, and water and nutrient cycles, allows scientists to begin with some good guesses about what are likely to be key elements of sustainable systems. Thus, the threat to tropical forests demands increased research in conservation, agriculture, and forestry but in no way obviates the need to understand the composition, structure, and dynamics of intact forests.

Although knowledge of tropical forests remains far from complete, accounts of the "state of the art" of tropical forest ecology can provide valuable summaries of widely dispersed information. Such accounts can also stimulate further research by providing baseline data, highlighting emergent patterns, and identifying controversies and gaps in scientific knowledge. A number of treatments of tropical rain forest ecology have focused on one or a few research themes at a variety of diverse sites (e.g., Richards 1952; Whitmore 1984; UNESCO 1978a; Golley 1983c; Prance and Lovejoy 1985; Longman and Jenik 1987; Jacobs 1988; Bawa and Hadley 1990; Gentry 1990b; Gómez-Pompa et al. 1991). Such treatments can take advantage of research strengths at selected sites and are especially valuable to researchers working in the subdisciplines of biology that are treated. It is difficult, however, to draw higher level comparisons and conclusions from such works because data on one aspect of a site may not be interpretable without information about its context (the rest of the ecosystem). This may be especially true because the complexity of biotic interactions in tropical areas makes it difficult to treat any one aspect in isolation. Two sites that are similar in climate, for example, may well differ in patterns of biotic diversity and, therefore, in the ways that these elements of diversity are organized into communities. Vegetation structure may be comparable at two sites but may support animal communities that are quite disparate.

In this volume we have chosen the alternative approach of providing a comprehensive overview of a single tropical rain forest—La Selva Biological Station. Located in the Atlantic lowlands of northeastern Costa Rica, La Selva is a research station and a 1500+ ha reserve of premontane rain forest. The facility is owned and administered by the Organization for Tropical Studies (OTS), a consortium of U.S., Costa Rican, and Puerto Rican universities and research institutions (table 1.1). La Selva has been studied by researchers and students for more than thirty years, and enough information has accumulated to permit a substantive treatment of the composition and ecology of this rain forest (although not without lacunae).

Comprehensive treatments of this sort potentially permit one to understand the interrelationships among the major ecosystem components: climate, geomorphology, plants, and animals. They are also subject to the idiosyncrasies related to the

Table 1.1 OTS Member Institutions, 1991–1992

University of Arizona
Arizona State University
Auburn University
University of California at Berkeley
University of California at Davis
University of California at Irvine
University of California at Los Angeles
Centro Agronómico Tropical de Investigación y Enseñanza
University of Chicago
City University of New York
University of Colorado
University of Connecticut
Cornell University
Universidad de Costa Rica
Duke University
Universidad Estatal a Distancia
University of Florida
Florida International University
University of Georgia
Harvard University
University of Hawaii
University of Illinois
Indiana University
University of Kansas
Louisiana State University
University of Maryland
University of Miami
University of Michigan
Michigan State University
University of Minnesota
University of Missouri-Columbia
University of Missouri-St. Louis
Museo Nacional de Costa Rica
Universidad Nacional
University of North Carolina
North Carolina State University
Ohio University
Ohio State University
Pennsylvania State University
University of Puerto Rico
Rutgers University
Smithsonian Institution
State University of New York at Stony Brook
Instituto Technológico
Texas A & M University
Tulane University
University of Utah
University of Washington
Washington University
University of Wisconsin, Madison
Yale University

particular history of research productivity at individual sites, with the result that they are rarely if ever strictly parallel (see chap. 26). Thus, the three volumes that are likely to be most often compared to the present treatment are, in fact, quite different, reflecting both different research traditions and decisions by the editorial teams. The remarkable and massive volume on El Verde rain forest, Puerto Rico (Odum 1970), was the result of an organized, well-funded, and comprehensive effort to survey the forest before and after exposing it to ionizing radiation in the mid-1960s. Even twenty years later, El

Verde may well be the most thoroughly known tropical forest, with particular strengths in floristics, vegetation structure and dynamics, nutrient cycling, and water relations. Leigh et al. (1982) chose to organize their valuable contribution on the ecology of Barro Colorado Island, Panama, around the strong seasonality of this forest and its effects on the biota. Both of these books include chapters that provide synthetic, site-specific overviews as well as others that are primary research reports. Jordan's (1989) synthetic treatment of the Amazonian rain forest at San Carlos on the Río Negro in southern Venezuela takes an ecosystem approach and emphasizes the stress associated with very low nutrient availability in this forest. Human use of this ecosystem is covered as well with chapters on slash-and-burn agriculture and secondary succession.

Contributors to the present volume provide a broad taxonomic and thematic synthesis of research at La Selva although the historical emphasis on plant biology and vegetation dynamics will be clear. They describe the climate and physical environment; survey taxonomic groups; summarize population and community level processes and interactions among species; and treat land use and applied ecology. The authors of these chapters synthesize what is known about La Selva in their subject areas but also identify gaps in knowledge. Some of these gaps are startling, and we hope that identification of them will stimulate future research. Authors place La Selva in context by comparing it to other research sites, emphasizing other tropical forest sites. Finally, authors identify areas for future research that are likely to be particularly fruitful or to yield information that is fundamental to understanding the forest at La Selva. Charting the future course of research is risky, and known paths (areas that have already received considerable attention) are likely to be emphasized. These authors, however, have perspectives on the general state of their fields as well as an understanding of research accomplishments at La Selva and thus are well positioned to take on the task.

Chapter 2 completes this introductory section of the book by providing information on La Selva's physical setting, history, and development, including both land acquisition and facilities. We have organized the chapters on research productivity at La Selva into five parts: (1) Abiotic Environment and Ecosystem processes; (2) The Plant Community; (3) The Animal Community; (4) Plant-Animal Interactions; and 5) La Selva's Human Environment. We introduce each part to highlight the major themes of the component chapters and note unresolved or controversial issues.

Section I summarizes what is known of the abiotic environment and ecosystem level processes at La Selva. Climate, geology, geomorphology, and hydrology are described in chapter 3, followed by a review of the soils at La Selva and of research on soil processes in chapter 4, and soil fertility, nutrient acquisition, and cycling in chapter 5.

Part 2 on the plant community examines vegetation structure, forest dynamics, and life-history processes among plants. It begins with a description of the vegetation and composition of the flora in chapter 6 and treats population-level phenomena of individual species in chapter 7. The fine-scale structure of the forest is described in chapter 8 with data on density, spatial patterns, growth, and mortality rates of trees.

The role of gaps in forest structure and dynamics is the subject of chapter 9. Plant physiological ecology is surveyed in chapter 10, including the responses of plants to variation in light, temperature, moisture, and soil conditions. Part 2 concludes with two chapters on plant reproductive biology: chapter 11 on phenology and chapter 12 on reproductive systems, including pollination mechanisms.

In chapters 13 to 18 in part 3 on the animal community, the available information is summarized on the six best-known animal taxa at La Selva: the five major vertebrate groups (fish, amphibians, reptiles, birds, and mammals) and butterflies, the best-known invertebrate group. In chapter 19 the coverage is extended to less well-known groups and the fauna of La Selva is placed into the perspectives offered by other topical sites and by ideas about the generation and maintenance of biodiversity. These taxonomically oriented chapters are followed by a synthesis of research on animal life histories and population biology at La Selva in chapter 20.

Reflecting the importance of plant-oriented research at La Selva, the section on interactions among species focuses in separate chapters on two major ways in which animals interact with plants: as purely exploitative herbivores in chapter 21 and in the more often mutualistic role as frugivores in chapter 22. The considerable work on pollination biology (which is mostly animal mediated at La Selva) is reviewed in part 2 on the plant community, chapter 12, because, in general, this research has emphasized plant reproduction rather than the behavior and foraging of animal pollinators.

Part 5 places La Selva Biological Station in its human context. Chapter 23 deals with the dramatic changes that have occurred in northeastern Costa Rica from 1960 to 1990, including patterns of settlement and land use. The chapter closes with an examination of what has become a major land-use system in the region (including the Cordillera Volcánica Central Biosphere Reserve of which La Selva and Braulio Carrillo National Park are both parts)—conservation. The agricultural development of the Atlantic lowlands region of Costa Rica (including the environs of La Selva) is examined in chapter 24 with attention to both subsistence and commercial (including export) agriculture. The forest industry in northeastern Costa Rica is described in the final chapter of this part, chapter 25, including deforestation, forest management, and plantation forestry. These last two chapters include surveys of research on these subjects at La Selva and recommendations for enhancing research productivity in agriculture and forestry at the research station.

Finally, Gordon Orians takes on the challenge of synthesizing a volume that in many ways constitutes an unfinished progress report. His contemplative analysis in chapter 26 makes thought-provoking reading.

This book is intended for a diverse audience, including those who know La Selva from personal experience, those who have worked at other tropical sites, and those whose tropical experience is vicarious. We have endeavored to make the contents accessible. We hope that scientists of whatever disciplinary specialty will be able to move among the chapters with ease and that the serious natural history tourist will also find much here that is comprehensible. For those who conduct research at La Selva the book will serve as a reference and a source of ideas for future projects. For the many who visit the station as students it will orient and instruct. The information presented here for La Selva will serve others as a source of data about one tropical rain forest and as a basis for comparisons with other sites.

We dedicate this book to the people of Costa Rica, who in every sense have made it possible: those who were among the founders of the Organization for Tropical Studies; those who were instrumental in safeguarding the station's biodiversity via its connection to Braulio Carrillo National Park; those who have worked at the research station in any capacity since its earliest days; and those who have never seen La Selva but who make its existence possible by being part of the national consensus for education and the pursuit of knowledge, and for conservation and the wise use of natural resources. May this book help to convey all that is special about Costa Rica, its people and biological wealth.

2

La Selva Biological Station

Lucinda A. McDade and Gary S. Hartshorn

In this chapter we place La Selva in its physical and biotic context. Our goals are to give readers enough background on the station to move to any other chapter in the book and to provide researchers and visitors with a framework for understanding their surroundings. We also present an overview of La Selva's history and development with a focus on those aspects that have had the most impact on research at the station (e.g., expansion of the reserve, enhancement of the station's facilities). We refer readers to appendix 1 for an analysis of research productivity at La Selva and to appendix 2 for discussion of the more pragmatic aspects of field station management (e.g., administration, governance). D. A. Clark (1988), D. B. Clark (1988, 1990) and Stiles and Clark (1989) provide useful overviews of La Selva Biological Station from differing perspectives. Stone (1988) provides a comprehensive historical treatment of the Organization for Tropical Studies (OTS) and of La Selva's role in the consortium. Pringle (1988) documents that part of La Selva's history that resulted in its connection to Braulio Carrillo National Park.

THE PHYSICAL AND BIOTIC SETTING

La Selva Biological Station is located at the confluence of the Sarapiquí and Puerto Viejo rivers in the cantón (county) of Sarapiquí, province of Heredia, Costa Rica (10° 26' N, 83° 59' W; figs. 2.1, 2.2). These rivers and their tributaries (Peje River on the west and the Sábalo-Esquina creeks on the east) provide natural boundaries on three sides. The entire reserve encompasses 1,536 ha (ca. 3,795 a. or 5.9 sq. mi.), of which more than 90% is located within these natural boundaries. The areas of La Selva that are outside the natural boundaries are the 100-ha La Guaria Annex, 4 ha on the East Bank, and La Flaminea, an area of about 40 ha extending from the East Bank to the Puerto Viejo–Horquetas Highway (fig. 2.2). These extensions of La Selva beyond the natural boundaries are relatively new and, as described, they exemplify OTS's efforts to improve access to the station, to increase the kinds of research that can be accommodated, and to provide an outreach center for the general public.

Geographically, the La Selva Biological Station is located precisely where the foothills of the central volcanic mountain chain of Costa Rica give way to the Caribbean coastal plain. At the northern end of the property, along the Río Puerto Viejo, the elevation is about 35 m above sea level. The relatively flat terraces near the river rapidly yield to steep hills that

reach 137 m elevation at the southwest corner of the station. These hills indicate what is to come for travelers venturing farther south. Braulio Carrillo National Park (BCNP) begins at La Selva's south border and climbs to nearly 3,000 m at the summit of Volcán Barva only 35 km away (linear distance) or about 60 km via the current La Selva–Barva trail. The history of the extension of BCNP to include La Selva is described later in this chapter.

Of La Selva's total 1,536 ha, about 55% is in primary forest, 7% is selectively logged primary forest, 11% is young secondary forest, 18% is early successional pasture, and 8% is abandoned plantations. An additional 0.5% is in areas that are managed for research and demonstration purposes (e.g., arboretum, successional strips), and slightly more than 1% is in developed areas. These vegetation types occur across a range of physical, edaphic, and historical conditions. Primary forest habitats, for example, extend from ridge tops to swamp forests on several soil types (see chaps. 4 and 6). This diversity of habitats provides land suitable for a comparable diversity of research purposes. In general, most of the primary forest is reserved for low-impact work, whereas areas in secondary forest, abandoned pasture, and plantations are available for manipulative research. Land-use designations ("zoning") are discussed under governance in appendix 2.

Environmental conditions at La Selva place it in Holdridge's tropical wet forest life zone (Hartshorn and Peralta [1988]; see also chap. 6, this vol.). The area receives about 4 m (157 in.) of rain annually. Perhaps more important than the absolute quantity of rain is the fact that the dry season is rarely long or severe (chap. 3). The forest is evergreen, but Newstrom et al. (chap. 11) demonstrate that phenological events in many plants are correlated with the seasonal pattern of rainfall. As is true of many tropical regions, the daily range in temperature (6°–12° C) far exceeds the range of diurnal averages over the year (<3° C). In terms of mean monthly average temperatures, August is the warmest month (27.1° C) and January is the coolest (24.7° C). Visitors to La Selva are surprised to find that nights may be quite cool (in part, owing to the narrowness of the Central American isthmus and, in part, to cold air drainage down the slopes of the high mountains to the south).

Whether La Selva's soils are representative of soil types that occur at other lowland tropical sites has received much discussion. Specifically, many Central American soils are of recent volcanic origin and are correspondingly relatively rich

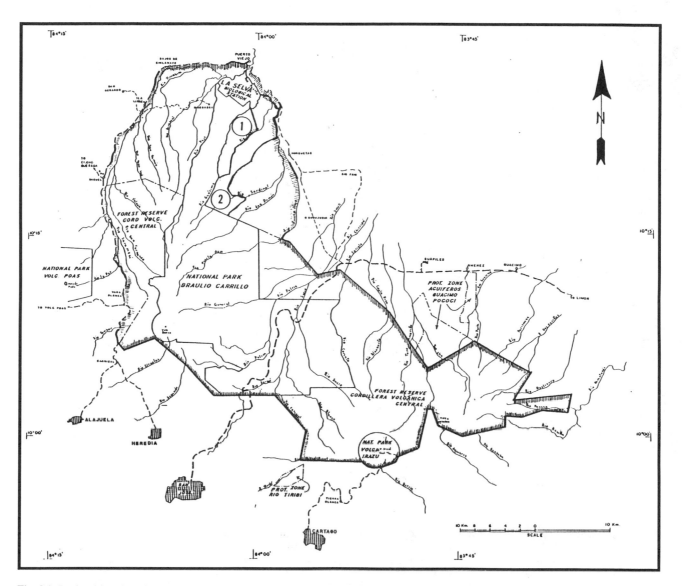

Fig. 2.1. Regional location of La Selva Biological Station, Braulio Carrillo National Park, and the Cordillera Volcánica Central Biosphere Reserve. 1 = Extension of Braulio Carrillo National Park declared April 1990. 2= Zona Protectora (Protected Zone) declared April 1990.

in nutrients although low in organic matter. In addition, significant portions of La Selva's soils are alluvial and, thus, relatively rich. In fact, soils at La Selva are highly variable, and Sollins et al. (chap. 4) indicate that they span much of the fertility range found elsewhere in the lowland wet tropics.

La Selva's rivers and creeks are important biological systems at the field station and also important determinants of the facility's ambiance. The reserve is bounded and drained by one of the major river systems of the Atlantic slope of Costa Rica. These streams offer opportunities for aquatic research (chap. 3) and for recreation, but they also present barriers. At the extreme, access to the station for the first twenty years of its existence was by boat only. The romantic aspects of this form of transportation rapidly pale against the harsh reality of transporting all supplies, building materials, and fuels by dugout. La Selva's rivers and creeks also mean that bridge building has always been an important aspect of reserve management.

High biological diversity and productivity are key elements

of the fascination that tropical wet forests hold for biologists and nature lovers. As documented throughout this volume, La Selva's biological resources are rich and largely intact. The vascular flora includes nearly two thousand species (nearly one-quarter of the country's total), of which more than four hundred are trees (chap. 6). We know very little yet about nonvascular plants and fungi. There are probably more than four thousand species of moths (D. Janzen and M. M. Chavarría pers. comm.) and perhaps five hundred species of butterflies (chap. 13). The remainder of the invertebrate fauna is clearly diverse but with few exceptions remains very poorly documented (chap. 19). The more than four hundred bird species (nearly one-half of the country's total avifauna) have made the site famous among ornithologists and their amateur colleagues. A notable exception to the litany of high diversity at La Selva is fish. La Selva's rivers contain an order of magnitude fewer species than the large, continental, Neotropical rivers (e.g., the Amazon and Orinoco rivers; chap. 14). The biological diversity of La Selva has been compared to that of

Fig. 2.2. La Selva Biological Station. Shading indicates land uses at the time of incorporation into the reserve or at the present in the case of Developed areas, the Rafael Chavarría Ecological Reserve and the Arboretum. Roman numerals identify the OTS permanent plots (see chaps. 6, 8). Small streams (Q. = quebrada) are indicated by solid lines. Trails are indicated by dashed lines, with acronyms as follows: AM = Avenida Marañon, CC = Camino Central, CCA = Camino Cantarrana, CCC = Camino Circular Cercano, CCL = Camino Circular Lejano, CEN = Camino Experimental Norte, CES = Camino Experimental Sur, LEP = Lindero El Peje, LOC = Lindero Occidental, LS = Lindero Sur, SAT = Sendero El Atajo, SCH = Sendero La Chanchera, SHA = Sendero Hartshorn, SHO = Sendero Holdridge, SJ = Sendero Jaguar, SLV = Sendero Las Vegas, SOC = Sendero Occidental, SOR = Sendero Oriental, SR = Sendero Ribereño, SSA = Sendero Sarapiquí, SSE = Sendero Sábalo-Esquina, SSO = Sendero Suroeste, SUA = Sendero Suampo, SUR = Sendero Surá. Refer to Fig. 2-3 for names used for specific areas of the La Selva reserve.

three other sites in the Neotropics (Barro Colorado Island, Panama; Cocha Cashú, Peru; the Manaus area, Brazil; Gentry 1990a). In general, total numbers of species in major groups are quite similar across these sites (allowing for differences in size and habitat diversity), but differences in component parts of that diversity suggest many fruitful avenues for research.

HISTORY OF LA SELVA

La Selva's history is important for both scientific and sociopolitical reasons. History is essential for understanding the present ecological and successional status of various parts of the reserve and, therefore, to interpret the results of current research correctly. It is also important to understand the sequence of human events that led to the protection of La Selva. As the world experiences massive destruction of natural ecosystems, any paths that lead to long-term protection merit examination.

For many years it was thought that La Selva's primary forests had never been disturbed. This idea required reevaluation upon the discovery of the pre-Columbian potsherds and worked stone along the main rivers (Quintanilla 1990) and of charcoal embedded in the soil essentially everywhere at La Selva that pits have been dug. Radiocarbon dating of four soil charcoal samples indicates fires about 1100 B.P. and 2400 B.P. (Horn and Sanford 1993). It is now clear that native Americans inhabited the alluvial terraces at La Selva beginning about 3000 B.P. (Quintanilla 1990). Most of La Selva's alluvial soils undoubtedly had a complex history of cultivation and abandonment cycles although the patch of primary forest in the Reserva Ecológica Rafael Chavarría (fig. 2.2) must have been undisturbed for an extended period. The presence of charcoal suggests that, perhaps more significantly, vegetation even far from the alluvial terraces may have been extensively and intentionally altered by periodic cutting and burning for shifting cultivation. Obviously, such information is of great interest to modern ecologists who have wondered whether a forest like La Selva's could ever regenerate following any sort of nonnatural disturbance. Research on the spatial distribution of trees of species favored by indigenous people and early settlers provides further evidence of human intervention at La Selva, probably in the form of selective cutting (D. B. Clark pers. comm.).

The Sarapiquí region was the principal route between the Caribbean coast and the Central Valley until the late nineteenth century. Voyagers poled their dugout boats up the San Juan and Sarapiquí rivers to El Muelle (a few kilometers north of Puerto Viejo) then traveled by mule up over the mountains to the Central Valley. In their 1853 journey through the Sarapiquí region, the German naturalists Moritz Wagner and Carl Scherzer (1856) comment on the small openings in the forest and rudimentary shacks along the overland route, including one unoccupied rancho at Rancho Quemado (site of present-day Puerto Viejo). As recently as 1953, the village of Puerto Viejo consisted of three small homes made of split palm trunks and thatched with palm leaves (L. R. Holdridge pers. comm.).

Throughout the region, population density remained very low until the 1970s with intensive land use restricted to the richer alluvial soils along the main rivers. In the 1970s and 1980s human populations increased rapidly and land-use patterns in La Selva's neighborhood changed accordingly (see chap. 23 and 24). La Selva's responses to these changes will be critical to its continued existence.

Farsighted tropical forester Dr. Leslie Holdridge acquired the core La Selva property (Old La Selva, fig. 2.3) in 1953 with the goal of putting into practice his ideas about the sustainable, nondestructive use of tropical wet forests. In keeping with his plans he named his new "farm" Finca La Selva (jungle farm). He was especially interested in experimenting with tree crops that could be planted without total clearing of the native forest. On alluvial soils near the Puerto Viejo River, Holdridge planted *Cordia alliodora* (laurel), a source of high-quality hardwood; *Bactris gasipaes* (peach palm, or pejibaye), a clonal palm that produces heart-of-palm and vitamin-rich fruits; cocoa under an overstory of native trees; and robusta coffee (*Coffea canephora*). Holdridge also established mixed plantations of cocoa, laurel, and pejibaye, in addition to small trials of other perennials. These plantings were mostly in areas that had been previously disturbed, leaving the primary forest intact. It is a tribute to Holdridge that, although his ideas found little acceptance in the 1960s, the search for sustainable land-use systems for the lowland tropics that attained emergency status some twenty years later has resulted in experimentation with many of the same concepts and crops.

Holdridge also used Finca La Selva as a weekend retreat. It was, indeed, a considerable retreat: after an interminable drive on a tortuous road from San José to Puerto Viejo, one boarded a dugout canoe for the final 4 km of the trip. It was also a rustic retreat: Holdridge built a two-story farmhouse on a bluff overlooking the Río Puerto Viejo (this building, now remodeled, is still in use as the River Station). There was no electricity, no phone, and few modern conveniences.

OTS bought La Selva from Leslie Holdridge in 1968 for fifty thousand dollars. The transfer included 587 ha of land (of which about 560 were in primary forest and the remainder in agricultural plantings and developed areas) and the limited

physical plant. OTS was a fledgling, five-year-old organization created with the goal of facilitating research and education in the tropics (Stone 1988). Prime among its activities were field courses in tropical biology. The first OTS course participants spent considerable time in classrooms in San José, but by 1968 these courses had developed a unique field-oriented format. With Holdridge's encouragement OTS courses had used La Selva as a site before it became OTS property. Holdridge had also encouraged research use of La Selva, and it had already been the site of important work on tropical forests by his students (e.g., Petriceks 1956; Budowski 1961) and on birds (Slud 1960).

During the OTS years, and in particular over the 1980s, La Selva became a major tropical field station. It is barely recognizable as Holdridge's Finca La Selva: access is easy, new buildings have been constructed, and the preserve has nearly tripled in size. Perhaps most significantly, La Selva is now the lowland terminus of a large national park that begins on the upper slopes of the volcanoes that surround the Central Valley. Facilities have improved, and the kinds and quantities of research conducted there have increased accordingly. The human environment both within the reserve and in its neighborhood has also changed rapidly. With station use now averaging more than forty-five persons per day, La Selva is no longer an isolated retreat. In the following sections we describe these changes and how OTS has planned and funded them. These changes have also required concomitant development of administration and governance at La Selva as described in appendix 2.

DEVELOPMENT OF LA SELVA

Land Acquisition

During the nearly twenty-five years that OTS has owned La Selva, six areas have been added to the core property purchased from Holdridge in 1968 (fig. 2.3). Altogether, the additions to La Selva have nearly tripled the size of the station from the 587 ha of Finca La Selva to the 1,536 ha of La Selva Biological Station. These additions are the result of at least five organizational goals: (1) OTS has sought to expand the reserve to natural boundaries to improve its protectability; (2) additional primary forest has been acquired to increase La Selva's viability as a conservation unit, and areas with disturbed habitats have been added to have suitable land for (3) research requiring intensive manipulation and for (4) facilities development (both of which are judged inappropriate for primary forest); and (5) certain pieces have been acquired because they improve access to the core of the reserve. We briefly describe each of the six additions to La Selva, pointing in each case to the goals involved.

Until the 1970s La Selva was a forest within a forest. Most of the Sarapiquí region was still forested, and basic issues about critical size and degree of isolation of preserved areas were just beginning to be raised. By the early 1970s, however, La Selva was threatened with becoming first a peninsula and then an island of forest surrounded by pasture and cultivated land. Precisely these concerns prompted the purchase of Annex A in 1970, adding 56 ha to the east flank and extending the reserve to the natural boundary formed by the Sábalo-Esquina creeks (fig. 2.3). La Selva researchers G. Hartshorn

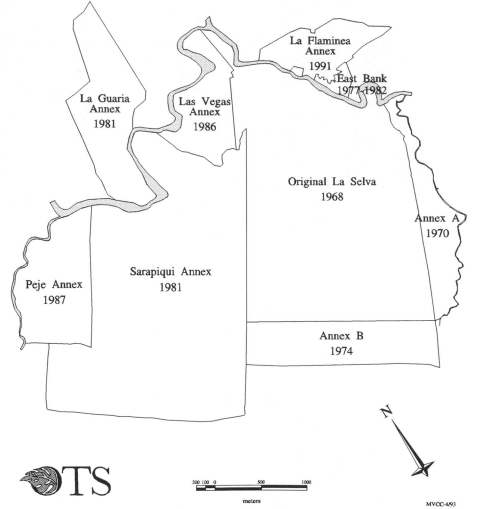

La Flaminea
Annex
1991

East Bank
1977-1982

La Guaria
Annex
1981

Las Vegas
Annex
1986

Original La Selva
1968

Annex A
1970

Peje Annex
1987

Sarapiqui Annex
1981

Annex B
1974

N

OTS

200 100 0   500   1000

meters

MVCC-4/93

Fig. 2.3. History of progressive acquisition of land at La Selva Biological Station. See text for further explanation. In 1993, OTS was still in process of acquiring parts of the La Flaminea Annex as a series of independent parcels; the boundaries shown reflect OTS's intentions.

and G. Stiles proposed that some of the early successional stages be maintained on Annex A by periodic cutting. Beginning in 1971 five .5-ha *successional strips* have been cut on a five-year rotation. The cutting of the oldest strip occurs in February, but the slash is not burned and there is no attempt to suppress vegetative sprouting. Although the regeneration is not representative of secondary succession following slash-and-burn agriculture, the successional strips have proven very useful for educational purposes, and many course projects conducted at La Selva use these 2.5 ha.

The 1974 purchase of Annex B (87 ha) was in response to the owner's plans to log this 300–400-m-wide strip of primary forest along the south border of La Selva (fig. 2.3). The successful fund-raising drive for this purchase only highlighted the increasing vulnerability of La Selva to insularity caused by deforestation of surrounding areas. The financial difficulties encountered by OTS during the mid-1970s (Stone 1988), however, delayed further expansion of La Selva for several years.

In the late 1970s OTS acquired about 4 ha of land on the East Bank of the Río Puerto Viejo immediately across the river from the station to improve access to the Puerto Viejo–Horquetas road, which is a short distance from the river (fig. 2.1). By 1980 an all-weather road was built from the highway to the East Bank, so that access to La Selva involved "only"

maneuvering down a muddy river bank, a short boat ride across the river, and a steep climb up to the buildings. With acceptable (easily acceptable, nonprimary forest that was not subject to flooding) building sites on the reserve side of the river already committed, it was decided to site further support facilities on the East Bank.

The addition of the Sarapiquí Annex in 1981 nearly doubled the size of the reserve. There were two pressing reasons for acquiring this piece of property. First, the owner was preparing to clear that portion that was still in good forest (about one-third of 631 ha). Second, for some time awareness had been growing within OTS of the need to acquire land suitable for manipulative research. With tropical rain forests disappearing rapidly, managers are reluctant to permit research that seriously damages even a small part of a conserved area. Still, certain critical questions can only be addressed by large-scale manipulative research. The Sarapiquí Annex with its secondary forests, abandoned pastures, and primary and high-graded forests was, thus, ideal for acquisition. This purchase extended La Selva's western boundary to the Sarapiquí River and included a strip of property (100 ha) on the north side of that river (figs. 2.2, 2.3). This strip extends from the river to the main highway between San José and Puerto Viejo. Now called La Guaria Annex, this strip was mostly in pasture with a few patches of secondary forest. It has proved important for access to the western portion of La Selva (via cable "ferry" over the Río Sarapiquí) and as a site for forestry and soils research (see chaps. 4, 25).

A 68-ha parcel of land located on the corner formed by the junction of the Puerto Viejo and Sarapiquí rivers (figs. 2.2, 2.3) was owned by the Hershey Chocolate Co. until 1986. With La Selva's destiny manifest and the experimental cocoa plantation apparently adding little to corporate coffers, the property was donated to OTS by Hershey. Ironically, Holdridge had split off the Las Vegas property when he originally bought La Selva and sold it to his associate J. Robert Hunter. It was Hunter who arranged the donation to OTS more than thirty years later. The Las Vegas Annex completed OTS ownership of land near the confluence of the two rivers; it was largely in cacao and pejibaye plantations (now abandoned), and its addition substantially increases land available for manipulative research. In 1992 John Ewel initiated a project on this site that seeks to understand ecosystem functions in simplified early successional communities.

Completing the story of additions to OTS holdings at La Selva disrupts the time line but has, perhaps, other organizational advantages. After the extension of Braulio Carrillo National Park to La Selva's south boundary (see below), one significant in-holding of 105 ha remained within the natural boundary formed by the Río Peje on the west (figs. 2.2, 2.3). This parcel was owned by a North American and was a likely target for colonization. Concerns about the prospect of a settlement on a poorly protected flank aside, La Selva badly needed land with degraded soils to use in forestry trials (that the Peje Annex has such soils indicates that it was, in fact, a poor target for colonization). The Peje Annex was purchased in 1987.

Finally, in 1993, OTS is negotiating the purchase of about 50 ha of land between the original 4 ha on the east bank and the Puerto Viejo–Horquetas highway (fig. 2.3). This strip is part of a much larger holding (La Flaminea) that is owned by the Costa Rican government for division and redistribution as small farms. OTS was able to convince the government that the piece immediately adjacent to La Selva was best used as an interface between the biological station and the public. Plans are being prepared to develop the site as a visitors' center with a variety of indoor and outdoor educational displays as well as picnic and recreational areas.

With this purchase OTS will accomplish its major goals of extending property lines to natural boundaries, protecting additional forest, adding substantial areas of disturbed habitats for manipulative research, and improving accessibility. As OTS developed La Selva into a modern field station, deforestation of adjacent areas increased the peninsularization of La Selva, and forest clearing south of the reserve threatened to isolate the still small area of protected forest. Thus, the campaign to extend Braulio Carrillo National Park to La Selva was fundamental to the survival of La Selva Biological Station as a viable conservation unit.

Connection to Braulio Carrillo National Park

In 1974 Hartshorn emphasized the increasing insularity of La Selva in a letter to the OTS Fund-Raising committee for Land Acquisition, "OTS cannot continue to buy up expensive little parcels of virgin forest to maintain continuity with the virgin forests extending from Volcán Barva and Volcán Cacho Negro [to La Selva]." With deforestation increasing in this frontier area of Costa Rica, it was clear that forested lands that were not protected would not remain forested for long. At that time OTS apparently could do little to prevent La Selva from becoming an island of forest in a sea of pasture, especially given the organization's financial condition (Stone 1988). As an island, without doubt the reserve would experience edge effects and species extinctions that would reduce the site's conservation usefulness and research potential.

A number of determined individuals looked toward the forest-covered slopes of the volcanoes to the south of La Selva and dreamed of a much larger conservation unit joining La Selva in the lowlands with the high mountains of the Cordillera Central. When the Costa Rican government established BCNP in the highlands in 1978, lines began to be drawn on maps proposing links between this 32,000-ha park and La Selva. Beyond OTS's proprietary interest in the ecological security of La Selva, an expanded park would be an exceptionally valuable conservation unit because it would protect one of the few unbroken transects of primary forest in Central America that spans an elevational gradient of this range (35 m to nearly 3,000 m). Increasing evidence that numerous bird species make regular seasonal migrations along this gradient (Pringle et al. 1984) provided additional strong support for the park extension from the perspective of conservation: How effective as conservation units are isolated highland and lowland reserves if elevationally migrating animals cannot survive because their essential seasonal movements are interrupted by modified landscapes in between?

Two successive presidential administrations of Costa Rica expressed support for the expanded park, but the country's economic problems in the 1980s precluded funding for the project. As an emergency measure, President Carazo designated the narrow corridor connecting La Selva to the Cordillera Central Forest Reserve (at about 900 m elevation) as a Zona Protectora (Protected Zone) in 1982. This designation allows landowners to maintain title and current use of the land but forbids intensification of land use (specifically deforestation). With this further stimulus an international consortium of conservation groups was formed to raise funds to purchase holdings within the *zona*. In this effort OTS was joined by the Nature Conservancy, the World Wildlife Fund, U.S., and the Costa Rican National Parks Foundation. Crucial momentum was provided by a one million dollar challenge grant from the John D. and Catherine T. MacArthur Foundation, the challenge being to match the donation dollar for dollar. Eventually more than two million dollars were raised from more than two thousand donations that ranged from a child's one dollar to MacArthur's one million dollars. The park extension was formally declared in 1986 by President Monge. It added 12,698 ha to BCNP for a total of 44,099 ha in the enlarged park.

International recognition of the value of the conservation unit came in 1988 when the United Nations Educational, Scientific, and Cultural Organization (UNESCO) recognized BCNP as the nucleus of the Cordillera Volcánica Central Biosphere Reserve (fig. 2.1). The reserve also includes two smaller national parks (Irazú and Poás Volcano national parks), a forest reserve, La Selva, and adjacent unprotected lands. The international system of biosphere reserves under

UNESCO's Man and the Biosphere Program (MAB) is intended to integrate conservation and sustainable management of natural resources. Butterfield (chap. 23) surveys the biosphere reserve concept and prospects for developing the Cordillera Volcánica Central Biosphere Reserve as an effective model.

The extension of BCNP and its recognition as a biosphere reserve did not, of course, solve all problems. Parks must be administered and guarded in perpetuity, and these activities are expensive. Ironically, funds for maintenance of parks are especially difficult to raise despite the fact that this sort of financial security is essential to protect the investment made in purchasing conservation areas. Furthermore, the issue of insularity remains. The park protects a large portion of Costa Rica's biodiversity in part because it includes many life zones across the elevational gradient. Each life zone, however, is represented by a relatively small area, and this is especially true of the lowland regions. Only 13% of the protected area is in the lowlands (below ca. 300 m, Hartshorn and Peralta [1988]) and the narrow panhandle of land that extends down to La Selva (fig. 2.1) is especially vulnerable to the effects of small area and peninsularization. The latter problem was dealt with in early 1990 when outgoing President Arias declared an extension of BCNP to the east (fig. 2.1). He also declared a Zona Protectora that will stabilize land use and protect the eastern lowland flank of BCNP, an active colonization zone (see chap. 23). The former problem has apparently been solved by a grant from the U.S. Agency for International Development (USAID). The Forest Resources for a Stable Environment Project (FORESTA), underway in 1991, will provide an endowment for management of BCNP as well as funds for a variety of sustainable development and natural resource management projects in areas surrounding the park.

The BCNP–La Selva project has secured a significant portion of Central America's biodiversity by effective cooperation among conservation organizations and between a tropical country rich in natural resources and a monetarily rich world community of concerned citizens. It has ensured a future for La Selva as part of a viable conservation unit and, in consequence, as an appropriate site for research that requires intact ecosystems.

Research Support

Into the Forest. After acquiring La Selva, OTS sought to make the forest accessible to researchers. Some 50 km of trails are maintained at La Selva in a pattern designed to provide access to all areas of the reserve as well as to special habitats. To prevent strandings and to help relocate special sightings, noteworthy plants, and projects, the trails are marked every 50 m with acronyms for their names and linear distances from the trailheads. Trails between the buildings within the developed areas and those intended for use by the general public are, for the most part, cement sidewalks. Within the forest, the most heavily traveled trails (those nearest the living quarters) as well as problem spots (wet areas, steep slopes) have been planked. The planks are covered with hardware cloth to increase traction. The most heavily used portions of the trail system are being converted to narrow cement sidewalks. Although this amenity may add to the comfort of casual visitors, the primary reason for the walkways is to avoid soil compac-

tion from the heavy traffic and trail widening that inevitably occurs when people go around muddy spots.

Developing a geographic reference system has long been one of OTS's highest priorities for La Selva. The original property was covered by a 200-m grid system that was extended to newly acquired areas when possible. OTS was less successful in securing topographic data for the reserve although topographic surveys were carried out for limited areas by researchers whose projects demanded these data. In 1991 these earlier survey efforts were superseded. A grant from the U.S. National Science Foundation (NSF) supported a complete survey of the reserve, resulting in a 50 m × 100 m grid over the entire property. Permanent monuments are at all grid points and where grid lines cross trails with x, y, and z coordinates, to decimeter accuracy, for each. The data are stored in a computer-based Geographic Information system (GIS), which is designed to present the information visually as maps. Other kinds of geographically referenced information can be added for long-term storage, for output as maps, and to facilitate decision making about the siting of projects. For example, the GIS will store information on soils (chap. 4), making it possible for researchers to locate their sites easily with respect to soil types. Once the locations of on-going research projects are entered, it will be easier to avoid conflicts among projects and to site new ones. Individual researchers can enter data (e.g., the location of troops of monkeys or of individuals of a given tree species) and generate maps for research and presentation purposes. In sum, these developments will dramatically improve the storage, retrieval, and comparison of geographically referenced data.

Biological Inventory and Database. Knowledge of the flora and fauna at a research site is clearly an invaluable tool for research. Without reliable taxonomic information, many kinds of research are difficult if not impossible, and intersite comparisons are hindered if the organisms involved cannot be identified. As detailed in parts 2 and 3 of this volume, progress toward a biological inventory of La Selva is very uneven. For some groups (e.g., vertebrates and trees) species lists are undoubtedly nearly complete. At the opposite extreme, nearly all invertebrate groups are very poorly known. Only in cases where a specialist on a given taxon has worked extensively at La Selva is good information available on what species are present (chap. 19). OTS has actively stimulated work toward a complete flora of La Selva via research funded by three NSF grants. In 1993, the organization is in the initial phases of a similar effort in conjunction with the Instituto Nacional de Biodiversidad (INBio) toward an anthropod fauna for the station (see chap. 19).

Research productivity at La Selva is stored as hard copies of all publications and theses that result from work at the station. The station also maintains copies of OTS coursebooks that contain reports on nearly thirty years of research projects conducted by more than two thousand students. This literature (including the coursebook reports) is also accessible as a key-worded database file that can be searched by subject and taxonomic groups. In the late 1980s OTS adopted a policy on archiving significant sets of data from research at the station to safeguard such valuable data while protecting the rights of the researchers who collected them (see app. 2).

Logistics and Facilities. For the first ten years or so of La Selva's existence as an OTS field station, the pace of change was slow. Minor enhancements turned a weekend retreat into a rustic field station. Generators were added to provide dim lights and barely adequate refrigeration. A wing was added to the Holdridge farmhouse that provided private rooms upstairs for long-term researchers and a screened "laboratory" downstairs (with little beyond wooden table tops and cold running water). A cement block prefabricated schoolhouse was installed and equipped with air conditioning to provide lab space with a semicontrolled environment. For the first time, it was possible to bring temperature and humidity sensitive equipment to La Selva without its certain demise. Access continued to be an adventure: arrivals were timed to coincide with one of the thrice daily dugout canoe runs into town. From the muddy banks of the Río Puerto Viejo near town, visitors, luggage, equipment, supplies, and generator fuel were shuttled the remaining 4 km (about thirty minutes) to the station itself. There was no phone, and radio communication with the OTS office in San José was unreliable. Groups were housed in a single room crammed with wobbly "bunk cots." This bunkroom served researchers as well when the six tiny single rooms and three double rooms were occupied.

The pace of change accelerated in the late 1970s when the U.S. National Science Foundation responded favorably to a series of proposals from OTS Executive Director Donald Stone. These proposals pointed to heightened demand for the facility against the backdrop of the crises of tropical deforestation and ignorance of tropical ecosystems. The identification of La Selva by the National Research Council's Committee on Research Priorities in Tropical Biology as among the most appropriate sites for intensive development of research potential (NRC 1980) provided further justification for NSF support. Since 1977 NSF support for La Selva has been continuous and has directly funded most of the major facilities' enhancements. In improving access, accommodations, and research facilities at the site, OTS goals have been to remove impediments to use of the site by improving communications and access, to provide appropriate housing for both long-term researchers and short-term visitors, and to support the full spectrum of research and teaching that can reasonably take place at a field station.

On-line electricity arrived at La Selva in 1979, and a microwave telephone followed shortly thereafter. An all-weather road to the East Bank across from the reserve was constructed, and access was further simplified with the 1982 completion of a 100-m-long footbridge across the Río Puerto Viejo. A building boom greatly increased the number of buildings at La Selva in the 1980s. Duplex cabins were constructed to provide comfortable housing for long-term researchers. A two-story laboratory was completed in 1981 with air-conditioned space for labs and offices downstairs, ambient lab space in screened wings, and a large ambient meeting/classroom upstairs. A second phase of construction in the mid-1980s developed the East Bank as the center of residential life at La Selva. The new dining hall seats seventy-five, and two dormitory-style buildings accommodate twenty short-term visitors each. The East Bank siting of these new buildings reflects the lack of easily accessible space on already deforested land on the west bank. Other building projects include a new director's house, remodeling of the old Holdridge farmhouse for use by long-term researchers, remodeling of an old workshop as a public education center with displays and classroom space, construction of an office and nursery building at the La Guaria Annex to serve forestry projects, and remodeling of the original air-conditioned lab as a library, herbarium and researchers' lounge. A new chemical analytical lab was completed in 1989 and outfitted with a range of equipment. This laboratory has signaled a quantum leap in the kinds of research that can be done at La Selva. The latest building project, four two-bedroom units for long-term researchers with families, is scheduled for completion in mid-1992. These new units will enable La Selva to house young children adequately for the first time, making it possible for researchers to continue their field work even as they take on family responsibilities.

In charting the course of development at La Selva OTS's overarching goal has been to facilitate research by making the station appropriate for many kinds of projects, both in terms of habitats and facilities, and more comfortable for extended stays. Patterns of station use strongly suggest that OTS has been successful. In 1982 a total of 57 researchers had projects at La Selva, and research and educational use of the station totaled 7,320 person-days. By 1991 these measures of activity had increased to 257 researchers and more than 16,000 person-days. Appendix 1 summarizes the research trends at La Selva, examining both the quantity of publications and the subject areas investigated. The remainder of this book documents the productivity of these researchers.

LA SELVA AND ITS NEIGHBORHOOD

The growth and development that have occurred at La Selva during its two decades as a field station were paralleled by significant changes in its neighborhood. In 1968, when OTS acquired the property, its remoteness was nearly a drawback. The 100 km between San José and La Selva were traversed by a rough mountain road that took four hours in a four-wheel-drive vehicle and longer by bus. Puerto Viejo was a one-street frontier town that offered very little material support to the station. At this writing, the road is now paved to Puerto Viejo, and the same 100 km can be traversed in two hours. The highway that runs south from Puerto Viejo to Las Horquetas and on to the Braulio Carrillo highway will be completed by the mid-1990s and will reduce the travel time further. Puerto Viejo has grown considerably and supplies more of the station's growing needs. The population of the Sarapiquí region has increased and will, no doubt, continue to do so as a result of the large-scale expansion of the banana industry in the region and government settlement projects (chap. 23). These changes have required commensurate changes in the attitude that the field station takes toward the world beyond its borders.

The ease of access to La Selva and the growth of neighboring communities have added two groups of station users: local people and ecotourists from abroad who wish to make day visits. Beginning in the mid-1980s, OTS took steps to accommodate these groups more effectively via construction of a visitors' center, enhancement of the most heavily used trails for comfort and safety, preparation of trail guides, and hiring of additional staff. Ecotourists have always been welcome at La Selva for overnight visits so long as they do not compete with the highest priority users (researchers and educational groups, see app. 2). Improved roads into the Sarapiquí area, together with the development of tourist lodges in the vicinity, have made day or "walk on" visits by such groups feasible. Ideally, OTS would like to be certain that day visitors have a high-quality experience and that they learn something about La Selva and OTS while at the station. OTS would also like to ensure that local people benefit from the ecotourism boom in Costa Rica. These parallel concerns have resulted in the training of a group of local people as naturalist guides (Paaby et al. 1991). Ecotourist groups that visit the station as "walk-ons" are required to hire one of these guides to accompany them. The guides are not OTS staff, but their work at La Selva relieves pressures on staff time, enhances the quality of visits by ecotourists, and yields income for local families.

The Environmental Education and Community Relations Program was founded in 1985 with the twin goals of educating the public about conservation in general and the roles of research stations in particular and of winning community support for the station. The program's staff organize regular visits by school and community groups to La Selva and are frequent participants in community meetings and festivities. The visitor's center is central to handling these groups; it provides a meeting place and exhibits that are designed to convey the purpose of a biological research station and to illustrate the kinds of research that take place there. Most of these activities will be moved to the new public center that is planned for the La Flaminea property near the Puerto Viejo–Horquetas highway (fig. 2.3). The location of the new facility will offer a visual perspective of the entire biosphere reserve, presenting the opportunity for environmental education about the region's conservation areas and ecological problems as well as a vehicle for explaining OTS's role in the region and the country. The government's natural resource management agencies, in particular those participating in local projects, will be involved in planning and implementing displays and activities at the new facility, and it will be run largely by local residents. The goal is for the new center to serve as La Selva's window on and for the world (Weekes 1992).

CONCLUSION

OTS's primary objective at La Selva is to facilitate tropical research and education by providing an accessible field site complete with living and laboratory facilities. Over the OTS years, the improvement of each of these aspects of site quality has received high priority and the achievements have been significant as described herein. One of the most important developments was, no doubt, the extension of Braulio Carrillo National Park down to La Selva. This addition allayed fears about the ability of a small station to sustain a species-rich tropical rain forest. This step has no doubt improved the success rate of OTS proposals for funds to improve facilities at the station and of individual researchers' proposals as well. These enhancements have, in turn, resulted in rapid expansion of research and educational use of La Selva and marked diversification of the former (see app. 1). In particular, better lab facilities have brought projects that require relatively high-tech capabilities. Similarly, the acquisition of terrain suitable for manipulative work has considerably extended the scope of research at the station. Further expansion of the La Selva reserve seems unlikely although involvement in the Cordillera Volcánica Central Biosphere Reserve (see chap. 23) will bring a larger role in regional land management than the station's size alone would indicate.

La Selva's next two decades as a field station will certainly bring challenges that are as complex as those that arose during the first two decades although we expect these to be somewhat different. The challenge now is not to enlarge the reserve, but to protect La Selva and BCNP and to develop the Cordillera Volcánica Central Biosphere Reserve as a model of effective collaboration between scientists and communities for conservation and sustainable development. From improving access and enhancing facilities, the challenge has evolved to maintaining facilities while keeping station rates affordable. Recruitment of researchers has ceased to be a challenge and, in the near future, the challenge may instead be to decide among a surfeit of projects that seek siting at La Selva. Rather than attracting visitors to La Selva, the challenge now becomes how to regulate use by nonpriority groups to give them a high-quality visit while protecting scientific users from adverse impact. We look forward to the continued evolution of the station to meet those new challenges.

ACKNOWLEDGMENTS

We thank David Clark, Beverly Stone, Donald Stone, and our coeditors for careful and insightful comments on earlier drafts. David Clark and Beverly Stone graciously assisted in the location of numerous unpublished facts and figures.

PART I

ABIOTIC
ENVIRONMENT
AND
ECOSYSTEM
PROCESSES

COMMENTARY

Lucinda A. McDade

Part 1 provides information on the abiotic environment and ecosystem-level processes at La Selva. In chapter 3 Sanford et al. describe La Selva's physical environment, including climate, geomorphology, topography, and a section about the physical and biological characteristics of La Selva's extensive streams. In chapter 4 Sollins et al. present a detailed comparative treatment of La Selva's soils. They begin this chapter with a brief introduction to soil science and continue with a description of each of La Selva's soil consociations based substantively on a recent soil survey conducted by two of the chapter's authors, Sancho and Mata. Sollins et al. next survey progress on soils process research at La Selva (e.g., mineralization, physical structure, and chemistry). In chapter 5 Parker links the abiotic to the biotic in a systematic treatment of processes of nutrient cycling at La Selva. A number of common themes emerge clearly from a great deal of interrelated material in these three chapters.

La Selva's environment is diverse and variable, which should be reflected in the spatial patterns of the plant and animal communities. Key aspects of La Selva's climate are high rainfall (4 m annually, on average) and lack of extended dry periods (no month averages less than 100 mm of precipitation). Sanford et al., however, note that mean rainfall is probably less informative and less biologically important than unusual events. For example, a thirty-day period with no rainfall in 1983 almost certainly stressed the forest. Such rare events may well be more important than their low frequency would suggest. At the other extreme, anyone who spends more than a few months at La Selva is likely to witness one of the spectacular floods of the Río Puerto Viejo. Such floods strongly affect the riparian landscape and vegetation, and the lengthy periods of rainfall associated with them may affect everything from plant phenology and animal activity patterns to tree-fall rates.

The simplistic notion that La Selva's soils were of three types (recent alluvium, old alluvium, and residual) has given way to a kaleidoscope patchwork of some twenty-four soil consociations with specific histories (in nature and source of parent material) and characteristics. Sollins et al. plead for greater appreciation of the diversity of tropical soils, pointing out that only a small portion of tropical soils fit the stereotype of "red, infertile, hardening irreversibly upon clearing."

The streams at La Selva are also highly diverse chemically and biologically with spatial patterns related to water source, substrate, light levels, and amounts of autochthonous and allo-

chthonous input. A single stream can change from autotrophic to heterotrophic within a few meters as a result of canopy cover, substrate, and inputs. It has even been suggested that subsurface drainage from volcanoes to the south of La Selva may account for the extreme chemical characteristics of some streams (Pringle 1991; Pringle and Triska 1991a, 1991b). The complexity exhibited by La Selva's streams offers great research potential and opportunities for collaboration between aquatic and terrestrial ecologists working on nutrient-cycling patterns.

Soils should be viewed as fundamental to ecological processes, and projects related to spatial pattern should reference soil types. All of the authors in this section identify the recently completed soil survey as a major advance. Sollins et al. recommend that data on soils be treated as seriously as other kinds of basic data in the siting of projects. This is certainly feasible now that good soil maps exist. The Geographic Information System (GIS) that is being installed at La Selva (see chap. 2) will also greatly facilitate the archiving of geographically referenced data. All of these authors urge that researchers precisely locate their projects so that soil type can be treated as a potential variable to explain patterns.

The La Selva ecosystem, although highly variable, is relatively nutrient-rich compared to those in other tropical areas. Sanford et al. note that La Selva's streams are generally at the nutrient-rich end of the spectrum for tropical streams. As biological confirmation of their relative richness, there is no evidence that macronutrients limit primary productivity in La Selva's streams. In contrast, experiments demonstrate that micronutrients and light are often limiting. Similarly, Parker points to relatively high nitrogen levels in La Selva's soils and indicates that nutrients have not been shown to limit plant growth at La Selva. Phosphorous and potassium are likely to be the most critical nutrients, but in the understory low light levels probably restrict plant growth far more significantly (see chap. 9, 10).

What is not known is vast and continues to impede research progress. All of these authors lament the large gaps in basic knowledge of La Selva. In fact, the material in some sections of these chapters serves more to identify what is not known than to describe what is known (e.g., processes of nutrient retention and reallocation by plants in chap. 5).

Perhaps the most startling fact to emerge from Sanford et al.'s discussion of La Selva's climate is how few data are available. Meteorological records are nearly nonexistent except for a twenty-five-year record of precipitation, and most of the additional data that are available are not actually from La Selva but from an area just south of the station. This points to the fact that an organization like OTS that lacks institutional funding for research has difficulty setting its own research priorities (especially when they are expensive). As climate change emerges as one of the most critical research issues, the lack of baseline meteorological data becomes particularly poignant. The lack of meteorological data stymies research in a number of areas, and researchers often must begin projects by collecting baseline data. The same problem was encountered with efforts to intensify the 200-m grid system and to obtain a detailed topographic map of the station. Substantial improvements in both areas are reported in chapter 2.

The historical absence of good information on soils was a major hindrance to research on ecosystem processes. The recently completed soil survey by Sancho and Mata (1987) makes it possible for the first time clearly to reference research results to soil type. Unfortunately, some earlier research projects cannot now be clearly sited, and their results are less useful. Before Sancho and Mata's survey, a tendency existed to generalize from work on the relatively fertile alluvial terraces despite the fact that these areas represent only a small fraction of the soil diversity at La Selva. Parker further points to the absence of rather basic data on nutrient acquisition, cycling, and loss and identifies several reasons for it. Simple lack of research is clearly a significant part of the problem, but, as noted, the reduced usefulness of data from studies that cannot now be sited accurately is also partly responsible.

Research opportunities are essentially unlimited. In each chapter the authors identify numerous avenues for future research that are particularly important or likely to yield results of special significance. The following list includes those that come up repeatedly or that bridge the chapters.

> Data from long-term monitoring of meteorological conditions are invaluable. Every effort should be made to begin gathering more complete meteorological data as soon as possible.
>
> The availability of good soil data presents greatly enhanced opportunities to explore the consequences of these differences on nutrient cycling, on the streams that drain areas with different soil types, on plant community structure and composition, and on animal distributions, to name only a few possibilities.
>
> Nutrient transfer processes, both uptake from the soil and nutrient release from plants and animals, are very little studied but will be vital to understanding nutrient processing at the ecosystem level.
>
> It is critical to achieve a better understanding of the relationship between climate and the plant community. For example, evapotranspiration rate changes when a forest is cleared (chaps. 3, 5), which suggests that forests are very important regulators of the water cycle at the local level. Parker reports that a clear-cut area adjacent to La Selva experienced a 31% reduction in evapotranspiration over the first six months compared to intact forest. Clearly, large-scale cutting has the potential to alter the water cycle (and regional climate) significantly.
>
> Improved understanding of the effect of disturbance and different land uses on ecosystem characteristics and on climate is needed. Parker reports that small-scale disturbances do not seem to result in increased soil nutrient availability, whereas those at intermediate scales (500–2,500 m^2) result in a rapid increase in nutrient concentrations in the soil solution, increased soil water percolation, and losses of soil nutrients. Data are not available for large-scale disturbances nor has the successional process been studied with respect to changes in nutrient levels.
>
> Sollins et al. hint at the relationships between the plant community and soils. For example, long-lived trees certainly have the potential to change their substrates, and animals such as earthworms and peccaries may also be important. Are the relatively high nitrogen levels in La Selva's soils related to biological nitrogen fixation associated with leguminous trees? To what extent are soil characteristics related to these kinds of biotic influences?

As Sanford et al. indicate, much of the relevant research has been by researchers working independently and without overt efforts to collaborate or to design research that will together yield a larger picture. As a result, with many of these subjects, we are at the point of having just enough information to begin to ask the really exciting questions that will tie the La Selva ecosystem together. A number of recent achievements (e.g., the soil survey, installation of a new meteorological station, a topographic survey, establishment of the GIS) are likely to result in rapid research progress in these areas. We hope that by identifying both what is known and what remains to be learned, this section of the book will enhance this process by stimulating research and collaborative links between scientists with interests in these areas.

3

Climate, Geomorphology, and Aquatic Systems

Robert L. Sanford, Jr., Pia Paaby, Jeffrey C. Luvall, and Eugenie Phillips

Until recently, this review of the climate, geomorphology, and aquatic systems of the La Selva Biological Station would have been nearly impossible owing to the lack of research in these subject areas at La Selva. Indeed, if written in 1980, this chapter would have included the precipitation data and uninformed speculation on all other topics. Major research efforts, however, on climate (Luvall 1984; Parker 1985) and on aquatic systems (Paaby 1988; Pringle et al. 1984, 1986; Pringle 1990; Pringle and Triska 1991b; Phillips 1987; Stout 1982) have provided much of the data base for the review here. Additionally, a survey of the geomorphology of La Selva by Alvarado (1985) has provided enough information to stimulate considerable discussion on the interaction of geomorphological processes with soil formation, nutrient cycling, and aquatic systems.

Although research on ecosystem characteristics of La Selva is underway, it is far too early to synthesize that information. Most research discussed in this chapter has been conceived and conducted as independent, marginally related projects; hence, only an outline of ecosystem ecology can be developed for La Selva. Some subject areas are well developed (considering the relatively short time that they have been systematically queried), whereas other key areas (including geomorphology) have received little attention. Although the patchy array of ecosystem research does not yield a complete overview of ecosystem processes at La Selva, the priorities for future research can be better defined (see also chap. 5). More importantly, for the first time, integrated ecosystem research can be an objective of future studies.

CLIMATE

An Overview of Tropical Wet Forest Climate

Tropical wet forests occupy low-altitude zones within about 23°27′ latitude of the equator. Within this region development of tropical forests occurs under two climatic conditions: mean annual daily temperatures between 18° and 24° C, with more diurnal temperature variation than seasonal, and yearly precipitation totals exceeding 2,000 mm with no month receiving less than 100 mm. Forests in four main regions meet these criteria: the Amazon and Orinoco basins in South America; the Central American isthmus; the Congo, Niger, and Zambezi basins of central and western Africa; and the Indo-Malay-Borneo-New Guinea archipelago. These tropical wet

forest regions are roughly defined by the equatorial trough (also called the intertropical convergence zone, or ITCZ), which is the major synoptic feature that influences maritime air mass circulation and seasonal rainfall patterns at low latitudes. The ITCZ, historically known as "the doldrums," is the region where the northern and southern hemispheric trade winds converge in the equatorial trough. The equatorial low-pressure trough is a region of atmospheric pressure gradient/wind interaction between the subtropical high-pressure areas and the equatorial low-pressure areas (Riehl 1979). Tropical seasonality is determined by rainfall patterns that are caused by movement of the equatorial trough to the north and south of the equator (fig. 3.1). This generally results in a single annual peak of precipitation in most tropical regions and a tendency toward two peaks per year at or near the equator. Localized orographic effects also influence rainfall and temperature variability within the equatorial region.

Synoptic Meteorological Patterns Affecting Costa Rica

The precipitation patterns of Central America and Costa Rica are influenced primarily by three large-scale atmospheric circulation patterns: the equatorial low-pressure trough, easterly waves embedded in the trade winds, and the polar trough (Dunn and Miller 1964). Hurricanes are infrequent in the region. Only one hurricane struck the coast of Costa Rica (Martha, November 1969) during the one hundred years records have been kept (Coen 1983) although several others have come close.

The strength and position of the equatorial low-pressure trough determines the seasonal synoptic precipitation patterns of Central America and Costa Rica. During the dry season, or *verano* (summer), from December through April or May, the equatorial trough is south of Central America, reaching its

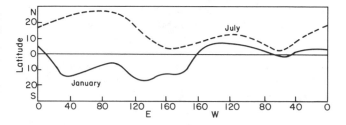

Fig. 3.1. Mean global position of the equatorial low-pressure trough (intertropical convergence zone) in January and July (Riehl 1979). Longitude at La Selva is 83° 59′ W.

Fig. 3.2. A GOES East meteorological satellite image showing the equatorial trough during the rainy season in Cosa Rica (June 24, 1988). The equatorial trough is defined by the band of clouds representing thunderstorms over Central America. Easterly waves imbedded in the trade winds can also be seen as large cloudy areas north of the equatorial trough.

southernmost position in February. During the wet season, or *invierno* (winter), which usually begins in May, the equatorial trough moves back over Central America to a position that is just north of Costa Rica (fig. 3.2). If the equatorial trough moves far enough north of Costa Rica, a short, often erratic dry period, or *veranillo* (little summer), may occur.

Easterly waves are a series of low-pressure troughs originating over West Africa and traveling via the trade winds from east to west (Dunn and Miller 1964; Riehl 1979; Barry and Chorley 1987). They are characterized by a leading edge of high pressure (divergence and subsidence of air masses) with fair weather followed by an area of low pressure (convergence and ascent of moist air masses), which causes an increase in cloud formation and precipitation. Easterly waves are common in the region from June through September when a wave passes over the eastern Caribbean about twice each week on average (Dunn and Miller 1964).

Other periods of continuous precipitation lasting several days (locally called *temporales*) occur from November through February and are caused by cold (originally polar)

air masses (*nortes*) from the North American continent. These trigger slow-moving depressions along the Central American coast with large areas of stratiform clouds and steady rain (Riehl 1979; Lydolph 1985; Portig 1976; Whiteside 1985).

Complex local variations in precipitation patterns are caused by the uplifting, cooling, and subsequent release of moisture as air masses pass over the Central American continental divide. The vegetation in Costa Rica reflects the effect mountains have on precipitation patterns. The moisture-laden northeastern trade winds (the most persistent winds in the region) support tropical moist forests on the Atlantic slope, whereas seasonally dry tropical forests are typical of the Pacific side.

The Climate at La Selva

Meteorological records of the climate at La Selva are limited primarily to long-term (twenty-five-year) daily precipitation records. Short-term meteorological measurements of solar radiation, net radiation, air temperature, and relative humidity have been collected by several researchers at La Selva. Precipitation data have been collected at Rafael's Point since late-1957; these data were not collected consistently on a daily basis, however, until late-1962. In April 1982 a new precipitation collection station was installed on the east bank knoll near the bridge over the Río Puerto Viejo known as the Bridge Station. The precipitation data from Rafael's Point are consistently lower than from the Bridge Station. The difference is probably because the edge of the forest canopy at Rafael's Point is less than 10 m from the collection station. We use data from both stations in our analysis. Because of the length of the record, the long-term analysis of precipitation data presented here is based on daily total precipitation records from 1963 to 1991 from Rafael's Point. We also report the average weekly rainfall based on the shorter, but more accurate, Bridge Station rainfall record from 1982 to 1991 (table 3.1).

The long-term average annual precipitation at La Selva is 3,962 mm, ranging from a low of 2,605 mm in 1983 to a high of 5,659 in 1970 (table 3.2). The monthly distribution of the precipitation is bimodal, with peaks of more than 400 mm per month occurring both in June–July and November–December. The period with least precipitation is February–April, and March is normally the driest month (table 3.2). The weekly pattern of precipitation, based on weekly averages of rainfall for nine years, reveals the striking onset of the wet season during the second week of May (fig. 3.3). The driest week of the year (March 26–April 1) and the wettest week of the year (August 20–26) receive 10 mm and 147 mm, respectively. Interestingly, the distribution of weekly precipitation shows little evidence of the veranillo season (but see chap. 11) although this may be the result of the relatively short duration of the Bridge Station collection record. In a general sense, the distribution of precipitation at La Selva is characteristic of the Atlantic side of Central America (Gramzow and Henry 1972; Portig 1976).

Precipitation patterns at La Selva are compared to other well-documented lowland tropical forest sites in figure 3.4. The San Carlos, Venezuela, lowland evergreen rain forest is slightly drier and lacks a distinct February–April dry period. From December through March both Ivory Coast and Barro Colorado Island lowland forest sites are much drier than La Selva. The peak precipitation period for these sites is June through August.

There are no long-term data of dew point temperatures at La Selva; air temperature, however, has been measured since 1984 at the Bridge Station (table 3.3). The average monthly air temperature is 25.8° C, which differs little among months. The maximum recorded daily air temperature is 36.6° C (3:00 P.M., May 3) and the minimum recorded nighttime air temperature is 16° C (3:00 A.M., March 13) (Luvall unpublished data). The diurnal range in air temperature of 6°–12° C greatly exceeds the average monthly change, which is <3° C.

Other micrometeorological measurements for La Selva include solar radiation, air temperature, and vapor density from August 1982 to August 1983 in a pasture area near the south boundary (Luvall 1984). Total solar radiation ranges from 17.5 MJ per day in October to 26.6 MJ per day in April (fig. 3.5). There is some seasonality in solar radiation because of the change in solar declination. Day-to-day change, however, can be greater than month-to-month change. Months of low

solar radiation have higher average monthly precipitation and, hence, persistent cloud cover. In July, for example, the sun's zenith angle is the greatest, but average daily total solar radiation is less than in the dry month of April. Little change is exhibited daily or monthly in total air moisture content of about 19×10^{-6} gm cm^{-3}. Vapor density deficits are important because they affect photosynthetic rates through effects on stomatal conductance. Deficits may change quickly, so average monthly values provide little information on the maximum deficits to which tropical rain forests are exposed. Deficits on the order of 10–12×10^{-6} gm cm^{-3}, however, are common during the dry season (Luvall 1984). Additionally, a series of atmospheric profiles was measured at La Selva in February 1988 in conjunction with a National Aeronautics and Space Administration) remote-sensing research project (Luvall unpublished data). Figure 3.6 illustrates a characteristic dry season atmospheric profile of air and dew point temperatures according to radiosonde measurements above La Selva. The morning profile indicates a moist air layer about 3,400 m thick, capped by dry, cold air, which is characteristic of tropical atmospheric profiles. The dry cap is caused by trade wind inversion, and it acts as a barrier to vertical development of precipitation during the dry season. During the rainy season, the dry cap is much higher in the atmosphere and much less pronounced (Riehl 1979; Whiteside 1985; Barry and Chorley 1987). The atmospheric profile changes seasonally in thickness of the moist air layer as well as in overall moisture content. Little if any change occurs in the air temperature profile (Riehl 1979; Whiteside 1985; Barry and Chorley 1987).

The Interaction of Tropical Rain Forest and Climate

Monthly averages of precipitation reveal little about the effect of precipitation extremes on tropical forests. It is more useful to examine the length of time between precipitation events (fig. 3.7). Precipitation of >5 mm per day was used as the threshold value for significant rainfall events. Lighter precipitation does not contribute significantly to soil moisture because 3 mm of rain is intercepted by the forest canopy and evaporated back into the atmosphere (Parker 1985). Thirty days was the maximum recorded period with total daily precipitation ≤5 mm (March–April, 1983; fig. 3.7). The long-term average for the driest months (March and April) is about twelve days. Thus, in some years the forest at La Selva may experience significant moisture stress in March and/or April.

A water budget was calculated for a primary forest adjacent to La Selva for the thirty-day March–April 1983 drought (Luvall 1984; Parker 1985) (table 3.4). Soil moisture for the first 70-cm depth was measured using tensiometers, and evapotranspiration was calculated using the Penman-Monteith equation. The water budget for the primary forest during that period indicated a 40% reduction in total soil moisture content in the upper 70 cm of soil (table 3.4 from Luvall [1984] and Parker [1985]). The forest probably would have experienced moisture stress during that period. Available soil water content, however, depends on soil type, and the rate at which the forest uses the available soil moisture (evapotranspiration) depends on forest type. Therefore, forests on the various soils at

Table 3.1. Weekly rainfall from the Bridge Station at La Selva, 1982–1991 (mm)

Week	1982	1983	1984	1985	1986	1987	1988	1989	1990	1991	Mean
1	—	59.7	188.3	26.9	64.1	41.3	56.8	126.6	119.3	47.3	81.1
2	—	141.8	27.6	18.5	24.2	17.0	4.4	69.5	30.0	24.1	39.7
3	—	64.0	28.2	4.7	24.8	26.1	3.2	118.0	187.4	4.0	51.2
4	—	23.7	20.7	30.0	29.6	117.8	150.3	17.8	151.9	13.4	61.7
5	—	3.2	21.2	5.5	98.5	43.9	203.2	10.2	47.7	83.1	57.4
6	—	1.3	70.7	50.9	0.0	22.2	118.1	44.1	8.5	15.7	36.8
7	—	42.6	1.2	143.1	41.5	3.3	74.0	16.3	7.5	22.5	39.1
8	—	71.6	43.5	26.9	10.1	16.3	16.1	48.8	27.5	24.2	31.7
9	—	92.5	155.2	34.8	57.3	25.1	23.9	39.6	7.6	116.3	61.4
10	—	27.5	22.8	47.0	34.5	4.0	40.0	68.0	76.4	11.0	36.8
11	—	46.4	18.9	11.0	3.6	0.0	34.3	3.9	162.2	0.0	31.1
12	—	0.0	31.0	0.0	105.5	0.0	125.0	7.9	50.2	11.5	36.8
13	—	0.0	0.0	3.9	13.9	25.3	19.4	4.0	0.0	27.7	10.5
14	12.3	14.0	4.1	0.5	29.4	90.6	6.8	77.9	0.0	63.8	29.9
15	10.9	1.1	0.0	57.1	9.5	73.5	33.9	85.8	19.2	27.0	31.8
16	4.6	0.0	66.9	11.3	82.5	28.2	4.9	29.0	4.7	18.0	25.0
17	46.6	1.0	1.1	0.0	84.2	9.1	0.0	14.9	7.6	6.5	17.1
18	9.2	92.5	7.8	17.4	119.3	12.9	4.6	14.2	41.4	63.3	38.3
19	15.4	272.0	132.0	76.3	19.0	18.7	190.0	0.0	168.3	189.1	108.1
20	32.2	99.5	55.3	31.0	11.0	59.4	51.7	16.8	218.1	26.2	60.1
21	33.7	24.3	84.0	64.3	56.6	123.0	131.7	174.2	82.3	28.5	80.3
22	53.7	80.5	63.7	69.2	0.0	18.7	109.3	74.5	50.0	131.4	65.1
23	71.2	33.4	60.5	124.7	80.2	64.1	44.7	169.1	306.3	79.8	103.4
24	137.6	143.0	78.0	84.1	74.0	206.0	28.4	105.3	55.8	189.9	110.2
25	79.9	17.8	120.2	287.3	180.9	19.8	81.8	125.2	19.4	48.7	98.1
26	300.4	74.4	77.8	76.9	200.2	23.7	59.9	59.4	158.0	91.4	112.2
27	60.0	229.3	36.8	119.2	97.1	52.9	34.0	115.9	106.2	24.4	87.6
28	262.3	44.8	21.8	46.1	123.3	133.0	67.1	111.8	39.5	179.2	102.9
29	188.0	230.5	148.3	57.3	92.9	111.4	142.4	115.8	40.2	182.3	130.9
30	301.5	98.8	28.8	68.4	109.5	35.3	107.0	122.7	114.4	265.5	125.2
31	197.7	108.2	60.6	103.1	163.8	169.4	16.5	86.1	66.6	140.0	111.2
32	108.6	77.8	174.4	117.4	75.3	70.5	149.8	78.0	169.4	242.5	126.4
33	105.0	52.2	141.6	87.5	118.0	133.9	19.8	13.2	123.0	151.1	94.5
34	249.4	81.8	41.8	130.1	165.6	121.5	203.3	120.0	248.9	110.3	147.3
35	87.8	156.7	65.9	44.9	159.5	81.0	126.8	34.5	64.5	121.1	94.3
36	54.8	39.8	44.9	194.9	258.1	56.3	112.4	75.4	90.8	67.0	99.4
37	213.9	86.1	67.4	59.4	97.6	127.0	43.4	58.9	58.3	195.1	100.7
38	74.6	48.4	26.3	75.4	131.1	20.2	83.8	170.5	93.2	185.5	90.9
39	133.5	70.7	92.5	14.4	24.1	48.4	40.0	185.2	52.0	43.6	70.4
40	197.2	40.6	61.5	60.3	218.0	61.9	173.8	8.6	47.3	—	96.6
41	117.6	127.0	208.3	17.0	127.1	103.6	16.2	61.1	103.7	—	98.0
42	134.0	55.9	15.9	27.9	47.3	29.0	236.9	20.0	57.7	—	69.4
43	344.8	216.1	138.0	94.4	6.4	146.0	37.0	34.2	119.8	—	126.3
44	63.7	115.5	125.3	48.4	20.5	58.6	6.2	183.9	70.5	—	77.0
45	251.0	39.9	61.5	175.7	20.8	11.0	38.2	25.2	150.7	—	86.0
46	109.0	21.1	164.6	85.3	112.4	84.2	81.9	54.7	59.6	—	85.9
47	15.0	30.9	117.5	16.4	71.0	308.5	72.4	151.4	139.9	—	102.6
48	16.0	26.2	7.5	14.2	7.3	3.2	99.8	161.1	101.2	—	48.5
49	48.5	30.4	97.0	77.0	165.8	53.2	138.1	47.3	98.7	—	84.0
50	231.7	11.4	72.2	51.3	18.9	41.9	162.2	36.0	86.9	—	79.2
51	124.4	42.2	10.1	13.9	22.3	190.2	268.0	15.6	84.2	—	85.7
52	50.6	38.6	161.4	25.6	8.9	49.4	39.8	105.2	4.2	—	53.7
TOTAL	4,548.3[a]	3,548.7	3,572.6	3,128.8	3,917.0	3,391.5	4,133.2	3,713.3	4,398.7	3,272.0	3,879.3[a]

[a]The total for this year is incomplete.

Table 3.2 Monthly rainfall in mm from Rafael's Point at La Selva, 1963–1991

Year	Jan	Feb	Mar	Apr	May	Jun	Jul	Aug	Sep	Oct	Nov	Dec	Total	Year
1963	245.4	148.3	339.9	552.5	331.5	350.3	318.0	257.0	454.2	489.0	603.5	502.7	4,592.3	1963
1964	131.1	38.7	137.4	119.4	351.5	209.6	415.0	344.9	303.3	327.2	367.3	148.5	2,891.2	1964
1965	528.1	236.5	200.9	49.0	401.3	408.9	851.0	486.7	283.7	360.7	280.7	535.9	4,623.4	1965
1966	399.0	268.7	324.4	372.4	621.5	303.5	357.4	409.4	366.0	345.9	338.8	667.5	4,774.5	1966
1967	383.5	99.3	103.4	417.3	498.3	554.7	559.3	625.9	367.5	324.9	624.1	392.2	4,950.4	1967
1968	454.7	116.6	307.1	386.6	568.2	318.8	450.9	255.0	372.4	516.6	277.4	394.2	4,418.5	1968
1969	194.1	106.2	134.1	128.0	250.2	519.7	311.4	479.6	357.9	400.1	678.9	644.9	4,205.1	1969
1970	196.3	477.3	209.8	502.2	274.1	449.3	340.6	415.5	701.3	293.9	506.5	1292.6	5,659.4	1970
1971	484.3	119.6	169.7	174.7	193.0	492.6	438.5	207.0	287.4	368.3	227.5	158.1	3,320.7	1971
1972	588.1	183.6	77.0	211.7	281.4	473.3	686.5	376.0	402.4	257.8	286.9	366.9	4,191.6	1972
1973	260.0	137.1	37.2	81.6	497.2	364.2	403.1	263.4	136.7	311.5	548.6	261.1	3,301.7	1973
1974	256.4	184.8	116.2	168.7	256.0	428.1	542.0	402.4	187.1	323.7	348.2	303.8	3,517.4	1974
1975	122.0	98.3	170.7	167.2	130.2	529.6	225.2	428.9	176.6	340.3	755.6	544.4	3,689.0	1975
1976	319.3	150.4	115.6	121.5	415.3	347.5	1,065.8	485.5	737.7	163.4	364.8	195.1	4,481.9	1976
1977	134.2	117.4	153.9	115.4	171.4	695.2	919.4	422.1	345.5	350.5	294.2	164.1	3,883.3	1977
1978	139.8	271.9	215.6	109.0	374.4	469.0	708.9	555.0	182.6	359.4	339.7	388.8	4,114.1	1978
1979	195.1	150.8	204.1	747.1	252.1	451.5	274.3	673.5	245.3	433.5	537.0	420.5	4,584.8	1979
1980	262.4	171.3	30.7	193.5	239.6	591.8	510.7	393.2	432.9	264.7	549.9	774.8	4,415.5	1980
1981	481.6	279.7	207.3	317.0	330.0	386.5	299.9	454.7	242.2	375.0	842.8	427.8	4,644.5	1981
1982	68.7	128.6	62.8	83.1	154.7	496.5	772.2	547.8	518.2	715.3	386.9	498.6	4,433.4	1982
1983	264.0	126.1	117.6	8.8	447.8	287.8	631.7	282.7	251.8	487.6	85.1	107.3	3,098.3	1983
1984	229.6	175.2	119.0	61.9	262.0	267.4	235.7	360.6	203.9	426.8	340.6	322.9	3,005.6	1984
1985	64.4	188.8	64.5	53.6	124.6	553.9	264.1	414.2	310.5	188.6	248.1	130.0	2,605.3	1985
1986	167.8	60.4	181.5	203.4	140.0	477.0	389.1	533.5	545.8	384.9	173.3	187.9	3,444.6	1986
1987	215.9	39.5	16.3	185.3	203.9	294.3	325.6	444.2	280.2	362.6	381.5	298.8	3,048.1	1987
1988	347.7	244.3	194.8	43.6	417.9	180.6	336.8	457.0	249.3	424.3	227.1	571.6	3,695.0	1988
1989	290.6	107.4	90.9	173.0	268.1	430.5	429.1	276.5	442.6	201.8	340.6	246.6	3,297.7	1989
1990	452.4	73.1	252.7	23.7	496.5	503.2	287.0	685.4	252.5	296.7	473.4	246.5	4,043.1	1990
1991	55.7	201.5	53.1	98.7	400.2	337.3	590.4	619.1	360.8	—	—	—	—	1991
Mean	273.5	162.1	152.0	202.4	322.5	419.7	480.7	433.0	344.8	360.5	408.2	399.7	3961.8	
Standard deviation	144.1	88.0	84.5	173.3	133.2	116.7	215.0	124.6	142.7	108.2	176.9	244.7	723.2	

Record Wet and Dry Months

	Jan	Feb	Mar	Apr	May	Jun	Jul	Aug	Sep	Oct	Nov	Dec	Total	
Maximum	588.1	477.3	339.9	747.1	621.5	695.2	1,065.8	685.4	737.7	715.3	842.8	1,292.6	5,659.4	
Year	1972	1970	1963	1979	1966	1977	1976	1990	1976	1982	1981	1970	1970	
Minimum	55.7	38.7	16.3	8.8	124.6	180.6	225.2	207.0	136.7	163.4	85.1	107.3	2,605.3	
Year	1991	1964	1987	1983	1985	1988	1975	1971	1973	1976	1983	1983	1985	

Table 3.3 Mean monthly maximum, minimum, and average air temperatures for La Selva Bridge Station, 1984–1991

Month	Mean Monthly Maximum	Mean Monthly Minimum	Mean Monthly Average
January	29.7	19.8	24.7
February	30.0	19.8	24.9
March	30.7	19.8	25.2
April	31.1	20.5	25.8
May	31.0	21.1	26.0
June	30.7	21.5	26.1
July	30.0	21.7	25.9
August	30.3	23.9	27.1
September	31.1	21.8	26.5
October	30.7	21.5	26.1
November	30.5	21.2	25.9
December	29.9	20.4	25.1

La Selva could have different water budgets during the same period of drought with more or less moisture stress.

The key variable for estimating tropical rain forest hydrologic budgets is evapotranspiration (ET). (For a more complete description of this approach see appendix 1 and Shuttleworth et al. 1984; Luvall 1984; and Monteith 1973.) Luvall (1984) found that about 47% of total annual precipitation is returned to the atmosphere by evapotranspiration from the primary forest at La Selva. Estimates for the forests of the upper Río Negro region and the eastern and central Amazon basin range from 48% to 52% (Jordan and Heuveldop 1981; Marques et al. 1977). Tropical rain forests can be usefully ranked based on ET rates because ET rates depend on both energy inputs and forest type (table 3.5). Although the variety of techniques and time frames used make it difficult to compare directly the values in table 3.5, the reported ET rates from

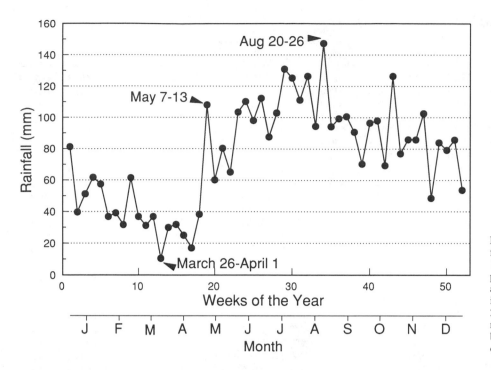

Fig. 3.3. Average weekly precipitation from the Bridge Station at La Selva, 1982–1991. The lowest average weekly precipitation occurs March 26–April 1, and the greatest average weekly precipitation is during the week of August 20–26. For the nine-year precipitation record at the bridge site, the wet season begins rather abruptly during the second week of May.

tropical rain forests generally range from about 2.7 to 6.0 mm per day.

Removal of the tropical forest canopy alters the hydrologic cycle in two significant ways: by reducing interception and decreasing transpiration. Under identical conditions water intercepted by the forest canopy will evaporate first at a rate two to five times faster than water is transpired (Rutter 1963, 1967; Stewart 1977; Moore 1976; Singh and Szeicz 1979; Murphy and Knoerr 1975). Luvall (1984) found that 35% of the total evapotranspiration for a primary forest near La Selva occurred as evaporation directly from the forest canopy. Evaporation rates equaled 30% of the total evapotranspiration in an Amazon rain forest (Shuttleworth et al. 1984). During the rainy season, when two of every three days has measurable precipitation, the intercepted water loss is an important component of the hydrologic cycle. Removing tropical forest destroys the vegetation that intercepts precipitation and evaporation becomes more rapid.

The second alteration of the hydrologic cycle occurs when the transpiring surface area is reduced and less moisture is removed from the soil. In the study near La Selva by Luvall (1984) and Parker (1985), a 31% reduction in ET occurred in a clear-cut during the first six months compared to an adjacent intact primary forest. Even though regrowth from sprouts was rapid, evaporation from the soil surface was a significant component of water loss. During the dry season, soil moisture in the clear-cut was significantly greater, a pattern that continued for three years after the cut (G. G. Parker pers. comm.).

Summary and Research Needs. Climatic seasonality at La Selva is determined by rainfall patterns that are caused by the position of the equatorial trough. Precipitation peaks occur in June–August and October–December. A short dry season usually occurs in February–April. Intervals between daily rainfall

totals of >5 mm average about twelve days during the dry season. A period of thirty days without rain (>5 mm/day), however, was recorded in 1983, and such extreme events may have a strong effect on ecosystem function. The climate at La Selva is notable for its lack of variability in air temperature, relative humidity, and solar radiation throughout the year. Diurnal change in air temperature of about 6°–12° C is much greater than the monthly change of <3° C. Variability in solar radiation throughout the year mostly depends on cloud cover and not on change in the sun's zenith angle. ET is an important component in the hydrologic cycle of La Selva where approximately 47% of the total precipitation is returned to the atmosphere from intact forest. Forest clearing strikingly reduced ET.

Clearly, a great deal remains to be learned about La Selva's climate and its impact on the biota. Ideally, long-term micrometeorological measurements of solar radiation, air temperature, relative humidity, wind speed and direction, and soil moisture should be collected. These data should be taken at least hourly and should be archived with OTS; they would be used by a wide variety of researchers at La Selva and elsewhere. Substantial progress along these lines has been achieved with installation of a class 2 meteorological station at La Selva in early 1992.

Little is known about the impact of meteorological phenomena on the forest at La Selva. For example, soil moisture can be significantly reduced during the dry season. No research at La Selva, however, relates soil moisture status to some expression of moisture stress in the forest canopy. Such research might address any number of important questions: Does reduced soil moisture affect primary productivity, trigger flowering or leaf abscision, or reduce the ability of individuals of a particular species to compete?

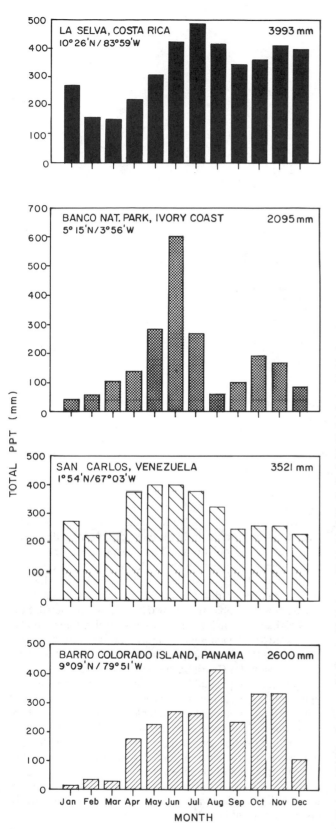

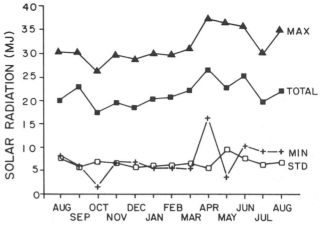

Fig. 3.5. Average daily solar radiation totals by month for La Selva, August 1982–August 1983 (Luvall 1984). "Max" and "min" refer to monthly daily maxima and minima; "total" refers to daily totals averaged for the month.

GEOMORPHOLOGY

Geomorphology is most simply defined as the study of land forms (Ritter 1978). Because biologists seek different information from the landscape, geomorphology should be the basis for integration of diverse research results. This is especially true for tropical environments where biodiversity and land-form diversity are complex. The regional geomorphology of the Caribbean lowlands of northern Costa Rica is complex and poorly understood. Near the Caribbean, Pleistocene alluvial deposits are common with occasional, highly weathered Tertiary volcanic hills (Madrigal 1970). The lower slopes of the continental divide are partially colluvial-alluvial in origin. At La Selva most of the terrain is formed by ancient (Lower Pleistocene) lava flows and (rarely) Quaternary Plio-Pleistocene mud flows (lahars). Exceptions are the alluvial terraces near the Puerto Viejo and Sarapiquí rivers. The ancient mud flows include materials of considerable heterogeneity in size and origin, including andesitic and basaltic boulders that are buried in a clayey/sandy matrix (Alvarado 1985). The lava flows are either andesitic or basaltic in origin. (See chapter 5 for detailed information about soil formation on the mud flows and terraces).

La Selva Geomorphology

The geomorphology and geology of La Selva has been the subject of one brief survey study (Alvarado 1985). Alvarado defines five geomorphologic units for La Selva (fig. 3.8): riverine alluvial deposits of Holocene and late Pleistocene origin; swamp alluvium, also of Holocene origin; Esquina andesites (EA); El Salto basalt-andesite (ESBA); and Taconazo basic andesites (TBA). These last three units are thought to be ancient lava flows. The alluvial deposits are tentatively divided into five terraces, T1–T5. Terraces 2, 3, and 4 are the best defined and most extensive and are discussed in detail in chapter 4.

Beginning several hundred meters inland from the rivers are the three lava flow areas, each with a distinct petrochemical signature. Two of these areas (EA and ESBA) are large and extend beyond the south boundary of La Selva, whereas

Fig. 3.4. Monthly precipitation patterns from selected tropical forest sites. La Selva, Costa Rica, 1963–1987; Ivory Coast, Africa, 1935–1973 (Bernard-Reversat et al. 1978); San Carlos, Venezuela, 1950–1958 and 1970–1978 (Jordan and Heuveldop 1981); Barro Colorado Island, Panama, 1977–1980 (Dietrich et al. 1982).

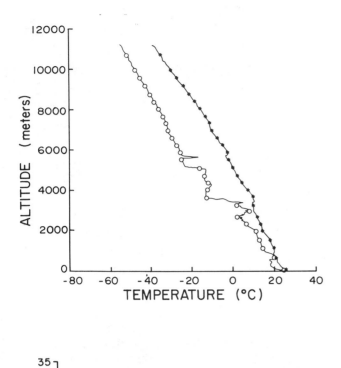

Fig. 3.6. Morning radiosonde profiles of air and dew point temperatures from La Selva during February 1988. Air temperatures are represented by * and dew point temperatures by *o*.

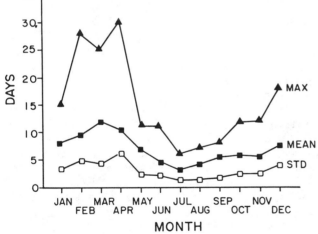

Fig. 3.7. Amount of time (days) between daily precipitation totals of precipitation >5 mm at La Selva, 1963 to 1987.

the third (TBA) is very small. ESBA is a pyroxene andesite with olivene and considerable quantities of sometimes large (3-cm) phenocrystic plagioclase inclusions. The petrochemistry of boulders from the ESBA flow reveals relatively low amounts of TiO_2, P_2O_5, Ba, and Zr but large amounts of Al_2O_3, FeO, MnO, MgO, CaO, Sr, and Cr in comparison with EA boulders. In contrast, boulders from EA are richer in Ba, Zr, and Cu but poorer in Al_2O_3, FeO, MgO, MnO, CaO, Cr, Ni, and Sc. EA is an andesite with microphenocrystic plagioclase inclusions and no olivene. The petrochemistry of TBA boulders is quite similar to ESBA except for P_2O_5 and MgO, which are present in amounts relatively similar to EA boulders. TBA is basic andesite that is somewhat unusual with its large pyroxene phenocrystal inclusions, previously known only from the Zurquí hills 40 km south of La Selva.

Topography

The historical lack of a high resolution (1:10,000 scale or better) surface map of La Selva has caused problems for many research efforts, particularly for those dealing with dispersed animal or plant populations and for ecosystem studies. A survey attempt to acquire a high-resolution topographic map of La Selva was largely unsuccessful (Petriceks 1956). More recently, as part of the acquisition effort for the Sarapiquí Annex, a topographic survey of 640 ha was completed. Surveys and accompanying maps of six La Selva trails are also completed (D. Clark pers. comm.), and an accurate, high-resolution (1:1,000 scale) topographic map exists for 34 ha of La Selva in La Guaria Annex (P. Sollins pers. comm.). A comprehensive topographic survey of La Selva was completed in late 1991. The survey grid was 50 m × 100 m and excluded only the three permanent research plots (16 ha total, see chap. 8). The data from the survey are now available on the geographic information system (GIS) at La Selva. For the first time a contour map of all of La Selva has been generated (fig. 3.9). Interpolation of the survey points by the GIS system also provides a surface feature map that dramatically illustrates the extent of the alluvial terraces and the topography of the highly dissected, ancient lava flows (fig. 3.10).

Research Needs. The single survey of geomorphology at La Selva is a good initial effort but provides only a coarsely resolved basis for integration of research results. The Alvarado

Table 3.4 Precipitation, interception, evapotranspiration, and net precipitation at La Selva Biological Station, August 1982–August 1983

Year	Month	Precipitation[a] (mm)	Interception[b] (mm)	Evapotranspiration[c] (mm)	Soil Water[d] (mm)	Net Precipitation[e] (mm)
1982	Aug	296	44	90	609	252
	Sep	502	74	217	594	427
	Oct	725	110	178	612	615
	Nov	409	61	211	607	348
	Dec	422	62	204	504	360
1983	Jan	282	39	143	604	243
	Feb	165	21	118	518	144
	Mar	127	18	180	588	109
	Apr	21	20	99	351	1
	May	508	77	188	531	431
	Jun	248	34	212	586	214
	Jul	655	100	184	644	555
	Aug	356	50	190	628	306
Summary		4,716	710	2,213	7,376	4,006
Average		363	55	170	567	308
Standard deviation		196	29	42	74	168
Maximum		725	110	217	644	615
Minimum		21	18	90	351	1

[a]Precipitation was collected on an event basis.
[b]Interception was calculated from regression of measured interception and precipitation.
[c]Evapotranspiration was calculated daily with the Penman-Monteith equation.
[d]Soil water was measured weekly using tensiometers.
[e]Net precipitation is precipitation minus interception.

Table 3.5 Average daily evapotranspiration from some tropical rain forests

Area	Forest Type[a]	Rate (mm/day)	Technique
(1) Costa Rica La Selva	lowland rain forest	5.9	Penman-Monteith
(2) Venezuela	montane rain forest	2.7	energy budget
(3) Amazon	lowland rain forest	2.7	atmospheric water vapor flux
(4) Amazon	lowland rain forest	4.0	Penman
(5) Amazon	lowland rain forest	3.5	eddy correlation
		5.1	Penman
(6) Amazon	lowland rain forest	6.0	tritiated water
(7) Kenya	montane rain forest	4.6	Penman
(8) East Africa	bamboo forest	2.9	gauged wier
	montane rain forest	4.1	gauged wier
(9) Panama	lowland rain forest	3.6	gauged wier
(10) West Java	lowland rain forest	4.1	water balance

Sources: (1) Luvall (1984); (2) Grimm and Fassbender (1981); (3) Marques et al. (1977); (4) Villa Nova et al. (1976); (5) Shuttleworth et al. (1984); (6) Jordan and Heuveldop (1981); (7) Dagg and Blackie (1970); (8) McCulloch et al. (1964); Pereira (1967); (9) Dietrich et al. (1982); (10) Calder et al. (1986).

[a]Forest type according to source references.

study should be built upon by work on fluvial land forms, drainage basin morphometry and hydrology, and water erosion on slopes. Combined with a comprehensive topographic survey and data on chemical weathering (see chap. 5), this information would provide a physical basis for integrating ecosystem research at La Selva.

Lack of an accurate topographic map of La Selva has limited many research efforts from evolutionary studies of particular organisms to ecosystem studies of nutrient cycling. The recently completed topographic map of La Selva will enhance the integration of terrestrial and aquatic research as well as significantly augment efforts to interpret results across disciplinary boundaries. Finally, the topographic map in the GIS system at La Selva provides researchers in all fields with a means for spatially explicit research in lowland tropical forests.

AQUATIC SYSTEMS

La Selva is drained by several streams that are tributaries of the rivers that bound the reserve (the Puerto Viejo and Sarapiquí, fig. 3.11). Most have their origins above La Selva at elevations of 150–500 m in Braulio Carrillo National Park. Many of these streams flow through essentially unbroken forest, whereas parts of others run through areas that are disturbed to varying degrees. These watershed characteristics, together with source water chemistry and the typical changes in streams between headwaters and mouth, yield a diversity of aquatic systems and rich opportunities for comparative research. The lower portions of La Selva's streams, near their confluences with the Puerto Viejo River, also differ in their behavior at flood stage. The Surá and Salto can flow backward, whereas the Sábalo-Esquina experiences spates.

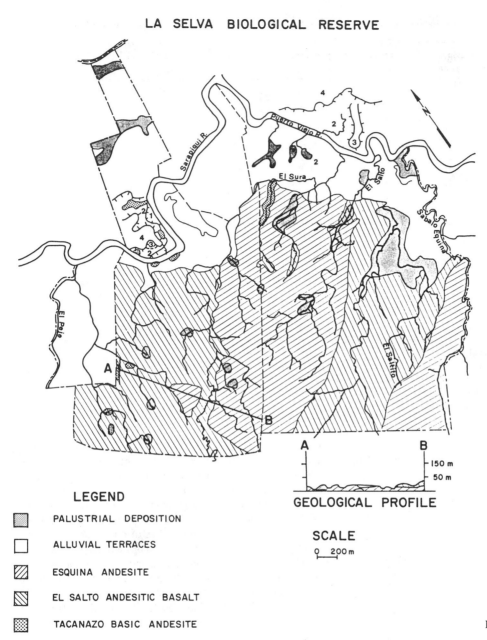

LA SELVA BIOLOGICAL RESERVE

LEGEND

▨ PALUSTRIAL DEPOSITION

☐ ALLUVIAL TERRACES

▨ ESQUINA ANDESITE

▨ EL SALTO ANDESITIC BASALT

▨ TACANAZO BASIC ANDESITE

GEOLOGICAL PROFILE

SCALE
0   200 m

Fig. 3.8. Geological map of La Selva.

La Selva's streams and their plant and animal life have been the subject of several theses and of ongoing research (e.g., Burcham 1985; Paaby 1988; Phillips 1987; Pringle et al. 1986; Pringle 1990, 1991; Stout 1978, 1980, 1982). From patterns emerging from this research we identify what appear to be exceptions to these generalities and point to promising avenues for future research.

In general, the streams of La Selva are narrow with rocky bottoms and have low pH and low conductivity near their source springs. They are wider, silty, circumneutral in pH, and high in conductivity in downstream areas (e.g., table 3.6). Several of La Selva's streams run through swampy areas that have little stable substrate (fig. 3.11; table 3.6). Stream waters have comparatively high concentrations of total and soluble P, NO_3, and NH_4 (table 3.6) and are generally richer in these nutrients (particularly P) in downstream areas. N:P ratios,

however, are generally lower than the suggested optimum for algal growth of 16:1 (Redfield 1958).

Exceptions to these overall patterns of stream chemistry point to local source waters that are affected by physicochemical differences at the landscape level. For example, portions of the Salto that flow through swampy areas have high P and NO_3 concentrations, reflecting substantial inputs from saturated terrain (table 3.6). Similarly, the Surá has very high P concentrations in the downstream portions (140 and 99 ppb, dry and wet season respectively, station 7, fig. 3.11), suggesting an external source somewhere in the watershed. High P values in Carapa Creek just before it joins the Surá (221 ppb, station 6, fig. 3.11) implicates this stream's watershed. Pringle (1991; Pringle and Triska 1991a, 1991b) hypothesizes that these pronounced changes in water chemistry are the result of groundwater inputs along geologic faults, with the

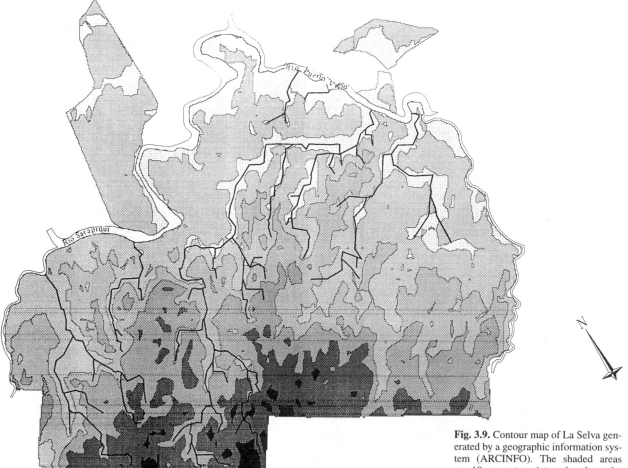

Fig. 3.9. Contour map of La Selva generated by a geographic information system (ARCINFO). The shaded areas are 10-m contour intervals where the lightest shade represents 20–29 m elevation and the darkest shade is 120–129 m elevation. Streams are shown as solid black lines.

groundwater coming from higher on the slopes of the central volcanoes of Costa Rica. Alternative explanations include locally distinct geological formations under the unusual watersheds (e.g., locally restricted sedimentary rock).

Where stream beds are rocky with emergent boulders, reophytes such as *Dicranopygium* (Cyclanthaceae) and *Cuphea* (Lythraceae) grow on the rocks. Primarily because of dense forest canopy over these generally small forest streams, productivity is likely to be largely heterotrophic via allochthonous input except where streams flow through natural or human-made open areas. For example, where the Surá flows through abandoned pasture, large mats of algae grow on the available substrata, suggesting that these portions of the stream may be autotrophic. Changes from heterotrophy to autotrophy probably occur sequentially over short segments of the stream where openings in the riparian canopy occur. Periphyton (attached algae, 90% of which are diatoms; Paaby [1988]) are responsible for most of the autochthonous productivity. The periphyton community is diverse (about seventy species identified to date) and capable of rapidly colonizing available substrate. Growth of periphyton in the Salto is limited by light and micronutrients, with supplemental macronutrients (N, P, N+P+Si, Si) having no effect on growth

(Pringle et al. 1986). Similarly, fertilization with N, P, N+P, Mo, and N+Mo failed to stimulate the growth of periphyton in the Saltito (Paaby unpublished data).

The aquatic insects of La Selva have been studied by J. Stout (1978, 1979b, 1981b) and R. Stout (1982), with emphasis on Naucoridae (Hemiptera). Stout (1981a) contrasted migration patterns and behavioral response of these insects to the spates of the Sábalo-Esquina with those to the backfloods of the Surá. Abundances and distributions of two species of bugs were unaffected by backflooding, but adults of both species were washed out by severe spates and population levels remained low for as long as one year following especially severe flooding. Research on the fish of La Selva's streams is surveyed by Bussing (chap. 14). Burcham's (1985) work on fish communities confirms differences among La Selva's streams in patterns of productivity. In forested streams, nearly half of the fish species ate terrestrial insects, and most of the remaining species were omnivorous. In contrast, the fish communities from a more open stream included 50% diatom specialists. There is, thus, at least preliminary evidence that communities of aquatic animals at La Selva are influenced by the spatial and temporal heterogeneity of habitats provided by these streams.

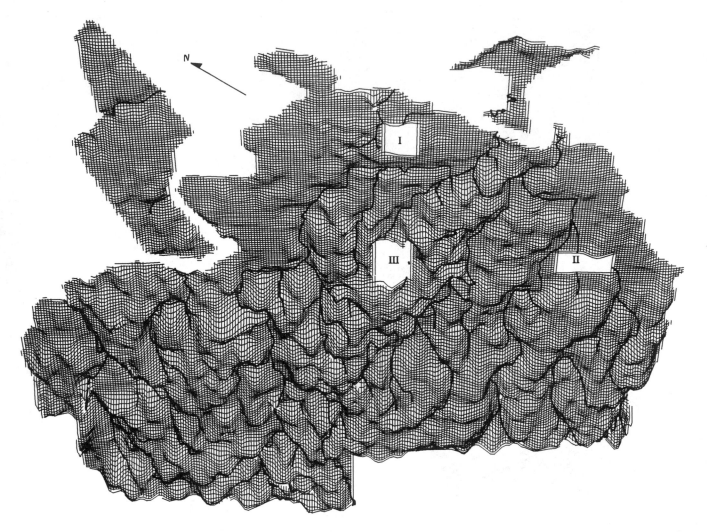

Fig. 3.10. Relief map of La Selva generated by a geographic information system that interpolated data from the topographic survey. Note that most of the alluvial terraces (flat area in the center foreground) are near the confluence of the Sarapiquí and the Puerto Viejo rivers.

Comparisons to Other Sites

Comparisons should draw upon long-term studies of aquatic systems that are integrated with terrestrial work to yield an understanding of patterns at the landscape level. Such a synthesis is not yet possible for La Selva, and data for other tropical sites are incomplete as well. The following should, therefore, be viewed as an initial glimpse at overall patterns.

The high nitrate concentrations found in streams at La Selva are very similar to those found in the Bombak River in Malaysia (Bishop 1973). Higher values, however, can be found in the Gambia River in Africa, in rivers in the Antilles, and even in some temperate rivers (table 3.7). In contrast to temperate landscapes, where N is often limiting in terrestrial systems but abundant in aquatic systems, lowland tropical landscapes (both terrestrial and aquatic) often have high N values (Gower 1987; Cushing et al. 1980). At La Selva both terrestrial and aquatic systems have high N values. Phosphorus, which is believed to cycle more efficiently than N in tropical forests (Vitousek and Sanford 1986), is found in surprisingly high concentrations in many tropical rivers. Phosphorus

levels in La Selva's streams are higher than those reported for Malaysia, Papua New Guinea, the Antilles, and Venezuela and higher than those commonly found in the Amazon basin (table 3.7). Overall, nutrient concentrations fall in the upper part of the range for tropical areas, with P levels many times higher than those considered by Bishop (1973) as polluted (>25 ug/l).

Research Needs

In light of the established focus on streamwater chemistry, one key area of research concerns nutrient dynamics in drainage areas, particularly swamps. Swamps may act as nutrient sinks (Yarbro 1983) or sources (Brinson et al. 1983) and, thus, influence forest nutrient dynamics and stream processes. The spiral "tightness" (nutrient retention time per unit of stream length) is closely related to sediment properties and biological activity within sediments and depends highly on the presence and distribution of stream primary producers (Elwood et al. 1983). Almost all streams at La Selva expel gas when bottom sediments are disturbed, suggesting anoxic stream bottoms.

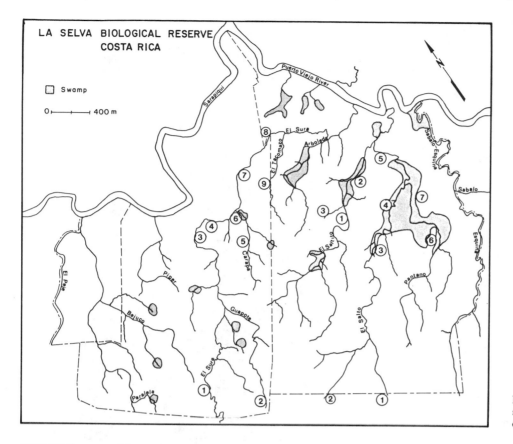

Fig. 3.11. Streams and numbered sampling sites at La Selva Biological Station, Costa Rica.

Table 3.6 Summary of the physical and chemical characteristics of El Salto stream and its main tributary, the Pantano stream

Site	Channel Width (m)	Stream Bottom	Season	n	Conduct (μS/cm$_2$)	Phosphate (ppb)[c]	Nitrate (ppb)	TP[a] (ppb)	N:P[b] (molar)
1	14–16	large rocks, boulders	wet	2	24	30	131	27	11:1
2	3–4	silt, mud	wet	2	15	12	24	9	6:1
3	10–12	rocks, pebbles, silt	dry	2	268	—	—	195	—
			wet	24	104[d]	62	101	93	2:1
4[e]	11–12	few rocks, sand, silt, mud	dry	2	150	122	—	134	—
			wet	2	167	107	63	87	2:1
4[f]			wet	2	420	—	—	360	—
4[g]			dry	4	480	125	98	232	1:1
4[h]			dry	2	—	282	122	266	1:1
4[h]			dry	2	—	627	105	333	0.7:1
5	15–16	mud, debris	dry	2	241	—	—	135	—
			wet	2	149[d]	76	55	112	1:1
6	3–4	silt, mud	dry	4	58	n.d.[i]	155	5	69:1
7	5–6	silt, mud	dry	2	98	n.d.	89	17	12:1
			wet	2	53	23	51	34	3:1

Sources: Site 3, dry season data (Phillips 1987); Site 3, wet season data (Pringle et al. 1986; Phillips 1987); all other site data (Paaby 1988).
Note: For location of sample sites see fig. 3.9.
[a]TP = total phosphorus
[b]N:P = nitrate-N μM:total-P μM. TP is used because acid hydrolysis eliminates the turbidity that interferes with the analysis for soluble-P.
[c]ppb = parts per billion (μg/l)
[d]n = 11
[e]El Salto main stem
[f]Dripping water from steep banks of stream channel
[g]Surface water flowing from swamp
[h]Trickle tributaries, samples taken during heavy rain.
[i]n.d. = none detectable

Table 3.7 Physical and chemical characteristics of different water courses in various tropical streams

Area	pH	Conductivity (μS/cm)	Alkalinity (mg/l)	Nitrate (μg/l)	Phosphate (μg/l)	TP (μg/l)	Source
Amazon region							
Rain forest streams	4.5	10	1.1			10	Furch 1984
Rio Solimões (nutrient rich area)	6.9	57	6.7			105	Furch 1984
Headwaters	4.0–6.6			0–150	0–110		Sioli 1975
Main channel	4.2–5.5			0–200	0–50		Sioli 1975
Venezuela							
Caura R.				63	2.48	20.47	Lewis 1986
Orituco R.	8.11		42.2	108	4.35		Cressa & Senior 1987
Limón R.	7.8	71	29.6			45	Lewis & Weibesahn 1976
Antilles							
St. Vincent Island	8.0–8.4			135–163	5–10, 19–40[a]		Harrison & Raukin 1976
Guadaloupe Island	circum neutral	41–43		500–2,200	16–68		Starmuhlner & Therezien 1982
Sri Lanka (Ceylon)							
Mountain streams	5.8–6.1	18–44.4		89–235	30–128		Starmuhlner 1984
Papua New Guinea							
Creeks	7.4	157	44				Petr 1983
Purari River	7.4–7.8	109–141	52–66	25–61.4	32	358	Petr 1983
Africa							
Gambia River	6.9–7.4	32–77		182–1652	10–54		Lesack et al. 1984
Others					78–230		Golterman 1975
Malaysia							
Gombak River	6.5–7.1	30–41	13–18	70–300	10–25		Bishop 1973
Tributaries	6.1–7.1	17–36	6.8–18.4	70–160	10–55		Bishop 1973
Costa Rica							
Q. El Surá upstream	3.8–5.8[c]	20–23	0–0.6	129–222	1–30	3–45	Paaby 1988
Q. El Surá downstream	6.3–7.4	116–255	39–76	78–178	99–160	86–195	Paaby 1988
La Selva (except Surá)	4.8–7.7	46–280	14.7–52.5	36–163	12–140	25–134	Paaby 1988

[a]Volcanic soils, St. Vincent Island.
[b]Latitude and climate similar to La Selva.
[c]One data point, rain event.

Therefore, anaerobic processes such as denitrification may be important mechanisms of N loss from the system and deserve further attention.

The apportioning of autotrophy compared to heterotrophy should be a focus of future research in stream productivity. The streams of La Selva are low-order streams, strongly suggesting that heterotrophy dominates biological activity. La Selva streams, however, are not homogeneous throughout their runs; they flow under dense shade and through open areas and gaps. This canopy heterogeneity imparts many microenvironments for aquatic primary producers that build the important autochthonous energy inputs to streams. Additionally, as Stout (1980) points out, La Selva streams are subject to large amounts of organic input from the drainage area. This C source is used mainly by shredding aquatic insects, suggesting that considerable quantities of insects should be found in these streams.

Stout and Vandermeer (1975) measured species richness of aquatic insects in comparable temperate and tropical streams (including the Quebradas Sábalo and Esquina at La Selva Biological Station). Previous studies had reported no differences in species richness between midlatitude and tropical streams (Patrick 1964; Patrick 1966), suggesting that aquatic insects are an exception to the rule that species richness is greater in tropical latitudes (MacArthur 1972). Stout and Vandermeer

(1975) suggested that the conclusions in previous reports were in error because of "inadequate sampling" and report that species diversity among aquatic insects is, indeed, significantly higher for tropical than midlatitude streams.

The aquatic habitats of La Selva comprise a very heterogeneous system. Variability in riparian vegetation, swamps along stream runs, and variation in allochthonous matter are ideal for the study of land-water interactions, nutrient spiraling, and stream productivity.

ACKNOWLEDGMENTS

We thank G. Hartshorn, H. Hespenheide, S. Horn, E. Leigh, S. Molden, K. Pringle, W. Reiners, J. Stout, several anonymous reviewers, and, especially, Lucinda McDade for comments that improved earlier versions of this chapter. We also thank M. Nelson for help drafting the figures and G. Jedlovec for the Geostationary Operational Environmental Satellite (GOES) image photographs. K. McElwain, B. Techau, and D. Jepsen worked with the original draft and with four subsequent revisions.

APPENDIX 3.1

The evaporation of water (ET) from the forest canopy can be expressed either as latent heat flux (LE, Wm^{-2}) for use in energy budgets or as a depth of water (ET, mm) for use in hydrologic budgets. To understand how and why ET is an important term in the hydrologic cycle in tropical rain forests and why removing the forest alters the hydrologic cycle, it is useful to first define ET in terms of the thermal energy budget (Wm^{-2}) of a forest canopy:

$$Q^* = LE + H + G \qquad (1)$$

where:

Q^* = the net radiation of the forest canopy
LE = the latent heat flux
H = the sensible heat flux
G = the canopy and soil heat storage

Values of Q^* for tropical rain forest canopies are generally about 70%–85% of solar radiation (Luvall 1984; Shuttleworth et al. 1984). Only a small fraction of this energy is used in photosynthesis, and the remainder is partitioned among LE, H, and G. In tropical rain forests LE is the most significant term, usually accounting for more than 70% of Q^* (Luvall 1984; Shuttleworth et al. 1984).

The evaporation of water from forest canopies is very difficult to estimate because it requires the measurement of many environmental factors of a forest canopy including net radiation, air temperature, vapor density deficits, wind, precipitation amount, and history, available soil moisture, canopy leaf area index, and plant stomata response. The measurement of these factors for a primary tropical forest canopy usually involves construction of towers above the canopy for routine micrometeorological measurements and poreometer measurements of stomata resistance. One common approach to measuring ET is the Penman-Monteith equation (Monteith 1973) which combines environmental factors with plant factors:

$$ET = \frac{s(Q^* - G) + (d\,p\,V/Ra)}{s + c(1 + Rc/Ra)}\,L \qquad (2)$$

where:

ET = evapotranspiration
L = latent heat of vaporization for water
s = slope of the saturation vapor pressure relationship
Q = net radiation flux
G = energy flux (storage to the ground or canopy)
d = density of air
p = specific heat of air
V = vapor density deficit of the air
c = psychrometric constant
Ra = aerodynamic resistance of the canopy
Rc = canopy resistance = stomatal resistance/leaf area index

4

Soils and Soil Process Research

Phillip Sollins, Freddy Sancho M., Rafael Mata Ch.,
and Robert L. Sanford, Jr.

Soils, together with correlative factors such as climate and topographic position, strongly influence plant growth, survival, and reproductive success. These processes determine, in turn, the composition and productivity of rain forest communities (e.g., Ashton 1964, 1976b; Webb 1969; Austin et al. 1972; Baillie et al. 1987) as well as the opportunities and constraints for agriculture and forestry (e.g., Sánchez 1976; Uehara and Gillman 1981). Thus, an adequate understanding of soil processes and properties is a prerequisite to any study of natural or managed terrestrial ecosystems.

Tropical soils are often lumped into a single category: red, infertile, and hardening irreversibly upon clearing. Nothing could be farther from the truth. The soils of the lowland humid tropics are as diverse as those of any other region. Some are, indeed, red and infertile. Some do harden irreversibly upon drying although these account for only about 7% of the tropical landmass (Sánchez 1976). But significant areas are occupied by young alluvial soils that are among the most fertile in the world. And young fertile soils are the rule throughout the volcanically active parts of the tropics where fresh, rapidly weathering mineral material is added constantly to the land surface. La Selva is, in certain regards, a microcosm of humid tropical soils. Although unusual in their physical properties, La Selva's soils span a large portion of the range in soil chemical properties found anywhere in the humid tropics.

Much information about the soils of the humid tropics has accumulated over the past few decades. To organize and use this information efficiently requires a soil classification system. Several are in use today in the tropics. Soil scientists in France, Brazil, and Australia have developed systems to use locally. A system developed by the Food and Agriculture Organization (FAO) has gained some international acceptance but provides only two hierarchical levels; thus, degrees of similarity among soils are difficult to judge. The only fully hierarchical classification system used widely is Soil Taxonomy (USDA Soil Survey Staff 1975, 1987, 1990). This system is used extensively within Costa Rica and elsewhere in Latin America and is used here.

Soil Taxonomy classifies soil mainly on the basis of diagnostic horizons (distinctive layers in the soil) and climate. Other criteria are used but vary widely with soil type and are too complex to describe here. In all, three soil orders are represented at La Selva: Entisols (young mineral soils lacking a B horizon), Inceptisols (relatively youthful soils with undistinguished B horizons), and Ultisols (well-weathered soils with an accumulation of clay in the B horizon). For those interested in learning more about Soil Taxonomy, an excellent text is available (Buol et al. 1980) in addition to the primary descriptions (Soil Survey Staff 1975, 1987, 1990).

An important feature of Soil Taxonomy is that the names themselves convey important information about the properties of the soils and their positions within the taxonomic hierarchy. For example, one of La Selva's soils is classified as an Oxic Humitropept. The *ept* in this name indicates that the soil is in the order Inceptisol. The appearance of *trop* in the name tells one further that the soil occurs in the tropics. The *humi* indicates that it contains large amounts of organic matter and the *oxi* that it has weathered to the point where its capacity to retain cations is quite low. Other terms relevant to La Selva's soils include *fluv* (for fluvial), *aqu* and *aquic* (for wet), *lithic* (for stony or in contact with rock), *psamm* (for sandy), *eu* (meaning rich in base cations), *dys* (for poor in base cations), and *histic* (for an accumulation of mineral-free organic matter). The term *andic* (or *and*) comes from the Japanese word for black and refers to soils, generally of volcanic origin, with distinctive claylike minerals that give rise to unusual properties such as high organic-matter content and water-holding capacity. Lastly, the term *typic* means simply "typical." Sollins et al. (1988) discuss in more detail the ecological information that can be inferred from the names used in Soil Taxonomy.

Soil Taxonomy differs fundamentally from the Linnaean system of biological nomenclature and classification. Because of biological processes, organisms can be grouped into well-defined units, called species, with relatively few intergrades among them. Soils, however, grade continuously from one type to another. Because of this continuity, the borders between even the highest-level taxa (orders) must be drawn arbitrarily, which means that the correction of a small error in laboratory results or a slight change in sampling location on the landscape may shift a soil into a different order. Although sometimes disconcerting to biological taxonomists, who are accustomed to clearer distinctions, at least between higher taxa, the fact that soils span a continuum is an inevitable and challenging feature of soil classification.

Soil survey and classification organize information on the chemistry and physics of soils, which can then be used to predict nutrient availability and biological activity, especially as they change in response to weather and to land-use practices. In this chapter we first describe the current state of knowledge concerning soil survey and classification at La Selva. Next we describe what is known of the chemical and physical properties of La Selva's soils. Last we place La Selva's soils in the context of tropical soils worldwide. Studies of soil biology and nutrient cycling and of the effects of soil processes on plant communities at La Selva are discussed elsewhere (chapter 5).

This chapter is intended to serve soil scientists and other researchers interested in working at La Selva and in comparing La Selva with other sites worldwide. It is impractical to try here to explain the basic concepts and terminology of an entire field of study (soil science plus large portions of geology and geomorphology). We have, however, attempted to define those terms that do not appear in standard dictionaries and whose meaning we feel is critical to an overall understanding of La Selva's soils and to write in a way that allows a reader to benefit from reading the chapter even if unfamiliar with the more technical terms. We would be most pleased if this chapter, coupled with a desire to research tropical terrestrial ecosystems, prompted some readers to dig deeper into the field. For those, we recommend general introductions to soils and nutrient cycling by Brady (1984) and Schlesinger (1991), more specialized texts and monographs by Sánchez (1976) and Uehara and Gillman (1981), and a particularly insightful essay on "Macrovariability of Soils of the Tropics" by Van Wambeke and Dudal (1978).

SOIL FORMATION AT LA SELVA

Soil scientists recognize five major factors of soil formation (Jenny 1941): climate, vegetation, topographic position, parent material, and soil age. La Selva spans only some 120 m of relief; consequently, climate is unlikely to have played a major role in creating soil variability at the site. Vegetation may have been an important influence, especially in the primary forest where canopy emergents persist for centuries and may create distinctive soil conditions under their canopies, but such effects have not yet been documented at La Selva. Topographic position is important locally at La Selva—distinctive soils have developed in poorly drained areas and along the small V-shaped streams—but, in general, parent material and soil age account for most of the differences between soil types at La Selva.

The nature of the parent material and soil age are determined by the nature and timing of the geomorphic processes that create the land surface. At La Selva, all parent materials resulted originally from volcanic activity. Ashfall, lahars (volcanic mud flows), and lava flows, three geomorphic processes unique to volcanic landscapes, have been important on the east slope of the Cordillera Central. At La Selva specifically, however, there is no evidence of ashfall. This lack is not surprising because the site lies well upwind of Costa Rica's volcanoes. (For example, as one moves westward from La Selva, the limit of subaerial ashfall is encountered about 2 km northeast of San Miguel toward La Virgen along the main highway from San José to Puerto Viejo [fig. 4.1].) Lahars, too, have played only a minor role in building La Selva's land surface. (Large areas of lahars are in evidence, however, east of the town of Río Frío toward Guápiles and to the north of the road between Chilamate and San Miguel [see fig. 4.1]) Lava flows, in contrast, have played a major role. Descending from the Cordille-

ran volcanos periodically during the last two million years, they have, at one time or another, covered almost the entire La Selva land surface. At the higher elevations, the flows have weathered slowly to form the so-called residual soils that we see now. At lower elevations, the lava flows were later covered over by alluvial and colluvial deposits.

Alluvial and colluvial processes operate in volcanic landscapes just as they do elsewhere. Lying atop the lava flows throughout the lower elevations at La Selva is alluvium deposited during overflow of the Río Sarapiquí and Río Puerto Viejo (fig. 4.1). These deposits can be presumed to span a fairly narrow range of mineralogy and chemical composition because all the material derives from flows and ejecta of the Cordilleran volcanoes and was well mixed during transport. Texture of the original deposits, however, appears to have varied considerably, depending on factors such as stream gradient and distance from the stream channel at the time of deposit. Because finer deposits weather faster than coarse ones, such textural differences have played a major role in determing the current state of the soils at La Selva.

Age is the other dominant factor determining soil properties at La Selva, especially for the alluvial soils. The alluvium was deposited as terraces, with the youngest occupying the lowest slope positions. The oldest terraces are widely regarded as Pleistocene in age, but dates are lacking. The lowest terraces flood periodically and contain an especially complex mosaic of soils because their properties are still so strongly affected by even small differences in age and original texture. The weathering sequence in this volcanic alluvium parallels in many ways those observed on lahars and volcanic ash deposits. Sancho and Mata (1987) referred to the terrace landscape as fluvio-volcanic to emphasize this point.

EARLY SOIL SURVEY AT LA SELVA

In the earliest reference to the soils of La Selva Petriceks (1956) divided Finca La Selva into red-brown lateritic soils occupying 85% of the area (613 ha), swamp soils (38 ha), and recent alluvial soils (35 ha).

In a second study researchers from the University of Washington distinguished three major landforms: two alluvial terraces (a lower and an upper) and an area of older soils toward the back of the property that they believed had formed in lava flows (Bourgeois et al. 1972). Modal profiles were described for nine soil series, all Entisols or Inceptisols, and samples were analyzed chemically.

In 1981 the Organization for Tropical Studies (OTS) began a conscious effort to stimulate ecosystems studies at La Selva. As part of these efforts, several investigators (G. Parker, M. Huston, P. Werner, P. Sollins, and P. Vitousek) carried out informal soil survey and classification in the areas where they worked. Such work set the stage for the first comprehensive survey of the soils of La Selva.

CURRENT SOIL SURVEY

In 1987 F. Sancho and R. Mata of the University of Costa Rica, under contract to R. Sanford and J. Denslow, surveyed the entire La Selva property. They described profiles by horizon based on standard soil survey methods (FAO 1968) at a

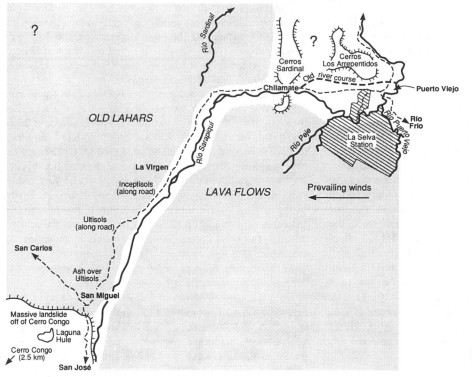

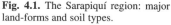

Fig. 4.1. The Sarapiquí region: major land-forms and soil types.

density of twenty-five observations per km² with a combination of pits (to 1.5 m depth) and auger holes (to 1.2 m depth). They then sampled the pits for subsequent chemical and physical analysis (table 4.1) and classified the soils according to the original version of Soil Taxonomy (USDA Soil Survey Staff 1975). Later, in 1989, two mapping units were resampled and reclassified according to the revised Soil Taxonomy (USDA Soil Survey Staff 1987, 1990).

Methods
Moisture content at −0.033 and −1.5 MPa was measured with a pressure plate on field moist, reconstructed cores and expressed on a unit dry-weight basis. Bulk density was measured with a coring device designed to minimize compaction. Parallel sampling of a subset of the pits with bulk density cans gave identical results. For all other analyses, soil was air dried and sieved (2 mm). Texture was analyzed by the Bouyoucos method after the soil had been treated with hydrogen peroxide to remove organic matter and dispersed by shaking for two hours with sodium hexametaphosphate. Designation of textural classes followed the rules of USDA Soil Survey Staff (1975). Soil pH was measured in water and potassium chloride (1 M) at a soil:solution ratio of 1:2.5. The pH in sodium fluoride was measured after two minutes. Organic matter (OM) content was measured by the Walkley-Black method. Oxalate-extractable aluminum and iron were determined by shaking 0.4 g of soil for four hours in the dark in 40 ml of extracting solution (mixture of ammonium oxalate and oxalic acid, 0.2 M with respect to oxalate, with proportions adjusted to give pH 3). Five drops of superfloc (0.4%) were added, the suspension was centrifuged (2,000 rpm, five minutes) and filtered (Whatman 42). The aluminum and iron concentrations in the supernatant were determined by atomic absorption spectrometry.

Cation exchange capacity (CEC) was measured with ammonium acetate (1 M) at pH 7 (USDA Soil Conservation Service 1972); the extract was analyzed for bases by atomic absorption spectrometry. For exchangeable acidity, samples were shaken for ten minutes in potassium chloride (1 M), filtered and titrated with sodium hydroxide (0.01 M). In the results shown in table 4.1, effective CEC (ECEC) was calculated as the sum of calcium, magnesium, potassium, and exchangeable acidity. Base saturation was then computed as the ratio of Ca + Mg + K to ECEC.

Phosphorus data in table 4.1 refer to surface soil (0- to 15-cm depth) samples gathered by Sancho and Mata during their 1987 survey. Samples were collected from all pits and auger holes, extracted on-site with dilute acid ammonium fluoride, and analyzed for phosphorus colorimetrically (Vitousek and Denslow 1987). Not all extracts were analyzed.

On the basis of profile observations and soils analyses, Sancho and Mata (1987) mapped most of La Selva at a scale of 1:10,000. (The Peje Annex was surveyed but no map was prepared because no base map was available.) In their overall arrangement of soil types, they recognized the same basic landscape units identified by Bourgeois et al. (1972)—older soils developed from lava flows toward the back of the property, alluvial terraces along the major rivers (Sarapiquí, Puerto Viejo, and Peje), and depositional zones in low-lying areas along creeks. They identified a total of twenty-three consociations and one complex (fig. 4.2). (Consociations are mapping units in which a single soil type occupies at least 75% of the area; similar but different soil types comprise the remaining area. Soil complexes group two or more dissimilar soil types that occur together in regular and repeating patterns too intricate to warrant separation at the scale of the soil survey.)

In what follows, we discuss each consociation and the single complex, emphasizing how and why they differ. We

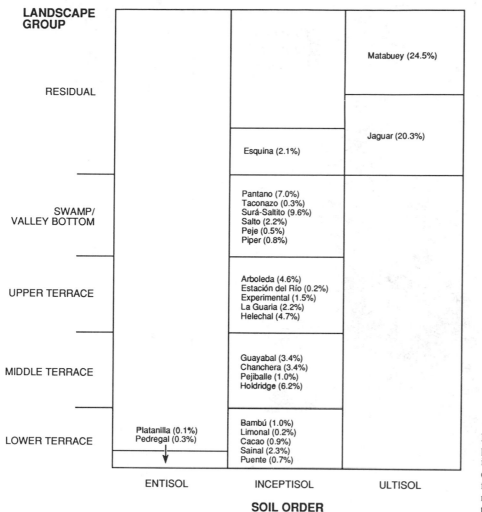

LANDSCAPE GROUP

Matabuey (24.5%)

RESIDUAL

Jaguar (20.3%)

Esquina (2.1%)

**SWAMP/
VALLEY BOTTOM**

Pantano (7.0%)
Taconazo (0.3%)
Surá-Saltito (9.6%)
Salto (2.2%)
Peje (0.5%)
Piper (0.8%)

UPPER TERRACE

Arboleda (4.6%)
Estación del Río (0.2%)
Experimental (1.5%)
La Guaria (2.2%)
Helechal (4.7%)

MIDDLE TERRACE

Guayabal (3.4%)
Chanchera (3.4%)
Pejiballe (1.0%)
Holdridge (6.2%)

LOWER TERRACE

Platanilla (0.1%)
Pedregal (0.3%)

Bambú (1.0%)
Limonal (0.2%)
Cacao (0.9%)
Sainal (2.3%)
Puente (0.7%)

ENTISOL   INCEPTISOL   ULTISOL

SOIL ORDER

Fig. 4.2. Soil consociations and complexes at La Selva. Area of each block is proportional to logarithm of area occupied by the mapping unit. Values are for La Selva Biological Station only; many of these units occur also outside the station.

describe areal extent, parent material, landform, and topography as well as salient chemical and physical characteristics. Elevations, where known, come from a 1982 survey along major trail systems of La Selva (D. A. Clark pers. comm.) or from a 1984 survey of a portion of the La Guaria Annex. Sites of major research projects are located with respect to the mapping units on which they occur.

The boundaries, profile descriptions, and physical/chemical data presented here supersede those in the report by Sancho and Mata (1987). In the years since Sancho and Mata completed their survey, several boundaries and relationships between consociations have been better defined. Topographic information has been unearthed that was not available to Sancho and Mata at the time of their survey. The Helechal consociation, in particular, has been resampled and its boundaries redefined. Material in this chapter thus represents the state of knowledge as of summer 1991, but boundaries will continue to change and new mapping units may be defined as more researchers become familiar with the soils of La Selva. (In late 1992 a topographic survey was completed that will, no doubt, substantially alter mapping unit boundaries.)

The Residual Soils
Three lava flows—the Esquina (andesitic), Salto (andesitic/basaltic) and Taconazo (andesitic/basaltic) (Alvarado 1985)—

give rise to the major "residual" soil consociations of La Selva (here, "residual" means simply "derived from lava flows"). Sancho and Mata (1987) recognized one mapping unit (the Matabuey consociation) on Salto lava and three on Esquina lava: the Arboleda at the lowest elevations near the upper river terrace, the Esquina on a lightly scarped area in the extreme southern corner of La Selva, and the Jaguar on the remaining strongly undulating to lightly scarped terrain. The Arboleda soils are more likely very old alluvial soils. In addition, we now recognize that the Matabuey consociation, as presently defined, derives from both Salto and Taconazo lavas.

The Matabuey (bushmaster, the largest of La Selva's poisonous snakes) consociation is the most widespread of La Selva's soils (fig. 4.3). It occupies the hilly upland portions of La Selva except toward the center of Old La Selva, where it is replaced by the Jaguar unit. To the west it extends to (and presumably beyond) the Río Peje. Chemically, the Matabuey soils are strongly acid (see table 4.1), organic matter rich, and highly leached, with a low degree of base saturation (30%) and a fairly large amount of exchangeable acidity. The increase in measured clay content between A and B horizons meets the criterion for an argillic horizon; thus, the soils were classified as Ultisols rather than Inceptisols. None of La Selva's Ultisols, however, show well-developed clay skins. Depth to C horizon varies. On steep slopes near the rivers, saprolite

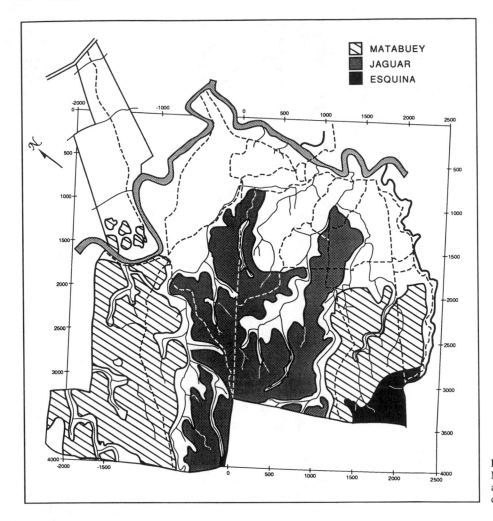

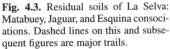

Fig. 4.3. Residual soils of La Selva: Matabuey, Jaguar, and Esquina consociations. Dashed lines on this and subsequent figures are major trails.

(highly weathered rock) outcrops at the surface. On ridgetops, however, B-horizon thickness can exceed 8.5 m (Parker 1985), an indication of the advanced stage of weathering of this soil. The majority of the unit was classified as a Typic Tropohumult because of its organic-matter content; inclusions are mainly Typic Dystropepts. Most of the TRIALS forestry plantations (see chapter 25) were established on hilly areas of Matabuey soils surrounded by younger soils derived from alluvial deposits. Studies by Luvall (1984) and Parker (1985) were conducted on Matabuey soils just outside the southern boundary of the Sarapiquí Annex.

The Matabuey unit, as described by Sancho and Mata (1987), occurs also near the Río Sarapiquí upstream from the ford and cable-car crossing, where it forms discontinuous cliffs along both sides of the river. It also occurs in hilly portions of the La Guaria Annex where the Taconazo lava flow was not (or is no longer) covered by upper-terrace deposits from which derive the soils of the Helechal consociation. Because the relationship between the Matabuey and Helechal units is complex, discussion is postponed to the end of this section after both units have been described.

The Jaguar consociation is the second most extensive unit at La Selva. Developed from the Esquina lava flow (andesitic), it occupies the central portion of the property between the Quebradas Salto and Surá see (fig. 4.3) and is bounded on

both sides by Arboleda consociation. Sancho and Mata (1987) found no chemical or physical differences between the Jaguar and Matabuey soils although they did note differences in color (see table 4.1). The two were separated mainly to maintain correspondence between lava flow and soil consociation.

Vitousek and Denslow (1987) compared P levels and plant growth on what they called "lava-flow-1" and "lava-flow-2" soils at La Selva. The "lava-flow-1" soil in their table 1 corresponds to the Jaguar consociation (plus Arboleda), their "lava-flow-2" soil to the Matabuey. For their plant bioassays (their table 2), lava-flow 1 corresponds again with Jaguar, but their lava-flow-2 soil came actually from a site within the Arboleda consociation. (Their old alluvial soil is in the Holdridge consociation.) All gaps used by Denslow et al. (1990) for outplanting experiments fell within the Jaguar consociation.

The Esquina (corner; name of creek) consociation is restricted to the extreme southeastern corner of La Selva (fig. 4.3) on the Esquina lava flow (andesitic). Except possibly for higher organic matter content, Sancho and Mata (1987) found no chemical differences between the Esquina unit and the Jaguar or Matabuey. Although the Esquina and Matabuey units derive from the same lava flow, the Esquina soils lack argillic horizons (for unknown reasons) and were, thus, classified by Sancho and Mata (1987) as Andic Humitropepts.

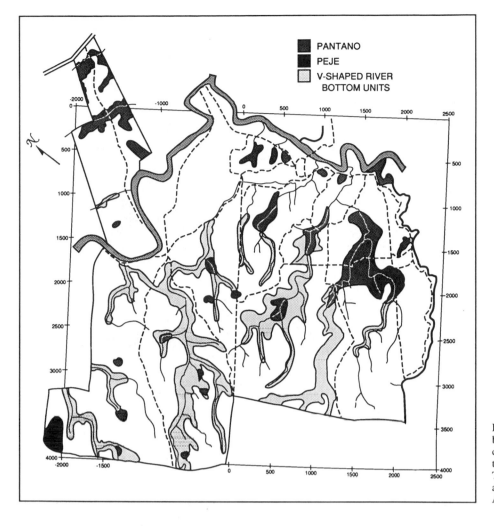

Fig. 4.4. Soils of the swamps and valley bottoms: Pantano complex, conso-ciations of the V-shaped valley bottoms (Surá-Saltito, Salto, Piper, and Taconazo), and the Peje consoci-ation (mainly off the map on the Peje Annex).

PANTANO
PEJE
V-SHAPED RIVER BOTTOM UNITS

Soils of the Valley Bottoms and Swamps

Throughout the hilly to undulating landscape that has developed on the lava flows and upper terraces, streams have formed small V-shaped valleys with small depositional zones at the bottom of each. Soils along the Quebradas Salto, Taconazo, and Piper are distinctive; the rest are combined into a single consociation (Surá-Saltito) named for the two largest streams along which this unit occurs (fig. 4.4).

The Surá-Saltito consociation, which accounts for the majority of the valley bottom soils, is extremely acid (pH 4.0–4.2), OM-rich, and very base poor. The soils are clayey and of heavy texture, colors are brown to yellowish-brown, and drainage is moderate. Rocks outcrop on 5%-10% of the area mainly near the Quebrada Surá. Generally the B horizon extends below 1.2 m, but in places a C horizon was encountered at <1.2–m depth that consists of saprolite derived from weathered lava. The dominant soils of this unit are Typic Humitropepts; inclusions are mainly in Lithic and Aquic subgroups.

The Salto consociation is found only along the Quebrada El Salto, upstream of the area where the creek traverses soils of the Pantano complex (see below). Colors are dark brown to dark yellow brown, and drainage is poor to moderate. Surface soil is acid, very base poor, and rich in OM. The C horizon, consisting of rock and saprolite, is usually encountered at <50

cm depth, and rocks are abundant throughout the profile. The dominant soil is a Lithic Humitropept.

The Taconazo consociation is restricted to the margins of the Quebrada El Taconazo along the western border of the original La Selva property. The Taconazo unit along with portions of the Matabuey derive from Taconazo lava, the oldest of La Selva's parent materials (Alvarado 1985). Stones are abundant in the upper parts of the profile. Surface soil is strongly acid and base-poor. Gray colors below a 22-cm depth indicate poor drainage; in fact, the water table rises in places to within 50 cm of the soil surface. This soil yielded quite high levels of extractable phosphorus (table 4.1). The dominant soil is a Typic Tropaquept.

Soils of the Piper consociation are moderately well drained and stone free. The soils are notably low in clay, thus, very readily worked. The surface soil is much less acid (pH 5.2) than in the Surá/Saltito, Salto, or Taconazo units. Unlike the other valley-bottom soils, those of the Piper unit are rich in bases and organic matter and, thus, presumably quite fertile. The unit floods periodically. The dominant soil is a Fluventic Humitropept.

The Peje consociation occupies an aggrading, U-shaped flood plain along the Río Peje at the extreme western boundary of the Peje Annex. Unfortunately, no base map was available for the Peje Annex at the time of the soil survey, so the

Table 4.1 Chemical and physical properties of La Selva soils

Mapping unit	Profile No.	Classification	Horizon	Depth (cm)	Munsell Color	pH H₂O	pH KCl	pH NaF	Organic Matter (%)	Ca	Mg	K	KCl Acidity	Sum of Bases	ECEC	CEC (NH₄OAc)	Base Saturation[b] (%)	P[c] (mg/kg)	Water Content −.3 MPa	Water Content −1.5 MPa	Bulk Density (Mg/m³)	Clay (%)
												(cmol(+)/kg)							(% by weight)		
Residual units																						
Matabuey[a]		Typic Tropohumult	A	0–18	7.5 YR 3/4	5.0	4.0	8.7	7.13	0.59	0.46	0.35	3.15	1.40	4.55	28.0	30.8	0.6	46.3	33.0	0.80	41
			Bt1	35–88	7.5 YR 4/4	4.8	4.2	9.7	2.70	0.53	0.21	0.19	1.85	0.93	2.78	20.1	33.4	—	38.1	31.2	0.84	62
Jaguar[a]		Typic Tropohumult	A	0–19	7.5 YR 3/4	4.5	4.2	8.6	6.75	0.54	0.53	0.29	3.50	1.35	4.85	29.0	27.9	0.7	39.8	32.8	0.69	40
Esquina	164	Andic Humitropept	Bt1	31–72	7.5 YR 4/4	5.2	4.6	9.7	1.40	0.33	0.14	0.13	1.90	0.60	2.50	21.3	23.8	—	39.5	31.0	0.86	60
			A	0–26	7.5 YR 3/4	4.3	3.7	8.0	11.93	0.79	0.92	0.42	4.10	2.13	6.23	42.6	34.2	1.7	45.9	29.8	0.66	31
			Bw1	54–117	7.5 YR 3/4	5.3	4.3	9.2	0.54	0.16	0.13	0.08	1.80	0.37	2.17	22.4	17.1	—	40.0	33.7	0.85	56
Valley bottom and swamp units																						
Surá-Saltito[a]		Typic Humitropept	A	0–19	10 YR 3/4	4.2	4.0	8.0	10.74	0.63	0.38	0.29	3.67	1.30	4.97	33.0	26.2	9.7	46.9	36.6	0.57	42
Salto	165	Lithic Humitropept	Bw1	38–99	7.5 YR 4/4	4.9	4.4	9.6	2.06	0.52	0.19	0.08	2.17	0.79	2.96	23.1	26.7	—	42.4	32.5	0.84	64
			A	0–12	10 YR 4/6	4.7	4.2	9.3	6.16	0.59	0.42	0.36	2.30	1.37	3.67	25.8	37.3	15.0	34.0	26.9	N.A.	19
Taconazo	127	Typic Tropaquept	Bw	12–45	10 YR 4/8	5.3	4.5	9.7	1.81	0.18	0.21	0.12	1.20	0.51	1.71	19.0	29.8	—	34.0	28.4	N.A.	42
			A	0–12	10 YR 4/2	4.4	4.1	9.3	3.02	1.30	0.79	0.36	2.50	2.45	4.95	32.4	49.5	3.6	52.9	45.7	0.44	11
Piper	126	Typic Humitropept	Bw1	12–22	7.5 YR 5/2	5.1	4.3	10.2	4.09	0.73	0.25	0.15	1.70	1.13	2.83	36.5	39.9	—	37.2	33.9	0.76	36
			A11	0–16	10 YR 3/3	5.2	4.7	9.8	9.38	11.50	2.67	0.71	0.50	14.88	15.38	38.6	96.7	6.4	45.2	32.6	0.72	4
Peje	167	Typic Humitropept	Bw1	35–108	10 YR 4/4	6.2	5.2	9.8	1.68	12.30	3.30	0.20	0.40	15.80	16.20	29.8	97.5	—	39.4	25.9	0.69	6
			A11	0–30	10 YR 3/3	5.5	4.6	9.5	5.70	8.00	3.92	0.71	0.50	12.63	13.13	35.0	96.2	7.4	49.3	30.9	0.82	10
Pantano	89	Histic Tropaquept	Bw	59–94	10 YR 4/3	6.0	5.0	9.5	1.21	11.40	3.86	0.13	0.30	15.39	15.69	33.0	98.1	—	44.1	29.0	0.74	20
			Ag	0–15	10 YR 3/1	4.9	4.2	9.5	10.36	1.98	0.71	0.51	1.60	3.20	4.80	37.1	66.7	—	56.1	34.7	N.A.	13
Pantano	145	Typic Tropaquept	Bg	15–100	10 YR 5/1	5.0	4.2	9.7	10.68	2.60	0.83	0.41	1.50	3.84	5.34	35.0	71.9	—	53.7	34.0	N.A.	23
			A	0–18	10 YR 3/2	4.9	4.3	8.8	7.64	3.38	1.67	0.56	1.10	5.61	6.71	29.2	83.6	—	43.4	30.8	0.45	33
			Bw1	18–55	10 YR 4/3	5.4	4.6	9.3	1.68	2.22	1.63	0.42	0.70	4.27	4.97	22.3	85.9	—	39.9	29.9	0.86	61
Units of the Río Sarapiquí terraces																						
Platanilla	119	Typic Tropopsamment	A	0–28	10 YR 3/2	5.3	5.2	9.4	4.89	11.25	2.84	1.73	0.30	15.82	16.12	25.3	98.1	4.7	31.0	17.6	0.84	4
Cacao	114	Fluventic "Andic" Eutropept	A	0–20	10 YR 3/2	6.2	5.2	9.7	5.36	13.00	2.54	0.92	0.20	16.46	16.66	23.0	98.8	—	37.5	26.2	0.81	5
			Bw	20–35	10 YR 3/3	6.9	5.4	9.9	0.74	9.25	2.04	1.89	0.30	13.18	13.48	18.8	97.8	—	42.1	10.1	0.89	10

Soil	No.	Classification	Horizon	Depth	Color																
Limonal[a]		Fluventic Eutropept/	A	0–10	10 YR 3/2	6.1	4.6	9.1	4.57	10.25	4.52	0.95	0.63	15.73	16.35	22.1	96.2	26.5	33.8	18.7	1.01	7
		Typic Tropopsamment/ Fluvaquentic Eutropept	Bw/ C1	30–45	10 YR 3/3	6.4	4.8	9.4	1.01	9.52	3.68	0.33	0.70	13.53	14.23	20.6	95.1	—	28.0	19.0	1.10	9
Bambú	115	Andic Eutropept	A11	0–19	10 YR 3/3	6.6	5.6	10.2	2.55	4.43	1.58	2.30	0.40	8.31	8.71	31.7	95.4	0.2	43.7	32.6	0.76	10
			Bw1	29–95	10 YR 3/3	7.0	5.3	9.8	0.00	11.50	3.54	0.46	0.20	15.50	15.70	30.0	98.7	—	56.3	27.8	0.78	10
Guayabal	116	Andic Dystropept	A11	0–16	10 YR 3/3	4.6	4.3	9.1	6.63	1.15	0.63	0.41	1.00	2.19	3.19	19.2	68.7	1.4	37.7	31.2	0.83	25
Chanchera	113	Andic Dystropept	Bw	48–97	10 YR 3/4	5.6	5.2	9.8	0.27	0.90	0.21	0.10	0.20	1.21	1.41	17.1	85.8	—	53.1	43.4	0.71	54
			A	0–9	10 YR 3/3	4.3	—	8.4	7.71	1.10	0.88	0.46	3.60	2.44	6.04	24.2	40.4	7.9	45.6	40.5	0.82	22
Helechal[a]		Oxic Humitropept/ Oxic	Bw1	27–71	10 YR 4/3	4.9	4.1	9.9	0.94	0.75	0.21	0.26	3.80	1.22	5.02	26.5	24.3	—	39.8	27.4	0.67	45
			A	0–10	10 YR 3/4	4.3	3.9	8.7	9.15	0.50	0.27	0.18	2.72	0.96	3.68	29.3	26.0	—	39.8	33.0	0.76	38
		Oxic Humitropept/ Oxic Dystropept	Bw1	10–32	10 YR 4/4	4.4	4.1	9.1	3.20	0.29	0.09	0.19	1.63	0.58	2.20	19.3	26.2	—	35.1	30.7	0.93	46
La Guaria	88	Fluventic Dystropept	A	0–22	10 YR 3/3	4.6	4.2	9.3	4.38	0.83	0.75	1.09	1.50	2.65	4.15	21.3	63.9	4.5	36.6	27.8	0.91	33

Units of the R. Puerto Viejo terraces

Soil	No.	Classification	Horizon	Depth	Color																
Puente	151	Typic Dystrandept	A11	0–21	10 YR 2/2	6.4	5.6	10.1	12.06	19.75	4.42	2.49	0.30	26.66	26.96	51.0	98.9	3.7	54.6	36.0	0.47	4
Holdridge[a]		Andic Humitropept	Bw1	33–104	10 YR 4/3	6.6	5.0	10.1	1.68	4.34	3.29	0.35	0.30	7.98	8.28	34.0	96.4	—	62.6	33.4	0.76	19
			A	0–22	10 YR 4/3	4.7	4.3	9.3	8.91	1.34	0.81	0.44	2.47	2.59	5.06	19.9	51.2	9.8	41.4	30.0	0.68	28
Pejiballe	107	Typic Humitropept	Bw1	26–77	10 YR 4/4	5.3	4.5	10.0	1.50	0.62	0.40	0.16	1.87	1.18	3.05	23.2	38.8	—	37.4	27.3	0.76	40
			A11	0–26	10 YR 3/3	4.7	4.3	9.6	6.27	1.03	0.46	0.36	1.60	1.85	3.45	19.6	53.6	0.6	43.6	34.9	0.65	30
Experimental	112	Andic Humitropept	Bw1	43–94	10 YR 3/4	5.3	4.7	9.6	2.28	0.98	0.33	0.26	0.50	1.57	2.07	19.6	75.8	—	41.8	34.0	0.89	40
			A11	0–15	10 YR 3/3	4.3	3.8	8.4	13.20	1.35	0.58	0.51	3.40	2.44	5.84	18.2	41.8	0.3	46.3	34.1	0.76	35
Estación del Río	152	Typic Dystropept	Bw1	91–134	10 YR 3/4	4.9	4.2	9.8	0.00	1.00	0.38	0.46	2.20	1.84	4.04	18.6	45.5	—	34.4	28.1	0.95	50
			A	0–15	10 YR 4/3	4.3	4.1	8.5	7.57	0.54	0.71	0.33	3.50	1.58	5.08	32.0	31.1	15.5	40.5	33.1	0.65	34
Arboleda	128	Typic Humitropept	Bw1	32–106	10 YR 4/4	5.1	4.2	9.5	0.60	0.32	0.25	0.09	3.00	0.66	3.66	22.0	18.0	—	37.8	29.4	0.90	54
			A	0–14	10 YR 4/3	5.3	4.9	8.9	4.03	11.00	2.96	0.41	0.40	14.37	14.77	34.9	97.3	1.5	42.4	33.9	0.55	6
			Bw1	14–93	10 YR 4/4	5.2	4.3	9.8	1.34	1.38	0.38	0.15	1.00	1.91	2.91	14.0	65.6	—	31.2	29.9	0.82	23

[a]Mean of multiple soil pits.

[b]Base saturation = (sum of bases/ECEC) × 100.

[c]Acid ammonium fluoride extraction.

soils of this area were described by Sancho and Mata (1987) but not mapped. A small area of Peje consociation lies within the Vargas annex and, thus, appears on the soil map (fig. 4.4). The Peje consociation is distinctive in that depth to the C horizon averages only about 90 cm. The dominant soil is a Typic Humitropept whose surface horizon is quite similar chemically to those of the Piper and Arboleda units. The A horizon is moderately acid, and base content is high. The B horizon of the Arboleda, however, is substantially more acidic and base-poor than that of either the Piper or Peje units.

The Pantano (swamp) complex is found in La Selva's swampy depressions. High water tables in such areas have caused reducing conditions, and OM has accumulated. The soils, all Tropaquepts, are typically gleyed and mottled. Three soil types recur regularly in a predictable but extremely complex pattern apparently unrelated to soil age or parent material. Histic Tropaquepts occupy areas where the water table is often above the soil surface and organic matter has accumulated in especially large amounts. Where the water table remains lower, soils are Typic Tropaquepts. Lithic Tropaquepts have formed wherever the streams have cut the valley floor down to the underlying bedrock. The three types form a similar, highly interdigitated pattern throughout the swampy portions of the property and were, thus, grouped as a complex by Sancho and Mata (1987) rather than mapped separately.

The Río Sarapiquí Terraces

The lower terraces along the Río Sarapiquí appear to have been formed as the river downcut through old alluvium. The new material was deposited atop the old during overbank flow. Currently, the river is downcutting mainly through the Taconazo lava flow (and through buried soils derived from it). The Río Sarapiquí shows little evidence of meandering although one abandoned channel can be seen along the base of the Cerros Los Arrepentidos (see fig. 4.1).

The lowest terraces comprise an active floodplain with an exceedingly complex mosaic of soil types (fig. 4.5). All of the lowest terrace soils are base-rich with pHs in the 6's. The Pedregal (rocky land) consociation consists of skeletal soils (rich in coarse fragments) with no B horizon. The dominant soil is a Typic Tropofluvent. (It was originally designated a Fluventic Troporthent by Sancho and Mata [1987] on the assumption that the unweathered cobbles in the C horizon represented a lithic contact, but, in fact, a lithic contact must be continuous or nearly so [USDA Soil Survey Staff 1987]). Despite their highly skeletal nature, these soils support a forest of large stature, mainly in the vicinity of a gravel-mining operation that is steadily destroying it.

The Platanilla (Heliconia patch) consociation lies along the Río Sarapiquí on the western side of the Vargas annex (see fig. 4.5). Soils are sandy textured without gravel or rocks. The dominant soil type is a Typic Tropopsamment.

The Cacao (cocoa) consociation occupies mainly the now-abandoned cacao plantation at the junction of the Ríos Sarapiquí and Puerto Viejo (see fig. 4.5). A weakly developed cambic B horizon overlies a C horizon that shows striking discontinuities, evidence of repeated depositional cycles that may indicate alternation of flooding by the two rivers. Textures range from silt to sand; rocks and gravel are absent. Base content and pH are high as in all the lower-terrace soils. The

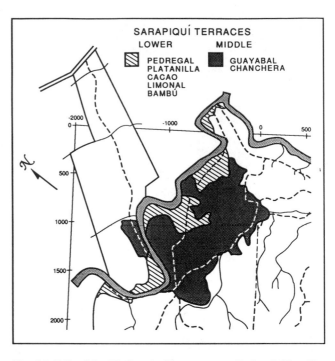

Fig. 4.5. Soils of the Río Sarapiquí lower terraces (Pedregal, Platanilla, Cacao, Limonal, and Bambú consociations) and middle terraces (Guayabal and Chanchera consociations).

dominant soil was designated a Fluventic "Andic" Eutropept by Sancho and Mata (1987) to emphasize its similarity to ash-derived soils.

The Limonal (grove of lime trees) consociation occupies the northern bank of the Río Sarapiquí at about 38- to 42-m elevation near the ford and cable crossing. Surface horizons, including a weakly developed (cambic) B, are sandy without gravel or rocks. A-horizon pH is only very slightly acid, and base-cation content is quite high (table 4.1). The C horizon shows evidence of distinct depositional events; in some areas stream-rounded cobbles are encountered at <1-m depth. The high sand content results in a bulk density that is unusually high by La Selva standards (1.0 g/cm³) and which distinguishes the Limonal from the Cacao consociation. The major soil type was classified as a Fluventic Eutropept by Sancho and Mata (1987) but could equally well be designated an "Andic" Tropofluvent. The lower-terrace plots of Robertson and Sollins occupy about one-third of this mapping unit, and the four soil pits upon which the previous description is based lie immediately adjacent to these plots.

Soils of the Bambú (bamboo) consociation occur only on the Las Vegas Annex along the Río Sarapiquí. The soils appear to differ from the Limonal only in that the B horizons of the Bambú are better and more deeply developed. Surface soil pH is 6.6 with a high pH (10.2) in sodium fluoride and low bulk density even in the B horizon (0.76 g/cm³). (The high NaF pH indicates the presence of large amounts of allophane [an amorphous gel-like clay mineral] or amorphous organo-alumino materials [USDA Soil Survey Staff 1987]. The low bulk density [<0.85 g/cm³], and pH [in water] >5 are corroborative.) The ECEC of this soil is quite low, however. Measured clay contents of 10% in the A and B horizons (table 4.1) could be an artifact caused by the presence of allophane. The dominant soil is an Andic Eutropept.

The middle terraces of the Río Sarapiquí have not flooded in recent memory; thus, these soils, although still quite fertile, are more deeply and extensively weathered than those of the lowest terrace. The Guayabal (guava grove) consociation lies on level terrain in the Sarapiquí Annex and is traversed by the Atajo and Las Vegas trails. Surface soil is moderately acid, and base content is moderate. A C horizon consisting of compacted alluvial sand underlies a deeply developed (cambic) B. A buried alluvial soil is encountered at about a 2-m depth. The dominant soil is an Andic Dystropept.

The Chanchera (pig barn) consociation occupies gently undulating land in the Sarapiquí Annex and is traversed by a trail of the same name. The surface Chanchera soil is more acid and base-poor than the Guayabal but still classifies as an Andic Dystropept. The B horizon is deep and well developed but contains abundant volcanic rocks in various stages of weathering.

The Helechal (fern patch) consociation occupies the highest terrace within the La Guaria Annex (fig. 4.6) and is the site of the upper-terrace plots of Robertson and Sollins. Relief is gently undulating; hills that protrude from this unit, especially near the Río Sarapiquí, create islands of much older soil (mapped currently as Matabuey). The following information is based on four pits, located adjacent to the Robertson-Sollins plots, that were described and sampled by Sancho and Mata in 1989. The Helechal surface soil (0.15 cm) is strongly acid (pH 4.4) and extremely base-poor (only 1.0 cmol (+)/kg total bases). Sodium fluoride pH is between 8.7 and 9.4, and bulk density is <0.85 g/cm³ throughout the top 1 m of profile. Levels of oxalate-extractable aluminum and iron are too low to meet the new definitions of Andic subgroups (USDA Soil

Survey Staff 1987). All four profiles meet the requirements for Oxic subgroups, based on clay contents calculated as 2.5 times moisture contents at −1.5 MPa. The soils classify, therefore, as Oxic Dystropepts or Oxic Humitropepts depending on organic matter content, which is highly variable (7–15 kg C/m² to a 1-m depth). A 2-ha portion of what was mapped originally by Sancho and Mata as Helechal consociation, lying to the south of the small stream that crosses Ave. Marañon (see fig. 4.5), is a relatively unweathered soil that classified as a Typic Hapludand upon further sampling in 1989 (F. Sancho and R. Mata unpublished data).

Described by Sancho and Mata (1987) as occupying a middle terrace, the Helechal, in fact, constitutes the uppermost terrace of the Río Sarapiquí. A topographic survey along Ave. Marañon and in the vicinity of the Robertson/Sollins plots shows that the Helechal unit ranges in elevation from 46 to 52 m, and lies some 1 to 10 m higher than the La Guaria consociation (described next). The sequence of deposits in which the Helechal soils have formed can be seen well in a roadcut along Ave. Marañon between locations 1000/−1340 and 1100/−1320 on the La Selva grid. Here a fine-textured surface deposit about 1-m thick, from which the Helechal surface soil derives, overlies a weakly cemented, well-weathered sand, which overlies, in turn, a buried profile developed from Taconazo lava. The cemented sand occurs at depth throughout the lower parts of the Río Sarapiquí valley and may represent the lahar (volcanic mud flow) from which the La Guaria soils have formed (described next). At least some of the Helechal soils (along with the La Guaria, described next) appear to have been used for mechanized rice production for about three years around 1960. Their extreme infertility may result in part from this land-use history in addition to natural processes of soil development.

The La Guaria (nearby bar and settlement named, in turn, after the national orchid of Costa Rica) is unique among La Selva's soils in that it derives from a lahar rather than an alluvial deposit or lava flow. Its exact age is unknown, but the lahar appears to be older than the upper-terrace deposits of the Río Sarapiquí but younger than the Taconazo lava. Angular to subangular well-weathered rocks of diverse composition are found embedded in a clay matrix in the subsoil at many locations; the angularity of the rocks indicates that they were deposited as a lahar rather than as alluvium, and their diverse composition rules out the possibility that they derive from a lava flow. Interestingly, the La Guaria land surface is lower than the Helechal, even though the latter is nearer the present river course, suggesting a depositional history that is, at the very least, complex.

La Guaria soils occupy nearly level land between the Helechal unit and the main road to Puerto Viejo (see fig. 4.6). Surface soil pH averaged 4.6, and bases summed to 2.5 cmol (+)/kg. The La Guaria consociation was classified as a Fluventic Dystropept by Sancho and Mata (1987) because organic matter content varied somewhat irregularly with depth and because a buried soil was found at 1.7-m depth in one of the two pits. Subsequent field work suggests that this pit was located in an area where Helechal soils lie atop the lahar, in which case the the La Guaria soils will need to be reclassified. The Canadian forestry trials and nearby Nelder plots (chap. 24) are located on this unit, as is one set of TRIALS plantings.

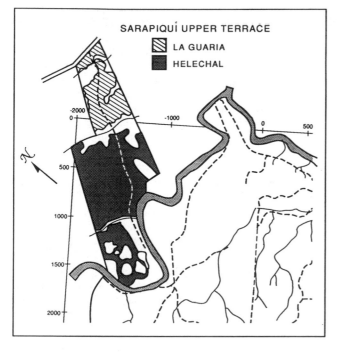

Fig. 4.6. Soils of the Río Sarapiquí upper terraces (Helechal and La Guaria consociations).

The Río Puerto Viejo Terraces

The Río Puerto Viejo has a gentler gradient than the Río Sara-piquí; thus, it has meandered more and deposited finer materi-als. At least four major terraces are evident near La Selva, of which only the lowest floods currently.

The Puente (bridge) consociation occupies generally flat land at about 29-m elevation near the mouths of the Quebradas Surá, Salto, Sábalo (fig. 4.7). Excellent profiles can be seen at the swimming hole below the bridge over the Quebrada El Salto. The entire profile is stone free but light textured. The high pHs in water and NaF (6.4 and 10.1) and low bulk density (0.47 g/cm³) of the A horizon suggest the presence of substan-tial amounts of allophane. A cambic B horizon with mottling extends below 1-m depth, suggesting moderately poor drain-age. The major soil type is a Typic Dystrandept; inclusions are mainly Aquic Dystrandepts. Low-lying areas along the Que-brada El Surá between the River Station and the laboratory complex, and extending upstream into the arboretum are cur-rently assigned to this unit purely on the basis of elevation.

The Sainal (a semi-invented word meaning area frequented by sainos, pigs or peccaries) consociation, as presently recog-nized, includes the level area bounded by the Cacao consocia-tion to the west and extending east to the slope break at about 120 m on the Sendero Occidental. The terrace lies at about 36 m elevation and floods at least occasionally. This consocia-tion was not described or sampled by Sancho and Mata (1987); their pit no. 107, labeled La Selva soil, is located in-stead within the Pejiballe consociation discussed later. (We have chosen not to use the name La Selva for any map unit to avoid difficulties in distinguishing between a soil from some-where at La Selva and a soil from a specific map unit named La Selva.) The only chemical data for the Sainal soils come from augered surface soil samples that gave a mean value for extractable phosphorus of 14.2 μg/g.

The Holdridge (trail named for Leslie Holdridge, founder of Finca La Selva) consociation occupies flat to gently undu-lating land on a middle terrace (36- to 38-m elevation) to the north and west of Quebrada Esquina (see fig. 4.7). Whether this and the Sainal mapping unit flood regularly is uncertain; both, however, did flood in 1970 (V. Chavarría pers. comm.). Soils are stone free and moderately weathered with intermedi-ate acidity and base saturation. The Río Puerto Viejo cuts through soils of this unit at Rafael's point, exposing many lay-ers of alluvium. Below these, beginning about 1 m above the dry-season water level, is a buried, highly weathered soil grading downward into Taconazo saprolite. The successional strips, OTS plots 4 and 5, and the well-drained portion of OTS plot 2 (M. Lieberman et al. 1985) are all on this mapping unit as are sites sampled by Robertson (1984), Sollins et al. (1984) and Werner (1984). The dominant soil is an Andic Humitropept.

The Pejiballe (a palm that produces edible fruit and heart) consociation occupies a middle terrace at about 42-m eleva-tion in the general vicinity of the laboratories and researcher cabins. This terrace has not flooded in historic times (I. Alva-rado and V. Chavarría pers. comm.). Soils near the dining hall and other facilities across the river may belong to this consoci-ation because they appear to be at the same elevation. A 10-m profile showing many depositional layers is kept continually exposed near the directors' house as the area is undercut

steadily by the Río Puerto Viejo. The surface soil is well drained and stone free with an A horizon that is strongly acid, base poor, and OM rich. Soils of this unit are less weathered than those of the Experimental, which it adjoins. It is similar to the Helechal and La Guaria on the Río Sarapiquí terraces but richer in OM and base metal cations. The dominant soil is a Typic Humitropept.

The Experimental (name of trail) consociation occupies undulating and moderately dissected upper-terrace land (40- to 50-m elevation) lying mainly to the northeast of the Que-brada El Surá. This unit occupies terrain more dissected than does the Helechal or La Guaria unit along the Río Sarapiquí or the Estación del Río farther upstream along the Río Puerto Viejo. Chemically, however, all these upper-terrace soils are similar. The dominant soil is an Andic Humitropept; inclu-sions are mainly Andic Dystropepts. The well-drained por-tions of OTS plot 1 (M. Lieberman et al. 1985) are located on this map unit.

The Estación del Río consociation (river station) occupies a small, gently undulating portion of the upper terrace at the site of the old field station and the Reserva Ecológica Rafael Chavarría (see fig. 4.7). The dominant soil is a Typic Dys-tropept.

The Arboleda (arboretum, in local usage) consociation forms a band of strongly undulating land just downslope from the Jaguar unit in the general vicinity of the Quebrada Ar-boleda (see fig. 4.3). The origin of the soils of this unit has been debated for several years. Although long viewed as lying well above the uppermost alluvial terrace and, thus, necessar-ily residual, these soils lack clay skins in the B horizon and are substantially less acid (pH 5.3 in the A horizon) and more base rich than the Jaguar or Matabuey soils. It now seems clear that this consociation occupies a very old river terrace, now so thoroughly dissected that it is barely recognizable as such. This reappraisal is based on the following reasoning. The highest parts of the Arboleda unit appear to correspond in elevation to the uppermost land surface to the north of the La Selva station along the main road between Puerto Viejo and Río Frío; cuts made through this surface during road relo-cation in 1989 revealed it as an old alluvial terrace now highly dissected. In addition well-weathered stream-rounded pebbles can be found exposed on trails to about 60 m elevation in nu-merous parts of what was mapped originally as Arboleda soils and in other parts of the property. The dominant soil of the Arboleda consociation was classified tentatively as a Typic Humitropept by Sancho and Mata (1987) because auger samples showed consistently dark colors, which suggest high OM levels. In fact, OM levels in the single soil pit sampled were fairly low (7.5 kg C/m²); thus, additional sampling is needed to classify these soils definitively.

Bioassays by Denslow et al. (1987) that show slight growth response to phosphorus and cations but not to nitrogen used an Arboleda soil. Of the natural tree-fall gaps studied by Vitousek and Denslow (1986), numbers 4 and 7 were on Arboleda soils; the rest were on Jaguar. OTS plot 3 (M. Lie-berman et al. 1985) is located on this map unit.

Major Unresolved Issues

Although work summarized here has given the first compre-hensive picture of La Selva's soils, it raises yet new questions

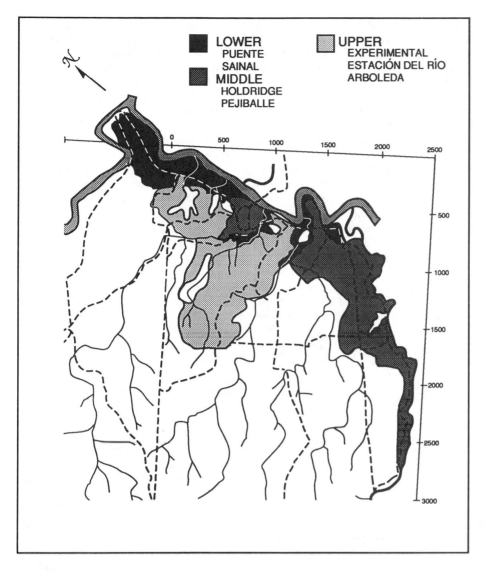

LOWER
PUENTE
SAINAL

MIDDLE
HOLDRIDGE
PEJIBALLE

UPPER
EXPERIMENTAL
ESTACIÓN DEL RÍO
ARBOLEDA

Fig. 4.7. Soils of the Río Puerto Viejo terraces: lower (Puente and Sainal consociations), middle (Holdridge and Pejiballe consociations), and upper (Experimental, Estación del Río, and Arboleda consociations).

about their genesis and classification. The age of these soils remains unknown. We really have no idea of the time it has taken a parent material of a given texture to weather to its current state in the warm, moist environment at La Selva.

The influence of parent material on the nature of La Selva's residual soils remains poorly understood. Vitousek and Denslow (1987) felt that the Esquina and Salto lavas have given rise to soils of different color and phosphorus availability. Sancho and Mata (1987), however, found few chemical and morphological differences between any of the residual soils. It must be noted that the geological survey of Alvarado (1985) is preliminary; thus, the boundary between Salto and Esquina lava must be regarded as tentative. Better mapping of the lava flows might yet show the corresponding soils to be quite different.

The correspondence between lava flow and soil type in the current mapping is further confused in that the Matabuey soils have developed in two rather different lava flows. Originally, Sancho and Mata (1987) described the Matabuey unit as developed solely from Salto lava. Subsequent field work has shown that, in the hilly areas immediately adjacent to the Río Sarapiquí including much of the La Guaria Annex and adjoin-

ing farms, the soils presently regarded as Matabuey consociation have developed from Taconazo lava, readily identifiable by the presence of large (>1 cm) feldspar crystals. (For example, a trench extending uphill from the parking area at the end of Ave. Marañon [1440, −1160 on the La Selva grid] uncovered Taconazo saprolite to about a 54-m elevation but revealed no sign of Salto lava overlying Taconazo.) Despite the difference in parent material, Sancho and Mata's data show no difference in chemistry or morphology between the two portions of the Matabuey unit.

The current mapping of hilly areas adjacent to the Río Sarapiquí is questionable. At present the hills are mapped as Matabuey and the surrounding areas as Helechal. In fact, the degree of soil development increases steadily as one moves upward toward the crests of these hills, presumably because the upslope positions are more stable. (For example, a trench at grid location 1440/−1160 in the La Guaria Annex showed that the B horizon increases in thickness from 40 cm to > 3 m and is increasingly reddish as one proceeds uphill. Numbers and size of iron nodules increase also.) Nonetheless, stream-rounded stones with obvious weathering rinds are found in the upper 50 cm of the profile, even at the tops of the highest hills (e.g., at a 60-m elevation throughout the adjoining property immediately southwest of the La Guaria Annex). Apparently, alluvium at one time covered this entire landscape then was removed to varying depths by fluvial and colluvial processes. At present, soils at the highest points are largely residual in origin, but those at lower positions on the landscape derive from some indeterminate mixture of residuum and old alluvium. Indeed, in some places highly weathered soil has moved downslope and now lies atop very young alluvium (e.g., location 1360, −1300 in the La Guaria Annex). Quite possibly the entire hilly area along the Río Sarapiquí (and, perhaps, the area south of the Río Puerto Viejo now mapped as Arboleda) would be mapped best as a complex.

Finally, the origin of the lahar that forms the parent material for the La Guaria soils and underlies at least portions of the Helechal consociation represents a major mystery. Given its degree of weathering, stratigraphic position, and position in the landscape, it seems possible that the lahar descended from the Cerros Los Arrepentidos rather than from the volcanoes of the cordillera, after the first of the lava flows that formed the landscape of Old La Selva but, clearly, well before the current cycle of deposition and downcutting by the Río Sarapiquí. Substantial field and laboratory work will be needed, however, to document fully its extent and origin.

SOILS PROCESS RESEARCH AT LA SELVA

Mineralogy

The mineralogy of a soil—that is, the nature and amount of clay-size mineral particles—is, perhaps, the single best correlate of its genesis, fertility, and physical structure. Unfortunately, mineralogy has been little studied at La Selva.

For the lower- and some middle-terrace soils (table 4.1), high pHs in sodium fluoride suggest the presence of allophane as an early weathering product of the volcanic parent materials. Werner (1984) described La Selva's soils as dominated by allophane, kaolinite (a weathered layer-silicate clay) and halloysite (a precursor of kaolinite) with quartz and gibbsite (aluminum hydroxide) present in similar, but presumably small, amounts. In fact, Werner's work dealt only with a middle-terrace consociation (Holdridge) and so agrees with current understanding.

For the one upper-terrace soil studied to date (Helechal at 0–15 cm depth), X-ray diffraction patterns indicate the presence of kaolinite, halloysite, and gibbsite (J. Borchers and J. Baham, Oregon State University, unpublished data). To see these patterns, it was necessary to remove iron sesquioxides, which precluded detection of goethite (a common iron oxide). Citrate/dithionite extractable iron averaged 10.5% (SE = 0.3%) in surface soil (0–10 cm depth) from this unit, however, suggesting a substantial accumulation of goethite (P. Sollins unpublished data).

On the basis of these admittedly meager data and patterns observed for volcanic soils elsewhere in the world, one can expect a certain weathering sequence at La Selva. Allophane should be abundant at early stages, converting later to halloysite, then to kaolinite, and lastly to gibbsite. Goethite should accumulate in large amounts throughout the weathering process. Some small amount of 2:1 layer-silicate clays must be present even in the most weathered La Selva soils to account for the fairly high levels of potassium chloride-exchangeable aluminum. It should be noted that most of these minerals are predominantly variable-charge; that is, the cation and anion exchange capacity (CEC and AEC) are affected strongly by the pH and ionic strength of the soil solution (Uehara and Gillman 1981; Sollins et al. 1988).

Physical Structure and Hydrology

Soil physical structure is important because it affects water-holding capacity (thus, plant growth), rates and pathways of water infiltration (thus, erosion and nutrient movement), and aeration and microbial activity (thus, turnover and transformations of carbon and nitrogen). The physical characteristics of La Selva soils are unusual and reflect their volcanic origin.

To compare La Selva soils with others, we used data provided by the USDA Soil Survey Staff (1975), selecting soils for which the temperature was isomesic or isothermic and the moisture regime was udic or perudic (table 4.2). Although most of the soils were from Puerto Rico and Hawaii, they may be fairly representative of the tropics because they were chosen by the Soil Survey Staff to represent a wide range of soil types. Comparison with a wider range of soil would be preferable but was not possible with the resources available.

As mentioned, La Selva's soils reflect their volcanic origin by their physical characteristics. Bulk density, for example, averages lower at La Selva than elsewhere (fig. 4.8a). Water-holding capacity (at −1.5 MPa) averages higher in La Selva soils, reflecting a somewhat different pore-size distribution than in soils elsewhere. Clay content tends to be lower at La Selva (fig. 4.8b), but the low clay content probably reflects the difficulty in dispersing the clay rather than any lack of clay per se. Although unusual by world standards, such physical characteristics are typical of volcanic soils (e.g., Warkentin and Maeda 1980; Sollins et al. 1983; Spycher et al. 1983; Russel and Ewel 1985).

Within La Selva, however, soil physical structure varies greatly. The weathered soils, such as the Ultisols Matabuey, Jaguar) and the more weathered Inceptisols (e.g., La Guaria, Helechal, Experimental, and Arboleda), have excellent structure. Because the soils are clay rich, fine pores are abundant and the soils retain large amounts of plant-available water. The clay is strongly microaggregated, however, with numerous large pores between the microaggregates. Water infiltrates readily into the soils through these large pores and drains freely. Thus, because of microaggregation, these soils behave like sandy soils in that they drain freely yet like clay soils in that they retain large amounts of plant-available water.

At La Selva, aggregation has been studied explicitly only on the Helechal consociation. Strickland et al. (1988) compared aggregation and aggregate stability of the dominant Helechal soil with that of five U.S. soils and found that the Helechal was the most extensively and stably aggregated. At least three factors could contribute to the extremely stable aggregation. Given the mineralogy of this consociation, binding by iron and aluminum oxides and hydrous oxides is probably important (see Tisdall and Oades 1982). Charge density is low because the soil pH is near the point of zero charge; thus, there is little tendency for the clay particles to repel one another (cf. Sollins et al. 1988). Last, the high OM content of the Helechal

Table 4.2 Chemical and physical properties of tropical soils worldwide.

Classification	Location	Land Use or Vegetation Type	Horizon	Depth (cm)	pH H$_2$O	pH KCl	Delta pH (H$_2$O-KCl)	Organic carbon (%)	K	Ca	Mg	Sum of Bases	CEC (NH$_4$OAc)	Water content (−1.5 MPa) (% by weight)	Bulk density (Mg/m³)	Clay (%)
										cmol(+)/kg						
Typic Acrustox	Brazil	Secondary forest	A1	0–10	5.0	4.2	0.8	3.24	0.15	0.42	0.32	0.91	15.0	24.0	—	71.6
			B1	30–65	5.0	4.7	0.3	1.35	0.01	0.02	0.02	0.06	6.0	24.1	—	72.3
Tropeptic Haplorthox	Puerto Rico	Pasture	Ap	0–15	5.6	5.0	0.6	4.30	0.90	8.30	1.50	10.80	20.5	31.9	0.96	77.1
			B21	15–33	4.7	4.6	0.1	1.75	0.50	1.80	0.20	2.50	10.6	33.6	1.07	76.8
Aeric Tropaqualf	Puerto Rico	Pasture	Ap	0–18	4.6	4.0	0.6	1.48	0.10	9.40	3.00	12.80	20.7	21.7	1.20	40.5
			B21t	30–43	4.8	3.5	1.3	0.54	0.10	5.10	4.40	10.30	18.5	22.9	1.29	48.4
Oxic Dystrandept	Maui	Pasture	Ap	0–20	6.2	5.4	0.8	8.87	1.70	17.50	6.30	25.80	53.7	41.0	0.70	—
			B21	20–41	6.6	5.8	0.8	3.51	1.80	11.80	5.40	19.00	40.6	49.1	0.70	—
Typic Eutrandept	Maui	Cultivation	Ap	0–23	6.7	5.8	0.9	5.81	6.10	35.80	11.70	54.00	61.1	35.2	0.87	23.7
			B2	41–63	7.4	6.4	1.0	2.58	0.30	42.70	12.30	61.30	63.1	43.2	0.76	13.9
Typic Hydrandept	Hawaii	Old field?	Ap	0–18	5.4	4.5	0.9	11.70	0.50	9.40	2.60	12.80	53.1	101.9	0.51	—
			B21	18–36	5.2	4.5	0.7	6.55	0.20	0.80	0.30	1.50	33.7	154.5	0.33	—
Typic Vitrandept	Hawaii	?	A1	0–10	6.6	5.5	1.1	5.27	1.30	7.50	1.90	10.90	27.5	19.4	0.78	—
			B2	10–20	6.5	5.6	0.9	2.12	0.30	1.70	0.30	2.40	18.6	23.7	0.75	—
Typic Dystropept	Puerto Rico	Old field?	Ap	0–15	4.2	3.5	0.7	2.28	0.30	1.60	0.70	2.60	18.7	22.7	1.20	40.2
			B	15–33	4.6	3.7	0.9	1.35	0.10	3.00	0.70	3.80	16.6	22.9	1.11	38.0
Typic Eutropept	Puerto Rico	Old field?	Ap2	3–15	5.0	3.7	1.3	1.77	0.40	13.20	9.40	23.20	35.0	22.5	1.26	32.0
			B	15–33	5.0	3.6	1.4	0.73	0.20	14.60	13.20	28.40	38.2	19.8	1.17	21.3
Typic Acrorthox	Mayaguez, PR	Old field?	A1	0–28	5.1	4.3	0.8	6.04	0.10	1.30	0.10	2.90	25.4	26.5	1.08	54.5
			B1	28–46	5.0	4.4	0.6	2.04	0.00	0.10	0.00	0.10	12.1	22.8	1.18	57.7
Haplic Acrorthox	Belem, Brazil	Evergreen tropical forest	A1	0–4	4.1	3.5	0.6	2.76	0.08	0.07	0.09	0.27	6.7	13.1	—	34.7
			B21	19–87	4.7	4.2	0.5	0.58	0.01	0.01	0.01	0.05	2.8	20.8	—	64.2

Table 4.2 (continued)

Classification	Location	Land Use or Vegetation Type	Horizon	Depth (cm)	pH H₂O	pH KCl	Delta pH (H₂O-KCl)	Organic carbon (%)	K	Ca	Mg	Sum of Bases	CEC (NH₄OAc)	Water content (−1.5 MPa) (% by weight)	Bulk density (Mg/m³)	Clay (%)
									cmol(+)/kg							
Tropeptic Eutrorthox	Vega Alta, PR	Pasture	Ap	0–20	5.6	4.9	0.7	2.73	0.60	8.10	2.10	10.90	15.1	26.7	1.22	73.1
			B21	20–46	5.4	4.7	0.7	0.94	0.60	4.20	1.60	6.50	9.5	30.1	1.25	81.2
Typic Gibbsiorthox	Kauai	Pasture	Ap1	0–38	5.2	4.2	1.0	3.88	0.10	0.50	0.60	1.30	14.6	30.9	1.12	—
			B21	48–76	5.0	5.1	−0.1	1.58	0.00	0.00	0.60	0.60	4.8	33.4	1.12	—
Tropeptic Umbriorthox	Kauai	Pasture	Ap	0–23	5.9	4.9	1.0	4.39	0.80	2.20	2.00	5.20	19.8	34.2	0.91	—
			B21	23–53	5.4	5.4	0.0	1.72	0.20	1.00	0.60	2.00	9.4	40.3	0.92	—
Typic Torrox	Honolulu, HI	Cultivated	Ap1	0–15	5.7	5.0	0.7	1.92	0.40	6.00	3.00	9.70	—	22.3	—	—
			B21	38–66	6.4	5.8	0.6	0.50	0.00	3.00	2.20	5.80	—	22.1	—	—
Oxic Plinthaquult	Toa Baja, PR	Cultivated	Ap	0–25	4.2	3.5	0.7	3.25	0.20	1.10	0.60	2.00	14.2	25.6	1.19	55.7
			B1t	25–33	4.2	3.2	1.0	1.18	0.10	2.60	0.40	3.20	12.7	32.2	1.29	76.4
Typic Tropohumult	Barranquitas, PR	Pasture	Ap	0–15	4.6	3.6	1.0	5.20	0.50	2.40	3.10	6.00	23.6	29.8	0.90	67.2
			B21t	15–28	4.5	3.5	1.0	2.01	0.10	0.10	1.90	2.10	19.9	34.1	1.02	81.8
Typic Tropudult	Puerto Rico	Pasture	Ap	0–15	4.7	3.7	0.0	2.40	1.60	2.10	0.90	4.70	14.1	20.7	—	47.9
			B21t	15–41	4.6	3.4	1.2	0.67	0.50	0.10	0.50	1.10	12.8	23.7	—	55.1
Aquentic Chromudert	Gurabo, PR	Orchard	Ap	0–18	4.8	4.0	0.8	1.96	0.40	5.80	11.00	17.60	25.5	22.4	1.20	47.7
			A12	18–38	5.3	4.4	0.9	1.17	0.20	9.90	17.10	28.20	26.9	26.1	1.15	61.4
Udic Chromustert	Lajas, PR	Old field	Ap	0–18	5.8	—	—	1.64	0.50	23.30	16.40	40.50	41.7	20.4	1.36	51.6
			A12	18–36	5.9	—	—	1.29	0.50	23.60	17.50	42.10	41.2	23.5	1.41	55.0

Source: All data based on compilation by USDA Soil Survey Staff (1975).

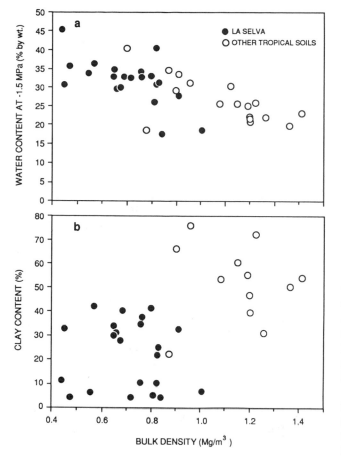

Fig. 4.8. Physical properties of La Selva soils and of other humid tropical soils. La Selva data are from table 4.1; data for other tropical soils taken from compilation by USDA Soil Survey Staff (1975); see text: (1) water-holding capacity at -1.5 MPa matric potential (dry weight basis) compared to bulk density; (2) clay content compared to bulk density.

is probably important, but this could be both a cause and an effect of the stable aggregation. Binding by allophane is probably important in the less-weathered soils at La Selva, but this has not yet been documented.

Because of strong aggregation, the texture of La Selva soils can be exceedingly difficult to measure. For example, field-moist Helechal surface soil shaken in sodium hexametophosphate solution gave a clay content of 73% (Strickland et al. 1988), whereas air-dried Helechal soil treated similarly yielded a clay content of only 38% (table 4.1). Sollins (1989) confirmed that air-drying of Helechal soil caused nearly irreversible aggregation of clay-sized particles.

Aggregation alters water flow through the soil by creating large pores (macropores) between the aggregates. Root decay also creates macropores as does the activity of soil animals. Dye-staining patterns for the Helechal soil give visual evidence of the extensive macropore network (Sollins and Radulovich 1988). A more quantitative technique (Radulovich et al. 1989) showed that, of the total porosity of 0.62 m³/m³, about 13% was the result of macropores (those pores draining between 0 and -3 kPa).

Macroporosity affects rates of nutrient leaching. To the extent that water flows through the macropores and bypasses the micropores of the soil matrix, the nutrients already in the soil matrix tend not to leach, and nutrients in litter leachate and throughfall tend not to enter the soil matrix (White 1985; Sollins 1989). Solute exchange between the rapidly moving by-pass flow and the relatively immobile micropore water within the soil matrix was studied on the Helechal consociation by Sollins and Radulovich (1988). Solute breakthrough curves showed that more than two pore volumes of $CaCl_2$ solution must flow through the soil before $CaCl_2$ stops diffusing into the fine pores. This result implies that equally large volumes are needed to leach nutrients from the fine pores. Mc-Voy (1985) and Seyfried and Rao (1987), working with a less-weathered Typic Dystropept under forest at Turrialba, Costa Rica, reported a similar, although somewhat less-pronounced, tendency for the soil to resist solute entry and leaching.

Measuring leaching rates is problematic in soil in which water drains mainly through macropores rather than through the soil matrix. Tension (porous-cup) lysimeters sample mainly the matrix water and give no indication of flow volumes; large zero-tension lysimeters collect the water actually draining the profile but are notoriously unreliable. Radulovich and Sollins (1987) improved water-collection efficiency of zero-tension lysimeters in the Helechal soil by using very large lysimeters and by pressing their rims upward into the soil, thus creating a saturated zone that promoted drainage into the lysimeter.

Because water drains rapidly through the macropores, it infiltrates readily into the soil instead of ponding on the soil surface or running off. On the Helechal soil, for example, initial infiltration rates averaged 3,900 mm/hr although they were highly variable (Radulovich and Sollins 1985). Erosion rates are, thus, low except on steep slopes and on trails where foot traffic has compacted the soils. The structure of this soil can be damaged, however; light foot traffic, for example, compacted the Helechal soil enough to decrease penetrability by 50%–70% (Radulovich and Sollins 1985). The lower-terrace soils contain less clay (table 4.1) and appear to have a less stable structure; lower-terrace sites maintained vegetation-free for several years have undergone considerable erosion (P. Sollins and G. P. Robertson, unpublished data).

Charge Chemistry, CEC and pH

As mentioned, La Selva's soils are rich in variable-charge constituents (certain clay minerals and organic matter) whose CEC and AEC depend on the pH and ionic strength of the soil solution (see Uehara and Gillman 1981; Sollins et al. 1988). Some indication of this effect can be seen by plotting actual CEC (calculated as the sum of exchangeable bases plus exchangeable acidity) against CEC as measured with ammonium acetate (1 M) at pH 7. Although the two measures correlate (fig. 4.9), the ammonium acetate method consistently overestimates actual CEC, sometimes by a factor of 6. There are two reasons: (1) CEC increases with pH and is, thus, much higher at pH 7 than at actual soil pH; (2) CEC increases with ionic strength and is, therefore, much higher in a 1 M solution than in the actual soil solution (about 10^{-3} for the Helechal soil; Sollins and Radulovich 1988). This striking effect of measurement technique on the value of a critical soil parameter should serve as a warning that the limitations of techniques must be understood when discussing results of soils

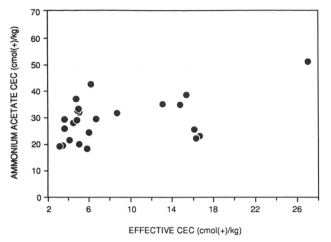

Fig. 4.9. CEC of La Selva soils as measured with molar ammonium acetate at pH 7 relative to actual CEC (sum of base plus acid cations). Data from table 4.1.

analyses, and that, in any case, methods must always be described carefully.

In terms of pH and cation levels, La Selva's soils can be grouped into two categories: Those with pH <5 tend to have low levels of exchangeable base-metal cations (K, Ca, Mg) and high levels of exchangeable acid-metal cations (H^+ and Al); those with pH >5 are more base rich and have lower levels of exchangeable acidity (fig. 4.10). Sum of bases and base saturation both increase with pH, but base saturation levels off at 100% above pH 5.0, which is to be expected because levels of exchangeable Al tend to be very low in the less-acid soils (fig. 4.10). La Selva's most base-poor soils (Helechal, Jaguar, and Matabuey) average 1.0–1.4 cmol (+)/kg in the A horizon (table 4.1), well above the lowest values reported elsewhere (table 4.2, see also the compilation by Lopes 1984). Whether cation levels influence productivity at La Selva is an important but unresolved question (see chap. 5).

CEC and pH are linked intimately in variable-charge soils (fig. 4.11); if pH changes seasonally or in response to management, so will CEC. Soil pH varies because of shifts in the balance between rates of processes that generate H^+ (sources) and rates of those that consume it (sinks). The major H^+ sources are plant nutrient uptake, root respiration, decomposition of organic matter to CO_2 and organic acids, and nitrification; weathering of soil minerals provides the largest H^+ sink although denitrification and nutrient release during decomposition can also be important (Van Breemen et al. 1983; Binkley and Richter 1987). There are limits to the extent that pH can shift because weathering (the major H^+ sink) proceeds at a rate roughly proportional to H^+ concentration in solution. Thus, for example, as pH drops because of increased nitrification, the higher H^+ concentrations cause weathering reactions to speed up, thereby increasing the rate at which H^+ is consumed.

Work on the Helechal consociation illustrates some of the large shifts in pH that can occur in response to disturbance. Ammonification rates are high in this nitrogen-rich soil, but vegetation takes up large proportions of this ammonium, and only a small fraction is nitrified. When plots were kept

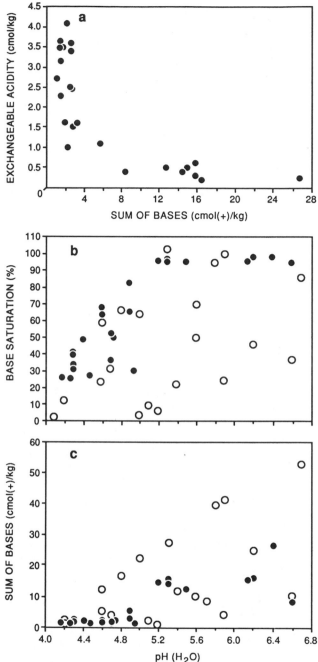

Fig. 4.10. Exchangeable base-metal and acid-metal cations in soils of La Selva (solid circles; data from table 4.1) and other tropical sites (open circles; data from table 4.2): (1) exchangeable acidity (acid cations) compared to sum of base cations; (2) base saturation compared to soil pH in water; (3) sum of base cations compared to soil pH in water.

vegetation-free even for a few months, pH dropped nearly one unit to below the point of zero charge (PZC) (Sollins et al. 1988; Robertson 1989). Laboratory experiments confirmed that the pH drop was the result of increased nitrification, triggered, in turn, because uptake had stopped while mineralization had increased (G. P. Robertson and P. Sollins unpublished data). These large pH drops are important because they mean that (1) base cation levels (K, Mg, Ca), already low, could

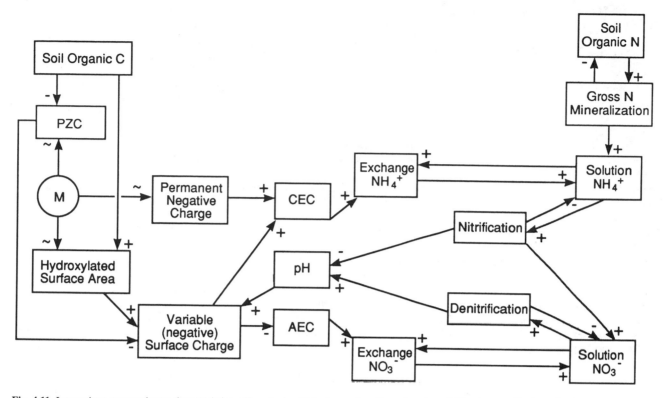

Fig. 4.11. Interactions among charge characteristics, pH, and microbial nitrogen transformations in variable-charge soils (from Sollins et al. 1988). A "+" indicates that the interaction increases the rate or value); a "−" indicates the opposite effect; a "~" indicates that effects are variable. PZC is point of zero charge.

drop even farther as the remaining cations are displaced from exchange sites; (2) CEC decreases, which reduces the ability of the soil to retain cations added in fertilizer or mulch; (3) AEC increases and may even exceed CEC, in which case ammonium will be more mobile than nitrate, the reverse of the situation in the soils from which most of our knowledge of nitrogen dynamics derives; and (4) problems of aluminum toxicity are exacerbated.

Soil pH may also significantly affect soil physical structure. As pH changes, the charge on clay surfaces changes, as does the tendency for the particles to repel each other (Uehara and Gillman 1981; Sollins 1989). From a practical standpoint, this suggests that tropical ecosystems may need to be managed to avoid large and rapid changes in soil pH with attendant effects on nutrient retention and mobility.

Anion Exchange and Adsorption
Anion and ligand exchange are also important processes in variable-charge soils. Anion exchange capacity increases as soil pH approaches the PZC. Because the pH approaches the PZC with increasing depth in the soil, nitrate tends to accumulate in the subsoil (Matson et al. 1987; Sollins et al. 1988). Nitrate accumulation is to be expected in La Selva subsoils, but only low levels of nitrate have been found to date (P. Sollins and G. P. Robertson, unpublished data).

Sulfate adsorption tends to increase with decreasing pH and with increasing levels of iron oxides and allophane (e.g., Uehara and Gillman 1981). Phosphate–extractable S averaged 1.1 cmol (−)/kg at 0- to 15-cm depth and 2.8 cmol (−)/kg at 15- to 30-cm depth (D. W. Johnson et al. 1979); soils used in

this study were taken from just north of OTS plot 1 and belong to the Sainal consociation (D. W. Johnson pers. comm, 1990).

Phosphorus adsorption is common in variable-charge soils and tends to be only weakly pH-dependent below pH 5.5 (Sánchez 1976; Sollins et al. 1988). Although no work has been done on mechanisms controlling phosphorus availability at La Selva, inferences are possible on the basis of the classification of La Selva's soils. Allophane, an extremely strongly phosphorus-sorbing mineral because of its large surface area (Uehara and Gillman 1981), is apparently still present in the less-weathered soils (lower-terrace and some middle-terrace soils); thus, high levels of phosphorus sorption are to be expected. Although allophane is apparently absent from the older soils, goethite appears to be abundant and also sorbs phosphorus strongly (see Sollins et al. 1988).

Vitousek and Denslow (1987) measured extractable phosphorus (acid ammonium fluoride) in surface soil sampled along trails throughout the old La Selva property and Sarapiquí Annex. They found that levels were highest in the lower-terrace soils, intermediate in the upper-terrace soils, and quite low in all the residual soils. Levels were always slightly higher in what are now called the Jaguar than in the Matabuey soils. The more extensive set of surface soil samples collected by Sancho and Mata in 1987 confirmed earlier findings about the three major groups of consociations at La Selva (fig. 4.12) but showed almost no difference between the Jaguar and Matabuey units (see table 4.1); results of phosphorus extraction by a second method (sodium bicarbonate at pH 8.5) showed much less clear patterns even between the lower- and upper-terrace consociations (see fig. 4.12). Bioassays (e.g., Denslow et al. 1987) suggest that phosphorus availability does, indeed, limit plant growth in many of La Selva's soils (see chap. 5).

Carbon and Nitrogen Levels

La Selva soils vary widely in carbon content. Carbon stores to 1-m depth range from 2 to 25 kg/m² (see table 4.1). By comparison, a set of forested soils in Puerto Rico spanning a wide range of life zones, forest types, soil groups, and topography ranged only from 6.5 to 10.0 kg C/m² (Weaver et al. 1987). At La Selva, all the most carbon-poor soils are sandy riverbank soils, and all the older soils tend to be carbon rich. The correlation between carbon content and soil age is not consistent, however; several of the young alluvial soils (e.g., the Andic Humitropepts of the Holdridge and Experimental units) are among the richest in carbon (>10 kg/m² to 1-m depth). Such high carbon levels likely result from the presence of allophane, which tends to bind or occlude the OM and, thereby, inhibit its decomposition (see Sollins et al. 1988). In the more highly weathered soils at La Selva, the high levels of iron and aluminum oxides and hydroxides (e.g., goethite, gibbsite) are probably responsible, at least in part, for the large amounts of organic mater. In the swamp and valley-bottom soils, organic matter may accumulate because anaerobic conditions hinder decomposition.

Studies suggest that the light fraction of the soil organic matter, mainly partially decomposed plant debris, accounts for only a small portion of the total soil carbon at La Selva. In samples (0–15-cm depth) from an Helechal soil, the light fraction accounted for only 2% of total soil carbon compared to 5%-50% for five temperate-zone soils spanning a wide range of physical structure and mineralogy (Strickland and Sollins 1987). The most likely explanation is that the light-fraction material decomposes more rapidly in tropical than temperate soils. Root turnover studies at La Selva (R. L. Sanford unpublished data) support this hypothesis in that they show very low levels of dead roots despite high rates of root death.

Along with the organic matter, nitrogen has also accumulated in large amounts, and phosphorus and cations rather than

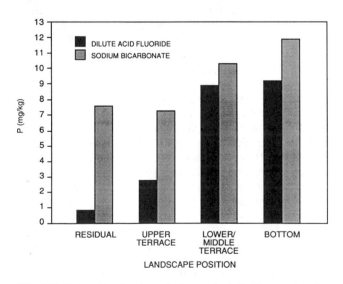

Fig. 4.12. Extractable phosphorus by two methods (acid ammonium fluoride and sodium bicarbonate) for the major groups of soil consociations at La Selva.

nitrogen tend to limit plant growth (see chap. 5). Why nitrogen has accumulated in such large amounts is not known. Nitrogen fixation by leguminous trees has often been suggested, but results to date are not definitive. With the large range in mineralogy and physical structure in La Selva's soils, the site offers an exceptional opportunity to study mechanisms of OM adsorption on mineral surfaces and the effects of such adsorption on soil organic mater (SOM) turnover and nitrogen mineralization.

RECOMMENDATIONS

Scientists studying tropical ecosystems would benefit greatly from closer collaboration with soil scientists and geologists both when designing studies and when interpreting results. Plant community ecologists, in particular, could use soil and landform classification to much greater advantage, especially when comparing their results with those of others. Several early studies of tropical plant communities (e.g., Ashton 1964) took full advantage of what was known of soils and geology of the region studied, but much recent work has not maintained this momentum. Unless differences among soils are recognized and understood, finding patterns in plant community parameters will be a slow process at best. In general, the lack of soils data for major sites of ecosystem research in the tropics (e.g., Los Tuxtlas, Mexico, and Barro Colorado Island, Panama) continues to make meaningful comparisons impossible.

Even at La Selva, ecosystem and plant community scientists could benefit from a better understanding of the differences in soil types within and among study sites. Probably the biggest single obstacle to progress at La Selva is the lack of topographic information and the inadequacy of the current geographic grid system. The lack of topographic information prevents accurate delineation of terrace boundaries and correlation of soil types between the Río Sarapiquí and Río Puerto Viejo terrace systems. Because grid points are widely spaced and inaccurately located, the sites of most previous studies cannot be pinpointed or assigned unequivocally to a soil mapping unit (see also comments by Vitousek and Denslow 1987). Work currently underway will greatly expand the present 200-m grid and will provide accurate elevations. The grid system will then form the basis for a geographic information system, and as the soil mapping units are redefined, sampling locations can be reassigned readily among soil types.

Methods of soil analysis can affect results markedly, especially in variable-charge soils where measurements at nonambient levels of pH or soil solution composition can grossly distort important soil parameters such as CEC (Uehara and Gillman 1981; Sollins et al. 1988). Those in the Tropical Soil Biology and Fertility Program have attempted to standardize many methods of soils analysis (Anderson and Ingram 1989). Deviation from such standard methods is often warranted but must be noted. Unless methods and locations are documented, results of soils analyses cannot be interpreted or generalized.

La Selva has seen almost no work on animals and plants as factors in soil formation. Soil pH does tend to be lower, for example, around *Pentaclethra* trees, presumably because the trees fix nitrogen, which is then nitrified (P. Sollins unpublished data). Leaf-cutter ants clearly affect soil properties at

La Selva, yet the effects of these animals, and the factors controlling their abundance and activity, remain unstudied. Areas where mammals and reptiles have burrowed or dug for food are also common. With a combination of pasture, second growth, and intact forest on a wide variety of soil types, La Selva provides an ideal setting for such studies.

Researchers have yet to take full advantage of the variability in chemical and physical properties of La Selva's soils. Early work focused on the soils nearest the station: Holdridge, Experimental, and Arboleda. As access to the property improved, emphasis shifted toward the residual soils, but the majority of the lower- and middle-terrace soils have remained little studied, in part because they no longer contain primary forest. As interest in soil chemistry, physics, and microbiology has increased at La Selva, more advantage has been taken of the striking contrasts among the alluvial soils (e.g., the PLOTS project of Robertson and Sollins). In addition, three projects have begun studies on the more highly weathered (presumably older) soils in the hills behind Puerto Viejo (W. A. Reiners and M. Keller, G. Gillman and P. Sollins, R. Butterfield and E. González). The Sarapiquí region, in general, provides exceptional opportunities to study effects of parent material, soil age, and colluvial and fluvial processes on soil genesis in a region where climate and mineralogy of the parent material hold nearly constant across a broad area.

ACKNOWLEDGMENTS

The initial soil survey by F. Sancho and R. Mata was funded by NSF grant no. BSR 83-06923 to J. Denslow and coordinated by R. L. Sanford and E. Newell. Supplemental funding was provided by NSF grant no. BSR86-05047 to P. Sollins and G. P. Robertson. The help of J. Tilley, G. Bracher, and N. Rudd in preparing tables and figures is gratefully acknowledged. We are especially indebted to the many people who contributed data, ideas, and critical review of the manuscript: E. Bornemisza, D. A. Clark, D. B. Clark, J. Denslow, G. Gillman, B. Haines, M. Hands, G. Hartshorn, E. Kramer, L. McDade, W. Melson, F. Montagnini, J. Núñez, R. Ortiz, C. Palm, G. Parker, W. Reiners, D. Richter, G. P. Robertson, C. Schnell, G. Uehara, P. Vitousek, and several anonymous reviewers.

5

Soil Fertility, Nutrient Acquisition, and Nutrient Cycling

Geoffrey G. Parker

Twenty-five years of soil, nutrient, and ecosystem research at the La Selva Biological Station have contributed substantially to the development of hypotheses of ecosystem processes in tropical rain forests but have yielded a fragmented body of detail on such processes at the La Selva site itself. La Selva soils are generally fertile by tropical standards but are exceedingly variable (see chap. 4). Information on La Selva forests suggests a correspondingly eutrophic nutrient cycle. Variation in trophic status among La Selva forests, undoubtedly very high, remains largely unexplored.

Observations at La Selva have advanced the understanding of (1) the "mobile ion" mechanism of soil leaching, used to predict soil and surface water responses to acidification in extratropical regions (e.g., D. W. Johnson et al. 1977, 1983); (2) the broad gradient in trophic status among tropical forests (e.g., Jordan and Herrera 1981; Jordan 1985; Vitousek and Sanford 1987); (3) specific consequences of substrate fertility on nutrient processing in forests, including the efficiency of nutrient use (Vitousek 1984); and (4) the short-term influence that biological activities may exert over the nature and degree of soil fertility, particularly in variable-charge soils (Sollins et al. 1988).

The soils of La Selva are now differentiated taxonomically, and their spatial variation is understood at a coarse scale of resolution. Few studies of plant performance, however, have been conducted within this mosaic of soil fertility. Reports on the processing of nutrients by plants are generally lacking. The relation between the distribution of species and/or communities and soil fertility and hydrology is sparsely understood for plants and almost unknown for animals.

La Selva is, nonetheless, uniquely situated to promote studies of the relationship between soil, plants, and ecosystem processes. The variation in substrate fertility at La Selva together with the background of detail on some aspects of nutrient processing and the theoretical framework relating soil properties and nutrient processing constitute a tremendous opportunity for productive investigations.

Background

In the late 1960s University of Washington researchers compared Atlantic slope wet tropical rain forest (La Selva) and seasonally deciduous Pacific slope forest (Finca Comelco in Guanacaste Province) with studies on pedology, litter fall, decomposition, and the chemistry of precipitation and soil water (e.g., Gessel et al. 1979, 1980; D. W. Johnson et al. 1975, 1977). At the outset they recognized the importance of variation in soil characteristics (e.g., Bourgeois et al. 1972) and initiated studies on widely differing soil types at both sites. Raich (1980a) (University of Florida) studied soil carbon budgets in young and old forests at La Selva. Carl Jordan and students (University of Georgia) linked micrometeorology, hydrological budgets, and soil solution chemistry in a study of the effects of disturbance at different spatial scales (Luvall 1984; Parker 1985). Pringle et al. (1986) studied nutrient limitations to algal growth in streams, and Stout (1980) described leaf decomposition in streams. Several short-term projects conducted at La Selva during Organization for Tropical Studies (OTS)-sponsored courses have dealt with aspects of nutrient processing (e.g., McColl 1970; Gower 1987).

In the 1980s investigators in several large projects have focused on a variety of processes in canopy gaps (see chap. 9), the biological modification of soil chemical properties (see chap. 4), and nutrient processing in streams (see chap. 3). Numerous aspects of soil fertility have been studied on related soils at higher elevation near Turrialba (e.g., Harcombe 1977; Ewel et al. 1982; Russel and Ewel 1985; Matson et al. 1987).

Despite the long history of related efforts, understanding of the acquisition, cycling, and loss of nutrients specific to the La Selva forest is far from complete. Several factors have contributed to this deficiency. First, despite the initial work of the University of Washington group, systematic description of nutrient processing for La Selva (with estimates of stocks and flows) has not been an objective of subsequent investigations. Furthermore, the locations of many studies have not been precisely referenced within the existing spatial variation in soil fertility. Accordingly, it can be difficult to ascertain on exactly which soil type some studies were conducted (see chap. 4). Often, studies undertaken at sites within more fertile but, generally, less extensive portions of La Selva have been assumed to characterize the entire forest. Consequently, nutrient processes in the La Selva forest are not only fragmentarily characterized but also poorly linked to the geography of soil fertility.

Framework

The following reviews a quarter-century of process studies on the availability, allocation and use, cycling and loss of major plant nutrients in both the intact and disturbed forests at the La Selva Biological Station. As a framework for integrating the various results of nutrient-plant studies at La Selva, I take the approach that the type and degree of nutrient processing depend on the fertility and hydrology of the substrate, ultimately on the mineralogy, chemistry, and physics of the soil (see also Trudgill 1977; Jordan and Herrera 1981; Vitousek and Sanford 1987; and Sollins et al. 1988 for further elaborations of this view). First, I review the nature and variation of soil fertility at La Selva. I treat soil chemical and physical properties relevant to the cycling of nutrients only briefly; an extended discussion is found in chapter 4. I summarize the available information on nutrient stocks and transfers, including movement of materials into, within, and out of the forest and discuss the effects of various disturbances on these processes. I compare La Selva with other lowland tropical rain forests with regard to nutrient processing and, finally, suggest some courses of promising research.

SOIL FERTILITY

La Selva soils are extensively classified in a large-scale survey (Sancho and Mata 1987) into twenty-four mapping units with a complex distribution (chap. 4). At a landscape scale of resolution, three dominant phases are in that spatial pattern. Roughly one-third of the property consists of comparatively fertile inceptisols (immature soils with poorly developed lower horizons) and some entisols (recent soils) of alluvial origin on three terraces of variable width (100 m–1 km) near the major rivers forming the northern and western borders of La Selva. Most of the remainder of the area in the steeper southern foothills contains relatively infertile, more acid ultisols (highly weathered soils with low stocks of basic cations) and some inceptisols derived from weathering in place of several lava flows (Alvarado 1982; chap. 3). In topographic depressions scattered within the dominant phases are numerous (more than two dozen exceed 1 ha) inclusions of swamp association soils (all inceptisols). I refer to these phases as alluvial, residual, and swamp, respectively.

On the whole, the soils of La Selva are weathered and often very deep, composed of predominantly clay-sized particles with low bulk density, relatively high organic matter content and porosity, and a stable aggregate structure. They generally retain large amounts of water yet readily permit water flow. Most of the ion exchange capacity of La Selva soils is variable; it may convert relatively rapidly between net negative (dominated by cation exchange capacity) and positive (anion exchange capacity), depending on soil pH and organic matter. They have a very large capacity for adsorbing sulfate and, particularly, phosphate ions from solution. There is no average soil condition at La Selva, however. Nearly all the soil properties show pronounced variation over several spatial scales. For example, the fertility of the La Selva forest soils ranges widely from moderate (some alluvial soils) to very infertile (some residual soils).

Similarly, La Selva cannot be represented by a single nutri-

ent cycle. Available comparisons between different substrates at La Selva indicate substantial differences in most major process rates. A nutrient cycle assembled from these disparate studies would not apply to any real situation. Documentation of nutrient cycles specific to the various La Selva forests requires a coordinated effort using a standard methodology focused on a unified set of processes in a location well referenced to the soil mosaic.

NUTRIENT ACQUISITION

Mineral ions become available in soluble form through input from outside the forest (atmospheric and fluvial deposition) and by weathering of minerals in bedrock and soil. For nitrogen, another atmospheric input is the fixation of dinitrogen gas by both free-living and mutualistic microbes. Weathering rates of primary or secondary minerals have not been studied at La Selva. Additional weathering, however, is unlikely to supply significant amounts of fresh mineral ions in the extensively weathered soils of La Selva (see chap. 4). Instead, atmospheric and fluvial deposition probably represent the major inputs.

Atmospheric Deposition

Wet Deposition. Rainfall at La Selva is extremely dilute (see table 5.1), but abundant precipitation (a 3,993 mm annual average at the Casa de Rafael station; see chap. 3) provides a substantial solute input. Deposition rates are low for inorganic nitrogen, phosphorus, and sulfur but relatively high for hydrogen ions, sodium, potassium, and chloride (Parker 1985). Ionic concentrations correlate inversely with the amount of precipitation and positively with the length of the preceding dry period: large storms generally produce more dilute rainwater than small ones, and events in the dry season (January–April) are more concentrated in solutes than rainy season storms.

The quality of rainwater at La Selva is strongly influenced by the Caribbean Sea, 50 km to the northeast and, for the majority of the year, upwind of the biological station (see chap. 3). The ionic composition is dominated by several major sea salts: sodium, chloride, potassium, and sulfate. La Selva precipitation resembles seawater in some respects. If all precipitation chloride is of marine origin, then more than 70% of the ionic mass in bulk precipitation at La Selva may be accounted for as a simple dilution of seawater (Parker 1985). The ionic mass in excess of the seawater contribution probably derives from local terrestrial sources.

The organic components of precipitation, often a major contributor to bulk nitrogen and phosphorus input (e.g., half of all phosphorus and two-thirds of all nitrogen in rain at Turrialba [Hendry et al. 1984]), have not been studied at La Selva. Precipitation concentrations of dissolved organic carbon averaged 5 mg C L^{-1} over several storms from June to August (Raich 1983a). The input of hydrogen ions in bulk precipitation (volume-weighted mean pH 5.2) compares with inputs in industrialized temperate zone regions, which have far less precipitation although at more acidic levels (pH 4.1–4.4) (Galloway et al. 1984). Industrial pollution is unlikely to contribute to precipitation acidity at La Selva. Organic acids (e.g.,

Table 5.1 Major ion chemistry for volume-unweighted bulk precipitation and streamwater grab samples, June 30, 1982–July 12, 1983.

Ion	Rainwater[a]		Streamwater[b]	
	Mean Concentration[c]	Standard Deviation[d]	Mean Concentration[c]	Standard Deviation[e]
pH	5.15	0.35	6.13	0.21
EC[f]	14.3	10.4	15.2	4.08
Ca	0.144	0.182	0.367	0.137
Mg	0.108	0.145	0.254	0.080
K	0.505	1.16	0.264	0.638
Na	0.754	0.997	1.23	0.110
NH_4-N	0.036	0.088	0.003	0.007
NO_3-N[g]	0.021	0.050	0.106	0.107
PO_4-P	0.009	0.021	0.0006	0.0029
Cl	1.79	1.55	2.38	0.757
SO_4-S	0.399	0.579	0.238	0.181

[a]Samples taken at the MAB project meteorological station.

[b]Samples taken from a small catchment draining into Quebrada El Bejuco, Sarapiquí Annex, Watershed 4.

[c]Concentrations in mg element L^{-1}, except electrical conductivity and pH.

[d]N = 42–49.

[e]N = 53.

[f]Electrical conductivity, $\mu S/cm$ at 25° C.

[g]$(NO_3 + NO_2)-N$.

Galloway et al. 1984), possibly with some strong mineral acids from volcanic emissions (see N. Johnson and Parnell 1986; Parnell 1986), are more likely to control hydrogen ion inputs.

Dry Deposition. Not all atmospheric deposition is sampled in bulk precipitation. The sedimentation of dust and other large particles, the impaction of fine aerosols, and the sorption of gases and vapors contribute atmospheric inputs between precipitation events (Galloway and Parker 1980). Such fluxes are collectively called dry deposition. No estimates of gas, vapor, aerosol, or dust input, however, have been reported for the La Selva forest, except for carbon dioxide uptake in net photosynthesis (see chap. 10).

Ambient concentrations of airborne substances in the atmosphere above and within the La Selva forest have not been characterized, but they are probably low by temperate zone standards. High and frequent rainfall (nearly two-thirds of all days have measurable precipitation, and prolonged dry periods are rare) tends to wash the lower troposphere, yielding low loads of gases, large aerosol particles, and dust (Lawson and Winchester 1979; Artaxo-Neto et al. 1982). The aerosol component, especially the fine fraction (<1 μm diameter), is likely to be somewhat enriched in carbon, potassium, sulfate, and phosphorus compared to free-tropospheric values as found in other forested regions in the humid tropics (Delmas et al. 1978; Crozat 1979; Lawson and Winchester 1979; Artaxo-Neto et al. 1982; Crutzen 1987; Harriss 1987, Talbot et al. 1988). Particulate deposition of marine salts and of combustion products from biomass burning is likely to occur episodically. Deposition of volcanic ash appears not to have been important historically (chap. 4).

The dry deposition flux of an airborne substance is often modeled as the product of its ambient air concentration and a transfer coefficient, the deposition velocity, which depends on the nature of the substance, the receiving surface, and atmospheric conditions. Deposition velocities of water-soluble or reactive gases (e.g., SO_2, O_3, NO_x) are high, but ambient concentrations in the lower tropical troposphere are generally low. Tropical rain forests appear to be potentially enormous sinks for such compounds (Kaplan et al. 1988), but actual dry deposition of gases and particles is probably low. Atmospheric inputs at La Selva are largely dominated by solutes delivered in rainfall.

Measurements of air quality above and within tropical rain forests are scarce. Exchanges of gases and vapors between the lower atmosphere and rain forests have been estimated only in the 1980s (in central Amazonia: Andreae and Andreae 1988; Kaplan et al. 1988; Zimmerman et al. 1988). Almost nothing is understood about atmosphere-canopy exchange of aerosols in rain forests.

Nitrogen Fixation. Several lines of evidence suggest that nitrogen fixation may be substantial at La Selva. Soil solution losses of inorganic forms of nitrogen greatly exceed inputs in bulk deposition under residual soils (Parker 1985). High soil nitrification potentials have been measured in both mature and successional forest areas (Robertson 1982, 1984; Vitousek and Denslow 1986; Robertson and Tiedje 1988; Vitousek and Matson 1988). Nitrate concentrations in interstitial water from the residual soils are among the highest reported in any forest (Parker 1985). Finally, leguminous forest trees are ubiquitous (18.7% of all stems and 44% of the total basal area, according to Gessel et al. [1979]) and often have roots nodulated with nitrogen-fixing bacteria (L. McHargue pers. comm.).

Biological nitrogen fixation has been assayed at La Selva only on epiphyllic communities and in termites. Bentley and Carpenter (1980, 1984) found that nitrogen fixed by leaf-surface microbes was quickly transferred to host leaves of *Welfia* palms. Some of the activity was associated with leaf-surface liverworts (Bien 1982). Additionally, the nitrogen fixation rate of gut-dwelling symbionts of *Nasutitermes* soldiers (Prestwich et al. 1980) and of whole colonies (Prestwich and Bentley 1981) was sufficient to replace whole colony nitrogen within one year or less. Extrapolation to areal rates is difficult, but these mechanisms could easily account for substantial nitrogen inputs. Forman (1975) estimated that cyanophilous lichens in a Colombian rain forest canopy alone might fix 1.5–8 kg N ha^{-1} yr^{-1}. Presumably, other components of the La Selva phyllosphere could contribute as much. Nitrogen fixation inputs to tropical rain forests have been estimated to range between 40 and 110 kg ha^{-1} yr^{-1} (Cole and Johnson 1979).

Many leguminous trees at La Selva, including the dominant *Pentaclethra macroloba,* have *Rhizobium*-nodulated roots (D. Janos, L. McHargue pers. comm.), but the extent of nodulation and the rate of nitrogen fixation are unknown. The nitrogenase activity of free-living soil nitrogen-fixers has also not been reported.

Fluvial Deposition
Riverborne sediments are deposited during episodic floods along the larger rivers, particularly on the lower terraces of

the Sarapiquí and Puerto Viejo. The spatial complexity of the soil taxonomic units in the alluvial regions of La Selva (Sancho and Mata 1987) is consistent with numerous historical episodes of flooding. Deposition by flooding is restricted to the vicinity (several hundred meters) of the rivers and lower reaches of the tributary streams where flow may reverse during high floods.

The rates of riverine inputs are unknown. Stratigraphic analyses and isotopic dating of the deep alluvia may indicate the frequencies of large floods and the magnitudes of inputs. Nutrient budgets for frequently flooded alluvial areas cannot be balanced without a quantification of fluvial deposition.

NUTRIENT TRANSFER PROCESSES

In the following account of plant-based nutrient cycling, I focus on three classes of transfer processes: the uptake of nutrients from the soil solution/exchange complex, the retention and reallocation of material in plant tissues, and nutrient return from plants to the available pool in the soil. The terminology follows Cole and Rapp (1981) and Waring and Schlesinger (1985).

Nutrient Release from Plants

Nutrients may be released from vegetation through herbivory, decomposition of plant parts (following tree fall, litter fall, and root sloughing), or through leaching by water of both above- and below-ground live tissues.

Litter Production. Litter-fall rates have been measured in different areas of the La Selva forest on several occasions: Gessel et al. (1979) report an annual total of 7.83 megagrams (Mg) dry matter per hectare, apparently on the old alluvial terrace. An estimate of 8.1 Mg ha^{-1} yr^{-1} was reported in Vitousek (1984). Litter production in a forest plot on residual (Matabuey) soil was estimated at 8.7 Mg ha^{-1} yr^{-1} (Parker unpublished data). In Parker's study monthly litter fall was distributed rather evenly throughout the year (fig. 5.1) and was not correlated with monthly rainfall. A recent study by R. Sanford (pers. comm.), however, found a positive relation between amounts of precipitation and litter fall in both alluvial and residual soils. Total litter fall (leaves, branches, and other tissues) measured on six plots throughout La Selva averaged 10.9 Mg ha^{-1} yr^{-1} (range 9.5–12.4 Mg ha^{-1} yr^{-1}). The differences between soil types were not great (R. Sanford pers. comm.). Most of these estimates are in the range of the value 8.5 Mg ha^{-1} yr^{-1} predicted for average annual temperature and precipitation conditions at La Selva by the equation of Brown and Lugo (1982, fig. 7),

$$L = 16.0 + 16.7 \log (T/P) - 6.5 (T/P),$$

where L is annual litter fall (Mg ha^{-1} yr^{-1}) and T/P is the ratio between annual mean temperature (°C) and precipitation (mm) multiplied by 100.

Reports of the nutrient composition of above- and below-ground litter are fewer than of the litter-fall rate. In old alluvial forest soils the potassium content of leaf litter fall totaled 32 kg ha^{-1} yr^{-1} (Gessel et al. 1979); nitrogen and calcium were 135 and 59 kg ha^{-1} yr^{-1}, respectively (Vitousek 1984). The nutrient content of litter of forests in six plots in several La

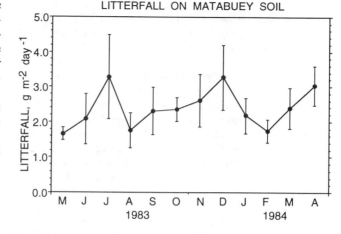

Fig. 5.1. Monthly leaf litter fall (means and standard errors) under intact rain forest on residual soils near the southern boundary of the La Selva Western Annex (ten samples). Annual litter fall was 8.73 Mg ha^{-1} yr^{-1} and total rainfall was 3,660 mm.

Selva soils has been measured by R. Sanford (pers. comm.), who found total phosphorus inputs lower on residual than on alluvial soils, in agreement with a trend in soil phosphorus concentrations. There was little trend in litter-fall nitrogen. Similarly, C. Jordan (pers. comm.) found leaves on the forest floor of a residual soil had lower concentrations of total phosphorus than did those on an older alluvial soil. Where differences in nutrient release from aboveground litter have been observed between plots on different soils, they tend to reflect contrasts in tissue concentrations rather than differences in the litter-fall rate.

The rate at which trees and branches die and become converted into litter (coarse woody litter fall, Vogt et al. [1986]) is unknown, despite an earlier emphasis on canopy gap creation through tree falls (Hartshorn 1980; Sanford et al. 1986; chap. 9). Decaying logs are conspicuous on the forest floor and, undoubtedly, account for much of the standing stock of litter in the La Selva forests. The chemical content of woody litter at La Selva has not been reported.

Production of below-ground litter has also been estimated by Sanford (pers. comm.), who found that fine root (<2 mm diameter) production in a residual soil plot (10.4 Mg ha^{-1} 10 months^{-1}) was more than double that in an alluvial soil site (4.6 Mg ha^{-1} 10 months^{-1}). Nitrogen and phosphorus concentrations of fine root litter were similar between the two plots. Thus, unlike the pattern for aboveground litter, inputs of nitrogen and phosphorus from roots were higher in the residual soil than in alluvial soil.

Arrested Litter. A striking feature of the La Selva forest is that a great deal of litter fall is intercepted and retained in the crowns of understory plants (and, probably, of epiphytes, too). Litter-intercepting plants commonly have palmoid (lamppost) growth-forms and have been referred to as detritophilic, or wastebasket, plants (D'Arcy 1973; W. Alverson pers. comm.). Conspicuous detritophiles in the La Selva understory are the geonomoid palm genera *Geonoma* and *Asterogyne* (Raich 1983b); *Quararibea pumila;* and saplings of various other species (W. Alverson pers. comm.). The arrested litter decomposes aboveground and, undoubtedly, enriches stemflow from and soil solution around such plants. Raich (1983b) suggests that the litter-trapping habit of *Asterogyne martiana* contributes to its success in the forest understory.

Litter Decomposition and Mineralization. Decomposition of both above- and below-ground litter has been inadequately studied at La Selva. Leaf litter decay is clearly very rapid: the standing stock of soil surface litter, although dependent on the season, is very low (3.30 Mg ha^{-1}; Cole and Johnson [1979]), far less than the annual rate of litter fall. Accordingly, the average residence time of identifiable litter is less than one year (0.4 years; Gessel et al. [1979]). Decay rates are undoubtedly species specific: the rate of dry mass loss from confined leaves of *Pentaclethra macroloba* and *Bursera simaruba* had exponential decay constants of 2.77 and 4.16 yr^{-1}, respectively (Gessel et al. 1979). The release rate of soluble minerals from aboveground litter (net mineralization) is likely to be correspondingly high, judging from the extremely high mineralization potentials reported by Robertson (1989) for nitrogen in surface horizons in various La Selva soils. Concerning the decay of coarse woody debris (Harmon et al. 1986) nothing has been reported for La Selva. Direct characterizations of litter decay processes for below-ground tissues are in progress (R. Sanford pers. comm.).

Earthworms likely play an important role in decomposition processes in some La Selva soils: their densities have been estimated to average about 200 m^{-2} in the upper 10 cm of soil (R. Sanford pers. comm.), one of the highest values reported in any tropical rain forest. Little is known, however, about the decomposing activities of specific organisms, such as earthworms, leaf-cutting ants, and termites, nor generally about the decomposer community, macro- and microinvertebrates, and microbes.

Leaching from Plants. Removal of soluble material from aboveground plant tissues by rainwater was measured for short periods by Johnson et al. (1977), McColl (1970), and Raich (1980a). These observations showed that throughfall and stemflow in alluvial soil forests have higher concentrations of base cations (Ca, Mg, Na, and K) and carbon, and higher pH than bulk precipitation outside the forest, a pattern consistent with most rain forests studied (Parker 1983). No annual budgets for the release of solutes from the forest canopy are available, nor has the leaching of aboveground tissue been compared between forests on different La Selva soils. Differences in stemflow quality between several species of trees suggest that the nutrient composition of stemflow water may be species specific (McColl 1970), but this possibility has not been systematically investigated. Solute leaching from below-ground tissues has also not been compared in forests on different soils at La Selva.

Uptake and Accumulation
The amounts of mineral nutrients taken up or incorporated by forest trees at La Selva have not been reported. Incorporation rates are often calculated as the rate of tissue production multiplied by chemical concentration in the new tissue (e.g., Sprugel 1984). Of these values only the aboveground biomass increment has been estimated for stems >10 cm dbh in the La Selva permanent plots (D. Lieberman et al. 1990). The various measurements necessary to construct an estimate of uptake and accumulation are laborious to obtain in diverse forests because species differences in growth, nutrient uptake, and incorporation require extensive sampling and chemical analyses.

Mycorrhizal Associations. Roots of the majority of woody plants in the La Selva forest form intimate associations with fungi (mycorrhizae). Most of the mycorrhizal trees form the vesicular-arbuscular type of endomycorrhiza (VAM) (Janos 1980a) though ectomycorrhizal associations (EM) are found in some species of the Caesalpiniaceae, Myrtaceae, and Nyctaginaceae (e.g., *Neea*). Janos (1980b) considers that the majority of mature forest species at La Selva exhibit poor development or complete lack of root hairs and are, thus, probably highly dependent on these associations ("obligately mycotrophic").

VAM provide a network of fungal hyphae invading the soil, which greatly extends the absorption surface of the plant rhizoplane and increases the uptake of scarce or immobile nutrients and water. The fungal association costs the plant some portion of its photosynthate, but the carbon cost has not been quantified. Under experimental conditions the presence of the fungal infection improves plant growth, water uptake, and seedling survival and may also enhance the nodulation of nitrogen-fixing plants (Janos 1980a). According to Janos (1983), mycorrhizae, furthermore, may promote the direct decomposition of litter. Whether or not mycorrhizae decompose organic matter, their presence enhances uptake of mineralized nutrients (e.g., Herrera et al. 1978; Stark and Jordan 1978) and may reduce the leaching of nutrients from the soil solution and exchange complex. The extent to which this "direct cycling" (Went and Stark 1968a, 1968b) contributes to the nearly "closed cycle of nutrients" in tropical rain forests suggested by Richards (1952) is unknown.

The presence of the VAM may influence the outcome of competitive interactions between plants, the composition of plant communities, and the trajectory of succession (Janos 1983). The type and degree of the interaction will depend on mineral availability and the likelihood of infection (which depends on the presence and density of spores). Plant dependence on the fungal association may be relatively unimportant in some habitats (soils of disturbed areas such as slash-and-burn clearings, waterlogged soils, or high elevation areas) and some species (crops and early successional plants). But for many of the habitats and most of the forest species at La Selva, mycorrhizae are likely an important determinant of many plant/soil interactions. Despite the large literature on the potential of mycorrhizal associations to increase the mineralization of organic matter and the uptake of phosphorus at low soil levels, however, the relevance of the association under field conditions in the La Selva forest is unknown.

Nutrient Redistribution. Within-plant movements of materials (e.g., in nutrient retranslocation from leaves before abscission) may be substantial but patterns of within-plant allocation of carbon and mineral nutrients have not been reported for La Selva.

Limiting Nutrients. No mineral nutrient has been shown to be limiting to terrestrial plant growth under field conditions at La Selva. The element most likely to be limiting, however, particularly in the residual foothills, is phosphorus. In acid soils generally (Sánchez 1976) and in volcanic tropical soils in particular (Parfitt 1980; Uehara and Gillman 1981; Bertsch et al. 1984; Sollins et al. 1988), soil phosphorus is largely occluded in complexes with aluminum and iron oxides and hydroxides. Other major fractions of phosphorus include those bound in soil organic matter or in the biomass of soil microbes (Hedley et al. 1982), but most of the total phosphorus is unavailable for plant uptake. Little phosphorus enters the system from deposition or is newly weathered from primary rock. Little is detected in soilwater or streamwater (Parker 1985; Luvall and Jordan 1984) although Pringle et al. (1986) found high concentrations of soluble orthophosphate in a third-order stream draining the residual hills (see chap. 3). Soil surface phosphorus concentrations (at 1–4 mg kg^{-1}; NH$_4$F extraction) in residual soils are considered to be in the deficient range (Bertsch and Cordero 1984). Plants grown on such soils are often deficient in foliar phosphorus (e.g., Bertsch et al. 1984) and grow more poorly than when in phosphorus-enriched media (Bertsch and Cordero 1984; Vitousek and Denslow 1987). Denslow et al. (1987) found growth of potted *Phytolacca rivinoides,* a nonmycotrophic successional species, responded to phosphorus amendments.

Potassium is probably the next most critical nutrient after phosphorus. Exchangeable potassium is extremely low at La Selva, particularly in the residual soils. Plants grown in such soils are occasionally deficient in foliar potassium (e.g., Bertsch et al. 1984). Denslow et al. (1987) found a reduction in growth of potted plants of several species of *Miconia* and *Piper* in a growth medium lacking added potassium compared to that in a complete nutrient treatment.

Potentially limiting nutrients in La Selva aquatic systems probably differ from those in soils. Neither phosphorus nor potassium appears to limit growth of attached algae in a third-order stream (Quebrada El Salto) draining the residual foothills. Short-term accumulations of chlorophyll-a on artificial substrates enriched with various nutrient combinations were most pronounced in the presence of micronutrient combinations (B, Co, EDTA, Fe, Mn, Mo, Zn; Pringle et al. [1986]). Nitrogen and phosphorus amendments neither yielded a significant growth response nor enhanced the effect of the micronutrients.

The chemistry of La Selva streams is as variable as that of its soils. Luvall and Jordan (1984) found solute concentrations in weekly samples of two adjacent streams in the Quebrada El Bejuco region of the Sarapiquí Annex differed consistently in nearly every parameter measured (table 5.2). The larger catchment (WS 3) drained a swampy area (C. Jordan pers. comm.), and its streamwater always had higher pH, conductivity, and concentrations of most ions than did streamwater from the smaller catchment (WS 4). A survey of stream reaches in the Quebrada El Bejuco vicinity suggested that the chemistry of WS 4 was the more typical (Parker unpublished data). The causes of such small-scale variation in water chemistry are poorly studied at La Selva (but see chap. 2). I suspect that differences in the pathway and residence time of water are more important than differences in vegetation or even sub-

strate geology (G. Alvarado pers. comm.) in these catchments.

PROCESSES OF NUTRIENT LOSS

Nutrients are lost from forest by volatilization to the atmosphere, percolation of soilwater to ground- and streamwater, and in particulate and dissolved losses from erosion. The majority of materiel lost from the system probably moves via percolating water and streamflow.

Gases lost from La Selva ecosystems include respired carbon dioxide, carbon monoxide, methane, and a wide variety of volatile hydrocarbons; ammonia, nitric oxide, nitrous oxide, and dinitrogen; and reduced sulfur compounds (hydrogen sulfide and others). Nitrous oxide losses from soils as a result of denitrification have been measured by Robertson and Tiedje (1988) at rates of 7.6–21 kg ha^{-1} yr^{-1}. Haines et al. (1987) report that various sulfur gases (particularly dimethyl sulfide and methyl mercaptan) are produced in litter or soil (B. Haines pers. comm.) or from roots of plants (Haines and McHargue 1986). The total sulfur emissions from these sources are on the order of 1 g ha^{-1} yr^{-1} or less. Burning of biomass during dry seasons is a potentially important loss of carbon, nitrogen, sulfur, and various other elements (in ash) in the region of the biological station. Although fires are not commonplace at La Selva currently, evidence of charcoal is in the soils (R. Sanford pers. comm.).

Lateral flow of water over the soil surface (overland flow) is very rare in La Selva forests as the intact soils have extremely high infiltration capacities (Parker 1985; Sollins and Radulovich 1988). Erosion is normally not high from intact soils of the La Selva region, unless infiltration is reduced, as when the soil surface is compacted though foot traffic (Radulovich and Sollins 1985). The rise in turbidity of the major rivers following large rains may be the result of erosive land use practices within their basins. Nitrate, chloride, and basic cations are the ions most affected (e.g., Parker 1985).

Table 5.2 Comparison of ion chemistry for grab samples of streamwater draining adjacent catchments, June 30, 1982–July 12, 1983.

Ion	Watershed 3[a] Mean Concentration[b]	Watershed 3[a] Standard Deviation[c]	Watershed 4[a] Mean Concentration[b]	Watershed 4[a] Standard Deviation[c]
pH	7.04	0.32	6.13	0.21
EC[d]	66.4	24.1	15.2	4.08
Ca	3.87	2.06	0.367	0.137
Mg	1.89	0.930	0.254	0.080
K	1.23	0.891	0.264	0.638
Na	3.15	1.34	1.23	0.110
NH$_4$–N	0.005	0.009	0.003	0.007
NO$_3$–N[e]	0.080	0.245	0.106	0.107
PO$_4$–P	0.063	0.045	0.0006	0.0029

[a]Watersheds in the vicinity of Quebrada El Bejuco, Sarapiquí Annex.
[b]Means are not flow-weighted. Concentrations in mg element L^{-1}, except electrical conductivity and pH.
[c]N = 53.
[d]Electrical conductivity, μS/cm at 25° C.
[e](NO$_3$ + NO$_2$)–N.

Solutes lost in percolating soil water may move in two distinct modes. Much of the solute flow is via slow percolation of soilwater in contact with the extensive surfaces of the fine-sized pores (micropore or matrix flow). Rapid movement of large quantities of water may occur in large soil pores and channels during large storms (macropore flow), acting analogously with a storm drain. Both modes of flow occur at La Selva: most of the soils have appreciable clay content and stable aggregates with abundant small and large pores (Sancho and Mata 1987; chap. 4). An extreme version of macropore flow (pipe flow) may be observed emerging from local roadcuts or in soil pits. As solute concentrations reflect the degree of contact between water and soil surfaces, macropore flow is likely to be more dilute than matrix flow. Sollins (1989) suggests that the diversion of water into macropores acts to reduce leaching of soil nutrients.

Partitioning soilwater transport between micropore and macropore flow is problematic, but Sollins and Radulovich (1988) report evidence of both sorts of water flow ("biphasic flow") in alluvial soils in the La Guaria Annex. Determination of the amounts of solutes moving by each mode of flow is even more difficult (Russel and Ewel 1985) and has not been satisfactorily resolved for any soil at La Selva (e.g., Parker 1985; chap. 4).

EFFECTS OF VEGETATION CHANGE

Effects of Disturbances

The type and scale of a disturbance influence its effects on nutrient processes. Intermediate-scale experimental clear-cutting (500 and 2,500 m²) of forests on residual soils results in rapid, short-term increases in nutrient concentrations in soil solutions, soil water percolation, and losses of soil nutrients (Parker 1985). With no additional disturbance, the large pulse of nutrients lost below the rooting zone in percolating water is transient, and concentrations return to predisturbance levels in less than two years. Small-scale disturbances, such as in natural or artificial treefall gaps (<100 m² of crown opening), do not result in increased soil nutrient availability (Vitousek and Denslow 1986) or solution losses (Parker 1985) compared to the intact forest. Thus, single treefall gaps do not appear to be sites of transient fertility in the residual soil forests, in contrast to the predictions of Anderson and Swift (1983) and Bazzaz (1983). Responses of nutrient processing in large disturbances, on the order of typical slash-and-burn clearings (several hectares or more in extent), have not been studied in the La Selva region.

Successional Effects

Carbon and mineral nutrient dynamics have been studied following forest cutting primarily on alluvial soils in Annex A. Most of the work has been conducted on plots of different ages since clearing along the Quebrada Esquina, not so much on the adjacent cyclically cleared strips (see chap. 2). Available soil nutrient cations and nitrate showed a depletion in the early (eight-to-sixteen-year) stages of a three-decade chronosequence, with temporary increase in organic matter, total nitrogen, ammonium-nitrogen, and phosphate-phosphorus; and longer-term increases in exchangeable iron, aluminum, and

hydrogen ions (Werner 1984). Robertson (1984) found that nitrogen mineralization and nitrification potentials were highest in the eight-year-old plot of the thirty-one-year sequence. Denitrification rates were lowest in the midsuccessional sites (Robertson and Tiedje 1988), probably because of nitrate limitation. In the first year after forest cutting there were large decreases in soil organic matter (Raich 1983b) although fine roots rapidly regrew to within 92% of the level of the mature forest (Raich 1980b).

Observations in forests on alluvial soils have dominated most studies of successional nutrient dynamics at La Selva, but Vitousek and Denslow (1986) examined nitrogen and phosphorus availability within treefall gaps on residual soil. They found net mineralization of nitrogen very high and of phosphorus relatively low in the upper soil within the crown zone of natural gaps. These values, however, were found not to be different from those of the intact forest.

COMPARISON WITH OTHER LOWLAND TROPICAL RAIN FORESTS

On the basis of a comparison of ecosystem parameters from seven lowland tropical rain forests, Jordan (1985) suggest that La Selva is among the more eutrophic of rain forests studied. The ecosystem parameters described here are structural characteristics whose lumped values indicate broad functional differences between systems. The data assembled by Jordan (1985) are tabulated in table 5.3. From this comparison, La Selva has a moderate- to large-statured forest (with a very low fraction of below-ground biomass), which produces large quantities of nutrient-rich, nonscleromorphic foliage that decomposes readily. Furthermore, it exhibits none of the extensive superficial root mat, canopy scavenging of precipitation nutrients (Jordan et al. 1979), and low soil solution concentrations observed on extremely poor substrates in blackwater regions (Amazon caatinga and oxisol forests). Furthermore, La Selva forests exhibit high availability and rapid interconversions of some nutrients, judging from the extremely high nitrogen mineralization, nitrification, and denitrification potentials measured. Its leaf production, however, is less than the very high values reported for forests on dolomite (high magnesium- and calcium-carbonate) soils in Darién, Panama (Golley et al. 1975).

These data on ecosystem attributes were subjected to a principal component analysis of the rain forest sites (information on some parameters was not available for La Selva; Jordan [1985]). On the highly significant first axis (table 5.4) that emerged from the analysis, La Selva is positioned close to the forest on the most fertile soils (Darién, Panama) but quite removed from forests on the most nutrient-poor sites (upper Venezuelan Amazon). Climatic conditions such as temperature and precipitation were not correlated with this axis. The suite of conditions promoting growth is, undoubtedly, a contributor to the complex gradient relating these sites.

The fertility suggested by Jordan's analysis cannot apply uniformly to the entire region: information available for La Selva came primarily, although not exclusively, from studies on the richer, alluvial soil sites. With more information, La Selva's position within the trophic gradient will be revised. Nonetheless, the overall nutrient regime of La Selva forests

Table 5.3 Ecosystem characteristics in tropical moist and rain forests

Parameter	Amazon Caatinga San Carlos, Venezuela	Oxisol Forest San Carlos, Venezuela	Lower Montane Rain Forest El Verde, Puerto Rico	Evergreen Forest Banco, Ivory Coast	Dipterocarp Forest Pasoh, Malaysia	Lowland Rain Forest La Selva, Costa Rica	Moist Forest, Darién, Panama
Root biomass (Mg ha^{-1})	132	56	72.3	49	20.5	14.4	11.2
Above-ground biomass (Mg ha^{-1})	268	264	228	513	475	382	326
Root:shoot ratio	0.49	0.21	0.32	0.10	0.04	0.04	0.03
Percentage roots in superficial mat	26	20	0[a]	0[a]	0[a]	0[a]	—
Specific leaf area (cm^2 g^{-1})	47	65	61	—	88	139	131–187
Leaf area index	5.1	6.4	6.6	—	7.3	—	10.6–22.4
Predicted leaf biomass (Mg ha^{-1})	10.8	9.8	10.8	—	8.3	—	10.4
Leaf litter production (Mg ha^{-1} yr^{-1})	4.95	5.87	5.47	8.19	6.30	7.83	11.3
Predicted leaf turnover time (yr)	2.2	1.7	2.0	—	1.3	—	0.9
Wood productivity (Mg ha^{-1} yr^{-1})	3.93	4.93	4.86	4.0	6.4	—	—
Leaf decomposition constant (k)	0.76	0.52	2.74	3.3	3.3	3.47	3.2
Biomass:phosphorus ratio in leaf litter fall	2,631	7,237	5,000	1,365	3,282	2,024	1,319
Biomass:nitrogen ratio in leaf litter fall	135	95	—	64	82	52	—

Source: From Jordan 1985.
[a]Approximately.

Table 5.4 Ranking of well-studied tropical rain forests along the first axis of a principal components analysis (centered and standardized) of major ecosystem attributes

Rain Forest Site	PCA Score
spodosol (San Carlos, Venezuela)	100.0
oxisol (San Carlos, Venezuela)	71.5
El Verde (Puerto Rico)	45.8
Banco (Ivory Coast)	15.7
Pasoh (Malaysia)	13.4
La Selva (Costa Rica)	3.2
Darién (Panama)	0.0

Source: From Jordan (1985, table III.5).

may be characterized as relatively eutrophic for tropical rain forests. In parallel with the variable fertility of the soil, however, a great deal of within-site variation in the trophic status of forests is likely across the La Selva region.

RESEARCH NEEDS

The demonstration of the variation of soil fertility at La Selva must now be extended with studies that explore the consequences of that pattern. First, previously undocumented but important aspects of carbon and nutrient processing in La Selva forests (e.g., biomass and productivity; nutrient stocks and fluxes) should be measured. Next, rates of specific components of nutrient cycling should be compared between forests on differing soils, successional states, or disturbance regimes. Finally, emphasis should shift into elucidation of mechanisms and limitations contributing to large-scale patterns.

The linkage between characteristics of nutrient cycling and variations in trophic status between La Selva forests is an appropriate major theme for future studies. Are there between-forest differences in (1) amounts of mineral nutrients taken up, retained, and released; (2) extent and rates of mineralization, nitrification, and denitrification; (3) dependence on mycorrhizal associations; and (4) efficiency of nutrient use? Several extant hypotheses could provide a framework for investigating soil/nutrient cycle linkages.

First, within-stand cycling of a limiting nutrient should be more efficient when the nutrient is in short supply. Concentrations of nutrients in biomass (particularly short-lived parts) should be lower on infertile soils ("nutrient-use efficiency;" Vitousek [1984]) than on rich ones. In oligotrophic situations, soil solution concentrations of important nutrients will be lower and the nutrient cycle ought generally to be more closed and conservative than in eutrophic situations.

Next, specific chemical differences between systems can affect major solute flows (by whatever means those differences arise). The character of soil solution anions and the capacity of the soil to regulate the mobility of those ions are hypothesized to control the type and degree of solution leaching (Nye and Greenland 1960; D. W. Johnson et al. 1975, 1979; Johnson and Cole 1980). The chemistry of soil solutions varies dramatically between La Selva soils: D. W. Johnson et al. (1977) found alluvial soilwater had high pH, with bicarbonate the dominate anion, whereas Parker (1985) found soil solutions in the residual soil were much more acid, with nitrate the anion in highest concentration. Because the mobility of the anion in the soil solution is a major determinant of the leaching of its associated cation, these differences in the makeup of the soil solution could affect regional solute balances, streamwater chemistry, and the response of soil fertility to disturbance.

Finally, in some La Selva soils, biological processes may directly affect, in the short term, the nature and capacity of the ion exchange complex, the foundation of soil fertility. When soil pH and/or organic matter content change in soils dominated by variable-charge minerals, exchange sites can be converted from a net negative surface charge to ones that are positive (Sollins et al. 1988). The concomitant shift between net cation and net anion exchange capacity could alter nutrient availability, the presence of some ions in soil solution (e.g., NH_4^+ versus NO_3^-), their availability for plant uptake, and their potential to be leached.

Unmeasured Stocks and Fluxes

Some investigations of gross stocks and flows of nutrients are currently underway, yet many aspects of the nutrient cycle have not been quantified. These include decomposition and release of minerals from boles, roots, and leaf litter; estimates of above- and below-ground biomass and elemental content, including nutrient standing stocks in canopy leaves and forest floor material; and, rates of carbon and element accumulation in wood, leaves, and roots. Associated processes of interest include the rates of material lost in erosion and in volatilization (CO, H_2S, N_2, and radiatively active "greenhouse" gases

CO_2, CH_4, and N_2O); acquisition of minerals through not only dry deposition of particulates and gases but also chemical weathering of primary and secondary minerals; and the hydrological balance of the forest. Estimation of these important features of material processing will provide the basis for meaningful comparisons between forests on different sites.

Regulative Processes

Continued attention should be paid to major factors controlling nutrient-cycling processes. Long-term plant responses are ultimately controlled by inherent soil fertility, site history, and degree of waterlogging; investigations of these should be extended.

The La Selva climate is nearly isothermal, with diurnal temperature variations far greater than the seasonal ones. Thus, the major short-term influences on plant responses will probably be the amount and timing of precipitation. The long-term precipitation record from the La Selva Biological Station provides a description of the duration of droughts and downpours. Rainfall is frequent at La Selva, but the intervals between rains heavy enough to penetrate the forest canopy (typically 3 mm or more) often exceed one week or more, depending on the season (see chap. 3). Moisture in the soil rooting zone exhibits marked seasonal variation (Luvall 1984; Parker 1985), and wilting of leaves of understory plants may be observed in dry periods. Furthermore, some aspects of nutrient availability depend on soil moisture; nitrate, for example, may accumulate in dry soil (Birch 1958).

Effects of Scale

The complexity of soil fertility and soil moisture regimes at La Selva has components at a variety of spatial scales from a regional altitudinal toposequence (Alvarado and Buol 1975; Marrs et al. 1988), to the hillslope catena, to the magnitude of individual trees at all life stages, to the size of individual soil aggregates. Studies of nutrient processing might be organized at each of these scales. The altitudinal transect from the alluvial zone of La Selva to the upper elevations of Braulio Carrillo National Park is a unique opportunity to study a very complex gradient of soil fertility. Indeed, initial studies of soil chemistry and mineralization along the altitudinal transect (Marrs et al. 1988) reveal the marked effects of elevation on regional soil fertility and nutrient availability. A systematic survey of stream chemistry would complement and extend the recently completed soil survey (Sancho and Mata 1987). The slopes of numerous small, steep catchments in the basalt hills provide for replicable observations of hillslope effects on soil development (Jenny 1941). Even smaller-scale variations in nutrient relations are evident from work on nutrient availability within the distinct regions of canopy gaps (crown, bare-soil, and trunk zones; Vitousek and Denslow [1986]).

Rainfall varies spatially as well as temporally. Incident precipitation peaks at midelevations along the elevational transect from La Selva to Braulio Carrillo National Park (Marrs et al. 1988). At a much smaller scale, the amount of precipitation input at the forest floor (throughfall) may vary dramatically over distances of only centimeters (fig. 5.2). Shielded, totally dry positions may be adjacent to marked drip points where amounts of throughfall may exceed incident precipita-

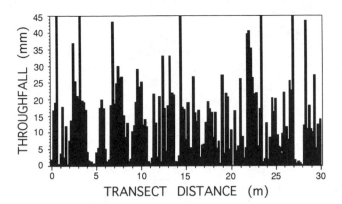

Fig. 5.2. Small-scale spatial variation (0.2 m resolution) in spatial distribution of throughfall under forest on alluvial soil for a single storm of February 11, 1988. Taken on a 30-m E-W transect at 25 m west of CES trail (800-m mark). Values greater than 45 mm indicate sampler overflow.

tion manyfold. Such short-scale variation could affect soil anaerobiosis, nutrient distribution, root growth, and seedling establishment. Other aspects of the hydrological balance are also likely to be spatially variable.

Species-fertility Interactions

Present understanding of nutrient processing and soil variation suggest some study of the reciprocal relation between particular plant species and the quality of the local substrate. There are consistent associations between some plants and soil types at La Selva (Hartshorn 1983b). A study of phytosociological associations conducted in the La Selva permanent plots by M. Lieberman et al. (1985) suggested that site elevation was an important factor in the makeup of the local woody plant community. To what extent is the observed distribution of plant species substrate-determined? Do individual plants exert distinct effects on the chemical, physical, or microbiological properties of soils within their influence?

Tropical tree species often exhibit marked differences in tissue chemistry (e.g., Rodin and Bazilevich 1967; Golley 1983a, 1983c), suggesting different strategies of acquisition and use as well as potential depletion/enrichment of local soils. At La Selva, for example, *Terminalia oblonga* (Surá) has heartwood high in calcium (Hartshorn 1983b), wood of *Dipteryx panamensis* is said to have very high concentrations of silica, and other species likely exhibit variation in aluminum levels. Soils under the dominant canopy species, *Pentaclethra macroloba* appear to have lower pH than in areas removed from this species (chap. 4). Soils under female *Trophis involucrata* individuals had significantly higher phosphorus concentrations compared with soils under the males (Cox 1981), but whether this represents a cause or an effect of the species presence is uncertain. Under *Simarouba amara* however, sexual differences in soil levels of available phosphorus (females had higher levels) are more consistent with gender-based effects on phosphorus availability than with sexual differences in site requirements (R. Sanford pers. comm.). To what extent do these species influence local nutrient processing? Investigations of species-specific tissue chemistry and local variation in both soil and soil solution chemistry could provide some answers.

The effects of animals on nutrient processing is poorly understood. Potentially important functions performed by animals include soil mixing and the promotion of aggregate structure (e.g., earthworms), redistribution of canopy tissues (leaf-cutting ants; Haines [1975]), methanogenesis from gut symbionts (wood-eating insects such as termites), production of readily decomposed frass and feces, and consumption and leaf area loss (sucking and chewing folivores; chap. 21).

More attention could be paid to nutrient processing in specific habitats. Essentially nothing is known of nutrient processing within the La Selva forest canopy, including nutrient allocation and redistribution; leaf biomass and turnover (and longevity); litter interception; and rates of herbivory. The boles of trees have several distinct habitats for plant growth, including the zones along which stemflow preferentially travels (colonized by some climbing plants and even some roots; Sanford [1987]). Swamp forests likely have distinct patterns of nutrient processing (because of prevalent anaerobic soil conditions) and, undoubtedly, influence downstream water quality (see chap. 3).

The ubiquity of mycorrhizal associations (especially VAM) in the La Selva forest has likely influenced the persistence, vigor, and productivity of forest trees under nutrient-limited conditions, but a great deal more remains to be learned about the importance of the symbioses. For example, how obligatory is mycotrophy across different soil types at La Selva? Which aspects of changes in community composition may be attributed to the presence or absence of VAM? What is the actual carbon cost of the symbiosis and under what conditions is it not cost effective to form the association? Can mycorrhizal exploitation of decomposing litter "tighten" the nutrient cycle?

Some aspects of plant-nutrient interactions may be addressed experimentally by altering substrate fertility, such as with amendments of fertilizer, lime, or enzyme inhibitors or with various types of organic matter manipulation (e.g., Harcombe 1977; D. Janos pers. comm.). Some nutrient manipulations are best carried out in the laboratory and greenhouse (Denslow et al. 1987), but other treatments could be undertaken in the field.

For forests at La Selva there is now detailed documentation of spatial variation in soil chemistry and physics; a long background of studies on selected aspects of nutrient cycling; numerous current efforts to understand some mechanisms of nutrient acquisition, transfer, and loss; and a variety of soil/nutrient cycling hypotheses appropriate to the La Selva forest ecosystems. Together these represent an unusual opportunity to probe the reciprocal relationships between plants and their substrate across complex patterns of soil fertility.

ACKNOWLEDGMENTS

This work was supported by The U.S. Man and the Biosphere Program (contract 81-888), the Mary Flagler Cary Charitable Trust, and the Smithsonian Environmental Research Program. I thank C. Jordan, J. Luvall, F. Montagnini, P. Sollins, W. Reiners, and several anonymous readers for helpful comments on an earlier draft and D. Janos, R. Sanford, and P. Sollins for useful discussions and access to unpublished work.

PART II

THE PLANT COMMUNITY: COMPOSITION, DYNAMICS, AND LIFE-HISTORY PROCESSES

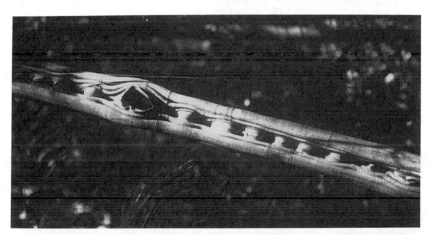

COMMENTARY

Kamaljit S. Bawa and Lucinda A. McDade

The plant community is the structuring element in images conjured by the phrase "tropical rain forest," and, indeed, it frames the experience for most visitors to rain forests. Plants define a forest; they are the physical and ecological matrix in which all other organisms exist; providing both substrate and energetic base for the consumer community. Knowledge of the plant community is, thus, essential to understand the workings of the forest and to research the other organisms that occur there. Historical factors have provided added impetus for research on plants at La Selva. As detailed by Hartshorn and Hammel (chap. 6), the station's original owner, Leslie Holdridge, was a plant ecologist. In the Organization for Tropical Studies (OTS) era, the first large-scale project at La Selva was the comparative ecosystem study that emphasized forest ecology. This emphasis has continued in recent years and it is not surprising that this section is the largest and, perhaps, the most complex in the book. Seven chapters in this section review the status of knowledge of the plant community at La Selva and three conceptually different aspects of the plant community there: vegetation and floristic patterns (chap. 6), forest dynamics (chaps. 7, 8, 9, and 10) and reproductive biology (chaps. 11, 12).

VEGETATION AND FLORISTIC PATTERNS

The characterization of local vegetation is the first step toward understanding the structure and dynamics of a forest. Knowledge of forest physiognomy and species composition at the landscape level provides background information for many other kinds of research. Hartshorn and Hammel (chap. 6) survey La Selva's forests in terms of vegetation types and floristic composition. The complex pattern of soil types (see chap. 4), together with rolling hills, streams, and swamps, makes the vegetation heterogeneous. Vegetation types and species composition in swamps are different from those on the well-drained ridges. Species composition also varies across soil types in patterns that are not yet completely understood. In addition to edaphic and topographic complexity, parts of La Selva have also been subjected to land uses of varying intensity and duration. Abandoned pastures and plantations, secondary forests of different ages, and various managed areas are included within the reserve.

Accelerating deforestation in the tropics has added urgency to efforts to inventory local and regional floras. The Neotropics is the world's richest region floristically (Gentry 1978b), and La Selva, with about fifteen hundred species of native vascular plants known to date, is a prime example of high species richness on a very small spatial scale. Even after an intensive plant-collecting effort extending over one decade, new species are still being added to the flora. Deforestation around La Selva has probably added species via dispersal of weedy species onto the station and, indeed, many of the recently recorded species for La Selva occur in second-growth habitats. Extinction is expected as another consequence of forest fragmentation or, in the case of La Selva, "peninsularization" (see chap. 2). Extinctions can only be documented by repeated sampling, however. For most tropical forests, including La Selva, even the first plant inventory has not been completed. Clearly, inventories to establish baseline data on the occurrence and relative abundances of major groups of organisms are urgently needed for protected areas in the tropics.

Hartshorn and Hammel also present data on the relative diversity of various life-forms among La Selva's plants. For each life-form two to three families have many more species present than the others. Such a pattern of dominance by certain families, especially for trees, has been noted for other tropical forests (Gentry 1988; Ashton 1988). It has been suggested that the species richness patterns of tropical rain forests cannot be fully understood until this issue of familial dominance is addressed (Ashton 1988).

Compared to other lowland rain forests in lower Central America and northwestern South America, La Selva is very rich in epiphytes, but it is otherwise similar floristically and in terms of life-forms. In the last section of chapter 6 Hartshorn and Hammel present a series of species vignettes to exemplify some of the more interesting floristic patterns, taxonomic problems, and reproductive systems among plants at La Selva. Many of these are amenable to investigation from systematic, evolutionary, and ecological perspectives and would make excellent research projects.

FOREST DYNAMICS: DEMOGRAPHY, PATTERNS OF TREE DISPERSION, GAP ECOLOGY, AND PHYSIOLOGICAL ECOLOGY

Understanding forest dynamics is fundamental to several aspects of rain forest ecology, including the successful management of tropical forests for human uses. Equally importantly, forest dynamics figure prominently in theories that seek to account for tropical diversity. The competing models of species diversification and maintenance predict certain patterns of demography, dispersion, and regeneration. These topics are addressed by the authors of the next several chapters, who approach the subject of forest dynamics from different and largely complementary perspectives. Clark (chap. 7) addresses forest dynamics as the sum of individual demographic patterns of the component species. Lieberman and Lieberman (chap. 8) examine patterns of density and dispersion among forest trees and then ask how these data may be successfully explained in terms of processes and how they mesh with the predictions of various models of species diversity. Denslow and Hartshorn (chap. 9) examine gap formation and filling as one of the primary ecological processes involved in forest regeneration and, perhaps, in the maintenance of high species diversity. The conclusions drawn in these three chapters are as informative for their differences as for their common points. In chapter 10, on plant physiological ecology, Fetcher, Oberbauer, and Chazdon provide the link between the population and community level patterns established in the previous three chapters and the individual plant responses to environmental heterogeneity that must ultimately account for these patterns.

As Clark points out, demographic studies in the tropics are beset with difficulties. Tropical forest plants are difficult to age, often occur at low densities or are inaccessible (e.g., epiphytes, lianas), and genets of lianas and some herbaceous perennials are hard to define. Despite these obstacles, a rich body of knowledge is accumulating. Clark summarizes patterns of growth, reproduction, and survival for a number of species, organizing these by life-form. She shows considerable spatial and temporal variation in these attributes within and among species, which is very often associated with microsite differences, notably light. Much of the available demographic information, however, is from the study of components of growth and reproduction in different species. Clark argues that comprehensive data for individual species are necessary and calls for long-term studies that include the major components of the life cycle. Furthermore, Clark pleads for an interdisciplinary approach to integrate morphology, physiology, and environmental factors with population dynamics. Because numerical changes are often accompanied by alterations in the genetic composition of populations, one might add that the integration of genetic and ecological approaches to tropical plant population biology is also long overdue.

Although Clark laments the paucity of demographic studies of lianas and epiphytes, the near absence of such work is not surprising given that some very fundamental aspects of the demography of much more accessible life-forms have yet to be addressed. Despite significant progress it will be some time before one can approach higher-level questions about plant demography at La Selva: What population processes account for the dominance of *Pentaclethra macroloba?* How is the demography of common species different from that of less-abundant species? How do differences in population density within and among species influence patterns of reproduction, recruitment, and genetic diversity?

Lieberman and Lieberman (chap. 8) analyze the structure of the forest at La Selva at a fine scale by focusing on the trees in 12.4 ha of permanent forest inventory plots. They provide data on basal area, intertree distances, size frequency distribu-

tions, foliage height, and canopy dimensions. These data lay the groundwork for comparisons of forest physiognomy among various tropical rain forests. The information on dispersion provides a firm base from which to evaluate various models of species diversity.

An interesting result from the Liebermans' work is that population densities of dioecious species at La Selva are significantly higher than those of monoecious or hermaphroditic species. Hubbell and Foster (1986c) noted a similar trend for dioecious species on Barro Colorado Island (BCI). It is not clear what mechanisms are responsible for this pattern, but it is intriguing that the mean population density factor of hermaphroditic species is exactly half that of the dioecious species. Could it be that some density-dependent factor maintains a certain minimum distance between seed-bearing individuals so that the density of seed-bearing trees is similar across sexual systems? Differences in density could also be a sampling artifact. Density is often related to the size of organisms, and it would be interesting to examine the density of dioecious and hermaphroditic species while controlling for tree size.

The Liebermans conclude their chapter with an emphasis on the dynamic nature of the forest and with the apt comment that the spatial dispersion of adults cannot be understood without investigating the processes that influence the distribution of seeds, seedlings, saplings, and juveniles. Data from long-term demographic studies are, thus, complementary to long-term data from permanent inventory plots.

Forest dynamics in naturally occurring and experimental gaps and the role that gaps play in determining forest structure are the subjects of chapter 9. Denslow and Hartshorn first present data on patterns of tree mortality, size and frequency of gaps, and the range of microclimatic conditions in gaps. Gaps are common at La Selva, with frequency inversely related to gap size. The environment within gaps (e.g., light, relative humidity, soil moisture) varies with gap size and changes quickly as gaps close. Denslow and Hartshorn next consider responses of plants to the altered environmental regime of gaps. The premise for much of the research on gap ecology at La Selva is that responses to environmental changes resulting from gap formation differ among species. It is very clear that plants in gaps respond quickly to the increased light that follows gap formation. Some differences among species in their responses to higher light levels in gaps have been documented, but by and large the differences have not been as pronounced as we might have supposed. Further, most species that have been studied have some degree of shade tolerance.

What is less clear is whether any significant nutrient effects exist in gaps at levels relevant to plants. It has been predicted that nutrient pulses associated with the decay of leaves and small twigs from the fallen tree's canopy would produce a nutrient flush important to establishing plants. On relatively nutrient-rich soils, such as those portions of La Selva where this research was conducted, there simply do not seem to be significant effects on nutrient levels as a result of gap formation (or even of artificial fertilization). As Denslow and Hartshorn point out, however, almost nothing is known of belowground processes, and it is possible that roots present in the crown area before gap formation quickly absorb any nutrients available. This suggests the intriguing idea that gaps may be as important to adjacent trees not directly affected by it as to those plants that occur in the gap proper.

A fascinating aspect of gap ecology that remains poorly understood is the relative role in gap filling of advance regeneration (juveniles already established in the understory) versus dispersal following gap formation. Variation among gaps related to size, season of formation, neighboring trees, and so forth, is likely to be considerable. As pointed out by Denslow and Hartshorn, a number of the features of lowland tropical forest (e.g., high diversity, low population densities, variable fruiting patterns) "contribute substantial noise" to the effort to understand which individuals fill which gaps and why.

At the heart of arguments about regeneration, dispersion, and community structure are differences in physiological responses among species to environmental heterogeneity in time and space. Fetcher, Oberbauer, and Chazdon review research in this area in chapter 10. The authors first document considerable spatial variation in light and, to a lesser extent, temperature and relative humidity, and then summarize research to date on the responses of plants to environmental variation in light. Although plants at La Selva are remarkably shade tolerant, including species commonly considered pioneer or shade intolerant, significant interspecific differences in photosynthetic and growth responses to high light conditions exist. The environment at La Selva is less extreme in availability of water and nutrients, but, as reviewed by Fletcher et al., the limited research suggests interspecific differences in plant response to drought conditions (which are not as rare at La Selva as might be expected; see chap. 3). Nutrient dynamics at La Selva may be strongly influenced by a high incidence of mycorrhizal and nodulating (nitrogen-fixing) species although, as noted, a great deal remains to be learned about below-ground processes.

Plant growth and biomass allocation in a sense integrate the physiological responses of plants over time. In general, plants grow faster in high light conditions and produce relatively more roots and smaller leaves than plants grown in the shade. Pioneer species generally grow faster in high light conditions than do late successional species. Interestingly, individuals of a number of species grew no better or were inhibited in full sun compared to partial shade. It emerges that plants at La Selva face a dilemma: they must deal with chronic low light conditions while being able to respond quickly to high light conditions. Studies of acclimatization to different light regimes suggest that pioneer or gap species are able to respond quickly to high light conditions in a number of ways (e.g., increased leaf production, photosynthetic adjustments). Research by King (1991), however, indicates that they do so at the cost of maintaining a high biomass in leaves even in deep shade. More shade-tolerant species are able to decrease relative biomass devoted to leaves under shaded conditions, perhaps enhancing survival.

Plant architecture and biomechanical properties are also reviewed here as key aspects of the response of plants to their environment. Morphological differences among understory palms can be understood in the context of biomechanics and crown construction in a light-limiting environment. Similarly, architectural differences among trees are related to growth

rates, trunk size, and canopy dimensions. Fast-growing pioneer species are constructed more precariously than slower-growing species. This work is especially interesting in light of the frequency of damage to plants in the understory from falling trees and other debris.

Fetcher et al. suggest that we might do well to think of shade tolerance and ability to respond to high light conditions not as end points of a single continuum but as distinct suites of traits. They also point to the need to undertake studies to enable separation of phenotypic plasticity from genetic variation. Progress has been made in understanding the differential growth rates among species in natural habitats, but the ecological and evolutionary significance of these responses remains largely unexplored.

Conclusions: Accord and Discord. The authors of chapters 6–9 agree on a number of salient features about the forest at La Selva Biological Station, not the least of which is that a great deal remains to be learned. Aspects of forest ecology that have received little attention (or that are recalcitrant to elucidation) include below-ground processes and how these relate to aboveground dynamics, patterns and timing of establishment in gaps, population genetic structure and its determinants, and canopy ecology.

It is clear that the forest at La Selva is dynamic. Whether measured in turnover rates within plant populations (Clark), in mortality rates of trees (Lieberman and Lieberman), or rates of gap generation (Denslow and Hartshorn), one comes to the conclusion that the "stable forest primeval" image is wholly inappropriate for La Selva. This means that the average square meter of forest floor lies within a gap every hundred years or so. It may be more productive to think of the forest as a matrix of gaps of differing age and regeneration histories rather than undisturbed or primary.

Partially as a result of this dynamism, life in the understory is fraught with danger as falling trees, branches, and litter cause high rates of death and damage to understory plants whether they are permanent denizens of this stratum or seedlings and saplings of canopy trees. Although light is generally thought to be the primary factor limiting plant well-being in the understory, this constant rain of detritus from above is certainly an additional hindrance. Many understory plants have tortured histories of growth after multiple knockdowns (McDade unpublished data). The architecture and biomechanical properties of understory plants (herbs and shrubs as well as juveniles of trees) merit further study in this context. If "damage from above" is as common as it appears, one might well expect ontogenetic changes in structure as saplings grow through this dangerous habitat.

The understory at La Selva is also a very dark place. Understory plants and juveniles of canopy species grow exceedingly slowly in this light-limited environment. In contrast, water and nutrients are rarely limiting. Increased appreciation for the importance of rare events in structuring communities, however, should stimulate research on the effects of the occasional severe dry seasons at La Selva.

Given the prominence of light limitation, it is not surprising that the "shade-tolerant versus intolerant" paradigm is a frequent theme and even an organizing concept in these chapters. Hartshorn's pioneering work categorized tree species at La Selva in terms of this rubric, and others have also differentiated pioneer and climax species based on differences in response to light (e.g., Whitmore 1989a). The authors of chapters in this section disagree on the usefulness of the paradigm although we suspect that they agree to a greater extent than is apparent. Clark argues that the paradigm has gotten far ahead of our knowledge of tropical forest plants. Given that we know exceedingly little about any aspect of the regeneration ecology of any but a very few species, she is correct almost by definition. Minimally, it seems clear that we are not dealing with two or three clear-cut and nonoverlapping groups of trees but with a continuum of regeneration requirements. For example, based on extensive growth and mortality data for the forty-five most common species on the permanent plots, the Liebermans have recognized four patterns of growth behavior. There is, however, considerable variability within categories as well as considerable overlap in growth rates and longevity between them.

Perhaps more significantly, as reviewed by Fetcher et al., we now know that essentially all of the species studied from La Selva—even those thought of as pioneer species—show remarkable shade tolerance and are able to survive and grow (slowly!) in the understory for extended periods. This notion is confirmed by the Liebermans' examination of spatial pattern among individuals of the "rapid growth–short life span" category. Although clumped distributions are expected if these are pioneer species that only regenerate in gaps, the results showed no evidence of clumping in two of three plots. Data from the long-term inventory plot on BCI (Hubbell and Foster 1986b) also give little indication of gap association. (See chapter 9, however, for caveats about the use of plant distributions as an adequate index of habitat requirements.) Plants at La Selva differ in their ability to respond to high light conditions and here the "pioneer" species perform largely as predicted by the paradigm, whereas "climax" species are often damaged by high light conditions.

An additional wrinkle here is the increasing evidence that individuals may vary ontogenetically in their requirements. Among the intriguing results of D. A. Clark and D. B. Clark (1992; see fig. 7.6) are significant differences among life stages of the same species in the light environment in which they occur. Similarly, D. A. Clark and D. B. Clark (1987a) showed that saplings of some species appear to lose their shade tolerance at a certain stage. Hubbell and Foster (1990a) also found that different life-history stages of the same species respond differentially to factors influencing regeneration. The concept of ontogenetic variation in shade tolerance must, therefore, be considered in any discussion of regeneration requirements of rain forest trees.

Considering the shade-tolerant versus intolerant paradigm in the context of this information, it is useful to reexamine what the theory actually sought to predict. Clearly, the portion of the paradigm that has been interpreted to predict that certain species *require* high light conditions to *establish* appears to be true of very few species, if any. Similarly, "shade intolerance" does not actually seem to exist if it is defined as inability to establish and survive in the dark of the forest understory. Although seedlings and saplings of most (all?) tree species can persist in the short term in this habitat, tree species differ in their performance in high light conditions. We do not yet know whether saplings of any species are capable of achieving canopy height without at some time experiencing enhanced light. Is it possible that some (all?) canopy tree species require relatively high light conditions at some phase(s) of the life cycle to reach the canopy? It is interesting that given the dynamic nature of the forest at La Selva, "tree saplings reaching the forest canopy have endured more than one episode of gap formation and filling" (Denslow and Hartshorn in chap. 9). The time has come to clarify the predictions of the paradigm and to test these appropriately, incorporating the fact that long-lived individuals may undergo ontogenetic transformations in light response just as in many other characteristics.

Understanding the reasons for high species diversity in tropical forests is surely one of the most fascinating challenges for biologists. It is, perhaps, a reflection of progress toward meeting this challenge that the authors of these chapters reach different conclusions based on work with the same species at the same site. Equilibrium models of species richness are based on competition and niche diversification with species exploiting different resources in spatially and temporally complex environments (Denslow 1987a and references therein). This model predicts nonrandom or patchy distributions of conspecific adults, reflecting the patchy distribution of areas of suitable habitat (Ashton 1969; Hubbell and Foster 1990a). In contrast, models attributing a role to processes such as predation predict less clumping than that initially generated by seed shadows (Janzen 1970; Connell 1971). It is hypothesized that species-specific predators and pathogens are abundant in the vicinity of adult plants and the probability of establishment should increase with distance from the parent tree.

Clark cites numerous studies at La Selva and elsewhere that have yielded results consistent with the Janzen-Connell model. Lieberman and Lieberman argue, however, that most of this research does not, in fact, actually test the model. They suggest that their finding of random dispersion of adults of most tree species at La Selva is inconsistent with both the predation and competition models of species coexistence, and they call for new paradigms for the contrasting patterns of species diversity in temperate and tropical areas. Denslow and Hartshorn warn that it is dangerous to draw conclusions about processes of diversity based on the distribution of adults and without comparisons across a wide spatial scale. They suggest that species are differentially adapted to the rich mosaic of microsites that occur in dynamic tropical forests and that they are differentiated with respect to regeneration requirements (see also Ashton 1969; Gómez-Pompa 1971; Grubb 1977b; Orians 1982). As discussed, however, the evidence for such differentiation is not clear-cut. In fact, Fetcher et al. provide evidence that individuals of essentially all species can persist in the forest understory, at least in the short term (and, thus, do *not* have differentiated niches?) while also suggesting that species differ in their responses to higher light conditions (and, therefore, *do* have differentiated niches?). An added complexity, as noted, is that individuals may differ ontogenetically in these traits.

Plant species may, of course, differentiate in response to other kinds of environmental variation beyond regeneration requirements. The authors in this section mention soil type in particular in this context, and there is limited evidence for a few tree species that distribution is related to soil. The availability of detailed soil data for La Selva (see chap. 4) makes it possible to address this question and should stimulate considerable research. It must also be kept in mind that the soil-plant interface is two-way. Long-lived canopy trees may well be capable of altering the soil on which they occur.

It is interesting that most of the discussion on the "species diversity problem" focuses on factors that *permit* the sympatric occurrence of many species. As important are the processes that *generate* this diversity. To frame the question correctly, one needs to know whether high species diversity is the result of the sympatric existence of many closely related species or of the coexistence of species representing numerous distinct phylogenetic lineages. The fact that the floras of many tropical areas are dominated by species-rich families and genera (see chap. 6) suggests the former, but careful phylogenetic studies will be necessary to resolve this issue. If species richness in tropical forests is related to the piling up of sister taxa in local areas, then it will be critical to understand speciation processes in the tropics. Despite considerable interest in this subject, few monographic studies of tropical forest plants have been conducted. Urgently needed are careful phylogenetic analyses, especially of species-rich (hence, difficult!) groups.

Biologists are clearly not yet approaching consensus on the issue of species diversity. Might it be more productive to avoid framing the issue as a dilemma to be resolved by accepting a single explanation and ruling out all others? Is it not, perhaps, more realistic to consider that species richness at La Selva might be maintained by a combination of equilibrium and nonequilibrium processes? Relatively high speciation and low extinction rates must be considered potential contributors as well. In this light, the Liebermans' call for new ideas seems wholly appropriate.

REPRODUCTIVE BIOLOGY

Plant reproduction is the subject of chapters 11 and 12. Phenological studies are an important component of plant population biology and community ecology, but as Newstrom et al. (chap. 11) emphasize, temporal patterns of growth and reproduction in plants also influence the structure and dynamics of animal communities. As noted by Newstrom et al., the strong seasonality characteristic of temperate ecosystems yields a preponderance of annual phenological cycles. In contrast, in tropical moist forests, seasonality is less pronounced and environmental conditions do not absolutely preclude biological activity at any time of the year. Are phenological events, therefore, less seasonal in these areas? Answering such questions is difficult because of the paucity of long-term phenological studies in the tropics. In addition, methodological and terminological differences make it difficult to compare the data that do exist for different forests.

Newstrom et al. contribute a new classification system for phenological events, using factors that they empirically found to be the most useful in their efforts to analyze the data from a twelve-year phenological study of trees at La Selva. Although in chapter 11 they emphasize flowering, the classification system is readily applicable to other phenological events (e.g., leafing, fruiting). These authors make a convincing case

for distinguishing levels of analysis in phenological studies. For example, in their study, individuals of the same species sometimes had different flowering patterns, and flowering patterns at the population level often were built from patterns that varied among individual plants. At the level of individual trees, Newstrom et al. indeed document that flowering is not predominantly annual. About half of 254 individual trees (representing 211 species) flowered in a subannual pattern, one-quarter flowered annually, and fewer than 10% flowered either continually or supraannually. One hopes that comparative data will soon be available from other tropical forests as well as for other phenological events and life-forms.

Newstrom et al. conclude with a discussion of the factors responsible for the timing of phenological events, both ultimate (selected) and proximate (environmental and physiological cues). A great deal remains to be learned about the determinants of phenology before generalizations can be made. Recurrent phenomena should be relatively easy to predict, yet despite a very large number of phenological studies no predictive models exist for the phenological behavior of tropical plants (Bawa et al. 1990). It will also be important to relate plant phenological patterns to those of their pollinators, seed dispersers, and herbivores. For example, an overwhelming majority of lowland tropical plants are pollinated by insects (Bawa 1990; see chap. 12). A notable problem in understanding the role of pollinators in influencing flowering patterns is the lack of information about temporal patterns of growth and reproduction in insects. Curiously, although entomologists are also interested in phenology (e.g., Wolda 1988), attempts toward a joint conceptual and empirical treatment of plant and insect phenology have been limited.

Kress and Beach (chap. 12) consider pollination mechanisms and sexual systems of plants. Interest in plant-pollinator interactions stems from several factors. At the species level these interactions provide rich material for the study of co-evolution; at the guild and community levels, for the study of community structure and organization; and at various taxonomic levels for the study of speciation (Bawa 1990). Pollinators also regulate gene flow via pollen and, together with plant mating systems and seed dispersal ecology, determine the patterns of genetic differentiation in plant populations. Thus, tropical lowland rain forests provide a unique opportunity to examine the effects of various mating patterns stemming from diverse pollination and sexual systems as well as the effects of patterns of plant dispersion on the spatial organization of genetic variability in populations.

The focus of chapter 12 is on differences in pollination mechanisms and sexual systems among plants occupying various vertical strata within the forest. Pollination by medium-sized to large bees is more common in the canopy, whereas hummingbirds and moths are more common pollinators of flowers of understory plants. Kress and Beach review the considerable body of knowledge accumulated on sexual systems of plants at La Selva. Although a number of studies have shown that most plants at La Selva are strongly outcrossed, there are significant differences between trees and understory species with respect to the incidence of self-incompatibility and dioecy. Kress and Beach note that the potential for inbreeding may be greater for herbaceous plants and small shrubs than for trees and, as a result, the genetic structure of populations in the two groups may be different. The limited genetic data support this point of view. The two understory species of small shrubs investigated so far exhibit low levels of genetic variation within populations (Sytsma and Schaal 1985; Heywood and Fleming 1986). In contrast, canopy trees are highly variable and appear to show little genetic differentiation between populations (Buckley et al. 1988; Hamrick and Loveless 1989).

Of all tropical sites, the most comprehensive data on plant reproductive systems have been amassed for La Selva. Considering, however, that the total number of native flowering plant species is estimated to exceed sixteen hundred, it will be many years before complete data are obtained. Stiles's work, referenced in chapter 12, is among the most complete studies of a group of plants serviced by the same pollen vectors, in this case, hummingbirds. Recent research on beetle pollination has underscored the importance of a common but neglected plant-pollinator interaction in the tropics. Additional work is required to document fully the distribution of these and other systems, especially those involving bees and moths, and to elucidate the role of plant-pollinator interactions in maintaining species diversity and community stability (Bawa 1990). Continuing studies of sexual systems, which are now being extended to the analysis of mating systems and genetic structure, should increase understanding of gene flow and microevolution in plant populations (O'Malley and Bawa 1987;

Hamrick and Loveless 1989). In addition, knowledge of population genetic structure will be immensely useful in management and conservation of forest genetic resources (Bawa and Ashton 1991; Hamrick et al. 1991). Efforts to domesticate tropical forest trees (chap. 25) also need to take genetic variation into account.

Overall, the authors of chapters 6–12 document progress in understanding plant community composition, structure, dynamics, and reproductive biology. It is obvious that only a beginning has been made and that many problems remain. Progress in many areas may come from an interdisciplinary approach designed to tackle selected problems comprehensively as emphasized by several authors. For example, Clark argues that research on regeneration and demography must be integrated with physiological ecology and reproductive biology, including phenology and genetics. Another recurrent theme in various chapters is the hope that better knowledge of community ecology and population biology will provide a sound basis for the management and conservation of tropical rain forest (see also Bawa et al. 1990). The realization of this hope, too, will require contributions from many disciplines. It will also require reorientation of basic research and a sharper definition of the unresolved critical issues in conservation and management of tropical forest resources so that these may be addressed through plant ecological research.

6

Vegetation Types and Floristic Patterns

Gary S. Hartshorn and Barry E. Hammel

An emphasis on plant research, particularly trees, has distinguished La Selva among tropical biological stations from the beginning. Dr. Leslie Holdridge, renowned forester and previous owner of the property (cf. Gómez and Savage 1983), was the most important figure in establishing this botanical focus. While on the forestry faculty at Inter-American Institute for Agricultural Sciences (IICA) (now Tropical Agricultural Research and Education Center [CATIE]), Turrialba, in the late 1950s and early 1960s, Holdridge brought student groups to La Selva and encouraged graduate students (e.g., Janis Petriceks [1956], Humberto Tasaico [1959], and Gerardo Budowski [1961]), to conduct thesis research there. During Holdridge's ownership and the early days of the Organization for Tropical Studies (OTS), the site also attracted temperate biologists who contributed information on La Selva's vegetation and flora.

The trees of La Selva traditionally have been better known than the smaller, more accessible herbs and shrubs. This bias is typical of many tropical forests where the economic importance of timber species has focused attention on the physically dominant growth-form. The number of tree species known from La Selva has increased as a result of recent collecting for the La Selva flora project (Hammel and Grayum 1982) but less dramatically than has the total number of plant species. By 1979 (when the flora project began) approximately 270 species of trees greater than 10 cm dbh were recorded among a total of 710 species of vascular plants (Hammel and Grayum 1982; Hartshorn 1983b). As of 1988, approximately 323 species of trees were known among 1,678 species of naturally occurring vascular plants. The total recorded flora more than doubled during this interval, whereas the tree flora increased by only 20%. Beyond floristics, trees are the focus of major research projects on demography, ecophysiology, breeding systems, phenology, pollination, seed dispersal, and mycorrhizal relations, among others. La Selva's primary forests are also the locale for vegetation research on stand structure, gap dynamics, floristic patterns, and dominance-diversity relations.

Our objectives in this chapter are to describe the physical appearance of La Selva with emphasis on vegetation types, successional patterns, forest structure, and dominance-diversity relations; to characterize the La Selva flora by summarizing available information on the numerical distribution of species among families, genera, and growth-forms, as well as the phytogeographic affinities of the flora; and to pro-

vide a close-up "snapshot" of a few species that exemplify some of the interesting floristic and biological patterns of La Selva.

LA SELVA VEGETATION

The natural vegetation of La Selva is classic tropical rain forest (*sensu* Richards 1952) characterized by a species-rich, multilayered community with impressive trees, lianas, epiphytes, and broad-leaved monocots. The diversity and abundance of these growth-forms distinguish La Selva from temperate forests and less humid, tropical forests. La Selva trees are not particularly tall, compared to other lowland forests such as that in Corcovado National Park (Hartshorn 1983b) or the dipterocarp forests of the tropical Far East (Whitmore 1984). The La Selva primary forests are distinctive because of the abundance of one leguminous tree species, *Pentaclethra macroloba* (Willd.) Ktze. Because *Pentaclethra* is so common, even tree diversity is not immediately impressive on a small scale. On the other hand, nearly every growth-form (from understory herb to canopy tree) has epiphytes and epiphylls (mosses and liverworts that grow on leaves). The abundance of climbing aroids, Cyclanthaceae, other trunk epiphytes, and canopy lianas also is visually impressive.

In different vegetation classification systems, the natural forest of La Selva is known as tropical wet forest (Holdridge 1947), tropical rain forest (Richards 1952), continuously wet tropical rain forest (Walter 1973), tropical ombrophilous lowland forest (Mueller-Dombois and Ellenberg 1974; UNESCO 1978b), tropical humid forest (Udvardy 1975), tropical lowland evergreen rain forest (Whitmore 1984), and tropical moist forest (Myers 1980). The latter—or an inverted form (moist tropical forest)—is used in a far more general sense than Holdridge's (1947) narrowly defined tropical moist forest.

An early ecological map of Costa Rica (Tosi 1969) indicates two life zones (*sensu* Holdridge 1967) for La Selva: tropical wet forest on the east side and basal belt (warm) transition to tropical premontane wet forest on the west side. The 4,000-mm isohyet is the division between these two life zones, and the long-term average annual rainfall for La Selva hovers around this value. A detailed forest typing study, however, maps the entire Puerto Viejo region (including all of La Selva) within the tropical wet forest life zone (Hartshorn and Peralta 1988). The restricted occurrence of a few tree species on the western side of old La Selva appears to be related to soil differences (see chap. 4) rather than a change in life zones.

In the hierarchical life zone classification system (Holdridge 1967) all of La Selva's vegetation types are wet atmospheric associations. Conditions are more humid than would be expected from average rainfall at La Selva because of moisture-laden trade winds (December–April) and cold air flow down the nearby slopes of the Cordillera Volcánica Central. The latter contributes to canopy condensation and drip almost every night at La Selva. Even during the dry season rarely more than one month passes with less than 100 mm of rainfall, resulting in the perhumid conditions that contribute to the abundance of epiphytes. Only in exceptional years (e.g., 1961, 1973) does a month have less than 100 mm of rain or three to four weeks go by without significant rainfall (see chap. 3).

VEGETATION TYPES

The present vegetation of La Selva is determined largely by two principal factors: physiographic position and recent land-use history. Ridges are sites richest in species, whereas poor drainage strongly influences vegetation type and reduces plant diversity (M. Lieberman et al. 1985). Pre-1968 conversion of La Selva forests to plantations of cacao (*Theobroma cacao* L., Sterculiaceae), laurel (*Cordia alliodora* (Ruíz & Pavón) DC., Boraginaceae), pejibaye (*Bactris gasipaes* H. B. & K., Arecaceae) and banana (*Musa paradisiaca* Linn., Musaceae), or pasture, plus some selective logging, have produced several easily recognized vegetation types (table 6.1). The purpose of this section is to describe briefly the major vegetation types on the principal land forms of greater La Selva, that is, the OTS land between the Puerto Viejo and Sarapiquí rivers. The structural physiographic descriptions used here predate the formal classification and mapping of La Selva's soils (see chap. 4). As a result, some of the inferred correlation between soils and vegetation may not be precise.

Primary Forest on Rolling Terrain

Primary forest on rolling terrain is the principal vegetation type and the most extensive land-form of La Selva (table 6.1); it occurs on the Residual Landscape Group (chap. 4). The primary forests of La Selva are dominated by *Pentaclethra macroloba,* which comprises about 40% of the timber volume (Bethel 1976) and has an importance value of 18%–23% (table 6.2). Large *Pentaclethra* trees usually form the base of the forest canopy at 30–35 m, with scattered individuals of

Table 6.1 Vegetation types of La Selva

Vegetation Type	Area (ha)	Percentage
Primary forest	848	55.2
Moderately high-graded forest	114	7.4
Secondary forest	162	10.6
Early successional pasture	272	17.7
Abandoned plantations	115	7.5
Aboretum and managed habitats	8	0.5
Developed areas	17	1.1
TOTAL	1,536	100.0

many other canopy species attaining heights of 40–55 m, which gives the canopy of this forest type a very irregular texture (fig. 8.7 in chap. 8). The vertical discontinuities are further accentuated, of course, by the frequent occurrence of tree falls (see chap. 9).

Subcanopy palms are a distinctive component of this vegetation type. *Welfia georgii* Wendland ex Burret, *Iriartea deltoidea* Ruíz & Pavón and *Socratea exorrhiza* (C. Martius) Wendland usually rank among the top five tree species in importance values (table 6.2). Although these tall palms do not attain large diameters (generally <30 cm dbh), they occur at high densities, which accounts for their high importance values. The abundance of subcanopy palms in the La Selva primary forest is accentuated by their large leaves (especially *Welfia*), long spines (*Astrocaryum confertum* Wendland ex Burret), high stilt roots (*Iriartea* and *Socratea*), and elegance (particularly *Euterpe macrospadix* Oersted). Typical subcanopy dicot species in this forest type include *Dendropanax arboreus* (L.) Dcne. & Planchon (Araliaceae), *Dystovomita paniculata* (J. D. Smith) Hammel (Clusiaceae), *Protium panamense* (Rose) I. M. Johnston and *P. pittieri* Porter (Burseraceae), and *Unonopsis pittieri* Safford (Annonaceae).

The understory of this vegetation type is, perhaps, the most difficult layer to define or describe. Nevertheless, a large suite of tree species attains sexual maturity at a small size (5 cm dbh or 5 m tall) and usually does not exceed 30 cm dbh or 15 m in height. Typical understory species include *Anaxagorea crassipetala* Hemsley (Annonaceae), *Cassipourea elliptica* (Sw.) Poir. (Rhizophoraceae), *Guarea rhopalocarpa* Radlk. (Meliaceae), *Rinorea deflexiflora* Barth (Violaceae) and *Symphonia globulifera* L.f. (Clusiaceae). *Symphonia globulifera* at La Selva is small compared to those on other sites.

A conspicuous set of woody species less than 5 m tall corresponds to Richards's (1952) shrub or D layer. It is characterized by treelets and colonial palms but not by bushy shrubs. Particularly abundant in this layer are the multiple-stemmed palms: *Geonoma congesta* Wendland ex Spruce; an undescribed *Bactris* sp.; and the dwarf, single-stemmed palms, *Asterogyne martiana* (H. Wendl.) H. Wendl. and *Geonoma cuneata* Wendland ex Spruce. Common dicot treelets in this vegetation type include *Carpotroche platyptera* Pittier (Flacourtiaceae), *Herrania purpurea* (Pittier) R. E. Schultes (Sterculiaceae), *Mabea occidentalis* Benth. (Euphorbiaceae), *Ocotea atirrensis* Mez & J. D. Smith ex Mez (Lauraceae), and *Miconia simplex* Tr. (Melastomataceae).

Most of the small (<1 m tall) plants are seedlings of trees and woody lianas rather than herbaceous plants. Nevertheless, some perennial herbs do occur on the primary forest floor, such as *Asplundia uncinata* Harling (Cyclanthaceae), *Besleria columneoides* Hanst. (Gesneriaceae), *Calathea cleistantha* Standley (Marantaceae), *Danaea wendlandii* Reichenb. F. (Marattiaceae), and *Spathiphyllum fulvovirens* Schott (Araceae).

Primary Forest on Old Terrace

Primary forest on old terrace covers appreciable areas in the northern part of La Selva, particularly along the Río Puerto Viejo (see fig. 2.2, chap. 2). This vegetation type occurs on the Upper Terrace Landscape Group (chap. 4). This relatively flat land-form (40–45 m elevation) appears to be a Pleistocene

Table 6.2 Stand characteristics of a 4-ha plot (3) in primary forest on rolling terrain, La Selva

Species	Density		Frequency		Basal Area		Importance
	Stems	Percentage	Subplots	Percentage	m²	Percentage	Percentage value
Pentaclethra macroloba	264	12.4	98	6.70	31.97	34.6	17.94
Socratea exorrhiza	218	10.3	86	5.88	2.66	2.88	6.36
Welfia georgii	163	7.71	85	5.81	3.25	3.52	5.68
Iriartea deltoidea	150	7.10	68	4.65	3.72	4.03	5.26
Protium pittieri	92	4.35	63	4.31	2.56	2.77	3.81
Warscewiczia coccinea	90	4.26	54	3.69	1.49	1.62	3.19
Euterpe macrospadix	69	3.27	39	2.67	0.76	0.83	2.25
Protium glabrum	46	2.18	33	2.26	1.40	1.52	1.99
Pourouma aspera	30	1.42	26	1.78	2.17	2.35	1.85
Virola sebifera	38	1.80	30	2.05	1.44	1.56	1.80
Subtotal top 10 spp.	1,160	54.9	582	39.8	51.42	55.7	50.13
TOTAL	2,113	100	1,462	100	92.34	100	100

Note: Subplots are 20 × 20 m. Plot 3 is on the Arboleda soil consociation, Río Puerto Viejo terrace.

terrace that is no longer subject to flooding. The best example of this vegetation type lies north of the Quebrada Surá and west of the laboratory complex. The Camino Experimental Sur (CES) trail traverses this vegetation type, including some good examples of the slow dissection of this land-form by small streams (e.g., the corners of plot 1). The central ridge in the Holdridge Arboretum, the Sendero Oriental (SOR) trail, Rafael's bluff, and the *Cordia* stand along the Sendero Holdridge (SHO) trail appear to represent smaller, more dissected remnants of, perhaps, the same old alluvial terrace.

As with the preceding vegetation type, *Pentaclethra* dominates the primary forest on this land-form (table 6.3). Occasional, exceptionally large emergents of *Dipteryx panamensis* (Pittier) Record (Fabaceae), *Hymenolobium mesoamericanum* Lima (Fabaceae) and *Sloanea latifolia* (L. C. Rich.) K. Schum (Elaeocarpaceae) characterize the canopy of this type of primary forest at La Selva. A notable absentee from this forest type is *Carapa nicaraguensis* C. DC. (Meliaceae), which occurs in primary forests in swamps and on hilly terrain of La Selva. *Warscewiczia coccinea* (DC.) Klotzsch (Rubiaceae) is an abundant subcanopy tree in this forest type; in contrast, *Iriartea* is absent from plot 1. *Capparis pittieri* Standley (Capparaceae) and the colonial palm *Bactris porschiana* Burret are prominent in the forest understory.

Primary Forest in Swamp

Primary forest in swamp is a minor but distinctive vegetation type in La Selva that occurs on the Swamp/Valley Bottom Landscape Group (chap. 4). It occurs most extensively along the middle section of the Quebrada El Salto, which includes about half of plot 2. Many other smaller patches of swamp forest occur along or near the numerous creeks of La Selva. Although these swamp forests will flood with 20–30 cm of water during very heavy rains (e.g., *temporales*), standing floodwaters usually do not remain for more than three to five days. A detailed analysis of the microtopography of the western part of plot 2 indicates four habitats: poorly drained firm sediments (62%); very poorly drained, soft sediments (23%); standing water (9%); and 5% with running water (M. Lieberman et al. 1985). Stem density and species presence are highly skewed to the firmer sediments (chap. 8).

Pentaclethra also dominates the swamp forest; other canopy species, however, comprise the top five importance values (table 6.4), contrasting with the situation on firm ground where subcanopy trees have high importance values. Characteristic canopy trees of the swamp forest include *Carapa nicaraguensis, Luehea seemannii* Triana & Planchon (Tiliaceae), *Otoba novogranatensis* Moldenke (Myristicaceae), *Pachira aquatica* Aublet (Bombacaceae), and *Pterocarpus officinalis* Jacq. (Fabaceae).

Although subcanopy tree species occur in this forest type, few (e.g., *Grias cauliflora* L. [Lecythidaceae], *Pithecellobium valerioi* [Britton & Rose] Standley [Fabaceae]) appear to be limited to swamp forest. Swamp forest understory does have several characteristic species: *Adelia triloba* (Muell.-Arg.) Hemsley (Euphorbiaceae); *Astrocaryum alatum* Loomis (Arecaceae); *Bactris longiseta* Wendland ex Hemsley (Arecaceae); *Chione costaricensis* Standley (Rubiaceae) and *Psychotria chagrensis* Standley (Rubiaceae).

Most swamp forest species also occur on the hilly, better-drained terrain of La Selva, but at much lower densities. Hartshorn (1972) attributes the occurrence of *Pentaclethra* on ridges, hills, and old terraces of La Selva to the lack of an effective dry season. Some typical swamp forest species (such as *Carapa*) are fairly common on the steep slopes south of La Selva (200–600 m elevation), probably also because of the lack of an effective dry season and, perhaps, higher average annual rainfall than at La Selva (Hartshorn and Peralta 1988).

Open Swamps

Open swamps occur in the northern half of La Selva on the Swamp/Valley Bottom Landscape Group (chap. 4) wherever drainage is extremely poor and there is standing water for several months of the year. Some of the open swamps of La Selva

Table 6.3 Stand characteristics of a 4.1-ha plot (1) in primary forest on old terrace, La Selva

Species	Density		Frequency		Basal Area		Importance
	Stems	Percentage	Subplots	Percentage	m²	Percentage	Percentage value
Pentaclethra macroloba	228	12.4	96	7.60	43.16	38.9	19.62
Welfia georgii	291	15.8	104	8.23	6.41	5.78	9.95
Socratea exorrhiza	203	11.1	88	6.96	2.84	2.56	6.86
Protium panamense	52	2.83	42	3.32	2.12	1.91	2.69
Warscewiczia coccinea	70	3.81	37	2.93	1.40	1.26	2.67
Laetia procera	29	1.58	25	1.98	4.73	4.26	2.60
Goethalsia meiantha	45	2.45	29	2.29	2.71	2.44	2.40
Dendropanax arboreus	35	1.91	32	2.53	2.89	2.60	2.35
Dipteryx panamensis	9	0.49	9	0.71	5.80	5.22	2.14
Casearia arborea	40	2.18	33	2.61	1.16	1.04	1.94
SUBTOTAL top 10 spp.	1,002	54.5	495	43.1	73.22	65.9	53.22
TOTAL 171 species	1,837	100	1,149	100	111.08	100	100

Note: Subplots are 20 × 20 m. The well-drained portions of plot 1 are on the experimental soil consociation, Río Puerto Viejo terrace. The poorly drained portions are on the Pantano consociation of the same terrace.

Table 6.4 Stand characteristics of a 4-ha plot (2) in primary forest in swamp, La Selva

Species	Density		Frequency		Basal Area		Importance
	Stems	Percentage	Subplots	Percentage	m²	Percentage	Percentage value
Pentaclethra macroloba	265	16.7	54	7.05	30.77	29.8	18.83
Carapa nicaraguensis	69	4.34	27	3.53	13.41	13.0	6.95
Iriartea deltoidea	133	8.37	35	4.57	3.38	3.27	5.40
Pterocarpus officinalis	33	2.08	20	2.61	6.71	6.49	3.73
Welfia georgii	76	4.78	30	3.92	1.70	1.64	3.45
Astrocaryum alatum	68	4.28	33	4.31	0.84	0.82	3.13
Goethalsia meiantha	36	2.26	20	2.61	3.50	3.38	2.75
Apeiba membranacea	22	1.38	15	1.96	4.16	4.03	2.46
Colubrina spinosa	39	2.45	21	2.74	0.49	0.47	1.89
Protium panamense	24	1.51	14	1.83	1.12	1.09	1.47
SUBTOTAL top 10 spp.	765	48.1	269	35.1	66.08	63.9	50.06
TOTAL	1,590	100	766	100	103.38	100	100

Note: Subplots are 20 × 20 m. The well-drained portion of plot 2 is on the Holdridge soil consociation, Río Puerto Viejo terrace. The poorly drained portion is on the Pantano consociation of the same terrace.

were caused by forest conversion for pasture (e.g., east of SHO 1800 in Annex A); however, most open swamps appear to be natural. Excellent examples of natural open swamps occur along the Cantarrana and CES trails as well as in the middle sections (1–2 km upstream from the main rivers) of the many creeks that traverse La Selva.

The low vegetation of open swamps is characterized by perennial herbs such as *Calathea lutea* (Aublet) Schultes (Marantaceae) and *Spathiphyllum friedrichsthalii* Schott (Araceae) and the subscandent shrubs *Acalypha diversifolia* Jacq. (Euphorbiaceae) and *Malvaviscus arboreus* Cav. (Malvaceae). The large herbs typically occur in the more open swamps, whereas shrubs are prominent in the more brushy, narrow streamside swamps. Low, brushy vegetation occurring naturally in open areas within a matrix of forest is known locally as *tacotal;* the term, however, is also applied elsewhere to the

thick vegetation of the first few years of succession that precede the *charral* (machete) stage. La Selva tacotales may contain a few isolated trees of *Cecropia obtusifolia* Bertol. (Cecropiaceae), *Pentaclethra,* and so forth, but the successional process appears to be greatly retarded in these open swamps. For example, the open swamp east of SHO 1800 was cleared for pasture in the early 1960s and purchased by OTS in 1970. Secondary forest has still not reclaimed the swamp after some twenty-five years.

Riparian
Riparian vegetation includes a number of tree species that appear to be restricted to this habitat on the Lowest Terrace Landscape Group (chap. 4) in La Selva. The large trees that gracefully arch over the Río Puerto Viejo are *Ficus insipida* Willd. (Moraceae), *chilamate,* and *Pithecellobium longifolium*

(Humb. & Bonpl.) Standley (Fabaceae), *sota caballo*. Figs of the taller chilamate trees provide the principal food of the *machaca* fish (see chap. 14). The shorter sota caballo trees often lean out over the river and seem remarkably resilient to bank erosion. *Cordia lucidula* I. M. Johnston (Boraginaceae), *Inga marginata* Willd. (Fabaceae), *Nectandra reticulata* Ruíz and Pavón (Lauraceae), and *Posoqueria latifolia* (Rudge) Roem. & Schult. (Rubiaceae) are mostly restricted to this habitat.

Several other species frequently occur along streams, but also occur in other alluvial habitats, such as *Inga ruiziana* G. Don. (Fabaceae), *Luehea seemannii* (Tiliaceae), and *Myrcia splendens* (Sw.) DC. (Myrtaceae). A number of species typical of higher elevations occur occasionally along the Puerto Viejo and Sarapiquí rivers in La Selva; these include *Acacia ruddiae* Janzen (Fabaceae), *Carludovica rotundifolia* Wendland ex Hooker F. (Cyclanthaceae), *Clusia minor* L. (Clusiaceae), *Cuphea utriculosa* Koehne (Lythraceae), and *Piper friedrichsthalii* C. DC. (Piperaceae).

Secondary Forest
Secondary forest now comprises substantial areas of La Selva because of OTS purchases of adjoining lands to add to Holdridge's original Finca La Selva. Approximately 200 ha of young secondary forests occur on the Residual Landscape Group (chap. 4) of the Sarapiquí Annex. These forests arose in the early 1970s after clear-cutting of the primary forest on hilly terrain and general failure to establish pasture. Now approximately fifteen years old, these young secondary forests have densely packed canopies dominated by *Cecropia insignis* Liebm. and *C. obtusifolia* (Cecropiaceae), *Goethalsia meiantha* (J. D. Smith) Burret (Tiliaceae), *Laetia procera* (Poepp.) Eichl. (Flacourtiaceae), and *Rollinia microsepala* Standley (Annonaceae).

Some of the secondary forests at the north end of the Sarapiquí Annex (e.g., along the Sendero Atajo [SAT] trail) are not as robust as those farther south. This lack of vigor appears to be related to repetitive cleaning of these pastures before abandonment; in fact, some patches of pasture still exist in this part of La Selva. The odoriferous stand of *Psidium guajava* L. (Myrtaceae), guava, (on the Middle Terrace Landscape Group, chap. 4) through which the SAT trail passes probably originated from cattle-dispersed seed from founder trees left in the pasture by the previous owner. Now that succession is proceeding, the guava trees do not appear to be spreading; however, the adult trees will persist, probably, for many years.

Some interesting stands of secondary forest occur along the SHO trail on the Middle Terrace Landscape Group (chap. 4) where pastures were abandoned between 1966 and 1968. A 1975 inventory documented the dominance of shade-intolerant gap species (Hartshorn 1983b). The proximity of primary forest probably contributed to the floristic richness of this stand of secondary forest. A substantial block of secondary forest occurs in the southeast section of the Sarapiquí Annex, where a swath 200–300 m wide of primary forest bordering the south boundary was cut in the early 1970s to protect the farm from squatters by the legally condoned method of demonstrating "use" or "improvement" of the land. Local informants concur that the felled forest was neither burned nor planted to grass. No studies have compared the resultant secondary forest to others with different postclearing fates.

Abandoned Plantations
Abandoned plantations are the most extensive vegetation type along the Puerto Viejo and Sarapiquí rivers, principally on the Upper and Middle Terrace Landscape Groups (chap. 4). Except for a small patch along the Sendero Occidental (SOC) trail near its juncture with Quebrada Leonel, no stands of undisturbed forest remain on this land-form. It appears, however, that almost all of the native tree species are present because so many trees were left for shade when the cacao plantations were established. Some exotic shade trees were also planted (*Dalbergia tucurensis* Donn. Smith and *Erythrina poeppigiana* O. F. Cook (both Fabaceae). Characteristic canopy tree species of this forest type include *Brosimum alicastrum* Sw. (Moraceae), *Bursera simaruba* (L.) Sarg. (Burseraceae), *Castilla elastica* Sessé (Moraceae) *Cordia alliodora* (Ruíz & Pavón) DC. (Boraginaceae), *Hernandia stenura* Standley (Hernandiaceae), *Hura crepitans* L. (Euphorbiaceae), *Terminalia oblonga* (Ruíz & Pavón) Steud. (Combretaceae), and *Zanthoxylum panamense* P. Wils. (Rutaceae).

Typical subcanopy tree species are *Alchornea costaricensis* Pax & K. Hoffmn. (Euphorbiaceae), *Bravaisia integerrima* (Spreng.) Standley (Acanthaceae), *Casearia corymbosa* Kunth (Flacourtiaceae), *Pleuranthodendron lindenii* (Turcz.) Sleumer (Flacourtiaceae), and *Simira maxonii* (Standley) Steyermark (Rubiaceae). Representative understory tree species include *Chamaedorea tepejilote* Liebmann (Arecaceae), *Guarea brevianthera* C. DC. (Meliaceae), *Lonchocarpus oliganthus* Hermann (Fabaceae), *Ocotea cernua* (Nees) Mez (Lauraceae), and *Pithecellobium catenatum* J. D. Smith (Fabaceae). Shrubs or treelets less than 10 cm dbh include *Ardisia opegraphia* Oersted (Myrsinaceae), *Cordia lucidula* I. M. Johnston, *Justicia aurea* Schlecht. (Acanthaceae), *Piper nudifolium* C. DC. (Piperaceae), and *Psychotria chiapensis* Standley (Rubiaceae). Common herbs in this habitat are *Bomarea obovata* Herb. (Amaryllidaceae), *Clidemia reitziana* Cogn. & Gleason ex Gleason (Melastomataceae), *Dieffenbachia longivaginata* Croat & Grayum ined. (Araceae), *Psychotria alfaroana* Standley, and *P. tonduzii* Standley (both Rubiaceae).

Managed Habitats
Managed habitats is a catch-all category that affects several vegetation types. Here, only two key areas are described—the Holdridge Arboretum and the successional strips. The Holdridge Arboretum is a 3.5-ha patch of former cacao plantation mostly on recent alluvium but with a long peninsula of old alluvial terrace. The cacao was removed in 1968, leaving the rich panoply of shade trees to start the arboretum with 661 trees representing 111 native species. Over succeeding years the planting of trees has raised the totals to about 1,200 trees of 250 native species. The herbaceous vegetation is cut three to four times a year, giving the arboretum a parklike attractiveness.

In 1970 a series of five successional strips was initiated in response to the growing isolation of La Selva's primary forest by surrounding pastures. The specific objective was to maintain representative patches of the early successional stages by cutting one strip (ca. 0.5 ha) each year on a five-year rotation. Fortuitously, the annual cut is made in February, which provokes a major fruiting of the colonizer *Phytolacca rivinoides* Kunth & Bouche (Phytolaccaceae) that begins in September just in time for the arrival of latitudinal migrant birds (see chap. 17).

The successional process on the strips may not be representative of the region. Because the slash is not burned, many woody and herbaceous species sprout copiously. Nevertheless, many classic pioneer species do colonize the strips. The cosmopolitan weed *Erechtites hieracifolia* (L.) DC. (Asteraceae) germinates in a few weeks and produces mature seeds within six weeks. Another early colonizer is the aforementioned *Phytolacca,* which barely persists into the second year. The pioneer trees *Cecropia obtusifolia* (Cecropiaceae), *Heliocarpus appendiculatus* Turcz. (Tiliaceae) and *Ochroma lagopus* Swartz (Bombacaceae) also get an early start from the seed bank. These vigorous pioneer trees easily become reproductive within the five-year rotation. Many other tree species regenerate by stump sprouts in the successional strips.

THE FLORA

The La Selva Flora project, begun in 1979 under the direction of Robert Wilbur (1986) and involving resident collectors for a total of approximately forty person-months through 1984, has produced about twenty thousand collections. The dramatic increase in number of species recorded from the area (from 710 to ca. 1,680; app. 3) was inevitable because previously the major herbaceous and epiphytic groups (e.g., ferns, Orchidaceae, Araceae) had been almost totally ignored. During this period the area of La Selva grew from 730 to ca. 1,400 hectares (see chap. 2), but the increased number of species was more the result of the all-inclusive nature of general collecting than to the increase in size of the property. Undoubtedly, future workers will find species not included in the eventually published flora, especially if intensive plot studies are undertaken in primary forest on the Sarapiquí Annex and near the southern boundary of the original property. For example, a second species of *Ormosia* (Fabaceae; Hartshorn, unpublished data) as well as two species of ferns new to the property (Grayum and Churchill 1987) were found during intensive plot inventories in 1985.

The herbaceous and weedy flora of secondary and human-maintained habitats (e.g., successional strips, laboratory clearings, abandoned pastures) is very dynamic (Grayum and Churchill 1987); species new to the flora periodically colonize these successional habitats while others disappear. For example, the grass *Pseudechinolaena polystachya* (Kunth) Stapf was abundant in the pejibaye grove in 1978 but has not been seen since (Grayum, pers. comm.; Judziewicz and Pohl 1984). *Trema micrantha* (L.) Blume (Ulmaceae) arrived on the East Bank with construction of the footbridge, where only *T. integerrima* (Beurl.) Standley was present previously. Despite these sources of plant species new to La Selva, we believe the La Selva tree list is nearly complete and that the overall flora, estimated to contain nearly two thousand species (Hammel and Grayum 1982; Hammel 1986i), will see no more dramatic increases.

Taxonomic Richness

In the discussion that follows we avoid the disparity of taxonomic opinion concerning the delimitation of fern families by treating the Polypodiaceae in the broad sense. This, along with seven other families of ferns or fern allies (Cyatheaceae, Gleicheniaceae, Lycopodiaceae, Hymenophyllaceae, Marattiaceae, Ophioglossaceae, and Selaginellaceae) is included in the pteridophyte group. We also treat the three groups of legumes as one family, Fabaceae. Three pairs of families, Cecropiaceae/Moraceae, Heliconiaceae/Musaceae, and Viscaceae/Loranthaceae (each of which is often treated as one family), are here treated as six families. At the familial level the La Selva flora contains 2 gymnosperm, 8 pteridophyte, and 122 flowering plant families for a total of 132 families of vascular plants. Among these, four (Balsaminaceae, Caprifoliaceae, Musaceae, Portulacaceae) are known only from cultivation or as escapees, and four (Brassicaceae, Caryophyllaceae, Loasaceae, Molluginaceae) are known only from ephemeral habitats (river sandbar) or from areas of human disturbance. One additional family, Papaveraceae (*Bocconia*), is known only from sterile collections or sightings; hence, it is not on the official list.

Of the 1,744 species based on herbarium specimens, 66 are cultivated (app. 3). Approximately 220 species are widespread adventives (referred to as exotics) typically associated with pastures, road clearings, and other large-scale human disturbances. The remaining 1,458 "natives" also include species of disturbed habitats such as tree-fall gaps and riverbanks, but these are species that seem likely to have been present before recent human disturbance. The "natives" are comparable to Foster and Hubbell's (1990) concept of the "forest flora" on Barro Colorado Island (BCI) in Panama, which includes 966 species. On BCI, an additional 241 more-or-less naturalized weedy species found in cleared areas and along the island's shoreline give a total of 1,207 (plus 65 cultivated) species now known from the island (Foster and Hubbell 1990a). Croat (1978) included another 140 mostly weed species, which had been reported by Standley (1933) but apparently no longer, or never did, occur on the island.

Although a few of the 220 "exotic" species (13% of the flora) at La Selva may have disappeared since collecting for the flora began, most are still present, and many are abundant and naturalized in disturbed habitats. As a result, figures 6.1 and 6.3 deal with a flora of 1,678 species, comparable to BCI's native and naturalized flora of 1,207 species. The rankings in the bar graphs (figs. 6.1 and 6.3), however, are based on the "natives" as defined previously. For example, because two-thirds of the grasses are introduced weeds, the family Poaceae (fig. 6.1) is ranked eighteenth even though in total number of species it is eighth (table 6.5).

Among the native flora of La Selva's natural habitats, the pteridophytes, Orchidaceae, Araceae, Rubiaceae, Melastomataceae, Fabaceae, Piperaceae, Moraceae, Euphorbiaceae, and Arecaceae rank highest in numbers of species (fig. 6.1; app. 3). Even when considered alone, "the broadly-circumscribed Polypodiaceae . . . reigns as the largest family in the La Selva vascular flora" (Grayum and Churchill 1987) with 127 of the 171 species of pteridophytes. Most of the top twenty families at La Selva are also among the top twenty at other lowland

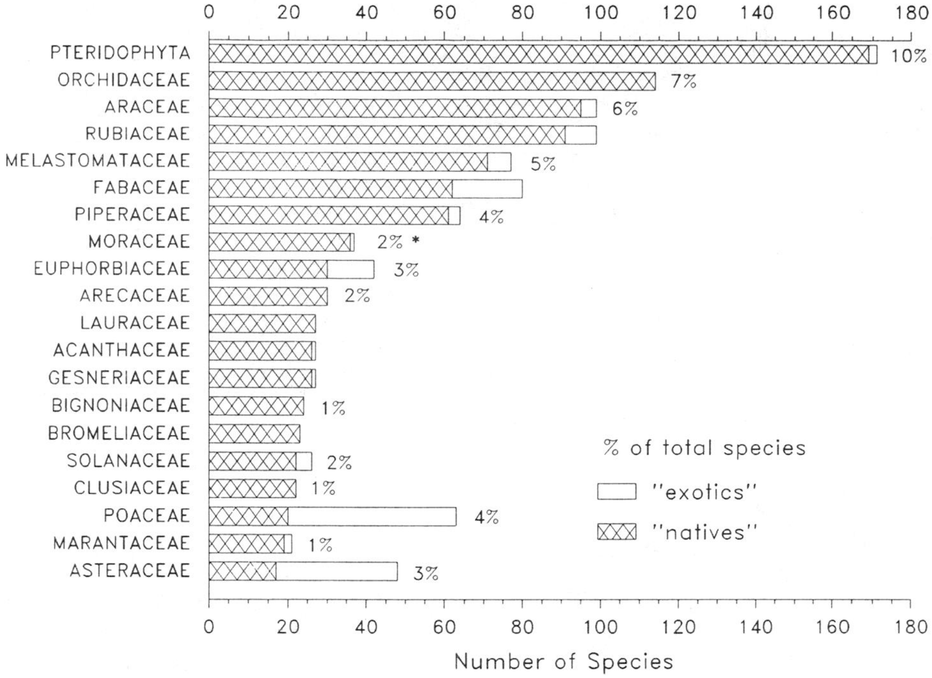

Fig. 6.1. Families with the most species at La Selva. Total flora includes 1,458 native and 220 widespread adventive (exotic) species; rankings, however, are based only on native species. *Includes six species of Cecropiaceae.

Neotropical sites (Gentry 1990b). Four groups (Fabaceae, pteridophytes, Rubiaceae, and Moraceae) are always among the top ten (table 6.5). In contrast, Araceae and Piperaceae are very rich at La Selva compared to most Central Amazonian forests, and Gesneriaceae and Marantaceae are diverse at La Selva but are not among the top twenty at other lowland Neotropical sites (Gentry 1990b). Palms are very conspicuous at La Selva, both in terms of abundance and richness; La Selva has nearly twice as many species as BCI, and the family ranks higher in species richness than at any other Neotropical site analyzed by Gentry (1990b).

Clearly, the greatest part of La Selva's floristic richness comes from a wealth of species in primarily lowland taxa, such as among the pteridophytes *Adiantum, Danaea, Selaginella, Tectaria, Thelypteris* subgenera *Goniopteris* and *Meniscium,* and *Trichomanes* (Grayum, pers. comm.). Another factor that may contribute to La Selva's richness, however, is its physiographic position at the transition zone between the extensive coastal plain of northeastern Costa Rica and the low but steep foothills of the Barva massif. This factor accounts for the presence of many species more typical of areas beyond the La Selva boundaries.

La Selva has a relatively large number of families that are more species-rich at higher (500–2,500 m) elevations, such as Aquifoliaceae, Clethraceae, Chloranthaceae, Ericaceae, Lauraceae, Magnoliaceae, Sapotaceae, and Symplocaceae. This pattern is further exemplified by the occurrence at La Selva of species typical of higher elevations, even in largely lowland families. Nearly all of the top twenty families have one or a few species at La Selva that are more abundant at elevations hundreds of meters higher, e.g., *Polybotrya alfredii* Brade (Pteridophyta), *Warrea costaricensis* Schlechter (Orchidaceae), *Anthurium lancifolium* Schott (Araceae), *Ladenbergia sericophylla* Standley (Rubiaceae), *Tibouchina longifolia* (Vahl) Baillon ex Cogn. (Melastomataceae), *Erythrina gib-*

bosa Cufodontis (Fabaceae), and *Pothomorphe umbellata* (L.) Miq. (Piperaceae). Likewise, numerous coastal-plain species just make it to La Selva along the lower, northern part of the property, but they do not occur in the dissected southern part of the property nor higher in the foothills. Gentry (1986) proposed that endemism and high species diversity in general, at least along the mountains in northern South America and southern Central America, are a result of such complex juxtaposition of different vegetation types.

The most diverse genera of La Selva plants (fig. 6.2) are typical of the wet lowlands of Central America as well as the base of the tropical Andes. *Piper* is among the three larger genera at La Selva, BCI, Río Palenque Science Center (RPSC) in Ecuador, and Manu in Amazonian Peru; however, it does not occur among the top twenty genera at Manaus in Brazil (Gentry 1990b). Three other genera (*Philodendron, Anthurium, Ficus*) are also among the ten most diverse at La Selva, BCI and RPSC, but not at Manaus (table 6.6). At La Selva most species in the top ten genera are treelets (*Piper, Psychotria, Miconia*), epiphytes (*Philodendron, Anthurium, Peperomia, Ficus*—in part), or terrestrial herbs (*Thelypteris*). This contrasts with the Manaus site where the large genera are mostly trees (table 6.6).

In contrast to the strong floristic differences between La Selva and Central Amazonia, the taxonomic similarities between the floras of La Selva and trans-Andean South America are probably associated with the relative proximity and similar geologic history of southern Central America and northwestern South America. Similarities in the distribution of growth-form categories among these sites may be more influenced by climatic regimes and soils.

Growth-forms

We present categories here more or less in the order of woody-canopy to herbaceous-understory, that is, trees, lianas, epi-

Table 6.5 Comparison of the ten richest families at three sites

Rank	La Selva		BCI		Manaus	
1	Pteridophytes	173	Fabaceae	112	Fabaceae	104
2	Orchidaceae	114	Pteridophytes	102	Sapotaceae	48
3	Araceae	99	Orchidaceae	90	Rubiaceae	42
4	Rubiaceae	99	Rubiaceae	66	Chrysobalanaceae	40
5	Fabaceae	80	Poaceae	63	Lauraceae	38
6	Melastomataceae	77	Araceae	46	Pteridophytes	37
7	Piperaceae	64	Asteraceae	42	Annonaceae	36
8	Poaceae	63	Moraceae	36	Moraceae	35
9	Asteraceae	48	Melastomataceae	35	Lecythidaceae	30
10	Euphorbiaceae	42	Piperaceae	32	Melastomataceae	26

Sources: La Selva data are from appendix 3, this volume. BCI and Manaus data are from Gentry 1990b.
Note: Includes native species and widespread adventives.

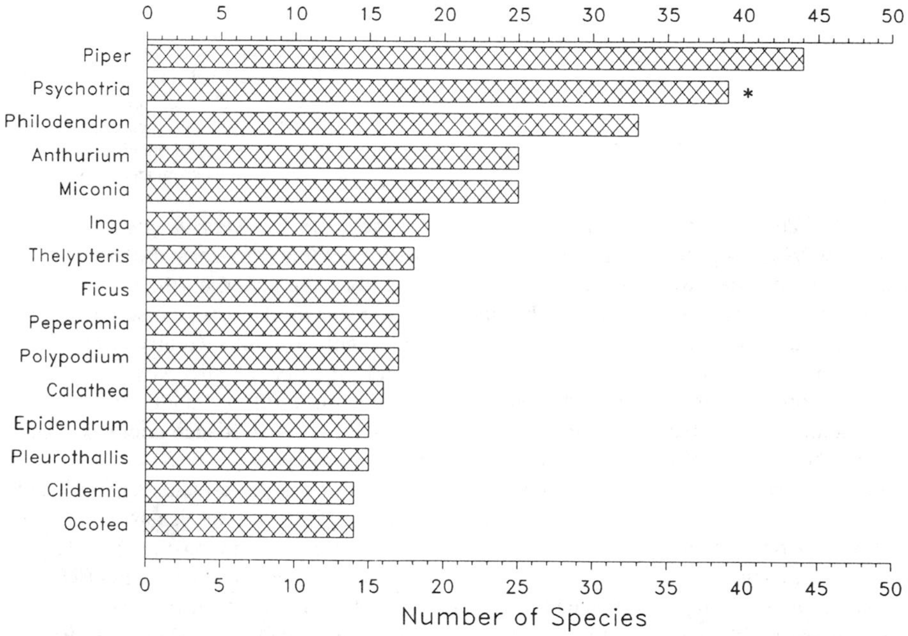

Fig. 6.2. Genera with the most species at La Selva, based only on native species. *Includes five species of *Cephaelis.*

phytes, shrubs, and herbs. This logical order should facilitate comparisons with data in other chapters where forest strata are discussed. Trees are defined as independent, mature individuals greater than 10 cm dbh or 5 m tall (Holdridge and Poveda 1976). Lianas are large, woody, mostly dicot vines of the canopy and subcanopy; nondicots include one gymnosperm *(Gnetum),* one aroid *(Heteropsis),* one palm *(Desmoncus),* and one bamboo *(Elytrostachys).* Epiphytes include strict epiphytes (orchids, bromeliads, *Anthurium* spp.), stranglers *(Ficus, Clusia),* more or less herbaceous, root-climbing plants (species of *Asplundia, Drymonia, Monstera,* and *Philodendron* that attain the subcanopy), and epiphytic parasites (Loranthaceae and Viscaceae). Mature woody plants (including many palms) smaller than the minimum tree size are placed in a shrub category, although many of these small, single-stemmed plants are more typically "treelets" than multistemmed shrubs. Herbaceous, low-growing vines and scandent plants of the understory are included in the herb category (including simple, erect herbs).

The percentage of trees, lianas, epiphytes, shrubs, and herbs at La Selva (fig. 6.3) appear to be typical for moist and

wet Neotropical forests (Gentry 1990b). La Selva, however, has a high percentage of epiphytes (23%) for lowland Neotropical forests. Understory shrubs and herbs comprise a higher percentage of dry Neotropical floras (nearly 75%; Gentry 1990b) than at La Selva (ca. 50%).

Most studies of tropical forests focus on trees, the visually and biomass-dominant growth-form, and seldom fail to mention that tropical areas harbor many more species of trees than temperate areas. La Selva has more than three hundred species of trees, whereas the whole state of Missouri has fewer than half as many. In a small area (440 acres) in Indiana, trees account for less than 8% of the flora (Yatskievych and Yatskievych 1987), whereas at La Selva they account for 20% of the species.

Researchers familiar with other tropical forests may be misled by the abundance of *Pentaclethra macroloba* to think that the tree flora is not very diverse. In fact, La Selva's primary forests are as rich in tree species as other Central American lowland forests (Hartshorn 1988). Not only are La Selva's forests dominated by a leguminous tree species, they also derive much species richness from the Fabaceae (fig. 6.4). High

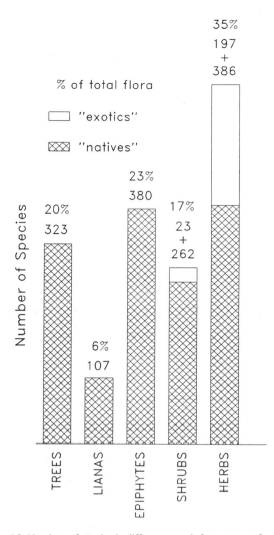

Fig. 6.3. Numbers of species in different growth-form types at La Selva. Herbs include herbaceous vines. Total flora includes 1,458 native and 220 widespread adventive (exotic) species; ranking, however, is based only on native species.

richness of leguminous trees is typical of lowland Neotropical forests, but the prominence of trees in the Lauraceae and Rubiaceae (exceeding those in Moraceae, even including Cecropiaceae) is unusual. This may be partly the result of a contribution from normally higher elevation species or local speciation in typically midelevation families.

The top-ranked liana families (Bignoniaceae, Fabaceae, Sapindaceae, and Malpighiaceae) are the same as for BCI and the lowland Neotropics in general (fig. 6.5). Although wet tropical forest like La Selva and the Colombian Chocó region appear to have fewer liana species than BCI, this may be exaggerated because lianas are the growth-form most difficult to collect and still the most poorly known at La Selva (Hammel and Grayum 1982). If wet forests are truly depauperate in liana species (not just lower percentage but fewer species) compared to drier forests, the explanation might be ecological displacement by epiphytes and/or stranglers.

La Selva is unique among lowland Neotropical floras recently analyzed (Gentry and Dodson 1987b; Gentry 1990b) in its richness of epiphytes (23% of all vascular plant species). Only Río Palenque Science Center in Ecuador with 21% of its species as epiphytes can compare with La Selva. Central Amazonian forests, as well as African and Asian forests, have a much less rich epiphyte flora. The pteridophytes and three families of monocots (Araceae, Bromeliaceae, and Orchidaceae) account for approximately 70% of epiphytic species richness at La Selva (fig. 6.6). These same four groups are the richest in epiphytic species on a worldwide basis, accounting for about 80% of all epiphytes (Kress 1986). Dicots, however, also contribute; Costa Rica has twice as many epiphytic dicot species as Java and four times the number reported for tropical West Africa (Burger 1980).

Most of the epiphytes at La Selva (primarily orchids and pteridophytes) are true epiphytes that germinate above the ground. About 25% of the epiphytic species are climbers (mostly *Philodendron*), and fewer than 10% are stranglers or other hemiepiphytes whose roots eventually reach the ground. Most epiphytes do not change growth-form as they mature; hemi-epiphytes, however, have aerial roots that descend to the ground.

Table 6.6 Comparison of the ten richest genera at three sites

Rank	La Selva		BCI		Manaus	
1	Piper	44	Psychotria	33	Licania	21
2	Psychotria	39	Piper	21	Inga	17
3	Philodendron	33	Inga	18	Protium	14
4	Anthurium	25	Ficus	16	Eschweilera	13
5	Miconia	25	Miconia	14	Swartzia	13
6	Inga	19	Epidendrum	13	Aniba	12
7	Thelypteris	18	Philodendron	13	Miconia	12
8	Ficus	17	Polypodium	13	Ocotea	11
9	Peperomia	17	Anthurium	12	Casearia	10
10	Polypodium	17	Cyperus	11	Couepia	10

Sources: La Selva data are from appendix 3, this volume. BCI and Manaus data are from Gentry 1990b.
Notes: Psychotria includes *Cephaelis.* Includes native species and widespread adventives in the totals.

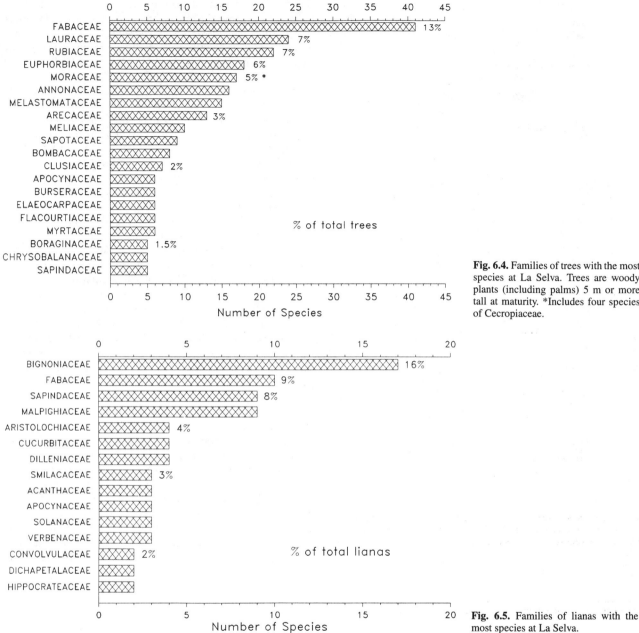

Fig. 6.4. Families of trees with the most species at La Selva. Trees are woody plants (including palms) 5 m or more tall at maturity. *Includes four species of Cecropiaceae.

Fig. 6.5. Families of lianas with the most species at La Selva.

Although many epiphytes have physical and physiological adaptations for short periods of drought, the lack of a definite dry season coupled with high rainfall (4,000 mm) at La Selva are the main factors that allow epiphytes to flourish. Similar reasoning may be partially responsible for the greater diversity of epiphytes in the west lowland Neotropics compared to similar areas of the Paleotropics, especially Africa. The Andean Neotropical region may have a longer history of uninterrupted, favorably wet, warm climate than other tropical regions. Longer and more regular droughts, perhaps, have restricted development of epiphyte diversity in the Paleotropics (Madison 1977; Gentry and Dodson 1987b). The endemic family Bromeliaceae also contributes to the greater diversity of epiphytes in Neotropical forests.

The shrub (+ treelet) flora includes many of the most di-

verse genera at La Selva. Melastomataceae, Piperaceae, and Rubiaceae with their three large genera (*Miconia, Piper,* and *Psychotria,* respectively) account for nearly half the richness of treelet species (fig. 6.7). A diverse small-tree layer seems to be typical of the wet lowlands of southern Central America and northwestern South America but apparently less typical of Central Amazonia and drier forests with poorer soils (cf. Gentry 1990b).

The emphasis on trees in tropical floras tends to obscure the even greater richness of herbs and herbaceous epiphytes in wet neotropical forests. These would still be the world's most species-rich plant communities even if the trees were omitted (Gentry and Dodson 1987a). At La Selva, as for most floras (tropical and temperate), herbs account for more species than any other growth-form. This is true whether one contrasts

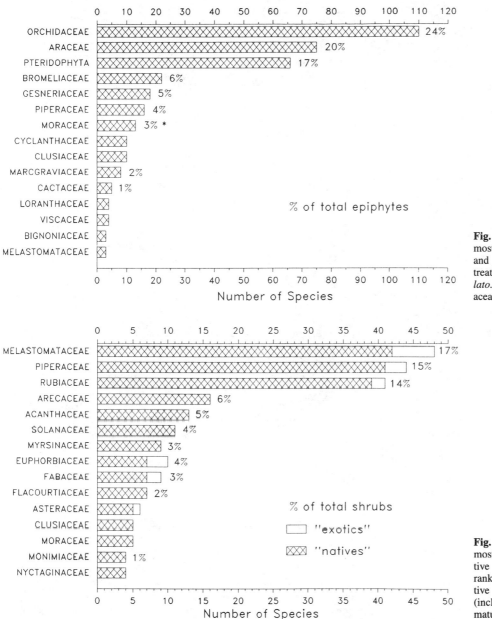

Fig. 6.6. Families of epiphytes with the most species at La Selva. Loranthaceae and Viscaceae are parasites, usually treated together as Loranthaceae *sensu lato*. *Includes two species of Cecropiaceae.

Fig. 6.7. Families of shrubs with the most species at La Selva, including native and naturalized exotic species; ranking, however, is based only on native species. Shrubs are woody plants (including palms) less than 5 m tall at maturity.

terrestrial herbs with other growth-forms (fig. 6.3) or all herbaceous plants with all woody plants. Considering only plants of natural habitats, epiphytes and terrestrial herbs are equal in numbers of species and each is more diverse than any of the other three categories.

The ranking of the most diverse families of terrestrial herbs (fig. 6.8) is somewhat confounded by the abundance of herbaceous "exotics." Even when paleo- and pantropical weeds are excluded, the pteridophytes at La Selva have more than twice as many species of terrestrial herbs as any single family of flowering plants and account for 17% of the herbaceous flora. The grasses at La Selva are quite rich, although more than half are Old World or pantropical weeds that inhabit abandoned pastures and other human-made clearings. Nevertheless, the native grass species of natural habitats are numerous, and the grass family is tied for third place among herbs with Araceae and Marantaceae. Grasses are also well-represented on BCI

(seventy-seven species; Croat 1978) or slightly more than at La Selva (sixty-two). Among the ten top-ranked herb families, the pteridophytes, grasses, composites, and sedges are also rich in temperate floras. Interestingly, the last three are the weediest families at La Selva and account for more than 40% of the exotic species. Among the tree and shrub floras (except for legumes) none of the ten most diverse families is also diverse in temperate floras.

Geographic Affinities and Endemism

Specific information on geographic distribution of plants at La Selva is available only for those few families already published (Atwood 1988; Grayum and Churchill 1989a, b; Hammel 1986e, f, g, h, i, j). The main theme of affinity with South America, however, has already been noted by Gentry (1978b; 1982a) for southern Central America, by Standley (1937) and Gómez (1982) for Costa Rica, and by Hammel and Grayum

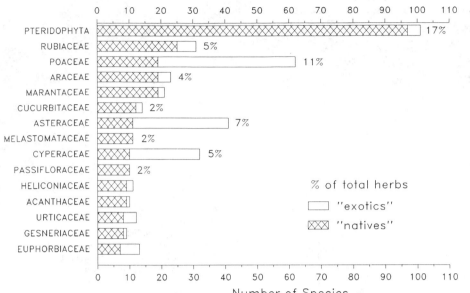

Fig. 6.8. Families of herbs with the most species at La Selva, including native and naturalized exotic species; ranking, however, is based only on native species.

(1982), and Hammel (1986a) specifically for La Selva. The pteridophytes of La Selva, although in general more wide-ranging than the seed plants, also show definite evidence of southern rather than northern affinities (Grayum and Churchill 1987). An analysis of six families (Cyclanthaceae, Marantaceae, Cecropiaceae, Clusiaceae, Lauraceae, and Moraceae) representative of the La Selva primary forests shows that roughly 85% of the species are known also from Panama or South America, whereas only about 45% (essentially the widespread species known from both Central and South America) occur also in Nicaragua or farther north. About 40% of the species known outside of Costa Rica are represented only from areas to the south, but less than 2% of the La Selva species occur only to the north (Hammel 1986a).

Recent work on the Nicaraguan flora has increased the number of "southern" species in the six families known from that country, but the number of "northern" species occurring at La Selva has not increased as a result of intensive collecting. These general trends are also true for several other large families (e.g., Rubiaceae, Araceae) for which manuscripts for the flora have been prepared.

This "pattern" of affinity with South America seems a logical outcome of any scenario of tectonic movement, vulcanism, original range, or recent dispersal; that is, dispersal has been predominantly from the nearest tropical continental source to this tropical isthmian region. The exceptional examples of West Indian or northern Central American affinities would require more interesting explanations, such as long-distance seed dispersal.

Almost 45% of the species in the group of six families appear to be endemic to Central America, and approximately 10% are endemic to Costa Rica. Likewise, among the more than fifty new species of flowering plants that have been described or are in manuscript from work at La Selva during the last ten years (table 6.7), few of them are narrowly restricted to the Sarapiquí region of Costa Rica. A much greater proportion may well prove to be restricted to the Caribbean lowlands of southern Central America. New species and endemism

are also infrequent among the La Selva fern flora (Grayum and Churchill 1987) as well as among the similarly wind-dispersed Orchidaceae (Atwood, pers. comm.).

Obviously, endemism varies from one family to another, but for the Caribbean lowlands of Costa Rica, 10% seems to be a reasonable estimate based on what we are learning about the distribution of species recently described from La Selva. Although this rate of endemism is not nearly as great as for tropical islands or some areas of Mediterranean climate (that have endemism levels as high as 90%), it is probably several times higher than for most temperate regions of high endemism (Gentry 1986).

SOME SPECIES DESCRIPTIONS

Although it is impossible in this overview to focus on many of the plant species at La Selva, we discuss several here to exemplify the interesting features and problems of this rich flora (see also Janzen 1983b for ca. forty additional taxa). Species are grouped according to four patterns that have emerged from our work on the La Selva flora.

Poorly Known Species

Poorly known species were reported first from La Selva, but have easily accessible populations or individuals at other sites. We assume that most of the disjunctions discussed next (La Selva/Panama, La Selva/South America) are only apparent and that the species will be found in intervening wet lowland forests once these areas are intensively explored.

Dicranostyles ampla Ducke (Convolvulaceae). This member of an otherwise South American genus is a large, woody liana with very small, white flowers and elliptical, Annonaceae-like leaves that are quite unlike those of a typical morning glory. It is common in the forest canopy on the ridges of plot 3. When collected for the La Selva flora project it was the first record for the genus in Central America (Hammel and Grayum 1982). An earlier collection from Santa Rita ridge in Panama has now been correctly placed in this species. In

Table 6.7 New species described or discovered in La Selva since 1979

Family	Genus	Species	Source
Acanthaceae	*Justicia*	1	McDade 1982
Acanthaceae	*Razisea*	1	McDade 1982
Annonaceae	*Annona*	1	Schatz in prep.
Annonaceae	*Cymbopetalum*	1	Schatz 1985
Annonaceae	*Unonopsis*	1	Schatz in prep.
Araceae	*Dieffenbachia*	4	Croat and Grayum in prep.
Araceae	*Homalomena*	1	Croat and Grayum in prep.
Araceae	*Monstera*	3	Croat and Grayum in prep.
Araceae	*Philodendron*	12	Croat and Grayum in prep.
Araceae	*Syngonium*	2	Croat and Grayum in prep.
Aristolochiaceae	*Aristolochia*	1	Barringer 1983
Bombacaceae	*Quararibea*	1	Alverson 1985
Clusiaceae	*Clusia*	1	Hammel 1986b
Clusiaceae	*Tovomitopsis*	1	Hammel 1986b
Cucurbitaceae	*Cayaponia*	1	Taylor in prep.
Cucurbitaceae	*Cionosicyos*	1	Taylor in prep.
Cyclanthaceae	*Asplundia*	2	Grayum and Hammel 1982
Cyclanthaceae	*Dicranopygium*	1	Hammel 1986d
Cyclanthaceae	*Sphaeradenia*	2	Grayum and Hammel 1982
Euphorbiaceae	*Dalechampia*	1	Armbruster 1984
Lauraceae	*Licaria*	1	Hammel 1986c
Lauraceae	*Nectandra*	1	Hammel 1986c
Lauraceae	*Ocotea*	1	Hammel 1986c
Lauraceae	*Phoebe*	1	Hammel 1986c
Malpighiaceae	*Lophanthera*	1	Anderson 1983
Melastomataceae	*Blakea*	2	Almeda in prep.
Melastomataceae	*Clidemia*	1	Almeda in press
Melastomataceae	*Miconia*	2	Almeda in prep.
Myrsinaceae	*Auriculardisia*	1	Lundell 1984
Rubiaceae	*Coussarea*	2	Taylor and Hammel in prep.
Rubiaceae	*Faramea*	1	Taylor and Hammel in prep.
Rubiaceae	*Hillia*	1	Taylor and Hammel in prep.
Rubiaceae	*Hoffmannia*	1	Dwyer in prep.
Rubiaceae	*Manettia*	1	Taylor and Hammel in prep.
Rubiaceae	*Psychotria*	1	Hamilton 1985
Rubiaceae	*Randia*	1	Dwyer in prep.
	TOTAL	58	

Costa Rica the species is still known only from La Selva. It is an example of how poorly known the canopy liana flora of the primary forest is in southern Central America.

Gnetum leyboldii Tul. (Gnetaceae) is another primary forest liana that is very poorly known in Central America. La Selva (the only Costa Rican locality for this species) is the northernmost record of the genus. The large woody stems can be recognized by their swollen nodes. The leaves are simple, opposite, and broadly elliptic. The oblong, cylindrical seeds resemble an enlarged date palm pit and are edible raw or toasted. This is one of the few gymnosperms of tropical lowland habitats. The only other plants with "naked seeds" from the lowlands of Costa Rica are *Zamia skinneri* Warscz. ex A. Dietr. and *Podocarpus guatemalensis* Standley.

Licania affinis Fritsch (Chrysobalanaceae) is a medium tree (15–20 m) previously known only from a few collections in eastern Panama and South America (Prance 1972a). This species is still known in Costa Rica only from La Selva, where it grows on a forested bluff over the Río Peje. This small patch of forest is not unlike cloud forest in aspect and is remarkable for the number of La Selva species that are known only from

there (e.g., *Alchornea latifolia* Swartz, Euphorbiaceae, and *Potalia amara* Aublet, Loganiaceae) as well as for the relatively large populations of species usually found at higher elevation (e.g., *Ladenbergia sericophylla* Standley, Rubiaceae, and *Anthurium lancifolium* Schott, Araceae). Ridgetops add much to the plant species richness of La Selva, perhaps especially because La Selva is an ecologically intermediate area, lying at the interface between mountains and extensive coastal lowlands.

Maranthes panamensis (Standley) Prance & White (Chrysobalanaceae). When this large tree was found at La Selva (Hammel and Grayum 1982), it was the first collection for Costa Rica of a genus previously known in the New World only from two collections in central Panama, where it was thought possibly to have been introduced (Prance 1972a). Since then it has been found in several areas of Costa Rica, including Tortuguero National Park, Braulio Carrillo National Park, and the Osa Peninsula. Its Neotropical presence is clearly natural, and it is now considered taxonomically distinct from the Old World species *M. corymbosa* Blume, with which it was originally identified (Prance 1972a). This species is quite rare on old La Selva, where it is known only from two individuals in ridge forest near plot 3 and along Quebrada Esquina at the southeast corner. It is more common in the central ridge forest of the Sarapiquí Annex. This phenomenon of high beta-diversity is typical of ridges in La Selva. Additional examples of ridgetop differences in abundance include *Geonoma deversa* (Poiteau) Kunth, *G. oxycarpa* C. Martius (both Arecaceae), and *Metaxya rostrata* (Kunth) C. Presl (Metaxyaceae).

Pariana parvispica R. Pohl (Poaceae) is an understory, bambusoid grass endemic to the Caribbean lowland forests of Costa Rica. It is apparently the northernmost species of a primarily South American genus and the only species in Costa Rica. It is unusual in the genus for its small and very inconspicuous inflorescences (borne on short leafless shoots sometimes hidden among leaf litter) with white anthers; other species have very conspicuous, relatively long inflorescences with bright yellow-orange anthers and are reportedly insect pollinated (Soderstrom and Calderon 1971). At La Selva the species apparently flowers throughout the year; flowers have been collected in March, May, September, and November. Judziewicz and Pohl (1984) provide a map to the known La Selva localities for this species. Seven other species of understory bambusoid grasses occur at La Selva, mostly restricted to primary forest.

Qualea paraensis Ducke (Vochysiaceae) was reported as restricted to South America, with a northern range extension into the Darién of eastern Panama (Gentry 1982b; Robyns 1967). Nevertheless, Costa Rican foresters have known a species of *Qualea* in the Osa Peninsula and the San Carlos/San Miguel area (Alajuela Province) for more than twenty years. The genus is now known to occur in southern Nicaragua. The fact that a particular species is common and sought after or recognized by loggers, foresters, or ecologists bears no necessary relation to how well known it is in the taxonomic/floristic literature or how well represented in herbaria. Only in the 1980s, with stimulus from floristic projects and more concerted efforts to collect trees, did *Qualea* specimens begin to confirm the distribution northwest of Panama. At La Selva, *Qualea* is known only from a few individuals on a ridge in primary forest on the Sarapiquí Annex. It also occurs on hills just north of Puerto Viejo and Chilamate.

Spachea correae Cuatrecasas & Croat (Malpighiaceae) is another species described from Panama and first collected in Costa Rica at La Selva. This large (150 cm dbh, 45 m tall) canopy tree is rare but is now known from several localities in the Caribbean lowlands of northern Costa Rica and southern Nicaragua and from the Osa Peninsula in the Pacific lowlands of southern Costa Rica. At least two other species of Malpighiaceae (*Byrsonima crispa* Adr. Jusseau and *Lophanthera hammelii* Anderson) and several species in other families such as *Amphidasya ambigua* (Standley) Standley (Rubiaceae), *Annona amazonica* R. E. Fries aff. (Annonaceae), *Hirtella lemsii* L. O. Williams & Prance (Chrysobalanaceae) and *Naucleopsis naga* Pittier (Moraceae) are presently known in Costa Rica only from this same restricted bimodal distribution. The absence of these wet forest species from the dry forests of northern Pacific Costa Rica (Guanacaste) is not surprising, but their absence from equally wet southern Caribbean Costa Rica most likely indicates inadequate exploration.

Species with Locally Distinct, Sympatric Forms

Several taxa occur as distinct species at La Selva, but they are treated as single species by specialists working with herbarium specimens. This common problem of local floras may have many different resolutions or explanations depending on the taxa: for example, ecotypic variation, overlap of ranges of usually disjunct extremes of variation within a species complex, poor herbarium representation of a particular species leading to the dismissal of that entity as an unusual form of a better-represented species, and different species concepts.

Dichorisandra hexandra (Aublet) Standley (Commelinaceae) is a common widespread species presenting a taxonomic problem not only at La Selva but also in many other parts of its range. Here it is restricted to areas of primary forest (e.g., along the back of the Camino Central Lejano [CCL], along Quebrada Esquina, plot 1) where it is a viney or scandent climber on tree trunks up to about 2–3 m or clambering over low shrubs. A second form is somewhat less common at La Selva and is restricted to more disturbed habitats on alluvial soils. This latter species appears to be an annual plant, or at least seasonally produces aboveground shoots, and forms an erect, branching herb with a shrublike habit. The two species also differ in characters of pubescence and flower color. Both have been grown from cuttings in the greenhouse where they maintain their distinctive habits. Gentry and Dodson (1987) reported this same species-pair problem for Río Palenque, Ecuador. However, according to one specialist in Commelinaceae, examination of herbarium material from the complete range of these species does not allow any clear-cut separation of the two forms (Faden pers. comm.). This is a paradox commonly faced by biologists who know well the species at a particular site; in many cases the insight of the field biologist is borne out when the herbarium taxonomist restudies the problem in view of character differences emphasized at the particular site. In others, perhaps two ends of a "Rassenkreis" are involved, with genetic continuity elsewhere in the range, whereas local sympatric forms behave as distinct species.

Hyeronima alchorneoides Allemão (Euphorbiaceae) is a large dioecious canopy tree that occurs on alluvial soils. At La Selva it is found especially along the SHO trail south of the successional strips. The distinctively colored, orangish-brown trunk is often more than 1 m in diameter, and the crown can extend up to 50 m. The species can be recognized by the bright yellow or reddish-orange senescent leaves beneath the tree. The leaves are subtended by foliose but saccate stipules, which can vary in size, shape, and presence or absence within an individual depending on the phenological status and age of the shoot; these variable stipules are the source of much taxonomic confusion in the genus. Because these features have been used to distinguish species, as many as five species have been included on lists from La Selva, where only two occur. As is often the case with large forest trees, too little material has been available for herbarium taxonomists to understand intraspecific variation. The second species, *H. oblonga* (Tul.) Muell.-Arg., is a much smaller tree with smaller, narrower leaves and occurs only along streams and on ridges in the southern part of the property.

Swartzia simplex (Swartz) Sprengel (Fabaceae) is a small tree with unusual compound leaves reduced to a single leaflet.

The very common 2–3-m shrub, which has always gone by this name at La Selva, appears to be restricted to the Caribbean lowlands of Costa Rica and is rare in herbaria. A larger tree (to 10 m) with larger leaves, more prominent venation, and longer stipules (and relatively rare at La Selva) has a wide Central American distribution. Although herbarium material of both species has been identified by specialists as *S. simplex,* researchers at La Selva have always considered the two as separate species, presuming that the rarer one was undescribed. Ironically, if a new name is needed, it will be for the locally common species.

Urera elata (Swartz) Grisebach (Urticaceae) is one of five *Urera* species at La Selva although Burger (1977c) reports only four species of *Urera* for the entire country. Judging from the apparent inadequacy of the current literature to accommodate variation seen in the field and from the apparent confusion in herbaria, the genus is in need of taxonomic revision. The few species involved (fewer than fifteen?) and small size of the plants suggest that the genus would be a good choice for a research project. *Urera elata* is one of the more common species in the genus at La Selva. It is a shrub 2–3 m tall, common in old secondary woods, especially along the SOC trail. A liana, *U. eggersii* Hieron. with similarly shaped leaves, is also common at La Selva. It can be distinguished from *U. elata* by its liana habit (to 15–20 m), reddish, exfoliating bark, and leaves with fewer, more distant lateral veins. The liana occurs from Mexico throughout Central America and into northern South America. As presented here the "problem" may seem solved; the resolution is the same as for this species pair at Río Palenque in Ecuador (Gentry pers. comm.). According to another interpretation, however, all Central and South American material in the complex is *U. eggersii,* and *U. elata* is restricted to the West Indies (R. Liesner pers. comm. and in herbarium). The other species at La Selva are *U. baccifera* (L.) Gaud., *U. caracasana* (Jacquin) Grisebach, and *U. laciniata* (Goudot) Weddell.

Widespread Species of Unusual Size at La Selva

This category overlaps somewhat with the previous pattern and may have similar explanations. A size-related phenomenon includes tree species that flower while small and in the understory but apparently produce fruits only when much larger, for example, *Licaria sarapiquensis* Hammel (Lauraceae). Other canopy and subcanopy tree species may produce flowers and fruits even while in the understory (e.g., *Clarisia biflora* Ruíz & Pavón, Moraceae) and some species of *Dichapetalum* (Dichapetalaceae) may flower as shrubs in the understory and later become lianas.

Symphonia globulifera Linn. (Clusiaceae) is a small (<10 m) understory tree at La Selva, but it is a canopy tree in coastal swamp forests (e.g., Tortuguero National Park) and at midelevations (800–2,000 m) in Braulio Carrillo National Park, as well as in parts of Central and South America. Flower color and leaf venation differences also seem to correlate with the size differences when comparing specimens from La Selva with those of certain other sites. A simple resolution of this problem by describing the La Selva form as a new species is confounded by the fact that a canopy tree can sometimes flower as a sapling in the understory. Careful examination of specimens from throughout Central America may reveal that

the La Selva plants represent either an extreme in a bimodal cline of variation or that two species, indeed, occur.

Plants with Unusual Morphology or Pollination

Asplundia uncinata Harling (Cyclanthaceae) has a dwarf palmlike habit and forms large patches in primary forests of northeastern Costa Rica. It was thought to be endemic to this area, but it is now known from southern Nicaragua and western Panama. In well-drained areas of La Selva it can be the dominant understory plant, replacing the palm *Asterogyne martiana* Wendland ex Burret, which is more abundant in poorly drained areas. The small, white unisexual flowers (male and female intermixed) are arranged on a cylindrical spadix, which is held erect on a short peduncle among the leaf bases. Flowering usually begins during the dry season and proceeds with the spathes opening under pressure of the expanding vermicellilike staminodia. The stage at which the staminodia are fully expanded and producing a very powerful, raspberry-mint fragrance, with hundreds of small weevils crawling about, is quite impressive but very ephemeral.

The tepals of male flowers of this and most species in the family exude a sticky resin, whose function and fate are unknown. The inflorescences are protogynous, and the weevils crawl through openings among the tightly packed staminate flowers into small spaces below them, where the glistening stigmas lie. It is not known at what point, or even if, the resin disappears, but it is exposed on the inner face of the openings through which the weevils crawl. Trigonid bees have been seen gathering pollen but could scarcely serve as pollinators (Schremmer 1982 for *Carludovica*) because they are too large to reach the receptive stigmas. Soon after the anthers dehisce, the staminate flowers fall and the infructescence slowly matures over a period of two to three months. The fruits are bright yellow-orange at maturity and are essentially dehiscent berries, with the apex of each fruit falling away to expose the juicy mass of seeds on the infructescence. The seeds are probably dispersed by bats, birds, and other vertebrates, who may remove the fruit caps to get at the flesh.

Bauhinia guianensis Aublet (Fabaceae) is a common liana in primary forest throughout the property. The large, undulate, ribbon-shaped trunk is very distinctive and gives rise to the common name "monkey's ladder" (see Fig. 7.4). The leaves are also quite distinctive, more or less heart-shaped but partially split down the middle, and, thus, resemble the wings of a butterfly. At the end of the dry season the elastically dehiscent pods can be heard exploding like popcorn and raining seeds to the forest floor. With the onset of the rainy season, the seeds readily germinate, and for a few months the whole forest, especially in gaps and along trails, is carpeted with seedlings.

Clusia flava Jacquin (Clusiaceae) is a large, sometimes strangling, dioecious epiphyte that is common on the trunks of trees left in pastures in the vicinity of Puerto Viejo. It has thick, leathery obovate leaves and viscous, cream-colored latex. The four pale yellow petals are very thick and rubbery and throughout anthesis remain erect and mostly overtopping the ovary or stamens. The flowers emit an unforgettable strong, sweet fragrance of artificial banana flavoring and attract small weevils and Nitidulidae. As in most species of *Clusia,* the seeds are covered with a bright orange aril. Oropendolas *(Psaricolius)* and honeycreepers *(Chlorophanes, Dacnis)* feed on these arils. Related species of *Clusia* at La Selva that are also beetle pollinated include *C. amazonica* Planchon & Triana (= *C. oedematopoidea* Maguire), *C. cylindrica* Hammel, *C. gracilis* Standley, and *C. quadrangula* Bartlett.

Clusiella elegans Planchon & Triana (Clusiaceae) is presumed to be closely related to *Clusia,* but that is problematic because *Clusiella* has fruits, seeds, and pollen strikingly different from anything else in the tribe (Hammel 1984). Costa Rican specimens are especially interesting because staminate plants have not yet been seen (Hammel 1986g). The implication is that in this part of its range the species may be apomictic. Unfortunately, although *Clusiella* is rather common at La Selva, it is a canopy epiphyte and difficult to study; it has been collected most often in tree and branch falls. The genus was first reported for Costa Rica from La Selva. The name *C. elegans* is used provisionally here; all of the Costa Rica collections and most from Panama are now recognized as a distinct species (Hammel unpublished data), which is now known from extreme northeastern Costa Rica (Limón Province near Cerro Coronel) to eastern Panama. Typical *C. elegans* is known from central Panama to northern South America.

Cybianthus schlimii (Hooker f.) Agostini (Myrsinaceae) has a plethora of synonyms: *C. spectabilis* (Standley) Agostini, *Ardisia spectabilis* Standley, *Correlliana spectabilis* (Standley) D'Arcy, *Weigeltia spectabilis* (Standley) Lundell. Placement in four different genera reflects the fact that it has a rather unusual morphology, a "palmoid" habit (see D'Arcy 1973). At La Selva it is an unbranched shrub about 1.5 m tall, with all of its oblong-spatulate, dentate and nearly epetiolate leaves clustered toward the top of the stem. This "trash bucket" of leaves is cluttered with fallen debris, spider webs, and, toward the base of the leaves, with decayed matter. In species with this habit, the trash bucket provides a potting medium for adventitious roots (Dressler 1981). The rather inconspicuous brownish flowers (ca. 4 mm in diameter) are borne in a narrow, axillary panicle about 15–20 cm long. Much more conspicuous are the swarms of shiny, metallic-green Euglossine bees invariably seen visiting these flowers.

Souroubea gilgii Richter ex Gilg & Wendland (Marcgraviaceae) is a bushy epiphyte with long, leafy branches that is common along rivers and on trees left in pastures near La Selva. The leaves are alternate, narrowly obovate, leathery, and glossy and, like many Marcgraviaceae, have glandular dots visible on the lower surface. The racemes of bright orange flowers appear May through July and produce a most extraordinary fragrance of honeydew melon and butter. The other common species of the family are heavily visited by hummingbirds as well as bees and wasps because of the large quantities of nectar present in the often pitcher-shaped nectaries. Others (e.g., species of *Marcgravia*) with less colorful flowers that are apparently not fragrant may be visited by bats. Nevertheless, most species appear to be autogamous (Bedell pers. comm.).

Future Research Needs

Considering the very dynamic nature of the floras of secondary and riparian habitats and of the epiphyte floras (most of the herbaceous flora), published floras of the humid tropics are always inherently "incomplete." Beyond this problem, tropical wet floras are so diverse, with so many species difficult to collect (large trees, canopy epiphytes, lianas), that even in an area as small as La Selva, overlooked species will continue to be discovered. Nevertheless, the La Selva flora, as presented in appendix 3, is probably 90% known, and its publication will make available abundant information about nearly all the commonly encountered species and facilitate further, more specific research.

Continued exploration of La Selva is important for pragmatic and theoretical reasons: to achieve the goal of a published flora that treats all species; to know how many species are packed into this small area; and to know what they are and how they are distributed on the property. With regard to total numbers, the more time that is taken to collect for a flora, the more misleading the published account might be. Weedy or other ephemeral species (probably including many herbaceous epiphytes) come and go. As arrivals are found, the list grows, but even if species are suspected of having disappeared, lists usually remain inflated and can result in misleading comparisons (Hammel and Grayum 1982).

Published florulas are extremely useful, especially in the tropics where so few exist; the Flora of Barro Colorado Island (Croat 1978) is a standard reference throughout Central America. Florula projects provide the data base and preliminary analysis for later revisionary work. More importantly, they allow identification of plants by other botanists and nonbotanical scientists for all types of ecological studies. Not only the final product but also the process can be an excellent vehicle for training students. Along these lines, the La Selva flora should be a great stimulus for further exploration of and comparison with the adjoining Braulio Carrillo extension.

A number of other related projects would further improve the botanical resource base at La Selva, among them, completing and updating the La Selva reference herbarium. This collection, housed in sealed wooden cabinets and apparently insectfree, is incomplete. Only certain taxa, e.g., Annonaceae, Clusiaceae, Melastomataceae, Moraceae, Rubiaceae, *Inga* (Fabaceae) are well represented. Often the most common species are not represented at all; during the intensive collecting phase collectors saw a greater need to keep a reference collection of rarer species for comparison with new collections. Making this collection more useful to researchers in general by including all common species and by updating the names should be a top priority for facilities improvement. Small fruit/seed, seedling, and pollen reference collections have been discussed and begun at various times at La Selva. All could provide very important tools for a variety of projects (e.g., studies of fruit dispersal, frugivory, seedling survival as related to isoenzyme diversity or pedigree, pollen dispersal, and pollen-stigma interactions).

The laboratories with modern equipment at La Selva could provide facilities for approaching such questions as How do many species within a genus coexist (e.g., *Piper, Psychotria, Philodendron*)? Why are there so many species of *Piper, Psychotria,* and *Philodendron* and only one each of *Sarcorhachis, Lasianthus, Homalomena,* respectively, in the same families? Are species pairs such as those including *Dichorisandra hex-*

andra, Swartzia simplex, and *Urera elata* more, or less, distinct genetically and morphologically different from each other than are other pairs of congeners?

It should be clear from this introduction that for a wet tropical forest, the flora of La Selva is relatively well-known but that many taxonomic problems remain (see app. 3). Of course, relative to temperate floras, La Selva is still poorly known. Lianas and large trees of primary forest (especially on the ridges) and all habit types of the poorly explored newer properties are almost certainly the repositories of species not yet found or poorly known. Habitats and growth-form groups should be the focus of efforts to complete the flora, a worthwhile goal that has still not been accomplished for any such forest. Undoubtedly, many additional species of herbs of secondary and other dynamic habitats would be found if these areas were systematically surveyed on a regular basis. Rather than providing data for discussing floristic similarities between areas, however, or elucidating the total number of species that exist in a given flora, a rigorous study of these weedy floras would be interesting precisely to document and analyze their dynamic nature. One might assume that any particular age plot in the artificially maintained successional strips would always have more or less the same set of species. This does not, however, seem to be the case. Are the factors purely stochastic and dependent on accidents of seed dispersal, or can dependent variables (e.g., season/weather, treatment, site) be defined?

The "native" forest flora is probably also dynamic although on a different time scale. One might assume that a particular habitat type would harbor the same set of species throughout the property. This does not seem to be the case, however. Ridge-top floras, at least, differ in a spatial sense and probably reflect the temporally dynamic nature of the species composition of primary forest.

Now that the soil survey of La Selva has been completed (see chap. 4), researchers obviously need a quantitative inventory of the vegetation of greater La Selva to correlate vegetation types with soil types. Given the importance of La Selva as a research site and the recent OTS initiatives in applied ecology, scientists need more comprehensive information on the distribution, composition, and structure of disturbed vegetation types. La Selva also offers excellent opportunities to monitor successional processes, including experimental manipulations on abandoned pastures and in plantations. Finally, the long-term studies of tree regeneration, survival, and growth (now covering twenty years) on the permanent inventory plots (chap. 8) must be continued.

7

Plant Demography

Deborah A. Clark

It has been variously stated that plant demography, as a corpus of knowledge, is a newly born branch of plant ecology. If so, tropical plant demography (and I would also add tropical plant population ecology) is still very much in gestation.

Sarukhán 1980

In 1992, twelve years after José Sarukhán made this evaluation, tropical plant demography is still in its infancy. Even at intensively studied sites such as La Selva, comprehensive information on population density, size structure, rates of growth and survival through ontogeny, reproductive rates, and how these vary in space and time do not exist for any species. This is scarcely surprising. The application of demographic concepts to plant populations is still a recent phenomenon, and tropical rain forest communities present special challenges to the plant demographer.

Why study plant demography in tropical rain forests? Understanding the mechanisms that maintain the exceptional species richness of these communities may well depend on analyses of life-history patterns of the component species. Plants are also the resource base for many of the remaining species. The demographic traits of a given plant species may be critical for the herbivores, pollinators, seed dispersers and predators, and other animals that use it. On an ecosystem level, the population processes of the larger plants strongly affect both net primary productivity and the physical structure of the forest. Finally, tropical rain forests include an exceptional richness of plant life-forms: emergent and canopy trees, lianas, hemiepiphytes, subcanopy palms, understory treelets, shrubs, dwarf palms, terrestrial and epiphytic herbs, and epiphylls. Analysis of the degree to which demographic diversity parallels life-form diversity will enhance overall understanding of plant population phenomena.

Over the 1980s plant demographic study intensified at La Selva and other tropical rain forest sites. Detailed research on a few species and several communitywide studies have provided insights but have barely scratched the surface. In this chapter I review the plant demographic research carried out at La Selva, relate the findings to data from other sites, and suggest lines to follow in this area of tropical research.

ASPECTS OF THE LA SELVA ECOSYSTEM AFFECTING PATTERNS OF PLANT DEMOGRAPHY

Diversity and Rarity

A first major challenge for plant demographers at La Selva is species richness. Most of the 1,458 naturally occurring vascular plant species presently known from La Selva (see chap. 6) occur in low numbers. For a group of six plant families, Hammel (1986b, 150) found about 75% of the species to be rare or uncommon ("known from a few individuals or from a few small populations at La Selva"). Seventy percent of the tree and liana species in 12.4 ha of La Selva primary forest (three plots on different soils) have overall densities of less than one individual larger than 10 cm dbh (diameter at breast height) per hectare (chap. 8). Rarity is a general characteristic of tropical rain forest plant communities (e.g., Poore 1968; Hubbell and Foster 1986c). Many species also have very few individuals in certain juvenile size classes (see references in D. A. Clark and Clark 1987a). Demographers have, therefore, tended to concentrate their studies on the more common plants, which may have population characteristics that differ from those of most rain forest plant species.

Temporal Variability

A second challenge is presented by year-to-year variability in weather patterns and plant behavior at La Selva. Although the climate is characterized by a "drier season" and a "wetter season" often interrupted by a "little dry season," these patterns are based on long-term averages. Among years, the large variation in rainfall in any given month (chap. 3) is paralleled by temporal changes in radiation and soil moisture. Such climatic variability can be reflected in changes in growth, survival, or reproduction of plants at La Selva. Significant year-to-year variation has been found in the diameter growth of canopy and emergent trees at the station (D. A. Clark and Clark in review). Clear evidence exists of variability in plant reproduction. Species at La Selva have been shown to vary in the number of reproductive episodes per year (Bullock et al. 1983) and in the total size of seed crops in different reproductive episodes (Bullock 1982; Bullock et al. 1983; McHargue and Hartshorn 1983; Young 1986a). Moreover, some species have multiyear intervals between reproductive episodes (Frankie, cited in Hartshorn 1972; Opler, Frankie, et al. 1980; D. A.

Clark and Clark 1987b; chap. 11). The joke that every field biologist's first year at a site is a "normal" year points to the misconceptions that can arise from short-term studies. Demographic studies of tropical rain forest plants should span multiple years (Stiles 1978c). Only in protected reserves such as La Selva is it possible to embark on long-term demographic study with some confidence that the study population will remain intact.

Spatial Heterogeneity
Even within a small reserve such as La Selva, the variability among habitats can have major impacts on plant demographic processes. La Selva encompasses at least four soil types that differ in nutrients such as phosphorus (Vitousek and Denslow 1987; Sancho and Mata 1987; chap. 4). Many plant species are absent from certain of these soils (e.g., many grasses; Judziewicz and Pohl 1984), and abundance of others varies with soil type. For example, two species of leguminous emergent trees, *Dipteryx panamensis* (Papilionoideae) and *Pithecellobium elegans* (Mimosoideae), have essentially reciprocal distributions within La Selva, with the former restricted to alluvial soils and the latter almost confined to poorer, residual soils (Clark and Clark unpublished data). Another emergent tree, *Carapa guianensis* (Meliaceae), is a dominant species in swamp (62 stems >10 cm dbh/ha), but is relatively rare on steep slopes with residual soils (9 stems/ha; McHargue and Hartshorn 1983). The floristic shift in trees across an elevational gradient within La Selva (M. Lieberman et al. 1985; chap. 8) may be the result of such soil-tree interactions. Significant intraspecific differences in plant growth may exist on the different soil types.

The age of vegetation varies greatly within La Selva. Patches within primary forest can be classified as "gap," "building phase," and "mature forest," depending on the extent of recuperation since a tree or branch fall (Whitmore 1984). Different successional stages following human disturbance are also present within the station. Across these gradients La Selva plants vary significantly in demographic characteristics (e.g., Bullock and Bawa 1981; Werner 1985; D. A. Clark and Clark 1987b; D. B. Clark and Clark 1987).

The Dynamic Nature of the La Selva Forest
The primary forest at La Selva is a scene of constant change (see chap. 9). Trees and large branches are falling to the ground, opening up new gaps and smashing smaller plants in the process. Smaller branches, bromeliads and other epiphytes, 6 m-long palm fronds, smaller leaves, and fruits fall constantly as well. The lifetime risk of suffering physical damage is, therefore, high for plants at La Selva. The morphology and capacity for resprouting of different species may be related to life history, with long-lived understory inhabitants being particularly well-equipped to minimize or repair such physical damage.

Another consequence of this dynamism is the uncertainty of environmental conditions at any microsite over time. Because most plants are sessile and, therefore, committed to their rooting site (lianas are the exception), they are likely to experience a major change in light and temperature if they live for many years. Estimates that La Selva's forest turnover rate is 118 years (Hartshorn 1978) and that 6% of the primary forest

is in young gaps at any time (Sanford et al. 1986) suggest that most shaded microsites will be exposed to higher light levels within a few decades at most. Similarly, microsite conditions in a new tree-fall gap can change rapidly because of regrowth in and around the gap. Fetcher et al. (1985) found that the temperature and humidity regime at 70 cm above the ground in a new gap became indistinguishable from that of adjacent understory after just two years. These environmental changes certainly affect the demography of plants in the lower levels of the forest (see Smith 1987).

Particular Challenges to the Plant Demographer in Tropical Rain Forest
Compared to temperate zone research, demographic studies of tropical rain forest plants are made much more difficult by the lack of methods to age individuals of most plant groups. Because no equivalent of winter exists (such as a reliable, pronounced dry season), the plants lack annual growth rings and annually demarcated shoot growth, tools of the temperate zone demographer (Bormann and Berlyn 1981). To age plants or to measure their growth rates the tropical rain forest researcher is reduced to brute force: careful remeasurement of individuals over time. For long-lived plants, age estimates require accumulating data on growth and survival of all size classes.

Another factor is the relative inaccessibility of many species. Lianas, epiphytes, and hemiepiphytes accomplish much or all of their growth and reproduction in the forest canopy. Although some demographic research has been accomplished by using climbing techniques (e.g., Putz 1984b), information on these plant groups is very limited.

The large proportion of dioecious plants also makes demographic study more difficult because data must be obtained for both sexes. For taxa with a long immature period, sexing individual plants can require very long-term observation (e.g., D. A. Clark and Clark 1987b). In addition, larger population samples are needed for estimates of reproductive output. The incidence of dioecy in La Selva's vascular plant flora is not known, but it is likely to exceed the value of 9% calculated by Croat (1978) for the tropical moist forest of Barro Colorado Island, Panama (BCI). Thirty percent of the species in six families at La Selva are dioecious (Hammel 1986b), and Bawa's (1979) estimate for the trees is 20%.

The Consequences of Human Impact on La Selva
With the recent discoveries of charcoal and human artifacts in soil cores in the center of La Selva (Horn and Sanford 1992), it is clear that the "primary forest" has been subject to human activity and to fire (perhaps anthropogenic) at least within the last 1,100 to 2,430 years (dates of carbon-dated samples of charcoal). Although the extent of human effects on the forest can only be speculated at this point, they may have been major. Anthropological study of indigenous inhabitants of a lowland wet forest in Panama has revealed extensive silvicultural activities; some plant species perceived to be of no value are removed by hunter/gatherers passing through forested areas, whereas others are planted in natural forest (Gordon 1982). One species often planted in *Dipteryx panamensis,* the most common emergent tree on La Selva alluvial soils (its seeds are edible and are a major attractant for wildlife).

In recent years the progressive peninsularization of La Selva is likely to have affected the demography of some plant species, particularly those present at low densities. Although the long-term ecological security of the station was greatly increased in 1986 by the extension of Braulio Carrillo National Park (chap. 2), La Selva is the tip of a forest peninsula surrounded by pasture. Aggressive species of secondary vegetation may be encroaching on the La Selva forest as Janzen (1983b) has reported for pioneer species in Santa Rosa National Park. Research has not been focused on this issue.

La Selva plants also may be affected by faunal change as a result of hunting. One large omnivorous mammal, the white-lipped peccary, has been extinct from the entire surrounding region for at least twenty years (E. López pers. comm.). This species may have been the dominant seed predator for some plant species, which may now experience greatly increased seed survival. The disturbances in the understory caused by the activities of large herds of white-lips no longer occur at La Selva; this may have altered the abundances and dynamics of understory plant populations. Baird's tapir, a large and potentially important herbivore, is present at La Selva, but tracks are observed only rarely; protection from hunting in the adjacent national park is expected to increase tapir activity in La Selva proper. The large fruit- and seed-eating birds, such as guans and curassows, and the smaller peccary (collared peccary) now appear to be recovering from past hunting pressure (pers. observation). Great green macaws, potentially important seed predators for some canopy trees, are nearly extirpated from the entire region, and are now sighted only in rare, small groups at La Selva. The decline of this nomadic bird species is probably more the result of habitat fragmentation than hunting pressure (Stiles et al. 1989). Observations by local residents (I. Alvarado D. pers. comm.) indicate that the scarlet macaw was extirpated from the region decades ago. Incursions by hunters into La Selva are now rare occurrences. The continued presence of all the large vertebrate predators has apparently prevented artificial increases in small- and medium-sized mammals. In contrast, on BCI, where the big predators have been lost, medium-sized mammals are more abundant, and seed and seedling survival are lower than on the adjacent mainland (DeSteven and Putz 1984; Sork 1987).

PLANT DEMOGRAPHY AT LA SELVA

Herbaceous Plants

Demographic information exists for few species of herbaceous plants at La Selva. Only one, a small terrestrial primary forest fern, *Danaea wendlandii* (Marattiaceae; fig. 7.1), has been investigated in detail (Sharpe 1988; Sharpe and Jernstedt 1990). The size distribution of individuals, size-specific rates of growth, survival, and reproduction, and relative performance of individuals originating from spores or from vegetative buds were studied during three years. Growth was slow in this abundant fern; leaf production rates were only 1.1 to 1.6 leaves per plant per year (populationwide means). Sharpe used total counts of nodes along stems of harvested plants combined with mean leaf production rates to estimate plant ages. For sterile and fertile adults mean age estimates were ten to thirteen years (maximum twenty-five years). Estimated

potential lifetime reproductive output of a *D. wendlandii* sporophyte was one or two sporophylls (spore-producing leaves) and up to nineteen ramets (vegetatively produced "offspring"). This approach includes assumptions that need to be evaluated. Size dependence and/or temporal variability in leaf production rates would produce errors in such age estimates as would disproportionate mortality of slower-growing plants. Given the information now available, it would be extremely interesting to use genetic markers to study the population consequences of this species' combination of sexual and vegetative reproduction.

For forest herbs an important demographic factor may be physical damage caused either by falling trees and litter or by animal activity. Approximately 25% of the *D. wendlandii* ferns in Sharpe's primary forest plots were completely covered by fallen trees or leaves during a two-year period (Sharpe 1988). Similarly, Kiew (1986) studied four species of herbs in a disturbed Malaysian rain forest and found evidence that all were subject to major physical damage (in five years the stem or apex was broken on one or more of the five plants in each species sample); all four species had the capacity to recover from such damage.

Based on his work on BCI, Smith (1987) proposed that the relatively low species richness of terrestrial herbs in the understory of tropical forests is the result of a high probability of local extinction of these species, in turn related to their particular life-history traits. Herbs in BCI's tropical moist forest appear tightly tied to changes in light availability in the understory. Growth and reproduction occur for the most part only in microsites that are strongly lit from a canopy opening above or nearby. As light levels decline, these plants decrease in size and may die if light conditions fail to improve before the plant's reserves are exhausted. Given the low light availability in La Selva's shaded understory sites (chap. 10), this scenario merits investigation.

Dieffenbachia longispatha (Araceae), a clonal terrestrial aroid of primary and secondary forest on alluvial soils at La Selva, flowered in all years of a four-year study (Young 1986a, 1990), but only 27% of the individual plants flowered in this period. Fruit set varied enormously among inflorescences: 3% to 90% of flowers produced fruits, and abortion of whole inflorescences was common. Similarly, Beach (1982) found that fewer than 5% of the mature-sized plants of *Cyclanthus bipartitus* (Cyclanthaceae) flowered in a two-year period although there was flowering in both years. Although no light measurements were made, these studies are consistent with Smith's (1987) hypothesis of light limitation for understory herbs.

Light was also implicated as limiting growth and reproduction in forest understory species of *Heliconia* (Heliconiaceae). Stiles (1975) observed that in habitats associated with primary forest, *Heliconia* abundance appeared to be linked to the availability of light. Multiple *Heliconia* clumps under forest canopy failed to flower in two consecutive blooming seasons. Stiles speculated that plants in shaded sites may be unable to allocate reserves simultaneously to growth and reproduction. The highest densities of *Heliconia* were in second-growth and open areas; the species of these habitats can very rapidly produce large clumps vegetatively. With their pivotal role as food plants for numerous hummingbird species (Stiles 1978c) and as larval food for hispine beetles (Strong 1982a), the *Heli-*

Fig. 7.1. The abundant understory fern *Danaea wendlandii* is subject to severe light limitation and a constant high risk of physical damage from falling debris (Sharpe 1988). Photo: J. Sharpe.

conia illustrate well the importance of plant demographic processes in community ecology.

Among epiphytes, groups such as orchids and bromeliads are of great interest demographically. When perched on the boles or branches of canopy trees, these plants are subject to dry, relatively high light conditions and to the risk of being carried to extremely different, lethal microsites if their branches or trees fall. Their aerial habitat, however, makes epiphytes very difficult to study, and estimates of population sizes, growth rates, mortality, and reproductive output are totally lacking.

Comparative demographic study of the most species-rich herbaceous groups would be a valuable approach to the understanding of relationships among coexisting, closely related taxa. The Araceae, rich in both species and life forms (herbaceous climbers, epiphytes, and terrestrial plants) is one such group. Currently under investigation (Horvitz in press; Le Corff 1992) are the interactions among demographic processes and forest light environments in the species-rich herb genus *Calathea* (Marantaceae).

Shrubs

Burger (1980) has pointed out a striking difference between the species richness patterns of trees and understory shrubs of lowland Neotropical rain forest: the high species richness of trees is from a large number of genera and families, whereas the many species of shrubs are concentrated in relatively few genera and families. At La Selva, as in other wet lowland forests in the Neotropics, these dominant shrub groups are Rubiaceae (e.g., *Psychotria,* forty-one species, Wilbur et al. 1991), Piperaceae (*Piper,* ca. forty-nine species, R. Marquis pers. comm.), and Melastomataceae (e.g., *Miconia,* twenty-eight species, Wilbur et al. 1991). The mechanisms accounting for the disproportionate species diversity in these shrub genera remain unknown. The demography of these species is of interest both because of this remarkable co-occurrence of congeners and because of the ecological importance of shrubs in understory (e.g., the effect of this guild on tree regeneration has been the focus of experimental work at La Selva; see chap. 9).

The best-studied shrub genus at La Selva is *Piper.* An intensive study included analysis of the species distributions within the reserve, comparative study of life-history patterns, and evaluation of the role of vegetation reproduction (Greig 1991). Using ordination techniques to analyze distribution patterns of forty-two *Piper* species at La Selva, Greig identified three major species groups: those characteristic of open, high-light habitats; those found in late secondary and primary forest on older alluvial and residual soils; and forest species restricted to recent alluvial soils. Highest species richness and density of *Piper* occurred in large gaps within primary forest. Major regeneration differences were related to the species groups. The forest species rarely reproduce by seed and, instead, heavily depend on various forms of vegetative reproduction (fig. 7.2); in contrast, the *Piper* species characteristic of young, open habitats principally reproduce by seed and show little capability for propagation by fragmentation. For a subset of five contrasting *Piper* species, Greig found that seed number and seed size were inversely correlated, with the species of open habitats characterized by far higher seed numbers and proportionately less seed predation.

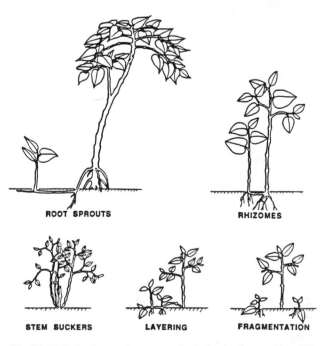

ROOT SPROUTS

RHIZOMES

STEM SUCKERS   LAYERING   FRAGMENTATION

Fig. 7.2. Methods of vegetative propagation in the shrub genus *Piper.* This is the dominant form of reproduction for most *Piper* species in late-successional or primary forest at La Selva (from Greig 1991, 101, with permission).

In a survey of sixteen *Piper* species in different habitats at La Selva, Gartner (1989) demonstrated the high incidence of physical damage to plants in forest microsites as well as a general capacity for vegetative growth after breakage from litter fall and other agents. In a study of the habitat affinities and densities of *Piper* at La Selva, Fleming and Maguire (1988) found that species richness and density were greater in second growth than in primary forest and that species richness in gaps was correlated with gap size. A detailed study of the impact of herbivory on *Piper arieianum* (Marquis 1984b, 1987) demonstrated that herbivory decreases both growth and reproduction and that the impact is greater on smaller plants (see chap. 15). Since 1980 Marquis has also followed the growth and survival of several La Selva *Piper* species in primary forest (Marquis 1988, unpublished data).

Other studies at La Selva have demonstrated major impacts of gaps on understory shrubs. Levey (1988b, 1990) found greatly increased fruit production by shrubs in the higher light conditions of tree-fall gaps. Experimentally planted cuttings of primary forest shrubs in the genera *Miconia* and *Piper* grew faster in gaps (Denslow et al. 1990); fertilization and shade-house experiments indicated that the shrubs' growth rates were limited by light rather than nutrients.

An Understory Cycad

Long-term demographic information has been obtained on the only cycad at La Selva, *Zamia skinneri* (Zamiaceae) (D. A. Clark and Clark 1987b; D. B. Clark and Clark 1988, D. B. Clark et al. 1992). This understory species is dioecious, with no noticeable vegetative dimorphism between the sexes. Population processes in primary and secondary forest have been documented through censuses of 180 individuals three to five times annually since 1980. In this species sexual expression is clearly related to plant size. Sexually active individuals are larger than nonreproductives, and reproductive females are larger than reproductive males (no sex changes have been observed). The frequency of production and size of cones increase with plant size for both sexes.

In the understory of primary forest, *Z. skinneri* shows the demographic consequences of extreme light limitation. Over the first six years of the study, 74% of the adult-sized plants in primary forest were reproductively inactive, and in only two years were both male and female cones produced in this population. The apparent adult sex ratio, based on plants reproductively active in the six years, was highly biased toward males (4:1). Only 5% of the adult-sized plants in the primary forest sample were active females, and none of these produced a cone more than once in the entire period. In the secondary forest, where light levels were higher (based on an index of the openness of surrounding canopy), more adult-sized plants produced cones (73%) and the sex ratio was not significantly different from 1:1. Canopy openness of microsites differed significantly among plants grouped by reproductive category (active females in the most open sites, nonreproductives in the most closed sites), even though the differences in openness were small. These patterns indicate that reproduction is strongly limited by light availability and that reproduction is more costly (more resource limited) for females than for males.

These cycads grow very slowly, and reproduction further

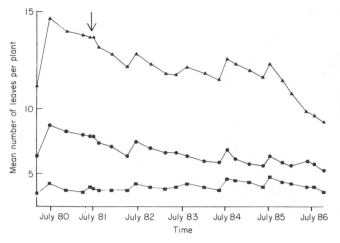

Fig. 7.3. The cost of reproduction in the cycad *Zamia skinneri* in primary forest at La Selva. Total leaf number declines subsequent to cone production with the effect lasting several years for female plants (first appearance of cones indicated by arrow); (triangles) active females in 1981; (circles) active males in 1981; (squares) nonreproductives in 1981 (taken from D. B. Clark and Clark 1988 with permission from the *Journal of Ecology*).

decreases growth (D. B. Clark and Clark 1988). For the entire sample, average annual leaf production was only 1.3 leaves per plant (*Z. skinneri* crowns consist of 1 to approximately 24 long-lived leaves; mean minimum time alive = 4.6 years; maximum ≥ 8.9 years; D. B. Clark et al. 1992). One year before cone production, male and female plants produce many new leaves; cone production is then followed by a decline in leaf number over the next several years (fig. 7.3). The longer and more severe depression in leaf production for females reflects their greater investment in reproduction. For both sexes, reduced growth after reproduction affects future reproductive potential, which depends on plant size. Although *Z. skinneri* may be an exceptionally "slow" species, it exemplifies well how conditions in the dark understory of La Selva's primary forest can limit growth and reproduction.

Palms

The palms are numerically and structurally an important group at La Selva. Smaller palms are prominent in the understory (Chazdon 1986c), and the larger species account for 25.5% of all stems greater than 10 cm dbh (D. Lieberman et al. 1985b). Their population processes, therefore, have important bearing on many aspects of the forest ecosystem.

Palms are of particular interest to tropical plant demographers because their growth form may permit inference of age and growth history. Many species produce clear, permanent leaf scars. Combined with estimates of rates of leaf production obtained by observation of plants of all sizes, these scars have been used to estimate past growth. Such inferred histories, however, can incorporate substantive error because of variability in growth rates (e.g., Lugo and Rivera Batlle 1987). Palm demographers have applied this technique in other tropical forests (Bullock 1980; Sarukhán 1980; Piñero et al. 1984; Lugo and Rivera Batlle 1987), but it has not yet been used to age palms at La Selva. Bullock (1986) measured internode dimensions of seedlings, juveniles, and adults of the large palm *Socratea exorrhiza* at La Selva. The variable patterns of

internode lengths along stems appeared to reflect periods of differential growth resulting from suppression and release, developmental changes, or reproductive events. Paul Rich (unpublished data) has carried out annual measurements of diameter, height, leaf number, and internode dimensions for *S. exorrhiza* and several other large palms at La Selva. His data will enable evaluation of internodes and leaf scars as indicators of growth history in these palms. As part of an ongoing study of the morphology and ecophysiology of understory palms, data are also being collected on leaf production and survival of *Geonoma cuneata*, *Geonoma congesta*, and *Asterogyne martiana* (see Chazdon 1986c).

Almost no data are available on reproductive output of palms at La Selva. Sarukhán (1980) cites unpublished data from La Selva of J. Vandermeer showing that seed production by *Welfia* palms increases with adult size as has been shown for the Mexican forest palm, *Astrocaryum mexicanum* (Piñero et al. 1982).

Palm seedlings are abundant at La Selva. Werner (1985) found that palm seedlings 30–100 cm tall cover 23% of the ground in primary forest, based on quadrats in Organization for Tropical Studies (OTS) plot 2 (see chap. 8 for descriptions of OTS plots 1–4). Vandermeer (1977) followed seedling dynamics of one species, *Welfia georgii*, in primary forest. In a seven- to eight-month period mortality was 38%. Physical factors (leaf and branch falls) accounted for 46% of the seedling deaths. Seedlings closer to adult *Welfia* showed significantly higher mortality. The large leaves of adult *Welfia* palms are an important source of mortality for small understory plants (Hartshorn 1972; Vandermeer 1977) and accounted for 14% of the seedling mortality in this study. Even when all litter-caused deaths are discounted, however, increased seedling mortality close to *Welfia* adults indicated that interseedling competition or natural enemies caused disproportionate losses of *Welfia* seedlings close to parent trees. Vandermeer's (1977) data on seedling distribution support this idea. In quadrats placed randomly within a 60 × 100 m plot, he found more than twice as many *Welfia* seedlings at greater than 3 m from the nearest *Welfia* adult than closer to adults despite the likelihood that most seed input is within 3 m of adults. Such analyses have not been made for seedlings of other La Selva palms.

Lianas

Woody climbers (lianas; fig. 7.4), one of the most characteristic plant groups of lowland tropical wet forest, have not been studied at La Selva. Because their reproductive structures are largely restricted to the forest canopy, many liana species at La Selva remain unidentified (B. E. Hammel pers. comm.). For no species is there an estimate of the number of reproductive individuals within the reserve. As shown by Putz (1984b) on BCI, individual lianas may loop over the canopy and between the canopy and the understory, interconnecting dozens of canopy trees. Because fallen stems often reroot (Putz 1984b), delimitation of individuals (genets) will require use of genetic markers.

At La Selva the lianas are structurally important, and they clearly compete with the crowns of canopy trees for light. Werner (1985) estimated that liana foliage accounted for 15% of the total canopy foliage in primary forest. Most canopy-level trees of seven canopy and emergent tree species at La

Fig. 7.4. The strikingly braided stem of the "monkey-ladder" vine (*Bauhinia guianensis,* Caesalpinoideae) makes it easily recognizable vegetatively; other morphologically distinguishable lianas at La Selva remain unidentified for lack of reproductive material. Photo: R. Marquis.

Selva were occupied by lianas; the basal area of lianas per tree increased with tree diameter (D. B. Clark and Clark 1990).

Very slow growth in diameter may be characteristic of the lianas as a group. Two *Doliocarpus* (Dilleniaceae) species at La Selva were projected to have average diameter growth of just 1 mm per year, with many individuals lacking diameter growth in the thirteen-year measurement period (Peralta, Hartshorn, et al. 1987). Lianas that attain large diameters may be of great age and may extend long distances through the forest (Putz 1984b). Because of these characteristics, lianas may experience significant genetic change over their lifetimes through somatic mutation.

Trees

The Existing Data. Although most plant demographic information from La Selva is for trees, there are detailed published studies of population processes for only a handful (table 7.1) of the more than 320 species (chap. 6). One major communitywide study has focused on long-term tree growth and mortality patterns in La Selva primary forest (D. Lieberman et al. 1985a, 1985b, 1990; M. Lieberman et al. 1985). In three forest inventory plots established in 1969–70 (OTS plots 1–3, 12.4 ha total), all woody stems greater than 10 cm in diameter (4,318 trees and lianas) have been remeasured three times over a total of nineteen years. In subsamples of the plots, seedlings (Peralta, Barquero, et al. 1987a) and saplings 2–10 cm diameter have also been repeatedly inventoried. Tree dynamics were also studied for two years in several successional stages and primary forest by Werner (1985). He measured survival, growth, and crown dimensions for all trees greater than 5 cm in diameter in plots of increasing age: five year (one of the La Selva successional plots); eight year (the "1973 Plot": succession following clearing in 1973 with treatments using burning, movement of litter, and/or application of methyl bromide); sixteen year (secondary forest east of Sendero Holdridge);

thirty-one year (a *Goethalsia meiantha* (Tiliaceae)-dominated stand west of Sendero Holdridge); and primary forest (1 ha in OTS plot 2). In OTS plot 1, Hartshorn (1972, 1975) intensively studied the population structure and dynamics of La Selva's dominant tree, *Pentaclethra macroloba,* and an associate canopy species, *Stryphnodendron excelsum* (both Mimosoideae), during one year. Long-term demographic and microsite data have been obtained for nine species of canopy and emergent trees at La Selva (table 7.1); the study, begun in 1983, is based on extensive population samples of all post-seedling life stages of these trees in more than 150 ha of primary forest (D. A. Clark and Clark 1987a, 1992; D. B. Clark and Clark 1987, 1990, 1991). Many shorter-term studies of trees have also contributed to the data base. This research has revealed a high turnover of trees in the La Selva forest and diverse life-history modes. Nevertheless, to understand tree population processes at La Selva will require detailed demographic and ecological study of many more species over the long term.

Inter- and Intraspecific Patterns of Tree Mortality. Communitywide mortality rates of La Selva trees are strikingly high. Over a 13-year interval (1970–1982), D. Lieberman et al. (1985b) found an exponential annual mortality rate of 2.03% for trees and lianas greater than 10 cm in diameter. In the following 2.5-year interval (D. Lieberman et al. 1990) mortality was 2.34%. These rates make La Selva one of the most dynamic tropical wet forests known. For tropical wet forests in Southeast Asia, Africa, and the Neotropics, most exponential annual mortality rates are less than 1.5% per year (summarized in Swaine, Lieberman, et al. 1987b). In notable contrast is the mortality measured in a 50-ha plot of tropical moist forest on BCI (Hubbell and Foster 1990b). In the period 1982–85, trees greater than 8 cm in diameter showed a mortality rate of 3.04% per year; this high rate (nearly three times those measured previously on BCI; Putz and Milton 1982;

Table 7.1 La Selva tree species for which detailed population studies have been published

Species (Family)	Study Period (years)	Most Recent Reference
Carapa guianensis (Meliaceae)	6	McHargue and Hartshorn 1983
Cecropia insignis (Cecropiaceae)	>4	D.A. Clark and Clark 1992
Cecropia obtusifolia (Cecropiaceae)	>4	D.A. Clark and Clark 1992
Compsoneura spruce (Myristicaceae)	2	Bullock 1982
Dipteryx panamensis (Papilionoideae)	>9	D.A. Clark and Clark 1992
Guarea rhopalocarpa (Meliaceae)	4	Bullock et al. 1983
Hyeronima alchorneoides (Euphorbiaceae)	>9	D.A. Clark and Clark 1992
Hymenolobium mesoamericanum (Papilionaceae)	>9	D.A. Clark and Clark 1992
Jacaratia dolichaula (Caricaceae)	5	Bullock and Bawa 1981
Lecythis ampla (Lecythidaceae)	>9	D.A. Clark and Clark 1992
Minquartia guianensis (Olacaceae)	>9	D.A. Clark and Clark 1992
Pentaclethra macroloba (Mimosoideae)	1	Hartshorn 1975
Pithecellobium elegans (Mimosoideae)	>9	D.A. Clark and Clark 1992
Simarouba amara (Simaroubaceae)	>4	D.A. Clark and Clark 1992

Lang and Knight 1983) appears to reflect extensive die-off related to the record "El Niño" drought of 1982–83. Long-term data bases are needed from La Selva and other sites to determine if such extreme tree mortality events also take place in tropical wet forests. Episodes of mortality as marked as that documented for BCI would have a major impact on ecosystem processes at all levels.

When all species in the OTS plots are treated together (trees >10 cm diameter), mortality rates do not vary with size (D. Lieberman et al. 1985b). A lack of size dependence in communitywide mortality rates has also been found in other tropical forest study plots when analyses are based on all stems greater than 10 cm in diameter (e.g., Malaysia, Manokaran and Kochummen 1987; Ghana, Swaine, Hall, et al. 1987). Although this is at first surprising, it is, perhaps, explained by the heterogeneity of the trees in each size class in such plotwide samples. The 10–30 cm diameter size class, for example, includes the very largest, perhaps senescent individuals of the smaller subcanopy species, many palms, and the vigorous "subadults" of canopy and emergent species.

Analysis of the relationship between mortality rates and ecological grouping (guild) is an interesting next step. For example, D. A. Clark and Clark (1992) found highly significant lower death rates than the communitywide values for six canopy and emergent species. For trees greater than 10 cm in diameter, this group had an exponential annual mortality rate of only 0.44%. In Malaysia, Manokaran and Kochummen (1987) also found significantly different mortality rates among tree guilds (stems >10 cm in diameter): emergents, 1.4% per year; understory trees, 2.6% per year; secondary species, 3.8% per year.

Although mortality rate and size were unrelated for all stems greater than 10 cm in diameter in the OTS plots, trees 2–10 cm in diameter had much higher death rates (4.6% per year) than the larger trees (D. Lieberman et al. 1985b). Comparative data are provided by Werner (1985) from OTS plot 2. Although his mortality rate for trees greater than 10 cm in diameter over two years was similar to that obtained by D. Lieberman et al. (1985b), he found much lower mortality rates than they did for the 2–10 cm in diameter class (1.6% com-

pared with 4.6% per year, respectively; χ^2, p <0.001). The difference may be explained by sampling (all stems in 1 ha compared with subsamples from 12.4 ha, respectively), or the thirteen-year interval of the latter study may have included an episode(s) of high mortality for this size class. Again, ecological groups may also differ in mortality rates in the 2–10 cm in diameter class.

Hartshorn (1972) proposed that trees are generally characterized by a sharp decline in mortality rate with increasing juvenile size (with pioneer species a possible exception). In the few data sets available from La Selva, mortality rates are indeed highest in the smallest sizes and rapidly drop with increasing sapling size (table 7.2). An exception, however, is the largest size class (>80 cm in diameter) of the dominant canopy species *Pentaclethra macroloba,* with 12% mortality during the one-year study. Analysis of the extensive data for *Pentaclethra* from the OTS plots (D. Lieberman et al. 1990) will reveal whether this suggestion of senescence at large diameters is meaningful or related to sampling error.

Growth Rates of La Selva Trees. Evaluation of tree growth rates under natural conditions is challenged by high variance across individuals, across size classes, and through time. Mean growth values are of limited utility. To identify the factors underlying this variance repeated measurements of individual trees must be coupled with data on microsite and climatic conditions and, ideally, with genetic markers. Tropical forest trees present additional complications. Because of the high frequency of buttressing and other bole anomalies, diameter often must be measured several meters above the ground (fig. 7.5). Shedding of bark, damage from falling trees, and bole irregularities caused by lianas or pathogens can all introduce error into measurements. Diameter can actually shrink because of short-term changes in the tree's water status (Schulz 1960). Finally, as demonstrated by Hazlett (1987) for three species of La Selva trees, diameter growth may follow marked seasonal rhythms. Height growth of tropical forest trees is affected both by high variance across individuals and size classes and by frequent stem and crown damage. Combined with the local rarity of most species, these factors ex-

Table 7.2 Relationship between size class and mortality rates for La Selva trees

	Pentaclethra macroloba[a]	
	Annual Mortality Rate (%/yr)	Number of Individuals (N)
Height Class (cm)		
0–50	52.0	(2,292)
50–100	27.7	(137)
100–150	11.4	(88)
150–200	14.0	(57)
200–250	2.8	(36)
250–300	8.1	(37)
Diameter (cm) (> 300 cm tall):		
2 < 5	2.6	(38)
5 < 10	0.0	(19)
10 < 20	2.7	(37)
20 < 40	2.4	(41)
40 < 60	1.9	(54)
60 < 80	2.3	(44)
> 80	12.0	(25)

	Other Species[b]					
	Annual Mortality Rate (N)					
	LA	MG	HM	DP	PE	HA
Diameter (cm)						
≤1	12.9 (62)	2.7 (59)	17.3 (10)	19.0 (79)	15.5 (26)	— (1)
1–4	1.5 (52)	0.9 (59)	2.4 (11)	0.0 (8)	2.4 (11)	3.3 (8)
4–10	1.1 (24)	0.0 (54)	0.0 (14)	0.0 (7)	1.7 (15)	0.0 (14)
10–20	0.0 (17)	0.0 (33)	0.0 (12)	0.0 (3)	0.0 (16)	0.0 (14)
20–30	0.0 (6)	0.0 (14)	0.0 (7)	0.0 (4)	0.0 (5)	0.0 (6)
30–50	0.0 (9)	1.6 (33)	0.0 (3)	0.0 (17)	0.0 (24)	0.0 (7)
50–70	0.0 (8)	1.5 (17)	0.0 (3)	0.0 (22)	1.8 (29)	0.0 (11)
70–100	0.0 (21)	— (1)	0.0 (5)	0.7 (36)	0.0 (28)	0.0 (18)
>100	0.0 (12)	— (0)	3.9 (7)	0.7 (38)	2.6 (10)	0.0 (20)

[a]Data for the canopy species *Pentaclethra macroloba* were taken over one year for all individuals (or subsamples) in 4 ha of OTS plot 1 (taken from Hartshorn 1972).

[b]Data for the remaining canopy and emergent species are for individuals more than 50 cm tall in an extensive sample of La Selva primary forest over a four-year period and were calculated as exponential annual mortality rates. LA: *Lecythis ampla;* MG: *Minquartia guianensis;* HM: *Hymenolobium mesoamericanum;* DP: *Dipteryx panamensis;* PE: *Pithecellobium elegans;* HA: *Hyeronima alchorneoides* (taken from D.A. Clark and Clark 1992).

plain why so little is known about patterns and determinants of growth rates in tropical forests. In recent studies, however, researchers have made a start toward understanding tree growth at La Selva.

M. Lieberman and Lieberman (1985) used data for the forty-five tree species represented by twelve or more individuals in the combined area of the OTS plots to simulate composite lifetime growth curves. For each species they randomly chose individuals from the sample of measured trees (partially stratified by tree size) and combined these individuals' thirteen-year growth increments to "grow" artificial trees from 10 cm in diameter to the maximum size observed in the sample. The procedure, repeated seventeen hundred times for each species, produced all lifetime growth curves that could be generated with the actual growth data available for individuals in a species sample. The maximum, median, and minimum growth rates estimated for each species were the slopes of the fastest, median, and slowest simulated growth curves. These growth estimates show several patterns. The substantial intraspecific differences between simulated maximum and minimum rates indicate high interindividual variance in diameter growth for most species. Of the forty-five species, only *Stryphnodendron excelsum* (Mimosoideae) showed a median (projected) growth rate of greater than 10 mm per year. Most species' simulated minimum growth rate was less than 0.5 mm per year, indicating that many individuals grew little or not at all in thirteen years. As a group the understory trees had the lowest projected median growth rates. Because diameter increments were only from trees greater than 10 cm in diameter, however, this observation may reflect the fact that the sample includes only the largest, perhaps senescent, individuals of these species. An important caveat regarding the growth simulations is that they assume no autocorrelation through time of an individual's growth rates. Such autocorrelation has been demonstrated in other studies of tropical tree growth (Swaine et al. 1987a; D. A. Clark and Clark 1992).

The first (1969–82) growth data from the OTS plots were based on diameter at breast height (1.3 m above the ground), including any buttresses or bole anomalies. Trees were mea-

Fig. 7.5. The high frequency of buttresses and other stem irregularities in La Selva trees makes climbing ladders a necessity for growth measurements. Photo: R. Marquis.

sured in 1982 both at breast height and above the buttresses (see fig. 7.5) but were measured only at breast height in 1969 (D. Lieberman et al. 1985a). These first growth data are, therefore, not directly comparable with data from studies based on bole increment (especially for trees with large buttresses, diameter size class and growth differ depending upon where they are measured). In the 1985 recensus (Lieberman et al. unpublished data), diameters were measured both at breast height and above the buttresses, thus providing the first data on bole growth for many La Selva trees (33% of stems >10 cm diameter were buttressed; chap. 8). The OTS plot diameter data from these and subsequent intervals will also enable assessment of the degree of growth autocorrelation.

To characterize growth of a given tree species the relation between size class and growth rates must be assessed. Although Richards cited the urgent need for such information in his 1952 classic on tropical rain forest, forty years later the phrase "extremely scanty" still accurately describes the data. For La Selva size-specific growth data are available for only a handful of species. The five emergent species studied by D. A. Clark and Clark (1992) show substantial size-dependent increases in median diameter growth rates, from less than 1 mm per year for smaller juveniles (<4 cm diameter) to 5–14 mm per year as adults. For Pentaclethra macroloba limited data have been published on growth at given sizes (Hartshorn 1972; Hazlett 1987). Hartshorn found mean one-year diameter increments of 3.0–3.7 mm per year for trees 10–80 cm in diameter. Much higher mean rates, however, were measured by Hazlett for two small samples of trees 20–40 cm in diameter (10.6 mm per year (N = 5) and 7.4 mm per year (N = 3)). For three trees greater than 1 m in diameter, Hartshorn found similarly high annual diameter increments (mean = 10.3 ± 7.6 mm). This apparent rapid growth in very large individuals contrasts with the higher mortality rate Hartshorn (1975) found for this size class and casts doubt on the idea of senescence in this species. The very large sample of measured Pentaclethra in the OTS plots (chap. 8) offers an opportunity to explore in detail the relationship between growth and size in this important species.

Growth rates of the same species can differ greatly between primary and younger forest and among successional stages. For Goethalsia meiantha (Tiliaceae), a species characteristic of secondary forest and gaps in primary forest at La Selva, Werner (1985) found a marked decline in diameter growth with stand age from 40–50 mm per year in five- and eight-year-old second growth to 0–10 mm per year in a thirty-one-year-old plot dominated by Goethalsia. Werner speculated that the growth decline was the result of increasing crowding of crowns over time with a resultant imbalance between crown volume and stem. For Pentaclethra Werner (1985) found notably high diameter growth rates (10–20 mm per year) of trees in the early stages of succession. Similar trends have been demonstrated for a canopy species in the Philippines, Parashorea malaanonan (Dipterocarpaceae)—significantly greater growth in full sun as well as a pronounced increase in growth rate with increasing tree size in primary forest (W. H. Brown, cited in Richards 1952).

Height growth of small saplings has been studied for a few La Selva species. For Pentaclethra macroloba Hartshorn (1972) documented very slow mean height growth (<10 cm per year) for saplings up to 1.5 m tall and a gradual increase to 30 cm per year for saplings 2.5 to 3.0 m tall, although growth was highly variable at all sizes. Limited data he obtained for four other canopy species also suggest slow increases in height growth in this size range. The six canopy and emergent species studied by D. A. Clark and Clark (1992) similarly showed very slow median rates of annual height growth as small saplings in primary forest.

Seedling growth rates in primary forest have been documented for one species, the emergent Dipteryx panamensis (D. B. Clark and Clark 1987). For the cohort of seedlings (N = 147) in a 1-ha plot, stem diameter actually decreased steadily in the first year after germination, probably because of the consumption of maternal reserves (at eleven months, nearly half the seedlings had failed to produce any leaves beyond the original pair; D. B. Clark and Clark 1985). From one to five years postgermination, average stem diameter increased at about 0.5 mm per year. The five surviving five-year-olds had 4–7 mm diameters and heights of 36–85 cm.

For such young and/or small trees, this very slow growth in primary forest strongly contrasts with growth rates in the open. Huston (1982) planted seeds of several La Selva primary forest species in former pasture on alluvial soil; at two years old, Pentaclethra saplings were nearly 5 m tall, and Dipteryx had reached 8 m. Clearly, most young trees in the primary forest are strongly suppressed. Given the dynamics of the forest canopy, juveniles of many species might be expected to pass through several periods of suppression; in some species, however, saplings may lose their growth potential after experiencing a period of zero or very slow growth.

To understand tree regeneration in tropical forests, detailed growth histories are needed of individuals in all size classes, across the spectrum of forest environments. Lacking such data, we are as yet ignorant of the patterns of growth that actually lead to attainment of adult size for nonpioneer tropical wet forest trees.

Gap-dependence and Shade Tolerance. The role of gaps in tropical tree regeneration has inspired a voluminous theoretical literature (see references in Denslow 1987b; D. A. Clark and Clark 1992). The central paradigm has been that tropical forest tree species are differentially dependent on gaps for regeneration. Although terminology varies greatly among authors (D. A. Clark and Clark 1987a), most have defined three general categories of species: those that can only regenerate in large gaps or clearings (often referred to as "pioneers"), those that cannot regenerate in the absence of a gap of small or moderate size (often called "gap" or "(shade-) intolerant" species), and those that do not require a gap for regeneration (often called "primary" or "shade-tolerant" species). These ideas have stimulated much interesting research, but the terminology is vague and does not lead to predictions testable with field data. How would one determine if a tree species requires gaps for regeneration and if there is specialization on a given range of gap sizes? One objective criterion is whether or not germination and/or establishment takes place only in high light. Swaine and Whitmore (1988) have proposed that tropical trees be divided into "pioneers" and "climax, or nonpioneer, species" on this basis. Most current concepts of gap dependence, however, refer more generally to an inability to grow and/or survive in nongap situations. Elucidation of the role of environmental factors in tropical tree regeneration will require quantitative data on the performance of individual trees in known microsite conditions.

At La Selva, where Hartshorn (1978, 1980) developed his seminal ideas about the importance of gaps, these issues have received much attention. Even now, however, nearly all classification of La Selva trees by regeneration mode is according to the "forester's (or ecologist's) eye." Hartshorn (1980) reported that he had classified 320 La Selva tree species as shade intolerant ("require gaps for successful regeneration") or shade tolerant, based on his familiarity with the forest and on observations of tree growth in the Holdridge (La Selva) Arboretum. He proposed that 71% of the canopy species, 48% of the subcanopy species, and 39% of the understory species were shade intolerant. Species lists and the data used to classify species were not published, but these figures are often cited as evidence of the importance of gaps for tree regeneration at La Selva and in Neotropical forests in general.

The relationship between juvenile tree distribution and gap size was studied at La Selva by Barton (1984). In twenty-three old gaps (openings with regrowth at least 2 m tall) he inventoried all individuals more than 1 m tall of eleven tree species. Gaps greater than 300 m² were classified as "large." Three of the six species he had classified as pioneers, *Ochroma lagopus* (Bombacaceae), *Heliocarpus appendiculatus* (Tiliaceae), and *Castilla elastica* (Moraceae), occurred only in large gaps as predicted. The remaining three species, *Cecropia obtusifolia* (Cecropiaceae), *Goethalsia meiantha,* and *Hampea appendiculata* (Malvaceae), occurred in gaps of both sizes. All species that he had classified as shade tolerant (*Socratea exorrhiza, Welfia georgii, Iriartea deltoidea, Pentaclethra macroloba,* and *Guarea* spp.) occurred in both gap classes but were present in proportionately more of the larger gaps. As Barton concludes, these data support the idea of gap-size specialization only for the three pioneers restricted to larger gaps; no species showed evidence of specializing in smaller gaps. A similar pattern was found by Brokaw ((1985a) in his six-year study of thirty tree-fall gaps on BCI. Although the density of pioneers was highly significantly correlated with gap size (20–532 m²), density of nonpioneers was unrelated to gap size.

Werner (1985) documented changes in forest composition along the successional gradient he studied on an old alluvial terrace at La Selva (Holdridge Association; see chap. 4). *Ochroma lagopus, Trema micrantha* (Ulmaceae), and *Heliocarpus appendiculatus* only occurred in the two youngest plots (five–eight year succession). *Goethalsia meiantha, Hampea appendiculata,* and *Pentaclethra macroloba* were present in all plots (successional stages and primary forest).

In the La Selva primary forest Brandani et al. (1988) located all tree seedlings and saplings in fifty-one gaps and former gaps ranging in age from one to sixteen years (ages greater than eight years were estimates). They divided each gap into bole, crown, and root zones and analyzed within-gap distributions of juveniles of the thirty most abundant species. Juveniles of many species were over- or underrepresented in one or more of the zones. The preferred zone shifted, depending on whether or not the gap-forming tree was a *Pentaclethra macroloba.* These patterns indicate the need for soil analysis in gaps formed by different species. The authors conclude that juvenile trees at La Selva may be strongly influenced by subtle soil differences within tree falls or by long-term local effects of canopy trees on soils.

Stimulated by Hartshorn's initial work, D. A. Clark and D. B. Clark focused on a diverse suite of nonpioneer species for their long-term study of life-history patterns in canopy and emergent trees. Two had been subjectively classified as shade tolerant (*Minquartia guianensis* and *Lecythis ampla*), two as exemplary gap species (*Dipteryx panamensis* and *Hyeronima alchorneoides*), and two had not been classified (*Pithecellobium elegans* and *Hymenolobium mesoamericanum*). In 1988 three "high light" species (*Cecropia obtusifolia, C. insignis,* and *Simarouba amara*) were added for comparison. Survivorship, growth, and microsite have been monitored annually for trees of all postseedling size classes. Findings from the first six years were used to assess the life histories of these species (D. A. Clark and Clark 1992). When compared in terms of juvenile microsites (crown position, an index of light availability; and forest phase, gap, building, mature), the nine species show four contrasting patterns (fig. 7.6). The groups able to establish in low light under mature canopy (*A* and *B*) differ in sapling morphology and in their sensitivity to low light as small saplings. Although Group *B* species showed a significant mortality response to slightly lower light, those of Group *A* did not. Two species (Group *C*) were strongly associated with gap or building phase sites as small saplings and as large poles. The two species of *Cecropia* (Group *D*) showed the strongest association with gap/building sites and higher crown illumination. Although these patterns indicate differences in regeneration modes, a striking finding of the study to date is the similarity among the nonpioneers (Groups *A–C*) in juvenile performance and some microsite characteristics. All showed growth responses to small increases in light, the capacity for substantial growth in high light, the ability to survive years without growth, and very low mortality at larger juvenile sizes. These life history patterns are complex and do not follow prevailing paradigms regarding expected trade-offs between growth capacity and the ability to withstand suppression.

For the remaining La Selva trees, there are no published data to demonstrate that a gap or clearing is required for germination or establishment. It is likely that many species can establish in the understory as has been shown for *Welfia georgii* (Vandermeer 1977) and *Carapa guianensis* (McHargue and Hartshorn 1983). If this pattern is typical, then colonization of newly formed gaps is not the major process determining gap occupancy as has been tacitly or explicitly assumed in the past (Hartshorn 1978, 1980; Orians 1982; Brandani et al. 1988). Instead, seedlings and saplings already in place would be expected to dominate in new gaps. In a Venezuelan Amazonian forest Uhl et al. (1988) studied regrowth in experimentally formed gaps. They found that nearly all young trees present after four years were from advance regeneration, not from seedlings establishing after the gap formed.

In their retrospective study of gaps at La Selva, Brandani et al. (1988) attempted to distinguish between juvenile trees that had been present before the gap formed and those that colonized as seeds or newly germinated seedlings after gap formation. Classified as survivors of gap formation were saplings that showed evidence of physical damage from debris or

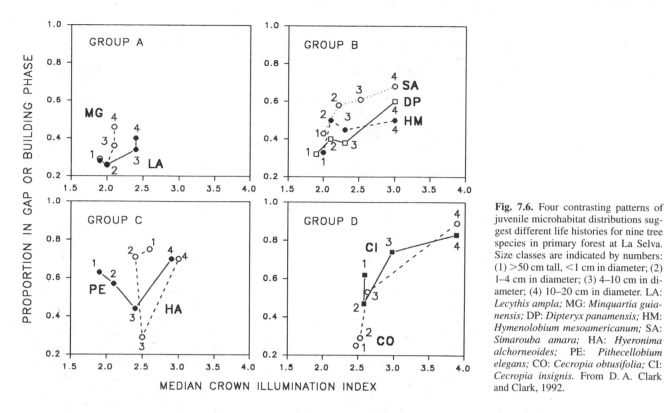

Fig. 7.6. Four contrasting patterns of juvenile microhabitat distributions suggest different life histories for nine tree species in primary forest at La Selva. Size classes are indicated by numbers: (1) >50 cm tall, <1 cm in diameter; (2) 1–4 cm in diameter; (3) 4–10 cm in diameter; (4) 10–20 cm in diameter. LA: *Lecythis ampla;* MG: *Minquartia guianensis;* DP: *Dipteryx panamensis;* HM: *Hymenolobium mesoamericanum;* SA: *Simarouba amara;* HA: *Hyeronima alchorneoides;* PE: *Pithecellobium elegans;* CO: *Cecropia obtusifolia;* CI: *Cecropia insignis.* From D. A. Clark and Clark, 1992.

that were considered too big to have grown up since the gap formed. They concluded that the great preponderance of gap occupants were colonizers. There is a problem, however, with this classification system: the older the gap, the less likely were juvenile trees to have been classified as survivors. Data presented in Orians (1982) from six of the gaps (ages given as 1.5–15 years) show a highly significant correlation between gap age and percentage of occupants classified as colonizers (R_s = .94, p = .01). The relative roles of colonization and advance regeneration in the filling of gaps at La Selva remains undetermined. Needed are data on many species' germination and establishment requirements and on the actual process of gap closure.

Focusing on a gap/nongap dichotomy raises several difficulties, including the diversity of microsites present both in gaps and in understory (D. A. Clark and Clark 1987a), the transitory nature of conditions within gaps (Fetcher et al. 1985), and the many currently used definitions of gaps (see D. A. Clark and Clark 1992). An alternative approach is to focus on specific environmental factors that influence regeneration. One factor that has been clearly tied to tree performance in the La Selva primary forest is crown light environment. For six species of canopy and emergent trees under long-term study (see D. A. Clark and Clark 1992), juvenile growth rates in primary forest are highly significantly correlated with different measures of light availability, even within very small ranges of light levels (Oberbauer et al. 1988, 1989; D. A. Clark and Clark 1992). When Werner (1985) analyzed the relationship of tree diameter growth to age of succession, tree size, crown dimensions, and light (crown position) for all stems combined, the multiple correlation coefficient of diameter growth with eleven variables was .54. The most important variable was light. Evaluation of growth response to available

light as a function of size class will provide one objective basis for characterizing tree species by regeneration mode.

Quantitative data on tree performance in measured microsite conditions are likely to change many "foresters' eye" perceptions of species' regeneration patterns. For example, *Pentaclethra macroloba* has been thought of as a shade-tolerant species, partially because of its abundant juveniles in the forest understory. Werner (1985) found this species to have the third highest diameter growth rate for 2–10-cm saplings among all 261 tree species occurring in his different-aged plots (only *Cecropia* and *Goethalsia* grew faster). As noted by Werner, *Pentaclethra*'s rapid growth in high light and abundance in early succession are characteristics usually associated with "gap species." At the end of his two-year study Werner (1985, 161) expressed doubt about the distinction between gap and primary forest species and wondered how *Pentaclethra* should be considered in this context. The complex life history patterns found for six canopy and emergent species at La Selva (D. A. Clark and Clark 1992) similarly highlight the problem with using the imprecise terminology of "gap dependence" and "tolerance."

Life Spans of La Selva Trees. No direct methods exist for aging individuals short of following them from germination (Bormann and Berlyn 1981). The short half-life of stems greater than 10 cm in diameter (thirty to thirty-four years; D. Lieberman et al. 1985b, 1990) and the rapid estimated forest turnover (118 years; Hartshorn 1978) at La Selva suggest that realized adult longevities of more than 100 to 200 years are rare for subcanopy and canopy trees. D. Lieberman et al. (1985a) used growth simulations for the forty-five most abundant tree species in the OTS plots to estimate potential longevity after reaching 10 cm in diameter. For each species, they estimated life span by dividing the maximum observed diameter in their sample by the estimated minimum growth rate. The greatest calculated longevity was 442 years (the emergent *Carapa guianensis*); the shortest for canopy trees was 78 years (*Goethalsia meiantha* and *Inga* cf. *pezizifera*). Because these values derive from the largest individuals observed, they depend on sample size (Median test: projected life span or maximum observed size compared with sample size for the twenty-one canopy species; $p = .02$). Not taken into account are the size-dependence of tree diameter growth and the probability that survival is related to growth rate. For these reasons these estimates only give rough bounds on the values to be expected.

For only seven La Selva tree species are there growth data from which life span up to the size 10 cm in diameter can be estimated. For *Pentaclethra macroloba* extrapolation from Hartshorn's (1972) 1-year data on mean growth rates at successive juvenile sizes leads to an estimate of 92-years for growth from seedling to 10 cm in diameter. For six canopy and emergent species (D. A. Clark and Clark 1992), similar calculations based on median annual growth rates of three size classes ($\geq$50 cm tall, $\leq$1 cm in diameter; >1–4 cm in diameter; >4–10 cm in diameter) in a 4-year period yield estimates of 98–183 years to reach 10 cm in diameter; estimates based on the maximum growth rates in these size classes, however, are much lower (17–39 years). It is possible that only the faster-growing saplings survive to subadult and adult sizes, especially given the high risk of physical damage for plants in the La Selva understory. Long-term data on growth and survival are needed to estimate satisfactorily time spent in these smaller size classes.

Structure and Stability of Tree Populations. The diverse size distributions shown by tropical trees have been used to infer regeneration processes. Classic "reverse J-shape" distributions (abundances highest in the smallest size classes and declining sharply with increasing size) were found in OTS plot 1 for *Pentaclethra macroloba* and a less-common canopy tree, *Stryphnodendron excelsum* (Hartshorn 1972). Such population structures have been thought to correspond to shade-tolerant, primary forest species (Knight 1975; Whitmore 1984). A strongly contrasting pattern was found in the same plot for *Dipteryx panamensis* (D. B. Clark and Clark 1987). Although the smallest size class was the most abundant, adults (trees >30 cm diameter) outnumbered 1–30 cm in diameter juveniles, and the 10–30 cm in diameter class was represented by only one tree in the 4 ha. Populations with very low numbers in all or some juvenile classes have been thought to represent either relict species (e.g., Knight 1975) or "gap" or "intolerant" species capable of rapid growth in gaps (Hartshorn 1980).

Growth and mortality rates in each size class must be factored into interpretations of patterns of population structure. Analysis of the adult size distribution, subadult growth rates, and mortality patterns of *Dipteryx* at La Selva demonstrated that this species is regenerating in situ in spite of the rarity of larger juveniles (D. B. Clark and Clark 1987). In a Nigerian moist forest Jones (1955) found that for some tree species with "deficient" size classes, the low numbers of individuals were compensated by relatively rapid growth at these sizes. A similar caveat was provided by Hubbell and Foster (1987a), who found on BCI that trees with apparently similar tolerance characteristics show variable population structures.

Hartshorn (1972, 1975) was first to apply life-table analysis to a tropical tree. From his one-year study of *Pentaclethra macroloba* in OTS plot 1, he had data on densities, growth and survival rates of seedlings to adults, and estimates of seed production and seed mortality rates. After interpolating the few missing growth and mortality data and estimating the relationship of seed production to adult size (he assumed a normal curve of seed production versus size class), he constructed a life table for *Pentaclethra*. Using standard matrix techniques, he calculated an intrinsic rate of increase of 1.002. This is the predicted ratio of population sizes between successive years, once the population structure has reached the theoretical stable stage distribution. A value so close to 1.0 indicated that the *Pentaclethra* population would be stable upon attainment of stable stage distribution. The actual size distribution in OTS plot 1, however, was very significantly different from the stable stage distribution (Kolmogorov-Smirnov one sample test; $p < .01$), indicating that this 4-ha "population" was still growing (there were proportionally many more seeds and seedlings than in a stable population). Limitations of this study were its short time span and the small study plot. It would be interesting now to reestimate the life-table parameters (to evaluate their stability over time) and to measure size distribution over a larger area ($\geq$10 ha) to see if the theoretical stable stage distribution is met at a larger scale.

Few life-table studies of other tropical trees exist. Piñero et al. (1984) constructed Leslie matrices for the understory rain forest palm, *Astrocaryum mexicanum,* at Los Tuxtlas, Mexico. Based on population samples from six forest plots, they calculated an intrinsic annual rate of increase of 1.0046. The observed population structure of *Astrocaryum* significantly differed from the stable stage distribution (the local population had not attained the structure indicated by its life-table characteristics). Bullock (1980) also applied life-table analysis to an African understory palm and calculated an intrinsic annual rate of increase of 1.0125. In a study of rain forest emergent trees in New Guinea, Enright (1982) found that the species with apparent high-light requirements for growth showed intrinsic annual rates of increase of 0.986–1.052 in three study plots; he interpreted this variation as a reflection of the spatial and temporal heterogeneity of regeneration for such species.

Major problems of interpretation of life-table analyses include the likelihood that most species show both temporal and spatial variation in life-table parameters; the usual finding that populations are not at the theoretical stable stage distribution; and the lack of any way to put confidence intervals around the projected intrinsic rate of population increase, the principal "product" of such studies. S. H. Bullock (pers. comm.) has suggested that perhaps the greatest use of life-table studies is to carry out sensitivity analysis, probing the relative effects of variation in different population parameters. Because of the considerable effort required to construct life tables for tropical plant populations, such studies should be carefully justified.

Reproduction. The most detailed studies of tree reproductive output at La Selva are of three dioecious understory species:

Jacaratia dolichaula (Bullock and Bawa 1981); *Compsoneura sprucei* (Bullock 1982); and *Guarea rhopalocarpa* (Bullock et al. 1983). Newstrom et al. (chap. 11) discuss flowering patterns in these species at the individual and population levels. For *Jacaratia,* females were larger than males, indicating that reproductive maturity is attained at a greater size (= age?) in females (no evidence exists of sex change in *Jacaratia*). For the other two species no such intersexual size difference was detected. In all three species production of flowers and fruits was significantly correlated with tree size. All species showed a marked skewness in the pattern of fecundity across individuals; in *Compsoneura,* for example, 20% of the individuals of each sex accounted for 73%–85% of the fecundity of that sex. As Bullock and Bawa (1981) pointed out, although this pattern results in marked domination of seed crops by a small proportion of adults in the short term, the relationship between fecundity and size means that individual trees are likely to shift in relative dominance as they grow. Shifts in reproductive dominance should increase heterogeneity in population genetic structure.

Males flowered more frequently than females in all three species. In *Compsoneura,* for example, although 52% of the males flowered three times during the study, only 13% of the females did so. The size dependence of reproduction and sex-dependent frequency of flowering indicate a clear cost of reproduction, which is greater for females. Similar patterns have been shown for a number of other dioecious species (see Piñero et al. [1982] and references in D. A. Clark and Clark [1987b]). The proportion of adults that flowered in each episode varied markedly for both *Compsoneura* and *Guarea;* this must result in significant genetic variability among seed crops.

Some fragmentary information exists on the reproductive output of canopy tree species at La Selva. Hartshorn (1972) inventoried flowers in the crown of a freshly felled adult *Pentaclethra macroloba* and estimated a total of slightly more than one million flowers. In Hartshorn's (1975) life-table matrix, the increase in seed production with increasing tree size is a hypothesized pattern, based on the assumption that proportional fecundity of increasing size classes is normally distributed. McHargue and Hartshorn (1983) obtained direct evidence of increasing fecundity with increasing size for the emergent tree *Carapa guianensis* by counting fruit valves under the crowns of five adults. Extrapolated seed production increased monotonically with tree diameter, from 754 seeds under a 20-cm diameter tree to 3,944 seeds under an individual 95 cm in diameter.

Age or size at first reproduction is largely environmentally determined for canopy trees. In primary forest at La Selva canopy species probably attain reproductive maturity at about 25–35 cm diameter, the approximate size at which individuals reach the canopy. The diameter of the smallest reproductively mature *Pentaclethra macroloba* in Hartshorn's (1972) study was 37 cm. The hypothesis that reproductive maturity is triggered by exposure of the crown to full sunlight is supported by very precocious reproduction in open sites. *Pentaclethra* has been observed to fruit at three years when planted in full sun (Huston 1982), and *Dipteryx* produced fruit in less than five years in similar conditions (personal observation).

Great variability in the size of successive seed crops and long interflowering intervals (see chap. 11) hinder the study of reproduction in tropical canopy trees. McHargue and Hartshorn (1983) found high interyear variability in seed production for *Carapa guianensis:* in one six-year period were two years when very few members of the population fruited and three years when nearly all adults had abundant fruit crops. Hartshorn (1972) cites unpublished data suggesting that *Stryphnodendron excelsum* populations reproduce at multiyear intervals. For a rare emergent species, *Hymenolobium mesoamericanum,* reproductive episodes are separated by several years (Clark and Clark unpublished data; M. Grayum unpublished data). Multiyear studies of reproduction at the population level are needed for a spectrum of La Selva species before general patterns can be inferred.

Seed and Seedling Demography. Although seed predation is known to have major impacts on tropical tree populations (Janzen 1983d), there have been few studies of this demographic aspect of La Selva trees. In their study of *Carapa guianensis* McHargue and Hartshorn (1983) found that 54%–96% of seeds had been removed from fruit valves under adult crowns. *Haplomys* rats were shown to eat *Carapa* seeds, and a lepidopteran *(Hypsipyla)* commonly infested seeds. McHargue and Hartshorn (1983) speculate that multiyear intervals between major fruit crops allow *Carapa* to "escape" intense seed predation through predator satiation (cf. Janzen 1971b).

In strong contrast *Pentaclethra macroloba* appears to suffer relatively low levels of seed destruction. Hartshorn (1972) estimated that parrots and squirrels destroy 5%, 3% are lost to insects predispersal, and 1% of seeds on the ground are infested by insects. Although he found 47% removal rates for seeds experimentally placed on the ground, many of these seeds may not have been eaten, owing to their high alkaloid and toxic amino acid content (Hartshorn 1972). Janzen (1970) has speculated that *Pentaclethra's* dominance in the wet Atlantic lowlands is the result of its relatively recent invasion of these areas with its normal complement of seed predators "left behind." Hartshorn (1975) tested this idea with his *Pentaclethra* life table. When he set seed survival at 5%, as if under intense seed predation, the resulting intrinsic rate of population increase dropped from 1.002 to 0.986. This rate would lead to a 13% population decrease in ten years.

Seedlings of La Selva trees experience high mortality rates. Peralta, Barquero, et al. (1987a) followed dynamics of tree and liana seedlings in multiple primary forest transects over a one-year period. The half-life of seedling cohorts was two to six months with an annual mortality rate of greater than 80%.

Densities of tree seedlings in the La Selva primary forest are fairly low. Average total densities recorded by Peralta, Barquero, et al. (1987a) for all tree and liana seedlings combined were only 5–15 per m², except locally just after germination of abundant crops. For *Dipteryx panamensis* densities of first-year seedlings in 1 ha of OTS plot 1 ranged from 0.15 per m² in the aggregations under parent trees to less than 0.05 per m² (D. A. Clark and Clark 1984). For the emergent tree *Nectandra ambigens* (Lauraceae) at Los Tuxtlas (Mexico), Córdova Casillas (1985) found much higher average seedling density (9.5 per m² in a 1.7-ha plot). On BCI, however, where mammalian seed and seedling predators are very abundant, predation on *Dipteryx panamensis* seeds and seedlings can result in almost no naturally occurring seedlings in the population (DeSteven and Putz 1984); on the adjacent mainland *Dipteryx* seedling densities are comparable to those at La Selva.

Dipteryx panamensis has an annual fruiting cycle (chap. 11), which produces an annual cycle of seedling abundance at La Selva (D. B. Clark and Clark 1987). Seedlings at high density can experience up to 90% mortality in the first year (Clark and Clark unpublished data). *Dipteryx* seedling mortality at La Selva is higher both at higher seedling density and closer to adult trees although partial correlation analysis indicates that density is the predominant factor. Although more than 30% of seven-month-old seedlings at low initial density (<0.05 per m²) survived two years, fewer than 5% of the high-density seedlings (0.11–0.15 per m²) lived this long (D. A. Clark and Clark 1984). This pattern of early mortality is consistent with the Janzen-Connell hypothesis that natural enemies of tropical trees increase interindividual spacing through seed predation or damage to seedlings. The relationships among seedling density, herbivory, and seedling mortality suggest that herbivores are responsible for this spacing process in *Dipteryx* at La Selva (D. B. Clark and Clark 1985).

In their study of seed dispersal in *Casearia corymbosa* (Flacourtiaceae), a tree on alluvial soils at La Selva, Howe and Primack (1975) found that seedlings left under adult crowns had much higher mortality than those carried to sites farther from adults; by seven to eight months after germination, all seedlings under adults had died (Howe 1977). Hartshorn (1972) presents seedling distribution data for one adult *Stryphnodendron excelsum* that suggest the same process. Many (but not all) of the studies of seed and juvenile mortality of tropical woody plants have produced evidence consistent with the Janzen-Connell spacing hypothesis (D. A. Clark and Clark 1984; Hubbell and Foster 1990a). Detailed studies of early mortality of more La Selva tree species would enable evaluation of the importance of this process in the community as a whole. Particular attention should also be given to the role of pathogens in mortality of seeds and seedlings; although this has not yet been studied at La Selva, Augspurger (1983c) has presented convincing evidence of the importance of pathogens in early mortality of a BCI tree species.

La Selva tree seedlings face a significant risk of death or severe damage from falling litter, branches, and trees, and the activities of vertebrates in the understory. Hartshorn (1972) found a one-year mortality rate of 52% for seedlings of *Pentaclethra macroloba,* and he estimated that 38% of the deaths were the result of litter fall. For *Dipteryx panamensis* seedlings seven to fifty-nine months old, D. B. Clark and Clark (1987) found that at least 16% of the mortality was from litter fall. Species-independent estimates of the risk of such physical damage to tree seedlings were obtained by monitoring physical damage to inert models of tree seedlings in two primary forest sites at La Selva (D. B. Clark and Clark 1989). More than 80% of the models were flattened, knocked over, ripped in pieces or dug up during one year. Of the total observed "mortality" at least 23% was attributable to litter fall, and at least 25% was attributable to digging and other activity

by vertebrates. Many tree seedlings at La Selva probably can withstand and repair such physical damage.

Studies in other Neotropical forests have identified similar patterns. Córdova Casillas (1985) attributed 18% of the seedling mortality of *Nectandra ambigens* at Los Tuxtlas to litter fall. On BCI Augspurger (1983c) found that litter fall and digging by mammals were significant mortality agents for seedlings of *Platypodium elegans* (Papilionoideae), and Aide (1987) estimated that at least 22%–47% of sapling mortality of the liana *Connarus turczaninowii* (Connaraceae) was caused by limb falls.

CONCLUSIONS AND FUTURE RESEARCH NEEDS

Plant demographic studies at La Selva have provided some detail on a few species and limited data on community dynamics. More questions have been raised than answered, however, and too few species have been studied in detail to generalize about the demographic properties of different plant guilds. No population studies have been made of any of La Selva's lianas, hemiepiphytes, or epiphytes, or of most herb and shrub species. For the trees, the best-studied plant group at La Selva, detailed demographic information is available for only a few of the 320+ species.

What can be concluded from the studies to date? First, in the shaded forest understory, plant growth and reproduction can be very slow. Long-term study will be needed to evaluate the population and individual level consequences of slow growth and low reproduction in the understory. Second, turnover within La Selva plant populations can be very high. Although on a forestwide basis the trees appear to balance high mortality with high recruitment (D. Lieberman et al. 1985b, 1990), how stable populations of individual plant species are within the station is not known. Third, the high rates of tree and litter fall at La Selva can result in high rates of physical damage to seedlings, saplings, and smaller trees; this damage is likely to greatly affect the demography of many plant species. Finally, at La Selva, as at other tropical rain forest sites, theory has outstripped the data base in the study of interactions between gaps and plant demography. Although the development of the "gap paradigm" has stimulated much recent field research, it has resulted in the proliferation of generalizations not backed up by field data. The effects of light availability, soil fertility, soil moisture, and plant/animal interactions on growth and survival through regeneration are not yet understood for any species at La Selva. Autecological studies are needed that link demographic performance of individuals in all size classes to environmental factors before many ideas about regeneration processes in tropical rain forest can be evaluated.

Future plant demographic research at La Selva should focus on a broad spectrum of life-forms and species. Concerted efforts should be made to evaluate levels of intraspecific demographic variation, both spatial and temporal. Nothing is yet known about the extent to which a species' population processes are affected by the marked differences in soil fertility and topography within La Selva. Equally interesting is the degree to which plant populations vary over time. Major seasonal or annual differences in rates of growth, mortality, or reproduction of dominant plant species could have wide-reaching impacts on forest processes; even in rare species, such temporal variability could cause major impacts on associated animals. This aspect of the demography of La Selva plants will only be detected with long-term studies.

Finally, given the daunting complexity of a forest such as La Selva, much progress will be made when plant demographers link efforts to those of workers in other fields by studying the same species. Study populations from on-going demographic projects on La Selva plants can serve as a major resource for research on topics such as the physiological correlates of plant life-history patterns or the relationship of sapling architecture to growth rates and survival in the understory. Although on-going plant population studies will greatly expand existing knowledge, the more than sixteen hundred spe-

cies of higher plants at La Selva provide a rich source of unanswered research questions for generations of tropical plant ecologists to come.

ACKNOWLEDGMENTS

In all phases of the development of this chapter David Clark provided key ideas and critiques. Important contributions were also made by several reviewers and the editors: Kamal Bawa, Steve Bullock, Julie Denslow, Jim Hamrick, Henry Hespenheide, Egbert Leigh, Gordon Orians, Richard Primack, and Francis Putz, and, especially, Lucinda McDade. I thank the Organization for Tropical Studies and the U.S. National Science Foundation (BSR 85-16371 and BSR89-18185) for support during the preparation of this chapter.

8

Patterns of Density and Dispersion of Forest Trees

Milton Lieberman and Diana Lieberman*

THE ISSUES

A consequence of the high species diversity that characterizes tropical forests is the low density of most tree species and the large expected distances between conspecific trees. The considerable distance between conspecifics was noted by Alfred Russel Wallace who in 1878 (65) wrote, "If the traveller notices a particular species and wishes to find more like it, he may often turn his eyes in vain in every direction. Trees of varied forms, dimensions and colours are around him, but he rarely sees any one of them repeated. Time after time he goes towards a tree which looks like the one he seeks, but a closer examination proves it to be distinct. He may at length, perhaps, meet with a second specimen half a mile off, or may fail altogether, till on another occasion he stumbles on one by accident." These impressions were shared and communicated by subsequent generations of biologists (Richards 1952; Owen 1971; Hubbell 1979).

The recognition that adult conspecific trees in most tropical forests are often far apart led to several studies to examine patterns of tree spatial dispersion. These studies (Pires et al. 1953; Ashton 1964, 1969; Poore 1968; Lang et al. 1971; Richards and Williamson 1975; Greig-Smith 1979; Hubbell 1979, 1980; D. Lieberman 1979; Forman and Hahn 1980; Rathbun 1980; Fleming and Heithaus 1981; Pemadasa and Gunatilleke 1981; Thorington et al. 1982; Newbery et al. 1986; Sterner et al. 1986) indicate that many tropical tree populations are randomly dispersed, others are clumped, and a very few are hyperdispersed.

Despite these findings, the premise that conspecific tropical trees were typically farther apart than would be expected by chance continued to be widely accepted. Several hypotheses were raised purporting to explain tropical tree spacing (Janzen 1970, 1971b, Connell 1971; and reviewed by Howe and Smallwood 1982), prominent among which were density-responsive or distance-responsive predation of seeds or seedlings, severe intraspecific competition for nutrients or light, and intraspecific allelopathy. Much of the ensuing research dealt with processes of density-dependent or distance-dependent mortality of seeds or seedlings in selected tree species.

The impetus for many of these studies was the notion that the postulated spacing mechanisms might offer an explanation for the coexistence of tree species in high-diversity assemblages, an extension of the ideas of Paine (1966) and others regarding predation and animal species diversity. Thus, the causal linkage between diversity and distance became convoluted: first, spacing between conspecifics was seen to be an expected consequence of high local diversity; then distance per se was considered to be necessary for the existence of diverse assemblages.

Rather than parse even finer the putative mechanisms leading to regular spacing between conspecific trees, we reexamine the original paradigm and ask whether actual measured distances between conspecifics are remarkable. That is, are the distances greater than would be predicted on the basis of population density alone? To test this, we carried out statistical analyses of spatial patterns of trees in mapped plots at La Selva.

Statistical techniques available for the analysis of spatial point patterns have advanced steadily in sophistication and power since the mid-1960s (cf. Ripley 1981; Diggle 1983). In contrast, biological insights into the meaning of pattern in tropical forests have been dominated by an older and, perhaps, stultifying *Weltanschauung*. Four characteristics that are shared by most (although by no means all) previous studies of tropical dispersion may account in part for the lack of recent discovery in this area.

1. Most research on dispersion in tropical trees has focused exclusively on species populations. Yet trees have attributes apart from species membership that may be relevant to questions of pattern and spacing. Among these are size, sexual system, growth and regeneration behavior, and density.

2. Dispersion studies have generally evaluated individual species in isolation; the context in which all of these trees exist—the forest—is often disregarded. Interpretation of pattern, however, becomes difficult if the surrounding forest matrix is ignored. On occasion, it is impossible to generate appropriate null models except by examining the forest as a whole (or a random sample thereof).

3. Measurements of spatial pattern have seldom been replicated in several sites; hence, one cannot ask whether patterns are idiosyncratic or are repeated with consistency from stand to stand. Lack of replication invites a posteriori explanation.

4. Rigorous tests of these ideas depend upon the use of extensive, communitywide data sets of the kind that have seldom been available in the past. Exceptionally complete data

* Original manuscript dated April 1987.

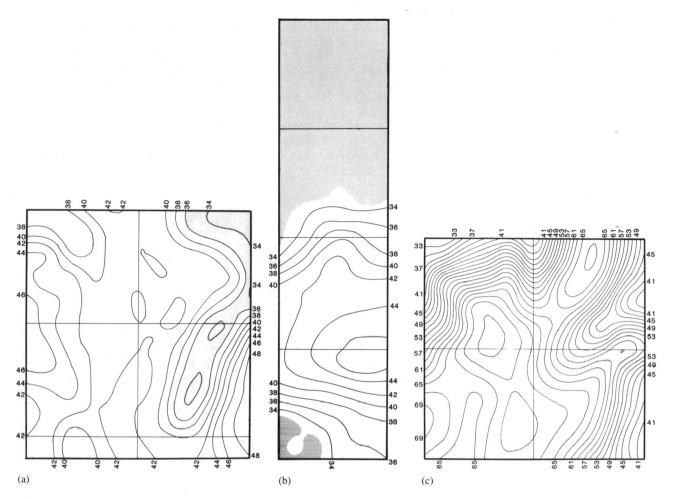

Fig. 8.1. Topographic maps of the three permanent plots. (*A*) Plot 1 (220 × 200 m); (*B*) Plot 2 (100 × 400 m); (*C*) Plot 3 (200 × 200 m). Contours are drawn at 2-m intervals of elevation. Swamp areas are shaded.

have now been collected at La Selva, permitting us to evaluate for the first time the most important hypotheses and speculations raised earlier.

In this chapter, we first report on the density and physical structure of the La Selva forest to provide a context for explanations of spacing among individuals. We then consider dispersion patterns of trees over several domains: the forest itself; species populations; size classes; and guilds (based on sexual system, growth behavior, and rarity). Finally, we compare results with those from other tropical forests and offer suggestions for future research.

SAMPLE PLOT DESCRIPTION

Three permanent inventory plots totalling 12.4 ha were established at La Selva in 1969–70 for a long-term study of tree growth rates and forest dynamics (fig. 8.1 A–C). Sites were chosen to represent contrasting substrates: plot 1 (4.4 ha) lies on an old alluvial terrace, plot 2 (4.0 ha) on old alluvium and colluvium, and plot 3 (4.0 ha) on residual soils overlying basalt parent material. The plots span an elevational range of 32–71 m above sea level and include steep hills, rolling terrain, plateaus, and swamp.

In 1982–83 a second complete plot inventory was con-

ducted; at this time a 10–m grid system was established in the plots to facilitate accurate mapping. All surviving tagged trees and lianas ≥10 cm in diameter at breast height (dbh) and all new recruits ≥10 cm dbh were measured to the nearest millimeter at breast height (and also above buttresses or stilt roots, if present), mapped to the nearest meter, and identified. Results presented in this chapter are generally based upon this second enumeration. Inventories have since been carried out in 1985 and 1989.

Subsamples randomly located within the permanent plots were used to estimate density of individuals in three younger life stages: saplings 2–10 cm dbh, seedlings ≤50 cm in height, and viable buried seeds in the soil seed bank. The tallies of seedlings and buried seeds included all tree and liana species capable of reaching 10 cm dbh at maturity. Saplings 2–10 cm dbh were enumerated in six 20 × 20 m subplots (two subplots in each plot). Seedlings were censused in forty-eight permanent transects 0.5 × 10 m in size; plots 1 and 3 have fifteen transects each, and plot 2 has eighteen. The abundance of germinable seeds in the soil seed bank was assessed from nineteen soil samples 0.5 × 0.5 m in area and 3 cm deep; six samples each were taken from plots 1 and 3, and seven were taken from plot 2. The soil collected from the plots was brought to a fully enclosed shade house, spread out in wooden flats, watered as needed, and occasionally remixed. Seedlings were identified and counted as they appeared until all germination ceased.

Previously published studies of the permanent plots have dealt with geology and soils (Bourgeois et al. 1972), population dynamics of the dominant species, *Pentaclethra macroloba* (Hartshorn 1972, 1975), floristic patterns (Hartshorn 1983b; M. Lieberman et al. 1985; D. Lieberman and Lieberman 1987; Peralta, Hartshorn, et al. 1987), tree mortality (D. Lieberman et al., 1985b, 1990; Peralta, Hartshorn, et al. 1987), stand turnover rates (Hartshorn 1978; D. Lieberman et al. 1985b, 1990; Peralta, Hartshorn, et al. 1987), tree growth and longevity (D. Lieberman et al. 1985a; D. Lieberman and Lieberman 1987; Peralta, Hartshorn, et al. 1987; M. Lieberman et al. 1988), seedling dynamics (D. Lieberman and Lieberman 1987; Peralta, Barquero, et al. 1987b; D. Lieberman et al. 1990); and forest canopy structure (M. Lieberman et al. 1989; M. Lieberman and Lieberman 1991).

FOREST STRUCTURE AND PHYSIOGNOMY

Stem Density

The total number of stems ≥ 10 cm dbh enumerated in 12.4 ha in the 1982–83 inventory was 5,530, a mean density of 446.0 individuals ha^{-1}. Four life-forms were included: dicot trees with 72.0% of stems (representing 241 species); arborescent palms with 25.5% (7 species); lianas with 2.4% (20 species); and tree ferns with 0.1% (1 species). Altogether, 269 species in 162 genera and 62 families were present. Relative abundances of species are shown in fig. 8.2. The most abundant species, *Pentaclethra macroloba*, contributes 14% of stems ≥ 10 cm dbh and 51% of individuals in the canopy.

Density in twelve 1-ha subplots ranged from 356 to 564 stems ha^{-1} (table 8.1). The lowest density was in the west-central hectare in the plot 2 swamp and is related to the presence of a meandering stream some 6 to 10 m wide; tree density throughout the plot 2 swamp diminishes progressively as drainage becomes poorer (M. Lieberman et al. 1985). The highest density was on the very steep slopes (30%–60%) in the northwest hectare of plot 3.

The mean number of saplings 2–10 cm dbh per 20 × 20 m subplot was 79.7, giving a projected sapling density of 1,992 ha^{-1}. Between June 1983 and June 1984, the mean seedling density within plots varied between 5 and 15 m^{-2} although for

short periods heavy germination increased the local density to more than 100 seedlings m^{-2} in some transects (D. Lieberman and Lieberman 1987; Peralta, Barquero, et al. 1987b). The mean density of buried tree and liana seeds in the samples was 232 m^{-2} although this varied markedly; mean density within plots was 102.7 m^{-2} in plot 1, 509.1 m^{-2} plot 2, and 36.7 m^{-2} in plot 3 (D. Lieberman and Lieberman 1987).

The size class distribution of trees and lianas in each of the three plots is shown in figure 8.3; individuals of all life stages are included from germinable seeds in the soil seed bank to the largest trees. The distribution indicates good stocking in all size classes except for the largest (>90 cm dbh). Density of these large trees ranges from 1.3 ha^{-1} in plot 3 to 3.4 ha^{-1} in plot 1. The dynamic nature of the La Selva forest has been implicated in the paucity of large trees (Hartshorn 1978). The log-log relationship between frequency and basal area among saplings and larger stems suggests that the abundance of small individuals may be limited by the preemption of space by larger individuals (M. Lieberman 1977).

Basal Area and Biomass

Plot basal area values were calculated directly using 1982 dbh measurements of all stems ≥ 10 cm dbh in the stand. Diameter above buttresses was used instead of dbh for individuals with buttresses or stilt roots rising above breast height. Basal area (mean ± S.E.) of trees in the plots is 24.685 ± 0.794 $m^2 ha^{-1}$. Values in individual 1-ha subplots ranged from 21.42 to 29.88 $m^2 ha^{-1}$ (table 8.1). Tabled values are somewhat lower than those reported previously for these plots; in earlier reports (D. Lieberman et al. 1985b, 1990; D. Lieberman and Lieberman 1987), we used the convention of averaging dbh and diameter above buttresses for stems with buttresses.

Biomass was estimated from basal area using the linear regression equation $y = 0.89\ x - 136$, where y is total aboveground dry biomass (kg) and x is basal area (cm^2). This relationship was developed from measurements of biomass and stem dimensions by the Departamento de Ingenieria Forestal, Instituto Tecnológico de Costa Rica under contract to NASA's Earth Resources Laboratory (A. T. Joyce unpublished data). The work was carried out in a tract of lowland forest a few kilometers southeast of La Selva. The one hundred trees harvested, dried, and weighed ranged in size from 3.7 to 116.0 cm dbh and represented the following abundant species: *Carapa guianensis, Pentaclethra macroloba, Protium panamensis, Protium pittieri, Tetragastris panamensis, Virola*

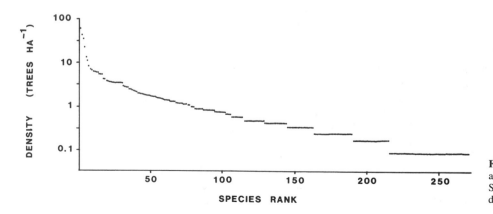

Fig. 8.2. Species abundance curve for all stems ≥ 10 cm dbh on 12.4 ha at La Selva; species ranked according to density.

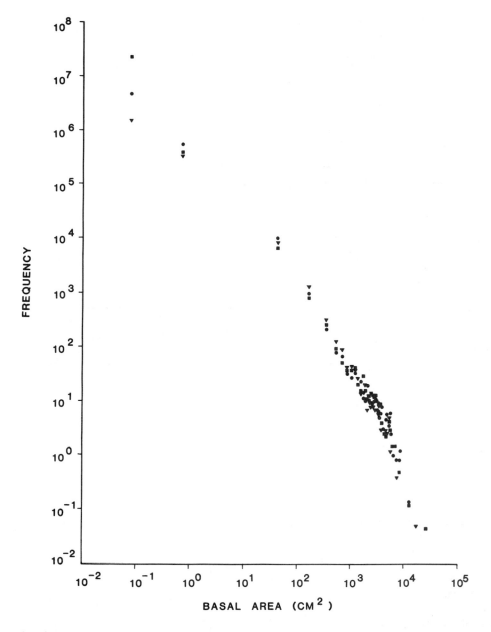

Fig. 8.3. Size class distribution of trees and lianas in each of the three permanent plots based upon 1982 inventory data. The three smallest size classes shown (in ascending order of size, from the left) are buried viable seeds, seedlings, and saplings. Diameter of stems was arbitrarily estimated as 0.1 cm for buried seeds and 0.3 cm for seedlings. Plot 1, ○ ; Plot 2, ■ ; Plot 3, ○ .

koschnyi, Virola sebifera, and *Vochysia ferruginea.* Most of these trees have wood of intermediate density.

The mean dry biomass of live trees in the La Selva permanent plots is calculated to be 221 T ha^{-1} (or 22.1 kg m^{-2}). Biomass in individual 1-ha subplots ranged from 192 metric tons to 268 metric tons (table 8.1). Basal area and biomass computations were based upon dbh unless the tree bole had buttresses or stilt-roots rising above breast height in which cases diameter above buttresses was used. A total of 32.9% of individuals ≥10 cm dbh at La Selva have buttresses although the proportion varies widely within and between plots (table 8.1). Mean water content (by weight) of trees, obtained by oven-drying of both wood and foliage samples, was approximately 50%.

Foliage Height and Canopy Structure
Measurements of total height and mean crown diameter were made for all individuals ≥10 cm dbh in a 20 × 360 m transect

in the vicinity of plot 3. Altogether 361 stems were encountered, of which 248 were dicot trees, 96 palms, 15 lianas, and 2 tree ferns. Because of difficulties in assessing the dimensions of liana crowns, lianas were omitted from the following analyses. Height of trees ranged from 4 to 40 m, with a mean and S.E. of 15.5 ± 0.35 m (fig. 8.4). Crown diameters ranged from 2 to 25 m, with a mean of 5.95 ± 0.17 m (fig. 8.5).

To examine the vertical distribution of foliage in the transect, the foregoing data were combined as follows. The cross-sectional area, or horizontal projection, of each crown was calculated from its diameter; these values were then summed within tree height classes (fig. 8.6). The median height of crowns (weighted by crown area) was 18 m with relative peaks at 17–18 m and 23–24 m. Fully 54% of the total cross-sectional area of crowns was contributed by trees between 16 and 26 m in height. The total crown area projected by all trees in the transect was 12,976 m², indicating an average of 1.8 crowns over any given point on the ground.

Table 8.1 Stand characteristics of the twelve 1-ha subplots within the three permanent plots

Plot	Hectare	Elevation (m)	Density, Stems (ha⁻¹)	Basal Area (m²)	Biomass (T)[a]	Buttressed Trees (%)
1	NW	36–47	422	29.15	260.73	16.6
	NE	34–42	408	23.70	212.00	43.1
	SW	40–47	438	24.13	215.90	12.8
	SE	34–48	414	23.81	212.97	35.5
2	W	33	405	29.98	268.12	27.2
	WC	32–39	356	21.64	193.60	18.8
	EC	33–45	447	27.69	247.66	31.3
	E	33–45	372	23.79	212.86	35.7
3	NW	33–71	564	23.97	214.47	35.3
	NE	39–67	554	23.65	211.58	38.6
	SW	57–69	476	21.42	191.67	46.4
	SE	41–69	519	23.28	208.30	44.9

Note: Subplots are mapped in fig. 8.1A–C. Plot 1 lies on an old alluvial terrace, plot 2 on alluvium and colluvium, and plot 3 on residual soils overlying basalt parent material. The western half of plot 2 is generally swampy.

[a]Calculated from basal area (see text for details).

The upper canopy at La Selva is quite uneven. The tallest trees (≥30–40 m) are relatively scarce; their crowns cover only 5.6% of the ground area and contribute only 3.1% of the total cross-sectional area of crowns. The roughness of the upper canopy is shown in a laser profiler track produced over La Selva by NASA's Airborne Oceanographic Lidar (fig. 8.7).

Changes with Time

Mortality of trees and lianas ≥10 cm dbh was measured from 1969 to 1982 (D. Lieberman et al. 1985a). During the thirteen-year period, 23.2% of the individuals died; the annual mortality rate, λ, calculated as the slope of $\log_e$ survivorship versus time, was 2.03% with all three plots combined. The expected stand half-life is 34.2 years. Mortality was highest in plot 3 (2.24% per year) and lowest in plot 1 (1.80% per year).

The observed turnover rate based upon stem mortality is considerably faster than turnover rates calculated previously on the basis of gap area formation (Hartshorn 1978). We estimate that at least half the tree mortality in the permanent plots occurs without creating new canopy gaps or contributing to existing ones (D. Lieberman et al. 1985b, 1990). Mortality was approximately equal to recruitment into the 10 cm dbh class. Both density and basal area changes over thirteen years were quite small in relation to total mortality and to differences among plots (D. Lieberman et al. 1985b, 1988; D. Lieberman and Lieberman 1987; Peralta, Hartshorn, et al. 1987c).

SPATIAL DISPERSION: ANALYSIS OVER SEVERAL DOMAINS

Methods

We analyzed dispersion patterns using the nearest-neighbor method of Clark and Evans (1954). The expected mean distance from a tree to its nearest neighbor, if the trees are distributed at random, is given by:

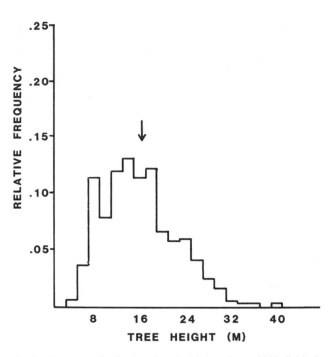

Fig. 8.4. Frequency distribution of tree height based upon 346 individuals ≥10 cm dbh in a 20 × 360 m transect. Arrow indicates mean.

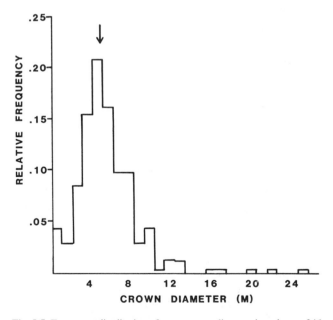

Fig. 8.5. Frequency distribution of mean crown diameter based upon 346 individuals ≥10 cm dbh in a 20 × 360 m transect. Arrow indicates mean.

$$\hat{r} = \frac{1}{2\sqrt{\rho}}$$

where ρ is the density of the trees in question. The index of dispersion R is the ratio of observed $(\bar{r})$ to expected $(\hat{r})$ mean distances. The magnitude of deviation from the null expectation of $R = 1.0$ can be statistically evaluated; values of R significantly smaller than 1.0 indicate clumping, whereas those significantly greater than 1.0 indicate regularity or hyperdispersion. To develop a table of critical values for statistical evaluation of *R*, we used computer simulation to generate expected distributions of *r̂*, drawing random coordinates over a wide range of sample sizes for each particular plot size and shape in this study.

The Forest

Dispersion of all trees ≥10 cm dbh was measured in each of four areas: plot 1 (4.4 ha), plot 2a (2-ha swamp), plot 2b (2-ha upland), and plot 3 (4 ha). Trees were significantly hyperdispersed in plots 1, 2b upland, and 3, and were randomly dispersed in plot 2a (table 8.2a). Plots were then subdivided into 1-ha subplots, and dispersion indices recalculated at the smaller scale (table 8.2b). Significant hyperdispersion was found in eight of the twelve hectares; random dispersion was found in the two eastern hectares in plot 1, the west-central (swamp) hectare in plot 2, and the northwest hectare in plot 3. The value of *R* was greater than 1.0 in ten of the twelve hectares.

Interpretation. Significant hyperdispersion indicates, by definition, that trees are spaced farther from their nearest neighbors than expected on the basis of their overall density. The observed intertree distances, however, are still quite small (see table 8.2), averaging from 2.14 to 2.85 m. These distances are far lower than the 5.95 m mean diameter of tree crowns (see fig. 8.5); this is physically possible only because tree crowns are not all at the same height but occupy a three-dimensional volume.

To interpret these results, we considered which subplots were randomly dispersed and which were hyperdispersed. Random dispersion in plots 1 and 2 was found in those subplots with the lowest density and basal area; in plot 3, however, random dispersion was found in the subplot with the highest density and basal area but with the steepest slopes. Hyperdispersion in the plots is, thus, associated with dense stands, except on the steepest terrain. We conclude that hyperdispersion of trees is a result of the close packing of tree crowns, which precludes the occurrence of very near neighbors ≥10 cm dbh.

Alternative hypotheses can be ruled out in this instance. Host-specific pathogens and herbivores, severe intraspecific competition for nutrients, and intraspecific allelopathy cannot be implicated because the pattern applies to all trees independent of species membership.

Populations

We measured dispersion within species populations in each of four areas: plot 1, plot 2a (swamp), plot 2b (upland), and plot 3. In each area all species with ten or more individuals per plot were included. Results of the nearest-neighbor analysis are shown in figure 8.8 (A–C). Of 104 populations belonging to 65 species, 83% (representing 59 species) were randomly dispersed, 5% (2 species) were significantly hyperdispersed, and 13% (13 species) were significantly clumped. Seven of the 13 species that showed significant clumping occurred in more than one plot. Only one of these was consistently clumped: *Cecropia obtusifolia,* a short-lived pioneer, was clumped in plots 1 and 2a. The two species showing signifi-

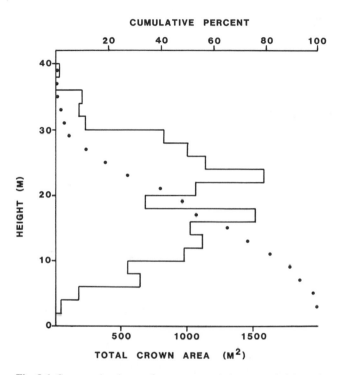

Fig. 8.6. Cross-sectional area of tree crowns relative to tree height, estimated from measurements of mean crown diameter and height of each tree ≥10 cm dbh on a 20 × 360 m transect. Histogram indicates sum of crown areas within height classes; curve shows cumulative percentage of total crown area as a function of height.

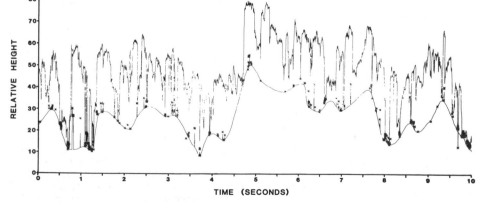

Fig. 8.7. Laser profiler track from La Selva with the flight path crossing generally west to east near plot 3. Dual trace shows canopy profile and ground topography; the ground contour is interpolated between occasional ground hits (*). One second (horizontal scale) is approximately equivalent to 100 m on the ground; the vertical scale is greatly exaggerated. The sensor used was NASA's Airborne Oceanographic Lidar, a pulsed Nd-YAG laser system, mounted in a P-3A turboprop aircraft and flown at 1,600 ft. (488 m) above terrain.

Table 8.2 Dispersion pattern of all trees ≥10 cm dbh in the permanent plots

Plot[a]	R	Hectare[b]	$\bar{r}$ (m)	R
1	1.054**	NW	2.61	1.073**
		NE	2.57	1.038
		SW	2.63	1.101**
		SE	2.45	0.995
2a Swamp	1.000	W	2.67	1.074**
		WC	2.54	0.956
2b Upland	1.080**	EC	2.57	1.087**
		E	2.85	1.097**
3	1.052**	NW	2.14	1.016
		NE	2.29	1.079**
		SW	2.41	1.052*
		SE	2.41	1.095**

Note: Nearest-neighbor distances calculated on the scale of plots or half-plots and 1-ha subplots. Density per subplot is reported in table 8.1. R = observed mean nearest-neighbor distance/expected mean nearest-neighbor distance; $\bar{r}$ = observed mean nearest-neighbor distance (m); * = $P < .05$; ** = $P < .01$.

[a]Plots or half-plots.
[b]One-hectare subplots.

cant hyperdispersion were *Pentaclethra macroloba* and *Welfia georgii. Pentaclethra* was hyperdispersed in plots 1, 2b, and 3 but randomly dispersed in plot 2a, whereas *Welfia* was hyperdispersed in plots 1 and 3 but randomly dispersed in plot 2b and clumped in plot 2a.

In order to consider pattern on a finer scale, we measured dispersion in *Pentaclethra* in each of the twelve 1-ha subplots. Significant hyperdispersion of *Pentaclethra* was found in the easternmost (upland) hectare in plot 2 and in three hectares in plot 3 (NW, SW, SE); dispersion was random in the other eight hectares (although R was greater than 1.0 in all twelve) (table 8.3). As in the all-tree analyses, *Pentaclethra* was hyperdispersed in those subplots having the highest population density.

Interpretation. Inspection of species distribution maps suggests that most of the clumped patterns arise in response to topographic heterogeneity within an area. Examples include *Rinorea pubipes,* which occurs in plot 3 on plateaus and ridges; *Warscewiczia coccinea,* which is locally abundant in the southwest portion of plot 1; *Cecropia obtusifolia,* which is found in plot 1 mainly in the swamp; and *Adelia triloba, Lonchocarpus oliganthus,* and *Pterocarpus officinalis,* among others, which are distributed within the plot 2a swamp in relation to a drainage mosaic. The presence of the stream and areas of standing water in the swamp makes part of the area unsuitable for tree growth, contributing to the number of clumped patterns in that area. These species distribution patterns have been analyzed in detail elsewhere (M. Lieberman et al. 1985; Peralta, Hartshorn, et al. 1987). A few clumped populations occur in patches or aggregates apparently independent of major topographic features. Such a distribution is seen in *Casearia arborea* and *Goethalsia meiantha* in plot 1 and may be related to establishment in certain kinds of

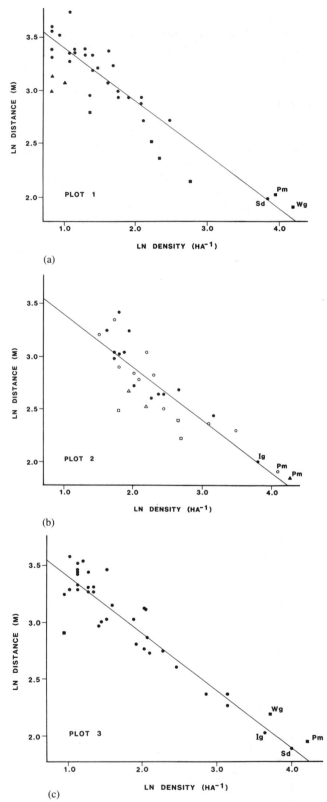

(a)

(b)

(c)

Fig. 8.8. Dispersion pattern of tree species in plot 1 (*A*), plot 2 (*B*), and plot 3 (*C*), as measured by mean nearest-neighbor distance (Clark and Evans 1954). Diagonal line indicates expected nearest-neighbor distance for randomly dispersed populations; values above the line are hyperdispersed, those below are clumped. Symbols: random dispersion, circles; nonrandom dispersion, triangles (*P* <.05) or squares (*P* <.01). Plot 2 was subdivided into swamp (2 ha), open symbols, and upland (2 ha), closed symbols. The most abundant species are *Pentaclethra macroloba, Pm; Welfia georgii, Wg; Socratea durissima, Sd;* and *Iriartea gigantea, Ig.*

Table 8.3 Dispersion pattern of *Pentaclethra macroloba* trees ≥10 cm dbh in the permanent plots

Plot[a]	R	Hectare[b]	$\bar{r}$ (m)	N	R
1	1.114**	NW	8.36	45	1.109
		NE	7.57	62	1.182
		SW	9.05	37	1.086
		SE	7.50	52	1.071
2a Swamp	1.043	W	6.12	71	1.024
		WC	7.99	48	1.095
2b Upland	1.104*	EC	6.80	63	1.070
		E	6.42	83	1.162*
3	1.145**	NW	7.08	72	1.194*
		NE	7.84	57	1.173
		SW	7.71	70	1.282**
		SE	7.40	65	1.183*

Note: Nearest-neighbor distances calculated on the scale of plots or half-plots and 1-ha subplots. R = observed mean nearest-neighbor distance/expected mean nearest-neighbor distance; $\bar{r}$ = observed mean nearest-neighbor distance (m); * = $P < .05$; ** = $P < .01$.

[a]Plots or half-plots.
[b]One-hectare subplots.

microsites; these species, however, were not clumped in other areas.

The hyperdispersion of *Pentaclethra macroloba* and *Welfia georgii* can only be interpreted in light of their density and crown characteristics. *Pentaclethra*, a broad-crowned mimosoid legume, and *Welfia*, a large-leaved geonomoid palm, are the two most abundant species in the permanent plots, contributing 13.7% and 9.6%, respectively, of the stems ≥10 cm dbh and 35.8% and 3.3%, respectively, of the basal area. *Pentaclethra* forms the base of the canopy at a height of 25–35 m (Hartshorn 1983a), whereas the subcanopy palm *Welfia* seldom exceeds 20 m (M. Lieberman et al. 1988).

The mean distance between *Pentaclethra* individuals and the nearest conspecific (plots 1, 2b, and 3) was 7.1 m, and the expected nearest neighbor distance was 6.3 m. These distances are less than the mean crown diameter of the species, 9.6 m (range 4–22 m, n = 48). We conclude that hyperdispersion of *Pentaclethra* stems results from dense packing of crowns as was seen in the forest as a whole.

In *Welfia* palms the mean distance to the nearest conspecific (plots 1 and 3) was 7.9 m; the expected distance was 7.0 m. These distances are comparable to the crown diameter of this species (mean 6.6 m, range 5–8 m, n = 29). Crown development in *Welfia* is generally completed early in life before substantial height growth is achieved, and the volume occupied by juvenile *Welfia* crowns is, therefore, sizable (Rich et al. 1986). This may limit the recruitment of nearby conspecifics. Further, falling *Welfia* fronds are known to contribute heavily to the mortality of seedlings and saplings of conspecifics (Vandermeer 1977) and other tree species at La Selva (Hartshorn 1972, 1975, 1983b). The continual shedding of these massive fronds hinders the establishment of neighbors beneath the crown of *Welfia*.

The two next most abundant species in the plots are the iriartoid palms *Socratea durissima* and *Iriartea gigantea*. Un-

like *Welfia*, these palms are stilt-rooted and quite tall (25–35 m) and have small crowns with fronds only 1.5–2 m long (Holdridge and Poveda 1976). Crowns of mature *Welfia* palms weigh as much as 250 kg, or ten times the weight of *Socratea* crowns, and six times that of *Iriartea* crowns in the same height class (Rich 1986). Both *Socratea* and *Iriartea* were randomly dispersed in all areas in which they occurred (fig. 8.8).

Spacing between Trees. The foregoing analyses pertain to distances between conspecifics and demonstrate intrapopulation patterns independent of the forest matrix. Another aspect of tree spacing involves the distance from a given tree to its nearest neighbor of any species. For each tree species with ten or more individuals (all plots combined), we measured the distance from each stem to its nearest neighbor, independent of species. Mean distance was then determined for each species (fig. 8.9). Values ranged from 1.3–3.2 m. For purposes of comparison the mean intertree distances, using all species, were 2.57 m (plot 1), 2.61 m (plot 2a), 2.70 m (plot 2b), and 2.31 m (plot 3).

The distribution of distance to nearest neighbor of any species did not differ significantly between *Welfia* palms (mean 2.63 m) and trees in general (Kolmogorov-Smirnov test, $P > .05$). In *Socratea* (mean 2.15 m) and *Iriartea* (mean 2.2 m) the distances were significantly smaller than those for trees in general (Kolmogorov-Smirnov test, $P < .05$), indicating that neighboring trees are closer to these two palms.

Size Classes

Although spatial dispersion is measured in trees as a static property, its origin is dynamic, arising from the dispersal of seeds followed by postdispersal attrition. To assess changes in pattern with increasing tree size, we examined the dispersion of individuals belonging to two size classes. Analyses were done in the twelve 1-ha subplots.

Empirical allometric data (Lieberman and Lieberman unpublished data) indicate that trees greater than 40 cm dbh may be presumed to have their crowns in the canopy, whereas trees less than 40 cm dbh are usually in the subcanopy or un-

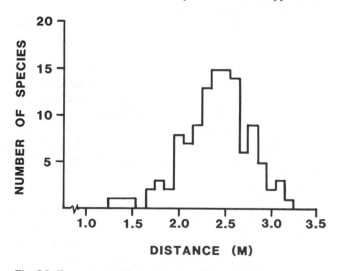

Fig. 8.9. Frequency distribution (number of species) of mean distance from a given individual (species x) to its nearest neighbor (any species). All species with ten or more members are included (N = 117).

Table 8.4 Dispersion patterns in canopy trees (all species) >40 cm dbh and a matched number of understory trees ≥10 cm dbh (all species)

Plot	Hectare	Trees >40 cm dbh		Trees 10 cm–X cm dbh		
		N	R	N	X(cm)	R
1	NW	60	1.156	60	11.9	1.073
	NE	69	1.089	71	11.6	1.043
	SW	51	1.188	52	11.2	1.188
	SE	59	1.237*	58	11.3	1.114
2a	W	80	1.072	82	11.7	1.122
Swamp	WC	48	0.945	47	11.2	1.094
2b	EC	81	1.108	80	12.1	1.181*
Upland	E	70	1.275**	71	12.3	0.921
3	NW	53	1.252*	52	10.5	1.096
	NE	48	1.146	47	10.6	1.030
	SW	53	1.168	55	10.7	0.998
	SE	56	1.181	55	10.6	1.065

Note: The upper size limit in the smaller trees is X. R = observed mean nearest-neighbor distance/expected mean nearest-neighbor distance; $\bar{r}$ = observed mean nearest-neighbor distance (m); * = P < .05; ** = P < .01. Large trees are significantly more regular in dispersion than small trees (P < .05).

derstory. We used this criterion to compare dispersion in canopy trees and a matched sample of the smallest understory trees.

For each 1-ha subplot all trees were ranked by dbh. All individuals greater than 40 cm were placed in the "canopy" group; starting with the smallest tree, an equal number of small trees (≥10 cm dbh) was then counted and assigned to the "understory" group. Spatial dispersion was measured within each group (table 8.4).

In general, the canopy and understory size classes tended toward hyperdispersion at the 1-ha scale; R was greater than 1.0 in eleven of the twelve hectares in the canopy tree samples, and in ten of the twelve hectares in the understory tree samples. Canopy trees are more regularly spaced than understory trees in the same 1-ha subplot (t-test for paired comparisons, one-tailed, P <.05).

A similar analysis was performed for *Pentaclethra macroloba* trees (table 8.5). All trees > 40 cm dbh were placed in the "canopy" group; the "understory" group included all trees ≥ 40 cm dbh. In general, *Pentaclethra* trees in the canopy tended toward hyperdispersion at the 1-ha scale, whereas the understory class did not. In canopy trees, R was greater than 1.0 in eleven of the twelve subplots. Canopy trees in *Pentaclethra* were more regularly spaced than understory trees in the same 1-ha subplot (t-test for paired comparisons, one-tailed, P <.01).

Interpretation. Two alternative processes could lead to the increased regularity of spacing, or repulsion, from smaller to larger size classes. The conventional explanation would invoke "Janzen-Connell spacing"—density-responsive or distance-responsive mortality of progeny mediated by host-specific seed predators (Janzen 1970), herbivores (Connell 1971), or pathogens (Augspurger 1983a, 1983c, 1984b; Aug-

spurger and Kelly 1984). Indeed, results such as these are often taken as prima facie evidence of Janzen-Connell spacing (D. A. Clark and Clark 1984).

We found comparable changes in dispersion as a function of size both in populations of *Pentaclethra macroloba* and in the all-tree analyses. Because the mechanism postulated for the Janzen-Connell process is host-specific, this process cannot be responsible for the results of the all-tree (null) analyses. Because we cannot distinguish between the null pattern and the pattern in *Pentaclethra,* we cannot conclude that Janzen-Connell spacing is implicated in either.

A more parsimonious explanation is progressive physical crowding of growing trees, more specifically crown-crown interactions among neighboring individuals. This explanation is particularly reasonable given the very short distances between neighbors (conspecific or otherwise) and is consistent with results reported in the previous section. Hyperdispersed patterns only occurred in the most densely packed populations. Crown contact is greatest among large trees that have large crowns positioned within the same general height range in the canopy. Understory trees have small crowns distributed over a much wider vertical range, and their crowns often pass through that range relatively quickly. The dynamics of crown spacing involves not only differential mortality or thinning but also differential growth based upon the size and distance of neighbors: smaller trees may remains suppressed, temporarily excluded from the next size class, until crown space becomes available by the death or senescence of a larger neighbor. Almost all tree species in the area are, in fact, capable of surviving lengthy periods in which diameter growth is nil (D. Lieberman et al. 1985a; Peralta, Hartshorn, et al. 1987). An increase in hyperdispersion with size is, thus, expected.

Guilds
Density and spatial dispersion were evaluated for tree species grouped by sexual system, growth behavior, and rarity.

Table 8.5 Dispersion patterns in *Pentaclethra macroloba* trees > 40 cm dbh and ≤ 40 cm dbh

Plot	Hectare	Trees >40 cm dbh		Trees <40 cm dbh	
		N	R	N	R
1	NW	29	1.267	16	1.028
	NE	39	1.096	25	0.958
	SW	21	1.026	16	1.013
	SE	34	1.149	18	0.753*
2a	W	40	1.112	31	1.019
Swamp	WC	22	1.070	26	1.143
2b	EC	26	0.961	37	1.124
Upland	E	38	1.199	45	1.075
3	NW	31	1.362**	41	1.013
	NE	27	1.308*	30	0.991
	SW	31	1.359**	39	1.096
	SE	34	1.270*	31	1.160

Note: R = observed mean nearest-neighbor distance/expected mean nearest-neighbor distance; $\bar{r}$ = observed mean nearest-neighbor distance (m); * = P < .05; ** = P < .01. Large trees are significantly more regular in dispersion than small trees (P < .01).

Sexual System. Bawa, Perry, et al. (1985) and Bawa, Bullock, et al. (1985) categorized 333 tree species at La Selva as hermaphroditic (65.5%), monoecious (11.4%), or dioecious (23.1%) on the basis of floral sexuality (see also chap. 12). Hermaphrodites have bisexual flowers, monoecious species produce male and female flowers on the same plant, and dioecious species bear male and female flowers on separate plants. Using these data we were able to categorize the sexual system of 77% of the species and 94% of the individuals in the permanent plots. Hermaphroditism is the commonest sexual system among both tree species and individuals. Within the plots 65.2% of species and 47.0% of individuals are hermaphrodites; 10.1% of species and 31.1% of individuals are monoecious; and 24.7% of species and 21.9% of individuals are dioecious. The disproportionate number of individuals in the monoecious category is due in part to the abundance of palms at La Selva; this family contributes only 2.6% of the species but more than 25% of the individuals ≥10 cm dbh (M. Lieberman et al. 1985).

Self-incompatibility was recorded by Bawa, Perry, et al. (1985) in 86% of hermaphroditic tree species tested, and the monoecious palms generally show asynchronous male and female phases; thus, an overwhelming majority of tree species at La Selva appear to be obligate outcrossers (Bawa and Opler 1975; Bawa 1979; Bawa, Perry, et al. 1985; see also chap. 12). It has been suggested that because the distance between mates can constrain reproductive success in outcrossing species, such species should tend not to occur at very low density (Ashton 1984; Bawa, Perry, et al. 1985; Hubbell and Foster 1986a). The density of potential mates may, in fact, be substantially lower than the density of conspecific individuals; the theoretical probability that the nearest (or any other) conspecific tree is a potential mate ranges from 1.0 in hermaphroditic species to 0.5 or less (depending upon the sex ratio) in dioecious species.

To test the prediction that population densities vary with sexual system, we compared the distribution of density (number of trees ≥ 10 cm dbh per species per plot) in hermaphroditic, monoecious, and dioecious species (fig. 8.10). The three groups differ significantly in density (Kolmogorov-Smirnov test, $P < .05$); the median density of hermaphrodites is 0.78 ha⁻¹, that of monoecious species is 1.0 ha⁻¹, and that of dioecious species is 1.6 ha⁻¹.

There is no significant association between sexual system and spatial dispersion pattern ($G = 7.34$, 4 d.f.; $.5 > P > .1$). Of fifty-one hermaphroditic populations (belonging to thirty-one species), 77% were randomly dispersed, 18% were hyperdispersed, and 6% were clumped; in seventeen monoecious populations (seven species), 765 were randomly dispersed, 12% were hyperdispersed, and 12% were clumped; and in thirty-two dioecious populations (nineteen species), 93.8% were randomly dispersed, none was hyperdispersed, and 6% were clumped.

Growth Behavior. Long-term growth rates of the forty-five most abundant tree species in the permanent plots were analyzed using growth simulation (M. Lieberman and Lieberman 1985), a statistical bootstrapping procedure. From these results (D. Lieberman et al. 1985a, D. Lieberman and Lieberman 1987), four main patterns of growth behavior were recognized: (1) slow maximum growth rates, rather constant growth and short projected maximum life spans (seen in understory species); (2) slow maximum growth rates, higher variability in growth, and long life spans (subcanopy species); (3) rapid maximum growth rates, highly variable growth, and long life spans (canopy and subcanopy species); and (4) very rapid maximum growth rates, little variability in growth, and very short life spans (canopy and subcanopy species). We hypothesized (D. Lieberman et al. 1985a) that these growth patterns might be related to general patterns of shade tolerance,

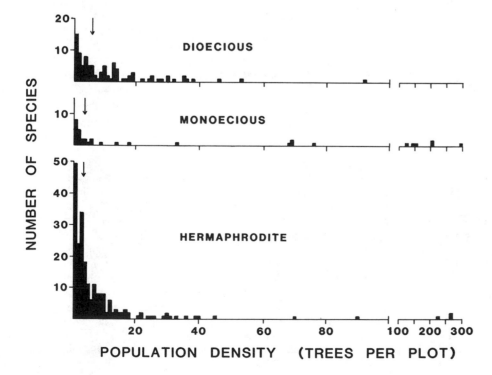

Fig. 8.10. Distribution of population density (trees per plot) for tree species categorized as hermaphroditic, monoecious, and dioecious. Arrows indicate medians. Density scale is condensed above one hundred trees per plot.

with fast growing, short-lived species being least tolerant of shade.

It has been frequently suggested that establishment of more shade-intolerant species should be confined to tree-fall gaps, transient canopy openings distributed as a mosaic throughout the forest (Grubb 1977b; Denslow 1980a; Hartshorn 1980; Brokaw 1982; see chap. 9). We examined the dispersion of trees in the fourth group mentioned to test whether the fastest-growing trees at La Selva show significant clumping in their adult distribution, that is, whether they tend to occur together, nearer to one another than would be expected by chance. The following species, pooled for analysis, were included in this group (in order of increasing maximum growth rate): *Casearia arborea, Ocotea hartshorniana, Stryphnodendron excelsum, Otoba novogranatensis, Inga coruscans, Inga* cf. *pezizifera, Goethalsia meiantha,* and *Hernandia didymanthera.* These trees (8 species, 312 individuals), were found to be randomly dispersed in plots 1 and 2 but showed significant clumping ($P <.05$) in plot 3.

We then considered the distance from every tree to its nearest neighbor of any species as an index of spacing around trees in this group; only young individuals (10–20 cm dbh) were used. Distances were compared in two groups: (1) all trees and their nearest neighbors of any species (the null distribution) and (2) fast-growing, short-lived trees and their nearest neighbors of any species. The two groups do not differ significantly (Kolmogorov-Smirnov test, $P >.05$); young, fast-growing trees are no more widely separated from their neighbors than are trees in general.

Rare Species. We analyzed the spatial dispersion of rare species within the permanent plots to test the hypothesis (Hubbell 1979; pers. comm.) that rare species might occur in an aggregated pattern. Species represented in a given plot by a single individual ≥10 cm dbh were pooled for analysis; plot 1 had 52 such species, plot 2 had 43, and plot 3 had 44. In all plots the dispersion of these rare trees was random ($P >.05$). We repeated the analysis using species represented by either one or two individuals, again pooling the species for analysis. This approach added 28 species to plot 1 (n = 108 individuals), 23 species to plot 2 (n = 89), and 22 species to plot 3 (n = 88). Again, the dispersion of rare trees as a group was random in all plots.

COMPARISONS WITH OTHER TROPICAL RAIN FORESTS

On global and regional scales the dispersion pattern within any tree species will be significantly clumped, the result of nonrandom distribution over the range of habitats represented. It is at smaller scales—10^4 to 10^5 m^2—that comparisons of dispersion patterns between populations are likely to be most instructive (Whitmore 1984; Newbery et al. 1986). Studies of dispersion in individual tree species are quite numerous, and indicate that many kinds of patterns occur. It is difficult to draw general inferences from these studies as scales of measurement, modes of analysis, hypotheses tested, rigor, and criteria for choice of species and inclusion of individuals vary. Because the studies lack replication, one cannot distinguish between species effects and site effects. Many of these papers

have been summarized by D. A. Clark and Clark (1984) and will not be considered here.

Several tropical forest studies have assessed dispersion in a large proportion of species within a given assemblage, and these constitute the most suitable basis for comparisons between sites. Table 8.6 presents the results of these studies. Adult tropical trees appear at the scales measured to be most often clumped or random and very seldom hyperdispersed. Clumping is attributed to inefficient seed dispersal (Poore 1968; Ashton 1969), vegetative reproduction by root suckers (D. Lieberman 1979), microhabitat preferences (Forman and Hahn 1980), or regeneration in treefall gaps (Newbery et al. 1986). Where size classes are compared, adult trees generally tend to show less clumping than juveniles (Lang et al. 1971; Hubbell 1979) although several exceptions are reported. An increase in regularity of spacing with increased density or tree size class has also been described in temperate zone stands (Laessle 1965; Christensen 1977; Yeaton 1978).

In an early review of spatial pattern in tropical trees, Greig-Smith (1969) reported that only one clear case of hyperdispersion was known to him (P. Greig-Smith and J. A. R. Anderson unpublished data), this being the largest size classes of *Shorea albida* in lowland peat forest in Sarawak, an atypical forest in which this species constitutes a very large proportion of the canopy. He proposes intraspecific competition as a possible source of the pattern but notes that this interpretation is unsubstantiated.

Spatial patterns in the La Selva permanent plots conform to general findings elsewhere in the tropics except for the hyperdispersion of La Selva's dominant species. The situation here may be similar to that described by Greig-Smith in Sarawak; in both cases, very dense packing of crowns of similar form and height appear to be involved.

CONCLUSIONS

Spatial dispersion statistics oversimplify forests in two particularly important ways. The two-dimensional pattern of tree locations is an abstraction of a geometrically complex three-dimensional structure. Stems have height as well as thickness; foliage is dispersed within crowns, which are themselves dispersed within a volume bounded by the soil surface and the top of the canopy; and the ground is often nonplanar. The static representation of points is an abstraction of a very dynamic system. Trees are dispersed as seeds, germinate, and die, causing fluctuations in the geometry (and hence spatial dispersion) of the forest. By neglecting these forest attributes, one may easily overlook or misinterpret a significant source of spatial pattern. Dispersion studies reported in this chapter indicate that the three-dimensional geometry of the forest and the dynamics thereof exert a major influence on the two-dimensional spatial pattern of trees at La Selva.

Trees ≥10 cm dbh, pooled for analysis independent of species tend to be hyperdispersed. The pattern is most pronounced in high-density stands. It is not seen on very steep terrain (where crowns may be staggered) nor in sites that are partly devoid of trees because of habitat unsuitability (e.g., swamps). There were no instances of significant clumping in the all-tree analyses.

The most abundant species in the plots (*Pentaclethra mac-*

Table 8.6 Summary of results of tropical forest dispersion studies

Site and Forest Type	Plot Area (ha)	Tree Density (ha^{-1})	Species Richness (Area Indicated)	Trees Analyzed	Results	Source
Brazil						
Terra firma forest, Pará	3.5	432.4 $\geq$ 10 cm dbh	179	6 most abundant spp.	6 random 0 clumped 0 hyperdispersed	(1)
Peninsular Malaysia						
Lowland dipterocarp forest, Jengka	20.6	134.4 $\geq$ 29 cm dbh	375	13 abundant spp.	6 random 7 clumped 0 hyperdispersed	(2)
Lowland dipterocarp forest, Jengka	20.6	134.4 $\geq$ 29 cm dbh	375	all trees	random	(2)
West Sarawak						
Mixed dipterocarp forest	2.9	376.5 $\geq$ 19 cm dbh	—	15 spp.	7 random 8 clumped 0 hyperdispersed	(3)
Panama						
Lowland moist forest, Barro Colorado I.	1.5	2712.8 $\geq$ 2.5 cm dbh	130	15 spp.	4 random 11 clumped 0 hyperdispersed	(4)
Lowland moist forest, Barro Colorado I.	1.5	2712.8 $\geq$ 2.5 cm dbh	130	size classes (spp. pooled)	> 20 cm dbh random < 20 cm dbh clumped	(4)
Lowland moist forest, Barro Colorado I.	1.5	2712.8 $\geq$ 2.5 cm dbh	130	all trees	clumped	(4)
Costa Rica						
Lowland dry forest, Guanacaste	13.4	1185.7 $\geq$ 2 cm dbh (260.8 $\geq$ 10 cm dbh)	135	all spp. with $\geq$ 2 individuals	17 random 44 clumped 0 hyperdispersed	(5)
Lowland dry forest, Guanacaste	13.4	1185.7 $\geq$ 2 cm dbh (260.8 $\geq$ 10 cm dbh)	135	juveniles vs. adults 30 spp.	adults less clumped than juveniles in 16 spp., more clumped in 5 spp., equally clumped in 9 spp.	(5)
Ghana						
Lowland dry forest, Pinkwae	0.36	2671.4 $\geq$ 3 m height	15	14 most abundant spp.	6 random 8 clumped 1 hyperdispersed	(6)
U.S. Virgin Islands						
Moist semievergreen forest, St. John	4.0	— $\geq$ 10 cm dbh	37	16 most abundant spp.	3 random 12 clumped 1 hyperdispersed	(7)
Panama						
Lowland moist forest, Barro Colorado I.	5.0	171.2 $\geq$ 19 cm dbh	112	63 spp.	37 random 26 clumped 0 hyperdispersed	(8)
Sarawak						
Kerangas (heath) forest, Sabal F.R.	19.2	836.6 $\geq$ 9.7 cm dbh	343	64 most abundant spp.	34 random 30 clumped 0 hyperdispersed	(9)
Costa Rica						
Lowland wet forest, La Selva	12.4	446.0 $\geq$ 10 cm dbh	269	104 populations (65 spp in 1–4 plots) with $\geq$ 10 individuals	85 random 13 clumped 5 hyperdispersed	(10)
Lowland wet forest, La Selva	12.4	446.0 $\geq$ 10 cm dbh	269	size classes (spp. pooled, & *Pentaclethra macroloba,* the canopy dominant)	canopy trees more hyperdispersed than understory trees	(10)

Table 8.6 (continued)

Site and Forest Type	Plot Area (ha)	Tree Density (ha^{-1})	Species Richness (Area Indicated)	Trees Analyzed	Results	Source
Lowland wet forest, La Selva	12.4	446.0 ≥ 10 cm dbh	269	sexual system guilds	various; no association between dispersion and sexual system	(10)
Lowland wet forest, La Selva	12.4	446.0 ≥ 10 cm dbh	269	fast growing, short-lived trees (spp. pooled)	random in 2 plots, clumped in 1 plot	(10)
Lowland wet forest, La Selva	12.4	446.0 ≥ 10 cm dbh	269	rare spp. (pooled)	random	(10)
Lowland wet forest, La Selva	12.4	446.0 ≥ 10 cm dbh	269	*Pentaclethra macroloba,* the canopy dominant	hyperdispersed	(10)
Lowland wet forest, La Selva	12.4	446.0 ≥ 10 cm dbh	269	all trees	hyperdispersed	(10)

Sources: (1) Pires et al. 1953; (2) Poore 1968; (3) Ashton 1969; (4) Lang et al. 1971, with additional analyses by Forman and Hahn 1980; (5) Hubbell 1979; (6) D. Lieberman 1979; (7) Forman and Hahn 1980; (8) Thorington et al. 1982; (9) Newbery et al. 1986; (10) this chapter.

roloba and *Welfia georgii*) both tend to be hyperdispersed. Again, the pattern is strongest where the species grow most densely. Observed mean distances between conspecifics in *Pentaclethra* and in *Welfia* are roughly equivalent to the mean crown diameter of the species. Dispersion in all other species is either random or significantly clumped. Clumped patterns are often idiosyncratic, occurring within a species at some sites and not others. Most clumping within species appears related to topographic heterogeneity within sites. Clearly, the great majority of species at La Selva are not hyperdispersed, that is, conspecifics are not farther apart than would be expected by chance. Large distances between conspecifics appear simply to result from the low relative abundance of the species.

In analyses of all trees independent of species and *Pentaclethra macroloba* hyperdispersion increases as a function of tree size. Trees in the canopy have, in nearly all cases, higher indices of dispersion than smaller trees.

Membership in guilds based upon sexual system did not influence dispersion although population densities differed among dioecious, monoecious, and hermaphroditic species. Rare species as a group did not depart from randomness in their dispersion. Fast-growing, short-lived trees (species pooled) were significantly clumped in one of the three plots; spacing around these trees (distance to nearest neighbor of any species), however, was no different than around trees in general. Thus, the real differences in growth behavior (D. Lieberman et al. 1985b, 1990; D. Lieberman and Lieberman 1987), physiology (chap. 10), and other aspects of the biology of fast-growing trees are not manifested in spatial dispersion.

Hyperdispersion within species (*Pentaclethra* and *Welfia*), hyperdispersion among all trees, and increasing regularity of spacing with size are all interpreted in terms of dense physical packing of tree crowns. Processes leading to this three-dimensional packing include differential growth or suppression of neighbors as well as differential mortality. Although the hyperdispersed pattern is demonstrable within species that occur at very high density, the pattern does not depend exclusively upon species membership but reflects dynamic processes within the forest as a whole.

RESEARCH NEEDS

Although it is technically quite easy to measure spatial pattern from stem maps, it is far more complicated to study the processes that produce the patterns. We still know little about how, where, and at what life stages attrition occurs or how species and stands differ in that regard.

Two approaches have been used to elucidate the development of pattern over time in tropical trees: different size classes have been compared at a single time (e.g., Hubbell 1979; Sterner et al. 1986) and individual cohorts of a given size class have been followed over time (e.g., Augspurger 1983a; D. A. Clark and Clark 1984; Connell et al. 1984). Most within-cohort studies have focused almost exclusively on seeds and seedlings, and very few have dealt with a substantial portion of the life span. Events affecting the dispersion of seedlings may have little or no bearing on the spatial pattern of adults, particularly where spatial scales are not comparable; results of seedling studies must, therefore, be extrapolated with caution. For example, D. A. Clark and Clark (1984) report that seedling densities in the emergent legume *Dipteryx panamensis* reach 0.15 m^{-2} around parents at germination, declining rapidly in a density-dependent fashion over the next two years, with no seedlings surviving within 8 m of the parental bole (directly beneath the parental crown). It is questionable whether such findings can be used to interpret dispersion in adults; mature *Dipteryx* trees occur at densities of only 2 ha^{-1}. Long-term studies of marked individuals in many size classes are needed, ideally including a sizable fraction of the species occurring at the site.

Because of high local diversity, many individual tropical

forest species are, necessarily, relatively rare, and tropical biologists have sought to explore the implications of that rarity. Rabinowitz (1981) noted that rarity at a site may arise for many reasons, involving the combination of geographic range, habitat specificity, and local population size of the species; most importantly, rarity in one area need not imply rarity elsewhere. Thus, it is dubious whether rare species should be treated as a single, monolithic guild. Information is vitally needed on the distribution and abundance of tropical tree species over their entire range (see Hall and Swaine 1981).

Density alone is insufficient to quantify effective population size in populations that are continuous over a large area. Effective population size is a function of both the local density and the variance of dispersal distance (Wright 1943, 1969), and it follows that a tree species with low density but efficient long-distance dispersal of pollen or seeds may have a much larger effective population size than one with equal density but shorter dispersal distances. Quantitative observations on pollen and seed movement in plant populations are difficult to make and are a particularly formidable challenge in dense and diverse tropical forest; yet an assessment of the evolutionary implications of low population density and large distances between conspecifics in diverse tropical assemblages may depend upon the acquisition of such data (Bawa 1979; see chap. 12).

Much of the necessary comparative information can best be obtained through long-term studies of marked individuals in permanent plots such as those at La Selva. Once such plots have been established, certain biases that may affect the direction of tropical ecological studies may be eliminated. Most notably, no a priori (and, often, unavoidably naive) decisions need be made about which of the multitude of species are most worthy of study; instead, all species are included, regardless of their rarity or abundance, ease of identification, physical attractiveness or charisma, taxonomic or trophic affinities, or occurrence in comfortable or unpleasant microhabitats. Further, parameters of interest can be compared among species. It is often difficult to evaluate results with rigor in the absence of a comparative frame of reference.

Comparisons between forests are also essential. We still understand little about how and why tropical forests differ both within and between regions although it is abundantly clear that they do (Swaine, Lieberman, et al. 1987). The relationships among environmental influences, physical structure, dynamics and floristic patterns are of particular interest. As quantitative data continue to accumulate from long-term investigations at La Selva and from parallel studies elsewhere and as more investigators are able to broaden their experience by visiting a variety of tropical forest sites, a clearer picture should emerge.

The local coexistence (stable or otherwise) of scores or hundreds of ecologically similar tree species is the single most obvious attribute of tropical forests. Studies of species dispersion patterns have traditionally been prompted by the quest for explanations of this diversity and its maintenance, a quest that has dominated tropical ecology for more than three decades. The focus on tropical diversity is understandable and can be traced to exclusive niche theory and its corollaries, a product of temperate zone ecology which renders the existence of high species richness paradoxical. Had the discipline begun with a tropical perspective, ecologists might now regard high diversity as the norm and consider that the comparatively depauperate assemblages of the temperate zone were in need of explanation; this latter problem might prove to be quite tractable in comparison. Even the incorporation of non-equilibrium components into community competition models—though clearly a step toward realism—has failed to explain coexistence in high richness species assemblages (Connell 1978, 1979; Hubbell 1979; Connell et al. 1984; Hubbell and Foster 1986a). We know of no compelling evidence, in fact, to suggest that processes of competitive exclusion have any influence on tropical forest diversity. The acceptance of this body of theory has led interpretations of dispersion data away from parsimony and may have diverted attention from many significant aspects of tropical ecology.

ACKNOWLEDGMENTS

We are grateful to Gary Hartshorn and Rodolfo Peralta for their collaboration in all aspects of the La Selva permanent plot studies; M. D. Swaine and the late J. B. Hall, whose thoughtful analyses of West African forests produced the intellectual underpinnings of this study; Michael Auerbach, Ross Kiester, D. McC. Newbery, and George Schatz for useful discussions on dispersion and its measurement; and Danilo Brenes, Gerardo Vega, Manuel Víquez, and María de los Angeles Molina for able assistance in the field. The Organization for Tropical Studies provided logistical support. Amos Bien, Gary Hartshorn, David Janos, and Rodolfo Peralta offered valuable comments on the manuscript. A. T. Joyce kindly made available unpublished data on harvested tree biomass. Portions of this study were carried out while M. L. was a National Research Council Senior Research Associate with the Earth Resources Laboratory, NASA. Field work was supported by National Science Foundation grants BSR-8117507 and BSR-8414968 and National Geographic Society grant 3132-85. Manuscript preparation was supported by NASA Grant NAGW-1033.

9

Tree-fall Gap Environments and Forest Dynamic Processes

Julie Sloan Denslow and Gary S. Hartshorn

Early studies of tropical rain forest structure distinguished high forest from the tangled patches of fast-growing saplings, lianas, and herbs characteristic of "disturbed" areas around tree falls (e.g., Richards 1952; Jones 1955; Beard 1955; Webb 1959). Although canopy gaps were acknowledged to be sites of tree seedling establishment and growth, shade-intolerant, "pioneer" species such as *Cecropia* (Moraceae), *Ochroma* (Bombacaceae), and *Trema* (Ulmaceae) were not considered legitimate components of the rain forest climax community. Current studies describe the canopy topography as an integral component of forest physiognomy (e.g., Torquebiau 1987; Oldeman 1983; Hubbell and Foster 1986a), and the environmental heterogeneity associated with gaps is now widely recognized as an important influence on seedling survival and growth rates of many rain forest trees and shrubs (Hartshorn 1980; Denslow 1980a; Brokaw 1985b; Whitmore 1978a, 1984). So-called pioneer species simply represent one end of a continuum of shade tolerance and light requirements among tropical forest trees (Brokaw 1985b; Denslow 1987b).

Research at La Selva has played an important role in advancing our understanding of the structure and dynamics of forest communities. Observations of gap formation rates in the Organization for Tropical Studies (OTS) plots (Hartshorn 1978, 1980; see chap. 6) helped stimulate the growing awareness of the role of tree-fall gaps in rain forest community ecology. Whereas earlier studies on rain forest dynamics had emphasized patterns associated with secondary succession following logging or agriculture (e.g., Budowski 1965; Ewel 1970, 1976), the La Selva studies focused research efforts on the natural regeneration processes occurring in tree-fall openings in primary forest (e.g., Hartshorn 1978, 1980; Vitousek and Denslow 1987; Chazdon and Fetcher 1984b, Sanford et al. 1986). This interest has also been pursued at other Neotropical sites: the Estación Biológica Los Tuxtlas, Mexico (e.g., Piñero and Sarukhán 1982a,b; Gómez-Pompa 1971; Gómez-Pompa et al. 1976; Gómez-Pompa and del Amo 1985; Martínez-Ramos 1985; Martínez-Ramos et al. 1988); the Smithsonian Tropical Research Institute at Barro Colorado Island (BCI), Panama (e.g., Augspurger 1984b; Brokaw 1982, 1985a, 1986, 1987; Hubbell and Foster 1986a, 1986b, 1987a, 1987b; Smith 1987); San Carlos de Río Negro, Venezuela (e.g., Uhl and Murphy 1981; Uhl 1982; Uhl et al. 1988); and

at Institut Françis de Recherche Scientifique pour le Développement en Coopération (Centre ORSTOM de Cayenne), French Guyana (e.g., Foresta et al. 1984; Charles-Dominique 1986).

The literature on ecological processes occurring in tropical tree-fall gaps has been extensively reviewed in several recent publications (Brokaw 1985b; Martínez-Ramos 1985; Denslow 1987b). We refer the reader to those papers for overviews of current directions in gap research. Because the environmental heterogeneity associated with tree-fall gaps figures importantly in the biology of so many forest organisms, many reviews in this volume also treat aspects of gap ecology, including plant demography (chap. 7), patterns of density and dispersion of forest trees (chap. 8), plant physiological ecology (chap. 10), frugivory (chap. 22), and plant-herbivore interactions (chap. 21). We do not repeat those discussions here but focus on the contribution of canopy gaps to forest structure, environmental diversity, and forest dynamics at La Selva and in other tropical forests.

GAP CHARACTERISTICS

Gap Formation

Canopy trees at La Selva snap, uproot, and die standing. Landslides are not known at La Selva, and Costa Rica is not within the hurricane belt (but see Boucher et al. 1990; Yih et al. 1991). Although no recent evidence of fire has been found at La Selva, widespread traces of charcoal in the soils suggest that the forest may have burned occasionally in pre-Columbian times (Sanford pers. comm.). D. Lieberman et al. (1985b) report that at least 26% of trees dying on the OTS plots between 1969 and 1982 died standing and at least 38% fell or were fallen upon. Another 37% died and decayed in the intervening time period, however, leaving no clues as to the causes of their deaths. Comparable figures for the percentage of trees dying standing in other Neotropical forests are 10% (San Carlos de Río Negro, Uhl 1982) and 14% (Barro Colorado Island, Putz and Milton 1982). In those forests the majority of trees larger than 15–20 cm dbh died as the trunks snapped, whereas 90% of the La Selva tree-fall gaps studied by Hartshorn (1980) were caused by uprooted trees.

In the La Selva primary forest, most tree-fall gaps occur in June–July and November–January, the wettest months of the year (Brandani et al. 1988). The peak in numbers of new tree falls observed during the rainy season on BCI also suggest that high winds and waterlogged soils are frequent contributing factors (Brokaw 1982). Uhl et al. (1988) note the importance of occasional high winds in creating large gaps in Venezuela. Data from the OTS plots at La Selva suggest that gap formation rates are also higher on alluvial or swamp soils than in the uplands at La Selva (Hartshorn 1978) although stand turnover rates are highest in the upland plot (D. Lieberman et al. 1985b). Among emergent trees at La Selva, lightning is an occasional source of mortality. Small trees are often buried under the debris of large individuals, contributing to a mortality rate among saplings (2–10 cm dbh) about double that of larger trees (>10 cm dbh, 2.03% per yr; D. Lieberman et al. 1985b). Although the possibility that some species are disproportionately more likely to be gap makers than others has not been directly addressed, Brandani et al. (1988) found *Pentaclethra macroloba* logs in 41% of fifty-one gaps formed in the OTS plots; *Pentaclethra* accounts for 13.6% of the stems larger than 10 cm dbh and 62% of the stems larger than 45 cm dbh in the permanent plots (Bethel 1976; Lieberman et al. 1985b).

Size-Frequency Distributions

In general, gap microclimatic conditions are a function of gap size (Schulz 1960; Chazdon and Fetcher 1984a; Denslow 1980a, chap. 10 this vol.). Although environmental variation within gaps is substantial, total incident radiation in the center of a tree-fall gap is a function of gap shape, orientation, and diameter; height of the surrounding forest; and sun angle. Soil and air temperatures, in turn, are functions of the duration of direct sun on the forest floor. Unfortunately, data are scarce on the size-frequency distribution of canopy gaps with respect to topography, soil characteristics, and vegetation history. Martínez-Ramos et al. (1988) used growth patterns of an understory palm to map and date gaps at Los Tuxtlas Biological Station, Mexico; they found a higher turnover rate on slopes there, which is likely the case at La Selva as well.

Sanford et al. (1986) interpreted aerial photographs (scale 1:2400) taken along two overflight transects over the 1400 m and 2400 m grid lines on Old La Selva. In figure 9.1A their data are compared with those from the old forest on BCI, Panama (Brokaw 1982), from the elfin forest at Monteverde, Costa Rica (Lawton and Putz 1988), and from mesic deciduous forest of the eastern United States (Runkle 1982). All of the latter surveys were made on the ground by monitoring new gap formation over several years in permanently marked transects or grids. On a gap area basis, La Selva is seen to have a smaller proportion of small gaps (<200 m²) and a larger proportion of very large gaps (<300 m²) than do the other forests studied. The relative paucity of small gaps at La Selva may be the result of the larger minimum gap size (40 m²) in the La Selva study.

The daily light availability, measured as photosynthetic photon flux density (PPFD), in a small gap in a forest of low canopy height is greater than in similar sized gaps in a tall forest. If gap size classes are relativized by mean height of the

forest (gap diameter/canopy height), the frequency distributions of functionally equivalent gaps (in terms of light environments) can be more directly compared. On this basis the gap environments in the La Selva and BCI forests may be similar (fig. 9.1B). Temperate forest gap environments may also be similar to those of the tropical forests although daily PPFD in the temperate gaps is more seasonally variable. The prevalence of small gap sizes in the Monteverde elfin forest is offset by the lower canopy there. Daily PFD in the Monteverde gaps is potentially equivalent to that in much larger gaps in taller, lowland forests. Gap-size associated variation in total incident radiation as well as differences between gap and understory light levels, however, will be considerably less under a climatic regime characterized by dense cloud cover and extensive periods of overcast skies (as in elfin forest) than in forests exposed to long periods of unobscured sun. Similarly, the extent of influence of canopy opening on light levels under adjacent intact canopy is a function of canopy height and latitude (Canham et al. 1990).

All of the tall, mesic forests for which we have data are characterized by many small gaps and few very large gaps. Because ingrowth from surrounding trees may close small or narrow gaps faster than large gaps (Runkle 1982), turnover times for small gaps may be higher than for large gaps (Martíanez-Ramos et al. 1988). We are likely underestimating the effect of small gaps on understory light environments. Branch and small tree gaps are a common, but ephemeral, source of increased light levels for understory trees and shrubs and for establishing saplings.

In contrast, gaps larger than 400 m² are a small proportion of all gaps formed but comprise a substantial proportion of total gap area (at least 21% of total gap area is in gaps larger than 400 m² at La Selva, Sanford et al. 1986). Large areas of young secondary vegetation within the forest occur where gaps have continued to enlarge as bordering trees fall. One of these is above the Sendero Circular Circano (CCC) south of its junction with the Sendero Suroeste (SSO) (fig. 2.2, chap. 2). This gap formed in July 1975 on a steep slope. Although young trees have established in its center, the edges of this gap continue to expand, probably because of unstable soils, altered wind currents, and structurally unbalanced trees. Another example is found on the western edge of the swamp south of Sendero Suampo; the first trees in this clearing fell in 1983.

An estimated 3.4% of Holdridge's original Finca La Selva is in open, successional vegetation that appears to be relatively permanent (Sanford et al. 1986). Although shade-intolerant tree species are present in these areas (e.g. *Ochroma lagopus, Cecropia obtusifolia*), poor drainage and waterlogging apparently impede the establishment of a continuous tree cover. One of the most familiar of these *tacotales* is on the Sendero Oriental (SOR) just west of the El Salto Bridge (fig. 2.2, chap. 2). Others occur along stream margins or swampy areas. Tacotales and enlarging gaps are important in the maintenance of shade-intolerant species (e.g., *Calathea, Heliconia*) in rain forest communities. Because canopy closure in large gaps is primarily from upward growth of establishing saplings, shade-intolerant saplings are able to overtop slower growing shade-tolerant competitors. Shade intolerant species persist longer in large gaps (Brokaw 1987), which may provide sufficient

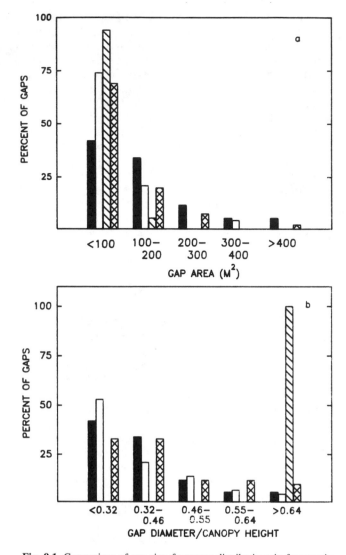

Fig. 9.1. Comparison of gap size–frequency distributions in four mesic forests. La Selva = filled bars; BCI old-growth forest = open bars; Monteverde elfin forest = diagonal hatching; Great Smoky Mountains National Park temperate deciduous forest = cross hatching. Gap size is presented as (A) area (m²) and (B) gap diameter as a proportion of the height of the surrounding canopy. La Selva (Sanford et al. 1986): minimum gap area = 40 m², maximum vegetation height in gaps = 5 m, forest canopy height = 30 to 40 m; Barro Colorado Island (Brokaw 1982): minimum gap area = 20 m², maximum vegetation height in gaps = 2 m, forest canopy height range = 23 to 30 m; Monteverde elfin forest (Lawton and Putz 1988): minimum gap area = 4 m², maximum vegetation height in gaps = 3 m, forest canopy height range = 5 to 15 m; Great Smoky Mountain National Park, United States (Runkle 1982): minimum gap area unspecified, maximum vegetation height in gaps = 10 to 20 m, forest canopy height range = 25 to 32 m.

time for these species to attain canopy stature and reproductive maturity.

Gap Habitats

Canopy openings are a major source of environmental heterogeneity at the forest floor. Tree and branch falls invariably result in some increase in light availability to the forest understory. In addition, deep soil layers may be brought to the surface in root pits and mounds (Hartshorn 1978; Putz et al.

1983; Putz 1983), and large quantities of woody and leafy litter are dropped with the tree crown. Otherwise, substantial areas of the gap floor in the vicinity of the fallen boles may remain relatively undisturbed. To these commonalities of gap structure must be added the considerable environmental variation associated with soil texture and composition: slope, exposure, and drainage; structure of the understory vegetation at the time of tree fall; fine root density (Sanford 1989); quantities of lianas brought down with the tree falls (Putz 1984a, 1984b); recent disturbance history of the site; and the activities of soil-foraging animals (e.g., agouti, paca, coati, peccary, armadillo, and leaf-cutting ants).

Light

Total incident radiation in gap centers is a function of gap size, geometry, exposure, and height of the surrounding forest. Total PFD measured at a height of 1 m on two to four sensors in the centers of each of four large (275–335 m²) clearings at La Selva were 8.6%–23.3% of full sunlight on sunny days. Comparable measurements in the forest understory were 0.4%–2.4% full sunlight and 2.8%–11.1% on gap-forest boundaries (Denslow et al. 1990). Chazdon and Fetcher (1984b) recorded similar values of 20%–35% in the center of a 400-m² clearing, 9% in a 200-m² clearing, and 1%–2% in the forest understory. These understory values are comparable to those measured in other rain forests (Pearcy 1983; Bjorkman and Ludlow 1972; Chazdon and Fetcher 1984a; Pearcy and Robichaux 1985). Gap area alone, however, is an imperfect predictor of light environments in the centers of gaps. Estimates of total incident sunlight from fisheye canopy photography are plotted against area in six mapped gaps in figure 9.2. Differences in gap geometry and orientation and the influence of other nearby gaps account for the lack of a significant linear correlation.

Spatial variation in light availability is likely greater at seedling height (<20 cm) than higher in the forest profile. The distribution and abundance of understory plants exerts considerable influence on light availability to establishing seedlings, both under intact forest canopy and following tree fall (Brokaw 1983; Denslow et al. 1991, but see Marquis et al. 1986). The understory vegetation at La Selva is not only well developed but also comprises a large number of low-growing, large-leaved herbs and shrubs, some of which occur in large aggregations: Cyclanthaceae (*Asplundia, Carludovica, Cyclanthus*); acaulescent and short-stemmed palms (*Geonoma, Asterogyne,* and saplings of larger species such as *Welfia, Iriartea,* and *Socratea*); Araceae (*Dieffenbachia*), Heliconiaceae (*Heliconia*), and Marantaceae (*Calathea*).

In addition to an increase in total incident radiation in gaps, the intensity of light received differs considerably between gap and understory environments. Plants in the centers of gaps are exposed to more bouts of high-intensity radiation for longer periods than are plants in the forest understory (Chazdon and Fetcher 1984b; chap. 10). Moreover, the low-intensity, diffuse light in the understory is deficient in photosynthetically active wavelengths (400–700 nm, Bjorkman and Ludlow 1972). Of the total incident radiation in the understory on a sunny day, however, 61%–77% is received in short-duration flecks of full sun (Chazdon and Fetcher 1984b).

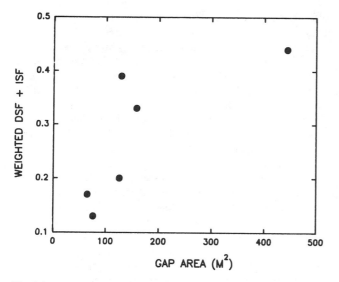

Fig. 9.2. Available light in the centers of six gaps as a function of gap area estimated from the crown projection of the surrounding trees. Available light is measured as the annual proportion of full sunlight estimated as the weighted average of direct site factor (direct beam radiation) and indirect site factor (reflected sky light) (Rich 1990). Means and standard errors of three to four points distributed along the center line of the gap.

On overcast days the light in both gap and understory environments is diffuse, shadows are less marked, and the differences between gap and understory light levels diminished (Chazdon and Fetcher 1984b). Thus, the magnitude of microclimatological differences across a gap-understory transition will depend, in part, on local patterns of clear and cloudy weather.

Light levels in the centers of gaps decrease with ingrowth from tree crowns and with the upward growth of saplings, shrubs, vines, and herbs within the gaps. This upward growth is more abundant and faster growing in large, bright gaps than in small, with the result that light environments at seedling heights in large and small gaps converge within the first year following tree fall (fig. 9.3). Light levels at 2 m remain higher in large gaps than in small following one year of regrowth. At 2 m above the ground, large gaps are, thus, characterized by both higher initial light levels and longer duration of enhanced light levels.

Fine-scaled data on foliage height distributions have shown that gaps are obconical rather than cylindrical, as often modeled, and that individual gaps that are discrete at ground level may coalesce with neighboring gaps higher in the forest profile (Hubbell and Foster 1986b). Thus, tall saplings are not only overtopped by fewer crowns than small saplings, they are more likely to be in or near a gap.

Soil Moisture and Nutrient Availability
Because the fall of a canopy tree produces a large volume of litter and because leaf and fine-stem litter decompose rapidly in many rain forest habitats (Anderson and Swift 1983; Jordan 1985), gaps have often been described as local hot spots of nutrient availability (e.g., Brokaw 1985b). Recent work at La Selva and elsewhere, however, has failed to document the predicted effects. Vitousek and Denslow (1986) took paired monthly samples of the upper 15 cm of soil in the crown zone

of four recent tree falls and adjacent forest understories on residual soils. Over the course of the year they detected no significant differences in either pool sizes or mineralization rates of nitrogen although there was some evidence of an early, ephemeral pulse in N mineralization rates. Recent studies have confirmed a short-lived peak in NO_3-N [nitrate-nitrogen] (but not NH_4-N [amonium-nitrogen]) about sixty days after tree fall (Sanford unpublished data). Extractable phosphorus in samples from gaps was consistently, although nonsignificantly, higher than in the understory samples (Vitousek and Denslow 1986). With the exception of lower nutrient availability in the exposed subsoils of the root zone, no differences in nitrogen and phosphorus availability were demonstrated among zones within gaps. Similarly, Parker (1985) detected no increase in solutes (NO_4-N, K, Na, Ca) in soilwater collected below 70 cm soil depth following the creation of six small (mean area = 97 m²) gaps on residual soils beyond La Selva's south boundary.

Several factors likely contribute to these results. Nitrogen availability in La Selva soils is high (Vitousek and Denslow

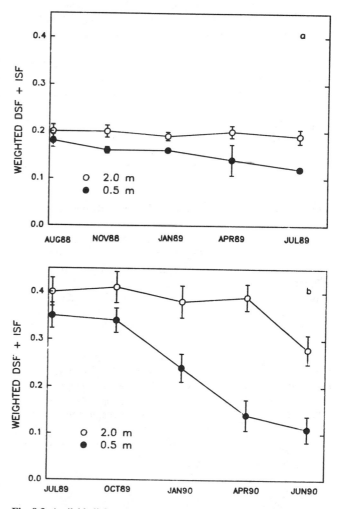

Fig. 9.3. Available light measured during the first year following tree fall in the centers of two gaps: (A) a small gap (approximately 126 m²) and (B) a large gap (approximately 611 m²). Available light is estimated as proportion full sunlight as in figure 9.2. Fish-eye photographs were taken at 0.5 m and 2.0 m above the ground.

1987) and may obscure any small increases from litter mineralization; the high adsorptive capacity of La Selva soils for P (see Vitousek and Denslow 1986, Uehara and Gillman 1981, chap. 4) may obscure any increase in phosphorus. In addition, intact root systems of surviving plants probably absorb a substantial portion of any increase in available nutrients. Although La Selva does not have a surface root mat characteristic of some tropical rain forests (Jordan 1985), the majority of fine roots occur in the upper soil layers (Raich 1980b; Parker 1985). Increased root growth of surviving plants in response to gap formation may quickly absorb early pulses of some nutrients. The hypothesis is consistent with Parker's (1985) observation of significant increases in soil water solutes below two large clear-cut areas (600 m² and 2,500 m²), (where there would have been extensive fine-root death) although no such effect was found in small gaps. Sanford (1989), however, found a lower fine-root biomass in the crown zone of a single two-year-old gap than in adjacent understory. Clearly, the dynamics of fine-root growth are an important but poorly understood component of rain forest ecosystems.

It is, moreover, difficult to make a strong case that a small increase in nitrogen or phosphorus availability in small tree-fall gaps is important to seedlings as they establish. Fertilization experiments in pots of intact residual soils at La Selva showed no evidence that *Miconia* and *Piper* shrubs were nitrogen limited (Denslow et al. 1987). Moreover, most La Selva trees and shrubs may not be strongly phosphorus limited, at least at light levels typical of forest understory and tree-fall gaps. Although laboratory analyses of residual La Selva soils suggest that they are deficient in phosphorus (Vitousek and Denslow 1987; Bourgeois et al. 1972), phosphorus fertilization produced no significant reponse in potted *Miconia* and *Piper* cuttings (Denslow et al. 1987). In addition, field fertilization experiments of seven species of *Miconia* and *Piper* planted as rooted cuttings into the crown zone of four recently formed gaps on residual soils produced no significant growth responses to a complete nutrient treatment (Denslow et al. 1990). Mycorrhizal associations in many La Selva species (Janos 1980), including *Miconia* and *Piper* (Denslow personal observation) may enhance phosphorus availability and account for the lack of response to phosphorus fertilization in these species. *Phytolacca rivinoides,* a nonmycorrhizal species, does show a strong growth reponse to added phosphorus and it may be limited to relatively nutrient-rich areas.

In contrast to nitrogen and phosphorus availability, soil moisture was consistently higher in gaps than in adjacent understory (Vitousek and Denslow 1986) as has been observed in gaps in other tropical forests (Becker et al. 1988; Uhl et al. 1988). Both higher throughfall and lower evaportranspiration rates in gaps may contribute to this pattern.

ECOSYSTEM PROCESSES

Forest Dynamics

Hartshorn's (1978) estimate of turnover time of 118 ± 27 years for La Selva is based on annual rates of gap formation and is similar to turnover times measured in other mesic tropical and temperate forests (BCI, Panama: 114 years in old-growth forests, Brokaw 1982; Monteverde, Costa Rica: 95 years in an elfin forest, Lawton and Putz 1988; Great Smoky Mountain National Park, USA.: about 100 years, Runkle 1982).

Stand half-life (calculated using a logarithmic model from Hartshorn's [1978] data) were estimated by D. Lieberman et al. (1990) as 53 to 92 years (mean = 77.3 years). This value corresponds to gap birth rates of 0.96% (range 0.71%–1.31%) of the land area each year. Tree mortality was measured on the same plots at La Selva at 2.03% per year, in the upper end of the 1%–2% range measured for other tropical forests and equivalent to a stand half-life of about 34 years (D. Lieberman et al. 1985b).

Disagreement in the two estimates of stand half-life for La Selva arises in part from the imperfect correspondence between tree mortality and gap formation (D. Lieberman et al. 1985b, 1990). Sizes of gaps created by the fall of single trees vary as a function of the size of the tree (Brokaw 1982). Gaps may be created also by the fall of multiple trees at a single time or over several years. Where such tree falls effectively overlap, the total area disturbed may be less than if the trees had fallen independently. Not all trees create gaps when they die; if a tree dies slowly over several years, lateral growth of adjacent crowns may fill the small openings. Likewise, larger trees may lose portions of their crowns, creating small gaps not associated with whole tree mortality.

Visitors to La Selva are impressed by the frequency of tree-fall clearings, and many hear large trees fall during heavy rains and even on windless days. The tree mortality data from the permanent plots suggest that the La Selva forest is one of the most dynamic measured although gap formation rates place it in the center of the range for a wide variety of temperate and tropical forests. The significance of differences in turnover times for forest structure is not clear, however. Certainly the details of gap processes differ strikingly between forest types as disparate as northern coniferous (Heinselman 1973) and tropical rain forests, although estimated turnover times are similar. The distribution and duration of the microclimatic conditions associated with tree falls are likely of more direct significance to forest composition than is the turnover time per se.

For a number of reasons gaps may play a more important functional role at La Selva than in many other systems studied to date. With the exception of the occasional simultaneous fall of several canopy trees, the La Selva forest does not receive major, periodic, canopy-opening events common in forests subject to hurricane, landslide, or fire. These cataclysmic events are not only major sources of tree mortality, but also trigger the germination and establishment of large cohorts of trees that influence forest structure for many years (e.g., Weaver 1989; Whitmore 1989b). Species establishing in the high light conditions following a major canopy-opening event will be primarily fast-growing and shade intolerant. For many years following large gap formation, tree-fall gaps will be small in this second-growth forest because large trees will be rare (Brokaw 1982; Lang and Knight 1983; Hubbell and Foster 1986a; Saldarriaga et al. 1988). Until natural mortality reduces the predominance of this cohort, gap processes will have little influence on forest structure within that patch. La Selva appears to be a relatively unusual example of a forest not strongly influenced by such large-scale, canopy-opening events.

In addition, the tall canopy, multiple layers of foliage, and a well-developed shrub and palm understory result in very low understory light levels. As a result seedling mortality is high and sapling growth is slow except where the canopy is opened by tree or branch falls. Processes occurring in gaps are, therefore, likely to be an important influence on forest structure at La Selva. In forests with less shaded understories, gaps may be less critical for seedling survival and sapling growth. Thus, gap processes may be less important in forests with lower or more open canopies or dominated by seasonally deciduous trees.

Plant Establishment and Growth

Accumulating data suggest that plant growth and reproduction are closely related to relatively small changes in light availability associated with small openings in the forest canopy (see D. A. Clark and Clark 1987a, chap. 10). That most plants grow faster and experience lower mortality rates under some degree of canopy opening than under an intact forest canopy has been observed in experimental plantings of *Miconia* and *Piper* cuttings in natural gap environments (Denslow et al. 1990), in marked individuals of naturally occuring tree seedlings (D. B. Clark and Clark 1985) and saplings (Piñero and Sarukhán 1982a; Brokaw 1985a; Chazdon 1985; D. A. Clark and Clark 1987a; DeSteven 1989) and in planted seedlings (Augspurger 1984a, 1984b, Augspurger and Kelly 1984; Popma and Bongers 1991). This pattern is most easily explained by more favorable carbon balances in high light environments although Augspurger (1984a, 1984b, and Augspurger and Kelly (1984) also demonstrated lower mortality rates from pathogens among seedlings planted into gaps. No species has yet been shown to grow best at the low light levels characteristic of the understory under intact canopy although mortality rates in the shade vary widely among species (Augspurger 1984a; D. A. Clark and Clark 1987a; Denslow et al. 1990).

Mortality rates are highest for seedlings and small saplings in all rain forest habitats (Hartshorn 1972; D. A. Clark and Clark 1984; D. Lieberman et al. 1985b). In the understory mortality is caused by falling leaves and branches (especially palm leaves) (Vandermeer 1977), fungal pathogens (Augspurger 1983c, 1984b; Augspurger and Kelly 1984), and herbivory (D. B. Clark and Clark 1985; Denslow et al. 1991; Marquis 1984a, 1984b and chap. 21) and is probably exacerbated by the expense of replacing leaf tissue in a light-limited environment. Denslow et al. (1991) found that mortality of *Inga* seedlings was significantly higher near understory palms than in the slightly higher understory light environments not affected by these palms. Plants growing under palms lost significantly more stem length to herbivores as well. Where such palms or other low-growing, large-leaved herbs are abundant, seedlings of woody plants are significantly less common (Denslow et al. 1991).

Although it is not yet clear whether plants growing in gaps experience different rates of herbivory (in percentage of tissue lost) than plants growing in the understory, the cost of herbivory in lost leaf-days of photosynthesis is probably considerably less for plants growing in gaps than for more heavily shaded understory individuals (Coley et al. 1985). Coley's (1983a) studies at BCI have shown that herbivory is higher on young leaf tissue than old and higher on high-light requiring pioneer species than on shade-tolerant species. Because both

new leaves and high-light requiring species are more common in gaps, the incidence of herbivory may also be higher there (see also Harrison 1987).

Reproductive output is also greater in plants growing in gaps than under an intact canopy. Trees and shrubs produce larger crops in longer fruiting seasons when in or adjacent to gaps (Levey 1988b; Piñero and Sarukhán 1982a). Some indication exists that both overall fruit removal rates and competition among fruiting plants for frugivores are higher in gaps than in the understory, probably because of a greater abundance of frugivorous birds there (Denslow et al. 1986; Denslow and Moermond 1982; Levey 1988b, chap. 22).

Communitywide, a large proportion of tree species at La Selva are thought to depend on some degree of canopy opening either for seed germination and seedling establishment or for sufficient growth to reach reproductive size. From observations of the growth responses of saplings to canopy opening Hartshorn (1978) estimated that approximately 75% of the 105 canopy tree species at La Selva depend on gaps for successful regeneration. These early impressions are generally being confirmed as more detailed species studies are completed.

Several authors have suggested heuristic categories of life-history patterns based on germination, establishment, and growth responses in different light environments (Bazzaz 1984; Bazzaz and Pickett 1980; Denslow 1980a, 1987b; Hartshorn 1980; Martínez-Ramos 1985; Uhl and Murphy 1981; Van Steenis 1958; Whitmore 1984). In most cases the individual categories are roughly comparable among authors and rely on a continuum in shade tolerance from high (climax, persistent, small-gap species) to low (pioneer, nomad [*sensum* Van Steenis 1958], large-gap, ruderal species). Bazzaz (1984) and Hartshorn (1980), among others, provide lists of attributes expected of species in the different categories. These suites of adaptive characteristics are also discussed in chapters 7 and 10.

Differences in light requirements among tree and shrub species have consequences for the ability of seedlings and saplings to survive and grow in gaps of different sizes. For example, some species at Barro Colorado Island are unable to persist in gaps smaller than 150 m^2 (Brokaw 1982, 1987). Although the dependence of these pioneer species on large gaps has been well demonstrated, the importance of gaps for more shade-tolerant species is less clear. Several recent studies have shown fine-scaled differences in growth patterns of ecologically similar, phylogenetically related species growing in forest habitats, including understory environments (Augspurger 1984a; Chazdon 1985; Pearcy and Calkin 1983; Pearcy and Robichaux 1985; Brokaw 1987; Denslow et al. 1990; see discussion in Denslow 1987b). Patterns of photosynthetic light responses, acclimation, and relative growth rates, however, are not always consistent with a priori designations of life-history strategies (Fetcher et al. 1987; Popma and Bongers 1991; Walters and Field 1987).

Distributions of seedlings and saplings may vary within gaps (Nuñez-Farfán and Dirzo 1988). Ellison et al. (in press) found significantly more Melastomatace seedlings in the disturbed soil of root zones than in other areas of gaps at La Selva. In older gaps, several species of saplings appear to be overrepresented in gap root zones, and the species compositions of root, bole, and gap zones differed from one another among gaps (Brandani et al. 1988). These data suggest the role that environmental patterns associated with tree falls may play in the establishment, growth, and reproduction of tropical rain forest plants. It remains to be shown whether such fine-scaled habitat differences importantly influence the distributions and abundances of canopy trees.

Seed Dispersal

Propagule dispersal is a critical and poorly studied component of plant regeneration. The seeds of most woody species at La Selva are ingested by animals and, thus, are likely to be widely dispersed from the parent plant (chap. 22). Denslow and Gómez (1990) found that less than 50% of the seed rain to new gaps could have been dispersed from parent plants within 50 m of the gap; the remainder had been dispersed longer distances. They found that rates of seed rain were spatially and temporally variable but that overall rates of seed rain were high (49 seeds per m² per month). Assuming soil seed stocks of 100 to 2,000 seeds per m² (Hopkins and Graham 1983; Putz 1983; Guevara S. 1986), these data suggest that turnover times of seeds in rain forest soils are short (two to forty-one months).

Nevertheless, the effectiveness of seed shadows may vary dramatically as a result of differences in seed size, crop size, number and behavior of dispersers, seed dormancy capacities, sources and rates of seed mortality, and germination requirements (Howe and Smallwood 1982). Some small-seeded species apparently require the levels and quality of light of large gaps for germination (e.g., *Cecropia, Heliocarpus, Ochroma, Trema;* Vázquez-Yánes and Orozco-Segovia 1984, 1985, 1987, 1990). Seedlings of these species recruit for a relatively short period soon after tree fall (Brokaw 1986, 1987; Uhl et al. 1988). As a low canopy forms over the ground, subsequent recruitment is increasingly restricted to more shade-tolerant species. Yet not all small-seeded species are shade intolerant. Understory herb and shrub species produce some of the smallest seeds at La Selva (e.g., *Miconia, Spathyphyllum, Piper*). Seeds of these genera were common components of the seed rain into four natural tree-fall clearings at La Selva (Denslow and Gómez 1990), but little is known of their seed germination requirements or dormancy capacities. Large-seeded species are typically uncommon in soil seedbanks; they germinate within a few months of dispersal, often triggered by the start of the rainy season rather than canopy opening (Garwood 1983; Brokaw 1982, 1986).

Long-term studies in Panama have shown that seedlings recruiting within the first few months following gap formation (including those in place at the time of tree fall) are more likely than later arrivals to survive to fill the canopy opening (Brokaw 1987). The importance of this advance recruitment has long been used by foresters to assess the prospects for regeneration following logging (Whitmore 1986). Gap and understory components are, thus, closely coupled in forest dynamic processes and those factors affecting seedling growth and survival in the understory are pertinent to any model describing the role of gaps in forest dynamic processes.

In addition, seed production is often highly variable within a species from year to year and from place to place. Chesson and Warner (1981) have proposed that the probability of propagule establishment varies among species from year to year. This temporal variation enhances the likelihood of coexis-

tence among species occupying similar habitats, producing in effect a temporal partitioning of establishment opportunities. Such a model is particularly apt for species, such as rain forest trees and shrubs, with long-lived adults and low and variable fruit production and seedling establishment. These factors, coupled with high species diversity and low population densities, contribute substantial noise to any attempt to determine which processes most strongly affect canopy composition of a forest.

TREE-FALL GAPS AND THE COEXISTENCE OF SPECIES

A central question, especially in species-rich communities, concerns the role of gaps in the maintenance of plant species diversity. Are tree and shrub species adapted to different microenvironments within the temporal and spatial mosaics that comprise forest understory and tree fall gaps? and Do competitive interactions among co-occurring (and differently adapted) species play an important role in the determination of canopy occupancy (e.g., Hartshorn 1978, 1980; Denslow 1980a; Brokaw 1985b; Orians 1982; Brandani et al. 1988)? Alternatively, are most canopy species habitat generalists? and Is canopy occupancy largely determined by the fortuitous presence of seedlings or saplings that can take advantage of increased light levels created when the gap is formed (e.g., Hubbell 1979; Hubbell and Foster 1986a, 1987b)? In the latter case neither differential adaptation to microhabitats nor competitive interactions among co-occurring species would importantly influence which sapling will fill the hole in the canopy left by a tree fall.

These hypotheses represent extremes in the importance attributed to competitive interactions in the determination of forest structure and composition. Hubbell and Foster (1986a, 1987b) suggest that, although biological interactions do occur, they are likely swamped by chance and historical events and, therefore, unlikely to influence species distributions or extinctions. Their model relies on processes such as immigration and speciation for the maintenance of within-community species diversity because, uninterrupted, drifting abundances reduce species diversity in the long term. Support for the Hubbell-Foster model has come primarily from data on the distribution patterns of trees and on the distributions of saplings with respect to canopy openings (Hubbell and Foster 1986a). With the exception of high light-requiring species whose saplings are significantly associated with canopy gaps (e.g., Brokaw 1987), most species show no gap association. As the authors point out, however, such data are inadequate to assess relative performances in gaps and in forest understory because growth and survival are not measured.

This observation highlights a common conceptual difficulty in studies of gaps and forest dynamics—the assumption that plant distribution is an adequate index of habitat requirements. As an indicator of soil characteristics or moisture requirements, plant distribution may reflect important environmental pattern, but the ephemerality of gap microenvironments severely jeopardizes its usefulness in that context. From accumulating data one can demonstrate that many tree and shrub species respond to canopy openings yet exhibit some degree of shade tolerance (e.g., chap. 7, 10; Denslow et al. 1990). Considering the small size of most gaps (Sanford et al. 1986) and the low incident light levels in those gaps (Chazdon and Fetcher 1984a), a degree of shade tolerance would seem to be necessary for the survival of most rain forest trees and shrubs. Moreover, it is likely that tree saplings reaching the forest canopy have endured more than one episode of gap formation and filling. Such a pattern has been observed in mesic deciduous forest of the eastern United States (Canham 1989) and is probably the case in tropical forests as well. As a consequence, one should expect to find saplings of shade-tolerant species randomly distributed among gap and nongap habitats although they may, nevertheless, depend on the light levels in gaps for substantial vegetative growth or reproduction.

Shade-intolerant species, however, do not survive canopy closure and their distributions in and out of gaps are better evidence of their habitat requirements. Nevertheless, Brokaw (1987) has shown that saplings of shade-intolerant species may become established in gaps in which they are not able to persist. Thus, the sizes of seedling or sapling classes may have little relation to the future abundance of a tree in the canopy (White et al. 1985).

The observed growth plasticity of some shade-tolerant tropical trees—the ability to both survive the low incident light levels of the forest understory and make rapid growth in tree fall gaps (noted in chap. 7) was, however, unexpected and emphasizes the complexity of growth responses in plants and the inadequacies of current categorization of plant growth patterns. In the final analysis, the relative growth responses of different species to forest microhabitats will be a more useful indicator of habitat requirements than will the distribution of individuals alone. Long-term, spatially replicated studies will be critical to successfully incorporate temporal and spatial variability into an understanding of forest dynamic processes.

The alternative model proposes that competitive interactions among saplings for canopy space importantly influence canopy composition. It does not propose that each species is restricted to a unique microhabitat or that competitive interactions for space are highly predictable in space or in time. The factors affecting seedling and sapling survival are small scaled and ephemeral. Their effects are likely to be manifest in probabilities of canopy occupancy over large spaces and long times.

If species do differ importantly in their abilities to establish and grow in the environmental mosaic associated with canopy openings, then the species compositions of forests characterized by different disturbance regimes should reflect those differences (Denslow 1980b, 1985). At one level this appears to be the case. Forests subject to large-scale disturbance are often characterized by the relative abundance of high-light requiring species (e.g., Denslow 1980b; Hubbell and Foster 1987b). The effects of disturbance regime on the relative abundances of more shade-tolerant species, however, are less well documented.

The most promising models for the development of sustained-use exploitation systems that maintain intact population and ecosystem processes in complex tropical forests will be based on an understanding of the natural regeneration processes and the relationship between canopy opening and the differential establishment and growth of species. One prototype of a sustained-yield logging operation for the Neotropics is being tested in the Palcazú Valley of Peru based on the successive cutting of strips wide enough to allow the regeneration of fast-growing, economically important tree species and narrow enough for establishment of seedlings from the bordering intact forest (Hartshorn 1989a, 1989b). In Costa Rica, Portico, S.A., is attempting to manage forests of caobillo (*Carapa nicaraguensis*), using a gap model that provides resources for *Carapa* regeneration and also maintains considerable diversity of nonexploited species as well (see chap. 25 for other examples). The artificial increase in the number of gaps in these managed forests will also provide an excellent opportunity to study the long-term consequences of a known disturbance pattern on plant and animal composition.

CONCLUSIONS AND RESEARCH NEEDS

Research on forest dynamics and ecological processes in gaps at La Selva and other rain forest research stations has focused attention on these small-scale disturbances as important sources of environmental heterogeneity on the forest floor. As data accumulates on growth patterns of species, it is apparent that variation in light availability is the preeminent factor affecting plant growth, at least at small gap sizes and in moist forest on relatively nutrient-rich soils. Although species responses vary under different patterns of light intensity and duration, most species likely have some degree of shade tolerance. It is also apparent that important qualitative and quantitative differences exist in microclimates, seedling demography, patterns of nutrient availability, and patterns of closure in gaps of different sizes and histories. One can only speculate, however, about the consequences of gap processes for larger-scale forest structure and composition.

Detailed studies of ecosystem processes associated with gap formation lag behind studies of population phenomena. Although studies at La Selva suggest that light is the primary limiting resource for seedling growth, below ground processes influencing soil moisture or nutrient availability will be relatively more important in dry forests or in forests on nutrient-poor substrates. Comparative studies in forests of different stature, phenology, and nutrient status will be important in elucidating large-scale patterns.

Researchers are also moving toward the incorporation of both small- and large-scale environmental heterogeneity into conceptual models of forest dynamics and processes affecting forest composition and processes (e.g., Horvitz and Schemske 1986; Álvarez-Buylla and García-Bárrios 1991). The environmental heterogeneity contained in greater La Selva and the contiguous Braulio Carrillo National Park ensure that La Selva will continue to be fertile ground for research into the dynamics and regeneration of tropical forests.

ACKNOWLEDGMENTS

We are grateful for many years of OTS encouragement and support, which, more than any other factors, have nurtured this and other research at La Selva. We also thank N. Brokaw, S.T.A. Pickett, J. Armesto, and P. Ashton for their thoughtful reviews of an earlier draft of the manuscript and NSF for its grants in support of our research on forest structure and dynamics at La Selva. This is contribution number 5 from the La Selva Gaps Project.

10

Physiological Ecology of Plants

Ned Fetcher, Steven F. Oberbauer, and Robin L. Chazdon

One objective of physiological ecology is to understand the controls of the distribution and abundance of plants by investigating their physiological responses to the environment. To this end, one studies patterns of resource use and allocation by individual plants in an ecological context. Because plant morphology is intimately related to physiology, one wishes to understand physiological and morphological traits in relation to microsite, growth, phenology, successional status, demography, and life-history patterns. Most research on physiological ecology of plants is based on three premises: species with similar ecological roles or growth forms should show similar physiological responses (Mooney et al. 1980); physiological and morphological plasticity within and among species should be correlated with the range of conditions experienced by these species (Bazzaz 1979, 1984); and to understand the physiological basis of a plant's performance under particular environmental conditions one must integrate structure and function at the whole-plant level.

The first premise underlies much of the research on plant physiological ecology at La Selva. It has been particularly central to studies of shade tolerance. Demographic studies and field observations indicate that trees can be classified according to apparent differences in shade tolerance (Brokaw 1985b; Hartshorn 1978; Knight 1975; Swaine and Whitmore 1988; Whitmore 1975). Species that grow quickly in open environments, often called *early successional* or *pioneer* species, are at one end of the spectrum. At the other are *late successional* or *shade-tolerant* species, characterized by seedling establishment in shaded environments and slow growth rates. Species exhibiting intermediate degrees of shade tolerance, called *gap* species, occur mainly in mature forest but appear to require a canopy gap to grow and reproduce (Whitmore 1975; Bazzaz 1984). This intermediate category is potentially the broadest because of the diversity of gap environments in mature forest. Although tropical ecologists widely accept the shade-tolerance continuum (but see chap. 7), it remains to be shown whether any tree can grow to the upper levels of the canopy without sometime receiving increased light associated with a gap. At La Selva the rate of canopy disturbance is high enough (Hartshorn 1978; D. Lieberman et al. 1985a, chap. 9) to ensure that nearly every tree is exposed to gap conditions at some point in its lifetime.

It has been proposed that species with similar degrees of shade tolerance have similar physiological responses (Bazzaz 1979, 1984). Intraspecific variation in physiological and mor-

phological traits, however, complicates discussion of relative shade tolerance among species. This variation can be partitioned into several components. First, variation results from genetic differences between individuals. Second, variation results from phenotypic plasticity or acclimatization exhibited when an individual is exposed to different environments. Such plasticity can be regarded as a heritable trait subject to natural selection (Bradshaw 1974). Third, individuals can respond differently to different environments (genotype-environment interaction, Falconer 1981). Finally, variation can result from ontogenetic stages (e.g., physiological differences between juveniles and adults). Studies at La Selva have focused on acclimatization and, to a lesser extent, on juvenile-adult comparisons.

One of the most important assumptions underlying research in physiological ecology is that changes in physiology or morphology that result from acclimatization are adaptive. Researchers seek to discover how the observed responses contribute to enhanced physiological performance. In many cases it would be desirable to test the assumption that acclimatization is adaptive by measuring components of fitness on the same individuals that are used for physiological measurements. Unfortunately, this approach is rarely feasible. In the following sections we place some observed responses in an adaptive framework, but we warn that critical tests to determine whether such responses are truly adaptive are lacking.

In this chapter we review current knowledge of the spatial and temporal heterogeneity of critical resources for plants (light and water) within and among habitats at La Selva. We then discuss plant responses to light, including changes in leaf morphology, photosynthesis, and leaf carbon budgets. After reviewing water and nutrient relations, we move to the whole plant and describe studies of growth, biomass allocation, and architecture.

PHYSICAL ENVIRONMENT

Previous studies of the physical environment of tropical forests concentrated on vertical profiles of radiation, temperature, and humidity (Baynton et al. 1965; Cachan and Duval 1963; Cachan 1963; Chiariello 1984; Evans 1939), and little was known about horizontal variation in these parameters. One of the main objectives of physiological research at La Selva has been to describe regimes of light, temperature, and humidity in different forest microenvironments. The focus has been on light because of its importance for photosynthesis and its extreme variability. Differences in temperature and humidity are less pronounced although significant variation exists (Fetcher et al. 1985).

Spatial Variation in Light Regime

The understory light regime is characterized by high temporal variability because of sunflecks—small, brightly lit patches of direct beam radiation penetrating the forest canopy (Pearcy 1983; Chazdon 1988; Chazdon and Pearcy 1991). Studies of the distribution of photosynthetic photon flux density (PPFD, the flux of photons between 400 and 700 nm wavelength per unit area), have confirmed the visual impression that light varies greatly over space and time. In table 10.1 we summarize average PPFD and total daily PPFD from several studies of light regimes at La Selva. Understory light levels are 1%–2% of that received in clearings (Chazdon and Fetcher 1984b; Oberbauer et al. 1988). PPFD in the understory is usually less than 10 μmol m^{-2} s^{-1} but is much higher in sunflecks (Chazdon and Fetcher 1984a, 1984b; Chazdon 1986b). A substantial fraction of total daily PPFD (10%–78%) comes during sunflecks (Chazdon 1986b).

The light regime in canopy gaps contrasts greatly with that of the understory (see also chap. 9). Compared to the understory, gaps are more heterogeneous in PPFD. Gaps vary greatly in size, from those caused by single branch falls to those caused by multiple tree falls. Total daily PPFD increases

with gap size as expected (table 10.1). Barton et al. (1989) found that total daily PPFD in the center of gaps increases linearly with gap area, with the slope of this relationship increasing with brightness of the day. Total daily PPFD also varies with location within gaps (Chazdon 1986b; Barton et al. 1989; Denslow et al. 1990). The edges of gaps receive less light than the center so that the edges of large gaps have total daily PPFD similar to the centers of small gaps (table 10.1).

In clearings PPFD is determined largely by the weather. On bright days with few clouds total daily PPFD may exceed 30 mol m^{-2} d^{-1} as compared to 10–20 mol m^{-2} d^{-1} for cloudy or overcast days (Chazdon and Fetcher 1984b). Sunny days at La Selva typically begin clear, but scattered clouds appear by midmorning and persist until nightfall. Clearings generally receive 20–30 mol m^{-2} on these days. A pyroheliograph was used to measure solar radiation in a clearing at La Selva from February 1983 to October 1984. Four patterns were recognized for which typical traces are shown in figure 10.1. Designating the dry season as January–April and the wet season as May–December, little difference is apparent in the frequency of overcast and partially cloudy days between seasons (fig. 10.2). The dry season had more sunny days, however, whereas the wet season had more "sun-cloud" days (i.e., days that began sunny but became dark in the afternoon, fig. 10.1D). This pattern was frequently associated with rain.

Little information is available about the vertical distribution of PPFD at La Selva. In the understory, light did not vary

Table 10.1 Photosynthetic photon flux density (PPFD), temperature at noon, vapor pressure deficit (VPD) at noon, and relative humidity (RH) at noon for several microhabitats at La Selva

	Average PPFD (mol m^{-2} s^{-1})	Total Daily PPFD (mol m^{-2} d^{-1})	Noon Temperature (°C)	Noon VPD (kPa)	Noon RH (%)
Understory					
<1.5 m	7.9, 5.9[a]	0.26, 0.34[a]	24.7, 23.5[b]	0.22, 0.28[b]	93, 90[b]
		0.26–0.33[c]			
>1.5 m		0.51[c]			
Crowns of Saplings					
Dipteryx panamensis		0.42[d]			
Lecythis ampla		0.31[d]			
Gap					
150 m^2		2.86[c]			
200 m^2	59.3[f]	2.56[e]			
		3.27[c]			
400 m^2	150, 145[a]	6.5, 6.3[a]	27.3, 25.2[b]	0.49, 0.46[b]	86, 86[b]
		5.83[c]			
Edge		1.08–1.46[c]			
		3.13[f]			
Clearing	530, 643[a]	22.9, 27.8[a]	31.0[g]	1.25[g]	72[g]
		30[d]			
		30.8[f]			
Canopy			27.0, 25.3[b]	0.92, 0.55[b]	74, 83[b]

[a]Chazdon and Fetcher 1984b; median in wet season, median in dry season.
[b]Fetcher et al. 1985; mean for April–October, mean for November–May.
[c]Chazdon 1986b.
[d]Oberbauer et al. 1988.
[e]Chazdon and Fetcher 1984b; median in dry season.
[f]Barton et al. 1989.
[g]Fetcher et al. 1985; mean for April–October.

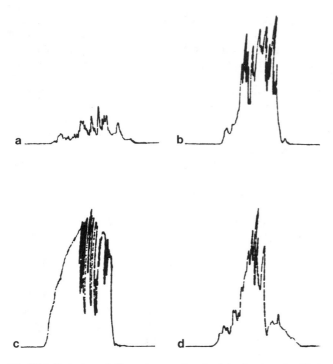

Fig. 10.1. Typical pyroheliograph traces from La Selva for days classified as overcast (*A*), partially cloudy (*B*), sunny (*C*), and sun-cloud (*D*).

between 0.7 m and 1.4 m above ground level (Chazdon 1986b). PPFD was significantly greater above 1.5 m than below that height (table 10.1). Oberbauer and Strain (1986) measured PPFD incident on leaves of *Pentaclethra macroloba* at three different levels: canopy, midcanopy, and understory. PPFD in the canopy reached 2,000 μmol m^{-2} s^{-1}, but 71% of the observations were less than 400 μmol m^{-2} s^{-1}. This difference occurred because the sensor was not above the canopy but within it and because of the high frequency of cloudy conditions. In the midcanopy PPFD was considerably less with 81% of the observations less than 100 μmol m^{-2} s^{-1}. In the understory 98% of the observations were less than 100 μmol m^{-2} s^{-1}.

Spectral composition of light can have profound effects on seed germination (Vázquez-Yánes and Orozco Segovia 1984) and plant growth (Smith 1982). Plant canopies alter spectral composition by selectively absorbing photosynthetically active radiation (400–700 nm) while allowing radiation in the near infrared (700–3,000 nm) to pass through. Thus, light under a plant canopy is enriched in wavelengths greater than 700 nm. Lee (1987) measured the spectral composition of light in clearings, gaps, sunflecks, and understory at La Selva and Barro Colorado Island, Panama (BCI). He found that the quantum ratio (R:FR) of red (658–662 nm) to far red (728–732 nm) radiation decreased as follows: clearing (mean R:FR = 1.22), gaps (0.90), sunflecks in the understory (0.87), and diffuse light understory (0.40). Gaps and sunflecks were more variable than clearing and understory in R:FR. Energy received between 400 and 700 nm as a fraction of total incoming shortwave radiation (300–1100 nm) declined in the same order: clearing (0.52), gap (0.40), sunflecks (0.36), and understory (0.17). Changes in spectral composition were strongly correlated with percent full sun PPFD received and

Lee (1987) was able to predict R:FR to within 0.2 units using the amount of instantaneous PPFD received at a site. He also found that the relationship between R:FR and PPFD differed between La Selva and BCI, possibly because of different atmospheric conditions (Lee 1987).

Spatial Variation in Temperature and Humidity

Fetcher et al. (1985) used hygrothermographs to monitor temperature and relative humidity at about two-week intervals for more than one year in a clearing, a 400-m^2 tree-fall gap, understory near the gap, and a platform located 30 m high in the canopy of a large tree. Temperature and relative humidity data were used to calculate vapor pressure deficit (VPD). VPD is the difference between actual pressure of water vapor in the atmosphere and saturated vapor pressure and is a measure of the gradient that drives transpiration and evaporation.

Daily fluctuations of temperature were greatest in the clearing where daytime temperatures exceeded 35.5°C during April–October (Fetcher et al. 1985). Nighttime temperatures at all microsites during this period averaged about 22°C, which is lower than for many tropical sites. In contrast to the clearing, understory temperatures rarely exceeded 27.5°C. The gap and canopy had similar mean daytime temperatures (table 10.1). The period from November to May was cooler than that from April to October; nighttime temperatures in December and January occasionally dropped below 17.5°C (Fetcher et al. 1985). Patterns of VPD differed in one important respect from patterns of temperature. VPDs in the canopy were considerably higher than in the gap because of greater air movement in the canopy (table 10.1). VPDs were lowest in the understory and highest in the clearing (table 10.1).

The canopy provides a different microenvironment from that in any of the ground level habitats, not only in light but also in temperature and humidity. It is cooler than clearings and less humid than gaps (Fetcher et al. 1985). Therefore, leaves in the canopy can be expected to have different energy budgets than leaves near the forest floor that lose less energy to convection and evaporation. Thus, plants such as trees and

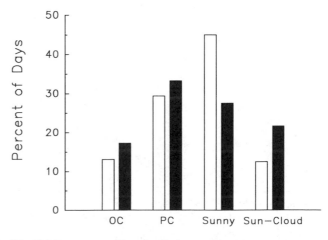

Fig. 10.2. Percentages of days classified as overcast (OC), partially cloudy (PC), sunny, and sun-cloud at La Selva during the dry (January–April, unshaded bars, N = 160) and wet seasons (May–December, shaded bars, N = 387).

lianas must adjust to drier environments over the course of their development.

PLANT RESPONSES TO LIGHT

Leaf Morphology

In general, leaf morphology changes with photosynthetic parameters as plants acclimate to high or low light. Much of the change in light-saturated photosynthetic rates that accompanies acclimation has been attributed to changes in the ratio of surface area of mesophyll to leaf area (Nobel et al. 1975) or specific leaf mass (mass per unit leaf area) (Chabot and Chabot 1977; Patterson et al. 1978). Therefore, plasticity in leaf morphology can be a useful index of a plant's capacity to adjust photosynthesis to differing light environments.

Although leaves of all species that have been studied at La Selva are thinnest in shade, the pattern of adjustment of leaf thickness differs between shade-tolerant and early successional species (table 10.2). As seedlings, shade-tolerant species appear to be less plastic with respect to leaf thickness than early successional species. Leaves of seedlings of shade-tolerant species grown in full sun were generally thinner than leaves of adults in the forest canopy (table 10.2). Apparently, these species (which rarely encounter full sun as juveniles) do not acquire the capability to make thicker leaves until later in life (Oberbauer et al. 1983). In contrast, leaves of early successional species that were grown in full sun were thicker than canopy leaves of adults of the same species, indicating that seedlings of these species have the capacity to respond quickly to high light conditions. Leaf thickness may be influenced by other factors (e.g., water relations, leaf energy budgets, mineral nutrition; Grubb 1977a). These may become more important as a tree matures, as suggested by the fact that adult leaves of early successional species were thinner than the leaves of seedlings grown in full sun (table 10.2).

Increases in the number and size of stomata may be necessary to achieve increased rates of gas exchange in high light. Stomatal densities were generally higher for seedlings grown in high light environments (e.g., Oberbauer and Strain 1985, 1986). For example, stomatal densities on the lower (abaxial) surface of *Heliocarpus appendiculatus* leaves were 164 per mm^2, 255 per mm^2, and 383 per mm^2 for plants grown in full shade, partial shade, and full sun, respectively; stomatal dimensions were unaffected (Fetcher et al. 1983). Leaf gas exchange capacity in well-lit environments may also be increased by facultative development of stomata on both leaf surfaces. Mean stomatal densities on the upper (adaxial) leaf surface were 28 per mm^2 and 55 per mm^2 in seedlings of *H. appendiculatus* grown in partial shade and full sun, respectively, but no adaxial stomata were developed in full shade. *Hampea appendiculata* also has facultative stomatal development on the adaxial leaf surface (N. Fetcher unpublished data).

Photosynthetic Responses to Light

Photosynthetic studies at La Selva have concentrated on acclimatization responses of seedlings of canopy trees and understory palms to differing light regimes. The major questions have been (1) Do species with apparent differences in toler-

Table 10.2 Leaf thickness (mm) of plants of three successional categories

| Species | Seedlings | | | Adults |
	FSH	PSH	FSU	CAN
Shade-tolerant canopy				
Carapa guianensis	0.139 a	0.147 a	0.215 b	0.266
Pentaclethra macroloba	0.076 a	0.118 b	0.161 c	0.149
Sloanea latifolia	0.137 a	0.156 b	0.182 c	0.237
Lecythis ampla	0.089 a	0.102 a	0.124 b	0.170
Gnetum leyboldii	0.148 a	0.187 b	0.217 c	0.306
Shade-tolerant gap				
Bursera simaruba	0.105 a	0.132 b	0.183 c	0.167
Cordia alliodora	0.112 a	0.146 b	0.181 c	0.166
Dipteryx panamensis	0.129 a	0.153 a	0.155 b	0.180
Goethalsia meiantha	0.083 a	0.102 b	0.125 c	0.140
Virola koschnyi	0.127 a	0.136 a	0.151 b	0.205
Early successional				
Hampea appendiculata	0.132 a	0.161 b	0.206 c	0.193
Heliocarpus appendiculatus	0.085 a	0.130 b	0.165 c	0.131
Ochroma lagopus	0.092 a	0.140 b	0.212 c	0.150
Cecropia obtusifolia				0.143

Notes: Leaves were collected from the canopies of trees in the field (CAN) and from seedlings of the same species grown in containers under three light regimes: full shade (FSH), partial shade (PSH), and full sun (FSU). Mean leaf thicknesses with the same letter are not significantly different ($p > 0.05$) by analysis of variance and Duncan's multiple range test; leaves taken from tree canopies were not included in the statistical test.

ance to shade also differ in their capacities to adjust to gradients of light? (2) Does the capacity for acclimatization contribute to increased carbon gain for plants growing in different environments? Considerable progress has been made on the first question, whereas only preliminary results are available for the second. Most photosynthetic work has been conducted with plants grown under shadehouse conditions (Oberbauer and Strain 1985; Fetcher et al. 1987) or with plants grown in growth chambers and temperature-controlled greenhouses (Oberbauer and Strain 1984). Although several studies are underway, few data have been published for photosynthesis under field conditions except for *Pentaclethra macroloba* and shrubs of the genera *Miconia* and *Piper*. Plants of several growth-forms remain to be studied (e.g., lianas, epiphytes, adult trees).

Characteristics of gas exchange have been measured for seedlings placed in full shade (2% of full sun), partial shade (20% of full sun), and full sun (Oberbauer and Strain 1985; Chazdon 1986b; Fetcher et al. 1987). Photosynthetic response to PPFD was measured for the most recently expanded leaf of two to eight individuals per species and treatment. Photosynthetic rates in full shade were very low because of low light intensity and low rates of light-saturated net photosynthesis. Thus, when an understory leaf receives high PPFD during a sunfleck, its photosynthetic rate will remain low even if it can respond instantaneously to increased light. The dynamics of leaf response to sudden changes in light intensity may impose further limitations on carbon dioxide uptake, depending on the degree of leaf activation and the length of the sunfleck

(Chazdon and Pearcy 1986a, 1986b). For seedlings of canopy trees grown in full shade, mean light-saturated photosynthetic rates range from 5.39 μmol CO_2 m^{-2} s^{-1} (1 μmol CO_2 m^{-2} s^{-1} = 0.273 μg C m^{-2} s^{-1}) for *Pentaclethra macroloba* to 2.17 μmol CO_2 m^{-2} s^{-1} for *Goethalsia meiantha* (Oberbauer and Strain 1985; Fetcher et al. 1987; N. Fetcher unpublished data). For the palms *Asterogyne martiana, Geonoma cuneata,* and *Genonoma congesta* mean light-saturated photosynthetic rates were 3.26, 3.79, and 3.08 μmol CO_2 m^{-2} s^{-1}, respectively (Chazdon 1986b). Photosynthesis was saturated at 200–400 μmol m^{-2} s^{-1} (PPFD) in both tree seedlings and palms.

Many gap and early successional species have low light-saturated photosynthetic rates when grown in full shade. Seedlings of four such tree species, *Ochroma lagopus, Cecropia obtusifolia, Hampea appendiculata,* and *Goethalsia meiantha,* had light-saturated photosynthetic rates at or below the median of 2.97 μmol m^{-2} s^{-1} for fifteen tree species studied by Fetcher et al. (1987; unpublished data). When these low rates are combined with the high respiration rates characteristic of these species (Fetcher et al. 1987), the resulting leaf carbon balance would be expected to be close to zero in understory conditions. Simulated leaf carbon budgets under understory light regimes, however, have shown no marked difference between early and late successional species (N. Fetcher unpublished data). These results may be explained by the increase in apparent quantum yield at low light for seedlings of early successional species when grown in full shade compared to quantum yields of the same species when grown in partial shade or full sun (Fetcher et al. 1987). Apparent quantum yield is the yield of photosynthesis in moles of carbon dioxide per mole of photons incident on the leaf.

Photosynthetic rates differed more markedly between seedlings of early and late successional species when these were grown at higher levels of light (Oberbauer and Strain 1984; Fetcher et al. 1987, unpublished data). Mean net photosynthetic rates of seedlings grown in partial shade ranged from 7.76 μmol CO_2 m^{-2} s^{-1} for *Heliocarpus appendiculatus* to 3.58 μmol CO_2 m^{-2} s^{-1} for *Pouteria standleyi.* When grown in full sun, mean net photosynthetic rates ranged from 10.71 μmol CO_2 m^{-2} s^{-1} for *Terminalia oblonga* to 2.90 μmol CO_2 m^{-2} s^{-1} for *Lecythis ampla.* For all species seedlings grown in full sun had rates equaling or exceeding those of seedlings grown in partial shade (Fetcher et al. 1987, unpublished data). Early successional species generally had the highest rates under both partial shade and full sun, suggesting that they could achieve higher rates of carbon acquisition and, consequently, more rapid growth in well-lit environments than late successional species. The difference in photosynthetic rate in high light accounts for part of the striking divergence in growth rates between early and late successional species in clearings and large gaps, but differences in construction costs of support tissue are also implicated.

Thus, differences in photosynthetic capacity between early and late successional species are more pronounced under high light conditions than under low light. Despite the apparent equality of photosynthetic rates in low light, late successional species may be better equipped to survive understory conditions because of differences in leaf structure. Leaves of many early successional species become thin and fragile when grown under full shade, whereas those of late successional species are often thicker and more robust (table 10.2).

Photoinhibition caused by overexposure to saturating PPFD (reviewed in Powles 1984) has been proposed as a mechanism for reduced photosynthetic capacity of tropical forest plants (Langenheim et al. 1984). It can cause reduced photosynthetic rates at high PPFD or reduced quantum yields of photosynthesis at low PPFD. Castro et al. (1991) found evidence for chronic photoinhibition in seedlings of *Dipteryx panamensis* grown in full sun but did not find strong indications of photoinhibition for four other species, including *Ochroma lagopus* and *Inga edulis.* Although interspecific variation in susceptibility to photoinhibition apparently exists, it may be less significant than other factors in determining a tree's response to high light environments. Fernández and Fetcher (1991) found little evidence of photoinhibition for saplings of eight tree species growing in forestry plantations at La Selva. Nevertheless, further research on the interaction of nutrient supply, leaf temperature, and water availability with photoinhibition is needed before it can be discarded as a mechanism limiting photosynthesis of tropical trees in clearings or canopies.

Field measurements of net photosynthetic rates for plants at La Selva are few although these are important to test the validity of measurements on plants grown under controlled conditions. Using a closed gas exchange system, Oberbauer and Strain (1986) measured photosynthetic rates for leaves of *Pentaclethra macroloba* in the understory, midcanopy, and canopy. Mean light-saturated photosynthetic rate for canopy leaves was 6.5 μmol CO_2 m^{-2} s^{-1} compared to 6.7 μmol CO_2 m^{-2} s^{-1} for seedling leaves developed in full sun (Oberbauer and Strain 1985). Mean light-saturated photosynthetic rate for potted seedlings grown in the understory was 5.0 μmol CO_2 m^{-2} s^{-1} compared to 5.4 μmol CO_2 m^{-2} s^{-1} for seedlings grown in full shade. For this species, at least, acclimatization potential is similar for leaves developed under similar light regimes in field and shadehouse. *Pentaclethra macroloba,* however, shows little change in light-saturated net photosynthetic rate as a function of light levels compared to other species (Fetcher et al. 1987; Oberbauer and Strain 1986) and further comparisons of field- and shadehouse-grown plants are needed.

Field measurements of photosynthesis have been used to determine the degree to which light availability limits growth of different species in gaps and in the understory. For shrubs of the genera *Piper* and *Miconia* growing in the center of a gap, light-saturated rates of photosynthesis were higher in individuals of species that typically occur in clearings as compared to understory species (Denslow et al. 1990). Relative stem growth rates were significantly greater in gap centers than in gap edges or understory microsites. Because fertilization did not significantly influence growth rate under any field conditions (see chap. 9), light is presumed to be the critical limiting resource for these species (Denslow et al. 1990). McDonald et al. (1991) measured photosynthetic rates of two shrubs and two trees in the genus *Miconia* and found that rates had doubled four months after a simulated gap was opened. No significant difference in photosynthetic rates was found, however, among species.

Plants of the tropical forest understory have been used as models in studies of adaptation to extreme shade because of their ability to carry out their entire life cycle under low light conditions (Björkman et al. 1972; Boardman 1977). At La Selva three types of understory plants have been investigated: palms (Chazdon 1985, 1986b), a cycad, *Zamia skinneri,* and a dicot herb, *Triolena hirsuta.* All three had low light-saturated rates when grown in full artificial shade, with palms having the highest rates (3–4 μmol CO_2 m^{-2} s^{-1}). Leaves of *Z. skinneri* had lower rates (2.4–3 μmol CO_2 m^{-2} s^{-1}) and *T. hirsuta* had the lowest rate (2 μmol CO_2 m^{-2} s^{-1}).

Leaves of many understory herbs in the tropics are unusually colored, with blue iridescence, velvety surface sheens, variegation, and red or purple undersides found in many species. These colorations are thought to be adaptations to extreme shade (Lee 1986; Lee et al. 1990). Red undersides, produced by an abaxial anthocyanin layer, are particularly common at La Selva where more than a dozen species possess the trait. Lee et al. (1979) hypothesized that the abaxial anthocyanin layer reflects unabsorbed light back into the leaf mesophyll for further absorption, thereby increasing photosynthetic efficiency at low light. *Triolena hirsuta* is of particular interest because the population is polymorphic for this trait. When grown under different levels of neutral shade, plants with the anthocyanin layer had higher chlorophyll content and light-saturated photosynthetic rates (fig. 10.3), but photosynthetic efficiencies at low light were similar to plants without anthocyanin (Oberbauer 1987). These results suggest that the color variants of *T. hirsuta* are physiologically different but are not consistent with the backscattering hypothesis. The prevalence of an abaxial anthocyanin layer at La Selva demands further work.

At this stage we can only partially resolve the issue of whether species with similar ecological roles have similar photosynthetic responses. For ecological groups that are also phylogenetic groups (e.g., understory palms or understory shrubs from genera such as *Piper* or *Miconia*) the answer seems to be yes. For ecological groups that assemble phylogenetically divergent taxa (e.g., shade-tolerant trees from different families) the answer seems to be no. For example, although both *Minquartia guianensis* (Olacaceae) and *Carapa guianensis* (Meliaceae) are considered shade tolerant, the latter exhibits considerable plasticity in light-saturated photosynthetic rate, whereas the former does not (Fetcher et al. 1987). This work, however, used seedlings and saplings; the photosynthetic characteristics of adults may converge. It should also be noted that although species may differ considerably in their abilities to respond to high-light environments, this difference is rarely expressed under light conditions prevailing on the forest floor at La Selva. In general, seedlings and saplings growing in the primary forest have leaf morphology and photosynthetic rates characteristic of plants growing in full shade (S. F. Oberbauer, unpublished data).

Leaf Carbon Budgets

A plant's carbon budget is the difference between carbon gained by photosynthesis and carbon lost through respiration, herbivory, and shedding of plant parts (Mooney 1972). Although plants can tolerate negative carbon budgets for short

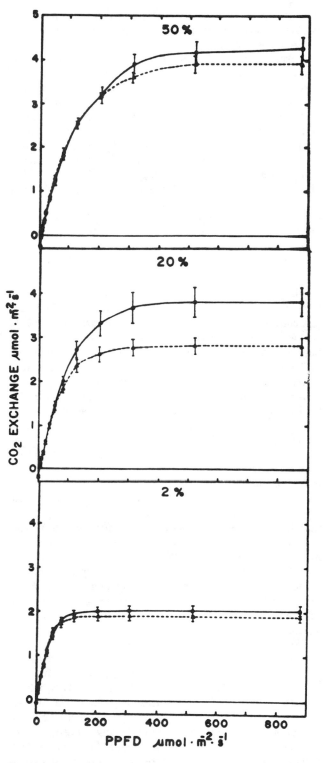

Fig. 10.3. Photosynthetic light responses of variants of *Triolena hirsuta* with (solid line) and without (dashed line) anthocyanin layer on the leaf underside. Plants were grown at 50%, 20%, and 2% of full sun. Values are mean (N = 6) ± one standard error.

periods by drawing on stored reserves, a positive carbon budget is necessary in the long term for survival, growth, and reproduction. The concept of carbon budget is useful to link short-term physiological measurements with these longer-term phenomena, which are more closely tied to plant fitness. It is nearly impossible to quantify all components of a plant's carbon budget because of methodological difficulties, particularly losses of carbon by roots. As a first step toward addressing whole plant carbon budget, however, one can examine individual leaves. Leaf carbon budget is daytime photosynthesis minus nighttime respiration; it provides an upper bound for estimated plant carbon gain. That is, if leaf carbon budget is negative, then whole plant carbon budget will surely be negative because of additional respiratory losses by stems and roots. If leaf carbon budget is positive, then whole plant carbon budget may or may not be positive, depending on respiratory and other losses.

Daily leaf carbon budgets have been calculated for palm seedlings grown in full shade (Chazdon 1986b) and for tree seedlings grown in full shade, partial shade, and full sun (Fetcher 1986; N. Fetcher, unpublished data). Both approaches assumed instantaneous photosynthetic response to changes in PPFD. This assumption may overestimate carbon budget in the understory because of time lags associated with activation of the photosynthetic apparatus after long periods in deep shade (Pearcy et al. 1985; Chazdon and Pearcy 1986a; Pearcy 1987). Pearcy (1983) found that two-thirds of the sunflecks in a rain forest in Hawaii were of less than 30 s in duration. If La Selva is similar, many sunflecks may be gone before the leaf becomes fully activated photosynthetically. A series of sunflecks, however, can induce a full photosynthetic response similar to that seen after exposure to continuous light (Chazdon and Pearcy 1986a). Carbon gain during brief sunflecks (5–20 s) may exceed that estimated from steady state rates (Chazdon and Pearcy 1986b). Because of complexities introduced by the dynamics of photosynthetic induction, carbon budgets calculated to date will require additional refinement before they can be taken as representative of understory conditions. Still, assuming that the dynamics of photosynthetic response are not very different among species, carbon budgets calculated using steady-state values should be useful for comparing species within the same habitat or the same species across habitats.

For leaves of the understory palms *Asterogyne martiana*, *Geonoma cuneata,* and *Geonoma congesta,* the light compensation point for positive daily carbon budget was 0.20 mol m^{-2} d^{-1} (Chazdon 1986b). Carbon budgets were highly correlated with total daily PPFD under diffuse light conditions but not so correlated with high levels of direct radiation because of saturation at low light levels. In extremely shaded environments (<10 μmol m^{-2} s^{-1} at midday) 40%–60% of the daily carbon budget is contributed during sunflecks (Chazdon 1986b).

Carbon budgets were calculated for leaves of tree seedlings of fifteen species ranging from pioneers to species of primary forest. Photosynthetic response to PPFD was measured for seedlings acclimatized to full shade, partial shade, or full sun (Fetcher et al. 1987). Carbon budgets were simulated for individual plants exposed to PPFD regimes measured in understory, gap center, or clearing microenvironments. Thus,

there were four sources of variation in leaf carbon budgets of tree seedlings: species, acclimatization, simulated PPFD regime, and variation among individuals. Of these sources the most important factor was microenvironment; differences among species were small compared to differences from the simulated PPFD regime. Differences in acclimatization produced differences in carbon budget that were intermediate between microenvironment and species effects (Fetcher 1986; N. Fetcher, unpublished data). Leaves developed in full shade had the highest carbon budgets in the simulated gap and simulated understory environments, whereas leaves developed in full sun generally had the highest carbon budgets in the simulated clearing environment (Fetcher 1986).

Analysis of leaf carbon gain is an initial step in developing an integrated picture of plant performance in different microhabitats. The work on tree seedlings has produced two major findings. First, microenvironmental variation clearly makes a stronger contribution to variation in carbon gain than does acclimatization or species differences in light response curves. Second, leaves developed in full shade have the greatest potential for carbon gain in most microsites at La Selva; only in the middle of large clearings are other types of leaves favored. Future analyses of carbon gain should incorporate the dynamics of photosynthetic response to light as well as response to other environmental factors.

WATER RELATIONS

Although tropical moist forests are usually associated with plentiful rainfall, moderate drought occurs at La Selva with some frequency. Rainfall during the driest month (February) averages 152 mm, but a calendar month with less than 50 mm total rainfall occurred during the dry season in about one-third of the years from 1961 to 1982 (see chap. 3). In 1980 rainfall in one eight-week period (February 15–April 14) totaled 50 mm. In 1983, an El Niño year, only 16.1 mm fell during a seven-week dry season period. For plants that normally encounter soils at or above field capacity during much of the year, even short periods without precipitation can be significant. Drought can be even more important for epiphytes that have little stored water available to the roots. Consequently, mechanisms for conserving or storing water, including Crassulacean acid metabolism (CAM) and CAM idling (Medina 1983; Ting et al. 1985), succulence, and sclerophyllous leaves, are most conspicuous in epiphytes.

The structure and composition of the forest at La Selva reflects the interaction between a moderate dry season (Medina 1983) and periodic drought. Seventy-five percent of the overstory species are evergreen (Frankie et al. 1974) and the dominant tree at La Selva, *Pentaclethra macroloba*, is confined to regions without an effective dry season (Hartshorn 1983a). Several of the emergent trees (e.g., *Lecythis ampla*, *Dipteryx panamensis, Pithecellobium elegans, Ceiba pentandra,* and *Hymenolobium mesoamericanum*) and a number of canopy species (e.g., *Bursera simaruba, Pterocarpus hayesii, Cordia alliodora*) are conspicuously leafless for up to several months during the dry season.

Water balance has been implicated in the control of phenology of leaves and flowers in tropical forests such as Costa Rican dry forests (Daubenmire 1972; Opler et al. 1976;

Borchert 1980; Reich and Borchert 1982), dry forest in Ghana (D. Lieberman 1982), dry forest in Mexico (Bullock and Solis-Magallanes 1990), and at Barro Colorado Island, Panama (Augspurger 1979; Rundel and Becker 1987; Wright and Cornejo 1990a), but the importance of water balance in controlling phenology has yet to be demonstrated clearly for trees at La Selva (Breitsprecher and Bethel 1990). On the basis of an irrigation experiment Wright and Cornejo (1990a) suggest that atmospheric humidity rather than soil moisture may regulate leaf fall at Barro Colorado Island. At La Selva Hazlett (1987) found that *Pentaclethra macroloba* had greatest stem expansion during the dry season, but two other species showed reductions in stem growth during the dry season. Extended dry periods at La Selva are usually accompanied by leaf drop from canopy trees. Flowering of many species commences following the start of the wet season. Phenological patterns of some deciduous species, however, do not follow the dry season closely. For example, after flowering and fruiting during the dry season, trees of *Cordia alliodora* may remain leafless for weeks or even months after the start of the wet season.

To investigate responses to drought of trees of different leaf types Oberbauer (1988) withheld water from seedlings of eight species of canopy trees ranging from shade-tolerant evergreens to evergreen pioneer species. Relative growth rates of *Pentaclethra macroloba* and *Dipteryx panamensis* were least affected by drought. The deciduous species, *Cedrela odorata* and *Bursera simaruba,* were affected most severely (table 10.3). Drought caused severe reduction in stomatal conductance and increased leaf abscission in these species (S. F. Oberbauer, unpublished data). Despite highly reduced growth, however, individuals of these deciduous species survived far longer than the other species under severe drought. Interestingly, seedlings of *Cordia alliodora* showed no tendency for deciduousness or stomatal closure under drought although adults are deciduous.

During extended dry periods plants of the forest floor show effects of water stress quickly. Plants of *Piper holdridgianum* are among the first to wilt in the understory and do so with a leaf water potential of −0.7 MPa (S. F. Oberbauer, unpublished data). (Water potential is the free energy of water in plants or soil relative to that of pure free water at the same temperature, which is defined as 0. More negative values mean that water is less available to the plant.) During extreme dry periods terrestrial cyclanths are no longer able to stay erect. This degree of water stress is by no means an annual occurrence but is likely to be important, particularly for seedlings that have recently germinated or that have shallow root systems because some species produce deep roots slowly following germination (Oberbauer 1990). Seedlings of *Hampea appendiculata* had predawn leaf water potentials as low as −1.2 MPa during the dry season of 1984 (S. F. Oberbauer, unpublished data). Oberbauer et al. (1987) found that seedlings and saplings of *Pentaclethra macroloba* had reduced stomatal conductances (0.7 and 1.2 mm s^{-1}, respectively) associated with lower predawn water potentials (−0.68 and −0.33 MPa, respectively) during the dry season. Pressure-volume analysis indicated that turgor was maintained but, undoubtedly, at the cost of reduced photosynthesis.

Dense vegetation protects plants growing in the understory

Table 10.3 Mean percent change in relative growth rate of controls for drought-stressed seedlings

Species	Level of Drought Stress	
	−0.7 MPa (Percent control)	−1.7 MPa (Percent control)
Pentaclethra macroloba	−3.9	−25.6
Dipteryx panamensis	−0.6	6.9
Pterocarpus hayesii	6.6	−21.1
Bursera simaruba	−14.9	−105.1
Cedrela mexicana	−10.7	−89.8
Cordia alliodora	−23.6	−29.1
Heliocarpus appendiculatus	−10.6	−47.3
Ochroma lagopus	−17.4	−77.5

Note: Mean percent change (N = 10) in relative growth rate of controls for seedlings of eight tree species drought stressed to −0.7 and −1.7 MPa predawn water potential.

and in gaps from the fairly severe irradiance and thermal climate at La Selva (Fetcher et al. 1985). As a result, evaporative demand varies widely even in the absence of pronounced drought. For example, vapor pressure deficits in the canopy can be as high as 2.0 kPa (Fetcher et al. 1985), and canopy air movement is one hundred times greater than in the understory (S. F. Oberbauer, unpublished data). PPFD in clearings and in the canopy can be as high as 2,400 μmol m^{-2} s^{-1}, and air temperatures in clearings can be even higher than those in the canopy. The understory, in contrast, presents little evaporative demand. Soil moisture varies less widely than evaporative demand. During low rainfall periods however, soil moisture under tall forest may be lower than in gaps (Vitousek and Denslow 1986). Together with the greater root density under tall forest than in gaps (Sanford 1989) this observation suggests that, during dry periods, canopy trees and other components of fully developed forest are using water more rapidly than the vegetation growing in gaps.

Available data indicate that leaf water potentials of canopy trees remain high even during the dry season. Both *Pentaclethra macroloba* (measured in 1982) and *Cordia alliodora* (measured in 1984) had midday and predawn leaf water potentials greater than −2.0 MPa and −0.4 MPa, respectively, during the dry season (Oberbauer et al. 1987, unpublished data). Leaf water potentials of canopy lianas are similar to those of canopy trees (S. F. Oberbauer, unpublished data), but water relations of lianas at La Selva have not been thoroughly examined. Investigation of liana water relations would undoubtedly prove very interesting; midday water potentials of two lianas, *Anomospermum* sp. and *Gnetum leyboldi,* growing in the same tree canopy were substantially different (−1.60 compared with −1.14 MPa, respectively; S. F. Oberbauer, unpublished data).

In clearings midday wilting of herbs and shrubs is frequent on sunny days as reported also for *Piper auritum* in Mexico (Chiariello et al. 1987). Midday stomatal closure occurred frequently for saplings of several tree species during the dry season at La Selva even though predawn water potentials were near −0.1 MPa and midday potentials exceeded −1.5 MPa (Oberbauer 1985). Similar midday stomatal closure was also reported for tree saplings at Barro Colorado Island, Panama

(Fetcher 1979). Midday wilting and stomatal closure in clearings and gaps are presumably linked to reduced photosynthetic rate although this has not been demonstrated for plants at La Selva.

During periods of average rainfall, leaf water potentials in the understory remain high. Midday leaf water potentials of woody species are maintained above -1.2 MPa and may be higher for herbaceous species (Oberbauer 1985; Oberbauer et al. 1987; S. F. Oberbauer, unpublished data).

Turgor potentials and mechanisms of turgor maintenance have not been intensely studied at La Selva. From the few data available it appears that understory plants and trees at La Selva lose turgor at water potentials higher than those reported for BCI (Robichaux et al. 1984). *Piper holdridgianum* loses turgor at -0.7 MPa, and turgor is lost in *Pentaclethra macroloba* at above -2.0 MPa (Oberbauer and Strain 1985; Oberbauer et al. 1987). In contrast, most trees examined at BCI lost turgor below -3.0 MPa, whereas most shrub and gap species lost turgor below -2.0 MPa (Robichaux et al. 1984; Rundel and Becker 1987). These differences presumably reflect adaptation to more extreme dry season conditions on BCI. Hawaiian species of *Dubautia* reach zero turgor between -0.9 and -1.6 MPa (Robichaux 1984). Only two species at La Selva have been tested for osmotic adjustment. Slight seasonal adjustment was found for *Simarouba amara* (Cordero 1988) and *Pentaclethra macroloba* (Oberbauer et al. 1987), but neither species exhibited osmotic adjustment across a light gradient.

Plants that become established in the understory and grow to the canopy must cross a steep gradient of evaporative demand (Givnish and Vermeij 1976; Chiariello 1984; Fetcher et al. 1985) that may require substantial adjustments in plant water relations. Oberbauer et al. (1987) examined leaf water relations in individuals of *P. macroloba* from understory to canopy and found little change in water potential with height other than the reduction in osmotic potential required to compensate for gravitational potential. Stomatal conductance, however, increased substantially from understory to canopy as stomatal densities doubled (Oberbauer and Strain 1986). Stomatal conductances were positively correlated with irradiance, absolute humidity deficit, and, in the understory, leaf water potential.

The difference in evaporative demand between different microenvironments at La Selva suggests that species that occupy such sites should differ in water relations (Bazzaz and Pickett 1980; Bazzaz 1984). Oberbauer (1985) compared a range of species from different stages of succession and found that early successional species had higher conductances than late successional species. No pattern of leaf water potential in relation to successional rank was apparent. A partial explanation for the lack of a clear pattern may be that spatial and temporal variation in water stress for plants rooted in the soil is small at a site that receives as much precipitation as La Selva. Much more variation in water stress is expected along the vertical gradient from forest floor to canopy. Interesting patterns of water use are likely for species of epiphytes and lianas that are found at different locations along this gradient.

MINERAL NUTRITION

Although soils are greatly diverse at La Selva, nitrogen availability generally is high (see chap. 4). Phosphorus availability is low except in recent alluvium (Vitousek and Denslow 1987). Patterns of soil type and species distribution present many questions, but plant nutrition has received relatively little attention to date.

In a study of variation in nutrition in *Pentaclethra macroloba* Oberbauer (1983) grew seedlings under three levels of irradiance in combination with added nitrogen (N), phosphorus (P), and N + P on old alluvial soil inoculated with *Rhizobium* and mycorrhizae. Growth of *Pentaclethra macroloba* in 20% of full sun and full sun was significantly improved with nitrogen addition mainly because of an increased production of leaf area. Photosynthesis and respiration were significantly increased by nitrogen supplementation, with photosynthesis affected more than respiration. Phosphorus had a positive effect on growth in full sun but a slightly negative effect in 20% full sun. A significant interaction between nitrogen and phosphorus was found only in full sun. Nitrogen significantly decreased nodulation in full sun and 20% full sun. No nodules were produced in any treatments at 2% full sun.

Other studies of plant nutrition at La Selva investigated whether nutrient changes after tree falls could influence the vegetation that fills gaps. As reviewed by Denslow and Hartshorn (chap. 9), experiments with potted and field-grown plants have provided little evidence for nutrient limitation of growth of either gap or understory shrub species. Of eight species tested only *Phytolacca rivinoides* responded significantly to fertilization. This species, the only nonmycorrhizal species studied, showed enhanced growth with phosphorous and N + P supplementation, which supports the idea that mycorrhizae are important for acquisition of phosphorus and that their absence results in reduced growth from phosphorus deficiency. *Phytolacca rivinoides* also appeared to be limited by potassium (K) in some La Selva soils (Denslow et al. 1987).

The importance of mycorrhizae and symbiotic nitrogen fixation has been the focus of considerable work at La Selva (Janos 1977, 1980b; Bentley and Carpenter 1980, 1984; chap. 4). Most of the woody plants at La Selva are mycorrhizal and, under experimental conditions, these associations generally enhance plant growth and survival (chap. 4; see also McHargue 1981). Almost nothing is known, however, about the relevance of mycorrhizae under field conditions. It is apparent that some species cannot grow without mycorrhizae and that species with light seeds that colonize disturbed habitats are least likely to be dependent on mycorrhizae (Janos 1980b).

Considering the relatively high availability of N in La Selva soils, a surprisingly large number of nitrogen-fixing legumes are found there (Gutschick 1981), including the canopy dominant, *Pentaclethra macroloba*. Of forty-five indigenous species of woody legumes at La Selva that were examined for nodules 82% were nodulated (three of six Caesalpinaceae, twenty-two of twenty-two Mimosaceae, and twelve of seventeen Fabaceae; McHargue 1981). McHargue (1987) tested *Pithecellobium catenatum* for the effect of nodulation on growth. Plants inoculated with *Rhizobium* had 16% greater shoot biomass and 24% smaller root to shoot ratios than uninfected plants. McHargue (1987) also found that five species of nodulated legumes showed greater growth and reduced nodulation with increasing levels of nitrogen.

Despite the large number of nodulated species, few legumes growing in the understory are nodulated, presumably because low light limits the supply of photosynthate to support nitrogen-fixing symbionts. Nodulation is usually increased by transfer to high light. In some cases, nodulation is also limited by phosphorus; phosphorus supplementation promoted nodulation in individual plants of normally nodulating species that had not otherwise formed nodules (L. A. McHargue pers. comm.). The presence of mycorrhizae, which should increase phosphorus, may also enhance nodulation (L. A. McHargue, unpublished data).

Nonsymbiotic nitrogen fixation is also important. Bentley and co-workers demonstrated that epiphyllous Cyanobacteria fix significant amounts of nitrogen at La Selva and that the fixed nitrogen is incorporated into leaves of the substrate plant at a rate potentially accounting for 10%–25% of leaf nitrogen (Bentley and Carpenter 1980, 1984).

GROWTH AND BIOMASS ALLOCATION

Carbon budgets, as mentioned, are useful to integrate physiological information and infer probable responses of plants to changes in their environment. Another useful method is to measure plant growth directly (McGraw and Wulff 1983). This approach has two advantages: first, plant growth integrates environmental effects on physiological processes such as photosynthesis and water relations that are measured on short time scales; second, many studies have shown that components of fitness such as survival, competitive ability, and reproductive output are strongly correlated with plant size (McGraw and Wulff 1983). Measurements of plant growth can confirm conclusions based on studies at other levels and, at the same time, permit extrapolation of the results of those studies to population and community levels.

Three types of growth studies have been pursued at La Selva. In one, naturally established plants were measured nondestructively in the field (see chap. 7). In the second, plant growth was analyzed in containers under more or less controlled conditions (Evans 1972). This technique permits estimation of relative growth rate (rate of accumulation of biomass per unit biomass), which can be partitioned into specific leaf mass (mass of leaves/unit leaf area), leaf area ratio (mass of leaves/unit mass of whole plant), and unit leaf rate (rate of accumulation of biomass/unit leaf area). In the third type of growth study plants were sown or transplanted into the field.

Growth Rates of Trees

We refer readers to chapters 7 and 8 for detailed discussion of the growth of trees at La Selva. Here we reiterate only the patterns that are most fundamental to plant physiological ecology.

Long-term growth data from the OTS permanent inventory plots collected by the Liebermans and colleagues reveal a range in median diameter growth rates of 0.35 mm yr^{-1} (*Anaxagorea crassipetala*) to 13.41 mm yr^{-1} (*Stryphnodendron excelsum*). The mean median growth for all species of 2.65 mm yr^{-1} was similar to rates estimated for similarly sized trees from BCI (Lang and Knight 1983), Ghana (Swaine and Hall 1986), and Sarawak (Primack et al. 1985). Four main patterns of tree growth were distinguished using data from the plots to simulate growth and longevity (see chap. 8): (1) understory species with slow maximum growth rates and short life spans; (2) subcanopy shade-tolerant trees with growth

rates similar to understory trees but significantly longer life spans; (3) canopy and subcanopy species that tolerate shade but respond opportunistically to increased light levels with relatively rapid maximum growth rates and long life spans; and (4) shade-intolerant canopy and subcanopy trees with fast maximum growth rates and short life spans. Variation in growth rates was higher among species in groups 2 and 3 than among those in groups 1 and 4; species in the latter groups have relatively short life spans and tend to occupy a narrow range of light conditions (D. Lieberman et al. 1985a).

Long-term studies of selected canopy tree species (D. B. Clark and Clark 1984; D. A. Clark and Clark 1987; chap 7) have demonstrated that saplings of all species studied survive and grow in the forest understory, contrary to initial expectations that individuals of at least four species would grow only in gaps. For saplings of all species the predominant direction of incident light was lateral rather than overhead. Differences were marked among larger individuals (10–20 cm dbh) in light regime (see fig. 7.6), suggesting that these species differ in shade tolerance as individuals reach larger size classes.

Effects of Light Regime on Growth of Seedlings and Allocation of Biomass

Observations of tree regeneration at La Selva stimulated a series of investigations on the effects of light environment on early seedling growth and biomass allocation (Fetcher et al. 1983; Oberbauer and Strain 1985; Oberbauer and Donnelly 1986; Fetcher et al. 1987). Seedlings germinated from locally collected seed were grown in shade houses or full sun in a clearing. These growth experiments were designed to examine responses to a sudden environmental shift, such as the abrupt increase in light availability following a tree or branch fall. Measurements of photosynthesis, leaf conductance, and leaf morphology accompanied the growth analyses (see previous sections).

Seedlings generally responded positively to increased light levels; in four of six species studied total biomass was lowest in plants that remained in full shade (2% full sun) for the duration of the experiment (Fetcher et al. 1987). In all species, however, growth was similar or inhibited in plants maintained in full sun for the duration of the experiment when compared to plants maintained in partial shade (20% full sun). When seedlings of *Dipteryx panamensis*, *Virola koschnyi*, and *Pentaclethra macroloba* were transferred from full shade to full sun, leaves became necrotic and soon abscised, causing a marked reduction in plant growth rate (Oberbauer and Strain 1985; Fetcher et al. 1987). Seedlings of *Cordia alliodora* and *Heliocarpus appendiculatus* also lost leaves during this transfer but rapidly recovered because of high rates of leaf production. Seedlings grown in partial shade did not suffer loss of leaves when transferred to full sun. Seedlings of *Pentaclethra macroloba* grown in full sun and partial shade did not differ in final biomass, and switching between these two treatments had no significant effect on biomass (Oberbauer and Strain 1985; Fetcher et al. 1987).

Seedlings of the early successional species *Heliocarpus appendiculatus* and *Cordia alliodora* responded quickly to changes in environmental conditions (Fetcher et al. 1987). In contrast, dry weights of seedlings of *D. panamensis*, *V. koschnyi*, and *P. macroloba* (species of mature forest) responded less quickly (Oberbauer and Strain 1985; Fetcher et al. 1987).

These species did not adjust completely to the new environment during the experiment (six to ten weeks). *Hampea appendiculata* also showed incomplete adjustment to the new environment during the experimental time frame. Although this species is typically found in small gaps and secondary forest, seedling growth was strongly inhibited when they were transferred to full sun early in development (Fetcher et al. 1987).

Patterns of biomass allocation of tree seedlings also changed with light regime during growth. Leaf area ratios of *Pentaclethra macroloba* seedlings were inversely related to light levels and were affected by the environment in which seedlings were grown before being transferred as well as by the new environment (Oberbauer and Strain 1985). Seedlings of *Dipteryx panamensis* and *Heliocarpus appendiculatus* also had decreased leaf area ratio with increased light, but only in *H. appendiculatus* was leaf area ratio affected by the previous environment as well (Fetcher et al. 1983). For these three species, root: shoot ratios generally were positively correlated with light during growth (Fetcher et al. 1983; Oberbauer and Strain 1985).

Analysis of the growth of seedlings of six tree species of different successional rank showed that, under full sun conditions, both relative growth rate and unit leaf rate were highest in early successional species (*Ochroma lagopus, Heliocarpus appendiculatus, Cordia alliodora*) compared to the shade-tolerant primary forest species *Pentaclethra macroloba* and *Brosimum alicastrum* (Oberbauer and Donnelly 1986). Seedlings of *Terminalia oblonga* had lower unit leaf rates and relative growth rates than expected based on the early successional rank of this species. Leaf area ratio and specific leaf area decreased along a successional gradient from early to late, whereas root: shoot ratio increased (Oberbauer and Donnelly 1986).

Huston (1982) planted seeds of *Hampea appendiculata, Dipteryx panamensis, Pentaclethra macroloba,* and *Carapa guianensis* in cleared pasture using three experimental treatments: monospecific compared with mixed stands, full sun compared with partial shade (30% full sun), and fertilized compared with unfertilized plots. In monospecific stands neither shading nor fertilization affected growth. In unfertilized mixed stands in full sun the early successional species *H. appendiculata* had the highest mean relative growth rate, followed by *D. panamensis, P. macroloba,* and *C. guianensis,* which are larger, slower-growing species of primary forest canopy. The same ranking of relative growth rate was obtained in mixed stands in partial shade where rates were initially higher than in full sun but later declined to between 90% and 34% of rates in full sun. Fertilization produced higher growth rates for *H. appendiculata* and lower growth rates for the other species as the result of suppression by this species. Thus, changes in nutrient levels can have unexpected effects on plant growth depending on the relative capacity for growth of the species experiencing the change.

Acclimatization to Differing Light Regimes

Studies of growth at La Selva suggest that early successional species adjust more rapidly to changes in light regime, primarily because of higher rates of leaf production (Fetcher et al. 1987). This finding needs further study, however. Popma

and Bongers (1991) found little difference between seedlings of three Mexican species when they were switched between light regimes.

Mean relative growth rates of all species studied at La Selva were similar in partial shade and in full sun (N. Fetcher, unpublished data). Regeneration requirements of common tree species at La Selva may depend upon physiological tolerance of sun as well as shade. These results are consistent with studies of growth of tree seedlings from other rain forest areas (Coombe 1960; Coombe and Hadfield 1962; Lebron 1979; Fasehun and Audu 1980; Whitmore and Gong 1983; Augspurger 1984a).

Most species are capable of some degree of plastic adjustment to differing light regimes. The degree of plasticity appears to depend on the trait under consideration, the species, and the relative magnitude of the increase or decrease in light level. In a study of seedlings of five species, N. Fetcher (unpublished data) found that unit leaf rate and relative growth rate were more plastic in the pioneer species *Ochroma lagopus* than in three shade-tolerant species. Similarly, Denslow et al. (1990) found that relative growth rates of two light-demanding species of *Miconia* were more plastic than the shade-tolerant *Miconia gracilis*.

On the other hand, measures of carbon allocation such as leaf weight ratio appear to be more plastic in shade-tolerant species. King (1991) studied saplings of nine species and found that those associated with gaps did not decrease the fraction of biomass allocated to leaves in response to increasing shade, whereas more shade-tolerant species were able to do so. As a result, the gap-associated species experienced greater increases in relative growth rates with increasing light availability. N. Fetcher (unpublished data) found that leaf weight ratio appeared to be equally plastic for seedlings of early and late successional species.

The relative magnitude of the increase or decrease in light can also influence the type of plastic response. Fetcher et al. (1983, 1987) observed both physiological and morphological changes for plants transferred from full shade (2% of full sun) to partial shade (20% of full sun). Popma and Bongers (1988) found increases in unit leaf rate and relative growth rate after transferring seedlings grown in forest understory (0.9%–2.3% of full sun) to a small gap (2%–6% of full sun). They observed few significant morphological changes, however. McDonald and Strain (1991) showed that leaves of *Miconia* species grown under 5% of full sun showed some adjustment of photosynthesis when transferred to 20% and 34% of full sun. Further investigations on acclimatization should attempt to determine the amount of light needed for seedlings or saplings to produce leaves that differ morphologically from those produced in the forest understory. Results to date suggest a value between 6% and 20% of full sun. If there is variation in this threshold between species, it may prove to be a useful indicator of shade tolerance.

Growth and Leaf Production in Understory Species

Understory species of rain forests are notoriously slow-growing (Evans 1939), and long-term studies are required for accurate measurements of growth and leaf production. Leaf production in the dioecious understory cycad *Zamia skinneri* is strongly seasonal and about half of the plants produced only one leaf per year (D. B. Clark and Clark 1988). Growth and reproductive biology of *Z. skinneri* are reviewed by Clark in chapter 7.

Growth of understory palms at La Selva is also strongly light-limited. After twenty months of growth, potted seedlings of *Asterogyne martiana*, *Geonoma cuneata*, and *Geonoma congesta* grown in a gap-edge site had higher rates of leaf production, were taller, and had larger leaves than seedlings of similar age grown in a closed-canopy site (Chazdon 1986a). For naturally growing reproductive individuals of *A. martiana* and *G. cuneata* the size and number of inflorescences were significantly higher in gap-edge areas than under closed canopy (Chazdon 1984). Annual rates of leaf production increased with plant size in the single-stemmed species *A. martiana* and *G. cuneata*, but in *G. congesta*, a multiple-stemmed palm, juvenile stems had higher rates of leaf production and stem elongation than older stems (Chazdon 1984, 1992). Ramets of *G. congesta* in the age class with the highest frequency of reproduction exhibited the greatest decrease in both crown size and leaf size over a three-year period (Chazdon 1992). Although this pattern may indicate a significant cost of reproduction, patterns of the distribution of growth and light within clones of *Geonoma congesta* strongly suggest that older, taller ramets supplement the growth of sprouts and juvenile ramets (Chazdon 1986b, 1992). Among the three palm species studied, *A. martiana* had the highest annual rate of leaf production at all stages, whereas *G. cuneata* had the lowest. These results reflect differences in the number of leaves present as well as different intrinsic growth capacities (Chazdon 1984). Growth and reproduction of the palm *Astrocaryum mexicanum* in lowland rain forest of Veracruz, Mexico, are also highly dependent on light levels (Sarukhán et al. 1984). Rates of leaf and seed production were significantly higher in plants growing in gaps compared to those in closed-canopy forest sites (Piñero et al. 1982; Martínez-Ramos 1985).

Growth responses to leaf and ramet removal in the multiple-stemmed palm *Geonoma congesta* were examined at La Selva over a three-year period (Chazdon 1991a). Removal of all but one ramet or 50% of the leaves of experimental ramets did not significantly affect increments of growth of height or stem diameter of the remaining ramets. Defoliated ramets had a significantly higher annual rate of leaf production per initial number of leaves than nondefoliated ramets, whereas the removal of ramets had no effect on leaf production. Moreover, defoliation and removal of ramets had no effect on cumulative number of reproductive structures produced over a three-year period by individual ramets. These results suggest that, in multiple-stemmed as well as solitary understory palm species, stored reserves are mobilized to maintain or accelerate rates of leaf production and sexual reproduction after defoliation (Chazdon 1991a; Oyama and Mendoza 1990; Mendoza et al. 1987). Further research is needed to clarify the role of carbohydrate storage and mobilization in moderating effects of herbivory and physical damage on growth and reproduction of understory species.

In some ways experimental studies of growth and biomass allocation at La Selva confirm what we have learned from studies of photosynthesis and carbon gain; although the intrinsic capacities for growth of species may differ, microenvironment may be the most important predictor of plant performance. Differences in photosynthetic capacity among early and late successional species were more apparent when plants were grown under high light than under low light (Fetcher et al. 1987). Similarly, differences in growth responses among seedlings of these species were not readily apparent when they were grown under full shade. These results are further supported by in situ growth measurements of saplings of gap-dependent and shade-tolerant canopy trees that show similar growth and photosynthetic responses with variation in light availability (D. A. Clark and Clark 1987a, chap. 7). Early successional pioneer species, however, can respond faster to changes in light regime compared to shade-tolerant and gap-dependent species. It is, thus, important to compare the responses of species under a wide range of environmental conditions. Seedlings and saplings of canopy trees at La Selva do not appear to differ as much in their abilities to tolerate shade as in their abilities to grow rapidly in full sun.

Studies of plant growth at La Selva further suggest that plasticity in photosynthetic responses of leaves is related to successional status. In many species rates of leaf production and turnover increase with light level (Oberbauer and Strain 1985; Chazdon 1986b; Fetcher et al. 1987). Those species that have the highest intrinsic growth rates, the pioneers, should exhibit the highest rates of leaf turnover under full sun conditions. Such high rates of leaf turnover may facilitate rapid redistribution of nutrients to developing leaves. In pioneer species the distribution of nutrients to individual leaves may pose a major limitation to carbon gain. For plants under forest shade, however, the microenvironment of individual leaves poses the major limitation to carbon gain. Plant performance in the understory closely parallels light availability (Chazdon 1988). Shade-tolerant species, which initially establish in the understory, tend to exhibit little variation in photosynthetic capacity compared to pioneer species (Bazzaz 1984; Denslow 1987b; Chazdon and Field 1987; Fetcher et al. 1987). Even when differences exist among species, these seldom produce consistent differences in growth rate because they are overwhelmed by environmental heterogeneity.

PLANT ARCHITECTURE

Canopy Structure and Light Interception

The structure of the canopy of an individual plant significantly influences the interception of light and allocation of biomass. Chazdon (1985) assessed the consequences of leaf size, leaf display, and crown size for interception of light in the understory palm species *Asterogyne martiana* and *Geonoma cuneata* at La Selva. Reproductive plants of *A. martiana* had longer leaves, more leaves per plant, and greater total leaf area than *G. cuneata*. Chazdon (1985) calculated two components of efficiency of light interception: angular efficiency was calculated as the area of horizontally projected leaves divided by total leaf area, exposure efficiency was calculated as the projected crown area divided by the projected area of all the leaves in the crown. It measured the degree of self-shading within the canopy. Total efficiency was calculated as the product of the two components. Efficiency of light interception was higher in reproductive individuals of *G. cuneata* than in *A. martiana*, but seedlings and juveniles of the two species did not differ significantly in efficiency of light interception. The reduced efficiency of light interception in reproductive *A. martiana* was attributed to the presence of older, more pendant leaves (which reduced angular efficiency) within the crowns rather than to self-shading. both species exhibited a low degree of self-shading <20% of total leaf area), a consequence of the even distribution of leaves around the main axis and bifid leaf morphology.

Mechanics of Support for Leaves and Crowns

Leaf display requires significant investments in petioles, branches, stems, and roots. Because investments in these structures represent a metabolic cost to plants, minimizing the allocation of biomass to structural components should permit faster growth. The mechanical stability of leaves, branches, and entire trees will be compromised, however, if structural investments are too low. Recent studies at La Selva have explored relationships between biomass allocation to support structures, mechanical properties of support tissues, and growth rates of a variety of species.

In the understory palms *Asterogyne martiana* and *Geonoma cuneata* the costs of biomass for leaf support increased disproportionately with leaf and crown size (Chazdon 1986a). Mechanical stresses acting on petioles also increased markedly with leaf size, resulting in a lower safety factor for the largest leaves of both species (Chazdon 1986a; safety factor is the ratio of observed load to the load sufficient to break the petiole, branch, or stem of a plant). *Geonoma congesta* showed similar increases in petiolar stress with increases in leaf size, but the allocation of biomass to leaf support was lower in the divided leaves of reproductive stems than in the smaller, bifid leaves of juvenile stems. This pattern resulted from a shorter relative petiole length in divided compared with bifid leaves as well as differences in the position of center of mass of the leaves. Safety factors for leaves of reproductive plants were from three to four in all three species. The scaling of petiole dimensions with leaf size in these understory palms acted to maintain a constant degree of leaf curvature (elastic similarity) rather than to maintain a constant mechanical stress (Chazdon 1985, 1986a). Given these patterns of biomass allocation, maximum leaf size in these species appears to be subject to mechanical constraints.

Although palms lack a secondary cambium, increased requirements for support during growth can be met in three ways: (1) initiation of growth in height with a stem diameter that is sufficient for future growth; (2) increase in stem diameter by sustained cell expansion in primary tissue; (3) increase in stiffness and strength of stem tissue by sustained lignification (Rich 1986). In a study of mechanical architecture of six species of canopy and subcanopy palms at La Selva, Rich (1986) found appreciable secondary changes in the mechanical properties of stems that enable these palms to maintain a constant safety factor against mechanical failure while height increases. Furthermore, *Socratea exorrhiza, Iriartea gigantea,* and *Euterpe precatoria* showed increases in stem diameter with increases in stem height (Schatz et al. 1985; Rich et al. 1986). Relative to dicotyledonous trees of similar height, juveniles of palms that reach the canopy have stems that are overbuilt in diameter but underbuilt in mechanical properties (Rich 1986; Rich et al. 1986).

Stem growth and crown size of canopy palm species are also subject to mechanical constraints. *Iriartea gigantea* and *Welfia georgii* had similar maximum stem diameter and mechanical properties, but *Iriartea gigantea* was taller and had a smaller crown than *W. georgii* (Rich 1986). *Cryosophila albida* had the fastest rate of leaf production among the six palms studied but the slowest rate of increase in height (Rich 1986). *Socratea exorrhiza* and *Euterpe precatoria* had the highest rates of stem elongation and produced weaker stems. These results suggest that the rate of increase in height and stem strength are inversely related. The relationship among stem diameter, canopy size and weight, and intertree spacing in these palm species is discussed in chapter 8.

Within the genus *Geonoma* multiple-stemmed species have relatively narrower stems for crowns of a given height compared to solitary species (Chazdon 1991b), increasing their mechanical instability (King 1987). A further consequence of the narrower stem in clustered palms is the lower biomass invested in supporting crowns. Perhaps this is one explanation for the relatively high rates of leaf production in *Geonoma congesta* compared to those described for solitary rain forest understory palms from similar light environments (Chazdon 1984; Mendoza et al. 1987; Oyama 1990; Oyama and Mendoza 1990).

King (1987) determined stability safety factors for four understory palms and treelets, one species of treelet growing in a clearing, and saplings of two tree species growing under natural conditions at La Selva. Reproductive individuals of *Geonoma congesta, Faramea suerrensis,* and *Piper arieianum* in the understory had low mean safety factors of 2.0–2.6, whereas the mean safety factor for *Piper auritum* in a clearing was 4.3. Among the understory species *Asterogyne martiana* had the highest mean safety factor (3.6). Low safety factors permit plants to support their crowns at a greater height for a given stem biomass but also make these plants less stable. Mechanical instability is common in understory species (Busby et al. 1980; King 1987). Safety factors of saplings more than 2.5 m tall of *Pentaclethra macroloba* and *Pourouma aspera* were similar to those for the understory species and decreased significantly with height (King 1987).

Allometric relationships between height and stem diameter in dicotyledonous trees at la Selva further delineate the inverse relationship between rate of increase in height and stem strength. *Pentaclethra macroloba,* a long-lived canopy species, had a high margin of safety against mechanical failure based on the allometric relationship between height and stem diameter. In contrast, the faster-growing, shorter-lived *Pourouma aspera* appeared to have decreasing safety factors with increasing height (Rich et al. 1986).

CONCLUSIONS AND DIRECTIONS FOR FUTURE RESEARCH

The premise that species with similar ecological roles are physiologically similar appears to be supported although in a rather unusual way. Nearly all tree species appear to be shade tolerant to some extent, and few, if any, are found only in open clearings. Species characteristic of primary forest are extremely shade tolerant as seedlings. Even seedlings of pioneer tree species, such as *Ochroma lagopus* and *Heliocarpus appendiculatus,* have considerable tolerance to shade as shown by their survival in 2% of full sun. Detailed studies of one species, *Pentaclethra macroloba* (Oberbauer and Strain 1985, 1986; Oberbauer et al. 1987), and of the guild of understory palms (Chazdon 1985, 1986a, 1986b; Rich 1986) have shown that shade tolerance is a multifactored response to a light-deficient environment. Traits that appear to have adaptive value in low light environments are present at several levels of plant organization including leaf morphology, metabolic rates, biomass allocation, and plant architecture.

As one learns more about the ecology and physiology of various species, distinctions between successional categories blur. For example, schemes that seek to classify plants by successional status fail to account for ontogenetic shifts in shade tolerance. Some canopy tree species appear to lose the ability to survive in extreme shade at the sapling stage, whereas others retain it (D. A. Clark and Clark 1987a). Understory plants remain shade tolerant throughout their life spans. At La Selva shade tolerance seems to be the norm. Perhaps the next focus of research should be how shade intolerance develops in saplings of some species.

Although tolerance of shade and of high light have been regarded as end points on a single physiological continuum, it may be more useful for future studies to consider them as distinct entities. By separating these two phenomena one may better resolve the contradiction posed by species such as *Pentaclethra macroloba* that can tolerate deep shade and also grow rapidly in well-lit conditions. The physiological and morphological traits that characterize shade tolerance have been well studied (Boardman 1977), whereas traits that confer resistance to photoinhibition are just beginning to come under scrutiny (Powles 1984; Demmig et al. 1988).

Among environmental factors light appears to be the most important in limiting growth, at least for the seedlings, saplings, and understory plants studied to date. Light regimes in various microhabitats have been well characterized except that more information is needed on the distribution and dynamics of sunflecks. Less is known about other environmental factors, such as temperature, humidity, soil water, and nutrients, but the existing data suggest that these factors limit plant growth only sporadically.

Whether variation among species in capacity for acclimatization is the result of greater phenotypic plasticity of individuals or of greater genetic variation for response to light levels in some populations is not known. We suspect it is the former, but the latter possibility should be tested if only to establish how much variation in physiological traits exists among individuals. Plasticity in response to environmental variation other than light has been less well studied, but work suggests that early successional species are more capable of rapid responses to increased nutrients (Huston 1982; Denslow et al. 1987).

Plants growing in rain forest habitats are faced with an enormous variety of dynamically changing conditions. Indeed, the challenge facing plants in tropical rain forests may not be how to cope with the stress imposed by an extreme environment but how to maximize fitness when confronted by spatial and temporal heterogeneity. The quantity and quality of light, soil nutrients, herbivory, pathogens, and physical damage can all vary simultaneously to influence growth, survival, and fecundity. We need more studies of photosynthetic dynamics, the effects of water balance on leaf loss, and whole-plant integration (including clonal integration). To be most useful, these investigations should consider variation over more than one temporal or spatial scale.

The variable environment may select for a rather conservative suite of physiological and morphological traits. For example, most species have relatively low photosynthetic rates and can tolerate a wide range of environments. We have found no new physiological pathways or records in photosynthetic performance although this may change as research expands to include groups that are poorly represented in the temperate zone such as lianas and epiphytes. Instead, we have acquired a new appreciation for environmental uncertainty and the need to incorporate it into our thinking.

Further studies are required to link physiological ecology to plant/animal interactions and to applied research on tree regeneration and soil fertility at La Selva. Physiological research has much to offer in collaboration with these fields. Examples include the effect of defoliation on photosynthetic rates, responses to nutrients and light in trees that are potentially useful for reforestation, and effects of environmental heterogeneity on reproduction.

The challenge before investigators is twofold: to understand the physiological and morphological basis for differential growth responses within and among species in natural habitats and to assess the ecological and evolutionary significance of these responses. As articulated in this chapter, some progress has been made in addressing the first part of this challenge, but the second remains uncharted territory.

ACKNOWLEDGMENTS

We thank C. Field, B. Strain, L. McDade, B. Osmond, J. Voltzow, and three anonymous reviewers for their comments. Much of the research that is summarized here was conducted with support from the National Science Foundation, the Jesse Smith Noyes Foundation (through OTS), and the United States Department of Agriculture Competitive Grants Program.

11

Diversity of Long-term Flowering Patterns

L. E. Newstrom, G. W. Frankie, H. G. Baker, and R. K. Colwell

Phenology is the study of the timing of recurring natural phenomena, such as periodic biological events, in relation to climatic changes (Caprio 1967; Stearns 1974). It is an important aspect of population biology because it impinges on so many features of each species and on the dynamics of interspecific interactions in competition, herbivory, pollination, and frugivory. Temporal changes in plant resources profoundly affect animals that use plants, for example, butterflies and moths, which oviposit on plants and consume leaves as larvae and floral resources as adults. Cycles of plant growth and reproduction are crucial to understanding the functioning of ecosystems (Lieth 1974) and processes of primary production (Sarmiento and Monasterio 1983). Thus, a critical requirement for monitoring, managing, and conserving ecosystems is a better understanding of phenological mechanisms. For recent reviews of aspects of phenology see Baker et al. (1983), Bawa (1983), Rathcke and Lacey (1985), and Primack (1985b).

In the lowland tropical rain forest the two most significant features of phenology are that continual warmth and moisture foster growth all year long (Richards 1952) and that phenological patterns are more diverse than in any other ecosystem (Sarmiento and Monasterio 1983). In temperate forests winter inhibits growth so that annual seasonal cycles are predominant if not essential in all species. In tropical dry forests plants also experience strong seasonal synchrony (Frankie et al. 1974a; Sarmiento and Monasterio 1983; Reich and Borchert 1984); but in lowland tropical rain forests no severe seasons markedly restrict or synchronize plant growth. Yet even in such relatively uniform and favorable climatic conditions discrete cycles do occur in vegetative growth in most species, especially trees (Borchert 1973, 1978, 1991; Hallé et al. 1978; Breitsprecher and Bethel 1990). These cycles have different frequencies that may be regular or irregular. When dealing with phenology in the aseasonal tropics, we are confronted with several questions: Are reproductive patterns mainly continuous or discrete? Are they regular or irregular? What factors influence the evolution of reproductive patterns and what climatic factors, if any, control reproductive cycles? To answer these questions, phenological patterns need to be described and quantified.

Timing of flower bud induction and anthesis (flower opening) is fundamental to the timing of fruit and seed production. In contrast, temporal patterns of leafing may differ radically from flowering (Tomlinson, 1980; Newstrom et al. 1991). Fruiting patterns may differ slightly from flowering patterns

because, although flowering always precedes fruiting, fruits may sometimes fail to set or may have prolonged maturation times (Frankie et al. 1974a; Janzen 1978b; Stephenson 1981; Levey 1987a). For example, two *Spondias* species in Panama flower simultaneously but present mature fruits to dispersers at different times (Croat 1974). Conversely, various dipterocarp species in Malaysia mature fruits simultaneously after flowering sequentially (Ashton 1988; Ashton et al. 1988). In addition, fruiting may directly affect time of flowering through internal correlative factors (Bernier 1988), particularly in alternate bearing species (Monselise and Goldschmidt 1982). We will discuss here only anthesis stage flowering patterns; leafing and fruiting patterns from our project will be reported elsewhere. For a review of fruiting at La Selva see chapter 16.

Flowering can be viewed at different time scales. Some phenological investigations focus on daily phenomena, such as hour of anthesis (Primack 1985a), some on yearly occurrences, such as dates and duration of flowering, and others on long-term behavior over a span of more than ten years. Individually tagged trees with long-term records provide sufficient data to determine predictability of flowering, but little attention has been given to this aspect of tropical phenology (see Hallé et al. 1978; Longman 1985).

Approaches to flowering phenology have incorporated several different levels of analysis. Most surveys of tropical flowering have emphasized community-level questions with only brief descriptions of individual-level patterns (e.g., Koelmeyer 1959; Medway 1972; Croat 1969, 1975; Dieterlen 1978; Cruz Alencar et al. 1979; Putz 1979; Van Schaik 1986; Koptur et al. 1988). Individual-level flowering patterns have been analyzed in more detail in population-level studies but mostly for short-term time spans (e.g., Augspurger 1980, 1981; Bullock and Bawa 1981; Bullock et al. 1983; Borchert 1980; Reich and Borchert 1982; Bullock 1982) or, if long-term, for only a few species (e.g., Milton et al. 1982; Piñero and Sarukhán 1982a; D. A. Clark and Clark 1987b; Michaloud 1988). Although short-term population studies maximize data on variation among individuals (at the cost of temporal variation), long-term synoptic surveys of many species maximize data on temporal variation within a few individuals (at the cost of characterizing population behavior). Thus, these two types of studies are complementary (see Individual and Population Patterns).

In this chapter we describe flowering patterns from the La

Selva Phenology Project, a long-term synoptic survey of phenology in lowland tropical rain forest trees at La Selva Biological Station. We first present a conceptual framework that classifies long-term flowering patterns, distinguishes levels of analysis, and interprets seasonality (see the summary in table 11.1). After describing four major flowering patterns, we present the proportions of each pattern as a flowering profile for the tree community level. To consider ultimate factors that may influence flowering patterns we discuss aspects of guild-level patterns, fig-wasp pollination, pollinator foraging strategies, and plant sexual systems. Finally, we briefly comment on proximate factors that may play a role in the control of flowering in lowland tropical rain forest trees.

DESCRIPTION OF FLOWERING PATTERNS

The La Selva Phenology Project

Baker and Frankie's phenological research began in 1968 in both lowland tropical rain forest at La Selva Biological Station and tropical dry forest in Guanacaste, Costa Rica, as part of an ecosystem study to compare and contrast the two types of forests. At La Selva Frankie and Baker extended the data collection for twelve consecutive years on phenology of 457 marked trees belonging to more than 200 species. The first trees marked in 1969 for the long-term survey formed the beginning of the arboretum at La Selva. Frankie et al. (1974a) and Opler, Frankie, et al. (1976, 1980) reported results from both ecosystems for the first three years of the project and

Table 11.1 Summary of conceptual framework for classifying flowering patterns

Variables to Describe Patterns	
Frequency	Number of cycles per year with respect to on/off phases
Regularity	Variability in lengths of cycles and phases
Duration	Length of time in each cycle or phase
Amplitude	Intensity or quantity of flowering
Date	Month or season of year in which flowering occurs
Synchrony	Simultaneous occurrence of a particular event

Main Classes of Flowering (based on frequency)	
Continual	Always in flower but may have brief gaps
Sub-annual	Irregular multiple flowering phases per year
Annual	One major flowering phase per year
	Subclasses: brief, intermediate, extended
Supra-annual	Multiyear cycles of flowering
	Subclasses: alternate, rare

Levels of Analysis	Seasons at La Selva
Flower	Main dry ~ January to May
Inflorescence	Early wet ~ May to September
Branch	Veranillo ~ September or October
Branch complexes	Late wet ~ October to January
Whole plant	Features of Multispecies Patterns
Population	Gaps (nonflowering phases)
Guild	Overlap among species
Community	Amplitude fluctuations
	Permutations of species

Newstrom et al. (1991) reported preliminary results for the long-term (twelve years) project.

To analyze the long-term data we first characterized the diversity of flowering patterns to provide a conceptual framework for quantitative analyses. We used frequency and regularity as criteria because they reflect predictability of flowering. Other advantages of this new classification are that it has an explicit time scale, applies to other phenological events (e.g., leafing and fruiting), and pertains to different levels of analysis. Further analyses of the long-term data (to be reported elsewhere) treat relationships among patterns of leafing, flowering, and fruiting; predictability of flowering in relation to rainfall distribution; and associations among phenological patterns and canopy position, plant sexual systems, pollinators, and dispersers.

Methods. Leafing, budding, flowering, and fruiting events in 457 tagged trees were observed monthly with the aid of binoculars from January 1969 through March 1981. Amount of flowering was recorded relative to the typical crop for each tree in three categories: no, light, or heavy flowering. Newly accessible species and trees were added to the study after construction of a trail system. Because only trees competent to flower were included, it was often impossible to locate more than a single tree of rare species. Sample sizes, therefore, consisted of four to six trees for each of eleven species, two to three trees for each of seventy-eight species, and one tree for each of the remaining species.

To classify flowering patterns we used only trees with ≥ five years of data and excluded palms (which require separate analysis). We grouped 254 trees from 211 species of 121 genera and 59 families into four major patterns (and 46 trees into intermediate classes that will be described elsewhere).

To investigate seasonality at La Selva we analyzed rainfall distribution from 1969 through 1980 using a seasonal decomposition time series analysis based on the Lowess curve smoothing technique (seasonal decomposition function [SABL] in S program, Becker et al. [1988]). This analysis partitioned daily millimeters of rain into long-term trend, seasonal, and irregular (residual) components (see Cleveland et al. 1982; Cleveland 1985).

Classification Systems

Criteria for Classification. One of the challenges in tropical phenology is how to summarize patterns without losing the character of the temporal sequence. Our approach, therefore, has been to categorize flowering patterns graphically (see Tufte 1983; Cleveland 1985). To elucidate the problem, reports of five-month flowering cycles in some figs (Koelmeyer 1959), and seven- to ten-month cycles in *Delonix regia* and *Lagerstroemia speciosa* (Richards 1952) lack a measure of variance and, therefore, do not adequately reveal the degree of irregularity in cycle lengths. To our knowledge regular nonannual flowering patterns with low variance have not been clearly documented with long-term, individual-level data. Of the many different variables used to describe and classify phenological patterns (see Bawa 1983; Rathcke and Lacey 1985; Primack 1985b), we found the most useful to be: **frequency, duration, amplitude, date,** and **synchrony** and have added **regularity.** Because tropical phenology terms have been inconsistently used and often vaguely defined, see app. 11.1 for our definitions of selected terms.

Previous Classifications. Three main types of classification have been commonly used for tropical flowering patterns, excluding classifications that incorporate additional non-flowering criteria that we do not discuss here, such as leafing (Reich and Borchert 1984) or life-form (Sarmiento and Monasterio 1983). In the first type of classification categories have been based primarily on **duration** of flowering: for example, *seasonal* or *extended* (Frankie et al. 1974a), *mass* or *steady state* (Augspurger 1983b), and *mass* or *extended* (Bawa 1983). Flowering duration has particular significance for the development of pollinators' search images for nectar and pollen. These classifications, however, lack a firm distinction between categories and have no explicit time or amplitude scales. For example, *mass* has been used not only for flowering that lasts for a few days or weeks (Augspurger 1980; Appanah and Chan 1981) but also for many weeks or months (Perry and Starrett 1980; Appanah 1982). In addition, *mass* (and related terms such as *general* and *gregarious* flowering) has referred to high amplitude flowering with multiyear cycles that are synchronized among many species (Janzen 1978b; Silvertown 1980; Ashton 1988). Furthermore, categories based on duration alone do not indicate how often flowering occurs; for example, the *extended* category of Frankie et al. (1974a) included flowering of several different frequencies.

In the second type of classification derived mainly from herbarium studies (e.g., Croat 1969, 1975, 1978; Dieterlen 1978) and short-term surveys (e.g., Tomlinson 1980) categories have been based primarily on **date,** such as month or season of flowering. Date of flowering is crucial for research on the interdependence of plant and animal life cycles and the effects of climate on phenology. Defining categories, however, on date alone does not indicate frequency and may, therefore, imply an annual cycle, which may be true for most temperate flowering but not for tropical flowering.

In the third type of classification categories have been based on a combination of several criteria. Gentry's (1974) classification, the most widely used, distinguished five categories based on several criteria: duration, timing, amplitude, and population-level synchrony. These categories, first described for the Bignoniaceae, reveal the temporal character of patterns more completely than any other classification. The terms have become complicated in the literature, however, because they have been applied to patterns that meet some but not all of the criteria originally specified by Gentry. Moreover, Gentry's categories do not encompass all the variation found in our data for La Selva.

Classification Based on Frequency and Regularity. We have classified flowering patterns based on **frequency** and **regularity**, thereby incorporating an explicit time scale and an indication of predictability. Janzen's (1978b) classification for seed production patterns used frequency, but regularity has not been used previously. The frequency criterion, defined as the number of **cycles** per year (with each cycle comprising a flowering and a nonflowering **phase**) refers to on/off cycles of flowering rather than amplitude fluctuations of light and heavy flowering. Because our data suggest that yearly climatic

events influence flowering, we used an annual time scale for four frequency classes (fig. 11.1):

1. **continual** with flowering interrupted by only a few short breaks;
2. **sub-annual** with irregular multiple cycles in most years;
3. **annual** with only one major cycle per year; and
4. **supra-annual** with flowering in multiyear cycles.

The regularity criterion indicates the amount of variation over time in cycle lengths and component phases. Figure 11.2 shows two main classes: regular, with similar phase and cycle durations; and irregular, with variable phase and cycle durations. Annual cycles were the most regular, whereas sub-annual cycles were the most irregular (fig. 11.3).

Our multidimensional classification system uses other criteria, such as duration, amplitude, and date, to distinguish subdivisions of the main classes. The presence of intermediates between classes emphasizes that flowering frequency in our data formed a continuum. Thus, we intend this classification to be a convenient "cognitive map" for seeing variation in flowering and not a replacement for past classifications or a rigid categorization of a process that is in essence dynamic.

Levels of Analysis. In temperate phenology analysis level has not been an issue because patterns tend to be similar at every level; in tropical phenology we do not have this luxury. Phenological patterns can be analyzed hierarchically (see table 11.1). In flowering, for example, the lowest level of analysis concerns the timing of anthesis among flowers within an individual inflorescence, which may be synchronous or asynchronous, regular or sporadic. The longevity of each flower also varies among species although the flowers of most tropical plants last only one day or part of a day (Frankie et al. 1983; Bawa 1983; Primack 1985a). Likewise, flowering rate (flowers per day per inflorescence), flower longevity, and total number of buds per inflorescence together determine inflorescence longevity—the time over which an individual inflorescence produces flowers.

At the next level, flowering among inflorescences or branches of inflorescences on an individual plant has its own pattern quite independent of patterns among flowers within inflorescences. In agroforestry Akunda and Huxley (1990) distinguished this level as branch phenology. Asynchronous flowering in different branches of the same tree, *manifold flowering,* is quite common in lowland tropical rain forests (Richards 1952; Koelmeyer 1959; Medway 1972; Frankie et al. 1974a; Borchert 1978). The succeeding level focuses on phenological patterns among plants within a population, and the fine level treats phenological patterns among different species in the same assemblage.

Although most data document patterns within and between populations, the complete phenological hierarchy can be illustrated with information available for the understory, largely hummingbird-pollinated tree, *Hamelia patens* (Rubiaceae) (R. K. Colwell and S. Naeem unpublished data). At La Selva individual flowers of this plant open between midnight and 2:00 A.M. although anther dehiscence and initiation of nectar flow do not occur until about 6:00 A.M. when the first hummingbirds visit. Each flower ceases to produce nectar by midday and usually falls from the plant by mid-afternoon.

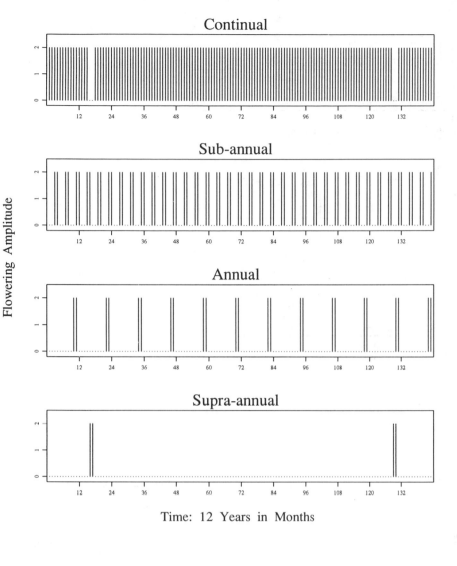

Time: 12 Years in Months

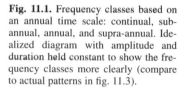

Fig. 11.1. Frequency classes based on an annual time scale: continual, sub-annual, annual, and supra-annual. Idealized diagram with amplitude and duration held constant to show the frequency classes more clearly (compare to actual patterns in fig. 11.3).

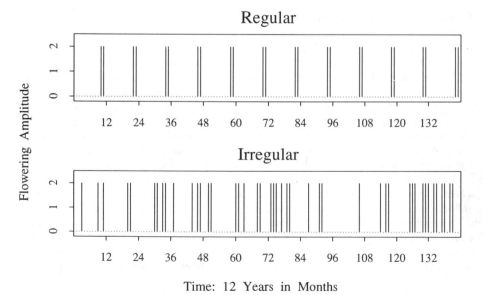

Time: 12 Years in Months

Fig. 11.2. Regularity classes: regular and irregular patterns based on variability in cycle length. Idealized diagram with amplitude held constant to show the contrast between the patterns (compare to actual patterns in fig. 11.3). In the regular pattern flowering and nonflowering phases are of similar duration; in the irregular pattern flowering and nonflowering phases are both highly variable.

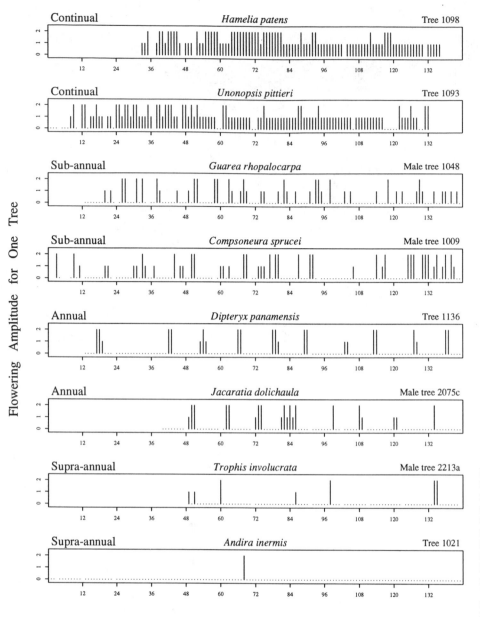

Fig. 11.3. Frequency and regularity of flowering. Time series graphs of monthly flowering in individual trees of eight species in the lowland tropical rain forest at La Selva Biological Station. Sub-annual flowering patterns are irregular, annual are regular, but data are inadequate to determine regularity in supra-annual patterns. Each panel represents one tree from 1969 through 1980. Amplitude categories: 1 = light flowering; 2 = heavy flowering; dots on X axis = no flowering; blanks = missing data.

At the level of the individual inflorescence, once flowering begins, an inflorescence produces from zero to five flowers each day. The distribution of number of flowers open per day per inflorescence does not differ significantly from the corresponding Poisson distribution ($p > .82$, Kolmogorov-Smirnov test, two-tailed) indicating random variation about the mean value of about 1.5 flowers per day (fig. 11.4). The total number of flowers produced during the life of an inflorescence is quite variable, ranging from about thirty (three weeks of flowering) to more than one hundred (ten weeks of flowering).

At the level of the individual, each *H. patens* tree bears from one to several dozen flowering inflorescences at a time, with no evidence of synchrony in the initiation of flowering among different inflorescences on the same plant. Three trees in our study at La Selva flowered continually; but trees in the tropical dry forest of Guanacaste (Opler, Frankie, et al. 1980)

and in Trinidad, W. I. (Colwell 1986), flowered only during the wet season in an annual pattern. Thus, at the species-level, *H. patens* comprises at least two different flowering patterns.

Similar variation in type of flowering pattern from one habitat or ecosystem to another has been documented for many other species (Frankie 1974a; Borchert 1980; Huxley 1983; Mure 1986). Consequently, Borchert (1980, 1983) has emphasized that we cannot characterize a species as having a "normal" flowering pattern. Even in the same habitat, phenological patterns can differ depending on level of analysis. For instance, a population may comprise both sub-annual and supra-annual patterns at the individual level as in *Ficus natalensis* (Moraceae) in the tropical rain forest in Makoku, Gabon (Michaloud 1988). In this population the summation of all the individual patterns produces a continual flowering pattern at the population level. A flowering profile of the population

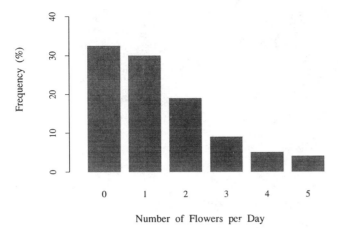

Fig. 11.4. Number of flowers open per day inflorescence in *Hamelia patens* at La Selva Biological Station. Data are pooled from six infloresences, each on a different tree, each followed from first to last flower between October and December 1986 (R. K. Colwell and S. Naeem unpublished data).

shows the relative proportions of each of the individual level flowering patterns, for example, percentage of sub-annual and supra-annual patterns in a *F. natalensis* population. In the tropics, patterns may differ at any level of analysis, but in the temperate zone, patterns tend to be annual at all levels.

Seasonality. When describing phenological patterns, we reserve the term **seasonal** to mean the association of an event with a recognizable climatic season no matter what the frequency. We also distinguish seasonal trends for on/off cycles from those based on amplitude fluctuations. To elucidate, sub-annual patterns show seasonal trends at the individual level for both a higher likelihood of flowering (on/off cycles) and a prevalence of heavier flowering (amplitude fluctuations). In our data annual flowering patterns were always seasonal, but some supra-annual flowering patterns had seasonal associations and others did not. Seasonal amplitude differences may be quite large as in the continual population-level flowering of *Piper arieianum* (Piperaceae), a shrub at La Selva, in which major population peaks in February contributed from 62% to 76% of the total annual fruit yield (Marquis 1988).

Seasons at La Selva have been categorized in two ways. Precipitation peaks in June–July and November–December and is lowest in March (chap. 3). One interpretation considered two wet and two dry seasons (Frankie 1974a), but other authors have excluded the less pronounced and unpredictable second dry season (Stiles 1978c). To investigate this question we used a seasonal decomposition time series analysis because twenty-five-year averages and short-term time series do not show the variation in seasons over time. The analysis partitioned twelve years of daily rainfall (1969 through 1980) into long-term trend, seasonal, and residual (irregular) components; these sum to the actual rainfall. The seasonal component revealed a regular progression of four seasons each year: a four- to five-month drier season from January to May, a three- to four-month wet season from May to September, a one-month drier season in September or October, followed by a two- to three-month wet season from October to January (see third panel from the top in fig. 11.5). The one-month dry

season, the *veranillo,* varied considerably in amplitude (almost disappeared in 1972 and 1973) yet was always relatively drier. Because pronounced veranillos may affect flowering, we recognize these four seasons: (1) main dry, (2) early wet, (3) veranillo, and (4) late wet.

The regularity or severity of these seasons from the plant's perspective, however, remains unknown. The variation in the residual was extremely high: a range of 1,200 mm compared to only 400 mm in seasonal and 300 mm in long-term components (see scale bars to the right in fig. 11.5). How rainfall distribution relates to long-term flowering patterns requires more detailed analysis.

Individual and Population Patterns
In this section we describe characteristic features of four major flowering patterns at two levels of analysis, individual and population, using our long-term individual data and other relevant short-term population studies from La Selva.

Continual Flowering. In continual flowering patterns, also called *ever-flowering* (Corner in Richards 1952; Sarmiento and Monasterio 1983), *ever-blooming* (Sachs 1960) and *continuous* (Stiles 1978c), trees flower almost all the time, so predictability is high. This class of flowering ranges from strictly **continuous** flowering, meaning a constant uninterrupted succession, to **continual** flowering that has brief but infrequent interruptions (see Fowler 1965). Truly continuous flowering rarely occurs at the individual level (only a few trees in our data) but is more common at higher levels of analysis, such as the guild or community.

To explain, one individual of *Unonopsis pittieri* (Annonaceae) had continual flowering with several short breaks as shown in figures 11.3 and 11.6. Similarly, two *Hamelia patens* trees (not shown) had continual flowering each with only one four-month break from December to March, but a third tree (shown in figs. 11.3, 11.6) may have flowered continuously depending on whether missing data represents flowering or not. In *H. patens* all three trees showed the same seasonal association with a predominance of heavy flowering in the early wet season and light flowering during the main dry season (fig. 11.7). At the population level seasonal maxima in flowering in this species are quite marked at La Selva (F. G. Stiles and S. Naeem pers. comm.; Levey 1987a).

Sub-annual Flowering. Sub-annual patterns, the least predictable, have multiple flowering phases in most years with highly irregular cycle lengths. Similar patterns have been called *multiple-bang* (Gentry 1974), *episodic* (Bullock et al. 1983), *intermittent* (Berg 1989), and *periodic* (Haber and Frankie 1989). Three features characterize the sub-annual pattern: variable durations in both flowering and nonflowering phases, variable number of cycles per year, and variable dates of flowering onset, which can occur at any time of year.

Bullock et al. (1983) demonstrated the complexities of **sub-annual** flowering in an intensive short-term population study of *Guarea rhopalocarpa* (Meliaceae), an understory dioecious tree at La Selva. Our data corroborate the general pattern for long-term flowering at the individual level: in three trees, flowering phase durations varied from one to four months, non flowering phase from one to nine months, and

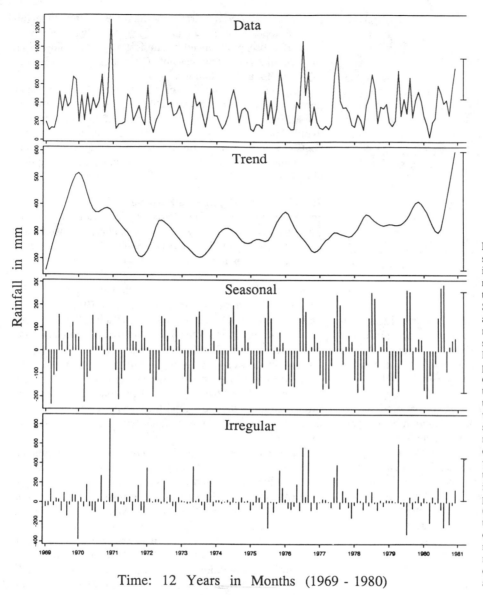

Time: 12 Years in Months (1969 - 1980)

Fig. 11.5. Seasonal decomposition time series analysis (Lowess curve smoothing in SABL, S program) of daily rainfall for twelve years (from 1969 through 1980) at La Selva Biological Station. The top panel (data) shows the actual distribution of total rainfall in millimeters. The second panel (trend) shows a numerical description of the long-term trend in rainfall. The third panel (seasonal) shows the seasonal oscillations above and below the center of the data (1 bar = 1 month). Bars below the line indicate the two drier seasons with a four- to five-month main dry season (~ January to May) and a one-month minor dry season (~ September or October). The minor dry season, or veranillo, is brief, variable in amplitude, and unpredictable (almost nonexistent in 1972 and 1973). The bottom panel (irregular) displays the residual (actual rainfall minus the long-term trend component minus the seasonal oscillations). The vertical lines on the right of each panel indicate the same magnitude, revealing the relative scale of each panel (graph from Newstrom et al. 1991).

the number of cycles varied from one to four per year (figs. 11.3, 11.6). Over the twelve years of the study, each of the three trees flowered in all but one month; this nonflowering month, however, was May for tree 1020, June for tree 1149, and December for tree 1048 (fig. 11.7). Flowering was more likely to occur in both drier seasons when heavier flowering was also predominant (fig. 11.7).

The population-level flowering pattern in *G. rhopalocarpa* also had a highly irregular sub-annual pattern (approximately five flowering phases per year for 119 trees according to Bullock et al. 1983). Population flowering peaked in September and October and different combinations of trees participated in each population-level flowering phase.

In a second example of sub-annual flowering, in the understory dioecious tree *Composoneura sprucei* (Myristicaceae), individuals flowered less frequently than *G. rhopalocarpa*. In our data one female and two males trees had **sub-annual** patterns, but a second female tree had an **annual** pattern (fig. 11.6). Variability in phase durations, number of cycles per year, and dates of flowering prevailed in the three

sub-annually flowering trees. Months in which certain individuals never flowered were April and June, and times of seasonal maxima were similar to those of *G. rhopalocarpa* (fig. 11.7). In contrast, the annually flowering tree flowered regularly only once each year between July and January (fig.11.6).

The population-level pattern in *C. sprucei* appeared to have two major flowering phases per year but only eighteen months of data were available (Bullock 1982). As with *G. rhopalocarpa*, different combinations of trees participated in each population flowering phase. Male trees flowered more frequently than female trees, implying that female trees may be more likely to have a long-term annual pattern than males in this species.

Annual Flowering. Annual patterns, the most predictable, have only one major flowering phase, which regularly repeats at the same time every year. In our data these patterns can be subdivided according to duration of flowering phase because, unlike sub-annual patterns, the duration tends to be consistent from year to year and can be arbitrarily divided into brief (<

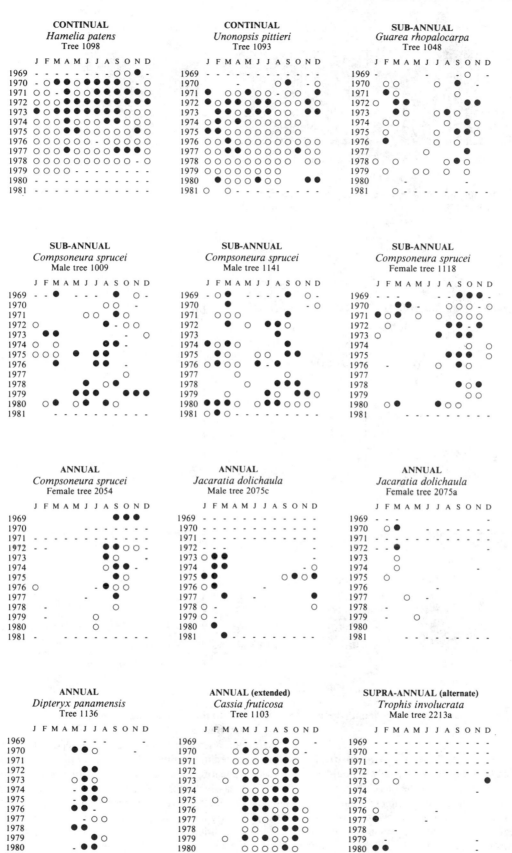

Fig. 11.6. Duration and date of flowering. Matrix graphs (year by month) of flowering for twelve years from 1969 to 1981 in trees of eight species at La Selva Biological Station. Each graph represents flowering in one tree; ● = heavy flowering; ○ = light flowering; blank = no flowering; — = missing data.

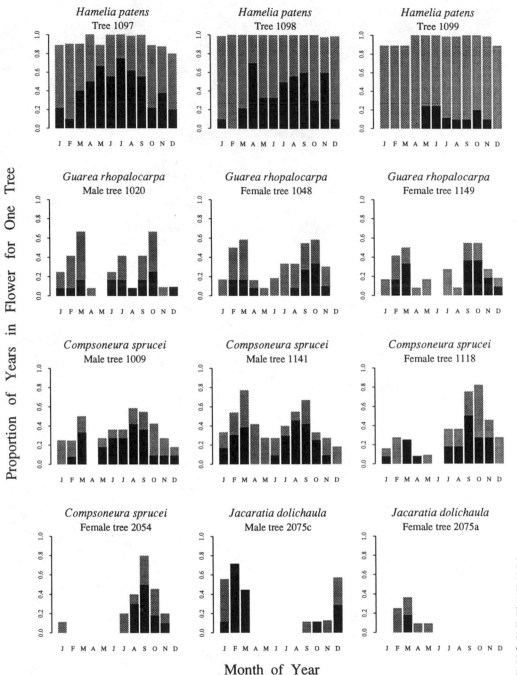

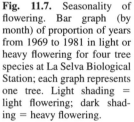

Fig. 11.7. Seasonality of flowering. Bar graph (by month) of proportion of years from 1969 to 1981 in light or heavy flowering for four tree species at La Selva Biological Station; each graph represents one tree. Light shading = light flowering; dark shading = heavy flowering.

one month), intermediate (one to five months), and extended (> five months).

Subclasses according to Duration. **Annual brief** flowering patterns, called *big bang* by Gentry (1974), could not be documented in our data because of the one-month census interval. At least one species at La Selva, *Tabebuia guayacan* (Bignoniaceae), had brief flowering, but we do not know its frequency or regularity (G. Frankie, pers. observation).

Annual intermediate patterns, called *cornucopia* by Gentry (1974) and *seasonal* by (Frankie et al. 1974a), are common. For example, in *Dipteryx panamensis* (Fabaceae), a hermaphroditic emergent tree, two trees flowered each year for

two to three months in the early wet season (fig. 11.6). Perry and Starrett (1980) documented annual flowering at the population level in this species. They reported that each tree fluctuated in amplitude slightly out of phase with other trees while appearing to be in full flower at all times, a pattern they called "massive blooming deception." Problems with time scale for the term *mass* have been mentioned, but *D. panamensis* clearly has a flowering pattern qualitatively different from annual brief patterns in which flowering lasts for only a few weeks (e.g., *Hybanthus prunifolius,* which flower less than ten days in the tropical moist forest at Barro Colorado Island, Panama [Augspurger 1980], and trees of several *Shorea* species, which flower less than twenty-five days in the tropical rain forest in Malaysia [Appanah and Chan 1981; Appanah 1985; Ashton et al. 1988]).

Some species have **annual intermediate** patterns at the individual level that combine at the population level to form an **annual extended** pattern. For example, in the understory dioecious tree, *Jacaratia dolichaula* (Caricaceae), the population flowering phase (including buds) extended 9 months from October to June, but most female trees flowered for less than 3.5 months and most males slightly longer than that (Bullock and Bawa 1981; and fig. 11.6). The extended population flowering phase may be attributed mainly to male trees that had earlier and more variable onset times than females (Bullock and Bawa 1981; and fig. 11.6). Nevertheless, peak flowering at the population level, both in numbers of trees and numbers of flowers per tree, occurred in February during the main dry season for both males and females (Bullock and Bawa 1981). The male tree in our long-term data flowered most heavily in February and the female tree in March (fig. 11.7).

Although annual extended patterns do exist at the individual level, they are not common. Only four trees from three species in our study showed this pattern. For example, one tree of *Cassia fruticosa* (Caesalpinaceae) flowered for ≥ six months beginning in the early wet season (fig. 11.6).

Types of Flowering Phases. Annual flowering patterns have a diversity of amplitude curves. First, sporadic precocious or tardy off-peak bursts of flowering may accompany the major flowering phase at the individual level (fig. 11.8 *A, B*) as in *Cassia fruticosa* in 1973, 1975, and 1979 (fig. 11.6). Augspurger (1982) differentiated between major and minor flowering phases at the population level.

Second, amplitude curves may have two or more flowering peaks (fig. 11.8 *C*). For instance, in *Jacaratia dolichaula*, five male trees had bimodal flowering phases (Bullock and Bawa 1981) and one male tree in our study had a multimodal flowering phase in 1975 (fig. 11.6). Similarly, in the tropical dry forest of Guanacaste *Cochlospermum vitifolium* trees flowered annually in the dry season with individual-level, multimodal amplitude peaks during each flowering phase (K. S. Bawa pers. comm.)

Third, at the individual level, on/off minicycles may be included within a major flowering phase (fig. 11.8 *D*). For example, *C. fruticosa* had short "pauses" within the annual flowering phase in 1972 and 1978 (fig. 11.6). We consider this pattern to be annual because the minicycles are embedded within a regular cycle of seasonal flowering that has an overall annual frequency. *Melicytus ramiflorus* (Violaceae) has a similar pattern in temperate New Zealand (Powlesland et al. 1985).

Finally, the shape of the curve may range from symmetric (fig. 11.8 *E*) to skewed (fig. 11.8 f). At the population level, *J. dolichaula* had a symmetric amplitude curve for the number of male trees in flower but a positively skewed curve for female trees (Bullock and Bawa 1981). Bawa (1983) and Rathcke and Lacey (1985) have reviewed the significance of the shape of amplitude curves in flowering.

Supra-annual Flowering. In supra-annual patterns, flowering cycles have multiyear time spans, but regularity and sea-

sonal associations of these patterns cannot be determined with only twelve years of data. In the alternate type of supra-annual pattern flowering occurs on average every other year or every few years and usually occurs in the same season. For example, one female *Jacaratia dolichaula* tree, flowered annually for three years and then flowered every other year for four years. In *Trophis involucrata* (Moraceae), an understory dioecious tree at La Selva, one male tree (shown in fig. 11.3, 11.6) flowered {2 years on, 1 off, 2 on, 2 off, 2 on}, whereas a second male tree (not shown) flowered {3 years on, 1 off, 1 on, 1 off, 4 on}. Yet a third tree, a female, flowered annually, missing only one of eight years.

In the rare supra-annual pattern flowering is very infrequent. For example, two trees of *Andira inermis* (Fabaceae) flowered only once in twelve years, but one tree flowered three times in nine years (fig. 11.9). All three trees flowered in August 1974. Because flowering phases in this species tend to be brief (< one month), more frequent flowering may have been missed by the one-month census interval. *Ceiba pentandra* (Bombacaceae) also flowered supra-annually at La Selva within the forest but more frequently at the margins of the forest (G. W. Frankie pers. observation).

At the population level, over a six-year time span, the understory dioecious *Zamia skinneri* (Cycadaceae) produced cones three times in secondary forest but only twice in primary forest at La Selva (D. A. Clark and Clark 1987b; chap. 7). Cone production in this species represents a seasonal supra-annual pattern with population-level synchrony, a distinction first made by Janzen (1978b).

Community-level Patterns

Overall Pattern. At La Selva the individual level flowering patterns for the tree community level (viz., continual, subannual, annual, and supra-annual) sum to form a continuous community level pattern with seasonal peaks. In other words, the forest is never without some individuals of some species in flower (Frankie et al. 1974a). We do not report analyses for the community pattern from the full twelve-year data here, but in earlier work Frankie et al. (1974a) showed that more species flowered in the main dry season, at the beginning of the early wet season, and close to the veranillo, whereas fewest species flowered in the late wet season. Lowland tropical rain forests typically have continuous flowering patterns at the community level (Richards 1952; Whitmore 1984; Longman and Jenik 1987) and usually have regular annual seasonal peaks in flowering (Koelmeyer 1959; Medway 1972; Dieterlen 1978; Wong 1983; Van Schaik 1986). In Malaysia, however, one of the more aseasonal forests showed no regular seasonal peaks in flowering (Putz 1979) but different sampling and analytical methods make comparisons among studies inconclusive.

Flowering Profiles. What proportions of each type of pattern occur in the tree community at La Selva? Because relatively uniform and favorable conditions prevail in the lowland tropical rain forest, do more trees have continual than annual flowering patterns? Based on our twelve-year data, the flowering profile at the tree community level showed the opposite, 29% of the trees had annual flowering patterns and only 7%

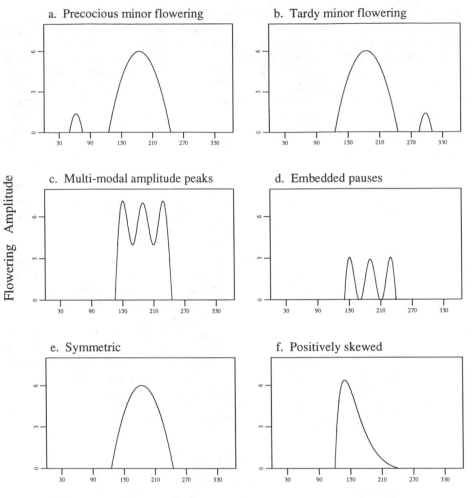

a. Precocious minor flowering

b. Tardy minor flowering

c. Multi-modal amplitude peaks

d. Embedded pauses

e. Symmetric

f. Positively skewed

Flowering Amplitude

Time: One Year in Days

Fig. 11.8. Types of flowering phases according to shape of amplitude curve in annual flowering patterns. (*A*) Major flowering phase with precocious bursts of minor flowering. (*B*) Major flowering phase with tardy bursts of minor flowering. (*C*) Multimodal major flowering phase with multiple peaks. (*D*) Pauses or on/off minicycles embedded in major flowering phase. (*E*) Symmetric major flowering phase. (*F*) Positively skewed major flowering phase.

had continual (fig. 11.10). The most predominant pattern, however, was the irregular sub-annual pattern with 55% of the trees. Supra-annual patterns were not common with only 9% of the trees. (We calculated the flowering profiles for number of trees and also for number of species, but the appropriate unit of analysis is trees because population patterns for each species cannot be characterized with our data. In any case, the results were similar).

Striking differences in climate occur between understory and canopy habitats in lowland tropical rain forests (Richards 1952; chap. 12). Are the flowering profiles likewise different for each strata? For example, Frankie et al. (1974a) found more *extended* flowering patterns in subcanopy than canopy trees. Preliminary assessment of 190 trees in our classification indicated that canopy trees may have a lower proportion of sub-annual and supra-annual flowering patterns than subcanopy trees and almost no continual patterns (fig. 11.10) (see Newstrom et al. 1991). The differences between canopy and subcanopy were significant (Poisson regression, change in scaled deviance = 6.79, 1 df, $p \leq 0.01$, Generalized Linear Interactive Modeling System [GLIM] program; Payne 1986). These strata differences in flowering may be associated with stratification of pollination systems observed by Frankie

(1975) and Bawa, Bullock, et al. (1985) at La Selva (see also chap. 12).

We lack information for life-forms other than trees to make up the complete plant community picture for La Selva (cf. Croat 1978; Dieterlen 1978). Similar quantified flowering profiles at the tree community level have not previously been reported for other forests, but we expect that Malaysian lowland tropical rain forests will have a much higher proportion of supra-annual patterns because of the predominance of dipterocarp trees (Appanah 1985, 1990; Ashton 1988; Yap and Chan 1990).

INTERPRETATION OF FLOWERING PATTERNS

Interpretations of tropical phenological patterns have often confused proximate and ultimate factors (Borchert 1983; Bawa 1983). Proximate factors explain how patterns operate physically, whereas ultimate factors address how patterns are selected in certain species. To elucidate, a night-blooming species may open flowers because of water or chemical changes as a result of light or temperature differences, but the adaptive significance of this strategy lies in the nocturnal foraging habits of its pollinators, such as hawk moths. Proximate

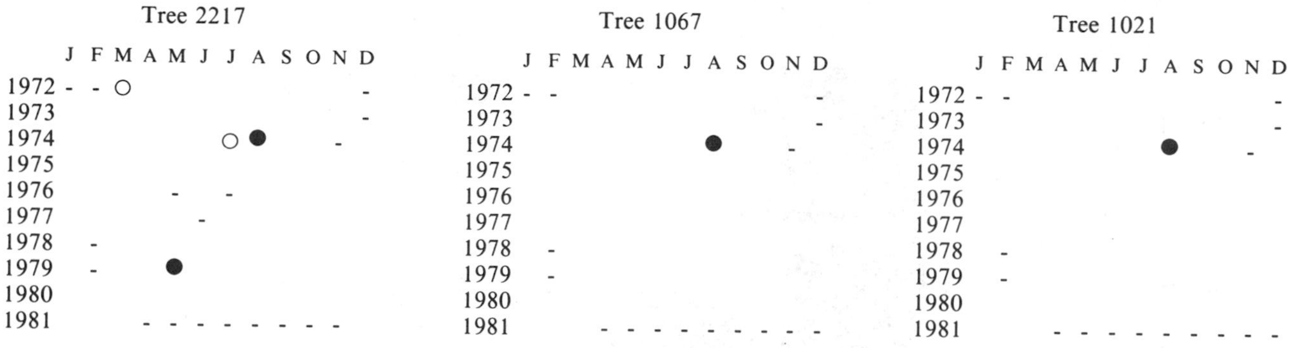

Tree 2217

```
            J F M A M J  J  A S O N D
     1972  - -  O                     -
     1973                             -
     1974              O ●            -
     1975
     1976            -    -
     1977              -
     1978  -
     1979  -        ●
     1980
     1981       - - - - - - -
```

Tree 1067

```
            J F M A M J  J  A S O N D
     1972  - -                        -
     1973                             -
     1974              ●   -
     1975
     1976
     1977
     1978  -
     1979  -
     1980
     1981       - - - - - - - -
```

Tree 1021

```
            J F M A M J  J  A S O N D
     1972  - -                        -
     1973                             -
     1974              ●   -
     1975
     1976
     1977
     1978  -
     1979  -
     1980
     1981       - - - - - - - -
```

Fig. 11.9. Duration and date of flowering in supra-annual patterns. Matrix graphs (year by month) of flowering from 1972 to 1981 in three trees of *Andira inermis* at La Selva Biological Station. All three trees flowered in August 1974, suggesting the possibility that some threshold level of an infrequent cue may have been responsible. Each graph represents flowering in one tree; ● = heavy flowering; O = light flowering; blank = no flowering; — = missing data.

and ultimate factors, thus, require different types of analysis and explanation. For example, to consider ultimate level questions one must relate the phenological patterns of plant reproduction to other constituents of the biota, both plants and animals. The more we know about plant breeding and sexual systems, the better we shall be able to appreciate links between reproductive strategies of plants and the evolution of their phenological patterns.

Ultimate Factors

Pollination systems and pollinator foraging patterns have a major role in the evolution of flowering patterns (Bawa 1983; Feinsinger 1983; Frankie and Haber 1983; Frankie et al. 1983; Rathcke and Lacey 1985). For example, flowering of *Trophis involucrata* in the main dry season and at the driest time of day at La Selva may have adaptive significance for wind pollination in this species (Bawa and Crisp 1980). In another example fewest species bloom in the late wet season at La Selva when selection against flowering may result from frequent day-long heavy rainfall that inhibits pollinator activity and damages unprotected pollen and flowers (I. Baker pers. comm.). In contrast, showers tend to be confined to afternoons in the early wet season, permitting pollinators and flowers to function during the morning. In some species, however, selection for events other than flowering and pollination could also be important (e.g., fruit maturation during time of highest resource availability [Janzen 1978b; Augspurger 1982] or highest seed disperser availability [Snow 1965, 1971; Janzen 1967; Bawa 1983]). Selection may also be strong for optimum germination time, escape from predators, optimum feeding time for larvae of pollinators, and many other life-history aspects (Bawa 1983; Rathcke and Lacey 1985; Primack 1985b, 1987; Aide 1988).

Pollination and flowering can be viewed as a two-way interaction: flowering time affects the population dynamics of pollinators, whereas pollinator activity affects plant reproductive success, mating patterns, and gene flow. That is, pollinators develop foraging strategies and life cycles in response to timing and abundance of plant resources, and, conversely, flowering strategies are successful because of pollinator availability and activities in the context of guild and community-level phenomena.

The type of foraging strategy influences gene flow (e.g., territorial foraging restricts wide crossing between plants and encourages sib crossing and selfing [Baker 1973]). Pollinators, however, are not necessarily locked into particular foraging strategies. Some hummingbirds, for example, switch from territorial to traplining foraging, depending on the abundance and composition of flowering species in the community (Feinsinger and Colwell 1978). Factors that influence the evolution of flowering and pollinator strategies represent a complex multicausal network at the community level. Consequently, no direction of causality is implied in the ensuing discussion.

Guild-level Patterns: Competition for and Maintenance of Pollinators. One of the most important levels of analysis for examining the evolution of pollination systems and flowering patterns is the guild level. Interpreting guild-level patterns using our classification reveals four important features of multispecies flowering patterns: gaps, overlap, amplitude fluctuations, and permutations. Stiles (1978c) has observed two contrasting guild-level flowering patterns related to hummingbird pollination at La Selva. The first pattern, a *multispecies staggered sequence,* has a regular succession of annual flowering species and is remarkably common in both temperate and tropical areas (Heithaus 1974). At La Selva, such sequences occurred not only among hermit hummingbird-pollinated plants (fig. 5 in Stiles 1978c) but also among beetle-pollinated plants (H. Young 1986a, 1990, and pers. comm.). In other forests, this flowering pattern occurred among medium- to large-bee plants (Frankie 1975; Frankie et al. 1983; Appanah 1985), hawk moth–pollinated plants (Haber and Frankie 1989), and bat-pollinated plants (Baker 1973). The same type of pattern occurs in fruiting (Snow 1965; Smythe 1970) and within some taxonomic groups (Gentry 1974; Appanah 1985; Ashton 1988), to mention only a few examples from the literature. The second guild-level pattern, the *multispecies mixed sequence,* has a variety of different flowering patterns that overlap in a disorderly succession and appears to be most common in the tropics. To elucidate, the guild-level pattern for flowering of nonhermit hummingbird-pollinated plants has a mixture of continual, sub-annual, and annual flowering patterns (fig. 4 in Stiles 1978c).

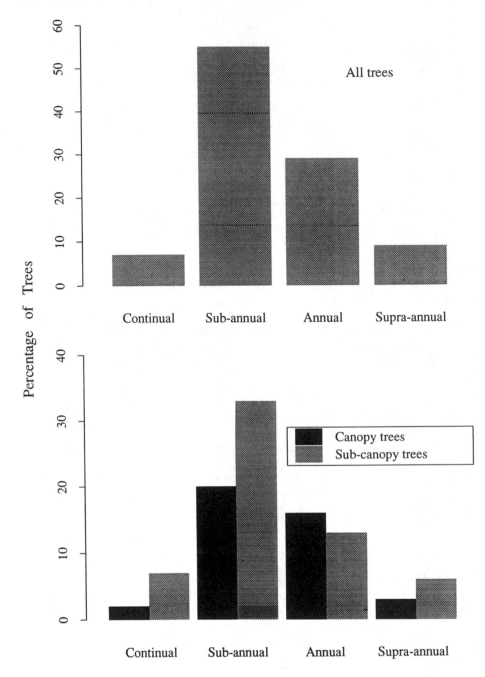

Fig. 11.10. Flowering profiles for (*A*) all trees (*N* = 254) and (*B*) canopy (*N* = 76) and subcanopy trees (*N* = 114) showing the percentage of trees with the four major flowering patterns at La Selva Biological Station. The number of trees is the appropriate unit of analysis but results for the number of species were similar. We excluded forty-six trees with intermediate patterns, all trees with less than five years of data, and all trees of Palmae (graph from Newstrom et al. 1991).

Gaps. In the tropics, guild-level flowering patterns often have few or no gaps (i.e., nonflowering phases). In the staggered sequence of the hermit hummingbird-pollinated species, only one brief gap occurred in four years of continual flowering, whereas in the mixed sequence of nonhermit hummingbird-pollinated species no gaps occurred (Stiles 1978c). During guild-level gaps pollinators may be dormant (e.g., beetles at La Selva) or migrate out of the area (e.g., some hummingbirds at La Selva). Guild-level flowering phases may be prolonged, lasting for all or most of the year (as in continuous patterns or annual extended patterns), or short, lasting for only a portion of the year (as in annual intermediate or brief patterns (e.g., only seven weeks in arctic zones [P. G. Kevan, pers. comm.]).

Overlap. The degree of overlap in guild-level patterns varies, but in general, mixed sequences have extensive overlap, whereas staggered sequences may have minimal overlap. Several authors have proposed that nonoverlapping staggered sequences evolved to reduce competition for pollinators or dispersers (see references in Baker 1973), an idea that has generated much controversy in phenology (e.g., Heithaus 1974; Feinsinger 1983; Pleasants 1983; Rathcke 1983; Primack 1985b; Rathcke and Lacey 1985; Wheelwright 1985a; and Armbruster 1986). There are two separate issues in the controversy: the appropriate statistical test for overlap and the evidence needed to demonstrate that competition for pollinators may be responsible for evolution of the sequence. The sequence of six species of thrip-pollinated *Shorea* in the tropical rain forest of Malaysia did have some overlap (Ashton et al. 1988), but the sequence of ten species of hermit hummingbird-pollinated plants did not have overlap (Stiles 1977). Poole and Rathcke (1979) challenged the existence of non-overlap in the hummingbird example on statistical grounds, but Cole (1981), Stiles (1979c, 1985c), and Pleasants (1990) provided further statistical arguments for it.

The use of null or other models requires careful construction of biologically realistic hypotheses and appropriate statistical approaches to avoid type II errors (see Fleming and Partridge 1984; Armbruster 1986; Ashton et al. 1988). Pleasants (1990) reviewed the statistical problems in testing a pattern for segregation in resource space, commonly referred to as competitive displacement. He concluded that, for floral phenologies, the guild-level parameter "mean pairwise overlap" (as used by Pleasants 1980; Cole 1981; and Ashton et al. 1988) has a higher capacity to recognize competitive displacement than other previously used parameters.

Competition for pollinators as a selective force in phenological strategies, however, is more complex than the accurate framing and testing of statistical hypotheses. Evidence that pollinators have been competing, or limiting plant reproduction, has been reviewed by Waser (1983) and Rathcke and Lacey (1985). Interspecific pollen transfer as a cause for divergence in flowering dates is supported, but no strong evidence exists to support competition for pollinators leading to reduced visitation, decreased pollen donation, or decreased seed set (Rathcke and Lacey 1985). Moreover, coincidence of flowering dates does not necessarily imply sharing of pollinators. Even if species broadly overlap in flowering time, their manner and timing of resource presentation could differ significantly, allowing them to avoid competition for the same pollinators. Studies reported by Frankie and Haber (1983) and Frankie et al. (1983) on large bees provide evidence for this possibility. We may conclude that nonoverlapping staggered sequences are a biological reality but the underlying ultimate causes for the pattern remain a complex issue that requires further investigation.

Amplitude Fluctuations. Flowering amplitude at the guild level tends to fluctuate in concert with changes in pollinator numbers, activity, or needs (see the review by Rathcke and Lacey 1985). Cycles of scarcity in food supply are critical to animal populations and may even determine the carrying capacity of a community (Terborgh 1986b). Seasonality in animal populations and abundance of plant resources have received much attention (e.g., Foster, 1982b; Milton et al. 1982; Windsor et al. 1989; Terborgh 1986b; Gautier-Hion and Michaloud 1989). At La Selva Stiles (1977, 1978c, 1980a, 1985c) found that the main dry season peak of flowering by hummingbird food plants coincides with the bird's breeding season and that the second flowering peak in the early wet season coincides with the peak of moult. In the late wet season, the time of scarcest flowering, hummingbirds' weight and fat reserves decrease and some species migrate out of La Selva while other species visit flowers not well adapted for hummingbird pollination.

A second example, with a supra-annual cycle rather than an annual one, has been studied in Malaysian dipterocarp forests (Chan and Appanah 1980; Appanah and Chan 1981; Appanah 1985; Yap and Chan 1990; La Frankie and Chan 1991).

After many years of sporadic light flowering, a sequence of six *Shorea* species burst into full, high-intensity flowering. At this time the short-lived thrip pollinators have an exponential population explosion. At other times different species that flower more frequently at low or intermediate amplitudes maintain the thrips at low population levels.

Permutations. A fourth feature of guild-level flowering patterns is that the same permutation of plant species may repeat each year. In the *Shorea*-thrip example six species repeated the same permutation in high-intensity flowering years (Ashton et al. 1988; La Frankie and Chan 1991). In the hermit hummingbird example ten species repeated almost the same permutation with only three species changing position slightly in four years (see fig. 5 in Stiles 1978c). In temperate phenology multispecies sequences shift together to earlier or later flowering dates in Spain, indicating that yearly climatic shifts affect different species in the same direction and to a similar extent, thus preserving the order (Arroyo 1990). We lack information on this aspect of tropical multispecies sequences, but in any multispecies sequence with a predominance of sub-annual patterns, the permutations each year would not be in the same order.

Sub-annual Flowering and Fig Pollination. One of the most interesting associations between pollinators and flowering patterns occurs in the fig-wasp pollination system because the obligate mutualistic relationship means that flowering time has critical consequences for pollinator survival. More than 750 species of figs (*Ficus*, Moraceae) are pollinated by species-specific wasps (Agaonidae, Hymenoptera) that live and reproduce inside the fig syconia, the structure enclosing both flowers and fruits. The wasps fly from their natal syconia to a different receptive syconia but apparently do not survive free-living for more than one week (Kjellberg and Maurice 1989; Bronstein 1989).

A reinterpretation of fig phenology using our classification shows at least two main types of flowering patterns: (1) sub-annual and supra-annual flowering at the individual level combined into a continual pattern at the population level in at least fifteen monoecious tropical species (see citations in table 1 of Bronstein et al. 1990; Milton et al. 1982; Michaloud 1988; Windsor et al. 1989; Berg 1989) and (2) a specialized annual flowering pattern at both the individual and population levels in the dioecious temperate *Ficus carica* (Valdeyron and Lloyd 1979; Kjellberg et al. 1987).

Although space does not permit full exploration of the intricate links among phenology, pollination, and plant sexual systems in figs (especially for the complex temperate pattern in *F. carica*), several points are important in considering monoecious tropical figs. First, at the individual level selection may favor sub-annual patterns over continual if high within-crown synchrony enhances the power of the presumed olfactory cue that syconia may produce to attract wasps (Janzen 1979; Bronstein 1989; Frank 1989). Synchronized flowering within a tree also promotes outcrossing (Janzen 1979). Second, because flowering is protogynous (i.e., female flowers open before male flowers), the population must be asynchronous for temporal coincidence of male and female flowering phases, which is essential for pollination (Bronstein et al.

1990). Sub-annual flowering produces highly asynchronous populations because of the irregular dates of onset among trees. Third, local wasp survival depends on continual availability of newly receptive female phase trees (Janzen 1979; Bronstein et al. 1990), so the number of participating trees throughout the year is crucial (see Bronstein et al. 1990). The population as a whole can maintain a continual flowering pattern with sub-annually flowering trees because the onset dates are scattered throughout the year although higher population-level amplitude may occur in certain seasons (see Milton et al. 1982; Windsor et al. 1989). Thus, a sub-annual flowering pattern at the individual level with continual flowering at the population level has adaptive value for fig-wasp pollination in monoecious tropical figs. Other selective forces, however, may promote asynchrony within trees (see Janzen 1979; Bronstein 1989; Frank 1989; Kjellberg et al. 1987; Kjellberg and Maurice 1989), which would lead to longer duration flowering phases or, ultimately, a continual flowering pattern at the individual level.

The fig-wasp system is species specific, hence, the guild level is equivalent to the population level. Other species with sub-annual flowering at the individual level appear to have sub-annual rather than continual population-level flowering patterns (e.g., *Guarea rhopalocarpa* (moth pollinated) (Bullock et al. 1983) and *Compsoneura sprucei* (thrip pollinated) (Bullock 1982), both dioecious species. For plants that do not have species specific pollinators, we need to know how the population-level patterns combine at the guild level to understand the relation between flowering patterns and pollinators.

Foraging Strategies. For pollinators the duration of flowering is important because of the time needed to develop search images. Two flowering patterns classified with respect to duration have been associated with two different pollinator foraging strategies: *extended* flowering and traplining pollinators (Janzen 1971a; Baker 1973; Gentry 1974; Frankie et al. 1974b; Frankie 1975; Stiles 1975; Feinsinger and Colwell 1978) and *mass* flowering and opportunistic pollinators (Baker 1973; Gentry 1974; Frankie 1974b; Augspurger 1980, 1983b; Baker et al. 1983; Bawa 1983; Primack 1985b; Rathcke and Lacey 1985). These two flowering patterns cannot be reinterpreted using our classification because the frequency and regularity associated with them have not been well documented. *Extended* flowering, generally characterized by long duration flowering phases with low amplitude and low synchrony, could include continual, sub-annual or annual extended cycles, whereas *mass* flowering, generally characterized by short duration flowering phases with high amplitude and high synchrony, could include sub-annual, annual brief, or supra-annual cycles. How these patterns combine at the guild level has received little attention, but the flowering patterns each have adaptive value at the individual and population levels.

Extended Flowering and Trapliners. Extended types of flowering have been reported as most common in the lowland tropical rain forest understory (Frankie et al. 1974a; chap. 12) but generally not in tropical dry forests (D. Janzen pers. comm.; Frankie 1975) or in temperate forests (D. Janzen pers. comm.; Gentry 1974). The association with trapliners relies on low amplitude (a few flowers per day) and widely dispersed plant populations so that the pollinators forage from plant to plant over long distances. Euglossine bees (Janzen 1971a), *Heliconius* butterflies (Gilbert 1975), some hummingbirds (Stiles 1975; Feinsinger and Colwell 1978), and bats (Baker 1973; Baker et al. 1983; Bawa 1983) are known to forage in this manner. Long duration flowering provides resources for pollinators for most or all of the year and plants can presumably afford to stay in flower because of the low intensity. Plants pollinated by trapliners must produce enough floral resources to merit pollinator visits, yet not so much that pollinators do not move to another plant (Janzen 1971a; Heinrich and Raven 1972). When plants become clumped, pollinators no longer need to engage in circuit foraging. For example, at La Selva, dispersed *H. patens* trees in the forest had traplining visits by hummingbirds, but the clumped trees in the *pejibaye* grove were defended by territorial hummingbirds (R. Colwell pers. observation).

Bawa (1983) and Rathcke and Lacey (1985) have suggested that *extended* flowering has several advantages for a plant's reproductive success. Long duration of flowering may result in lower risk of reproductive failure because of bad weather or lack of pollinators. It may also provide potential control of the relative investment in flowers and fruits to match short-term patterns of resource availability. The low amplitude of flowering reduces geitonogamy (pollination between different flowers on the same plant). Asynchrony between individuals and long duration of flowering results in high diversity of matings over time for outcrossing plants.

Mass Flowering and Opportunistic Pollinators. Mass flowering patterns have been reported as common in lowland tropical rain forest canopies and in tropical dry forests during the dry season (Frankie et al. 1974a; Frankie 1975) but not in temperate forests (Gentry 1974). Problems with the time scale of *mass* flowering have been discussed previously. Here we use the term to mean a pattern with a large display of flowers opening simultaneously with such brief (less than one month) duration that pollination depends on rapid response by pollinators to visual or olfactory cues. Flowers in leafless tropical dry forest trees (Janzen 1967a) or the canopy of lowland tropical rain forest (Frankie 1975; Bawa 1983) have enhanced visibility over long distances.

A sudden massive display of flowering in trees depends upon the availability of preformed dormant buds that open on cue (Opler et al. 1976). Plants with this pattern usually attract a large number of pollinators from different taxonomic groups, such as bees, wasps, and flies. For example, up to fifteen thousand bees representing seventy species were recorded visiting six *Andira inermis* trees in tropical dry forest of Guanacaste (Frankie et al. 1976). The abundance of flowers encourages foragers to remain concentrated within a plant, but pollinators do move between plants because of aggressive, competitive interactions (Frankie and Baker 1974; Frankie et al. 1976), to avoid predators (Gentry 1978b), or possibly to follow changes in timing of nectar rewards between plants (Frankie and Haber 1983; Frankie et al. 1983). Even the low rate of pollinator movements typical among *A. inermis* trees produced abundant fruit set (Frankie et al. 1976).

Selection for brief blooming durations may be related to

competition for pollinators and minimizing interspecific gene flow (Bawa 1983). Plants rare in space or time should flower "massively" because a small number of flowers will not attract pollinators (Augspurger 1982; Bawa 1983), whereas clumped or continually flowering plants do not need large individual displays. Outliers in space or time also may not escape seed predation, which may also select for mass flowering (Janzen 1978b; Silvertown 1980). The briefer the flowering, the more important high within-population synchrony will be for outcrossing plants (Bawa 1983; Rathcke and Lacey 1985). Many briefly flowering plants, however, are able to self (Bawa 1983). We must, therefore, consider breeding system, nature and timing of rewards, and dispersion of the plant population when interpreting the effects of pollinator foraging on flowering patterns.

Plant Sexual Systems. In flowering phenology male and female functions may be separated in time at any level of analysis. Temporal separation of male and female functions from the intrafloral level up to, but not exceeding, the whole plant level are referred to as dichogamy (Lloyd and Webb 1986), whereas separation of male and female functions between plants is referred to as dioecy (Bawa 1980a). In dioecious species consistent differences in flowering patterns between male and female plants have important consequences for demographic characteristics of plants (see chap. 7). Bawa (1983) explained these intersexual differences in flowering patterns for both dioecious and monoecious species in terms of sexual selection theory, that is, males, with energetically inexpensive gametes, optimize quantity of matings, whereas females, with expensive gametes and parental investment costs, optimize quality of matings (see references in Bawa 1983).

At the individual level male trees flower more frequently than females in species with sub-annual patterns as in *Guarea rhopalocarpa* (Bullock et al. 1983) and *Composoneura sprucei* (Bullock 1982) and annual patterns as in *Jacaratia dolichaula* (Bullock and Bawa 1981). This translates at the population level into a higher number of male trees in flower at any given time. The difference may result in females belonging to a different flowering frequency class than males; for example, sub-annual patterns in most males and alternate patterns in many females of *C. sprucei* or annual patterns in most males and alternate patterns in many females of *J. dolichaula*. The tendency for alternate bearing in females may reflect the cost of fruit production or correlative influences from fruits (Bullock and Bawa 1981; Bawa 1983; Monselise and Goldschmidt 1982).

In general, males tend to have more reproductive units and more rewards for pollinators than females (Bawa 1983). In *G. rhopalocarpa*, *C. sprucei*, and *J. dolichaula* male trees had higher intensity of flowering than females at the individual level. For both annual and sub-annual patterns in these three species flowering phases in males last much longer than in females. In addition, for the supra-annual pattern in *Zamia skinneri*, the duration of male pollen release was longer than that of female cone receptivity at both the individual and population levels (D. A. Clark and Clark 1987b).

Seasonal patterns with annual or supra-annual frequencies have an option that does not exist in sub-annual patterns—the control of onset date for earlier flowering in males. In *J.*

dolichaula the population-level pattern has a male phase (with only male trees flowering) that precedes a hermaphroditic phase (with both male and female trees flowering). Population-level supra-annual cone production of *Zamia skinneri* had the same pattern as do many other species cited in Lloyd and Webb (1977) and Bawa (1983). Such earlier male reproduction confers an advantage if early entrainment of pollinators leads to increased chances of pollen deposition when females become receptive (Bullock and Bawa 1981; D. A. Clark and Clark 1987b).

Proximate Factors
At the mechanistic level, flowering patterns are interpreted with respect to internal (endogenous) and external (environmental) factors. In the relatively equable climate of the lowland tropical rain forest what factors might control annual flowering so that some species flower in the early wet season *(Dipteryx panamensis)*, others in the main dry season *(Trophis involucrata)*, and still others in the transition between seasons? In addition, some species flower more frequently in the lowland tropical rain forest than in tropical dry forest, whereas other species show the reverse pattern. From studies of flowering physiology we know that control mechanisms are basically different in different species and that the flowering process (floral evocation, initiation, and development) proceeds to completion only when the appropriate balance and timing of essential factors are achieved (Bernier 1988). Furthermore, at any stage in the many interrelated steps leading to a flowering event, both internal and external factors can be summed, substituted, or overridden (Bernier et al. 1981a, 1981b; Kinet et al. 1985; Bernier 1988). For example, in *P. arieianum* flowering was delayed in response to leaf loss only at off-peak times (Marquis 1988), suggesting that conditions promoting flowering were stronger at peak times. The nutrient diversion hypothesis (Sachs 1977; Sachs and Hackett 1983) and the multifactorial model for control of flowering (Bernier et al. 1981b; Bernier 1988) provide unifying theories for flowering through diverse alternative pathways incorporating many different promotive and inhibitory processes that take place concurrently in various plant parts.

Flowering patterns in tropical trees are thought to be primarily the result of endogenous factors (Hallé et al. 1978; Borchert 1983, 1991). In several tropical dry forest species the internal functional state of a tree may counteract external conditions, resulting, for example in leafing and flowering in the dry season or cessation of growth in the wet season (Reich and Borchert 1982, 1984). Tree size also affects flowering patterns (Borchert 1978, 1980, 1991). Several authors have reviewed the many different internal and external factors thought to play a role in tropical phenology (Richards 1952; Huxley and Van Eck 1974; Whitmore 1984; Longman 1985; Longman and Jenik 1987; Borchert 1983, 1991). We need much more observational and experimental information, particularly at the individual level, before any factors can be unequivocally identified as important for lowland tropical rain forest species. It is beyond the scope of this chapter to discuss all potentially important factors so we have limited our comments to suggestions about the potential role of a few factors in three different flowering patterns at La Selva.

Continual Flowering. One of the striking intraspecific changes in flowering pattern from one ecosystem to another occurs in *Hamelia patens,* which flowered continually at La Selva but annually in the tropical dry forest in Guanacaste. Thus, in drier seasons, this species flowers less intensely in the lowland tropical rain forest and not at all in the tropical dry forest. This suggests that lack of moisture inhibits flowering in this species and that conditions during the main dry season at La Selva are sufficiently dry to affect flowering intensity but not to inhibit flowering altogether.

Sub-annual Flowering. Almost no information exists on mechanisms that may control irregular sub-annual flowering patterns. Species with this flowering pattern may be viewed as similar to continually flowering species except that they have longer and more frequent nonflowering phases. Thus, the pattern may be primarily the result of inhibiting factors such as constraints from branch growth patterns (tree architecture) or the effects of internal interactions similar to those proposed in Borchert's (1978, 1991) model in which internal changes in water economy of the tree have an important role. The most interesting feature of sub-annual flowering is that, at the individual level, a number of trees never flowered in a certain month. This result suggests that some inhibiting factor reoccurs annually, but the months differ for different trees of the same species (e.g., May, June, and December in *Guarea rhopalocarpa* and April and June in *Compsoneura sprucei*). In both these species different trees participated in different population-level flowering phases. In particular, in *G. rhopalocarpa,* the entire population seldom or never flowered simultaneously, but the entire population frequently went out of flower simultaneously (Bullock et al. 1983). In general, less-frequent and less-intense flowering occurred in the wetter seasons in these two species. Perhaps, excess rainfall or low light conditions during these seasons produced synchronized nonflowering phases at the population level.

Supra-annual Flowering. Monselise and Goldschmidt (1982) have reviewed the mechanisms proposed to explain alternate flowering cycles in horticultural species (e.g., correlative influence of fruits as carbohydrate sinks preventing the next season's floral bud induction). In rare supra-annual patterns the infrequency of flowering may be the result of the occurrences of a critical level of some cue needed for floral bud induction. The simultaneous flowering in August 1974 in three *Andira inermis* trees suggests such a mechanism.

Hydroperiodicity (dry/wet cycles) may have a role as a cue for floral bud induction in some supra-annually flowering species. Trees of both *A. inermis* and *Ceiba pentandra* flowered less frequently (rare supra-annually) at La Selva and more frequently (alternate supra-annually) in tropical dry forest of Guanacaste (G. W. Frankie, personal observation). In Africa *C. pentandra* flowered regularly every year in dry forest and supra-annually in wet forest (Baker 1965). *Shorea* species in Malaysia demonstrate a similar pattern with trees flowering most often and regularly in drier forests, supra-annually in wet forests, and scarcely at all in swamp forests (Yap and Chan 1990). Experimental evidence for xeroinduction (an absolute requirement for water stress to initiate floral buds) in lowland tropical rain forest exists for the herbaceous *Geophila renaris*

(Bronchart 1963; Bernier 1981a), but this mechanism has not been investigated for lowland tropical rain forest trees. In tropical dry and moist forests and in tropical agriculture the importance of moisture stress and other related internal and external factors have been investigated for shrubs and trees (e.g., Alvim 1960; Daubenmire 1972; Opler et al. 1976; Alvim and Alvim 1978; Augspurger 1980, 1981, 1982, 1983b; Borchert 1980, 1983, 1991; Reich and Borchert 1982, 1984; Wright and Cornejo 1990a, 1990b; Wright 1991).

Other factors may play a significant role in supra-annual flowering. For example, in swamp forests (or in waterlogged soil during wet seasons), growth and, therefore, flowering may be inhibited by poor soil aeration preventing activity in the root system (Oldeman in Alvim and Alvim 1978:462). Temperature changes have also been suggested to influence supra-annual flowering (Ashton et al. 1988), but we need to distinguish among factors in terms of which stages of the flowering process they are likely to affect (e.g., floral bud induction compared with bud break [see Borchert 1983]). The diverse stages of flowering do not react similarly to the same external and internal variables (Bernier 1988).

Supra-annual patterns, however, may be primarily the consequence of insufficient levels of some promoting factor. One of the most important environmental factors promoting flower production is the "light integral" (the integral of irradiance X time) (Vince-Prue 1984; Kinet et al. 1985). D. A. Clark and Clark (1987b) proposed that lack of a critical level of resource accumulation because of limiting factors such as light could account for the low reproductive frequency (approximately every four years [see chap. 7]) in *Zamia skinneri* in the La Selva primary forest. Abundance of light may also explain why *C. pentandra* flowered more frequently at forest margins than in closed forest at La Selva and why *C. pentandra* and *A. inermis* flowered more frequently in more open dry forests. Greater frequency and amplitude of flowering have been found in higher light conditions in other species at La Selva (Bullock et al. 1983; Chazdon 1986c; Marquis 1988) and in other tropical forests (Ng 1977; Barthélémy 1986; Van Schaik 1986; S. J. Wright and C. P. Van Schaik, pers. comm.). Large gap, small gap, and understory specialists have different reproductive patterns (Denslow 1980a), and some species do not flower at all until light increases (Hartshorn 1980). Changes in flowering patterns under different light conditions as a function of cloud cover, gap opening, and canopy closure merit further investigation.

The interaction of external factors, such as light intensity, cloud cover, and rainfall, in natural experiments makes these factors difficult to investigate (Longman and Jenik 1987; Wright and Cornejo 1990a, 1990b). Studies comparing flowering patterns in the same species across habitats or ecosystems (e.g., Borchert 1980; Reich and Borchert 1982; Huxley 1983; Akunda and Huxley 1990) and experimental manipulation of variables in the greenhouse (e.g., Longman 1985; Longman and Jenik 1987) or in the field (e.g., Wright and Cornejo 1990a, 1990b) provide important insights for potential roles of the different internal and external factors. More information on long-term individual-level flowering patterns will significantly contribute to this severely understudied topic in tropical flowering phenology.

CONCLUSIONS

Long-term flowering patterns, as classified here at the individual level according to frequency and regularity, reveal a tremendous diversity and unpredictability in flowering of lowland tropical rain forest trees at La Selva. Irregular sub-annual flowering, the most common pattern at the individual level, occurs in more than half the trees in our study, whereas regular annual patterns are the next most common (one-third of the trees) and continual and supra-annual are not as well represented (each less than 10%). Three important characteristics of the sub-annual flowering pattern are variable flowering and nonflowering phase durations, irregular multiple flowering cycles per year, and unpredictable onset dates at any time of year. Even with such high irregularity, distinct seasonal trends typically occur in this pattern. What proximate mechanisms might govern such an irregular pattern is an open question, but inhibitory factors or constraints resulting from tree architecture may be important.

The evolution of sub-annual flowering patterns is an equally unexplored question. Except for the tropical monoecious fig-wasp pollination system, data is lacking on the organization of individual-level sub-annual flowering patterns into population level and guild level patterns. Interpreting guild-level flowering patterns using our classification system shows four distinctive features for multispecies patterns: gaps, overlap, amplitude fluctuations, and permutations. At La Selva two guild-level patterns have been distinguished. In the *multispecies staggered pattern* members of the guild flower annually in an ordered, staggered sequence. In contrast, the *multispecies mixed pattern* comprises a disordered mixed sequence of species with continual, sub-annual, and annual flowering patterns.

The field of tropical phenology has not yet had the benefit of standardized quantitative methods and terminology. The complexity of tropical flowering patterns discourages reliance on conventional methods because averages and other statistical measures often lose the character of the temporal sequence, especially for irregular patterns. Graphical and other analytical approaches (e.g., time series, curve smoothing) may produce new insights. The concept of separate levels of analysis, each with a flowering profile showing the proportions of component patterns that, when added together, make up the overall pattern provides another tool for addressing the complexity.

One of the most important tasks in tropical phenology is to develop a common language to communicate concepts and to render interpretations of data more compatible among investigators. We suggest incorporation of explicit time and amplitude scales, precise definitions for terms and classifications, and clear differentiation of patterns at each level of analysis (see summary table 11.1). More quantitative and standardized results will lead to a better understanding of the highly irregular and unpredictable flowering cycles such as the poorly known sub-annual pattern, that we find to be so common in the lowland tropical rain forest at La Selva.

ACKNOWLEDGMENTS

We dedicate this chapter to the memory of Irene Baker, who was a constant source of inspiration and encouragement to us. We thank J. Frankie, R. Echeveria, C. Esquivel, G. Hartshorn, and F. G. Stiles for assistance in recording observations, monitoring data collection, and verifying species names. OTS and NSF funded data collection. Funding for data entry and analyses was provided by OTS, NSF, and the Department of Entomology at the University of California, Berkeley. J. Barthell assisted with data entry and verification. G. Casterline, T. Porco, S. Jacobson, and P. Spector kindly provided consultation for statistical and graphical methods. We are grateful to five anonymous reviewers for helpful criticism of the manuscript. In addition, we thank K. S. Bawa, R. Borchert, E. O. Guerrant, P. Hall, S. Koptur, D. G. Lloyd, L. McDade, S. Naeem, G. Orians, J. Rosenthal, and F. G. Stiles for comments on various stages or portions of the manuscript.

APPENDIX 11.1: TERMINOLOGY

Tropical phenology lacks explicitly defined and consistently used terms for describing patterns. Many terms have multiple connotations (e.g., *periodic* and *cycle*), and a number of terms are used to indicate the same phenomenon (e.g., sub-annual flowering patterns have been called *episodic, intermittent, periodic,* and *multiple bang*). It is important to distinguish between two of the many connotations of *periodic*: the phenomenon of cyclic behavior in which some activity alternates with a state of rest, whether regular or irregular, as used in tree growth literature and only that class of cyclic behavior that has regular cycle lengths or a *period* as used in chronobiology. Sweeney (1987) defines a cycle in chronobiology as a sequence of events repeating in the same order and at the same time interval, a meaning borrowed from the description of oscillations in physics. The corresponding terms *frequency* (number of cycles per unit time) and *period* (1/frequency) are determined by spectral analyses (Sweeney 1987; Platt and Denman 1975). In tropical phenology Putz (1979) first used this type of analysis for periodicities of flowering peaks at the community level in Malaysia.

In our classification, to incorporate information on the irregularity of the temporal sequences in flowering patterns, we have defined a cycle as a sequence of "on" and "off" phases (such as flowering and nonflowering) that may have equal or variable cycle lengths and frequency as the number of cycles per year. Whether the concept of a "variable periodicity" is useful remains an open semantic question (cf. Newstrom et al. 1991), but to avoid confusion, we prefer to use the terms *period, periodic,* and *periodicity* as they are used in physics and chronobiology (the same sequence of events reoccurring at the same time interval). For example, in our classification, the annual pattern has regular cycle lengths (one year) and the sub-annual pattern has irregular cycle lengths (most less than one year). In chronobiology the annual pattern would have a period of one year, but the sub-annual pattern appears to be *aperiodic*. Spectral analysis of the sub-annual pattern may reveal a tendency for bimodal peaks per year in flowering amplitude. Further confusion arises from the use of *periodic* as well as *episodic* to mean a rare or unusual event in a long-term time span. For this reason we have avoided using *periodic* (Haber and Frankie 1989) and *episodic* (Bullock et al. 1983) to refer to the sub-annual flowering pattern.

Variables for describing flowering patterns are not necessarily independent; a study of their interrelationship is needed. For instance, high synchrony usually accompanies high amplitude and short duration (Augspurger 1983b). To avoid proliferation of new terms we use the same terms for any phenological event (leafing, flowering, and fruiting) and for patterns at all levels of analysis; therefore, leafing and fruiting can be substituted for flowering in the following definitions.

> **Amplitude.** The quantity of activity, or intensity of response, such as number of flowers on a tree, or flowering trees in a population, or species in a guild or community. A flowering phase may have multiple flowering amplitude peaks (fig. 11.8C) that can be measured or estimated qualitatively or quantitatively (e.g., Frankie et al. 1974a; Opler, Frankie, et al. 1980; Augspurger 1983b). Counting number of inflorescences, however, may be practical only in understory plants (e.g., Bullock et al. 1983). The problem of resolution arises here as broader categories obscure more subtle fluctuations.

> **Cycle.** A repeating sequence of "on" and "off" phases, such as flowering and nonflowering. Each repetition of the cycle may be variable in length (as in sub-annual flowering) or nearly equivalent (as in annual flowering). For some annual flowering patterns a larger cycle of one major flowering phase per year contains brief "pauses" of nonflowering within in, which we refer to as embedded minicycles in the larger cycle or *pulsed annual* (fig. 11.8D). Monocarpic plants do not have repeating cycles at the individual level but only one reproductive cycle per lifetime (e.g., *Tachigalia versicolor* [Foster 1977] and some bamboo species [Janzen 1976b]).

> **Date.** The calendar time in days, months, or seasons of the year when the event occurs; it is an important reference point for relating plant phenology to external abiotic or biotic cycles. Date of flowering can be measured for the first onset, or maximum, modal, or last flowering, each reflecting different properties of the plant's reproductive effort (Primack 1985b).

> **Duration.** The length of time a unit remains in a given phase or cycle. Categories of duration in our classification are **brief** (less than one month, **intermediate** (more than one month and less than five months), and **extended** (more than five months). The problem of resolu-

tion arises in that a long census interval will overestimate the duration of each phase and underestimate the number of cycles per unit time. Daily observations are usually impractical, however, and a two-week or one-month census interval is commonly used.

> **Frequency.** The number of cycles of "on" and "off" phases per unit time. We defined four arbitrary frequency classes based on an annual time scale (figs. 11.1, 11.3): **continual** = always in flower with no or few brief interruptions; **sub-annual** = > one cycle per year; **annual** = one cycle per year; and **supra-annual** = multiyear cycles. Within the continual class, a **continuous** pattern means that there are no interruptions at all, but this is not common at the individual and population levels. Within the supra-annual class, more frequent flowering patterns with one to three years of flowering interspersed with one to three years of nonflowering belong to the subclass **alternate** and infrequent flowering patterns to the subclass **rare**.

> **Phase.** Any defined portion of a complete cycle, such as flowering or nonflowering. The nonflowering phase at the population or guild levels can also be referred to as a gap. In many phenological studies, the term *episode* refers to the flowering phase and *interval* to the nonflowering phase.

> **Regularity.** The variability in length of cycles or phases. Regular patterns, such as annual flowering, have consistent cycle lengths and duration of component phases, whereas irregular patterns, such as sub-annual flowering, have variable durations of both the cycle and component phases. Predictability analyses of Colwell (1974), Raveh and Tapiero (1980), and Stearns (1981) may provide a useful tool to quantify this variable (see Putz 1979).

> **Seasonal.** The temporal association of an event with a recognizable climatic season. Although an annual pattern is always seasonal in our data, the term *annual* refers to frequency (one cycle per year), not to seasonality. Supra-annual patterns may or may not have seasonal associations.

> **Synchrony.** The simultaneous occurrence of the same event in most or all of the units being considered (e.g., flowers on a tree or flowering trees in a population). Different quantitative measures of synchrony have been used (Primack 1980; Augspurger 1983b; Lack 1982; Wright 1991).

12

Flowering Plant Reproductive Systems

W. John Kress and James H. Beach

La Selva Biological Station has had a notably rich history of research on plant reproductive biology. Work on floral biology at La Selva began in 1968 with multiyear community surveys of tree and shrub phenology (Frankie et al. 1974; Opler, Baker, et al. 1980; chap. 11). Bawa and co-workers initiated an extended research program in 1976, using both descriptive and experimental techniques to elucidate flowering phenology, sexuality, pollinators, and breeding systems of many canopy and subcanopy species. From the mid-1970s on many additional species-level studies of plants found in all forest strata have been carried out at La Selva. As a result, considerable data exist on pollinator relationships and sexual and breeding systems of La Selva plants.

Researchers addressed several broad conceptual issues in the ecology and evolution of plant reproductive systems, and their work provides foci for organizing the available information for La Selva and comparing it to what is known about other tropical forests. First, based on the observation that wet tropical forests are characterized by high species diversity and low population densities, two speciation models for tropical trees were proposed: inbreeding followed by sympatric speciation (Fedorov 1966) and outcrossing coupled with allopatric divergence (Ashton 1969). To address this debate research was initiated at La Selva to inventory sexual and incompatibility systems of forest trees (Bawa 1979; Bawa, Bullock, et al. 1985). These studies demonstrated a high incidence of obligate outcrossing among trees at La Selva, which supported the allopatric model of species divergence. Studies on the actual genetic population structure and mating patterns of trees, however, have only just begun (Bawa and O'Malley 1987; O'Malley and Bawa 1987).

A second major conceptual issue addressed by studies at La Selva concerns interactions between plants and their pollinators. Early qualitative reports of pollinator relationships have served as a point of departure for quantitative and experimental approaches to problems at the community and species levels. At the community level phenomena such as the distribution of pollination systems as related to vertical stratification of the forest (Bawa, Bullock, et al. 1985) and the temporal organization of flowering among plants that share the same pollinator (Stiles 1975, 1978c) have been studied. At the species level research has been directed toward understanding the ecology and evolution of different plant-pollinator associations: small bees and dioecious plants (Bawa and Beach 1981;

Beach 1981); hummingbirds and heliconias (Stiles 1975; Kress 1983a, 1983b); and beetles and aroids (Young 1986a, 1986b, 1988), palms (Beach 1984; Henderson 1984), and cyclanths (Beach 1982).

Other studies have examined ecological constraints on the evolution of various pollination modes and breeding systems, the diversity of incompatibility systems, and the systematic distribution and diversification of reproductive systems in specific plant groups. Most of the studies conducted at La Selva can be interpreted in the context of the adaptive significance of reproductive traits that promote outcrossing.

In light of these conceptual issues we assemble and summarize here the available information on floral biology of trees, shrubs, and herbs at La Selva. Pollinators, sexual systems, and incompatibility systems for many flowering plant taxa at La Selva are compiled in appendix 12.1. For canopy and subcanopy species, data are largely from Bawa, Bullock, et al. (1985). Excluded from appendix 12.1, however, are taxa listed by those authors for which pollinator data are not available. Information on understory plants comes from a number of sources as indicated. Some of these data are from unpublished observations by the authors. These are anecdotal observations that have not been studied in-depth; they are included for completeness and as indicators of taxa or life-forms in need of more research. Historical emphasis has been on the biology of trees at La Selva with the result that shrubs and herbs have been less well studied. We have compensated here by placing considerable emphasis on these latter life-forms.

The patterns elucidated for La Selva are then compared to data on reproductive biology that are available for other tropical forests. Although limited, community-level work at other sites permits a number of generalizations about the reproductive biology of tropical plants. These comparisons, in turn, point clearly to gaps in knowledge and lead to a discussion of key areas for future research.

FOREST STRATIFICATION

To elucidate the diversity of plant reproductive systems at La Selva, we present our overview in the context of spatial distribution in the forest. Plants at La Selva can be categorized into canopy, subcanopy, and understory strata, depending on their vertical position at reproductive maturity. Herbs, shrubs, and small trees less than 5m high make up the understory. Trees more than 5m tall but of less than canopy stature at maturity are considered subcanopy. The canopy includes those trees flowering at the top of the forest, including emergent species. Because some species begin flowering in the subcanopy but eventually reach the upper stories of the forest, it is sometimes more appropriate to refer to the canopy and subcanopy together as the *upper stratum*. Shrubs and herbs in the understory then comprise the *lower stratum*. Although we initially divided our data on reproductive systems into canopy, subcanopy, and understory, it became apparent that most differences are between the upper and lower strata. We also recognize that some large shrubs and small trees provide a gradual intergradation of the lower and upper strata that complicates any interpretation of vertical stratification. Yet even if the assignment of some taxa to a particular stratum is difficult, vertical stratification in tropical forests appears to affect the evolution of plant and animal morphology and life-history traits (Smith 1973) and of plant reproductive systems as demonstrated here.

Trees at La Selva comprise a group of about 320 species belonging to 60 families. The abundance of these species varies from sixty individuals/ha ($\geq$ 10 cm dbh) for the dominant *Pentaclethra macroloba* (Willd.) Kuntze, to taxa known only from one individual (D. Lieberman and Lieberman 1987). In addition, the trees range in stature from 5 m to more than 50 m. Fabaceae, one of the most important families of Neotropical forests, is the most diverse family of trees at La Selva (table 12.1). Legumes are also abundant in a number of individuals at La Selva, second only to the family Arecaceae in number of stems greater than 10 cm dbh per hectare (D. Lieberman and Lieberman 1987). Other important families of trees at La Selva are Lauraceae, Rubiaceae, Euphorbiaceae, Melastomataceae, Meliaceae, Moraceae, Annonaceae, Sapotaceae, and Bombacaceae (table 12.1; see also fig. 6.4). In the context of high diversity and variation in population density and size of individual trees, it is worth noting that research on the reproductive ecology of La Selva trees has emphasized common and accessible species belonging to major taxonomic families. A small amount is known about many species as the result of broadly targeted community surveys, but the reproductive biology of very few trees has been examined in comprehensive detail (e.g., *Dipteryx panamensis;* Perry and Starrett 1980).

Unlike the upper level of the tropical forest at La Selva, which is dominated by dicotyledons, monocotyledons are a much more significant component of the understory. Even though some palms and epiphytic orchids, bromeliads, and aroids are present in the canopy and subcanopy, these families as well as members of the herbaceous Cyclanthaceae, Commelinaceae, Costaceae, Heliconiaceae, Marantaceae, Poaceae, and Zingiberaceae are very diverse in the understory (table 12.1). In addition, many dicotyledonous shrubs and small trees in the families Acanthaceae, Annonaceae, Gesneriaceae, Melastomataceae, Moraceae, Myrsinaceae, Piperaceae, Rubiaceae, and Solanaceae are found in the lower stratum of the forest (table 12.1). A current estimate of the numbers of taxa in the understory at La Selva (excluding epiphytes) is 365 species in 154 genera and 62 families (M. H. Grayum, Kress, Hammel, and Schatz unpublished data). About one hundred of these species are rare (only one or two

individuals are known at La Selva), leaving 265 species in 129 genera and 59 families that account for most of the individuals occuring in the understory. The accessibility of understory plants for observation and manipulation has allowed more in-depth studies of the floral biology of taxa in this stratum than in the canopy, for example, *Aphelandra* (McDade 1984, 1985), *Cyclanthus* (Beach 1982), *Dieffenbachia* (Young 1986a, 1986b, 1988), and *Heliconia* (Stiles 1975; Kress 1983a, 1983b).

POLLINATION SYSTEMS

As a result of both community-wide inventories and taxon specific studies, a great deal is known about the diversity and distribution of pollinators among plant taxa at La Selva (app. 12.1; table 12.2). To establish with certainty the pollinator of a plant species requires considerable effort to demonstrate the actual transfer within or between flowers of viable pollen. Most studies do not fulfill this requirement. For this reason some controversy exists over which animals are simply visitors to flowers and which are pollinators. Without bogging down in this controversy, we have tried to include well-documented reports as well as logical inferences when identifying pollinators. For example, extrapolations have been made to La Selva species from studies carried out on the same species at other wet forest sites or on congeneric species sharing the same floral characteristics. In all cases these inferences are clearly marked in the appendix with literature citations or as preliminary observations in need of further documentation. We have recognized the limitations of these data in drawing our conclusions. Where more than one flower visitor is important in the pollination of a species, several are listed. In such cases the first is considered the most important and is used in the compilation of summary tables. The class "small diverse insects" is used where a single most important pollinator could not be identified (see Bawa, Bullock, et al. 1985). In many cases the syndrome probably does represent a generalized pollination system, but the number of taxa in this category is likely to decrease when additional critical observations are made.

Upper Stratum

Bawa and co-workers (Bawa, Bullock, et al. 1985) surveyed the pollination systems of La Selva trees. Trees of both the canopy and subcanopy were predominantly pollinated by bees; when pooled across size classes this system comprised 41% and 38%, respectively, of the pollination systems in those strata (table 12.2). "Small diverse insects" make up the next most common category of pollinators in the canopy (27%) and subcanopy (14%). A significant proportion of species listed as having pollinators of this class may prove to be primarily bee pollinated. Thus, bees may ultimately be found to pollinate as many as 60% of canopy and subcanopy tree species. Moths, the next most common pollinators of tree species at La Selva, have been observed on six species of Meliaceae, four Rubiaceae, and two Caricaceae (app. 12.1). Other syndromes are less common and are scattered among various families. Vertebrates (bats and birds) are relatively rare as pollinators in the upper stratum. No canopy trees are known to be wind pollinated. With the exception of the subcanopy species *Trophis*

Table 12.1 Reproductive systems of the largest flowering plant families in the canopy, subcanopy, and understory at La Selva

Taxa (Genera/Species)	Sexual System[a]	Breeding System[a]	Pollination System[a]
Canopy and Subcanopy			
Monocotyledonae			
Arecaceae (7/8)	M	— (PA/PG)	bees/beetles
Dicotyledonae			
Fabaceae (15/42)	H	SI (PA)	bees
Lauraceae (5/20)	H	—	—
Rubiaceae (12/18)	S/H	SI (SH)	moths/bees
Euphorbiaceae (9/11)	M/D	—	diverse insects
Melastomataceae (3/10)	H	—	bees
Meliaceae (4/9)	D		moths
Moraceae (5/8)	M	— (PG)	wasps
Annonaceae (5/7)	H	SC (PG)	beetles
Sapotaceae (1/7)	H	—	—
Bombacaceae (5/6)	H	SI (SH)	bats
Understory			
Monocotyledonae			
Araceae (7/21)	M/H	— (PG)	beetles/euglossine bees
Marantaceae (4/18)	H	SC (SH)	euglossine bees
Arecaceae (7/17)	M	— (PG/PA)	beetles/bees
Poaceae (7/10)	M	—	—
Heliconiaceae (1/7)	H	SC (SH)	hummingbirds
Costaceae (1/7)	H	SC (SH)	euglossine bees/hummingbirds
Cyclanthaceae (4/6)	M	— (PG)	beetles
Commelinaceae (4/4)	H	—	—
Zingiberaceae (1/4)	H	SC (SH)	hummingbirds
Dicotyledonae			
Rubiaceae (17/52)	S/H	SI (SH)	bees/butterflies/moths
Melastomataceae (8/48)	H	—	bees
Piperaceae (1/33)	H	—	—
Acanthaceae (7/14)	H	SC (PG)	hummingbirds
Annonaceae (6/10)	H	SC (PG)	beetles
Solanaceae (5/8)	H	— (SH)	bees
Myrsinaceae (2/7)	H	SC	bees
Moraceae (5/6)	D		wind
Urticaceae (3/5)	D		wind?
Gesneriaceae (2/5)	H	SC (PA)	hummingbirds

Notes: Number of taxa is vouchered species of primary forest plants only, including emergent and gap species. Excluded are epiphytes, vines, lianas, very rare taxa, and those species only known from forest edges or other human-disturbed areas. D = dioecious; H = hermaphroditic; M = monoecious; PA = protandrous; PG = protogynous; S = distylous; SC = self-compatible; SH = simultaneously hermaphroditic (flowers); SI = self-incompatible; — = no data.

[a]Predominant systems in the family.

racemosa (Moraceae), all of the known wind-pollinated taxa at La Selva (six species) are understory plants.

The largest families of trees at La Selva have a diversity of pollinators including bees, beetles, moths, wasps, and bats (table 12.1). In the Fabaceae, La Selva's largest family of trees, most species are pollinated by bees (Arroyo 1981) although many other modes exist (e.g., moth pollination in *Pithecellobium*, Bawa, Bullock, et al. 1985; bird pollination in *Erythrina*, Stiles 1978c; Neill 1987; bat pollination in the liana *Mucuna*, Dobat 1985; beetle pollination in *Pentaclethra*, J. H. Beach unpublished data). Protected nectar and pollen and the trip floral mechanism of papilionoid legumes are thought to be specializations for bee pollination. The Lauraceae and Sapotaceae stand out as important tree families at La Selva; very little is known of their reproductive biology.

The only comprehensive account of the reproductive biology of a canopy tree at La Selva is for the legume *Dipteryx panamensis* (Pitt.) Record (Perry and Starrett 1980), the second most abundant leguminous tree in the primary forest. The flowers of *Dipteryx* are hemaphroditic and function for one day. Mature trees produce a large floral display consisting of an estimated 0.6 to 1.2 million pink lavender flowers during the flowering season. The flowers open at dawn and produce a strong fragrance. An average six microliters of nectar per flower is produced primarily during the first two hours of anthesis with a mean concentration of 34% sucrose equivalents. Perry and Starrett (1980) collected twenty species of bees at the flowers. Several species were nectar robbers that carried few or no *Dipteryx* pollen grains during their interfloral foraging. Other visitors to the flowers included flies, humming-

Table 12.2 Pollination systems of flowering plant species at La Selva

| | Forest Stratum | | | | | | | |
| | Understory | | Subcanopy | | Canopy | | All Strata | |
Pollination System	(N)	(%)	(N)	(%)	(N)	(%)	(N)	(%)
Bee, medium to large[a]	33	21.9	15	20.3	19	37.3	67	24.3
Hummingbird	36	23.8	3	4.1	2	3.9	41	14.9
Bee, small[b]	24	15.9	13	17.6	2	3.9	39	14.1
Beetle	25	16.6	9	12.2	1	2.0	35	12.7
Small diverse insects[c]	7	4.6	10	13.5	14	27.4	31	11.2
Moth	6	4.0	10	13.5	6	11.8	22	8.0
Butterfly	7	4.6	3	4.1	2	3.9	12	4.3
Bat	2	1.3	6	8.1	2	3.9	10	3.6
Wasp	0	0.0	4	5.4	3	5.9	7	2.5
Wind	6	4.0	1	1.4	0	0.0	7	2.5
Fly	5	3.3	0	0.0	0	0.0	5	1.8
TOTAL species	151	100	74	100	51	100	276	100

Note: See appendix 12.1 for references by species.

[a]Category includes mostly Anthophoridae.

[b]Category includes mostly Meliponinini, Halictidae, and Megachillidae.

[c]Category includes small Hymenoptera, Diptera, Coleoptera, Lepidoptera, and Thysanoptera.

birds, and butterflies, but these visitors all failed to operate the floral mechanism that brings the stigma and anthers into contact with the animal. Their analysis of quantitative data on flower visitors at *Dipteryx* verifies the difficulty of identifying a single most important pollinator for a species, particularly in biotically diverse habitats such as La Selva. Although bees appear to be the main pollinators of *Dipteryx,* they found considerable variance among flower visitors in the pollen loads carried and in their efficacy as pollinators.

The Arecaceae are well-represented in both the upper and lower strata and have been the focus of a fair amount of recent research at La Selva and at other sites (Uhl and Moore 1977; Beach 1984; Henderson 1986). At various heights within the subcanopy, *Welfia, Euterpe, Iriartea,* and *Prestoea* are all pollinated by a heterogeneous set of small bees, wasps, and flies. *Welfia* probably has the most restricted suite of pollinating bee species. The low subcanopy palm *Cryosophila* is beetle pollinated. In general, beetle pollination in the upper stratum is probably much more common than we have been able to document because of the limited number of observations of epiphytic Cyclanthaceae and Araceae.

Lower Stratum

Most of the information accumulated on plant reproductive systems in the forest understory has been the result of studies of individual plant taxa. The only surveys of understory plants across taxonomic families are those of Stiles (1978a, 1978b) and Grove (1985). Stiles's comprehensive work on hummingbird-pollinated taxa provided a detailed picture of the diversity of plants serviced by this pollinator guild at La Selva. Grove's survey included ten understory species in five families that were visited by hummingbirds and euglossine bees. Her studies encompassed many aspects of floral biology in addition to pollinator behavior. Additional systematic studies on specific plant families or genera (e.g., Acanthaceae, Annonaceae, Are-

caceae, Gesneriaceae, Heliconiaceae, Marantaceae) have been indispensible in providing a clear picture of the spectrum of pollinators in the understory.

Understory taxa are visited by a diversity of invertebrates and vertebrates. The four largest categories of pollinators (table 12.2) are hummingbirds (thirty-six plant species, or 24%, of the total species studied), medium to large bees (thirty-three plant species, or 22%), beetles (twenty-five plant species, or 17%), and small bees (twenty-four plant species, or 16%). If considered at the level of family, thirteen of the nineteen largest understory families at La Selva (table 12.1) have members that are visited by one or more of these pollinators. Other pollinators are present in much lower percentages in the understory and no other single vector accounts for pollination of more than 5% of the understory flora that has been studied (table 12.2).

Hummingbirds, large bees, and beetles are long-distance fliers that travel considerable distances through the forest during foraging (Janzen 1971a). Ninety-four plant species (62% of the total) and nine of the largest plant families in the understory are visited almost exclusively by these long-distance pollinators: hummingbirds and/or euglossine bees in Marantaceae (Kennedy 1978; Schemske and Horvitz 1984; Grove 1985), Costaceae (Schemske 1981, 1983; Grove 1985), Zingiberaceae (Grove 1985), Gesneriaceae (Wiehler 1983; Grove 1985), Acanthaceae (McDade 1985) and Heliconiaceae (Stiles 1975; Kress 1983b); beetles in Cyclanthaceae (Beach 1982; G. E. Schatz unpublished data) and Annonaceae (Schatz 1987); and beetles or medium to large bees in Araceae (Young, 1986b). These families include both herbaceous and woody species.

The largest family of plants found in the understory, the Rubiaceae, exhibits the greatest diversity of pollination systems. Bees, butterflies, moths, and hummingbirds are all known to visit species in this family (app. 12.1). Conspicuously absent pollinator categories in the Rubiaceae are euglossine bees, beetles, and bats.

Although substantive experimental evidence is absent for most reported cases, wind pollination is found in three lower stratum families: Moraceae (*Sorocea* and *Trophis*), Urticaceae (*Myriocarpa* and, probably, other genera), and Arecaceae (*Chamaedorea*). Although the palm family has been historically and anecdotally perceived as predominantly wind pollinated, work at La Selva has demonstrated that this conjecture is largely fallacious. In the lower (and upper) stratum at La Selva, with the exception of *Chamaedorea,* animal pollination, involving highly specialized plant-pollinator relationships, is the rule.

Very little has been published about the reproductive biology of six of the largest understory families at La Selva: Commelinaceae, Melastomataceae, Myrsinaceae, Piperaceae, Poaceae, and Solanaceae (table 12.1). For Melastomataceae at other sites Renner (1986, 1989 for Amazonia) and Lumer (1980, 1983 for Monteverde, Costa Rica) report visitation by bees of various sizes (and, in one genus, rodents). Other evidence suggests that apomixis[1] may be prevalent in certain gen-

---

[1] *Apomixis* is asexual reproduction; *agamospermy* is a form of apomixis involving reproduction by means of seeds, which are produced by a variety of asexual processes and are not the result of the typical sexual mechanism (meiotic reduction followed by syngamic fusion). See Frankel and Galun (1977) for discussion.

era (*Clidemia, Leandra, Miconia;* Renner 1986, 1989; Sobrevila and Arroya 1982). In Piperaceae pollination by small bees, small flies, and small beetles (Semple 1974) has been described for some taxa. Our observations suggest that bees are visitors to many Myrsinaceae and Solanaceae as well.

MATING PATTERNS

Mating patterns in plants are determined and influenced by many different factors, for example, population density, phenology, sexual system, and physiological self-incompatibility. We now know enough about the diversity and distribution of sexual and self-incompatibility systems in all strata of the forest at La Selva to test some of the hypotheses discussed earlier on mating patterns (extent and distribution of inbreeding compared with outcrossing mechanisms) and the degree of heterogeneity present in tropical plant populations.

The constraints imposed by certain sexual systems, for example, monoecy and dioecy, upon self-fertilization and inbreeding in flowering plants have been indentified in research on both temperate and tropical plant species (Bawa and Beach 1981). In addition to selective pressures for outcrossing, ecological factors, such as pollinator behavior and morphology, that influence the evolution of various sexual systems in angiosperms have been particularly addressed by studies at La Selva (Bawa 1980b; Beach and Bawa 1980; Beach 1981).

Physiological self-incompatibility is known to be common among flowering plants in the temperate zones (de Nettancourt 1977). Information on the extent of self-incompatibility in tropical plants has only become available in the last decade, much of it from studies conducted at La Selva. The metabolic shutdown of the pollen grain or pollen tube after self-pollination effectively prevents self-fertilization and reduces inbreeding. Although self-incompatibility can fail because of physiological stress or as the result of tissue damage, experimental pollinations in the field under natural conditions demonstrate that "leakage" in self-incompatibility is rare in tropical trees (Bawa, Bullock, et al. 1985; table 12.3).

Isozyme electrophoresis has been a valuable tool of geneticists and ecologists investigating the population genetic structure of temperate zone plants and animals, but quantitative estimates of genetic variation and levels of outcrossing present in tropical plant populations are few (see Hamrick and Loveless 1986, 1989; Murawski and Hamrick 1990; Murawski et al. 1990). To fully understand mating patterns in tropical forest plants, quantitative studies on genetic variation will be an important complement to current knowledge of pollination, sexual, and incompatibility systems in these taxa.

Upper Stratum

Floral Sexuality. With respect to floral sexuality, the data of Bawa, Bullock, et al. (1985) coupled with several other reports (J. H. Beach and W. J. Kress unpublished data) indicate that about two-thirds of the trees and shrubs occurring in the canopy and subcanopy produce hermaphroditic flowers (table 12.3). Five distylous[2] subcanopy species have been studied,

Table 12.3 Sexual systems of flowering plant species at La Selva

| Sexual System | Forest Stratum | | | | | | All Strata | |
| | Understory | | Subcanopy | | Canopy | | | |
	(N)	(%)	(N)	(%)	(N)	(%)	(N)	(%)
Hermaphroditic	168	74.7[a]	124	67.4[b]	64	65.3	356	70.2
Monoecious	35	15.5	18	9.8	10	10.2	63	12.4
Dioecious	22	9.8	42	22.8	24	24.5	88	17.4
TOTAL species	225	100	184	100	98	100	507	100

Sources: Data primarily derived from the 372 species cited in appendix 12.1 plus 88 subcanopy and 47 canopy species enumerated by Bawa, Perry, et al. (1985; not included in app. 12.1). Some of the species listed in that publication were reclassified according to the definitions of strata followed here.
[a]Including fourteen distylous species.
[b]Including five distylous species.

but more heterostylous species are likely to be added to the overall count when additional field observations are undertaken.

Monoecism in the canopy and subcanopy is relatively uncommon (twenty-eight species; table 12.3) and is concentrated in three families: Arecaceae, Moraceae, and Euphorbiaceae. Studies on epiphytic members of the Cyclanthaceae and Araceae will probably increase the number of monoecious species in the upper stratum (J. H. Beach unpublished data). All of the upper stratum palms at La Selva are monoecious with separate male and female flowers in the same inflorescences. The temporal ordering of gender, however, splits the palm species into two reproductive groups. In the upper stratum, *Welfia, Euterpe,* and *Iriartea* are protandrous, whereas *Astrocaryum* and *Cryosophila* are protogynous.

Dioecious species comprise 24% and 23%, respectively, of the canopy and subcanopy tree floras (Bawa, Bullock, et al. 1985; table 12.3). Work on the reproductive ecology of La Selva's dioecious species has contributed significantly to the debate about the evolution of dioecism (Bawa 1980b; Beach and Bawa 1980; Beach 1981). Reproduction of dioecious taxa has been examined in more detail than any other sexual system of trees at La Selva (Beach and Bawa 1980 on *Coussarea;* Bawa and Crisp 1980 on *Trophis;* Bawa 1980b; Bullock and Bawa 1981 on *Jacaratia;* Bullock 1982 on *Compsoneura;* Bullock, et al. 1983 on *Guarea;* see chap. 11).

Self-incompatibility. Among the hermaphrodite trees at La Selva the combined frequency of self-incompatibility in canopy and subcanopy taxa is 88% ($n = 17$; table 12.4). For the monoecious Araceae and Moraceae, self-incompatibility has not been tested and is irrelevant as an outcrossing mechanism because pollination in these families either includes discrete male and female flowering phases (Henderson 1986) or an intricate insect life-history relationship that precludes the occurrence of self-pollination (Wiebes 1979). No breeding

[2] *Distyly* is the production of two distinct kinds of flowers on different individuals of plants of the same species. Flowers of the long-styled, or *pin,* morph have relatively long styles and short stamens, whereas those of the short-styled, or *thrum,* morph have relatively long stamens and

short styles. Distyly is one kind of *heterostyly; tristyly* also exists, involving three different floral morphologies among plants of the same species. These sexual systems are generally interpreted as mechanisms to promote disassortative pollination (between flower types), which results in increased levels of cross-pollination.

Table 12.4 Breeding systems of flowering plant species at La Selva

| | Forest Stratum | | | | | | | |
| | Understory | | Subcanopy | | Canopy | | All Strata | |
Breeding System	(N)	(%)	(N)	(%)	(N)	(%)	(N)	(%)
Self-compatible	25	65.8	1	9.1	2	25.0	28	49.1
Self-incompatible	13	34.2	10	90.9	6	75.0	29	50.9
TOTAL species	38	100	11	100	8	100	57	100

Note: Data include only La Selva species for which experimental information is available. See appendix 12.1 for references by species.

system data are available for major elements of the tree flora, including the Lauraceae, Melastomataceae, and Sapotaceae.

Among families of canopy trees at La Selva the Fabaceae have received the most attention. As an example, canopy-level experimental pollinations in *Dipteryx* resulted in 2% fruit set by self-pollinated flowers compared to 23% of cross-pollinated controls (Perry and Starrett 1980). Self-incompatibility is widespread among legumes and is generally thought to be gametophytically[3] controlled (de Nettancourt 1977; Arroyo 1981) although other incompatibility mechanisms have been suggested (Kenrick and Knox 1985; Seavey and Bawa 1986). Four of five additional leguminous trees tested at La Selva (*Lonchocarpus, Pentaclethra, Pithecellobium,* and *Swartzia*) exhibit self-incompatibility as well (Bawa, Perry, et al. 1985).

Quantitative Estimates of Outcrossing. The investigations of Bawa and co-workers described have shown that most of the trees of the La Selva forest are obligate outcrossers. Inbreeding, however, because of gene exchange among closely related conspecifics, may be greater than one would predict based on the presence of mechanisms that promote outcrossing. O'Malley and Bawa (1987) have conducted the only study of the mating system of a tropical tree species at La Selva. Using starch gel electrophoresis, they demonstrated that in *Pithecellobium pedicellare* (DC.) Benth. (Fabaceae), an obligate outcrosser, the outcrossing rate was high (similar to that in self-sterile temperature conifers) but that the potential for inbreeding was not negligible. The departure from the mixed mating model for two of four enzyme loci sampled suggests that the population of *P. pedicellare* may be subdivided. The high outcrossing rate also suggests, however, that the breeding population of these widely dispersed canopy trees is large. This investigation is significant because it confirmed the predicted mating system of a species that had been inferred from observed pollination, breeding, and sexual systems. This method can now be applied to other taxa at La Selva and other tropical sites.

[3] *Gametophytic self-incompatibility* is a physiological rejection response of flower tissues to the gene products of a developing male gametophyte or pollen tube. Inhibition of pollen tube growth typically occurs in the tissues of the style. *Sporophytic self-incompatibility* is a rejection response of the receptive tissues of the flower to the genetic identity of the sporophyte parent of the pollen grain. Inhibition in these cases usually occurs on the surface of the stigma.

Lower Stratum

Floral Sexuality. Seventy-five percent (168 species; table 12.3) of the understory species at La Selva possess hermaphroditic flowers. The main exceptions to hermaphroditism are the monoecious, often protogynous[4] Arecaceae (Schmid 1970; Bullock 1981; Beach 1984; Henderson 1984), Araceae (Young 1986a, 1986b, unpublished data), and Cyclanthaceae (Beach 1982; Hammel 1986f), and the dioecious wind-pollinated Moraceae and Urticaceae (Burger 1977b, 1977c; Bawa, Bullock, et al. 1985; Bawa and Crisp 1980; Hammel 1986j). Protandry has been recorded in some monoecious Arecaceae (J. H. Beach unpublished data) and hermaphroditic Gesneriaceae (Grove 1985) and Rubiaceae (McDade 1986, pers. comm.). The Annonaceae are special in possessing large, hermaphroditic, protogynous flowers (Schatz 1985, 1987). The only distylous taxa are found in the Rubiaceae.

Self-incompatibility. Of the thirty-eight understory species that have been tested at La Selva, 66% (twenty-five species; table 12.4) are self-compatible. These self-compatible species are distributed among twelve different families (Araceae, Cannaceae, Costaceae, Heliconiaceae, Marantaceae, Zingiberaceae, Acanthaceae, Annonaceae, Gesneriaceae, Loganiaceae, Myrsinaceae, and Rubiaceae). Members of the Rubiaceae, a family of woody plants found mostly in the canopy and subcanopy, account for nearly all the reported cases of self-incompatibility (Bawa and Beach 1983) in the understory. It should be recognized that most of the taxa tested for self-compatibility in the understory (excluding the Rubiaceae) have large conspicuous flowers that are visited by large bees or hummingbirds.

In the studies of understory plants at La Selva both pollen tube growth in styles and/or seed set were assessed after controlled pollinations to determine the extent of self-rejection in a species. In many of the self-compatible taxa self-pollinations resulted in lower numbers of pollen tubes per style or seeds per fruit than cross-pollinations (Heliconiaceae, Kress 1983b; Acanthaceae, Costaceae, Gesneriaceae, Marantaceae, Zingiberaceae, Grove 1985; Acanthaceae, McDade 1985). Thus, even though these tests indicate that the species are self-compatible, cross-pollinations are on the average more successful than self-pollinations. Reduced fecundity after selfing may be the result of either the presence of partial self-incompatibility systems (where pollen tube growth is slower) or inbreeding depression expressed in early embryo or endosperm development. The heterotic model of gametophytic self-incompatibility (Mulcahy and Mulcahy 1983), in which both pre- and postfertilization rejection phenomena are attributable to a single polygenic system, may also explain these observations of the self-compatibility response in understory taxa (McDade 1985).

Bawa and Beach (1983) examined self-incompatibility phenomena in fourteen species of La Selva Rubiaceae found in both the understory and subcanopy. The significance of their results merits discussion here. Nine of ten distylous

[4] *Dichogamy* is temporal separation of the sexual phases in bisexual or perfect flowers. *Protandrous* flowers shed pollen some hours or even days before they become receptive as females. *Protogynous* flowers have stigmas that become receptive some hours or days before pollen is shed.

species of shrubs and small trees were found to be self-incompatible as indicated by the physiological rejection of self- and intramorph pollen and by patterns of fruit set. Each of four monomorphic species tested also proved to be self-incompatible. The site of pollen tube inhibition varied within the distylous species. In short styles pollen tube inhibition occurred at or near the stigmatic surface, the characteristic rejection response of distylous species with sporophytic self-incompatibility (de Nettancourt 1977). Inhibition of intramorph pollen tubes in long style flowers occurred deep in the style as is typical for gametophytic self-incompatibility. These results suggest that more than one class of incompatibility system may operate within a single species, raising the possibility that the distylous, sporophytically determined incompatibility may overlay and mask an evolutionarily older, gametophytic system within the same plant (Bawa and Beach 1983).

The study of breeding systems and sexuality of La Selva Rubiaceae has advanced knowledge not only of the distribution of self-incompatibility in the family but also of the mechanisms of mate selection in tropical forest populations. In addition to sporophytic (stigmatic rejection) and gametophytic (stylar rejection) systems of self-incompatibility in La Selva species, *Warszewiczia*, a subcanopy species, shows "late-acting self-incompatibility" (pre- or postzygotic rejection in the ovary; Seavey and Bawa 1986), a third, less understood type of pollen genome rejection that may be more widespread in tropical species than previously believed. Further field and laboratory studies on the physiology and behavior of pollen and pollen tubes in Rubiaceae and other families at La Selva will increase understanding of the evolutionary ecology of self-incompatibility mechanisms in tropical communities.

PLANT-POLLINATOR INTERACTIONS

Associations between Reproductive Components

Certain components of plant reproductive systems (e.g., sexual systems, phenological patterns) tend to be nonrandomly associated with specific pollinator classes (e.g., Gentry 1974; Frankie 1975; Opler, Baker, et al. 1980; Bawa and Beach 1981; Bawa 1983; Frankie and Haber 1983). From the data summarized here associations are apparent between breeding, sexual, and pollination systems of both woody and herbaceous plants distributed throughout the forest strata at La Selva (table 12.5). Hermaphroditic flowers are associated with a number of different pollinator types. The majority of plant species in both the upper and lower strata pollinated by small and large bees (67% and 93% of the species, respectively), butterflies (100% of the species), hummingbirds (100% of the species), and bats (80% of the species) possess perfect flowers. Distylous taxa are associated to a much lesser extent with some of these pollinators, including bees (only 4 of 106 plant species), hummingbirds (2 of 41 plant species), moths (4 of 22 plant species), and, most commonly, butterflies (7 of 12 plant species). Sixty-two percent (21 species) of the plant species pollinated by beetles are monoecious (mostly Araceae, Arecaceae, and Cyclanthaceae), but 13 species (38%) of beetle-pollinated plants have hermaphroditic flowers. Ten of these latter 13 species are members of the family Annonaceae that are protogynous and, hence, "temporally monoecious."

Dioecious taxa are most commonly pollinated by small diverse insects, some by moths and the wind. Specifically, 20 species (46%) of the dioecious taxa from all three strata involve generalist insect pollination systems. These 20 species are spread over eight families and 13 genera. In contrast, only 3% of the hermaphroditic species have generalist insect pollinators. A distinct relationship between dioecism and anemophily is indicated by the fact that six of seven wind-pollinated species at La Selva are dioecious trees.

Although strong ecological correlations exist between some pollination and sexual systems in tropical wet forest plants, associations between breeding systems and other reproductive traits are less marked. Self-incompatibility systems are associated with nearly all the pollinator types found at La Selva (except wind, wasps, and flies), that is, bats, settling moths, hawk moths, butterflies, bees of all sizes, beetles, and "small diverse insects" (table 12.5). It is striking, however, that only three of the twenty hummingbird-pollinated plants tested are self-incompatible in contrast to fourteen of the sixteen species pollinated by moths and butterflies. This relationship suggests that some significant associations may become apparent as more species are tested for self-compatibility.

Heliconias and Hummingbirds

Reproductive biology of the genus *Heliconia*, a common and conspicuous element in the understory, has received considerable attention during the last twenty years at La Selva (Linhart 1973; Stiles 1975, 1978c, 1979b; Kress 1983a, 1983b) and at other sites in the tropics (McDade 1983; Dobkin 1984, 1987; Kress 1985). The relationship between heliconias and hummingbirds represents a well-documented case of tropical plant-pollinator interactions.

Hummingbirds are the primary and, probably, exclusive pollinators of Neotropical *Heliconia*, which typically possess brightly colored inflorescences with long, tubular, diurnal, odorless, hermaphroditic flowers producing considerable amounts of nectar (Stiles 1975, 1979b). Neotropical species of *Heliconia* fall into two ecological groups based on the type of hummingbird pollinator (Linhart 1973; Stiles 1975, 1979b). The first and largest group consists of species that form small clumps in closed forest habitats, have long curved floral tubes, and produce few flowers each day with abundant or concentrated nectar. Members of this group are pollinated by hermit hummingbirds (Phaethorninae), which do not defend territories but forage over long distances between widely spaced plants. The second, smaller group includes species that grow in large, monoclonal stands at forest edges or in open habitats, have short, straight perianth tubes, and produce many flowers daily with relatively smaller amounts of dilute nectar. This localized and long-lasting nectar source provides a dependable food supply for pollinating nonhermit hummingbirds (Trochilinae), which establish and defend territories around these plants. Each of these groups has representatives at La Selva (Stiles 1975).

Physiological self-incompatibility is rare in Central American *Heliconia*. In a study of nineteen species (including seven species at La Selva), pollen tube growth in styles and fruit set were used to determine the extent of self-incompatibility (Kress 1983b). Responses to self-pollination ranged from total self-rejection in one species to full self-compatibility in

Table 12.5 Relationships between pollination systems, sexual systems, and breeding systems of La Selva flowering plant species

Pollination System[a]	Sexual System				Breeding System	
	Hermaphroditic[b] (N = 165)[c]	Dioecious (N = 44)	Monoecious (N = 50)	Distylous (N = 17)	SC[d] (N = 29)	SI[e] (N = 29)
Bee, medium to large (N = 67 species)	61	3	2	1	4	5
Hummingbird (N = 41)	39	0	0	2	17	3
Bee, small (N = 39)	23	5	8	3	4	2
Beetle (N = 34)	13	1	21	0	2	1
Small diverse insects (N = 31)	5	20	6	0	0	2
Moth (N = 22)	8	9	1	4	0	7
Butterfly (N = 12)	5	0	0	7	2	7
Bat (N = 10)	8	0	2	0	0	2
Wind (N = 7)	0	6	1	0	0	0
Wasp (N = 7)	0	0	7	0	0	0
Fly (N = 5)	3	0	2	0	0	0

Note: Data include 276 of the 372 species in appendix 12.1 for which both pollinator type and sexual system are known.
[a]Pollination systems are ordered by decreasing overall abundance in the flora (see table 12.2).
[b]Excluding distylous taxa.
[c]Number of species.
[d]Self-compatibility.
[e]Self-incompatibility.

the majority of taxa tested. Autogamy in excess of 25% was detected in only five species; most plants require a probing visitor for pollination and seed set. Separation of the stigma and anthers within a flower is less than 2 mm and probably has little effect on outcrossing. Dichogamy does not appear to play any part in the breeding system. Thus, physiological and morphological specializations for outcrossing are few in *Heliconia*.

Observations of the interactions between plants and pollinators suggest that daily phenological patterns of flower production in conjunction with foraging patterns of the hummingbirds significantly influence the level of inbreeding that occurs in a population. In species visited by nonhermit, territorial birds pollen movement between plants appears to be limited. In contrast, pollen flow and, presumably, degree of outcrossing are significantly greater in species that are visited by traplining birds. These patterns of pollen flow were demonstrated in *Heliconia* at La Selva by using histological marker dyes to track pollinator movement within and between plants of traplining and territorial species (Linhart 1973). Using fluorescent powders, researchers identified little pollen flow between territories in nonhermit-pollinated *H. imbricata* at La Selva although territories do occasionally break down and pollen is exchanged between them (Ray 1982; Ray, Stiles, and Kress unpublished data). Investigations on the actual levels of genetic variation in populations of *Heliconia* pollinated by long-distance and territorial foraging birds are currently under way (Kress and Roesel 1991).

The syndrome of hermaphroditic flowers, self-compatibility, low daily output of flowers per plant, and pollination by long-distance foragers (e.g., hermit hummingbirds, euglossine bees) is not uncommon in understory species at La Selva. A survey based on published accounts, information from taxonomic monographs, and unpublished observations

by the authors identifies at least five additional families, which are conspicuous components of the understory, that possess this combination of traits: Acanthaceae, Costaceae, Gesneriaceae, Marantaceae, and Zingiberaceae (table 12.1; app. 12.1). This suite of reproductive traits has apparently evolved multiple times in tropical angiosperms.

CONCLUSIONS

Pollinators and other components of plant reproductive systems have figured prominently in theoretical population biology as determinants of mating patterns, genetic differentiation, and speciation in tropical trees (Ashton 1969; Fedorov 1966; Frankie 1976; Janzen 1971a). The organization of plant-pollinator relationships at the community level in tropical wet forest has also been investigated (e.g., Feinsinger 1976, 1978). Although many questions in plant population biology and community ecology may be addressed most effectively through long-term community analyses (e.g., Feinsinger et al. 1987) or by using molecular approaches (e.g., Hamrick and Loveless 1986, 1989; O'Malley and Bawa 1987), the broadly based surveys conducted at La Selva have provided important insights into the role of pollinators in the evolution of rain forest plants.

The information summarized in this chapter on the diversity and distribution of sexual, breeding, and pollination systems of forest plants at La Selva provides baseline data for understanding some of the ecological constraints on the evolution of plant reproductive modes. New reports for understory and subcanopy species brought together here provide evidence in support of the vertical stratification of reproductive features (see Bawa, Bullock, et al. 1985). In addition, comparisons to other tropical forest systems can now be made.

Forest Stratification

The distribution of sexual systems among forest strata at La Selva is roughly homogeneous (table 12.3). A slightly greater proportion of taxa in the understory layer is either hermaphroditic or monoecious, whereas dioecy is more common in the canopy and subcanopy. The most striking difference between strata in reproductive modes is the predominance of self-incompatibility systems in most of the hermaphroditic taxa of the upper forest level (table 12.4). Self-incompatibility is rare in understory taxa except in one family: ten of the thirteen self-incompatible understory species studied are woody Rubiaceae. If the rubiaceous shrubs and trees are excluded, the disparity in breeding systems between the forest strata becomes even more striking with 89% self-compatible species in the understory.

Bees are the most common pollinators in all strata of the tropical forest (table 12.2). Medium to large bees are especially important as visitors in the canopy, whereas small bee pollinators are almost as prevalent as large bees in the subcanopy and understory. Hummingbirds comprise the largest category of floral visitors in the understory but are relatively unimportant as pollinators of trees. Beetles contribute significantly as pollinators in the understory and subcanopy but not in the canopy. "Small diverse insects" (small bees, flies, beetles, butterflies, moths, and thrips) are the second-largest category of pollinators in the canopy, are less dominant in the subcanopy, and are relatively unimportant in the understory. Of the seven wind-pollinated species at La Selva, all but one are understory plants. Only one of the sixty-eight dioecious species found in the upper stratum is wind pollinated (*Trophis racemosa;* Bawa and Crisp 1980).

Reproductive systems that enhance outcrossing appear to predominate in both canopy and understory plants at La Selva. In the upper strata obligate outcrossing is ensured by the prevalence of dioecism or self-incompatibility barriers. In understory plants, despite the absence of physiological barriers to self-fertilization, cross-pollination may be common because of the prevalence of at least two reproductive syndromes: (1) low daily flower production coupled with reliable, long-distance pollinators (e.g., euglossine bees, hummingbirds) and (2) monoecy, protogyny, and beetle pollination.

There are some obvious exceptions to these generalizations about the spatial stratification of reproductive systems. For example, some canopy members of the genus *Erythrina* (Fabaceae) are self-compatible and pollinated by hummingbirds (Neill 1987), both traits characteristic of understory species. Members of the Rubiaceae are aberrant in the understory because they possess self-incompatibility systems (Bawa and Beach 1983). Piperaceae, which are also very abundant in the understory at La Selva but have not been carefully studied, may also prove to be an exception (Gómez-Pompa and Vásquez-Yánes 1974).

The overrepresentation of monocotyledons in the understory could, in part, account for the stratification of reproductive systems. The diversity of reproductive systems in the understory is comparable, however, in both subclasses of angiosperms (table 12.1). The evolutionary convergence of similar pollinator and breeding systems in such diverse families as the Heliconiaceae (monocots) and Acanthaceae (dicots) is a good example of the independent origin of corresponding reproductive systems in each of the two angiosperm subclasses.

Stratification of plant floral features could also be the result of stratification of the pollen vectors themselves. It is known that hermit hummingbirds primarily occur in the forest understory (Stiles 1978a; Stiles and Wolf 1979). If their preferred habitat is determined by factors other than nutritional requirements, then the prevalence in the understory of plant species with extended flowering periods that are visited by these birds is partly explained. The same is not true of large bees (e.g., euglossines), however, which are important pollinators in all levels of the forest (Frankie et al. 1983). More species in all strata need to be tested to confirm the observed stratification in self-incompatibility systems.

Environmental factors, especially light intensity, that affect the growth patterns of plants may be the primary driving force in the evolution of flowering phenological patterns (see chaps. 10, 11). The high light environment of the canopy may support photosynthetic levels high enough to produce the massive floral displays common in these species. A sustained low level of flower and fruit production exhibited by many understory plants is one of the main reproductive strategies that has evolved in these habitats with reduced light. It is not clear if solitary, reliable pollinators, such as hummingbirds, beetles, and euglossine bees coevolved with these understory plants (Janzen 1971a; Stiles and Wolf 1979; Feinsinger 1983). Once reliable long-distance pollinators were available to these extended bloomers in the understory, however, the pollen flow patterns that resulted from their foraging behavior may have reduced the selective pressures for the evolution of physiological or morphological mechanisms that promote outcrossing. The production of a few flowers per plant per day may decrease the probability of geitonogamy[5] enough to reduce natural selection for other outcrossing systems.

Comparisons with Other Tropical Forests

Pollination Systems. Many ecological and evolutionary studies on pollination systems of tropical forest species have been undertaken in recent years (for a general discussion see Baker et al. 1983). La Selva, however, is the only tropical forest site for which there is a systematic enumeration of pollination systems for a large portion of the taxa. Thus, quantitative, community-level comparisons of La Selva with other sites are difficult to make. Still, enough is known from other sites to comment.

One would expect a priori, because of climatic and taxonomic differences, that high-altitude forests would have markedly different pollinator faunas than that of La Selva. In the forests of the Jamaican Blue Mountains Tanner (1982) found that forty-six of fifty-six tree species (82%) belonged to the generalist insect pollination syndrome, six species were presumed to be pollinated by wind, and birds and moths were each the probable pollinators of two species. A majority of tree and shrub species in a Venezuelan cloud forest were also observed to be pollinated by small insects (Sobrevila and Arroyo 1982). Although these studies were based on limited samples of woody plants, small insects (especially bees) seem to be relatively more important in montane forests than at La Selva where other pollinators (particularly medium to large

[5] *Geitonogamy* is the transfer of pollen between flowers on the same plant. Barring somatic mutations, its genetic consequences are the same as *autogamy,* or transfer of pollen to the stigma of the same flower.

bees) play a proportionately larger role in the pollination of trees of the canopy and subcanopy (table 12.2). Little is known about the vertical stratification of pollinators in these montane sites or for any other Neotropical forest community.

Recent research in Paleotropical forests has addressed specific ecological and evolutionary aspects of plant reproduction, but no quantitative information exists on the distribution of pollinators for any Old World community comparable to what is available for La Selva. Appanah (1981, 1990), following Ashton et al. (1977), qualitatively described the pollination systems of forest plants in Peninsular Malaysia. In a classification similar to ours he categorized species based on the vertical position of flowering in the community. He circumscribed four groups: (1) canopy and emergent trees with flowers presented above the canopy, (2) plants with flowers borne within and below the canopy, (3) species that are cauliflorous, and (4) flowering plants in the lowermost understory level of shrubs and herbs. The majority of species in the first category have large, conspicuous flowers that are visited by large bees, bats, or sunbirds. Plants flowering in the subcanopy, including many dioecious taxa, possess small unspecialized flowers, a trait characteristic of generalist small insect pollination. Cauliflorous species, which range in habit from canopy trees to understory species, comprise a heterogeneous group, with bird, bat, bee, beetle, and other insect flower visitors. The species of the fourth group, the equivalent of our understory class, also show a diversity of floral biology. Flower size and morphology in this group range from small and simple to large and complex (e.g., *Rafflesia;* Beaman et al. 1988). Pollinators in the Malaysian forest understory include flies, bees, wasps, beetles, butterflies, and moths.

In the Malaysian forest canopy a group of flowering trees stands in marked contrast to the common mode of pollination by large animals. These taxa are members of the Dipterocarpaceae, including *Hopea, Shorea, Vatica,* and *Dipterocarpus,* and possess a specialized pollination syndrome consisting of minute flowers, an absence of nectar rewards, and nocturnal anthesis. Pollination by thrips (Thysanoptera) has been described for special species of *Shorea* at the Pasoh Forest study site in Peninsular Malaysia (Chan and Appanah 1980; Appanah and Chan 1981). Thrips oviposit on the flowers and flower buds of the trees, and the emergent larvae feed on the petal tissue. After eight days, the larvae metamorphose into adults that feed on mature floral tissues and pollen. The flowers function for one day, reaching maximum receptivity during the night. The following morning, the corolla and staminal tissues are shed and thrips ride the floral parts to the ground. The thrips subsequently ascend to the canopy the following evening and are presumed to effect cross-pollination as they are attracted to the sweet odors of new flowers that are beginning anthesis. It is suspected that in Southeast Asian forests many dipterocarp trees may depend on these pollinators. In contrast, Thysanoptera have been implicated in the pollination of only one Neotropical tree, *Compsoneura* (Myristicaceae) at La Selva (Bawa, Bullock, et al. 1985).

As in Neotropical forests, wind pollination has been reported infrequently in Paleotropical forests. Only one of 760 species of upper stratum trees in a rain forest in Brunei (Borneo) was reported as wind pollinated (Ashton 1969). Experience from La Selva has shown, however, that the syndrome is easily overlooked and, if present, is often found in taxa of the lower stratum.

Mating Patterns. In addition to the work at La Selva the frequency and distribution of sexual systems has been tabulated from several other tropical sites (table 12.6). The most complete and comparable enumerations are those of Croat (1979) for the flora of Barro Colorado Island (BCI), Panama, and Bullock (1985) for the flora of the Chamela Biological Station in Jalisco, Mexico.

The distribution of sexual systems in the BCI flora corresponds closely to that at La Selva (table 12.6). This similarity may, in part, be the result of overlap in the taxonomic composition of the two floras or, more particularly, the similar frequency distribution of phanerogamic life-forms (Bullock 1985). The vertical stratification of sexual systems in the BCI flora parallels that of La Selva in the increased frequency of dioecism from the understory to the canopy strata (Croat 1979).

Bullock's (1985) study of the sexuality patterns of the lowland Chamela deciduous forest corresponds remarkably to that found at La Selva and, to a lesser extent, to the patterns in the dry, deciduous forest in Guanacaste, Costa Rica (Bawa and Opler 1975; table 12.6). Bullock also found an increasing proportion of dioecious species from lower to higher forest strata.

At higher elevation, Sobrevila and Arroyo (1982) and Tanner (1982) report similar frequencies for hermaphroditism and dioecism within montane forest sites in Venezuela and Jamaica, respectively (table 12.6). In the Venezuelan study a significantly greater proportion of dioecious species was also detected among trees and shrubs (31%; 38% for trees alone) than among herbs and vines (3% dioecious species).

In general, the distribution of sexual systems is fairly uniform across Neotropical sites (table 12.6), especially considering the substantial differences in sample size and taxonomic composition of the various samples. Further study of the significance of sexuality patterns in tropical forests will clearly require an analysis of the effect of the taxonomic component on the distribution of sexual systems.

Relatively little information is available on the diversity of sexual systems in tropical Asian forests. No community surveys of plant sexuality have as yet been published. In a mixed dipterocarp forest in central Sarawak Ashton (1969) reported that 26% of 711 tree species were dioecious. This proportion is higher than that found at any lowland Neotropical site.

The frequency of self-incompatibility among hermaphrodite plants shows considerably more variation among sites than does the distribution of sexual systems (table 12.6). This variation may owe, in part, to the limited samples, nonrandom selection of taxa, and inconsistency in the plant life-forms tested among studies. For example, the incidence of incompatibility systems at La Selva (51% self-incompatible) differs from that at the Guanacaste site (79% self-incompatible). The discrepancy disappears when the comparison is restricted to the same life-form. The Guanacaste sample included only trees, and the comparable number for trees at La Selva is 84% self-compatible.

Self-compatibility among hermaphrodites in the Jamaican montane forests (88%) is the highest for any tropical site sampled thus far. Tanner (1982) suggested that the high inci-

Table 12.6 The distribution of sexual and breeding systems in Neotropical forests

Site (Reference)	Sample (Taxa)	Sexual System[a]	Breeding System (Sample)[b]
La Selva Biological Station Premontane wet forest (this chap.)	507 spp. (angiosperms: trees, shrubs, herbs)	H 70% M 12% D 17%	SC 49% SI 51% (57 spp.)
Barro Colorado Island, Panama Moist forest (Croat 1979)	1,212 spp. (phanerogams: all life-forms)	H 76% M 11% D 9% X 4%	no species tested
Chamela Biological Station Jalisco, Mexico Deciduous forest (Bullock 1985)	708 spp. (angiosperms: all life-forms)	H 70% M 13% D 12% X 5%	SC 24% SI 76% (33 spp.)
Guanacaste, Costa Rica Semideciduous dry forest (Bawa & Opler 1975)	130 spp. (angiosperms: trees only)	H 68% M 10% D 22%	SC 21% SI 79% (34 spp.)
Altos de Pipe, Venezuela Humid premontane cloud forest (Sobrevila & Arroyo 1982)	75 spp. (angiosperms: all life-forms)	H 84% M[c] D 16%	SC 56% SI 44% (25 spp.)
Blue Mountains, Jamaica Montane rain forest (Tanner 1982)	56 spp. (phanerogams: trees only)	H 68% M 11% D 21%	SC 88% SI 12% (8 spp.)

[a]H = hermaphroditic; M = monoecious; D = dioecious; X = mixed sexuality patterns, including polygamy (with bisexual and unisexual flowers on the same or different individuals of the same species), gynomonoecy (production of unisexual female and bisexual flowers) and andromonoecy (production of unisexual male and bisexual flowers); SC = self-compatible; SI = self-incompatible.

[b]Number of species tested for self-incompatibility.

[c]Monoecious species were not distinguished from hermaphrodites in the study.

dence of self-compatility may be evolutionarily related to the scarcity of pollinators in a cool and wet mountain climate. This hypothesis was supported by results from another montane forest, the Altos de Pipe site in Venezuela (table 12.6), which has the second highest percentage of self-compatible hermaphrodites for a tropical flora.

The distribution of incompatibility systems in tropical deciduous forest is relatively uniform across the three sites studied (table 12.6), with self-incompatibility three to four times as common as self-compatibility. These percentages contrast with the much higher percentage of self-compatibility (49%) in the wet forest at La Selva and in cloud forests. As noted earlier, mostly woody taxa were sampled at the dry sites. If only the woody taxa at La Selva are considered, the incidence of self-compatibility is much lower (17%).

As with sexual systems, data on the prevalence of incompatibility systems in tropical Asian forests are scarce. In Malaysia, Chan (1981) showed that self-incompatibility is present in a number of species in the Dipterocarpaceae, especially in the thrips-pollinated genus *Shorea*. Although some species of *Shorea, Hopea,* and *Dipterocarpus* appeared to be self-compatible, other evidence suggests that at least some of these taxa may be apomictic. In addition to the Dipterocarpaceae experimental pollinations have shown that several Malaysian forest species of Sapindaceae and Bombacaceae are self-incompatible (Ha et al. 1988b). If these limited data are representive of tropical Asian forests as a whole, the prevalence of self-incompability systems may be similar to that found in Neotropical trees.

Apomixis, although not widespread, has been reported in

both Neotropical and Paleotropical trees. At La Selva, agamospermy has not been found in any woody or herbaceous taxa although very few species have been carefully studied. Sobrevila and Arroyo (1982) provide indirect evidence for agamospermy in *Clidemia* and *Miconia* (Melastomataceae) at their Venezuelan cloud forest site. Renner (1986, 1989) also reports apomixis in other Neotropical members of this family. Among Old World taxa, Kaur et al. (1978, 1986) found evidence for apomixis in five species of polyploid *Shorea* and *Hopea* (Dipterocarpaceae). Cytological and embryological evidence for agamospermy in four Malaysian dioecious species of *Garcinia* (Clusiaceae) has also been recently published (Ha et al. 1988a). The extent and potential significance of apomixis in tropical trees in both the Neotropics and Palaeotropics remains to be explored.

The study by O'Malley and Bawa (1987) in which outcrossing rates were estimated for one tree species at La Selva is one of the few such studies conducted in tropical forests. Hamrick and Loveless (1986, 1989) have summarized the limited genetic data on tropical species based on isozyme electrophoresis and have provided some preliminary information for twenty-nine woody plant taxa at BCI, Panama. They conclude that tropical tree species may exhibit equal or greater amounts of genetic variation than temperate zone tree species. Their sample also indicates that understory and subcanopy shrubs and trees are more genetically variable than canopy trees (in percentage of polymorphic loci, number of alleles per locus, and heterozygosity). In contrast, Heywood and Fleming (1986) demonstrated relatively low levels of genetic variation in three understory species of the genus *Piper* at Parque Nacional Santa Rosa (a Costa Rican dry forest). In another investigation of Neotropical understory taxa Linhart and colleagues (1987) studied the effects of different hummingbird foraging behaviors on the level of inbreeding in two members of the Acanthaceae at Monteverde (a Costa Rican cloud forest reserve). Their results showed that the level of inbreeding was lower in the species pollinated by long-distance foragers despite the absence of any self-incompatibility system.

In the Paleotropics Gan and co-workers (Gan et al. 1977; Gan et al. 1981; Ashton et al. 1984) have analysed levels of genetic variation in some Malaysian members of the Dipterocarpaceae and Sapindaceae in an effort to understand population genetics and systematic relationships. Their results showed high levels of within-population genetic variation in most of the taxa they tested, suggesting that outbreeding predominates. Low levels of variation in one species, however, were attributed to possible self-fertilization. As more results from genetic research on both Old and New World tropical plants become available, they will provide quantitative tests of predictions about plant mating systems derived from observations of pollinators, sexual systems, and incompatibility systems.

Future Research

Pollination, breeding, and sexual systems have been studied in a great diversity of taxa in many different plant families at La Selva during the last twenty years. Some families, such as the Rubiaceae, Acanthaceae, Heliconiaceae, Arecaceae, and Annonaceae, have received much attention. Other families, including the Melastomataceae, Piperaceae, Solanaceae, Poaceae, and Commelinaceae, are nearly unstudied at La Selva. The reproductive biology of epiphytes and vines, two of the most important life-forms contributing to the great diversity of tropical forests (Gentry and Dodson 1987b; chap. 6), has been almost completely ignored (Ackerman 1986). It is striking that no observations from La Selva have been published on reproduction in the Orchidaceae, the largest flowering plant family in the flora. Similarly, minimal information exists on the reproductive biology of epiphytic members of the Cyclanthaceae and Araceae. New data on these unstudied taxa (especially the epiphytes) may significantly change understanding of the distribution of reproductive systems in the forest, which is woefully incomplete without this information.

Research is also needed to address the impact of pollinator life-history traits on plant reproductive features. One example, described previously, of a well-studied system at La Selva is the interaction between hermit hummingbirds and widely dispersed understory taxa producing few flowers per day (e.g., heliconias, gingers, gesneriads; Stiles 1975, 1978c; Grove 1985). Another system is the guild of protogynous, nocturnally flowering, often monoecious taxa (e.g., Arecaceae, Araceae, Cyclanthaceae, Annonaceae) pollinated by beetles at La Selva (Beach 1982, 1984; Young 1986a, 1986b, 1988; Schatz 1987, unpublished data). An understanding of the life histories of other pollinators may provide insights into the selective factors that have determined floral evolution.

Many of the data on reproductive systems of La Selva plants (app. 12.1) were originally collected to address conceptual issues regarding the evolution of tropical plant populations, especially hypotheses on plant-pollinator coevolution

and models of speciation. We have now confirmed that most of the tree species at La Selva, and probably other rainforest habitats as well, are obligate outcrossers. These results support Ashton's (1969) contention that wide outcrossing and, presumably, allopatric speciation best explain the origin of the great diversity of tropical tree species. Only a few investigations, however, have actually documented the mating system of tropical species (O'Malley and Bawa 1987; O'Malley et al. 1988). It is imperative that additional studies on genetic variation and outcrossing rates of taxa in all strata of the forest at La Selva are executed to provide a more thorough picture of the spectrum of mating systems found in tropical plants (Hamrick 1987).

More importantly, it must be emphasized that although outcrossing may be widespread no studies have as yet addressed the actual mechanisms of speciation in tropical forest taxa. Investigations of the processes of speciation are difficult and complex (White 1978). Convincing experiments to demonstrate speciation processes will require not only information on the ecology and population genetics of species but testable hypotheses on phylogenetic history as well. The identification of sister group relationships by cladistic analyses and the determination of genetic population structure and mating patterns using modern molecular techniques have the potential to provide unambiguous case studies of population divergence. Likewise, phylogenetically based studies of character evolution at the species level will provide insight on the degree of coevolution of plant-pollinator relationships that cannot be achieved by ecological correlations alone. Clearly, parallel investigations by population biologists, ecologists, and systematists will be necessary to address the as yet unsolved controversy on models of speciation in tropical plants.

Information on plant mating systems is central to an understanding of the evolutionary mechanisms that account for the origin and maintenance of high species diversity in tropical forests. It is equally important for sound practices of forest conservation (O'Malley and Bawa 1987). Data on the size of breeding populations and the foraging ranges of potential pollinators and dispersal agents are critical if tropical forest reserves are to be wisely managed to maintain species diversity. An understanding of the complex interactions that take place among organisms will provide a sound scientific basis for determining the size, shape, and composition of managed forests if they are to fulfill their conservation roles for more than a few years.

ACKNOWLEDGMENTS

We thank B. Alverson, T. Antonio, K. Bawa, S. Bullock, D. and D. Clark, J. Denslow, P. Feinsinger, K. Grove, L. McDade, J. Miller, S. Renner, C. Roesel, G. Schatz, C. Sobrevila, and H. Young for helpful discussions, comments on the manuscript, and unpublished data. B. Hammel, M. Grayum, and W. Burger were especially cooperative in the identification of plant taxa at La Selva. This research was funded in part by grants from the National Science Foundation to W. J. K. (BSR-8706524) and J. H. B (BSR-8503378), the National Geographic Society (NGS-3824-88), and the Neotropical Lowlands Research Program of the Smithsonian Institution.

APPENDIX 12.1 Checklist of the reproductive biology of the flowering plants of La Selva.

Taxon	Habitat Classes[a]	Sexual System[b]	Breeding System[c]	Pollinator Classes[d]	Source[e]
Monocotyledonae					
Amaryllidaceae					
Crinum erubescens	U	H		hawk moth	K&B upbl
Araceae					
Anthurium ochranthum	U	H		medium bee	W&D 1976
A. trisectum	U	H		medium bee	W&D 1976
Dieffenbachia beachiana	U	M		beetle	K&B upbl
D. hammelii	U	M		beetle	K&B upbl
D. longispatha	U	M		beetle	Yng 1986a, 1986b, 1988
D. longivaginata	U	M		beetle	K&B upbl
D. oerstedii	U	M		beetle	Val 1983
D. seguina	U	M		beetle	K&B upbl
Dracontium pittieri	U	H		fly	K&B upbl
Philodendron grandipes	U	M		beetle	Yng 1986a
Spathiphyllum friedrichsthalii	U	H	SC	medium bee	M&A 1986; W&D 1976
S. fulvovirens	U	H		euglossine bee	W&D 1976
S. laeve	U	H		euglossine bee	W&D 1976
S. phryniifolium	U	H		euglossine bee	W&D 1976
S. wendlandii	U	H		euglossine bee	W&D 1976
Arecaceae					
Asterogyne martiana	U	M		fly	K&B upbl; Sch 1970; U&M 1977
Astrocaryum alatum	S	M		beetle	Bw, Bck et al. 1985; Bck 1981
Bactris longiseta	U	M		beetle	Bw, Bck et al. 1985
B. porschiana	U	M		beetle	Bch 1984
B. wendlandiana	U	M		beetle	Bck 1981
B. sp. nov.	U	M		beetle	Bck 1981
Calyptrogyne sarapiquensis	U	M		bat	K&B upbl
Chamaedorea exorrhiza	U	D		wind	Bw, Bck et al. 1985
C. warscewiczii	U	D		wind	K&B upbl
Cryosophila albida	S	M		beetle	Bw, Bck et al 1985; Bck 1981; Hnd 1984
Euterpe macrospadix	S	M		small bee	K&B upbl
Geonoma congesta	U	M		small bee	K&B upbl
G. cuneata	U	M		fly	K&B upbl
G. ferruginea	U	M		small bee	K&B upbl
G. interrupta	U	M		small bee	K&B upbl
G. longevaginata	U	M		small bee	K&B upbl
Iriartea gigantea	S	M		medium bee	Bw, Bck et al. 1985; Bck 1981
Prestoea decurrens	S	M		small bee	Bw, Bck et al. 1985; Bck 1981
Socratea durissima	S	M		beetle/bee	Bw, Bck et al. 1985; Bck 1981
Synechanthus warscewiczianus	U	M		small diverse insects	K&B upbl; Hnd 1986
Welfia georgii	S	M		medium bee	Bw, Bck et al. 1985; K&B upbl
Bromeliaceae					
Aechmea magdalenae	U	H		hummingbird	K&B upbl; Stl 1978c
A. mariae-reginae	S	H		hummingbird	K&B upbl
Guzmania lingulata	S	H		hummingbird	K&B upbl
Cannaceae					
Canna tuerckheimii	U	H	SC	hummingbird	K&B upbl; Stl 1978c
Commelinaceae					
Campelia zanonia	U	H			K&B upbl
Dichorisandra hexandra	U	H			K&B upbl
Floscopa robusta	U	H			K&B upbl
Tripogandra serrulata	U	H			K&B upbl
Zebrina huehueteca	U	H			K&B upbl
Costaceae					
Costus guanaiensis	U	H		euglossine bee	K&B upbl; Smk 1981, 1983
C. malortieanus	U	H	SC	hummingbird	Grv 1985; Stl 1978c

APPENDIX 12.1 (continued)

Taxon	Habitat Classes[a]	Sexual System[b]	Breeding System[c]	Pollinator Classes[d]	Source[e]
C. pulverulentus	U	H	SC	euglossine bee hummingbird	Grv 1985; Stl 1978c
C. scaber	U	H		hummingbird	K&B upbl; Maas 1977
Cyclanthaceae					
Asplundia costaricensis	U	M		beetle	K&B upbl; Hmm 1986f
A. sleeperae	U	M		beetle	K&B upbl; Hmm 1986f
A. uncinata	U	M		beetle	K&B upbl; Hmm 1986f
Carludovica sulcata	U	M		beetle	K&B upbl; Hmm 1986f
Cyclanthus bipartitus	U	M		beetle	Bch 1982; Hmm 1986f
Dicranophygium umbrophilum	U	M		beetle	K&B upbl; Hmm 1986f
D. wedelii	U	M		beetle	K&B upbl; Hmm 1986f
Haemodoraceae					
Xiphidium caeruleum	U	H		small bee	Bmn 1983; K&B upbl
Heliconiaceae					
Heliconia imbricata	U	H	SC	hummingbird	K&B upbl; Stl 1975
H. irrasa	U	H	SC	hummingbird	K&B upbl; Stl 1975
H. latispatha	U	H	SC	hummingbird	K&B upbl; Stl 1975
H. mariae	U	H	SC	hummingbird	K&B upbl; Stl 1975
H. mathiasiae	U	H	SI	hummingbird	K&B upbl; Stl 1975
H. pogonantha	U	H	SC	hummingbird	K&B upbl; Stl 1975
H. sarapiquensis	U	H	SC	hummingbird	K&B upbl; Stl 1975
H. trichocarpa	U	H	SC	hummingbird	K&B upbl
H. umbrophila	U	H	SC	hummingbird	K&B upbl; Stl 1975
H. wagneriana	U	H	SC	hummingbird	K&B upbl; Stl 1975
Marantaceae					
Calathea cleistantha	U	H		euglossine bee	Kdy 1978; K&B upbl; S&H 1984
C. cuneata	U	H		euglossine bee	Kdy 1978; K&B upbl; S&H 1984
C. gymnocarpa	U	H		euglossine bee	Kdy 1978; K&B upbl; S&H 1984
C. inocephala	U	H		euglossine bee	Kdy 1978; K&B upbl; S&H 1984
C. insignis	U	H		euglossine bee hummingbird	Kdy 1978; K&B upbl; Stl 1978c
C. lasiostachya	U	H	SC	euglossine bee	Grv 1985; K&B upbl
C. lutea	U	H		euglossine bee hummingbird	Kdy 1978; K&B upbl; Stl 1978c
C. marantifolia	U	H	SC	euglossine bee hummingbird	Grv 1985; K&B upbl; Stl 1978c
C. micans	U	H		euglossine bee	Kdy 1978; K&B upbl; S&H 1984
C. similis	U	H		euglossine bee	Kdy 1978; K&B upbl; S&H 1984
C. venusta	U	H		euglossine bee	Kdy 1978; K&B upbl; S&H 1984
C. warscewiczii	U	H		euglossine bee	Kdy 1978; K&B upbl; S&H 1984
Ischnosiphon inflatus	U	H		euglossine bee	Hmm 1986i; K&B upbl; S&H 1984
Pleiostachya pruinosa	U	H		euglossine bee	Hmm 1986i; K&B upbl; S&H 1984
Poaceae					
Cryptochloa concinna	U	M			Pohl 1980
Orthoclada laxa	U	H			Pohl 1980
Pariana parviflora	U	M		small diverse insects	Pohl 1980; S&C 1971
Pharus latifolius	U	M			Pohl 1980
Zingiberaceae					
Renealmia alpina	U	H		hummingbird	K&B upbl; Stl 1978c
R. cernua	U	H	SC	hummingbird	Grv 1985; Stl 1978c
R. pluriplicata	U	H	SC	hummingbird	Grv 1985; Stl 1978c
Dicotyledonae					
Acanthaceae					
Aphelandra aurantiaca	U	H		hummingbird	Grv 1987
A. dolichantha	U	H		fly	K&B upbl
A. storkii	U	H	SC	hummingbird	Grv 1987; McD 1985; M&K 1980

APPENDIX 12.1 (continued)

Taxon	Habitat Classes[a]	Sexual System[b]	Breeding System[c]	Pollinator Classes[d]	Source[e]
Bravaisia integerrima	C	H		medium bee	Bw, Bck et al. 1985
Herpetacanthus panamensis	U	H		medium bee	K&B upbl
Justicia aurea	U	H		hummingbird	C&W 1981; M&K 1980; Stl 1978c
J. sarapiquensis	U	H		hummingbird	K&B upbl; Hmm, p.c.
Odontonema callistachyum	U	H		hummingbird	K&B upbl; Stl 1978c
Razisea spicata	U	H		hummingbird	K&B upbl; Stl 1978c
R. wilburii	U	H	SC	hummingbird	Grv 1985
Ruellia metallica	U	H			K&B upbl
Amaranthaceae					
Alternanthera costaricensis	U	H			K&B upbl
Anacardiaceae					
Spondias radlkoferi	C	D		small diverse insects	Bw, Bck et al. 1985
Annonaceae					
Anaxagorea crassipetala	U	H	SC	small beetle	Bw, Pry et al. 1985a
A. phaeocarpa	S	H		beetle	K&B upbl
Annona montana	S	H		scarab beetle	Bw, Bck et al. 1985
Cymbopetalum costaricense	U	H		scarab beetle	Bw, Bck et al. 1985; Stz 1985
C. torulosum ("C.sp." of Bw, Bck et al. 1985; Bw, Pry et al. 1985)	U	H	SC	scarab beetle	Bw, Bck et al. 1985; Bw, Pry et al. 1985; K&B upbl; Stz 1985
Desmopsis microcarpa	U	H		beetle	Bw, Bck et al. 1985
Guatteria aeruginosa	U	H		beetle	Bw, Bck et al. 1985
G. diospyroides ("G. inuncta" of Bw, Bck et al. 1985)	U	H		beetle	K&B upbl
Sapranthus viridiflorus	S	H		beetle	K&B upbl
Xylopia bocatorena	S	H		beetle	K&B upbl
Apocynaceae					
Lacmellea panamensis	C	H		large bee	Bw, Bck et al. 1985
Stemmadenia donnell-smithii	S	H		euglossine bee	Bw, Bck et al. 1985; Jnz 1971a
S. robinsonii	S	H		euglossine bee	Jnz 1971a
Tabernaemontana arborea	S	H		moth	Bw, Bck et al. 1985
Aquifoliaceae					
Ilex sp.	C	D		small diverse insects	Bw, Bck et al. 1985
Araliaceae					
Dendropanax arboreus	C	D		small diverse insects	Bw, Bck et al. 1985
D. stenodontus	U	D		small diverse insects	Bw, Bck et al. 1985
Asteraceae					
Clibadium pittieri	U	M			McV 1984
Lasianthaea fruticosa	U	H			McV 1984
Begoniaceae					
Begonia semiovata	U	M		small bee	K&B upbl
Bignoniaceae					
Schlegelia sulfurea	C	H		hummingbird	Hmm, p.c.
Tabebuia guayacan	C	H		medium to large bee	Bw, Bck et al. 1985
T. rosea	C	H		medium to large bee	Bw, Bck et al. 1985
Bombacaceae					
Ceiba pentandra	C	H		bat	Bw, Bck et al. 1985
Matisia ochrocalyx	S	H		bat	Alv, p.c.
Ochroma pyramidale	C	H	SI	bat	Bw, Bck et al. 1985; Bw, Pry et al. 1985
Pachira aquatica	S	H		bat	Bw, Bck et al. 1985
Quararibea asterolepis ("Q. sp. A" of Bw, Bck et al. 1985)	S	H		moth	Bw, Bck et al. 1985

APPENDIX 12.1 (continued)

Taxon	Habitat Classes[a]	Sexual System[b]	Breeding System[c]	Pollinator Classes[d]	Source[e]
Q. parvifolia	S	H		bat	Alv upbl
Q. pumila ("Q. sp. B" of Bawa, Bullock, et al. 1985)	U	H		euglossine bee	Alv 1984, upbl
Boraginaceae					
Cordia alliodora	C	H		moth	Bw, Bck et al. 1985
C. linnaei ("C. lineata" of Bawa, Perry, et al. 1985)	U	S			J. Mil, p.c.
C. lucidula	S	H			Bw, Pry et al. 1985
C. porcata ("C. nitida" of Bawa, Bullock, et al. 1985; Bawa, Perry, et al. 1985)	S	H	SI	butterfly	Bw, Bck et al. 1985; Bw, Pry et al. 1985;
Burseraceae					
Bursera simaruba	C	D		small diverse insects	Bw, Bck et al. 1985
Protium costaricense	S	D		small diverse insects	Bw, Bck et al. 1985
P. sp. ("P. glabrum" of Bawa, Bullock, et al. 1985)	S	D		small diverse insects	Bw, Bck et al. 1985
Tetragastris panamensis	S	D		small diverse insects	Bw, Bck et al. 1985
Cactaceae					
Epiphyllum sp.	S	H		bat	Hmm, p.c.
Capparidaceae					
Capparis pittieri	S	H	SI	bat or moth	Bw, Bck et al. 1985; Bw, Pry et al. 1985
Caricaceae					
Jacaratia dolichaula	S	D		hawk moth	Bw 1980b, Bw, Bck et al. 1985
J. spinosa ("J. costaricensis" of Bawa, Bullock, et al. 1985)	C	D		hawk moth	Bw, Bck et al. 1985
Chloranthaceae					
Hedyosmum scaberrimum	S	D			Brg 1977a
Clusiaceae					
Clusia flava	S	D		beetle	Hmm, p.c.
C. minor	S	D		large bee	Hmm, p.c.
C. uvitana	S	D		large bee	Hmm, p.c.
C. valerii	S	D		large bee	Hmm, p.c.
Symphonia globulifera	S	H		hummingbird	Bw, Bck et al. 1985; K&B upbl; Hmm, 1986g
Tovomitopsis glauca ("T. sp. A" of Bawa, Bullock, et al. 1985)	U	D		small diverse insects	Bw, Bck et al. 1985; Hmm 1986g, upbl
T. nicaraguensis ("T. sp. B" of Bawa, Bullock, et al. 1985)	U	D		small diverse insects	Bw, Bck et al. 1985; Hmm 1986g
T. silvicola	U	D			Hmm 1986g
Vismia bilbergiana	S	S			Hmm 1986g
Combretaceae					
Terminalia amazonia	C	H		medium bee	Bw, Bck et al. 1985
Chrysobalanaceae					
Hirtella lemsii	S	H			K&B upbl
Dichapetalaceae					
Dichapetalum axillare	S	H		small bee	Prc 1972b
D. donnell-smithii	S	H		small to medium bee	Bw, Bck et al. 1985; K&B upbl

APPENDIX 12.1 (continued)

Taxon	Habitat Classes[a]	Sexual System[b]	Breeding System[c]	Pollinator Classes[d]	Source[e]
Euphorbiaceae					
Acalypha diversifolia	U	M			K&B upbl
Adelia triloba	S	D		small diverse insects	Bw, Bck et al. 1985
Croton schiedeanus	S	M		small diverse insects	Bw, Bck et al. 1985
Hyeronima oblonga	C	D		small diverse insects	Bw, Bck et al. 1985
Mabea occidentalis	U	M		bat or marsupial	Stn 1981, 1983
Sapium jamaicense	C	M		small diverse insects	Bw, Bck et al. 1985
Fabaceae					
Andira inermis	S	H		medium bee	Bw, Bck et al. 1985
Cassia fruticosa	U	H		large bee	Bw, Bck et al. 1985
Dalbergia tucurensis	C	H		medium bee	Bw, Bck et al. 1985
Dipteryx panamensis	C	H	SI	large bee	Bw, Bck et al. 1985; Bw, Pry et al. 1985; P&S 1980
Dussia cuscatlantica	C	H		medium to large bee	Bw, Bck et al. 1985
D. macroprophyllata	C	H		medium to large bee	Bw, Bck et al. 1985
Erythrina cochleata	C	H		hummingbird	Bw, Bck et al. 1985; Nel 1987; Stl 1978c
E. gibbosa ("E. sp." of Bawa, Bullock, et al. 1985)	U	H		hummingbird	Bw, Bck et al. 1985
Hymenolobium pulcherrimum	C	H	SC	medium to large bee	Bw, Bck et al. 1985; Bw, Pry et al. 1985
Lonchocarpus oliganthus	S	H	SI	small bee	Bw, Bck et al. 1985; Bw, Pry et al. 1985
L. pentaphyllus	S	H		small to medium bee	Bw, Bck et al. 1985
L. velutinus	C	H		small to medium bee	Bw, Bck et al. 1985
Pentaclethra macroloba	C	H	SI	small beetle	Bw, Pry et al. 1985; K&B upbl
Pithecellobium gigantifolium	U	H	SI	butterfly/moth	Bw, Bck et al. 1985; Bw, Pry et al. 1985
P. longifolium	C	H		butterfly/moth	Hmm, p.c.
Pterocarpus officinalis	C	H		medium bee	Bw, Bck et al. 1985
Stryphnodendron excelsum	C	H		small to medium bee	Bw, Bck et al. 1985
Swartzia cubensis	C	H		large bee	Bw, Bck et al. 1985
S. simplex	S	H	SI	large bee	Bw, Bck et al. 1985; Bw, Pry et al. 1985
Flacourtiaceae					
Carpotroche platyptera	U	D			Slm 1980
Casearea arborea	S	H	SI	small diverse insects	Bw, Bck et al. 1985; Bw, Pry et al. 1985
C. corymbosa	S	H	SI		Bw, Pry et al. 1985
Hasseltia floribunda	S	H		small diverse insects	Bw, Bck et al. 1985
Laetia procera	C	H		medium bee	Bw, Bck et al. 1985
Gesneriaceae					
Besleria columneoides	U	H	SC	hummingbird	Grv 1985; Stl 1978c
B. robusta	U	H		hummingbird	Grv 1987
Columnea purpurata	U	H		hummingbird	Grv 1987
Episcia lilacina	U	H		medium bee	Whl 1983
Hernandiaceae					
Hernandia sternura	C	M		small diverse insects	Bw, Bck et al. 1985
Lacistemataceae					
Lozania pittieri	S	H			K&B upbl
Lamiaceae					
Scutellaria purpurascens	U	H			K&B upbl
Lauraceae					
Ocotea atirrensis	U	H			Hmm 1986h
O. dendrodaphne	U	H			Hmm 1986h
O. tenera	U	H		beetle	Bw, Bck et al. 1985; Hmm 1986h

APPENDIX 12.1 (continued)

Taxon	Habitat Classes[a]	Sexual System[b]	Breeding System[c]	Pollinator Classes[d]	Source[e]
("O. sp." of Bawa, Bullock, et al. 1985)					
Lecythidaceae					
Eschweilera calyculata	S	H		large bee	Bw, Bck et al. 1985
Lecythis ampla	C	H		medium to large bee	Bw, Bck et al. 1985
Loganiaceae					
Spigelia humboldtiana	U	H	SI	small bee	Grv 1985
Magnoliaceae					
Talauma sambuensis	S	H		beetle	Gbb et al. 1977; K&B upbl
Malpighiaceae					
Bunchosia macrophylla	U	H		small to medium bee	Bw, Bck et al. 1985
B. ocellata	U	H			K&B upbl
Byrsonima crispa	C	H		medium to large bee	Bw, Bck et al. 1985
("B. aerugo" of Bawa, Bullock, et al. 1985)					
Malvaceae					
Hampea appendiculata	S	D		small to medium bee	Bw, Bck et al. 1985
Malvaviscus arboreus	U	H		hummingbird	Bw, Bck et al. 1985; Stl 1978c
Pavonia fruticosa	U	H		hummingbird	K&B upbl; M&D 1984; Szm 1981
P. rosea	U	H		hummingbird	K&B upbl; M&D 1984; Szm 1981
Melastomataceae					
Aciotis rostellata	U	H			K&B upbl
Clidemia densiflora	U	H		small bee	Rnn 1986, p.c.
C. japurensis	U	H		small bee	Rnn, 1986, 1989
C. ombrophila	U	H		small bee	Rnn 1986, p.c.
C. reitziana	U	H		small bee	Rnn 1986, p.c.
C. septuplinervia	U	H		small bee	Rnn 1986, p.c.
Conostegia bracteata	U	H			K&B upbl
C. icosandra	S	H			K&B upbl
C. micrantha	S	H			K&B upbl
C. montana	S	H			K&B upbl
C. rufescens	S	H			K&B upbl
C. setosa	S	H			K&B upbl
Henriettea tuberculosa	S	H			K&B upbl
Leandra dichotoma	U	H			K&B upbl
L. granatensis	U	H			K&B upbl
Miconia affinis	S	H		medium bee	Bw, Bck et al. 1985; K&B upbl
M. ampla	S	H			K&B upbl
M. appendiculata	S	H			K&B upbl
M. barbinervis	U	H			K&B upbl
M. bracteata	U	H			K&B upbl
M. brenesii	U	H			K&B upbl
M. centrodesma	U	H			K&B upbl
M. dorsiloba	U	H			K&B upbl
M. elata	S	H			K&B upbl
M. gracilis	U	H			K&B upbl
M. impetiolaris	S	H			K&B upbl
M. longiflora	S	H			K&B upbl
M. nervosa	S	H			K&B upbl
M. simplex	U	H			K&B upbl
Ossaea laxivenula	U	H			K&B upbl
O. macrophylla	U	H			K&B upbl
O. micrantha	U	H			K&B upbl
O. robusta	U	H			K&B upbl
Triolena hirsuta	U	H			K&B upbl

APPENDIX 12.1 (continued)

Taxon	Habitat Classes[a]	Sexual System[b]	Breeding System[c]	Pollinator Classes[d]	Source[e]
Meliaceae					
Cedrela odorata	C	M		moth	Bw, Bck et al. 1985
Guarea glabra ("G. bullata" of Bawa, Bullock, et al. 1985)	S	D		moth	Bw, Bck et al. 1985
G. grandiflora ("G. grandifolia" of Bawa, Bullock, et al. 1985)	C	D		moth	Bw, Bck et al. 1985
G. guidonia	S	D		moth	Bw, Bck et al. 1985
G. macropetala	C	D		moth	Bw, Bck et al. 1985
G. rhopalocarpa	S	D		moth	Bw, Bck et al. 1985; Bck, Bch & Bw 1983
Trichilia septentrionalis	S	D		small diverse insects	Bw, Bck et al. 1985
Monimiaceae					
Siparuna nicaraguensis	U	D			Ant, p.c.
S. pauciflora	S	D			Ant, p.c.
S. tonduziana	U	D			Ant, p.c.
Moraceae					
Brosimum alicastrum	C	D		small diverse insects	Bw, Bck et al. 1985; Hmm 1986j
Ficus brevibracteata	S	M		wasp	Bw, Bck et al. 1985; Hmm 1986j
F. dugandii	C	M		wasp	Bw, Bck et al. 1985; Hmm 1986j
F. glaucescens ("F. sp." of Bawa, Bullock, et al. 1985)	S	M		wasp	Bw, Bck et al. 1985; Hmm 1986j
F. insipida	C	M		wasp	Bw, Bck et al. 1985; Hmm 1986j
F. maxima	S	M		wasp	Bw, Bck et al. 1985; Hmm 1986j
F. tonduzii	S	M		wasp	Bw, Bck et al. 1985; Hmm 1986j
F. velutina ("F. trigonata" of Bawa, Bullock, et al. 1985)	C	M		wasp	Bw, Bck et al. 1985; Hmm 1986j
Maquira costaricana	U	D			Brg 1977b; Hmm 1986j
Perebea angustifolia	U	D			Brg 1977b; Hmm 1986j
Sorocea pubivena	U	D		wind	Bw, Bck et al. 1985; Hmm 1986j
S. trophoides	U	D		wind	Bw, Bck et al. 1985; Hmm 1986j
Trophis involucrata	U	D		wind	Bw, Bck et al. 1985; B&C 1980; Hmm 1986j
T. racemosa	S	D		wind	Hmm, p.c.
Myristicaceae					
Compsoneura sprucei	U	D		thrips	Bw, Bck et al. 1985
Virola koschnyi	C	D		small diverse insects	Bw, Bck et al. 1985
Myrsinaceae					
Auriculardisia auriculata ("Ardisia auriculata" of Bawa, Bullock, et al. 1985)	U	H		small bee	Bw, Bck et al. 1985
A. nigropunctata ("Ardisia nigropunctata" of Bawa, Bullock, et al. 1985)	U	H	SC	small bee	Bw, Bck et al. 1985; Bw, Pry et al. 1985
A. wedelii	U	H		small bee	K&B upbl
Weigeltia spectabilis	U	H		euglossine bee	K&B upbl; Hmm, p.c.
Myrtaceae					
Eugenia sp.	S	H			K&B upbl
Myrcia splendens	U	H			K&B upbl
Nyctaginaceae					
Neea psychotrioides	U	D			K&B upbl
N. sp.	U	D			K&B upbl

APPENDIX 12.1 (continued)

Taxon	Habitat Classes[a]	Sexual System[b]	Breeding System[c]	Pollinator Classes[d]	Source[e]
Ochnaceae					
Cespedesia macrophylla	C	H		medium to large bee	Bw, Bck et al. 1985
Ouratea lucens	U	H		medium bee	Bw, Bck et al. 1985
O. sp.	S	H			K&B upbl
Piperaceae					
Piper aequale	U	H			Brg 1971; Smp 1974
P. arieianum	U	H			Brg 1971; Smp 1974
P. augustum	U	H			Brg 1971; Smp 1974
P. auritum	U	H			Brg 1971; Smp 1974
P. biolleyi	U	H			Brg 1971; Smp 1974
P. biseriatum	U	H			Brg 1971; Smp 1974
P. carrilloanum	U	H			Brg 1971; Smp 1974
P. cenocladum	U	H			Brg 1971; Smp 1974
P. colonense	U	H			Brg 1971; Smp 1974
P. culebranum	U	H			Brg 1971; Smp 1974
P. garagaranum	U	H			Brg 1971; Smp 1974
P. glabrescens	U	H			Brg 1971; Smp 1974
P. hispidum	U	H			Brg 1971; Smp 1974
P. holdridgeianum	U	H			Brg 1971; Smp 1974
P. imperiale	U	H			Brg 1971; Smp 1974
P. melanocladum	U	H			Brg 1971; Smp 1974
P. peracuminatum	U	H			Brg 1971; Smp 1974
P. phytolaccaefolium	U	H			Brg 1971; Smp 1974
P. reticulatum	U	H			Brg 1971; Smp 1974
P. sancti-felicis	U	H			Brg 1971; Smp 1974
P. sinugaudens	U	H			Brg 1971; Smp 1974
P. tonduzii	U	H			Brg 1971; Smp 1974
P. umbricola	U	H			Brg 1971; Smp 1974
P. urophyllum	U	H			Brg 1971; Smp 1974
P. urostachyum	U	H			Brg 1971; Smp 1974
Rhamnaceae					
Colubrina spinosa	S	H		small diverse insects	Bw, Bck et al. 1985
Rhizophoraceae					
Cassipourea elliptica	S	H		small bee	Bw, Bck et al. 1985; K&B upbl
Rubiaceae					
Bertiera guianensis	U	H		small bee	K&B upbl
Cephaelis elata	U	S	SI	hummingbird	B&B 1983; Bw, Bck et al. 1985; Stl 1978c
C. glomerata	U	S		butterfly	K&B upbl
C. tomentosa	U	S		hummingbird	K&B upbl
Coussarea impetiolaris	S	S		moth	K&B upbl
C. talamancana ("C. sp. A" of Bawa, Bullock, et al. 1985)	U	D		small bee	B&B 1981; Bw, Bck et al. 1985; K&B upbl
C. sp. nov. (Beach 1467)	S	S	SI	moth	B&B 1983; Bw, Pry et al. 1985; K&B upbl
Faramea suerrensis	U	S	SI	butterfly	B&B 1983; K&B upbl
F. talamancarum	U	S		butterfly	Bw, Bck et al. 1985; K&B upbl
F. sp. (Beach 1477) ("F. occidentalis" of Bawa, Bullock, et al. 1985)	S	S	SI	butterfly	B&B 1983; Bw, Bck et al. 1985; K&B upbl
Genipa americana	S	D		small to medium bee	K&B upbl
Geophila cordifolia	U	H	SC	small bee	B&B 1983; K&B upbl
Gonzalagunia bracteosa	U	H		medium bee	K&B upbl
Hamelia patens	U	H	SI	hummingbird	B&B 1983; Bw, Bck et al. 1985; Bw, Pry et al. 1985; Stl 1978c
H. xerocarpa	U	H	SI	large bee	B&B 1983; Bw, Bck et al. 1985; Bw, Pry et al. 1985

APPENDIX 12.1 (continued)

Taxon	Habitat Classes[a]	Sexual System[b]	Breeding System[c]	Pollinator Classes[d]	Source[e]
Hoffmania liesneriana	U	H			K&B upbl
Pentagonia donnell-smithii	U	H		hummingbird	K&B upbl; Stl 1978c; McD 1986
Posoqueria grandiflora	U	H	SI	hawk moth	B&B 1983; Bw, Bck et al. 1985
P. latifolia	U	H	SI	hawk moth	B&B 1983; Bw, Bck et al. 1985; Bch 1983
Psychotria acuminata	U	S	SC	butterfly/small bee	B&B 1983; K&B upbl
P. brachiata	U	H		small diverse insects	K&B upbl
P. chagrensis	U	S		small bee	K&B upbl
P. chiapensis	U	S	SI	hawk moth	B&B 1983; Bw, Bck et al. 1985; K&B upbl
P. eurycarpa ("Coussarea sp." Beach 1424 of Bawa, Perry, et al. 1985)	S	S	SI	moth	B&B 1983; Bw, Pry et al. 1985; K&B upbl
P. graciliflora	U	S		small bee	K&B upbl
P. marginata	U	S		small bee	K&B upbl
P. officinalis	U	S	SI	butterfly/small bee	B&B 1983; K&B upbl
P. orchidearum	S	H	SC	small bee	B&B 1983; K&B upbl
P. suerrensis	U	S	SI	butterfly	B&B 1983; K&B upbl
Randia armata	U	D		hawk moth	Bw, Bck et al. 1985
R. genipoides (R. grandiflora of Bawa, Bullock, et al. 1985)	S	D		hawk moth	Bw, Bck et al. 1985
Rudgea cornifolia	U	S	SI	medium bee	B&B 1983; Bw, Bck et al. 1985
Simira maxonii	S	H		small to medium bee	Bw, Bck et al. 1985
Warszewiczia coccinea	S	H	SI	butterfly hummingbird	B&B 1983; Bw, Bck et al. 1985
Rutaceae					
Zanthoxylum mayanum	C	D		small diverse insects	Bw, Bck et al. 1985
Z. panamense	C	D		small diverse insects	Bw, Bck et al. 1985
Sabiaceae					
Meliosma donnell-smithii	S	H		medium bee	Bw, Bck et al. 1985; Hmm p.c.
Sapindaceae					
Cupania guatemalensis ("C. cf. cinerea" of Bawa, Bullock, et al. 1985)	U	M		small bee	Bw, Bck et al. 1985
Simaroubaceae					
Simarouba amara	C	D		small diverse insects	Bw, Bck et al. 1985
Solanaceae					
Cestrum megalophyllum	U	H		hawk moth	Bw, Bck et al. 1985
C. racemosum	S	H			K&B upbl
Cyphomandra hartwegii	S	H		large bee	Bw, Bck et al. 1985
Lycianthes sanctaeclarae	S	H		large bee	K&B upbl
Markea neurantha	S	H		bat	K&B upbl; Voss et al. 1980
Solanum arboreum	U	H		medium bee	Hbr 1983
S. argenteum	S	H		medium bee	Bw, Bck et al. 1985
S. rugosum	U	H		medium bee	Bw, Bck et al. 1985
Witheringia cuneata	U	H			K&B upbl
Sterculiaceae					
Herrania purpurea	U	H		fly	K&B upbl
Tiliaceae					
Apeiba membranacea	C	H		large bee	Bw, Bck et al. 1985
Goethalsia meiantha	C	H	SI	small diverse insects	Bw, Bck et al. 1985; Bw, Pry et al. 1985
Heliocarpus appendiculatus	S	D		small diverse insects	Bw, Bck et al. 1985
Luehea seemanii	C	H	SI	moth	Bw, Bck et al. 1985; Bw, Pry et al. 1985

APPENDIX 12.1 (continued)

Taxon	Habitat Classes[a]	Sexual System[b]	Breeding System[c]	Pollinator Classes[d]	Source[e]
Ulmaceae					
Trema micrantha	C	M		small diverse insects	Bw, Bck et al. 1985
Urticaceae					
Myriocarpa longipes	U	M		wind	Brg 1977c
Pilea diversissima	U	D			Brg 1977c
Urera caracasana	U	D			Brg 1977c
U. elata	U	D			Brg 1977c
Verbenaceae					
Aegiphila falcata	S	D		small bee	Bw, Bck et al. 1985
A. panamensis	S	D		small bee	Bw, Bck et al. 1985
Vitex cooperi	C	H	SI	medium bee	Bw, Bck et al. 1985; Bw, Pry et al. 1985
Violaceae					
Gloeospermum diversipetalum	U	H			K&B upbl
Rinorea deflexiflora ("R. cf. pubipes" of Bawa, Bullock, et al. 1985)	U	H		small bee	Bw, Bck et al. 1985
Vochysiaceae					
Vochysia ferruginea	C	H	SC	butterfly/bee	Bw, Bck et al. 1985; Bw, Pry et al. 1985
V. hondurensis	C	H		medium to large bee	Bw, Bck et al. 1985

Sources: Data derived from Bawa, Bullock, et al. (1985) for canopy and subcanopy species and from the references cited under "Source" for understory plants. Woody taxa listed by Bawa, Perry et al. (1985) for which pollinator information are lacking are excluded here, and the reader is referred to that publication for the data.

[a]Habitat Classes: U = understory (herbs, shrubs, and small trees to about 5 m in height); S = subcanopy (trees taller than 5 m but less than canopy stature at maturity; includes also some epiphytes normally flowering in the subcanopy); C = canopy (emergent and canopy trees). Native secondary and riparian forest trees are also included. Cauliflorous species are included in the stratum where flowers are displayed.

[b]Sexual Systems: H = hermaphroditic; M = monoecious; S = distylous; D = dioecious. Data are from published reports or from direct observation by the authors.

[c]Breeding Systems: SC = self-compatible; SI = self-incompatible. Data based on published reports or doctoral dissertations as referenced under "Source."

[d]Pollinator Classes: based largely on published reports from La Selva. Included also are data from unpublished observations by the authors or, in a few cases, from inference from well-documented studies of the same species or congeneric species sharing the same floral characteristics at other wet forest sites (as referenced under "Source"). These latter data should be considered preliminary and in need of further documentation. Where more than one class of flower visitor is important in the pollination of a species, these are listed. In such cases the first is considered the most important and is used to compile the summary tables. The class "small diverse insects" is used where a single most important pollinator could not be identified (see Bawa, Bullock, et al. 1985).

[e]Sources: references from which data were obtained.

Abbreviations:

Alv = Alverson	Hmm = Hammel	Rnn = Renner
Ant = Antonio	Hdn = Henderson	S&C = Soderstrom & Calderon
B&B = Bawa & Beach	J.Mil = J. Miller	S&H = Schemske & Horvitz
B&C = Bawa & Crisp	Jnz = Janzen	Sch = Schmid
Bch = Beach	K&B upbl = Kress & Beach unpublished	Slm = Sleumer
Bck = Bullock	Kdy = Kennedy	Smk = Schemske
Bmn = Buchmann	M&A = Montalvo & Ackerman	Smp = Semple
Brg = Burger	M&D = McDade & Davidar	Stl = Stiles
Bw = Bawa	M&K = McDade & Kinsman	Stn = Steiner
Bck, Bch & Bw = Bullock, Beach & Bawa	Maas = Maas	Stz = Schatz
Bw, Bck et al. = Bawa, Bullock et al.	McD = McDade	Szm = Sazima
Bw, Pry et al. = Bawa, Perry et al.	McV = McVaugh	U&M = Uhl & Moore
C&W = Corbet & Willmer	Nel = Neill	Val = Valerio
Gbb = Gibbs	P&S = Perry & Starrett	Voss = Voss
Grv = Grove	p.c. = personal communication	W&D = Williams & Dressler
Hbr = Haber	Pohl = Pohl	Whl = Wiehler
	Prc = Prance	Yng = Young
	Pry et al. = Perry et al.	

PART III

THE ANIMAL COMMUNITY

COMMENTARY

Henry A. Hespenheide

Natural history studies in Costa Rica have a tradition that has been broadly chronicled by Gómez and Savage (1983) and, more narrowly, for butterflies by DeVries (1987). Both accounts include contributions from indigenous and adoptive Costa Ricans and visiting scientists from the United States and European countries. Although Costa Rica did not figure extensively in the Biología Centrali-Americana project sponsored by Godman and Salvin (1879–1915), subsequent biological research there has developed rapidly. It is probable that a greater proportion of its total biota is known than that for any other tropical American country, although particular taxa may be better known elsewhere. La Selva Biological Station has played a growing part in that development by providing a base for the variety of faunal studies reviewed here.

The growth in knowledge of the biota of any area typically follows a pattern in which vascular plants and higher vertebrates are the first groups studied, then lower vertebrates, and, more gradually, invertebrates. Among invertebrates, butterflies and certain families of beetles—groups popular with both professional and nonprofessional collectors—are studied first and others at less predictable intervals. In that regard, La Selva's history has been typical. The first faunal inventory was Slud's (1960) study of the birds, and by the time of Janzen's (1983b) *Costa Rican Natural History,* lists could be presented for all of the terrestrial vertebrate classes. In the latter complication 1,190 species were listed in the four classes of terrestrial vertebrates known from Costa Rica, of which 509, or about 45% of the total, were known or expected from La Selva. Although *Costa Rican Natural History* was selective rather than complete, at least 79 of La Selva's vertebrate species (15%) were, in turn, well enough known to have separate species accounts written about them. Among invertebrates, a few groups were known adequately at that time to have species lists for Costa Rica and La Selva: namely, short-horned grasshoppers (Orthoptera, Acridoidea), sphingid moths (Lepidoptera, Sphingidae), and certain families of butterflies. In just nine years after the *Costa Rican Natural History* appeared, guides to Costa Rican freshwater fish (Bussing 1987), birds (Stiles et al., 1989), and butterflies (DeVries 1987) were published, and the faunal lists for La Selva have been added to and revised considerably. In addition, the terrestrial vertebrates of La Selva have been considered as part of a more general comparison of four lowland Neotropical rain forest sites (Gentry 1990a).

In this section we review in turn the vertebrate groups known from La Selva as well as the butterflies. For each group authors briefly summarize what is known about the fauna both taxonomically and ecologically, compare this knowledge with what is known from other tropical sites, and then indicate directions for future research at La Selva. After the reviews of individual taxa, I will compare them as part of a general discussion of the fauna. I will emphasize the poor state of knowledge of tropical insects and discuss the general issue of tropical species richness and its relation to climate, history, and ecological diversity. La Selva is one of the few well-documented examples of the species richness (*biodiversity*) of wet tropical forests, and comparisons can be made with nearby areas that are more seasonal (e.g., Barro Colorado Island) or that have had different historical biogeographic influences (e.g., Guanacaste).

In the concluding chapter authors Braker and Greene review the autecological studies that have been conducted on animals at La Selva. Intensive studies of individual species have been relatively few, and these authors make a special plea for long-term studies coordinated in such a way that they can be compared with those conducted at other tropical sites. They point to the surprising diversity of mating systems and patterns of reproduction in what is thought of as a relatively constant and predictable environment (see also Vitt 1986) and urge that more life-history studies be undertaken. They also note the importance of predation and the special role that frogs may play in maintaining the diversity and populations of large predators.

13

Patterns of Butterfly Diversity and Promising Topics in Natural History and Ecology

Philip J. DeVries

Two of the most frequent questions asked of any biologist with expertise in a particular group of organisms are How many species occur at such and such a site? and How does species richness compare with such and such a site? Often a site is judged a priori by the answers to these questions: high species richness is equated with a good site because organismal biologists are susceptible to the seduction of high diversity. Although uninteresting on its own, a species list is basic information about any site, and such lists are used extensively for site comparisons and decision making in conservation biology. To the specialist a species list is an indicator of what is known and how many more life histories remain to be discovered and a possible framework for exploring the evolutionary patterns of diversity (see Mound and Waloff 1978; May 1988). The lists compiled for many organisms at La Selva are surging with species, and for this reason most biologists consider it a "good site."

By comparing the butterfly species richness of La Selva to that of other tropical temperate sites and highlighting some aspects of butterfly biology, I wish to impress upon prospective researchers that even though a multitude of butterfly species are known from La Selva (and the species list will continue to lengthen) the general biology and ecology of the vast majority of all tropical butterfly species remains unknown. Accordingly, this chapter is divided into two general sections. In the first I describe the patterns observed in comparing species lists by subfamilies and augment broader comparisons made elsewhere (DeVries 1983, 1987). Although I am fully aware that sampling among sites is not equal, geographic areas and climatic regimes differ, and evolutionary and ecological histories are not equivalent among sites, such comparisons do provide a context for La Selva species richness and a taxonomic background from which to sample butterfly biology. In the second section of this chapter I cover selected aspects of tropical butterfly diversity and suggest areas that may be fruitful for field biologists working at La Selva and other tropical sites. Readers interested in population biology, ecology, natural history, and evolution of tropical butterflies will find useful summaries in Boggs (1981b), Boggs and Gilbert (1979), Boggs et al. (1981), Chai (1990), Denno and Donnelly (1981), DeVries (1987), Douglas (1986), Singer and Mandrachia (1982), Smiley (1978a, 1978b, 1985b), Silberglied 1984; Vane-Wright and Ackery (1989), Young (1979, 1982), and

references therein. Some of these studies (Boggs, Denno and Donnelly, DeVries, Smiley, and Young) were conducted at or near La Selva. For recent developments in butterfly systematics see Ackery (1988), Ackery and Vane-Wright (1984),- DeVries et al. (1985), Harvey (1987, 1991), Kitching (1985), Miller (1987, 1988), and Scoble (1987). Although butterflies are comparatively well known among insects and systematic studies are on the upswing, there is still no resolved phylogeny for the group as a whole.

THE COMPARATIVE DIVERSITY OF BUTTERFLIES AT LA SELVA

In general, species richness of Costa Rican butterflies is highest in lowland and midelevational rain forest areas and falls off with increasing elevation and/or decreasing rainfall (DeVries 1987). Therefore, because of its geographical location, La Selva is rich in butterfly species. As in other groups of animals and plants, the bulk of the butterfly fauna is composed of species that range widely along the Atlantic lowlands of Central America. At least 204 species of Papilionidae, Pieridae, and Nymphalidae have been reported from La Selva (DeVries 1983, unpublished data), and with very incomplete records I estimate that at a minimum 50 species of Riodinidae and 50 Lycaenidae will be found there. (Given the inverse relationship of body size and number of species [May 1978] and my own recent experience with riodinids, I tend to agree with R. Robbins [pers. comm.], who suggests that the La Selva fauna contains at least twice this many species in each of the latter two families.) Should some industrious biologist eventually tabulate all of the La Selva skippers (Hesperiidae), my guess is that the total fauna could increase by at least another 175 species. A conservative total estimate of the butterflies and skippers occurring at La Selva would, thus, be 479 species, a rich fauna by any standard although not nearly as rich as the 1,200 species estimated to occur at the larger Tambopata Reserve, Peru (Lamas 1981, 1985b). When the butterfly fauna of Parque Braulio Carrillo is eventually determined and the fauna of La Selva placed in context, these figures should increase substantially and make this area an even "better" site for butterflies.

To place the butterfly fauna of La Selva in a comparative context, the total estimate of butterflies and skippers indicates that La Selva has two hundred more species (60%) than the entire United States east of the Great Plains (data from Opler and Krizek 1984). For just the Papilionidae, Pieridae, and Nymphalidae La Selva has seventy-eight (62%) more species than the eastern United States. In short, even though the area covered by the eastern United States is several orders of magnitude larger than that of La Selva, one effect latitude has on diversity seems clear: persistent freezing winters are not kind to butterflies.

By comparing the species counts of the Papilionidae, Pieridae, and Nymphalidae at La Selva to those at other sites that differ in habitat and climate (Parque Nacional Corcovado, Parque Nacional Santa Rosa [before the development of the Guanacaste National Park in 1987], and Barro Colorado Island, Panama], a pattern emerges that, although not very surprising, is useful. The data for Barro Colorado Island are derived from Huntington (1932) and DeVries (unpublished data), and all other sites are from DeVries (1983, 1987, and unpublished data). Comparison among these faunas demonstrates a well-established pattern seen in many insect, animal, and plant groups: overall species richness is greatest in areas of high rainfall, low elevation, and low latitude (table 13.1). For butterflies this trend is most evident in nymphalids and, as noted later, the riodinids. It is likely that if all of Guanacaste National Park, Braulio Carrillo, and the lands in the BCI National Monument were included, the estimates for the tropical sites would be greater and further increase the contrast between temperate and tropical sites, illustrating the contribution of elevational gradients and topographic diversity over relatively small geographic scales.

When the contribution of each family to total species richness is examined (table 13.1) an interesting pattern emerges. The Papilionidae comprise between 9% and 10% of the total species at all sites. The contribution of Nymphalidae to the butterfly species count increases toward the tropics: 65% in the eastern United States, 73% at Santa Rosa, 79% at BCI, and 80% at Corcovado and La Selva, scarcely a surprise. The Pieridae, however, show an increase in species richness from 10% to 26% of total species from tropical to temperate sites. Why the pierid species richness varies in this manner is unknown, but it may be related to larval host plants and adult microhabitat preference. For example, almost all species in the Coliadinae and many Pierinae use weedy, herbaceous and woody legumes, or Cruciferae as larval host plants, and the adults inhabit sunny, open areas. It seems likely that the abundance of potential pierid host plants increases from rain forest to dry forest to temperate areas. The increase in pierid species from wet to drier sites may reflect the influence of human disturbance as it creates habitats for both host plants and butterflies.

A comparison of taxa commonly shared among all fives sites (table 13.2) shows, not surprisingly, greater overlap between those sites having similar climatic characteristics and geographic proximity. Hence, La Selva shares more taxa with Corcovado and Barro Colorado Island than it does with the United States, and so forth. Finally, of the entire 367 species used in the comparison, 98 are confined entirely to the U.S. sites, and only 19 species are shared in common between all fives sites (table 13.2). Of these 19 wide-ranging taxa, all are weedy species that occur in open second growth, and many

Table 13.1 Minimum estimates of total species per site in three butterfly families

Families	Total Species				
	LS	CV	BCI	SR	US
Papilionidae	20	18	12	13	11
Pieridae	20	18	17	21	33
Dismorphiinae	4	1	0	0	0
Pierinae	4	5	5	5	11
Coliadinae	12	12	12	16	22
Nymphalidae	164	138	107	91	82
Charaxinae	18	11	11	12	3
Apaturinae	2	0	1	2	2
Libytheinae	1	1	1	1	1
Nymphalinae	47	46	31	35	37
Acraeinae	1	1	0	0	0
Heliconiinae	15	16	11	12	3
Melitaeinae	10	6	4	9	12
Danainae	4	4	3	4	3
Ithomiinae	24	20	12	5	0
Morphinae	7	5	6	1	0
Brassolinae	9	8	9	4	0
Satyrinae	26	20	18	6	21
TOTAL	204	174	136	125	126

Sources: DeVries (1983, 1987, unpublished data); Huntington (1932); Opler and Krizek (1984).

Note: Comparison codes: LS = La Selva; SR = Santa Rosa National Park; CV = Corcovado National Park; BCI = Barro Colorado Island; US = eastern United States.

are known to migrate (e.g., *Battus polydamas, Papilio cresphontes, Ascia* spp., *Eurema* spp., *Phoebis* spp., *Junonia evarete, Dryas iulia, Marpesia chiron, Cissia hermes*). As the tropical faunas become better known it will be interesting to see whether the lycaenids and riodinids shared among these sites are also migratory and/or species common in second-growth.

THE LA SELVA BUTTERFLY FAUNA

From the comparisons presented here it is obvious that La Selva contains a very diverse butterfly fauna. With such an impressive array of species it is rather sobering to realize that fewer than 7% of the combined papilionid, pierid, and nymphalid species have been studied in any detail (only the Heliconiinae), and no lycaenid or riodinid species at La Selva has been studied. In this section I focus on patterns associated with communities of tropical butterflies that have received little or no attention.

Influences on Diversity

Many factors interact to influence the distribution and diversity of butterfly faunas (Shapiro 1975; Gilbert and Singer 1973, 1975; Vane-Wright 1978; Chew and Robbins 1984). Because butterflies depend upon specific plants as larval hosts, it follows that site differences in butterfly species richness and abundance may, in part, reflect host-plant diversity (Shapiro 1975; Benson 1978; Haber 1978; Gilbert and Smiley 1978; Lawton 1978; Strong et al. 1984). For example, in Costa

Table 13.2 Number of butterfly species shared among sites

Family	LS CV	LS SR	LS BCI	LS US	CV SR	CV BCI	CV US	SR BCI	SR US	BCI US	All
Papilionidae	12	6	8	3	8	8	3	6	3	3	2
Pieridae	13	9	12	8	14	10	8	13	12	9	7
Dismorphiinae	1	0	0	0	0	0	0	0	0	0	0
Pierinae	3	1	2	0	3	3	1	3	2	1	1
Coliadinae	9	8	10	8	11	7	7	10	10	8	6
Nymphalidae	106	56	81	14	59	69	13	52	16	10	10
Charaxinae	6	6	9	0	7	8	0	8	1	0	0
Apaturinae	0	1	0	0	0	0	0	1	0	0	0
Libytheinae	1	1	1	0	1	1	0	1	0	0	0
Nymphalinae	37	22	23	6	23	19	5	20	5	3	3
Acraeinae	0	0	0	0	0	0	0	0	0	0	0
Heliconiinae	12	9	10	3	8	7	3	7	3	2	2
Melitaeinae	4	2	2	1	4	3	1	2	3	1	1
Danainae	4	4	3	3	4	3	3	3	3	3	3
Ithomiinae	15	4	11	0	4	8	0	3	0	0	0
Morphinae	5	1	6	0	1	5	0	1	0	0	0
Brassolinae	6	3	4	0	4	5	0	3	0	0	0
Satyrinae	16	3	12	1	3	10	1	3	1	1	1
TOTALS	131	71	101	25	81	87	24	71	31	22	19

Sources: DeVries (1983, 1987, unpublished data); Huntington (1932); Opler and Krizek (1984).

Note: Comparison codes: LS = La Selva; SR = Santa Rosa National Park; CV = Corcovado National Park; BCI = Barro Colorado Island; US = eastern United States; All = All sites.

Rica the area around Rincón de Osa that adjoins Corcovado National Park to the north has an ithomiine richness and abundance as high, or higher than La Selva. In contrast, the Corcovado basin itself has many fewer ithomiine species. One striking difference between these areas is that diversity of understory Solanaceae is lowest in the basin and comparable between La Selva and Rincón (see Haber 1978). Other likely examples of host limitation are the low ithomiine richness at Santa Rosa, which correlates with low diversity of Solanaceae; *Dulcedo polita* (Satyrinae), which is rare or absent at all other sites but is found commonly near stands of *Welfia* palms at La Selva; the abundance of *Heliconius* individuals at Corcovado (but not La Selva) resulting from the abundance of *Passiflora* (L. Gilbert pers. comm.); and populations of *Kricogonia lyside* (Pieridae) localized in Guanacaste near its host plant, *Guaiacum officionale*. In some cases, however, the measurement of abundance may be misleading. For example, caterpillars of the riodinid *Thisbe irenea* may be abundant throughout the year on patches of its host plant, but the adults are very rarely seen and, hence, would be considered rare (DeVries unpublished data). Finally, it should be noted that both insect abundance and diversity may be low in the face of high host abundance and diversity (see Scriber 1973; Strong et al. 1984; Miller 1987), indicating that a variety of factors work in concert as determinants of herbivore diversity. Excepting the genus *Heliconius* (Benson 1978; Smiley 1978b; Denno and Donnelly 1981; Gilbert 1984), the effects of host density and diversity and adult and larval resources have not been studied for populations of tropical butterflies. The La Selva

butterfly fauna would be an excellent place to begin such studies.

Seasonality

Insect ecologists have long been interested in seasonality (as a correlate of species richness) and diapause (see reviews of Danilevskii 1965; Riedl 1983; Tauber et al. 1986). Most studies, models, and theories treating the evolution of insect seasonality are based on temperate examples, and tropical species are considered "special cases." Without belaboring the point, it is obvious to all tropical field biologists that tropical insects show seasonality (ask any *campesino*) and that tropical insects have different seasonal constraints from temperate ones. It also should be obvious that seasonality in tropical insects deserves further study. Considering the patterns of species richness in the tropics it seems likely that seasonal responses in temperate insects represent special cases of species evolving from tropical ancestors. The migrations of the monarch butterfly (*Danaus plexippus*), the snout butterfly (*Libytheana carinenta*), the painted lady (*Vanessa cardui*), and *Kricogonia lyside* all appear to be spectacular examples of tropical species "intent" on colonizing the temperate zones, and more subtle examples surely exist among other species. In other words, perhaps researchers should begin modeling temperate insect seasonality after tropical examples.

Studies of seasonality in tropical butterflies are rare and have concentrated on the population dynamic effects of a few taxa (Smiley 1978a; Young 1979; Gilbert 1984). Three types of data on tropical butterflies could be gathered that would greatly increase what is known of the evolution of insect seasonality. First, life-history records of species should be chosen so that data will ultimately be comparable to data taken in other areas. Researchers could use a simple form with phylogenetically related taxa and take monthly (or weekly) censuses of the presence or absence of adults, their reproductive condition, presence and abundance of caterpillars, host use, and presence or absence of a diapause in egg, larval, or pupal stages. If done as a comparison among microhabitats, such a monitoring study could also be applied to developing empirical and theoretical bases for conservation biology. A second avenue of research would be comparative physiology of those species showing diapause and those that do not. The documentation of lipid accumulation seems a likely place to begin (e.g., Pullen 1987). Third would be comparative studies on intraspecific variation in seasonality and diapause for species having broad geographic or elevational ranges. The transect from La Selva to the high elevations in Braulio Carrillo would be an excellent place from which to draw material. With field and laboratory manipulations added, it should be possible to assess the effects of predators, parasitoids, food availability and quality, temperature, light, and moisture on seasonality of butterflies. Were data like these available for the sites listed here (tables 13.1 and 13.2), it would have been possible to make ecological and evolutionary, rather than strictly faunistic, comparisons.

Vertical Stratification

Recent interest in how to estimate total insect species richness (May 1988; Gaston 1991) in large part stems directly from sampling done in the tropical forest canopy (e.g., Erwin 1982, 1990). Knowing this, it is scarcely surprising that substantial fractions of the total butterfly fauna at many tropical sites are found mainly in the forest canopy, not in the understory where the majority of biologists work. Regardless of how many species of butterflies there are, documenting how these species partition a forest habitat seems fundamental to understanding tropical butterfly biology. Vertical stratification of butterflies (the occurrence of different species at differing levels of the forest), however, has been studied very little. Perhaps the best-known study is that of Papageorgis (1975), who showed that some mimetic groups of butterflies (heliconiines, ithomiines, and their mimics) in a Peruvian forest are stratified at different levels by color pattern. She suggested that these butterflies are stratified in response to predator pressures at different levels of the forest for mimetic resemblance. Although this pattern is not at all obvious in Central American heliconiine species (L. Gilbert pers. comm.; pers. observation), stratification does very clearly occur in a guild known as fruit-feeding nymphalids. The fruit-feeding nymphalid species derive most or all of their adult nutrition by feeding on the juices of rotting fruits, sap flows in trees, carrion, or dung. A preliminary study using stratified traps baited with rotting fruit showed that fruit-feeding nymphalid butterflies at La Selva (all palatable nonmimics) are stratified between the canopy and the understory by size, color pattern, species and subfamily, and all of these characteristics are correlated with phylogenetic affinity (DeVries 1978b). Interestingly, the vertical stratification by these butterflies is most easily explained by the light level

preferences of the different species, not by where fleshy, rotting fruits are expected to occur.

Given the history of canopy work done at La Selva and the growing interest in vertical stratification of insects, it is surprising that so little work has been done at La Selva, particularly on butterflies. A long-term trapping study of fruit-feeding nymphalids at La Selva comparing canopy and understory captures could provide more accurate estimates of species richness and data on how forest structure and seasonality influence insect stratification and population dynamics. Three reasons argue that this method of data collection would be pertinent to conservation biology and the design and management of nature reserves: the trapping technique is simple; it is replicable; and it allows mark/release/recapture studies to be conducted on a selected subset of forest butterflies. A long-term study using these methods is currently being conducted with my colleagues in southwestern Costa Rica at the Las Alturas Biological Station. Regretfully, no similar study exists at La Selva to allow a comparison between the two sites with their distinct forest types. For future biologists who may have complete access to the forest canopy and wish to perform mark/release/recapture studies, the problem will remain of how to study those flower-feeding species that must be sampled with nets or observed with binoculars (papilionids, pierids, lycaenids, and riodinids).

Mimicry and Other Defenses

An important component of all tropical butterfly faunas is the suite of mimetic species. The study of mimicry embraces, among other things, the evolution of warning coloration, ecological chemistry, genetics, and predator-prey interactions. Of the seven commonly recognized Batesian and Müllerian butterfly mimicry complexes in Costa Rica (see DeVries 1987), six occur at La Selva, and all of these have figured prominently in mimicry studies conducted elsewhere (see Gilbert 1983; Brower 1984; Turner 1984 for summaries and reviews). Although the fauna is rich in models and potential mimics, no study has concentrated on butterfly mimicry at La Selva.

A tropical perspective of butterfly mimicry and defenses against predators suggests a greater complexity than that of the classic monarch and viceroy mimicry system of the temperate zone. Indeed, Ritland & Brower (1991) show that even this paradigm is more complex than thought previously—they are Müllerian mimics! Although there are excellent fundamental theories and empirical studies on the acquisition of noxious chemicals and mimicry in butterflies, researchers are just beginning to appreciate the mimetic diversity inherent in tropical systems. Five areas of study, all of which can be carried out at La Selva, seem especially promising for advancing understanding of Neotropical butterfly mimicry.

Acquisition. The exemplar aposematic models for Neotropical mimicry complexes are *Heliconius* and ithomiine butterflies, yet recent studies have changed previous assumptions of how these butterflies acquire their distasteful properties. *Heliconius* butterflies are apparently unpalatable not because their caterpillars feed on toxin rich *Passiflora* host plants but because they synthesize cyanogenic compounds as adults through assimilation of nitrogenous products from pollen

(Nahrstedt and Davis 1981, 1983). Viewed in a different manner, this result implicitly suggests that model *Heliconius* butterflies are palatable automimics when they emerge from the pupa and Müllerian mimics sometime after feeding on pollen. Although ithomiine and danaine butterflies may acquire chemical defenses from larval feeding, they have been demonstrated to augment their unpalatability to predators through adult feeding on plants containing pyrrolizidine alkaloids, and males may transfer these acquired chemicals to females through mating (Boppre 1984, 1990; Brown 1984). The fact that distasteful properties can be gained without feeding on a host plant rich in secondary compounds suggests that researchers need to revise their thinking about palatability in butterflies. Systematic surveys of palatability in all butterfly groups could prove very interesting indeed and may well broaden our horizons in bad taste.

Palatable and Unpalatable Local Faunas. In spite of the theoretical consideration given the evolution of mimicry (e.g., Endler 1986a; Guilford 1986; Mallet and Singer 1987), the dynamics of tropical Müllerian and Batesian faunas and the suite of predators that select for mimicry are poorly understood and rarely studied empirically. Using a large number of species from local butterfly fauna, Chai (1986, 1987, 1990) conducted the first large-scale Neotropical study in Parque Nacional Corcovado, Costa Rica. In this study he probed the gustatory and visual responses of a specialist predator (the rufous-tailed jacamar, Galbulidae) to the community characteristic of butterflies, testing the ideas of Bates (1862), Müller (1879), Poulton (1887, 1908), and Turner (1984) about what characteristics unpalatable models should exhibit. Chai showed that to assess the palatability of butterflies in a community jacamars use a suite of behavioral and phenotypic characteristics of which color pattern is only one character cue. Further, some pierids and nymphalids previously thought to be palatable Batesian mimics are not eaten by jacamars. Contrarily, some heliconiines are palatable to jacamars, which suggests that palatability cannot be judged entirely upon phylogenetic affinity and highlights the "palatability spectrum" initially described by Brower for monarch butterflies and later amplified by Turner (1984). Because not all birds or lizards are specialists on butterflies but clearly do exert selection on coloration, behavior, and shape (Swynerton 1919), one now needs to know the responses of a range of butterfly predators to understand selection on butterfly communities as a whole.

Thermoregulatory and Aerodynamic Conditions. Using the palatability data gathered by Chai (1987), two studies (Chai and Srygley 1990; Srygley and Chai 1990b) have broken new ground in the physiological ecology of butterflies by comparing the behavioral, physiological, and microhabitat characteristics of a subset of palatable and unpalatable butterflies. In this study they begin to characterize the differences in thermal requirements of many species of Corcovado butterflies and their physiological traits. Perhaps most significantly, in this work they demonstrate how much can be learned about the ecology of mimicry with a minimum of equipment. Their hypothesis that palatable butterflies require higher operating temperatures to fly swiftly out of harm's way and, hence,

should spend more time basking, clearly deserves more study. Although methods for studying the physiology and biomechanics of animal flight can be complicated (see Ellington 1984), considerable data can also be gathered in the field with few items of equipment. For example, with a portable video camera R. Dudley (pers. comm) has documented considerable kinematic variation in the aerodynamic mechanisms that affect forward flight, flight behaviors, and maneuverability in a spectrum of morphologically diverse Panamanian butterfly species. Recent field studies (Chai 1990; DeVries and Dudley 1990; Dudley and DeVries 1990; Srygley and Chai 1990a) suggest that simple thermocouple probes, videocameras, and biomechanical techniques are useful tools when exploring the physiological constraints on migration, mimetic species, and the effect of morphological variation on predator evasion in butterflies.

Unexplored Mimicry Groups. Although nymphalids, pierids, and papilionids provide fertile ground for experimental and theoretical work on mimicry, obvious mimetic complexes involving lycaenid and riodinid butterflies have been ignored. The neglect of these groups is without doubt because of their size; if riodinid and lycaenid butterflies were larger, many species would have long ago been considered aposematic. Two examples of riodinid-lycaenid mimicry complexes that occur at La Selva, and elsewhere, deserve study. One completely unexplored complex includes species in the riodinid genera *Esthemopsis, Brachyglenis, Mesene,* and *Symmachia* and moths in the notodontid subfamily Dioptinae, especially the genera *Josia, Phaeochlaena,* and *Erbessa* (J. Miller, pers. comm; see also Bates 1859; Seitz, 1916–1920). These taxa all fly in the same forested areas, often at the same time of day. On the basis of their warning coloration, they are likely to represent Batesian and Müllerian mimicry complexes. Another probable Batesian and Müllerian complex includes the lycaenid genus *Eumaeus,* the riodinid genera *Uranesis, Hades,* and *Lyropteryx,* the nymphalids *Diaethria* and *Persisama,* and various pericopid and dioptid moths. In this example *Eumaeus* and the pericopids are clearly unpalatable. Upon capture many day-flying pericopid moths emit a froth from the thorax that smells of cyanide. In addition to the adults reflexively bleeding a liquid from the thorax, all stages of the life cycle of *Eumaeus* species appear to be chemically well defended (Bowers and Larin 1989; DeVries 1977, unpublished data; Rothschild et al. 1986), adults are refused by jacamars (P. Chai pers. comm.), and insectivorous rodents will not eat caterpillars or pupae (pers. observation). Any observations on mimicry and palatability in these systems would be new and interesting.

Caterpillar Mimicry. In general, a greater portion of a butterfly's offspring are killed as caterpillars than as adults (Smiley 1985b). It is clear that butterfly caterpillars may be chemically and physically defended (see review in Brower 1984), but caterpillar mimicry is an important but unexplored area of butterfly biology. Janzen (1988, pers. comm.) has noted the similarities between nymphalid butterfly caterpillars and urticating saturniid moth caterpillars and suggested this and other examples as apparent cases of mimicry. My own recent work with riodinids indicates that their caterpillars may strongly resemble members of the Papilionidae, Arctiidae, Lasiocampidae, Noctuidae, and Megalopygidae, groups that have well-developed defenses (fig. 13.1). No systematic studies of the defenses, relative palatability, or mimicry of butterfly caterpillars have been done nor have specific predators been identified against which such defenses may be directed. Simple lists of caterpillar species accepted and refused by predators (as compiled by herpetologists, ornithologists, mammalogists, and entomologists who maintain captive organisms) would go a long way toward documenting and understanding mimicry and larval defenses.

Biologists at La Selva can make major contributions toward understanding the evolution and maintenance of butterfly mimicry in adults and caterpillars by employing field methods like those of Marshall and Poulton (1902), Swynnerton (1919), Swynnerton (in Carpenter 1942), and Chai (1986). Work that documents how insectivorous birds, lizards, and insects react to unpalatable and mimetic butterflies should include experiments with caterpillars and pupae as well as adults. A study that includes all life stages would be an instant classic.

THE RICHNESS IN MINIATURE

The families Lycaenidae and Riodinidae comprise small, seldom-observed butterflies. Of the two groups, the "hairstreaks," or lycaenids, are most familiar to the inhabitants of La Selva, especially the white and black striped *Arawacus lincoides* that occurs in second growth and along trails. Many lycaenids customarily rub their hindwings together when at rest to show the peculiar color patterns on the distal portion of the hindwing to advantage. These together form a false head image that is employed as a defense against predators (Robbins 1981). Although no other group of butterflies can match the riodinids in diversity of color patterns and wing shapes—they may look like members of several butterfly or moth families—the riodinids are the least frequently observed of all butterflies. Many species land on the underside of leaves with their wings open, others land on top of leaves and have jerky movements like a clockwork toy, and most of the riodinids are active either in the early morning or late afternoon. As a group, they are most often mistaken by the casual entomologist as "odd looking" lycaenids.

Inconspicuous as the lycaenids and riodinids may be, the biology of these lilliputians is unparalleled elsewhere in the butterflies. The Lycaenidae and Riodinidae account for more than 40% of all the world's butterfly species (Vane-Wright 1978; Robbins 1982), a substantial portion of which is Neotropical. Lycaenid and riodinids are the exceptions to the traditional wisdom of butterfly and host plant relationships: as a group their caterpillars have a spectacularly wide phytophagous diet breadth (within and among species) and may include carnivory or tropholactic feeding by unrelated species (Farquarson 1922; Ehrlich and Raven 1964; Clark and Dickson 1971; Cottrell 1984; Pierce 1984, 1987; Thomas et. al. 1989). Given such characteristics it seems evident that understanding these butterflies is important to understanding insect diversity and evolution of diet. One of the most intriguing life-history traits of lycaenids and riodinids, however, is myrmecophily—their habit of forming symbioses with ants.

Lycaenid and riodinid caterpillars may possess adaptations for forming, mediating, and maintaining symbioses with ants that include secretory organs that produce food secretions for ants, organs that produce semiochemicals, sound-producing organs that attract ants (fig. 13.1), and a suite of behavioral and morphological characters for living with ants (reviewed in Cottrell 1984; DeVries 1988a, 1990, 1991a, 1991b). In the past, general studies have typically combined the riodinids and lycaenids under the family Lycaenidae (Atsatt 1981; Pierce 1985, 1987). Treated in this way, understanding of riodinid evolution and ecology has, thus, been inferred from what is known of lycaenids. When considered individually, however, a number of differences between the two groups are evident: (1) caterpillar morphology suggests that the trait of forming symbioses with ants in butterflies has evolved at least twice—once in the lycaenids and once in the riodinids; (2) lycaenid species richness is uniformly high in all tropical regions, but of the twelve hundred or so named riodinid taxa, more than 92% occur only in the Neotropics; (3) myrmecophily is widespread in the Lycaenidae but relatively uncommon in the Riodinidae; (4) riodinid myrmecophily evolved only in the Neotropics; and (5) myrmecophilous riodinid taxa tend to feed on plants with extrafloral nectaries, whereas the non-myrmecophilous taxa tend to use plants without extrafloral nectaries. For further discussion see DeVries (1991b).

A Suite of Interactions

The basis of many ant-plant symbioses on the one hand, and of riodinid and lycaenid caterpillar-ant symbioses on the other, is, in effect, the same: ants provide protection from enemies in exchange for secretions rich in amino acids and sugars (Cottrell 1984; DeVries 1988a, 1991c; DeVries and Baker 1989; Huxley and Cutler 1991; Pierce et al. 1987). Moreover, evidence suggests that co-occuring plants, myrmecophilous butterfly caterpillars, and other insects share the same suite of protective ant taxa (DeVries 1991c). The interactions between plants and insects that benefit from protective ants demonstrate how common species interactions may be more complex than thought previously.

Studies of ant-herbivore interactions (e.g., Buckley 1983; Horvitz and Schemske 1984), in addition to observations indicating that ant taxa may be shared by co-occuring plants and insect herbivores (DeVries 1991c), suggest that the evolution of insect myrmecophily has allowed these herbivores to invade and exploit ant-plant symbioses (DeVries 1991b; DeVries and Baker 1988). For example, the caterpillars of the riodinid *Thisbe irenea* are found commonly from Belize to Panama (including the La Selva area) on saplings of several *Croton* species (Euphorbiaceae). Like most butterflies, *Thisbe* caterpillars feed mainly on leaf tissue. In addition to leaf tissue, however, drinking extrafloral nectar contributes significantly to caterpillar growth, a habit that is typical of many myrmecophilous riodinids (DeVries 1991b; DeVries and Baker 1988). In this system caterpillars are protected against enemies while they feed on plant tissue by the same ants that deter herbivores from the plant. Moreover, such riodinids may typically add insult to herbivory (fig. 13.1) because they use the basis of the plant-ant symbiosis (extrafloral nectar) for their own growth benefit (DeVries 1991b, 1991c). Consideration of the many other myrmecophilous insects and plants

Fig. 13.1. Riodinid caterpillars. *A.* The nonmyrmecophilous caterpillar of *Ancyluris inca.* The body is clothed in a dense covering of setae and resembles a moth caterpillar in the family Arctiidae. *B.* A *Camponotus sericiventris* ant feeding on amino acid secretions produced by an eversible tentacle nectary organ (arrow) on a caterpillar of *Synargis mycone. C.* A C. sericiventris ant showing an "alarm" reaction to the presumed semiochemicals produced by the everted anterior tentacle organ (arrow) on caterpillar of *Synargis mycone. D.* Caterpillar of *Nymphidium mantus* drinking at an extrafloral nectary (arrow) while being tended by an *Azteca* sp. ant, who, in turn, is feeding from the secretion produced by the caterpillar. Photos from color transparencies (copyright, P. J. DeVries).

that occur in the tropics paints an overall picture of an even larger and more complex suite of interactions than is generally thought.

The details of ant-caterpillar-plant interactions in most lycaenid and riodinid systems are largely unexplored. Yet these systems offer a wealth of theoretical and empirical problems that include diet breadth, predator-prey interactions, herbivory, insect communication, phylogenetics, and the evolution of symbioses. Simple field observations and experiments can furnish insights into the ecology and evolution of butterfly-ant symbioses and provide clues to how and why particular riodinid-ant and lycaenid-ant symbioses are maintained. Given the numbers of plants bearing extrafloral nectaries in the Neotropics and the abundance of herbivores (some myrmecophilous) that feed on them, La Selva seems an ideal place to explore the various levels of myrmecophilous insects and their host plants.

FINALE

La Selva contains a rich and unstudied butterfly fauna that can be used as a tool to generate questions and explore ecological and evolutionary patterns of rain forest insects. If a study considers the interplay between a single butterfly species, its host plants, and the animals that interact with them, the result is information valuable to understanding community ecology. When the entire butterfly fauna of La Selva and its plant animal interactions are considered, a picture of staggering ecological and evolutionary complexity results.

Regardless of what subjects or semantics fashionably fuel the literature of ecology and evolution, the fact remains that systematics and natural history are its cornerstones. Meaningful comparisons of species diversity, seasonality, unpalatability, or mutualisms depend upon knowing, not guessing, what species occur where and when and what they do for a living. Although studying natural history has been thought of at times as a quaint pastime for amateurs, it should be borne in mind that "natural history" is twentieth-century organismal biology and continues to provide the new data to be used by present and future biologists. To appreciate its role in the development of ecology and evolution one need only consider that solid, taxonomically based, natural history observations will stand as fact and be used profitably one hundred years from their dates of publication.

The study of butterflies has been instrumental in the development of an impressive array of theoretical and empirical subdisciplines in biology: biogeography, chemical communication, coevolution, ecological chemistry, genetics, host selection, mimicry, mutualisms, sexual selection, and population biology. Of these disciplines, it is notable that many had their origins in studies of tropical butterfly species. If this history is any predictor, La Selva-based quantitative natural history studies on butterflies coupled with experimental manipulations and a healthy appreciation for the richness of the La Selva butterfly fauna will prove to be science of lasting value.

ACKNOWLEDGMENTS

This manuscript is an outgrowth of arguments, comments, correspondence, and discussions provided over the years by T. M. Aide, R. Brown, J. J. Bull, P. Chai, T. Dee Lite, D. Feener, K. Gaston, L. E. Gilbert, H. Greene, N. Greig, D. Grimaldi, D. Janzen, I. J. Kitching, R. Lande, E. G. Leigh, J. Longino, M. C. Singer, and R. I. Vane-Wright. For field-related help thanks go to A. Bien, I. A. Chacon, J. Gamboa, T. Gutierrez, N. Greig, D. H. Janzen, R. Marquis, J. May, D. Perry, F. G. Stiles, the 1979–1980 denizens of Finca La Selva, and the 1985–86 denizens of Barro Colorado Island, Panama. The manuscript was improved by comments from R. Dudley, T. Emmel, N. Greig, D. Harvey, E. G. Leigh, J. Longino, and R. Robbins. Over the years portions of my work have been supported by a Fulbright Hayes fellowship, a Smithsonian predoctoral and postdoctoral fellowship, the Museo Nacional de Costa Rica, the University of Texas, Stanford University, and a fellowship from the MacArthur Foundation. I thank these institutions for allowing me to study tropical butterflies. I dedicate this chapter to the inspired diversity of Art Blakey, Clifford Brown, Miles Davis, Thelonius Sphere Monk, and Bud Powell.

14

Ecological Aspects of the Fish Community

William A. Bussing

Published accounts of Central American freshwater fishes have dealt primarily with systematics and geographic distribution. Numerous collections made at La Selva Biological Station since 1959 have provided material for new species descriptions (Bussing 1963, 1967, 1979) as well as systematic treatments involving certain other local species (Bussing 1976b, 1980, 1987; Bussing and Martin 1975). During 1962 I carried out an ecological study of the fishes in the La Selva region, primarily in the Rio Puerto Viejo but also in unique adjacent habitats (Bussing in press). More recently, Burcham (1988) compared the trophic structure and distribution of the fish communities in a forest stream and in a pasture stream at La Selva. Through the years several fish studies of one to three days duration have been carried out by students attending various Organization for Tropical Studies (OTS) courses (Bergmark 1985; Melcher et al. 1983; Miller et al. 1977; Pratt 1977; Rodríguez-M. 1980; Romero 1981); these latter reports and an unpublished paper by Vaux et al. (1984) have been included in the review presented here.

THE ENVIRONMENT

Although the Puerto Viejo, a tributary of the Río San Juan drainage on the Atlantic slope of Costa Rica, is the most conspicuous aquatic environment at the La Selva Biological Station, numerous tributary streams, swamps, and at least one spring provide a variety of fish habitats. The bed of the main river channel of the Puerto Viejo consists principally of soft clay although fallen logs and bends in the river create a variety of current and substrate conditions from oxbow backwaters to rocky riffles, each favoring distinctive fish associations. The average width of the Puerto Viejo at La Selva is 40 m, and the average depth is 2 m at low water. What appear to be slow-flowing stretches on the Puerto Viejo reach speeds up to 50 cm/s and riffles reach a maximum velocity of 150 cm/s. Burcham (1988) found velocities of 27 to 46 cm/sec. at several sites on the El Surá and Sábalo streams, two affluents of the Puerto Viejo (figs. 2.2, 3–9). Average annual rainfall in the area is about 4 m and flooding (fifteen to thirty times per year) may occur at anytime of the year. Stout (1982) cites depths during flooding up to 9 m in the Surá and Sábalo streams and levels 12 m above normal on the Puerto Viejo at La Selva (Stout 1981a). She recorded the maximum velocity on the Surá at 120 cm/s and estimated much greater velocities for the Sábalo during spates. Fish habitats are constantly being

modified by flood-created changes, and, even along the Surá, Melcher et al. (1983) found erosion and slumping at curves in spite of few spates in seven years.

Turbidity was reduced during low water periods in February–April and September but remained high in the Puerto Viejo during much of the remainder of 1962. Other physico-chemical characteristics of the water were quite stable. Annual water temperature variation in the Río Puerto Viejo during 1962 was 6.5° C (19.5° C on January 3–4 to 26° C on June 14). Air temperatures reached a high of 31° C during several days in June and a low of 17° C on February 12. Burcham (1988) found higher average temperatures in two streams and greater variation in the Sábalo than in the Surá during 1985. The concentration of dissolved gases and minerals remained relatively constant, fluctuating slightly with changes in rainfall. The aquatic environments can be characterized as heterotrophic, containing soft water, that is, low in mineral content and of circumneutral pH.

THE ICHTHYOFAUNA

Forty-three species of fishes are known from La Selva Biological Station (Bussing in press), and a few additional species may be added in time (app. 4). The richest diversity of fishes in Costa Rica is found in the Río San Juan drainage, which contains at least fifty-four species, some of which may augment the La Selva ichthyofauna from time to time. The greatest diversity of this system is probably found downstream from the confluence of the Puerto Viejo with the Sarapiquí as the Río San Juan is approached. Certain large river or lacustrine fishes, such as *Atractosteus tropicus, Dorosoma chavesi, Cichlasoma citrinellum,* and *C. managuense* have not been recorded from La Selva but are present in the larger river downstream.

Although four species new to science were discovered in the last twenty-six years in the La Selva area, none are restricted to the region, and it is unlikely that additional undescribed species will be taken there in the future. Nearly all genera at the station are represented by a single species. The most notable exception is *Cichlasoma* with eight species although these belong to four groups, sometimes considered sections (Regan 1906–1908) or even subgenera or genera (Fernández-Y. 1969; Loiselle 1985). This can be considered a diverse fish community for the generally depauperate Central American region. A similar site in the Usumacinta drainage of southern Mexico and Guatemala would presumably present a somewhat richer fauna (Miller 1976). In contrast, Myers (1947) found small (4-ft. wide, 2-ft. deep) lowland streams in the Amazon basin with 60 to 100 species of fishes, so an Amazonian site with aquatic habitats comparable to La Selva's would certainly contain more than 100 fish species. Saul (1975) collected 101 species at a site in upper Amazonian Ecuador. He sampled a variety of habitats during a six-month period in a deforested region.

The La Selva fishes belong to the San Juan Fish Province and are principally derived from migrants that probably arrived in Middle America during the Late Cretaceous or Paleocene from South America (Bussing 1976a). The only northern element reaching the province is the gar, *Atractosteus* which reaches the San Juan basin but is not yet recorded from La Selva. Certain groups, such as the Poeciliidae and Cichlidae, have evolved several genera and species endemic to the region. The other dominant family, Characidae, contains genera of which all but one, *Bramocharax,* are much more species-rich in South America. Frequently, the distributions of the individual species in a given genus are contiguous and nearly or completely nonoverlapping. The fact that these distribution patterns are the same for numerous unrelated groups of fishes suggests that speciation may have been caused by a series of vicariance events in the past rather than by competitive exclusion. Central American representatives of genera also found in South America are usually ecologically generalized, eurytopic forms, perhaps in response to the less stable and geographically less diverse isthmian environment.

PRINCIPAL LA SELVA FISH FAMILIES

The majority of the La Selva fishes belong to three large families: the Characidae, Poeciliidae, and Cichlidae. The characids are small (<10 cm except *Brycon*), silvery schooling fishes and are active at all levels of the water column. *Astyanax* and *Bryconamericus* are omnivores, both consuming roughly one-half insects and one-half plant material (table 14.1). *Astyanax,* however, is ubiquitous, whereas the very similar *Bryconamericus* is found almost exclusively in small streams. Other characids use a wide variety of food sources. Most characids are oviparous egg scatterers although terrestrial spawning of *Brycon petrosus* alongside Panamanian streams was reported by Kramer (1978), and Janet Burcham (pers. comm.) noted a similar behavior for *B. guatemalensis* at La Selva. Spawning periods are seasonal or possibly continuous for some species and unknown for others. Eggs hatch quickly (one or two days), and as far as is known, none of the local characids provide parental care. On April 17, 1962, plankton collections revealed large numbers of characid fry being swept down the Puerto Viejo during low water conditions. These presumably represent the result of massive spawning of either *Astyanax* or *Brycon*) the two most locally abundant species of local characids.

Poeciliids are small (<5 cm) viviparous fishes usually found close to shore. Many insectivorous species swim at or near the surface, whereas detritus and sediment feeders browse constantly on the bottom or other substrates. These latter forms apparently use diatoms, foraminifera, and, perhaps, bacteria that abound in the sediments on rocks, other

substrates, and debris. As in *Astyanax* and *Bryconamericus,* habitat partitioning is especially apparent among poeciliids belonging to the same trophic guilds. For example, three insectivores are seldom syntopic: (1) *Brachyrhaphis parismina*—Puerto Viejo and the Sábalo, a partially unshaded stream; (2) *B. holdridgei*—spring pool of El Surá and oxbow backwaters only; and (3) *Priapichthys*—forest streams such as El Surá. One sediment feeder, *Phallichthys tico,* is found exclusively in stagnant oxbow backwaters. Two other sediment feeders are most abundant in separate microhabitats: (1) *Poecilia*—shallows of low to moderate current velocity in the Puerto Viejo and the sunlit Sábalo, absent from the Surá, and (2) *Phallichthys amates*—in Puerto Viejo and streams including the Surá, always in slack water situations. *Alfaro* and *B. parismina* breed during all months. Individuals produce large broods at well-separated intervals. *Poecilia* and *Neoheterandria* reproduce with apparent peak periods during most (perhaps all) months.

Members of the family Cichlidae, known locally as mojarras, are spiny-rayed fishes active near the bottom and close to rock or debris refuges. The guapote, *Cichlasoma dovii,* may reach 50 cm in length, but other species are of small to medium size (10–30 cm). *Cichlasoma dovii* and *C. loisellei* are piscivorous and generalized predators, respectively. *Cichlasoma tuba* eats fig leaves, liverworts, mosses, and fruit. Romero (1981) conducted a series of experiments to determine that *C. tuba* actually ate the petioles rather than leaves of the fig, *Ficus glabrata.* Other mojarras ingest considerable detritus and sediment along with variable amounts of aquatic insects and plant matter. Most cichlid species live close to the shores of the wide Puerto Viejo and in smaller streams (Vaux et al. 1984; Burcham 1985). Two species, however, *C. loisellei* and *Herotilapia* were found only in oxbow backwaters or stagnant inlets of the Puerto Viejo. Ripe specimens of the following were collected in all or nearly all months of the year: *C. dovii, C. tuba, C. septemfasciatum,* and *Neetroplus. Cichlasoma alfari* apparently spawns at two peaks, January–February and August–September. Insufficient material prevents generalizations regarding ecology and seasonality of reproduction in other species of Puerto Viejo cichlids.

HABITAT PARTITIONING

The La Selva waters generally do not teem with fishes. Bergmark (1985) in a study of the role of fish in the diet of Puerto Viejo residents concluded that fish biomass production is low. At one time or another all La Selva fishes occur in the Puerto Viejo although the bulk of the populations of some species may inhabit certain adjacent habitats. Rodríguez (1980) correlated high fish species diversity with increasing structural diversity of the substrate in the Puerto Viejo. Likewise, Gosse (1963) found that the Congo River, with its diversity of shorelines and islands, contains far more species than its small affluents although certain species are apparently restricted to the smaller streams. Burcham (1988) revealed striking differences in the fish community structure between two La Selva streams, the Sábalo and El Surá. Her sampling sites in each stream were specifically chosen to represent extreme situations: a stream surrounded by pasture (a section of the Sábalo) and the heavily canopied forest stream, the Surá. Twenty-six

Table 14.1 Ecological characteristics of selected species of La Selva fishes

Species	Trophic Level (Adults)	Habitat	Reproduction Type	Reproduction Frequency
Characidae				
Astyanax fasciatus	omnivore	midwater	egg scatterers	May–December
Brycon guatemalensis	macrophytes	midwater	egg scatterers	December–July
Pimelodidae				
Rhamdia nicaraguensis	aquatic insects	benthic	egg scatterers	?
Poeciliidae				
Alfaro cultratus	terrestrial insects	near surface	viviparous	continuous
Brachyrhaphis parismina	terrestrial insects	substrate	viviparous	continuous
Neoheferandria umbratilis	detritivore sediment	substrate	viviparous	intermittent
Atherinidae				
Atherinella hubbsi	aquatic insects	midwater	eggs in plants	December–July
Cichlidae				
Cichlasoma dovii	piscivore	benthic	adhesive eggs	February–September
C. alfari	aquatic insects	benthic	adhesive eggs	2 peaks
Heterotilapia multispinosa	detritivore	benthic	adhesive eggs	?
Neetroplus nematopus	omnivore	substrate	adhesive eggs	?
Mugilidae				
Agonostomis monticola	predator	rapids	catadromous	May–August
Joturus pichardi	algivore	rapids	catadromous	December–January
Eleotridae				
Gobiomorus dormitor	predator	benthic	adhesive eggs	?
Gobiidae				
Awaous tijasica	aquatic insects	benthic, sand	adhesive eggs	?
Sicydium altum	algivore	benthic, rock	adhesive eggs	?

fish species were found in the Sábalo compared to nineteen species at the Surá site. Total numbers of individuals were also far greater in the Sábalo. Repeated sampling in each stream yielded 2,266 specimens of the algivore *Poecilia* in the Sábalo, whereas the species was never collected in the Surá. Working in Mexico, Darnell (1953, 1962) found that the distribution of the related *Poecilia sphenops* was determined largely by bottom type, which was, in turn, related to food supply. Other species common to both streams at La Selva revealed this same disparity, being far more abundant in the Sábalo apparently because of the more autochthonous nature of food resources in this exposed pasture stream (Burcham 1988).

Few Central American river systems are extensive in size, and, probably, none have the habitat diversity of the Amazon, for example. In addition, the lower Central American land mass is geologically relatively young, so many species may not have had time to reach the region. These factors alone may explain the relative paucity of the freshwater ichthyofauna compared to that in rivers of similar size in the Amazon basin. Unpredictable extreme fluctuations in water level also may affect fish diversity. This often brief but sudden flooding scours the mostly soft clay substrate, constantly alters habitats, and allows little development of aquatic vegetation. A pool rich in decaying leaf litter may suddenly be transformed into a sterile sand-bottomed stretch. New logs are swept downriver to form new habitats of relatively little stability. Conversely, the predictable long-term annual flood cycle of the Amazon, for example, may be beneficial to fish that are able to take advantage of vast flood plains for reproductive and nutritional purposes (Goulding 1980).

TROPHIC PARTITIONING

The La Selva fish fauna includes a considerable variety of food specialists (table 14.1). Occasionally one sees schools of large machacas, *Brycon*, under chilamate trees, *Ficus glabrata*, in the Puerto Viejo or tributaries. These fishes frequently gobble whole figs an instant after they hit the water. Miller et al. (1977) noted this behavior and determined that *F. glabrata* seeds found in the lower intestine of machacas germinated within one week. Pratt (1977) commented on the intense fish activity during the time of greatest fruit fall, also found seeds, fruits, and flowers of *Ficus* in the digestive tracts of *Brycon* and determined that the seeds germinated in five days, evidence that these fish are dispersal agents for *F. glabrata*.

Numerous other La Selva fishes are insectivores: *Gymnotus, Rhamdia* spp., *Rivulus, Gobiesox*, and *Atherinella*. Large marine migrants such as *Carcharhinus, Tarpon, Pomadasys* and *Centropomus* are known to be piscivorous and apparently forage in the rivers where they may remain for extended periods. Other specialized species include the algivore *Carlana*, the piscivore *Bramocharax*, and *Roeboides*, which usually eats fish scales but is largely an insectivore at La Selva.

As pointed out earlier, where two species seems to have nearly identical food habits, numerous examples show that they occupy different habitats, thus reducing direct competition. Although food categories of similar species overlap considerably, a closer look suggests that actual *species* of animal or plant foods are often quite different. La Selva fishes show a high correlation of morphology (dentition and gut/length ratio) to food habits (Bussing in press).

Although La Selva fish consume far more aquatic than terrestrial organisms, the ultimate energy source for most aquatic anthropods is probably terrestrial vegetation. A few species consume considerable autochthonous vegetation in the form of filamentous algae: *Joturus, Sicydium, Poecilia,* and *Carlana.* The latter three may account for little biomass although *Joturus,* especially in former times under reduced fishing pressure must have taken advantage of the abundant algae in swift waters.

Tropical freshwater fishes in other parts of the world show similarities to the La Selva species in the manner in which they partition food resources. Lowe-McConnell (1975) compared the diet of fishes on three tropical continents and found that allochthonous vegetable and, especially, insect materials play a huge role; sediment and detritus are important food sources for certain specialists; and considerable numbers and variety of piscivorous fishes are typically present. Fish communities with similar trophic structures were found by Matthes (1964) in streams in the Congo basin and Bishop (1973) in a small Malayan river. Quite different conclusions were reached by Knöppel (1970) based on his examination of stomach contents of fishes in small Amazonian rain forest streams. He concluded that food specialists were not present; stomach contents of various families were rather uniform; dentition and intestinal ratios were not correlated to food habits; location, season, and age of fish species did not greatly affect food habits; and fishes found their food in the whole living space regardless of their microhabitat. These general conclusions certainly do not apply to the La Selva fish fauna nor to numerous other tropical fish communities that have been studied. Roberts (1972) pointed out that some of Knöppel's conclusions are not borne out by statements and data in the main text of his paper although he does concur that there are fewer narrow tropic specialists in small streams.

REPRODUCTION

As noted, some fish species at La Selva, especially poeciliids and cichlids, reproduce continuously, whereas others, such as *Neoheterandria* (Poeciliidae), appear to reproduce intermittently throughout the year (table 14.1). Others, such as *Astyanax* (Characidae), *Atherinella* (Atherinidae), and *Cichlasoma dovii* (Cichlidae), are probably seasonal spawners in the Puerto Viejo, sexually active about half the year. *Cichlasoma alfari* was sexually active during two peaks in 1962, January–February and August–September, which were low rainfall months in that year. Ripe specimens of the catadromous *Agonostomus* (Mugilidae) were found from May to August; the similar *Joturus* is said to migrate downstream during December to January. Additional reproductive data for some of these species indicates that breeding seasons are not necessarily the same at other localities or in lentic environments. A similar spectrum of reproductive strategies is found in other tropical rivers (Lowe-McConnell 1975) although the long migrations characteristic of some Amazonian fishes (Goulding, 1980) are rare and, perhaps, restricted to *Brycon,* which moves to minor tributaries of the Puerto Viejo to spawn.

It should be clear that although the list of fish species that occur at La Selva is reasonably complete, knowledge of their life-history patterns is superficial, and the same can be said of Central American fishes in general. Growth rates, competition, predation pressures, community organization, biotic versus physical factors in relation to ecological distribution, and many other aspects of fish biology remain to be studied. The manageable size of the fish fauna and excellent working conditions at La Selva should make it an important site for research on fish biology. Toward understanding the ecological and historical factors that determine community composition, it will be especially interesting to compare La Selva's low biomass and relatively depauperate fauna to the richer faunas in the mainland South American drainages.

15

Amphibian Diversity and Natural History

Maureen A. Donnelly

The composition of the La Selva herpetofauna is well known (Duellman 1990), largely owing to the efforts of Jay M. Savage, his students, and associates. Frogs and toads are diverse, abundant, and occur in most habitats within the station, whereas caecilians and salamanders form minor components of the fauna. Although the taxonomic composition of the La Selva amphibian fauna is well known, long-term ecological studies on amphibians have only recently been conducted at this site, and others will be necessary before generalizations about the La Selva assemblage can be made that will allow rigorous comparisons to other lowland Neotropical amphibian assemblages.

One of the most striking features of amphibian biology is the wide array of reproductive modes exhibited by the three living orders (Gymnophiona, Caudata, and Anura), and although few ecological data exist for the La Selva amphibians, the reproductive mode characteristic of each species is known. In this chapter I describe the composition of the fauna, compare the La Selva fauna to five other Neotropical lowland faunas, review reproductive modes exhibited by La Selva amphibians, review studies that have been completed at La Selva, and propose future research on this diverse group of animals.

COMPOSITION OF THE AMPHIBIAN FAUNA

The forty-eight amphibian species known from La Selva represent nine families. The anurans dominate, accounting for 92% of the amphibian fauna (app. 5). The number of amphibian species known from La Selva (ca. 15 km²) is slightly higher than that known from West Virginia, a state approximately equal to the size of Costa Rica (ca. 55,000 km²). Fourteen anuran species and twenty-eight salamander species are known from West Virginia (Green and Pauley 1987). Thus, although the diversity of salamanders is higher in that state than it is at La Selva, the diversity of anurans is much lower. The La Selva amphibians are part of a widespread, tropical lowland assemblage that ranges from the Atlantic lowlands of Tamaulipas, Mexico, south to central Panama, and in isolated areas of the Pacific lowlands of Guatemala, the Golfo Dulce region of southwestern Costa Rica, and in extreme western Panama. The historical factors associated with the evolution of the fauna have been described in detail by Savage (1982). Savage examined distributional patterns, using generalized tracks and areas of endemism. Guyer (1990) used the method of Simberloff and Connor (1979) to analyze distributional

data for eight Neotropical herpetofaunas, including La Selva. Guyer's analysis confirmed the integrity of the eastern Caribbean lowland track described by Savage. Historically, a generalized tropical herpetofauna extended from 40° N southward in Paleocene times. At the end of the Paleocene, the presumed connection between North and South America was broken and Middle America was isolated from South America. Climatic and physiographic changes isolated the Central American component of the Old Northern unit in Middle America in the Oligocene. Orogenic events in the Oligocene and the Miocene separated eastern and western lowland areas in Mesoamerica, and the uplift of lower Central America closed the Panamanian Portal in the Pliocene. Subsequent to closure of the portal, dispersal occurred in both directions, but most of the South American groups have only reached Panama (Savage 1982). The fifteen genera of amphibians known from La Selva include representatives from three of the four historical source units (Old Northern, Young Northern, Middle America, South American) identified by Savage (1982), and are largely derived from the Middle American and South American source units (table 15.1).

In his treatment of the Costa Rican herpetofauna, Savage (1980) recognized nine geographic zones within the republic. In this system, La Selva lies in the Northeastern zone and its fauna is typical of the northeastern lowlands and adjacent foothills. Savage and Villa (1986) indicated the general geographic distribution for all Costa Rican species. The Northeastern zone is the most diverse of the nine, with seventy-seven species of amphibians (three caecilians, eleven salamanders, and sixty-three frogs and toads). The La Selva fauna contains 62% of the amphibian species known from the Northeastern zone. This percentage is incredibly high given that the elevational range of that zone, as defined by Savage and Villa (1986), extends from sea level to 1,500 m. The species from this zone that are not known from La Selva are probably restricted to montane regions and are not likely to be encountered at La Selva (J. M. Savage, pers. comm.).

Gymnophiona

One caecilian, *Gymnopis multiplicata,* is known from La Selva. This species is elongate, limbless, essentially blind, and fossorial. It is the most widespread caecilian in Costa Rica and occurs from sea level to 1,400 m. Individuals can be collected in forests under rocks, logs, burrowed into soil, under leaf litter, and in other moist microsites (M. H. Wake 1983).

Table 15.1 Principal historic source units of the La Selva amphibian genera

Source Unit		
Old Northern	Middle American	South American
Bolitoglossa	Gymnopis	Leptodactylus
Oedipina	Gastrophryne	Dendrobates
Rana	Bufo	Phyllobates
	Eleutherodactylus	Centrolenella
	Agalychnis	Scinax
	Hyla	
	Smilisca	

Source: Sensu Savage 1982.

Individuals emerge to the surface of the forest floor at night during heavy rains, presumably to forage. The diet consists of a variety of invertebrates that are ground dwelling or burrowing forms. Like all caecilians, *G. multiplicata* has a pair of protrusible tentacles located on the side of the head between the nostril and eye. The tentacle lies immediately anterior to the eye in *Gymnopis,* whereas it is located closer to the nostril in other Costa Rican caecilians. The tentacles are thought to conduct airborne particles to the olfactory epithelium and they are moved continually by muscles and nerves that are associated with the eyes in other vertebrates (M. H. Wake 1983). Because of their secretive habits, very little is known about caecilian ecology.

Caudata

The salamanders known from La Selva belong to the tribe Bolitoglossini of the family Plethodontidae. This family of lungless salamanders has undergone an extensive radiation in Mexico and Central America (D. B. Wake and Elias 1983; D. B. Wake and Lynch 1976). Hendrickson (1986) presented a vicariant biogeographic hypothesis that explained distribution patterns in the tribe Bolitoglossini. The observed areas of salamander endemism correlate with the geologic history of "suspect terranes," or crustal fragments, accreted to a continental craton (a stable horizontal rock formation of the earth's crust that is the principal part of a continent and its continental shelf). The area cladograms suggested by this hypothesis are congruent with bolitoglossine phylogeny (Hendrickson 1986). In Costa Rica salamander species diversity is highest in the Cordilleras (Central, Tilarán, and Talamanca; Savage and Villa 1986). Although eleven salamander species are known from the Northeastern zone, only three have been recorded from La Selva (app. 5); a fourth species, *Bolitoglossa arborescandens,* may occur there (J. M. Savage pers. comm.) because it has been seen at nearby Finca El Bejuco in Chilamate (pers. observation). Salamanders are relatively uncommon at La Selva and are typically found at night, perched on low vegetation or in the leaf litter. The two species of *Oedipina* are fossorial or occur under wet moss mats or rotted logs (Brame 1968). None of the three species has been studied in any detail at any site, and little is known of their ecology.

Anura

Anuran species diversity increases with decreasing latitude and the faunas of lowland Neotropical forests are among the richest known (Duellman 1988; Duellman and Trueb 1986). La Selva has a large anuran fauna, and these animals dominate the amphibian fauna in number of individuals, number of species, variety of habitats used, and variety of reproductive modes. Three families, Leptodactylidae, Hylidae, and Centrolenidae, account for 80% of the anurans at La Selva. The species-rich and diverse genus *Eleutherodactylus* accounts for 32% of the total anuran fauna (app. 5). Within Costa Rica two of the forty-four species of anurans are restricted to the Northeastern zone of Costa Rica (*Gastrophryne pictiventris* and *Agalychnis saltator*). The remaining taxa are found either in the northeastern and southeastern lowlands, northeastern and southwestern lowlands, in all four lowland regions, or are widely distributed throughout the country (Savage and Villa 1986). Although the composition of the La Selva anuran fauna has been well documented (Scott et al. 1983), three species (*Centrolenella pulverata, C. granulosa,* and *Smilisca sordida*) were collected for the first time at La Selva during twenty-four months of continuous monitoring (1982–83, Guyer and Donnelly unpublished data). These additions represent a 7% increase in amphibian species richness. Two canopy-inhabiting species (*Anotheca spinosa* and *Hyla miliaria*) and a toad (*Bufo melanochloris*) are known from elsewhere in the Northeastern zone and may occur at La Selva (J. M. Savage pers. comm.).

Anurans occur in several habitats at La Selva; there is a large leaf-litter assemblage (S. Lieberman 1986), a swamp-breeding assemblage (Donnelly and Guyer unpublished data), and a riparian assemblage. Although most La Selva anurans are nocturnal, several species are diurnal, and some are active during both day and night (S. Lieberman 1986; Miyamoto 1982; Slowinski et al. 1987). Miyamoto (1982) showed that juveniles of some species of *Eleutherodactylus* are active diurnally, whereas adults are active nocturnally.

Maiorana (1976) hypothesized that Neotropical amphibian faunas are "equitable" and that although many species coexist, none of them dominates numerically. This hypothesis lacks empirical support. In fact, most Neotropical amphibian faunas, including that of La Selva, exhibit high species richness with a few very abundant species and many uncommon ones (Duellman 1978, 1990; S. Lieberman 1986; Scott 1976).

COMPARISON OF LA SELVA TO OTHER NEOTROPICAL LOWLAND SITES

The taxonomic compositions of five other lowland Neotropical amphibian faunas have been relatively well-documented. Here I compare the La Selva amphibian fauna to those of Tortuguero National Park, also located in Savage's (1980) Northeastern zone; the Osa Peninsula in southwestern Costa Rica (Rincón de Osa and Corcovado National Park); Barro Colorado Island (BCI), Panama; Los Tuxtlas Biological Station in Veracruz, Mexico; and, finally, Santa Cecilia, Ecuador. These sites vary markedly in size, which may hinder meaningful comparisons among them. Most of the research on amphibians has been concentrated on 750 ha of "old" La Selva. Tortuguero National Park includes 19,000 ha; Corcovado National Park extends over 41,800 ha; BCI is 1,500 ha; Los Tuxtlas includes 700 ha; and the Santa Cecilia site is 300 ha.

The three orders of amphibians occur in approximately the

same proportions, and anurans dominate the fauna (in number of species) at all five sites (table 15.2). Three species of caecilians are known from Santa Cecilia, whereas one caecilian occurs at the other sites (table 15.2). Salamander species diversity (number) is highest at Los Tuxtlas, where the Caudata account for 14% of the amphibians (table 15.2). Wake and Lynch (1976) described distribution, ecology, and evolutionary history of tropical American plethodontids. Based on distributional data, they recognized thirteen regions in tropical America; Los Tuxtlas occurs in Region 4 (southern Sierra Madre Oriental, from 22° N to the Isthmus of Tehuantepec), La Selva, Tortuguero, the Osa, and BCI occur in Region 9 (lower Central America), and Santa Cecilia occurs in Region 12. Region 4 has more genera (seven) than the other regions and it harbors a rich salamander fauna (thirty-one species, twenty-four endemics). Wake and Lynch (1976) hypothesized that three of the thirteen regions, including Regions 4 and 9, were major evolutionary centers for the tropical lungless salamanders because these areas are characterized by high species diversity and high levels of endemism.

Nine amphibian families are present at La Selva, Tortuguero, the Osa Peninsula, and Los Tuxtlas; eight are known from BCI; and ten are represented at Santa Cecilia (table 15.2). The Neotropical caecilians from these five sites belong to the family Caeciliidae; the salamanders belong to the family Plethodontidae; and the anurans represent nine different families (table 15.2). The frog families Bufonidae, Leptodactylidae, Ranidae, Centrolenidae, and Hylidae are represented at all six sites, and three families (Leptodactylidae, Hylidae, and Centrolenidae) account for more than 50% of species at all six sites. The frog family, Rhinophrynidae, is represented

only at Los Tuxtlas, and the family Pipidae is represented only at Santa Cecilia (table 15.2). The families Microhylidae and Dendrobatidae occur at five of the six sites (table 15.2).

At the species level, analysis of overall similarity of the amphibian faunas among the six Neotropical sites (a total of 195 species at all sites) indicates that La Selva forms a cluster with the other Costa Rican sites (Tortuguero and the Osa Peninsula). The Costa Rican sites are clustered with BCI. Los Tuxtlas and Santa Cecilia fall outside of this lower Central American cluster (fig. 15.1; app. 5). The results from this analysis agree with those reported for the reptiles (chap. 16) and are consistent with the biogeographic history of the Middle American herpetofauna (Savage 1982) and the distributional analysis of Guyer (1990).

La Selva and Tortuguero

Both the La Selva and Tortuguero sites contain taxa belonging to nine amphibian families (table 15.2). At the species level the sites share twenty-nine taxa in common (60% of the La Selva total, 88% of the Tortuguero total). This is not surprising in that both are in the same geographic province (Savage and Villa 1986). Fieldwork by Peter N. Lahanas (unpublished data) has increased knowledge of species composition at Tortuguero, and additional taxa that occur at La Selva doubtless will be found there. At both sites the leptodactylids, especially *Eleutherodactylus,* and hylids dominate the community to account for more than half of the total amphibian fauna. La Selva has a more diverse assemblage of centrolenid taxa (six species), whereas Tortuguero is characterized by more

Table 15.2 Number of species in amphibian families and orders from six Neotropical sites

Taxon	Site[a]					
	TU	TO	LS	OSA	BCI	SCA
Caeciliidae	1	1	1	1	1	3
Gymnophiona	1	1	1	1	1	3
TOTAL	(2)[b]	(3)	(2)	(2)	(3)	(3)
Plethodontidae	6	2	3	4	2	2
Caudata	6	2	3	4	2	2
TOTAL	(14)	(6)	(6)	(8)	(6)	(2)
Bufonidae	3	5	3	4	2	4
Centrolenidae	1	1	6	7	3	3
Dendrobatidae	0	2	2	5	2	5
Hylidae	15	9	13	11	9	37
Leptodactylidae	11	10	16	14	12	25
Microhylidae	3	1	1	1	0	5
Pipidae	0	0	0	0	0	1
Ranidae	2	2	3	1	2	1
Rhinophrynidae	1	0	0	0	0	0
Anura	36	30	44	43	30	81
TOTAL	(84)	(91)	(92)	(90)	(91)	(95)
TOTAL Amphibians	43	33	48	48	33	86

[a]The sites are indicated by the following abbreviations: TU = Los Tuxtlas; TO = Tortuguero; LS = La Selva; OSA = Osa Peninsula; BCI = Barro Colorado Island; and SCA = Santa Cecilia.

[b]Values in parentheses are percentages of the total fauna in each order.

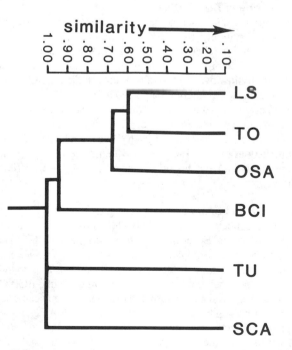

Fig. 15.1. Dendrogram based on similarity of amphibian faunas at six Neotropical sites. Similarity is based on Jaccard's Index (of overall dissimilarity where zero indicates maximum similarity and one indicates maximum dissimilarity [Pimentel and Smith 1986]), and sites were clustered with the unweighted pair group method (UPGMA) (Pimentel and Smith 1986). The sites are indicated by the following abbreviations: LS = La Selva; TO = Tortuguero; OSA = Osa Peninsula; BCI = Barro Colorado Island; TU = Los Tuxtlas; and SCA = Santa Cecilia.

bufonid species (five). The remaining families at both sites contribute less than 10% to the total fauna.

La Selva and the Osa Peninsula

The La Selva and Osa Peninsula lowland sites occur on opposite coasts, are separated by the Cordilleras, and are subject to different climatic conditions (Coen 1983). Forty-eight species are known from each site, and they share twenty-nine species (56% of the La Selva and Osa totals). The dendrobatid fauna at the Osa is more diverse (five species in three genera) than the faunas at La Selva and Tortuguero (two species in two genera).

La Selva and Barro Colorado Island

The Panama site is drier and more seasonal than the La Selva site, and this may influence amphibian species diversity. Eight families of amphibians occur on BCI, all of which occur at La Selva (table 15.2). Although the familial compositions are similar, the species occurring at the two sites are different. Only sixteen amphibian species in five families occur at both La Selva and BCI. The fauna of BCI has been studied since the island was formed in the early part of this century (Myers and Rand 1969; Rand and Myers 1990). Within the past thirty years at least one species (*Rana warschewitschii*) has become extinct on the island (Myers and Rand 1969; Rand and Myers 1990). During the last decade, one species (*Hyla rufitela*) experienced a short-lived population explosion (Rand et al. 1983), but its numbers have declined since 1983 (Rand and Myers 1990).

La Selva and Los Tuxtlas

Los Tuxtlas clusters outside of the other Middle American sites, and only eight species of anurans in four families are common to La Selva and Los Tuxtlas (19% of the Los Tuxtlas total and 17% of the La Selva total). Four of these eight species are hylids, two are leptodactylids, one is a ranid, and the other is the wide-ranging toad, *Bufo marinus*. The salamander fauna at Los Tuxtlas is richer than at La Selva although both regions are considered to be centers of tropical plethodontid diversification (Wake and Lynch 1976). The lack of overall similarity at the specific level between these two sites is also consistent with the historical scenario proposed by Savage (1982).

La Selva and Santa Cecilia

Only two species of amphibians known from La Selva occur at Santa Cecilia, Ecuador (Crump 1974; Duellman 1978). The two shared taxa, *Bufo marinus* and *Leptodactylus pentadactylus*, are widespread in the Neotropics and are also found at three of the other sites (Tortuguero, Osa Peninsula, BCI). The dissimilarity of these two lowland faunas at the specific level is not surprising because the Central and South American taxa apparently evolved in isolation after the separation of Middle America from the South American continent during the early Cenozoic (Savage 1982).

REPRODUCTIVE MODES OF THE LA SELVA AMPHIBIANS

Jameson (1955, 105) once said of a frog's perspective, "if it is not small enough to eat nor large enough to eat you, and does not put up a squawk about it, mate with it." This simplistic view of amphibian reproductive behavior has stimulated a tremendous amount of research during the ensuing decades, and it is now known that amphibians exhibit a greater variety of reproductive modes and life-history strategies than any other terrestrial vertebrate class. Several reproductive trends are exhibited in all three orders of amphibians, including changes from external to internal fertilization, large to small clutch size, aquatic to terrestrial oviposition sites, simple to elaborate mate recognition and courtship patterns, the development of parental care, a shift from free-living aquatic larval stage to direct development of the eggs, and a shift from oviparity through ovoviviparity to viviparity (M. H. Wake 1977).

Reproductive mode, as defined by Salthe and Duellman (1973), combines site of development with mode of development. The mode of development includes rate and duration of development, size and stage at hatching, and nature of parental care. Twenty-nine modes of reproduction have been described for anurans, seven for salamanders, and four for caecilians (Duellman and Trueb 1986). Rather than contribute to the proliferation of systems to classify amphibian reproduction, I herein use the Duellman-Trueb system (1986), making minor modifications as necessary to accommodate the diversity among La Selva's amphibians.

The generalized mode of amphibian reproduction involves deposition of eggs in water and development into free-swimming aquatic larvae that later metamorphose into miniature adults; this mode is characteristic of most temperate zone amphibians. In the continental United States, for example, amphibians exhibit four reproductive modes, and 90% of all species oviposit in water (Duellman and Trueb 1986). In contrast, 67% of the La Selva species are characterized by modifications of this generalized mode. La Selva amphibians have eight reproductive modes ranging from the aforementioned generalized mode to viviparity (app. 5).

Caecilian Reproduction

Gymnopis multiplicata is viviparous (mode V, app. 5). Although behavioral and ecological data relative to courtship, mate selection, and demography are lacking, the morphological studies of M. H. Wake have been invaluable toward understanding this secretive, Pantropical group. All caecilians are presumed to have internal fertilization (Duellman and Trueb 1986). The posterior portion of the male cloaca is modified into a phallodeum (intromittent organ) that has a species-specific morphology (M. H. Wake 1977). The male inserts this reproductive organ into the female's vent during copulation (Duellman and Trueb 1986). Ovarian clutch size in *G. multiplicata* ranges from twenty-five to thirty eggs, and brood size ranges from two to nine. Breeding behavior in many species is seasonal; breeding begins with the onset of the rainy season (M. H. Wake 1977).

The fertilized eggs are retained in the oviducts; after the yolk supply is exhausted the embryos hatch in situ (within the oviduct). Upon hatching, a species-specific fetal dentition develops. The fetal teeth differ in number, arrangement, and crown morphology from adult teeth. Fetal caecilians feed on material produced by the oviducal glands of the female, resulting in a placental form of maternal nutrition. The rasping action of the fetal teeth stimulates proliferation of and secretion by glandular cells to ensure production of the fetal food supply (M. H. Wake 1977, 1983).

Aquatic Eggs and Aquatic Larvae

Sixteen species in four families (Microhylidae, Bufonidae, Hylidae, and Ranidae) have a generalized mode of reproduction and oviposit in water (anuran modes 1 and 2 of Duellman and Trueb 1986). *Gastrophryne pictiventris* breeds in shallow, temporary pools that form in low-lying areas of the forest following heavy rains (Donnelly, de Sá, et al. 1990). Reproductive behavior of the three bufonid species is not well studied at La Selva, but all three oviposit in temporary ponds formed during the rainy season. It has been suggested that use of such temporary sites provides larvae with refuge from aquatic predators (Heyer et al. 1975), but larvae in these sites may be subject to desiccation. The largest anuran known from the reserve, *Bufo marinus,* usually oviposits in the shallow margins of permanent and temporary bodies of water that are typically located in the open. Clutch size in females is large, ranging from five thousand to twenty-five thousand eggs (Zug 1983). *Bufo haematiticus* congregates in pools along forest streams and rivers to breed. Males call from beneath loose boulders (Scott 1983a), but frequency and timing of breeding behavior have not been determined.

The three ranids from La Selva also exhibit the generalized mode of reproduction although data from La Selva are lacking for these species. I have encountered *Rana vaillanti* in the "Research Swamp" (ca. line 250 on the Camino Experimental Sur); but did not observe calling, amplexus, or oviposition. In other parts of its range males of this species call from floating vegetation or in water near pond margins (Greding 1972, 1976). During oviposition, *R. warschewitschii* attaches its eggs to the undersides of rocks in streams (Scott and Limerick 1983). *Rana taylori* has been recorded from La Selva, but it is uncommon. Elsewhere this species breeds in large, semipermanent or permanent bodies of water.

Nine species of hylids in three genera exhibit the generalized reproductive mode (app. 5), and eight of the nine species breed in large temporary ponds that form in the forest at the onset of the rainy season (Donnelly and Guyer unpublished data). *Smilisca baudinii* usually calls in monospecific choruses in small temporary pools, frequently in pastures near the station (Duellman 1970; pers. observation). *Smilisca puma* calls from small puddles within forests but also has been taken from pools in open fields nearby. *Smilisca sordida* breeds in pools associated with small creeks (anuran mode 2 of Duellman and Trueb 1986) during the dry season when stream volume and flow rate are minimal (Duellman 1970; Scott and Limerick 1983). *Scinax [Ololygon] elaeochroa* is an "explosive" breeder (sensu Wells 1977b), and breeding individuals form large aggregations immediately after the large ponds fill in the early wet season. Breeding activity begins at dusk and continues for at least forty-eight hours (Donnelly and Guyer unpublished data). *Scinax boulengeri* and *Hyla phlebodes* are swamp breeders encountered throughout the rainy season in low densities. *Hyla rufitela* occurs sporadically in the Research Swamp, but reproduction has not been observed. This frog oviposits in shallow forest ponds elsewhere in its range (Duellman 1970).

Eggs in Foam Nest and Aquatic Larvae

The two species of *Leptodactylus* from La Selva deposit eggs in foam nests (anuran mode 8 of Duellman and Trueb 1986). Males construct nests during amplexus by kicking semen, mucus, air, eggs, and water into a foamy mass. Both species breed in shallow water on the periphery of larger water bodies (Gregory 1983; Scott 1983b; pers. observation). The tadpoles enter the water upon hatching. *Leptodactylus pentadactylus* tadpoles are relatively large and prey upon tadpoles of other species (Heyer et al. 1975).

Terrestrial Eggs and Aquatic Larvae

Deposition of terrestrial eggs and parental transport of tadpoles to water is characteristic of the two dendrobatids at La Selva (anuran mode 14 of Duellman and Trueb 1986). In *Dendrobates pumilio* courtship is prolonged and eggs and sperm are deposited on leaves while adults are in vent-to-vent contact (Limerick 1980). Newly hatched tadpoles are transported to aquatic rearing sites by one of the parents (the "nurse frog"). Male *Phyllobates lugubris* carry up to six tadpoles at a time (Donnelly, Guyer, et al. 1990), whereas in *D. pumilio* females function as nurse frogs and transport one or two tadpoles to the axils of bromeliads (pers. observation). In *D. pumilio* females return to the axils containing their larvae and deposit nutritive eggs upon which the developing larvae feed (Weygoldt 1980).

Terrestrial Eggs and Direct Development

Seventeen amphibians at La Selva oviposit terrestrial eggs that undergo direct development (anuran mode 17 and salamander mode IID of Duellman and Trueb 1986). All members of the family Plethodontidae and the genus *Eleutherodactylus* at La Selva are characterized by this reproductive mode. It has been suggested that direct development has been important in the radiation of both groups in the New World tropics (Duellman and Trueb 1986; Savage 1975; Wake and Lynch 1976). In these species clutch size is small, and the eggs contain substantial amounts of yolk. Oviposition occurs in moist microsites, and clutch attendance has been described for many species.

All plethodontids have internal fertilization. During courtship, males deposit a sperm packet (spermatophore) on the substrate that is picked up by the cloacal lips of the females. The sperm are stored in the female spermatheca (located in the roof of the cloaca) that is the homologue of the male pelvic gland. The cap of the spermatophore releases spermatozoa that arrange themselves in whorls in the spermathecal tubules where they remain until ovulation. At the time of ovulation, sperm are expelled by contraction of muscles surrounding the spermathecal tubules and fertilize the eggs as they enter the cloaca (Duellman and Trueb 1986).

Female plethodontids usually guard egg clutches although male attendance has been reported (Salthe and Mecham 1974; Vial 1968). Clutch attendance is thought to contribute to increased survival of the young by protecting against predation, preventing or reducing fungal infection, and enhancing aeration via agitation of the eggs and embryos (Duellman and Trueb 1986).

Internal fertilization has been described for two species of *Eleutherodactylus* (Townsend et al. 1981), and parental attendance of clutches has been observed (Drewry 1970; Hayes 1985; Townsend et al. 1984) but not at La Selva. Internal fertilization and parental clutch attendance may be widespread in these terrestrial frogs but, unfortunately, scant data exist for Neotropical mainland populations, and data are not available for the fourteen species from La Selva.

Arboreal Eggs and Aquatic Tadpoles

In their classification of anuran reproductive modes Duellman and Trueb (1986) recognize three modes of arboreal egg deposition: arboreal eggs that hatch into tadpoles that drop into ponds or streams (mode 18); arboreal eggs that hatch into tadpoles that drop into water-filled cavities in trees (mode 19); and arboreal eggs that hatch into froglets (mode 20). Ten of the La Selva anurans oviposit arboreally (anuran mode 18 of Duellman and Trueb 1986), but the tadpoles develop in three different aquatic habitats, so I recognize these as variants of reproductive mode 18.

Agalychnis callidryas, A. saltator, and *Hyla ebraccata* place their eggs on vegetation over large, still bodies of water. The eggs develop into tadpoles in this aerial site, and they fall into the water upon hatching to complete metamorphosis (my mode 18A). Although the eggs are not susceptible to aquatic predators, some eggs are attacked by parasitic flies that oviposit in the egg mass (Villa 1977, 1980, 1984; Villa and Townsend 1983; pers. observation) or are infected by fungi (Villa 1979). Two snake species, *Leptodeira septentrionalis* and *Imantodes inornatus,* are predators on these eggs and are frequently encountered on rainy nights moving through swamp vegetation (pers. observation). Wendy Roberts (Museum of Vertebrate Zoology, University of California) is currently investigating egg survival in the hylid frogs that deposit arboreal eggs at La Selva. She is following clutches from oviposition to hatching (or clutch failure), identifying sources of mortality, and conducting experimental manipulations to determine the evolutionary advantages of this reproductive strategy.

A modification of this reproductive mode is characteristic of one species at La Selva, *Agalychnis calcarifer* (my mode 18B). This poorly known species occurs in Costa Rica and Panama and is the sister-taxon of *A. craspedopus,* a species that occurs in South America (Duellman 1970; Hoogmoed and Cadle 1991). *Agalychnis calcarifer* males form choruses in the understory of primary forest away from bodies of standing water (Marquis et al. 1986). A clutch of *A. calcarifer* eggs was found on the upper surface of a *Philodendron* leaf overhanging a very small pool in a depression along the trunk of a fallen tree at La Selva (Donnelly et al. 1987). The appearance of the egg mass of *A. calcarifer* resembled that of the other two species of *Agalychnis,* but the locations of the clutch and the chorus were different. Duellman (1970) described a similar oviposition site for a clutch of *A. calcarifer* from the Darien of Panama. Wendy Roberts has gathered detailed data recently on courtship and oviposition for this species at La Selva. Clutches of the South American species, *A. craspedopus,* were found on leaves overhanging water-filled hollow logs in Peru, and males called from perches in the forest (Hoogmoed and Cadle 1991). I recognize the reproductive mode of *A. calcarifer* as a modification of Duellman and Trueb's mode 18 because the oviposition site and site of tadpole development differs from other *Agalychnis* species at La Selva.

All members of the family Centrolenidae oviposit on vegetation over high-gradient streams (my mode 18C). Some glass frog species deposit eggs on the upper surface of leaves and others place them on the lower surfaces. Egg color varies with oviposition site. Eggs that are deposited on upper surfaces are usually dark (black or brown), whereas those placed beneath the leaf are white or light green (McDiarmid 1983). Clutch size is small and tadpoles frequently hatch during heavy rains. This facultative hatch may increase the probability that tadpoles will reach the stream when they drop and it may make hatchlings less vulnerable to predation when the water is turbid. After the tadpoles enter the stream they become bright red and bury themselves in the leaf litter and debris in quiet eddies (McDiarmid 1983).

Comparison of Anuran Reproductive Modes among Sites.

The greatest diversity of anuran reproductive modes is seen in the Neotropics; in South America, twenty-one reproductive modes are known but fourteen are restricted to tropical regions (Duellman and Trueb 1986). Thirteen reproductive modes characterize the anurans that occur in the six lowland sites described earlier (table 15.3). The amphibians of Tortuguero and Los Tuxtlas are characterized by seven of the thirteen modes, those at La Selva, the Osa Peninsula, and BCI by eight modes, and those at Santa Cecilia by eleven of the thirteen modes (table 15.3). Comparisons of this type are complicated because the sites vary climatically. Similarity or difference in the number of modes may also be related to phylogenetic constraints. Although the generalized modes (modes 1 and 2 of Duellman and Trueb 1986) characterize some frogs and toads at all of the six sites, more than one-half of the anurans at each site exhibit more specialized modes (table 15.3), in marked contrast to temperate areas.

REVIEW OF RESEARCH CONDUCTED WITH LA SELVA AMPHIBIANS

Amphibians are relatively small, can be locally abundant, exhibit complex social behaviors and, in many Neotropical sites, several species are active year-round. For these reasons, amphibians are amenable for studies of community ecology, population biology, behavior, and physiology (Duellman and Trueb 1986; Ryan 1985; Wells 1977a, 1977b). Research on amphibians at La Selva has, for the most part, not yet taken advantage of these traits. Work has been generally short term and has focused on individual species, comparisons between sympatric species, and composition and dynamics of two assemblages: the leaf-litter community and the swamp-breeding assemblage of frogs.

Ecological Studies of *Dendrobates pumilio*

The brightly colored, diurnal, poison frog, *Dendrobates pumilio,* is the only amphibian that has been studied in any detail at La Selva. Kitasako (1967) was one of the first to study the behavior and population structure of the strawberry poison frog at the reserve. He used mark-recapture techniques to estimate frog densities of 5 and 6.25 individuals per 100 m² on his study plots. He described, but did not quantify, feeding behavior, calling behavior, courtship, male-male aggression, tadpole transport, dispersion patterns, and homing behavior. Bunnell (1973) found that males were distributed evenly in

Table 15.3 Number of anuran species by reproductive modes at six Neotropical lowland forest sites

Mode No.	Site					
	TU	TO	LS	OSA	BCI	SCA
Aquatic Eggs and Larvae						
1 In lentic water	17	14	15	11	9	34
2 In lotic water	2	1	1	2	1	0
3 Early stages in small natural/constructed basin	0	0	0	1	0	0
4 In treeholes/aerial plants	2	0	0	0	0	1
Eggs on Dorsum of Aquatic Female						
11 Direct development	0	0	0	0	0	1
Eggs in Foam Nest						
8 Foam nest on pond	3	1	2	5	3	8
22 Nest in terrestrial burrow	0	0	0	0	0	1
Terrestrial Eggs						
14 Aquatic larvae carried to water by parent	0	2	2	5	3	8
17 Direct development	8	9	14	9	9	15
Arboreal Eggs and Aquatic Larvae						
18A In lentic water	3	2	3	3	2	11
18B In very small pools	0	0	1	0	1	0
18C In lotic water	1	1	6	7	3	3
Eggs Carried by Adult						
27 Direct development	0	0	0	0	0	1

Source: Reproductive modes from Duellman and Trueb (1986).

Note: TU = Los Tuxtlas; TO = Tortuguero; LS = La Selva; OSA = Osa Peninsula; BCI = Barro Colorado Island; and SCA = Santa Cecilia.

the forest and showed almost no lability in their location during her eight-day study. She concluded that male vocalizations played an important role in territorial behavior by maintaining intermale distance.

Limerick (1980) described courtship and oviposition behavior of *D. pumilio*. She observed five complete courtship sequences and portions of six others. Courtship is prolonged (bout time = 10–98 min.), is initiated when a female approaches a calling male, and results in sperm and egg deposition in the leaf litter. The clutch size in this species is small ($\bar{X}$ = 3.5 eggs).

McVey et al. (1981) reported on territoriality and homing behavior in *D. pumilio*. They censused two plots in the arboretum and conducted displacement experiments to measure homing capabilities. Females were spatially more clumped than males and were found most frequently near bromeliad-containing trees or in association with males. Nearest-neighbor distances were measured between calling males ($\bar{X}$ = 4.8 m), and 23% of the marked males remained near their initial capture site for one month. Females and males successfully returned to original capture sites following displacements of 12 m and 20 m. McVey et al. (1981) concluded that homing behavior in females was related to female egg attendance and/or tadpole transport, and they suggested that males defend territories containing calling perches and oviposition sites.

I conducted a fifteen-month investigation on the autecology of this species with a focus on territoriality (Donnelly 1987, 1989a, 1989b, 1991). Territoriality is best understood in the context of intraspecific competition for a limited resource.

Because most anurans defend resources used for reproduction, I conducted a manipulative field experiment to determine whether males were defending such resources (Donnelly 1987). I increased the availability of oviposition sites (leaf litter) and tadpole-rearing sites (bromeliads) and predicted that if either or both of these resources were the object of defense, male density would increase on treatment plots relative to unmanipulated control plots. The addition of leaf litter did not result in increased male density, but the addition of bromeliads did increase density of adult males (Donnelly 1989a). These results indicated that bromeliad availability was limited, and I concluded that males defend access to bromeliads. The density of adult females also increased on bromeliad-supplemented plots. I examined four demographic parameters (number of immigrants, number of recruits, number of emigrants, and number of survivors) for response to supplementation of reproductive resources. The number of males surviving on bromeliad-supplemented plots was significantly higher than on litter addition or control plots. Bromeliad addition also resulted in increased numbers of female immigrants on treatment plots relative to control plots. Home-range size, dispersion, and spatiotemporal overlap were not affected by the addition of reproductive resources (Donnelly 1989b). The observed increase in density on bromeliad addition plots was the result of additional males and females using unoccupied space. I concluded that the addition of bromeliads enhanced the plots so that additional animals were able to use space that was unsuitable before the resource was added.

Philip Robakiewicz (University of Connecticut at Storrs) conducted ecophysiological experiments with *D. pumilio* and found that treadmill performance varies incredibly among males. No correlation exists between stamina and body size, but stamina and the ability to defend a territory are associated (pers. comm.).

Dendrobates pumilio is abundant and occurs year-round at La Selva. I examined a sample of 468 frogs collected at the reserve in the early 1970s to determine seasonality in reproduction and age-structure (Donnelly 1989c). Preserved frogs were measured snout-to-vent (SVL) and were sexed by dissection. Individuals were assigned to one of three age classes (adult males, adult females, juveniles) based on condition of the reproductive organs and development of male secondary sexual characteristics. The minimum size at which I observed sexual maturity in males (expanded vasa deferentia and vocal slits) and females (expanded oviducts and pigmented ovarian eggs) was 19 mm SVL. These data are useful for estimating age in the field. The sex of adult-sized individuals can be determined by examining the throat (males have brown throats). Males, females, and juveniles were captured in all months and there were no obvious seasonal trends in abundance. Frogs in the smallest size class (recently metamorphosed individuals), however, appeared in a pulsed fashion during the wet season approximately two months after the onset of the rains. I observed a similar pattern of recruitment during 1982–83 (Donnelly 1989a). This pattern contrasts with that reported for another dendrobatid, *Colostethus nubicola*, on BCI (Toft et al. 1982). *Colostethus nubicola* was more abundant, and juveniles were sighted more frequently in the dry season than in the wet. Both of these patterns differ from that reported for the South American dendrobatid *Epipedobates parvulus* (Crump 1974). Crump concluded that this species was an aseasonal (continuous) breeder because gravid females and juveniles were collected during every month of the year. Seasonal reproduction in *D. pumilio* at La Selva was suggested by the pulsed appearance of juveniles in the smallest size class, seasonal changes in oviduct size, and seasonal changes in ovarian complement (the number of eggs present in the ovary at any one time).

Leaf-litter Amphibian Community

The leaf-litter amphibians have received considerable attention, and these organisms are also regularly sampled during field problems by Organization for Tropical Studies (OTS) course participants. Many amphibians that dwell in the leaf litter are characterized by derived reproductive modes, and they do not migrate to breeding ponds. As a consequence, juveniles, subadults, and adults occur together in time and space. The number of amphibian species in the leaf-litter assemblage varies from fifteen (Heinen 1992) to twenty-two (S. Lieberman 1986), and the assemblage is dominated by *Eleutherodactylus* species (table 15.4). Scott (1976) studied this assemblage by sampling nineteen litter plots (7.6 × 7.6 m) located in primary forest in 1970 and 1971. S. Lieberman (1986) worked with samples taken from ninety plots (8 × 8 m) and from pitfall traps distributed along transect lines in primary forest and abandoned cacao groves (1973–74). Heinen sampled amphibians in 5 × 5 m plots in primary forest (*N* = 25) and two cacao groves (*N* = 25 in each grove). The times since these groves were abandoned were twenty-five and five years. Duellman (1990, app. 24.1) indicated habitat use for La Selva species, presumably based on his own experience with these organisms.

In the first such study for any group of vertebrates Scott (1976) compared New and Old World tropical leaf-litter communities. La Selva, along with a site on the Osa Peninsula,

showed a much greater abundance of amphibians and reptiles than sites in Borneo and the Philippines. The average density for La Selva and the Osa site was 17.1 individuals per 100 m², whereas the density in Borneo was 1.5 per 100 m². Scott (1976) surmised that these patterns owed either to basic differences in the composition of the community or differences in the structure and function of these tropical ecosystems. Scott (1976) eliminated abundance of snake predators, forest structure, and species richness as explanations for these differences and suggested that reduced litter production in Bornean forests accounted for the lower population densities. Scott's data on the seasonality of litter fall at both La Selva and the Osa site indicated that litter depth was greatest during the dry season. He suggested that litter patchiness at both sites increased during the wet season and that the abundance of amphibians and reptiles was affected by changes in this resource base. Inger (1980) cited two recent estimates of leaf-litter production in Malayan and southern Thai forests that refute Scott's ideas. Inger (1980) hypothesized that the reduced density of leaf-litter amphibians in the Indo-Malayan region is associated with mast fruiting by the dipterocarps that dominate these forests and results from decreased food availability for these insectivorous, secondary consumers during nonmast years.

S. Lieberman (1986) compared differences between primary forest and a secondary forest (an abandoned cacao grove) in species composition, species abundance, diurnal activity patterns, and seasonality of environmental variables to determine how these variables interact to influence patterns of seasonality in the composition and abundance of the herpetofauna. Species diversity (number) was significantly higher in the forest than in the cacao; twenty-two species of amphibians were collected in the forest, whereas only fourteen were collected in cacao plots. Sixteen of the twenty-two forest species were represented by fewer than ten specimens; eight of the fourteen species of amphibians collected in the cacao were represented by fewer than ten specimens. *Eleutherodactylus bransfordii* and *D. pumilio* were the first and second most abundant species in both habitats, accounting for 35% and 24% of the total collections, respectively.

S. Lieberman (1986) discovered a seasonal pattern in the volume of leaf litter fall, with litter volume highest at the juncture of the wet and dry seasons (generally April–May) and lowest at the end of the wet season (generally December). Litter depth was positively correlated with number of species and number of individuals collected in each study plot. Fauth et al. (1989) also found a positive correlation between amphibian and reptile species richness and litter depth. Although the cacao groves were less complex vegetatively than the undisturbed primary forest, S. Lieberman found that litter was significantly deeper in the cacao and that significantly more anurans were collected there than in primary forest. Seasonal differences in the composition and abundance of the amphibian fauna were significant. Abundance of animals in the leaf-litter assemblage was highest during the dry season and declined during the wet season. S. Lieberman (1986) suggested that the seasonal variation in density of amphibians resulted from different recruitment patterns and responses to changes in litter arthropod populations. She further suggested that strong seasonality in both environmental factors and amphib-

Table 15.4 Composition of the leaf-litter amphibian fauna at La Selva

Taxon	Source	Taxon	Source
Oedipina sp.	L, H	E. cruentus	L, H
O. pseudouniformis	S, D	E. diastema	S, L, H
O. uniformis	S, D	E. fitzingeri	L, D, H
Bolitoglossa colonnea	D	E. gollmeri	L
Gastrophryne pictiventris	L, D, H	E. mimus	ALL
Bufo coniferus	D	E. noblei	S, L, D
B. haematiticus	S, L, D	E. ridens	ALL
B. marinus	D	E. talamancae	S, L, H
Centrolenella albomaculata	S	Leptodactylus melanonotus	D
C. prosoblepon	S, L	L. pentadactylus	L
Eleutherodactylus altae	S, L	Hyla rufitela	L
E. biporcatus	ALL	Scinax elaeochroa	L
E. bransfordii	ALL	Dendrobates pumilio	ALL
E. caryophyllaceus	L, H	Phyllobates lugubris	L, D
E. cerasinus	ALL		
E. crassidigitus	ALL	Rana warschewitschii	S, D, H

Notes: The following abbreviations are used to indicate the authority: S = Scott (1976); L = S. Lieberman (1986); D = Duellman (1990); H = Heinen (1992); ALL = all four authorities. Scott, Lieberman, and Heinen collected their samples in leaf-litter plots at La Selva. See the text for a full explanation of their sampling methods.

ian and reptile population densities dispelled the notion that this wet lowland tropical forest was aseasonal.

Anuran Feeding Patterns. Three studies of feeding patterns of leaf-litter anurans at La Selva have been completed. Limerick (1976) compared the feeding patterns of the two most abundant anurans in the La Selva leaf-litter assemblage, *D. pumilio* and *E. bransfordii*. S. Lieberman (1986) studied feeding in the leaf-litter community, and I examined how feeding in *D. pumilio* was influenced by sex and age (Donnelly 1991).

Limerick (1976) found that the diet of *D. pumilio* consisted of several species of small ants and mites ($\bar{X}$ number of items/ stomach = 57.9; $\bar{X}$ prey length = 1.42 mm); these two prey types accounted for 98% of the total prey removed from the stomachs. *Eleutherodactylus bransfordii* ate fewer and larger items than *D. pumilio* ($\bar{X}$ number of items/ stomach = 3.0; $\bar{X}$ prey length = 3.84 mm). *Eleutherodactylus bransfordii* consumed fourteen categories of food items; ants accounted for 30% of the prey items, spiders 11%, mites 10%, and coleopterans 10%. The diet of *E. bransfordii* was diverse (H' = 0.98; H' = $\Sigma p_i \log_{10} p_i$) and even (J' = 0.86; J' = H'/$\log_{10}$k); the diet of *D. pumilio* was less diverse (H' = 0.33) and less even (J' = 0.32) than the diet of *E. bransfordii*.

S. Lieberman (1986) examined stomach contents from individuals of eighteen amphibian species to determine how food resources were used and if differential use of arthropod prey by amphibians (and reptiles) played a role in structuring the community. She also compared the proportions of items taken by frogs with the proportions of insects captured in pitfall traps to determine whether amphibian (and reptile) species specialized on certain prey types. The results suggested that leaf-litter amphibians used food resources differently; some taxa consumed prey in roughly the same proportions as they occur in the leaf litter (generalists); some concentrated on small, hard-bodied ants and mites; and others concentrated on large, soft-bodied orthopterans. These categories correspond to the "guilds" identified by Toft (1980) for an assemblage of leaf-litter anurans in Peru. Toft (1980) found that body morphology, foraging strategy, and antipredator tactics covaried with diet. Ant specialists (dendrobatids and bufonids) had narrow mouths, were active foragers on small, hard-bodied prey and were protected by skin toxins. Cryptically colored ant avoiders were sit-and-wait predators on mobile, soft-bodied arthropods. Toft (1981) found that feeding patterns displayed by leaf-litter frogs in Panama reflected the same continuum (ant specialists to ant avoiders). Four of eighteen leaf-litter amphibians at La Selva (*Bufo haematiticus, Gastrophryne pictiventris, Dendrobates pumilio,* and *Eleutherodactylus diastema*) were ant specialists or ant and mite specialists. The results for the La Selva community differ somewhat from the results reported by Toft (1980, 1981) for the leaf-litter anurans in Peru and Panama. Toft described *Eleutherodactylus* species as being ant avoiders, but at La Selva it appears that the species of *Eleutherodactylus* have evolved several different feeding strategies, including specialization on ants, coleopterans, spiders, or orthopterans (S. Lieberman 1986).

Toft (1985) suggested that ant specialization was a derived trait in the family Dendrobatidae, and S. Lieberman (1986) described *D. pumilio* as an ant and mite specialist based on a small sample (two frogs per month, one from forest and one from abandoned cacao, for a total of twenty-six frogs). Dendrobatids do not exhibit the extreme sexual size dimorphism typical of most other frogs, but they are behaviorally dimorphic. Behavioral differences between males and females might affect total foraging time and the type of prey encountered. Toft (1980, 1981) demonstrated that prey size increased with frog size, and it might be expected that diet composition would change as body size and gape increase with age. Previous studies of feeding in Neotropical frogs have not determined if feeding patterns were affected by age or sex. I examined stomach contents from 326 frogs collected monthly for fifteen months from the Las Vegas Annex at La Selva to see if feeding in *D. pumilio* was influenced by age- or sex-related behavioral differences (Donnelly 1991). *Dendrobates pumilio* consumed ants, mites, and several types of other prey in all months of the study, and consumption patterns varied temporally for males, females, and juveniles. Mites were important in the diet of juveniles, and females consumed significantly more ants than males and juveniles. All three age-sex groups consumed a variety of other prey types in a pulsed fashion, and I concluded that frogs responded to naturally occurring fluctuations in prey availability (Donnelly 1991). Other types of prey may be more nutritious or easier to digest than ants and mites, and frogs may, therefore, prefer other prey when it is available. The results from my study indicated that placement of a species along a continuum with ant specialists on one extreme and ant avoiders on the other depends on when the sample is taken, the age structure of the sample, and the sex ratio of adults in the sample.

Other Studies

Geographic Patterns of Intraspecific Variation in *Eleutherodactylus bransfordii.* Savage and Emerson (1970) compared color and structural variation in Central American frogs allied to *E. bransfordii*. Forty-two Costa Rican populations, including one at La Selva, were examined. Their analysis delineated six basic dorsal color patterns, four distinct ridging patterns, and thirteen modifying color characters that are superimposed at random on the dorsal color pattern. The La Selva population of *E. bransfordii* represented nineteen of the twenty-three aforementioned characters; in a random sample of sixty La Selva animals they found fifty-one distinct phenotypes based on different combinations of the characters. Their analysis of coloration and ridging patterns identified four nonoverlapping morphs, three of which are found in the La Selva population. They found no geographical correlates of color variation, no differences associated with sex, and they suggested that variation represents a balanced polymorphism that may have evolved in response to the mosaic of color and textures that characterizes the litter of the forest floor.

Predation on Anurans by Invertebrates. Fritz et al. (1981) conducted experiments comparing attack frequency on *D. pumilio* (N = 23) and species of *Eleutherodactylus* (N = 23) by the large ponerine ant *Paraponera clavata*. *Paraponera clavata* accepted (bit) twenty of the *Eleutherodactylus*, rejected two, and attacked then rejected one. Only one dart-poison frog was accepted, thirteen were rejected, and nine were attacked and rejected. When *P. clavata* attacked and rejected *D. pumilio,* it wiped its mouthparts on its forelegs or on the bark of trees; this behavior was not observed when the ant attacked and rejected species of *Eleutherodactylus.* Szelistowski (1985) examined predation by the ctenid spider, *Cupiennius coccineus,* on *D. pumilio* (N = 20) and species of *Eleutherodactylus* (N = 20). Twelve of the poison-dart frogs were attacked and rejected; eleven of the rain frogs were attacked and consumed by the spiders. Although *D. pumilio* was not acceptable to *C. coccineus* as food, there was not a significant difference in the proportion of *Dendrobates* and *Eleutherodactylus* that were attacked, and Szelistowski (1985) suggested that the spider used chemosensory cues rather than visual cues in discriminating between these frogs after the initial attack.

Habitat Use in *Eleutherodactylus.* Miyamoto (1982) examined vertical habitat use by frogs of the genus *Eleutherodactylus* at La Selva and Monteverde. Three species occurred at both sites: *Eleutherodactylus fitzingeri, E. gollmeri,* and *E. rugulosus.* Three patterns of space/time use were detected at both sites: diurnal and ground-active, nocturnal and ground-active, and nocturnal and arboreal. Six of the twelve La Selva species studied by Miyamoto (1982) were at least partially diurnal and ground-active. Only one species, *E. diastema,* was exclusively nocturnal and arboreal. This species has been collected in leaf-litter plots as part of three studies (see table 15.4), and it is interesting that Miyamoto never saw it on the ground in leaf litter. Miyamoto (1982) found that *Eleutherodactylus* males move from the leaf litter to arboreal calling sites at night, and the individuals of *E. diastema* found by other workers in the leaf litter may have been males.

Swamp-breeding Frogs. Craig Guyer and I have studied the phenology of reproduction in an assemblage of swamp-breeding hylid frogs. Our data (unpublished) were collected during a fifteen-month study (1982–83), and they suggest that reproduction in this assemblage is seasonal. Aichinger (1987) reported seasonal reproduction for an assemblage of frogs in Peru, and Gascon (1991) observed similar patterns for frogs in Brazil. The patterns at La Selva, Peru, and Brazil differ from the aseasonal pattern of reproduction observed at Santa Cecilia, Ecuador (Crump 1974; Duellman 1978). The diversity of reproductively active frogs at La Selva is highest at the onset of the rainy season because both the explosive breeders and the prolonged breeders are present in swamps as soon as they fill. Recruitment of recently metamorphosed individuals was pulsed, and the appearance of small frogs was associated with drying of the swamp (Donnelly and Guyer unpublished data).

CONCLUSIONS AND SUGGESTIONS FOR FUTURE RESEARCH

Research on La Selva amphibians has focused on the ecology and behavior of one species, *D. pumilio,* and one assemblage, leaf-litter inhabitants, but La Selva's amphibians are diverse and provide ecological and evolutionary biologists with numerous research opportunities. Caecilians and salamanders are uncommon or rare at La Selva, poorly studied throughout

their ranges, and any information on these taxa would add to knowledge of their biology. Duellman (1978) has suggested that because of the greater variety of structural habitat and the generally more equitable climatic conditions, amphibians in tropical wet forests have exploited a wider array of reproductive strategies than in temperate and more seasonal tropical sites. Crump (1974) suggested that diversity in reproductive strategies allows the coexistence of a large number of species by increasing the number of available breeding sites. Quantitative data on the influence of diversity of breeding sites on the number of species at a breeding pond are largely lacking for La Selva. The members of the family Hylidae exhibit four reproductive modes at La Selva, and individuals of several species use the same pond for reproduction. Despite this ideal situation for comparative studies of reproductive biology and population dynamics, such studies are lacking.

Many of the derived reproductive modes are associated with increased levels of parental care as exhibited by members of the Centrolenidae, Dendrobatidae, Plethodontidae, and Leptodactylidae. Comparative studies of parental care at La Selva could relate this behavior to reproductive output, egg and larval survivorship, and population size and structure. In addition to parental care, predator avoidance strategies, foraging behavior, and mate selection have not been studied.

The most glaring lack of basic biological information concerns frogs of the genus *Eleutherodactylus.* This diverse group of frogs occurs in several habitats at La Selva and clearly is important in the flow of energy in this forest. These frogs are predators on arthropods (S. Lieberman 1986), and they, in turn, are prey items for many vertebrates (Greene et al. unpublished data). No detailed ecological study has been completed for any of the fourteen species that occur at La Selva. Investigations focusing on these species are essential to understand the overall functioning of the ecosystem, particularly predator-prey systems.

Although detailed ecological data have been collected on the ubiquitous *Dendrobates pumilio,* no studies of mate selection or parental care have been conducted. Most of the research involving this species has been conducted in secondary forest situations. Population densities seem to be lower in primary forest than in secondary forest sites (pers. observation), but few empirical data exist to support this observation. If densities are lower in forest than in secondary sites, social interactions, feeding patterns, and population structure may also differ. Comparative demographic and behavioral studies in undisturbed and disturbed forests would add to the large body of information regarding this species.

The riparian anuran assemblage has not been studied at La Selva possibly because of low population densities along the larger streams of "old" La Selva. The larger *quebradas* in the reserve harbor healthy fish populations, and the stream-breeding anurans may use smaller streams for reproductive purposes. It is possible that this group of amphibians exists in higher densities in the southern part of the reserve where the relief is more pronounced, but data are lacking.

Although Guyer and I studied one swamp-breeding assemblage, our data are descriptive and preliminary. Phenological patterns of adults and juveniles need to be examined more quantitatively in the future. Although we have observed seasonality in the adults and juveniles, we know nothing of the dynamics of the larval community. These dynamics should be quantified and the trophic characteristics of the component taxa should be investigated. Patterns observed at La Selva could be compared with those observed by Gascon (1991) in Brazil.

Ecophysiological studies are difficult to complete in most tropical field sites because of equipment requirements, but a few have been conducted on some Neotropical amphibians. The laboratory facilities at La Selva have improved enough that studies requiring air-conditioned environments (e.g., physiology) are now feasible. Taigen and Pough (1985) pointed out the complex correlations among morphology, foraging strategies, diet, predator avoidance, reproduction, water loss, and metabolism in anurans. The La Selva amphibians would provide ecophysiologists with excellent opportunities to determine how these traits are correlated and how they vary among syntopic species.

Wells and his associates have conducted several studies of anuran calling behavior in Panama (Schwartz and Wells 1983, 1984; Wells 1988; Wells and Bard 1987). The species studied in Panama also occur at La Selva, but calling behavior has not been examined for the La Selva species. Calling energetics have also been examined for some Neotropical frog species (Bucher et al. 1982; Ryan 1985), but similar studies have not been conducted at La Selva.

Recently, considerable attention has focused on the apparent widespread decline in amphibian populations (Barinaga 1990; Borchelt 1990; Phillips 1990; Vitt et al. 1990; D. B. Wake 1991). It is difficult to evaluate the generality of this phenomenon because many observations are anecdotal and long-term data on amphibian populations are rare (especially for tropical taxa). Pechmann et al. (1991) presented data on population size for four amphibian species based on twelve years of continuous sampling. They concluded that these species were not declining at a protected study site in South Carolina. They stressed that long-term monitoring efforts at several sites are necessary to distinguish between naturally occurring fluctuations and severe declines. However, the decline of the Golden Toad, a Costa Rican cloud forest endemic, has now been well documented (Crump et al. 1992). The situation at La Selva offers herpetologists an excellent opportunity to conduct such long-term studies because the La Selva amphibians are protected, and baseline data on species composition, relative abundance, and density exist for several species.

The La Selva amphibians are an integral part of the forest ecosystem. The current state of knowledge regarding these animals poses more questions than it answers. These organisms have not been studied in any detail (with the exception of *Dendrobates pumilio*), and they are suitable for addressing a wealth of conceptual areas from natural history to population genetics to ecophysiology. Because of my general interest in population ecology of tropical organisms and my concerns regarding the status of amphibians populations in the tropics, I argue strongly for long-term studies of marked populations both among habitats and among sites in the Neotropics. Long-term demographic studies, like those of Turner (1960) and Pechmann et al. (1991), provide baseline information on naturally occurring cycles of abundance. These studies also provide insights into the types of factors, both biotic and abiotic, that may function in regulating density. Such studies are prerequisite to understanding the role of amphibians in the overall functioning in this and other tropical lowland forests.

ACKNOWLEDGMENTS

I thank J. M. Savage for introducing me to the Neotropics and the La Selva herpetofauna. I am grateful to the Organization for Tropical Studies for logistic support in Costa Rica and for funding my research at La Selva with a Jessie Smith Noyes fellowship. J. M. Savage and R. W. McDiarmid provided me with a species list for the Osa Peninsula (Rincón de Osa and surrounding areas). P. N. Lahanas provided me with a species list for Tortuguero and additions to the species list for the Osa Peninsula (Corcovado National Park). R. Vogt provided me with a list of the amphibians and reptiles of the Los Tuxtlas Biological Station. J. T. Heinen sent me a copy of a manuscript that was in press dealing with the leaf-litter herpetofauna at La Selva. I thank B. I. Crother, M. L. Crump, C. Guyer, D. M. Hillis, C. C. Horvitz, S. Kirkpatrick, J. M. Savage, and two anonymous reviewers for commenting on this manuscript. This is contribution no. 264 from the University of Miami Department of Biology Program in Tropical Biology, Ecology, and Behavior. Final revisions to this document (1988–92) were made while I was a Boschenstein Research Fellow in the Department of Herpetology and Ichthyology at the American Museum of Natural History.

16

The Reptile Fauna: Diversity and Ecology

Craig Guyer

Ecological research on Neotropical reptiles has developed slowly since the mid-1950s, and studies at La Selva have contributed significantly to knowledge of these animals. In this chapter I focus on three questions: What species comprise La Selva's reptile fauna? How does La Selva's fauna compare with that in other Neotropical sites? and How do ecological interactions of these reptiles compare with those at other Neotropical sites? I highlight the importance of La Selva as a site of reptile diversity and suggest avenues for future research.

SPECIES COMPOSITION

La Selva has served as an important collecting locality since the late 1950s when L. Holdridge owned the property and encouraged its use by biologists. Because of its long and relatively continuous use as a study site, the herpetofauna at La Selva is one of the best known among lowland wet forest sites in the Neotropics. A. H. Brame, J. DeAbate, W. E. Duellman, A. G. Kluge, R. J. Lavenberg, D. C. Robinson, J. M. Savage, N. J. Scott, E. H. Taylor, and J. Villa, were among the pioneers who sampled the area of Costa Rica that includes La Selva. Extensive collections are housed at the University of Costa Rica, University of Kansas, Museum of Vertebrate Zoology (University of California at Berkeley), and the University of Miami. These collections have been invaluable in documenting Neotropical reptile diversity, in determining evolutionary relationships among reptile groups (e.g., Lieb 1981; Presch 1970; Werman 1986), and in studying the biogeographic history of lower Central America (e.g., Savage 1966, 1982; Scott 1969). Additionally, these collections have provided a vital data base for assessing the ecological roles of reptiles at La Selva and similar sites (Greene 1988; Guyer 1986; S. Lieberman 1986; Scott 1976).

Scott et al. (1983) presented a species list for La Selva reptiles that summarized more than thirty years of fieldwork by various researchers. It included four turtles, one crocodilian, twenty-three lizards, and forty-seven snakes. One species (*Leptophis mexicanus*) was apparently listed incorrectly for La Selva because no voucher or other supportive material exists. Since 1982 this list has been updated by nearly continuous monitoring of the fauna by long-term researchers. During this time all of the species previously recorded for La Selva were observed except two species of lizard, three snakes, and one turtle (app. 6). The unsighted turtle and snakes are rare (known from only a few specimens) and secretive and may

well still occur at La Selva. The lizard species (*Basiliscus vittatus* and *Ameiva quadrilineata*) are more common in open habitats near the Caribbean coast, and both are conspicuous when present. The lack of recent observations of these lizards indicates that they are now absent from La Selva. It is questionable, however, whether these two species ever occurred there as viable populations.

In addition to confirming most of the species previously recorded for La Selva, the updated list includes one species of crocodilian, two lizards, and eight snakes not found previously (app. 6). This represents a 14% increase in species richness, assuming that all species on the Scott et al. (1983) list are, in fact, still present. Many of the additions are secretive animals that likely always have been present at La Selva and do not represent recent invaders from other areas. Three species (*Lampropeltis triangulum*, *Leptodeira annulata*, and *Oxybelis aeneus*), however are large, conspicuous, and widespread and apparently have escaped detection until recently despite these characteristics. Despite many years of active collecting of reptiles at La Selva, a few species will undoubtedly be discovered in the future. The fauna of lizards, turtles, and crocodilians will, probably, not be altered substantially, but snake species will continue to be added because of their secretive habits and lower population densities.

Although few data on population densities exist, careful reading of the species list suggests that reptile diversity has remained stable since the 1960s. The rapid degradation of forest surrounding La Selva appears to have had little immediate effect on species composition of reptiles. Many of the species are characterized by small, sedentary, and abundant individuals, indicating that the current La Selva property may be large enough to conserve viable populations. The larger snakes (e.g., *Lachesis muta*, *Boa constrictor*, *Clelia clelia*), however, may be represented by only a few individuals on La Selva proper (H. W. Greene pers. comm.). This is true of *Crocodylus acutus* and perhaps of *Caiman crocodylus* as well. The continuity between La Selva and the much larger Braulio Carrillo National Park to the south may prove vital for conserving these species (chap. 2).

Although the La Selva herpetofauna is well known, additional long-term monitoring will be necessary to determine whether altered patterns of land use near La Selva are affecting species richness. Long-term data also are needed to determine whether population densities are being maintained for species that are potentially sensitive to human disturbance.

Candidates for such studies are provided by Braker and Greene (chap. 20).

FAUNAL COMPARISONS WITH OTHER NEOTROPICAL SITES

Lists of reptile species are available from nine other Neotropical sites (sites and sources are listed in table 16.1). All except one represent localities of sizes similar to La Selva (the Iquitos sample included a much larger area; Dixon and Soini 1975, 1977), and all have been sampled intensively for at least one year or less intensively over many years. I have used these lists to compare faunal diversity and composition from these sites to that of La Selva. The ten localities represent six lowland, one midelevation, and two highland wet forests, as well as one lowland dry forest (table 16.1). When comparing the composition of the La Selva reptile fauna to that of other Neotropical sites, I analyzed snakes, lizards (including amphisbaenians), and turtles and crocodilians separately.

The eighty-seven species of reptiles found at La Selva represent a rich fauna compared with published lists from other Neotropical sites. This, undoubtedly, reflects the paucity of published lists from Amazonia. The number of species present at La Selva, however, is identical to the number of species reported by Duellman (1978) from Santa Cecilia, an Amazonian site with the greatest species richness of any Neotropical site of comparable size (table 16.1). The lizard faunas of all sites contain roughly equivalent proportions of species among families except for a notably high diversity of Gymnophthalmidae in Amazonia (table 16.2). This represents the major difference between La Selva and Santa Cecilia. Gymnophthalmids are small, attenuate lizards that inhabit leaf-litter, rotting logs, bromeliads, and loose soil (Duellman 1978); all genera in this family are members of Savage's (1966, 1982) South American historical element. Among snakes, all Neotropical faunas are dominated by the family Colubridae. No differences occur among snake faunas at the family level, and the La Selva fauna is comparable in diversity to the rich fauna at Santa Cecilia (table 16.2). The diversity of La Selva reptiles is greater than that at upland sites (Cerro de la Muerte and the Monteverde Preserve) or the dry site (Cañas; table 16.2). This

faunal richness results from an increase in numbers of species in all families rather than a shift in the contribution of any one family.

In comparing faunas at the generic level, I have used Savage's (1982) classification of biogeographic groups. For all orders of reptiles, the La Selva fauna consists principally of genera that have distributions centered in Central America (table 16.3). This is also true of the other Central American sites. Not surprisingly, the two South American sites (Santa Cecilia and Iquitos) are composed of genera whose distributions are centered in South America with only a few centered in Central America.

At the species level analysis of overall similarity of the reptile fauna among the Neotropical sites in table 16.1 yields similar results for snakes, lizards, and turtles plus crocodilians (fig. 16.1). All three analyses indicate that La Selva clusters tightly with the three other lowland wet or moist forest sites in Central America (Rincón de Osa, Tortuguero, and Barro Colorado Island [BCI]). These lowland sites cluster next with midelevation wet forest sites in Central America represented by Las Cruces (1,500 m). These wet forest sites are next most similar to the dry forest site (Cañas), and this cluster is approximately as similar to the highland wet forest sites (Monteverde plus Cerro de la Muerte) as it is to the South American sites (Santa Cecilia plus Iquitos). Thus, the species composition of the La Selva reptile fauna is most similar to other lowland and midelevation wet forest sites in Central America and is least similar to highland and South American lowland faunas.

Similarity among reptile faunas, however, is not a simple function of decreasing distance. Some sites that are nearer to La Selva (e.g., Cañas) have faunas that are quite different, whereas others that are farther away (e.g., BCI) are more similar to La Selva. Of sites in Central America, La Selva is least similar to the highland sites of Cerro de la Muerte and Monteverde. This pattern suggests that elevation and associated climatic changes are important in determining faunal composition as well as geographic proximity. Unlike some birds and insects (chap. 17, Janzen 1983e), none of La Selva's reptiles appear to be elevational migrants.

The analyses of faunal composition among Neotropical

Table 16.1 Numbers of species in five orders of reptiles at ten Neotropical sites

Site	Turtles	Crocodiles	Lizards	Amphisbaenians	Snakes	Total	Life Zone	Source
LS	4	2	25	0	56	87	Lowland wet	Scott et al. 1983
BCI	5	2	21	1	39	68	Lowland wet	Myers and Rand 1969
								Savage and McDiarmid
OSA	2	2	23	0	43	70	Lowland wet	pers. comm.
TORT	5	2	22	0	30	59	Lowland wet	Lahanas pers. comm.
LC	0	0	16	0	23	39	Midelevation wet	Scott et al. 1983
CAN	3	2	17	0	35	57	Lowland dry	Scott et al. 1983
								Hayes and Pounds pers.
MONT	0	0	9	0	26	35	Highland wet	comm.
CERRO	0	0	3	0	6	9	Highland wet	Scott et al. 1983
SC	6	2	27	1	51	87	Lowland wet	Duellman 1978
IQUI	9	4	40	2	87	142	Lowland wet	Dixon and Soini 1975, 1977

Note: Site abbreviations are La Selva (LS); Barro Colorado Island (BCI); Rincón de Osa (OSA); Tortuguero (TORT); Las Cruces (LC); Cañas (CAN); Monteverde Preserve (MONT); Cerro de la Muerte (CERRO); Santa Cecilia (SC); Iquitos (IQUI).

Table 16.2 Number of species in each lizard and snake family for ten Neotropical sites

Family	Site									
	LS	BCI	OSA	TORT	LC	CAN	MONT	CERRO	SC	IQUI
Lizards										
Anguidae	3	0	0	2	0	0	1	1	0	0
Iguanidae	13	12	11	12	9	7	7	2	10	16
Gekkonidae	4	4	3	3	1	3	0	0	3	6
Gymnophthalmidae	0	1	3	0	2	1	0	0	10	14
Teiidae	2	2	3	3	1	2	0	0	3	3
Scincidae	2	1	2	1	2	3	1	0	1	1
Xantusiidae	1	1	1	1	0	0	0	0	0	0
TOTAL	25	21	23	22	15	16	9	3	27	50
Snakes										
Aniliidae	0	0	0	0	0	1	0	0	1	1
Anomalepidae	0	2	0	0	0	0	0	0	0	0
Boidae	2	3	2	1	0	2	0	0	5	5
Colubridae	46	30	35	22	21	27	24	5	36	65
Leptotyphlopidae	0	0	0	0	0	1	0	0	0	0
Micruridae	3	2	2	2	0	1	1	0	5	8
Tropidophiidae	1	0	0	0	0	0	0	0	0	0
Typhlopidae	0	0	0	0	0	0	1	0	0	2
Viperidae	4	2	4	3	2	3	2	1	4	6
TOTAL	56	39	43	28	23	35	28	6	51	87

Note: Site abbreviations are La Selva (LS); Barro Colorado Island (BCI); Rincón de Osa (OSA); Tortuguero (TORT); Las Cruces (LC); Cañas (CAN); Monteverde Preserve (MONT); Cerro de la Muerte (CERRO); Santa Cecilia (SC); Iquitos (IQUI).

sites indicate that reptiles at La Selva are most similar to those on other Central American lowland sites. These sites cluster at a level of faunal similarity comparable to that of the two Amazonian sites (fig. 16.1). These findings support placement of the La Selva reptile fauna in Savage's (1966) Eastern Central American Lowland Herpetofauna. This biogeographic unit has a wide distribution, ranging from the Atlantic lowlands of Tamaulipas, Mexico, to central Panama and including the Osa Peninsula of Costa Rica. Until recently, this distribution was more or less continuous. Increased human disturbance, however, has splintered reptile habitats in this region. Loss of lowland forest increases the importance of La Selva as a site for reptile research because it has a well-known and diverse assemblage of species that is apparently maintaining its richness and is representative of a once more wide-ranging fauna.

The analyses reflect the biogeographic history of the La Selva reptile fauna. When compared at the species level, La Selva (and other Central American lowland wet forest sites) are quite dissimilar from lowland wet forest sites in Amazonian South America. This finding conforms to Savage's (1966, 1982) biogeographic hypotheses that the Central American herpetofauna represents a unit of endemism and that there has been a long separation of this land mass from South America. The phenetic analyses used to characterize the La Selva reptile fauna and compare it to that on other Neotropical sites, however, are not sufficient to test the complex vicariance and dispersal model proposed by Savage (1982) to explain the biogeography of Central America. The appropriate data are cladograms of relationships among taxa from these sites with

which area cladograms can be constructed. Unfortunately, these are available for very few groups that occur at La Selva (Cadle 1982; Presch 1970; Werman 1986). Only with much additional systematic work will the historical relationships of La Selva to other Central American and South American sites be clarified.

ECOLOGICAL COMPARISONS WITH OTHER NEOTROPICAL SITES

Studies of reptile faunal composition are necessary preludes to more detailed analyses of factors involved in structuring communities. Assemblages of reptiles inhabiting the leaf-litter community are among the best investigated of tropical reptilian groups. In these studies seasonal patterns of change in weather, forest structure, and food availability have been used to explain cycles of reptile abundance.

The leaf-litter community depends upon leaf fall from forest trees. In many Central American forests leaf fall peaks during the dry season, leading to accumulation of litter at that time (Daubenmire 1972; Ewel 1976; Frankie et al. 1974a; Kunkel-Westphal and Kunkel 1979). At other sites litter builds to peak levels early in the wet season (Lambert et al. 1980; Toft and Levings 1977). After accumulating, leaf litter is reduced gradually throughout the rest of the year as a consequence of decreased leaf drop, decomposition by microorganisms, and the mechanical action of rainfall (Levings and Windsor 1982). Thus, leaf-litter abundance varies in a cyclic and potentially predictable fashion even in the moist tropics. Leaf litter provides habitat for a diverse invertebrate fauna

Table 16.3 Biogeographic distributions of reptile genera from ten Neotropical sites

Reptile Genera	Biogeographic Unit[a]	Biogeographic Distribution[b] Site[c]									
		LS	TORT	OSA	BCI	LC	CAN	MONT	CERRO	SC	IQUI
Snakes	WT	.10	.09	.09	.10	.05	.17	.15	.00	.06	.10
	SA	.26	.09	.30	.32	.28	.10	.20	.00	.64	.63
	CA	.62	.77	.61	.58	.61	.53	.55	.75	.30	.27
	NA	.02	.05	.00	.00	.05	.20	.10	.25	.00	.00
	TOTAL	39	22	33	31	18	30	20	4	33	41
Lizards	WT	.07	.09	.06	.07	.10	.07	.20	.00	.05	.04
	SA	.14	.09	.31	.36	.40	.20	.20	.00	.86	.81
	CA	.72	.73	.56	.57	.40	.53	.40	.33	.09	.15
	NA	.07	.09	.06	.00	.10	.20	.20	.67	.00	.00
	TOTAL	14	11	16	14	10	15	5	3	22	26
Turtles and Crocodiles	WT	.00	.00	.00	.00	.00	.00	.00	.00	.00	.11
	SA	.25	.43	.25	.29	.00	.20	.00	.00	.89	.78
	CA	.75	.43	.50	.42	.00	.60	.00	.00	.11	.11
	NA	.00	.14	.25	.29	.00	.20	.00	.00	.00	.00
	TOTAL	4	7	4	7	0	5	0	0	9	9

[a]Biogeographic units (from Savage 1983) are widespread Neotropical (WT); distribution centered in South America (SA); distribution centered in Central America (CA); distribution centered in North America (NA).

[b]Data are proportions of total number of genera at each site that belong to each biogeographic unit.

[c]Site abbreviations are La Selva (LS); Barro Colorado Island (BCI); Rincón de Osa (OSA); Tortuguero (TORT); Las Cruces (LC); Cañas (CAN); Monteverde Preserve (MONT); Cerro de la Muerte (CERRO); Santa Cecilia (SC); Iquitos (IQUI).

whose abundance is correlated with the seasonal pattern of leaf-litter deposition (Buskirk and Buskirk 1976; Janzen 1973c) or with leaf-litter depth (Levings and Windsor 1982; S. Lieberman and Dock 1982; Toft and Levings 1977; Wolda 1978b). Additionally, cyclic changes in litter depth result in altered humidity levels that could affect the quality of refuges and nest sites as well as alter the numbers of parasites (e.g., mites and ticks). The number of vertebrate predators, in turn, is correlated with the seasonal abundance of invertebrates (Janzen and Schoener 1968; S. Lieberman 1986; Skutch 1950; Toft 1980b). Thus, for a variety of sites in the Central American tropics correlative evidence links seasonal change in leaf fall and litter accumulation with the number of leaf-litter arthropods and their predators. Lizards are among the most abundant vertebrate components of this predator assemblage. Some lizards are known to alter densities of arthropods (Pacala and Roughgarden 1984; Schoener and Spiller 1987), indicating that reptiles can be important as regulators of prey abundance.

Although the phenological cycles described suggest a consistent pattern, extensive data on patterns of rainfall, leaf-litter depth, insect abundance, and reptile abundance are available from only two sites, BCI and La Selva. At both sites leaf-litter accumulates during the dry season when numerous canopy tree species drop their leaves (Frankie et al. 1974a; Levings and Windsor 1982). Responding to this alteration in habitat availability, litter arthropods at BCI begin increasing in density by late in the dry season but do not reach peak abundance until early in the wet season (Smythe 1970; Toft and Levings 1977; Willis 1976). At this time increased moisture provides

an environment conducive to rapid litter decomposition (Levings and Windsor 1982). By the end of the wet season continued decomposition together with low input of new litter reduces leaf-litter depth and arthropod abundance. In this regard BCI differs from La Selva where increased habitat availability results in peak densities of leaf-litter arthropods late in the dry season (S. Lieberman and Dock 1982). At La Selva precipitation is greater during all seasons than at BCI (see Guyer 1988a; Rand and Rand 1982; and chap. 3 for rainfall summaries). Therefore, the more intense dry season at BCI may have correspondingly stronger effects on reptiles. Leaf-litter arthropod populations at La Selva decline during the wet season as litter decomposes. If food limits population density of predators of these leaf-litter arthropods, then predator densities should be correlated with arthropod density, peaking in the early wet season (BCI) or late dry season (La Selva). This prediction holds for at least some arthropod predators at both sites (S. Lieberman 1986; Toft 1980b).

Lizards in the genus *Norops* (*Anolis* of other authors; Guyer and Savage 1986) have served as model species within the reptile leaf-litter communities at La Selva and BCI. Detailed studies of *Norops limifrons,* an important arthropod predator at BCI, indicate that population density is regulated by factors affecting rates of egg production and egg mortality (Andrews et al. 1983). The observations that growth rate (Andrews et al. 1983), body mass (Wright 1979), and fecundity (Andrews and Rand 1974; Sexton et al. 1963, 1971) are reduced during the dry season food shortage are all consistent with the hypothesis that food availability can limit population density. Manipulations similar to the elegant experiments of Stamps and Tanaka

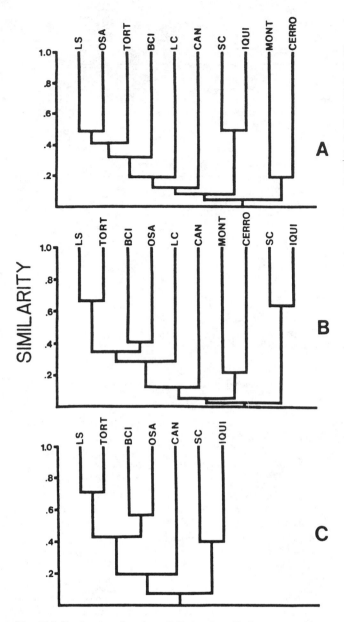

Fig. 16.1. Dendrograms based on similarity of reptile faunas at ten Neotropical sites. Similarity is based on Jaccard Index and clustering is based on the unweighted pair group method (Pimentel and Smith 1985). Analyses presented separately for snakes (*A*), lizards (*B*), and turtles and crocodilians (*C*). Site abbreviations are La Selva (LS); Barro Colorado Island (BCI); Rincón de Osa (OSA); Tortuguero (TORT); Las Cruces (LC); Cañas (CAN); Monteverde Preserve (MONT); Cerro de la Muerte (CERRO); Santa Cecilia (SC); Iquitos (IQUI).

(1981), however, are needed to determine the influence of cyclic changes in water availability on abundance of *Norops limifrons*. Because moisture does not appear to limit egg survival during dry months at BCI (Andrews et al. 1983), reduced water availability may not affect *N. limifrons* by dehydrating eggs or encouraging other sources of mortality. Instead, the dry season may reduce the availability of oviposition sites.

At La Selva population studies have centered on *Norops humilis*. Serial samples of *N. humilis* indicate a distinctive cycle of abundance of this species that appears to be consistent from year to year (Guyer 1986, 1988a) and that coincides

with cycles of arthropod abundance (S. Lieberman and Dock 1982). Experimental enhancement of food availability causes increased population density of adults, demonstrating that food can regulate lizard numbers (Guyer 1988a). Population size increases rapidly because of increased juvenile and adult male immigration and increased survival of adult females (Guyer 1988a). This alteration of population density is associated with growth of males, increased rate of egg deposition, and increased home range overlap (Guyer 1988b). Thus, males may respond to seasonal changes in food availability by altering growth, whereas females may respond by altering reproductive output.

Data for *N. humilis* at La Selva indicate that cyclic changes in arthropod abundance can cause cyclic changes in lizard abundance. The generation time for this species is six months (Guyer 1986), allowing for rapid demographic responses to food availability. Data documenting this response were collected over an area similar to that under a single forest tree (Guyer 1988a). Because Neotropical tree species differ in timing of leaf drop (Frankie et al. 1974a) and individuals of particular tree species can be widely dispersed (D. B. Clark and Clark 1987), patches of thick leaf litter and abundant arthropods in primary forest may be widely spaced. Therefore, increased population size should result from demographic changes within these sites rather than from movements of individuals from one site to another. Thus, the area under single forest trees appears to be an appropriate spatial scale for documenting cycles of abundance of La Selva's leaf-litter reptiles. These cycles should differ among such sites depending on the tree species, an observation that would be obscured if samples from different areas were pooled (e.g., Fitch 1973).

In two recent studies researchers have compared the physiological performance of lizards at La Selva with performances of lizards at higher elevations and latitudes (Van Berkum 1986, 1988). Because sprint speed appears important in foraging ability, escape from predation, and perhaps social interactions, this trait was chosen to assess one possible aspect of fitness among lizards from different sites, including La Selva. As ectotherms, lizards must cope with variation in environmental temperatures. Research by Van Berkum tested the common assumption that lizards at tropical sites with less-variable temperature regimes (like La Selva) should have less-variable temperature tolerances relative to north temperate sites or tropical sites with more variable temperatures (e.g., dry forest or higher elevation sites). This assumption predicts that lizards at sites with less-variable temperatures should (1) be able to locomote over a narrower range of temperatures (tolerance range), (2) perform daily activities over a narrower range of temperatures (variability in field body temperature), and (3) sprint at near maximal velocities over a narrower range of temperatures (performance breadth). When tropical lizards (from many sites, including La Selva) were compared to temperate forms, a consistent pattern of narrower tolerance range for tropical species was observed (Van Berkum 1988). No reduction in variability of field body temperatures was demonstrated for tropical lizards relative to temperate forms. Only a comparison of tropical *Ameiva* (from La Selva) and temperate *Cnemidophorus* supported a narrowing of performance breadth among tropical lizards.

Comparisons of tropical anoles indicate that species at La

Selva exhibit narrower variability in field body temperatures (Van Berkum 1988) and narrower performance breadth (Van Berkum 1986, 1988) relative to anoles at other sites. These patterns suggest that ectotherms at La Selva experience less-variable environmental temperatures than at Santa Rosa or Monteverde and temperate sites and that selection might have narrowed the range of temperatures over which La Selva reptiles perform well. This suggestion conforms to the arguments of Janzen (1967b) relative to the effects of the thermal environment on distribution and diversity of tropical ectotherms.

A final avenue of reptile research at La Selva attempts to establish the role that this group plays as predators upon other vertebrates. Among predators at La Selva snakes are notable for their abundance and diversity. In fact, the predators of La Selva and Manú, Peru, appear to differ from other assemblages described by Greene (1988) in that snakes comprise more than half of the vertebrate predator species. Thus, although the number of vertebrate predators at La Selva is impressively large, most of them are secretive serpents that are easily overlooked relative to sites with visually more conspicuous predators (e.g., mammalian carnivores of African savannahs [Greene 1988]). La Selva's snakes can be divided into several foraging guilds: frog-eaters, predators on other snakes, bird-eaters, and mammal-eaters. More than half of La Selva's snake fauna belongs to the frog-eating guild. Therefore, much of La Selva's vertebrate predator diversity is associated with an abundant and diverse frog fauna (Greene 1988), a correlation documented by Arnold (1972) for snakes in general.

CONCLUSIONS AND FUTURE RESEARCH

In this chapter I summarize the avenues that research on reptiles has taken at La Selva. This work has resulted in a fairly complete faunal list, description of patterns of population and community dynamics based upon detailed study of only a few species inhabiting the leaf litter, and development of hypotheses relative to how this group interacts with its environment as predators and as potential prey. Clearly, much observational and experimental work remains to be done at La Selva.

Basic natural history data are lacking for nearly all species of reptiles at La Selva. Information gathered on any species is likely to improve understanding of population and community organization of this group of vertebrates. The work of Greene and Santana (1983) articulates the potential for such studies. By determining the foraging pattern of the bushmaster (*Lachesis muta*) these investigators documented that this snake eats primarily spiny rats (*Proechimys*). This rodent is an important dispersal agent for *Welfia* (Vandermeer 1979), an abundant subcanopy palm. Thus, long-term observation of foraging by predators such as *L. muta* may ultimately lead to a more precise understanding of how reptiles affect forest dynamics by regulating seed dispersal agents. Other reptiles at La Selva may serve as effective controls on prey populations and might be amenable to experimental manipulations similar to those performed at other tropical sites (Pacala and Roughgarden 1984; Schoener and Spiller 1987).

Additionally, natural history studies should help regional conservation efforts. For example, studies of reptiles at La Selva have largely overlooked chelonians and crocodilians, despite the fact that *Caiman crocodylus* and *Rhinoclemmys*

funerea are commonly observed and, apparently are good candidates for population studies. Because these species are impacted by human activities in the Puerto Viejo region, observations at a less disturbed site like La Selva could have important management and conservation implications.

Reptile abundance within the leaf litter at La Selva peaks late in the dry season rather than in the wet season as it does at the more seasonal BCI. Documentation of factors responsible for this difference is an important avenue for future research. Much of what is known about cycles of abundance of leaf-litter reptiles comes from detailed studies of only two anoline lizards (*N. humilis* and *N. limifrons*). Although anoles have been used as model vertebrates (Schoener 1983), data on other genera are needed. Abundant lizard species such as *Sphenomorphus cherrei* and *Lepidoblepharis xanthostigma* are attractive candidates for such studies at La Selva.

Examination of demographic variation at La Selva should allow testing and refining of theories of reptile life-history evolution, a major focus of research on north temperate lizard populations. Long-term data on demographic variation within and among lizard populations coupled with monitoring of resource availability and predator intensity are crucial for examination of these theories (Tinkle and Dunham 1986). Except for the pioneering efforts to monitor *Norops limifrons* on BCI (Andrews et al. 1983; Andrews and Rand 1982), few relevant data are available from mainland Neotropical lizards and none are available from La Selva. Long-term population studies of lizards (and other vertebrates) are potentially important at La Selva because this site is contiguous with a protected elevational transect. Thus, demographic variability of lizard populations could be studied among different elevational sites as well as among years. Such studies should allow assessment of the role of elevational patterns of thermal sensitivity (Van Berkum 1986, 1988) in shaping life-history evolution.

Needed also at La Selva are community-level studies of reptiles. As with population studies, most of the literature related to community organization among reptiles has dealt with lizards. This is especially true of Caribbean anoles, for which predictions of community structure (Schoener 1977), biogeography (Roughgarden et al. 1983), and morphology (Moermond 1979) have been generated. Similar studies of mainland anoles have been attempted recently (Pounds 1988) and should be expanded to include La Selva. Work by Andrews (1979) near Limón, Talbot (1979) at La Selva, and Corn (1981) near Río Frío suggest that the competitive interactions thought to be important in shaping island anole communities may be less intense for mainland anoles. Caging experiments, however, similar to ones performed on some Caribbean island anoles (Roughgarden et al. 1983), are needed to determine whether potential anole competitors at mainland sites affect each others' growth and reproduction. Additionally, studies on the role of predation in regulating population and community structure of anoles are needed at La Selva. This factor has been hypothesized to be important by several investigators (Andrews 1979; Guyer 1988a; Talbot 1979) but has received almost no detailed study.

An additional avenue of community-level research is examination of the foraging ecology of the snake assemblage at La Selva. This assemblage is among the richest of published Neotropical faunas, but the secretive nature of these predators makes them difficult subjects for study. Several species, however, are abundant enough to make such a study feasible (e.g., *Bothrops asper, Leptodeira septentrionalis, Leptophis depressirostris, Sibon nebulata,* and *Rhadinaea decorata*). Data on foraging habits indicate that much of the diversity of La Selva's snake fauna is accounted for by a frog-eating guild (Greene 1988). An understanding of the organization of this assemblage awaits study of the seasonal activity of snakes relative to temperature, rainfall, and food availability because these factors have been hypothesized to be important in other tropical snake assemblages (Henderson and Hoevers 1977; Henderson et al. 1978a, 1978b; Silva et al. 1985; Vitt and Vangilder 1983).

La Selva can also play a key role in biogeographic research. Such studies should incorporate tests of ecological explanations of faunal distribution and composition. Attention needs to be focused on determining species composition among habitats at La Selva so that analyses of alpha and beta diversity can be attempted. By relating habitat specialization to geographic distribution, hypotheses explaining why some species of reptiles have broader geographic distributions than others might be generated. Species might be widespread because they perform well in a variety of habitats or because their habitats are widely distributed. Additionally, these data might indicate whether differences in diversity among Neotropical sites result from differences in numbers of habitat types, from differences in specialization among component species, or from biogeographic causes.

On a broader scale, two general hypotheses of Savage (1966, 1982) for the history of Central American reptiles need testing: that members of the Central American herpetofauna will display predictable evolutionary relationships with taxa in South America, North America, and the Greater Antilles and that highland taxa will display sister-taxon relationships with lowland taxa. These tests require phylogenetic analyses of reptiles that occur throughout the New World so that cladograms can be converted into hypotheses of relationships of Savage's (1966, 1982) biogeographic areas. Although phylogenies are available for a few reptile groups at La Selva, much

systematic work remains to be done. Because anoline lizards have been used extensively in population and community investigations at La Selva and other Neotropical sites, a phylogenetic study of these lizards would be especially valuable. Such a study will allow an assessment of the contribution of phylogeny to patterns of morphology, habitat use, and behavior observed for the anole community at La Selva.

The La Selva reptile fauna is rich, well-documented, and representative of the eastern Central American lowland herpetofauna as defined by Savage (1966). The relatively complete faunal list as well as the opportunities afforded by its species richness, biogeographic location, and apparent stability in the face of nearby human disturbance indicate that La Selva should remain an important site for research on Neotropical reptile ecology and evolution.

ACKNOWLEDGMENTS

My deepest appreciation is extended to Jay M. Savage for the encouragement, advice, and friendship that resulted in my interest in tropical ecology. S. M. Hermann, L. McDade, M. A. Donnelly, and H. Hespenheide deserve special thanks for improving ideas presented in this chapter as well as the writing itself. Others who read and helped improve this chapter were H. W. Greene, R. W. McDiarmid, N. J. Scott, L. J. Vitt, and two anonymous reviewers. The refinement of the reptile list for La Selva was aided by several researchers, especially M. A. Donnelly, M. Fogden, H. W. Greene, and M. Santana. For access to species lists I thank P. N. Lahanas (Tortuguero), J. M. Savage, and R. W. McDiarmid (Rincón de Osa), and M. P. Hayes, J. A. Pounds, and W. Timmerman (Monteverde Preserve). Many of the thoughts presented here were formulated while I was a Jesse Smith Noyes predoctoral fellow through OTS. Thanks go to the office staff of OTS as well as to David and Deborah Clark, the La Selva Biological Station directors, who provided countless courtesies during my months at La Selva. Support for completing this work was provided by the Department of Zoology and Wildlife Science and the Agricultural Experiment Station at Auburn University.

17

Birds: Ecology, Behavior, and Taxonomic Affinities

Douglas J. Levey and F. Gary Stiles

With more than four hundred species, the avifauna of La Selva is the richest recorded for any Middle American site. It is also one of the most intensively studied, beginning with the year-long study of Slud in 1957–58 (Slud 1960). In the 1960s Skutch studied breeding of several species, and Stiles and L. L. Wolf began their long-term studies on hummingbirds. Other studies during this period included those of Linhart (1973) and Orians (1969a). During the 1970s and 1980s the pace of ornithological work at La Selva quickened, with intensive studies of hummingbirds (Wolf et al. 1972; Stiles 1975, 1978b, 1978c, 1979a, 1980a; Stiles and Wolf 1974, 1979), puffbirds (Sherry and McDade 1982), flycatchers (Sherry 1984), and frugivorous birds (Denslow and Moermond 1982; Moermond and Denslow 1983, 1985; Levey 1987b, c, 1988a; Loiselle 1987, 1988; Moermond et al. 1987; Blake and Loiselle 1991). La Selva has also provided important data on the role of latitudinal and altitudinal migrants in tropical avifaunas (Stiles 1980a, 1980b, 1985a; Loiselle 1987, 1988; Levey 1988a; Loiselle and Blake 1991). Nevertheless, even after thirty years of ornithological work, we lack a thorough understanding of the La Selva avifauna. We still know very little about the basic biology of many species, especially the large, rare, or canopy species. Annual variation in size and structure of the avifauna is also only beginning to be understood. Even a variable as simple as the number of species is only a minimum estimate: witness Slud's (1960, 63) assertion that he had "found all the species to within 1% of the total." The species list has since grown by 24% (table 17.1; app. 7).

In this chapter we will summarize what is and is not known about the La Selva avifauna. A similar overview with a different perspective is provided by Blake et al. (1991). We necessarily take a broad approach, concentrating on the community and population levels. By doing so we hope to provide both a general introduction to the birds and a catalyst for future work.

COMPOSITION AND AFFINITIES OF LA SELVA'S AVIFAUNA

By 1991 411 species of birds have been recorded from La Selva or its immediately adjacent areas; 256 species are known or strongly suspected to breed there, and at least 12 others breed in the general region (table 17.2). This gives La Selva the richest recorded avifauna of any Middle American site (Slud 1976; Stiles 1983a). In part, this diversity simply reflects the fact that many competent observers have worked

extensively at La Selva, recording rare and accidental species that would probably be missed in a less well-studied area (e.g., *Morphnus guianensis, Empidonax albigularis, Sterna fuscata*). We believe, however, that the richness of La Selva's avifauna principally reflects its favorable climate, geographic position, and diverse habitats.

Wet, mildly seasonal climates and complex evergreen vegetation, as found at La Selva, favor higher bird species richness than do more highly seasonal climates and deciduous vegetation (Orians 1969a; Willis and Eisenmann 1979; Stiles 1983a; Slud 1976). Geographically, La Selva's position on the Caribbean slope near the southern end of the Middle American isthmus places it in a belt of evergreen forest that extends continuously to South America, source of the world's richest tropical avifauna. Thus, the La Selva avifauna is considerably richer (although with fewer endemic forms) than that of the ecologically isolated, but climatically similar Osa Peninsula (Stiles 1983a). La Selva's position at the base of the Cordillera Central allows its avifauna to be further enriched, both seasonally and irregularly, by thirty-five species of highland birds (table 17.2). Finally, its position less than 50 km from the Caribbean coast, along which the migration routes of many northern-breeding species are concentrated, helps account for the occurrence of more than eighty such species (table 17.2).

The importance of habitat diversity can scarcely be overestimated. With the addition of the Las Vegas Annex, including an extensive area of riverine scrub along the Río Sarapiquí (the periodic flooding of which maintains this area in a state of early secondary succession), La Selva now has nearly all of the natural terrestrial habitats of the Caribbean lowlands of Costa Rica, even though certain aquatic habitats are lacking (extensive marshes or lakes, *Raphia* palm swamps, estuaries, and seacoast). With the addition to the station of human-induced second growth of various ages and origins, and the adjacent agricultural and pastoral areas, it should be no surprise that nearly half of Costa Rica's avifauna has been recorded at La Selva.

As might be expected from La Selva's geographic position, the affinities of its breeding avifauna are overwhelmingly with South America (table 17.3). Over 85% of the species include South America in their breeding distributions, and nearly 60% are widespread on both sides of the Andes. By contrast, only 4% breed north of the extreme southern United States, and only 30% extend north of the limits of wet tropical forest in southern Mexico. Most of the species extending well into

Table 17.1 Additions to the La Selva list of avifauna

Number of Species	Holdridge et al.[a] 1957	Slud 1960	Stiles 1977[b]	Levey and Stiles[c] This Study
Total number of species	247	331	389	411
Number of species added over previous total		84	58	22
Percent change from previous total		34	18	6

[a]As reported in Slud 1960.
[b]Unpublished list distributed by the Organization for Tropical Studies.
[c]Additional species are listed in appendix 7.

Table 17.2 Elements of the La Selva avifauna by seasonal and breeding status

Avifauna	Large Land and Water Birds[a]	Higher Nonpasserines[b]	Suboscine Passerines[c]	Oscine Passerines[d]	Total Species
Tropical Residents: Breeding in Costa Rica					
Permanent resident, definitely breeds at La Selva	29	68	70	63	230
Seasonal resident, definitely breeds at La Selva	2	0	2	1	5
Irregular visitor or resident, breeding recorded	7	3	2	1	13
Present year-round, breeds on Caribbean slope, possibly at La Selva	8	0	0	0	8
Total breeding avifauna					256
Sporadic visitor, any time of year; probably does not breed at La Selva, but does so on Caribbean slope	7	2	0	4	12
Vagrant or stray, definitely nonbreeding	9	6	5	1	20
Regular altitudinal migrant, all years	0	2	3	3	8
Irregular altitudinal migrant, some years only	0	6	2	4	12
Altitudinal wanderer on regular, even daily basis	3	3	0	0	6
Altitudinal vagrant or stray	2	2	3	2	9
Altitudinal component					35
Long-distance Migrants: Breeding in North America					
Seasonal (mostly winter) resident	10	2	3	26	41
Regular spring or fall transient only	3	4	5	19	31
Irregular or sporadic vagrant	6	3	1	4	13
"Nearctic" component					86
TOTALS	86	101	96	128	411

[a]Families Tinamidae through Laridae.
[b]Columbidae through Picidae.
[c]Dendrocolaptidae through Tyrannidae.
[d]Hirundinidae through Ploceidae (or Fringillidae).

North America are waterbirds, whose breeding requirements are linked to particular types of water bodies, irrespective of the surrounding vegetation type (Stiles 1983a). Most of the other species extending into North America are associated with second growth, open areas, or human settlements (e.g., *Troglodytes aedon, Sturnella magna*). No true forest-interior bird of the La Selva avifauna breeds north of southern Mexico. A further indication of the southern affinities of the breeding avifauna is the fact that nearly half of the species (12 of 25) at or near their southern limit at La Selva are replaced by very closely related congeners immediately to the south (e.g., *Dysithamnus striaticeps* by *D. puncticeps, Campephilus guatemalensis* by *C. melanoleucos*); most are members of superspecies complexes. Several others (e.g., *Aphanotriccus capitalis*) are Central American endemics with close relatives farther south.

Table 17.3 Northern and southern limits of breeding range of species that breed at La Selva

Families	Northern Limits				Southern Limits			
	Widespread in NA	S U.S., N and C Mexico	S Mexico to N Honduras	SE Honduras, Nicaragua, N Costa Rica	C Costa Rica, W Panama	C and E Panama	N and W SA (Transmontane)	Widespread in SA
Large land and water birds[a] (N = 46 spp.)	7	13	18	8	2	0	9	35
Higher nonpasserines[b] (N = 71)	1	20	27	23	6	2	21	42
Suboscine passerines[c] (N = 74)	0	13	33	28	6	0	21	47
Oscine passerines[d] (N = 65)	2	19	23	21	11	8	19	27
TOTAL (N = 286)	10	65	101	80	25	10	70	151

[a]Families Tinamidae through Laridae.
[b]Columbidae through Picidae.
[c]Dendrocolaptidae through Tyrannidae.
[d]Hirundinidae through Ploceidae (or Fringillidae).

COMPARISON OF LA SELVA WITH OTHER NEOTROPICAL SITES

Here we compare the breeding avifauna of La Selva with those of six other Neotropical sites (data from Smithe 1966; Stiles 1983a, unpublished data; Rodríguez 1982; Leck et al. 1980; Pearson 1972; Pearson et al. 1977; Willis and Eisenmann 1979; also see Karr et al. 1991 for further comparisons with other sites). To control the effects of humidity, altitude, and temperature on avifauna composition, we selected sites in the lowland moist forests of the northern Neotropics. Tikal, Guatemala, is near the northern extreme of tropical wet forests. Rincón de Osa is an isolated enclave of humid forest on the Pacific slope of Costa Rica. Barro Colorado Island (BCI) is in central Panama (we used the total resident list, ignoring extinctions). Los Katíos is in the Chocó region of Colombia, at the northern end of transmontane wet lowland forest in South America and near the source area of the Central American avifauna. Río Palenque, Ecuador, is at the southern end of South American transmontane forest and Limoncocha, Ecuador, is an Amazonian site similar to Río Palenque but on the opposite side of the Andes. Precise areas of each site are not available. Although area certainly affects the number of recorded species, we feel areas surveyed are sufficiently large to yield a good estimate of the complete local avifauna.

At each site we divided the avifauna into sixteen groups of ecologically similar taxa. This method of classification is obviously simplistic; boundaries between ecological groups seldom lie strictly along taxonomic lines. Nevertheless, the classification system makes possible a comparison of the representation of taxonomically restricted ecological groups in the absence of detailed data on each species. We tallied species and genera in the groups (app. 17.1) and calculated the proportion held in common with La Selva (table 17.4).

Among those compared the most similar avifauna to La Selva's is Rincón de Osa as might be expected by its proximity. Pacific slope species and replacement of some Atlantic slope species with congeneric endemics account for most of the difference with La Selva. To the south, the avifaunas of BCI, Los Katíos, and Río Palenque are progressively less similar to La Selva. Even at the far end of the transmontane avifauna, Río Palenque, more than half the species, and three-quarters of the genera, are represented at La Selva. At the northern end of the transmontane avifauna, Tikal, approximately the same proportions of La Selva taxa are found. In this area many north Central American endemics replace La Selva species and species that occur in the highlands of Costa Rica (e.g., furnariids) are common in the lowlands.

Species richness is much greater across the Andes at Limoncocha in the Amazon. Groups that show large increases relative to sites such as La Selva are most notably suboscine insectivores (antbirds, woodcreepers) and, to a lesser extent, nonpasserine frugivores (pigeons, parrots, trogons, barbets), suboscine frugivores (manakins, cotingas), and nine-primaried oscine frugivores (especially tanagers).

Note that the coefficients of similarity for both species and genera are higher for comparisons between Río Palenque and La Selva (0.56, 0.79, respectively) than between Río Palenque and Limoncocha (0.45, 0.78, respectively). This reaffirms that Central America, from southern Mexico to western Ecuador, constitutes an avifaunal unit, whose richness increases south from Mexico to a peak in the Chocó of Colombia, then declines towards Ecuador (Haffer 1985). La Selva, in southern Central America, shows higher diversity than sites to the north. To the south diversity probably changes little until central Panama (where it is lower for climatic reasons), then increases through eastern Panama to the Chocó. The increases in richness as one approaches the Chocó, or the Amazon, are the result of not only many new groups but also an increase in species within widespread Neotropical genera (e.g., *Myrmotherula, Automolus, Myiozetetes, Tangara*). Often, only one species in each of these genera reaches the northern extreme of the gradient, Tikal, which has the fewest species per genus (table 17.4). La Selva falls along the middle of the gra-

Table 17.4 Proportion of breeding species and genera at six Neotropical sites that are also found at La Selva, mean number of species/genus, and total number of species at each site

Breeding Species	Tikal Guatemala	La Selva Costa Rica	Osa Costa Rica	BCI Panama	Los Katíos Colombia	Río Palenque Ecuador	Limoncocha Ecuador
Species	0.536		0.718	0.693	0.597	0.564	0.354
Genera	0.740		0.889	0.857	0.798	0.791	0.707
Species/genera[a]	1.20	1.28	1.23	1.27	1.34	1.33	1.47
Total number species	199	256	240	217	296	255	383

Source: Data from appendix 17.1.

[a]Mean number.

dient but has more species per genus than any other similar Central American site for which data are available.

DIET CATEGORIES, FORAGING ECOLOGY, AND GUILDS

Stiles (1983a) classified La Selva's resident avifauna into seven broad diet categories. To avoid a heterogeneous "omnivore" category he placed some species partly in one category, partly in another. Representation in these diet categories follows in order of decreasing percentage of avifauna: (1) small arthropods 39%; (2) fruits and large seeds 27%; (3) large arthropods and small reptiles/amphibians 13%; (4) larger vertebrates 9%; (5) nectar 7%; (6) small seeds 4%; and (7) carrion 1%. The frugivore and nectarivore groups are more prominent at La Selva than at drier lowland sites, the carnivore group less so. With increasing elevation in humid life zones, frugivores and nectarivores increase proportionately, whereas consumers of large arthropods and small reptiles/amphibians decline drastically (Stiles 1983a; see also Orians 1969a; Terborgh 1977).

Within a given diet category, morphological features of species can be related to characteristics of the particular food types they exploit to determine how and to what extent morphology constrains foraging mode and food choice. At La Selva such studies have focused on frugivores (Moermond and Denslow 1983, 1985; Levey et al. 1984; Moermond et al. 1986; Levey 1987b), insectivores (Sherry and McDade 1982; Sherry 1984), and nectarivores (Wolf et al. 1972; Stiles 1975). For all these groups results indicate that size and bill dimensions constrain food choice, but other morphological features related to flying, perching, and reaching ability may also play major roles in foraging. Bill dimensions appear to be most important in hummingbirds and least important in those frugivores that thoroughly mandibulate fruits.

Species that feed in similar ways and habitats constitute guilds (Root 1967). Several guilds at La Selva have been examined for species interactions, patterns of resource partitioning, and community structure (sallying and hovering insectivores [Sherry 1984]; understory frugivores [Moermond and Denslow 1985]; hummingbirds [Stiles 1975, 1980a]). Studies of resource availability for frugivores and nectarivores have revealed important seasonal and supra-annual fluctuations in abundance (Frankie et al. 1974a; Stiles 1975, 1978c, 1980a, 1985a; Opler, Frankie, et al. 1980; Denslow et al. 1986; Levey 1988a; Blake and Loiselle 1991; Loiselle and Blake 1991). In hummingbirds interference competition (territoriality and dominance interactions) appears to be im-

portant in determining resource use at some flowers in some seasons. Interference, however, does not seem to characterize understory frugivores (Moermond et al. 1986). Different microhabitat use and seasonal movements between habitats are likely important mechanisms facilitating coexistence in many species (Levey 1988b; Stiles and Levey 1988).

In nearly all groups numerous foraging specialists offer interesting opportunities for comparative studies with other avifaunas. *Cacicus uropygialis* and *Myrmotherula fulviventris* spend much time investigating rolled leaves (especially palm fronds, in *Cacicus*) (Stiles pers. observation; see also Greenberg 1987). *Colonia colonus* specializes on *Trigona* bees, and *Terenotriccus erythrurus* feeds predominantly on fleeing homopterans (Sherry 1984). Some hummingbirds fulfill their protein requirements by hunting spiders, either taking them from webs (*Phaethornis* spp.) or directly from vegetation (jumping spiders, *Threnetes ruckeri*) (F. G. Stiles and H. Hespenheide unpublished data). *Accipiter superciliosus* apparently specializes on hummingbirds by traplining regular perching sites (Stiles 1978b). Even among small fruit-eating birds, which have been considered generalists (Leck 1971; Snow and Snow 1971; McKey 1975), some species have restricted diets. For example, *Euphonia gouldi* feeds largely on Anthuriums, and *Mionectes oleaginea* consumes mostly small arillate fruits (e.g., *Siparuna, Clusia,* and *Calathea;* Loiselle and Blake 1990, Levey unpublished data).

HABITAT USE

Patterns of habitat use play an important role in structuring tropical bird communities (Karr 1971; Schemske and Brokaw 1981; Karr and Freemark 1983; Stiles 1980b; Terborgh 1985a). Because habitats within the forest are not static over space or time, neither is the composition of the avifauna at a particular location. Thus, an understanding of avifaunal dynamics requires an examination of habitat requirements. What species are typical of each habitat, and what are their characteristic traits?

Here we take a structural approach, classifying four habitats by the position of the foliage-air interface (FAI). This interface is likely an important habitat parameter to most birds because it is where net primary productivity is highest and, hence, where insects, fruit, and nectar are most likely to be concentrated. *Canopy* occurs where the FAI is horizontal and high enough to allow a distinct zone underneath it, the *forest interior. Forest edge* occurs where the FAI is inclined, such as along streams or light gaps. *Nonforest* habitats have a horizontal FAI and no understory. We emphasize that the objective of the FAI classification is to place birds in their structural habitat, not to describe what each bird does or needs there (Stiles 1980b, 1983a; Terborgh 1985a). In particular, it does not exclude the possibility that certain additional features (e.g., small streams, vine tangles, presence of army ants) within a habitat type could be critical for a given species.

Frugivores and nectarivores are well represented among the canopy inhabitants (Loiselle 1987, 1988). Notably absent or poorly represented are wrens, antbirds, and manakins. Large flocks, often joined by numerous long-distance migrants, move noisily over large areas of the canopy. When they encounter a gap or watercourse, they frequently follow the FAI down near the ground. Indeed, canopy species can often be found in nonforest habitat. Few of these species, however, would be likely to survive without forest (Stiles 1985a). Characteristic species include *Ramphastos* spp., *Carpodectes nitidus*, *Ornithion brunneicapillum*, *Gymnostinops montezumae*, and *Pitylus grossus*.

Forest edge habitats, including tree-fall gaps, contain more species than any other habitat type (Stiles 1980b, 1983a). Many of these species also inhabit the canopy (e.g., *Caryothraustes poliogaster* or *Cacicus uropygialis*) and follow its contour. Nearer the ground, dense tangles of lianas or new secondary growth attract small insectivores, including *Thryothorus thoracicus* and *Myrmeciza exsul*. In gaps or along streams high densities of fruits and flowers often occur. Some understory species are especially characteristic of such habitats (Stiles 1975; Levey 1988b). They include *Arremon aurantiirostris*, *Mitrospingus cassinii*, *Manacus candei*, and *Phaethornis superciliosus*.

Forest interior species are predominantly insectivorous. Wrens, antbirds, woodcreepers, and some tanagers are common. Relatively few interior species are exclusively so (e.g., *Geotrygon veraguensis*, *Hylopezus perspicillatus*, *Platyrinchus coronatus*); most venture into gaps and some rise to the canopy. Very rarely, however, are forest interior species found in nonforest habitats. Given this dependence on forest, their typically sedentary nature, their low population densities, and their restricted diets, forest interior birds are probably the most ecologically inflexible species and are thought to be the most susceptible to extinction after deforestation (Willis 1974; Karr 1982; Sherry 1984; Stiles 1985a).

Nonforest habitats tend to have more granivores than other habitats (Karr 1971). The most characteristic attribute of nonforest species, however, is their colonizing ability (the ability to find and occupy disjunct patches of habitat). These birds usually move between patches of second growth via the canopy (e.g., *Myiozetetes granadensis*, *Ramphocelus passerinii*, *Turdus grayi*, *Saltator maximus*), but some (particularly *Sporophila aurita* and *Oryzoborus funereus*) are regular (albeit infrequent) in forest interior. Their colonizing ability suggests that they were once restricted to smaller, more ephemeral patches of second growth and required dispersal to find new areas of second growth as former habitat matured. Thus, their occasional occurrence inside the La Selva forest may not be the result of encroaching pasture lands but may instead reflect their historical pattern of low abundance and scattered distribution in the second growth of tree-fall gaps and waterways.

MIGRANTS

Approximately 25% of La Selva's land bird species are migrants. Most of these (approximately 18%) breed in the temperate zone (we term them *long-distance migrants*); the remainder are altitudinal migrants. In addition, some resident species (in particular, *Pipra mentalis* and *Mionectes oleagineus*) show dramatic seasonal changes in local abundance as nonresident individuals apparently enter or leave La Selva (Slud 1960; Levey 1988a; Loiselle and Blake 1991). On a smaller scale many resident individuals shift habitats in response to changing resource abundance. This especially characterizes frugivores and nectarivores (Stiles 1975, 1980a; Levey 1988a; Blake and Loiselle 1991). Together, these movements create a complex and dynamic avifauna.

We briefly discuss two most obvious types of migrants at La Selva, long-distance and altitudinal. We emphasize that the distinction between them is not as pronounced as typically assumed. Their many ecological parallels and taxonomic ties demonstrate that they represent the same phenomenon on different scales. This observation leads to the hypothesis that the evolution of migration to and from the Neotropics originated from seasonal movements of birds within the Neotropics; taxa of altitudinal and intratropical migrants were most prone to develop long-distance migration (Levey and Stiles 1992). This helps explain why most of the principal groups of North American migrants are drawn from frugivorous and nectarivorous taxa and why such prominent Neotropical groups as antbirds, ovenbirds, and woodcreepers are not found in North America (Levey and Stiles 1992). It also means that the current emphasis on the conservation of long-distance migrants is lacking a valuable perspective. The ecology of these birds cannot be fully understood outside the context of the tropical communities from which they were drawn and still form an important part. Understanding migration at all scales is necessary to understand migration at the largest scale.

Long-distance Migrants

Northern-breeding migrants are completely absent from La Selva for only two months, June and July (fig. 17.1). They reach their highest density in September and October on their way to wintering grounds farther south. Characteristic of this period are impressive numbers of *Catharus ustulatus*, *Cathartes aura*, *Contopus virens*, and *Buteo swainsonii* (Loiselle and Blake 1991). Migrants present from November to February are mostly winter residents. These species typically shun the forest interior, appearing more often in nonforest habitats than do permanent residents (Stiles 1980b; Blake and Loiselle 1991). In the forest migrants usually appear in the canopy with a flock (Loiselle 1987). Notable exceptions are *Empidonax flaviventris*, *E. virescens*, *Hylocichla mustelina*, and *Oporornis formosus*, which commonly occur alone, probably on individual territories, in the understory. Spring migration, March and April, produces a second but smaller peak in number of migrant species (fig. 17.1; see also Blake and Loiselle 1991). Many species seen in the fall apparently return to North America via the coast. Species that return via La Selva in the spring are much less conspicuous and leave more quickly than in the fall.

Of La Selva's oscine species, 21% are migrants that spend

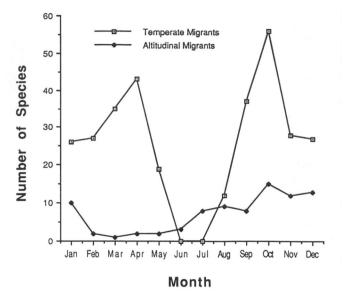

Fig. 17.1. Temporal distribution of temperate and altitudinal migrants at La Selva. (Data from Slud [1960]; Stiles [1980a]; Levey [1988c]; and unpublished data.)

more than five months at La Selva (table 17.2). These species are not simply "visitors" but form an integral part of the La Selva avifauna as is true with migrants at many other Neotropical sites (Keast and Morton 1980; Rappole et al. 1983; Greenberg 1984; Karr et al. 1991). Their interaction with permanent residents and role in community structure remain largely unexplored.

Several species of long-distance migrants at La Selva do not breed in the temperate zone. These species (e.g., *Myiodynastes luteiventris*) breed elsewhere in Costa Rica or northern Central America, "winter" in South America and, like temperate migrants, pass through La Selva during migration (Morton 1977). Some individuals may stay several months or even breed (e.g., *Vireo flavoviridis, Legatus leucophaius*) at or near La Selva.

Altitudinal Migrants

Most altitudinal migrants appear at La Selva late in the rainy season (July–December; fig. 17.1; also see Loiselle and Blake 1991). By midJanuary, the start of the dry season, they begin to return to higher elevations where they breed from approximately March to May (Skutch 1950, 1967b; Stiles 1985b; Rosselli 1989; Loiselle and Blake 1991). Migrant hummingbirds display a slightly different pattern, probably nesting at higher elevations in August–September to January–February and moving downslope in April–August (Stiles 1980a, 1983a, 1985a). One species, *Thalurania colombica*, breeds at La Selva and migrates upslope during its nonbreeding season. In general, altitudinal migrants are more forest dependent than are long-distance migrants (Stiles 1985a; Blake and Loiselle 1991).

Presumably, altitudinal migrations occur as birds track resources that change asynchronously in abundance at different altitudes (Feinsinger 1980; Stiles 1985c). If so, frugivores and nectarivores might be especially prone to altitudinal movement. Fruits and flowers tend to occur more sporadically and

in discrete patches for shorter times than insects. Indeed, among land bird species that are not permanent La Selva residents but are found in the adjacent highlands, significantly more frugivorous and nectarivorous species migrate into La Selva than species from three other dietary groups (fig. 17.2; $\chi^2 = 18.5$, d.f. $= 4$, $p < 0.01$; data from Stiles 1985a). Their appearance at La Selva coincides respectively with peaks of fruit and flower availability (Stiles 1978c, 1980a; Levey 1988a; Loiselle and Blake 1991; Blake and Loiselle 1991). This is also a period of low flower abundance at 500 and 1,000 m (Stiles 1985c, unpublished data) and low fruit abundance at 500 m (Loiselle and Blake 1991; Rosselli 1989). These patterns suggest that frugivores and nectarivores track fruit and flower abundance along an elevational gradient. Many more years of data at a number of sites are needed to confirm such tracking.

An alternative explanation for altitudinal migration is that highland birds are driven to lower elevations because of adverse seasonal conditions (e.g., high winds, heavy rains, and/or cold temperatures) that increase energy demands while decreasing foraging time (Wetmore 1926; Ramos 1983). Indeed, many tropical understory birds appear sensitive to even microclimatic changes (Karr and Freemark 1983). Even so, we doubt that adverse weather conditions directly cause altitudinal migration to and from La Selva. Such weather would most likely affect insectivores (which must forage longer for prey) more than frugivores or nectarivores (which are less time-limited in their foraging). Yet the frugivores and nectarivores are the most frequent migrants. In fact, some nectarivores even migrate *uphill* in the wet season. Seasonal fluctuations in abundance of birds that consume different resources are not correlated, suggesting that these movements are more related to asynchronous cycles of resource abundance than to seasonal abiotic factors (Blake and Loiselle 1991).

BREEDING AND MOLTING CYCLES

For more than 90% of the breeding birds of La Selva it is possible to estimate the nesting season fairly precisely (table

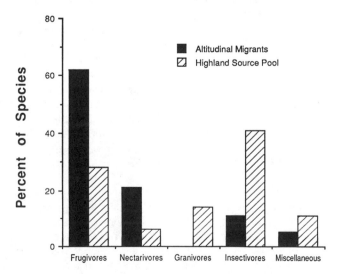

Fig. 17.2. Guild composition of altitudinal migrants at La Selva and their source pool, Caribbean slope highland species that are not permanent residents at La Selva. (Data from Slud [1964] and Stiles [1983a].)

Table 17.5 Breeding seasons of the La Selva avifauna

Avifauna	J	F	M	A	M	J	J	A	S	O	N	D	Total Species
					Number Species Known or Suspected to Breed Each Month								
Tinamous-galliformes	2	3	5	6	6	6	5	3	2	2	1	2	7
Hawks-falcon-vultures	8	17	16	16	13	6	2	1	0	0	3	3	19
Waterbirds	3	2	3	5	10	10	12	12	12	8	5	3	13
Pigeons	2	6	8	8	8	8	7	6	3	1	0	0	8
Parrots	2	5	8	8	6	5	1	0	0	0	0	0	8
Cuckoos	1	2	2	2	2	3	3	3	1	1	1	0	3
Owls	1	4	4	4	4	1	0	0	0	0	0	0	4
Swifts and nightjars	1	2	4	5	6	6	2	1	0	0	0	0	6
Hummingbirds	10	13	13	11	10	5	1	1	0	2	3	5	14
Trogons, toucans	3	5	7	7	7	4	0	0	0	0	0	0	7
Kingfishers	2	4	4	2	1	1	0	0	0	0	0	1	4
Motmots, jacamars, puffbirds	0	1	7	7	8	3	0	0	0	0	0	0	8
Woodpeckers	1	4	6	6	6	1	0	1	1	1	1	1	7
Woodcreepers, ovenbirds, antbirds	3	6	15	22	25	24	18	11	6	1	0	1	27
Cotingas, manakins, tityras	0	1	6	8	8	8	6	2	0	0	0	0	9
Flycatchers	1	6	23	29	29	27	10	2	0	0	1	1	30
Swallows, wrens, sylviids, warblers, vireos	1	4	12	18	21	16	9	5	1	0	0	1	22
Jays, icterids	3	5	7	10	10	10	5	3	0	0	0	0	10
Tanagers, finches	1	5	13	28	31	28	19	11	7	3	0	0	31
TOTAL	45	95	163	202	211	172	100	62	33	19	15	18	237

Sources: Data from Skutch (1954, 1960a, 1969); Stiles et al. (1989); Stiles and Levey (unpublished data).

Note: Breeding season was determined from four types of data: observation of active nests, observation of breeding behavior (e.g., adults carrying nesting material), presence of cloacal protuberances or brood patches on mist-netted birds, and molt (which usually follows closely upon breeding) in netted birds. Data from all years were combined. This method probably overestimates the length of the breeding season for any one year because "early" and "late" years are doubtless included in the sample.

17.5). This estimate is derived not only from direct observations of nesting but also from indirect evidence such as adults carrying nesting material, food, or fecal pellets, presence of juveniles, and presence of active brood patches on mist-netted birds. For the avifauna as a whole, a broad peak of nesting occurs from March through June, and a pronounced low point characterizes the last three months of the year (table 17.5). A similar breeding cycle has been recorded for birds throughout Central America (Skutch 1950). This general pattern, however, obscures much interesting variation, which evidently relates to habitat, diet, nesting habits, or taxonomic affinities as the following examples will demonstrate.

Perhaps the most obvious example of a habitat-related constraint is found in the waterbirds, which as a group show peak breeding activity in the middle of the wet season (May–September). Some second-growth species have much longer breeding seasons than do their forest-dwelling relatives. Examples include *Columbina talpacoti, Amazilia tzacatl, Synallaxis brachyura, Troglodytes aedon,* and *Tangara larvata.* Examples of diet-related patterns include the early breeding seasons of hummingbirds (peak January–May) and raptors (peak February–May) and the late nesting of most granivores (peak May–August). These periods of peak reproduction are, presumably, tied to resource flushes. The influence of nesting habits is seen in the earlier nesting of almost all hole-nesting birds (peak March–May) as opposed to open-nesting species (peak April–June). We do not know the basis of this difference.

Table 17.6 Length of breeding seasons among the La Selva avifauna

	2[a]	3[a]	4	5	6	7	8	9	10+	Total
	Number of Months Breeding Recorded									
Water and large land birds	0	3	16	8	7	2	3	1	2	42
Higher nonpasserines	0	10	25	14	8	3	3	3	1	67
Suboscine passerines	1	13	21	18	6	3	1	2	1	66
Oscine passerines	3	13	14	14	12	3	0	1	1	61
TOTAL	4	39	76	54	33	11	7	7	5	237

Sources: Data from Skutch (1954, 1960, 1969); Stiles et al. (1989); Stiles and Levey (unpublished data).

Note: For methods see table 17.5. Families grouped as in table 17.1.

[a]These are probably underestimates in many cases because of lack of data.

On a population level, the breeding season of most La Selva birds lasts three to six months, with a mode at four months (table 17.6). Longer breeding seasons characterize several groups, however, notably tinamous (eight to nine months), pigeons (most six months or more), and several small waterbirds (e.g., *Jacana*), which breed most of the year when permanent water is available but breed much more seasonally in areas such as Guanacaste. Occasionally, individuals breed outside the species' normal breeding season (e.g., *Ramphocelus* in November; fig. 17.3).

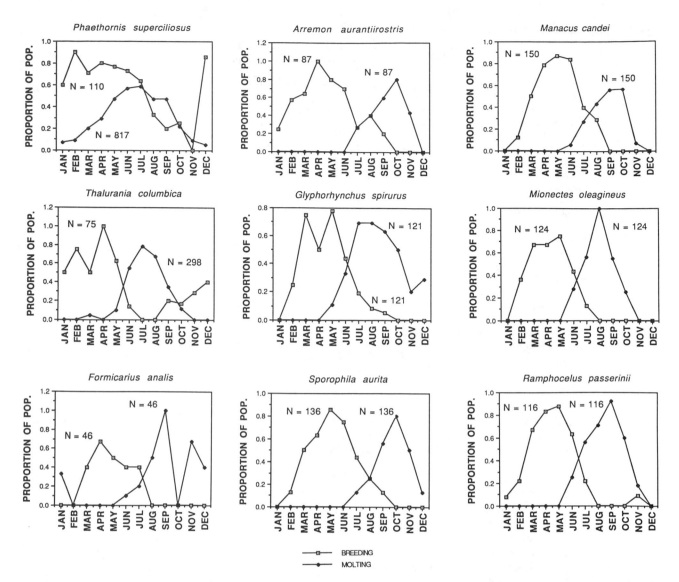

Fig. 17.3. Breeding and molting seasons: some representative patterns for species for which we have large sample sizes of adults (mist net data and collected specimens). (Breeding proportions of hummingbirds are all from specimens. This may lead to overestimation of breeding season because males have enlarged testes when females do not have enlarged ovaries. Enlarged testes may be more related to territorial defense than breeding.)

In contrast, it is very unusual to find adults molting out of season; molting season is generally more sharply defined than breeding season (fig. 17.3). The apparent flexibility of the nesting season and inflexibility of the molting season suggest that breeding and molt may be controlled differently at the physiological level. Molt may obey relatively regular cues such as photoperiod, whereas breeding may be tied to rainfall patterns and resource levels, which vary more from year to year (see Stiles 1980a). In most species, molt follows breeding with little or no overlap. Overlap at the population level is frequent because of the asynchrony of breeding between individuals, which in part reflects high rates of nest predation and repeated nesting attempts (fig. 17.3). Breeding and active molt at the individual level, however, is much less frequent and occurs primarily in species with protracted molt or long breeding seasons (e.g., pigeons, tinamous, cuckoos). Protracted molt reduces daily energy expenditures and a long breeding season increases breeding asynchrony, resulting in a three- to

four- month period of molt-breeding overlap at the population level (e.g., *Glyphorhynchus spirurus,* fig. 17.3). In a few species overlap between molting and breeding may be substantial. Individuals of *Phaethornis superciliosus,* for example, molt very slowly at the same time each year, evidently determined by their calendar date of hatching (Stiles and Wolf 1974). Their slow molt spaces out energetic demands, thereby permitting broad overlap of molting and breeding (Foster 1975). Still, females of this species often suspend molt during a breeding attempt. In general, the relationship between molting and breeding remains very poorly understood in tropical birds (Fogden 1972; Foster 1975). Studies using marked individuals followed through several seasons are sorely needed.

Breeding and molting cycles of La Selva birds have been related to the food resources most explicitly for hummingbirds (Stiles 1975, 1978c, 1979a, 1980a). The two major peaks of flower availability in the year correspond closely to hummingbird breeding and molting periods. In frugivores the pattern is less clear. Fruits are most abundant during the mid-to-late rainy season, August–December (Frankie et al. 1974a, Opler, Frankie, et al. 1980; Denslow et al. 1986; Levey 1988a; Blake and Loiselle 1991; Loiselle and Blake 1991). Frugivore numbers peak during this period principally because of an influx of frugivorous migrants. Breeding of the most common frugivores (*Manacus candei, Pipra mentalis, Mionectes oleagineus,* and *Ramphocelus passerinii*), however, occurs during a relative fruit low, March–June (fig. 17.3; Levey 1988a, Loiselle and Blake 1991). In contrast, breeding of the two most common frugivores at a Panamanian site was closely tied to the period of peak fruit abundance (Worthington 1982). Breeding of frugivores at La Selva may be timed so that young are fledged when fruit abundance is increasing (nestlings are fed few fruits; Morton 1973). Alternatively, breeding may coincide with peak insect rather than fruit abundance (Levey 1988a) because protein is a critical resource for egg production and nesting (Romanoff and Romanoff 1949; Fogden 1972). Unfortunately, seasonality of insects has received little study at La Selva. In the only quantitative study, S. Lieberman (1986) found a peak of leaf-litter arthropod abundance at the end of the dry season/beginning of wet season. Otherwise, we can only cite casual observation and conventional wisdom as suggesting a peak of insect availability in the Neotropics at the start of the rainy season (Wolda 1978a; Janzen 1983e, Stiles 1983a). The generality and magnitude of this peak is unknown; it is likely more pronounced in drier, more seasonal sites than La Selva. Nevertheless, we note that most insectivorous birds at La Selva breed during this time: 61% of insectivores that were captured with brood patches were netted in April–June (Levey unpublished data; see also fig. 17.3). Whether their breeding activity is ultimately related to insect abundance or to other factors (e.g., relative scarcity of altitudinal and long-distance migrants) awaits further study.

FLOCKS

Single- and multispecies flocks are a conspicuous feature of most Neotropical avifaunas, particularly in humid areas (Buskirk et al. 1972; Buskirk 1976; Munn and Terborgh 1979; Stiles 1983a; Munn 1985; Powell 1985). Single-species flocks probably represent family groups. In mixed flocks, species can be classified into many groups based on their social role in flock integration and behavior outside flocks (e.g., Moynihan 1962; Munn and Terborgh 1979; Powell 1979). Because almost nothing is known about multispecies flocks at La Selva, we distinguish only between *nuclear* and *attendant* species. Nuclear species are intraspecifically gregarious, usually noisy, and often dull-colored (Moynihan 1962; Buskirk 1976). Their frequent calls seem to rally the flocks when they form early in the morning and continue to attract individuals to the flock throughout the day (Munn and Terborgh 1979; Powell 1979). Nuclear species also appear to direct the flock's movement (Moynihan 1962; Munn and Terborgh 1979; Gradwohl and Greenberg 1980). Although only one nuclear species is necessary to form a flock, two to four species frequently form the core of activity around which the flock centers.

Attendant species, which join and travel with nuclear species, may be regularly or sporadically associated with flocks. Attendants are typically active arboreal feeders, such as frugi-

vores and insectivores, and they usually occur as territorial individuals or pairs (they are inter- but not intraspecifically gregarious). In contrast, sentinel (sit-and-wait foragers) and terrestrial species (with the exception of some Formicariidae) are overwhelmingly nonsocial and seldom join flocks (Buskirk 1976). Many attendant species may join a flock only briefly (perhaps while it is in their territory), whereas others may stay with the flock for most of the day.

In Peru (Munn and Terborgh 1979) and Panama (Gradwohl and Greenberg 1980) multispecies flocks contain one pair each of many species whose territories largely overlap (or coincide). Membership in the flock is restricted to one such pair per species and defense of the jointly held flock territory is communal. Multispecific territoriality has not been documented at La Selva but based on our observations of *Myrmotherula/Dysithamnus* flocks that encounter each other, it is likely to occur (Levey and Stiles pers. observation, unpublished data).

A continuum exists between single-species and mixed flocks. Single-species flocks may be almost never (e.g., *Ramphastos, Gymnostinops*), occasionally (e.g., *Mitrospingus, Habia*), or regularly (e.g., *Cacicus*) joined and followed by others. Frugivorous species predominate among single-species flocks both in the understory (e.g., *Mitrospingus*) and canopy (e.g., *Querula*). Probably all passerines at La Selva (except swallows and sentinel foragers) regularly or occasionally join flocks. Appendix 17.2 is a partial list of nuclear and regular attendant species in La Selva flocks. This list is based on our impressions; data are sorely lacking. Although the flock types we suggest are "typical" of La Selva, we stress that flock composition is not as fixed and regular as such a classification scheme implies.

RECENT CHANGES AND FUTURE TRENDS

When Slud first inventoried its birds, La Selva was surrounded by primary forest (see plate 9 in Slud 1960). By the early 1970s the northern boundary and parts of the eastern and western boundaries abutted pasture. Today, La Selva is at the tip of a forested peninsula in a sea of pasture. Stiles (1985a) predicted that if La Selva were to become an isolated island, it would lose thirty-six to thirty-seven species within twenty-five years. Fortunately, the extension of Parque Nacional Braulio Carrillo to La Selva's south boundary ensures that it will not become an island of forest. Nonetheless, obvious changes have occurred in the populations of some species as surrounding forest has been cut. Detailed observations of the avifauna over thirty years allow an analysis of changes in species composition.

Approximately 18% of La Selva's species have been recorded fewer than five times in thirty years. Extinction is difficult to document among such rare species, but we suspect at least five may now be locally extinct (*Harpia harpyja, Ara macao, Daptrius americanus, Chloroceryle inda,* and *Icterus mesomelas*). These species are very diverse in diet and habitat requirements. Perhaps more meaningful is an analysis of more common species that have shown marked changes in abundance since the 1960s (app. 7); species either much more or less abundant than Slud (1960) reported.

Interior forest birds should be most sensitive to encroaching deforestation because of their sedentary habits and small population sizes (Willis 1974; Karr 1982; Stiles 1985a). Forest edge and canopy species are quite mobile and probably more tolerant of adjacent forest clearings than forest interior species. Nonforest species, of course, should increase as forest is cut.

These patterns are only partially confirmed by our data (app. 7; table 17.7). For both increasing and decreasing species, the distribution among habitat types is significantly different from the distribution displayed by the entire community (χ^2 = 44.7 and 8.6, respectively; d.f. = 3, $p < 0.05$). As expected, increasing species are predominantly nonforest species and, as a corollary, no nonforest species are decreasing in abundance. Decreasing species include large proportions of forest interior, canopy, and edge species, but interior and canopy species are declining especially quickly (compare their percent representation as decreasing species and in the entire avifauna, table 17.7). Surprisingly, forest canopy species are decreasing faster than forest interior species. Most of the canopy species, and some of the interior species, are large (e.g., various raptors, macaws). Canopy frugivores, nectarivores, and granivores (e.g., *Ara*, *Selenidera*) that are intolerant of nonforested areas may well require larger areas than previously suspected because they travel long distances in search of patchily distributed food. Indeed, the tropical forest canopy is known for its "boom or bust" resources (Frankie et al. 1974a; Greenberg 1981; Loiselle 1988), and the species dependent upon these resources must, therefore, be under strong selection for mobility so that they can move between asynchronous resource patches. Perhaps remaining forest like La Selva are too small to always contain enough flowering or fruiting trees and too widely scattered to make movement between them feasible.

The remaining species that have decreased in abundance once occurred in La Selva's northern "tip" (Slud 1960; Stiles unpublished data) but are now found almost exclusively near the southern boundary (e.g., *Jacamerops, Selenidera, Thamnistes, Myiobius, Lanio, Chlorothraupis, Phaenostictus*). This shift suggests that they have been sensitive to the "peninsula effect," a decrease in species richness from the base to the tip of a peninsula.

As upland forest around the Braulio Carrillo extension is

cleared, future changes in the La Selva avifauna are inevitable. Although the deforestation is unfortunate, it will, nonetheless, provide valuable information for conservation biologists. Indeed, as demonstrated, we still cannot predict which species are most likely to change in abundance. It is, therefore, critical to keep quantitative records of the La Selva avifauna in the ensuing years.

FUTURE RESEARCH

As we have stressed throughout this chapter, relatively little is known about La Selva birds. Future research topics are unlimited, and almost any study is certain to uncover new and important information about the avifauna. For those interested in frugivores, nectarivores, and insectivores (especially flycatchers), a good data base exists from which to identify new questions and frame hypotheses. Although much remains to be done even with these relatively intensively studied guilds, some guilds very well represented at La Selva have received little or no study. Examples include understory gleaning insectivores (antwrens, wrens), trunk and branch foragers (woodpeckers, woodcreepers), small insectivores (greenlets, gnatcatchers), ant-followers (antbirds, woodcreepers), and bird-hawks (forest falcons, accipiters).

Autecological studies are currently rather unfashionable. This is unfortunate because the more general, comparative studies often rely on autecological data. La Selva's avifauna offers any number of common, observable species whose behavior is almost unstudied. In fact, the only La Selva species whose foraging, courtship, mating, and other behaviors have been studied in detail is *Phaethornis superciliosus* (Stiles and Wolf 1979).

On a community level we know almost nothing about the abundance and distribution of birds except the most common frugivores and nectarivores. Censusing and territory mapping of all species would yield valuable information on community structure and organization that could then be used to predict which species are most vulnerable to local extinction. An excellent example of such a study is Terborgh et al. (1990).

Our understanding of La Selva (and Neotropical!) birds is primitive as one can see from the topics in need of research: the relationship between breeding and molt phenology, nest predation, habitat selection, role of migrants, bioenergetics, ecophysiology, demography and long-term population dynamics, foraging ecology, mating systems and intraspecific social organization (and how these relate to foraging ecology), ecomorphology, song, plant-animal interactions, and effects of spatial and temporal variation in resource abundance. This list is far from exhaustive and is not meant to suggest priorities for future work. Instead, it serves to emphasize that even in some of the most extensively studied taxa and in one of the most intensely studied Neotropical sites, our level of understanding is still primitive.

Table 17.7 Characteristic habitats of species showing marked changes in abundance

Status	Forest Interior	Forest Canopy	Forest Edge	Nonforest
Increasing species	0.5 (3%)	1 (6%)	1.5 (9%)	14 (82%)
Decreasing or extinct species	9.5 (28%)	11.5 (34%)	11 (32%)	1 (3%)
All species	77 (19%)	87 (22%)	157 (39%)	77 (19%)

Source: Data from appendix 7.

Note: Habitat classification by Stiles (1983a). Species classified in two habitat types were counted as one-half in each. *Cochlearius cochlearius* and *Heliornis fulica* were not included in the analysis because their habitats did not match the four habitats considered.

Appendix 17.1 Breeding avifaunas of Neotropical wet lowland sites (number of species/number of genera)

Category	Tikal Guatemala Residents	Tikal Guatemala Common with LS	La Selva Costa Rica Residents	Rincón de Osa Costa Rica Residents	Rincón de Osa Costa Rica Common with LS	Barro Colorado Island Panama Residents	Barro Colorado Island Panama Common with LS	Los Katíos, Chocó Colombia Residents	Los Katíos, Chocó Colombia Common with LS	Río Palenque Ecuador Residents	Río Palenque Ecuador Common with LS	Limoncocha Ecuador Residents	Limoncocha Ecuador Common with LS
Large frugivores granivores	9/8	5/6	8/7	6/6	5/6	6/6	5/6	6/6	4/6	6/5	4/5	10/7	2/5
Waterbirds	13/10	11/10	19/16	18/16	17/16	17/14	14/13	34/26	21/19	11/9	8/8	33/27	15/16
Bird of prey	25/20	22/19	28/22	35/29	27/25	24/19	22/18	31/26	19/17	24/19	21/19	36/27	21/22
Nonpasserine frugivores	20/14	14/13	23/14	22/14	17/14	20/12	15/11	28/14	15/13	26/17	7/12	36/19	6/13
Nonpasserine aerial insectivores	5/5	2/4	6/6	7/6	5/5	6/5	5/5	7/7	4/5	6/5	4/5	10/8	6/5
Nonpasserine nectarivores	10/8	5/4	14/12	17/14	11/11	12/10	10/10	17/13	11/10	15/12	11/9	16/13	6/9
Nonpasserine insectivores nonwoodpecker	8/8	5/5	12/11	7/7	5/6	12/9	7/7	19/14	10/9	11/8	8/7	24/17	6/10
Nonpasserine insectivores: woodpeckers	7/7	5/5	7/6	7/6	3/5	5/4	3/4	8/7	3/5	8/7	3/6	11/8	1/6
Suboscine insectivores	15/14	8/12	29/24	27/24	20/22	24/19	20/19	34/25	19/19	30/25	20/20	64/40	12/18
Suboscine frugivores	9/6	7/6	12/10	12/9	7/7	12/9	9/8	11/8	7/7	12/9	7/8	19/15	3/6
Suboscine sallying insectivores	26/23	16/19	31/26	30/25	22/21	31/26	20/21	33/26	16/18	40/31	19/20	42/30	12/17
Oscine aerial insectivores	2/2	1/2	3/3	3/3	3/3	2/2	2/2	4/4	2/3	5/4	2/2	5/5	2/3
Oscine insectivores/ frugivores	10/7	6/5	6/5	6/4	5/4	6/5	4/5	6/5	3/3	4/3	0/2	5/3	1/3
Oscine gleaning insectivores	6/5	3/4	12/9	7/5	4/5	9/7	6/7	12/8	6/8	11/8	8/8	10/9	4/8
Nine-primaried Oscines: Insectivores/ omnivores	9/8	6/6	14/11	10/10	9/9	7/7	5/7	16/11	6/8	10/9	7/8	17/12	2/6
Nine-primaried Oscines: Frugivores/ granivores	25/20	12/15	32/18	26/18	18/17	27/17	7/16	30/21	19/18	36/21	15/16	45/24	13/18
Totals:													
All Species	199/165	122/135	256/200	240/196	178/176	217/171	164/169	296/221	165/168	255/192	144/155	383/264	113/164
Land Species Only	186/155	111/125	237/184	222/180	161/160	200/157	142/141	262/195	144/149	244/183	136/147	350/237	97/149

Sources: Data from Smithe (1966); Stiles (1983a, unpublished data); Rodriguez (1982); Leck et al. (1980); Pearson (1972); Pearson et al. (1977); and Willis and Eisenmann (1979).

A. Forest Understory
    1. Ground Flocks
        Nucleus: *Cyphorhinus*
        Attendants: *Henicorhina, Arremon, Microbates, Automolus*
    2. Antwren-Antvireo
        Nucleus: *Myrmotherula, Microrhopias, Dysithamnus, Hylophilus ochraceiceps*
        Attendants: *Microbates, Cyphorhinus, Microcerculus, Automolus, Xenops, Rhynchocyclus, Hyloctistes, Myiobius, Glyphorhynchus,* other woodcreepers
B. Understory to Middle Level
    3. Tanager Flocks
        Nucleus: *Tachyphonus delattrii, Chlorothraupis carmioli* (together or separately)
        Attendants: *Tachyphonus luctuosus, Rhytipterna, Lanio, Euphonia gouldi, Arremon, Cyanocompsa, Tolmomyias assimilis, Xenops,* various woodcreepers and flycatchers, sometimes *Monasa*
C. Upper Middle Level and Canopy
    4. Cacique Flocks
        Nucleus: *Cacicus uropygialis*
        Attendants: *Rhytipterna, Lipaugus, Trogon massena, Harpagus, Monasa,* numerous woodcreepers, flycatchers, sometimes *Electron*
    5. Greenlet-Honeycreeper Flocks
        Nucleus: *Hylophilus decurtatus, Cyanerpes lucidus*
        Attendant: *Polioptila, Euphonia gouldi, Chlorophanes, Dacnis, Vireolanius,* various flycatchers (esp. *Ornithion, Tolmomyias assimilis*), warblers, tanagers and woodcreepers

6. Grosbeak Flocks
    Nucleus: *Caryothraustes poliogaster*
    Attendants: *Vireolanius, Chlorophanes, Tityra,* various tanagers and flycatchers, sometimes trogons, warblers and woodpeckers
D. Second-growth and Edge
    7. Tanager-Saltator Flocks
        Nucleus: *Ramphocelus passerinii*
        Attendants: *Saltator* spp., *Ramphocelus sanguinolentus, Tangara larvata,* other tanagers and some flycatchers, warblers
    8. Flycatcher Flocks
        Nucleus: *Myiozetetes granadensis,* occasionally *Coryphotriccus*
        Attendants: *M. similis, Megarhynchus,* other flycatchers, tanagers and sometimes warblers and vireos
    9. Seed-eater Flocks
        Nucleus: *Sporophila aurita, Volatina jacarina, Tiaris olivacea*
        Attendants: *Oryzoborus* spp., *Passerina cyanea,* occasionally ground-doves, tanagers, etc.
E. Aerial Flocks
    10. Swift Flocks
         Nucleus: *Chaetura cinereiventris,* sometimes *Streptoprocne zonaris*
         Attendants: *Panyptila cayennensis, Cypseloides* spp., occasionally *Progne*
F. Ant-following Flocks[a]
    11. Antbird Flocks
         Nucleus: *Gymnopithys, Phaenostictus, Hylophylax*
         Attendants: *Dendrocolaptes certhia, Dendrocincla fuliginosa, Micrastur ruficollis, Attila, Neomorphus* (rare), *Gymnocichla* and *Habia* in second-growth

[a]A special case, as the ants are the attracting force, not the birds themselves (Willis and Oniki 1978).

18

The Mammal Fauna

Robert M. Timm

La Selva has been a field station of the Organization for Tropical Studies (OTS) more than a quarter-century. Habitat and wildlife have been protected or managed since the mid-1950s when Leslie R. Holdridge created Finca La Selva. In this chapter I explore the nature and diversity of the mammalian fauna of La Selva by integrating what is known about this fauna from the published literature in the context of my own impressions and experience. I present a complete list of the mammals that are known to occur at La Selva with estimates of their abundance in appendix 8 and discuss here the distributional patterns of these species. I also discuss research published on mammals at La Selva and briefly review some of the more significant research findings. I compare this fauna briefly with other tropical faunas, discuss the impact of human activities on mammalian distributions and abundance in this region, and note some of the major gaps in knowledge as suggestions for future research.

THE MAMMALIAN FAUNA OF COSTA RICA AND LA SELVA

Costa Rica is one of the few countries in the Western Hemisphere in which the entire mammalian fauna that was present at the time of European settlement is still largely extant. Good populations of most species can be found within the country; populations of a few species, however, have been reduced to sizes that can no longer be considered healthy and viable. Of the approximately 116 species of mammals originally found at La Selva (app. 8), perhaps only two species, the giant anteater and the white-lipped peccary, have been locally extirpated. Both species may have disappeared from La Selva since the 1970s (Timm et al. 1989). Although occasionally present, jaguars are now only rarely observed at La Selva and do not have a viable breeding population there. Thus, La Selva retains more than 98% of its component species.

The mammalian fauna of La Selva historically included 5 marsupials, 65+ bats, 3 (possibly 4) primates, 7 edentates, 1 rabbit, 3 squirrels, at least 9 long-tailed rats and mice (families Heteromyidae, Muridae, and Echimyidae), 1 pocket gopher, 1 porcupine, 1 paca, 1 agouti, 5 mustelids, 4 procyonids, 5 cats, 2 peccaries, 2 deer, and 1 tapir (app. 8). These 116 species represent 57% of the species of mammals found in the country (app. 8). This list cannot be considered a complete enumeration of all mammal species that have occurred or currently do occur at La Selva, however. Two species of bats, *Pteronotus*

davyi and *Chrotopterus auritus,* recently were added to the faunal list and, undoubtedly, additional species of bats and rodents will be found there (Timm et al. 1989). With the exception of these orders it is likely that the list of La Selva's mammals is complete.

This list of La Selva mammals differs from that of Wilson (1983) and unpublished lists that have been circulated over the years in that some species have been added and several species now suspected not to occur there have been deleted. Identifications of specimens in collections have been verified or corrected, and all published literature records have been evaluated. No species are included as "expected to occur in the area," a category that created an artificially high species count. I base estimates of relative abundance of species on my observations and those of other experienced individuals and on the logbook of mammal sightings kept at La Selva since 1979, which I critically evaluated. The 116 documented species are 21 fewer than Wilson's (1983) listing of 137 species known or expected to occur at La Selva; it is unlikely that these "expected" additional species occur in the Puerto Viejo region.

The majority of mammals found at La Selva are widespread species, typical of neo-tropical rain forests (app. 8). Most have a broad distribution throughout the tropical lowlands of Central America. Many species are even more widely distributed, including some with broad elevational ranges in Costa Rica (and elsewhere in Central America) and/or wide geographic distributions throughout the lowlands of Central and South America. Of the 116 species of mammals that occurred at La Selva, 58 (49%) are broadly distributed in the Neotropics, often occurring as far north as Mexico and ranging south through Central America and much of tropical South America; 35 (30%) occur in the northern tropics, generally being found from tropical Mexico through Central America and northern South America; 15 (13%) are restricted to Central America; 7 (6%) are very wide ranging, found from the United States through Central America and across most of South America; and 1 (1%) is widely distributed in North America, occurring as far south as central Panama (app. 8). Interestingly, the bat fauna of La Selva consists overwhelmingly of species that are either widely distributed in the tropics or are of northern Neotropical distribution (60 of 65 species, or 92%). La Selva's primates are all of northern Neotropical distribution. The edentates are all of either wide distribution in the tropics or of northern Neotropical distribution. The single species of rabbit found at La Selva has a wide-ranging distribution in the lowland tropics of Central and South America. The rodent fauna varies; some members are widespread in the tropics, whereas others are of northern Neotropical or Mesoamerican distribution. All of the smaller rodents, including the squirrels, pocket gopher, long-tailed rats and mice, and the porcupine, are either of northern tropical or Mesoamerican distribution, whereas agoutis and pacas are widely distributed in the tropics. The carnivore fauna is represented by many species whose ranges include much of the tropics but also includes species with more extensive nontropical distributions (7 of 15 species, or 47%).

In general, species that are common at La Selva are common throughout their ranges, and species that are rare at La Selva also are rare throughout their ranges. Notable exceptions in the bat fauna are Caribbean white bats, which are more abundant at La Selva than at any other site from which they are known and disk-winged bats, which are abundant at Tortuguero and elsewhere in the Caribbean lowlands but are rare at La Selva. Although this rarity is unexplained, it is not because of the absence of suitable roosting sites in the Puerto Viejo region. These tiny insectivorous bats roost only in the immature, rolled leaves of *Heliconia,* which are abundant at La Selva, and in which the bats are easily located if present.

Relatively few species of mammals are endemic to Costa Rica: six rodents, one carnivore, and, perhaps, two or three shrews. Most of these are species of mid- to high elevations and occur in restricted habitats. Of these endemics, only the pocket gopher, *Orthogeomys cherriei,* occurs at La Selva. *Orthogeomys cherriei* is found in a broad elevational band north and east of the Cordillera Central and north and west of the Cordillera de Tilarán in northern Costa Rica (Hafner and Hafner 1987). The Caribbean white bat is the only other mammal species at La Selva that has an extremely limited distribution. White bats are restricted to the Caribbean lowlands of Honduras, Nicaragua, Costa Rica, and extreme northwestern Panama (Timm 1982).

Three groups of mammals found elsewhere in the Neotropics are absent from the fauna of the La Selva region: shrews (Soricidae), coyotes and foxes (Canidae), and bears (Ursidae). A single genus of shrew, *Cryptotis,* is found in the Neotropics, and at least five species occur in Costa Rica, but all are restricted to the highlands (Woodman 1992). Canids do not occur in the Caribbean lowlands of Costa Rica although the family is widely distributed worldwide and occurs throughout North, Central, and South America. The bush dog, *Speothos venaticus,* does (did) occur in eastern Panama, including the Caribbean lowlands. Coyotes, *Canis latrans,* continue to expand their range in Costa Rica (Vaughan 1983; Monge-N. and Morera 1987) but have yet to reach the Caribbean lowlands. Bears have a disjunct distribution in the New World, being widely distributed in North America, absent in Central America, and with a single species, the spectacled bear (*Tremarctos ornatus*), occurring in the Andes of northern South America.

Historical biogeography of the Central and South American mammal faunas was reviewed by Hershkovitz (1972), Savage (1974, 1982), Marshall et al. (1982), and Rich and Rich (1983) and of Costa Rican rodents by McPherson (1985, 1986). An interesting review of mammalian ecology in Costa Rica is provided by Janzen and Wilson (1983); it includes in-

sightful discussions of the relatively low species diversity of terrestrial mammals, seasonal environmental stress, the largely frugivorous diet of many tropical "carnivores," and an overview of tropical bat radiation. They pointed out that the high species diversity of mammals seen per unit area in Costa Rica and elsewhere in Central America is the result of a dramatic increase in the number of bat species. Species richness of terrestrial mammals per unit area in tropical Central America is similar to that of temperate North America. Excellent summaries of the biology of many of the common mammals at La Selva and in Costa Rica may be found in *Costa Rican Natural History* (Janzen 1983b). Two extremely useful and nicely illustrated guides to Neotropical mammals, many of which occur at La Selva, were published. Eisenberg (1989) treats distribution and identification of the mammals of the northern Neotropics (defined as Panama, Colombia, Venezuela, and the Guyanas). Emmons and Feer (1990) is a field guide to Neotropical rain forest mammals and covers species occurring primarily below 1,000 m in elevation.

RESEARCH ON MAMMALS AT LA SELVA

Until the 1980s the northern Caribbean lowlands and midelevational slopes of Costa Rica received little attention from naturalists, particularly mammalogists. Goodwin (1946) discussed the distribution and natural history of the approximately 125 species of mammals then known or thought to occur in the country in his *Mammals of Costa Rica,* but he made little mention of the northeastern lowlands and cited no specimens from the region. Even today very few specimens of mammals from this region reside in scientific collections, which hinders efforts to better understand the fauna.

It was not until the 1960s that studies on mammals began at La Selva. The first published mention of mammals there and in the surrounding region was by Paul Slud (1960, 76) in his classic study of the avifauna. In the 1960s a few papers were published on rare species of mammals found at La Selva. As OTS developed La Selva into a working biological station, its popularity with researchers grew. In the 1970s and 1980s numerous publications on a wide array of subjects appeared on mammals at La Selva. These studies can be broadly categorized into four main areas: distribution and systematics, ecology and natural history, community structure, and mammal-plant and mammal-insect interactions.

Papers focusing on geographic distributions and systematics often include morphometric data, taxonomic notes, and valuable natural history information, especially on reproduction. Several distributional papers deal exclusively with bats (Casebeer et al. 1963; Starrett and Casebeer 1968; Gardner et al. 1970; LaVal 1977), and one deals with the possible occurrence at La Selva of night monkeys of the genus *Aotus* (Timm 1988). Wilson's (1983) checklist of mammals at the OTS field sites was the first modern assessment of mammalian distributions in Costa Rica, and as such gives researchers an important baseline. Two recent reports review the fauna of this region (Timm et al. 1989; Wilson 1990).

Timm et al. (1989) conducted a faunal survey of the elevational transect from 35 m to 2,600 m that encompasses La Selva and Parque Nacional Braulio Carrillo to Volcán Barva. We documented that at least 141 species of mammals occur in the region, including several species new to the area's fauna. Additionally, we review the systematics, distribution, and natural history of these species and provide new biological information on each. This is the first in-depth study of an elevational transect in the Neotropics, and the first comprehensive review of the mammalian fauna for any region of Costa Rica since Goodwin's (1946) *Mammals of Costa Rica.*

Publications from La Selva in the category of ecology and natural history of mammals include studies on natural history, behavior, general ecology, and evolutionary ecology. Emphasized here are specific ecological problems as well as in-depth autecological studies. Interestingly, most of the twenty-odd published studies in this category have been on bats. In addition, Fleming has published several reports of his studies on rodents; Greene (1989) provides interesting observations of aggressive interactions between male three-toed sloths; and Fishkind and Sussman (1987) provide a preliminary survey of primate densities at La Selva and the adjacent zona protectora (now Parque Nacional Braulio Carrillo), suggesting that six to ten groups of howler monkeys, two to four groups of spider monkeys, and two to three groups of white-faced capuchins are in the La Selva reserve. Two important studies, both on primates, are underway at La Selva in 1992. Amy Fishkind Campbell is studying white-faced capuchins, and Kathryn Stoner is working on howler monkeys. These studies represent the first modern autecological work to be undertaken on primates in the Caribbean wet forests of Costa Rica and as such will provide refined estimates of densities and valuable comparative data.

Several species of bats, including some of the most poorly known and those considered to be extremely rare, have been studied at La Selva. Until it was rediscovered at La Selva in 1961, *Ectophylla alba* (fig. 18.1) was among the least known of all mammals: only two poorly preserved specimens existed, one collected in the late 1880s and from Honduras and one in the early 1890s from Nicaragua (see Timm et al. 1989), with no associated biological data (Casebeer et al. 1963; Timm 1982). Several aspects of its biology have been investigated at La Selva, including its construction of roost "tents" by modifying leaves (Greenbaum et al. 1975; Timm and Mortimer 1976; LaVal and Fitch 1977; Timm 1982; Timm and Kermott 1982; Brooke 1987a, 1989). White bats create tents by severing the lateral veins and interconnected tissues on both *Heliconia* and *Calathea* leaves from near the base to the tip of the leaf, causing the sides of the leaf to droop down. The roosting bats hang from the midribs of the leaves and are protected from both predators and the elements by their tents (fig. 18.1). White bats roost only under these altered leaves, using them as diurnal roosts, maternity roosts, and night-feeding roosts. *Ectophylla alba* is now known from thirteen localities, but only at La Selva and in the adjoining Parque Nacional Braulio Carrillo is there assurance that it and its habitat will be protected. Other studies on specific aspects of the ecology of bats at La Selva focus on social organization and foraging in emballonurid bats (Bradbury and Vehrencamp 1976b), tent construction and social organization in *Vampyressa nymphaea* (Brooke 1987b), prey selection and foliage gleaning by *Micronycteris megalotis* (LaVal and LaVal 1980), reproduction during the dry season (Mares and Wilson 1971), the influence of human-made trails on foraging by frugivorous bats (Pal-

Fig. 18.1. Photograph of a colony of four white bats, *Ectophylla alba,* roosting in a tent cut from a *Heliconia* leaf. The bats typically hang curled in tight clusters from the midrib. Details of the cut side veins and interconnected tissues may be seen along the midrib of the leaf. The holes in the leaf were made by the bats' claws. Tents such as this one may be used for several weeks. *Ectophylla* roosts only under leaves that it modifies as tents.

meirim and Etheridge 1985), the influence of body size on diet and habitat in *Carollia* (Fleming 1991), ecology and systematics of tent-making bats (Timm 1984, 1985, 1987; Timm and Clauson 1990), foraging of vampire bats (Young 1971a), and ecology of *Saccopteryx bilineata* (Young 1972c; but see 1975).

In the early 1970s Theodore H. Fleming began a series of classic studies on the population ecology of Desmarest's spiny pocket mouse, *Heteromys desmarestianus* (see Fleming 1973, 1983; Fleming and Brown 1975; and additional references cited). He investigated population dynamics (Fleming 1974a), social organization (Fleming 1974b), growth and development (Fleming 1977a), and experimental responses of animals to manipulated food and water availability (Fleming 1977b). These studies were among the first in-depth on any species of small tropical rodent. As such, they provide an important basis for comparisons with temperate rodents and some of the first baseline data on tropical species. Fleming's work is especially important in elucidating the ecology of and roles played by both *H. desmarestianus* and *Liomys salvini* as "key industry" species in tropical forests. As these heteromyids are often the most abundant small mammals in the community, they serve as both major seed predators and seed dispersers and are important prey items for a wide variety of carnivores.

At La Selva Fleming found *Heteromys* to be the most abundant small terrestrial mammal present. Densities ranged from nine to eighteen per hectare and the population was stable over the two-year study. Both males and females bred throughout the year although reproductive activity declined markedly during May and June. Litter size averaged 3.1 young, and females produced up to five litters per year. *Heteromys* is perhaps the dominant granivorous vertebrate of the forest floor. Fleming found spiny pocket mice to feed extensively on seeds of the palms *Socratea durissima* and *Welfia georgii,* and we observed them to feed on the palms *Euterpe macrospadix, Geonoma* sp., and *Iriartea gigantea* and on *Meliosma* sp. (Sabiaceae) (Timm et al. 1989). Based on extensive trapping, Fleming suggested that *Heteromys* does not have mutually exclusive home ranges or territories. My live trapping of this species at Monteverde certainly supports that conclusion. There are, however, tremendous population fluctuations of spiny pocket mice at Monteverde (pers. observation).

Publications that focus on mammalian community structure at La Selva are exclusively concerned with bats (Findley 1976; Findley and Wilson 1983; LaVal and Fitch 1977). One of the first comparative studies of tropical bats was by LaVal and Fitch (1977), who compared the structure, movements and reproduction of the complex bat communities at La Selva, La Pacífica (tropical dry forest), and Monteverde (premontane moist and premontane wet forests). They found the highest species diversity of bats at La Selva; much of this diversity was due to insectivorous bats, both foliage-gleaning and aerial feeders. The three sites were similar in species diversity of nectar- and pollen-feeding bats. Most tropical bats reproduce seasonally. Bats at La Pacífica, with its sharply delineated wet and dry season, have the shortest and most sharply delineated reproductive seasons. A longer reproductive season was typical of bats at La Selva, and Monteverde was intermediate between the two (LaVal and Fitch 1977).

It is often assumed that the extreme species richness observed in a large, complex fauna like that of tropical bats can occur only if species are restricted to narrow and mutually exclusive feeding niches. Findley (1976), however, compared the bat fauna of La Selva to those at other tropical and temperate sites and demonstrated that although temperate faunas do exhibit greater rarity or even absence of ecologically distinctive taxa compared to tropical faunas, tropical and temperate faunas do not differ significantly in species packing or degree of niche overlap among component species.

Findley and Wilson (1983) demonstrate that species density of New World frugivorous bats is significantly greater than tropical African frugivorous bats. Capture rates with standard mist nets showed that the absolute numbers of species captured, as well as the numbers of individuals within species, are higher in the Neotropics. They did, however, observe a compensatory trend in biomass: the larger frugivorous pteropid bats of Africa (mean body mass of 52 grams) occurred in roughly the same total biomass density as the more numerous but smaller New World frugivorous phyllostomids at La Selva (mean body mass of 18 grams). It should be noted that the difference in size between pteropid bats and microchiropterans is also a phylogenetic one.

One of the advantages of a heavily used biological station such as La Selva is that biologists with diverse interests and skills are able to exchange ideas and expertise. Such a stimulating atmosphere encourages cross-disciplinary collaboration, such as the very fruitful research on mammal-plant interactions, including studies on seed dispersal by mammals (Vandermeer et al. 1979), seed hoarding (an extremely uncommon behavior among Neotropical mammals) by spiny pocket mice (Vandermeer 1979), the effect of predation on seeds and seedlings by mammals (Denslow and Moermond 1982; McHargue and Hartshorn 1983), bat pollination of flowers (Voss et al. 1980), the role of bats in dispersing *Piper* (O'Donnell 1989), and the alteration of leaf shape by bats to produce diurnal roosting structures (Foster and Timm 1976; Brooke 1987a, 1987b; Timm 1987).

Two important studies of mammal-insect interactions have been completed at La Selva. An apparently phoretic relationship between *Heteromys desmarestianus* and a newly discovered species of tineid moth (*Amydria selvini*) has been reported (Davis et al. 1986). Females of this moth were found only on spiny pocket mice; the male of the species is unknown. *Amydria selvini* has been found only at La Selva although I have searched unsuccessfully for it elsewhere in Costa Rica at higher elevations. The tight association between female moths and *Heteromys* suggests that the life cycle of the moth is associated with the nesting biology of the rodents, but this has yet to be demonstrated. A parallel association is found between three-toed sloths and sloth moths. The sloths at La Selva have large populations of these phoretic/ectoparasitic moths, but they have not been studied there. A review of the complex parasite fauna of sloths and its interesting biological relationships, based primarily on research conducted in Panama and Brazil, is provided by Waage and Best (1985). Three species of batflies of the family Streblidae that are host-specific, blood-feeding ectoparasites on *Carollia perspicillata* were investigated at La Selva by Fritz (1983). He found that the life cycles of the batflies are tightly synchronized with the life cycles of their hosts.

HISTORY OF KNOWLEDGE OF LA SELVA MAMMALS

Biologists in a wide array of fields have long been aware of the tremendous diversity of organisms found in the tropics. With the recent awareness of the plight of tropical rain forests, there has been a parallel scientific awakening and interest in this diversity. To understand the structure and nature of tropical rain forests biologists have attempted to characterize, compare, and contrast the diversity of particular organisms at and between sites.

Critical to these types of studies is the historical component of time and how one's knowledge of the biota at a particular site increases with time. The time component is quite complex and includes two aspects: the percentage of the total fauna available for sampling at the time of the study and the percentage of the total fauna known to the investigator(s). Incomplete tabulations might result from local extinctions before investigation, from species that are rare or difficult to capture or observe, or simply from lack of resources for adequate sampling.

Because such an excellent record of the mammalian fauna at La Selva now exists, it is useful to examine the rate of increase in knowledge of what species of mammals occur at La Selva—a species discovery curve of the fauna.

The question I address here is How long did it take to reach the current level of knowledge? and the database I use is the published literature. A review of the literature on research at La Selva contains more than forty-five primary references covering the thirty-one-year period from 1960 to 1990.

The mammalian fauna of La Selva, as recently as the 1960s, consisted of at least 116 species (app. 8). The discovery curve illustrates the considerable time it took to acquire this knowledge as measured by published reports in the literature (fig. 18.2). Only 16% of the fauna had been identified by 1970; 84% was identified by 1980. It was as late as 1986, how-

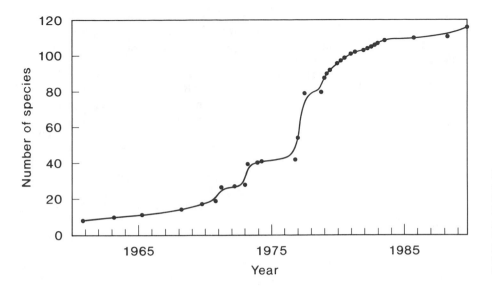

Fig. 18.2. Cumulative species discovery curve for mammal species known from the La Selva Biological Station. Data points are plotted cumulatively by year and represent the number of species added to the fauna based upon the published literature. The X axis is the date of publication of the primary reference to the nearest quarter year.

ever, before 95% of the species known to be present were documented there. Thus, a full twenty-six years elapsed between the first reports and documentation of 95% of the mammalian biodiversity.

This time lapse at such a well-studied site makes it clear that investigators need to exercise extreme caution in making comparisons between sites, especially in the tropics. Without a sufficient database and historical perspective, the comparisons made could be more misleading than insightful. For instance, sites such as La Selva and BCI both approach having 100% of the mammalian diversity identified. Most other Neotropical sites have been studied much less intensively and for a much shorter time and a much lower percentage of their faunas is likely known. Direct comparisons between La Selva or BCI and these other sites could, thus, produce spurious or misleading results.

LA SELVA AND OTHER TROPICAL SITES

With the previous caveats in mind comparisons between the mammalian fauna at La Selva and other tropical sites can be useful in understanding the structure and complexity of tropical ecosystems on a broad level. Published faunal lists of mammals at Neotropical sites are few, and attempts to make intersite comparisons of tropical mammal faunas have been hampered by lack of data on presence and abundance of species (see Eisenberg and Thorington 1973; Emmons 1984; Bourlière 1989).

Barro Colorado Island (BCI) in central Panama is the only Neotropical site that has been studied intensively for a longer period than La Selva; the first mammalogists visited the area in the late 1910s. La Selva and BCI are in close geographical proximity and have considerable faunal overlap. Barro Colorado Island was isolated as an island when the Río Chagres was dammed to form the central waterway of Panama Canal in 1914. Naturalists first visited BCI in 1916, and it was declared a reserve and tropical field station in 1923. Although there are numerous differences between La Selva and BCI in climate, geology, and the resulting forest, one significant difference now affecting the mammals of these two lowland tropical sites is the fact that BCI has been an island for seven

decades, whereas La Selva lies on the mainland. Additionally, BCI has no streams that flow year-round.

BCI has been the site of excellent long-term studies on several species of mammals, including white-faced capuchins, howler monkeys, sloths, red-tailed squirrels (*Sciurus granatensis*), agoutis, and coatis (see Leigh et al. 1982, and references therein). No studies, however, have been published about the La Selva and BCI mammal faunas that are directly parallel, making ecological comparisons between the two difficult. As with La Selva, several species of mammals have benefited from the reduction and/or elimination of predators by humans. The larger cats and raptors, especially harpy eagles, are absent or rare on BCI.

A comparison of the numbers of species represented in each order of mammals in all of Costa Rica, at La Selva, and at BCI shows that the La Selva and BCI faunas are extremely similar (table 18.1). La Selva and BCI have about the same number of marsupials (five and six, respectively), neither has any insectivores, and La Selva has more species of bats than BCI (sixty-six compared with fifty-six species). BCI may have only one additional species of primate (crested bare-faced tamarins, *Saguinus oedipus*); the two sites have the same number of edentates, rabbits, artiodactyls, and perissodactyls; La Selva, however, appears to have twice as many carnivores (fourteen compared to seven) and 60% more rodents (sixteen compared to ten). As with the La Selva fauna, much of the BCI fauna has a widespread distribution in the tropics, which accounts for the fact that the two sites share many species.

Just how representative the fauna and densities of mammals currently occurring on BCI is of what one would expect for a pristine lowland forest in central Panama has been the subject of much debate. Glanz (1982, 1990) documented both the historical and recent changes in abundance of terrestrial mammals on BCI, including the extinction of pumas (*Felis concolor*) and white-lipped peccaries, and the apparent tenfold increase of agoutis and squirrels, among others. He did, however, suggest that many species might just appear to be more abundant because they are less wary now as a result of protection from hunting.

The fauna of La Selva is much more terrestrial than the faunas found in the greater Amazon Basin. Faunas of season-

Table 18.1 Comparison of mammalian faunal diversity at the ordinal level between Costa Rica in total, La Selva, and Barro Colorado Island

Order	Costa Rica	La Selva	BCI
Marsupialia	8	5	6
Insectivora	5	0	0
Chiroptera	105	65	56[a]
Primates	5	3	5
Edentata	7	7	7
Lagomorpha	3	1	1
Rodentia	45	16	10
Carnivora	22	14	7
Artiodactyla	4	4	4
Perissodactyla	1	1	1
TOTALS	205	116	97

[a]Handley et al. 1991.

ally flooded forests such as those of Manaus are typically composed of a greater proportion of arboreal species. Although arboreal marsupials, primates, rodents, and carnivores are present at La Selva, the number of arboreal species in each group is considerably fewer than seen at Amazonian sites. The paucity of arboreal species is not a reflection of the forest but of geography; most of the truly arboreal species are of southern origin.

Terborgh (1988, 1990b) provided thought-provoking, controversial essays on the importance of large carnivores in maintaining biological diversity of both plants and animals in the Neotropics. He noted that BCI, an island with few of its larger predators left, has populations of agoutis, pacas, and coatis that are ten times higher than those observed at Cocha Cashú (with its predator populations intact) in Amazonian Peru and that populations of smaller prey species, such as cottontail rabbits, armadillos, and opossums, are from two to ten times greater on BCI. He suggests that at such high densities, these species act as significant seed predators and that the artificially elevated abundances of these mammals that occur on BCI may have had a major impact upon the structure of the forest. Although interesting, these sorts of comparisons between sites are largely untestable given the multitude of factors involved. Forest type differs from site to site, and the history of land use for each of the sites is quite different. For example, Manaus (Brazil) has notoriously poor soils and a strong dry season, both of which may contribute to the depauperate mammal fauna and low numbers of individuals there (Malcolm 1990). Manu (Peru) has thirteen species of primates, many of which occupy a squirrel-like niche, perhaps contributing to its paucity of squirrels. There is little acceptance among wildlife managers of the notion that a single predator species can actually control or regulate a prey species' population. Elimination of many or all predator species coupled with major habitat changes, however, may impact prey species' populations.

IMPACT OF HUMAN ACTIVITIES

Several species of mammals at La Selva undoubtedly have increased in abundance as human activities adjacent to the reserve opened the forest and as the reserve expanded to include secondary forest and other disturbed areas. Conversely, hunting pressure and forest destruction have a direct negative impact upon the number of species and, often, densities of mammals.

Subsistence farming, especially when based on small family garden plots surrounded by forest, has a positive effect on species diversity and abundance of certain marsupials, bats, and small to medium-sized rodents. Crops and their associated insect pests provide a dense and readily available source of food for animals. Additionally, opening up of the forest increases the edge effect, or ecotone habitat, creating a rich habitat for many species. Species that have undoubtedly increased in abundance because of human activity in the region include *Didelphis marsupialis, Glossophaga soricina, Carollia perspicillata, Artibeus jamaicensis, Dasyprocta punctata,* and, possibly, *Sylvilagus brasiliensis.* All are generalists within their particular feeding niches and are typically forest-edge species. For example, although *Glossophaga soricina* is a nectarivorous bat, it uses a broader range of plant species than do other nectar-feeding bats.

Agoutis and squirrels are considerably more abundant at La Selva than in large tracts of lowland rain forest in the Amazon Basin (pers. observation). Agoutis and, probably, other small to medium-sized rodents may have also benefited from the systematic killing of predators by humans. An interesting discussion of predators in tropical ecosystems, with special reference to La Selva, is presented by Greene (1988). He identifies one hundred species of vertebrates as predators at La Selva and independently concurs in attributing the high population densities of many rodents currently observed at La Selva and BCI to removal or reduction of predator populations by humans.

The three common species of primates at La Selva, *Ateles geoffroyi, Alouatta palliata,* and *Cebus capucinus,* all appear to be more abundant now than they were in the 1960s and early 1970s. Primate populations throughout much of the Caribbean lowlands of Central America were decimated by an epidemic of mosquito-borne yellow fever during the early 1950s (see Fishkind and Sussman 1987, and references therein). Although data are not available for the La Selva region, one assumes that this epidemic reduced primate populations in the reserve. Far fewer primates were observed at La Selva between the late 1960s and early 1980s than in similar tracts of primary forest elsewhere in the Neotropics (pers. observation). Although Milton (1982) suggests that the howler monkey population on BCI had rebounded from the yellow fever epidemic by 1970, that recovery was apparently not as rapid in the Sarapiquí region, perhaps because of other factors, including hunting. Primate populations at La Selva, however, have certainly rebounded during the late 1980s. Capuchin, howler, and spider monkeys can now be seen almost daily and often in large groups. K. Stoner (pers. comm.) estimated in 1990 that seven to twelve groups of howler monkeys, three to six groups of spider monkeys, and four to seven groups of white-faced capuchins were on the greater La Selva property. Howler monkeys are the most abundant. These results agree with a preliminary 10-day survey conducted at La Selva in 1986 by Fishkind and Sussman (1987) although Stoner's numbers for 1990 are higher.

This increase in primate abundance may also be owed, in

part, to a decrease in the abundance of natural predators, especially harpy eagles (*Harpia harpyja*). Cebid monkeys of the genera *Alouatta, Ateles,* and *Cebus* constituted more than one-third of the prey consumed by a pair of harpy eagles in Guyana; two- and three-toed sloths, opossums, and agoutis also were important prey items (Izor 1985). Harpy eagles were rare in the Puerto Viejo region as early as the late 1950s (Slud 1960) and are now absent.

Primates, as well as many other species, undoubtedly benefit from complete protection from hunting at La Selva. We are, perhaps, witnessing a period of fluctuation in numbers of many predatory and prey species because of the changing degrees of hunting pressure by humans. Because human populations increased in the Puerto Viejo region in the 1940s and 1950s before the establishment of La Selva as a reserve, one would assume that hunting pressure increased dramatically. Now that La Selva is well protected, hunting pressure has been effectively eliminated from the reserve although poaching continues in Braulio Carrillo as at La Selva into the early 1980s. It will be interesting to observe the population responses of predators and the larger prey species. Complicating the ability to detect such changes is the fact that individuals of many species become less wary when not hunted and, thus, are more easily observed, which gives the false impression of higher abundance even though numbers may not have changed.

Historically, families in the Sarapiquí region relied heavily upon local wildlife as a source of protein. Tapirs, white-lipped and collared peccaries, and pacas are highly prized meats and are among the largest mammals of the region. Medium-sized mammals such as agoutis, monkeys, and squirrels (*Sciurus* spp.) were hunted to a lesser extent. Tapirs and pacas are now uncommon at La Selva and in the general vicinity even where adequate habitat remains. White-lipped peccaries are extirpated from La Selva; no sightings have been confirmed on the property for several years although Slud (1960) found them to be common there in the late 1950s. In the past few years, however, the population of collared peccaries has increased dramatically.

The recent extirpation of white-lipped peccaries from La Selva may be affecting the nature of the forest there. Peccaries are both major seed dispersers and seed predators, and the activities of large herds could greatly affect the forest plants. Herd sizes of fifty or more individuals are commonly reported for white-lipped peccaries where they are not heavily hunted, and they tend to concentrate their activities in areas of favorite food resources.

Large- and medium-sized mammalian predators, including jaguars, ocelots, tayras, and coatis, have been persecuted by humans in the Puerto Viejo region as they have been throughout the human-inhabited tropical lowlands. These animals were killed to protect livestock and crops as well as for their pelts. Other predators whose populations have been reduced in the Puerto Viejo region, as well as throughout their ranges, include bushmasters (*Lachesis muta*) and hawks and eagles that prey on medium-sized rodents and primates. It is likely that the other three species of cats and the one river otter have also been affected by hunting and habitat destruction. Since Costa Rica's 1975 ratification of the Convention on International Trade in Endangered Species (CITES), the Costa Rican

government has effectively controlled illegal trade in wildlife. Previously, cat skins were openly sold in markets. Cat populations continue to be seriously threatened by both habitat destruction and hunting. The other species of carnivores are, perhaps, nearly as abundant now as in the past where suitable habitat exists, but systematic studies are lacking.

Giant anteaters probably are extirpated in this region and throughout most of the country. Originally, giant anteaters were found from Belize and Guatemala throughout the lowlands of Central America and tropical South America. Little is known about their original distribution and abundance in Costa Rica. In the late 1870s Alston (1879–82) reported that giant anteaters were rare in Costa Rica and confined to the low, hot forest lands near the coast. There have been no reports of *Myrmecophaga* from this region in recent years, and one presumes that they are now only found in the most remote regions of the country (Timm et al. 1989).

La Selva, even with its connection, Parque Nacional Braulio Carrillo, may be only minimally large enough to support populations of predatory bats of the subfamily Phyllostominae. The eighteen species of the phyllostomines that I report here for La Selva probably represent all the species expected to occur there. Some of these large predatory species are encountered in much higher densities elsewhere, in larger tracts of pristine forest than currently exist at La Selva (pers. observation). Large populations of predaceous bats were reported from La Selva by LaVal and Fitch (1977). Although all of these species are still present in the reserve, most are not now encountered as frequently as in the past (pers. observation). Apparently, this decline is the result of the destruction of much of the forest that surrounded La Selva and provided more extensive habitat for these bats. As with the large predatory cats, these bats may be good "indicator" species in that they may be among the first components to disappear as a result of human disturbance (fig. 18.3). Thus, the connection of La Selva to the recently expanded Parque Nacional Braulio Carrillo and the continued effort to expand Braulio Carrillo are vitally important to the maintenance of populations of these highly vagile, predatory bats, as well as other mammals.

Human impact in this region undoubtedly has altered the abundance of many species of mammals; some have increased in abundance, whereas many have decreased in abundance. With the exception of giant anteaters and white-lipped peccaries, all species of mammals present at La Selva before the arrival and colonization of the region by Europeans have survived there. La Selva and the adjoining Parque Nacional Braulio Carrillo are large enough to provide suitable habitat for most species of mammals but probably not for larger, highly vagile species such as giant anteaters and white-lipped peccaries. Every effort should be made to preserve as much of the forest surrounding La Selva and Braulio Carrillo as possible.

SUGGESTIONS FOR FUTURE RESEARCH

A review of the literature on mammals at the La Selva Biological Station includes more than forty-five original published contributions. Most deal with ecology and natural history and interactions between mammals and other biotic components of the environment. Although more has been published about mammals at La Selva than at other Neotropical sites, except

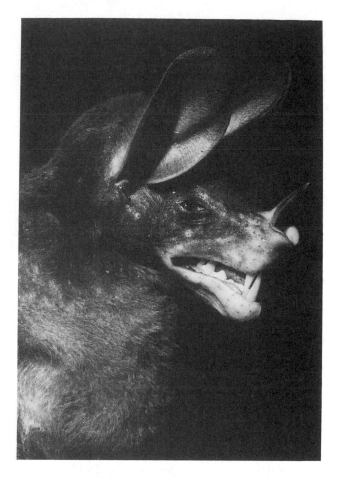

Fig. 18.3. The carnivorous bat *Vampyrum spectrum* has a large body and is found only in low numbers. They are monogamous and have only a single young per year with extended parental care. Like the terrestrial predators, it is among the first mammal species to disappear with fragmentation of rain forests.

BCI much remains to be learned about this complex fauna and its component species. When reviewing these research reports, it is ironic that one must conclude that more questions have been posed than answered. Interestingly, much more is known about the larger mammals on BCI and even at Manu in Peru than at La Selva.

Nonmammalogists might wonder why bats have been studied so much at La Selva (and elsewhere in the Neotropics) and rodents or other groups have been studied so little, in contrast to the voluminous literature on temperate rodents. Perhaps, part of the answer lies in the relative abundance of bats in the tropics and their relative ease of capture with mist nets. Most mammal species are nocturnal, wary of humans, and have excellent hearing and vision. Many of them (especially marsupials, primates, edentates, rodents, and carnivores) are partially or wholly arboreal in the tropics, adding to the difficulty of capture and study. The environment at La Selva has a three-dimensional complexity much greater than that of temperate sites or even other tropical sites. These difficulties, along with frequent heavy rainfall, render direct observation, radiotelemetry, and most of the standard mark-recapture techniques employed by temperate mammalogists difficult. Although most of the mammal research at La Selva has centered on bats, at-

tempts have been made to study other mammals. Some of these were unsuccessful, perhaps for some of the reasons mentioned above.

A spectacular tropical forest such as that found at La Selva presents unlimited opportunities for future study. This work needs to be undertaken at several different levels, as outlined next.

Faunal Surveys Coupled with Systematic Studies of the Mammal Species

Species of mammals undoubtedly occur at La Selva that have not yet been recorded. These previously undetected species probably include more than one species of bat and one or more species of long-tailed rodents. This lack of the most basic knowledge of what species are present is also true for all other Neotropical sites. Researchers quite literally do not have a complete listing of the mammal fauna of any New World tropical site. In addition, knowledge of many of the small mammals is so rudimentary that undoubtedly more than one good biological species may be included under a single name. This confusion is almost certainly true for *Oryzomys* (*Oligoryzomys*) *fulvescens*. Additionally, even though Costa Rica is one of the most intensively studied countries in the Neotropics, species of mammals new to science continue to be discovered there. Understanding of tropical forest ecosystems and the ability to make meaningful comparisons between La Selva and other sites are hampered by incomplete knowledge of the fauna.

Habitat Requirements, Life History Strategies, and Reproductive Modes

Essentially nothing is known about the habitat (and other) requirements of most species of mammals in the Neotropics, especially the small and medium-sized species that make up most of the fauna. Given the now widespread attention to the biodiversity crisis in the tropics, it is critical that biologists have a better understanding of the ecology of tropical mammals. Detailed knowledge of this sort will be critical for the proper management of species and wildlife reserves in the future. One of the most basic questions is How large an area does a given species need to maintain a viable population? The answer has direct implications for the size and shapes of reserves.

Population Biology and Community Ecology of Neotropical Mammals

Knowledge of the structure and dynamics of populations will be critical to understanding how communities are organized and how tropical systems differ from temperate systems and to understand all aspects of evolutionary biology theory. Studies on rates of reproduction, litter sizes, survival rates, and longevity are all feasible and will provide considerable insights.

Research on mammals in the tropics is entering an exciting phase. Investigators are now perfecting the techniques needed to work with these animals. La Selva's diverse fauna provides biologists with numerous opportunities to explore ecological and evolutionary questions. The field facilities at La Selva are superb, encouraging interactions among scientists and providing excellent access to a rich fauna.

ACKNOWLEDGMENTS

I thank the Organization for Tropical Studies, especially R. Butterfield, D. and D. Clark, L. A. McDade, C. E. Schnell, and D. E. Stone for making my visits to La Selva over the years possible, pleasant, and productive. R. K. LaVal and D. E. Wilson assisted in developing the abundance indexes provided in appendix 8, and I am grateful for their input. S. E. Abrams, S. D. Anderson, A. P. Brooke, B. L. Clauson, and K. E. Stoner provided constructive review and insightful comments on earlier drafts of the manuscript. Special thanks are due my wife, Barbara Clauson, who has ably assisted me with my research over the years, and for providing both of the photographs used herein.

# 19

An Overview of Faunal Studies

Henry A. Hespenheide

The six groups of animals reviewed in chapters 13–18 include two different phyla whose members inhabit both terrestrial and aquatic environments. Finding common themes among these six taxa or among any other combination of six animal taxa would be difficult. Each group has a taxon-specific evolutionary history and tradition of research (e.g., at La Selva, frugivory by birds versus breeding modes of frogs). The "highest" animals are structurally more complex than plants, but the major difference among animals from plants that frustrates generalization is their ecological diversity. Although plants and animals occupy the same physical environment, the trophic diversity of animals is decisively greater. Evolutionarily, it matters considerably whether an animal is herbivore, frugivore, or carnivore and even whether the herbivore is an insect or a vertebrate, whereas plants live in a trophically simple world of more or less sunlight and moisture.

Diversity is the central issue of tropical biology. In the sections that follow I first argue that ignorance of insect diversity, compared to knowledge of vertebrates, is pervasive and constrains research. For the groups that are known the reviews show that the directions taken in further research diverge once faunistic studies are relatively complete. Comparisons of La Selva with other sites help put into perspective the influences that the evolutionary history of an area and its climate have on diversity, which, in turn, determine the kinds of research undertaken to study it. Such comparisons are crude because of artifacts that result from differences between sites (site areas, variety of habitats, amount of prior or subsequent disturbances) and idiosyncrasies in the traditions of research done there. Using such comparisons, a review of patterns in latitudinal gradients in species richness shows that the relative importance of climate and biogeographic history differs between and within taxa. Finally, comparisons with other sites with longer histories of study than La Selva's show that changes in adjacent land-use patterns will bring changes to the fauna and, by implication, to faunal research at La Selva.

TAXONOMIC BIAS: SOME SPECIES ARE MORE EQUAL THAN OTHERS

As stated in the introduction to this section, the fauna of La Selva is very unevenly known, biased toward vertebrate groups and a few more "popular" invertebrate groups. All of the terrestrial vertebrate groups had been inventoried both for Costa Rica and La Selva when *Costa Rican Natural History*

was published (Janzen 1983b), whereas few families or subfamilies of insects had been so inventoried. About 15% of La Selva's vertebrates were treated individually in biological accounts, whereas only 4% of the 284 species of insects from groups that were adequately known to be listed had such accounts. Although the vertebrate lists have been revised since 1983, additions have taken the form of new distribution records for previously described species (see chaps. 15, 16) and, in the case of birds, usually for ecologically less important nonresidents (chap. 17). The number of mammals has actually decreased after critical pruning of the list by Timm (chap. 18). Only among freshwater fish (chap. 14) have vertebrate species been recently described from the La Selva area. Although none of the lists is considered by its authors to be complete, it is unlikely that more than 10% of La Selva's resident or breeding vertebrate species remain to be discovered and extremely unlikely that even 1% are unknown to science.

The minor changes to be expected in knowledge of vertebrates contrast strongly with major deficiencies in knowledge about most insect groups. Taxonomic knowledge of insects at La Selva varies from moderately complete (the larger butterflies; DeVries, chap. 13, 1987) to largely unknown (parasitic wasps). From data in table 19.1, based largely on unpublished surveys of a few isolated families or portions thereof, one can see that large numbers of La Selva's insects are yet undescribed—10%–85%, depending on the taxon. For comparison and corroboration, analyses of standardized samples of insects in adjacent Panama show that 42% of its 164 species of roaches are undescribed (Wolda et al. 1983), 89% of 295 species of Psocoptera (Broadhead 1983), and between 35% and 48% of Trichoptera species (McElravy et al. 1981). For large groups such as weevils among the beetles and parasitic wasps the proportions of undescribed species are probably close to those for Panamanian Psocoptera. Although the Flora of La Selva project has been underway for eight years (see chap. 6), only in 1992 has a general inventory of insects been funded and synoptic collections begun of a number of groups in addition to those few treated in 1983 (table 19.1 and Hespenheide, unpublished data). In addition, collections from La Selva are beginning to be treated in general revisionary studies (e.g., Menke 1988; Wibmer 1989). Because of the intrinsic scientific interest of tropical organisms and the important ecological roles of insects, there is general agreement that such inventories are desperately needed in tropical areas (Janzen and Adams 1985) before human destruction of environments leads

Table 19.1 Numbers of species and state of knowledge of certain insect groups at La Selva compared with those at Barro Colorado Island, Panama.

	La Selva			Shared		Barro Colorado Island		
Taxon	Known	Named	(%)	N	(%LS)	Known	Named	(%)
Odonata	90	—	—	—		90	—	
Lepidoptera, 3 families	204	204	(100)	101	(49)	136	136	(100)
Diptera, Asilidae	92	35	(38)	19	(27)	73	43–46	(59–63)
Coleoptera								
Cerambycidae	231	200	(87)	—		—		
Buprestidae								
subfamily Agrilinae	15	5	(33)	5	(33)	22	7	(32)
subfamily Trachyinae	96	44	(46)	22	(23)	81	43	(53)
both subfamilies	111	49	(44)	27	(24)	103	50	(49)
Curculionidae subfamily Zygopinae								
tribe Piazurini	19	11	(58)	12	(63)	23	14	(61)
tribe Zygopini	121	43	(35)	45	(37)	71	26	(37)
tribes Lechriopini and Lobotrachelini	94	13	(14)	11	(12)	30	10	(33)
all tribes	234	67	(29)	68	(29)	124	50	(40)
Curculionidae subfamily Tachygoninae all genera	18	4	(22)	5	(28)	11	3	(27)

Sources: Asilidae—E. M. Fisher unpublished data, Fisher and Hespenheide 1990; Buprestidae, Curculionidae—Hespenheide unpublished data; Cerambycidae—F. T. Hovore unpublished data; Odonata—Paulson 1985; Lepidoptera—DeVries chap. 13.

to extinction of large numbers of species there. Costa Rica has recognized that need as evidenced by the on-going work based at the Museo Nacional and by the formation of an Instituto Nacional de Biodiversidad (INBio). In cooperation with IN-Bio the arthropod fauna of La Selva is currently under study by a group of entomologists headed by R. Colwell and J. T. Longino. The project (ALAS), funded by the National Science Foundation and the U.S. Agency for International Development, will assess general arthropod diversity by a variety of ecological sampling methods as well as study several focal taxa more intensely.

Apart from nematodes and most mites (but cf. Colwell 1986) the insect fauna is, thus, the largest, most basic area of ignorance about La Selva in particular and tropical forests in general, if only because they are the group with the largest number of species (cf. tables 19.1, 19.2). Erwin and collaborators (Erwin 1982, 1983b; Erwin and Scott 1980) have extrapolated from samples of canopy insects collected by insecticide fogging to estimate that the Neotropical insect fauna may run to tens of millions of species and those of individual sites such as La Selva may run to tens of thousands. Although Southwood et al. (1982) have shown that canopy fogging collects only about 40% of a plant's herbivore fauna, Erwin's estimates of numbers of host-specific herbivores and assumption that most plant species have similar numbers of such herbivores probably still err in the direction of being too high (Stork 1988). Very simply, if only one million species are described currently and 30 million are predicted to exist (Erwin 1982), then the proportion of undescribed species in samples should be 97%. Although high, the proportions are not that high at La Selva (table 19.1) or at the Amazonian site of Tambopata (Fisher 1985; Paulson 1985; Lamas 1985a; Wilkerson and Fairchild 1985). Even so, it is clear that the majority of insect species are largely unknown taxonomically, not to mention biologically. Because of the pivotal ecological position of insects as the primary herbivores and important mutualists of plants (chap. 21), as predators and parasitoids of other insects,

and as a major diet item of many vertebrates (chap. 20), the need for their systematic study is paramount. Lack of basic taxonomic knowledge of insects seriously constrains a variety of plant and vertebrate ecological studies and will continue to do so for the foreseeable future.

AFTER FAUNISTICS, WHAT?

In groups such as birds there is a historical sequence in which levels of knowledge about the organisms are added: a given species is first named taxonomically and its distribution is determined; then its life history is studied (reproductive phenology, behavior, population growth); and, finally, its role in the community is ascertained (comparative studies, delimitation of the niche, energetics, coevolution). From the foregoing sections it is clear that La Selva animal groups differ considerably in their position in this sequence: Vertebrates are completely described taxonomically and largely inventoried.

Table 19.2 Similarity of the terrestrial vertebrate fauna of La Selva compared with two Neotropical sites (numbers of species)

Group	Guanacaste Costa Rica	Shared	La Selva Costa Rica	Shared	BCI Panama
Mammals (total)	112[a]	74	116[b]	62	96[b]
Chiroptera	63	43	65	30	55
All other	49	32	51	32	41
Birds					
Palo Verde	126[c]	58	236[c]	135	199[d]
Santa Rosa	135[c]	61			
Reptiles	55[e]	17	86[f]	39	68[g]
Amphibians	23[e]	11	48[h]	16	33[g]
Butterflies[i]	125	71	204	101	136

[a]Wilson 1983; [b]Timm chapter 18 and pers. comm.; [c]Stiles 1983b, all figures for birds are for resident breeding species only; [d]Willis and Eisenmann 1979; [e]Cañas area, Scott et al. 1983; [f]Guyer chapter 16; [g]Myers and Rand 1969; [h]Donnelly chapter 15 [i]DeVries chapter 13.

Moreover, certain vertebrates, such as some bats and hummingbirds, have been extensively studied in terms of their population and community ecology as well as their coevolved relationships with other components of the biota. By contrast, insects are almost unknown at all levels.

Animals function in several major ways in tropical plant communities: as exploitative herbivores (chap. 21); as mutualists with plants in the roles of pollinators (chap. 12), dispersing frugivores (chap. 22), or protectors; and as decomposers, predators, and parasites of each other (chap. 20). As reflected in the chapters of this volume, a disproportionate amount of research at La Selva has been on plants rather than on animals. Except for birds and some species of the other terrestrial vertebrates, most ecological research on animals has yet to be done. Kinds of research that involve or require knowledge about insects are limited because of ignorance of insect taxonomy and basic biology. Patterns of research funding and the styles of research esteemed by biologists themselves have recently favored conceptually oriented population and community studies at the expense of taxonomy and descriptive life histories, thus creating an intellectual gridlock because, for example, no one can identify insects. One fascinating correlate of this situation is that a significant fraction of insect taxonomy is done as intellectual moonlighting by professional biologists with conceptual alter egos or by individuals whose professional success is independent of their taxonomic studies.

Several themes recur in the chapters on the vertebrate taxa, but most frequently mentioned is the need for long-term monitoring of populations of single species. To the present these have been limited essentially to Ph.D. dissertation studies, probably because of the demands on senior North American researchers by funding agencies and promotional evaluation committees for sustained, annual research productivity. Intensive autecological or life-history studies have been conducted on only one or two species in each of the vertebrate taxa: among amphibians, *Dendrobates pumilio* and *Eleutherodactylus bransfordii;* reptiles, *Norops humilis;* birds, *Phaethornis superciliosus;* mammals, *Heteromys desmarestianus* and *Ectophylla alba.* For various reasons species-rich groups such as snakes and the frog genus *Eleutherodactylus* are very poorly known. The studies of Clark and Clark on *Zamia* (reviewed, chap. 7) have shown that insect herbivory is important in an irregular, superannual manner, an observation that could be made only because the observers were continuously in situ. Other irregular and unpredictable events are possibly ecologically crucial, such as population irruptions in insects that cause local defoliations or failure of such "keystone" resources (Terborgh 1986) as fruit crops that are known to occur in other tropical areas in anecdotal cases and may be important at La Selva. Donnelly, Guyer, and, especially, DeVries argue that basic, long-term natural history studies will be valuable for many species, especially for those that have no temperate-zone equivalent. The long-term environmental monitoring program at Barro Colorado Island (BCI) would be an excellent general model for La Selva but would require a larger resident scientific staff.

Except for leaf-litter amphibian and reptile communities and for reproduction, migration, and molt in birds, seasonality of animals at La Selva is more poorly known than at BCI although partly for the obvious reason that La Selva is less sea-sonal climatically. Seasonal reproduction is undoubtedly tied to seasonal changes in resources; seasonal patterns are known for flowering and fruiting (chap. 12) but nearly unknown for insects. Fish reproduction and movements must be tied to rainfall and episodes of flooding (chap. 14) as are those for aquatic insects (chap. 20).

Predation is the biotic factor that has been repeatedly alleged to be of greater importance in the tropics and suggested as a cause of enhanced tropical diversity. A number of studies of marine taxa or environments have demonstrated a latitudinal gradient in predation rates, with higher levels of predation in tropical areas (Bakus 1969, 1974; Bakus and Green 1974; Gaines and Lubchenko 1982; Palmer 1979; Vermeij 1978; Vermeij and Vail 1978). Direct data for terrestrial environments are lacking although predation has been invoked as a factor in promoting the high species richness of tropical animals (Elton 1973; Maiorana 1976; Moynihan 1971), as well as plants (chap. 7). At La Selva the large number of species of frog-eating snakes is related to the large number of species of frogs, for example. Participation in mimicry complexes is more frequent with decreasing latitude (Hespenheide 1986, 1992), as are taxa of leaf-mining beetles with hidden eggs (Hespenheide 1991), and both may be the consequence of higher predation intensity in tropical terrestrial environments although actual predation intensities remain to be measured. Even descriptive studies of diets have been relatively infrequent except of individual species or of small ecological assemblages or guilds (chap. 20).

Because of the extent of such phenomena as mutualism and mimicry in tropical wet forests, forms of coevolution in addition to simple exploitative herbivory and parasitism (e.g., the ecological monophagy of Smiley [1978b] and "three trophic level" studies of Price et al. [1980] offer many opportunities for research. In addition to insect-plant mutualisms, vertebrate frugivore studies are a strong tradition at La Selva for both birds and mammals (chap. 22). Gilbert (1983) has said that in tropical forests practically every diurnally active and conspicuous arthropod participates in some form of mimicry and argues convincingly for studying mimicry. Only a few descriptive studies have been published from La Selva (Hespenheide 1984, 1992) although many mimicry complexes involving large numbers of insect species are common there and have not yet been described (Masner 1988).

High species richness permits comparative studies essential to such evolutionary ecological research as reproductive strategies (chaps. 14, 15; also Vitt 1986). High species richness, however, makes community-level research forbiddingly complex. Stream-dwelling fish (chapter 14), leaf-litter amphibians and reptiles (chaps. 15, 16) and guilds of birds (chap. 17) have been the subject of matter of the first attempts at such studies but are only a beginning.

COMPARISON WITH OTHER SITES

As information about the fauna of La Selva accumulates, one wishes to know how typical La Selva is compared with other Neotropical or Old World tropical sites. One quickly discovers, however, that possible comparisons are limited. Taxonomic studies usually favor samples from a variety of localities rather than intensive study at a single one. The obvious choices for repeated studies are field stations like La Selva, but these are few in tropical areas. In Central America the oldest such station is Barro Colorado Island in Panama although information about a number of other areas has been developing rapidly: in Costa Rica at Monteverde and in Santa Rosa and Corcovado National Parks; in Mexico at Los Tuxtlas in Vera Cruz and Chamela in Jalisco. South American sites available for comparison vary with the taxon under consideration because, except for terrestrial vertebrates, none has been inventoried for all groups (Gentry 1990a). Overall, the few sites are widely separated geographically and, therefore, also biogeographically and climatically. For reasons that one might classify as historic the reserves associated with these stations differ in area and in range of altitudes and habitats, both physiographic and successional. All of these factors make comparisons difficult. Nevertheless, in number of species, La Selva's fauna varies from relatively poor (fish) to the richest in Central America (breeding land birds) but is poorer in most groups than are Amazonian localities.

Barro Colorado Island is the obvious choice for intensive comparisons because it is a lowland forest that is relatively close geographically and has a long history of biological research. A volume summarizing research at that site, *The Ecology of a Tropical Forest* (Leigh et al. 1982), focuses largely on seasonality, with more than two-thirds of the chapters in the book devoted to some aspect of that topic. Other "traditions" of research that emerge from the volume include studies of particular mammalian species, especially monkeys, army ants, and associated organisms, and the effect of food availability on animal populations and reproduction. Although vertebrate faunas have been inventoried to assess the importance of insularity, for example, as a cause of extinction (Myers and Rand 1969; Willis and Eisenmann 1979; Glanz 1982), they have been eschewed, even discouraged in the case of invertebrates, in favor of ecological studies. As a consequence, La Selva is already probably as well known faunistically as BCI, with the irony that BCI is part of the Smithsonian Institution, which includes the largest systematics institution in the United States.

In table 19.2 I have compared the terrestrial vertebrate fauna of La Selva with those of BCI and Guanacaste Province (or sites therein) of Costa Rica. Although BCI has higher rainfall than Guanacaste, the sites are similar in having more severe dry seasons than La Selva. On the other hand, the faunas of BCI and La Selva are apparently more similar in being derived historically from a South American fauna, either from recent (birds and mammals) or older (fish, amphibians, reptiles) immigrants, whereas that of Guanacaste includes a significant North American element (analyses for plants, Gómez 1982; reptiles and amphibians, Savage 1982; birds, Haffer 1985; butterflies, DeVries 1987). In table 19.2 I show that the faunas of La Selva and BCI are larger, more similar in size, and share a greater proportion of species. That of Guanacaste is generally much smaller (except for mammals, which are listed for the entire province rather than a single reserve associated with a research station) and shares a smaller proportion of species with La Selva, even though the Costa Rican sites are geographically much closer than either is to BCI. Despite the greater faunal similarity the differences between La Selva and BCI are still considerable: compared to BCI's generally

smaller fauna, shared species constitute two-thirds of the mammals and birds to about one-half of the amphibians and reptiles. In table 19.1 I corroborate this for insects: the faunas of asilid flies, buprestid beetles and zygopine weevils of La Selva and BCI are similar in size, somewhat similar in the proportions of undescribed species, yet share surprisingly few species (20%–35%); that is, they are *not* very similar at the species level. To lump these two sites as "lowland tropical forest" is to ignore profound differences in the composition of their biotas in addition to functional (phenological) ones imposed by differences in the severity of the dry season. Studying one site will not allow researchers to know the other.

As mentioned previously and in the sections on vertebrate taxa, La Selva has a smaller fauna than Amazonian localities. For insects, the Tambopata Reserved Zone of 5,500 ha in southern Peru (Erwin 1985) has more species of Odonata (Paulson 1985), Asilidae (Fisher 1985), Tabanidae (Wilkerson and Fairchild 1985), and Cicindelidae (Pearson 1985) than any other site known in the Neotropics, with about 50% more species than La Selva for the few groups for which both faunas are known; only the Sphingidae have fewer species reported from Tambopata than from La Selva (Lamas 1985). Neither site, however, has been exhaustively collected, and few ecological studies have been made at Tambopata. In the only comparison known to me of insect faunas of New and Old World tropical localities, DeVries (1987) shows Costa Rica's butterfly fauna to be larger than those of Liberia or Malaysia despite their much larger areas.

LATITUDINAL DIVERSITY AND COMPOSITION

La Selva's popularity as a research site derives in part from the high diversity of its biota. Greater diversity in tropical areas compared with temperate ones is documented for Central America and Costa Rica in the few groups for which relatively complete inventories have been prepared, primarily birds (also MacArthur 1969) and mammals. More careful analysis has shown that greater topical diversity does not occur uniformly in all taxa or in all environments that are geographically tropical. For example, Wilson (1974) showed that although mammals increase overall in numbers of species with latitude, most of the increase is due bats. Among mammals other than bats rodents decrease in diversity with decreasing latitude, whereas primates, edentates, and marsupials increase. Dominance of a taxon by a single subgroup may, thus, obscure contradictory patterns in smaller subgroups. Among La Selva's other vertebrate groups amphibians are dominated by frogs and reptiles by snakes, whereas freshwater fish and birds are not dominated by single taxa. DeVries (chap. 13) shows that butterflies increase overall in species number from temperate to tropical regions although the family Pieridae actually decreases.

One can see a similarly complex pattern in the various groups of beetles within the family Buprestidae (fig. 19.1; Hespenheide 1991). The combined diversity of the Agrilinae and Trachyinae increases overall toward the equator from temperate North America, but the two groups behave differently: the wood-boring Agrilinae peak in the subtropics and then decline again toward the equator, whereas the leaf miners (most Trachyinae) sharply increase in the tropics and dominate the family there. Even within the Trachyinae there are changes in dominance among geographic regions at the generic level. Insect distributions may well be more sensitive to climatic differences than vertebrates, and insects may, thus, be better organisms for studying patterns of diversity.

As discussed, one may distinguish a generally less diverse subtropical biota occupying the northwestern Pacific (Guanacaste) coast of Costa Rica with its stronger dry season from a more diverse tropical biota occupying the Caribbean and southwestern Pacific coasts with its greater and less seasonal rainfall (Gómez 1982; Savage 1982; Stiles 1983a; DeVries 1987). The differences in the distribution, diversity, and composition of subtropical and tropical biotas affect the latitudinal gradient in diversity and may be significant enough to have engendered, for example, a controversy over patterns in the diversity of tropical parasitoid Hymenoptera. Using the family Ichneumonidae as an example, Janzen (1981c) proposed that there was no greater and, perhaps, a lower diversity of parasitoids in the Neotropics than in temperate regions. This suggestion was based on sweep sampling (Janzen and Pond 1981), rearings of bruchid beetles (Janzen 1980a) and an analysis of published distributions (Janzen 1981c). Moreover, the pattern seems to be paralleled by the Ichneumonidae of tropical Africa (Owen and Owen 1974) and Australia (Gauld 1986). My data from La Selva for the Chalcididae (Hespenheide 1979), however, indicate diversity for that family of parasitoids is much higher than for any temperate area. Subsequent collections at La Selva indicate that at least some other parasitoid taxa are as disproportionately rich as the Chalcididae although the question cannot be resolved until inventories of insects at both La Selva and a site such as Santa Rosa in Guanacaste are relatively complete.

It may be that the Ichneumonidae will show a "subtropical depression" in diversity similar to those of the Membracidae (Wood and Olmstead 1984) and gall-forming insects (Price 1991). If so, work in areas as close geographically as Guanacaste and La Selva within Costa Rica could lead to very different conclusions about such characteristics of the "tropics" as diversity, seasonality, or coevolutionary relationships. For example, DeVries (1987) reports a lower frequency of mimicry in Guanacaste than in the Atlantic lowlands.

As shown, the Buprestidae show a well-developed subtropical fauna of wood borers compared to a well-developed tropical fauna of leaf miners, a pattern also reflected in faunal differences between Guanacaste and La Selva. In these cases it is not just the subtropical dry-wet seasonality of the Guanacaste region that would differentiate it from wetter and less seasonal La Selva but also its historical biogeographic relationships with subtropical Mexico compared to La Selva's relationships with South America (Gómez 1982; Savage 1982). Central Panama is climatically more seasonal than La Selva but also shows biogeographic relationships with tropical South America and is much more diverse than Guanacaste. This comparison suggests that both history and climate are important in determining the composition and size of a local fauna, that one cannot, therefore, study ecology or biogeography in isolation from the other, and that comparisons of biotas must be made in cognizance of both.

TROPICAL FAUNAL DIVERSITY AND THE FUTURE OF LA SELVA

Tropical faunas have been of special interest to naturalists, both for documenting the greater numbers of species ("diversity") at lower latitudes and explaining it. Those who have analyzed diversity (Soule 1986) point out that one can talk about diversity as having three different components of varying spatial scale: a "within-habitat" component, a "between-habitat" component, and a "geographic" component (the α-, β-, and γ-diversity of Cody 1986). A variety of models have been proposed to explain geographic patterns in biological diversity, but these have not always been careful to specify the scale being discussed. This problem arises in the vertebrate chapters in this volume when comparing numbers of species between sites that differ in area or contain different habitats because of gradients in elevation or inundation. Although Cody (1986) has been one of the few to attempt to partition these components of diversity, it is clear that, in comparison to temperate localities, La Selva has higher numbers of species at the

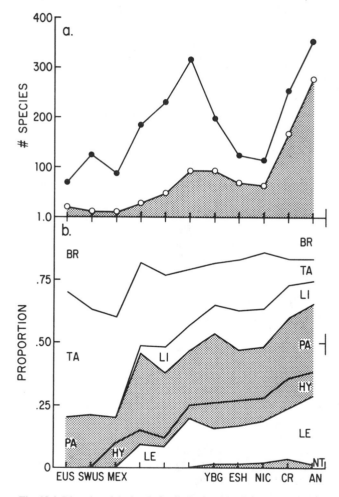

Fig. 19.1. Diversity of the beetle family Buprestidae in North and Central America. *(A)* Numbers of species in the subfamilies Agrilinae (wood borers, open area) and Trachyinae (leaf miners, shaded area). *(B)* Proportions of species in different genera of leaf miners (BR = *Brachys*, TA = *Taphrocerus*, LI = *Lius*, PA = *Pachyschelus*, HY = *Hylaeogena*, LE = *Leiopleura*, NT = *Neotrachys*). Geographic zones: EUS = Eastern United States; SWUS = Southwestern United States; MEX = Mexico; YBG = Yucatán Peninsula, Belize, and Guatemala; ESH = El Salvador and Honduras; NI = Nicaragua; CR = Costa Rica; PA = Panama.

within-habitat scale and Costa Rica is more diverse at the geographic scale. The adjacent Braulio Carillo National Park offers a chance to study the between-habitat component in an elevational transect with La Selva at the lower end.

It is, perhaps, trivially true that the number of species in an area is determined by a balance between processes that increase numbers of species (immigration from an adjacent area, speciation) and those that decrease numbers of species (local or global extinction). Ecologists have argued whether this balance is in equilibrium or not; it is probably in equilibrium on a within-habitat scale and, probably, is never in equilibrium at the geographic scale. Because ecological resources—energy and/or nutrients—are not unlimited, speciation and/or immigration effectively creates rarity as limited resources are divided more finely among species populations. The interesting questions for a given habitat/area is at what level of rarity extinction becomes likely and what factors determine how small populations can be before the probability of extinction becomes significant. This is a question crucial for research stations like La Selva, which are becoming islands of forest in the midst of agriculture. Barro Colorado Island has experienced diminution of its fauna since 1923 by extinction (birds, 24%—Willis 1974; Willis and Eisenmann 1979; Karr 1982; reptiles and amphibians, 3%–9%—Myers and Rand 1969; mammals, 7%—Glanz 1982) partially because of loss of early successional habitats and partly from area effects. La Selva has lost two mammals and others may be in danger (chap. 18); several reptiles may have been lost (chap. 16); and five birds have been extirpated (chap. 17).

Under natural conditions diversity is probably largely owing to the "tolerance" of rarity whether at the scale of a single habitat or of a geographic area. Isolation of field stations like La Selva and BCI by habitat destruction both removes possible sources of immigration and reduces area itself. Because total population size depends on the amount of habitat and the areal density of species, rare (low density) species are at increased risk when area diminishes. If local species number is a dynamic balance between low rates of extinction and immigration, habitat insularization will preclude "normal" extinctions from being balanced by recolonization. Levey and Stiles already report declines in abundances of forest bird species at La Selva, which are hypothesized to be the result of ecological peninsularization. Lovejoy et al. (1986) are attempting to determine experimentally the effect of insularization on bird faunas.

Extinction depends on rarity and rarity must, in turn, depend on the amount of resources and how finely they are divided (the "capacity" and allocation rules of Brown [1981]).

The amount of resources, especially energy, available to the community is influenced by the variability of the physical environment; allocation is based on the abundance, size, and trophic status of the organisms present. When analyzing the difference in the numbers of species of forest birds between an Amazonian community (207 species) and the eastern U.S. deciduous forest community (40 species), Terborgh (1980) concluded that 51% of the increase was the result of increased resources in tropical forests (34% from guilds not represented in the temperate zone, 17% from broader guild niches) and 49% from differences in allocation of resources (by tighter packing of the species). Findley (1976), however, found that differences in numbers of bats between temperate and tropical communities included only differences in resources and not in their allocation. Resources and their allocation are not independent because, for example, forest structure both regulates levels of photosynthesis and affects its subsequent allocation. Consideration of capacity and allocation are important to the problem of extinction at field stations like La Selva. It is interesting that species lost are often extreme specialists (antbirds; Willis and Eisenmann 1979) or the largest members of guilds (e.g., the giant anteater and Harpy Eagle at La Selva). Gilbert (1991) argues for studies of the ecology of rarity in view of the large contribution of relatively rare species to tropical diversity.

The future of research on the fauna of La Selva has immense potential, the significance of which has been anticipated by a small number of pioneering studies. The current fashion of testing evolutionary ecological hypotheses will, undoubtedly, continue, but the ecology of La Selva is the ecology of diversity, which can only begin to be understood when the extent of that diversity is understood; for example, in careful basic descriptive life-history and natural history studies of focal taxa (Gilbert 1991) or of defined subcommunities (Marquis 1991b). Most animal species other than vertebrates are undescribed. Study of the physiological ecology of La Selva animals has scarcely begun, and the importance of seasonality is largely unknown. As the plants that form the physical matrix and energetic foundation for animal communities become known taxonomically and ecologically, the animals that feed on, pollinate, and disperse them—and that also interact with one another—can be placed in a firm, clear context.

Behind these opportunities, however, is the prospect, or risk, that La Selva's fauna will change and diminish as species are lost with little chance for their natural replacement. The next twenty-five years should prove exciting ones for the study of the fauna of La Selva, with important implications for all of biology.

20

Population Biology: Life Histories, Abundance, Demography, and Predator-Prey Interactions

H. Elizabeth Braker and Harry W. Greene

Populations are assemblages of individuals among which there is substantial genetic continuity. They can be characterized by birth rates, death rates, age structures, and other parameters not applicable to single organisms or to communities. Tropical animals long have attracted the attention of population biologists, and, although usually based on short-term studies, several tentative generalities were emerging at the time La Selva became a research station. Early authors stressed reduced seasonality, habitat complexity, low population abundances, and high rates of predation as affecting life-history attributes of tropical organisms more strongly than their temperate counterparts (e.g., Dobzhansky 1950; Cody 1966; Paine 1966; MacArthur and Wilson 1967; MacArthur 1972). During twenty-five years of research at La Selva some of these generalities have been confirmed, whereas new insights into some aspects of animal population biology have challenged others.

La Selva has been and continues to be an important site for population biology. As of 1991, it appears relatively undisturbed in comparison to much of Central America. To some extent disturbance and its effects have been monitored (e.g., local extinction of an eagle, *Harpia harpyja,* and the white-lipped peccary, *Tayassu pecari*). La Selva is part of one of the largest preserved elevational corridors in Central America and, hence, can teach much about local movements and home ranges of animal species. It represents an important Central American sample of Neotropical biota for comparisons with South American sites and is one of few available study sites for the geographically more restricted and distinctive Middle American lowland component of the Neotropical biota. Most importantly, a database now exists on which to build future population-level research.

Our specific goals are to assess the scope of animal population biology at La Selva during its first quarter-century as a research station, to evaluate that work in the context of several conceptual questions, to compare animal population biology research at La Selva with that at other tropical sites and with accomplishments in the field in general, and to suggest topics for emphasis in future research. By focusing on the broader context for research projects and on gaps in our knowledge we hope to inspire a reassessment of research priorities among individual biologists and funding agencies.

LIFE HISTORIES

Theoreticians have argued that tropical animals should exhibit life-history patterns different from those of temperate animals. Dobzhansky (1950) suggested that populations of organisms occupying seasonal and/or severe environments might be limited primarily by density-independent factors, whereas populations in aseasonal or benign environments such as the tropics might be controlled largely by interspecific and intraspecific interactions. MacArthur and Wilson (1967) and MacArthur (1972) explored predictions about life histories of animals in aseasonal environments, and Pianka (1970) discussed the *r-K* dichotomy, with attention to tropical-temperate comparisons. Although the concepts of *r* and *K* selection are oversimplifications, they remain useful in a comparative sense (Stearns 1976; Boyce 1984). In this section we discuss studies of animal life histories at La Selva with particular attention to how they compare and contrast with life histories of animals from other tropical and extratropical sites. We emphasize interactions related to breeding and mating because of their direct link to population dynamics.

Mating Systems

Ecological constraints on sexual selection and resultant patterns of social organization affect population parameters by influencing mating structure, age-specific reproductive schedules, and dispersal (Emlen and Oring 1977; Partridge and Endler 1987). Predictions derived from early life-history models include the idea that animals in less seasonal environments should allocate a high proportion of energy to raising each offspring and should exhibit mating systems allowing them to do so (Williams 1966; Orians 1969b; Tinkle 1969; Cody 1971; Stearns 1976). These ideas have been well developed for avian mating systems. In an early discussion of factors leading to reduced clutch size in tropical environments Skutch (1949) suggested that high predation rates on nests may select for a clutch size smaller than that which parents would otherwise be able to feed. Orians (1969b) noted that where clutch size is strongly influenced by factors other than the number of offspring that can be supported polygynous and promiscuous mating systems should be more prevalent and suggested that such situations might occur in species in which adults feed on low-energy food sources such as pulpy fruits and nectar, which necessitate more frequent trips to the nest. Snow (1971) documented that polygynous and promiscuous mating systems do occur in birds that eat primarily fruit and nectar, dietary items available for tropical residents year-round.

We surveyed work on La Selva's animals to assess what types of mating systems predominate and whether mating systems are more diverse than for temperate relatives. Mating systems described for La Selva insects, birds, mammals, and amphibians range from monogamy to various types of polygyny to promiscuity (table 20.1). Mating systems may have a number of potential indirect or direct effects on population parameters (table 20.1). For example, mating systems that lead to high variance in mating success by one sex (e.g., leks where one or a few males achieve most matings) probably affect population genetic structure (Falconer 1981). We compared mating systems of La Selva species with mating systems of closely related taxa having temperate or seasonal tropical distributions (table 20.1). Ten of twelve species cited have a relative (member of the same genus or family) with a similar mating system. Thus, the available evidence does not indicate an evolutionary trend to polygyny or promiscuity in La Selva animals; indeed, for the species surveyed, mating system may be a phylogenetically conservative trait. A comparative study of mating systems of taxa from La Selva that have sister taxa in temperate localities might help resolve this question.

Parental Care

Parental care is a component of the life history because guarding of young from natural enemies can increase offspring survivorship. The allocation of reproductive energy to offspring care has been studied at La Selva in several insect and frog species. We exclude birds and mammals from this discussion because parental care is typical in these groups and because a comparative study has not been undertaken at La Selva.

Among membracid treehopper species the degree of parental care varied (Wood 1984). *Umbonia crassicornis* females limited parental care to insertion of egg masses into a branch of a host tree and to guarding eggs until they hatched. Even this relatively low level of investment can be important: egg masses survived better with the female present than when she was removed (Wood 1974, 1976, 1983). Female *Entylia bactriana* not only protected eggs and first instar nymphs from predators but attracted protective ants to the egg mass by means of honeydew secretions; after the eggs hatched, the ants tended and defended the treehopper nymphs (Wood 1977, 1983). The percentage of La Selva treehopper species with some form of parental care is about the same as in tropical high elevation and dry forest sites (Wood 1984). Parental care is only one aspect of treehopper life history and should be seen as part of a suite of correlated life-history characteristics, such as tendency to form aggregations, host specificity, and interaction with ant mutualists (Wood 1984).

Parental care is well developed within the aculeate and parasitic Hymenoptera (Wilson 1971). At La Selva females of the sphecid wasp *Trypoxylon superbum* built and provisioned nests as do females of other *Trypoxylon* species. In *T. superbum,* however, males stayed at the nest and guarded the larvae from predatory ants, a significant source of mortality in cavity-nesting wasps (Coville and Griswold 1984). The only sphecid wasps known to be social are in the genus *Microstigmus.* In one *Microstigmus* species at La Selva males and more than one female cooperated in construction, maintenance, and defense of a single nest (Matthews 1983).

In frogs parental care is widely distributed among fourteen of twenty families and appears to be most common in terrestrially breeding tropical species (McDiarmid 1978). At La Selva frog species vary in degree of parental care (chap. 15). Little or no parental care occurs in *Bufo hematiticus,* which deposits large egg clutches in pools and streams (Scott et al. 1983). At the other extreme is the complex form of parental care found in *Dendrobates pumilio.* Adults transport larvae to rearing sites in bromeliads and feed tadpoles by laying unfertilized eggs in the bromeliads (Limerick 1980; Weygoldt 1980; Donnelly 1987, chap. 15).

Parental care probably exists in many other species at La Selva but has yet to be studied in any great detail. The degree to which energy is devoted by one or both parents to caring for offspring must be quantified and its contributions to current and future parental fitness assessed not only for animal populations at La Selva but for their close relatives living in more seasonal environments. Until such studies are undertaken, it will not be possible to use La Selva organisms to test the assertion (e.g., Pianka 1970) that organisms found in less seasonal environments allocate a greater proportion of parental resources to offspring in the form of parental care than do organisms of highly seasonal habitats.

Cooperative Breeding

Cooperative breeding is characterized by the normal presence of helpers at some or all nests (Brown 1987) and is a life-history phenomenon when considered in its demographic context. It is probably very common among tropical birds and has now been documented for seventeen Neotropical bird families (Brown 1987; Skutch 1987). Skutch's early studies, some conducted at or near La Selva (Skutch 1935a, 1935b; 1961, 1987), were instrumental in the recognition of cooperative breeding as an important behavioral and ecological phenomenon.

Cooperative breeding has been postulated to evolve under conditions of population saturation in a given habitat, that is, the same kind of conditions postulated to occur in "K-selected" species in aseasonal tropical environments (Woolfenden and Fitzpatrick 1984; Brown 1987). Another, not mutually exclusive hypothesis, is that cooperative breeding and other complex social systems may arise from flock associations formed for maintenance of long-term feeding territories by resident insectivores and omnivores (Stiles 1983a; Brown 1987; Powell 1989).

In table 20.2 we list bird species for which cooperative breeding has been confirmed at La Selva and species occurring at La Selva known to breed cooperatively at ecologically similar sites. Roles of nest attendants for each species are also shown. La Selva is the site of the first description of cooperative breeding for four species (table 20.2). Representatives of nine other bird families are known to breed cooperatively elsewhere, and their breeding habits at La Selva should be investigated (Picidae, Columbidae, Psitaccidae, Apodidae, Furnariidae, Tyrannidae, Pipridae, Hirundinidae, Muscicapidae, and Emberiziidae; Skutch 1987; Brown 1987).

Table 20.1 Mating systems of La Selva animals

Mating System	Behavioral Attributes[a]	Known or Inferred Effects on Population	Similar[b]	Sources
Monogamy				
Ramphastos swainsonii[c] (Aves: Ramphastidae)[d]	Stable pair bond, defense of permanent, resource-based territories	Stable population size	Y	1, 2, 3, 4
Peropteryx kappleri & *Saccopteryx leptura* (Mammalia: Emballonuridae)	Roost in pairs, forage widely over canopy	Low rates of dispersal from natal area	N	5, 6
Norops limifrons (Reptilia: Iguanidae)	Mutual displays, long-lasting pair bonds	Stable population size	Y	7, 8, 9, 10
Resource Defense Polygyny				
Dendrobates pumilio (Amphibia: Dendrobatidae)	Males defend oviposition sites	Females select males defending sites	Y	11, 12
Thalurania furcata (Aves, Trochilidae)	Males defend flower clumps	High variance in male reproductive success	Y	13, 14, 15
Female Defense Polygyny				
Vampyressa nymphaea (Mammalia: Phyllostomidae)	Males roost with, defend harem in leaf "tent"	High variance in male reproductive success	Y	16, 17
Microtylopteryx hebardi (Insecta: Acrididae)	Males guard mated females	High variance in male reproductive success	Y	18, 19
Male Dominance Polygyny, explosive breeding assemblage				
Agalychnis calcarifer	Breeding limited in time, concentrated in space	High variance in male reproducive success?	Y	20, 21
Male Dominance Polygyny, leks				
Phaethornis superciliosus (Aves: Trochilidae)	Breeding concentrated on leks over long season	High variance in male reproductive success?	Y	14, 22
Manacus candei & *Pipra mentalis* (Aves: Pipridae)	Mating on leks, long season	Not studied intensively at La Selva	Y	23, 24, 25
Promiscuity				
Heteromys desmarestianaus (Mammalia, Heteromyidae)	Intraspecifically tolerant, overlapping home ranges	Low dispersal rates of young	P	26, 27, 28
Mionectes oleaginea (Aves, Tyrannidae)	Males call, attract females	Females disperse, raise young alone	Y	29, 30

Sources: (1) Skutch 1971; (2) Skutch 1976; (3) Howe 1983; (4) Skutch 1960b; (5) Bradbury and Vehrencamp 1976a; (6) Bradbury and Vehrencamp 197b; (7) Talbot 1979; (8) Guyer 1988a; (9) Guyer 1988b; (10) Andrews and Rand 1982; (11) Donnelly 1987; (12) Crump 1974; (13) Stiles 1975; (14) Stiles 1983a; (15) Stiles and Wolf 1970; (16) Brooke 1987b; (17) Williams 1986; (18) Braker 1986; (19) Riede 1987; (20) Marquis et. al 1986; (21) Pyburn 1970; (22) Stiles and Wolf 1979; (23) Breitswich 1982; (24) Gill 1979; (25) Foster 1985; (26) Fleming 1974a; (27) Fleming 1974b; (28) Jones 1988; (29) Sherry 1983a; (30) Briskie and Sealy 1987.

[a]Behavioral attributes known or thought to have an effect on population parameters for each species.

[b]Indicates whether the mating system described for the La Selva population has been described for a close relative occupying a seasonal habitat (seasonally dry tropical or temperate). Y = yes, N = no, P = possible

[c]Species

[d]Class: Family.

Cooperative breeding can take a variety of forms, most of which occur among La Selva birds. At the simplest level it may involve prenesting cooperation as in some manakin species (Foster 1985). Perhaps most commonly, attendants at nests only feed nestlings, reported by Skutch (1961) for the plain-colored tanager at La Selva. In some species, for example, the black-faced grosbeak at La Selva, attendants may defend young as well as feed them (Skutch 1972, 1987; Moermond 1981). Elsewhere in Costa Rica, subordinate groove-billed anis engage in all breeding activities, including laying eggs (Vehrencamp 1977, 1978, 1983; Skutch 1959, 1972, 1987), but the breeding biology of anis near La Selva is un-

known. Nest attendants take a variety of roles at La Selva (table 20.2), but their contribution to nesting success has not been studied nor has the relative nesting success of individuals breeding in a communal fashion versus individuals breeding in pairs. Description of breeding systems and quantification of reproductive success would help elucidate their roles in population dynamics of La Selva avian species.

Allocation of Energy to Growth and Reproduction

Theoretical treatments of life-history evolution assume a causal relationship between life-history components so that an increase in energy allocated to one component of the life history will be associated with a decrease in another (Gadgil and Bossert 1970; Reznick 1983; Stearns 1989). Trade-offs between life-history attributes may be assessed by means of within-taxon comparisons. At La Selva such comparisons are possible only for the Lepidoptera, for which life-history data are relatively abundant. Still, the wide variation in life-history attributes among the few La Selva butterfly and moth species studied delineates some problems in generalizing about life-history patterns (table 20.3). We examined clutch size (number of eggs laid per reproductive bout), egg weight (reflecting the amount of resource allocated by each female to each of her offspring), larval development time (reflecting growth rates), female lifetime fecundity, and total development time from egg to adult (roughly equivalent to age at first reproduction); all vary over a wide range of values among butterflies at La Selva (table 20.3). Clutch size is not a useful parameter for examining life-history trade-offs in butterflies because eggs are laid either singly or in large clusters (Stamp 1980; Chew and Robbins 1984).

If trade-offs exist among life-history parameters, one would expect systematic relationships among them. Rank correlations between pairs of life-history traits (egg weight versus lifetime fecundity, larval, and total developmental times and lifetime fecundity versus larval and total developmental times) for eight species, however, were not significant (table 20.3). That is, there is no evidence for this set of species that the observed parameters bear consistent relationships with one another. The sample size, however, is extremely small. A more useful comparison would examine trends in life-history

parameters within a single lepidopteran family across a climatic or latitudinal gradient. To our knowledge a large-scale comparison of this kind has not been undertaken, but Boggs's (1981a) work on heliconiine butterflies at La Selva exemplifies this approach.

Allocation of nutrient and energy resources to growth and reproduction interacts with other aspects of an organisms's life history and environmental constraints to determine reproductive strategy (Tinkle 1969). Heliconiine butterflies are an excellent group in which to study this interaction; species vary in life-history traits (including life span, number of matings, and egg production) as well as in habitat and nutritional requirements. Boggs (1981a) compared potential reproductive effort among three species: *Heliconius cydno* from La Selva, *H. charitonia* from nearby successional habitat, and *Dryas julia* from Texas. Allocation of larval nutrient resources to reproductive reserves at metamorphosis (ANM) could be predicted by overall adult nutrient intake and reproductive output. As would be predicted by allocation theory, the species from forest habitat at La Selva (*H. cydno*) had the smallest ANM, lowest total reproductive output, and obtained a large proportion of the nitrogen budget during the adult phase.

Comparison with Relatives of More Seasonal Habitats
A key prediction of early life-history models is that animals in less seasonal environments should be characterized by lower total reproductive effort and higher survival rates than relatives in seasonal habitats (Pianka 1970). Life histories of two La Selva vertebrates have been compared to those of close

Table 20.2 Bird species at LaSelva known or suspected to breed cooperatively

Species	Site[a]	Eggs per Nest	Number Female Breeders	Total Attendants	Retention[b]	Role(s) of Attendants[c]	Sources
Monasa morphoeus Bucconidae	LS	2–3	1	4	Yes	Brooding, feeding	3, 4, 13
Mitrospingus cassinii Emberizidae	LS	2	1	3–7	Yes	Feeding	4, 10
Tangara inornata Emberizidae	LS	2	1	4	Unknown	Feeding	4, 10
Caryothraustes poliogaster Emberizidae	LS	3	1	3–5	Yes?	Feeding, defense	4, 12
Crotophaga sulcirostris Cuculidae	GU	<20	1–5	<15	Yes	Nesting, reproduction, incubation, brooding, feeding, defense	1, 2, 4
Pteroglossus torquata Ramphastidae	BCI	3?	1	5	Yes?	Feeding	4, 5
Querula purpurata Cotingidae	TR	1	1	4	Unknown	Feeding, defense	4, 6
Cyanocorax morio Corvidae	MV, LC	2–3	1	<7	Yes	Nesting, feeding, defense	4, 7, 8, 9, 10
Campylorhynchus zonatus Troglodytidae	GU, LC	3–5	1	≥4	Yes?	Nesting, feeding	4, 8
Tangara larvata Emberizidae	BCI, SCR	2	1	1–4	Yes?	Feeding	4, 11

Sources: (1) Skutch 1959; (2) Vehrencamp 1977, 1978, 1983; (3) Skutch 1972; (4) Skutch 1987; (5) Skutch 1960b; (6) Snow 1971, 1976; (7) Skutch 1935b; (8) Skutch 1960b; (9) Lawton 1983; Lawton and Guindon 1981; (10) Skutch 1961; (11) Skutch 1969; (12) Moermond 1981; (13) Sherry 1983b.

[a]Location of research that documented cooperative breeding. GU = Guanacaste, Costa Rica; LS = La Selva; BCI = Barro Colorado Island, Panama; TR = Trinidad; LC = unspecified site, lowland Caribbean Costa Rica; MV = Monteverde, Costa Rica; SCR = unspecified site in southern Costa Rica.

[b]Retention of breeding group into nonbreeding season.

[c]Role of nest attendants (other than dominant or primary pair).

Table 20.3. Life-history attributes of Lepidoptera at La Selva

| Species | Clutch Size (No. Eggs) | Egg Weight (mg) | Lifetime Fecundity (No. Eggs) | Female Development Time | | Female Maximum Reproductive Life Span (days) |
				Larvae (days)	Total (days)	
Automeris phrynon (Saturniidae, Hemileucinae)	275	1.50	275	82	140	5
Parides arcus (Papilionidae, Papilioninae)	5	0.72	253	17	42	60
Parides childrenae (Papilionidae, Papilioninae)	6	0.71	212	17	42	68
Perrhybris lypera (Pieridae, Pierinae)	18	0.10	379	15	34	85
Siproeta steneles (Nymphalidae, Nymphalinae)	1	0.67	268	23	36	110
Anartia fatima (Nymphalidae, Nymphalinae)	1	0.08	521	16	28	50
Heliconius cydno (Nymphalidae, Heliconinae)	1	0.76[a]	217	12.7	23.3	38
Heliconius erato (Nymphalidae, Heliconinae)	1	0.70[a]	—	13.7	23.7	—

Notes: For simplicity variance measures reported in original literature are not included in the table. Results of Spearman rank correlations with P-values adjusted for multiple comparisons: egg weight versus lifetime fecundity, $p > 0.20$; egg weight versus larval development time, $p > 0.50$; egg weight versus total development time, $p > 0.50$; lifetime fecundity versus larval development time, $p > 0.50$; lifetime fecundity versus total developmental time, $p > 0.05$.

[a]Minimum estimate of egg weight based on reported egg diameters, assuming a spherical egg with density equal to that of water ($1 mg/mm^3$).

relatives from more seasonal habitats. The mouse *Heteromys desmarestianus* had a mean litter size of 3.06 at La Selva with about nine offspring born during a ten-month breeding season. The altricial young grew slowly, and females reached sexual maturity at about 7.9 months. The minimum annual probability of survival was 31% for males and 21% for females. In contrast, *Liomys salvinii,* a heteromyid inhabiting Costa Rican dry forests, had a larger mean litter size (3.78) but a shorter breeding season (seven months), producing fewer per capita offspring per year (mean of 6.8). Young *L. salvinii* are precocial, reaching sexual maturity in three to four months (Fleming 1974a, 1974b, 1977a). Annual survival rate for *L. salvinii* was 18%. Although total production of young was similar in the two species, the La Selva species had higher annual survivorship and a greater age at first reproduction than its dry forest relative (Fleming 1977a).

Members of the anoline lizard genus *Norops* (*Anolis* of other authors; see chap. 16) are common in leaf litter at La Selva and at Barro Colorado Island (BCI), Panama. Subtle differences in demography between species at each site suggest an effect of seasonal environments on life-history parameters. Female *N. humilis* at La Selva deposited eggs all year for a maximum of thirty-six eggs per year (Guyer 1986; 1988a, 1988b). A gap in recruitment of juveniles occurred during the dry seasons, however, and may have been an important cause of fluctuations in adult numbers (Guyer 1988a). In the more seasonal forest on BCI, female *N. limifrons* deposited approximately the same total number of eggs (forty per year) but at a reduced rate during the dry season (Andrews and Rand 1974, 1982). Eggs deposited during the dry season apparently suffered high rates of mortality (Andrews 1988). These data suggest that the BCI species possesses more "r-selected" traits (slightly more eggs with higher egg mortality) than does the La Selva species, but a definitive answer will require an in-

depth comparison of long-term demographic data, such as survivorship of sex and age classes during dry and wet seasons, ideally for a species such as *N. limifrons* that occurs at both sites.

ABUNDANCE IN SPACE AND TIME

La Selva is a spatially and temporally heterogeneous habitat. On the largest spatial scale habitats are differentiated by slope and soil characteristics as well as disturbance history. On a smaller scale microclimatic differences or priority effects may induce significant patchiness in resources used by animals (e.g., Hartshorn 1983b; Denslow 1980a, 1987b; Barton 1984; Fetcher et al. 1985; Brandani et al. 1988). The three-dimensional nature of the La Selva forest creates a marked vertical structural component. Seasonal and yearly fluctuations in resource abundance also occur (e.g., Frankie et al. 1974a; Stiles 1978c; Stiles and Wolf 1979; Levey 1988a, chaps. 11 and 22). In this section we examine spatial and temporal patterns of animal abundances at La Selva.

Patterns of Spatial Use
The amount of space used by individuals within a population may be indicative of resource availability in the habitat. A change in density could be accommodated by an alteration in patterns of spatial use. Two studies conducted at La Selva have investigated the use of space by members of a population.

Space-use patterns appear to be independent of population density in *Dentrobates pumilio* (Donnelly 1989b). Supplementation of a key reproductive resource (bromeliads) resulted in a density increase on experimental plots. No effect on size of home ranges, the degree of spatial overlap between individuals, or dispersion patterns was apparent. These data indicate that the increase in density was a result of new individuals being recruited into the population and using habitat that was formerly unoccupied by *D. pumilio.*

A similar result was obtained by Guyer (1988b), who investigated the effect of supplemental feeding on spatial use patterns of *Norops humilis.* Although overall density increased after food supplementation, the size of individual home ranges did not change as a result of food supplementation. Rather, individuals tended to overlap more with individuals of other age-sex groups in their use of space. The results of both studies indicate that individuals of La Selva populations use space in complex ways and that an experimental approach is necessary to elucidate factors responsible for observed patterns of spatial use by individuals.

Patterns of Microhabitat Use

One possible mechanism to maintain high tropical species diversity is fine interspecific partitioning of microhabitats (e.g., MacArthur et al. 1966; Orians 1969a; Karr and Roth 1971). The degree to which an organism is a habitat or microhabitat specialist can affect population structure through demographic phenomena such as vagility, dispersal, degree of inbreeding, and population size.

Many resident birds at La Selva are specific in microhabitat use with a mean of 1.8 habitat categories (of four defined by Stiles 1983) used by each species. The greatest number of resident bird species used forest edge habitats, followed by forest canopy, nonforest, and forest interior. Canopy and edge habitats shared a large number of species, probably because there is not a sharp interface between them (Stiles 1983a). Birds of second-growth and forest edge habitats were found more often in light gaps than in forest understory, whereas canopy and forest interior species rarely foraged in light gaps (Levey 1988b).

Evidence from other taxa for habitat specificity is sketchy but suggestive. Several bat species specialized in foraging microhabitat: *Rhynchonycteris naso* foraged mostly over rivers and *Peropteryx kappleri* in the subcanopy of mature forest (Bradbury and Vehrencamp 1976a). Of thirty-one acridid grasshopper species occurring at La Selva, 36% inhabit primarily forest edge and light gaps; 32%, forest canopy; 16%, forest floor or understory; 7% each, grassland and young second growth; and 3%, marshes (Rowell 1978; H. E. Braker, pers. observation). The butterfly *Heliconius erato* is a habitat specialist found only in relatively open successional areas (Smiley 1978a).

A few La Selva species appear to be more generalized in microhabitat use. Two bat species, *Saccopteryx leptura* and *S. bilineata,* foraged above rivers, above and in the canopy, and in open habitats (Bradbury and Vehrencamp 1976a). *Heliconius cydno* was abundant in forest gaps, along edges, and in the canopy (Smiley 1978a). Some microhabitats intergrade, such as light gaps and building phase forest, and species that typically use edge habitats may appear to be generalized in habitat use. For example, leks of the long-tailed hermit hummingbird, *Phaethornis superciliosus,* were usually located near clumps of flowering *Heliconia* plants at the edges of light gaps (Stiles 1975, 1983a).

The vertical dimension is an important component of microhabitat structure in many tropical forests (Pittendrigh 1950; Orians 1969a; Elton 1973; Papageorgis 1975; Wolda

1979; Erwin 1983a; Rees 1983; Sutton 1983). At La Selva some species appear to have affinities for particular strata. In nymphalid butterflies that feed on fruit as adults, individuals of two subfamilies were found almost exclusively in the canopy and those of two other subfamilies exclusively in the understory (DeVries 1988b). Among avian species, the La Selva forest canopy was dominated by frugivores and parrots, distinct from the suite of insectivores and frugivores found in adjacent understory (Loiselle 1988). Two species of lizard differed in height of perches occupied and in foraging habitat: *Norops limifrons* captured 81% of its prey arboreally and 19% in the leaf litter, whereas the reverse was true for *N. humilis* (Talbot 1979). Four leaf-nesting ant species used vegetation of different heights (Black 1987).

Some La Selva animals, however, use more than a single vertical layer. *Eleutherodactylus* frogs were active during the day in the leaf litter but moved up into the shrub layer at night. Males of several species used vegetation of slightly different height for nocturnal perching and calling (Miyamoto 1982). Of ten cockroach species studied, three were consistently found within the leaf litter, four migrated from understory diurnal perches to leaf litter to feed nocturnally, and three perched and foraged in understory (Schal 1982; Schal and Bell 1986). For mobile species like butterflies actual height above the ground may be much less important in determining microhabitat use than light levels. A breakdown of stratification in habitat use may, therefore, occur at forest edges and in light gaps (DeVries 1988b).

Seasonal Patterns

Although La Selva has a less seasonal environment than some other tropical forests, there are well-documented seasonal differences in resources important to animals. Rainfall, leaf flush, onset of flowering, and fruit ripening all occur seasonally (Frankie et al. 1974a; Stiles 1978c; Loiselle 1987a; Levey 1988a; chaps. 11, 22), and La Selva animal populations respond to temporal and spatial variation in resources to varying degrees.

Seasonal Variation in Abundance. A few studies at La Selva have documented seasonal variation in breeding population size. Five species of hermit hummingbirds were most abundant throughout the dry season and into the early wet season. Some of these numerical fluctuations, however, may be the result of differences in habitat use between seasons and/or migration into or out of La Selva (Stiles 1980a; and following sections). The cicada *Fidicina sericans* emerged in alternate dry seasons, causing sharp peaks in abundance of adults (Young 1972a). Varying rates of immigration, emigration, and survival resulted in significant temporal variation in abundance of adult *Dendrobates pumilio* (Donnelly 1989a). Although yearly mean population size remained relatively constant, densities of *Norops humilis* fluctuated over an order of magnitude seasonally. Densities on control plots ranged from three to twenty-five individuals per hectare with a yearly mean of 10.4 individuals per hectare (Guyer 1988a). Possible mechanisms causing seasonal differences in adult population size are seasonal reproduction, seasonal differences in recruitment and mortality, generation times less than one year, and/or a quiescent stage (such as insect diapause).

Some La Selva populations appear to vary little seasonally. Groups of twelve to fourteen adult *Microtylopteryx hebardi* grasshoppers occurred throughout the year in light gaps (Braker 1991). The cicada *Fidicina mannifera* emerged at low rates year-round (Young 1972a). Apparently, constant population size could result from overlapping generations or generation times greater than one year.

Seasonal Breeding. Both continuous (throughout a yearly cycle) and seasonal reproduction occur among La Selva animals. Although seasonal increases or decreases in birth rates may occur, adults in breeding condition have been found throughout the year in *Dendrobates pumilio* (Donnelly 1989a), *Norops humilis* (Guyer 1986), the vampire bat *Desmodus rotundus* (Janzen and Wilson 1983), the grasshopper *Microtylopteryx hebardi* (Braker 1991), two species of limnocorid bugs (Stout 1982), and *Heliconius* butterflies (Smiley 1978a; Boggs 1979).

Some La Selva species reproduce seasonally, apparently in concert with flushes in resource availability and/or periods of low climatic stress. A peak in reproduction occurred for twelve La Selva hummingbird species during the late dry season (February through April). This peak coincided with a yearly peak in blooming of hummingbird food plants, but the birds did not breed during a second blooming peak in the early wet season (Stiles 1980a). The number of pregnant female *Heteromys desmarestianus* peaked in the early dry season. The young born during this time would probably be weaned three to four weeks later during one of two yearly peaks in fruit fall (Fleming 1974a, 1977a). Bat species at La Selva may track patterns of resource availability with the period of weaning of young occurring at the most favorable times of year. Insectivorous bats bred once per year, early in the rainy season, whereas frugivorous and nectarivorous bats bred twice per year, at the beginning of the rainy season and late in the rainy season (Mares and Wilson 1971; Bradbury and Vehrencamp 1977a, 1977b; Wilson 1979; Janzen and Wilson 1983). At least four species of La Selva frogs are explosive breeders. Reproductive activity is confined to the onset of the rainy season when temporary marshes fill, providing breeding sites and larval habitat (M. A. Donnelly and C. Guyer, pers. comm.).

Seasonal Movements. The composition of the animal species assemblage is dynamic; many animals move seasonally within La Selva or in and out of the area. Such changes occur at several scales, ranging from small (between habitats) to medium (elevational migration) to large (latitudinal migration). Medium- and large-scale movements by La Selva avian species are discussed in detail elsewhere (chap. 17); here we highlight movements with potential population-level consequences.

Small-scale movements, or local changes in habitat, are undertaken by several hummingbirds (Stiles 1980a). Declines in hummingbird numbers in one habitat were often correlated with an increase in another; these shifts usually corresponded to local flowering peaks. Numbers of *Thalurania furcata* declined in mature forest in the early wet season and increased along edges and in second growth during the annual blooming period of *Heliconia imbricata*. This pattern is substantiated by observations of marked individuals that shifted habitat use seasonally (Stiles 1980a). Other animals, particularly those with specific breeding sites (e.g., swamp-breeding frogs) or resource requirements (e.g., frugivorous birds, Loiselle 1988) may also show seasonal shifts in local habitat use.

Medium-scale seasonal movements (in this case, elevational migrations) are probably very common among mobile La Selva species, especially those that depend on resources that shift seasonally in abundance (for a review of elevational migration in birds see chap. 17). Although many elevational migrants are thought to use La Selva only during their nonbreeding season (Blake et al. 1990), several species may have more complex patterns of seasonal habitat shifts. The white-tailed manakin, *Pipra mentalis,* is more abundant during the nonbreeding than the breeding season at La Selva, indicating that a resident population breeds at La Selva and is augmented during the nonbreeding season by elevational migrants (Loiselle 1987a; Levey 1988a). In the white-ruffed manakin, *Corapipo leucorrhoa,* sex and age categories differ in elevational migration patterns. This species breeds in Braulio Carillo National Park; immature males remain there during the dry season while females and adult males migrate to La Selva (Loiselle 1987a). If the La Selva-Braulio Carillo corridor is similar to other elevational gradients in Costa Rica, it is probable that many butterfly and moth species undergo seasonal elevational migrations (DeVries 1987; Janzen 1983). The effects of these seasonal elevational migrations of populations on gene flow and population structure have yet to be determined.

Long distance migrants engage in large-scale seasonal movements; most species do not breed at La Selva but interact with local animals as competitors, predators, or mutualists (Greenberg 1980, 1986, chap. 17). Eighty-five species in the La Selva avifauna are latitudinal migrants (Loiselle 1987b; chap. 17). A few species of other taxa also pass through La Selva during long distance migrations. Thousands of individuals of the uraniid moth *Urania fulgens* pass through La Selva during some, but not all, years (Smith 1972, 1983). A large migration occurred in 1978 (Young 1979a); another in 1987 (H. E. Braker pers. observation). The destination of these moths is so far unknown.

We are only beginning to appreciate the complexity of the spatially and temporally variable mosaic of animal abundances at La Selva. With the exception of recent studies on avian populations few quantitative data are available on seasonal abundance patterns. Such data will be critical to understanding population dynamics at La Selva and for comparing them with animal populations in other tropical forests.

DEMOGRAPHY AND LIMITING FACTORS

The relative significance of factors regulating natural populations has been controversial among ecologists for many years. Dobzhansky (1950) suggested that density-dependent factors, such as interspecific interactions, might be important in regulating tropical animal populations, whereas density-independent factors like weather should be more important limiting factors in temperate populations. Others have suggested that tropical populations should have relatively constant densities near carrying capacity (Elton 1958, 1973; Pianka 1976). Recent studies, however, have demonstrated large fluctuations in population densities of some insects in both seasonal and aseasonal tropical forests (Wolda 1978a, 1979, 1982). A continuum of factors is likely to impinge on any one species at any one time, and the relative importance of such factors can be elucidated only by carefully designed experimental manipulations. Several long-term studies of dynamics of animal populations have been conducted at La Selva. Early workers used observational approaches to track population characteristics over time, whereas recent efforts have manipulated environmental factors to measure their effects on demography. Surprisingly few studies have reported one very basic demographic parameter, population density, underscoring the lack of even this most basic information for most organisms at La Selva.

More detail is available for a few species. Densities of *Heteromys desmarestianus* ranged from seven to eighteen individuals per hectare, with little seasonal fluctuation. Densities of both adult and juvenile *Dendrobates pumilio* varied over time (Donnelly 1989a). For adults, population size varied owing to immigration, emigration, recruitment, and survivorship. For juveniles, only the last three factors were important (Donnelly 1989a).

Most species of *Heliconius* butterflies maintain low density, stable local populations (Smiley 1978a; C. A. Boggs, pers. comm.). Forest-inhabiting species (e.g., *H. cydno*) had lower population sizes (0.3–1.5 individuals/ha) than species occupying successional habitats (*H. hecale* and *H. erato;* 1.4–5.0 individuals/ha). Numbers of *H. sara* fluctuated dramatically, however, during the dry season (from ca. 100 individuals/ha in December to 10 individuals/ha in February [Smiley 1978a]).

Factors Limiting Animal Populations at La Selva
Climatic factors are partially responsible for differential mortality of two predaceous naucorid water bug species (Stout 1978, 1981a, 1982, 1983). In the fast-flowing Quebrada Sábalo, both species incurred severe population losses during the wet season; in the low-gradient Quebrada Surá, bug population densities were not as severely affected by occasional back-flooding during the rainy season. Dispersal and recruitment characteristics of the two species were also related to environmental conditions: densities of nymphal *Limnocoris insularis* were higher in the back-flooding Surá than in the Sábalo, probably because this species is vestigially winged and must colonize new habitats or recolonize old ones by swimming upstream. Adult population densities of *Cryphocricos latus* were similar in both streams; *C. latus* has a winged adult stage and colonizes upstream habitats by flying. Stout's detailed studies of these naucorids indicate that climatic factors are important in population structure; other factors that potentially may limit bug densities remain uninvestigated.

Evidence is accumulating from La Selva to support the view that predation plays an important role in limiting tropical animal populations and, therefore, in structuring tropical animal communities (Skutch 1949, 1967a, 1985; Klopfer and MacArthur 1961; Margalef 1968; Cody 1966, 1971; Ricklefs 1969; Strong 1983). Studies of the interactions of hispine

beetles and their *Heliconia* food plants, many conducted at La Selva, have been influential in the development of theories of dynamics of ecological communities (reviewed in Strong 1983; Strong et al. 1984; chap. 21). One important feature of the biology of hispine populations is the high mortality of immature stages, owing principally to parasitoids and predators. At La Selva parasitic chalcidoid wasps caused mortality of 35%–49% of the eggs of *Cephaloleia consanguinea* (Morrison and Strong 1981) and up to 75% of the pupae of *Chelobasis bicolor* (Strong 1982a). Although a key factor or life-table analysis has not been performed for these beetles, evidence indicates that juvenile mortality caused by parasites is extremely important in maintaining populations at low densities, more so than other biotic factors (Strong 1983).

Mortality of *Heliconius* butterfly larvae is more than 95% at La Selva, mainly because of the activities of predatory ants and parasitic wasps. Rates of predation varied between habitats with *Passiflora* vines in edge habitats receiving significantly more ant visitation than those in other habitats. Similarly, egg parasitism rates varied between habitats with 53% of eggs on vines in successional habitats parasitized compared to none in forest habitats (Smiley 1978a).

Although army ants have not been studied intensively at La Selva, they play an important role in regulating arthropod abundance in other Neotropical forests (Wilson 1971; Franks 1982; Franks and Bossert 1983). In one short-term study, arthropod abundance and species richness were reduced following the passage of foraging army ant swarms through the La Selva understory. The number of individual isopods was reduced by 44% and of chelonethids by 71% although numbers of other taxa did not change (Otis et al. 1986).

Predation by a chironomoid midge (*Pentaneura*) reduced density of larvae of two mosquito species in *Heliconia* bracts. In fact, the midge completely excluded one species from the bracts, even though the mosquito apparently can survive and grow in bract fluid. The effects of the midge on mosquito populations depended on the presence of a copepod, which served as an abundant alternative prey, thereby maintaining midge density even if mosquitoes were extirpated (Naeem 1988).

If a population is limited by the availability of a key resource (e.g., food, shelter, access to mates), it should show a density-dependent response to altered abundance of the critical resource. Several studies at La Selva have taken the approach of manipulating a key resource to assess its effects on population dynamics.

The importance of food in limiting population growth of *Norops humilis* was studied experimentally by Guyer (1988a, 1988b). Food supplementation did not affect individual home range size but did increase growth rates of males, egg production rates, and home range overlap of some sex/age groups. Manipulation of food resulted in an approximate doubling of population density with significant increases in numbers of hatchlings, immigrants, and survivors. The increase was temporary, however, because recruitment did not increase and loss of juveniles to death or emigration increased. The density of adults in the population increased, but this was attributed to increased survival and immigration (Guyer 1988a, 1988b). Guyer (1988b) suggested that experimental food supplementation resulted in increased competition for food and resultant high mortality and proposed that seasonal flushes of arthropod food resources could trigger similar demographic responses in their predators.

In contrast to *Norops,* hispine beetles at La Selva are probably not food-limited (Strong 1981, 1982a). These beetles eat rolled leaves of *Heliconia* (Strong 1977b). Beetle densities on host plants were typically low (20%–40% of rolled leaves are unoccupied by hispines). Leaves occupied by beetles received relatively little damage: over the first sixty days of development of twelve new *H. imbricata* leaves, two species of hispines removed less than 0.116% of leaf area (Strong 1983). Quality of the leaf material does not seem to be involved; fertilization of host plants did not lead to increased growth rates of hispines (Auerbach and Strong 1981).

Densities of the territorial frog, *Dendrobates pumilio,* were limited by the availability of a key reproductive resource (Donnelly 1989a, 1989b). Both male and female densities increased on experimental plots with addition of bromeliads, used for rearing tadpoles. Addition of leaf litter, used for oviposition, did not lead to density changes. The demographic mechanisms responsible for increased density on plots with bromeliad additions were increased survival of males and increased immigration, recruitment, and survival of females (Donnelly 1989a).

The small number of demographic studies at La Selva suggest a few tentative conclusions with three preliminary caveats. First, some populations undergo regular seasonal fluctuations. Investigators who sample annually but not continuously might happen to sample populations during the same phase of a seasonal cycle and, thus, assume a nearly constant population size, whereas sampling during different phases of a density cycle would yield very different conclusions. A second caveat is that many populations at La Selva exhibit spatial variation in density. In the three-dimensional habitat mosaic of the La Selva forest number of individuals per hectare may be a misleading measure of overall density. Many species are found in high density in light gaps but not in primary forest understory (e.g., *Microtylopteryx hebardi* grasshoppers, Braker 1991; *Heliconius sappho* butterflies, Smiley 1978a; and several bird species, Stiles 1983a, Levey 1988b). Reporting densities per unit of habitat area would provide a clearer picture of overall abundance of these animals. Finally, most studies at La Selva have been short-term ones. Only a few have tracked population densities of any animal for even one year (Fleming 1977a; Smiley 1978a; Stiles 1980a; S. Liebermann 1986; Donnelly 1989a, 1989b; Guyer 1988a, 1988b); to our knowledge, no study has closely followed a population for more than two years. Any conclusions about the dynamics of tropical animal populations are, therefore, preliminary at best.

Despite these caveats, several factors have been found to limit densities of animal populations at La Selva. These include both abiotic (e.g., flooding) and biotic factors (e.g., predators and parasitoids, competition for food or reproductive resources). Several factors, however, undoubtedly interact to determine the density of any particular species at any time. Construction of life tables and key factor analyses would help identify those factors having the greatest effect on populations, and additional experimental manipulations would provide insight into their relative importance.

PREDATOR-PREY INTERACTIONS

Knowledge of predator dietary patterns and of the way predators interact with prey populations is essential for understanding the nature of tropical food webs and community structure (Skutch 1949, 1967a; Klopfer and MacArthur 1961; Paine 1966; Ricklefs 1969; Janzen 1970; Connell 1971; Strong 1983; Terborgh 1988). We first review studies of diets of predators on arthropods and vertebrates at La Selva and then discuss the role of predators in structuring higher order interactions within La Selva animal communities.

Predators of Arthropods

A large assemblage of predators on arthropods exists at La Selva. Using species lists, published descriptions of food habits, and our own experience, we inferred the degree of insectivory among La Selva vertebrates. At least 48 mammal species (including 35 bat species) eat mostly arthropods, and 18 other mammals (including 5 marsupials) probably include a large proportion of arthropods in their omnivorous diets (Janzen and Wilson 1983; Robinson and Redford 1986). For birds, comparable figures are 115 and 30 (Stiles 1983a); for reptiles, 19 and 5 (Scott et al. 1983); for amphibians, 36 and 4 (Scott et al. 1983).

Predators on arthropods at La Selva range from narrow dietary specialists to broad generalists. The sphecid wasp *Trypoxylon xanthandrum* provisions nests with spiders of a single genus. The size distribution of prey spiders was smaller than that of congeneric spiders collected from the field, indicating that the wasps were strongly selective of both taxon and size of prey (Coville and Griswold 1983). The assassin bug *Salyavata variegata* is highly specialized for capturing worker termites at their arboreal paper carton nests (McMahan 1982, 1983). Two of sixteen species of tyrannid flycatchers are consistent narrow specialists (Sherry 1982, 1984). For example, 94% of the diet of *Terenotriccus erythrurus* was Homoptera (mostly fulgoroids).

Generalist feeders on arthropods include both polyphagous species that consistently forage on a variety of prey taxa and opportunistic species that exploit patchily available food. Some species of flycatchers in Sherry's (1984) study exemplify the former strategy, such as *Myiornis atricapillus,* which feeds on cryptic, immobile, widely distributed arthropods. Other flycatcher species, particularly overwintering species like *Contopus borealis,* feed opportunistically on aggregated, ephemerally abundant food resources (Sherry 1984). The foliage-gleaning phyllostomid bat *Micronycteris megalotis* is also a dietary opportunist. In September–October 40% of species consumed by *M. megalotis* were Lepidoptera, but during December and January, less than 5% were lepidopterans, and proportionally more orthopterans and coleopterans were eaten (LaVal and LaVal 1980).

Predators of Vertebrates

Approximately one hundred species of snakes, birds, and mammalian carnivores eat vertebrates at La Selva, and these are the subject of an ongoing team study (Greene 1988). The goals are to assemble autecological profiles for as many species as possible and to use them in aggregate to answer some community-level questions. Two species exemplify some preliminary findings.

The jaguar (*Panthera onca*) is a large, Neotropical cat. It eats a variety of vertebrates: predominantly capybaras (a large rodent) at a swamp site in Brazil (Schaller and Vasconcelos 1978); turtles and crocodilians in primary forest in Peru (Emmons 1987); and armadillos in second growth in Belize (Rabinowitz and Nottingham 1986). Our preliminary data indicate that jaguars at La Selva eat a wide range of vertebrate prey, with sloths comprising almost half of the diet. Green iguanas are important but only during the dry season when females come to the ground to lay eggs (Greene, Braker, and Santana unpublished data). These findings strengthen the generalization that jaguars are opportunistic predators, emphasizing locally abundant prey types (Emmons 1987). They also suggest that these cats might play a larger role than previously suspected in tropical ecosystem function through their effects on the behavior and population dynamics of large arboreal folivores (Greene 1988; Terborgh 1988; see fig. 20.1).

The bushmaster (*Lachesis muta*) is a large viper (2–3 m total length) that is restricted to Neotropical lowland wet forests. Examination of stomach contents and scats showed that at all ages and throughout its range, this snake feeds largely on rodents of the genus *Proechymys*. Radiotelemetrically facilitated observations at La Selva indicate that these snakes lie in wait for prey for days to weeks and abruptly move to new hunting sites following prey capture. At La Selva hunting sites usually are close to a palm (*Welfia georgii*), the fruits of which are eaten by *Proechymys* rats (Greene 1986; Greene and Santana 1983). The response of rodents and bushmasters to seasonality of fruit fall in *Welfia* (Vandermeer et al. 1979) remains to be studied.

From these sorts of data patterns are emerging regarding potential interspecific interactions among vertebrate predators.

1. Most predator feeding guilds contain members of each class of vertebrate, so a "community" study of only raptors or only snakes would miss numerous instances of diet overlap. For example, the double-tooth kite (ca. 100 g) probably overlaps more with the jaguar ($\geq$ 75 kg) in diet than with most other raptors because they both eat green iguanas, even though different iguana life stages may be eaten by each species (Greene 1988).

2. Although some tropical predators have narrow diets, many of them are opportunistic generalists. Eyelash pitvipers (*Bothriechis schlegelii*) eat frogs, lizards, birds, oppossums, bats, and rodents. Crane hawks (*Geranospiza caerulescans*) take whatever they locate in bromeliad axils and tree holes, including orthopterans, frogs, lizards, snakes, nestling birds, and bats.

3. Several tropical raptors feed largely or entirely on reptiles, especially snakes. The snake fauna is rich (at least fifty-six species; chap. 16), and about half of the species feed largely or entirely on frogs. The latitudinal gradient in predator species richness is caused mainly by an increase in snakes and raptors toward the equator, and that, in turn, probably rests in part on the high abundance and species richness of tropical frogs.

COMPARATIVE RESEARCH EMPHASES

The dynamics of population processes and relative importance of limiting factors at La Selva and other sites cannot be compared directly because few studies have used comparable methods and closely related organisms. We know of no comparisons of La Selva animal populations with those of other tropical lowland wet forests, and only Fleming (1974a, 1974b, 1977a) and Bradbury and Vehrencamp (1976a, 1976b, 1977a, 1977b) compared attributes of La Selva populations with close relatives in seasonal dry forest. The volume edited by Gentry (1990a) highlights the lack of parallel studies at La Selva and other Neotropical sites, making explicit comparisons impossible.

We compared topics emphasized in population biology studies at La Selva and at other tropical sites with emphasis on these topics in the general ecological literature for a five-year period. We surveyed the journals *Ecology* and *Biotropica* for the years 1982–86 for articles concerned with animal populations and classified them as life histories, patterns of abundance in space and time, demography, and predator-prey interactions. We included only those studies that explicitly involved population phenomena: studies of competition were included if population densities were reported but were excluded if only mechanisms or outcomes were stated; behavioral ecological studies were included if it was clear that the behavior in question had a direct and specific effect on some aspect of population biology (studies on mating systems were included, those on foraging behavior excluded). Approximately two hundred papers published in *Ecology*, sixty in *Biotropica*, and results from seventy studies conducted at La Selva and published in other journals were included in the analysis.

Scarcely 15% of papers meeting the above criteria in *Ecology* during the five-year period were conducted in tropical

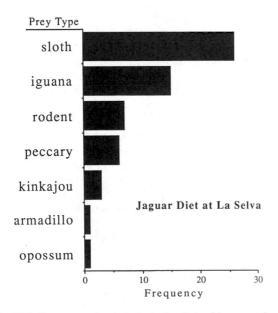

Fig. 20.1. Frequency of prey types in the diets of jaguars at La Selva based on analysis of feces.

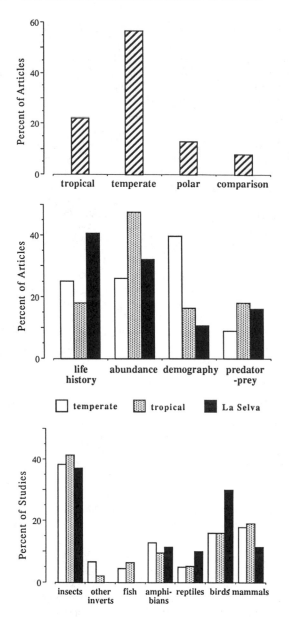

Fig. 20.2. Trends in animal population ecology, 1982–1988. Top: Percentage of population biology articles in the journal *Ecology* based on research at tropical, temperate, and arctic latitudes, and latitudinal comparisons (*N* = 200). Center: Percentage of animal population biology studies done in the temperate zone (open bars), in the tropics (hatched bars), and at La Selva (shaded bars) devoted to topics covered in this review. Bottom: Population-level studies on various animal taxa in the temperate zone, in the tropics, and at La Selva. Symbols as in B.

areas (fig. 20.2A). During the five-year period a majority of ecological studies in the temperate zone concerned life histories and factors limiting the abundance of organisms in space and time. Tropical population ecologists have studied similar topics but have emphasized seasonal patterns of abundance and movement (fig. 20.2B). Proportionally more attention has been paid at La Selva to life histories and interactions of predators and prey (fig. 20.2B). Temperate and tropical population biologists have focused on insects and birds (fig. 20.2C) with fewer studies on other invertebrate taxa or fish. At La Selva,

taxa have been studied in about the same proportion as at other sites except for an even greater frequency of bird studies and a paucity of studies on fish and noninsect invertebrates (Fig. 20.2C). Research emphases at La Selva have been similar to those at other tropical and extratropical sites but with greater emphasis on aspects of life-history phenomena and on particular taxa.

CONCLUSIONS AND FUTURE RESEARCH NEEDS

Early research at La Selva included in-depth descriptive studies of life histories and feeding ecology, whereas more recent studies have examined factors limiting populations and determining interactions. One can now ask whether studies done at La Selva support, refute, or have no bearing on generalizations about expected attributes of tropical populations.

We found limited support for traditional predictions of life-history theory for tropical animals. Some La Selva animals (e.g., *Heteromys desmarestianus, Heliconius cydno*) have significant energy investment by adults in offspring, are long-lived, and have relatively low and stable population densities, especially when compared to relatives from seasonal habitats. Other animal species at La Selva, however, have high reproductive rates, little parental care, and short life spans. In contrast to the idea that tropical populations are stable over time and in space, we find evidence for a large degree of spatial and temporal variability in most La Selva animal populations studied (e.g., *Dendrobates pumilio*). A variety of factors, both biotic and climatic, contribute to limiting animal populations. Predation appears to be a key element in structuring La Selva communities. Despite these advances, much basic information is still lacking on animal population biology. We suggest several focal areas for future research.

Except for a handful of studies, most information on animal life histories at La Selva is hidden within introductory paragraphs or confined to descriptions of one or a few individuals. Animal population biologists cannot ignore the need for empirically strong, descriptive natural history studies, especially as a prelude to design of studies on factors important in determining life histories (Greene 1986; Strong et al. 1983). New observational technologies (video recording, computer-assisted information gathering, and long-range radiotelemetry) can be coupled with increasingly sophisticated methods of analysis derived from population genetics and multivariate statistics.

Refinements of classical comparative methods in biology have proved useful for assessing differences among and within taxa (e.g., Brooks and McLennan 1991). With few exceptions (Fleming 1974a, 1974b, 1977a; Boggs 1981a; Braker 1989a) studies on animal life histories at La Selva have not included between-species or between-site comparisons.

It is increasingly clear that variation in life-history parameters may play an important role in population structure (Reznick and Endler 1982; Reznick 1983; Endler 1986b; Dunham et al. 1988), yet no published study has examined population variation in life-history traits for a La Selva animal. Such research could profitably focus on abundant or easily observable organisms about which a fair amount of life-history information is already available (e.g., *Anartia, Heliconius, Micro-*

tylopteryx, Norops, Dendrobates, Thalurania, Phaethornis, Heteromys).

Key species, representing different taxa and trophic groups, should be monitored on a continuous basis. Knowledge of long-term population trends is essential for understanding ecosystem function and for management (May and Seger 1986; Soulé and Kohm 1989). Low population densities, habitat complexity, and the vagaries of obtaining consistent funding for population research may make such studies difficult. Current concern over the population consequences of habitat reduction and insularization of reserves, however, makes demographic data on tropical forest animals of critical importance. The absence of such data for La Selva animals is particularly glaring when compared with the large and growing body of information available on tree populations at La Selva (chap. 7).

Several species are of particular interest in this regard. For some animals, as noted, a substantial research effort at La Selva has already resulted in a strong basis for future studies. Other species have been studied in detail at other Neotropical lowland sites and, a comparison with La Selva populations would be worthwhile (e.g., leafcutter ants, howler monkeys, two- and three-toed sloths). Still other species deserve study because of their potential roles in community structure as "keystone mutualists" (e.g., *Heliconius,* orchid bees, hummingbirds); as important prey species (e.g., agoutis, sloths, *Eleutherodactylus* frogs); or as predators (e.g., army ants, antbirds, leaf-litter herpetofauna).

There is a pressing need for demographic studies coupled with simultaneous measurement of resource availability and intensity of predation. We need more quantitative, experimental studies, such as those of Guyer (1988a, 1988b) and Donnelly (1989a, 1989b) that demonstrate mechanisms affecting demographic trends.

A few studies have focused on single members of the leaf-litter fauna and have contributed significantly to understanding of factors important in limiting populations at La Selva (Donnelly 1989a, 1989b; Guyer 1988a, 1988b). In one study S. Liebermann (1986) reported overall abundance of litter herpetofauna. Leaf-litter organisms are, for the most part, easily captured and observed, can be used in replicated experiments, and may be of great importance in ecosystem function. We urge additional use of leaf-litter organisms in population-level studies.

If population dynamics and community structure are indeed different in the tropics than in the temperate zone, research efforts should be directed both to trophic and taxonomic groups unique to Neotropical communities and to groups that occur in both temperate and tropical ecosystems. Examples of the former at La Selva are onychophorans and caecilians, of the latter, tettigoniid orthopterans and anoline lizards.

We encourage animal population biologists to take a broad-scale view of the Braulio Carillo–La Selva corridor, such as is being done for birds (Stiles 1983a, 1988; Blake et al. 1991; Loiselle 1987a). The ecology and population biology of elevational migrants and mobile species cannot be understood by studying them in only part of their ranges. For some common species traditional techniques (long-term, direct observations of marked individuals) could be combined with newer, technologically facilitated approaches (e.g., radiotelemetry, DNA fingerprinting).

Habitat and microhabitat use at La Selva should be quantified; we encourage future workers to refer precisely to microhabitat utilization. Many species at La Selva have narrow habitat requirements, but some habitat generalists also occur. We lack even a crude general picture of microhabitat requirements of La Selva animals. Comparisons of habitat use by a taxonomic group or a foraging guild at La Selva with relatives or ecologically similar animals in a more seasonal forest (e.g., Greenberg 1979; Thiollay 1988) should prove useful. The significance of microhabitat specificity for population structure should be addressed. Important areas of study include movement patterns and dispersal rates of animals with restricted habitat requirements, the mechanisms by which mates are found, and effects on both gene flow and population structure.

The studies reviewed provide initial evidence that predators may have large effects on prey populations and on higher order, community-level interactions at La Selva. In future studies, prey items should be identified at the greatest possible level of resolution because imprecise identification can lead to inaccurate estimates of diet breadth and overlap (Greene and Jaksic 1983) and does not provide information on the interactions of predator and prey populations. In most studies of insectivore diets at La Selva prey has been identified only to order or suborder (e.g., LaVal and LaVal 1980; Sherry 1982, 1984). Our present level of taxonomic sophistication permits identification of La Selva insect prey only to the generic or familial level (chap. 19), which underscores the need for continued faunistic work and the deposition of voucher specimens for ecological studies.

We concur with Otis et al. (1986), who emphasized the need for quantitative data on the impact of foraging by army ants and their followers and extend their plea to include the activities of other key predators, such as other ants, parasitic wasps, insectivorous birds, and some snakes. Studies such as those of Gradwohl and Greenberg (1982) that document the effect of a particular insectivore species on its prey would be particularly useful.

La Selva will become increasingly isolated from other forest, and land use in surrounding areas will continue to change dramatically. OTS will be forced to decide to what extent management of animal populations is necessary to maintain integrity of the reserve. Detailed, long-term studies on population biology will be crucial for designing and implementing such efforts.

PART IV

PLANT-ANIMAL
INTERACTIONS

COMMENTARY

Henry A. Hespenheide

The interactions between plants and animals have become a major focus of ecology in recent years. In this section authors review research on the ways in which animals exploit plants as herbivores and on their more frequently mutualistic relationship of frugivory. Of these two kinds of interactions, the latter has a long tradition of research at La Selva; the former is ecologically more pervasive but has received less study for reasons I describe. Pollination biology, the other important mutualistic plant-animal interaction was treated in the section on the plant community primarily because research at La Selva has concentrated largely on the effects of pollination on plant reproduction and evolution and has infrequently been concerned with the biology of the pollinators. In that respect studies of pollination that focus on animal pollinators form a major area of research yet to be developed at La Selva.

The study of herbivory at La Selva has largely been the study of herbivory by insects and has involved only a few systems in any depth: *Heliconius* butterflies feeding on *Passiflora*, hispine beetles on *Heliconia*, various insects on *Piper*, and acridoid grasshoppers on various plants. Conceptually, this research has primarily considered levels of host specificity of the herbivores, the possible occurrence of competition, and the effects of herbivory on plant fitness, with the minor theme of protective mutualism between ants and plants.

Although much has been accomplished, it is also clear that a vast amount of research remains to be done, both in view of the number of species of plants and of plant-feeding insects and of the neglect of mammalian herbivory. Marquis and Braker make a particular plea for extensive study of the natural history of herbivores: life histories, host relationships, and the relative effects of herbivores on the defenses and phenology of their plant hosts. Except, perhaps, for *Piper arieianum*, no complete herbivore fauna has been compiled for any La Selva plant. Unfortunately, because of the appallingly incomplete knowledge of the insect fauna, such inventories cannot be made with the expectation that even as many as half of the species would be known to science. Although La Selva is climatically less seasonal than such localities as Barro Colorado Island (BCI), herbivore faunas and their effects are clearly seasonal in ways yet to be studied. With such major systems as seed predators and leaf-cutting ants still to be comprehensively investigated there, the study of herbivory is clearly an area of great research potential at La Selva.

As indicated, the study of frugivory has been one of the stronger traditions in vertebrate research at La Selva, easily justified in terms of the proportion of plant species that produce animal-dispersed fruits and the number of vertebrates depending primarily or partially on this resource. Studies at La Selva have focused on birds and, therefore, lack a parallel with the extensive work on frugivory by mammals at BCI. Results from work at La Selva have blurred distinctions between "bird fruits" and "bat fruits," brought into question generalizations about patterns of fruiting, and failed to support the generalist-specialist hypothesis of frugivores and fruit types; that is, they have shown that nature is complex. Although these studies have not identified figs or any other plant as a "keystone" resource, they have shown that tree-fall gaps represent resource "hotspots" that have important implications for both birds and plants. Fruit resources exhibit wet season peaks and dry season troughs in availability and are associated with local, elevational, and larger-scale movements of frugivores. Despite the mutual benefits of frugivory to plants and birds, *coevolution* occurs nearly always in the *diffuse* sense of reciprocal adaptation and not, or rarely, in the restricted sense of an exclusive relationship between two species, which is probably not surprising in a community with hundreds of species of both frugivores and fruit-bearing plants.

In chapters 21 and 22 authors point to the need for long-term studies. The fact that La Selva is less seasonal in rainfall protects it from the pulses of productivity and insect activity that occur at more seasonal sites like BCI. As Marquis and Braker point out, there are no conspicuous examples of defoliation at La Selva similar to those documented for Barro Colorado Island. Many years of observations of particular systems, however, such as the liana *Byttneria aculeata* and its associated herbivores, suggest that striking year-to-year differences may occur. Some of these appear to be the result of year-to-year differences in rainfall, but not all are so obvious. Anyone with an evolutionary point of view should not be surprised that nature is dynamic. Long-term monitoring of a variety of biotic systems is required to establish the baselines against which shorter-term studies can be compared. As Levey et al. point out, more than nine years of fruiting data have been collected at La Selva in the past eighteen years, albeit on a limited scale, and comparable long-term data should be collected for insects and vertebrates. The thoughtful design and implementation of such a monitoring program could be the centerpiece of the next decades of research at La Selva.

21

Plant-Herbivore Interactions: Diversity, Specificity, and Impact

Robert J. Marquis and H. Elizabeth Braker

The empirical basis for our present understanding of plant-herbivore interactions comes largely from studies in temperate regions. This bias exists despite the assertion that herbivores represent a stronger selective force for their host plants in tropical communities than in temperate habitats (Doutt 1960; MacArthur 1969; Baker 1970; Levin 1976). The high plant species diversity characteristic of tropical communities has been attributed to spatial patterns of seed and seedling mortality resulting from herbivory (Janzen 1970; Connell 1971; see also pt. 2 introduction). Natural selection associated with herbivore attack is hypothesized to cause greater interspecific allelochemical diversity in tropical than temperate communities (Southwood 1977; Janzen 1983d). Geographic variation in selection pressure associated with herbivory is also thought to have produced clines in secondary chemical composition with a greater number and toxicity of compounds per plant species at lower latitudes and lower elevations (Levin 1976, 1978; Moody 1978; Levin and York 1978; Acosta-Solis 1979; Langenheim 1984; Coley and Aide 1991; but see McCoy 1978). In general, the data necessary to evaluate these hypotheses are lacking. Researchers are now just beginning to estimate herbivore abundance, diversity, and degree of specialization on individual host plants of tropical communities, the allelochemical composition of those plants, and the impact of herbivores on both individual plants and plant populations.

In this chapter we review current understanding of the factors known to control herbivore populations at La Selva and the degree of host plant specialization of those herbivores, determinants of differential damage both within and among plant species, and, finally, the impact of herbivory on individual plants and plant populations. Research at La Selva has contributed significantly to present understanding of tropical plant-herbivore interactions, particularly factors controlling insect population levels and the selective influence of herbivores on plant populations. In other areas, such as the importance of seed predators for plant population dynamics, biologists have limited information on the species involved and considerably less knowledge of the nature of their interactions. Our review reflects this variance in depth of study. We have attempted to be comprehensive to present the range of studies that have been conducted and to stimulate future studies in areas in which research has been particularly limited.

HERBIVORE FAUNA AT LA SELVA

In this section we review the herbivore species known to occur at La Selva. We provide a preliminary classification of their feeding habits based on published studies conducted at La Selva and other Neotropical sites and corroborated by our own field observations and those of colleagues.

Vertebrates

In table 21.1 we summarize the dietary habits of all vertebrate species known or suspected to feed on plant material at La Selva. "Plant material" includes leaves, stems, roots, flowers, and seeds. The vertebrate fauna can be grouped as obligate herbivores that feed primarily on plants year-round (plant material comprising greater than 50% of the annual diet) and opportunistic herbivores that feed only occasionally or seasonally on plants (plant material comprising less than 50% of the diet) (table 21.1). We do not include those species that are primarily frugivorous or consume pollen and nectar (feed mutualistically). These dietary patterns are considered in chaps. 22 and 12, respectively. Two primarily herbivorous mammal species, *Tayassu pecari* and *Tapirus bairdii,* were recently extant in the La Selva area and are now missing or essentially so owing to hunting (see chap. 18).

With the exception of *Heteromys desmarestianus,* a rodent seed predator, the feeding ecology of vertebrate herbivores at La Selva has not been studied (table 21.1). Studies have been conducted on the majority of these species either on Barro Colorado Island (BCI) in Panama or in the dry deciduous forest of Costa Rica, both more seasonal environments than La Selva. Ecological factors, such as plant species composition, predator abundance, seasonal production of fruit and leaves, and climatic factors (e.g., precipitation), differ greatly between these sites and La Selva (e.g., Frankie et al. 1974; Opler, Frankie, et al. 1980; Coen 1983; Dietrich et al. 1982; Foster 1982). These factors are likely to influence foraging behavior of vertebrate herbivores (e.g., Edwards 1983), and the results from work at such sites may not be valid for La Selva.

Invertebrates

At least 171 different insect families in nine orders and two mite families are known or expected to feed on plants at La

Table 21.1 Diets of vertebrate herbivores at La Selva

Classification[a]	Species[b]	Diet[c]	Site[d]	References
Reptilia				
Testudinata				
Emydidae	Rhinoclemmys annulata	HB	PA	Mmr 1971
	R. funerea	HG/HB	PA	M&L 1971; Ern 1983
Squamata				
Iguanidae	Basiliscus plumifrons	OF	GUA	VD 1983; S&L 1983
	B. vittatus	OF	GUA	S&L 1983
	Iguana iguana	HB	BCI	Rand 1978
	Polychrus gutturosus	IO	BR	V&L 1981
Mammalia				
Marsupialia				
Didelphidae	Didelphis marsupialis	FO	VE	O'Cn 1979
	Philander oppossum	IO	—	Wlk 1975
	Caluromys derbianus	FO	—	Wlk 1975
Primates				
Cebidae	Alouatta palliata	HB	BCI	Mlt 1979, 1980
			GUA	Gln 1981, 1983
	Ateles geoffroyi	FH	BCI	H&H 1969
	Cebus capucinus	FH	BCI	H&H 1969
Edentata				
Bradypodidae	Bradypus variegatus	HB	BCI	M&S 1975, 1978
	Choloepus hoffmanni	HB	BCI	M&S 1975, 1978
Lagomorpha				
Leporidae	Sylvilagus brasiliensis	HB	T	Chp 1983
Rodentia				
Sciuridae	Sciurus granatensis	FG	BCI	H&T 1978; Hny 1983
	S. variegatoides	?	—	Hny 1983
Heteromyidae	Heteromys desmarestianus	FG	LS, GUA	Flm 1974, 1983
Erethizontidae	Coendou mexicanum	HB	BCI, GUA	B&G, p.o. Jnz 1983a
Dasyproctidae	Dasyprocta punctata	FG	BCI	Smt 1978, 1983
	Agouti paca	FG	BCI	Smt 1978, 1983
Echimyidae	Hoplomys gymnurus	FG	LS	Van et al. 1979; Van 1983
	Proechimys semispinosus	FG	LS	Flm 1973, 1974a
Muridae	Oryzomys spp.	FG	PA	Wlk 1975
	Nyctomys sumichrasti	FG	—	Flm 1970, Grd 1983
Carnivora				
Mustelidae	Eira barbara	OF	CR	Jnz 1983c
Procyonidae	Bassaricyon gabbii	FO	—	Wlk 1975
	Nasua narica	OF	PA	End 1935; Kfm 1962, 1983
	Potos flavus	OF	—	Wlk 1975
Artiodactyla				
Tayassuidae	Dicotyles tajacu	FO/HB	GUA, PE, OP	Wlk 1975
Cervidae	Tayassu tajacu	FO/HB	GUA	Jnz 1983g
	Odocoileus virginiana	HB	M	Gll et al. 1978
	Mazama americana	FH	—	Wlk 1975
Perissodactyla				
Tapiridae	Tapirus bairdii	HB	GUA	Jnz 1983h
Aves				
Tinamiformes				
Tinamidae		FO	CR	Stl et al. 1989
Galliformes				
Phasianidae		FG	CR	Stl et al. 1989; Lpd 1977
Columbiformes				
Columbidae		FG	CR	Stl et al. 1989
Psittaciformes				
Psittacidae		FG	CR	Stl et al. 1989
Passiformes				
Fringillidae		FG	LS	Chap. 22; Mrt 1978

Table 21.1 (continued)

ªClassification: class, order, family.

ᵇSpecies list based on Scott and Limerick (1983), Stiles (1983), and Wilson (1983).

ᶜKey to dietary categories for herbivorous La Selva vertebrates (modified from Robinson and Redford 1986): Herbivore-grazer (HG) = > 50% grasses; herbivore-browser (HB) = > 50% leaves and twigs; frugivore-herbivore (FH) = > 50% fruits, rest green plant material; frugivore-granivore (FG) = Seeds and fruits eaten in varying proportions; frugivore-omnivore (FO) = > 50% fruits, remainder insects, vertebrates; omnivore-frugivore (OF) = > 50% invertebrates and vertebrates, some flowers and fruits; insectivore-omnivore (IO) = > 50% invertebrates. A diet category in parentheses means that the food item is eaten occasionally or opportunistically. ? = diet unknown.

ᵈKey to site locations: BCI = Barro Colorado Island, Panama; BR = Brazil; CR = in Costa Rica but neither Guanacaste nor OSA; GUA = Guanacaste, Costa Rica; M = Durango, Mexico; OP = Osa Peninsula, Costa Rica; PA = in Panama but not on BCI; PE = Peru; VE = Venezuela. Abbreviations:

B&G = H.E. Braker & H. Greene	Hny = Heany	p.o. = personal observation
Chp = Chapman	Jnz = Janzen	Rand = Rand
End = Enders	Kfm = Kaufmann	S&L = Scott & Limerick
Ern = Ernst	Lpd = Leopold	Smt = Smythe
Flm = Fleming	M&L = Moll & Legler	Stl = Stiles
Gll = Gallina	M&S = Montgomery & Sundquist	V&L = Vitt & Lacher
Gln = Glander	Mlt = Milton	Van = Vandermeer
Grd = Gardner	Mmr = Mittermeier	VD = Van Devander
H&H = Hladik & Hladik	Mrt = Morton	Wlk = Walker
H&T = Heany & Thorington	O'Cn = O'Connell	

Selva. A classification of the feeding modes in families of La Selva herbivorous arthropods is presented in appendix 21.1. It should be noted that many nonarthropod invertebrates also feed on plants. For example, herbivorous nematodes are abundant and diverse in tropical forests where some species feed on roots and meristematic tissue (Coleman 1970). The effects of nematodes on growth and yield in agroecosystems have been well documented (Freckman and Caswell 1985). We are not aware of any studies of the effects of nematodes on plant growth and/or fitness in tropical forest systems and, thus, have excluded them from the appendix.

HERBIVORE FAUNAL DIVERSITY

Determinants of Local Diversity

Local herbivore species richness for a host plant species is a function of both local and regional processes. Investigation in temperate systems indicates that positive correlations exist between numbers of insect species associated with a plant species, both locally and regionally, and geographical range and size of the host plant (Opler 1974; Lawton and Schröder 1977; Strong and Levin 1979; Cornell 1985). Research at La Selva suggests that these patterns also hold for tropical systems. Both host plant size and abundance influence the diversity of herbivore faunas for three different plant-herbivore systems at La Selva: hispine beetles and their host plants in the Zingiberales; *Heliconius* butterflies and flea beetles (Chrysomelidae), which feed on *Passiflora* vines (Passifloraceae); and geometrid moths and curculionid weevils, which feed on *Piper* (Piperaceae). In Costa Rica the total number of hispine beetle species associated with a given species in the Zingiberales is greater for plant species with larger ranges (Strong 1977). At La Selva local hispine species diversity on a plant species is positively correlated with plant abundance, plant size, and the number of congeneric plant species present (Strong 1977a). Local hispine diversity is also related to length of the growing season (Strong 1977b). In Costa Rica, in general, shorter growing seasons limit the herbi-

vore fauna to those species that have shorter developmental times, whereas both rapidly and slowly developing hispine beetles occur at La Selva (Strong 1977b). The relative importance of each of these factors in determining regional and local abundance is not known.

At La Selva local herbivore diversity (flea beetles and *Heliconius* butterflies combined) on *Passiflora* is positively correlated with local host plant abundance but not with total host plant range (Gilbert and Smiley 1978). In addition, herbivore diversity is also associated with habitat; second-growth and light-gap species of *Passiflora* have more diverse herbivore faunas than do understory species (Gilbert and Smiley 1978). Local species diversity of *Heliconius* butterflies is correlated with local abundance of *Passiflora* species (Gilbert and Smiley 1978).

Among *Piper* species at La Selva those that are more common and characterized by larger individuals are attacked by significantly more weevil (Curculionidae: Baridinae and Rhynchophorinae) and geometrid moth (Geometridae: Larentiinae) species than are host species whose individuals are rare or small (Marquis 1991b). Neither habitat (light gap compared with understory), soil type, or leaf pubescence level are significant predictors of *Piper* herbivore diversities.

Tropical Compared with Temperate Local Diversity

To understand the role of herbivores as selective agents for host plant traits it is important to know the diversity and abundance of herbivore species involved and their relative impact on plant fitness. In tropical habitats one rarely knows the herbivores of a given plant species, let alone their impact. Even studies of temperate systems have tended to focus on the most conspicuous herbivores (Marquis 1991a). Published descriptions of the herbivore faunas of tropical plants are few (Janzen 1977a; Auerbach and Strong 1981; Schupp 1986; Thomas 1990).

Diversity of herbivore faunas varies greatly among *Piper* species at La Selva (Marquis 1991b). For example, at least ninety-five different species representing six insect orders and twenty-five families feed on aboveground parts of *Piper ariei-*

anum at La Selva; fifty-six consume leaf tissue, six feed on reproductive parts, and thirty-three are sap feeders (Marquis 1991b). In contrast, only twenty species are known to feed on *P. holdridgeianum,* which is similar in size, habitat, and abundance to *P. arieianum.* For comparison Hespenheide (1985) reports eighteen herbivore species from five guilds feeding on the liana *Byttneria aculeata* (Sterculiaceae) at La Selva. Local diversity of herbivore species associated with *P. arieianum* is 2.5 to 11 times higher than reported for temperate plant species of comparable size (Marquis 1991b). An important point here is that herbivore diversity is not necessarily related to impact: nine times as many leaf-chewing species are associated with *P. arieianum* as with *P. holdridgeianum,* but average damage levels do not differ significantly between these two host species (Marquis 1991b).

Specialization with Respect to Host Plant
Knowledge of the degree to which herbivorous insects specialize on their host plants is critical for understanding life-history patterns of these insects (Futuyma 1976), their impact on plant populations (Thomas 1990), and their importance as selective agents for plant defenses (Feeny 1976; Rhoades and Cates 1976). Evidence has been presented both for and against the hypothesis that tropical forest herbivorous insects are more specialized than those of temperate communities (Connell 1971; Janzen 1973a, 1977a; Scriber 1973; Beaver 1979; Price 1980, 1991; Young 1982; Wood and Olmstead 1984). In this section we review patterns of host plant use for three taxa of La Selva herbivores. Conclusions are preliminary as available data on host use patterns are limited (Price 1991). We adopt the following definitions: *monophagous* species feed on one host plant species; *oligophagous* species feed only on congeners ("narrow oligophages") or on several genera in the same family ("broad oligophages"); *polyphagous* species eat host plants from several, unrelated taxa.

Butterflies
Butterflies at La Selva are primarily monophagous. Mean number of host plant species for 97 of 145 La Selva butterfly species in the Papilionidae, Pieridae, and Nymphalidae (DeVries 1985, 1987) is 1.38 (standard deviation (SD) = 0.39). Most members of the family Papilionidae, and of the nymphalid subfamilies Morphinae and Ithomiinae, are monophagous. Larger numbers of host plant species are eaten by butterflies in the Brassolinae, Danainae, and Heliconiinae. For example, the mean number of host plant species per heliconiine species is 2.5 (SD = 3.1); this value includes six monophagous species that are found on one *Passiflora* species each, and four oligophagous species that feed on four to five *Passiflora* species (Smiley 1978b, 1978a, 1985a). In the laboratory larval growth rates on different *Passiflora* species do not correlate with female oviposition patterns (Smiley 1978b). These observations suggest that the mechanism maintaining host plant use patterns may be ecological factors (the "ecological monophagy" of Smiley 1978a; see also Marquis 1991a) and/or by phylogeny (Futuyma 1991; Mitter et al. 1991) rather than by selection per se for biochemical specialization to the food plant.

We used data for La Selva butterflies (DeVries 1985) to test whether butterflies in general differ in their host specificity in temperate compared with tropical regions. In figure 21.1A we show the number of species in each dietary category (as defined previously) for La Selva Nymphalidae, Papilionidae, and Pieridae based only on host plant records from La Selva and for the same set of butterfly species, but including host records from all of Costa Rica (DeVries 1985). We compare these patterns to patterns for four geographically limited areas in the United States (Georgia, New York, Indiana, and southern California, data from Forbes 1960; Emmel and Emmel 1973; Harris 1972; Shull 1987) and for Great Britain (Heath et al. 1984). The nymphalid subfamilies Ithomiinae, Heliconiinae, Morphinae, and Brassolinae were omitted from the analysis because they are almost strictly tropical. Analysis showed that diet type is not independent of site (G-test, $p < 0.005$): compared to butterflies from each of the five temperate sites, La Selva butterfly species are more host specific than temperate butterflies whether based on host plant records only from La Selva or from all of Costa Rica. If data from the four omitted nymphalid subfamilies (Gilbert 1969; Benson et al. 1975; Smiley 1985a, 1985b; DeVries 1985) are included in the analysis, the pattern of greater specificity for the La Selva butterflies does not change.

Treehoppers (Homoptera: Membracidae)
Treehopper adults and nymphs are sapsuckers, tapping into the phloem of unhardened stems, mostly of shrubs and trees. Lowland tropical treehoppers are more generalized in host plant use than are temperate species or tropical high-elevation species (Wood and Olmstead 1984). Based on work with *Enchenopa binotata* complex, Wood and co-workers (Wood 1980; Guttman et al. 1981; Wood and Guttman 1981, 1983) explained this pattern by proposing a scenario for colonization of temperate habitats and hosts by an originally tropical polyphagous species. In response to diverse host phenologies in more severe climatic conditions membracid host races formed, each race (and, eventually, species) specializing on a different host species (Wood and Keese 1990). Wood and co-workers emphasize the importance of host growth phenology because the availability of unhardened twigs determines whether a plant can be attacked. In temperate North America twig expansion is asynchronous among species and limited to the spring, thus selecting for specialization by membracids (Wood et al. 1990). In contrast, in the tropics unhardened shoots of woody plants are available year-round and individuals of any one host species are relatively scarce, making specialization unlikely. This model for sympatric speciation in temperate membracids provides a mechanism for an increase in the degree of specialization from tropical to temperate habitats. In the case of *E. binotata* a tropical polyphagous species is hypothesized to have given rise to a temperate complex of cryptic and host-specific species. At La Selva most treehoppers are characterized by polyphagy, large nymphal aggregations, and frequent mutualistic association with ants (Wood 1982, 1983, 1984). Wood and Olmstead (1984) cautioned that generalizations are tentative because the ecology and host associations of most tropical membracids have not been thoroughly investigated.

Acridid Grasshoppers
The feeding biology of acridoid grasshoppers (families Acrididae and Romaleidae) is known more thoroughly for La Selva than for any other Neotropical site (Rowell 1978, 1983a,

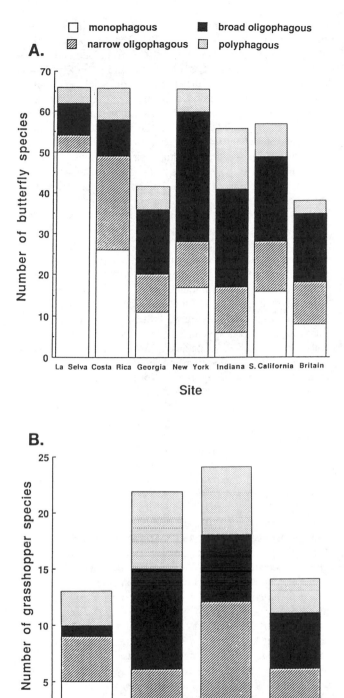

Fig. 21.1. Temperate compared with tropical diet patterns for pierid, papilionid, and nymphalid butterflies and acridid grasshoppers. See the text for definition of diet categories and data sources. *A.* Diet patterns for La Selva butterfly species based on La Selva host plant records (site, La Selva) and those for all of Costa Rica (site, Costa Rica) compared to five temperate sites (sites, Georgia, Indiana, New York, S. California, Britain). *B.* Diet patterns of grasshoppers at La Selva and three North American sites.

1983b, 1983c, 1983d, 1985a, 1985b, 1987; Waltz 1984; Braker 1986, 1989a, 1989b, 1991; Braker and Chazdon 1992). Grasshopper diet breadths at La Selva range from monophagy to extreme polyphagy with these feeding patterns following subfamilial lines. Herbs, shrubs, and trees are all eaten. Groups characterized by narrow diets include the Proctolabinae (seven La Selva species), which are monophagous or narrowly oligophagous on host plants in the Solanaceae and Nyctaginaceae, and one species of Copiocerinae at La Selva that feeds on understory geonomoid and bactroid palms. Members of the subfamily Ommatolampinae and the family Romaleidae tend to be oligophagous or polyphagous. The more host-restricted grasshopper species at La Selva oviposit in or on their host plants, whereas those with broader diets lay eggs in the soil (Braker 1989a, 1989b). Dietary patterns of several La Selva grasshopper species have been investigated in some detail.

Differences in local host abundance may affect the degree of polyphagy in members of the ommatolampine genus *Rhacicreagra* (Rowell 1985a, 1985b). In the wild *R. nothra* at La Selva eats a few species from several families, primarily Urticaceae and Asteraceae, and, more rarely, Amaranthaceae, Phytolaccaceae, and Poaceae. In captivity *R. nothra* accepts not only these plants but also the host plants of any of ten other allopatric Costa Rican *Rhacicreagra* species. Rowell (1985b) proposed that a "generic spectrum" of acceptable plants exists for *Rhacicreagra,* including members of seven plant families, and that a given species' diet is determined mainly by local availability.

Local host plant abundance is one factor determining diet patterns in the ommatolampine *Microtylopteryx hebardi* (Braker 1986, 1991). During a three-year period at La Selva more than four hundred sampled individuals were observed feeding on fifty-two plant species in nine monocot and seven dicot families. Most of these records, however, were from four plant families: Heliconiaceae (44%), Arecaceae (20%), Marantaceae (13%), and Araceae (10%). Individuals varied in diet breadth: some were monophagous when sampled over several months, whereas others fed on as many as six plant species in one hour. Individual "specialists" did not all use the same food plant. This variation was associated with several factors, including differences in local patch composition, movement rates associated with reproductive behavior, and home range size. Oviposition requirements influence plant species eaten by the females because only a few, relatively rare species are used as oviposition sites. A female's choice of oviposition site may influence the survivorship of her offspring; nymphs reared in the laboratory had differential survivorship on different diets (Braker 1986).

At least one La Selva grasshopper species is truly polyphagous. *Abracris* (*Osmilia*) *flavolineata* feeds on at least ten plant families in the wild (Rowell 1983a). In laboratory preference assays *Abracris* feeds indiscriminately on leaves of many woody and herbaceous plant species (Waltz 1984; Marquis and Braker 1987).

Figure 1B compares diet breadths of La Selva grasshoppers to those known from three north temperate desert or grassland sites (Joern 1979; Ueckert and Hansen 1971). More species are monophagous at La Selva than at the three temperate sites (G = 17.46; df = 9; *p* < 0.05). This conclusion is preliminary. The acridid species studied so far at La Selva are relatively common and inhabit successional areas such as light gaps and trail edges. We have little knowledge of the feeding habits of the large number of grasshopper species that occupy other habitats: forest floor, tree trunks and branches, and canopy (Rowell 1978).

REGULATION OF INSECT HERBIVORE POPULATIONS

Populations of organisms can be affected by density-dependent factors that regulate population numbers and density independent factors that disturb population growth processes. Density independent factors are often associated with climate. Because tropical environments have been viewed as less variable and, thus, more benign or predictable than temperate environments, more emphasis has been placed on density dependent, or biotic, factors (e.g., Dobzhansky 1950; Slobodkin and Sanders 1969; Baker 1970; Gilbert 1977, 1980; Janzen 1983d). Evidence presented below suggests that insect herbivore populations at La Selva are controlled by both density independent factors (specifically, rainfall) and density dependent factors (parasitism).

Rainfall

La Selva's climate is relatively aseasonal, but predictable variation in temperature and rainfall does occur (see chap. 3). Herbivores that feed externally on leaves (in contrast to leaf miners) of the understory shrub species, *Piper arieianum,* were fewest during periods of highest rainfall at La Selva (Marquis n.d.) despite the fact that leaf production (and, thus, resource renewal) is independent of rainfall (Marquis 1988). The basis for the relationship between rainfall and herbivore numbers is not known. It may be that extended rainfall affects flight time and oviposition behavior of adult Lepidoptera and/or causes larval mortality. Field observations suggest that extremely dry periods (e.g., at the end of the dry season) as well as exceptionally wet periods may lower abundance of both lepidopteran larvae and grasshoppers at La Selva (Braker 1986; Marquis and Braker, pers. observation). Thus, even at La Selva, where within-year climatic variation is relatively low, herbivore population levels are affected by climate (see also Janzen 1983e). At more seasonal tropical sites the effect of climate is even greater, with more arthropods during the wet season than the dry season (e.g., Janzen 1973c, 1985; Wolda 1978a, 1983; Windsor 1978; Frith and Frith 1985; but see Wong 1984).

Competition for Resources Compared with Parasitism

Competition between species for resources has been viewed as a major force structuring animal communities and influencing relative species abundance, the body size and shape of animals, and their use of resources (MacArthur 1972; Cody and Diamond 1975; May 1981). Competition between herbivores for a common resource (their host plant) can be manifested in two ways: actual depletion of that resource (e.g., Blakely and Dingle 1978) and through changes in the quality of uneaten tissue (e.g., induction of secondary compounds, changes in nutrient level) for herbivores colonizing previously damaged plants (e.g., Janzen 1973b; Faeth 1986; Karban et al. 1987).

Investigation of interspecific competition in the *Heliconia*-hispine beetle system at La Selva (Strong 1977a, 1977b, 1982a, 1982b; Strong and Wang 1977; Auerbach and Strong 1981; Gage and Strong 1981; Morrison and Strong 1981) was among the first to contradict the commonly held assumption that competition is a major determinant of community structure (see also Rathcke 1976). Rolled-leaf hispine beetles (Coleoptera: Chyrsomelidae) feed on the inner surface of rolled immature *Heliconia* (Heliconiaceae) leaves or, in a few species, on the top surface of mature leaves but under accumulated debris. Evidence suggests that present abundance, distribution, and feeding of hispine beetles are little affected by interspecific competitive interactions. Different hispine species co-occur on the same plant species (e.g., six species are associated with *H. latispatha* at La Selva) and show no evidence of agonistic behavior (Strong 1977a). Host plant species use patterns and feeding are not affected by the presence of other hispine species (Strong 1982, 1982b). Experimental addition of individuals and analysis of natural patterns of occurrence give no evidence that individuals avoid leaves that have been previously colonized either by conspecifics or by individuals of other species (Strong 1982a). In addition, because abundance of all species is low, the shared resource is superabundant. Hispine-caused damage averages only 1.5% of total leaf area (Strong 1983). Not all leaves or plants are colonized, but feeding trials show that noncolonized leaves are as acceptable as colonized leaves and that species do not segregate along gradients of light, host plant density, or vegetation type (Strong 1981). Finally, competition through changes in leaf palatibility because of past feeding has yet to be ruled out; an initial survey of *H. imbricata* and *H. latispatha,* however, found no evidence for presence of secondary compounds (Gage and Strong 1981).

Although competition in the past may have determined present larval use patterns, this hypothesis is extremely difficult to test (see Connell 1979; Strong 1983; Strong et al. 1984). If competition does take place in the herbivore communities associated with rolled leaves, it occurs only infrequently and contributes little to community composition and abundance of individual species. The same conclusion has been reached for communities associated with *Heliconia* floral bracts (Seifert and Seifert 1976, 1979) and leaf litter on top of *Heliconia* leaves (McCoy 1985). Instead, evidence suggests that parasitism may be a major factor in limiting rolled-leaf hispine beetle populations to levels below which food plant depletion does not occur. Parasitism on eggs and pupae account for 30%–70% and 90% of total mortality at each stage, respectively (Morrison and Strong 1981; Strong 1982a, 1983). Although no life table has been constructed, circumstantial evidence suggests that this level of parasitism is sufficient to control populations in this system. Strong (1983) notes that under heavy pesticide use plantations of bananas (a close relative of the Heliconiaceae) in Costa Rica suffer pest outbreaks. These outbreaks are associated with reduced parasitism rates on the pest herbivores. Rutilio et al. (1973) report a comparable example for *Rothschildia aroma* (Lepidoptera: Saturniidae) in El Salvador.

These results contrast strongly with what is known about seed-attacking beetles of dry forest plants of Costa Rica. In a thirteen-year study of specificity of host plant use by these beetles, Janzen (1980) found that 57% of the species studied had no parasitoids; those species with parasitoids were parasitized only infrequently and at low levels. Population size may instead be controlled by a combination of host plant defensive characters, including fruit and seed chemistry, fruit morphology, and phenology of fruit production (Janzen 1969b, 1978a, 1980a), and the presumably adverse effect of the long dry season (Janzen 1987a; Traveset 1991).

HERBIVORES AS REGULATORS OF PLANT POPULATIONS

Levels of Attack

An initial step toward understanding interspecific patterns of herbivore use of their host plants and the importance of herbivores as regulators of plant populations is to measure damage levels over a number of plant species. This has been done only rarely on a communitywide basis (leaf herbivory: Johnstone 1981; Coley 1983b; Dirzo 1984; Waltz 1984; Southwood et al. 1986; de la Cruz and Dirzo 1987; seed predation: Janzen 1980 and references therein; Lewinsohn 1991). Leaf herbivory may also impact plant growth, and both leaf herbivory and seed predation may affect plant fitness.

Leaf Herbivory. In the case of leaf herbivory it is important to know both what portion of a plant's canopy is missing at any one time because of herbivore damage and at what rate leaf area is removed.

An initial survey of interspecific variation in standing crop damage levels has been conducted at La Selva (Gonzalez et al. 1985). The results for forty-five different plant species demonstrate significant interspecific variation. Average damage per species sampled was 8.6% (SD = 34.8%, range = 0.5%–25.5%). No relationship existed between damage level and habitat. More intensive studies of individual plant species further support this pattern of high interspecific variation. Total herbivore damage to *Heliconia* spp. is invariably less than 2.5% of the total leaf area (Strong 1977b). In contrast, for individuals of five *Piper* species the average value was 15.9% (SD = 8.9%, range = 5.7%–19.8% per species) (Marquis unpublished data). In three understory geonomoid palms, mean percent leaf area missing per species (± 95% confidence intervals) ranged from 0.3% (± 0.01) to 5.4% (± 0.8) (Braker and Chazdon 1992).

Leaf damage also varies greatly within plant species at La Selva. In *Piper arieianum* estimates for twenty-five individuals ranged from 3% to just less than 50% of total leaf area missing (Marquis 1987). There was considerable intraspecific variation in damage to two *Neea* spp. (Nyctaginaceae), *Solanum eschylosum* (Solanaceae) (DeLapp 1979; Marquis 1987), and three fern species (*Polybotria cervina, Thelypteris turrialbae,* and *Adiantum obliquum;* Hendrix and Marquis 1983). The mean damage levels for each of the fern species (9.9%, 5.5%, 7.3%, respectively; overall mean = 7.8%) fell close to the midpoint of damage levels reported for angiosperm species at La Selva. These data, plus the fact that damage to these ferns was caused by a number of different herbivores, add further evidence (see also Balick et al. 1978; Auerbach and Hendrix 1980; Hendrix 1980) for the idea that ferns are not as free

from herbivore attack as has been previously suggested (Brues 1920; Ehrlich and Raven 1964; Regal 1977).

If herbivore pressure is greater in tropical than temperate habitats, then we would predict greater loss of leaf area per unit time in tropical habitats. Unfortunately, estimates of rates of leaf area loss are much less common than estimates of proportional leaf area missing. The average 8.6% damage level for the forty-five species sample of La Selva plants (Gonzalez et al. 1985) and 12.4% for a sample of seventy-seven species (Hartshorn 1973) are within the 7.0%–20.3% range given for other tropical forested habitats (Misra 1968; Odum and Ruíz-Reyes 1970; Edwards 1977; Adis et al. 1979; Coley 1982; Leigh and Smythe 1978; Proctor et al. 1983; Wint 1983; Dirzo 1984; de la Cruz and Dirzo 1987) and at the high end of the 1.8%–12.3% range for temperate habitats (Bray 1961, 1964; Reichle and Crossley 1967; Whittaker and Woodwell 1969; Gosz et al. 1972; Nielson 1978; Schowalter et al. 1981). In addition to methodological differences among studies (see Lowman 1984; Ohmart 1984; Landsberg and Ohmart 1989), a number of problems arise when making tropical-temperate comparisons. Leaf life spans (and, thus, periods of exposure to herbivores) vary greatly among sites. Values for herbivore damage at the community level are highly variable within temperate and tropical habitats (e.g., Putz and Holbrook 1988). Although, in general, plants in tropical communities may suffer more damage (Landsberg and Ohmart 1989; Coley and Aide 1991), one of the systems purportedly least susceptible to herbivore attack is the tropical white sand soil community (Janzen 1974; Jordan and Uhl 1978; but see Proctor et al. 1983). In contrast, the communities with the highest reported foliar damage are temperate *Eucalyptus* forests of Australia (Fox and Morrow 1983; Ohmart et al. 1983; Lowman 1984, 1986; Ohmart 1984; Morrow and Fox 1989).

Based on our experience, outbreaks of defoliating insects in which numbers of individuals of the same or different plant species lose 90% or more of their leaf area occur only rarely at La Selva. A problem in assessing damage at La Selva, particularly for tree species, is the high multilayered canopy, species heterogeneity, and potential for plants to refoliate quickly once damaged. Beyond data presented in table 21.2, we are not aware of outbreaks of insects at La Selva comparable to those reported from Barro Colorado Island in Panama (Wolda and Foster 1978; Wong et al. 1990), Estación Biológica Los Tuxtlas in Mexico (Dirzo 1987), Parque Nacional Guanacaste in the lowland dry forest of Costa Rica (Janzen 1981b, 1985, 1988), Parque Nacional Corcovado of Costa Rica (Janzen 1978a), or other tropical habitats (Gray 1972; Kenyi 1980).

Information on temporal patterns of folivory at La Selva comes from three studies. Marquis (1984b, 1987) measured damage for two years to *Piper arieianum,* an understory shrub at La Selva. Plants lost on average 1%–3% of their total leaf area during each 1.5–3 month census intervals (Marquis 1984b, 1987). Variation was extremely high, however: some plants lost no leaf area, others were totally defoliated by leaf-cutting ants (*Atta cephalotes*), still others lost up to 30% of their leaf area between censuses to herbivores other than *Atta.* The average daily loss of leaf material was 0.012%–0.080% (Marquis 1984a). Approximately one-half of the total damage that an average leaf of *P. arieianum* is likely to receive during

Table 21.2 Observed instances of heavy defoliation at La Selva other than by *Atta* ants

Plant Species	Insect Species	Number of Plants	Percent Defoliation	Habitat	Date	Reference
Dipteryx panamensis (Papilionaceae)	Unknown	Two canopy trees	50	Primary forest	April 1981	R.J. Marquis pers. observation
Ocotea sp. (Lauraceae)	Archaeoprepona demophon (Nymphalidae)	Two saplings	95	Understory primary forest	September 1987	R. J. Marquis pers. observation
Hamelia patens (Rubiaceae)	Unidentified sawfly sp. (Hymenoptera)	Five 2–4 m shrubs	100	Shaded arboretum	June 1981	R. J. Marquis pers. observation
Zamia skinneri (Zamiacae)	Eumaeus minyas (Lycaenidae)	Many plants	100	Primary, secondary forest	Various times	D. B. Clark and Clark 1991
Lecythis ampla (Lecythidaceae)	Unidentified attelabine weevil (Curculionidae)	Many canopy trees	75	Primary, secondary forest	June–July 1982	H. E. Braker & R. J. Marquis pers. observation
Dichopetalum sp. (Dichopetalaceae)	Phryganodes nr. croceipes (Pyralidae)	One 2.5 m vining shrub	100	Understory primary forest	June 1987	R.J. Marquis pers. observation

its lifetime occurs during leaf expansion (about eight weeks) (Marquis unpublished data).

Ernest (1989) found that both leaf age and habitat affect folivory rates in *Pentagonia donnell-smithii* (Rubiaceae), an understory treelet. Young leaves in secondary forest had twice the folivory rate (0.47% per day) as mature leaves in the same habitat, and thirty times more than mature leaves in the successional plots. Damage rate to young leaves was related to expansion rate: leaves on plants in the successional plots (0.18% loss per day) expanded more quickly than leaves of plants in secondary forest (0.47% loss per day). These intraspecific differences support the hypothesis of Aide and Londoño (1989) that rapid expansion of young leaves may reduce damage during this vulnerable stage.

Waltz (1984) enumerated damage to saplings of twelve tree species (eight shade-tolerant and four gap species) every one to two weeks from April through September 1975 and March through September 1976. Most leaves were not damaged from one census to the next (table 21.3). When leaves were damaged, however, young leaves lost more area than mature leaves in both shade-tolerant and gap species. Frequency of attack on mature leaves of a plant was positively related to plant size (total number of leaves) for ten of twelve species over both seasons. Frequency of attack on young leaves of a given plant was also positively correlated with plant size for five of ten species and with number of new leaves present on censused plants for six of eight species.

Both frequency of attack on an individual and amount of leaf area removed were related to its life-history class (Waltz 1984). Individuals of gap species were more frequently attacked and lost more leaf area per attack than did shade-tolerant species (table 21.3). This pattern held for both immature and mature leaves. In contrast, Coley (1982, 1983a) found significant differences in damage rates to mature leaves between gap and shade-tolerant species, but damage to immature leaves did not vary significantly between these two categories. Mean daily removal rates were 1.8 to 5.8 times higher in the Panamanian tree sample than for La Selva plants (Coley 1983a; see table 21.3). Differences between studies may, in part, owe to the fact that all plants enumerated in the Panama-

nian study were in light gaps, whereas those at La Selva were in both gaps and understory.

Fruit and Seed Predation. As noted in chapter 22, the distinctions between dispersers and predators is not clear cut. The disperser end of the spectrum is emphasized in that chapter, whereas here we focus on predation on fruits and seeds. Seed predators at La Selva represent a diverse assemblage of insects, birds, and mammals (table 21.4). Initial study of their

Table 21.3 Frequency of attack and mean loss rate to young and mature leaves

	Frequency of Attack (SD)[a]		Mean Loss Rate (SE)[b]	
	Immature Leaves	Mature Leaves	Immature Leaves	Mature Leaves
Shade tolerant				
La Selva	75.8 (4.9)	97.2 (0.7)	0.143 (0.025)	0.012 (0.002)
BCI (dry season)[c]	—	—	0.829 (0.142)	0.043 (0.038)
BCI (wet season)[c]	—	—	0.549 (0.134)	0.048 (0.048)
Gap				
La Selva	64.8 (11.8)	89.9 (9.0)	0.396 (0.150)	0.180 (0.119)
BCI (dry season)[c]	—	—	0.521 (0.368)	0.135 (0.167)
BCI (wet season)[c]	—	—	0.731 (0.187)	0.499 (0.167)

Note: Frequency of attack and mean loss rate to young and mature leaves of eight shade-tolerant and four gap tree species for two consecutive rainy seasons at La Selva (April–October) and thirteen shade-tolerant (persistent) and eight gap (pioneer) tree species for the dry and wet seasons on Barro Colorado Island, Panama (BCI).

[a]Percentage of observations with no damage (Waltz 1984, 116–117).

[b]Waltz (1984, 120) equates 1 cm^2 area loss per observation to approximately 4% leaf area loss per month, ignoring interspecific variation in leaf size.

[c]Coley 1982.

Table 21.4 Known seed predators of La Selva plant species

Plant Species	Plant Part	Predator	Reference
Carapa guianensis (Meliaceae)	Mature seeds	Tayussu tajacu, T. pecari (Artiodactyla) Dasyprocta punctata, Agouti paca (Rodentia)	McHargue & Hartshorn 1983
Phytolacca rivinoides (Phytolaccaceae)	Mature fruits	Rodents	Denslow & Moermond 1982
Psychotria pittieri, P. brachiata (Rubiaceae)	Mature fruits	Rodents	Denslow & Moermond 1982
Welfia georgii (Palmae)	Mature seeds	Heteromys desmarestianus (Rodentia)	Vandermeer et al. 1979
Jacaratia costaricensis (Caricaceae)	Green fruits	Parrots	Grove and Sauer 1977
Dipteryx panamensis (Fabaceae)	Green fruits	Ara ambigua (Psittacidae)	D. B. Clark pers. comm.
Pithecellobium macradenium (Mimosaceae)	Green fruits	Parrots	R. Butterfield pers. comm.
Stemmadenia donnell-smithii (Apocynaceae)	Green fruits	Pionus senilis (Psittacidae)	R. J. Marquis pers. observation
Pentaclethra macroloba (Mimosaceae)	Green fruits	Pionus senilus	Hartshorn 1972
Pentaclethra macroloba (Mimosaceae)	Cotyledons	Sesiid moth	Hartshorn 1983a
Stryphnodendron excelsum (Mimosaceae)	Green fruits	Pionus senilus	Howe & Primack 1975
Stryphnodendron excelsum	Seeds in pods	Acanthosceloides sp. (Bruchidae)	Hartshorn 1972; H. Hespenheide pers. comm.
Pithecellobium gigantifolium (Mimosaceae)	Seeds in pods	Eulechriops sp. (Curculionidae)	R. J. Marquis pers. observation
Simira maxonii (Rubiaceae)	Seeds in pods	Rosella sickingiae (Bruchidae)	Janzen & Wilson 1977
Piper spp. (Piperaceae)	Developing seeds	Numerous curculionid weevil spp.	Marquis 1988; Greig 1993
Various Piper spp.	Developing seeds	Sibaria englemani (Hemiptera)	Marquis 1988; Greig 1993
Various Piper spp.	Developing seeds	Hyalymenus pulcher (Hemiptera)	Greig 1993
Piper sancti-felicis	Developing seeds	Piasus eribricollis (Hemiptera)	Greig 1993
Carapa guianensis	Mature seeds	Hypsipyla ferrealis (Pyralidae)	McHargue & Hartshorn 1983

Note: Ordered by predators from mammals to insects.

effect on plant reproductive output demonstrates that seed attack is both intense, in the proportion of a plant's seed crop destroyed, and extremely variable within and among species (Hartshorn 1972; Janzen and Wilson 1977; McHargue and Hartshorn 1983; Schupp and Frost 1989; Greig 1993). Great green macaws *(Ara ambigua)* and various parrot species feed on green fruits of large-fruited species and are undeterred by fruits or seeds that are hard or resinous (Hartshorn 1972; Howe and Primack 1975; D. B. Clark pers. comm.; R. J. Marquis, pers. observation). Both parrots and rodents may repeatedly return to fruiting individuals and, in so doing, destroy a large portion of an individual's seed crop (Grove and Sauer 1977; Vandermeer et al. 1979; Denslow and Moermond 1982). In an experimental study of fruit removal on shrub species with small fleshy fruits Denslow and Moermond (1982) demonstrated that nocturnal disappearance of fruits (assumed to have been removed by rodents) represented 32%–73% of total fruit removal over a twenty-four-hour period. They hypothesized that infructescence morphology and orientation in

these plants is the result of conflicting selection pressures to make fruits inaccessible to seed-destroying rodents but still available to frugivorous birds, which pass seeds unharmed through their digestive tracts. A number of additional studies in which seed predation was not the major focus document the occurrence of high levels of attack on developing seeds and fruits (Grove 1985, *Calathea* spp. [Marantaceae]; Bullock et al. 1983, *Guarea rhopalocarpa* (Meliaceae); Howe 1977, *Casearia corymbosa* [Flacourtiaceae]).

When damage has been tallied for individual plants, high intraspecific variation in attack rates has been found (Janzen and Wilson 1977; Condon 1974; Vandermeer 1972; Boring 1979). For four of five species studied the portion of an individual plant's seed crop destroyed by insect predators depends on plant spacing: in both *Cassia grandis* (Condon 1974) and *Guarea rhopalocarpa* (Boring 1979) rates of attack were positively correlated with adult plant density. For pasture *Sida acuta* (Malvaceae), however, neither adult population density, average nearest neighbor distance, nor the number of seeds encountered per plot was correlated with percentage of seeds attacked (Vandermeer 1972). Finally, distance to adult trees and microhabitat may interact to determine whether a seed escapes predation: Schupp and Frost (1989) found that *Welfia georgii* seeds placed in understory at least 10 m away from an adult *Welfia*, were twice as likely to escape predation as seeds placed in light gaps (irrespective of the proximity to adults) and seeds placed under adult *Welfia* in closed canopy.

Knowledge of the population dynamics of insect seed predators at La Selva is limited to studies of the predispersal fauna of *Piper*. In this genus number of fruits available is correlated with number of attacking insects either positively (*P. arieianum*, Marquis 1988) or negatively (*P. urostachyum* and *P. sancti-felicis*, Greig 1993). Number of seed predators attacking *Piper arieianum* responds to the major peak of fruiting but not to the minor peak (Marquis 1988). As a result, Marquis (1988) proposed that plants of *P. arieianum* may escape seed predators by fruiting at times when they are absent. In contrast, Augspurger (1980, 1981) proposed that seed predators select for synchrony in *Hybanthus prunifolius;* plants forced to fruit out of synchrony with the rest of the population suffered greater seed predation than those fruiting in synchrony. Because seed predators of *Piper* fruits generally attack more than one plant species (Greig 1993), future research should include measurement of seasonal fruit abundance of all *Piper* species to relate overall resource base to population fluctuations of the insects.

With the exception of the rodent *Heteromys desmarestianus* (Fleming 1973, 1974a; Vandermeer 1979; Vandermeer et al. 1979), the host specificity, life cycle, and population dynamics of vertebrate seed predators at La Selva are unknown. *Heteromys desmarestianus* appears to be a generalist cache horder of seeds (Vandermeer 1979) that also feeds opportunistically on insects (Fleming 1973; Borges and Propper 1983). In laboratory trials it fed on mature seeds of eight species, representing six families and encompassing a large range of seed sizes (Vandermeer 1979; Denslow and Moermond 1982). Field-caught animals were carrying seeds of *Welfia georgii*, *Socratea durissima, Pentaclethra macroloba, Rauwolfia tetraphyllum, Virola sebifera*, a number of unidentified seeds, and one insect larva (Fleming 1974a).

Estimates of Impact

Seed Predation and Plant Abundance. Seed predators affect plant population dynamics by reducing the number of propagules available for recruitment to subsequent generations (Harper 1977). In a comparative study of population dynamics of two mature forest trees, *Pentaclethra macroloba* and *Stryphnodendron excelsum* (Leguminosae), at La Selva, Hartshorn (1972) found that seed predation intensity was inversely but perhaps spuriously related to adult plant abundance. *Stryphnodendron excelsum* is a rare member of the forest canopy, whereas *P. macroloba* is the most common tree species at La Selva, comprising 40% of stems more than 5 cm dbh. In the former species fruits "fall to the ground where some are assumed to be removed by vertebrate predators or dispersal agents, but density-responsive predation by bruchids probably destroys nearly all of the seeds remaining under the parent tree" (Hartshorn 1972, 78). In contrast, Hartshorn

(1972) estimated that just less than half (46%) of *P. macroloba* seeds escape predation. Predispersal predation was estimated at approximately 7%, and a field experiment showed that 47% of seeds on the ground are removed by vertebrates. Seeds of *P. macroloba* contain an unidentified alkaloid and an uncommon amino acid. The seeds are lethal to *Heteromys desmarestianus* in pure diet. It is unclear, therefore, how many of the moved seeds are actually consumed. Hartshorn (1983a) has hypothesized that the relatively high number of seeds that escape predation is one of the major factors accounting for the high adult abundance of *Pentaclethra* at La Selva (see also Janzen 1970; Connell and Lowman 1989). This system begs for further study.

For species of *Piper* the proportion of a plant's developing seeds destroyed by insects (Hemiptera and Coleoptera, table 21.4) depends on plant life history (Greig 1993). Late successional species (e.g., *P. urostachyum*) lose fewer total seeds to predispersal seed predators than do early successional species (e.g., *Piper sancti-felicis*). Because late successional species produce fewer seeds, these losses represent a greater proportion (up to 75% on average) of the seed crop than do the levels sustained by early successional species (5%–10%). Greig (1993) suggests that high seed predation rates on late successional *Piper* species may account for the almost complete absence of seedlings of these species at La Selva.

Leaf Herbivory and Adult Plant Fitness. A growing body of evidence indicates that leaf herbivores can be detrimental to plant fitness in natural systems (see reviews by Crawley 1983; Belsky 1986; Hendrix 1988; Whitham et al. 1991; Marquis 1992b). In part, because of the multitude of effects of folivory (Marquis 1992b) and the interplay between those effects and environmental conditions (e.g., Bentley and Whittaker 1979; Archer and Detling 1984; Dirzo 1984; Parker and Salzman 1985), opinions vary widely about the selective influence of herbivores for plant evolution (e.g., Owen and Wiegert 1976; Jermy 1984; Belsky 1986; McNaughton 1986; Whitham et al. 1991). For tropical forest ecosystems, despite the frequent assertion that herbivores are important (e.g., Baker 1970; Janzen 1970; Connell 1971; Levin 1976; Gilbert 1975, 1977, 1980; Hartshorn 1978) experimental data are limited (Janzen 1966, 1976a; Hartshorn 1972; Rockwood 1973; Bentley 1976; Augspurger 1980, 1981, 1984b; Schemske 1980; Becker 1983; Letourneau 1983; McHargue and Hartshorn 1983; Wright 1983; Augspurger and Kelly 1984; Dirzo 1984; Horvitz and Schemske 1984; Marquis 1984b, 1987, 1988, 1991a; McKey 1984; De Steven and Putz 1984; D. B. Clark and Clark 1985; Howe et al. 1985; Koptur 1985; Oliveira et al. 1987; Mendoza et al. 1987; Schupp 1986, 1988b, 1990; Sork 1987; Schupp and Frost 1988; Domínguez et al. 1989; Denslow et al. 1990; Howe 1990; Oyama and Mendoza 1990; Thomas 1990; Nichols-Orians 1991c; Vasconselos 1991; Greig 1993).

A long-term investigation of leaf herbivory as a selective force at La Selva included determining the effects of experimental herbivory on plant fitness and its relation to natural damage in the understory shrub *Piper arieianum* (Marquis 1984b, 1987, 1988, 1991a). Low damage (10% leaf area loss) resulted in significant decreases in growth for one year (Marquis 1991a). A single removal of 30% or more of leaf area reduced growth and seed production of smaller plants for two years (Marquis 1984b, 1987); larger plants recovered sooner. In addition, the viability of seeds produced by plants that were 50% defoliated decreased a significant 25% compared to plants that suffered less defoliation (Marquis 1984b, 1991b). The modifying influence of plant size on the effect of folivory underscores the importance of herbivory in this species: at the end of two years, smaller plants that had lost leaf area were still in the same size class and, thus, vulnerable to effects of equal magnitude after a repeated attack.

Removal of leaf area affected flowering time in *Piper arieianum*, but the effect depended upon the timing of damage (Marquis 1988). Low damage levels (10%–30% leaf area removed) caused one- to two-month delays in flowering time relative to the secondary flowering peak, which contributes 50%–65% less to total annual seed production than the main peak. Because cross-pollination is necessary to ensure high fruit set in *P. arieianum*, delayed flowering altered the portion of the population with which the defoliated plants overlapped in flowering time and, thus, their potential mates. In contrast, only complete defoliation before the major flowering peak of the species (February) resulted in a flowering delay. The results of these experiments suggest that degree of control of flowering time varies over the year.

Seedling Herbivory and Tree Species Diversity. Janzen (1970) and Connell (1971) independently proposed that the detrimental effects of herbivores on seeds and seedlings are of prime importance in the maintenance of high local species diversity in tropical forests. They hypothesized that herbivores decrease the probability of establishment of seedlings near adult plants through their effect as density-responsive predators on seeds and seedlings. The resultant spatial pattern of mortality favors the establishment of nonconspecifics near adult plants. In the views of Janzen (1970) and Connell (1971) two important phenomena result: spacing patterns are affected so that mean distances from adults to surviving juveniles are greater than mean seed dispersal distances; and the density of any individual species, regardless of that species's competitive ability, is low, allowing species of lower competitive ability to coexist (the compensatory mortality hypothesis, Connell 1978). D. A. Clark and Clark (1984) argue that to determine the influence of herbivores on tropical forest plant diversity it is necessary to study the dynamics of the processes involved and that single census studies of spacing patterns (e.g., Hubbell 1979; Fleming and Heithaus 1981; Sterner et al. 1986) are insufficient. D. A. Clark and Clark (1984) and D. B. Clark and Clark (1985) studied the establishment process in *Dipteryx panamensis* (Leguminosae), a canopy emergent at La Selva (see also Augspurger 1983a, 1983c; Augspurger and Kelly 1984; Connell et al. 1984; Howe et al. 1985; Schupp 1988b; reviews in Connell 1979; Connell et al. 1984; D. A. Clark and Clark 1984). Following heavy fruiting peaks in *D. panamensis* at La Selva, seeds and seedlings are relatively dense under adult trees (up to 0.15 per m²) although seeds are occasionally dispersed up to 30 m away from the nearest adult by birds and mammals (D. A. Clark and Clark 1984). Seedling mortality is high (up to 90%) during the first year. By following a cohort of seedlings in a 1-ha plot, D. A. Clark and Clark (1984) demonstrated that the mean distance of surviving seedlings from

adults increased over time as predicted by the Janzen-Connell model and that mortality was associated with herbivore damage (D. B. Clark and Clark 1985). A similar pattern was found for *Casearia corymbosa* (Flacourtiaceae, Howe and Primack 1975; Howe 1977), the only other tree species at La Selva in which seedling mortality patterns have been studied. In *D. panamensis* neighboring seedling density and not distance to the nearest adult accounted for most of the variance in seedling survivorship (D. A. Clark and Clark 1984). Seedling density was positively correlated with both incidence of apical meristem damage and leaf area loss; the latter was correlated with survivorship (D. B. Clark and Clark 1985).

The relative importance of herbivores in the maintenance of tropical tree diversity remains in question (Hubbell 1980; D. A. Clark and Clark 1984; Connell 1979; Connell et al. 1984; Howe et al. 1985). Reviews of available evidence show that density- and/or distance-responsive herbivory, although common, is not pervasive in tropical systems (Connell 1979; Connell et al. 1984; D. A. Clark and Clark 1984). Attack by herbivores (and pathogens, Augspurger 1984b) is only one postulated source of compensatory mortality and, in turn, is only one of a number of hypotheses to explain maintenance of diversity in tropical forests (Connell 1978). In *D. panamensis* at La Selva D. A. Clark and Clark (1984) were able to rule out nonherbivore sources of compensatory mortality, that is, adult interference and competition from neighboring conspecific seedlings. Herbivore damage to young seedlings in rain forests, however, does not necessarily affect their mortality; experimental defoliation of seedlings of two dipterocarp species in Malaysia, *Shorea leprosula* and *S. maxwelliana*, demonstrated that natural herbivory levels (mean area lost per year = 13% and 5%, respectively) did not affect mortality (Becker 1983). Thus, it is difficult to generalize the effects of a given level of herbivory on seedling establishment in tropical forests. Equivalent levels of herbivory may have very different consequences, depending on the influence of other biotic (Dirzo 1984; Parker and Salzman 1985) and abiotic factors (Augspurger 1983a, 1983c), and species-specific responses (Marquis 1992b).

SUSCEPTIBILITY OF PLANTS TO ATTACK

A major theme of plant-herbivore interactions concerns whether plants share defense arrays among species (and/or plant parts) and whether these defense patterns are based on differences in plant life history, macrohabitat, microhabitat preference, and individual plant parts. Two general types of defense, representing endpoints of a continuum, were emphasized initially. First, quantitative defenses, including compounds that reduce digestibility (e.g., tannins and phenolic resins: Mould and Robbins 1981; Robbins et al. 1987), are often associated with long-lived and large species (e.g., temperate *Quercus*). The quantity of these compounds in plant tissue is positively related to their deleterious effect on the herbivore. The species that contain these compounds are hypothesized to be more readily found by their herbivores and, thus, are "apparent" (Feeny 1976) or "predictable" (Rhoades and Cates 1976) to those insects. In contrast, "unapparent" or "unpredictable" plant species, exemplified by small, early successional or weedy species, are characterized by qualitative defenses. These compounds are present in small quantities but are strong deterrents and/or toxic to all but a few specialist insects. These plant species are unapparent as a result of their life-history traits (low abundance and/or short life cycle), which allow them to escape more readily from their herbivores in space and/or time than can apparent species.

Within an individual plant allocation of defense to a leaf has been predicted to reflect the value of that leaf (McKey 1974, 1979; Rhoades 1979; Mooney and Gulmon 1982). Leaves with longer life spans have greater lifetime return and, therefore, should be defended more heavily than leaves with shorter life spans.

Defense Regimes of La Selva Shrubs and Trees

To determine which characteristics account for within-plant and interspecific differences in damage levels Waltz (1984) measured a number of leaf physical and chemical factors on eight shade-tolerant and four gap plant species (table 21.3). The measured factors were total phenolics, condensed tannin, nonstructural carbohydrates, nitrogen, water content, and toughness of both immature and mature leaves. These factors have been shown to affect herbivore preference and feeding level in other plants (e.g., Scriber and Feeny 1979; Coley 1983a; McKey et al. 1981; Howard 1988; Marquis and Batzli 1989). The presence of tannins and other digestibility-reducing components such as lignins and fiber is often associated with a reduction in available nitrogen (for the effects of phenolic resins and tannins as toxins see Lincoln et al. 1982; Reese et al. 1982; Berenbaum 1984; Lindroth and Batzli 1984); therefore, "available nitrogen" was also calculated for each species as the ratio of leaf nitrogen content to the sum of condensed tannin concentration and a toughness value. These physical and chemical characteristics were then correlated with leaf life span, natural herbivory, and preference ratings of three generalist orthopteran herbivores at La Selva, *Abracis* (= *Osmilia*) *flavolineata* and *Leptomerinthoprora flavovittata* (Acrididae), and *Orophus conspersus* (Tettingoniidae). Waltz's original prediction was that gap species would escape attack more frequently (would be less apparent) than shade-tolerant species because the former are restricted to habitats that are unpredictable in time and space (see Hartshorn 1978). It was hypothesized that types and levels of defense should reflect these differences in life-history characteristics. The following patterns should be considered preliminary because the number of species sampled per life history type was small, and chemical analyses were not always on leaves from plants of the same habitat.

Mature leaves of all species were significantly tougher, lower in water, nitrogen, and condensed tannins, and higher in nonstructural carbohydrate content than new leaves (Waltz 1984). No significant differences existed between new leaves of shade-tolerant and gap species in any of the characters. In contrast, mature leaves of shade-tolerant species were, as predicted by the apparency theory, significantly tougher, lower in nitrogen, and lower in nonstructural carbohydrates than mature leaves of gap species. Mean total phenolic and condensed tannin content did not differ among new and mature leaves of gap and shade-tolerant species. Rather, differences were more closely related to growth form than to shade tolerance: both immature and mature leaves of canopy species had signifi-

cantly greater phenolic and condensed tannin content than did immature and mature leaves of subcanopy species. A difference in defensive regime between subcanopy and canopy tree species has been reported for only one other tropical forest (Coley 1983a; Waterman 1983).

Across all species studied by Waltz (1984) neither total phenolics nor condensed tannins increased with leaf life span. Across all species, however, toughness increased and available nitrogen decreased with life span, and both percentage of water and actual percentage of nitrogen decreased somewhat ($0.05 < p < 0.10$) with leaf life span. These latter results are consistent with the prediction that allocation to the defense of individual leaves should reflect their future value, assuming that value of a leaf is positively correlated with life span (Mooney and Gulmon 1982; McKey 1979; Coley et al. 1985) and that increased toughness is a deterrent to feeding (Coley 1983a). Fiber and polyphenol content have been found to increase with leaf life span for a sample of forty-one Panamanian tree species (Coley 1983a; Coley et al. 1985). Mean percentage ($\pm$ SE) of total phenolics (7.1 ± 1.1) in mature leaves for all species was comparable to those reported for other lowland tropical sites (Waterman 1983; Davies et al. 1988; Waterman et al. 1988), whereas condensed tannin values (3.7 ± 0.9) were lower than those reported for other sites but were similar to the mean of 2.36% reported for the twenty most common species at the "anamolously low" Kibale site of western Africa (Waterman 1983; Oates et al. 1990).

The characteristic of mature leaves that most strongly correlated with preference by all three orthopteran herbivores was available nitrogen, suggesting that the interplay between properties that reduce digestibility (tannins and toughness) and nitrogen content was important. Available nitrogen is an important correlate of species preference for a number of generalist herbivores: howler monkeys (Milton 1979), colobine monkeys (Waterman and Choo 1981; McKey et al. 1981), microtine rodents (Marquis and Batzli 1989). For both mature and immature leaves condensed tannin content was significantly negatively correlated with orthopteran preference but only among plants with greater than 2% dry weight tannin content. Condensed tannin concentration was also negatively correlated with natural losses to mature leaves only in those species with greater than 2% tannin by weight, again suggesting that tannins are important, but only above some threshold level. Condensed tannins also deter feeding by browsing ruminants in South Africa but only above 5% concentration (Cooper and Owen-Smith 1985). For natural herbivory at La Selva, only leaf toughness was correlated with observed differences in damage levels to mature leaves of all species. Toughness was also the most important correlate of amount of insect damage across all species for the sample of Panamanian trees (Coley 1983a) and for five species of Australian rain forest trees (Lowman and Box 1983). None of the measured characteristics was related to differences among La Selva species in natural damage levels to immature leaves.

Although gap species suffered greater rates of leaf area loss than shade-tolerant species to both mature and new leaves ($0.058 < p < 0.079$; see also Coley 1983b), Waltz (1984) found no evidence that gap species are less apparent to their herbivores than are shade-tolerant species. Although apparency theory would predict greater variance in attack rates among individuals of unapparent species (Feeny 1976), variation in attack rates among individuals of gap species was not greater than for those of shade-tolerant species.

In the only other reported test of the apparency theory using attack rates Coley (1983a) concluded that individuals of pioneer tree species are as easily found by their herbivores as individuals of persistent tree species. These results have led Coley and others (Coley 1987; Coley et al. 1985; Fox 1980), to question the value of the apparency hypothesis in explaining interspecific differences in defensive regimes. Local resource availability for plant growth (e.g., soil nutrient content: Janzen 1974; McKey et al. 1978) is related to both type and quantity of chemical compounds shared by entire communities of plants and to the associated herbivore fauna and abundance (but see Waterman 1983; Oates et al. 1990). To account for this latter pattern, Coley et al. (1985) proposed that resource availability in the environment to which a plant species is adapted is the major determinant of both amount and type of plant defense. Over evolutionary time a large allocation to defense is predicted for plants that grow in environments of low resource availability because resource limitation results in slow growth rate and, therefore, increases the impact of herbivory.

Based on comparison of phenolic concentrations in gap- and forest-adapted species of *Piper* (Piperaceae) and *Miconia* (Melastomataceae) and La Selva, Baldwin and Schultz (1988) argue that plant phylogeny must also be considered when trying to explain patterns of defense. They found that gap-adapted *Miconia* species were higher in leaf phenolics then forest-adapted *Miconia,* just the opposite predicted based on resource availability (Coley et al. 1985; see also Newberry and de Foresta 1985). No differences existed among gap- and forest-adapted *Piper* species. When Coley's (1983a) data were reexamined taking into account phylogeny, concentrations of phenolics in pioneer species were greater than those of persistent species within the same family (Baldwin and Schultz 1988). Thus, life history (Coley 1983a; Waltz 1984), phylogeny (Baldwin and Schultz 1988), and growth-form (Waltz 1984) all must be considered in analysis of interspecific differences in defense investment among plant species.

Ecological Factors
The susceptibility of plants to herbivore attack is often influenced substantially by the environment. These may involve changes in tissue quality as a function of soil nutrient levels (e.g., Gershenzon 1984), incident light levels (e.g., Langenheim et al. 1981), and past herbivory (e.g., Rhoades 1979; Carroll and Hoffman 1980; Schultz and Baldwin 1982; Faeth 1986; but see Fowler and Lawton 1985). These effects may influence the probability of attack by affecting leaf quality. The environment may also influence herbivore search and colonization patterns (Huffaker and Kennett 1959; Cromartie 1975; Atsatt and O'Dowd 1976), either as a result of the particular species composition and abundance of neighboring plants (Tavanainen and Root 1972; Root 1973; Atsatt and O'Dowd 1976; Kareiva 1983; Stanton 1983) or through an effect of the abiotic environment itself on the herbivore (Moore et al. 1988). The overall effect of the environment is, likely, complex. For example, Mueller and Wolf-Mueller (1991) demonstrated that epiphylls deter feeding by *Atta*

cephalotes on *Citrus paradisi* (Rutaceae) and *Cyclanthus bipartitus* (Cyclanthaceae) at La Selva, the first such report of deterrence by epiphylls. Thus, in addition to those factors mentioned, susceptibility of plants at La Selva to attack by *Atta* is likely determined by factors that influence colonization by epiphylls, the particular species composition of epiphylls that do colonize, and the interactive effect of epiphylls and the leaf on leaf quality (e.g., increased nitrogen concentration, Bentley and Carpenter 1984).

For many La Selva understory species herbivory levels are related to the habitat in which they are growing. For *Piper pseudobumbratum* populations along streams and in swampy habitats suffer significantly greater leaf damage than upland populations (Kelly et al. 1982). For the insect folivores of *Heliconia imbricata* that live under packs of fallen leaves that collect on living leaves (leaf-top herbivore guild) colonization is related to leaf pack moisture and microhabitat (McCoy 1984) so that only plants in gaps and along open trails are colonized. Preliminary experiments suggest that these leaf-top herbivores prefer tighter leaf packs, which are less susceptible to desiccation. Braker and Chazdon (1992) found a higher proportion of leaf area missing for individuals of three species of geonomoid palms (*Asterogyne martiana, Geonoma cuneata,* and *G. congesta*) growing in gaps than in understory. Individuals of *Piper urostachym* (Denslow et al. 1990), *P. sanctifelicis,* and *P. arieianum* (Marquis unpublished data) in gaps lose more leaf area than understory plants, whereas the opposite pattern holds for *Miconia barbinervis* (Denslow et al. 1990). For *P. arieianum* greater damage in gaps is associated with more herbivore species attacking gap than understory plants (R. J. Marquis unpublished data). Denslow et al. (1990) found a negative correlation between total phenolic concentration and folivory across three species of *Miconia* and four species of *Piper* planted in light gaps and neighboring understory.

Insect abundance parallels these differences in damage levels between plants in understory and gaps. Maleret et al. (1978) found that insect abundance was highest in a newly formed gap, intermediate in a gap in the "building stage" (see Whitmore 1984), and lowest in the understory of closed canopy forest. It is not known to what degree these patterns are related to differences in abiotic conditions or in plant species composition. Differential predation may also be important: Wrobel (1979) found lower abundance of predatory insects in open successional plots than in neighboring primary forest understory. Comparison of the effect of habitat on standing damage levels also must take into account the effects of habitat on leaf life span, which, in turn, determines the time that an individual leaf is exposed to herbivores (Marquis and Clark 1989).

Some portion of the observed intraspecific variation in leaf damage could result from soil nutrient effects on plant secondary chemistry (e.g., Gershenzon 1984). Nutrient availability varies substantially over the La Selva property (see chap. 4). When soil nutrient level was manipulated in three species each of *Piper* and *Miconia,* phenolic level was affected in five of six species (Denslow et al. 1987). In the majority of the cases, however, greater phenolic production was associated with increased soil nutrient level, opposite the result predicted if there were a trade-off between nutrient availability and production of carbon-based secondary compounds (Mooney and Gulmon 1982; Bryant et al. 1983; Coley et al. 1985). Manipulation of light level in three species of *Miconia* also yielded no evidence for a trade-off between growth and investment in phenolics (Denslow et al. 1990). In contrast, for seedlings of *Inga oerstediana,* Nichols-Orians (1991c) found a trade-off between growth and phenolic production when growth was nutrient-limited (on residual soil) but not when growth was light-limited (on alluvial soil). Thus, although support for the resource availability hypothesis of plant antiherbivore defense has come from studies in other areas (Bryant et al. 1987a, 1987b, 1989; Price et al. 1989), analysis of intraspecific variation in tannin chemistry of a number of shrub and tree species at La Selva provides only equivocal evidence for the existence of trade-offs between resource availability, growth, and allocation to defense (Denslow et al. 1987, 1990). Relevance of these kind of experiments for plant fitness depends on the relationship between observed changes in foliar chemistry and subsequent damage to the plant. Experiments with susceptibility of *Inga* seedlings to leaf-cutting ants suggest that condensed tannins are deterrents to *Atta* but only at threshold levels (Nichols-Orians 1991a, 1991c; Nichols-Orians and Schultz 1990). Denslow et al. (1990) discuss the difficulties of relating foliar chemistry measures to natural damage patterns.

Genetic Basis for Resistance

For herbivores to effect evolutionary changes in traits of their host plants both selection and a genetic response by the plant population to selection must occur. For *Piper arieianum* at La Selva intraspecific variation in damage levels is high and results in differences in growth and seed production. Using *P. arieianum* in a cloning experiment conducted over 3.5 years, Marquis (1990) demonstrated significant differences in leaf damage among genotypes. In some censuses certain herbivore species showed significant preferences for certain genotypes, but the most heavily damaged genotype was not necessarily the same for all insect species. Ranking of clones generally remained the same throughout the experiment. Finally, during a portion of the experiment a significant negative correlation existed between the amount of leaf damage and growth, suggesting that selection had occurred during the experiment. These results, along with data on the effects of experimental herbivory, suggest that leaf herbivores at La Selva can select for changes in defense characters in *P. arieianum.* Assuming that a significant portion of the clone effect is the result of additive genetic effects, one would expect a response to this selection. For only five other plant-herbivore systems are sufficient data available to demonstrate that herbivores have a potential selective impact on their host plants (Marquis 1992b).

MUTUALISMS WITH ANTS AND OTHER INSECTS AS DEFENSE

In mutualistic ant/plant associations plants provide some resource for predatory or parasitic insects, usually in the form of food (e.g., extrafloral nectar) and/or domicile (domatia). The ants use the plant as a foraging substrate, both feeding from extrafloral nectaries and capturing prey on the plant. Because these prey insects can be herbivores, the plant may de-

rive benefit from their removal (Bentley 1977). Additional benefit to the plant may include removal of vines and nutrient addition through breakdown of collected debris in domiciles (Benson 1985). Although most studies have concentrated on the ant fauna involved in these relationships, other Hymenoptera and Diptera that often visit extrafloral nectaries may be parasitic on herbivores (e.g., Hespenheide 1985; Koptur 1985).

Pearl Bodies

Pearl bodies are single or multicelled, lipid-rich structures that are produced on the leaves and shoots of many tropical plants; they are reported from fifty genera in nineteen tropical and subtropical families (O'Dowd 1982). The presence of pearl bodies is often associated with other ant-related features of the plant, such as extrafloral nectaries and domatia.

The role of pearl bodies as ant attractants has been investigated for two plant species at La Selva, *Ochroma pyramidale* (Bombacaeae) (O'Dowd 1979, 1980) and *Piper cenocladum* (Piperaceae) (Risch et al. 1977; Risch and Rickson 1981; Rickson and Risch 1984). *Ochroma* saplings grown in a greenhouse from seeds collected at La Selva produced an average of 402 pearl bodies per leaf (O'Dowd 1980). Pearl bodies, however, are not observed on *Ochroma* in the field, presumably because they are removed immediately after production (O'Dowd 1979). The rate of production of pearl bodies on leaves in the greenhouse saplings is greatest on those leaves that produce copious amounts of petiolar extrafloral nectar and are heavily patrolled by ants in the field. When ants were excluded from *Ochroma* saplings on the Osa Peninsula, those plants lost more leaf area than plants to which ants had free access (O'Dowd 1979). O'Dowd argues that pearl bodies (74.4% lipid per gm dry weight, comparable to the oil content of some seeds) and extrafloral nectaries function in concert to provide a high-energy reward to ants, thereby reinforcing the ants' fidelity to *Ochroma.* Absence of pearl bodies in the field on ant-patrolled *Ochroma* and other pearl-body producing species (Schupp and Feener 1991; R. J. Marquis pers. observation, N. Greig pers. comm.) suggests that this form of defense may be common but largely overlooked because pearl bodies are small and quickly removed by ants.

Pheidole bicornis ants are associated with the understory shrub *Piper cenocladum* at La Selva (Risch et al. 1977). The ants live in domiciles formed by the appressed margins of the petiole and collect pearl bodies that are 46% lipids by weight (Rickson and Risch 1984). Initially, the benefit to the plant in this relationship was unclear: the ants are small and unaggressive toward many phytophagous insects, and the plant is often damaged heavily by herbivores. Letourneau (1983), however, used ant exclusion to show that individuals of *P. fimbriulatum* on the Osa Peninsula, also inhabited by *P. bicornis,* experience reduced herbivory when inhabited by ants. Exclusion experiments with *P. cenocladum* have not been conducted. Risch and Rickson (1981) have shown that *P. cenocladum* responds to the short-term presence of ants (about forty days in this study) by increased production of pearl bodies. The bodies are produced in very low numbers (fifteen to twenty per petiole) when no ants are present, but a petiole containing ants produces about fifteen hundred food cells. This is the first report of an ant-plant that facultatively increases food production when ants are present.

Extrafloral Nectaries

Extrafloral nectaries (EFNs), nectar-secreting glands found outside of the flower, are a common feature of many tropical trees and tropical and temperate herbs (Bentley 1977). EFNs are hypothesized (e.g., Bentley 1977) as adaptations to attract ants, which, in turn, protect the plant against herbivores and/ or climbing vines. Although it has been suggested that EFNs are more frequent in tropical than in temperate areas (Bentley 1977), data allowing between-site comparisons of the distribution and abundance of EFNs are few. Data from La Selva primary forest understory do not confirm the prediction: the range in proportion of plants sampled that have EFNs is 0.009–0.08 (Chu 1977; K. Keeler pers. comm.). These values fall within the range reported for four different Nebraska habitats (0.0–0.083, Keeler 1980) and for high elevation Neotropical sites (0.0–0.27, Chu 1977; Keeler 1979). The La Selva successional plots have a high proportion of EFN plants compared to primary forest understory (0.022–0.292, K. Keeler, pers. comm.; H. A. Hespenheide unpublished data), but whether this holds for gaps is not known. Canopy and subcanopy trees and lianas have not been sampled at La Selva to determine if EFN abundance is similar to that reported for Costa Rican dry deciduous forest and riparian forest (0.8 and 0.1, respectively, Bentley 1976). Within successional stages at La Selva frequency of EFNs may vary with growth form; EFNs are more common on vine species than on herbs or woody plants in the first-year successional plot at La Selva (H. A. Hespenheide unpublished data).

In both short- and long-term studies at La Selva biologists have explored the ecological role of extrafloral nectary production. Results from short-term studies have been equivocal about the benefit of EFNs. Carroll (1969) simulated EFNs on bean seedlings at two planting densities and examined the plants after four days for the presence of meristematic tips. At low density seedlings with simulated EFNs had a higher percentage of shoot tips remaining than control plants. There was no difference between treated seedlings and control plants at high density. Fritz (1979) glued termites to plants with and without EFNs and found that these baits disappeared more quickly from EFN plants. Griswold (1977), however, found no difference in number of herbivores or in amount of seed predation among plants of *Costus nutans* (Zingiberaceae) with or without ants. Jones (1985) found that *Passiflora* branches with simulated EFNs had more termite baits removed than did controls, but there was no differences in removal rate between mature and new branches, each with simulated EFNs.

In contrast, results of long-term studies conducted at La Selva have generally shown a positive effect of EFN's on plant fitness parameters. In *Pentaclethra macroloba* nectar production at EFNs (located on the stem at the base of petioles) declines with age (Hartshorn 1972). Seedlings in short height classes that were attended by ants grew more quickly than did those from which ants were excluded.

The extrafloral nectaries of *Cassia fruticosa* (Caesalpinia-ceae) are located at the base of leaflets and on inflorescences. Deuth (1980) found that ant abundance on plants varies both spatially and temporally, leading to the prediction that protection might vary similarly. Although seedlings from which ants were excluded did not differ from controls in amount of damage, control plants on which ants were frequent visitors were significantly less damaged than control plants on which ant visitors were rare. Most importantly, no inflorescences of plants from which ants were excluded produced fruit, whereas those visited by ants matured varying numbers of fruits.

In his study of insect visitors to foliar EFNs of the vine *Byttneria aculeata* (Sterculiaceae) at La Selva Hespenheide (1985, unpublished data) emphasized the potential importance of visitors other than ants to EFNs, specifically predatory and parasitic flies and wasps. During the dry season ants, parasitoids, and flies are about equally frequent visitors to these nectaries. Twenty-four ant species were recorded on *Byttneria*, eight of which nested in hollow stems. The ants spent most of their time at EFNs, not in patrolling the plant. Many species of parasitic Hymenoptera visited the EFNs, and three of these were reared from herbivorous larvae found on *Byttneria*. Flies appeared to be nectar thieves; they were not observed performing any beneficial function for the plant. Hespenheide (1985) suggests that plants like *Byttneria* may benefit more from parasitic Hymenoptera visiting the nectaries than from ants.

Herbivore deterrence may be only one explanation for the evolution of plant/ant associations via EFNs. At La Selva ants of the genus *Crematogaster* build carton nests on tree trunks and collect fruits and extrafloral nectar of the epiphytic vine *Codonanthe crassifolia* (Gesneriaceae) (Kleinfeldt 1978). After eating the arils, the ants place the seeds in the carton nest, where the seeds germinate. Plants with ants grew more rapidly than plants without ant associates. Kleinfeldt did not observe any herbivore deterrence by *Crematogaster*. Instead, the hypothesized benefit to the plant in this association is through short-distance, but assured seed dispersal to a safe site. In turn, more rapid growth of seedlings emerging from the ant nests is presumed to result from increased nutrient availability. The benefit to the ant is a constant aril and nectar supply.

Plant Associations with Ants and Homoptera

An additional level of complexity in ant-plant associations is exemplified by *Ocotea atarense* (= *O. pedalifolia*) (Lauraceae) and its insect associates (Stout 1979a). Depressions along inner margins of the hollow stems are occupied by ants *Myrmelachista* sp. and two species of mealy bugs (Coccidae: *Dysmicoccus* sp.). Almost every plant examined (forty-nine/ fifty) contained ants that are active in and upon stems and on leaves. Ant activity is highest on young leaves from which experimentally placed insects and insect eggs are removed. Mealy bugs are found in stem depressions in apical portions of the plant. Because the mealy bugs are located inside the stems, Stout was not able to describe the interaction between ants and bugs, but collection of honeydew from Homoptera by ants has been well-documented (Way 1963). In such systems benefits to the plant may include protection from herbivores; to the ants, food and domicile; to the mealybugs, food, domicile, protection from predators, and, possibly, reduced competition from other herbivores. Ant-coccid associations are probably common in plant species with hollow stems (e.g., *Ceiba pentandra*, H. A. Hespenheide pers. comm.; *Cordia alliodora*, Carroll 1983; *Cecropia* spp., Janzen 1969a).

Other Plant/Ant Associations

At least sixteen species of plants occurring at La Selva are adapted for ant occupancy with specialized features such as domatia: *Ocotea atarense*, *O. dendrodaphne*, *O. nicaraguense* (Lauraceae); *Cordia alliodora* (Boraginaceae); *Cecropia obtusifolia* and *C. peltata* (Moraceae); *Codonanthe* (three spp.) (Gesneriaceae); *Clidemia setosa*, *C. pubescens*, *C.* spp., *Miconia* spp., *Triolena* spp., *Topobea* sp. (Melostomataceae); *Piper cenocladum* (Piperaceae); and *Acacia ruddiae* (Mimosaceae) (Wyatt 1974; Cort et al. 1977; Manning 1983; Short 1984; R. J. Marquis pers. observation; M. Grayum pers. comm.). Research on ant-plant relationships conducted so far at La Selva and other tropical sites (Janzen 1966, 1972; Rehr et al. 1973; Letourneau 1983; McKey 1974, 1984; Benson 1985; Koptur 1985; Schupp 1986) suggests wide variation in benefits received and costs incurred by both participants. It is unclear at this time what conditions affect the balance between costs and benefits (Beattie 1989; Cooper-Driver 1990).

FUTURE DIRECTIONS

We are beginning to accumulate the information necessary to generalize about tropical plant-herbivore systems and to determine whether those patterns are in accord with present theory. Initial predictions for tropical habitats based on knowledge in temperate systems have not always been confirmed. For example, tannin concentrations in immature leaves can be higher than in mature leaves (Coley 1983a; Nichols-Orians and Schultz 1990), short-lived tree species are not necessarily less apparent to their herbivores than long-lived tree species (Coley 1983a; Waltz 1984), and trade-offs do not necessarily exist between plant growth and phenolic production (Denslow et al. 1987, 1990; Nichols-Orians 1991c). More extensive investigation at tropical sites will continue to broaden the perspective on plants and their herbivores. In this section we suggest areas for future research at La Selva that are likely to add significantly to overall understanding of plant-herbivore interactions.

Leaf damage by herbivores at La Selva varies greatly both within (Marquis 1987) and among plant genera. Because of its high plant species diversity, the La Selva forest provides an opportunity for intrageneric comparisons of defense patterns and for studies of herbivore choice as related to plant life history. Further investigation will clarify life-history differences among La Selva tree species as well as the relative importance of leaf quality factors in determining interspecific differences in amount of damage. Intrageneric comparisons are possible for a number of species rich genera at La Selva (e.g., *Clidemia*, *Cordia*, *Guarea*, *Inga*, *Miconia*, *Ocotea*, *Passiflora*, *Piper*, *Pithecellobium*, *Psychotria*, and *Solanum*). Because members of these genera often occur in different habitats (e.g., gaps compared with understory) and represent different growth-forms, they provide an opportunity to investigate the relative importance of apparency (Feeny 1976; Rhoades and

Cates 1976) compared with resource availability (Coley et al. 1985) (or some combination of the two, Southwood et al. 1986) for herbivore damage and associated allocation to defense while controlling for phylogenetic differences (Baldwin and Schultz 1988).

The forest floor at La Selva is subject to an almost continual rain of seeds from the canopy above. Despite this high seed input seedling density is low; high concentrations of seedlings are rare, and when they do occur, are extremely short-lived events. What role do herbivores play as agents controlling seedling establishment and, thus, forest regeneration? Their influence may be as both pre- and postdispersal seed predators and as herbivores on newly establishing seedlings and older saplings. The effect of herbivores on seedling growth may be subtle: seedlings planted under understory palms and cyclanths grow slower and have a lower probability of survival than seedlings planted away from understory shade (Denslow et al. 1991). The growth differences associated with treatments owed mainly to differential herbivory and not, as expected, to differential light conditions (Denslow et al. 1991).

In light of the role proposed for seed predators in the maintenance of tropical tree species diversity and their demonstrated impact on distribution dynamics in some temperate plants (Louda 1982a, 1982b) general lack of knowledge of seed predators at La Selva is striking. Initial observation and investigation suggests that seed predation is pervasive across most, if not all, La Selva plant species and is likely to be a major factor limiting the number of propagules available for ensuing generations. At this time, we know the seed predators for only a few plant species, and have much less knowledge of their impact on plant reproduction and seedling establishment. In addition to their significance for theories of plant-herbivore interactions studies of seed predators could provide valuable information for the propagation of economically important tree species.

The evidence available at this time is contradictory about the degree to which herbivores determine the fates of the establishing seedlings in tropical forests. Leaf area loss in seedlings can be correlated with growth (D. B. Clark and Clark 1985) or may have no observable effect (Becker 1983). Shade-house studies of germination, establishment, and seedling growth of various La Selva tree species (R. J. Marquis unpublished data) demonstrate that in the absence of herbivores, seedling growth is markedly lower at light levels typical of understory compared to gap environments, but survivorship is almost independent of light level for at least some period. It appears then that seedlings of La Selva tree species are physiologically capable of tolerating the low light conditions of the understory. Yet the forest floor is not covered with seedlings. Because microenvironment and herbivory are likely to interact in their effect on establishment (Augspurger 1983c; Howe et al. 1985; Schupp 1988b; Turner 1989; Denslow et al. 1991), experiments that integrate the effects of herbivory, light, and falling debris (Clark, chap. 6) on seedling establishment would be extremely valuable. Transplanting seedlings to various locations may tell much about the role of herbivores as they affect plant distribution at La Selva. The expansion of Parque Nacional Braulio Carillo to La Selva's southern boundary provides additional potential for studying the importance of herbivores in limiting the elevational ranges of plants.

Through the experimental work on *Piper arieianum* we know how observed damage affects the plant fitness of one species at La Selva growing in one microhabitat (the understory). To what degree these results are applicable to other species in different habitats (particularly gaps) is not clear. Furthermore, the effects of herbivores may not be readily apparent. First, total tissue loss from a plant can be small (and, therefore, not readily observable) but extremely detrimental to growth and seed production (e.g., damage to shoot tips, D. B. Clark and Clark 1985; Marquis 1991a; flower buds and flowers, Bawa 1980a). For example, the role of herbivores in the fate of flowers is typically ignored but can be substantial: seed and pollen production by *Neea psychotrioides* at La Selva is reduced early in reproduction because of attack on flowers by gall-forming flies and chewing moth larvae, resulting in an average 39.1% and 6.5% destruction of male and female flowers, respectively (L. Wolfe unpublished data). Second, attacks that result in large amounts of leaf area loss are often episodic and unpredictable. For example, average standing damage levels to *Zamia skinneri* are 2% loss of leaf area or less, suggesting that herbivores have little importance in this system. When oviposition by the butterfly *Eumaeus minyas* is coordinated with leaf-flush, however, complete defoliation and even plant death can result (D. B. Clark and Clark 1991). Long-term studies are, thus, necessary. Inter- and intraspecific comparisons for different microhabitats (e.g., understory compared with light gaps, alluvial compared with residual soils) are needed to understand the interactions between resource base, allocation to defense, and the effect of herbivory.

The relative importance of factors that regulate insect herbivore population levels at La Selva is unknown. From studies of rolled-leaf hispine beetles, biologists suggest that parasitism is important, but at least for lepidopteran larvae at La Selva, parasitism rates vary greatly among species (R. J. Marquis pers. observation). Predation by birds controls the composition and numbers of insects in leaf packs in forest on Barro Colorado Island in Panama (Gradwohl and Greenburg 1982). By extrapolation avian predation can be expected to be an important regulator of herbivorous insect populations. Avian exclusion experiments such as the temperate zone studies by Holmes et al. (1979) and Marquis and Whelan (in press) could yield exciting results.

We are beginning to quantify the relative costs and benefits of defense against herbivores (Simms and Fritz 1990). For plants that attract ants as defense agents against herbivores it has been suggested that the cost of supporting ants may be high relative to the total energy budget of the plant (McKey 1984). Manipulation of plants that have mutualistic associations with ants provides an opportunity to determine the phenotypic cost of defense (e.g., through quantification of nectar production rates and constituents relative to the total carbon and nutrient budget of plants with extrafloral nectaries) and the benefit of defenses (through the effects of damage on fitness components for plants with and without ants). Domatia occur in at least six genera at La Selva and extrafloral nectaries are widespread.

We make a strong plea here for studies of the natural history of herbivore species that occur at La Selva, their host plants, and the secondary chemical constituents of those plants. How diverse are the herbivore faunas of individual plant species? How specific are herbivores in their use of plant species? What is the relative contribution of individual herbivore species to total damage, and in turn, resultant reduction in plant fitness? How do diversity and toxicity of secondary compounds relate to the diversity of herbivore species? Only for a few plant species and their herbivore faunas is the information necessary to begin to answer these questions available. Biologists know little about actual patterns of herbivore abundance, their host preferences and life cycles, let alone the factors that might control these processes. The impact of mammals as herbivores, their diet composition, and factors that affect that composition are unknown. Leaf-cutting ants, which may have the greatest impact of any single herbivore species at La Selva in some habitats, have been ignored except for a few behavioral studies (e.g., Anonymous 1973; Coyne 1974; Meagher and Damery Parrish 1975; Terwilliger 1981; Winnett-Murray 1981; Rudolph and Loudon 1986) and one set of studies of the interaction between host-plant chemistry and *Atta* preference (Nichols-Orians 1991a, 1991b, 1991c; Nichols-Orians and Schultz 1990). The best-known group of herbivores is the butterflies, for which we have only minimal natural history information for fewer than one-half of the species that occur at La Selva (DeVries 1985, 1987). Phytochemical studies at La Selva have been limited to the work of Gage and Strong (1981), Waltz (1984), Baldwin and Schultz (1987), Denslow et al. (1987, 1990), Nichols-Orians (1991a, 1991b, 1991c), and Downum et al. (1991). Only one secondary compound (safrole in *Piper auritum*, Marquis unpublished data) has been isolated and identified from a plant at La Selva. The first report (Downum et al. 1991) of phototoxins for seven plant families, based on work at La Selva, underscores the degree of ignorance about tropical plant secondary chemistry. Unique physical defenses also likely await discovery: for example, Wooten and Sun (1990) demonstrated that in *Heliconia wagneriana* active secretion of liquid into floral bracts reduces insect predation on developing ovaries by 17%. Initial secondary chemistry work at other tropical sites (Barro Colorado Island, Panama: Coley 1983a; Parque Nacional Guanacaste, Costa Rica: Janzen and Waterman 1984; Hubbell et al. 1984; Howard and Wiemer 1986; Howard 1987, 1988; four African forests: McKey et al. 1978, 1981; Gartlan et al. 1980; Waterman 1986; Ganzhorn 1988; Oates et al. 1990; Kakachi forest in south India: Oates et al. 1980; and two Malaysian forests—Davies et al. 1988; Waterman et al. 1988) will provide important bases for comparison of results from future investigation at La Selva.

The development of the Sarapiquí annex for forestry growth trials provides an excellent opportunity to integrate basic and applied aspects of plant-herbivore research. Initial work with *Hampea appendiculata* (Marquis and Clark 1989) suggests that trees in pastures may escape herbivores that normally attack them in forest settings (see also Janzen 1987b). We do not know whether lower damage is the result of low colonization rates of herbivores from the forest, differential predation and parasitism, habitat effects on leaf quality, or some combination of these factors. Tree density, distance from forest, and cultural practice all can be manipulated to determine their effects on herbivore species composition and abundance and resultant damage. Comparative study of the herbivore faunas in forest and plantation would be beneficial from both theoretical and applied points of view. Manipulation of herbivore abundance on trees and saplings under both managed and natural conditions would provide information about the effects of herbivores on growth of tropical trees, an area where information is largely anecdotal (Browne 1933, 1968; Kapur 1959; Gray 1972; UNESCO 1978b; Palmer et al. 1989).

CONCLUSIONS

The herbivore fauna of Estación Biológica La Selva is diverse in both vertebrate and invertebrate species. Little is known of the population dynamics and feeding biology of either group or of the impact of vertebrate herbivores. For those insect herbivores that have been studied parasitism, resource quality, and abiotic factors, such as rainfall, may control population levels. Degree of host plant specialization varies greatly both within and among major taxa of herbivorous insects. Based on available data, La Selva butterfly and grasshopper species appear to be more host specific than their temperate counterparts, whereas treehoppers at La Selva are less host specific than those in North America. Both host defensive regime and ecological factors affect the level of host specificity, but their relative importance is not known. The number of insect species found on any one plant species is correlated with local host abundance, habitat, overall host range, and number of related, co-occurring plant species.

The present information suggests that the insect herbivores of La Selva have important effects on both individual plant fitness and plant population dynamics. There is high variation in damage levels both among and within plant species, and corresponding diversity of herbivore faunas for any given plant species. Interspecific variation in damage levels is, in part, explained by differences in leaf quality factors, which are, in turn, related to plant life history and growth-form, supporting patterns already found in Panama. At the population level spacing patterns and actual species abundance may be influenced by the negative affects of herbivore damage on seedling establishment and early seedling growth. Within-species variation in damage level can be related to microhabitat, density of conspecifics, and individual plant genotype. At the individual plant level loss of leaf area to herbivores can result in longer-term decreases in plant reproductive output and, therefore, fitness. In many areas the ability to make meaningful comparisons between La Selva, other tropical sites, and temperate habitats is presently limited because of inadequate data from all areas.

ACKNOWLEDGMENTS

We thank C. Kelly, M. Berenbaum, J. Nitao, D. Strong, J. Schultz, S. Louda, P. Coley, and one anonymous reviewer for helpful discussion, criticism, and editing of the material presented here. Steven Passoa provided a number of useful literature references. We wish to express our thanks to OTS staff for their logistical and moral support throughout the years. Support for research during writing of this paper came from National Science Foundation grants BSR-8600207 and DEB 81-10197.

Appendix 21.1 Feeding habits of herbivorous arthropod families at La Selva

Taxon	Plant Part	La Selva Studies
Acarina		
Eriophyidae	LG = A,I	
Collembola		
Sminthuridae	LC = A,I	
Orthoptera		
Tetrigidae	LC = A,I	Marquis 1991b
Eumasticidae	LC = A,I	Marquis 1991b
Romaleidae	LC = A,I	
Acrididae	LC = A,I	Denno and Donnelly 1981; Waltz 1984; Rowell 1978, 1983a, 1983b, 1983c, 1983d, 1985a, 1985b; Braker 1986, 1989a, 1989b, 1991; Braker and Chazdon 1992; Marquis 1991b; Marquis and Braker 1987
Tettigoniidae	LC = A,I (LR = A,I)	Rentz 1975; Waltz 1984; Marquis 1991b
Gryllacrididae	LC = A,I	
Phasmotodea		
Phamastidae	LC = A,I	Marquis 1991b
Isoptera		
Termitidae	W = A,I	
Rhinotermitidae	W = A,I	
Thysanoptera		
Heterothripidae	FC = A,I LC = A,I	
Thripidae	FC = A,I FG = I LC = A,I LG = I M = A,I	
Hemiptera		
Heteroptera		
Miridae	CS = A,I	Greig 1993
Tingidae	CS = A,I	

Appendix 21.1 (continued)

Taxon	Plant Part	La Selva Studies
Lygaeidae	S = A,I LC = A,I	Ernest 1989
Largidae	CS = A,I	
Berytidae	CS = A,I	
Pyrrhocoridae	S = A,I	
Coreidae	CS = A,I	
Rhopalidae	CS = A,I	
Alydidae	CS = A,I	Greig 1993
Scutelleridae	CS = A,I	
Pentatomidae	S = A,I CS = A,I	Marquis 1988, 1991b; Greig 1993
Cydnidae	CS = A,I	
Plataspididae	CS = A,I	
Corimelaenidae	CS = A,I	
Homoptera		
Fulgoroidea	PH = A,I	Marquis 1991b
Cicadoidea	XY = A,I PH = A,I M = A R = I (CS = A,I)	Wood 1978, 1983; Wood and Olmstead 1984; Young 1972a, 1973a, 1976, 1980, 1983; Marquis 1991b
Psylloidea	PH = A,I SG = A,I (LG = I)	
Aleyroidea	PH = A,I	Marquis 1991b
Aphidoidea	PH = A,I (R,SG = I)	Marquis 1991b
Coccoidea	PH = A,I	Stout 1979a
Coleoptera		
Scarabaeidae	FC = A L = A M = A,I R = I	Young 1988a, 1988b
Buprestidae	LC = A,I LM = I SB = I SG = I W = I	Hespenheide 1983a, 1983b
Elateridae	FC = A LC = A SB = I W = I	
Cantharidae	FC = A	
Bostrichidae	M = A,I	
Rhizophagidae	R = A,I	
Languriidae	FC = A LC = A SB = I	D. B. Clark and Clark 1991
Phalacridae	FC = I	
Coccinellidae	(LC = A,I)	
Byturidae	FC = A,I	
Tenebrionidae	(S = A) (LC = A)	
Mordellidae	(SB = I)	
Pyrochroidae	FC = A LC = A	
Meloidae	FC = A LC = A	
Cerambycidae	FC = A W = I (SB = I)	
Chrysomelidae	FC = A LC = A,I LM = I SB = I PH = I	Strong 1977a, 1977b, 1983; Smiley 1982; Strong and Wang 1977; Strong and Auerbach 1981; Gilbert and Smiley 1978; Ernest 1989
Curculionidae	FC = A,I LC = A,I LR = I LG = I XY = I PH = A CS = I R = I	Hespenheide 1987; Marquis 1991b; Janzen and Wilson 1977; Crease and Trott 1982; Greig 1993
Brentidae	SG = I W = I	
Platypodidae	W = I R = I	
Scolytidae	SG = A,I	Sauer and Grove 1977
Bruchidae	S = I	Hartshorn 1972
Lepidoptera		
Micropterigidae	LC = I	
Eriocraniidae	LC = I	
Hepialidae	SB = I R = I	
Nepticulidae	LM = I R = I (LG = I)	
Opostegidae	LM = I	
Tischeriidae	LM = I	
Incurvariidae	LC = I S = I LM = I SM = I	
Heliozelidae	LM = I	
Tineidae	R = I	
Pyschidae	R = II	

Appendix 21.1 (continued)

Taxon	Plant Part	La Selva Studies
Lyonetiidae	LM = I (LC = I)	
Gracillariidae	LR = I LM = I	
Oecophoridae	SG = I LC = I LR = I	Marquis 1991b
Elachistidae	LM = I	
Coleophoridae	LM = I (LC = I)	
Momphidae	LC = I LM = I	
Cosmopterigidae	LC = I LM = I	
Gelechiidae	LC = I (LM = I)	
Epermeniidae	LM = I	
Carposinidae	SM = I R = I	
Glyphipterigidae	LR = I	
Plutellidae	LC = I	
Yponomeutidae	LC = I	
Douglasiidae	LM = I	
Heliodinidae	LM = I	
Sesiidae	S = I SM = I	Hartshorn 1983a
Cossidae	SB = I	
Tortricidae	LR = I (LM = I)	Marquis 1991a, 1991b
Uraniidae	LC = I	
Hesperiidae	LC = I	Marquis 1991a, 1991b
Papilionidae	LC = I	DeVries 1985, 1987
Pieridae	LC = I	DeVries 1985, 1987
Lycaenidae	LC = I	D. B. Clark and Clark 1991
Nymphalidae	LC = I	DeVries 1985, 1987; Marquis 1991b; Young 1971b, 1971c, 1971d, 1972b, 1973b, 1973c, 1974; Auerbach and Strong 1981; Andrews 1983; Ray and Andrews 1980
Zygaenidae	LC = I	
Megalopygidae	LC = I	Auerbach and Strong 1981
Dalceridae	LC = I	
Limacodidae	LC = I	Auerbach and Strong 1981
Pyralidae	LR = I (LM = I)	Marquis 1991a, 1991b
Thyrididae	S = I (FC = I	
Pterophoridae	LR = I SB = I	
Thyatiridae	LC = I	
Depranidae	LC = I	
Geometridae	LC = I	Marquis 1991b
Apatelodidae	LC = I	Marquis 1991b
Bombycidae	LC = I	
Lasiocampidae	LC = I	
Saturniidae	LC = I	Marquis 1984c
Sphingidae	LC = I	
Notodontidae	LC = I	
Dioptidae	LC = I	
Arctiidae	LC = I	
Lymantriidae	LC = I	
Noctuidae	LC = I	Marquis 1991b
Diptera		
Tipulidae	(LC = I)	
Bibionidae	(R = I)	
Cecidomyiidae	FG = I LM = I LG = I	Marquis 1991b
Syrphidae	LC = I	
Psillidae	SG = I	
Otitidae	LC = I	
Platystomidae	LC = I	
Tephritidae	LM = I (SG = I)	
Chloropidae	SB = I	
Agromyzidae	LM = I LG = I	
Anthomyzidae	LC = I LG = I M = I SB = I (LM = I)	
Opomyzidae	SB = I	
Anthomyiidae	(LM, R = I)	
Drosophilidae	LM = I XY = I	
Pantophthalmidae	XY = I	

Appendix 21.1 (continued)

Taxon	Plant Part	La Selva Studies
Hymenoptera		
Xyelidae	FC = I LC = I SB = I	
Pergidae	LC = I	
Argidae	LC = I LM = I	
Cimbicidae	LC = I	
Tenthredinidae	LC = I LM = I LG = I	
Siricidae	SB = I	
Cephidae	SB = I	
Agaonidae	S = I	
Eurytomidae	S = I SG = I	
Chalcidae	S = I	
Cynipidae	LG = I	
Megachilidae	LC = A, I[a]	
Anthophoridae	SB = A,I SG = A,I	
Formicidae	S = A,I FC = A,I[b]	Nichols-Orians 1991a, 1991b, 1991c;
	LC = A,I[b]	Nichols-Orians and Schultz 1990

[a]Nest cells area lined with cut leaves.

[b]Larvae of attine ants feed on fungus cultivated on cut plant material.

Note: Summarized from Needham et al. 1928; Brues 1946; Frost 1965; Borror et al. 1981; Crowson 1981; Strong et al. 1984. S = seeds; FC = flower chewer; FG = flower galler; LC = leaf chewer; LR = leaf roller or tier; LM = leaf miner; LG = leaf galler; M = meristem feeder; SB = stem borer; SG = stem galler; W = Wood borer; XY = xylem feeder; PH = phloem feeder; CS = sucks cell contents; R = root feeder; A = adult stage; I = Immature stage(s); ( ) = subset of family has this feeding habit.

22

Frugivory: An Overview

Douglas J. Levey, Timothy C. Moermond, and Julie Sloan Denslow

Seed dispersal by vertebrates may have been responsible for the spread and domination of angiosperms (Corner 1964; Regal 1977; Snow 1981; but see Herrera 1989). Its modern importance in tropical forests is reflected in the diversity and abundance of animal-dispersed plants. Several studies have estimated that more than 80% of tree and shrub species in tropical wet forests are animal dispersed (Frankie et al. 1974a; Croat 1975; Hilty 1980; Opler, Frankie, et al. 1980; Gentry 1982a; Stiles 1985b; Willson et al. 1989). These plants provide enough fruit to support large populations of fruit-eating birds and bats (Snow 1971; Snow and Snow 1971; Heithaus et al. 1975). In fact, fruit resources alone may account for approximately 20% of the difference between temperate and tropical avian species diversity (Karr 1971; Remsen 1985). Estimates of the tropical vertebrate biomass supported by fruits range from 50% (for birds) to 80% (birds plus mammals, Willis 1980; Terborgh 1986a; Janson and Emmons 1990).

The importance of tropical frugivory extends beyond its role in community structure. As a mechanism for seed dispersal, frugivory is hypothesized to be important in maintaining high species diversity of tropical plants (Janzen 1970). The advantages of dispersal include escape from high seed and seedling mortality near the parent (Janzen 1970; Connell 1971; D. A. Clark and Clark 1984), colonization of regeneration sites (Thompson and Willson 1978; Culver and Beattie 1978), and increased gene flow (Levin and Kerster 1974). Without dispersal agents fruiting plants may become prone to extinction (Howe 1977, 1984a; Temple 1977; D. A. Clark and Clark 1981). Nevertheless, little is known about the relative importance of selective forces that promote dispersal (Howe and Smallwood 1982; Howe 1986).

From an ecological perspective fruits offer an opportunity to monitor seasonal changes in abundance of an important food resource. Unlike other food items fruits are made for consumption. Hence, they are conspicuous and relatively easy to count. Documenting temporal variation in fruit abundance helps us understand population dynamics of frugivores. Currently, we know very little about what causes population fluctuations of any tropical vertebrate even though such fluctuations are common (Fogden 1972; Toft 1980b; Martin and Karr 1986b; chaps. 15, 16). Understanding these fluctuations is especially important in management of tropical forests. At a minimum one needs to estimate how much habitat is necessary to maintain frugivore populations. Failure to preserve a key fruiting plant species or disperser may have cascading effects through the community as other species are affected by the loss of a mutualistic link (Gilbert 1980; Janzen and Martin 1982; Howe 1984b; Terborgh 1986b).

From an evolutionary perspective fruits represent an interesting compromise between conflicting strategies (Snow 1971). The conflict arises because the aspects of seed dispersal that are beneficial to fruiting plants represent costs to frugivores who transport seeds without benefiting from their nutrient content. Similarly, what is beneficial to frugivores is a cost to fruiting plants, which must produce fruit pulp to attract seed dispersers. Examination of these costs and benefits allows one to explore the basis of plant-frugivore coevolution. Of particular interest is the question of how closely fruiting plants and frugivores are coevolved. Clearly, the interaction between these groups is mutualistic. However, we have yet to determine the level of interaction between the groups and the mechanisms underlying their coevolution (Howe 1984a; Futuyma and Slatkin 1983; Herrera 1986; Jordano 1987).

Here we present an overview of frugivory at La Selva. More general reviews of frugivory are provided by Howe (1986), Willson (1986), Fleming et al. (1987), Wheelwright (1988), and Jordano (1992). Our goal is to put La Selva work in the perspective of frugivory in general and tropical frugivory in particular. We begin with a general description of fruits and frugivores at La Selva and compare them with other tropical forests. Then we address the following questions: What factors influence fruit choice? What is the role of fruit handling in successful dispersal? How do spatial and temporal variation in fruit abundance affect frugivore populations? Given information on fruit choice, fruit handling, and spatial/temporal variation in bird and fruit abundance, what can one deduce about the coevolutionary patterns and constraints in fruit-frugivore interaction? In each section we present a brief review and then focus on what is known about the topic at La Selva. In this chapter we focus on avian frugivory only because it has received so much attention at La Selva.

GENERAL DESCRIPTION OF FRUITS AND FRUGIVORES

Fruits

We distinguish between fruits produced by shrubs and treelets and those produced by trees. This dichotomy follows similar divisions proposed by Smythe (1970), McKey (1975), and Janson (1983).

Fruits of Shrubs and Treelets. Ninety-five percent of 154 species of shrubs and treelets surveyed by Opler, Frankie, et al. (1980) at La Selva bore small fleshy fruits (Opler, Frankie, et al. 1980). Members of the Rubiaceae, Melastomataceae, Palmae, Piperaceae, and Gesneriaceae are found commonly in fruit in the understory and areas of second growth. Within the forest significantly more species and individuals are found to bear fruit in tree-fall gaps than in shaded understory (Marquis 1988; Levey 1988b, 1990). Three generalizations made about these small fruits are not well supported at La Selva. First, small-fruited species are usually envisioned as prolific fruit producers (Land 1963; Willis 1966; McKey 1975). Such plants often do produce large crops when in high light environments, but in the understory crop sizes are typically quite small, usually fewer than one hundred fruits per plant (Bullock 1982; Denslow et al. 1986; Murray 1987). For example, 69% of fifty fruiting plants found in fifty 100 m² quadrats had fewer than ten ripe fruits per plant per census period (Denslow et al. 1986). Similarly, small crop sizes have been observed in other tropical communities (Fleming 1981; Leighton and Leighton 1983; Terborgh 1983). Second, small-fruited species are often characterized as having numerous small seeds (Harper et al. 1970; McKey 1975). Exceptions at La Selva include three dominant families (Heliconiaceae, Arecaceae, and Rubiaceae), whose members all produce fruits with one-to-several large seeds. Third, fruits are thought to "advertise" their presence by being brightly colored (Snow 1971; Van der Pijl 1982). At La Selva the most common ripe fruit color among 101 understory species is very dark blue/black (table 22.1; see also Wheelwright and Janson 1985). Although the dark fruits may appear inconspicuous, their relative visibility (brightness; see Willson and Hoppes 1986) has yet to be determined. In most species with dark fruits contrasting colors provided by the infructescence or neighboring unripe fruit make the fruit display quite visible to the human eye (table 22.1).

Bat- and bird-consumed fruits display many differences (Van der Pijl 1982). Although these differences are pronounced, it is important to note they are not exclusive. Many birds, for example, take *Piper* fruits, and bats readily take "bird" fruits (Fleming et al. 1985; Dinerstein 1987; Palmeirim et al. 1989). Despite these exceptions the syndromes provide an interesting basis of comparison.

In general, fruits eaten by birds are small and spherical, are in brightly colored displays held close to foliage, and are without an odor. Common examples at La Selva are species of Melastomataceae and Rubiaceae. These watery fruits are typically low in proteins and lipids but high in carbohydrates (White 1974; Moermond and Denslow 1985); they tend to rot quickly. In contrast, other bird-dispersed species, such as those of *Heliconia*, produce dense, lipid-rich fruits that often remain on the plant for long periods with little sign of decay (White 1974; Moermond and Denslow 1985). Fruits dispersed by bats are typically green, odoriferous, and presented away from foliage. The most common examples at La Selva are in the Piperaceae.

Table 22.1 Fruit types and display characteristics of shrubs and treelets compared with trees at La Selva

Characteristics	Shrubs and Treelets[a] (No. Species)	Trees[b] (No. species)
Fruit type		
Arillate	27	23
Berry[c]	70	7
Other	4	1
Color of ripe fruit[d]		
Red/orange	24	12
Light blue	12	0
Dark blue/black	50	11
White	10	7
Green	5	1
Source of contrasting colors[e]		
Infructescence structure	21	16
Ripening fruits	44	6

Sources: Denslow et al. 1986; Levey 1988a.
Note: The sample of 132 species is based on censuses of all fruiting plants occurring in 50 quadrats (100 m²) and along 150 transects (12 m × 1 m).
[a]Height less than 10 m.
[b]Height more than 10 m.
[c]Functionally defined as a soft, watery fruit.
[d]Colors originally determined by reference to color swatches based on the Munsell system.
[e]Species without sharply contrasting colors not listed.

Tree Fruit. Ninety percent of the 185 tree species surveyed by Frankie et al. (1974a; see chapter 11) produce fleshy fruits. Seeds of approximately 50% of these species are probably dispersed by birds, 13% by bats, and the remainder by other mammals (Hartshorn 1978). Wind dispersal, which is almost absent among understory species (Opler, Frankie, et al. 1980), is also rare among trees (see also Gentry 1982a). Only 6% of the surveyed tree species are adapted for wind dispersal and 83% of these are canopy species (e.g., *Ceiba pentandra, Pterocarpus officinalis*).

Fruit production in the canopy is more seasonal than in the understory (Frankie et al. 1974a; Croat 1975; Opler, Frankie, et al. 1980). Species that produce small, berrylike fruits are rare; large-seeded arillate species predominate (table 22.1). Most of the large-seeded fruits are protected by a husk that opens when ripe to expose an arillate seed. The arils are typically white, orange, or red. Bright colors in other structures associated with infructescences are less common among trees than among understory plants, in which relatively inconspicuous fruits are associated with brightly colored accessory structures (table 22.1).

Examples of common trees with bird-dispersed seeds are *Protium* spp., *Virola* spp., and *Guarea* spp. Many species with especially large seeds (e.g., *Dipteryx panamensis, Lecythis ampla*) are dispersed by bats. Monkeys feed on numerous canopy fruits (e.g., *Welfia georgii, Dendropanax* spp.) but frequently drop, rather than disperse large seeds (Hladik and Hladik 1969). As in understory plants, canopy species are often dispersed by more than one type of disperser. Some species such as *Cecropia obtusifolia* and *Ficus* spp. are eaten and dispersed by almost all groups of frugivores.

Although wind dispersal is common among tropical forest

epiphytes and lianas (Croat 1975; Opler, Baker, et al. 1980; Gentry 1982a) many bear vertebrate-dispersed seeds (e.g., *Monstera, Syngonium* and other genera of Araceae). These fruits are often similar to those of understory shrubs (e.g., *Anthurium* spp.). Further, most canopy trees support many fleshy-fruited epiphytes (J. Denslow and D. Levey unpublished data). Although crop sizes are typically small, the abundance of epiphytes probably makes them an important component of the fruiting plant community.

Understory species are typified by a disproportionate number of berry-producing species, whereas canopy species are dominated by arillate fruits ($\chi^2 = 22.1$, d.f. = 1, $p < 0.001$; table 22.1). Similarly, color of ripe fruit and plant growth-form are not independent ($\chi^2 = 8.3$, d.f. = 3, $p < 0.05$; Table 22.1; White and Green categories combined). An examination of the residuals in the chi-square table reveals that most of the deviation from the expected distribution is the result of the relative abundance in the understory, and absence in the canopy, of light-blue fruits. Wheelwright and Janson (1985) found that species with brightly colored fruits were significantly more common in the understory in another Costa Rican forest. The evolutionary basis of such patterns remains largely unexplored (Willson and Whelan 1990; Cipollini and Levey 1991).

Frugivores

Here we define a frugivore as any organism whose diet includes more than 50% fruit. We stress that this definition, which follows Terborgh (1986a) and Fleming et al. (1987), is arbitrary and artificially imposes a dichotomy on a continuum of fruit dependency. Although many animals at La Selva eat fruit, probably none rely only on fruit. Animals that consume almost entirely fruit (e.g., some phyllostomid bats; T. Fleming pers. comm.) apparently must supplement their diets with insects, pollen, or other sources of protein (Morton 1973; Foster 1978). Note that under our definition, consumption of fruit is the only criterion by which frugivory is defined. We purposefully do not distinguish between frugivores that disperse seeds and those that destroy them because the distinction is usually unclear (Howe 1986). Research on the predation end of the spectrum (the destruction of fruit and seeds) is covered in chapter 21.

Fruit-eaters transcend taxonomic and trophic boundaries (table 22.2). Many carnivores will eat fruit. Swallow-tailed kites *(Elanoides forficatus),* for example, take fruit (Buskirk and Lechner 1978) and all Costa Rican Carnivora, except otter *(Lutra longicaudus)* and mountain lion *(Felis concolor),* "are known or alleged to consume large amounts of fruit" (Janzen and Wilson 1983, 430). Most granivores also depend on fruit if only for a source of seeds. Nearly all flycatchers will eat fruit (Traylor and Fitzpatrick 1982); wrens and vultures take fruit (McDiarmid et al. 1977; Willis 1980); and even some rails and antbirds are fruit-eaters (Skutch 1933; Morton 1973; Keeler-Wolf 1986; D. Levey pers. observation). In short, most vertebrates at La Selva probably take advantage of fruit resources. This is not surprising given that most fleshy fruits are conspicuous and provide an easily accessible energy source. What is surprising is that some mammals and birds (e.g., woodcreepers) apparently never eat fruit.

Here we briefly describe vertebrate frugivores at La Selva.

For discussion of invertebrate frugivores, which are most likely seed predators, refer to chapter 21. It is likely that all vertebrate frugivores disperse at least some seeds but their efficacy as dispersers varies tremendously (e.g., Levey 1987b). In the following sections we state when possible whether a given animal is likely to disperse seeds. We augment, where possible, descriptions of the behavior and role of frugivores that have received little study at La Selva with observations of the same species elsewhere in Central America.

Birds. Understory frugivores are extremely common at La Selva, accounting for approximately 50% of individual mist net captures (Levey 1988b; Blake and Loiselle 1991; D. Levey unpublished data). Approximately 28% of the resident species are frugivores (Stiles 1985b), and it is likely that all of these species supplement their diets with protein-rich food sources. We follow Moermond and Denslow (1985) in classifying these by the nonfruit portion of their diets, insects and seeds.

At La Selva most frugivorous species also eat insects. Only a few species seem to eat almost exclusively fruits. In particular, three manakins (*Pipra mentalis, Manacus candei,* and *Corapipo altera*) and one tanager *(Euphonia gouldi)* almost always have pulp or seeds in their feces (100% occurrence of pulp and/or seeds in more than 250 total samples; D. Levey unpublished data). All common La Selva tanagers are frequent fruit-eaters. They probably disperse most small (<2 mm) seeds but are likely to drop larger seeds under or near the parent plant (Levey 1987b). Large flycatchers (e.g., *Pitangus, Megarynchus, Myiodynastes, Myiozetetes* spp.) commonly consume fruits. In contrast, relatively few of the many smaller flycatcher species regularly take fruit (e.g., *Mionectes* spp., *Ornithion,* and *Zimmerius* do take fruits). Cotingas are primarily frugivorous (Snow 1982) but little is known about their feeding behavior at La Selva because they spend most of their time in the canopy where observation is difficult. Finally, species of Ramphastidae, Picidae, Trogonidae, Icterinae, and Turdinae are also commonly observed eating fruit. Although Turdinae are usually uncommon at La Selva, they may be important dispersers during fall migration when they are abundant. In October 1983, for example, three species of *Catharus* constituted 32% of all frugivore mist net captures and fecal analysis revealed that they were eating many fruits (D. Levey unpublished data).

The most common seed-eating frugivores at La Selva are emberizine finches (e.g., *Caryothraustes poliogaster* and *Arremon aurantiirostris*). These birds have strong bills and crack some seeds, but many seeds are dropped during fruit handling or ingested and defecated still intact (Levey 1986). Ground-feeding species such as tinamous, curassows, and ground- and quail-doves feed on fallen fruits, and most seeds are probably destroyed in the muscular gizzard. In the canopy pigeons and parrots are the most common seed-eating frugivores (Loiselle 1988). Pigeons (*Columba* spp.) eat both ripe and unripe fruit and are assumed to be primarily seed predators although their effect on seeds remains undetermined. Parrots usually eat large-seeded fruits (e.g., *Dendropanax, Guarea*), destroying the seeds and rarely ingesting pulp or aril. They also destroy many small seeds (e.g., *Ficus* spp., Janzen 1981a; Jordano 1983) but may defecate some intact (Fleming et al. 1985).

Table 22.2 Families of vertebrates at La Selva and their relative dependence on fruit, including seeds

Vertebrate Families	Total Species	Frugivory[a]	Seed Treatment[b]	Abundance[c]
Birds				
Tinamidae	3	+++	−	3
Cathartidae	3	+	?	2
Accipitridae	24	+	?	2
Cracidae	3	+++	+/−[d]	2
Aramidae	7	+	?	3
Columbidae	11	++++	+/−	4
Psittacidae	8	++++	−	4
Trogonidae	5	+++−++++	+	4
Motmotidae	2	+++	+	4
Ramphastidae	5	++++	+	3
Picidae	7	++	+	3
Cotingidae	13	+++	+	3
Pipridae	6	++++	+	5
Tyrannidae	41	+−+++	+	4
Hirundinidae	6	+	+	3
Corvidae	1	++	+	1
Mimidae	1	+++	+	2
Muscicapidae	8	++++	+	3
Vireonidae	8	+++	+	3
Emberizidae	85	+−++++[e]	+/−[f]	2–4[e]
Mammals				
Didelphidae	3	+++	+	3
Phyllostomidae	40	+−+++++[g]	+	1–5[g]
Cebidae	3	+++	+/−	3–4
Edentata	3	+	?	3
Sciuridae	4	+++	−	2
Heteromyidae	1	+−+++	−	4
Erethizontidae	1	++?	?	1
Dasyproctidae	2	+++	−	4
Echimyidae	2	+++?	−?	2
Muridae	5	+++?	−	2?
Mustelidae	4	+−+++	+	2
Procyonidae	2	++++	+	3
Felidae	5	++?	+	1
Tayassuidae	2	++++	−	2
Cervidae	1	+++?	−?	1
Tapiridae	1	+++	−	1
Reptiles and amphibians				
Emydidae	2	++?	+?	2
Iguanidae	13	++	?	3

Sources: Janzen 1983c; Macdonald 1984; T. Fleming pers. comm.; and D. Levey, T. Moermond pers. observations.

[a] + = fruit rarely included in the diet; ++ = fruit included occasionally; +++ = fruit included regularly; ++++ = fruit dominates the diet.

[b] + = probably disperses most seeds from fruits consumed; − = probably does not disperse most seeds from fruits consumed (but may, nonetheless, be an important disperser).

[c] 1 = very rare; 2 = uncommon; 3 = common; 4 = very common; 5 = abundant.

[d] Guans and Chachalacas defecate viable seeds. Curassows appear to destroy most ingested seeds (R. Buchholz pers. comm.).

[e] A diverse group in degree of frugivory and abundance at La Selva. Major subfamilies that are heavily frugivorous are Icterinae (twelve spp.), Thraupinae (twenty-one spp.), Cardinalinae (seven spp.) and Emberizinae (ten spp.).

[f] Seed treatment heavily dependent on seed size relative to bird size (see text).

[g] A diverse group in degree of frugivory and abundance at La Selva. Major subfamilies that are heavily frugivorous are Carolliinae (three spp.) and Stenodermatinae (thirteen spp.). Both of these subfamilies are generally abundant. One subfamily, Desmodontinae, takes no fruits.

Mammals. Until the late 1980s, mammalian frugivory at La Selva received little attention. A project begun in the late-1980s by A. Campbell (unpublished data), along with researchers at other Central American sites, suggests that fruit may be an important resource to most La Selva mammals.

Phyllostomid bats in the subfamilies Carolliinae and Stenodermatinae largely depend on fruit and are extremely common in Costa Rica and Panama (Fleming et al. 1972; Heithaus et al. 1975; Heithaus and Fleming 1978; Morrison 1978; Bonaccorso 1979; Fleming 1988). La Selva has three species of

Carollia and thirteen species in the Stenodermatinae, including the common Jamaican fruit bat (*Artibeus jamaicensis*). At other sites this species primarily feeds on canopy figs and has a large home range (Heithaus et al. 1975; Morrison 1978; Bonaccorso 1979). Other members of the Stenodermatinae are also canopy frugivores (Bonaccorso 1979). In contrast, *Carollia* spp. are understory frugivores that feed mainly on *Piper* spp. and have comparatively small home ranges (Heithaus et al. 1975; Heithaus and Fleming 1978; Bonaccorso 1979; Fleming 1981).

Fruits in the following genera are common at La Selva and frequently eaten by bats elsewhere: *Ficus, Piper, Spondias, Solanum, Dipteryx, Quararibea, Astrocaryum, Markea, Clusia,* and *Cecropia* (Heithaus et al. 1975; Heithaus and Fleming 1978; Bonaccorso 1979; Fleming and Heithaus 1981). Many of these fruits (e.g., *Ficus, Cecropia, Solanum, Piper*) are also taken by birds (Palmeirim et al. 1989). Curiously absent from most bat diets are fruits of Melastomataceae and Rubiaceae, both of which are primarily eaten by birds at La Selva.

Even highly frugivorous bat species depend on other food resources (Heithaus et al. 1975). Nevertheless, the correlation between bat breeding seasons and periods of high fruit abundance suggests that fruit is a critical resource to these bats (Fleming et al. 1972; Heithaus et al. 1975; Bonaccorso 1979; Dinerstein 1987).

Bats rarely ingest fruits in fruiting trees; instead they carry fruits to feeding sites or roosts where seeds are dropped or defecated in dense piles (Morrison 1978; Bonaccorso 1979; Fleming 1981; Fleming and Heithaus 1981). Seed and seedling mortality is high in these piles (Janzen 1971b). Some seeds, however, are undoubtedly dropped or defecated away from roosts (Morrison 1978; Fleming 1981; Fleming and Heithaus 1981).

Monkeys consume large quantities of fruit. Three primates occur at La Selva: white-faced Capuchins (*Cebus capucinus*), howler (*Alouatta palliata*), and spider (*Ateles geoffroyi*) monkeys. Howlers are known to consume fruits of numerous species at La Selva including *Ficus, Cecropia, Cordia, Brosimum, Spondias, Dendropanax,* and *Tetragastris* (Opler et al. 1975; Howe 1980; Milton 1980; Estrada et al. 1984; D. Levey pers. observation). On Barro Colorado Island, Panama, figs are their favored fruit; 36% of their feeding time was spent in fig trees (Milton 1980; see also Estrada and Coates-Estrada 1984). Spider monkeys are primarily frugivorous (Eisenberg 1983) and take a wide variety of fruits common at La Selva (e.g., *Ficus, Trichilia, Tetragastris, Welfia, Swartzea,* and *Virola;* Boucher 1981; Howe 1980, 1982; Eisenberg 1983). White-faced Capuchins often forage near the ground and, perhaps, include more smaller fruits in their diets. Oppenheimer (1982) and Freese (1977, 1983) report them feeding on *Neea, Siparuna, Ficus, Miconia,* and *Dendropanax.* At La Selva they feed heavily on *Welfia* (T. Fleming pers. comm.; D. Levey pers. observation). In general, monkeys are probably not good dispersers. They drop large seeds under the parent tree and defecate small seeds in dense clumps (Vandermeer et al. 1979; Howe 1980; Freese 1983; Estrada and Coates-Estrada 1984). Nevertheless, the huge number of seeds they ingest and defecate (Hladik and Hladik 1969) ensures that a few will be deposited in good sites for germination.

Among other La Selva mammals known to eat fruit are armadillos (*Dasypus novemcinctus,* Wetzel 1983), Tayras (*Eira barbara,* Brosset and Erard 1986, Janzen 1983c), coatis (*Nasua narica,* Kaufmann 1962, 1983; Opler et al. 1975; Jordano 1983), kinkajous (*Potos flavus,* Vandermeer et al. 1979), margays (*Felis wiedii,* Koford 1983) and most other Carnivora (Janzen and Wilson 1983). The three opossum species at La Selva may at times be important frugivores and seed dispersers (cf. Charles-Dominique et al. 1981), but little is known of their diets or abundance. Agoutis (*Dasyprocta punctata,* Smythe 1970; Vandermeer et al. 1979), squirrels (especially *Sciurus granatensis,* Heaney and Thorington 1978; Glanz et al. 1982; Heaney 1983), peccaries (*Tayassu tajacu,* Kiltie 1981b; Sowls 1983), and heteromyid rodents (especially *Heteromys desmarestianus,* Fleming 1974a) feed on seeds but also cache or defecate intact seeds (Janzen 1971b; Sowls 1983). Hence, these species are probably dispersers as well as seed predators (Smythe 1970, 1986; Janzen 1971b; Charles-Dominique et al. 1981). With the exception of bats and monkeys, however, extremely little is known about the effectiveness of mammalian seed dispersal.

Reptiles and Fish. Brown land turtles (*Rhinoclemmys annulata*) "relish" fruits (Ernst 1983, 416); the basilisk lizard (*Basiliscus basiliscus*) consumes *Ardisia, Cordia, Spondias, Ficus,* and *Sloanea* (Van Devender 1983); and iguanas (*Iguana iguana*) are also reported to eat fruit (Rand 1978). Some fish (e.g., *Brycon;* Characidae) apparently wait under fruiting *Ficus glabrata* trees to catch fruits that fall into the water; such fruits are found and consumed in seconds (chap. 14). Nothing is known about seed dispersal via these fish at La Selva but in the seasonally inundated forests of the Amazon, many fish clearly play an important role in seed dispersal (Gottsberger 1978; Goulding 1980). They may also be important dispersers of riparian species at La Selva.

FRUGIVORE COMMUNITIES: LA SELVA COMPARED WITH OTHER RAIN FORESTS

The taxonomic composition of assemblages of frugivores differs among tropical regions of the world. Such differences offer fertile ground for speculation and future work. A few such differences suggest particularly important evolutionary divergences in seed dispersal systems (see also Fleming et al. 1987; Gentry 1990a). Delineating the ultimate causes behind these differences is beyond the scope of this chapter; we refer the reader to Willson et al. (1989) for alternative hypotheses. Our comparisons will be limited to selected groups of birds and mammals because these groups are the best studied and also appear to be by far the most important seed-dispersers in all tropical wet forests (Gentry 1982a; Whitmore 1984; Gautier-Hion et al. 1985; Stiles 1985b; Howe 1986).

Birds

A great variety of birds feed on fruits in all major tropical forest regions. David Snow (1980, 1981) has provided a thorough Pantropical comparison of avian frugivores and their food plants. His comparison emphasizes the fruit-eaters that take large fleshy fruits. Each major tropical region has one or more groups of birds in this category, most drawn from widely disparate taxonomic groups. Cotingas (Cotingidae), toucans (Ramphastidae), and trogons (Trogonidae) are the major representatives of this group in the Neotropics. Hornbills (Bucerotidae) and fruit pigeons (*Ptilinopus, Ducula, Treron,* and related genera) are the predominant elements in the Paleotropics (e.g., Crome 1975; Leighton and Leighton 1983), and in New Guinea, many birds-of-paradise (Paradisaeidae) are similarly important (Beehler 1981; Pratt and Stiles 1985; Diamond 1986). Even some apparent similarities among regions may be illusory. The trogons of the Paleotropics do not appear to be as frugivorous as those of the Neotropics, particularly *Pharomachrus* spp. (Wheelwright 1983); the African species may eat fruit only very rarely (Brosset and Erard 1986). Nor should hornbills be taken as ecological equivalents of toucans. Some species are much larger than toucans, and many small hornbills are highly insectivorous (Kemp 1979; Brosset and Erard 1986). Among fruit pigeons, *Treron* spp. and *Columba* spp. are thought to be seed predators because of the action of their well-developed gizzards (Goodwin 1970; F. G. Stiles pers. comm.).

In all regions one can find a large number of small frugivore-insectivore species that consume small watery fruits (e.g., Thraupinae, Pycnonotidae, Muscicapidae). This category of fruit-eaters is by far the most species rich. The species richness of these birds varies among regions (see Karr 1976a, 1976b, 1980; Pearson 1977), with a particularly high diversity in the Neotropics and New Guinea. Despite the large number of species, this group contributes little to the biomass of tropical forest systems, especially in contrast to mammalian frugivores (Terborgh 1986a). Because of their small size and mixed diet, these frugivore-insectivores have received relatively little attention in most studies of frugivory. Yet they play an important role in seed dispersal in the Neotropics (Wheelwright et al. 1984; Loiselle and Blake 1990), and we suspect that they will be shown to be important components of Paleotropic dispersal systems also (e.g., Beehler 1981).

Data on possible regional differences in fruit-feeding behavior and fruit selection are almost totally lacking. A preliminary study of fruit-eating behavior of some central African fruit-eating birds (T. C. Moermond unpublished data) suggests that the small African species do not encompass the wide range of fruit-taking behavior described for small Neotropical fruit-eaters (see Moermond and Denslow 1985). For example, in the Neotropics manakins and tyrant flycatchers take fruit by hover gleaning, but in the Paleotropics no birds apparently forage in this manner. Lacking also in the Paleotropics are birds that crush fruits in their bills while eating, a common behavior among Neotropical tanagers and emberizids.

Mammals

The situation is much more complicated in mammals than in birds. We have divided them into seven major groups: small nocturnal rodents, terrestrial herbivores, carnivores, marsupials, squirrels, bats, and primates. Because the degree of frugivory and manner of seed handling of most mammals are poorly studied, mammalian groups undoubtedly include seed dispersers as well as seed predators and do not include all the mammals that take fruits.

Small, nocturnal rodents are often implicated as seed predators but may occasionally serve as seed dispersers through "sloppy" seed handling, seed caching, or, more rarely, through defecation of viable seeds. Although rodents are potential dispersers in every tropical region, studies of their fruit eating and potential seed dispersal are rare or nonexistent (but see Janzen 1986b). Nearly the same situation exists for the herbivores that feed on fallen seeds in almost every tropical wet forest (cf. Smythe 1986). This group includes the large caviomorph rodents, peccaries (Tayassuidae), and deer (Cervids) in the Neotropics (cf. species accounts in Janzen 1983b); pigs (Suidae), antelopes and chevrotains in the Afrotropics (Dubost 1984); pigs, mouse-deer (Tragulidae) and deer in the Asian tropics (Medway 1978, Payne et al. 1985); and wallabies (Macropodidae) in the Australian tropics (Strahan 1983). Tapirs (Tapiridae) in the Neotropics and Indo-Malaysian tropics, and elephants in African and Asian forests may also play important roles as dispersers (Alexandre 1978; Janzen 1983b, 1986b; Gautier-Hion et al. 1985). In addition, extinct mammals may have played an important role in all regions (e.g., Proboscideae, Ursidae, Driprodontidae; Janzen and Martin 1982, Pratt 1983; but see Howe 1985).

At La Selva we know that small rodents probably play a role in seed dispersal (Denslow and Moermond 1982; S. Hermann unpublished data). Large rodents such as agoutis are common but little studied (cf. Smythe 1978; Hallwachs 1986). Tapirs *(Tapirus bairdii)* are rare at La Selva, as is common for large mammals in forests near significant human populations.

Many carnivores have been noted to take occasional fruits; nevertheless, the species identified as regular frugivores are few. They include two mustelids (*Eira* and *Mephitis*) and three procyonids (*Potos, Bassaricyon,* and *Nasua*) in the Neotropics (Charles-Dominique et al. 1981; Janzen 1983b), and the palm civets (Paradoxurinae, Viverridae) in the Old World tropics (*Nandinia* for Africa [Charles-Dominique 1978] and *Paguma, Paradoxurus, Arctogalidia,* and *Arctictis* for Indo-Malaysia [Medway 1978; Payne et al. 1985]).

Charles-Dominique et al. (1981) present extensive data on diet and ranging patterns of five species of Didelphidae in Cayenne, French Guyana. Other Neotropical marsupials are poorly studied but are likely to consume fruit and disperse seeds. *Caluromys* spp., in particular, are heavily frugivorous and probably disperse many seeds (B. McNab pers. comm.). In New Guinea and Australia the phalangers (Phalanderidae) and many of the possums (Pseudocheiridae) eat fruits (Strahan 1983) but only as a minor part of the diet; most species are folivores (Smith and Hume 1984).

Squirrels (Sciuridae) are present in Neotropical, Afrotropical, and Indo-Malaysian forests. Although poorly studied, they appear to be primarily seed-predators (MacKinnon 1978; Emmons 1981; Glanz et al. 1982; Heaney 1983). Some species may contribute to seed dispersal of certain plants (Leighton and Leighton 1983; Becker and Wong 1985).

Bats are clearly important frugivores and seed dispersers (Fleming 1979, 1981, 1988; Marshall 1983; Charles-Dominique 1986; and others cited previously). The frugivorous bats of the Neotropics are small and in one family (Phyllostomidae) of the Microchiroptera. In contrast, the Microchiroptera of the Paleotropics are insectivorous or carnivorous (McNab 1971). In the Paleotropics frugivory in bats is generally confined to the large Megachiroptera (McNab 1971;

Medway 1978; Marshall 1983). In addition to size difference between the Old and New World frugivorous bats the two groups have apparently radiated differently with respect to foraging zones. Whereas Neotropical phyllostomids forage throughout the forest, Paleotropic pteropodids are generally canopy or forest edge feeders and do not commonly forage in the understory (Fleming et al. 1987).

Primates are often a dominant element among tropical forest frugivores. In particular, diverse assemblages of primates dominate the frugivore biomass in many Paleotropic communities (Terborgh 1986a). Neotropical rain forests have fewer, but often still quite numerous, species of primates (e.g., Panama [Hladik and Hladik 1969], Surinam [Mittermeier and Roosmalen 1981], Peru [Terborgh 1983; Janson and Emmons 1990]). An important difference between the frugivory system at La Selva, with only three primate species, and other rain forest sites is, thus, likely to be reduced significance of frugivory by primates. Indeed, "monkey fruits" appear much less common at La Selva than reported in Peru (Janson 1983; Terborgh 1983). Data to substantiate this impression, however, are lacking.

Primates are extremely variable in their effectiveness as dispersers (Howe 1980; Hladik and Hladik 1969; Gautier-Hion 1984; Terborgh 1983; Corlett and Lucas 1990). Old World primates are generally more folivorous than New World primates (Terborgh 1986a), but the degree of folivory compared to frugivory varies from region to region (Gautier-Hion 1983). The degree of folivory appears to be positively correlated with body weight in Malaysia (Chivers 1980), Gabon (Gautier-Hion et al. 1980), Panama (Hladik and Hladik 1969), and Peru (Terborgh 1983). Another difference between New and Old World primates is that New World primates are almost exclusively arboreal but many Old World primates are markedly terrestrial (Fleming et al. 1987).

It is particularly interesting to note that the New World Callitrichidae, which are all much smaller than the cebids or the cercopithids, are heavily insectivorous with many species also frugivorous (Terborgh 1983; Rylands 1984). Two *Saguinus* species in Peru appear to eat primarily small, sugary "bird" fruits, which are typically found in small, widely dispersed patches in the understory (Terborgh 1983; Garber 1986). These small, sometimes abundant primates, have no direct parallel in Africa or Asia. In Madagascar frugivorous lemurs have similar feeding patterns (Tattersall 1982). Although the Old World lorisids are in the same small size range, only a few are notably frugivorous (*Galago alleni* and *Perodicticus potto* in Africa [Charles-Dominique 1971, 1977] and *Nycticebus coucang* in Southeast Asia (Payne et al. 1985]). Their solitary, nocturnal behavior and relatively low population densities offer a very different picture from that of the diurnal and abundant tamarins and marmosets of the Neotropics.

FRUGIVORES AS SEED DISPERSERS

To understand the ecological and evolutionary importance of frugivores as seed dispersers, we need to know how dispersal affects plant fitness (Herrera 1986; Willson 1991; Jordano 1992). This is a difficult question especially because appropriate plant fitness estimators vary among plant species (Mur-

ray 1987). The ecological and evolutionary consequences of dispersal are beyond the scope of this chapter and have been reviewed elsewhere (Howe and Smallwood 1982; Beattie 1985; Howe 1986). Instead, we focus on one dimension of seed dispersal that we suggest is of universal importance— whether seeds are carried away from the parent plant. From this perspective seed dispersal has two important components. First, fruits must be chosen for consumption and then they must be handled in such a way that seeds are ingested and dispersed. Work at La Selva has focused on these components of dispersal among frugivorous birds.

Fruit Choice by Birds

Fruit and Fruit Display Attributes. Fruit choice by birds is influenced by a variety of fruit attributes: fruit size (Moermond and Denslow 1983; Wheelwright 1985b), seed size (Sorensen 1984), dry pulp mass (Johnson et al. 1985), pulp to seed ratio (Howe and Vande Kerckhove 1980; Herrera 1981), pulp taste (Sorensen 1983), sugar concentration (Levey 1987c), caloric content (Sorensen 1984; Jordano 1988), fruit color (Morden-Moore and Willson 1982; Willson and Thompson 1982), and ripeness (Moermond and Denslow 1983). In addition, other factors such as fruit crop size (Howe and De Steven 1979; Howe and Vande Kerckhove 1979; Stapanian 1982; Foster 1990), distance to nearest fruiting neighbors (Manasse and Howe 1983; Levey et al. 1984; Denslow 1987a; Sargent 1990) and nutritional content of other fruits (Jordano 1988) affect fruit choice on another level by influencing which fruiting plants are visited. How birds integrate these factors when choosing among fruits is complex and largely unknown (Martin 1985a; Moermond et al. 1987). We have been able, however, to examine the underlying decision-making process in fruit choice by offering captive birds pairwise choices of fruits on artificial infructescences (Moermond and Denslow 1983). These birds make consistent and transitive choices among fruits. This means that fruit choices are consistent with a maximization principle (McCleery 1978). Our working hypothesis is that this maximization principle is, in turn, based upon weighing costs and benefits associated with each fruit type. The spectrum of costs and benefits, and their relative importance, remains unclear but almost certainly includes fruit availability and bird morphology. Under a framework of cost-benefit analysis we discuss these factors and how they are weighed against each other in determining fruit choice in birds.

Fruit Availability. Fruit abundance, detectability, and accessibility are components of fruit availability. They define the context in which fruits are encountered by birds, and they influence fruit removal rates and frugivore abundance (Thompson and Willson 1978; Manasse and Howe 1983; Martin and Karr 1986b; Loiselle 1988; Loiselle and Blake 1990). If, as hypothesized previously, frugivores employ a cost-benefit approach in choosing fruits, then we would predict that fruit choice depends on context. A fruit preferred in one situation may not be taken when encountered in a different context.

We have examined how one component of fruit availability, accessibility, affects fruit choice (Moermond et al. 1986, 1987). Accessibility is a measure of how easily a fruit can be taken from an infructescence. It includes proximity of a fruit to a perch and perch diameter, angle, and flexibility. If a bird is given a choice between a fruit close to a perch and a fruit of the same species far from a perch, it chooses the closer, more accessible fruit (Denslow and Moermond 1982; Moermond and Denslow 1983). Similarly, birds will choose fruits adjacent to large, sturdy perches over identical fruits near small, flexible perches (Moermond et al. 1986).

What if both accessibility and fruit quality are varied by presenting a bird with a choice between a low-ranked, easily accessible fruit and a high-ranked, less-accessible fruit? Can birds simultaneously weigh both fruit quality and accessibility when selecting fruits? We have addressed this question by "titrating" fruit quality against accessibility. A bird is first presented with two equally accessible fruits, A and B. It usually shows a preference for one, say A, by taking it first. If A is then made less and less accessible over a series of trials, the bird usually switches its behavior and takes B before A (Moermond and Denslow 1983; Levey et al. 1984; Moermond et al. 1986). These switches are reversible and consistent among trials. Highly preferred fruits must be made very inaccessible before the less desirable but more accessible fruit is taken first. These results indicate that birds weigh fruit quality against accessibility in choosing among fruits. Thus, they appear to employ a cost-benefit assessment; costs of access are weighed against relative benefits of different fruits.

If birds are most likely to take easily accessible fruits, why are so many fruits placed on terminal infructescences where they are relatively inaccessible? Denslow and Moermond (1982) suggest that terminal infructescences reduce seed predation by arboreal rodents. Although fruit removal by birds may also be reduced, they argue that because rodents are less agile than birds, probability of predation by rodents decreases more rapidly than probability of dispersal by birds as fruits become more inaccessible. An alternative explanation of terminal fruit presentation may be that fruits are produced where flowers are placed, and flowers are placed to maximize pollination success.

Bird Morphology. Fruits that are accessible to individuals of one bird species may be relatively inaccessible to another because of differences in the birds' morphology and associated effects on foraging behavior. Avian frugivores at La Selva separate into two morphological groups (Moermond and Denslow 1985; Moermond et al. 1986; Levey 1987b). The first group is represented by the subfamilies Thraupinae and Emberizinae. Its members have relatively strong feet, narrow gapes, and long, stout bills. The second group contains Pipridae, Cotingidae, Trogonidae, and Tyrannidae. Members of this group generally have weak feet, wide gapes, short, flat bills, and short, broad and/or highly slotted wings. Whereas members of the first group usually reach for fruit from perches, members of the second group often take fruits during flight and can thereby remove fruits that would be difficult to reach from a perch. Consequently, perch structure is less important to them than to members of the first group (Moermond et al. 1986; D. Levey unpublished data). They tend to be more selective than birds in the first group (Moermond et al. 1986). This high selectivity may be related to the higher costs of taking fruit on the wing.

Integrating the factors responsible for fruit choice poses a challenging problem (Martin 1985a). In this section we have presented evidence that birds select fruits by balancing disparate costs and benefits. Alternative ideas need further attention (e.g., Worthington 1989; Loiselle and Blake 1990). We conclude that birds are selective but their choices depend on context. Because the contexts in which fruits are found are extremely variable, the costs and benefits associated with taking particular fruits will also vary, resulting in a dynamic fruit selection process. This conclusion yields two insights: first, little opportunity for specialization and coevolution exists because fruit choice by birds depends heavily on variables that are beyond the control of individual fruiting plants; second, the wide diet breadth of fruit-eating birds and overlap among species (Snow 1962a, 1962b; Leck 1971; Snow and Snow 1971; Worthington 1982; Wheelwright et al. 1984; Moermond and Denslow 1985) should not necessarily be interpreted as a lack of discrimination by birds. Instead, it is a likely consequence of the complex decisions underlying fruit selection. An alternative hypothesis is offered by Worthington and Olberg (1990).

Handling of Fruits and Seeds

Handling fruit and seeds has important consequences for both the bird and fruiting plant. From a frugivore's perspective seeds represent useless bulk. Yet they often constitute much of each fruit and if not processed efficiently could result in high costs of ballast or gut volume displacement. From a plant's perspective seed processing is critical because it determines how effectively seeds are dispersed. For example, Howe (1977) and Howe and Primack (1975) documented different methods of seed handling by birds feeding on *Casearia*. Some species carried seeds away from the parent tree, whereas others deposited them below the tree where survivorship was lower than for dispersed seeds (see also Augspurger 1984b; D. A. Clark and Clark 1984; Howe et al. 1985).

At one extreme of seed-handling techniques are granivores. It is important to note that although these species are usually assumed to be seed predators, many may also be seed dispersers; the two groups are not distinct (Janzen 1971b; Levey and Byrne in press). For example, at La Selva, two species of emberizid finches consume and defecate many small seeds in viable condition. In fact, these finches may be better dispersers of small seeds than tanagers and manakins because the seeds have longer passage times and are distributed among more defecations (Levey 1986).

Even among those frugivores that are unlikely to digest seeds, there are important differences in fruit and seed-handling techniques. Differences between the two morphological groups mentioned earlier are especially evident (Moermond and Denslow 1985; Moermond et al. 1986). The group with strong bills (henceforth, "mashers") thoroughly mandibulate fruits, often extracting and dropping seeds, especially large seeds (Levey 1987b). Birds with weak bills (henceforth, "gulpers") and wide gapes usually swallow fruit whole, ingesting all seeds regardless of size. Mashers also have higher ingestion rates (number of fruits eaten per feeding bout), longer handling times, are more sensitive to taste cues, take a wider range of fruit sizes than gulpers of equivalent size, and can suction drink (Moermond 1983; Levey 1987b, 1987c).

Large seeds are handled efficiently by mashers because

minimal cost is incurred in dropping them. Cost to the plant, however, is high because most seeds are dropped near the parent where relatively few will survive (Howe and Primack 1975; D. A. Clark and Clark 1984). Even though gulpers ingest large seeds, the seeds are still processed rapidly. In contrast to small seeds, which are defecated by both mashers and gulpers, large seeds are regurgitated (Levey 1987b). Regurgitation is a much faster method of voiding seeds than defecation (Johnson et al. 1985; Sorensen 1984; Levey 1986, 1987b). In most cases seeds are defecated within thirty minutes and regurgitated within fifteen minutes. Both mashers and gulpers can internally separate seeds from fruit pulp and pass the seeds before the pulp (Levey 1986; Levey and Grajal 1991).

VARIATION IN FRUIT AND FRUGIVORE ABUNDANCE

Spatial Variation

At La Selva production of fleshy fruits by shrubs and treelets is higher in secondary than mature forest (Opler, Frankie, et al. 1980; Blake and Loiselle 1991). Abundance of understory bird-dispersed fruits is also significantly higher in early second growth than in tree-fall gaps and significantly higher in gaps than in adjacent understory (Denslow et al. 1986; Levey 1988a; see also Martin 1985b; Blake and Loiselle 1991). These differences are primarily the result of higher densities and crop sizes in the higher light environments (Bullock 1982; Marquis 1988; Levey 1990; Blake and Loiselle 1991).

The diversity of fruiting plant species also varies across habitats. Opler, Frankie, et al. (1980) recorded more fruiting species in mature than secondary forest (ninety-one and forty-three species, respectively), but this pattern is confounded by a greater sampling effort in mature forest. Other studies have documented a higher species diversity of plants with fruit in second growth and gaps than in mature forest understory (Denslow et al. 1986; Levey 1988b; Blake and Loiselle 1991).

Spatial variation in fruit abundance apparently affects frugivore abundance; fruit-eating bird abundance mirrors fruit abundance (Levey 1988b; Blake and Loiselle 1991; Loiselle and Blake 1991). The concentration of fruits and frugivores in tropical tree-fall gaps suggests that gaps may be "hotspots" of frugivore activity and seed dispersal (Willson et al. 1982; Herrera 1985b; Levey 1988a, 1990). This conclusion has several important implications: First, the disparity in fruit crop sizes between gap and understory conspecifics (Marquis 1988; Levey 1990) suggests that shrubs may produce a large proportion of their lifetime output of fruit during the relatively short time they are in a gap (Levey 1990). Second, although gaps represent rare opportunities for understory plants to produce large crops and attract many dispersers (Denslow and Moermond 1982; Blake and Hoppes 1986; Martin and Karr 1986a), competition with neighboring plants for dispersers is probably severe because frugivores are most selective under such circumstances (Levey et al. 1984). Thus, pronounced neighborhood effects (Manasse and Howe 1983; Herrera 1986; Sargent 1990) may be typical of gaps. Third, the abundance of frugivores and high rates of fruit removal in gaps (Thompson and Willson 1978; Denslow and Moermond 1982;

Piper 1986) coupled with rapid seed-passage times (Levey 1986; Murray 1988; Worthington 1989) suggest that frugivores do not select fruits and disperse seeds randomly within tropical forests (Murray 1988; Schupp et al. 1989; Willson and Crome 1989; see also Hoppes 1988; Malmborg and Willson 1988). Instead, we hypothesize that they are most likely to take fruits from and disperse seeds into (or near) gaps. The consequences of such dispersal are complex and remain largely unresolved (e.g., Augspurger 1984b; Augspurger and Kelly 1984; Dirzo and Dominguez 1986; Webb and Willson 1985; Willson 1988; Schupp 1988b; Schupp and Frost 1989).

Temporal Variation

Fruit abundance at La Selva has been monitored for more than eleven years over a twenty-year period (Frankie et al., 1974a; Opler, Frankie, et al., 1980; Denslow et al. 1986; Levey 1988b; Loiselle and Blake 1991). To compare fruiting phenology at La Selva to BCI, another tropical site with almost equally extensive data, we standardized by dividing each monthly value into the range of monthly values reported (see the figure legend for sources). If data for more than one understory habitat or plant form were reported in a study, fruiting activity was calculated for each and then averaged across months.

In general, there is a single peak in fruit abundance in September–October and a period of relative fruit scarcity during the dry season (January–March; fig. 22.1). This single-peak pattern contrasts with double peaks of annual fruit abundance reported elsewhere in Costa Rica, Panama, and Peru (Frankie et al. 1974a, Croat 1975, Opler, Frankie, et al. 1980; Foster 1982b; Janson and Emmons 1990; fig. 22.1). Even during the period of relative fruit scarcity, however, ripe fruit is not uncommon. Frankie et al. (1974a, 902) concluded that "a continual source of food was . . . provided to resident frugivorous animals" throughout their three-year study.

The magnitude of seasonal shifts in fruit abundance depends on habitat (fig. 22.1). Individuals in the canopy, second growth, and gaps tend to display large, synchronous changes in fruiting activity (Frankie et al. 1974a; Levey 1988b). Understory fruit production, on the other hand, is much less seasonal (Frankie et al. 1974a; Opler, Frankie, et al. 1980; Levey 1988b). Because understory species tend to produce fruit over brief periods (Opler, Frankie, et al. 1980) the aseasonality of understory fruit abundance is apparently caused by many short, nonoverlapping fruiting episodes. Still, the understory is an area of relative fruit scarcity; most sites are without fruit for at least four months of each year (Denslow et al. 1986). Despite large communitywide fluctuations in fruit abundance in gaps, individuals of at least several species produce fruit nearly continuously in gaps, whereas conspecifics in the understory do not (Bullock 1982; Marquis 1988; Levey 1990).

The abundance of fruit-eating birds also displays seasonal fluctuations (Loiselle 1987a, 1988; Levey 1988b; Loiselle and Blake 1991). Large numbers of frugivores, mostly temperate and altitudinal migrants, start to appear in October–November when fruits are abundant. Frugivore abundance remains high through January–February and then drops during the dry season period of relative fruit scarcity. These concurrent shifts in frugivore abundance suggest that many frugivorous birds track changes in fruit availability through altitudinal migration

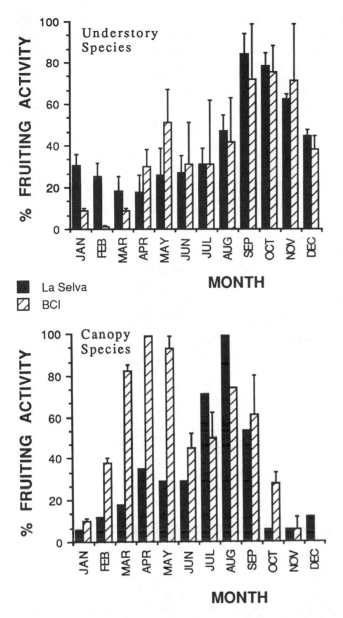

Fig. 22.1. Seasonal patterns of (top) understory and (bottom) canopy fruiting activity on Barro Colorado Island and La Selva. Fruiting activity reflects either the number of individuals in fruit per area or proportion of community bearing fruit, depending upon the data base. Bars are standard errors (each data base contributed only once per habitat to their calculation). (Data from Frankie et al. 1974a; Croat 1975; Opler, Frankie, et al. 1980; Foster 1982b; Denslow et al. 1986; Loiselle 1987a; Levey 1988b.)

(Rosselli 1989; Loiselle and Blake 1991). Movements of birds, bats, and primates at other tropical sites also appear tied to seasonal shifts in fruit abundance (Crome 1975; Marshall 1983; Leighton and Leighton 1983; Innis 1989; Te Boekhurst et al. 1990).

Temporal variation in frugivore abundance is likely to affect fruit removal rates. *Hamelia patens,* for example, produces ripe fruit continuously and experiences dramatic changes in disperser abundance (Leck 1972; Levey 1987a). During some periods its fruits rot on the tree, whereas at other times they are removed while still unripe. Fruits of *Hamelia* ripen more rapidly during periods of rapid fruit removal

(Levey 1987a). Denslow and Moermond (1982) reported higher fruit removal rates during a season of fruit scarcity than during a season of fruit abundance (see also Leck 1972). Similarly, Vandermeer et al. (1979) documented highest fruit removal rates of *Welfia georgii* from individuals that fruited slightly out of synchrony with the rest of the population. Croat (1974) and Snow (1965) argue that some species may shift their fruiting periods in response to changes in frugivore feeding activity. The extent to which such shifts are a cause or an effect of correlated shifts in frugivore activity, however, remains unclear.

Breeding activity and mortality patterns in many types of frugivores have been linked to seasonal changes in fruit availability (Kaufmann 1962; Fleming 1981; Smythe 1978; Bonaccorso 1979; Glanz et al. 1982; Oppenheimer 1982; Smythe et al. 1982; Russell 1982; Terborgh 1983). Widespread famine among frugivores during fruit shortages has been reported elsewhere in Central and South America (Foster 1977, 1982a; Terborgh 1986a) but has not yet been reported at La Selva. During the dry season of 1981, however, we observed fifteen species of frugivorous birds eating unripe or partially ripe fruits. Some species were using unusual foraging methods (e.g., *Ramphocelus* tanagers hopping on the ground), and we noticed more aggressive interactions at fruiting trees than usual (D. Levey unpublished data). Foster (1982a) estimates that severe fruit shortages occur on Barro Colorado Island approximately every ten years following unusually wet dry seasons. Indeed, January and February of 1981 were unusually wet—over a twenty-seven-year period only three Januaries and two Februaries were wetter (chap. 3).

During periods of extremely low fruit availability, the few species in fruit may be responsible for maintaining the entire frugivore community. Figs are often identified as such "keystone" mutualists (Leighton and Leighton 1983; Terborgh 1983, 1986b). At La Selva, we have not noticed any such keystone species. Figs, in particular, do not appear to play as important a role as elsewhere (see also Gautier-Hion and Michaloud 1989). Howe (1984b) suggested, however, that *Casearia corymbosa* may be a critical resource for *Tityra semifasciata* during times of fruit scarcity at La Selva.

FRUIT-FRUGIVORE COEVOLUTION

The importance of tropical fruits as resources for birds was well documented in the 1960s (Snow 1962a, 1962b; Land 1963; Willis 1966; Leck 1969) but a theoretical framework to explain the diversity of fruits and frugivores was lacking. Van der Pijl (1982) noted several syndromes among fruiting plants. Fruits commonly eaten by bats, for example, were different in predictable ways from those eaten by birds. These syndromes seemed to be clear evidence of coevolution, but the underlying mechanisms generating them were unclear. The first explanation (McKey 1975) was based on Snow's (1971) work and was later expanded by Howe and Estabrook (1977). It emphasized a continuum of evolutionary and ecological dependence between species of birds and fruiting plants and focused attention on the end points of the continuum.

At one end of the hypothesized continuum were closely interdependent *specialists.* Plants in this group produced relatively small crops of large-seeded, nutritious fruits and attracted a few species of uncommon but "reliable" dispersers. These birds were thought to eat only *specialist* fruits and disperse their seeds to favorable germination sites. At the other end of the continuum were *generalist* plants, producing large crops of small-seeded fruits. Because these fruits contained little more than sugar-water and seeds, birds feeding on them were necessarily generalists, taking a wide variety of fruits and insects. Hence, the birds were seen as opportunistic foragers that displayed little selectivity among the many species of fruits they ate.

The generalist-specialist hypothesis has generally lacked both theoretical and empirical support (Wheelwright and Orians 1982; Howe 1984a; Moermond and Denslow 1985; Herrera 1986). We review the difficulties encountered by the theory and suggest a current view of fruit-frugivore coevolution that is consistent with the results of recent studies.

Detailed studies of foraging behavior by frugivores quickly revealed many inconsistencies between predicted and observed patterns. First, generalist frugivores do not feed on small fruits opportunistically or haphazardly. Instead, they are highly selective in choosing among fruits and apparently weigh disparate costs and benefits of different fruits found in various contexts (Moermond and Denslow 1983; Moermond et al. 1986, 1987; Loiselle and Blake 1990). Second, large specialized birds do not rely exclusively on a few specialized fruits but, in fact, often take many species of fruits over a wide range of sizes, including Melastomataceae and Moraceae (refs. in Moermond and Denslow 1985; Wheelwright 1985b). Third, small generalist birds are not low-quality dispersers. Viable seeds have been found in the feces of almost all tanagers and manakins and even supposed seed predators (e.g., *Arremon*) captured in mist nets at La Selva (D. Levey unpublished data). Small birds, in general, (and tanagers, in particular) are usually active and, hence, are likely to disperse seeds away from the parent plant. Furthermore, they appear to be "reliable"; color-banded individuals return regularly to specific fruiting trees (Pratt 1984; D. Levey pers. observation). Fourth, "specialist" frugivores often do not provide effective dispersal. Trogons at La Selva and Monteverde (Costa Rica) frequently regurgitate seeds under or near the parent tree (Wheelwright 1983); oilbirds deposit many seeds in their roosting caves (Snow 1962c); and toucans may often be fruit "thieves" (Howe 1977). Fifth, it is often inaccurate to label a bird species as a high- or low-quality disperser because how well it disperses seeds may not be an inherent attribute of the bird but instead depend upon seed size, pulp texture, or other fruit traits (Levey 1987b). A tanager, for example, may be an excellent disperser of small seeds but may seldom ingest large seeds even though it feeds on large-seeded fruits. Finally, species interactions may change dramatically from site to site, a phenomenon not accounted for by the theory. Similar species of *Casearia* are dispersed by specialists at La Selva but by generalists elsewhere (Howe 1977; Howe and Vande Kerckhove 1979). These observations suggest that the theoretical framework proposed by McKey (1975) and Howe and Estabrook (1977) does not adequately describe tropical fruit-frugivore interactions.

How closely are fruiting plants and frugivores coevolved? We do not expect tight, species-specific coevolution for several reasons. Plants have little control over where frugivores deposit their seeds. Because birds are "rewarded" for taking fruits rather than dispersing seeds, there is no direct mechanism for the evolution of dispersal to especially good germination sites (Wheelwright and Orians 1982; Howe 1984a). In addition, spatial and temporal variation in dispersal effectiveness and fruit supply may often be high enough to swamp evolution of species-specific interdependence (Howe and Smallwood 1982; Howe 1984a; Herrera 1985a, 1986; Denslow et al. 1986). Dynamic fruit-frugivore associations such as those at La Selva preclude close coevolution. If there are no clear or consistent differences among dispersers in their effectiveness, there is little chance of selection for specialization with the best disperser. Finally, genetic constraints on quantitative traits are little understood but may hinder the evolution of tight fruit-frugivore mutualisms (Howe 1984a; Herrera 1985a, 1986).

For these reasons coevolution in the strict sense (mutual shifts in gene frequencies in two interacting populations) is undoubtedly rare between plants and fruit-eating birds. It is simply unclear how either group would benefit from species-specific interactions (Wheelwright and Orians 1982). The few examples of close coevolution that apparently do exist beg further study. These examples nearly all involve mistletoes (Wetmore 1914; Ridley 1930; Ali 1931; Sutton 1951; Doctors van Leeuwen 1954; Parker 1981; Davidar 1983; Reid 1986; Restrepo 1987). In this case sticky viscin clinging to each seed often results in birds wiping their vents against branches to dislodge the seeds, thereby "planting" it on the branch, the best potential germination site for a mistletoe. This may represent one of the very few cases where a plant in part determines its dispersal site, thereby satisfying one of the seldom met criteria for coevolution toward a more specific fruit-frugivore interaction (Wheelwright and Orians 1982).

On a very general level the syndromes described by Ridley (1930) and Van der Pijl (1982) do exist although their distinctness is blurred (see Howe 1986; Fleming et al. 1987). Differences between fruits typically consumed by birds, bats, rodents and monkeys are well-documented (Van der Pijl 1982; Janson 1983; Gautier-Hion et al. 1985; see also Willson et al. 1989; Jordano 1992). Within these groups are still further divisions of consumers and fruit types. At La Selva, for example, *Euphonia gouldi* eats mostly fruits of epiphytic plants (especially *Anthurium* spp.) and *Mionectes oleagineus* is clearly a specialist on arillate fruits (Loiselle and Blake 1990; D. Levey unpublished data). Clearly, patterns exist among frugivores of who eats what fruits. The fact that proposed theoretical constructs (McKey 1975; Howe and Estabrook 1977) are inadequate should not detract attention from these patterns; they still need to be explained. Indeed, fruit syndromes suggest that differences among dispersers have been important in the evolution of fruit attributes (Janson 1983; Gautier-Hion et al. 1985; Levey 1987b; Levey and Grajal 1991) and that reciprocal selection pressures are likely among groups of plants and groups of frugivores. Thus, seed-dispersal systems exemplify "diffuse" coevolution (Janzen 1980b) where evolutionary changes result from interactions between groups of species rather than between individual species.

The species groups responsible for the diffuse nature of

fruit-frugivore coevolution remain unclear. To delineate them one must keep a historical perspective (Janzen and Martin 1982; Howe 1985). More importantly, biologists need a more thorough understanding of how plant fitness and demography are tied to dispersal by different frugivores (Howe 1984a, 1989; Jordano 1987). If, as suggested earlier, dispersal of seeds away from a parent plant is an important and ubiquitous selective force, then the masher-gulper dichotomy may represent a level of diffuse coevolution between birds and fruiting plants. Because mashers drop most large seeds, large-seeded plants may have methods of increasing the probability of gulpers feeding on their fruits while simultaneously decreasing the possibility of mashers doing so. Predicted mechanisms for such selectivity have been described, tested, and generally supported (Levey 1987d; Rosselli and Stiles in press).

AVENUES FOR FUTURE RESEARCH

Nonavian Frugivory

Little is known about nonavian frugivory at La Selva. In particular, bats are an important but poorly understood group of frugivores and seed dispersers at La Selva that have only recently received attention (T. Fleming unpublished data). Given the wealth of detailed information on diets, foraging patterns, and annual cycles of frugivorous bats in more seasonal Neotropical sites (Fleming et al. 1972; Heithaus et al. 1975; Morrison 1978; Bonaccorso 1979; Foresta et al. 1984; Fleming 1988), work on frugivorous bats at La Selva would provide important comparative data. In addition, several similarities and differences between frugivorous bats and birds merit further attention. First, radiotelemetry studies of bats and birds in Cayenne suggest that dispersal of seeds by bats is more homogeneous than by birds (Foresta et al. 1984). Second, bats may display altitudinal migrations similar to those observed in frugivorous birds (see Marshall 1983; Bonaccorso and Humphrey 1984; Dinerstein 1987). Third, the seed-handling techniques of mashing and gulping described for birds appear also characteristic of Neotropical bats (Fleming 1986; Bonaccorso and Gush 1987).

Invertebrate frugivores have been almost universally overlooked (Janzen 1977b; Herrera 1982, 1984). Yet from a coevolutionary perspective the interactions between these fruit-consumers and their hosts are probably more intricate than between vertebrate frugivores and fruiting plants (Thompson 1982; Herrera 1984, 1986; Sallabanks and Courtney 1992). The interaction between pathogens and fruiting plants is especially interesting because it may directly affect the probability of a vertebrate frugivore feeding on a fruit and dispersing its seeds (Borowicz 1988; Jordano 1989; Buchholz and Levey 1990). On the one hand, secondary compounds that protect ripe fruits against nonmutualist frugivores are also liable to discourage potential seed dispersers. On the other hand, seed dispersers may sometimes prefer insect-damaged fruit because the insects increase the amount of available protein (Redford et al. 1984; Piper 1986; Drew 1988). In any case, the large proportion of damaged fruits on many plants strongly suggests that pathogens may be an important selective force for many fruit traits and, consequently, influence seed dispersal.

The Role of Spatial and Temporal Variation

Gaps at La Selva host high concentrations of fruiting shrubs and foraging frugivores and, thus, can be considered arenas of fruit-frugivore interactions. Yet researchers know little about the consequences of such intense, localized activity. Do birds and bats that feed in gaps bring seeds from other gaps? Are seeds that they take deposited nonrandomly in gaps? and if so, What are the consequences for seedling survival? Because gaps tend to enlarge (Hubbell and Foster 1986b; Runkle and Yetter 1987; Young and Hubbell 1991), if frugivores commonly deposit seeds around the perimeter of gaps (Hoppes 1988), then seed dispersal of many plants may be "directed" to an area of high potential for seedling establishment (Levey 1988a; Malmborg and Willson 1988; Schupp et al. 1989).

This mechanism for seedling recruitment into gaps presently lacks supporting data, but we feel it is likely important at La Selva. In other tropical forests, however, it probably does not operate. For example, in a Panamanian forest neither fruit-eating birds nor fruiting plants are prevalent in gaps (Schemske and Brokaw 1981; Willson et al. 1982). Such basic differences between geographically proximate forests remain unexplained.

As emphasized earlier, the impact of fruit seasonality on frugivore populations is poorly understood. Of particular interest is the extent to which annual cycles of frugivores are tied to fruit abundance and how closely frugivores track changes in fruit availability. Tracking may occur on a small scale (habitat shifts; Karr and Freemark 1983; Blake and Loiselle 1991) or on a large scale (altitudinal and latitudinal migration; Morton 1977; Levey 1988b; Rosselli 1989; Loiselle and Blake 1991; Levey and Stiles 1992 chap. 17). The degree to which frugivores can and do track fruits has important implications for conservation because large areas of contiguous forest may be necessary to maintain viable populations. How much forest will sustain viable populations of highly transient species is, however, unknown. For management of more sedentary species it is important to identify "keystone" species, if, indeed, they exist (DeSteven and Putz 1984; Terborgh 1986b; Gautier-Hion and Michaloud 1989). At La Selva we suggest that tree-fall gaps, which always seem to support relatively high levels of fruits, may be more important than specific keystone species in maintaining frugivores through periods of fruit scarcity (Levey 1990).

Integrating Multiple Factors at the Level of the Plant Species

Perhaps the most challenging task is framing frugivory and seed dispersal in the context of the many other components of a plant's life history (Schupp et al. 1989). A first step is to determine the relationship between fitness and patterns of seed dispersal; researchers need to move beyond studies of frugivore behavior and link such studies with plant demography (Howe 1989; Jordano 1992). How, for example, do seed deposition patterns influence seed and seedling survival (Howe et al. 1985; Fleming 1988; Schupp 1988a; Willson and Whelan 1990)? Only a handful of studies at La Selva have focused on postdispersal seed fates (Schupp and Frost 1989; Loiselle 1990; Byrne and Levey in press; Levey and Byrne in press). This is a critical area of research because both demographic and experimental studies have demonstrated that predation on seeds and seedlings can be the major limitation to recruitment in some tropical species (Sarukhán 1978; Augspurger 1984b; Howe et al. 1985; Sork 1987; Schupp 1990).

Seed dispersal also needs to be integrated with the many other factors affecting plant fitness. For example, how do gap dependence, pollination, herbivory, and seed predation influence fruit and seed attributes and, hence, dispersal success? Large seeds may be required for vigorous, shade-tolerant seedlings but may also increase risk of predation and decrease dispersibility (Snow 1971; Howe and Richter 1982). Similarly, fruiting phenology may be dictated by flowering phenology and, thus, constrained by pollinator abundance and/or weather rather than tied solely to disperser abundance. Hence, examining only seed dispersal or frugivory may yield conclusions that are misleading when placed in the broader context of the plant's autecology and evolution.

CONCLUSIONS

Fruiting plants and frugivores are a critical part of the La Selva ecosystem; most dicots rely on frugivores for seed dispersal, and most vertebrates rely on fruit resources for at least part of their energetic requirements. The central role of frugivory is apparently typical of other tropical wet forests as well, but the types of dominant fruits and frugivores vary widely. At La Selva small fruits are especially common, and the birds that feed on them have received much attention. These birds are selective in their choice of fruits, yet their choices are context dependent and their fruit-handling techniques vary in important ways. Thus, the major dispersers of a given plant species are often unpredictable. This type of fruit-frugivore interaction precludes close coevolution. Instead, the important selective pressures probably occur between groups of species, such as those determined by fruit type or fruit-handling technique.

Spatial and temporal heterogeneity in fruit and frugivore abundance emphasize the importance of long-term and large-scale studies. There are obviously times and places of intense frugivore and fruiting activity. Yet the impact of such variation is unclear, especially for the most common frugivores (small birds and bats) and fruiting plants (shrubs, treelets, lianas, epiphytes, and herbaceous plants).

ACKNOWLEDGMENTS

We thank T. Fleming, B. McNab, A. Worthington, and two anonymous reviewers for insightful comments on the manuscript. The Chapman Fund (American Museum of Natural History), National Science Foundation (NSF DEB 79-10991), Noyes Foundation, and the Organization for Tropical Studies made our research at La Selva possible. Thanks also go to G. Kiltie for typing the manuscript. Manuscript preparation was supported by the University of Wisconsin, Zoology Department, and an Archie Carr Postdoctoral Fellowship from the Department of Zoology at the University of Florida.

PART V

LA SELVA'S HUMAN ENVIRONMENT

COMMENTARY

Lucinda A. McDade

In part 5 of this book we step beyond La Selva's borders to place the station in the context of the region and its people and to explore how La Selva's mandate in research and education can be applied to human problems. Part 5 includes three chapters. In chapter 23, Butterfield addresses land use in the Sarapiquí region with emphasis on colonization patterns and conservation. Montagnini treats agriculture in chapter 24, and Butterfield presents an overview of forestry and related research in chapter 25. They characterize the development and current needs of these sectors and go on to address the role that La Selva, as a biological research station, might have in meeting these needs.

Butterfield begins chapter 23 with a historical overview of colonization in the region with particular attention to the current phase in which the government is parceling out large underused holdings to a new wave of settlers. The significance of this process is confirmed by the fact that about 50% of the inhabitants of Sarapiquí Cantón live in government-sponsored settlements and almost 40% of the area of the cantón is occupied by such projects (fig. 23.2). Clearly, the fate of these settlements and of the individuals who live in them is a key to the future of Sarapiquí. Butterfield does not paint an optimistic picture; she indicates that the settlers often confront small plots with poor soils and inadequate extension services. The result may be a segment of the population that is unable to meet even subsistence levels. With the agricultural frontier closed in Costa Rica, options are few for these settlers to relocate if their current farming efforts are not successful. It is a volatile situation and one to be watched.

The second section of Butterfield's chapter on land use deals with conservation. Colonization and conservation may at first seem oddly matched in a single chapter. When one realizes, however, that the Sarapiquí region is largely a mosaic of colonization projects and conserved areas, it is clear that these are appropriately combined here. These are the two main land uses in the region and their counterposition poses both problems and potential. Butterfield documents the variety of public and private reserves in the Sarapiquí. She also cautions that these reserves cannot be viewed as permanently safeguarded given the pace of change in the region.

Colonization and conservation come together in the third section of this chapter, which covers the Cordillera Volcánica Central Biosphere Reserve. The United Nations' biosphere reserve concept seeks to expand the notion of conservation units to include landscapes modified for sustainable use by humans in addition to strictly preserved areas. A large part of Sarapiquí Cantón (including La Selva) is in the Cordillera Volcánica Central Biosphere Reserve, and Butterfield points to the challenges involved in making this biosphere reserve more than idealized lines on a map. Specifically, she argues that La Selva and the Organization for Tropical Studies (OTS) can ill afford *not* to play significant roles in the effort to turn the reserve into a model of sustainable development and conservation for the tropics.

Montagnini (chap. 24) treats agriculture in the region, beginning with the historical development of farming in the Atlantic lowlands of Costa Rica. In taking this approach Montagnini points out that historical factors are frequently more significant determinants of present patterns of land use than are ecological conditions. Subsistence farms, cattle ranches, and commercial plantations (largely bananas, some cocoa) have been and continue to be the region's primary agricultural systems. Large cattle ranches developed in the region largely through buy-outs of small holders who had found the land unsuited for sustained agriculture. Ironically, this process is now coming full cycle as unproductive cattle ranches are being parceled out to small farmers by the Costa Rican government as described by Butterfield in chapter 23. Montagnini discusses patterns of land use in the region as a function of size and nature of farms. She describes some of the more promising ideas for intensification of agricultural productivity with emphasis on innovative crops. When discussing problems associated with enhancing productivity of both crops and livestock Montagnini identifies ecological and institutional constraints and argues that both must be addressed to improve living standards and protect remaining forested areas.

Montagnini closes by reviewing the accomplishments in agricultural research at La Selva and recommending increased OTS roles in this field. She feels that courses in agroecology are a large step in the right direction, especially because many researchers at La Selva have historically been alumni of OTS courses. She directly challenges researchers at La Selva to undertake projects that are relevant to the difficult situations faced by small farmers in the lowland wet tropics and to develop collaborative research with local scientists and institutions.

In the final chapter of part 5, Butterfield provides an overview of the status of forests in Costa Rica with emphasis on the region around La Selva. Although Costa Rica is a global leader in relative land area under protection, it also has one of the region's highest rates of deforestation outside of those areas. In just a few years essentially all of the forest outside of reserves will be gone. Butterfield identifies causes of deforestation and the likely social and economic impacts of the demise of accessible forest resources. The exhaustion of the forest resource may well put pressure on protected forests, especially as the price of wood products rises and the national balance of payments is adversely affected by wood imports. This projection lends special urgency to strengthening public support for parks and reserves and to increasing forest production by improving management of remaining forests for sustained yield and reforesting cut-over areas that are unsuited for farming.

After briefly characterizing the timber industry in Costa Rica Butterfield explores reforestation and forest management efforts as potential aids to alleviate the coming wood crisis. She surveys the status of projects in each of these areas and describes research necessary to enhance the contribution of each of these sectors.

The fundamental problem in reforestation is the lack of data on appropriate species, which has led to reliance on a few exotic species that have been well researched in other parts of the tropics. Research needs for natural forest management are extensive because experience in this field is very limited, particularly in the high-diversity forests characteristic of this region of the country. Although it is probably too late for such techniques to be widely applied in Costa Rica, they may be used in more extensively forested regions of the world.

Turning to research relevant to forestry at La Selva, Butterfield points out that forest research has always been important at La Selva with work on trees continuing to be a research focus. Much of this work has direct implications for forest management, including tree demography and physiological ecology, phenology, and forest regeneration. Especially useful are long-term plots that have been recensused periodically since 1968 (see chaps. 7, 8). The data from such projects have already been used by the Portico commercial venture in sustainable forest management to estimate expected growth rates of *Carapa* and, therefore, cutting cycles (R. Peralta pers. comm.).

Several projects designed to provide data useful for reforestation and plantation projects were established at La Selva in the 1980s with strong emphasis on native species. Butterfield presents some of the initial results of these projects: many native species are easy to propagate (despite assumptions to the contrary), and several natives have outstanding growth rates and characteristics (compared to the widely used exotics). Because native species have the additional advantage of hosting other members of the native flora and fauna, their use in reforestation is also a positive step toward conserving biodiversity. Butterfield urges OTS to play an active role in promoting forestry research as part of its ongoing institutional focus on tropical research, education, and the wise use of natural resources. She also challenges individual researchers at La Selva to consider the implications of their work for natural forest management and reforestation.

The common themes in these chapters are several. A region that was remote and sparsely populated until a dozen years ago is now witnessing the simultaneous closing of Costa Rica's agriculture frontier and the exhaustion of the forest resources that once seemed inexhaustible. Colonists from other parts of the country who receive small parcels of (largely) abandoned pasture land have little experience farming in the humid tropics and have received little guidance from extension agents. The area's exploitable forest resources are being rapidly consumed, but wood remains a primary building material and source of fuel. In the early 1990s an additional new development, the extensive expansion of the banana industry in the region east of La Selva, is turning Puerto Viejo into something of a boom town. Its ecological, economic, and sociological impacts can only be glimpsed at present. Against this complex socioeconomic and political backdrop are numerous private reserves as well as forest reserves and a large national park. There is certainly much grist for the pessimists' mill here. Optimists, however, will hope that the dynamic mix of people and circumstances in La Selva's environs will yield the intellectual and physical energy needed to discover and implement the creative solutions that are necessary to solve these complex problems.

It seems likely that, had this book been written ten years ago, this entire section would not have been included. In contrast, one can predict that a future version of this book with reports on research progress during La Selva's next twenty to thirty years would have many more pages devoted to these topics. It is also possible that these topics might not appear as a separate section of the book but rather as integral parts of many of the component chapters. A section in a chapter on plant reproductive biology, for example, might describe the pollination ecology of key tree species for reforestation and the co-plantings necessary to maintain these pollinators. Similarly, the chapter on forest dynamics might include sections on succession after large-scale disturbance and on reforestation. Much is to be gained in taking such an integrated approach, not the least of which is more information about the subjects we study. For example, Marquis points out in chapter 21 that a great deal more is known about the role of nematodes as herbivores in agricultural systems than in natural forests, but little transfer of this information has occurred. The treatment of these topics clearly depends upon the extent to which individual researchers are able to leap the chasm that has separated the traditional fields of "applied" ecology (agriculture and forestry) from "basic" ecology. Equally clearly, the successful bridging of the chasm will be of increasing importance to La Selva, OTS, and Costa Rica in the future.

23

The Regional Context: Land Colonization and Conservation in Sarapiquí

Rebecca P. Butterfield

When Leslie Holdridge bought Finca La Selva in 1953, it was located in the wilds of frontier Costa Rica. The Sarapiquí region was still relatively isolated and undeveloped when the Organization for Tropical Studies (OTS) acquired the property in 1968 (Stone 1988). Since then, the area has undergone dramatic changes, the most notable being the elimination of large tracts of tropical forest to establish cattle ranches. Recent LANDSAT images demonstrate the extent to which La Selva has become ringed by pastures (Sader and Joyce 1988): it remains as the tip of a peninsular forest sanctuary delimited by rivers (see also chap. 2). Since the 1950s the colonization of Sarapiquí has been continuous; land conflicts in the area, however, have escalated since the early 1980s. Until the mid-1980s, La Selva worried about homesteaders at its back boundary; it now has colonists living at its front door. The transformation of the region from a largely forested open frontier to a mostly agricultural, rapidly developing zone with high potential for conflict between interest groups has profoundly altered the context in which La Selva exists. A series of new socioeconomic, political, and environmental problems has emerged, offering new opportunities for research, training, and service as well as new challenges that OTS must be prepared to address.

In this chapter I sketch the historical changes in land-use patterns in Sarapiquí from the 1950s to the 1990s. I trace the development of the region from a frontier outpost with subsistence agriculture through the expansion of commercial cattle farms. I discuss the role of government-sponsored land reform projects in changing land-use and land-settlement patterns and identify opportunities created by the advent of new conservation units in the zone. I then present concepts of resource management with recommendations for the future role of La Selva in a dynamic and challenging environment: the Cordillera Volcánica Central Biosphere Reserve and the surrounding Atlantic lowlands of Sarapiquí.

THE AGRICULTURAL FRONTIER

The last forty years have seen a qualitative change in the character of Costa Rica's frontier communities. The Sarapiquí region, named for its principal river (fig. 23.1), provides a microcosm of the processes that have occurred in many frontier regions of Costa Rica and other tropical countries. Three his-

torical processes have influenced the development of the region in the years since La Selva was founded. (1) The growth of population and of commercial agriculture elsewhere in Costa Rica has led to land colonization, penetration of new roads, and the rapid development of formerly isolated hinterlands like Sarapiquí. (2) The expansion of government influence in the region has linked the countryside to national institutions that provide a variety of services. In particular, the process of land colonization, which was formerly rather spontaneous, has come increasingly under the control of the Instituto de Desarrollo Agrario (Institute of Agrarian Development [IDA]) and is, thus, fostered and regulated by the central government. (3) With the closing of the agricultural frontier and the rising awareness of the need to conserve some forests, unoccupied lands are no longer available for independent colonization. Wilderness homesteading has given way to organized invasions of underused land, institutionalized land reform, and the redistribution of already occupied land.

In the 1950s La Selva was located in one of the last frontiers of Costa Rica. Sarapiquí was characterized by its isolation and the predominance of subsistence agriculture. Although mule trails and the Sarapiquí River (navigable below Puerto Viejo) had made the region an important gateway to Europe during the nineteenth century, as late as 1950 no road for motor vehicles penetrated the region (Fernández 1985). The road from San José via Vara Blanca reached as far north as Cariblanco (fig. 23.1). From there settlers entered on horseback (F. Madrigal pers. comm.).

The subsistence farms of early settlers produced staple foods, such as rice, beans, corn, bananas, coffee, and pork. Some income was generated by collecting rubber from wild trees *(Castilla elastica),* extracting logs of *Cedrela odorata,* and selling dried river fish, bobo *(Joturus pichardi)* (I. Alvarado, F. Madrigal pers. comm.).

In the early-1950s a dirt road was pushed through to Puerto Viejo. Products could then be sent to San José via a grueling fifteen-hour trip. Puerto Viejo became an entrepôt between San José and the lower Río San Juan; products from Barra del Colorado and the Río San Juan were paddled upstream to Puerto Viejo where they continued via truck to San José. The lowland river settlers between Puerto Viejo and Nicaragua sold bananas, plantains, cacao, and copra. Puerto Viejo in the early 1950s boasted two country stores and several houses, or

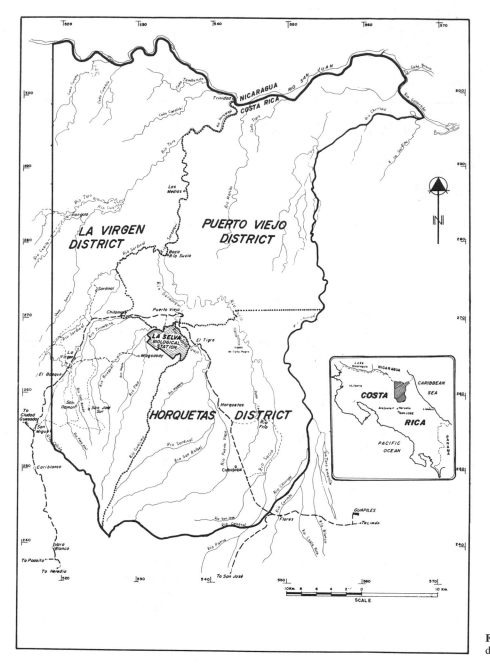

Fig. 23.1. Sarapiquí Cantón and districts.

ranchos, roofed with palm thatch (F. Madrigal, V. Chavarría pers. comm.).

Between 1952 and 1960 mechanized rice farms were established between La Virgen and Puerto Viejo along the Sarapiquí floodplain. The grain was shipped to a drying plant at Barranca near the Pacific port of Puntarenas. The land yielded two harvests per year until weeds became a serious problem and cultivation was abandoned (F. Madrigal pers. comm.; Pierce 1992).

By the 1960s there was regular bus service to San José (seven to nine hours) and a ferry across the Sarapiquí River that was met by a microbus from Las Horquetas. The Standard Fruit Company moved to Río Frío in 1967 to develop intensive banana plantations. They improved transportation infrastructure in the region with a road connecting Guápiles with Río

Frío and Puerto Viejo. The banana plantations also offered a new source of employment in the area.

The 1970s were boom years for Sarapiquí. Local governance was granted in 1970 when Sarapiquí was declared a cantón (county) with three districts: Horquetas, Puerto Viejo, and La Virgen (fig. 23.1). Shortly thereafter, a bridge was constructed over the Sarapiquí River to improve access to the Las Horquetas and Río Frío region. During the early 1970s Puerto Viejo started to receive government services: connection to the national telephone network, an agricultural high school, and a branch of the national bank. Permanent medical services through a Social Security clinic were not available until the 1980s. By 1984 two lawyers, a dentist, a doctor, and an agricultural extension agent had offices in Puerto Viejo (V. Chavarría pers. comm.).

The opening of the agricultural frontier in Sarapiquí is synonymous with deforestation of the region. Forest clearing was the traditional method of not only claiming untitled land but also demonstrating use of owned land; both were sanctioned by Costa Rican law. To initiate a land claim some land had to be cleared of forest and put into agricultural production. The cheapest way to keep land cleared and occupied to defend one's claim is by grazing cattle. Besides the traditional use of grazing to maintain land claims, the cattle export boom that occurred throughout Central America in the 1970s provided an export market for beef and improved prices on the local market as well. Credit for ranching was readily available (Leonard 1987) and helped support the establishment of large ranches in Sarapiquí. Land speculation combined with attractive beef markets have significantly altered land use in the area. Not surprisingly, the predominant land use in Sarapiquí is pasture. The most significant changes in land use from 1963 to 1983 (fig. 23.2) were the expansion of pasture and the shrinking of forested areas. This process of conversion of forests to pasture is well documented for Central America in general (Hartshorn et al. 1982; Leonard 1987) and for the Sarapiquí region (table 23.1). Within Sarapiquí the districts showing the greatest increase in cattle in the 1973–83 period were Horquetas and Puerto Viejo with 182% and 178% relative increase, respectively.

Frontier Colonization

For much of the nineteenth and early-twentieth centuries the Costa Rican government tried, with varying degrees of success, to populate its isolated frontier regions. Although much of the settlement of Sarapiquí was spontaneous, the government made several early attempts to promote settlement in this part of the Atlantic lowlands. Before 1962 several colonization projects were initiated in an effort to develop the region. The largest of these projects granted more than 20,000 ha to a group known as Acción Nacional de Trabajo in 1942. The homesteading area was a large triangular piece extending from the confluence of the Sarapiquí and Sucio rivers upstream to the present boundary of the Cordillera Volcánica Central Forest Reserve. It included La Selva and the entire area of what is now known as the Magsasay sector of the Braulio Carrillo National Park (Zona Protectora La Selva). Rectangular parcels of 100 ha were granted by the national government to homesteaders who cleared the forest and en-

Table 23.1 Increase in cattle and pasture in Sarapiquí by district and for the cantón as a whole, 1973 to 1983

Region	1973 No. Cattle (Head)	1973 Pasture (ha)	1983 No. Cattle (Head)	1983 Pasture (ha)	Relative Increase (Head %)
Puerto Viejo	7,834	6,466	21,743	21,635	178
La Virgen	14,138	16,035	16,371	18,956	16
Horquetas	8,001	9,584	22,588	20,053	182
Sarapiquí	29,973	32,085	60,702	60,644	103

Sources: DGEC, Censo Agropecuario 1974 and 1984a.

gaged in agriculture (González 1988). Few homesteaders managed to secure their land claims, and the project failed because the area was so remote.

Two penal colonies, Magsasay and El Plástico, were established near La Selva in the late-1950s. Magsasay, located near the south boundary of La Selva, had a working sawmill until the late-1960s (Gardella 1981; G. Hartshorn pers. comm.). The El Plástico colony, along the eastern border of Braulio Carrillo park at about 500 m elevation, cleared forest to create small patches of pasture. Both colonies have since been abandoned and incorporated into conservation units.

In 1962 Instituto de Tierras y Colonización (ITCO) (Land Colonization Institute, now IDA) was formed to promote and regulate colonization projects in frontier areas. The central government wanted to develop and colonize unused land and provide an outlet for the growing population of the central valley. During its first four years ITCO formed colonies of farmers on unexploited forest land (IDA 1987). In the mid-1960s when banana plantations were being established near Río Frío, ITCO claimed large tracts of land in the same area. They built roads and bridges to open new lands for agricultural development. Improved access to previously remote regions stimulated a large migration of land-poor farmers from other areas of Costa Rica. A near land rush was started in Sarapiquí, beyond the control of the government, as the new settlers raced to carve farms from virgin forest (M. Castro pers. comm.; CODESA 1983).

Because of limited transportation infrastructure and associated problems of marketing agricultural produce, the region had a very low population density before the colonization boom. In the two decades since 1963 the population of Sarapiquí increased 289% from 4,856 to 18,909 inhabitants (Dirección General de Estadísticas y Censos 1966 and 1984b), reflecting this wave of immigration. Many of the settlers were landless farmers from Guanacaste, San Carlos, and the west side of the central valley (Barahona 1980; UCR 1984). Others were wealthier individuals, some living in the central valley, who sought land for speculation and to establish cattle ranches. This second group usually obtained land either by buying out the original homesteaders once their claims were recognized or by hiring surrogates to work the homesteads they established.

Between 1973 and 1983 the number of farms in Sarapiquí increased 183% (table 23.2). The largest increase in numbers of farms occurred in Horquetas (443%), the site of the first ITCO projects and the last district to develop road access. In

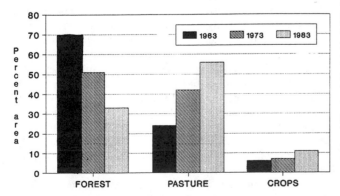

Fig. 23.2. Land use changes in Sarapiquí, 1963–83. Source: Censo Agropecuario.

Table 23.2 Changes in total number of farms in Sarapiquí by district and for the cantón as a whole, 1963 to 1983

Region	Total Number of Farms			Relative Increase 1973 to 1983 (%)
	1963	1973	1983	
Puerto Viejo	—	195	493	153
La Virgen	—	342	578	69
Horquetas	—	172	934	443
Sarapiquí	577	709	2,005	183

Sources: DGEC, Censo Agropecuario 1965, 1974, and 1984.

Table 23.3 Farm ownership in Sarapiquí by district and for the cantón as a whole, 1973 to 1983

Region	Farmer-Owned				Absentee Owner			
	Percent Total Farms		Percent Total Area		Percent Total Farms		Percent Total Area	
	1973	1983	1973	1983	1973	1983	1973	1983
Puerto Viejo	92	—	77	—	8	—	23	—
La Virgen	80	—	45	—	20	—	55	—
Horquetas	83	—	28	—	17	—	72	—
Sarapiquí	84	80	47	43	16	20	53	57

Sources: DGEC Censo Agropecuario 1974 and 1984.

the 1990s Sarapiquí is an area with many small owner-operated farms, a direct result of colonization projects involving individual farmers. More than 50% of its territory, however, is in large farms with absentee owners (table 23.3). These larger farms are mainly cattle ranching operations.

The Closing of the Frontier and the Rise of Governmental Land Reform

In the 1980s the agricultural frontier closed; unclaimed forest land in Sarapiquí, indeed, in Costa Rica as a whole, no longer exists. The few remaining forests outside of reserves are being rapidly destroyed. Land colonization has shifted from the homesteading of unclaimed virgin forest to squatter invasions of underused, abandoned, or absentee-owned cattle farms.

Since the 1980s, the focus of IDA has shifted from colonization projects to land redistribution programs. Identification of farms for purchase is often initiated by a land invasion. Politically well-connected landowners have also been implicated in using their influence either with government or with organized peasant groups to include their farms in the land redistribution program, thus converting a failing cattle ranch into cash. Once a farm has been identified by whatever means IDA buys and subdivides it into small (ca. 10 ha) parcels. Unfortunately, IDA does not evaluate the land-use capability of these properties, with the consequence that many IDA settlements are on worn-out, degraded land unsuitable for intensive, small-holder agriculture (Hartshorn et al. 1982).

Sarapiquí is one of the most active colonization zones in the country. Land invasions have been so frequent in Sarapiquí during the last decade that IDA cannot keep up with land purchases and development of new settlements (M. Castro pers. comm.). Considerable land in Sarapiquí has been acquired and settled by IDA (fig. 23.3). Of the entire cantón, 38% of the area is in IDA settlements. Of the district of Horquetas, 80% of the territory is in IDA settlements, whereas Puerto Viejo has the smallest land area (10%) in colonization projects (table 23.4). Although population data are estimates, it appears that close to 50% of the population of Sarapiquí now lives on IDA settlements.

Land use patterns on the smaller IDA farms differ significantly from those in the region as a whole (fig. 23.4). The IDA farmers devote nearly 60% of their land to subsistence crops (beans, rice, and manioc) and perennials (cacao and plantains), whereas the entire canton is dominated by cattle grazing with less than 20% of the area in crops. IDA farms are

too small for extensive cattle grazing although many colonists keep dairy cows (UCR 1984).

Land titles in IDA settlements are not granted until a claim has been occupied for ten years, so many farmers do not have titles to their farms. Untitled land does not qualify for bank agricultural credits for crop investment. Generally poor soils coupled with inadequate agricultural extension services are contributing to a rapidly growing class of subsistence or below-subsistence farmers in the Sarapiquí region. This situation, in turn, leads to increased dependence on off-farm employment (Schelhas 1991).

The marginality of these farms is underlined by the active land market in untitled IDA parcels. Some colonizers, unable to eke a living from their farms, sell their plots to other hopeful farmers although it is technically illegal to do so (Schelhas 1991). Several plots in an IDA settlement near Chilamate have been sold six to eight times each although no one had legal title (F. Madrigal pers. comm.). There are no concrete data on the turnover of IDA parcels, but it does appear that some IDA "farmers" make their living from real estate rather than crops (Brooijmans 1988). The role of speculation as a motivating force in land invasions merits further research. Whatever the impetus, the trend of subdividing pasture land into smaller farms will continue.

New roads and government services established during the 1960s and 1970s enabled landless farmers from other regions of Costa Rica to homestead new farms on unclaimed wilderness land. Rising international beef prices and abundant credit for cattle coupled with increasing land values also contributed to the land rush as entrepreneurs sought to make a profit from cattle and land. As virgin forest land became scarce, colonists began to occupy underused or abandoned cattle farms. The government land reform policy recognizes colonists' rights and has sanctioned the purchase of large cattle ranches for redistribution to small farmers. The absentee-owned cattle ranches formed during the 1960s and 1970s are now being divided into substantially smaller plots. The new class of subsistence farmers who occupy such lands face a marginal existence unless adequate land management techniques can be developed with the supporting infrastructure and extension services to present alternatives to a cycle of land clearing and speculative transactions. La Selva is located in the middle of a dynamic region with rapidly changing patterns of land use and settlement. The stabilization of such patterns depends on

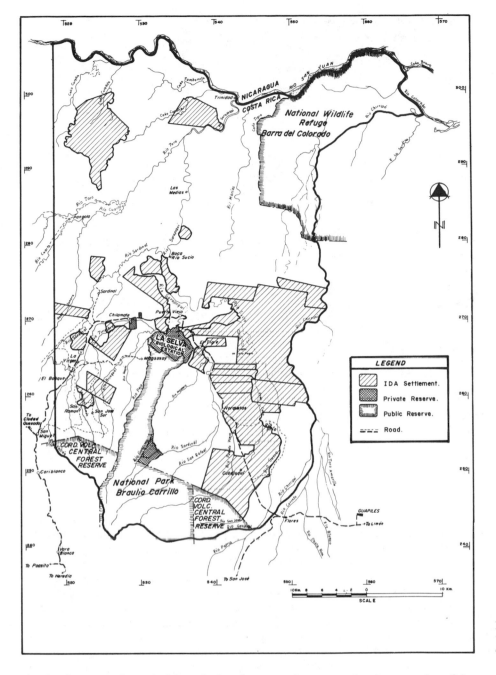

Fig. 23.3. IDA settlements and conservation units in Sarapiquí. Source: Sección de Inventario de Tierras, IDA.

the development of sustainable agricultural systems that can provide a livelihood for the increasing number of small farmers in the region.

CONSERVATION IN SARAPIQUÍ

Ironically, as deforestation continues unabated in Sarapiquí, new lands are being protected in both private and public conservation units. It is likely that these areas will soon be the only remaining examples of the once extensive forests of the region. There is no guarantee, however, that social, economic, and political pressures to gain access to these areas for the expansion of agricultural land and the procurement of forest products will not, in the long run, succeed. The challenge of conservation is not only the preservation of critical areas but

also the protection of those areas for the future. The Sarapiquí region reflects the situation of the country as a whole—rapid deforestation in an area with growing conservation reserves (see chap. 25).

Finca La Selva became the first forest preserve in Sarapiquí when Dr. Holdridge realized, with amazing foresight, that his farm was worth more as forest than as a commercial plantation. Since then, private and governmental efforts have protected an additional 15,017 ha of forest in Sarapiquí. The centerpiece of conservation in the region is the La Selva–Braulio Carrillo National Park complex that connects two important areas extending from 30 to 2900 m elevation. A corridor linking La Selva with Braulio Carrillo National Park was seen as essential to ensure the survival of altitudinally migrating species (see chap. 2; Pringle et al. 1984; Pierce 1992).

Table 23.4 IDA Settlements in Sarapiquí by District and for the cantón as a whole, 1963 to 1987

Region	Total (ha)	As Percentage of Total Land Area	Number Families Settled	As Percentage of Total Population
Puerto Viejo	4,593	10	192	19
La Virgen	13,994	34	476	43
Horquetas	29,278	80	1,374	53
Sarapiquí	47,865	38	2,042	43

Source: Unpublished data: Sección de Inventario de Tierras, IDA, San José.

Note: The figures for total population are estimates only. Two studies in the Río Frío area (ITCO 1980; Ureña 1983) have identified average family size on IDA settlements as five to six people. To be conservative the number of families settled has been multiplied by four people. Data on the number of families settled on 6,132 ha in IDA settlements are not available. This could potentially add an additional 613 families to the total.

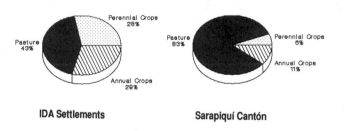

Fig. 23.4. Proportions of pasture and crops in IDA settlements and Sarapiquí Cantón.

Another large conservation unit in the region is Barra del Colorado Wildlife Refuge (fig. 23.3), established in 1985. The refuge incorporates 92,000 ha extending from the Nicaraguan border south toward Tortuguero National Park. The area is characterized by high rainfall and contains lowland flood plains, marshes, and swamps. It is an important refuge for the endangered manatee (*Trichechus manatus*) and for several sport fish (e.g., gar [*Atractosteus tropicus*] and tarpon). The idea of connecting the La Selva–Braulio Carrillo complex with Barra del Colorado has been discussed (Cifuentes 1983) but has received little serious attention.

The expansion of Braulio Carrillo National Park to La Selva has served as a catalyst for private conservation efforts in the region. Rara Avis, a 650-ha private forest reserve, is located on the east side of Braulio Carrillo National Park near Las Horquetas. Its goal is to promote ecotourism and sustained yield forestry as a model for alternatives to cattle ranching. Adjacent to Rara Avis is Selva Tica (570 ha), whose owners are interested in experimenting with natural forest management. Selva Verde (181 ha), a tourist lodge located in Chilamate to the west of La Selva, specializes in bird watching and ecotourist groups. El Bejuco, across the road from Selva Verde, has attracted small numbers of scientists and students to its modest forest reserve (21 ha).

An analysis of the conservation effort in Sarapiquí and in other parts of Costa Rica reveals that since the economic crisis of Costa Rica in the early 1980s the impetus for protecting forested areas has come from external sources with foreign financing. La Selva, Rara Avis, Selva Tica, Selva Verde, and El Bejuco are all foreign owned or financed. The extension of Braulio Carrillo National Park to La Selva received enthusiastic support from the Costa Rican government but was financed through the fund-raising efforts of international organizations

(Pringle 1988). Urban conservation groups, both foreign and national, support these reserves, but national consensus for conservation is lacking. Local support for these areas has, for the most part, not been cultivated.

What will be the fate of these protected areas, both public and private, as a growing population presses for more land and as diminishing timber resources raise the value of protected trees? Costa Rica has been a model for Latin America in setting aside land for conservation, but it has done so thus far without having to choose between livelihoods for rural citizens and conservation. As unoccupied land disappears and the population density increases, the choices become more difficult and more politically charged. Powerful national interests may be persuasive in opening access to park lands whether for mining, logging, or energy generation. Foreign-owned private reserves may be even more vulnerable to development pressures because they may not be viewed as part of the national patrimony but as valuable resources in the hands of foreigners. Several years of political maneuvering over a proposed expropriation of OTS's Wilson Botanical Garden in southern Costa Rica is a case in point. Although there is no cause for immediate alarm, such action is not outside the realm of possibility.

If the protection of conservation units, both private and public, is subject to social, economic, and political pressures, then it stands to reason that the conservation units have to demonstrate their socioeconomic value and develop a local and national constituency that believes that the protection of these areas over the long term is in their interest. The concept of a park or reserve that excludes use by humans and ignores the larger context in which it is situated is no longer viable. Protected areas necessarily involve people, especially those living in close contact with them. Support for natural resource conservation usually comes from the more affluent sectors of society and urban areas; the people with the highest potential impact on parks, however, are more often their immediate neighbors. In the tropics those neighbors are likely to be poor rural families with limited economic alternatives. If rural people become convinced that conservation is in their direct interest and that they can benefit from the wise use of natural resources, park protection is facilitated. This philosophy is incorporated in the concept of a Biosphere Reserve.

Cordillera Volcánica Central Biosphere Reserve
UNESCO, under its Man and the Biosphere (MAB) program, established international biosphere reserves as a model that

merges conservation and park protection with sustainable development. A biosphere reserve extends beyond park boundaries to incorporate local communities in a regional approach to conservation and natural resource management. This expanded concept of a conservation unit incorporates humans and manipulations by humans as part of the overall management objectives. Parks benefit people, and people, in turn, protect parks.

In 1988 UNESCO recognized the Cordillera Volcánica Central Biosphere Reserve in Costa Rica (fig. 23.5). The action plan for Biosphere Reserves from MAB outlines the reserve concept (UNESCO 1984). The function and location of each zone for the Cordillera Volcánica Central is identified.

1. *Core zone* of protected undisturbed habitat with endemism and genetic richness or unique natural features of exceptional and/or scientific interest. Basic research and monitoring are conducted in the core area to provide baseline data on natural systems. The core zone incorporates Braulio Carrillo and Irazú National Parks and parts of the La Selva Biological Station.

2. *Buffer zone* providing protection for the core area. Some manipulation of the environment is allowed, such as forestry and wildlife management activities. It includes areas suitable for experimental manipulations that have low environmental impact. The Cordillera Volcánica Central Forest Reserve forms a buffer around the upper portion of the conservation

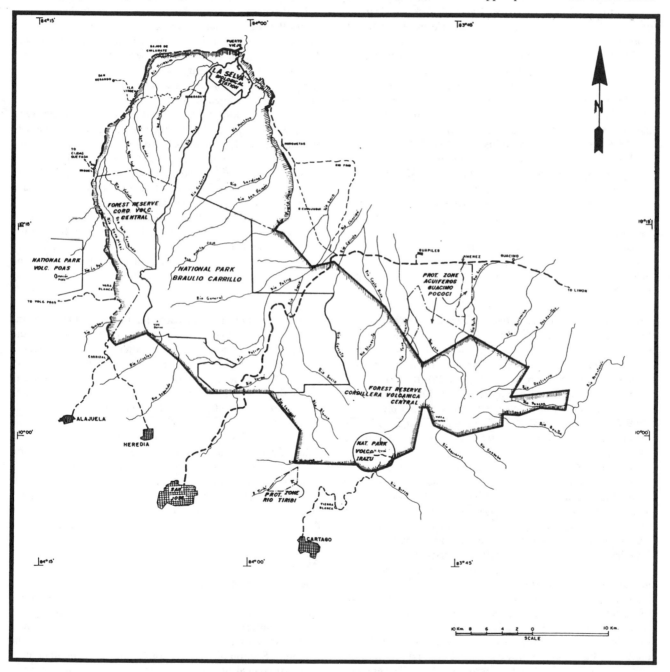

Fig. 23.5. The Cordillera Volcánica Central Biosphere Reserve, Costa Rica.

nucleus while the area around Las Horquetas that is still in primary forest is the focus of a new sustained-yield forestry project. Experimental lands at La Selva fall into this category.

3. *Restoration zone* where research and development are focused on restoring degraded landscapes to stable condition. Reforestation and reducing intensity of land use are among the methods used to reverse environmental degradation.

4. *Ecodevelopment zone* to serve as a model of long-term sustainable land-use practices. This includes traditional land-use practices and/or new crops and methods for stable agricultural development. These last two zones are found along the western boundary of the reserve (the Sarapiquí Highway), from Vara Blanca to Puerto Viejo. Lands extending from La Virgen toward Braulio Carrillo National Park are degraded pastures that need to be restored toward a more sustainable land use.

The four different land-use zones of the biosphere reserve concept are also found within La Selva, which, therefore, serves as a microcosm of the larger reserve. Within La Selva are absolute protected areas, areas of minimal environmental manipulation, and, in the new annexes, large areas for reforestation projects and for research on disturbed sites.

Since the first biosphere reserve was founded in 1976, hundreds of reserves have joined the worldwide system. Unfortunately, owing to the lack of national and UNESCO resources, many remain reserves only on paper with little progress toward achieving the program's goals. Biosphere reserve status, although internationally recognized, does not carry any legal sanctions. It is an innovative concept but one that requires international and national support and resources to become a reality. Furthermore, most biosphere reserves are simply coincident with previously established parks and do not include more than the "protected" core zone.

The Cordillera Volcánica Central Biosphere Reserve could become a model biosphere reserve because of two important factors: the location of a well-equipped, operating, research station (La Selva) within its boundaries and an endowment ($22 million, USAID) to manage the reserve with an emphasis on conservation and forestry (see chap. 25). If OTS takes an active role in the research, educational, and managerial functions of the reserve, La Selva will gain a higher profile as a contributing member of the local community and worldwide recognition for its role in the success of an international biosphere reserve.

The Man and the Biosphere Program has identified priority areas for research, including both basic and applied research within the core area and comparative research between natural and managed areas. La Selva has been outstanding in generating basic research; the baseline data already collected there and the scientific expeditions into Braulio Carrillo National Park lay the foundation for a better understanding of the ecological processes in the core area of the reserve. The area of greatest neglect in biosphere reserves, however, is the human-altered landscape. La Selva, with its recently acquired annexes of pasture and secondary forest land, is now in a position to develop research and educational programs that address this gap. Priority MAB research topics relevant to La Selva include soil biological processes, succession and regeneration, and restoration of degraded ecosystems, all of which coincide with ongoing research at La Selva.

Nevertheless, research alone does not affect land-use prac-

tices. A key to the success of the reserve will be the development of alternative sustainable land-use practices that will bring long-term benefits to neighboring communities. Stabilizing land use within the reserve and raising the standard of living for the local population decrease the necessity for marginal activities such as hunting, develop local support for the reserve, and stop the cycle of land clearing and degradation.

CONCLUSION

Without doubt, the years of ecological research conducted at La Selva have contributed to a better understanding of the natural processes of the tropical rain forest. At the same time social and economic change has led to large-scale deforestation within the region, thus affecting La Selva in a fundamental and permanent way. The old days of isolation in a jungle wilderness have given way to the increasing demands and expectations of a tropical research station located in a zone facing complex issues of land tenure and land use. Not only does the future of the rain forest depend on a broader perspective (taking account of human activities) but the future of La Selva depends on OTS's ability to become an active and contributing member of the national and regional community that forms the socioeconomic and political context of La Selva.

Since the 1950s the extensive forests of Sarapiquí have been largely converted to agriculture, primarily pasture. Much of this pasture is now being subdivided into small farms by the government land reform program. The small plots, poor soils, and occupants who lack land titles hinder the stability of these settlements. Following the model of the biosphere reserve concept, research and extension on land restoration and sustainable agricultural and forestry practices could bring long-term benefits to local residents and, indirectly, enhance park protection.

A broadening of the research facilitated by OTS into the natural resource management and agricultural fields could play a significant role in offering alternatives to La Selva's neighbors for improved land-use practices and better livelihoods. OTS has more than twenty-five years of experience in facilitating tropical research. The growth of OTS to more than fifty member institutions and the expansion of experimental lands at La Selva provide both the intellectual resources and physical facilities to take on the challenge of entering new fields of research aimed at improving land use in the tropics. La Selva is no longer hidden in the hinterlands of an isolated frontier; it has been pushed to the forefront of a dynamic region, and it has the potential, given the expertise and experience that OTS represents, to become a model of sustainable development and conservation for the humid tropics.

ACKNOWLEDGMENTS

I thank Charles Schnell, David Kauck, and Lucinda McDade for critical reviews and comments on earlier drafts of the manuscript. I also thank the following people for sharing their time and information with me: Isaías Alvarado, foreman of La Selva Biological Station; Miguel Castro, chief of the Land Inventory Section of IDA; Victor Chavarría, accountant at La Selva Biological Station; Gary Hartshorn, long-time resident of Costa Rica and researcher at La Selva; and Franco Madrigal, Sarapiquí resident and avid conservationist.

24

Agricultural Systems in the La Selva Region

Florencia Montagnini

A variety of land uses representative of the Costa Rican Atlantic lowlands can be observed in the vicinity of the La Selva Biological Station, including pastures, tree plantations, home gardens, and extensive monocultures in addition to primary and secondary forests. This diversity is amenable to comparative studies of the function of agroecosystems, including evaluation of their productivity and sustainability under local conditions.

In Costa Rica the agricultural sector accounts for approximately 17.9% of the gross national product, for 28% of total employment, and for 72.3% of export income (Banco Central de Costa Rica [BCCR], Costa Rican Central Bank, 1988). The principal agricultural exports are coffee, bananas, beef, sugar, rice, and ornamentals (BCCR 1988). Costa Rica has the ecological conditions and technical capability to produce all of its own food but has not done so because of crop preferences (e.g., farmers prefer to grow coffee and other traditional crops); low yields (e.g., cacao is often imported to satisfy local demand); high costs (e.g., weed control in rice); cheaper imports (e.g., wheat, apples, grapes); and lack of incentives.

In Costa Rica most regional development programs are promoting agricultural diversification to decrease farmers' dependence on external markets, cushion against fluctuating crop yields, and increase exports. For example, when international coffee prices are low (as in 1990), farmers who have a more diverse production system (e.g., coffee with citrus) are in a better position than those who grow coffee in monocultures. Additionally, those who have planted trees may be able to sell timber at good prices during years of low crop yields. More diverse agricultural production would make the country more independent of fluctuating international prices of export products.

The Atlantic lowlands of Costa Rica include the Atlantic and Northern regions, which are administrative divisions used by the Government Office for Planning (SEPSA 1982). The Atlantic Region is a 9,756 km² area covering the eastern portion of the Atlantic lowlands, including the whole province of Limón and part of Heredia (DGEC 1987). In Heredia the Atlantic Region includes the Horquetas District up to the Sarapiquí River (fig. 23.1, chap. 23). All of La Selva except La Guaria Annex is in this region. The Northern Region includes the Sarapiquí, Puerto Viejo and La Virgen districts near La Selva and extends northwest to include the counties of San Carlos, Guatuso, and Los Chiles, and the San Isidro de Peñas Blancas and Río Cuarto districts (SEPSA 1982). In this chapter I refer to the Atlantic lowlands as a whole and emphasize the Atlantic Region near La Selva (the watershed of the Sarapiquí River and the Horquetas District in Heredia Province).

The Atlantic Region is 19.1% of the total surface of Costa Rica, but the area in farms (owned by farmers or agricultural enterprises) comprises only 9% of the country's total, which is the lowest among the country's regions. Of the total area in farms 48% is not under active cultivation and remains under forest cover; only 49.4% is in agricultural use, again the lowest value for the entire country (Flores Silva 1987). The region has until recently been so strongly dedicated to banana and cacao cultivation that other crops have received little attention. In the 1990s the region produces more than 70% of the country's total production of bananas along with 62.5% of coconuts and 20.5% of plantains but only 4% of basic grains and 5% of beef cattle. The region generates a large proportion of foreign income because bananas are grown mostly for export (Flores Silva 1987). The level of economic development in the region, however, does not reflect its significance for the country's export income. Agricultural diversification could contribute to the development of a stronger regional economy, offering options beyond employment generated by the banana companies.

In the Atlantic Region agricultural diversification has already begun to respond to the opening of new markets and to improved transportation. The paving of the road between San José and Puerto Viejo de Sarapiquí, which was completed in early 1986, significantly stimulated agriculture, forestry, and other economic activities in the region. Diversification has involved expansion of macadamia, citrus, fish culture, dwarf coconut, and other crops (Flores Silva 1987). Many of these species were chosen based on market value and presumed adaptability to the climatic conditions of the lowlands but without much previous in situ evaluation or experimentation with management practices. Most of these activities are too recent for evaluation of their economic success or ecological impact. If the region is to play a more significant role in the whole country's development, much more needs to be done to develop and implement economically productive and ecologically sustainable agricultural and forestry systems.

A number of local and international institutions such as Ministerio de Agricultura ([MAG], Ministry of Agriculture), Dirección General Forestal ([DGF] Forest Service), Junta de Administración Portuaria y de Desarrollo Económico de la Vertiente Atlántica ([JAPDEVA] Council for Harbour Administration and Economic Development of the Atlantic Lowlands), Instituto de Desarrollo Agrario ([IDA] Institute for

Agrarian Development), University of Costa Rica (UCR), Centro Agronómico Tropical de Investigación y Enseñanza ([CATIE] Tropical Agriculture Research and Training Center), Escuela Agropecuaria para la Región Tropical Húmeda ([EARTH] College of Agriculture for Humid Tropical Regions) are operating in the Atlantic lowlands. Although their objectives differ, the overall goal is to develop alternative, ecologically sound land use systems. La Selva has begun to take part in this effort, but it can have a much more significant role both in evaluating current practices and in making scientifically based recommendations for better land-use practices.

In this chapter I survey traditional and more innovative land-use systems in the La Selva region, and emphasize opportunities for improvements, summarize Organization for Tropical Studies (OTS) involvement in research on the ecological basis of sustainable and productive land-use systems, and present recommendations for future agroecological research in the region.

ECOLOGICAL CONDITIONS OF THE ATLANTIC REGION

La Selva Biological Station is in the Atlantic Region of Costa Rica at its western border with the Northern Region. Except for the northern portion of the Talamanca mountain range, a small portion of the Cordillera Central, and small, extinct volcanoes, the region lies below 100 m elevation. The average slope is 6° (11%) in the upper parts, and 1°–2° (1%–3%) in the lower areas. Because of the flat terrain, swamps and areas with poor drainage are common. The general geology consists principally of recent alluvial deposits and lahars (Madrigal and Rojas 1980). The indigenous vegetation in the region was tropical rain forest.

The Atlantic lowlands are crossed by the only rivers in Costa Rica that are navigable for any distance from the coast: the San Juan and its tributaries, the San Carlos, and the Sarapiquí. Because of the influence of the trade winds and strong convection forces, the region receives abundant precipitation, with annual means from 2,000 to 9,000 mm. In many years there is no pronounced dry season, and farmers generally need to adapt to an excess rather than to a scarcity of water. Heavy rainfall and heat made the area inhospitable to European settlers, and significant settlement occurred only in the twentieth century.

In a general soil fertility survey of Costa Rica the Sarapiquí District was diagnosed as having problems of low fertility because of low pH (<5.5), high aluminum saturation (10%–50%), low cation exchange capacity (<5 meq/100g), low calcium (<4 meq/100 g), low magnesium (<1 meq/100 g), and low extractable phosphorous (<10 μg/g) (Bertsch 1986). Levels of micronutrients and potassium were considered adequate for most agricultural crops. Nitrogen is not generally included as part of routine soil tests because levels of soil total nitrogen are not a good indicator of nitrogen availability, and nitrogen demand by crops is usually measured with fertilizer experiments. Nitrogen fertilizers are heavily used in agriculture in the Atlantic lowlands, indicating that low nitrogen may be an additional factor in the region's generally low fertility (E. Bornemisza pers. comm.). Soils of the Sarapiquí area, thus, have low fertility and potential problems with acidity and high aluminum saturation. Fifty percent of the Río Frío region has been placed in class 3 land-use capability (Tosi 1972), which corresponds to conditions of moderate to low soil fertility, strong acidity, and moderate to poor drainage. These characteristics restrict the possibilities for crop species (UCR 1984). Research on soils at La Selva is summarized in chapter four. La Selva's soils share many features with other soils of the Costa Rican Atlantic lowlands as well as with other regions of the lowland humid tropics; thus, soils research at La Selva has broad regional as well as local significance.

TRADITIONAL LAND USE IN THE ATLANTIC LOWLANDS

The present land-use pattern in the Atlantic Region is complex and is not simply related to the ecological conditions that favor each activity. Land-use patterns can be better understood by examining the history of agricultural colonization in the region as well as the socioeconomic and ecological factors influencing agricultural production. In this section I examine the traditional land uses in pre-Hispanic times and the changes resulting from colonization by Hispano-American settlers in the late-1800s to 1900s.

Pre-Columbian Period

Archaeological investigations carried out in the Atlantic lowlands reveal a cultural complex similar to that in northern South America with dietary habits and ceramics most like those of people inhabiting the coastal Caribbean plains of Colombia (Hall 1984). It is likely that a single humid tropical forest culture extended from the Amazon and Orinoco river basins along the Caribbean coast to Honduras (Snarskis 1975, 1976).

Agricultural practices in the Atlantic lowlands in pre-Columbian times apparently had a very strong South American influence: most of the crops grown in the Atlantic lowlands were native to tropical forest zones in the northern part of South America (Hall 1984). These include tuber crops, such as cassava (*Manihot esculenta*) and tiquisque (*Xanthosoma violaceum*). Cultivation of pejibaye (peach palm, *Bactris gasipaes*) spread from the Orinoco valley. Cacao (*Theobroma cacao*) was also apparently introduced from South America, and its pods were used in religious ceremonies in the Amazon and Orinoco river basins (Stone 1977). It was generally believed that corn was introduced to the region from Mexico and Guatemala. Corn with South American characteristics, dated A.D. 100–300, however, has been discovered in excavations in the Atlantic lowlands (Snarskis 1975), suggesting that the variety grown here was of South American origin. Storage of corn must have been a problem because of the very humid climate of the region. In contrast, root tubers such as cassava produce year-round and store well in the ground, from which they can be harvested as needed.

During pre-Columbian times land-use patterns allowed for the regeneration of most forests. The relatively low population density permitted long periods of forest regeneration following slash and burn agriculture (Hall 1984). It is only during the twentieth century that settlers of European descent accelerated the deforestation process.

Agricultural Colonization

Settlement of the Atlantic lowlands resulted from three principal phenomena: spontaneous migrations from other regions of the country; the establishment of banana plantations by U.S. companies; and government colonization policies such as the establishment of prison colonies in the 1950s to 1960s and the more recent settlement projects of IDA. These different sources of colonization have determined land-use and tenure systems, agricultural practices, and the market economies of the region. These historical patterns also influence the kinds of changes that are feasible as well as the mechanisms appropriate for implementing change.

The Early Settlements: Subsistence Agriculture. After three centuries of European settlements largely restricted to the Central Valley, colonists began moving out in all directions about one hundred years ago. These early settlers were few and faced isolation from the Central Valley because of rugged mountains and a nearly total lack of roads and marketing potential for their products (Sandner 1961).

During this period forest cutting by individual pioneers or small groups of settlers with machetes and axes removed the largest trees, leaving stumps and understory seedlings to dry for burning. Commercial wood exploitation was very limited because of inadequate transportation and the absence of sawmills, markets, and technologies suited for tropical forest species. Small amounts of wood were used by settlers for construction and for firewood; a few stems were sold to local industries for furniture and construction. Commercially undesirable species were burned or left to decompose (Tosi 1971, 1974). Sandner (1959) estimated that 60%–80% of the area deforested between 1860 and 1960 was initially used for shifting agriculture. Settlers could buy public lands at low prices and obtain property title once the land was under cultivation. In practice, however, many farmers settled without claiming property titles (Hall 1984).

The small subsistence farms operated very much like indigenous agriculture. After two to three years of cultivation, fields were abandoned and new areas were cleared. On subsistence farms, traditional home gardens were planted to meet household needs. These gardens grew basic foodstuffs: rice, corn, plantains, beans, "robusta" coffee (adapted to low elevations), sugar cane, and various fruits and vegetables. These small farms also had chickens, pigs, and, frequently, a few cows. Some of the products were occasionally marketed to obtain cash for other needs. With relatively low population pressure, shifting agriculture was a viable alternative for the early colonizers. In many areas of the Atlantic lowlands, however, ecological conditions were not suitable for agriculture; colonists found that yields were too low for sustained cultivation, and settlers often sold their land to speculators and moved to other areas (Lambert 1969).

The Beginnings of Extensive Cattle Ranching. Large holdings in the area were acquired in most cases through buy-outs of small settlers as described. On large properties extensive beef cattle ranching was the most feasible land use before the development of modern transportation (Sandner 1961) because live steers could be transported to markets without significant deterioration. Efforts to establish plantations in in-

accessible regions such as the Sarapiquí valley often failed because of lack of railroads or paved roads for transportation of relatively fragile agricultural products to potential markets (León 1943).

The initiation of beef exports to United States markets in 1957 provided a major new economic incentive to the cattle industry in Costa Rica. Ranchers began to direct their efforts toward the colonization zones, and in one decade (1963–1973) the area in pasture in Costa Rica increased 62% (Parsons 1983). The center for ranching in Costa Rica traditionally has been Guanacaste Province because of its long dry season. With strong economic incentives after 1957 landscapes in Valle del General and Coto Brus, Turrúbares, Puriscal, Parrita, San Carlos, Sarapiquí, and Arenal became increasingly dominated by cattle pastures. The more humid areas of the Atlantic lowlands were spared from cattle encroachment, and only in the mid-1960s were cattle ranches established in this area.

In the late 1950s many farmers in the Atlantic lowlands planted rice on the flat areas of their farms. This "rice fever" lasted until the mid-1960s, when most farmers stopped growing rice because of the difficulty in controlling weeds after two to three years of cultivation. During the "rice fever" areas not suited for rice were dedicated to beef cattle. With higher beef prices in the mid-1960s ranching expanded to occupy the old rice fields.

Although these conditions promoted the establishment of ranches in the Atlantic lowlands, the region is not well-suited ecologically for extensive cattle raising, and management practices were inadequate. There were very few permanent workers on cattle ranches, and displaced farmers made up most of the labor force. Natural grasses were very unproductive, and exotic pasture species with higher yields were introduced only about fifteen years ago. As a result, productivity of cattle ranching was very low and extensive areas were degraded by inappropriate management practices and overgrazing (Parsons 1983).

Plantation Agriculture: The Beginnings of High-input Commercial Production. In the late nineteenth and early twentieth centuries, a number of plantations with similar characteristics were established in the Costa Rican Atlantic Region as well as in several other areas of the lowland humid Latin American tropics. Extensive virgin lands were then still available for agriculture, and governments exerted little control over the activities of multinational companies (Casey 1979). In the Atlantic lowlands of Costa Rica, banana plantations were started by the United Fruit Company (United Brands) in the Limón area in the late 1870s. Limón was the principal region for banana cultivation until the mid-1930s. Banana production declined after 1913 because of low prices, high incidence of fungal diseases (Sigatoka and Panama diseases caused by *Micosphaerella musicola* and *Fusarium oxysporum,* respectively), soil degradation, and labor unrest. By 1942 banana exports from Limón had ceased, and the region underwent severe economic depression. Alternative crops were planted (cacao, rubber, subsistence food crops) in the 1940s and early 1950s.

In the early 1950s considerable migration from other regions of Costa Rica to the Atlantic lowlands took place with settlers occupying new land as well as areas abandoned by

United Fruit. As the railroad system was expanded to Río Frío and Estrella, Standard Fruit Company, a subsidiary of Dole Fresh Fruit Company, installed banana plantations in 1956 in the areas near Limón previously occupied by United Fruit as well as in the Río Frío zone just east of La Selva. In the 1990s Standard Fruit expanded its operations to areas northeast of La Selva along the Río Sucio. United Fruit has moved its operations to the Pacific Coast where it grows mostly African oil palm although recently it has come back to the Atlantic Region to reinitiate banana production. In the mid-1960s the Costa Rican government gave incentives for banana production ("crédito bananero") to stimulate the establishment of plantations by local companies. By 1979 Costa Rican companies produced approximately 60% of bananas in the Atlantic Region although they still market their produce through Standard Fruit.

The banana companies established their plantations on the most fertile, well-drained riparian soils. In the early days of banana plantations (late 1800s) with intensive land use and inappropriate soil management these good soils deteriorated after seven to ten years of monoculture and had to be fallowed for about ten years before they could be put back into banana cultivation (Casey 1979). Immigrants from Jamaica and Costa Rican migrants from other regions of Costa Rica worked in the plantations. Independent banana producers also grew bananas on farms ranging from less than 10 ha to more than 20,000 ha and sold their produce to United Fruit (Casey 1979). Local farmers also had home gardens with subsistence crops, apart from bananas. Jamaican immigrants introduced Afro-Caribbean subsistence production systems, including polycultures with bread fruit, citrus, plantains, cassava, yams, and sweet potatoes for food and cacao and coconuts as cash crops. While the banana industry flourished, however, subsistence agriculture did not meet local demand for foodstuffs; most of the products for consumption by banana workers and local farmers were either imported or brought from other regions of the country (Jones and Morrison 1952). This was beneficial for the banana companies, who sold those products in their own stores.

In contrast to areas colonized by individual or group settlers, plantations were established from the start as highly specialized commercial agriculture. Standard Fruit Company grows "grand Cavendish" and "dwarf Cavendish" varieties of bananas, which are exported to the United States and Europe through Limón. Bananas are an ecologically demanding species; they require high temperature, high humidity, and abundant soil nutrients. They are a typical large-scale plantation crop because of high risk of crop failures from natural catastrophes and the high costs of site preparation, drainage, and labor administration. Banana cultivation requires intense use of agrochemicals: fertilizers are applied monthly (N, K, Mg, and micronutrients), and nematocides are applied four times per year. Fungicides and insecticides are also applied frequently. Fungicide applications are aerial; fertilizer and nematocide applications are manual. Some areas around Río Frío were abandoned by Standard Fruit because the high incidence of fungal diseases and nematode damage required such intensive application of agrochemicals that banana cultivation was no longer profitable.

In contrast to bananas, cacao has played a minor role in the development of commercial plantations in Costa Rica. In the Atlantic lowlands cacao was a major export crop in the 1940s and early 1950s (between the banana boom periods) as in the nineteenth century. With the reestablishment of the banana industry, however, cacao took a secondary position. In fact, cacao does best in areas with an extended dry season and the Atlantic lowlands are not ideal for the crop because of high annual rainfall and lack of a well-defined dry season. The strong rains and high humidity favor the spread of fungal diseases (e.g. *Monilia roreri*), delay or impede ripening of fruits, and complicate their collection and processing. The northern plains (especially Upala) are ecologically more appropriate.

In the Atlantic Region of Costa Rica cacao is mostly grown on farms managed by individuals or families with far less advanced technology than is used in bananas. In the 1990s in the Atlantic lowlands most cacao is grown on land abandoned from banana plantations or in remnants of forest. Cacao has a highly fluctuating price on the international market, and interest in its cultivation follows these price cycles. Because of high risks of crop failure and fluctuating prices, the Costa Rican Ministry of Agriculture recommends that farmers who grow cacao should diversify their production with coconuts, tuber crops (cassava, tiquisque), and spices (black pepper, ginger).

Plantation agriculture had and still has a strong socioeconomic impact in Costa Rica. As foreign companies converted land, installed drainage and transportation systems, and built ports and living facilities, they contributed to development of large areas. Until recently, however, most revenue from these enterprises left the country, yielding little benefit to the Costa Rican economy (Casey 1979). The establishment of plantations resulted in a rapid transformation of the forested landscape into fruit monocultures, but these also declined very rapidly when the banana industry was hit by fungal diseases and economic depressions. Hundreds of hectares of secondary forest developed in the areas abandoned by United Fruit in the early 1900s.

Despite past and present problems, bananas are still seen as a key crop for regional economic development owing to relatively favorable conditions in the Atlantic lowlands and the employment opportunities offered. In the 1990s the banana companies contribute much more to the local and national economy than in the past. They hire a higher proportion of local technical and administrative personnel, and the government imposes a tax on exported bananas and on company revenues (Gaceta Oficial, Government Public Document, 1978).

In the late-1950s to 1980s the construction of good roads eliminated the principal barrier to development of commercial agriculture in the Atlantic lowlands. An ongoing secondary colonization process began on land already partially deforested and populated (see chap. 23).

LAND USE IN THE ATLANTIC REGION AND FUTURE PROSPECTS

Land-use patterns in the Atlantic lowlands of Costa Rica changed over the 1980s in response to increased population, the improvement of roads, and a general trend in the country to diversify agricultural production. These changes, however, have often outstripped technical expertise and the capacity of agricultural extension agents. For example, technical problems in cacao management have restricted its expansion while enthusiasm for the cultivation of a promising ornamental, *Dracaena* spp., oversaturated the market. Thus, the transition from subsistence agriculture and extensive cattle raising to more intensive land-use systems is not complete: large areas (up to 70% of the area in farms in many districts) are still ranches. In spite of low productivity per hectare a cultural predilection for ranching sustains the practice. In many cases land is deforested and fenced, and cattle are installed just to claim the land. Additionally, lack of adequate markets and infrastructure has limited the production of food crops for local consumption (González Vega et al. 1970). In spite of better communications and more and better roads, the market structure is still not adequate to foster the production of many crops for local consumption or for export.

Land-use patterns in the Atlantic lowlands are changing continuously in response to changes in markets, agricultural technology, and economic policies. Between 1963 and 1982 the area in farms tripled, and the area in pastures increased severalfold while the forested area decreased sharply (table 24.1). Among agricultural crops, perennial crops predominate, mostly reflecting the large areas in bananas (table 24.1). I present the principal agricultural practices in the Atlantic Region, their main constraints, and potential for improvement next.

Agriculture

In the general statistics for the most commonly grown crops in the Atlantic Region presented in table 24.2 the area in cacao may be an overestimate because many plantations have been abandoned owing to fungal diseases (*Monilia*). The most frequent annual crops comprise a mix of subsistence and commercial crops (table 24.2). The data in table 24.2 are for the entire Atlantic Region, including farms of all sizes. Most farms in the region are small (table 24.3): 68% of farms are smaller than 20 ha and 19% are smaller than 4 ha. Farms less than 20 ha, however, cover only 16% of the total area. Medium (20–200 ha) and large farms (>200 ha) dominate. These areas are primarily devoted to bananas.

A different picture of crop preferences emerges if small farms are examined separately. In a survey of recently established IDA settlements (farms <20 ha) in the Río Frío region

Table 24.2 Principal crops in the Atlantic Region, 1973

Crops	Total Area (ha)	Farms (No.)
Perennial		
Bananas	20,698	802
Cacao	17,224	1,935
Plantains	1,551	664
Coconuts	940	781
Coffee	485	343
Annual		
Corn	5,245	1,532
Rice	753	474
Cassava	566	553
Beans	171	189
Sugarcane	146	122

Source: DGEC 1974, in Van Sluys et al. (1989).

Table 24.3 Farm size distribution in the Atlantic Region, 1984

Farm Size (ha)	Distribution				
	Number	Percentage of Total	Area (ha)	Percentage of Total	Mean Size (ha)
<4	1,754	19	3,400	1	2
4–20	4,445	49	43,000	15	10
21–200	2,577	29	125,100	44	49
>200	252	3	286,200	40	455
Total	9,028	100	286,200	100	32

Source: Preliminary results of 1984 Agricultural Census in Van Sluys et al. 1989.

near La Selva (UCR 1984) 63% of the land owned by these farmers was under cultivation; the rest was unused principally because of inadequate economic resources (table 24.4). Of the land under cultivation 39% was planted to annual crops and 23% to perennials, and 38% was used for cattle. Small farmers grew a mix of annual and perennial subsistence crops, and more than half of these farmers combined cattle with crops. Cattle were raised for market and local consumption. Although most crops were grown in monoculture, some examples of crop combinations were cacao and cassava, corn and beans, plantains and corn, plantains and beans, and plantains with cassava and cacao.

Perennial Crops. *Bananas* are largely produced and exclusively marketed by foreign companies; the Costa Rican companies sell their produce through Standard Fruit. The Costa Rican Association for Banana Production (ASBANA) supports local producers with technical research and extension. As noted, bananas are not a good crop for small farmers owing to high production costs. Commercial banana plantations, however, are expanding in the region.

Plantains are grown by small farmers for home consumption, whereas the produce of midsized farms is sold to local markets, principally San José; small amounts are exported. In general, both inputs and productivity of plantains are lower than bananas. In the early-1980s plantains began to be af-

Table 24.1 Land use in the Atlantic Region, 1963–1982

Land-use Category	Total Hectares Covered		
	1963	1973	1982
Annual crops	—	ca. 76,000	19,000
Perennial crops	39,100	44,400	78,300
Pasture	35,000	71,800	232,900
Area in farms	205,200	140,800	—
Nonagricultural land (mostly forest)	767,500	728,200	352,500

Sources: Dirección General de Estadísticas y Censos ([DGEC] National Office of Statistics and Census) 1974 and 1975; Unidad Regional de Asistencia Técnica ([IDA/RUTA] Reginal Unit for Technical Assistance) 1984 in Van Sluys et al. 1989.

Table 24.4 Frequency of crops in the Río Frío region

Crop	Cultivated Area[a] (% of total)	Farmers[b] (% of total)
Corn	20	68
Plantains	18	38
Cacao	14	30
Beans	8	64
Cassava	7	57
Aroids	7	45
Pineapple	7	30
Rice	6	30
Bananas	5	11
Pejibaye	4	11
Coffee	2	11
Fruit crops	2	8

Source: Univerity of Costa Rica 1984.
[a]Percentage of total area cultivated by small farmers.
[b]Percentage of farmers who grow them.

fected by the Sigatoka disease, and cultivation was drastically reduced (Van Sluys et al. 1989).

Cacao was the principal crop on small and midsized farms, especially in the southern part of the Atlantic Region, until 1978 when the *Monilia roreri* fungus became a serious problem. *Monilia* attacks only the pods; it can be controlled by planting resistant hybrids, eliminating the infected fruits, and reducing shade (intense shade creates a humid microclimate that favors the spread of the fungus). These measures are costly, however, and many plantations were instead abandoned. In the 1980s, there was renewed interest in cacao cultivation and new government incentives (MIDEPLAN 1984).

Coconuts are grown for oil and fresh fruit. In the Río Frío area coconuts are grown for local consumption. Most coconut plantations are along the Caribbean coast north and south of Limón. Plantations of dwarf coconuts can be found elsewhere, but these are generally small and show low productivity (Rojas 1978).

Coffee is cultivated primarily near Volcán Turrialba and along the Siquirres-Turrialba road. Most areas of the Atlantic Region are too low in elevation for optimal productivity of coffee. In lower areas conditions are better suited for other crops such as macadamia.

Macadamia and pejibaye are new to the Atlantic Region, and both are promising. At present, they are attractive only to large-scale farmers who have access to export markets. There is, however, a small national market for pejibaye fruit and heart of palm.

Annual Crops. *Corn* is a subsistence crop grown by small farmers throughout the Atlantic Region. On midsized farms, especially near Guácimo and Cariari, it is an important crop. The produce is sold to Consejo Nacional de Producción, ([CNP] National Production Council) at subsidized prices. Corn is not very profitable, apparently because of the high costs of chemical weeding, insect control, tillage, harvesting, and transportation.

Rice is grown as a commercial crop on midsized farms, especially in the areas around Batáan-Matina (NNW of

Limón). Its production is highly mechanized and costly as a result.

Cassava is grown by small farmers for subsistence and on midsized farms for the export market, especially to the United States.

Most Promising Crops for Agricultural Expansion. Among the more innovative crops the cultivation of aroids with edible tubers, including white tiquisque (*Xanthosoma sagitifolium*), red tiquisque (*X. violaceum*), and malanga (*Colocasia esculenta*) is very promising for export (especially to the United States where demand for them is increasing among the growing Hispano-American population). These crops are well adapted to the region, but they are highly susceptible to viral diseases. Because crops are propagated vegetatively, viral diseases are easily spread when new plantings are established. Investigators at the University of Costa Rica (UCR) are working to produce virus-resistant strains.

Ornamentals, principally foliage, ferns, itabo (*Yucca elephantipes*), and others are grown mainly for export to the United States; their cultivation is expanding after the opening of additional export markets (Japan, Europe). Ornamentals require intensive management and are, thus, an important source of employment.

At the UCR experimental farm in Río Frío, the adzuko bean (*Vigna angularis*) is under study; it has edible seeds and good yields, can be grown year-round and is disease tolerant. The lack of acceptance by consumers accustomed to black beans may limit its expansion.

Peach palm (pejibaye, *Bactris gasipaes*) cultivation for heart of palm was begun by Industrias de Desarrollo Agropecuario ([INDACO] Institute of Agricultural Development) for local industry and for export. This company at present absorbs most of the local crop. The species is well suited to the Atlantic Region and maybe an interesting alternative for cash income as the market expands.

Among fruit trees, citrus appears most promising. Cultivation of oranges for export to the United States is expanding, and a juice-processing plant (for oranges, pineapple, and other fruits) will soon be operating in the Sarapiquí area, built by Tico-Fruiti, a private company with Florida producers as partners.

The cultivation of bamboo (*Bambusa guadua*) for home construction and furniture is being promoted in the Atlantic Region through a project recently begun with financing from the Dutch government and the United Nations Development Program and technical assistance from IDA, DGF, and MAG.

Other crops that have been introduced or improved by UCR are fruit trees such as mamón chino (*Litchi chinensis*), carambola (*Averrhoa carambola*), water apple (*Eugenia malaccensis*), messina lemon (*Citrus aurantifolia*), and grapefruit (*Citrus grandis*). These should be well suited to the region, and they merit exploration. Examples of other innovative crops include chilis (*Capsicum* spp.), passion fruit (*Passiflora* spp.), papaya (*Carica papaya*), cardamom (*Elletaria cardamomo*), achiote (*Bixa orellana*), and medicinal plants mainly for the Costa Rican market for herb teas.

Ecological Constraints on Agricultural Production. Agricultural production in the Atlantic Region has a low yield/

cost ratio owing to problems with weed control, pests and diseases, soil drainage and fertility, and labor administration. The ecological problems are aggravated by the high costs of agricultural inputs, lack of credit, and high interest rates (Van Sluys et al. 1989).

Although the climate is relatively uniform, occasional dry periods or excessive rains may impede growth or ripening of crops. In some years, as in 1985, hurricanes may considerably damage bananas (Van Sluys et al. 1989). Drainage problems occur in large areas, limiting crop choices, increasing costs, and interfering with management practices.

Weeds are a major problem for annual crops; as mentioned, high costs of weed control in rice ended the "rice fever" in the Sarapiquí valley in the mid-1960s. Weed control is labor intensive and involves the use of toxic and expensive herbicides that may contaminate soil. Pests and diseases cause considerable losses (e.g., rotting of corn and cacao) and may increase production costs (e.g., *Monilia* control in cacao). Problems may be serious enough to discourage the cultivation of certain crops as was the case with Panama disease in bananas and *Monilia* in cacao. A few management recommendations exist to solve pest and disease problems, including selection of more favorable areas (drier areas for plantains), manipulating the microclimate (reducing shade in cacao plantations), and timing production (planting corn so that it ripens during a relatively dry period).

Management problems are related to soil characteristics (e.g., bad drainage, steep slopes) and climate (e.g., excessive rains). Problems of low soil fertility will probably increase with increasing pressure on the land and lack of appropriate practices to maintain soil fertility (Van Sluys et al. 1989).

Infrastructure and Institutional Constraints. Constraints include financing, access to agricultural inputs, availability and transfer of agricultural technology, and physical access to markets (Van Sluys et al. 1989). Many small farmers have little access to credit. Land titles are needed to obtain credit, and interest rates are high. The establishment of perennial crops, in particular, requires high initial investments and many years of interest payments before any benefits accrue. Labor costs are generally high because the large banana companies in the region pay relatively good salaries. Many farmers need to work outside the farm to supplement their incomes (Van Sluys et al. 1989).

Both farmers and government employees in charge of agricultural extension point to a lack of technical information on agricultural crops. Most of the available information comes from outside the region, which means that local experience with many crops is very limited. The foreign banana companies have developed their own technology, but they are not willing to share it with local producers. Even when local technology exists, its transfer is often inadequate. For example, as mentioned, the technology to control *Monilia* disease in cacao consists of planting resistant hybrids, pruning trees, reducing shade, and eliminating infected pods. Many farmers have heard about these techniques, a few know how to apply them, but the majority are not using these recommendations. Apparently, both research and extension are not effectively reaching farmers (Van Sluys et al. 1989).

Access to good markets is another constraint to agricultural production. For some products, such as corn and rice, CNP guarantees the purchasing price, but it is not certain how long this policy can be sustained by the government. For other products (e.g., plantains) there is a free market. The market for export crops, such as bananas, roots and tubers, ornamentals, and macadamia, is controlled by foreign companies, which generally have foreign headquarters. Small farmers, then, have access to these markets only through foreign companies, if at all.

Farmers need to improve this situation and have the potential to do so, but few farmers have access to capital, inputs, technology, and markets. The potential of alternative crops such as ornamentals, cardamom, macadamia, and pejibaye will not be realized as long as these constraints remain.

Livestock
Of the large increases in pasture areas in the Atlantic Region since the 1970s (table 24.1) only a small portion is used for intensive milk production. Most is used for extensive beef production with productivity values per hectare lower than many crops. The principal farming systems that include cattle are

Farms with dual purpose cattle (meat and milk) producing milk for consumption on the farm. Most such farms are small and also grow annual subsistence crops (corn, beans), as well as some perennial crops (cacao, plantains). Some may also have pigs and poultry. Sometimes part of the family's income is from sources outside the farm. Generally, productivity is very low; lack of cash and credit are the principal constraints to improvement.

Farms with dual purpose cattle producing milk for market. Dairy products are marketed locally and can generate a regular income that can be invested in improved pastures and stock and mineral supplements for cows. The most common races of cattle are Brahman, Indo-Brasil, Holstein-Frisian, and Jersey.

Specialized dairy farms. Generally, these are larger farms (up to 100 ha), with modern dairies and substantial investment in pastures, feed supplements, and veterinary care for cows. Products are generally sold commercially through large dairy companies (Borden, Dos Pinos).

Beef cattle ranches. These ranches either raise calves for sale to other ranches at about 150 kg, or eight months, or buy calves at this age and raise them to 500 kg (2–2.5 more years). Both operations occasionally take place on the same farm. Ranching at this scale requires improved pastures on fertile, well-drained soils. Such ranches often occupy recently cleared land in areas where land is still cheap.

Farms with pigs. A few specialized farms raise pigs, using banana residues and waste from dairy farms as feed.

Among the IDA farmers of Río Frío raising dairy cows for milk and cheese production generally occupies more than one-third of the area in farms, both at the subsistence level (fewer than four cows) and at the commercial level. Dairy farming has increased because many farmers in the IDA settlements came from the San Carlos valley (Northern Region in the Atlantic lowlands) and are familiar with this activity.

Dairy farming is seen as a promising activity in spite of its requirements: technological knowledge, high investments, and reliable markets (UCR 1984). CATIE and IDA promoted dairy farming in the region through credits, technical advice, and marketing facilities offered by the CATIE Dairy Farming Program in the 1970s to early 1980s; CATIE has ceased these activities.

Despite institutional efforts to promote technological improvements in dairy farming innovations are still not widespread among farmers. Mostly native grasses are used in pastures although they support only low production. Introduced grasses, such as *Cynodon nlemfuensis, Brachiaria decumbens,* and *Pennisetum purpureum,* are considered by UCR researchers to be most promising in yield and nutritional value for dairy cattle in the region.

In contrast to dairy cattle farming beef production is an extensive and almost exclusively commercial activity. Most beef cattle in the region are range fed, but even the more productive exotic pasture grass species are low in protein and fiber (Parsons 1983). Because of this dependence on pastures, most ranches are large and the mean stocking rate is only one animal per hectare. As a result, ranching generates few employment opportunities and promotes migration to cities. The ecological impacts of expansion of beef cattle ranching have been incalculable, including serious problems of soil deterioration (Parsons 1976). Excessive grazing may result in soil compaction, which exacerbates drainage problems and leads to soil erosion on steep slopes. With poor management productivity is low and after a few years new areas are cleared and put into pasture. Forest regeneration on abandoned pastures is often slow because of soil deterioration and distance from seed sources in large cleared areas.

Livestock Management Problems. The principal technical problems of dairy and beef cattle raising are high production costs that are not compensated by high prices, dependence on grasses of low productivity and quality, little integration with agriculture (e.g., crop residues and manure are underused as feed and fertilizer, respectively), and high incidence of cattle diseases, which are often related to the hot, humid climate and muddy pastures (Van Sluys et al. 1989).

Institutional Problems with Cattle Production. The problems encountered in livestock production mirror those encountered in agriculture (Van Sluys et al. 1989). Extension and veterinary services are few and, therefore, ranchers find little useful advice. Ranchers have little control of market institutions, which have been poorly organized (e.g., until recently, beef cattle were sold in Alajuela, and small farmers generally had to sell to middlemen). Large producers depend on the international market and confront fluctuating prices for their products. Whether selling to middlemen or directly to dairy companies, farmers must organize transportation themselves. There is little opportunity to expand the national market for dairy products because most such products are too expensive for the majority of the population to buy.

Dairy and beef cattle ranches can be most feasibly improved by intensifying management rather than by expanding onto marginal lands. Key strategies are improving pastures by planting more productive fodder species, using rotational grazing practices to avoid overgrazing and decrease risks of soil compaction and nutrient depletion, and adopting agrosilvopastoral systems that include fodder trees in pastures and in living fences. CATIE's livestock project is experimenting with these techniques near Guápiles and at the Los Diamantes Experiment Station. These more intensive practices can now be seen on a few farms in the vicinity of La Selva. If these practices are financially feasible (i.e., if increased productivity pays for the additional costs of intensive management), they may represent viable alternatives for management of ranches in the region.

FUTURE DIRECTIONS IN AGROECOSYSTEM RESEARCH IN THE ATLANTIC REGION

Apparently, there is still no clear consensus as to the most profitable agricultural activities in the Atlantic Region. Agriculture is developing quickly and in different directions depending on the initiatives of individual land owners, investors in commercial agriculture and forestry, and the national and foreign institutions dedicated to research, education, and development. Many production systems are economically and ecologically sound (e.g., cultivation of pejibaye, use of living fences in farms, intensive dairy farming). Other systems have problems but are still practiced because they are traditional in the region (e.g., cacao and rice production, extensive cattle raising). In many cases the availability of capital is the principal factor influencing the choice of a system (e.g., bananas, macadamia, and large-scale production of ornamentals require considerable capital investment). Farmers are experimenting with production systems according to their own interests (e.g., with techniques for growing macadamia). National and international institutions (MAG, UCR, CATIE) are investigating and promoting production systems for specific sectors (generally, the small farmers of the region).

Any organized effort to promote sound agricultural development should start by dividing the region into subregions of relatively homogeneous ecological and socioeconomic conditions (e.g., Horquetas, Río Frío, Guápiles, the Caribbean coast south or north of Limón). Within these subregions the sectors to which efforts are directed need to be clearly defined (small-, medium-, or large-scale farmers). Land-use capability must be assessed for each subregion, considering slope, soils, and other production factors to determine appropriate land uses (intensive agriculture, cattle, forestry, or agroforestry). Finally, research and extension should begin by examining existing ecologically and economically sound practices and should concentrate on improving them and on designing and promoting new systems if needed.

Agricultural diversification should supply farmers' needs for subsistence crops and cash by promoting cultivation of home gardens and of cash crops for local markets and for export. Agricultural research should focus on crops that are (or are expected to be) most profitable or those that contribute strongly to subsistence nutrition. In a small country such as Costa Rica it is possible to saturate markets quickly, with negative results for otherwise very promising crops such as passion fruit and chilis. Research should be concentrated on crops that are accepted by farmers, such as pejibaye and corn; those that are ecologically appropriate (macadamia, cardamom, pejibaye); those with excellent market potential even though they may have some technical problems (rice, plantains); and the traditional subsistence crops (cassava, beans, bread fruit, tiquisque, fruit trees, among many).

Agricultural diversification at both farmer and regional levels can also be attained by using agroforestry practices, ideally, combining subsistence crops with more marketable or export species. Agroforestry systems may successfully combine timber trees with cacao, rubber, coconuts, and ornamentals. Many annual crops, however, are shade intolerant and, thus, are not suited for agroforestry. Trees and crops may be profitably combined in the early stages of tree plantations (first to second year, depending on the species and planting distance) when trees do not provide much shade. Agroforestry practices should decrease the need for weeding, and trees may have positive effects on soil properties (Montagnini 1990a). Many areas in the region should be better suited for trees than for conventional or mechanized agriculture because of steep slopes, low fertility, and problems with soil compaction. On individual farms land that is not suited for agriculture can be used to grow trees in association with crops. Besides, many agroforestry practices require less labor than intensive agriculture and allow for more flexibility in management and timing of harvest (Kapp 1989). The following agroforestry systems already are present in the Atlantic Region (Kapp 1989):

Cacao, coffee, and other perennial crops with native trees for shade, many of which are valuable timber species (for a complete list see Kapp 1989)

Pastures with native trees (same as those with perennial crops)

Annual crops with native trees

Crops or pastures in alternating rotation with tree fallows

Living fences (*Gliricidia sepium, Diphysa robinioides, Bursera simaruba,* and others)

Mixed home gardens with food crops and many fruit and timber trees

Opportunities to improve these systems include

The promotion of better spatial designs (e.g., planting in lines rather than haphazardly to facilitate management practices)

Selection of tree species more appropriate for combinations with crops (eucalypts and pines, very much used in the region, are not well suited for combinations with crops; *Gmelina,* also common in the region, and many native trees are better)

Choice of shade-tolerant crop or pasture species (cassava and other root crops are more shade-tolerant than corn or rice, among forage species legumes are more shade tolerant than grasses)

Developing detailed management schemes (e.g., timing pruning and thinning to provide more light and increase inputs in the form of residues from pruning to the soil when crops can most use these resources).

To design improved systems that will be accepted by farmers the socioeconomic and institutional constraints described in the previous sections must also be considered.

OTS INVOLVEMENT IN EDUCATION AND RESEARCH ON SUSTAINABLE LAND-USE SYSTEMS IN THE ATLANTIC LOWLANDS

The involvement of OTS in agricultural research is a very recent phenomenon that began with small projects by participants in OTS agricultural ecology courses inaugurated in 1985. OTS offered its first agroecology course in Spanish in 1988, an outstanding promotion of agroecology education in Latin America. Research by course participants consists of projects completed in a few days although some students return for longer-term research. Topics examined are varied: herbivory, soil chemistry and microbiology, mycorrhizae, nitrogen fixation, crop ecophysiology, general surveys of socioeconomic aspects of land-use patterns, crop preferences, and land-use alternatives. Course projects also orient theses and other long-term projects. Most importantly, agroecologically oriented courses can influence the philosophy of those already involved in agricultural or forestry research, education, or practice toward greater concern for environmental issues and a more integrative, holistic approach to agriculture and forestry.

Increased interaction between OTS and Costa Rican institutions involved in agricultural and forestry research, training, and extension is desirable. Students in course projects can take advantage of data and recommendations resulting from local agricultural and forestry projects, and, in turn, provide data and innovative ideas toward solving specific problems. OTS research fellowships for students are making a good contribution by funding projects by Latin American scientists at La Selva. OTS might additionally strengthen ties between foreign researchers and local scientists by helping to identify local faculty to serve as advisers to foreign students working on applied problems at La Selva. Through its connections with its Costa Rican member institutions OTS can also facilitate collaborative research involving local and foreign scientists. Especially when one works on applied research, local expertise is valuable to ensure that the project is well-grounded in social, political, and institutional realities.

Other educational efforts can be implemented through OTS's Environmental Education Program. Initially focused on ecology and conservation, this program has recently expanded into agroecology and forestry through interaction with forestry projects at La Selva.

RECOMMENDATIONS FOR AGROECOLOGICAL RESEARCH AT LA SELVA

As noted, La Selva has soils that are representative of the Atlantic Region of Costa Rica as well as of many lowland areas in tropical America. Thus, research on soil fertility, soil/plant interactions, soil chemistry, and microbiology have potential for wide applicability. Research on management of traditional and more innovative agroecosystems could be applicable to other areas in Latin America as evidenced by the similarity of traditional cropping systems across broad areas of the lowland wet tropics.

Agroecological research at La Selva should focus on the most promising land-use systems for the economic benefit of farmers and the country, including subsistence crops. Existing forestry plots at La Selva can be used by OTS researchers and students in collaboration with the principal investigators of the projects to study management techniques, impact of trees on soils, and agroforestry uses of the species planted.

Buffer zones for protected areas offer interesting opportunities for agroecological research at La Selva, both on station property (La Guaria Annex, El Peje, La Flaminea) and elsewhere in the Atlantic lowlands (e.g., via participation in the recently funded Forest Resources for a Stable Environment Project [FORESTA]. This type of research fits well with OTS goals and is also becoming an important aspect of tropical ecological research as new areas are protected and surrounding lands need to be managed in ways that foster protection of the core. Agroforestry, management of secondary forests, and development of ecologically sound land-use systems are a frequent component of management plans for buffer zones in areas of tropical rain forest.

Whenever possible, research should be conducted in collaboration with local institutions (MAG, DGF, UCR, CATIE). Much research is currently undertaken by local institutions to design innovative land-use systems. For example, F. Bertsch and V. Vega (UCR) are conducting soil fertility research at the UCR experimental farm in Río Frío as part of a network of studies in the lowland humid tropics coordinated by North Carolina State University (Bertsch and Vega 1990). The experiment, which is in its early stages, consists of continuous cropping of a rice (Oryza sativa)-cowpea (Vigna unguiculata)-mucuna (Styzolobium spp., a legume fallow) sequence. Researchers from Wageningen Agricultural University (Holland) have been working at CATIE since 1986 in collaboration with MAG's "Atlantic Zone Programme" (Atlantic Zone Programme 1987). The objectives are to contribute to sustainable development in the Atlantic zone of Central America and Panama through research and training activities. Research focuses on production systems and on regional land-use planning. Other projects at CATIE could be of interest to La Selva researchers (Nitrogen-fixing Tree Project, agroforestry, wildlands management, and others). OTS researchers and course participants could take advantage of such experiments and projects to visit, exchange ideas, and, potentially, to initiate collaborative research.

Likewise, researchers involved in agroecological work at La Selva should be aware of the efforts of local institutions in the same discipline and encouraged to interact actively with their colleagues by sharing results and by collaborating on mutual projects whenever appropriate. This sort of interaction could be encouraged by organizing joint activities, such as workshops, seminars, and field trips. La Selva already facilitates local research by providing cheaper station fees for Costa Ricans. The opportunities at La Selva could be better publicized in local universities and other institutions by more frequent communication and by organizing seminars by La Selva researchers on subjects of common interest. Local researchers could supply background data and help to define priorities and goals. This type of interaction will magnify the contribution by OTS students and researchers at local and national levels.

Biologists in many projects at La Selva already interact with local researchers and employ local technicians and students, and more direct involvement of local scientists and collaborators is a highly desired next step. For example, data from local institutions that are concerned with development and with socioeconomic aspects of agriculture and forestry should be useful for defining goals and subjects of study for La Selva researchers. OTS can play a more significant role in its host country by building upon its educational and research programs and, especially, by capitalizing upon its pivotal position to improve linkages between U.S. and Latin American scientists and the general public.

ACKNOWLEDGMENTS

F. Bertsch, E. Bornemisza, L. McDade, F. Sancho, C. Schnell, and P. Sollins reviewed the manuscript and provided useful information and ideas. B. Auer and R. Rheingans edited the final version. Many other people offered ideas and constructive criticism in informal conversations and discussions on land-use systems in the Atlantic Region.

25

Forestry in Costa Rica: Status, Research Priorities, and the Role of La Selva Biological Station

Rebecca P. Butterfield

In this chapter I provide an overview of forestry in Costa Rica and place forestry research conducted at La Selva within the context of the country's forestry problem: the social and economic factors that contribute to deforestation in Costa Rica; the relation of the Costa Rican wood industry to forest resources with an emphasis on the region surrounding La Selva Biological Station; past and present forest development projects in Sarapiquí along with some of the obstacles to natural forest and plantation management; and related research at La Selva and nearby areas in both natural forest management and plantation forestry.

Costa Rica has an excellent and well-deserved international reputation for conservation. It also has one of the highest relative rates of deforestation in Central America (Leonard 1987). As late as 1943, forests occupied 70%–75% of the country (Sader and Joyce 1988; Garita 1989). A 1987 survey by the Dirección General Forestal ([DGF] General Forestry Directorate [DGF]) revealed that only 29% of the country remained forested. Of the existing forest cover 66% is within protected areas (parks, wildlife and forest reserves, and protected zones). A variety of activities are allowed within protected areas depending upon their legal status. National parks are largely government owned and protected although some private in-holdings remain to be bought out. No extractive or exploitive practices are allowed inside national parks. Protection zones and forest and wildlife reserves are less well protected, with severe squatter pressure in some areas (Hartshorn et al. 1982; Kauck and Tosi 1989). Legally, forest exploitation is permitted within forest reserves although many occupy upper watershed areas and should be protected for their hydrologic functions (Hartshorn et al. 1982; Garita 1989). Logging or agricultural practices could lead to severe erosion. The DGF exercises little control over the forest reserves; 60% to 90% of the land is privately held and conversion of forest to other uses is permissible (Rodríguez and Vargas 1988). Because the government does not own the land, it cannot manage the forests in these reserves. Forested areas outside of protected areas are privately owned and controlled; at current rates of deforestation, the remaining 504,000 ha of unprotected forest in the country will disappear by the end of the twentieth century.

The DGF divides the country into eight forestry regions. Sarapiquí and La Selva Biological Station are located in the Northern zone (fig. 25.1). Although it contains only 15% of the country's forests, the Northern zone has the largest area (148,610 ha) of unprotected forest (table 25.1), and it is this privately owned forest mass that will be the focus of land clearing and wood extraction during the 1990s.

CAUSES OF DEFORESTATION

Government policies greatly influence land-use patterns. The way in which forests are used depends, in part, on how governments perceive the resource. In Latin America forests have provided land to a growing population for agricultural expansion. Homesteading and the settling of forested areas have been promoted in the past as parts of national development strategies. Although incentive structures vary from country to country, Central American countries have generally had similar policies that seek to expand the agricultural land base. Matamoros (1987) has identified several major contributing factors to deforestation in Costa Rica: expansion of the agricultural frontier; increased population coupled with land concentration; homesteading and forest conversion; national and international bank credit policies; and illegal logging.

Expansion of the agricultural frontier has been at the expense of forests (see chap. 23). Forest clearing has been the traditional method of claiming land. It demonstrated that the land was being used and, under Costa Rican law, was considered an improvement that raised land value (Hartshorn et al. 1984; Porras and Villarreal 1985). Although the homesteading legislation has changed, the tradition persists.

Increased population coupled with land concentration (larger farms owned by fewer people) has pushed people to marginal areas (areas with steep terrain, high rainfall, and poor soils). Although much of this land is not suitable for permanent cultivation and should be maintained as forest to protect soils and watersheds, it is often cleared for agriculture. More than 50% of the land managed by the Instituto de Desarrollo Agrario ([IDA] Institute of Agrarian Development) has low agricultural potential and is better suited to forestry than farming (Matamoros 1987).

Although logging is not the leading cause of deforestation in Costa Rica, it is interrelated with homesteading and the conversion of forest to agriculture and pasture. Loggers are often the first to open roads into new areas; they extract some

Fig. 25.1. Costa Rican forestry regions. Forestry regions: 1 = Dry Pacific; 2 = Northern zone; 3 = Central Pacific; 4 = Western Central Valley; 5 = Atlantic zone; 6 = Central Region; 7 = Southern Pacific; 8 = Eastern Central Valley.

timber and farmers move in to develop farms and cattle pastures. Where roads are poor or nonexistent, farmers fell trees and leave them to rot (Keogh 1984). The DGF estimates that only 10% of the total volume of felled trees is actually extracted for timber (Segleau 1985). The unabated clearing of forests for agriculture has allowed the Costa Rican forest industry to depend on land clearing rather than managed forests or plantations for its supply of raw material.

National and international bank credit policies have stimulated the expansion of subsidies for certain land uses, especially cattle raising. Support for the cattle industry and beef exports was seen as a way to diversify Costa Rica's exports and to earn foreign exchange. Since the 1960s most of the forest cleared in Costa Rica has been converted to pasture, often with the aid of bank loans (Place 1981; Annis 1990). Similar credit incentives do not exist for maintaining land in forest cover (Kauck and Tosi 1989).

Costa Rica has good forestry laws but the lack of enforcement has allowed substantial areas of forest to be cut illegally. A permit must be obtained from the DGF to cut any tree larger than 10 cm in diameter whether for land clearing or wood harvesting. The meager resources of the DGF has limited enforcement to inspection stations with limited hours to check logging trucks for permits. Farmers clearing forest for agricultural purposes do not bother to get permits, and many unscrupulous loggers move cargos during the night or on weekends when the DGF is not working.

If current rates of forest clearing continue, the use of forested land to settle a growing population will no longer be an option. At present growth rates Costa Rica's population will double in the next twenty-eight years to nearly six million people (Population Reference Bureau 1990). As a consequence, there will be increased pressure on existing large

farms for redistribution through land reform and, potentially, on protected areas to provide new agricultural land (see chap. 23). Although logging has not been the prime motive for forest clearing, the timber industry will also suffer the consequences of exhaustion of unprotected forest reserves and may press for access to protected areas.

THE COSTA RICAN TIMBER INDUSTRY

Historically, the wood industry has been able to supply itself with raw materials without investing in future sources of timber by simply following the agricultural frontier. Few sawmills own forested tracts of land or manage land for timber production. Most land within forest reserves is privately owned; government logging concessions within reserves, therefore, do not exist (Rodríguez and Vargas 1988). Sawmills purchase logs at minimal prices from farmers as they are converting forest to agricultural use. Because farmers earn little from timber sales, they have little incentive to replant or to manage natural forest. The industry depends, in effect, on the selective mining of forests for logs of choice species without ensuring a future supply of wood from the same land.

The industry has also been notoriously wasteful. It is estimated that only 54% of the volume felled commercially is actually transported to sawmills. Of this, another 50% is lost in processing. An average of 50 m³ of timber is cut per hectare on commercially harvested land to provide only 12 m³ per hectare of processed wood (Fundación Neotrópica 1988; Flores 1985).

Logging is currently concentrated in the Northern and Atlantic zones of the country. In 1988, 75% of the country's industrial wood volume originated in these regions with nearly one-half (163,725 m³) cut in the Northern zone (based on log-

Table 25.1 Forest cover in Costa Rica by region, 1987

Forestry Region	Forest Area Protected (ha)	(%)	Forest Area Unprotected (ha)	(%)	Total Forest (ha)	Total National Forest (%)
Northern zone	70,050	7	148,610	30	218,660	15
Atlantic zone	385,350	40	104,230	21	489,580	33
Southern Pacific	267,337	28	133,403	27	400,740	27
Eastern Central Valley	159,309	16	35,171	7	194,480	13
Western Central Valley	39,920		31,040	6	70,960	5
Dry Pacific	41,400	4	40,920	8	83,320	6
Central Pacific	4,120	0.5	6,700	1	10,820	1
Central Region	5,120	0.5	3,260	0.5	8,380	1
TOTAL	972,606	100	503,334	100	1,475,940	100

Sources: Cartographic and Topographic Unit of the DGF. Dry Pacific based on 1982 LANDSAT images; Southern Pacific based on 1979 information, including the southern part of the Atlantic zone. All other data taken from 1987 LANDSAT images (Garita 1989).

ging permits, DGF 1989). This figure, however, should probably be at least doubled to include estimated uncontrolled and illegal logging.[1]

Critical wood shortages are predicted to begin in 1995 resulting in a potential annual wood import bill of $350 million (Flores 1985). This shortage will exacerbate an already troubled balance of payments situation. In 1990 Costa Rica was still a net exporter of wood and wood products. Exports, mostly to the United States and Canada, consisted of $21.8 million in furniture, doors, and wooden handicrafts. Imports, more than half from the United States and Nicaragua, included $3.4 million in crates, sawn timber, and plywood (DGF 1990).

By the end of the 1990s nearly all of the remaining forests will be inside private and public reserves (Flores 1985). As noted, rising wood scarcity can be expected to increase the pressure for access to protected forests. To safeguard forest reserves and parks, steps must be taken immediately to institute management of primary and secondary forests and to replant cutover lands and pastures for future wood production. Incentive structures and laws will have to be changed to encourage the development of a sustainable forest industry if Costa Rica does not want to depend upon wood imports. The challenge to scientists is to develop viable options, both biological and economic, for the rational exploitation of plantations and natural forests that can ease Costa Rica's forestry crisis and that can serve as models for other countries facing similar problems.

REFORESTATION: STATUS AND RESEARCH NEEDS

Status

Reforestation in Costa Rica was negligible before 1979 and had grown to 9,257 ha planted in 1989 (DGF 1990). Consider-

ing that approximately 50,000 to 60,000 ha are deforested in Costa Rica annually and that imminent wood shortages are projected, reforestation remains woefully inadequate. An estimated 37,000 ha of commercial plantations must be established annually from 1990 through 1998 to avoid large-scale wood importation (Flores 1985). Low prices for standing timber have discouraged tree planting. Several types of government incentives have become available since 1979 to encourage reforestation (Pérez 1989). Funds for reforestation continue to be limited, however, and cannot meet demand. In 1989 the extension department of the DGF received applications to plant 6,000 ha under its fiscal incentive program but had funds for only 3,000 ha (G. Canet pers. comm.). Under 1992 conditions the existing fiscal incentives for planting will not generate sufficient plantation area to offset the loss of forest and will not meet the growing internal demand for wood and wood by-products.

The two primary incentive programs are the Certificado de Abonos Forestales ([CAF] Certified Forestry Bonds) and the Fondo de Desarrollo Forestal ([FDF] Forestry Development Fund). The former, in effect since 1986, pays landowners about U.S. $1,000 per hectare in bonds over a five-year period to cover 100% of the costs of the first five years of plantation establishment and maintenance. These bonds are available to people with land titles and have been used mostly by large landowners. The CAF are not loans and no repayment is required. The FDF was created in 1989 for smaller farmers who often do not hold title to their lands. They are paid, through a local farmer organization, about $644 per hectare or circa 70% of the costs of plantation establishment and maintenance for a five-year period. The small farmer must repay this loan by giving 30% of the income from the final harvest back to the farmer organization as part of a revolving forestry fund. Both programs recognize private ownership of trees that are planted (OET/DGF 1990; Pérez 1989).

Although incentives are a positive step in promoting reforestation, they also create problems. Past abuses of the program include clearing of primary forest to establish tree plantations. Certain investment companies have also planted trees to capture reforestation credits and then sold shares or small plots without further replanting or maintenance of the plantations

[1] In 1986 the DGF approved a total of 468,748 m³ of wood for extraction, whereas a forest industry census in the same year revealed that 827,786 m³ was consumed (DGF 1988, 1989).

(Rodríguez and Vargas 1988). The goal of these projects is to profit from tax credits, not wood production.

A serious problem is the limited number of species approved for planting under these programs. Only species on the official DGF list for the particular forestry region can be planted. The DGF develops the list based on growth and yield tables for species in plantations. The lack of growth data for native species has precluded their use in incentive-supported reforestation. The approved list for the Northern zone includes *Eucalyptus deglupta, Gmelina arborea, Tectona grandis,* and *Cordia alliodora.* Of these *C. alliodora* is the only species native to the area although more than one hundred native species are currently being used in local sawmills (DGF 1988a).

Over the 1980s a total of 4,845 ha have been reforested in the Northern zone (DGF 1989). Officially, this equals one hectare planted for every ten hectares cleared although unofficially it is less than one to twenty. The most commonly used species for reforestation in the area are *Cordia alliodora* (69%) followed by *Gmelina arborea* (11%), *Pinus caribaea* (8%), and *Eucalyptus* spp. (4%) (DGF 1989).

Tectona grandis recently replaced *Pinus caribaea* on the official species list because pines exhibited poor form and growth. The wood of the eucalypt is acceptable for poles and pulp, but the tree does not grow well on infertile sites (Butterfield 1990). Most reforestation in the Northern zone is with *Cordia* although it also grows well only on fertile soils. It is often planted on sites to which it is not well adapted, resulting in high mortality rates and slow growth (Camacho 1981). Wood production from these plantations (69% of the total area planted) will be extremely low.

Despite its problems plantation forestry has an important role to play in Costa Rica. Large areas of forest have been cleared, creating agricultural lands of low productivity. This land may be better suited to intensive forestry, especially given the poor soils and high rainfall of the Northern zone. Tropical plantations are noted for high productivity with annual growth increments ranging from 35 m³ per hectare per year for *Gmelina arborea* in Brazil for fiber to 14 m³ per hectare per year for valuable hardwoods such as *Swietenia macrophylla* in Fiji (Evans 1982). Plantations can produce five to fifteen times more commercially usable wood per hectare than natural forest. Using estimates from Flores (1985) for total commercial wood volume harvested per hectare in an unmanaged mature forest in the Northern zone, the average forest growth rate of commercial species is 2.3 to 1.5 m³ per hectare.[2] Plantation wood is more uniform in quality and size, which facilitates harvesting and use by processors. By intensifying wood production in plantations pressure on existing natural forests for timber can be alleviated.

Research Needs

Forestry research in Costa Rica has, for the most part, been oriented toward commercial reforestation and plantation management. In 1946 Centro Agronómico Tropical de Investigación y Enseñanza (CATIE) initiated its first experiments to test species for use in plantations. Since then, more than 250 tree species have been introduced at CATIE in the arboretum

and in species trials (Combe and Gewald 1979). Larger plantations have been established throughout the country emphasizing *Pinus caribaea* and *Eucalyptus deglupta.* In 1965 the United Nations Food and Agriculture Organization (FAO) and Instituto de Tierras y Colonización ([ITCO] Land Colonization Institute, now IDA) established experimental plots throughout Costa Rica; the forest research department of the Ministry of Agriculture (MAG) established additional plots in the 1970s (Camacho 1981). These plots are now monitored by the DGF although their limited research budget inhibits adequate data collection and analysis. In total, sixty sites were planted with seventy species including twelve sites in the Atlantic zone and two in the Northern zone. A total of thirty-seven species were tested in these two zones of which seventeen were indigenous to the area. The best performing species based on results analyzed in 1981 included *Gmelina arborea, Pinus caribaea, Tectona grandis, Eucalyptus deglupta, Cordia alliodora, Stryphondendron microstachyum,* and *Jacaranda copaia*[3] (Camacho 1981). Although many of the DGF plots have been abandoned, follow-up research would yield useful information.

The heavy emphasis on exotic species for reforestation in Costa Rica results from adoption of techniques and species widely tested in other parts of the tropics. During the decade of the Green Revolution (1960s) foresters identified a few extremely fast-growing species for maximum production of wood in intensively managed plantations. During the 1960s CATIE introduced many of these promising exotics to Costa Rica. The planting of *Pinus caribaea, Eucalyptus* spp., *Gmelina arborea,* and *Leucaena leucocephala* around the world owes in part to the large data banks available on these species. Data on tree growth under plantation conditions in Costa Rica for these exotic species far exceed those available for native species.

Focusing on maximum wood production under highly favorable conditions has reduced the relevance of this research to real reforestation needs in Costa Rica today. Farmers seldom dedicate their best bottomland soils to trees; they plant uplands and marginal and degraded areas instead. The commercial operations tend to do the same or, if larger companies, to plant over a wide variety of edaphic conditions. The international "wonder species" may well offer high productivity when on fertile sites (although comparative studies have not been done), but the first decade of experience with reforestation in Costa Rica shows that they often perform suboptimally under the environmental conditions in which reforestation most often takes place (Rodríguez and Vargas 1988; Butterfield 1990; OET/DGF 1990).

Clearly, a wider range of species must be officially incorporated into reforestation programs. The diverse ecological zones of the country (twelve zones, according to the Holdridge Life Zone System), each with numerous soil types and many microsite variations, require a wide range of species to optimize productivity by matching trees to sites (Hartshorn et al. 1982). Native species that have been neglected in forestry programs but have a local market, provide food and habitat

---

[2] Estimate based on 45 m³ per hectare of harvested wood divided by a twenty- to thirty-year rotation.

[3] The last two species were never approved for reforestation by the DGF probably because of the lack of growth and yield tables. *Stryphnodendron microstachyum* published as *S. excelsum* in Camacho 1981.

for wildlife, and are well adapted to prevailing site conditions should be tested and compared to the exotic species currently being used.

One limitation on the use of native species in reforestation is the lack of commercial sources of seed. This constraint is especially problematic given that natural stands of valuable timber species are dwindling rapidly. It will be increasingly difficult to locate seed trees with superior form and vigorous growth. For example, *Tectona grandis* seed[4] in Costa Rica is considered to be of low genetic quality because the trees used for seeds are those with poor form left over from earlier harvesting. The development of protected stands or seed orchards will be essential to supply quality seed in the future. An additional complication is that many tropical seeds lose their viability rapidly. Research on seed storage techniques for native species has been minimal (e.g., Blance et al. 1991).

NATURAL FOREST MANAGEMENT: STATUS AND RESEARCH NEEDS

Status

The field of sustained yield-natural forest management in the tropics is still in its infancy despite a long history and increasing interest in recent years (Gómez-Pompa and Burley 1991; Buschbacher 1990). Research conducted in Asia, Africa, and South America has produced models (with varying degrees of success) for forest management and harvesting applicable under specific ecological conditions and for certain species (Buschbacher 1990). These models have yet to be tested in Central America to ascertain which management system, or variation thereof, would be most appropriate ecologically and economically. Although there has been little forest management experience in Costa Rica, several large development projects near La Selva have included natural forest management components.

In 1979 USAID financed a natural resource management project (CORENA, GOCR/AID 0–32). One component of the CORENA Project was to be sustained-yield, integrated forest management in Sarapiquí (specifically an area bounded by the Puerto Viejo and Guácimo rivers in the Las Horquetas area). The project proposed to integrate wood product industrialization with timber production from natural forest management. Ideally, forest owners would see a greater return by managing their forests through direct profit-sharing links with the wood-processing industry. Extensive inventories were conducted in the area's primary forests and 100 ha of forest were put under three different experimental harvesting/silvicultural regimes to identify management techniques that optimize natural forest regeneration. The treatments were (1) clearcutting and (2) felling all trees with a diameter greater than 40 cm and (3) greater than 60 cm (DGF 1988b). New roads were built into the experimental forestry areas and a portable sawmill and charcoal kiln were installed near Las Horquetas. The project also constructed a gasification plant to convert wood residues to electricity. Unfortunately, few data emerged from the project. USAID discontinued funding in 1984 as a result more

of administrative problems than difficulties with silvicultural aspects of the project (Alpizar et al. 1985). Follow-up measurements were not made in the experimental management areas, and it is impossible to assess the relative success of the three treatments. The sawmill and gasification plant were subsequently removed.

Recently, a similar project (Forest Resources for a Stable Environment [FORESTA], managed by Fundación para el Desarrollo de la Cordillera Volcánica Central [FUNDECOR]) has been developed by USAID and the Costa Rican Government to work in buffer zones around Braulio Carrillo National Park (including La Selva). Overall, the project has three principal components: (1) strengthening national parks (Irazú, Braulio Carrillo, and Poás) through infrastructure improvement and an endowment to defray operating costs; (2) natural forest management in three forested buffer zones around Braulio Carrillo park; and (3) farm forestry (agroforestry and reforestation). The project began in 1991 with headquarters in Puerto Viejo, Sarapiquí. Managers of the seven-year project seek to alter economic structures that lead to inappropriate use of forest resources by providing incentives and technical assistance to landowners to protect and manage their forests.

The project is risky in that the goal is to develop large-scale forest management while essential technical information is sketchy at best. Research and silvicultural experimentation will need to go hand in hand with forest management plans to develop working models for sustainable forest management. The farm forestry component also must be developed in conjunction with applied research to study appropriate species, agroforestry combinations, and management techniques for the successful implementation of farm forestry.

The only active commercial forest management project in the country is located northeast of La Selva in swamp forests bordering Tortuguero National Park. The project is managed by Portico, S.A., a private company that exports "royal mahogany" doors made from *Carapa nicaraguensis* (Meliaceae). In 1986 the company began managing 5,000 ha on a fifteen-year cycle to harvest *Carapa*. Trees more than 70 cm in diameter are harvested by methods designed to create forest canopy openings no larger than 1,000 m². More than 40 ha of permanent inventory plots have been established to monitor regeneration and growth in both managed and undisturbed forest. The company has established a small research department and encourages scientists to work with them to improve forest management practices (R. Peralta pers. comm.).

Although natural forest management has received international attention as a way to provide timber and maintain large areas under forest cover, its usefulness in Costa Rica is limited. Only 5% of the national territory that is still in forest is suited for such management and, as noted, that area is shrinking rapidly (Garita 1989). As primary forests disappear, secondary forests might prove to be a more important land base for future forest management. Reduced beef prices or cattle subsidies could conceivably produce a rapid increase in secondary forest as less-productive pastures were abandoned. No data exist, however, on the rate of pasture abandonment or the area in secondary forests. An additional difficulty with natural forest management, especially given the urgency of Costa Rica's wood crisis, is that it is a low-impact, extensive use of forests. Yields will, therefore, be low, and the volume of tim-

---

[4] Teak is not native, but it exemplifies the kind of problems that can occur if seed sources are not selected with care and properly managed.

ber that can be harvested in any one year will be small. It will be best suited to the production of high-value and value-added products.

Research Needs

Costa Rica is endowed with species-rich forests that include many valuable hardwoods, such as *Swietenia macrophylla* (mahogany), *Dalbergia retusa* (rosewood), and *Cedrela odorata* (Spanish cedar).[5] The Costa Rican biota contains elements from both North and South America (Holdridge and Poveda 1975; Janzen 1983b). The natural ranges of some tree species extend as far north as Mexico and as far south as Brazil. Other species have a more restricted range and are limited to Costa Rica and its immediate neighbors to the north or south (Holdridge and Poveda 1975). The 1,500 ha reserve at La Selva contains 323 tree species (chap. 6).

Ironically, forest management for timber is hampered by high diversity. Part of the immense waste in forest clearing is the result of the small number of species extracted for commercial purposes. Selective logging of a few species degrades large areas because choice species occur at low densities. Conversely, clear-cutting leaves many "noncommercial" trees on the ground. As preferred species become scarce, more species are used in local sawmills. During the 1950s only one species was logged in the Atlantic lowlands, *Cedrela odorata,* but now mills use up to one hundred species (DGF 1988a). Handling and marketing little-known timber remains a problem, however, and applied research is needed to extend the range of species that can be used. This would include, for example, techniques in wood treatment and drying to enable use of softer, less rot-resistant woods in general construction.

Primary and secondary forest management requires knowledge of species' responses to different silvicultural treatments (e.g., regeneration and growth). The most extensive research in this area in Latin America has been in Surinam (de Graf 1986). Each ecological zone has unique features that may require specific management techniques, and these will also vary among species. For instance, a high-light-demanding species will differ substantially from a shade-tolerant hardwood in environmental requirements. Information is needed on the response of commercial species and competing vegetation to changes in light, water, and nutrient resources that occur after forest disturbance (e.g., selective cutting). Little is known about how timber species respond to these changes and how best to encourage the growth of individuals of commercial species without damaging the forest ecosystem.

The greatest need in natural forest management research is for permanent study plots under different silvicultural treatments. Although interest in forest management projects in Costa Rica is growing, the establishment of permanent plots under different silvicultural regimes is, generally, not contemplated as part of these projects. The short time frame of most development projects is not conducive to the long-term studies needed to learn to manage long-lived trees successfully. In

this context the potential value of collaborative forestry research at biological stations like La Selva becomes evident.

RESEARCH AT LA SELVA AND ENVIRONS IN FOREST ECOLOGY AND PLANTATION FORESTRY

Forest Ecology Research

Much of the natural history and ecosystem level research at La Selva could be relevant to management of natural forests for specific ends, commercial or noncommercial. Historically, this research has not been focused on applied questions, but the basic knowledge generated constitutes a major resource. It seems increasingly likely that this body of knowledge will be explored for its value in forest management and plantation forestry. In recent years many forest ecology researchers have expressed interest in bringing their research to bear on topics relevant to forest management (pers. observation).

For a lowland tropical forest La Selva has a rich and well-known tree flora that provides an important foundation for developing forestry research (see chap. 6). In addition, several long-term databases (see chaps. 7, 8, 11) offer valuable information on forest ecology. The recently completed soil survey of La Selva also provides useful baseline data for forestry research (see chap. 4). Interestingly, some of the earliest studies conducted at La Selva included forest management (Petriceks 1956) and forest succession (Budowski 1961). In the late 1960s, as part of a comparative ecosystem study, the University of Washington established long-term forest inventory plots at La Selva and Palo Verde. Data continue to be collected for these plots at La Selva (see chap. 8). They are among the oldest established forest inventory plots in Central America and represent a tremendous source of information on natural forest inventory and growth. In 1985 and 1986 Hartshorn and Peralta established a transect of single-hectare, long-term forest inventory plots at elevational intervals of 250 m from La Selva to the top of Volcán Barva in Braulio Carrillo National Park (2,900 m). These plots complement those at La Selva and offer opportunities for altitudinal comparisons of species diversity and forest composition in forests not dominated by *Pentaclethra macroloba.*

Other basic research with potential for guiding future work in forest management includes tree demography and natural regeneration studies (see chap. 7) and phenology (see chap. 11). The work on tree-fall gaps (see chap. 9) sets a theoretical base for more applied research. Although this study has not been oriented toward commercial species, knowledge of the function of gaps and the mechanisms of forest regeneration within them can be applied to the development of silvicultural systems. Research about the effects of light on regeneration, tree growth, and species composition also provides data for silviculture treatments (see chap. 10). Especially useful are studies of optimal light levels for peak photosynthetic activity of commercial tree species (Fetcher et al. 1987). Direct seeding trials under different light and fertilization regimes have also been conducted (Huston 1982).

In the Northern zone, CATIE has initiated a natural forest management research project at two lowland forest sites: high-graded primary forest near the Corinto River, Guápiles,

---

[5] Costa Rica is about the size of West Virginia but has a tree flora of about 1,400 species in 487 genera and 116 families as compared to about 865 tree species native to North America north of Mexico (Zamora 1989; Brockman 1968).

and secondary forest at La Tirimbina near La Virgen, Sarapiquí. Permanent plots have been established in secondary forests of one, fourteen and twenty-five years of age at the La Tirimbina site. Silvicultural treatments that included liberation thinnings of commercially desirable species in the understory began in 1990. The plots will be monitored to ascertain effects on regeneration and growth of light-demanding and shade-tolerant timber species (Finnegan and Sabogal 1988).

Basic research conducted at La Selva and elsewhere provides a foundation for understanding forest dynamics. Communication is needed, however, between foresters and ecologists to facilitate the sharing of relevant information and the development of future research efforts. Foresters need to synthesize existing ecological research to develop working hypotheses to guide natural forest management. Ecologists have the opportunity to make their work applicable to forestry problems by consulting with foresters to identify information gaps and species of mutual interest. For example, the lack of information on the autecology of commercial timber species has been identified as one of the most critical obstacles to natural forest management (Gómez-Pompa and Burley 1991).

Plantation Forestry Research

In 1981 OTS acquired lands suitable for manipulation at La Selva in a purchase that nearly doubled the size of the station (see chap. 2). In 1983 the DGF established the first experimental forestry plantations at La Selva to test four species, planted at 2 m × 2 m spacing in the new La Guaria Annex. Two species, *Dipteryx panamensis* and *Goethalsia meiantha*, were established, using transplanted natural forest seedlings that did not survive. The other two species, *Pseudobombax septenatum* and *Gmelina arborea*, were planted from nursery stock and have shown good results at seven years of age with basal area growth of 6.2 and 8.0 m² per hectare per year, respectively (c. Espinoza unpublished data).

In 1984, through its Central American Fuelwood Project, CATIE planted two replicates of twelve promising fuelwood species in a Nelder design on the La Guaria Annex.[6] Nelder plots are designed to test simultaneously species adaptability and the effect of different spacing regimes (fig. 25.2). Species showing the best initial growth and survival were *Gmelina arborea* and *Acacia mangium*. Many of the *A. mangium*, however, fell over after three years because of root rot or damage to roots by pocket gophers (*Orthogeomys* spp; for additional information on these widespread pests see Sisk and Vaughan 1984). Additionally, most of the *A. mangium* cut during a fourth-year thinning were riddled with termites (pers. observation). Species that had the highest mortality and poorest growth were *Pithecellobium saman*, *Guazuma ulmifolia*, and *Cordia alliodora*. Unfortunately, the student who established these plots never finished the analysis, and neither CATIE nor OTS have data. In 1988 the first replicate was heavily thinned, destroying the Nedler design and leaving only the largest trees

of good form. The second replicate was lightly thinned by eliminating every other row. The plots are now abandoned.

In 1985 OTS collaborated with the DGF to establish a demonstration plot of fourteen native timber species on the La Guaria Annex (table 25.2). A complete randomized block design was used with five replicates. Each plot contains 7 × 7 trees at 2 m spacing. Total height and dbh (diameter at breast height, 1.3 m) were measured every six months on the inside 5 × 5 trees. A third-year analysis of growth data has identified *Vochysia guatemalensis*, *Stryphondendron microstachyum*, and *Terminalia ivorensis*[7] as the most outstanding species in survival and growth with basal areas of 17.9, 17.6, and 15.7 m² per hectare, respectively (table 25.2). *Vochysia guatemalensis* and *T. ivorensis* have uniformly straight stems, whereas *S. microstachyum* has a more irregular form. Although slower growing, several other species appear promising owing to their uniformity, straight stems, and higher wood value (*Vochysia ferruginea*, *Hyeronima alchornoedies*, *Dipteryx panamensis*, *Calophyllum brasiliense*). It is interesting to compare data for these promising species with data for the species most widely used for reforestation in the Northern and Atlantic zones, *Cordia alliodora*, which had a basal area of 2.4 m² per hectare at three years of age (Espinoza and Butterfield 1990). At four and six and one half years of age all of the species mentioned (except *C. alliodora*) were thinned, removing 50% of the stems from each plot in each thinning. Growth is now measured annually.

Soil fertility, nutrient cycling, and root development are being studied in this plantation on seven species. Results indicate that soil organic matter increased significantly in the first 15 cm under the tree plantations. Soil under both *S. microstachyum* and *V. ferruginea* had more than 6% organic matter, exceeding the 4.8% for a neighboring pasture and approaching the 7.5% for a secondary forest site. Increased organic matter content was linked to increased soil base content. *Stryphnodendron microstachyum* also exhibited high nitrogen recycling potential and promise as an agroforestry species (Montagnini et al. 1989, 1990).

In 1990 CATIE and the Central American and Mexican Coniferous Resources Cooperative based at North Carolina State University (CAMCORE) established a 4-ha provenance trial of *Vochysia guatemalensis* in the La Guaria Annex. The trial contains trees from six provenances: four from Costa Rica, one from Guatemala, one from Honduras, plus a local control from Sarapiquí; forty-eight half-sib families are included. Data will be collected for ten years after which the superior provenances will be kept to develop a seed orchard (CATIE 1990).

The TRIALS Project. The interest in native species generated by the plantation just described led to an OTS/DGF proposal to screen a large number of little-known tropical species for reforestation. This project, known as TRIALS, began in 1987. TRIALS has two major components: screening to identify species appropriate for large-scale plantation trials and long-term monitoring of tree/site interactions to identify spe-

---

[6] Species tested include *Acacia mangium*, *Calliandra calothyrsus*, * *Casuarina cunninghamiana*, *C. equisetifolia*, *Cordia alliodora*,* *Dipteryx panamensis*,* *Eucalyptus camadulensis*, *E. deglupta*, *E. grandis*, *Gmelina arborea*, *Guazuma ulmifolia*, and *Pithecellobium saman** (* indicates species native to Costa Rica, Torres 1985).

[7] Planting stock was purchased from a local nursery as the native *Terminalia oblong;* subsequently, the species was identified as *T. ivorensis* from West Africa.

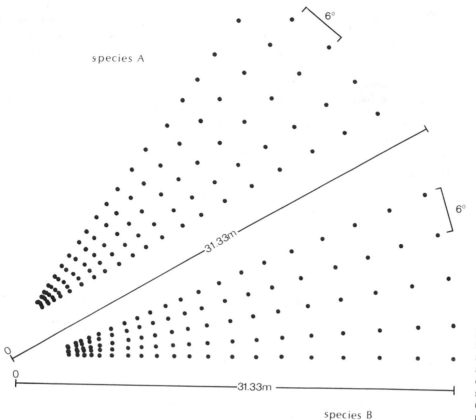

species A

6°

6°

31.33m

O

0

31.33m

species B

Fig. 25.2. Nelder design (two species shown here). Species are planted in arcs that connect to form a circle. Each arc or circle contains the same number of trees, which are at increasingly wider spacing with distance from the center (Torres 1985). Reprinted by permission.

cies with potential for rehabilitating marginal land. Additionally, data on seeds and nursery production for native timber species have been collected and used in programs to train local foresters and farmers in nursery management of native species. TRIALS researchers have also initiated pilot plantings of promising species on private farms and compiled information on previous experiences with native species in the region (OET/DGF 1990).

Screening trials have incorporated eighty-four timber species in replicated plots on 7.3 ha at four contrasting sites within La Selva. Emphasis is on native species with widely used exotics included for comparative purposes (64% are indigenous to the region and 83% to Costa Rica; see app. 25.1). Seeds of native species have been studied for their physical and biological characteristics, including fresh seed germination rates, seed weight, fruiting cycles and seed abundance, and seed collection techniques (González 1991). Most of the native species have proven surprisingly easy to propagate by seed, provided that seed is fresh.

A complete randomized block design is used with single tree plots. Each block, containing one tree of each species, is replicated twenty-four times per site. Blocks are contiguous with 3m × 3m spacing between all trees. Block sizes vary with the number of species being planted. See figure 25.3 for an example of the plantation layout.

Although it would have been ideal to plant all species at the same time, this was logistically impossible because seeds were, for the most part, not commercially available. Seed collection was necessarily opportunistic and techniques for ger-

minating and handling seedlings were developed as needed. Once seedlings of a minimum of eight species were ready for planting, sites were prepared and planted. Planting occurred only during the wet season (June–January).

The single tree plot design has several advantages: (1) it is a cost effective means to test a large number of species quickly; (2) border effects and the effects of one species dominating another are distributed randomly across the site and among all species; (3) because border effects are randomized, all trees are measured and there is a 1:1 tree planted to tree measured ratio; and (4) in highly variable sites such as hills and tree-invaded pastures, locating individuals of a species randomly over the site insures that it is thoroughly tested and helps to eliminate false results from microsite variations.

Sites were chosen to reflect common soil and topographic types within Sarapiquí. Emphasis was placed on marginal or degraded lands that are unsuited for agriculture or long-term grazing and, as such, are possible future targets for reforestation and land restoration programs. The four sites are pasture or abandoned pasture land. The parent material is of volcanic origin. Site 1 is the most fertile; it is flat and located on a weathered old alluvial terrace with Inceptisols (see chap. 4 for soil types and definitions). The pasture area has been abandoned since 1982. The other three sites are located on hills with residual soils that are highly acidic and heavily leached with very low base saturation (30%). These soils have been classified as Ultisols. Sites 2 and 3 are located in a shrub and tree-invaded pasture that has been abandoned since 1982. The second site was cleared of all vegetation before planting, cre-

Table 25.2 Third-year results for fourteen native hardwood species tested at La Selva Biological Station

Species	DBH (cm)	CV (%)	Height (m)	CV (%)	Basal Area (m²/ha)	(N)
Vochysia guatemalensis	10.1	24	8.0	12	17.7	105
Stryphnodendron microstachyum	9.8	41	6.5	29	17.6	101
Terminalia ivorensis[a]	8.7	34	8.2	26	15.7	120
Vochysia ferruginea	7.3	28	5.3	14	8.9	99
Hyeronima alchorneoides	6.2	40	5.3	34	7.9	109
Tabebuia rosea	5.1	27	4.5	20	5.4	123
Lonchocarpus ferrugineus	4.4	34	4.4	31	3.9	117
Dipteryx panamensis	4.3	49	5.5	42	3.5	99
Pterocarpus officinalis	4.4	52	4.7	37	3.3	86
Dalbergia tucurensis	3.6	31	4.7	22	2.6	118
Calophyllum brasiliense	3.7	38	4.2	25	2.5	101
Cordia alliodora	5.0	46	4.5	38	2.4	51
Schizolobium parahybum	3.7	67	3.0	60	1.8	60
Vitex cooperi	2.3	59	2.7	35	1.1	105

Source: Espinoza and Butterfield 1990 and unpublished data.

Note: DBH = diameter at breast height (1.3 m); CV = coefficient of variation; N = number of survivors from the original 125 trees planted.

[a]Native to West Africa, all others are indigenous.

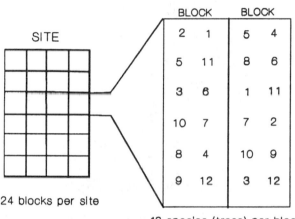

Fig. 25.3. Experimental design used for TRIALS Project: complete randomized blocks with single tree plots.

ating a full sun plantation. In site 3 the shrub and tree cover was maintained to lightly shade seedlings. Site 4 is in a recently abandoned pasture that had been in use for more than twenty-five years; it is considered the most degraded of the four sites.

Preliminary results from the species screening trials are presented in Table 25.3. Total heights for thirty-two species were summed and then ranked for each site. (The sum of heights is a measure of both growth and survivorship. See appendix 25.1 for species used in the ranking.) Additional columns in table 25.3 give number of survivors (of twenty-four trees planted per site) and average height of survivors for each species. The top ten species for sites 1 and 2 are presented in table 25.3. Although these lists have several species in common, differences reflect differing soil types. In general, the exotics grew well on site 1, which has relatively fertile, alluvial soils. The exotics, however, exhibited high mortality on site 2 with poorer upland soils. This mortality resulted, in part,

from higher herbivory. Two of the upland sites are interspersed with secondary forest, which shelters large nests of leaf-cutter ants. The large number of native species that exhibited good growth on the poor site indicates the potential usefulness of these species for reforestation of marginal land. The striking differences between these lists indicates that even within La Selva's limited range of site conditions, appropriate species for reforestation can vary widely (Butterfield 1990).

In 1988 a 12-ha plantation was established on the Peje Annex as part of the TRIALS project to monitor the effects of trees on degraded soils. Soil testing before planting measured pH, organic matter, total nitrogen, total carbon, cations, cation exchange capacity, extractable phosphorous, and bulk density. Five nitrogen-fixing (*Acacia mangium, Inga edulis, Pentaclethra macroloba, Pithecellobium macradenium, Stryphnodendron microstachyum*) and six nonnitrogen-fixing species (*Gmelina arborea, Hyeronima alchorneodies, Pinus tecunumanii, Virola koschnyi, Vochysia ferruginea, V. guatemalensis*) were planted in plots of 0.25 ha with four replicates. Species were selected for characteristics perceived to affect soil-building processes (high and low litter-fall, phosphorous or aluminum uptake, rooting depth) as well as for their potential for reforestation and wood production. Permanent inventory subplots have been established within each plot to monitor tree growth and for future silvicultural manipulation. Researchers are encouraged to conduct related studies in these plantations; they were established as a multipurpose resource for research on native trees (OTS 1987).

Overall, many of the species used in TRIALS have adapted well to pasture conditions and have demonstrated rapid growth rates (>2 m per year in height). The wide array of species incorporated into TRIALS is providing basic information for the production and planting of seedlings of many multiple-use species. Species that may have poor form or slow growth rates that would eliminate them from commercial plantations may, nevertheless, be useful for home gardens, agroforestry systems, or restoration projects.

Table 25.3 Top ten species ranked by height sums at two years of age for two contrasting sites at La Selva Biological Station

Site	Rank	Species	Origin	Height Sums	N	Average Height (m)
Site 1[a]	1	Eucalyptus deglupta	b	171.6	24	7.2
	2	Terminalia ivorensis	b	167.5	22	7.6
	3	Eucalyptus camaldulensis	b	128.9	21	6.1
	4	Jacaranda copaia	a	116.9	23	5.1
	5	Vochysia guatemalensis	a	104.4	23	4.5
	6	Dalbergia tucurensis	a	101.5	22	4.6
	7	Hevea brasiliensis	b	99.8	18	5.5
	8	Pinus tecunumanii	b	99.1	24	4.1
	9	Pinus caribaea var. hondurensis	b	97.0	24	4.0
	10	Laetia procera	a	96.7	24	4.0
Site 2[b]	1	Ochroma pyramidale	a	222.7	20	11.1
	2	Hevea brasiliensis	b	133.9	23	5.8
	3	Inga edulis	a	125.8	24	5.2
	4	Jacaran dacopaia	a	111.1	18	6.2
	5	Laetia procera	a	106.8	24	4.5
	6	Vochysia guatemalensis	a	103.7	24	4.3
	7	Dalbergia tucurensis	a	97.4	23	4.2
	8	Pinus tecunumanii	b	83.0	20	4.2
	9	Rollinia microsepala	a	82.3	20	4.1
	10	Gliricidia sepium	a	69.6	23	3.0

Note: Origin: a = species native to Costa Rica; b = exotic species. N = number of surviving trees. Average height is for surviving trees.
[a]Site 1: La Guaria Annex, with relatively fertile Inceptisols. Four of the top ten species are native.
[b]Site 2: Sarapiquí Annex, with relatively poor Ultisols. Eight of the top ten species are native.

Pilot plantings in 1989–1990 have established 16.5 hectares (sixty-six plots) of experimental plantations on private farms of the most promising native species based on results to date (*Vochysia guatemalensis, V. ferruginea, Stryphnodendron microstachyum, Hyeronima alchorneodies,* and *Jacaranda copaia;* González 1990). Species are planted in 0.25-ha plots and replicated across farms to incorporate different site types within Sarapiquí. Data will be taken on these plots for a minimum of five years.

The objective of the second phase of research (OTS/ITCR TRIALS II) is to develop several of the most promising native species for large-scale reforestation within the region. The popularity of many of the species with local farmers and high demand for seedlings has oriented the project toward seed tree conservation and provenance/progeny trials that can later be converted into seed production areas. Research priorities within the project include seed storage tests, genetic selection of fast-growing trees of good form, and improved nursery production. Growth and yield data will be generated through existing project and farmer plots located throughout the region.

The first regional interinstitutional working group has been organized for the Northern zone to coordinate experimental designs, data collection, and information dissemination between research, education, and development agencies and NGOs (nongovernmental agencies) interested in the development of native timber species. Members include OTS, ITCR (Instituto Tecnológico de Costa Rica), DGF (Seed Bank), FUNDECOR, EARTH, COSEFORMA (Cooperación en los sectores forestal y maderero), and CATIE (Genetic Improvement program). Seed trees of ten priority native species for the Northern zone (*Cordia alliodora, Calophyllum brasiliense, Dipteryx panamensis, Hyeronima alchorneoides,*

Stryphnodendron microstachyum, Terminalia amazonia, Virola koschnyi, Vochysia guatemalensis, V. ferruginea, and *Zanthoxylum mayanum*) are being identified and protected from Upala to Guápiles by COSEFORMA, OTS, ITCR, and FUNDECOR. Provenance/progeny (seedlings kept separate by mother tree) trials of two to four species will be established in 1993/94 on private farms and at collaborating institutions (ITCR, OTS, EARTH).

Yale University has initiated an additional forestry research program at La Selva to build upon TRIALS results. Promising species identified through TRIALS have been incorporated into monoculture and four-species polyculture plantations to investigate species/site interactions. Studies will include changes in soil properties, microclimate, nutrient availability, and nutrient cycling. A total of twelve native species have been incorporated into three mixed-tree plantations at 2 × 2 m spacing (Montagnini 1990b). The first plantation was established in mid-1991 with *Jacaranda copaia, Calophyllum brasiliense, Stryphnodendron microstachyum,* and *Vochysia guatemalensis.* Another was established in 1992 with *Albizia guachapele, Dipteryx panamensis, Terminalia amazonia,* and *Virola koschnyi.* The last plantation, established in 1992/93, contains *Genipa americana, Hyeronima alchorneiodes, Pithecellobium elegans,* and *Vochysia ferruginea.*

Because the Costa Rican government is preparing to spend millions of dollars on reforestation in the near future, the development of technical information on appropriate species for timber production, soil improvement, and land restoration is critical. Reforestation and plantation research at La Selva has laid the groundwork by developing basic information on promising species and extensive experimental plantations. So little is known about the management and ecology of native timber species in the tropics that the research agenda is broad and urgent.

CONCLUSIONS

Accelerating deforestation rates throughout the tropics have caused global concern for the lack of stewardship of forest resources and underlined the need for forest protection and management. The emerging forestry profession in tropical countries urgently requires information on which to base management decisions for the rational use of tropical forests. Although one century of research has provided data for development of management systems in the temperate zones, most of these models are not directly applicable to the tropics. The socioeconomic and silvicultural conditions faced by tropical foresters are sufficiently different to require new models.

Increased demands on the rapidly diminishing forests of Costa Rica point to the urgency of implementing sustainable management practices in both natural forest and plantation forestry. Land managers and development agencies cannot wait for the results of long-term research; thus, forest management and related research must be done simultaneously. Researchers need to collaborate with ongoing development projects, tailoring research goals to development needs to provide the scientific basis for a more rational use of land and forest resources.

Although there has been more than thirty years of forest ecology research at La Selva, the knowledge gained has not been widely employed toward more overtly applied research agendas nor communicated to tropical foresters in the field. More collaboration is needed to bridge the gap between basic ecological research and its application. One way to accomplish this goal is for ecologists to participate in development projects. Although scientists have expressed interest in such arrangements, development agencies are slow to recognize the need to fund research in conjunction with development activities. Scientists are often remiss in not perceiving the implications of their research for development purposes or in not including some "applied" goals within their "pure" research objectives.

Reforestation and plantation research at La Selva have explicitly been directed at development needs. The results of these projects will be useful only to the extent that they are disseminated to policymakers, extensionists, and development agencies. The growth data on native species have been of interest to the DGF, which has already allowed several "new" native species to be planted experimentally within the region, using government fiscal incentives. Regional workshops to disseminate results and publications in Spanish have also helped get the information to those who need it. Much more is needed.

OTS has been successful in promoting tropical research. As part of its strategy, OTS has not only provided facilities but also used courses and fellowships to promote interest in a particular field of research (tropical biology and, more recently, agroecology). It could use the same model to promote research in forestry and natural resource management. The human and scientific resources of OTS, through its fifty member institutions, can have a positive impact by focusing research efforts toward sustainable forest development. A tropical forest research station cannot do less than strive to provide the scientific basis for the rational use of the forest.

ACKNOWLEDGMENTS

En especial se agradezca a la Dirección General Forestal, quien inició la primera plantación experimental en La Selva y quien continua tomar un papel collaborador en la investigación forestal en La Selva. Plantation research and community forestry outreach by OTS have been supported with grants from the John D. and Catherine T. MacArthur Foundation, H. John Heinz III Charitable Trust, World Wildlife Fund, Andrew J. Mellon Foundation, Weyerhaeuser Company Foundation, and the Canadian Embassy. The author thanks L. McDade, G. Minnick, C. Schnell, R. Kellison, and J. Denslow for comments on earlier drafts of the manuscript as well as Gilbert Canet, director Desarrollo Campesino DGF, and Rodolfo Peralta, director of research, Portico S.A., for their time and information.

Appendix 25.1 Species tested in the OTS/DGF species screening trials at La Selva Biological Station

Species Name	Family
Exotic Species widely tested in the lowland humid tropics	
Araucaria hunsteinii[a]	Araucaricaceae
Acacia mangium[b]	Mimosaceae
Erythrina poeppigiana[a]	Papilionaceae
Eucalyptus camaldulensis[a]	Myrtaceae
Eucalytpus deglupta[a]	Myrtaceae
Eucalyptus saligna[a]	Myrtaceae
Gmelina arborea	Verbenaceae
Hevea brasiliensis[a]	Euphorbiaceae
Pinus caribaea var. hondurensis[a]	Pinaceae
Pinus tecunumanii[a]	Pinaceae
Tectona grandis	Verbenaceae
Terminalia ivorensis[a]	Combretaceae
Terminalia superba[b]	Combretaceae

Appendix 25.1 (continued)

Species Name	Family
Species indigenous to the region	
Brosimum lactecens	Moraceae
Calophyllum brasiliense	Clusiaceae
Carapa guianensis	Meliaceae
Castilla elastica[a]	Moraceae
Cedrela odorata[a] (= mexicana)	Meliaceae
Cordia alliodora[a]	Boraginaceae
Cordia bicolor	Boraginaceae
Cordia megalantha	Boraginaceae
Dalbergia tucurensis[a]	Papilionaceae
Dipteryx panamensis[a]	Papilionaceae
Genipa americana[a]	Rubiaceae
Goethalsia meiantha	Tiliaceae
Hampea appendiculata[a]	Malvaceae
Hernandia didymanthera	Hernandiaceae
Hyeronima alchorneoides[c]	Euphorbiaceae
Hymenolobium mesoamericanum (= pulcherrimum)	Papilionaceae

Appendix 25.1 (continued)

Species Name	Family
Ilex skutchii	Aquifoliaceae
Inga coruscans	Mimosaceae
Inga edulis[a]	Mimosaceae
Inga longispica	Mimosaceae
Inga thibaudiana	Mimosaceae
Jacaranda copaia[a]	Bignoniaceae
Lacmellea panamensis[a]	Apocynaceae
Laetia procera[a]	Flacourtiaceae
Lecythis ampla (= costaricensus)	Lecythidaceae
Miconia multispicata	Melastomataceae
Minquartia guianensis	Olacaceae
Nectandra kunthiana	Lauraceae
Nectandra membranaceae	Lauraceae
Ochroma pyramidale[a] (= lagopus)	Bombacaceae
Ormosia macrocalyx	Papilionaceae
Otoba novogranatensis (= Dialyanthera otoba)	Myristicaceae
Pentaclethra macroloba[a]	Mimosaceae
Pithecellobium elegans (= pedicellare)	Mimosaceae
Pithecellobium macradenium	Mimosaceae
Pouteria spp.	Sapotaceae
Pterocarpus rohrii	Papilionaceae
Rollinia microsepala[a]	Annonaceae
Sclerolobium guianensis	Caesalpiniaceae
Simarouba amara	Simaroubaceae
Stryphnodendron microstachyum[a] (= excelsum)	Mimosaceae
Tabebuia guayacan	Bignoniaceae
Terminalia amazonia[b]	Combretaceae
Terminalia oblonga (= chiriquensis, lucida)	Combretaceae
Virola koschyni[a]	Myristicaceae
Vochysia allenii	Vochysiaceae
Vochysia ferruginea[a]	Vochysiaceae
Vochysia guatemalensis[a] (= hondurensis)	Vochysiaceae

Appendix 25.1 (continued)

Species Name	Family
Zanthoxylum mayanum	Rutaceae
Species native to other parts of Costa Rica	
Albizia guachapele (= Pseudosamanea guachapele)	Mimosaceae
Bombacopsis quinata	Bombacaceae
Brosimum utile	Moraceae
Dalbergia retusa[a]	Papilionaceae
Dilodendron costaricense (= Dipterodendron costaricense)	Sapindaceae
Enterolobium cyclocarpum[a]	Mimosaceae
Gliricidia sepium[a]	Papilionaceae
Myroxylon balsamum	Papilionaceae
Ocotea spp.	Lauraceae
Ormosia spp.	Papilionaceae
Pithecellobium arboreum	Mimosaceae
Pithecellobium idiopodum (= pseudotamarindus)	Mimosaceae
Pithecellobium saman[a]	Mimosaceae
Pseudobombax septenatum	Bombacaceae
Simarouba glauca	Simaroubaceae
Tabebuia rosea	Bignoniaceae
Little-known exotic species	
Guazuma crinita[a] (from Peruvian Amazon)	Sterculiaceae

Note: Species names in parentheses are those used traditionally at La Selva for taxa that have been revised nomenclaturally.

[a]One of thirty-two species planted at four sites at La Selva Biological Station used for ranking in table 25.3.

[b]Two provenances.

[c]As H. oblonga in Espinoza and Butterfield 1990.

26

Prospects for a Comparative Tropical Ecology

Gordon H. Orians

In a synthesis chapter one should provide some integration of previous material that is not, indeed cannot be, captured in individual chapters. I could delve into many of the themes touched upon in previous chapters, or I could select a few themes for special emphasis. I chose the latter option because several themes capture most key aspects of the rich and varied research that has been carried out at La Selva. I first assess the contributions of research at La Selva to understanding of the causes of tropical species richness patterns. Because biologists have been extensively preoccupied with the large number of species in tropical ecosystems, this theme pervades many chapters of this book even though individual authors hold different perspectives on it and its appropriateness as a dominant theme in tropical biology. I then narrow my focus to individual species and examine the adaptations of organisms to tropical climates, soils, and biological interactions as revealed by studies at La Selva. The range of variation in climate and soils is less at La Selva than in the tropics as a whole, but a great deal has been learned at La Selva about adaptations of organisms in many different taxa to climatic and edaphic conditions in tropical wet forests. This information is intrinsically interesting, and it also forms the basis for broader comparisons with attributes of organisms in tropical environments where dry seasons are longer and more severe and where soils are more impoverished.

This approach leads naturally into a third theme—comparative tropical ecology. Many basic ecological and evolutionary questions, most of them as yet poorly answered, are comparative. What are the adaptations of organisms to increasing length of dry season? How do those adaptations influence patterns of community organization? To what extent can patterns in ecological communities be predicted on the basis of knowledge of the physical environment? Do similar communities evolve in different parts of the world with similar climates even though the biota had very different starting points? Given that community convergence is incomplete, are there components in which convergence is more striking? What types of information must be gathered to provide a basis for an insightful series of comparisons of tropical forests worldwide?

This book is both a progress report and a look into the future. Therefore, I end the chapter with some suggestions about the challenges currently facing tropical ecology and how research stations like La Selva can play creative roles in meeting those challenges. Predicting the future is always risky, but if researchers plan for the future they may be less surprised and less controlled by it than if they march into it unthinkingly.

SPECIES RICHNESS

The great richness of species in the lands where winter never comes has dominated the attention of biologists since the first European scientists visited the tropics. The striking latitudinal gradients in species richness of most taxa strongly invite, if not compel, speculation concerning their causes (Pianka 1966). Theories assign different weights to two components whose relative rates determine species richness: the rate of speciation and the rate of extinction. One theory maintains that, during most periods of geological history, speciation rates exceeded extinction rates. If so, the number of species would increase with time except during intervals of mass extinctions. In this view the tropics are thought to have more species because of the long time over which species have accumulated. At higher latitudes, however, the vicissitudes of the Pleistocene caused massive extinctions and shifts in species ranges from which the temperate zone is still recovering. According to this theory, the temperate zone might eventually approach the tropics in species richness if environmental conditions persisted relatively unchanged for sufficient time.

Most theories, however, have invoked current ecological interactions to explain patterns in species richness. For example, some theorists have suggested that benign tropical climates permit the survival of more species because only a few can tolerate the more rigorous conditions of higher latitudes. Benign climates are also postulated to cause the reliable presence of a wider variety of resources throughout the year. This dependability might allow more species to persist, both because niches can be smaller and because more resources are present. Alternatively, tropical climates are thought to be favorable for predators so that control of prey is more effective, resulting in fewer opportunities for competitive exclusion. The postulated result is more niche overlap than in the temperate zone. The fact that there are so many theories, not all of which are compatible, indicates how much biologists have yet to learn about the factors influencing species richness anywhere, even in the comparatively well-studied temperate regions. Nor is there reason to believe that any one theory is universally applicable.

Imposed on latitudinal patterns of species richness are

equally interesting patterns associated with local to regional climatic and soil differences and, for animals, differences in the structural complexity of plant communities. Nowhere in the world is the number of bird species in grasslands as large as it is in adjacent vegetation dominated by trees. Tropical plant species richness is, in general, inversely correlated with the length of the dry season and with elevation. Disturbance also influences species richness, but disturbance comes in many forms. It can either increase or decrease species richness depending on its type, intensity, frequency, extent, and duration.

Plant Species Richness

Neotropical forests are all similar floristically. The same families of plants dominate the four intensively studied Neotropical forests sites (Gentry 1990a). South American sites are richer in tree species than Central American sites, but Central American forests may be richer in epiphytes and understory herbs and shrubs (Gentry 1990a). Thus, 2–4 ha plots at La Selva have about 100 tree species larger than 10 cm dbh (Hartshorn 1983b), and 1-ha plots on BCI have about 93 tree species larger than 10 cm dbh (Foster and Hubbell 1990). In contrast, in upper Amazonia between 155 and 283 tree species have diameters larger than 10 cm dbh (Gentry 1988). The Manaus area, despite its long dry season and poor soils, has 179 tree species larger than 15 cm dbh in a 1-ha plot (Prance *et al.* 1976). The causes of this striking difference are unknown, but Gentry (1990a) speculates that the existence of a greater range of habitat types and the evolution of species adapted specifically to all those habitats create greater opportunities for "habitat spillover" in South America than in Central America where soils are uniformly richer.

Plants provide both the physical structure for most terrestrial communities and the resources that support all other species. The environmental resources that support plant growth—sunlight, water, and mineral nutrients—are universally required. Plants do differ in how much light they require for growth, their drought tolerance, and the minimal amounts of nutrients they require to sustain growth. Those differences, however, appear inadequate to explain plant species richness. Such differences fall far short of explaining the richness of La Selva where the vascular flora contains as many as two thousand species, more than four hundred of which are trees.

One promising approach to explaining the coexistence of plants competing for similar resources has been developed by David Tilman (1982, 1988). He suggests that the winners are those species able to reduce essential resources to lower levels than their competitors. Thus, if renewal rates of critical resources are slow enough that competing plants can maintain them at quite low levels, slight differences in nutrient availability could favor different species of plants. Relatively constant balances between resource depletion and resource renewal could be widespread in tropical forests although such balances have never been demonstrated empirically. There is, however, ample evidence that shifts in species abundances accompany minor changes in soil type in many different tropical regions (Ashton 1964, 1976b; Austin et al. 1972; Baillie et al. 1987).

Relationships between resource depletion and regeneration

are disrupted by such events as tree falls that provide inputs of some nutrients and temporarily reduce or increase the rate of their withdrawal. The mosaic of conditions within gaps has been postulated as an important contributor to tropical tree species richness (Brokaw 1985b; Denslow 1980a; Hartshorn 1978; Orians, 1982). On the one hand, the distribution of species regenerating in gaps suggests that patchiness on this scale may be important (Brandani et al. 1988). On the other hand, the postulated nutrient differences between bole and crown zones of gaps have not been detected at La Selva (Vitousek and Denslow 1986), nor is the concentration of fine root biomass different in crown zones than in other parts of gaps (Sanford 1989). Nonetheless, comparative studies, in which a variety of techniques are used to measure carbon and nutrient processing and cycling by forests on different soils in different successional states and in different gap zones are needed to resolve major uncertainties. The importance of slight differences in soil fertility on plant species distributions on infertile soils receives empirical support from the fact that extremely species-rich plant communities are found on ancient, highly infertile soils in the fynbos of South Africa and the impoverished ancient soils of western Australia (Whittaker 1977; Naveh and Whittaker 1979; Rice and Westoby 1983).

Better comparative studies in the tropics are needed because the most convincing evidence of the role of soil fertility on plant species richness comes from extratropical regions. Both La Selva and Barro Colorado Island occur on relatively rich, volcanic tropical soils. Much poorer soils are found in the Manaus region and other areas in South America with ancient soils. Within La Selva the three major soil types do differ in fertility (chap. 4), but these differences span only a modest fraction of the range found in tropical soils. Because a good soil map has become available only recently, comparative studies of ecological processes and products on different soil types have only begun at La Selva. Indeed, the lack of such comparative studies is a striking feature of the current volume. This is a promising subject for future research. As I will document, results from Manaus and Cocha Cashu support that soil infertility strongly influences the structure and functioning of ecological systems (Gentry 1990a).

The role of tree-fall gaps in maintaining tropical tree species richness has attracted a great deal of attention in recent years. Much work, both theoretical and empirical, has been carried out at La Selva. Most empirical research has attempted to infer the nature of regeneration processes in gaps from measurements of species compositions at single moments in time. Researchers, however, may be approaching the limits of what can be learned from such analyses of distributions of seedlings and saplings in gaps. Additional insights may require long-term studies of seedling survival so that spatial and temporal mortality patterns are demonstrated. The current gaps project at La Selva, in which distributions of understory plants have been mapped before the predetermined felling of trees, is an important new dimension to gap studies. This procedure, however, produces gaps without root zones. Studies of seed inputs to gaps have compared gaps with the adjacent undisturbed forest, but researchers lack data on distributions of seed inputs in relation to gap zone and distance from the edge of the gap (Augspurger and Franson 1988). Inputs of bird-dispersed seeds should be higher at the margins of gaps because birds defecate mostly while perched rather than in flight. Indeed, densities of seedlings of gap-requiring species are, at times, higher on the edges than in the center of gaps (Popma et al. 1988). Unfortunately, no one knows how these input patterns affect distributions of trees when they reach the canopy.

The influence of predators on the distributions of tropical plants was stimulated by the early work of Janzen (1970) and Connell (1971, 1978). They suggested that mortality rates of seeds and seedlings close to parent plants were so high that successful recruitment was much more likely at some distance from the parent. The original theories, although they were stimulated by empirical observations, suffered from imprecision concerning spatial scales. Janzen's original model is quite compatible with clumped, random, or hyperdispersed distributions of tree species, depending on the scale at which the processes operate and the scale at which measurements are made. Therefore, most earlier "tests" of that theory were inappropriate, but they did demonstrate that, at scales of square kilometers, most tropical tree species have clumped distributions (Hubbell 1979; Thorington et al. 1982).

Attempts to demonstrate the role of herbivores on distributions of tree species at La Selva have yielded equivocal results. As demonstrated by the Liebermans (chap. 8), if herbivores are preventing successful regeneration of trees of most species close to parent trees, the distance over which they are effective is very short at La Selva. Distances appear to be too short for predation to favor the coexistence of tree species, even though it may have subtle effects on local distribution patterns. Mortality rates from predators do decline with distance for at least some species. At La Selva, seedlings of *Dipteryx panamensis,* which initially are relatively dense under adult trees, suffer higher mortality rates from herbivore damage than seedlings farther from adult plants (D. A. Clark and Clark, 1984; D. B. Clark and Clark 1985). On BCI, saplings whose nearest neighboring canopy tree is conspecific grow more slowly and have higher mortality rates than saplings whose nearest canopy neighbor is heterospecific (Hubbell and Foster 1990b). Nonetheless, dispersal may be powerful enough to override predation as the dominant force affecting small-scale distribution patterns. Measures of the relative strengths of these counterbalancing forces are needed to determine the causes of intraspecific spacing patterns.

Because of the sensitivity of measures of clumping patterns to the scale at which measurements are made, more studies are needed of distribution patterns on multiple scales similar to those provided by Baillie et al. (1987) in Sarawak. Now that a good soils map of La Selva is available, measures at different spatial scales can use information on soils and changes in topography in the design of sampling locations and procedures. Clumped distributions of tree species are typically found when sampling units vary in soils and topography. Much attention will have to be given to determining appropriate null distributions. Nonrandom distributions are certain to be generated by structured patterns of seed inputs, gap formation, and predation. Thus, simple random null models are of little value in attempts to untangle the complex web of interacting factors that determine tree distribution patterns.

Animal Species Richness

The fauna at La Selva is poorly known. More than 4,000 species of moths and about 500 butterflies probably exist, but one cannot estimate even orders of magnitude for most other invertebrate taxa. The vertebrates, which are quite thoroughly sampled, have the richness expected of a wet tropical forest site in Central America. La Selva has fewer species of anurans than the intensively studied South American sites but as many species of lizards and snakes (Duellman 1990). In birds, however, La Selva is less rich than some Amazonian Basin forests. South American forests support more resident species of birds than Central American forests (244 and 251 species for La Selva and central Panama, respectively, compared to 300 and 332 species for Manaus and Cocha Cashu, respectively). Taxonomic diversity of forest birds at the generic level is as much as 20% lower in Central than in South America, but avian diversities are comparable at the familial level (Karr, Robinson, et al. 1990). These patterns, which reflect the greater regional avian species richness in South America compared to Central America, indicate that regional processes strongly influence local species richness (Ricklefs 1987).

Invertebrate species richness should be positively correlated with plant species richness because many invertebrate species are monophagous or oligophagous. For some groups, however, especially vertebrates, structural features of the environment rather than number of plant species exert the most powerful influences on species richness patterns. Why these larger animals do not view the world in a fine-grained way is intuitively obvious. It is less obvious, however, why species richness among some animal groups is inversely related to plant species richness. The most striking example comes from the wet-dry tropical regions of northern Australia where the plant-species-poor and regularly burned *Eucalyptus* woodlands have much higher numbers of ant, termite, and lizard species than adjacent monsoonal forests that are less often burned and have many more plant species (Braithwaite 1987; Braithwaite et al. 1988; Anderson pers. comm.). The converse, more intuitively reasonable pattern is found in the fauna of the Manaus area where poor soils and low productivity are associated with fewer species and lower densities of animals than in areas with richer soils (Lovejoy and Bierregaard 1990).

The causes of these surprising correlations are unknown, but the taxa involved are all primarily terrestrial ones that exploit resources gathered on or near the soil surface. Braithwaite (1987) suggests that the high frequency of fires in tropical *Eucalyptus* woodlands creates a mosaic of conditions to which different lizard species are adapted. The importance of ants and termites as processors of plant remains is inversely correlated with soil fertility and, hence, the quality of litter (Morton and James 1988). The dominance of these insects, in turn, favors lizards because they are better adapted for exploiting those prey than are birds and mammals. Birds and mammals have higher daily energy requirements and poorer abilities to survive long periods of low resource availability than do lizards.

Research in arid regions has demonstrated the importance of plant architecture on the number of species of arthropods that live and feed upon the tissues of different plant species (Lawton and Schroeder 1977; Moran 1980; Strong and Levin

1979). Larger and structurally more complex plants have a greater variety of tissues upon which herbivores can feed. In addition, more complex plants offer a greater variety of hiding places and more substrates to mimic. Exiting data from deserts suggest that the role of escape space from predators is more important than the variety of tissue types available (Schultz et al. 1977). The same patterns may exist in tropical forests but the relative roles of feeding ecology and escape from predators in generating those patterns is still unknown.

Whereas it is widely accepted that tropical herbivorous insects are more specialized dietarily than their temperate counterparts, this generalization, like so many about the tropics, is based on very scanty evidence (Beaver 1979). Data gathered at La Selva indicate that butterfly and grasshopper species are more host specific than their temperate counterparts, whereas treehoppers are less host specific than North American species (chap. 21). There is little information on how the number of species of arthropods feeding upon tropical plant species is related to their geographical distribution and abundance. Data from La Selva suggest that the number of insects associated with any one plant species is positively correlated with both local host abundance and overall host range. For some cultivated crops, such as cacao and sugarcane (Strong 1974; Strong et al. 1977), species richness of herbivores eating them is positively correlated with hectares of the crop in the region. Interestingly, this correlation holds both for regions where the crop is native and those where it is relatively recently introduced, suggesting that such patterns may develop fairly rapidly. Gathering such information for a large number of species of plants at La Selva will be a formidable task. Intensive studies of a few genera whose species differ widely in geographical range, number of habitats used, and local population densities are needed to determine if repeatable patterns exist of species richness of tropical herbivorous arthropods.

The numbers of species in some groups of predatory and parasitic arthropods are positively correlated with the species richness of their prey (Price 1980). For other groups such relationships do not hold. Insectivorous bird species richness is poorly correlated with plant species richness on small scales, even though species richness of their prey is positively correlated with plant species richness. For these groups general structure of the vegetation is more important because it influences the number of viable foraging modes in a given vegetation (MacArthur 1971; Karr and Roth 1971; Cody 1975).

The many gaps in knowledge of species richness and its determinants, combined with what researchers already know about patterns and causes, suggests that a search for broad intertaxonomic generalizations is unlikely to be successful. Nonetheless, prospects for a good understanding of patterns of species richness for specific taxonomic groups appear very good. La Selva is an excellent place for continuing studies of patterns and determinants of those patterns of species richness because the base of knowledge of organisms of many taxa is already extensive, and La Selva's biota is very rich. Projects now underway on patterns of arthropod species richness at La Selva should show how, and perhaps why, arthropod species are associated with plant species. Further studies of gap dynamics and relationships between plant species and soil types will throw further light on how so many species of woody plants can exist in a single tropical forest.

ADAPTATIONS OF INDIVIDUAL SPECIES

Species richness is not the direct object of natural selection. Rather, it is the by-product of interactions among species whose traits *are* influenced directly by natural selection. Therefore, the study of adaptations of species and how they are influenced by their physical and biological environments is an essential component of tropical ecology. Studies of individual species and genera have played prominent roles at La Selva. Data gathered from these studies form the basis for interpreting patterns at higher levels of ecological organization. Little progress, however, has been made in interpreting community-level patterns as functions of underlying species adaptations and interactions.

Life-history Traits

Life-history theory has emerged as a major component of "microecological" studies, and plant life-history traits have received considerable attention during the late-1970s and early-1980s (Bawa 1979, 1983; Bawa and Beach 1981; Bawa and Opler 1975; Beach 1981; Willson 1983; Willson and Burley 1983; Charnov 1982; Westoby and Rice 1982). Key ideas in recent theoretical developments have emerged from the recognition that the total fitness of hermaphroditic plants is a function of their contributions as both male and female parents and that selection normally favors equal investment in offspring of both sexes (Fisher 1929). Much work has been stimulated at La Selva by the recognition that the alternation of generations in plant life histories and the dependence of the gametophyte on the sporophyte among gymnosperms and angiosperms provide excellent conditions for intergenerational conflict over allocation of resources.

Extensive research on plant breeding biology at La Selva has revealed that most forest trees, even rare ones, are strongly outcrossed. These studies have also demonstrated relationships between breeding systems and modes of pollination and seed dispersal (chap. 12). Recognition of these patterns is stimulating development of new hypotheses and new field tests concerning the evolution of the complex and varied breeding systems of plants.

The high incidence of dioecy among tropical woody plants has attracted special attention (Bawa 1980a; Croat 1978; Hammel 1986a). At least 17% of the total vascular plant flora of La Selva is dioecious (chap. 12). This is a paradox in a species-rich forest where so many species are rare, making it especially difficult for individuals to exchange gametes. For a given population density the density of potential mates is approximately half as great for a dioecious species as for a hermaphroditic one. Therefore, other things being equal, one would expect fewer dioecious species in tropical regions than in temperate ones, especially if pollinators are frequently limiting. Clearly, other things are not equal, but it is not clear which factors are the most unequal. For only a few of the best studied dioecious tropical species do biologists know how effectiveness of pollen donation and seed set vary as a function of the isolation of individual plants, their age, and the size of their reproductive investment in any given year. These measurements are needed to determine the conditions under which a unisexual genotype can invade a population of hermaphrodites.

Most research on the physiological ecology of plants at La Selva has been based on the premises that species with similar ecological roles or similar growth-forms have similar physiological responses and metabolic rates and that physiological and morphological plasticity within and among species is correlated with the range of conditions experienced by those species (chap. 10). The value of these premises for organizing research is clear. Nonetheless, strict adherence to them may inhibit creative research to determine the degree to which plants of similar growth-forms living in the same environments use resources differently and to investigate how these differences contribute to interspecific relationships in, and composition of, plant communities. Biologists need to determine which are source and which are sink environments for plant species (Pulliam 1988; Pulliam and Danielson 1991). Without this information they cannot determine which habitat types are essential for the survival of the species. Knowledge of population dynamics in variable environments may also help them understand the evolution of some traits of species. For example, a plant may be found growing in a variety of environments but, if successful reproduction occurs in only one or a few of them, selection is unlikely to produce adaptations to the full range of conditions in those habitats.

Dioecy dominates the animal world of La Selva, but how these individuals associate with one another for reproduction is highly varied. Until recently, ornithologists believed that most birds were monogamous with a small number of polygynous species in a few families and some environments, such as marshes (Verner and Willson 1966; Lack 1968). The advent of color banding, combined with long-term studies, has revealed a high incidence of communally breeding species in tropical and subtropical latitudes (Brown 1987). Indeed, more communally breeding species are known from Australia and Africa than the rest of the world combined, even though avifaunas on other continents are better studied. The proportion of species known to be communal breeders is still low (158 of 5,274 species of passerines), but the number is rising steadily. Existing evidence indicates that communal breeding is especially prevalent among sedentary species that saturate their habitats so that young individuals have very limited independent breeding opportunities (Brown 1987). These conditions may be widespread among tropical forest birds, but, primarily because of the difficulties of working with those species compared to species of more open habitats, relatively little is known about breeding systems of birds in tropical wet forests. Details of the breeding organization of most avian species at La Selva remain to be determined. If, as currently appears to be the case, communal breeding is rarer among forest species than among those of savannahs, theories of the conditions favoring such breeding systems may need to be modified. Alternatively, habitat saturation among tropical forest birds may be rarer than currently judged.

It is becoming increasingly clear that many tropical organisms make seasonal movements to special habitats where they survive better during unfavorable periods. Viability of populations in large regions may, in fact, depend upon the existence of key small areas (Janzen 1984, 1986a). Individuals that survive unfavorable periods in those refuges spread out and colonize large areas during subsequent favorable periods. Recognition of the importance of seasonal migration to many species at La Selva was an important stimulus to the efforts to link the reserve to Braulio Carrillo National Park. This corridor connects La Selva and the mid- and high-elevation forests on the slopes of Volcán Barva. Researchers know which birds use this corridor for seasonal movements, but seasonal movements of insects, which may be even more widespread, are mostly undescribed. Data on seasonal movements of insects into and out of the tropical dry forests of Guanacaste (Janzen 1984) suggest that many insect species migrate seasonally to the Caribbean slope of Costa Rica. Thus, where a species can live depends upon patterns of different habitat types on several different spatial scales from local to regional. Because different species respond to different scales, distribution patterns are the result of a complex series of overlapping patterns whose influences are as yet only dimly perceived.

Demography of Tropical Organisms

The study of the demography of tropical organisms is in its infancy, but information on tree mortality from long-term study plots in Neotropical forests is considerable. Annual tree mortality averages about 1%–2% (Hartshorn 1990). Hubbell and Foster (1990b) attribute the high mortality rate (3%) on BCI to the long and severe dry season in 1983 associated with El Niño. Lower mortalities were reported during the years before the drought. Except on BCI, no size-specific mortality patterns have been detected among tropical trees (Hartshorn 1990).

Historically, La Selva has had a paucity of scientists in long-term residence and a corresponding paucity of long-term demographic studies of plants and animals. Long-term studies of tropical plants being carried out at La Selva in the 1990s (D. A. Clark and Clark 1989b), hold great potential, but existing data are insufficient for anything but tentative suggestions about demographic traits of tropical plants. Growth and reproductive rates of plants in shaded understory are very low, but this is not surprising. Physical damage from litter fall is high, and many species have very high turnover rates. Moreover, plant demography in the tropics, as elsewhere, has been severely hindered by researchers' inability to determine the parents of the readily countable seedlings and saplings. Seed shadows have been measured for isolated trees where sources of seed contamination were unlikely, but in most locations even the female parent of a seedling cannot be determined reliably on the basis of its physical location. Therefore, scientists have no data on relative genetic contributions of different individuals, either as male or as female parents, for any tropical plant. New molecular methods, such as DNA "fingerprinting," now offer the possibility of determining the parents of young plants. Such data can provide a sound basis for constructing life tables and determining comparative reproductive success of individuals in plant populations.

Information on reproductive success, both as a male and a female parent, as a function of degree of isolation of individuals, their age, and their reproductive investments in current and past bouts of reproduction is being gathered for tropical plants at La Selva and elsewhere (Bullock et al. 1983; D. B. Clark and Clark 1988). The considerable body of evidence indicating that many individuals flower irregularly, even though the species may flower every year, suggests that plants may store energy for considerable periods and then produce much larger seed crops than would be possible on a single year's surplus. The extent of this pattern and the relationship between reproductive success and the size of the reproductive effort need to be determined if scientists are to understand the complex spatial and temporal structure of plant populations.

Existing knowledge of tropical animal demography is also limited. Among birds, the best known group, breeding seasons are, on average, longer, clutch sizes are smaller, and nest survival rates are generally lower than in temperate environments (Cody 1966; Ricklefs 1966). Annual production of offspring per breeding female is not known for more than a handful of species. Annual survival rates of adult birds are higher in the tropics than in temperate areas, but data are few and ecologists do not agree on the generality of the patterns (Karr, Nichols, et al. 1990). Thus, basic demographic differences between temperate and tropical species of even the better-studied taxa are still uncertain.

Biologists are equally ignorant of the comparative degree of fluctuations in population densities among tropical and temperate species. Population densities of *Anolis limifrons,* an abundant lizard on BCI, fluctuate more than reported for any temperate zone lizard (Andrews and Rand 1989), and insect populations appear to fluctuate as much on BCI as they do in the temperate zone (Wolda 1978a). Densities of territorial adults in some BCI antwrens are nearly constant among years (Greenberg and Gradwohl 1986). The simple expectation that tropical populations are more stable than temperate ones evidently is only partly correct, and one cannot as yet predict for which taxa and under what conditions it is true.

Patchy Environments

Most populations in nature are metapopulations. Occupied patches are separated by stretches of unsuitable habitat across which individuals disperse. Immigrants enrich the genetic resources of local populations and recolonize patches from which the population has been extirpated. Searching individuals encounter habitats of varying quality some of which are already occupied and some of which are vacant. Choice mechanisms evolve in relation to population pressures, proportions of habitats that are suitable, difficulties of searching, and risks of mortality while searching (Levins 1962; Lande 1987).

Temperatures evidently varied substantially in the tropics during the Pleistocene. Precipitation patterns also alternated between wetter and drier periods during which forests expanded and contracted in opposition to savannas and other more open vegetation types. Debate continues over the locations of forest refugia and how important they were for speciation and current patterns of tropical species richness (Diamond and Hamilton 1980, Prance 1982, Simpson and Haffer 1978). Nonetheless, there is almost universal agreement that the Pleistocene was a time of great change in the tropics and that current lowland tropical climates may have been eliminated during glacial periods when temperatures dropped as much as 4°–5° C.

Today, the distribution of tropical forests is again changing rapidly, this time as a result of human activity. Vast forest areas are being reduced to small, isolated patches. Viability of populations of many species under these changing circumstances is uncertain. A few large vertebrates have been lost from La Selva, but the current ecological community is re-markably complete. Large predators and some other species have also been lost from BCI. Because loss of large mammals and raptors is one of the first consequences of human penetration of tropical forest regions, many tropical studies have been carried out in areas where this important community component is lacking. For this reason it is especially important to monitor mesopredator populations at La Selva now when large mammals are present so that data will be available should the top carnivores disappear in the future.

In addition, tropical climates are expected to change, both as a direct result of loss of forest cover and associated changes in evapotranspiration and albedo and indirectly as a result of global climatic warming. If tropical climates change, populations of many species may confront relatively unsuitable conditions. This is especially true for species with long-lived individuals, such as trees, in which the current adults may have grown up under conditions significantly different from those under which they then are living and under which their offspring must compete. To anticipate such changes, it is important to initiate studies of species at the boundaries of their ranges, especially those at the lower, warmer boundaries, because these are the most likely to become unsuitable in the near future. Projected rates of climatic change are high enough that demographic consequences should be detectable within a few decades. This is, of course, longer than the duration of standard studies, but stations such as La Selva, where projects can continue for many decades, should play special roles in research on the consequences of climatic change and habitat fragmentation.

Thus, although the body of information on the population biology of tropical species is growing, only a small fraction of species have been examined even superficially. The genetic structure of most tropical populations is unknown. Biologists cannot yet offer comparative generalizations about temperate and tropical populations within or between taxonomic groups nor can we say how demography within taxa varies in the tropics in relation to climate, soil, and topography. More information on population biology is needed both for basic theoretical reasons and as a basis for development of management schemes for species whose viability is being threatened by the habitat changes that are proceeding so rapidly in tropical regions today.

COMPARATIVE TROPICAL ECOLOGY

Tropical climates differ from one another primarily in the length and severity of the dry season. The dry season is short and variable at La Selva. Therefore, La Selva investigators have been less concerned with the effects of the tropical dry season on traits of individual species and ecosystem dynamics than have investigators at the three other intensively studied Neotropical forest sites: Barro Colorado Island (BCI) in Panama (Leigh et al. 1982), Manaus, Brazil, and Cocha Cashu, Peru (Gentry 1990a), or in the tropical dry forests of the Guanacaste lowlands in northwestern Costa Rica. The populations of many frugivores on BCI are limited by availability of food during those periods when few species are ripening fruits. Populations of large frugivores at Cocha Cashu appear to depend upon the presence of a few species of palms and laurels during the period of greatest fruit scarcity (Terborgh 1986a).

Interestingly, heavy rains during the dry season on BCI cause many plants to fail to flower and fruit and may, thus, be more devastating to frugivorous animals than unusually long dry seasons (Foster 1982a). Variations in the severity of the dry season at La Selva may have similar effects, but these are yet to be investigated. It would also be interesting to determine if some species are absent at BCI because the dry season is not as reliably dry as it is in some other areas. Comparisons between the Neotropics and Paleotropics might be especially informative. In the New World, the major mountain ranges are oriented north-south, roughly at right angles to the direction of movement of air masses. The deep penetration of cool air from both the north and the south into the Neotropics is probably favored by this configuration of mountains. In the Old World, in contrast, most of the mountain ranges are oriented east-west. They, consequently, produce fewer well-marked rain shadows, and they do not create conditions favorable to the penetration of cold air masses into tropical regions.

The influences of the El Niño Southern Oscillation (ENSO) are particularly strong in Asia, resulting in periodic droughts that are far more severe than those afflicting the Neotropics. This provides opportunities to study community structure in regions where extreme events differ in severity and frequency. Ecologists increasingly believe that such events are very important. Unfortunately, the study of rare events is difficult because they do not happen during most investigations. Nonetheless, such events should be studied when they do occur. More thought needs to be given to the kinds of data that should be gathered during normal years to facilitate comparisons with abnormal years.

At longer time scales the climate of the earth is changing. During the past fifty years the climate of BCI has been drying at an average rate of about 1 cm per year (Rand and Rand 1982). La Selva also may be getting drier although rainfall data over the past twenty-five years do not reveal such changes. Global climatic warming is now predicted to result from current patterns of combustion of fossil fuels and tropical deforestation. All current global climatic models predict that warming will be more pronounced at higher latitudes than in tropical regions. Nonetheless, even slight warming or changes in rainfall patterns in the tropics could have major effects on ecosystem processes. Under current climates, photosynthetic rates of plants are suppressed during the middle part of the day and rates of nocturnal respiration are high. Increases in temperature may further depress photosynthesis and increase respiration, resulting in lowered net primary productivity of tropical ecosystems. The consequences of these changes for all components of tropical forests could be substantial. Now is the time to initiate the research necessary to help scientists predict how tropical forests are likely to respond to climate warming. One obvious possibility is to study tropical forests that, for a variety of reasons, are hotter than average today.

Interest in convergent evolution of species, communities, and ecosystems was first stimulated by the global exploration of European plant ecologists and plant geographers in the middle of the last century. Since then, ecologists have been aware that vegetation is similar in different areas of the world with similar climates even though the dominant organisms are only distantly related taxonomically. Convergent evolution occurs when features of organisms or their assemblages come to resemble other organisms or assemblages more closely than was the case among ancestral states. Community-level convergence depends in part on adaptive trends among component species, but such features as community structure and species richness can converge in the absence of convergence among the component species. Conversely, individual species could converge even though community properties fail to do so.

Convergence can be driven by both the physical environment and by biological interactions. Convergence is expected because available energy is limited and its allocation to feeding, locomotion, defense, growth, reproduction, and maintenance may be driven by biological interactions such as competition, predation, and parasitism and by abiotic factors such as patterns of water availability and by the nearly universal temperature dependence of physiological processes. Therefore, if climates are similar, responses to the physical environment may be similar. Thus, the nature and intensity of biological interactions might well evolve in similar directions despite major differences in the ancestral traits of the interacting species. Convergence, however, is not an all-or-nothing phenomenon, and many different traits of organisms and their assemblages can potentially converge. Moreover, because multicellular life is probably monophyletic, complete independence of the biotas in different parts of the world does not exist. Unfortunately, climates are not exactly the same in any two regions of the world either.

For these reasons and because it is difficult to predict which features of organisms and communities are most likely to converge, demonstration of ecological convergence is difficult. The relevant time spans are too long for experimental manipulations to yield much useful information. Potential convergence needs to be studied at a number of spatial scales because species richness might converge, say, at the beta scale but not at the gamma scale (Cody 1975).

Most intensive studies of community convergence have focused on extratropical communities, particularly hot deserts and Mediterranean vegetation (Orians and Solbrig 1977; Mooney 1977). Mediterranean communities are excellent for this purpose because they prevail in five well-separated regions of the globe whose biotas are as distinct as can be found anywhere. The structural simplicity of those communities makes field investigations relatively easy. Hot deserts have some of the advantages of Mediterranean ecosystems, but they have been less isolated from one another historically. The most intensively studied pair, the North American Sonoran Desert and the Argentine Monte, share many species and genera.

Tropical forests offer substantial and, as yet, mostly unexploited opportunities for comparative studies of convergence. The tropical forests of America, Africa, and Asia have been long separated from one another. There are many pantropical plant genera and families, but many taxa are restricted to one continent, and the faunas of the three major wet forest areas are quite distinct. Climates can be matched quite closely although this is more difficult than would appear at first glance.

Effective studies of convergence, or its absence, require careful selection of study sites as well as comparable research methods so that detected differences reflect real differences in the ecological communities rather than methodological inconsistencies. Agreement is needed on the major questions to be addressed so that investigators at different sites study similar issues. For example, major differences in the questions asked and types of data gathered severely limit the comparisons that can be made currently between La Selva, BCI, Manaus, and Cocha Cashu. Because scientists are interested in both convergence and how communities change along gradients in the physical environment, we need to address comparable methods and research questions at additional sites.

Features of plant communities first attracted the attention of globe-trotting European biologists. The gross similarities evident among Mediterranean climate plants and tropical wet forest plants in different parts of the world have been extended by more detailed studies of ecophysiology (Mooney and Dunn 1970). Extensive data also exist on how various features of plant morphology vary with increasing severity of the tropical dry season and with the decreasing temperatures and higher humidities accompanying the ascent of tropical mountains (Tanner and Kapos 1982). Less clear is how those features of plants that evolve in response to interactions with mutualists and predators compare in communities with similar climates and along climatic gradients. Certain patterns are evident, however. Among those features directly responsive to the physical environment, such as leaf size and shape, tropical plants are very similar on all continents, as was noted in the nineteenth century by European plant geographers. In addition, most tropical wet forest plants depend upon animals for transfer of pollen (chap. 12).

Fruit dispersal systems appear to differ remarkably within and among continents. Neotropical forest trees and shrubs depend heavily on birds for seed dispersal. Frugivorous birds constitute large fractions of both species and avian biomasses in most Neotropical forests (Terborgh 1986a, 1990a). At La Selva 95% of treelets and shrubs and 90% of trees produce fleshy fruits (chap. 22), most of which are dispersed by birds. In contrast, primarily frugivorous birds constituted only about 11% of captures and 22% of the biomass of all birds banded at Manaus (Bierregaard 1990). Trees with mammal-dispersed fruits are more abundant in the recent floodplain forests at Cocha Cashu than in other Neotropical forests, including those in the Manaus area (R. B. Foster 1990). This situation is more similar to that found in Africa where mammals, up to the size of elephants, are important seed dispersers. Frugivorous birds are a much less conspicuous part of African forest communities (Moermond and Denslow 1985). The reduced importance of large mammals in the Neotropics may be a relatively recent phenomenon (Janzen and Martin 1982), but the richness of Neotropical frugivorous birds is, nonetheless, impressive. Southeast Asian forests are noted for the infrequent mast fruiting of dominant species, particularly in the Dipterocarpaceae, for which wind is the primary dispersal agent (Janzen 1974). These variations in structure of seed dispersal communities fit the more general pattern of less convergence in those components of community structure that are driven primarily by biological interactions than in those driven by direct responses to the physical environment (Orians and Paine 1983).

Determining why tropical forest trees are nonconvergent in their relationships with seed dispersal agents will be a difficult task. The climates of monsoonal Asia differ from those elsewhere, and the strong influences of ENSO result in periodic droughts of a magnitude greater than those currently experienced in other tropical regions. Is that difference, and the associated higher frequency of large-scale fires, pivotal to the evolution of fruiting traits of southeast Asian trees, or are biologists witnessing a historical artifact caused by dominance of Dipterocarpaceae? The same reproductive traits dominate fire-prone Australian forest trees, suggesting that more than historical accident is involved. Fires favor mass fruiting, storage of seeds in tough capsules, and heat-triggered germination, leading, in turn, to greater importance of predators capable of penetrating the tough capsules and organisms, such as ants, that are able to survive fires and then quickly harvest the rain of seeds that follows (O'Dowd and Gill 1984).

Plant species richnesses have been compared at several different spatial scales on different continents. In South America the patterns of species turnovers are especially interesting at locations such as Cocha Cashu where many different soil conditions are found in relatively small areas (Terborgh 1990a). The extremely species-rich floras on ancient, impoverished soils in the southern continents are the result of typical, high local species richnesses combined with unusually high rates of species turnovers across habitat and geographical gradients (Rice and Westoby 1983). Data are still too few to indicate whether small-scale tree species richness in tropical wet forests is comparable on different continents or how richness changes with increasing length of the dry season. Increasing aridity is, in general, associated with decreasing plant species richness.

Comparisons of animal species on different continents have tended to have a strong taxonomic bias. This owes in part to the necessary skills of field observers but also to the belief that intrataxon interactions are stronger and more decisive for community structure patterns than are intertaxon interactions. An alternative approach is to focus on animal guilds, groups of species that exploit a common resource base by similar means (Root 1967). Most guild studies, however, have focused on subdivisions of a single taxon, for example, foliage-gleaning birds, rather than expansions across taxa. How well intertaxon guilds can function as objects of study is not clear. For example, mammals and insects are the dominant folivores in tropical forests, but it is not evident how to study interactions between the two groups or how they influence one another's abundance. It is even more difficult to establish that some species in one taxon are absent because particular species in the other taxon are present.

Experimental studies of the seed-eating guild of the Sonoran Desert have revealed that ants and rodents exert strong reciprocal influences, but it is yet to be demonstrated that the presence of any one species depends upon the presence or absence of one or more members of the other taxonomic group (Brown and Davidson 1977; Brown et al. 1979). Moreover, the time frames for evaluating species viability are much longer than those of most ecological experiments. Development of methods for conceptualizing and experimentally studying such intertaxon problems is one of the greatest needs of contemporary community ecology. At La Selva, where the vegetation is dominated by long-lived perennials, experimental manipulations, even if they can be accomplished, may not yield results in reasonable time frames. Some as yet unimagined shortcuts are needed to provide answers more quickly.

Studies of convergence of folivore communities are almost nonexistent for any ecological community. Lawton (1982) has examined herbivore communities on bracken fern in different parts of the world, finding little evidence of convergence. The insect herbivores of no tropical tree, shrub, or herb species are well enough known to permit meaningful interspecific comparisons. Some comparative data would be very valuable, but the species for study must be chosen carefully. Given the difficulties of studying tropical plants, especially trees, a solid comparative data base will accumulate very slowly.

Although community structure, especially species richness and guild composition, has attracted the bulk of the attention of tropical ecologists, interest in tropical ecosystem research is increasing. Studies of patterns of productivity, energy flow, and nutrient cycling have been stimulated both by the basic need to understand how major ecosystem processes function under conditions of continuously high temperatures and abundant moisture and by the dismal success of much tropical agriculture. Early tropical plant community studies made extensive use of information on the soils and geology of the focal regions (Ashton 1964), but during the 1970s and 1980s attention has been primarily elsewhere. As a result, many aspects of nutrient cycles, such as decomposition and release of minerals from dead plants, rates of growth and accumulation of elements in vegetation, loss of materials through erosion, volatilization, and leaching, and hydrological balances are poorly known for tropical forests (chap. 5). The availability of good soils maps and of sites suitable for manipulative research suggests that ecosystem studies will feature prominently in future books about La Selva.

The science of comparative tropical ecology is still in its infancy, but some of the directions in which it needs to develop are already clear. The influences of climate and soils on ecosystem dynamics under tropical conditions are basic to understanding structure and dynamics of guilds of organisms exploiting the resources available in tropical ecological communities. If truly strong convergences or parallelisms exist among ecosystems in different tropical regions, these physical environmental conditions are likely to have been primary causes. Existing evidence, however, suggests that once one moves away from direct evolutionary responses to the physical environment, the details of history in different regions leave a strong imprint on the structure of ecological communities. Determining the relative contributions of unique historical events and widespread selective pressures on the features of today's ecological communities is the major challenge of comparative tropical ecology.

THE FUTURE

The future of tropical biological research will be full of surprises as individual investigators pursue their own ideas and as unexpected results stimulate unanticipated observations and experiments. This eclectic aspect of research should be encouraged. Nonetheless, understanding of tropical ecology is severely limited not only by scanty data but also by the absence of comparable studies in different tropical regions. The extensive studies at El Yunque in Puerto Rico (Odum 1970) and at Pasoh Forest Reserve in Malaysian dipterocarp forest (Leigh 1978) emphasized productivity, nutrient budgets, and cycling of nutrients. Studies at BCI have emphasized how climate affects interactions between vegetation, herbivores, and carnivores. Researchers at La Selva have tended to emphasize biological diversity and life-history traits, and, more recently, productivity. Studies at Cocha Cashú have emphasized community ecology of birds and mammals, interactions between plants and vertebrate herbivores, and vegetation succession (Terborgh 1990a). All of these themes are of great intrinsic interest, but differences in research emphases prevent researchers from making many interesting comparisons among even the best-known tropical sites.

To remedy this deficiency will require some international agreements about methods and common research questions that should be addressed at a number of major field stations. Some important comparative themes require large research teams. Graduate student education, however, is not well served by assigning students to specific pieces of large projects whose scope and objectives have been determined previously by other people. Further, such projects are not well suited to stations that have only small resident scientific staffs. Coordinating large projects without inhibiting the creativity of individual investigators is difficult.

Because of these difficulties, it is important to determine which types of questions can be pursued adequately at specific field stations and biogeographic regions and which truly require intercontinental comparisons. Uniformity is not needed for the former, whereas it is essential for the latter. Answers to basic questions concerning relationships between ecosystem productivity and soil fertility, litter quantity, quality, and decomposition rates, patterns of species richness at all scales, and effects of increasing length of dry season, for example, require comparable data from several continents and several sites within continents. Researchers from different continents need to work together to develop a research agenda designed to provide a powerful basis for a truly comparative tropical forest ecology.

A part of such a research agenda must be the gathering of detailed data on the physical environment. The value of the long-term climatic data from BCI is already evident. It is unfortunate that adequate climatic data have not been collected at La Selva. Funding for such routine monitoring is always more difficult to obtain than support for innovative research projects, but interpreting research results continues to be severely constrained by the lack of long-term climatic data. Funding agencies need to be convinced of the importance of such information and to allocate their support accordingly. Better climatic data combined with data on soil moisture status and moisture stress in the forest canopy will allow the important questions posed in chapter 3 to be addressed. How does reduced soil moisture affect primary productivity, flowering, leaf abcission, and competitive relations among species? How much does the structure of current communities depend on particular combinations of events in the past as opposed to reflecting average weather conditions?

Another urgent need is for tropical research to become more immediately relevant to conservation biology. The rate of destruction of tropical forests is so high that existing stations are destined to become islands in seas of exploited landscapes. Scientists at those stations will inevitably document some of the consequences of isolation of their study sites, but research at tropical stations has a larger role to play in conservation biology. In the United States, parks and reserves were established primarily in uninhabited wilderness areas. In wealthy temperate societies there is sufficient demand for recreation and open space and sufficient lands are available for intensive agriculture and forestry that parks and reserves can be removed from exploitation. In most tropical countries, however, eliminating large areas from the productive exploitation base is far more difficult. Ways must be found to provide substantial economic benefits from those lands while still preserving most of the ecological processes and species found in parks and reserves. Potential economic benefits from "extractive reserves," in which animals and plant products such as fruits, nuts, latexes and resins, are harvested in sustained ways that do not alter basic forest functioning, are in need of much study. Research is also needed on ways to increase income and moderate the negative impacts from "ecotourism," a rapidly growing industry in some tropical countries. Ecotourism, which already plays an important role at La Selva, is certain to increase in importance in the 1990s.

The concept of a park that serves both conservation and human support needs has been developed in UNESCO's Biosphere Reserve Network, a major program of the Man and the Biosphere Program (MAB). Although the MAB model has been observed in the breach more than in the observance, the basic model, which includes an unexploited core area, surrounded by a buffer zone in which activities that do not endanger the core are permitted, is a sound one. The buffer zone is, in turn, surrounded by a transition zone with more extensive exploitation. The Costa Rican government has adopted this model in its plans for a series of megareserves (R. Gámez pers. comm.). Research is urgently needed at and near park boundaries to determine the most important outside influences and how they are expressed. Studies are needed of patterns of reproductive success of species at different distances from forest boundaries. Sites where different types of agricultural practices exist adjacent to reserves are particularly valuable for study. There are rich possibilities for creating specific types of boundaries at research stations. Such studies should include both the effects of the exploited area on ecological processes in the adjacent forest and the effects of organisms moving out of the forest on adjacent exploited lands.

Studies at research stations can provide valuable information on the effects of loss of specific components of forest ecosystems and information on how those components might be retained in areas where human activity is widespread. Field stations can function as sites where management techniques for "supersaturation" can be developed, that is, where one learns how to intervene to maintain more species in the systems than would survive if natural ecological processes were allowed to unfold. For example, interventions might include artificially increasing invasion rates so that local extinction rates are reduced. This might not be appropriate at La Selva, but the surrounding areas offer possibilities for such interventions. To the extent that research stations like La Selva provide such information, they will be perceived as valuable to the people of the countries in which they are located, thereby in-

creasing their acceptance, probable longevity, and genuine usefulness.

Field stations also have a major role to play in the developing science of restoration ecology. Understandably, most attention of conservationists has been directed toward saving and setting aside areas that have not yet been destroyed. In many regions, however, so much habitat has been lost that species-rich ecosystems must be restored on degraded lands. Little is known about how to do that although the "Trials" Project at La Selva is providing basic information on the performances of trees in abandoned pastures. Studies are needed of vegetation succession on different soils in areas that have been farmed for varying lengths of time, and in sites more or less isolated from sources of seeds of colonizing plants. Field stations such as La Selva are unusually well situated for such studies because the rich base of information about the natural forest can guide the planning of insightful experiments and the interpretation of the results.

La Selva is also poised for increased involvement in the application of remote sensing technologies to analysis of vegetation patterns, physiological states of plants, and documentation of patterns of habitat alterations. Progress in the use of these technologies depends, in part, on ability to "ground truth" interpretations of remotely sensed images. Because the technology of remote sensing is advancing more rapidly than ability to use and interpret the data, the need for field studies in biologically well-known areas is great. This type of study, which has just begun at La Selva, is certain to increase substantially in the near future as biologists become more familiar with these technologies and come to appreciate their potential usefulness for helping answer important biological questions.

Prospects for the future of rain forest ecology are simultaneously promising and depressing. The quantity and quality of tropical researchers is better than at any time in the past, and the wealth of methods at their disposal is steadily improving. Questions that could not be answered a decade ago, can now be approached with some promise of success. At the same time the rate of loss of tropical forests threatens the persistence of the systems biologists are attempting to understand and the long-term viability of the major field stations upon which so much of tropical biology depends. As is often the case, the outcome depends on the comparative rates of key processes. The rate of knowledge accumulation is partly under the control of scientists. The major influences on other key factors have traditionally resided elsewhere, but scientists are increasingly stepping into the fray, accepting the view that their knowledge of tropical ecological systems and their understanding of the importance of tropical forests for the future requires involvement in the political processes that will determine the fates of those forests. Some of this involvement lies in the traditional educational roles of scientists but expanded to target nontraditional audiences, such as bureaucrats and politicians, whose decisions have such far-reaching environmental consequences. Important although education may be, the short period during which the future of tropical forests will be determined generates a sense of urgency that propels an increasing number of scientists, including many La Selva researchers, into more direct political action.

LA SELVA

APPENDIXES

APPENDIX 1
Patterns of Research Productivity 1951–1991

Lucinda A. McDade and Kamaljit S. Bawa

Research productivity at La Selva has been clearly documented in this book as information now available about many aspects of the ecosystem. In this appendix we take an entirely different approach to examining research productivity. We begin by analyzing the pattern of research publications from La Selva in terms of numbers and subject areas treated over time. After summarizing the trends suggested by this analysis, we discuss some of the factors responsible for these patterns. Finally, we look briefly and in general terms at what the future might bring.

ANALYSIS OF RESEARCH PATTERNS

To understand the development of research at La Selva we have analyzed the bibliography compiled by OTS of published papers and theses that are based largely or entirely on research at La Selva. Our goal was to examine the changes in quantity of research as expressed by numbers of papers published over time and in the subject matter treated. Thus, we counted papers and assigned each title to one of seven thematic categories designed for our purposes as follows:

Systematic Biology includes species descriptions, floristics, faunistics, species lists, and other research with explicit taxonomic goals: comparative biology, phylogenetics, character analysis.

The *General Ecology* category includes research devoted to understanding the biology of individual species: life histories, demography, behavior, and genetics. Papers placed in this category may cover more than one species but treat the species separately. In contrast, *Interspecific Ecology* encompasses research focused on the ecological interactions between species, including pollination biology, frugivory, predation, and herbivory; papers on coevolution (in the broad sense) are placed in this category. The General Ecology and Interspecific Ecology categories were initially combined, but this resulted in a very large and heterogeneous grouping. Further, these appear to be two rather distinct research traditions with a somewhat different history of development at La Selva. Recognizing Interspecific Ecology as a separate category leaves General Ecology still somewhat heterogeneous. In a more detailed analysis or one that sought to unravel the numerous intellectual threads within ecology one would wish to divide this category more finely. Any analysis of this nature, however, will inevitably have one category that is something of a catch-all.

Community Ecology includes papers treating the structure, dynamics, and organization in space and time of plant and animal communities. Papers on photosynthesis, water relations, metabolism, nutrient and energy budgets, and physiological adaptations are grouped under *Physiological Ecology*. *Ecosystem Ecology* encompasses work on soils, nutrient cycling, energy flow, and physical features of the habitat. Finally, *Applied Ecology and Conservation* assembles papers on forestry, agroecosystems, and conservation of natural resources.

This analysis is necessarily crude in several respects. The bibliography compiled by OTS is, no doubt, incomplete. There is no reason to believe, however, that the incompleteness of the subject matter has a pattern. The categories that we recognize are generally those used traditionally in biology, but, clearly, there are other ways to partition the papers and other categories that one might wish to recognize. Distinctions between our categories (or any others) are somewhat arbitrary, and although a given paper might include subject matter in more than one category, each was assigned to only a single most-appropriate category. In general, it was quite straightforward to place papers in one of these categories. The significant lag time between the conduct of research and publication must also be kept in mind. For example, the first papers on physiological ecology were published in 1980, but reflect work begun in the late 1970s. Similarly, the impact of the new analytical laboratory (see chap. 2) on research productivity is not yet (1991) apparent from publications based at La Selva.

The results of this analysis are presented in table A1.1. Because of the small number of papers published in the early years, these are pooled as 1951–65 and 1966–68. Beginning with 1969, the publications are grouped in two-year intervals. As noted, the data for 1983 do not include contributions to *Costa Rican Natural History*, edited by D. Janzen. Some sixty titles from this volume treat organisms that occur at La Selva. Most of these are species accounts rather than primary research publications. We include data for 1991 in the analysis of subject matter of research but do not total the publications for that year because the list is quite incomplete.

Two patterns are immediately obvious. First, the overall quantity of research at La Selva has increased continuously from barely a single paper per year in the early 1960s to about seventy papers per year in the late 1980s. Second, research has diversified considerably over this same period. Initially,

Table A.1 Publications based on research at La Selva Biological Station, 1951–1991

Years	Systematic Biology (No.)	(%)	General Ecology (No.)	(%)	Interspecific Ecology (No.)	(%)	Community Ecology (No.)	(%)	Physiological Ecology (No.)	(%)	Ecosytems Ecology (No.)	(%)	Conservation Applied Ecology (No.)	(%)	Total	No./Year
1951–65	13	76	0	0	0	0	4	24	0	0	0	0	0	0	17	1.1
1966–68	9	45	8	40	1	5	2	10	0	0	0	0	0	0	20	6.7
1969–70	9	39	8	35	2	9	2	9	0	0	1	4	1	4	23	11.5
1971–72	8	24	15	44	2	6	4	12	0	0	1	3	4	12	34	17.0
1973–74	13	34	14	37	3	8	6	16	0	0	0	0	2	5	38	14.0
1975–76	6	13	16	36	13	29	6	13	0	0	2	4	2	4	45	22.5
1977–78	2	4	9	20	18	39	10	22	0	0	2	4	5	11	46	23.0
1979–80	7	11	19	29	14	22	13	20	2	3	8	12	2	3	65	32.5
1981–82	19	22	25	29	21	24	10	11	1	1	4	5	7	8	87	43.5
1983–84[a]	20	20	24	24	17	17	8	8	11	11	8	8	12	12	100	50.0
1985–86	31	24	27	21	26	20	13	10	14	11	11	8	7	5	129	64.5
1987–88	20	14	49	35	13	9	16	11	11	8	18	13	14	10	141	70.5
1989–90	20	14	36	25	18	12	26	18	10	7	15	10	20	14	145	72.5
1991[b]	10	18	9	17	7	13	10	18	3	6	7	13	8	15	54	

[a]The data for 1983 do not include articles from *Costa Rican Natural History* (D. H. Janzen, ed.). Sixty-one of these are related to research at La Selva. The vast majority would be categorized as General Ecology (42), followed by Systematic Biology (8), Interspecific Ecology (6), and Community Ecology (4).

[b]The data for 1991 are incomplete; the numbers of publications should be disregarded, but their distribution among subject areas is unlikely to change dramatically.

most research was concentrated in the fields of systematic biology, species ecology, and community ecology. As late as 1973–74, publications in these fields accounted for more than 80% of total research productivity. Although these three subject areas continue to account for about half of all publications, productivity in other areas has diversified the research mix. In the following paragraphs, we identify what appear to be the most noteworthy trends in this diversification.

Research productivity at La Selva in the field of systematic biology is bimodal. Most of the early papers were systematically oriented. Relatively few papers on systematics were published in the mid-1970s, which is correlated with and, perhaps, reflects a disdain within the OTS community for making systematic collections, and a prohibition on collecting at La Selva (see app. 2). Since the early 1980s, papers on systematics have accounted for 15%-20% of total annual publications. Many of these are related to the Flora of La Selva project (see chap. 6), but animals have received considerable attention as well. In particular, the increase in the number of papers on insect systematics was significant in the 1980s.

By the mid-1970s, research on the ecology of interspecific interactions contributed significantly to total research productivity. Since that time, about 20% of the papers published each year have been in this category although there appears to be a trend toward relatively fewer publications on interspecific ecology in recent years. At La Selva this field has emphasized plant-animal interactions (e.g., pollination ecology, herbivory, frugivory) and was, no doubt, stimulated by the seminal work by D. Janzen and others in the late 1960s and early 1970s.

Research in physiological ecology was not established at La Selva until the late 1970s with papers on this work first published in 1980. The vast majority of this work is on plants (see chap. 10).

Ecosystems ecology received an early boost at La Selva from the large-scale, NSF-funded project that set out to compare wet and dry forest ecosystems (see Stone 1988). This project foundered on the rocks of administrative and logistic difficulties, and ecosystems-level research did not increase markedly again until the late 1970s. Publications in this field now regularly constitute about 10% of the total papers published each year.

Conservation and applied ecology began at La Selva with publications in forestry. The gradual increase in papers in these fields in part reflects increased interest within OTS in applied ecology. The forestry trials project (see chap. 25) and OTS courses in applied ecology (begun in the mid-1980s) have contributed significantly. Increased concern about the loss of tropical forest and species has also resulted in publication of a number of conservation-oriented papers in the scientific literature and in the semipopular press.

THE INGREDIENTS FOR RESEARCH PRODUCTIVITY

Research at La Selva has increased in quantity and diversified greatly since the 1950s, when it was first used by researchers, and since 1968, when OTS acquired the station and reserve. The reasons for these developments are many and complex. The factors responsible for the two kinds of changes (more and more diverse research) are at least potentially distinguishable. In this section we attempt to identify some of the reasons for the growth of research at La Selva with particular emphasis on those that might be of value to those trying to establish research stations elsewhere.

There are more researchers, more projects, and more publications from La Selva now than at any time in the past. As in all human endeavors, there are temporal changes in the popularity of different kinds of research, and tropical research has become increasingly popular in the late-twentieth century. The reasons for this are as complex and as intertwined as the reasons for La Selva's growing popularity as a research site in the tropics. The intrinsic fascination of high-diversity forests with complex biotic interactions has added appeal given the late twentieth century improvements in travel and medical care (prophylaxis and treatment of tropical diseases) have made the pursuit of this fascination less risky. In the late twentieth century rapid destruction of tropical ecosystems and organisms is an additional stimulus for many to undertake tropical research.

Given increased interest in tropical research, La Selva is an attractive site for a number of reasons, including its biotic diversity and relatively easy access from San José and from the United States and Europe. As documented in chapter 2, OTS has worked hard to build upon these intrinsic merits. Ensuring that the station continues to attract the core group of population- and community-oriented ecologists and, in fact, attracts more of them has been consistently of high priority. Land has been acquired on a number of occasions to protect more adequately the biodiversity housed at La Selva. Most significantly, OTS was instrumental in extending Braulio Carrillo National Park down to La Selva (see chap. 2). More and better living conditions and easier communication with the outside world have made the station attractive to a larger group of researchers, not just the extremely hardy. Although the extremely hardy sometimes complain about this trend, that these improvements bring more researchers is irrefutable.

Some of the developments sought by OTS have focused rather explicitly on increasing the scope of research that can be accommodated at La Selva. For example, land has been acquired to enhance habitat diversity at the station, and facilities have been improved to make the station suitable for a broad spectrum of research, including projects that require air-conditioned space, well-equipped laboratories, and computers. Of course, many of the developments that have occurred at La Selva have enhanced both the number of researchers and diversity of projects. For example, improved access makes it easier for more scientists to come and for them to bring more sophisticated gear with them.

OTS's role in stimulating research at La Selva has mostly taken the form of facilitating research by enhancing the site, as described here and in chapter 2. However, on occasion, OTS has taken a more active role in promoting research in areas deemed appropriate for the facility and important for its development as a research site. In the early 1980s, for example, OTS wished to stimulate ecosystem-level research at La Selva. The organization, therefore, convened a number of meetings that led to the development of research projects dealing with forest gaps (see chap. 9) and soil chemistry (see chap. 4) and, ultimately, to numerous publications in these areas.

Research at La Selva has been enhanced by OTS's other programs, in particular by graduate education and research

fellowships. OTS courses began to use La Selva regularly in 1968 for stays up to several weeks. In the 1990s, dozens of graduate students visit La Selva each year as course participants, many of whom are actively seeking a thesis or dissertation project. With courses offered in both basic and applied ecology (since 1985), the education program recruits research in a variety of fields. Of equal importance is OTS's program of research grants that provides limited funding (field expenses only) for projects by graduate students and young researchers in Costa Rica. Indicative of the importance of these programs is that of about eighty graduate theses based on research conducted at La Selva between 1965 and 1991, more than half (forty-seven) were by course alumni. Of the sixty-six theses completed since 1976, thirty-nine were by course alumni and thirty-four were funded partly or fully by OTS. These programs have certainly increased the number of research projects sited at La Selva, but OTS has not sought to direct the subject matter of this research. It is clear, however, that by initiating courses in applied ecology, OTS has stimulated diversification of research at La Selva (and its other research stations).

OTS is not a research institution with a staff of scientists, nor does it have significant funds to support research by others. Thus, although OTS has successfully competed for a series of grants from the U.S. National Science Foundation for enhancements and core support of La Selva, the researchers who have come to take advantage of these improvements are working on their own projects and have their own sources of financial support (the exception here is OTS's program of small grants for graduate students already described). Much of the research accomplished at La Selva has been funded by NSF grants to individual researchers. These projects are subject to rigorous peer review, and low success rates mean that only those judged to be the very best are funded. Reliance by researchers upon external sources of funding has, therefore, meant that the pattern of diversification of research at La Selva has been influenced by trends in various subdisciplines of biology as well as by planning by the organization and individual researchers.

We would be remiss not to point out that the experience at La Selva clearly demonstrates that individual scientists are exceedingly important in generating research momentum. Especially early in the development of a site, use by scientists who are productive and have significant numbers of students and postdoctorates can identify that site as "good" for research and set clear research traditions. The generation of basic information about a site paves the way for future projects, but scientists also "recruit" to a site for sociological reasons.

At least three strong, organismally centered research traditions are apparent from the history of research productivity at La Selva: birds, amphibians and reptiles, and trees. Together, these three traditions account for more than one-third of all publications from La Selva. In the initial stage of development of each of these research traditions, investigators conducted survey work that described the biota and made it possible to identify the organisms: P. Slud (birds); E. H. Taylor (amphibians and reptiles); L. Holdridge, G. Hartshorn, and others (trees). In the next phase, one or a few researchers developed long-term ecological research programs that built upon this earlier work: F. G. Stiles (birds); J. Savage and students (amphibians and reptiles); K. Bawa, G. Frankie, and others (trees). By producing basic information and by raising unanswered questions, this work, in turn, paved the way for the next generation of researchers: J. Blake, D. Levey, and B. Loiselle (birds, frugivory); M. Donnelly, H. Greene, and C. Guyer (amphibians and reptiles; population, reproductive and behavioral ecology); D. and D. Clark, J. Denslow, and collaborators, D. and M. Lieberman, R. Peralta (trees, forest dynamics, demography).

Based on experience from La Selva we suggest that five factors be considered fundamental to stimulating research at new field sites:

1. Facilities enhancement and habitat diversification attract researchers to a site and diversify the subjects of their projects.
2. Ancillary programs (e.g., courses, site visits, advisory committees) that bring people to the facility (even if ostensibly for other reasons) are good for research productivity, especially if these include scientists (or their mentors) who are likely to be looking for research sites.
3. Grants, even small ones, for research are very effective in attracting researchers, especially graduate students who do the sort of long-term research that is highly productive of data.
4. The accomplishment of basic systematic and descriptive work in a given field sets the stage for research in more derived areas by subsequent investigators.
5. Individual scientists are very important. Those researchers who produce works that identify a site as a fruitful place to work recruit future researchers to that site, especially if they are effective mentors for younger colleagues.

THE FUTURE OF RESEARCH AT LA SELVA

In La Selva's thirty years as a research station we have witnessed vast and rapid change in the tropics, including La Selva's neighborhood (see chap. 23). The assault on the environment and natural resources worldwide has brought the issues of conservation and management of natural resources to the forefront. The transformation is not confined to the biological world but also includes social, cultural, and economic changes. Against this complex and rapidly changing backdrop we expect to see continued evolution of the research mix and of the researcher community at La Selva.

A significant part of the research undertaken will, undoubtedly, continue to reflect the fundamental quest for knowledge about tropical forest. As deforestation continues, there will simply be fewer places where this kind of work can be done, and those that remain will receive increased attention from those wishing to study "pristine" ecosystems. At the same time we expect that future analyses of research productivity at La Selva will show relative increases in publications with direct relevance for conservation and for the well-being of tropical people. During the 1980s it became clear to many researchers that the creative solutions that are necessary to solve the problems of resource depletion in the tropics will come not from either traditional "camp" of pure or applied research but from a synthesis of these two approaches. Increasingly, ecologists, agronomists, foresters, anthropologists, and social scientists are willing to meet each other halfway. A research station, such as La Selva, that offers access to a wide diversity of natural and human-altered habitats is a natural location for such research.

Biodiversity and global change have become the watchwords of the 1990s. As efforts are launched to collect baseline data and to monitor future changes, La Selva is well situated to play a role as representative of Central American lowland rain forest. Although such data will be intrinsically interesting and exceedingly useful, it is unfortunate that abuse of the planet by humans is the impetus to their collection.

The 1980s saw increased participation of Costa Rican researchers at La Selva, and we hope that this trend will continue. Scientists from other Latin American countries should also be encouraged to work at La Selva, especially those from countries that lack wet tropical habitats and those who are interested in broadly comparative studies. La Selva should play a central role in the development of effective links between biological field stations and researchers in developing countries with emphasis on Latin America. At the same time we expect to see increased work at La Selva by scientists from developed countries beyond the United States. The resultant highly international mix of researchers should prove a fertile source of ideas, projects, and collaborative efforts.

One result of the patterns mentioned will be an increasing number of projects that include multidisciplinary and multinational teams of researchers as co-principal investigators. This process will be driven, in part, by the sense of responsibility that many scientists from developed countries feel to work collaboratively with Latin American colleagues and, in part, by the special informality of a research station that encourages interactions.

We envisage La Selva's research future as further diversified in several dimensions, and we anticipate a quite different book synthesizing the next several decades of research at La Selva. We predict that the authors of chapters in such a future volume will be more thoroughly international, including greater Costa Rican representation. The volume is likely to summarize substantial progress along the research themes that are emphasized in this book and that reflect continued fascination with pristine tropical ecosystems. At the same time we expect the future volume to include substantive reports on the conservation and management of tropical resources and on the sustainable use of tropical lands for human purposes. Comparative ecology of tropical forests is likely to figure more prominently, in part a product of the frustration that many authors express here at the near absence of parallel studies of the sort that permit regional synthesis of issues in tropical ecology. We also expect to see the results of research efforts designed to monitor anthropogenic changes in the physical environment and in biodiversity. Although it is tempting to speculate about what these last results will indicate, we will leave the future to be determined by the independent and concerted actions of all of us.

Administration and Governance

Lucinda A. McDade

La Selva Biological Station is owned and operated by the Organization for Tropical Studies (OTS), a consortium of U.S. and Latin American universities and research institutions. Its mission is to serve the scientific community broadly, according equal priority to research and education. La Selva's user community is, thus, diverse and places many potentially conflicting demands on the station. Further, the intensity of research station life and the identification that many researchers come to feel for a site lead to strongly held opinions. These factors argue for an administrative structure that is open to input yet able to make decisions efficiently. An additional complicating factor at La Selva is the rapid development that occurred during the 1980s. Decisions about land acquisition and new facilities (see chap. 2) are implicitly or explicitly decisions about priorities and future directions, and these are often controversial.

In this appendix, I present an overview of the development of the administrative structure for La Selva and of the present system of governance. The latter section includes examples of the kinds of issues that are confronted at research stations and how these have been handled at La Selva. My intent is to present enough contextual information to make clear the framework in which various policies were made and to be frank about issues that remain controversial. I hope that this appendix will be useful to those who face similar management issues elsewhere and that it will stimulate interchange among research station managers.

ADMINISTRATION

During La Selva's early years most administrative tasks were handled at one of the OTS offices (San José and North America), including the purchase of food, reservations, billing, and fund raising. In fact, for the first several years no OTS staff were in residence at La Selva. The researcher who had been in residence the longest and/or was the most gregarious became de facto responsible for meeting newcomers and orienting them to the station and reserve. By the mid-1970s, however, it was clear that OTS could not afford to depend upon volunteers for station management. The first series of station managers were graduate students who were carrying out research at La Selva and who received room and board in exchange for half-time services. As OTS prepared for major

development of facilities at La Selva in the late 1970s, it was apparent that the planned projects would require close professional supervision. Accordingly, the station manager position was upgraded to a full-time, Ph.D.-level station directorship in 1980.

In subsequent years staff have been added as administrative tasks have increased in *scope* (as a number of administrative functions have been moved to the research station and new facilities have been added) and in *volume* (as use of the station has increased more than fivefold since 1980). Staff at the station how handle reception of visitors, billing, purchasing, book and recordkeeping. As facilities and station use have increased, various parts of what was formerly a single job have become large enough to require specially designated persons. Eventually, new positions had to be created to supervise the growing staff. Just as administrative functions quickly outstripped the capacity of the station director, so, too, has the demand for scientific expertise grown well beyond what can be handled by one position. Increased station use means more researchers, educational groups, and ecotourists who need biological orientation to the site. The establishment of an environmental education program at La Selva in the mid-1980s further increased demands on staff time. To meet these needs two station biologist positions were added in 1987. These individuals supervise the environmental education program and orient newly arriving groups and researchers. In the late 1980s OTS committed itself to provision of research support at a level that requires a concommitant level of technical support staff. For example, the analytical lab, with its stock of OTS-owned equipment, is overseen by a laboratory manager. Similarly, long-term curation of the Geographic Information System (GIS) and databases at the station require staff with appropriate expertise.

La Selva's administrative staff has increased from a single half-time person in 1979 to twelve full-time in early 1992. The evolutionary process here has been largely empirical. Models that can be emulated are generally lacking or only partially parallel to La Selva's particular circumstances. For example, the Smithsonian Tropical Research Institute's Barro Colorado Island facility is a very well-developed research station that serves in many ways as a model for others. Its administrative structure, however, includes a staff of Ph.D.-level scientists, and it is, therefore, not comparable to La Selva's. La

Selva is also something of a special case in that it enjoys the administrative and fund-raising support of the OTS offices in San José and the United States.

La Selva's experience suggests at least three generalities that are likely to be relevant elsewhere. First, in situ intellectual leadership is fundamental for substantive progress. It is very difficult to imagine that the developments that occurred at La Selva throughout the 1980s could have taken place without a very high level of resident leadership. Second, if the general public visits a research station in any significant numbers, it is highly advisable to have designated staff to handle them. To do otherwise is to run the risk of unhappy visitors, overextended staff, and any of the many potential catastrophes associated with naive people at research stations. Third, enhancing resources for research also requires enhanced staff. When the multiple user phenomenon is coupled with the rigorous environment of the rain forest, the risks of damage to sensitive (and often expensive) equipment are compounded, and responsible oversight is essential. It is more difficult to generalize about other kinds of administrative roles because they are likely to be handled differently across sites. One additional generalization might be suggested, in the form of a caveat. Although handling the hotel functions (reservations, billing) off-site is superficially attractive, it is fraught with problems. It is difficult enough to verify the arrival and departure dates of people and extract payment from them in person, much less from distant offices and after a lapse of time.

GOVERNANCE

When operating La Selva and especially when setting policy for use of the station's resources, OTS has always relied heavily on independent (nonstaff) advice. This approach was vital in the early years when the facility lacked professional staff, but it has been just as critical more recently as increased usage of the station has resulted in complex and potentially controversial issues that require arbitration.

During La Selva's earliest years as a research station the executive committee of OTS played a significant role in guiding the development of the facility, setting regulations, and approving research projects. (The executive committee consists of officers and members-at-large who are elected from the organization's Board of Directors.) G. Hartshorn, in his capacity as a long-time researcher at La Selva, served as special advisor to the committee for years. In the face of increased use of the station and the prospect of rapid development and expansion of facilities, a formal La Selva Advisory Committee (LSAC) was constituted in 1976. The committee brings together an international group of scientists who represent diverse areas of expertise and who have worked at La Selva or have managerial experience at other field facilities. Formally, LSAC makes recommendations to the president of OTS on all matters of governance, policy, and land use at La Selva. Once accepted by the board of directors, LSAC's recommendations become operating policy for the station.

Perhaps the most significant single step in the evolution of governance at the station is the "Master Plan for Development and Administration" of the station. Many of the regulations for work at La Selva were previously recorded only as minutes of OTS Executive Committee meetings or in correspondence among committee members, the executive director of OTS, and individual researchers. Beginning with the preparation of operational plans for the 1980–81 period, the various unwritten rules that applied to La Selva were examined and codified or rejected in favor of modified regulations. After four years of research and writing by the station codirectors, extensive input from numerous individuals, and several revisions, the first version of the master plan was approved by the board of directors in 1984. It was revised and reratified in 1988. The plan is an open document that states OTS's management and development policies. It provides a stable framework within which researchers and educators can plan their use of the station and its facilities. The document has proved to be adaptable, and sections are modified and new policies added as necessary.

The issues, large and small, that must be dealt with at a research station are legion. The master plan in 1992 includes regulations dealing with zoning, marking of plants and animals, use of machetes, handling of poisonous snakes, and canopy access, to name but a very few. The complete La Selva master plan is available on request from OTS (in Spanish or English). I examine a few examples of the kinds of issues that the plan handles to indicate the scope of the document and the issues that research station managers must face.

Collecting

Many biologists hold strong opinions about collecting. Historically, La Selva has experienced a phase of essentially no restrictions on collecting, followed by a period of nearly absolute prohibition of collecting, followed by the present policy of permitting carefully regulated collecting. Interestingly, the middle phase prohibition of collection was, in part, a reaction to perceived abuses by earlier collectors. Recognition is now widespread that sensitive collecting, guided by research goals, is fundamental to increasing knowledge of tropical organisms. All recent collecting programs by specialists (e.g., Greene et al. collecting reptiles, the Flora project team) have produced new records for La Selva (and often new species).

The policy in 1992 on collecting first urges researchers to take specimens from disturbed areas outside of La Selva whenever possible. Collecting within the station requires a clear link to research goals and complete compliance with Costa Rican laws about collecting (including deposition of voucher specimens in the national collections). Special regulations apply to certain groups of organisms and life stages; the criteria are (1) how much is known about the organisms in question (the better known a group, the more stringent the requirements for additional collecting), (2) fecundity (ability to recover from collecting), and (3) possible ulterior motives for collecting (e.g., commercial value). Collecting is also restricted to protect long-term research projects for which unperturbed population dynamics are required. Examples of this policy include the special permission and justification required to collect butterflies that are of commercial value. Collection of vertebrates must be justified by research goals that cannot be addressed in other ways and by evidence that the populations in question can recover. Seed collecting must take into account the size of the total seed crop to avoid having negative impact on plant reproduction. Seeds may not be collected in areas where detailed demographic studies of selected species (e.g., the Clarks' work; see chap. 7) or of the entire tree community (e.g., the work in the OTS plots; see chap. 8) are underway. The station director is empowered to make most exceptions to the policy on collecting; the LSAC serves as final arbiter in difficult cases.

Zoning

Ideally, a research station would accommodate all types of field research. Given the state of tropical rain forests in the last decade of the twentieth century, however, many feel that certain kinds of research are too destructive to merit conduct in protected primary forest. Fortunately, the most destructive types of research usually do not require pristine forest as the starting point. Further, the recent series of land acquisitions at La Selva has yielded a site with habitats suitable for most kinds of research. This has not always been the case, and some research projects had to be sited elsewhere during the 1970s and early 1980s when La Selva's land holdings were almost exclusively in primary forest.

In the current zoning scheme set forth in the master plan research intensity has been matched with available habitats. Five zones are recognized. The *Rafael Chavarría Ecological Reserve* (fig. 2.2) is an absolute preserve to perpetuate the esthetic value of this relatively undisturbed portion of the forest. It is not available for research except with special permission. *Low Impact Areas* (including most of Old La Selva) are available for all research except *major* manipulations (defined for these purposes as at a scale greater than 100 m²). *High Impact Areas* are available for projects requiring larger-scale habitat manipulations. Within this category are eleven ranked parcels of land; those ranked highly are to be used only when areas with lower rank are not available or are unsuitable. Two areas of primary forest are in this category, but they are ranked tenth and eleventh and are to be used for highly manipulative work only when absolutely required. *Habitat Management Areas* are sites managed by OTS for demonstration or logistical purposes (e.g., arboretum, successional strips, trails). These sites are available for use by compatible research projects. The fifth land use category, *Developed Areas,* includes all building sites and their surrounding open areas. Again, research is encouraged in these areas, but site maintenance and safety take priority over research needs.

Exclusive Use

Some kinds of research are highly susceptible to disturbance. Even occasional pedestrian traffic, for example, can trample seedlings and bias studies of plant regeneration. As is often the case, the value of protecting such studies is offset by the detriments of denying other researchers access to these areas. In each case, pros and cons must be weighed, and it is precisely here that LSAC is extremely valuable. Thus far, requests for exclusive use have been few and carefully justified, and the ratio of "locked up" to open areas remains low.

Long-term Maintenance

Projects that require long-term maintenance most often are "institutional"-level projects, and the decision to undertake them is made at this level and in recognition of the future costs involved. Examples at La Selva include the arboretum, which serves as a living reference collection of native tree species,

and the successional strips, which are cut on a five-year rotation and provide easily accessible second-growth vegetation suitable for numerous projects. Individual researchers occasionally plan large-scale projects that offer considerable institutional benefits provided they are maintained over the long term. Only rarely will the researcher have funding beyond setup of the project or be in a financial position to guarantee maintenance in perpetuity. In such cases LSAC must make reasonable estimates of the long-term costs and then weigh these against long-term institutional benefits as a basis for recommending the project's approval.

Exotic Species

Despite concerns about the impact that introduced species can have on native species and ecosystems, often quite good reasons exist for the introduction of exotic species or genotypes. For example, La Selva is developing as a center for testing tree species suitable for reforestation in tropical wet lowlands. It would be difficult to fulfill this role by using only tree species and genotypes that are native to La Selva. Undoubtedly, many useful species will prove to be natives, but this could not be clearly demonstrated if the project could not grow widely used exotic species for comparison.

The primary concerns about exotics are that they may escape into natural habitats within La Selva; escape and become pests in the surrounding region; and invalidate genetic studies at La Selva by introducing foreign genotypes. The current policy on exotics prohibits the introduction of animals (including release of confiscated wildlife); introduced animals must be caged and removed following completion of the research. The policy requires evaluation of each proposed plant introduction in the context of what is known about the species's reproductive biology and careful monitoring of the introduced plants for evidence of reproduction. Research projects are required to bear the expense incurred in monitoring introductions for escapees and for eradication of spreading plants. The policy also restricts introduction to areas outside of the primary forest.

Data Management

Among the sensitive issues that must be dealt with at a research facility that is used by numerous individuals is the archiving of data. Although the proprietary rights of the individuals who collected the data must be recognized, there are also legitimate reasons for ensuring that such data sets do not "die" with the completion of an individual's research career. Clearly, OTS, as La Selva's owner, has the right to ensure that data collected there are widely available, and it also has the responsibility of protecting the right of researchers to take credit for their data. The 1988 version of the master plan for La Selva for the first time included explicit policy on archiving data sets. Investigators who have major research projects at La Selva are required to archive their data in a way acceptable to OTS, and OTS agrees in return to safeguard such data from release to unauthorized individuals.

Establishing Usage Limits

Setting usage limits reflects the potential conflict between maximizing use of the station and conserving its biological resources. No data are available on which to base such limits, and one might question whether such data are realistically obtainable. The issue is further complicated by the fact that different kinds of visitors impact differently on the reserve. The station is probably less affected by a dozen people who walk along a cement sidewalk once than by a researcher who repeatedly samples from sites off the main trails, but most comparisons are not as clear. Usage limits will also depend upon where these various kinds of visitors will be within the reserve. Although there is some confidence that the number of visitors to the primary forest (especially those who can be expected to get off the trails) should be limited, it is not clear that managed areas or those in secondary forest require as much protection. In fact, such protection would restrict the ability of these areas to serve, for example, as demonstration plots.

At remote facilities, available housing is an added consideration (including adequacy of the potable water and sewage disposal systems) and acceptable levels of perceived crowding therein. In the past, carrying capacity at La Selva has been set as much by available beds and chairs in the dining room as by educated guesses about what the forest can bear.

In the 1990s the usage limit for the *reserve* is set at sixty-five per day, with this total comprising as many as thirty-five researchers and thirty students or other group visitors. There is no usage limit for areas of the reserve that are not in primary forest. As many as seventy-five may be in residence at the station (this limit is set according to beds and dining hall capacity) provided that at least ten of these individuals have projects outside of the primary forest. These regulations are expressly provisional, and they are reexamined annually.

Priority for Station Use
OTS identifies research and education as the station's twin priorities. Researchers are accommodated on a first-come–first-served basis up to the limit of thirty-five. Those with ongoing, long-term projects at the station receive special consideration for the ten excess-capacity slots at times of peak usage. The thirty group slots are a bit more difficult to administrate because of competition for these slots from natural history tour groups. Ecotourists are welcome users of the station as long as they do not displace the higher-priority users. As commercial enterprises, tour companies are able to make reservations and pay deposits years in advance, whereas the academic courses that have priority for the group slots generally operate within a much shorter time frame. The current solution to this problem is to close La Selva to natural history groups during the peak course seasons (January–March and June–August) until OTS courses have made their reservations and other university groups that have used the station in the past year have been contacted concerning their intentions. Once these courses are scheduled for station use, the remaining dates are made available to natural history groups.

Day Visitors
Until recently, staying overnight at the station was, for all intents and purposes, the only way to visit La Selva. The recent improvement of the highway to Puerto Viejo and the development of tourist-style accommodations in the vicinity means that day visits are now feasible. These had previously been prohibited because of issues of control of *walk-ons,* liability, and impact on the forest. In response to increased demand to permit day visits OTS is experimenting with different systems for admitting walk-ons. Day visitation of natural history visitors is currently permitted by groups (up to twenty persons per day) provided that they are accompanied by an OTS-trained naturalist. The intention is to use these local naturalists as an interface between La Selva and the general public in a way that is mutually beneficial. The training program for these naturalists is described by Paaby et al. (1991).

Future Development
While retaining the ability to react opportunistically to unanticipated circumstances, a major facility like La Selva must have a responsible plan for its future direction. Within an integrated plan for development and use of the entire site, the more specific categories that require such planning include accommodations and amenities, research facilities and equipment, site development, and logistics. Questions of what (specific development proposed), when (priorities), where (siting), and how (resources: human, financial, and other) must then be addressed for each category. The master plan includes a list of high-priority projects for short-term development (five-year range), and a more general presentation of the issues of longer-term facilities development.

The previous sections are designed to illustrate the scope of governance at La Selva, most of which is handled by the station's Master Plan. It should be stressed that the plan is intended to evolve along with the research station. Future versions will, no doubt, modify current policy and make policy in new areas. The highly interactive process used in formulating and modifying the plan, however, is well established.

CONCLUSION
I have given only a hint of the kinds of issues and controversies that have faced and will continue to face OTS in administering La Selva in the future. To confront these challenges OTS has developed a system of administration and governance at the station that operates effectively and, most importantly, is flexible enough to deal with problems as they arise. Particularly vital is the role of the La Selva Advisory Committee as a panel of individuals with diverse expertise that is not aligned with any one point of view or any one group of users. As the issues at La Selva become more specialized (e.g., hazardous waste, forestry research), more specialized advice may be needed, perhaps in the form of ad hoc committees assembled to take on a particular issue. In fact, precedent exists within OTS for seeking this kind of advice from one-time committees set up to make recommendations on timely subjects.

The increased demand for use of the station is bringing a number of new issues into sharp focus. In early 1992 the station is full for significant parts of the year, raising the question of whether OTS's first-come–first-served policy should be reevaluated. Research users are concerned that small-scale projects may be "squeezed out" as the number of "big science" projects at La Selva increases. If several large projects with long-term funding each bring five or six researchers to the facility, correspondingly fewer spaces are available for individual researchers. One might anticipate considerable pressure in the near future for a reservation system that recognizes these issues, for reexamination of usage limits, and for increasing residential capacity at the station.

Demand for services from the general public will also increase. Completion of the Puerto Viejo-Horquetas Highway that lies just beyond the entrance to La Selva will certainly accelerate this trend. The increased pressure for access to the station from nonscientists is likely to increase debate about the proper role of these visitors at a biological station. There is certainly space for discussion about whether and to what extent research stations like La Selva can be everything to all people without losing their research and educational foci. The new Visitors Center on the La Flaminea property will play a key role; it is vital that it be designed to provide a satisfying experience for casual visitors, thus reserving the station's core to serve priority users.

Although the exact dimensions of La Selva's future cannot be foreseen, it is clear that demand for access to the station's rain forest and facilities will continue to grow. Apparent also is that this demand will come from an increasingly diverse audience of scientists and nonscientists alike. Consequently, management issues can only be expected to increase in both number and complexity. There is every reason to believe that the policy-making for La Selva will continue to be a lively aspect of research station management. Exciting and challenging times are ahead for La Selva!

APPENDIX 3

Vascular Plants: An Interim Checklist

Robert L. Wilbur and Collaborators

My "collaborators" are receiving much less than their deserved credit for they are full authors of their contributions, and I can only hope that I have not misrepresented the results of their study in this skeletal report. Their names and their taxonomic contributions to this checklist are as follows: Frank Almeda (Melastomataceae), W. S. Alverson (Bombacaceae), W. R. Anderson (Malphigiaceae), John T. Atwood (Orchidaceae), Thomas Croat and Michael H. Grayum (Araceae), Michael H. Grayum and Hugh W. Churchill (Lycopodiophyta and Polypodiophyta), Barry E. Hammel (Cecropiaceae, Guttiferae, Lauraceae, Moraceae, *Psychotria* of the Rubiaceae, Cyclanthaceae, Marantaceae), Emmet J. Judziewicz and Richard W. Pohl (Gramineae), W. John Kress (Heliconiaceae), Lucinda A. McDade (Acanthaceae), John M. MacDougal (Passifloraceae), George E. Schatz (Annonaceae), Damon A. Smith (Elaeocarpaceae), Charlotte M. Taylor (Rubiaceae), and Grady L. Webster and Michael J. Huft (Euphorbiaceae). A special thanks is due Dr. Michael Grayum, this project's first collector at La Selva, who has continuously advised me and tactfully corrected many of my mistakes.

The checklist is presented in alphabetical sequence of the families under the following groupings:

    I. Pteridophytic plants
       A. Lycopodiophyta
       B. Polypodiophyta
    II. Gymnosperms
    III. Flowering plants
       A. Monocots
       B. Dicots

Synonyms of certain species are included in parentheses, especially those preferred by some specialists or those in common use elsewhere or found in widely used treatments of Central American plants. Names in brackets are used to a limited extent for some familiar names no longer deemed correct (at least by the author of the treatment) to refer to the name employed in this checklist.

There is no intention to publish the name of any new taxa in this checklist. Numerous species are listed as new (*sp. nov.*), and many are given epithets of convenience and indicated as *ined*. These names, of course, are not validly published and have no nomenclatural standing except, I suppose, regrettably as *nomina nuda*. There seems to be no good alternative because some of these binomials have gained currency at La Selva, and it is not mnemonically helpful to designate such species by either letter or number.

The precise identities of other species are less certain, and this uncertainty is indicated by listing the binomial that seems closest with *vel aff.* added to indicate the systematist's basic unease together with confidence in the approximate identity. More and better material or additional study we hope will resolve the uncertainty.

A question mark following the binomial expresses uncertainty about the correctness of the identification. A question mark (?) appears before the specific epithet of certain species. These have often been suggested on good authority to be at La Selva and sometimes even are known to be there but lack a voucher seen by me or the author of the taxon. In other cases their reported presence seems unlikely.

I hope readers will send corrections and additions to me so that the published *Vascular Flora of La Selva* can profit from the more specialized knowledge and familiarity of others.

I. PTERIDOPHYTIC PLANTS
A. LYCOPODIOPHYTA
LYCOPODIACEAE
Lycopodium L.
    cernuum L.
       (= *Lepidotis cernua* (L.) Beauv.; *Lycopodiella cernua* (L.) Pichi-Serm.)
    dichaeoides Maxon
       (=*Huperzia dichaeoides* (Maxon) Holub; *Urostachys dichaeoides* (Maxon) Herter ex Nessel)
    dichotomum Jacq.
       (= *Huperzia dichotoma* (Jacq.) Trev. St. Léon; *Urostachys dichotomus* (Jacq.) Herter)
    linifolium L.
       (= *Huperzia linifolia* (L.) Trev. St. Léon; *Urostachys linifolia* (L.) Herter)

SELAGINELLACEAE
Selaginella Beauv.
    anceps K. Presl
    arthritica Alston
    atirrensis Hieron.
    bombycina Spring
    eurynota A. Braun
    flagellata Spring

oaxacana Spring
silvestris Aspl.
umbrosa Lemaire ex Hieron.

B. POLYPODIOPHYTA (= FILICOPHYTA)
CYATHEACEAE
Alsophila R. Br.
  cuspidata (Kunze) Conant
    (= *Nephelea cuspidata* (Kunze) R. Tryon)
  firma (Baker) Conant
    (= *Nephelea mexicana* (Schldl. & Cham.)
    R. Tryon)
Cyathea J. E. Sm.
  microdonta (Desv.) Domin
    (= *Trichipteris microdonta* (Desv.) R. Tryon)
  multiflora J. E. Sm.
    (= *Hemitelia multiflora* (J. E. Sm.) R. Br.)
  schiedeana (Presl) Domin
    (= *Trichipteris schiedeana* (Presl) Domin
  trichiata (Maxon) Domin
    (= *Trichipteris ursina* (Maxon) R. Tryon)
  ursina (Maxon) Lellinger
    (= *Cyathea ursina* (Maxon) R. Tryon)

GLEICHENIACEAE
Dicranopteris Bernh.
  pectinata (Willd.) Underw.
Gleichenia J. E. Sm.
  bifida (Willd.) Sprengel
    (= *Sticherus bifidus* (Willd.) Ching)

HYMENOPHYLLACEAE
Hymenophyllum J. E. Sm.
  brevifrons Kunze
  hirsutum (L.) Sw.
    (= *Sphaerocionium hirsutum* (L.) Presl)
  maxonii Christ ex C. Morton
  polyanthos (Sw.) Sw.
    (= *Mecodium polyanthos* (Sw.) Copel.)
Trichomanes L.
  angustifrons (Fée) W. Boer
  ankersii Parker ex Hook. & Grev.
  botryoides Kaulf.
  collariatum Bosch
  crispum L.
  curtii Rosenst.
  diaphanum HBK.
  diversifrons (Bory) Mett. ex Sadeb.
  ekmanii W. Boer
  elegans L. C. Rich.
  godmanii Hook.
  krausii Hook. & Grev.
  membranaceum L.
  pinnatum Hedw.
  rigidum Sw.
  tuerckheimii Christ
Danaea J. E. Sm.
  cuspidata Liebm.
  elliptica J. E. Sm.
  grandifolia Underw.

nodosa (L.) J. E. Sm.
wendlandii Reichenb. f.

METAXYACEAE
Metaxya K. Presl
  rostrata (HBK.) K. Presl

OPHIOGLOSSACEAE
Ophioglossum L.
  reticulatum L.

POLYPODIACEAE
Adiantum L.
  latifolium Lam.
  macrophyllum Sw.
  obliquum Willd.
  petiolatum Desv.
  seemannii Hook.
  tetraphyllum Humb. & Bonpl. ex. Willd.
    (= *A. fructuosum* Poeppig ex Sprengel)
  wilsonii Hook.
Anetium (Kunze) Splitg.
  citrifolium (L.) Splitg.
Antrophyum Kaulf.
  cajenense (Desv.) Sprengel
    (= *Polytaenium cajenense* (Desv.) Benedict)
  lanceolatum (L.) Kaulf.
    (= *Polytaenium feei* (Shaffner ex Fée) Maxon)
  ?lineatum (Sw.) Kaulf. [presence highly doubtful in
  spite of a voucher.]
Asplenium L.
  abscissum Willd.
  cirrhatum L. C. Rich. ex. Willd.
    (= *A. radicans* L. var. *cirrhatum* (L. C. Rich. ex
    Willd.) Rosenst.)
  cuspidatum Lam.
    (= *A. auritum* Sw.—the pinnate extreme.)
  falcinellum Maxon
  formosum Willd.
  holophlebium J. G. Baker
  otites Link
  pteropus Kaulf.
  riparium Leibm.
    (= *A. repandulum* auctt., non Kunze)
  serra Langds. & Fischer
  serratum L.
Blechnum L.
  gracile Kaulf.
    (= *B. fraxineum* Willd., *B. lellingeranum* L. D.
    Gómez)
  occidentale L.
  polypodiodes Raddi
Bolbitis Schott
  aliena (Sw.) Alston
  nicotianifolia (Sw.) Alston
  portoricensis (Sprengel) Hennipman
[Campyloneurum—see Polypodium]
Ctenitis (C. Chr.) C. Chr.
  sloanei (Poeppig ex Sprengel) Morton
  subincisa (Willd.) Ching

(= *Megalastrum subincisum* (Willd.) A. R. Sm. &
  R. C. Moran)
Cyclopeltis J. Sm.
  semicordata (Sw.) J. Sm.
Dennstaedtia T. Moore
  bipinnata (Cav.) Maxon
  cicutaria (Sw.) T. Moore
  obtusifolia (Willd.) T. Moore
Dicranoglossum J. Sm.
  panamensis (C. Chr.) L. D. Gómez
Didymochlaena Desv.
  truncatula (Sw.) J. Sm.
Diplazium Sw.
  cristatum (Desr.) Alston
  grandifolium (Sw.) Sw.
  ingens Christ
  lindbergii (Mett.) Christ
  lonchophyllum Kunze
  macrophyllum Desv.
  pactile Lellinger
  striatastrum Lellinger
Elaphoglossum Schott ex J. Sm.
  amygdalifolium (Mett.) Christ
  correae Mickel
    (= *E. "palmense,"* not Christ)
  grayumii Mickel
    (= *"E. sp. nov. ined."* in *Selbyana* 11: 94.1989)
  herminieri (Bory & Fée) T. Moore
  latifolium (Sw.) J. Sm.
  peltatum (Sw.) Urban
    (= *Peltapteris peltatum* (Sw.) C. Morton;
    Rhipidopteris peltatum (Sw.) Link)
Grammitis Sw.
  linearifolia (Desv.) Steudel
    (= *Cochlidium linearifolium* (Desv.) Maxon
    ex C. Chr.)
  serrulata (Sw.) Sw.
    (= *Cochlidium serrulatum* (Sw.) Bishop,
    Xiphopteris serrulata (Sw.) Kaulf.)
Hecistopteris J. Sm.
  pumila (Sprengel) J. Sm.
Hemidictyum K. Presl
  marginatum (L.) K. Presl
    (= *Diplazium marginatum* (L.) Diels; *D. limbatum*
    (Willd.) Proctor)
Hemionitis L.
  palmata L.
Hypolepis Bernh.
  hostilis (Kunze) K. Presl
  repens (L.) K. Presl
Lastreopsis Ching
  exculta (Mett.) Tindale ssp. guatemalensis (Baker)
    Tindale
    (= *L. chontalensis* (E. Fourn.) Lellinger)
Lindsaea Dryand. ex J. E. Sm.
  lancea (L.) Bedd.
  quadrangularis Raddi ssp. subulata K. U. Kramer
Lomariopsis Fée
  fendleri D. C. Eaton
    (= *L. vestita* E. Fourn.)
  japurensis (Martius) J. Sm.

Lonchitis L.
  hirsuta L.
[Megalastrum-see *Ctenitis subincisa*]
Nephrolepis Schott
  biserrata (Sw.) Schott
  multiflora (Roxb.) Jarrett ex Morton
  pendula (Raddi) J. Sm.
    (= *N. cordifolia* sensu many Central American
    authors)
  rivularis (Vahl) Mett. ex Krug
Oleandra Cav.
  articulata (Sw.) K. Presl
    (= *O. nodosa* (Willd.) K. Presl)
[Peltapteris Link
  Peltata (Sw.) Morton; see *Elaphoglossum peltatum*
  (Sw.) Urban]
[Phlebodium—see Polypodium]
Pityrogramma Link
  calomelanos (L.) Link
  ?tartarea (Cav.) Maxon [expected but not yet
    collected]
Polybotrya Humb. & Bonpl. ex Willd.
  alfredii Brade
  caudata Kunze
    (= *P. villosula* Christ)
  cervina (L.) Kaulf.
    (= *Olfersia cervina* (L.) Kunze)
  osmundacea Humb. & Bonpl. ex Willd.
Polypodium L.
  angustifolium Sw.
    (= *Campyloneurum angustifolium* (Sw.) Fée)
  ciliatum Willd.
    (= *Microgramma reptans* (Cav.) A.R. Sm.)
  crassifolium L.
    (= *Niphidium crassifolium* (L.) Lellinger)
  dissimlie L.
    (= *P. chnoodes* Sprengel)
  dulce Poiret
    (= *P. sororium* Humb. & Bonpl. ex Willd.)
  furfuraceum Schldl. & Cham.
  hygrometricum Splitg.
    (= *Pecluma hygrometrica* (Splitg.) M. G. Price)
  latum (T. Moore) T. Moore ex Sodiro
    (= *Campyloneuron latum* T. Moore)
  loriciforme Rosenst.
  lycopodioides L.
    (= *Microgramma lycopodioides* (L.) Copel.)
  maritimum Hieron.
  occultum Christ
    (= *Campyloneurum occultum* (Christ) L. D.
    Gómez)
  pectinatum L.
    (= *Pecluma pectinata* (L.) M. G. Price)
  percussum Cav.
    (= *Pleopeltis percussa* (Cav.) Hook. & Grev.,
    Microgramma percussa (Cav.) Soto)
  pseudoaureum Cav.
    (= *Phlebodium pseudoaureum* (Cav.) Lellinger)
[sororium Humbl. & Bonpl. ex Willd.—see P. dulce
  Poiret]
  sphenodes Kunze ex Klotzsch

(= *Campyloneurum sphenodes* Kunze ex Klotzsch) Fée)
  triseriale Sw.
Pseudocolysis L. D. Gómez
  bradeorum (Rosenst.) L. D. Gómez
    (= *Polypodium bradeorum* Rosenst., *P. colysoides* Maxon & Copel. ex Copel.)
Pteris L.
  altissima Poiret
  plumula Desv.
  propinqua J. Agardh
  pungens Willd.
    (= *P. quadriaurita* Retz.)
  tripartita Sw.
Saccoloma Kaulf.
  inaequale (Kunze) Mett.
Salpichlaena J. Sm.
  volubilis (Kaulf.) J. Sm.
    (= *Blechnum volubile* Kaulf.)
Stigmatopteris C. Chr.
  longicaudata (Liebm.) C. Chr.
  sordida (Maxon) C. Chr.
Tectaria Cav.
  athyrioides (Baker) C. Chr.
    (= *T. rheosora* (Baker) C. Chr.)
  brauniana (Karsten) C. Chr.
    (= *T. neotropica* L. D. Gómez)
  draconoptera (D. C. Eaton) Copel.
    (= *T. myriosora* (Christ) C. Chr.)
  heracleifolia (Willd.) Underw.
  incisa Cav.
  mexicana (Fée) C. Morton
  nicotianifolia (Baker) C. Chr.
    (= *T. euryloba* (Christ) Maxon)
  plantaginea (Jacq.) Maxon
  rivalis (Mett. ex Kuhn) C. Chr.
  rufovillosa (Rosenst.) C. Chr.
Thelypteris Schmidel
  angustifolia (Willd.) Proctor
    (= *Meniscium angustifolia* Willd.)
  balbisii (Sprengel) Ching
  biolleyi (Christ) Proctor
    (= *Goniopteris biolleyi* (Christ) Pichi-Serm.)
  curta (Christ) C. Reed
  decussata (L.) Proctor var. costaricensis A. R. Sm.
    (= *Glaphyropteris decussata* (L.) Fée)
  dentata (Forssk.) E. St. John
    (= *Cyclosorus dentatus* (Forssk.) Ching, *Christella dentata* (Forssk.) Brownsey & Jermy)
  francoana (Fourn.) C. Reed
    (= *Goniopteris francoana* (Fourn.) A. & D. Löve)
  ghiesbreghtii (Hook.) C. Morton
    (= *Goniopteris mollis* Fée)
  gigantea (Mett.) R. Tryon
    (= *Meniscium giganteum* Mett.)
  hispidula (Decne.) C. Reed
    (= *T. quadrangularis* (Fée) Schelpe, *Cyclosorus quadrangularis* (Fée) Tardieu-Blot, *Christella hispidula* (Decne.) Holttum)
  leprieurii (Hook.) R. Tryon
  lingulata (C. Chr.) C. Morton

  nicaraguensis (E. Fourn.) C. Morton
  poiteana (Bory) Proctor
    (= *Goniopteris poiteana* (Bory) Ching)
  resinifera (Desv.) Proctor
    (= *Th. pseudosancta* (C. Chr.) C. Reed, *Amauropelta resinifera* (Desv.) Pichi-Serm., *T. panamensis* (K. Presl) E. St. John)
  serrata (Cav.) Alston
    (= *Meniscium serratum* Cav.)
  torresiana (Gaudich.) Alston
    (= *Macrothelypteris torresiana* (Gaudich.) Ching)
  urbanii (Sodiro) A. R. Sm.
    (= *T. leucophlebia* (Christ) C. Reed; *Goniopteris leucophlebia* (Christ) Ching)
  villana L. D. Gómez
    (incl. *Th. pseudoaspidoides* L. D. Gómez)
Vittaria J. E. Sm.
  costata Kunze
    (= *Ananthacorus angustifolius* (Sw.) Underw. & Maxon)
  lineata (L.) J. E. Sm.
  stipitata Kunze

II. THE GYMNOSPERMS
CYADACEAE
Zamia L.
  skinneri Warscz. ex A. Dietr.
    (= *Aulacophyllum skinneri* (Warscz. ex A. Dietr.) Regel)

GNETACEAE
Gnetum L.
  leyboldii Tul.

III. THE FLOWERING PLANTS
A. MONOCOTS
AMARYLLIDACEAE
Bomarea Mirbel
  obovata Herb.
    (= *B. chontalensis* Seemann)
Crinum L.
  erubescens L. f. ex Aiton
Hymenocallis Salisb.
  littoralis (Jacq.) Salisb.

ARACEAE
Alocasia (Schott) G. Don
  cucullata (Lour.) Schott
  macrorrhizos (L.) Schott
  plumbea van Houtte
    (= *A. indica* Schott var. *metallica* Schott)
Anthurium Schott
  acutangulum Engler
  austin-smithii Croat & R. Baker
  bakeri Hook. f.
  bradeanum Croat & Grayum
  clavigerum Poeppig
  clidemioides Standley
  consobrinum Schott
    (= *A. cuneatissimum* (Engler) Croat, *A. consobrinum* var. *cuneatissimum* Engler)

cuspidatum Masters
flexile Schott
formosum Schott
friedrichsthallii Schott
gracile (Rudge) Lindley
interruptum Sodiro
lancifolium Schott
llanoense Croat ssp. oblongispicum Croat & Grayum
    ined.
ochranthum K. Koch
pentaphyllum (Aublet) G. Don
    var. pentaphyllym
    var. bombacifolium (Schott) Madison
ramonense Engler ex K. Krause
ravenii Croat & R. Baker
scandens (Aublet) Engler
spathiphyllum N. E. Br.
subsignatum Schott
trinerve Miq.
trisectum Sodiro
upalaense Croat & R. Baker
Caladium Vent.
    bicolor (Dryander ex Aiton) Vent.
Colocasia Schott
    esculenta (L.) Schott
Dieffenbachia Schott
    beachiana Croat & Grayum *ined.*
    concinna Croat & Grayum *ined.*
    hammelii Croat & Grayum *ined.*
    longispatha Engler & K. Krause
    longivaginata Croat & Grayum *ined.*
    oerstedii Schott
    seguine (L.) Schott
Dracontium L.
    gigas (Seemann) Engler
Heteropsis Kunth
    oblongifolia Kunth
Homalomena Schott ex Schott & Engler
    hammelii Croat & Grayum *ined.*
Monstera Adans.
    adansonii Schott var. laniata (Schott) Madison
    costaricensis (Engler & Krause) Croat & Grayum
    diversifolia Croat & Grayum *ined.*
    glaucescens Croat & Grayum *ined.*
    pittieri Engler
    skutchii Croat & Grayum *ined.*
    spruceana (Schott) Engler
    standleyana Bunting
    tenuis K. Koch
    tuberculata Lundell var. brevinodum (Standley &
    L. O. Williams) Madison
Philodendron Schott
    alliodorum Croat & Grayum
    angustilobum Croat & Grayum *ined.*
    aromaticum Croat & Grayum *ined.*
    aurantiifolium Schott
    bakeri Croat & Grayum *ined.*
    brevispathum Schott
    brunneicaulum Croat & Grayum *ined.*
    cretosum Croat & Grayum *ined.*

davidsonii Croat
ensifolium Croat & Grayum
fragrantissimum (Hook.) G. Don
grandipes K. Krause
hederaceum (Jacq.) Schott
herbaceum Croat & Grayum
inaequilaterum Liebm.
jodaviscanum Bunting
ligulatum Schott
opacum Croat & Grayum
platypetiolatum Madison
pterotum K. Koch
radiatum Schott
rhodoaxis Bunting
rigidifolium K. Krause
rothschuhianum (Engler) Croat & Grayum
sagittifolium Liebm.
scandens K. Koch & Sello
schottii K. Koch
    (= *P. talamancae* Engler)
sulcatum K. Krause
tenue K. Koch
tripartitum (Jacq.) Schott
wendlandii Schott
Rhodospatha Poeppig
    pellucida Croat & Grayum *ined.*
    wendlandii Schott ex Engler
Spathiphyllum Schott
    friedrichsthalii Schott
    fulvovirens Schott
    laeve Engler
    phryniifolium Schott
    wendlandii Schott
Stenospermation Schott
    angustifolium Hemsley
    marantifolium Hemsley
Syngonium Schott
    birdseyanum Croat & Grayum *ined.*
    macrophyllum Engler
    podophyllum Schott ssp. peliocladum (Schott) Croat
    (= *S. peliocladum* Schott)
    rayi Croat & Grayum *ined.*
    schottianum H. A. Wendl. ex Schott
    standleyanum Bunting
    triphyllum Birdsey ex Bunting
Urospatha Schott
    grandis Schott
    (= *U. tonduzii* Engler)
Xanthosoma Schott
    jacquinii Schott
    mexicanum Liebm.
    nigrum (Vellozo) Stellfeld
    undipes (K. Koch) K. Koch

[ARECACEAE—see Palmae]

BROMELIACEAE
Aechmea Ruiz & Pavón
    angustifolia Poeppig & Endl.
    magdalenae (André) André ex Baker

mariae-reginae H. A. Wendl.
mexicana Baker
nudicaulis (L.) Griseb.
pubescens Baker
Ananas P. Mill.
    comosus (L.) Merr. [cultivated]
Androlepis Brongn. ex Houllet
    skinneri Brongn. ex Houllet
        (= *A. donnell-smithii* (Baker) Mez)
Catopsis Griseb.
    juncifolia Mez & Wercklé ex Mez
    morreniana Mez
    sessiliflora (Ruiz & Pavón) Mez
Guzmania Ruiz & Pavón
    erythrolepis Brongn. ex Planchon
    lingulata (L.) Mez
    monostachia (L.) Rusby ex Mez
    scherzeriana Mez
Pitcairnia L'Her.
    atrorubens (Beer) Baker
    wendlandii Baker
Tillandsia L.
    anceps Lodd.
    bulbosa Hook.
    excelsa Griseb.
    festucoides Brongn. ex Mez
    leiboldiana Schldl.
    monadclpha (E. Morren) Baker
    pruinosa Sw.
    usneoides (L.) L.
    venusta Mez & Wercklé
Vriesea Lindl.
    gladioliflora (H. A. Wendl.) Antoine
    heliconioides (HBK.) Hook. ex Walp.
    kupperiana Süssenguth
    ringens (Griseb.) Harms

CANNACEAE
    Canna L.
        tuerckheimii Kraenzlin

COMMELINACEAE
    Campelia L. C. Rich.
    zanonia (L.) HBK.
        (= *Tradescantia zanonia* (L.) Sw.)
Cochliostema Lemaire
    odoratissimum Lemaire
Commelina L.
    diffusa Burm. f.
Dichorisandra J. C. Mikan
    hexandra (Aublet) Standley
Floscopa Lour.
    robusta (Seub.) C. B. Clarke
Tripogandra Raf.
    serrulata (Vahl) Handlos
Zebrina Schnizlein
    heuhueteca Standley & Steyerm.
        (= *Tradescantia huehueteca* (Standley & Steyerm.)
        D. Hunt)

?pendula Schnizlein
    (= *Tradescantia zebrina* Hort. ex Bosse)

COSTACEAE
    Costus L.
        barbatus Süssenguth
        bracteatus Rowlee
        laevis Ruiz & Pavón
        lima L.
        malortieanus H. A. Wendl.
        nitidus Maas
        pictus D. Don?
        pulverulentus K. Presl
        scaber Ruiz & Pavón

CYCLANTHACEAE
    Asplundia Harling
        euryspatha Harling
        ferruginea Grayum & Hammel
        longitepala Harling
            (= *A. antioquiae* sensu Hammel, *Selbyana* 9:197.
            1986.)
        multistaminata Harling
        sleeperae Grayum & Hammel
        uncinata Harling
        utilis (Oersted) Harling
            (= *A. costaricensis* Harling)
        vagans Harling
    Carludovica Ruiz & Pavón
        rotundifolia H. A. Wendl. ex Hook. f.
        sulcata Hammel
    Chorigyne
        pendula (Hammel) R. Eriksson
            (= *Sphaeradenia pendula* Hammel)
    Cyclanthus Poit.
        bipartitus Poit.
    Dicranopygium Harling
        umbrophila Hammel
        wedelii Harling
    Evodianthus Oersted
        funifer (Poit.) Lindman
    Ludovia Pers.
        integrifolia (Woodson) Harling
            (= *Carludovica integrifolia* Woodson)
    Sphaeradenia Harling
        carrilloana Grayum & Hammel

CYPERACEAE
    Calyptrocarya Nees
        glomerulata (Brongn.) Urban
        poeppigiana Kunth
    Cyperus L.
        compressus L.
        costaricensis J. Gómez L.
            (= *C. amplus* G. Tucker)
        flavescens L.
        laxus Lam.
            (= *C. tolucensis* HBK.)
        luzulae (L.) Retz.
        miliifolius Poeppig & Kunth

niger Ruiz & Pavón
odoratus L.
simplex HBK.
surinamensis Rottb.
tenuis Sw.
thyrsiflorus Jungh.
Dichromena Michaux
ciliata Pers.
(= *Rhynchospora ciliata* (Pers.) Kükenthal;
R. nervosa ssp. *ciliata* (Pers.) T. Koyama)
radicans Schldl. & Cham.
(= *Rhynchospora radicans* (Schldl. & Cham.)
Pfeiffer, *Dichromena microcepala* Bertero ex
Sprengel, *Rhynchospora radicans* ssp.
microcephala (Bertero ex Sprengel) W. W.
Thomas)
Eleocharis R. Br.
elegans (HBK.) Roemer & Schultes
nervata Svenson
retroflexa (Poiret) Urban
Fimbristylis Vahl
dichotoma (L.) Vahl
miliacea (L.) Vahl
Hypolytrum Pers.
schraderianum Nees
(= *H. nicaraguense* Liebm.; *H. longifolium* ssp.
nicaraguense (Liebm.) Koyama)
Kyllinga Rottb.
brevifolia Rottb.
(= *Cyperus brevifolius* (Rottb.) Endl. ex Hassk.)
pumila Michaux
(= *Cyperus tenuifolius* (Steudel) Dandy,
K. tenuifolia Steudel, *Cyperus densicaespitosus*
Mattf. & Kükenth.)
Mapania Aublet
assimilis T. Koyama
Rhynchospora Vahl
contracta (Nees) Raynal
(= *R. micrantha* Vahl)
corymbosa (L.) Britton
polyphylla Vahl
Scleria Berg
latifolia Sw.
macrophylla J. S. & C. Presl
microcarpa Nees
pterota K. Presl
secans (L.) Urban
setuloso-ciliata Boeck.

DIOSCOREACEAE
Dioscorea L.
bulbifera L.
polygonoides Humb. & Bonpl. ex Willd.
urophylla Hemsley

GRAMINEAE (= Poaceae)
Acroceras Stapf
zizanioides (HBK.) Dandy
(= *Panicum zizanioides* HBK., *Acroceras*
oryzoides (Sw.) Stapf)

Andropogon L.
bicornis L.
Anthephora Schreber
hermaphroditica (L.) Kuntze
Arundinella Raddi
berteroniana (Schultes) Hitchc. & Chase
Axonopus Beauv.
compressus (Sw.) Beauv.
Bambusa Schreber
vulgaris Schrader ex H. A. Wendl. [cultivated near
River Station]
Brachiaria Griseb.
fasciculata (Sw.) Parodi
(= *Panicum fasciculatum* (Sw.) Parodi)
mutica (Forsskal) Stapf
(= *Panicum purpurascens* Raddi)
ruziziensis Germain & Evrard
Chusquea Kunth
simpliciflora Munro
Coix L.
lachryma-jobi L.
Cryptochloa Swallen
concinna (Hook. f.) Swallen
Cynodon L. C. Rich
dactylon (L.) Pers.
Dendrocalamus Nees
sp. indet.
Digitaria Heister ex Fabr.
ciliaris (Retz.) Koeler
(= *D. adscendens* (HBK.) Henrard)
setigera Roth ex Roemer & Schultes
violascens Link
Echinochloa Beauv.
polystachya (HBK.) Hitchc.
Eleusine J. Gaertn.
indica (L.) J. Gaertn.
Elytrostachys McClure
clavigera McClure
Eragrostis Wolf
acutiflora (HBK.) Nees
Gynerium Willd. ex Beauv.
sagittatum (Aublet) Beauv.
Homolepis Chase
aturensis (HBK.) Chase
Hymenachne P. Beauv.
amplexicaulis (Rudge) Nees
Hyparrhenia Andersson ex Fourn.
rufa (Nees) Stapf
Ichnanthus Beauv.
nemorosus (Sw.) Doell
pallens (Sw.) Munro ex Benth.
(= *I. axillaris* (Nees) Hitchc. & Chase)
Ischaemum L.
timorense Kunth
Ixophorus Schldl.
unisetus (K. Presl) Schldl.
Lasiacis (Griseb.) Hitchc.
oaxacensis (Steud.) Hitchc. var. oaxacensis
scabrior Hitchc.
Lithachne Beauv.
pauciflora (Sw.) Beauv. ex Poiret

Melinis Beauv.
  minutiflora Beauv.
Olyra L.
  latifolia L.
Oplismenus Beauv.
  burmannii (Retz.) Beauv.
Orthoclada Beauv.
  laxa (L. C. Rich.) Beauv.
Oryza L.
  latifolia Desv.
Panicum L.
  boliviense Hackel
  [fasciculatum Sw.—see *Brachiaria fasciculata*
    (Sw.) Parodi]
  grande Hitchc. & Chase
  hirsutum Sw.
  laxum Sw.
  maximum Jacq.
  pilosum Sw.
    var. pilosum
    var. lancifolium (Griseb. ex Hitchc.) R. Pohl.
  polygonatum Schrader ex Schultes
  pulchellum Raddi
  [purpurascens Raddi—see *Brachiaria mutica*
    (Forssk.) Stapf]
  stagnatile Hitch. & Chase
  stoloniferum Poiret var. *major* (Trinius) Kunth
    (= *P. frondescens* G. F. W. Meyer)
  tricanthum Nees
  trichoides Sw.
Pariana Aublet
  parvispica R. Pohl
Paspalum L.
  conjugatum Bergius
  decumbens Sw.
  fasciculatum Willd.
  orbiculatum Poiret
  paniculatum L.
  plicatulum Michaux
  virgatum L.
Pennisetum L. C. Rich. ex Pers.
  purpureum Schumann
Pharus P. Br.
  latifolius L.
  virescens Doell
  vittatus Lemaire
    (= *P. cornutus* Hackel)
Pseudechinolaena Stapf
  polystachya (HBK.) Stapf
Saccharum L.
  officinarum L. [cultivated]
Sacciolepis Nash
  indica (L.) Chase
Setaria Beauv.
  geniculata (Lam.) Beauv.
  paniculifera (Steud.) E. Fourn.
Sporobolus R. Br.
  indicus (L.) R. Br.
    (= *S. poiretii* (Roemer & Schultes)
    Hitchc.)
  jacquemontii Kunth

Streptochaeta Schrader ex Nees
  sodiroana Hackel

HAEMODORACEAE
  Xiphidium Aublet
    coeruleum Aublet

HELICONIACEAE
  Heliconia L.
    curtispatha Petersen
    imbricata (Kuntze) Baker
    irrasa Lane ex R. R. Sm. ssp. undulata Daniels &
      Stiles var. undulata
    latispatha Benth.
    mariae Hook. f.
    mathiasiae Daniels & Stiles
      (= *H. vaginalis* Benth. ssp. *mathiasiae* (Daniels
      & Stiles) L. Andersson)
    pogonantha Cuf. var. pogonantha
    sarapiquensis Daniels & Stiles
    trichocarpa Daniels & Stiles var. trichocarpa
    umbrophila Daniels & Stiles
    vaginalis Benth.
      (= *H. deflexa* Daniels & Stiles)
    wagneriana O. Petersen

MARANTACEAE
  Calathea G. F. W. Meyer
    cleistantha Standl.
    crotalifera S. Watson
      (= *C. insignis* Petersen)
    donnell-smithii K. Schumann
    foliosa Rowlee ex Woodson
    gloriana H. Kennedy ined.
    gymnocarpa H. Kennedy
    hammelii H. Kennedy *ined.*
    inocephala (Kuntze) H. Kennedy & D. Nicholson
    lasiostachya J. E. Sm.
    leucostachys Hook. f.
    lutea (Aublet) Schultes
    marantifolia Standley
    micans (Mathieu) Koern.
    nitidifolia H. Kennedy
      (= *C. cuneata* sensu Hammel)
    pittieri K. Schumann
      (= *C. elegans* sensu Hammel)
    similis H. Kennedy
    venusta H. Kennedy
    warscewiczii (Mathieu) Koern.
  Ctenanthe
    villosa H. Kennedy ined.
  Ischnosiphon Koern.
    elegans Standley
    inflatus L. Andersson
  Pleiostachya K. Schumann
    leiostachya (J. D. Sm.) Hammel
    pruinosa (Regel) Schumann
  Stromanthe Sonder
    palustris H. Kennedy *ined.*

MUSACEAE (and see Heliconiaceae)
  Musa L.
    ?acuminata Colla [cultivated]
      (= *M. sapientum* L.)

ORCHIDACEAE
  Beloglottis Schlechter
    subpandurata (Ames & Schweinf.) Garay
  Brassia R. Br.
    caudata (L.) Lindley
  Bulbophyllum Thouars
    aristatum (Reichenb. f.) Hemsley
  Campylocentrum Benth.
    fasciola (Lindley) Cogn.
    micranthum (Lindley) Rolfe
  Catasetum L. C. Rich. ex Kunth
    "most likely either *C. maculatum* Kunth or *C. integerrimum* Hook." fide John Atwood but plant sterile.
  Cycnoches Lindley
    egertonianum Bateman
  Dichaea Lindley
    costaricensis Schlechter *vel aff.*
    panamensis Lindley
    standleyi Ames *vel aff.*
    trulla Reichenb. f.
    tuerckheimii Schlechter
  Dryadella Luer
    pusiola (Reichenb. f.) Luer
      (= *Masdevallia pusiola* Reichenb. f.)
    simula (Reichenb. f.) Luer
      (= *Masdevallia simula* Reichenb. f.)
  Elleanthus K. Presl
    cyanocephalus (Reichenb. f.) Reichenb. f.
    graminifolius (Barb.-Rodr.) Lojtnant
  Encyclia Hook.
    abbreviata (Schlechter) Dressler
    chacaoensis (Reichenb. f.) Dressler & G. Pollard
    fragrans (Sw.) Lemée
    pygmaea (Hook.) Dressler
  Epidendrum L.
    [boothii (Lindley) L. O. Williams—see *Nidema*]
    ciliare L.
    cristatum Ruiz & Pavón
    difforme Jacq.
      (= *Neolehmannia difformis* (Jacq.) Pabst)
    hawkesii A. H. Heller
    hunterianum Schlechter
    isomerum Schlechter
    nocturnum Jacq.
    octomerioides Schlechter
    physodes Reichenb. f.
    rigidum Jacq.
    schlecterianum Ames
      (= *Nanodes discolor* Lindley)
    sculptum Reichenb. f.
      (= *Epidendrum colonense* Ames)
    stangeanum Reichenb. f.
    strobiliferum Reichenb. f.
    trialatum Hagsater

Erythrodes Blume
  purpurea (Ames) Ames
  tuerckheimii (Schlechter) Ames
    (= *Aspidogyne tuerckheimii* (Schlechter) Garay)
Gongora Ruiz & Pavón
  unicolor Schlechter
Hexisea Lindley
  imbricata (Lindley) Reichenb. f.
Ionopsis HBK.
  satyrioides (Sw.) Reichenb. f.
  utricularioides (Sw.) Lindley
Jacquiniella Schlechter
  globosa (Jacq.) Schlechter
Kegeliella Mansfeld
  atropilosa L. O. Williams & A. H. Heller
Leochilus Knowles & Westcott
  labiatus (Sw.) Kuntze
Lepanthes Sw.
  blepharistes Reichenb. f.
Masdevallia Ruiz & Pavón
  tubuliflora Ames
Maxillaria Ruiz & Pavón
  aciantha Reichenb. f.
  acutifolia Lindley
  confusa Ames & Schweinf.
  endresii Reichenb. f.
  friedrichsthallii Reichenb. f.
  fulgens (Reichenb. f.) L. O. Williams
    (= *Ornithidium fulgens* Reichenb. f.)
  hedwigae Hamer & Dodson
  lueri Dodson
  neglecta (Schlechter) L. O. Williams
  nicaraguensis (Hamer & Garay) J. Atwood
    (= *Neourbania nicaraguensis* Hamer & Garay)
  tenuifolia Lindley
  uncata Lindley
  xylobiiflora Schlechter
Mendoncella A. D. Hawkes
  grandiflora (A. Rich.) A. D. Hawkes
    (= *Zygopetalum grandifolium* (A. Rich.) Benth. & Hook. ex Hemsley)
Myoxanthus Poeppig & Endl.
  scandens (Ames) Luer
  uncinatus (Fawc.) Luer
    (= *Pleurothallis uncinatus* Fawcett)
[Nanodes Lindley
  discolor Lindley—see *Epidendrum schlechterianum* Ames)]
[Neourbania Fawc. & Rendle—see *Maxillaria*]
Nidema Britton & Millsp.
  boothii (Lindley) Schlechter
    (= *Epidendrum boothii* (Lindley) L. O. Williams)
Octomeria R. Br.
  graminifolia (L.) R. Br.?
Oncidium Sw.
  [pusillum (L.) Reichenb. f.—see *Psygmorchis pusilla* (L.) Dodson & Dressler]
  [kramerianum Reichenb. f.—see *Psychopsis kramerianum* (Reichenb. f.) H. G. Jones]
  stenotis Reichenb. f.

Palmorchis Barb.-Rodr.
   powellii (Ames) Schweinf. & Correll
   silvicola L. O. Williams
Platystele Schlechter
   lancilabris (Reichenb. f.) Schlechter
Pleurothallis R. Br.
   acrisepala Ames & Schweinf.
   aristata Hook.
   biglandulosa Schlechter *vel aff.*
   brighamii S. Watson
   corniculata (Sw.) Lindley
   glandulosa Ames
   grobyi Batem. ex Lindley
   guanacastensis Ames & Schweinf.
   lewisii Ames
   microphylla A. Rich. & Galeotii
   pantasmi Reichenb. f.
   periodica Ames
   phyllocardioides Schlechter
   pruinosa Lindley
   [uncinata Fawc.—see *Myoxanthus uncinatus*
     (Fawc.) Luer]
   verecunda Schlechter
Polystachya Hook.
   foliosa (Lindley) Reichenb. f.
   masayensis Reichenb. f.
Psychopsis Raf.
   krameriana (Reichenb. f.) H. G. Jones
     (= *Oncidium kramerianum* Reichenb. f.)
Psygmorchis Dodson & Dressler
   pusilla (L.) Dodson & Dressler
     (= *Oncidium pusillum* (L.) Reichenb. f.)
Reichenbachanthus Barb.-Rodr.
   reflexus (Lindley) Brade *vel aff.*
Rodriquezia Ruiz & Pavón
   compacta Schlechter
Scaphosepalum Pfitzer
   microdactylum Rolfe
Scaphyglottis Poeppig & Endl.
   gracilis (Schlechter) Schlechter
   graminifolia (Ruiz & Pavón) Poeppig & Endl.
   huebneri Schlechter
   longicaulis S. Watson
   minutiflora Ames & Correll
   prolifera Cogn.
Sobralia Ruiz & Pavón
   decora Bateman
   fragrans Lindley
   mucronata Ames
   powellii Schlechter
Stanhopea Frost ex Hook.
   ?sp. indet.
Stelis Sw.
   cleistogama Schlechter
   endresii Reichenb. f.
   parvula Lindley
   purpurea (Ruiz & Pavón) Willd.
   superbiens Lindley
Trichosalpinx Luer
   blaisdellii (S. Watson) Luer

     (= *Pleurothallis blaisdellii* S. Watson)
   foliata (Griseb.) Luer
     (= *Pleurothallis foliata* Griseb.)
   orbicularis (Lindley) Luer
     (= *Pleurothallis orbicularis* (Lindley) Lindley)
Trigonidium Lindley
   egertonianum Bateman ex Lindley
   riopalenquense Dodson
Vanilla P. Mill.
   pauciflora Dressler
Warrea Lindley
   costaricensis Schlechter

PALMAE
Asterogyne H. A. Wendl. ex Hook. f.
   martiana H. A. Wendl. ex Burret
Astrocaryum G. F. W. Meyer
   alatum Loomis
   confertum H. A. Wendl. ex Burret
     (= *A. standleyanum* L. H. Bailey)
Bactris Jacq. ex Scop.
   gasipaes HBK. [cultivated]
     (= *Gulielma gasipaes* (HBK.) L. H. Bailey;
     G. utilis Oersted)
   hondurensis Standley
     (= *B. wendlandiana* Burret, *B. paula* L. H. Bailey)
   longiseta H. A. Wendl. ex Hemsley
     (= *B. fusca* Oersted)
   porschiana Burret *sp. nov.* fide H. E. Moore
Calyptrogyne H. A. Wendl.
   ghiesbreghtiana (Linden & H. A. Wendl.) H. A.
     Wendl.
     (= *C. sarapiquensis* H. A. Wendl. ex Burret)
Chamaedorea Willd.
   deckeriana (Klotzsch) Hemsley
   pinnatifrons (Jacq.) Oersted
   selvae Hodel
   tepejilote Liebm.
     (= *C. exorrhiza* H. A. Wendl. ex Guillaumin)
Cocos L.
   nucifera L. [cultivated dooryard trees]
Crysophila Blume
   warscewiczii (H. A. Wendl.) Bartlett
     (= *C. albida* Bartlett)
Desmoncus Martius
   costaricensis (Kuntze) Burret
Elaeis Jacq.
   guineensis Jacq. [cultivated & naturalized]
Euterpe Martius
   macrospadix Oersted
Geonoma Willd.
   congesta H. A. Wendl. ex Spruce
   cuneata H. A. Wendl. ex Spruce
   deversa (Poit.) Kunth
   ferruginea H. A. Wendl. ex Spruce
     (= *G. microspadix* H. A. Wendl. ex Spruce,
     G. microstachys H. A. Wendl. ex Burret,
     (= *G. versiformis* H. A. Wendl.
     ex Spruce)
   interrupta (Ruiz & Pavón) Martius

longevaginata H. A. Wendl. ex Spruce
oxycarpa Martius
Iriartea Ruiz & Pavón
deltoidea Ruiz & Pavón
(= *I. gigantea* H. A. Wendl. ex Burret)
Pholidostachys H. A. Wendl. ex Hook f.
pulchra H. A. Wendl. ex Hemsley
Prestoea Hook. f.
decurrens (H. A. Wendl.) H. E. Moore
Reinhardtia Liebm.
gracilis (H. A. Wendl.) Burret var. rostrata
(Burret) H. E. Moore
simplex (H. A. Wendl.) Burret
Socratea Karsten
durissima (Oersted) H. A. Wendl.
(= *S. exorrhiza* (Martius) H. A. Wendl.)
Synechanthus H. A. Wendl.
fibrosus (H. A. Wendl.) H. A. Wendl.
warscewiczianus H. A. Wendl.
Welfia H. A. Wendl.
georgii H. A. Wendl. ex Burret

POACEAE—see Gramineae

PONTEDERIACEAE
Eichhornia Kunth
diversifolia (Vahl) Urban
Heteranthera Ruiz & Pavón
reniformis Ruiz & Pavón

SMILACACEAE
Smilax L.
domingensis Willd.
mollis Humb. & Bonpl. ex Willd.
vanilliodora Apt

ZINGIBERACEAE
Alpinia Roxb.
purpurata (Veill.) Schumann [cultivated]
[Costus—see Costaceae]
Hedychium Koenig
coronarium Koenig
Renealmia L.
alpina (Rottb.)Maas
cernua (Sw.) Macbr.
concinna Standley
costaricensis Standley
pluriplicata Maas

B. DICOTS

ACANTHACEAE
Aphelandra P. Br.
aurantiaca (Schiedw.) Lindley
dolichantha J. D. Sm.
golfodulcensis McDade x sinclairiana Nees
storkii Leonard
Barleria L.
micans Nees

Blechum P. Br.
B. brownei Juss.
(= *B. pyramidatum* (Lam.) Urban)
Bravaisia A. DC.
integerrima (Sprengel) Standley
Dicliptera Juss.
podocephala J. D. Sm.
Herpetacanthus Nees
panamensis Leonard
(= *Standleyacanthus costaricensis* Leonard)
Hygrophila R. Br.
costata Nees
(= *H. guianensis* Nees)
Justicia L.
aurea Schldl.
comata (L.) Lam.
pectoralis Jacq.
refractifolia (Kuntze) Leonard
sarapiquense McDade
trichotoma (Kuntze) Leonard
Louteridium S. Watson
costaricensis Radlk. & J. D. Sm.
[Mendoncia Vellozo ex Vandelli—see Mendonciaceae]
Odontonema Endl.
cuspidatum (Nees) Kuntze [cultivated]
tubiforme (Schldl. & Cham.) Kuntze
(= *O. callistachyum* (Schldl. & Cham.) Kuntze)
Razisea Oersted
spicata Oersted
villosa Gómez-L. & Hammel *ined.*
wilburii McDade
Ruellia L.
biolleyi Lindau
metallica Leonard
palustris Durkee
tubiflora HBK. var. tetrasticantha (Lindau) Leonard
Thunbergia Retz.
alata Bojer ex Sims
erecta (Benth.) T. Anderson

ACTINIDIACEAE
Saurauia Willd.
yasicae Loesener

[AIZOACEAE—see Molluginaceae]

AMARANTHACEAE
Achyranthes L.
aspera L.
Alternanthera Forssk.
[costaricensis Kuntze—see *Jamesbondia*]
laguroides (Standley) Standley
sessilis (L.) R. Br.
tenella Colla
(= *A. ficoidea* sensu authors)
Amaranthus L.
spinosus L.
Chamissoa HBK.
altissima (Jacq.) HBK.

Cyathula Lour.
  achyranthoides (HBK.) Moq.
  prostrata (L.) Blume
Gomphrena L.
  serrata L.
    (= *G. decumbens* Jacq.)
Iresine P. Br.
  diffusa Humb. & Bonpl. ex Willd.
    (= *I. celosia* L.)
Jamesbondia Mears *ined.*
  costaricensis (Kuntze) Mears *ined.*
    (= *Altenanthera costaricensis* Kuntze)

ANACARDIACEAE
Anacardium L.
  occidentale L. [cultivated]
Mangifera L.
  indica L. [cultivated]
Spondias L.
  mombin L.
  purpurea L.
  radlkoferi J. D. Sm.
Tapirira Aublet
  guianensis Aublet

ANNONACEAE
Anaxagorea St.-Hil.
  crassipetala Hemsley
  phaeocarpa Martius
    (= *A. costaricensis* R. E. Fries)
Annona L.
  amazonica R. E. Fries *vel aff.*
  montana Macf.
  muricata L. [cultivated]
Cananga (DC.) Hook. f. & Thomson
  odorata (Lam.) Hook. f. & Thomson [cultivated]
Cymbopetalum Benth.
  costaricensis (J. D. Sm.) R. E. Fries
  torulosum G. E. Schatz
Desmopsis Safford
  microcarpa R. E. Fries
    (= *D. glabrata* Schery)
  schippii Standley
    (= *D. dolichopetala* R. E. Fries, *D. brevipes* R. E. Fries)
Guatteria Ruiz & Pavón
  aeruginosa Standley
  diospyriodes Baillon
    (= *G. diospyroides* subsp. *honduriensis* R. E. Fries, *G. inuncta* R. E. Fries, *G. platypetala* R. E. Fries)
  recurvisepala R. E. Fries
Rollinia St.-Hil.
  microsepala Standley
Sapranthus Seemann
  viridiflorus G. E. Schatz *ined.*
Unonopsis R. E. Fries
  hammelii G. E. Schatz *ined.*
  pittieri Safford
    (= *U. schippii* R. E. Fries)

Xylopia L.
  bocatorena Schery
  sericophylla Standley & L. O. Williams

[APIACEAE—see Umbelliferae]

APOCYNACEAE
Allamanda L.
  cathartica L. [cultivated]
Allomarkgrafia Woodson
  plumeriaeflora Woodson
Aspidosperma Martius & Zucc.
  megalocarpon Muell.Arg.
    (= *A. cruentum* Woodson)
Forsteronia G. W. F. Meyer
  viridescens S. F. Blake
Lacmellea Karsten
  panamensis (Woodson) Markgraf
Mandevilla Lindley
  hirsuta (L. C. Rich.) Schumann
  villosa (Miers) Woodson
Mesechites Muell.Arg.
  trifida (Jacq.) Muell.Arg.
Odontadenia Benth.
  grandiflora (G. W. F. Meyer) Miq.
  sp. indet.
Prestonia R. Br.
  exserta (A. DC.) Standlcy
  portobellensis (Beurl.) Woodson
  sp. nov.?
Rauvolfia L.
  purpurascens Standley
Stemmadenia Benth.
  donnell-smithii (Rose) Woodson
  robinsonii Woodson
Tabernaemontana L.
  amygdalifolia Jacq.
  arborea Rose
    (= *Peschiera arborea* (Rose) Markgraf)

AQUIFOLIACEAE
Ilex L.
  skutchii G. Edwin ex T. Dudley *ined.*

ARALIACEAE
Dendropanax Decne. & Planchon
  arboreus (L.) Decne. & Planchon
  gonatopodus (J. D. Sm.) A. C. Sm.
  stenodontus (Standley) A. C. Sm.
Oreopanax Decne. & Planchon
  capitatus (Jacq.) Decne. & Planchon
Schefflera J. R. & J. G. Forster
  sphaerocoma (Benth.) Harms
  systyla (J. D. Sm.) Viguier

ARISTOLOCHIACEAE
Aristolochia L.
  constricta Griseb.
  cruenta Barringer

grandiflora Sw.
pilosa HBK.
translucida Pfeifer

ASCLEPIADACEAE
Asclepias L.
  curassavica L.
Cynanchum L.
  spp. indet.
Fischeria DC.
  panamensis Spellman
Gonolobus Michaux
  albomarginatus (Pittier) Woodson
  denticulatus (Vahl) W. D. Stevens *ined.*
Matelea Aublet
  spp. indet.
Tassadia Decne.
  obovata Decne.
    (= *Cynanchium apocynellum* (Gleason &
    Moldenke) Spellman)

[ASTERACEAE—see Compositae]

BALSAMINACEAE
Impatiens L.
  balsamina L.
  wallerana Hook. f.

BEGONIACEAE
Begonia L.
  fischeri Schrank
  glabra Aublet
  multinervia Liebm.
  semiovata Liebm.
  sericoneura Liebm.
  sp. indet.

BIGNONIACEAE
Amphilophium Kunth
  paniculatum (L.) HBK.
Amphitecma Miers
  kennedyi (A. Gentry) A. Gentry
    (= *Dendrosicus kennedyi* A. Gentry)
Anemopaegma Martius ex Meissner
  chrysoleucum (HBK.) Sandw.
  orbiculatum (Jacq.) DC.
Arrabidaea DC.
  chica (Humb. & Bonpl.) Verlot
  patellifera (Schldl.) Sandw.
  verrucosa (Standley) A. Gentry
    (= *Scobinaria verrucosa* (Standley) Seibert)
Callichlamys Miq.
  latifolia (L. C. Rich.) Schumann
Clytostoma Miers ex Bureau
  binatum (Thunb.) Sandw.
Distictella Kuntze
  magnoliifolia (HBK.) Sandw.
Jacaranda Juss.
  copaia (Aublet) D. Don

Macfadyena A. DC.
  ?uncata (Andr.) Sprague & Sandw.
Mansoa DC.
  standleyi (Steyerm.) A. Gentry
    (= *Pachyptera standleyi* (Steyerm.) A. Gentry)
Martinella Baillon
  obovata (HBK.) Bureau & Schumann
Mussatia Bureau ex Baillon
  hyacinthina (Standley) Sandw.
[Pachyptera DC. ex Meissner—see Mansoa]
Paragonia Bureau
  pyramidata (L. C. Rich.) Bureau
Pithecoctenium Martius ex Meissner
  crucigerum (L.) A. Gentry
    (= *P. echinatum* (Jacq.) Baillon)
Schlegelia Miq.
  fastigiata Schery
  nicaraguensis Standley
  parviflora (Oersted) Monachino
[Schobinaria Seibert—see *Arrabidaea*]
Stizophyllum Miers
  inequilaterum Bureau & Schumann
Tabebuia B. A. Gomes ex DC.
  chrysantha (Jacq.) G. Nicholson
  ?guayacan (Seemann) Hemsley
  ?rosea (Bertol.) DC.

BOMBACACEAE
Ceiba P. Mill.
  pentandra (L.) Gaertner
[Hampea—see under Malvaceae]
Matisia Humb. & Bonpl.
  bracteolosa Ducke
    (= *Quararibea bracteolosa* (Ducke) Schumann)
  ochrocalyx K. Salp.
    (= *Quararibea ochrocalyx* (Schumann) Vischer
Ochroma Sw.
  pyramidale (Cav. ex Lam.) Urban
    (= *O. lagopus* (Cav. ex Lam.) Urban
Pachira Aublet
  aquatica Aublet
Quararibea Aublet
  asterolepis Pittier
  parvifolia Standley
  pumila Alverson

BORAGINACEAE
Cordia L.
  alliodora (Ruiz & Pavón) Oken
  bicolor A. DC.
  bifurcata Roemer & Schultes
  collococca L.
  cymosa (J. D. Sm.) Standley
  dwyeri Nowicke
  lucidula I. M. Johnston
  megalantha S. F. Blake
  panamensis Riley
  porcata Nowicke
  protracta I. M. Johnston

Heliotropium L.
  indicum L.
Tournefortia L.
  angustiflora Ruiz & Pavón
  glabra L.

BURSERACEAE
  Bursera Jacq. ex L.
    simaruba (L.) Sarg.
  Protium Burm. f.
    costaricense (Rose) Engler
      (= *P. correae* D. Porter)
    panamense (Rose) I. M. Johnston
    pittieri (Rose) Engler
      (= *Tetragastris tomentosa* D. Porter)
  Tetragastris Gaertner
    panamensis (Engler) Kuntze

CACTACEAE
  Disocactus Lindley
    ramulosus (Salm-Dyck) Kimnach
      (= *Rhipsalis ramulosa* (Salm-Dyck) Pfeiffer)
  Epiphyllum Haw.
    macropterum (Lemaire) Britton & Rose
    pittieri (Weber) Britton & Rose
  Rhipsalis J. Gaertn.
    baccifera (J. S. Mill.) Stearn
      (= *R. cassutha* J. Gaertner)
  Weberocereus Britton & Rose
    tunilla (Weber) Britton & Rose
      (= *Cereus tunilla* Weber)

CAMPANULACEAE
  Centropogon K. Presl
    coccineus (Hook.) Regel ex B. D. Jackson
  Hippobroma G. Don
    longiflora (L.) G. Don

[CAPPARIDACEAE—see Capparaceae]

CAPPARACEAE
  Capparis L.
    discolor J. D. Sm.
    frondosa Jacq.
    pittieri Standley
  Cleome L.
    longipes Lamb. ex DC.
    pilosa Benth.

CAPRIFOLIACEAE
  Sambucus L.
    canadensis L. var. laciniata A. Gray [cultivated]

CARICACEAE
  Carica L.
    mexicana (A. DC.) L. O. Williams
    papaya L. [cultivated]
  Jacaratia DC.
    dolichaula (J. D. Sm.) Woodson

spinosa (Aublet) DC.
  (= *J. costaricensis* I. M. Johnston)

CARYOPHYLLACEAE
  Drymaria Willd. ex Schultes
    cordata (L.) Willd. ex Roemer & Schultes

CECROPIACEAE
  Cecropia Loefl.
    insignis Liebm.
    obtusifolia Bertol.
  Coussapoa Aublet
    nymphaeifolia Standley
    villosa Poeppig & Endl.
      (= *C. panamensis* Pittier)
  Pourouma Aublet
    bicolor Martius subsp. scobina (Benoist) C. C. Berg &
      van Heusden
      (= *P. aspera* Trecul)
    minor Benoist
      (= *P. umbellifera* W. Burger)

CELASTRACEAE
  Crossopetalum P. Br.
    uragoga (Jacq.) Kuntze

CHLORANTHACEAE
  Hedyosmum Sw.
    scaberrimum Standley

CHRYSOBALANACEAE
  Couepia Aublet
    polyandra (Kunth) Rose
  Hirtella L.
    lemsii L. O. Williams & Prance
  Licania Aublet
    affinis Fritsch
    glabriflora Prance
    hypoleuca Benth.
    platypus (Hemsley) Fritsch [cultivated]
  Maranthes Blume
    panamensis (Standley) Prance & F. White
      (= *Couepia panamensis* Standley)

CLETHRACEAE
  Clethra L.
    lanata Martens & Galeotti

[CLUSIACEAE—see Guttiferae]

COMBRETACEAE
  Combretum Loefl.
    laxum Jacq.
  Terminalia L.
    amazonia (J. F. Gmelin) Exell
      (= *T. obovata* (Ruiz & Pavón) Steudel)
    bucidoides Standley & L. O. Williams
    oblonga (Ruiz & Pavón) Steudel
      (= *T. chiriquensis* Pittier)

COMPOSITAE
Acmella L. C. Rich. ex Pers.
  radicans (Jacq.) R. Jansen
Ageratum L.
  houstonianum P. Mill.
Bidens L.
  pilosa L.
Calea L.
  urticifolia (P. Mill.) DC.
Chaptalia Vent.
  nutans (L.) Polak.
Chromolaena DC.
  odorata (L.) R. King & H. Robinson
    (= *Eupatorium odoratum* L.)
Clibadium L.
  asperum (Aublet) DC.
  pittieri Greenm.
  surinamense L.
Conyza Less.
  apurensis HBK.
  bonariensis (L.) Cronq.
    (= *Erigeron bonariensis* L.)
Critonia P. Br.
  morifolia (P. Mill.) King & H. Robinson
    (= *Eupatorium morifolium* P. Mill.)
Eclipta L.
  prostrata (L.) L.
    (= *E. alba* (L.) Hassk.)
Elephantopus L.
  mollis HBK.
Eleutheranthera Poit. ex Bosc
  ruderalis (Sw.) Schultz-Bip.
Emilia Cass.
  fosbergii D. Nicolson
  sonchifolia (L.) DC.
Erechtites Raf.
  hieracifolia (L.) Raf. ex DC. cacalioides (Fisher ex
    Sprengel) Griseb.
Fleischmannia Schultz-Bip.
  microstemon (Cass.) R. King & H. Robinson
    (= *Eupatorium microstemon* Cass.)
  sideritidis (Benth.) R. King & H. Robinson
Galinsoga Ruiz & Pavón
  rivularis Poeppig & Endl.
Hebeclinium DC.
  macrophyllum (L.) DC.
    (= *Eupatorium macrophyllum* L.)
Heterocondylus R. King & H. Robinson
  vitalbae (DC.) R. King & H. Robinson
    (= *Eupatorium vitalbae* DC.)
Jaegeria HBK.
  hirta (Lagasca) Less.
Koanophyllon Arruda da Camara
  hylonoma (B. L. Robinson) R. King & H. Robinson
Lasianthaea DC.
  fruticosa (L.) K. M. Becker
    (= *Zexmenia frutescens* (P. Mill.) S. F. Blake)
Melampodium L.
  costaricense Stuessy
  divaricatum (L. C. Rich.) DC.

Melanthera Rohr
  aspera (Jacq.) Small
Mikania Willd.
  gonzalezii B. L. Robinson & Greenman
  guaco Humb. & Bonpl.
  hookeriana DC.
  leiostachya Benth.
  micrantha HBK.
  pittieri B. L. Robinson
  tonduzii B. L. Robinson
  vitifolia DC.
Neurolaena R. Br.
  lobata (L.) R. Br.
Piptocarpha Hook. & Arn.
  poeppigiana (DC.) Baker
    (= *P. chontalense* Banker, *P. costaricensis* Klatt)
Pseudelephantopus Rohr
  spicatus (Juss.) Vahl
  spiralis (Less.) Cronq.
Schistocarpha Less.
  eupatorioides (Fenzl) Kuntze
    (= *S. oppositifolia* (Kuntze) Rydb.)
Sinclairia Hook. & Arn.
  polyantha (Klatt) Rydb.
    (= *Liabum polyanthum* Klatt)
Spiracantha HBK.
  cornifolia HBK.
Struchium P. Br.
  sparganophorus (L.) Kuntze
Synadrella Gaertner
  nodiflora (L.) Gaertner
Vernonia Schreber
  brachiata Benth.
  cinerea (L.) Less.
  patens HBK.
Zexmenia LaLlave & Lexarza
  virgulta Klatt

CONNARACEAE
Rourea Aublet
  glabra HBK.
  suerrensis J. D. Sm.

CONVOLVULACEAE
Dicranostyles Benth.
  ampla Ducke
Ipomoea L.
  alba L.
    (= *I. bona-nox* L., *Calonyction bona-nox* (L.)
    Bojer, *Covolvulus aculeatus* L.)
  batatas (L.) Poiret
  batatoides Choisy
  phillomega (Vell.) House
  squamosa Choisy
  tiliacea (Willd.) Choisy
  trifida (HBK.) G. Don
Maripa Aublet
  nicaraguensis Hemsley
Merremia Dennst.

tuberosa (L.) Rendle
umbellata (L.) H. Hallier

CRUCIFERAE
Cardamine L.
flaccida Cham. & Schldl.

CUCURBITACEAE
Cayaponia Manso
glandulosa (Poeppig & Endl.) Cogn.
Cionosicys Benth.
macranthus (Pittier) C. Jeffreys *vel aff.*
Cucumis L.
melo L.
Cyclanthera Schrader
multifoliola Cogn.
Elateriopsis A. Ernst
oerstedii (Cogn.) Pittier
Fevillea L.
cordifolia L.
Frantzia Pittier
villosa Wunderlin
(= *Sechium villosum* (Wunderlin) C. Jeffrey)
Gurania (Schldl.) Cogn.
costaricensis Cogn.
makoyana (Lemaire) Cogn.
Melothria L.
dulcis Wunderlin
pendula L.
scabra Naudin
Momordica L.
charantia L.
Polyclathra Bertol.
cucumerina Bertol.
Psiguria Necker ex Arn.
bignoniacea (Poeppig & Endl.) Wunderlin
warscewiczii (Hook. f.) Wunderlin
(= *Anguria warscewiczii* Hook. f.)
Selysia Cogn.
prunifera (Poeppig & Endl.) Cogn.
Sicydium Schldl.
thamnifolium (HBK.) Cogn.

DICHAPETALACEAE

Dichapetalum DuPetit-Thouars
axillare Woodson
rugosum (Vahl) Prance

DILLENIACEAE
Davilla Vandelli
nitida (Vahl) Kubitzki
(= *D. multiflora* (DC.) St.-Hil.)
Doliocarpus Rolander
dentatus (Aublet) Standley
Pinzona Martius & Zucc.
coriacea Martius & Zucc.
(= *Doliocarpus coriaceus* (Martius & Zucc.) Gilg)
[*Saurauia*—see Actinidiaceae]

Tetracera L.
portobellensis Buerl.
(= *T. sessiliflora* Triana & Planchon)

ELAEOCARPACEAE
Sloanea L.
geniculata D. Sm. *ined.*
latifolia (L. C. Rich.) Schumann
medusula Schumann & Pittier
meianthera J. D. Sm.
rugosa D. Sm. *ined.*
tuerckheimii J. D. Sm.

ERICACEAE
Satyria Klotzsch
elongata A. C. Sm.
Sphyrospermum Poeppig & Endl.
buxifolium Poeppig & Endl.

ERYTHROXYLACEAE
Erythroxylum P. Br.
fimbriatum Peyr.
macrophyllum Cav.
(= *E. lucidulum* HBK.)

EUPHORBIACEAE
Acalypha L.
apodanthes Standley & L. O. Williams
arvensis Poeppig
costaricensis (Kuntze) Knobl. ex Pax & K. Hoffm.
diversifolia Jacq.
macrostachya Jacq.
radinostachya J. D. Sm.
?villosa Jacq.
Adelia L.
triloba (Muell.Arg.) Hemsley
Alchornea Sw.
costaricensis Pax & K. Hoffm.
latifolia Sw.
Alchorneopsis Muell.Arg.
floribunda (Benth.) Muell.Arg.
Caperonia St.-Hil.
palustris (L.) St.-Hil.
Chamaesyce S. F. Gray
hirta (L.) Millsp.
(= *Euphorbia hirta* L.)
hypericifolia (L.) Millsp.
(= *Euphorbia hypericifolia* L.)
hyssopifolia (L.) Small
(= *Euphorbia hyssopifolia* L.)
thymifolia (L.) Millsp.
(= *Euphorbia thymifolia* L.)
Conceveiba Aublet
pleiostemona J. D. Sm.
(= *Veconcibea pleiostemona* (J. D. Sm.) Pax & O. Hoffm.)
Croton L.
billbergianus Muell.Arg.
brevipes Pax
panamensis (Klotzsch) Muell.Arg.

schiedeanus Schldl.
smithianus Croizat
trinitatis Millsp.
Dalechampia L.
    dioscoreifolia Poeppig
    websteri Armbruster
Drypetes Vahl
    standleyi Webster
Hevea Aublet
    brasiliensis (Willd. ex A. Juss.) Muell.Arg. [cultivated]
Hura L.
    crepitans L.
Hyeronima Allemao
    alchorneoides Allemao
        (= *H. laxiflora* (Tul.) Muell.Arg.)
    oblonga (Tul.) Muell.Arg.
Jatropha L.
    gossypiifolia L. [cultivated]
    integerrima Jacq. [cultivated]
Mabea Aublet
    occidentalis Benth.
Manihot P. Mill.
    brachyloba Muell.Arg.
    esculenta Crantz
Omphalea L.
    diandra L.
Pera Mutis
    arborea Mutis
Phyllanthus L.
    amarus Schumann & Thonn.
    caribaeus Urban
    carolinensis Walter
    urinaria L.
Plukenetia L.
    volubilis L.
Poinsettia Graham
    heterophylla (L.) Klotzsch & Garcke
        (= *Euphorbia heterophylla* L.)
Richeria Vahl
    dressleri G. L. Webster *vel aff.*
Ricinus L.
    communis L.
Sapium P. Br.
    aucuparium Jacq.
        (= *S. jamaicense* Sw.)
    oligoneurum Schumann & Pittier
Tetrorchidium Poeppig
    euryphyllum Standley

[FABACEAE—see Leguminosae]

FLACOURTIACEAE
Carpotroche Endl.
    platyptera Pittier
        (incl. *C. crassiramea* Pittier & *C. glaucescens*
        Pittier)
Casearia Jacq.
    arborea (L. C. Rich.) Urban
    commersoniana Cambess.
    coronata Standley & L. O. Williams

    corymbosa HBK.
    sylvestris Sw. var. sylvestris
Hasseltia HBK.
    floribunda HBK.
[Lacistema Sw.—see Lacistemaceae]
Laetia Loefl. ex L.
    procera (Poeppig) Eichl.
[Lozania Mutis—see Lacistemaceae]
Pleuranthodendron L. O. Williams
    lindenii (Turcz.) Sleumer
        (= *P. mexicana* (A. Gray) L. O. Williams)
Ryania Vahl
    speciosa Vahl var. panamensis Monachino
Xylosma J. G. Forster
    chlorantha J. D. Sm.

GENTIANACEAE
Lisianthius P. Br.
    skinneri (Hemsley) Kuntze
Voyria Aublet
    ?tenella Hook.

GESNERIACEAE
Besleria L.
    columneoides Hanst.
    macropoda J. D. Sm.
    pauciflora Rusby
    robusta J. D. Sm.
    "viridiflora" *sp. nov. ined.*
Chrysothemis Decne.
    friedrichsthaliana (Hanst.) H. E. Moore
        (= *Tussacia friedrichsthaliana* Hanst.)
Codonanthe (Martius) Hanst.
    crassifolia (Focke) C. Morton
    macradenia J. D. Sm.
    uleana Fritsch
Columnea L.
    consanguinea Hanst.
        (= *Dalbergaria consanguinea* (Hanst.) Wiehler)
    linearis Oersted
    nicaraguensis Oersted
    "pectinatosepala" *sp. nov. ined.*
    purpurata Hanst.
    sanguinolenta (Klotzsch) Hanst.
        (= *Trichantha sanguinolenta* (Klotzsch ex Oersted)
        Wiehler)
Diastema Benth.
    hispidum (DC.) Fritsch
Drymonia Martius
    alloplectoides Hanst.
    "calycina" *sp. nov. ined.*
    coriacea (Oersted ex Hanst.) Wiehler
        (= *Alloplectus coriaceus* (Oersted ex Hanst.)
        Hanst.)
    macrophylla (Oersted) H. E. Moore
    multiflora (Oersted ex Hanst.) Wiehler
    pilifera Wiehler
    "sarapiquiensis" *sp. nov. ined.*
    serrulata (Jacq.) Martius
    stenophylla (J. D. Sm.) H. E. Moore

"submarginalis" *sp. nov. ined.*
warscewicziana Hanst.
Episcia Martius
lilacina Hanst.
Napeanthus Gardner
apodemus J. D. Sm.
bracteatus C. Morton
(= *N. bicolor* (L. O. Williams) Barringer)
Paradrymonia Hanst.
decurrens (C. Morton) Wiehler
(= *Centrosolenia decurrens* C. Morton, *Episcia decurrens* (C. Morton) Leeuw.)

GUTTIFERAE
Calophyllum L.
brasiliense Cambess. var. rekoi (Standley) Standley
(= *C. rekoi* Standley)
Clusia L.
amazonica Planchon & Triana
(= *C. oedematopoidea* Maguire)
cylindrica Hammel
flava Jacq.
gracilis Standley
minor L.
penduliflora Engler
quadrangula Bartlett
stenophylla Standley
uvitana Pittier
(= *C. erectistigma* Maguire)
valerii Standley
Clusiella Planchon & Triana
elegans Planchon & Triana
Dystovomita (Engler) D'Arcy
paniculata (J. D. Sm.) Hammel
(= *Tovomita pittieri* Engler, *D. pittier* (Engler) D'Arcy)
Garcinia L.
intermedia (Pittier) Hammel
(= *Rheedia intermedia* Pittier)
Marila Sw.
laxiflora Rubsy
(= *M. verapazensis* J. D. Sm.)
[Rheedia L.—see *Garcinia*]
Symphonia L.f.
globulifera L.f.
Tovomita Aublet
weddelliana Planchon & Triana
Tovomitopsis Planchon & Triana
glauca Oersted ex Planchon & Triana
nicaraguensis Oersted ex Planchon & Triana
silvicola Hammel
Vismia Vandelli
bilbergiana Beurl.
macrophylla HBK.
panamensis Duchass. & Walp.

HERNANDIACEAE
Hernandia L.
didymantha J. D. Sm.
stenura Standley

HIPPOCRATEACEAE
Salacia L.
?megistophylla Standley
Tontelea Aublet
hondurensis A. C. Sm.

HUMIRIACEAE
Sacoglottis Martius
trichogyna Cuatrecasas
Vantanea Aublet
barbourii Standley

ICACINACEAE
Calatola Standley
costaricensis Standley

LABIATAE
Hyptis Jacq.
brevipes Poit.
capitata Jacq.
obtusiflora Presl ex Benth.
suaveolens (L.) Poit.
verticillata Jacq.
vilis Kunth & Bouché *vel aff.*
Marsypianthes Martius ex Benth.
chamaedrys (Vahl) Kuntze
Scutellaria L.
purpurascens Sw.
Solenostemon Thonn.
scutellarioides (L.) Codd.
(= *Coleus blumei* Benth.)

LACISTEMACEAE
Lacistema Sw.
aggregatum (Berg.) Rusby
Lozania Mutis
pittieri (S. F. Blake) L. B. Sm.
(= *L. pedicellata* (Standley) L. B. Sm.)

[LAMIACEAE—see Labiatae]

LAURACEAE
Beilschmiedia Nees
anay (S. F. Blake) Kosterm.
sulcata (Ruiz & Pavón) Kosterm.
(incl. *B. mexicana* sensu Hammel)
Licaria Aublet
sarapiquensis Hammel
triandra (Sw.) Kosterm.
sp. nov. "A" Burger & van der Werff
Nectandra Rolander ex Rottb.
belizensis (Lundell) C. K. Allen
cissiflora Nees
(= *N. paulii* C. K. Allen)
hypoleuca Hammel
kunthiana (Nees) Kosterm.
(= *Ocotea cooperi* C. K. Allen, *Rhodostemonodaphne kunthiana* (Nees) J. Rohwer)
latifolia (HBK.) Mez
(= *N. purpurea* sensu Hammel)

membranacea (Sw.) Griseb.
    (= *N. standleyi* C. K. Allen)
reticulata (Ruiz & Pavón) Mez
    (= *N. mollis* (HBK.) Nees)
Ocotea Aublet
  atirrensis Mez & J. D. Sm. ex Mez
  babosa C. K. Allen
  bijuga (Rottb.) Bernardi *vel aff.*
  caracasana (Nees) Mez *vel aff.*
  cernua (Nees) Mez
    (= *O. caudata* (Nees) Mez and *O. killipii*
    A. C. Sm.)
  dendrodaphne Mez
  endresiana Mez
  floribunda (Sw.) Mez
    (= *O. wachenheimii* Benoist)
  hartshorniana Hammel
  insularis (Meissn.) Mez
    (= *O. ira* Mez & Pittier ex Mez)
  leucoxylon (Sw.) Lanessan
    (= *O. laseriana* C. K. Allen, *O. lenticellata*
    Lundell)
  meziana C. K. Allen
  mollifolia Mez & Pittier ex Mez
  nicaraguensis Mez
  tenera Mez & J. D. Sm. ez Mez
Persea P. Mill.
  americana P. Mill. [cultivated and escaping]
  rigens C. K. Allen
  silvatica van der Werff
Phoebe Nees
  chavarriana Hammel
    (= *Cinnamomum chavarrianum* (Hammel)
    Kosterm.)

LECYTHIDACEAE
Eschweilera Martius ex DC.
  calyculata Pittier
  costaricensis S. Mori
Grias L.
  cauliflora L.
    (= *G. fendleri Seemann*)
Lecythis Loefl.
  ampla Miers

LEGUMINOSAE—CAESALPINIOIDEAE
Bauhinia L.
  guianensis Aublet
Cassia L.
  [alata L.—see *Senna alata* (L.) Roxb.]
  [fruticosa P. Mill.—see *Senna fruticosa* (P. Mill.)
    Irwin & Barneby]
  grandis L. f. [cultivated]
  [leiophylla sensu auth.—see *Senna cobanensis*
    (Britton & Rose) Irwin & Barneby]
  [obtusifolia L.—see *Senna obtusifolia* (L.) Irwin &
    Barneby]
Macrolobium Schreber
  costaricense W. Burger

Sclerolobium Vogel
  costaricense Zomora & Poveda
Senna P. Mill.
  alata (L.) Roxb.
    (= *Cassia alata* L.)
  cobanensis (Britton & Rose) Irwin & Barneby
    (= *Cassia leiophylla* sensu authors)
  fruticosa (P. Mill.) Irwin & Barneby
  leiophylla (Vogel) Irwin & Barneby
  obtusifolia (L.) Irwin & Barneby

LEGUMINOSAE—MIMOSOIDEAE
Acacia P. Mill.
  hayesii Benth.
  ruddiae D. Janzen
  tenuifolia (L.) Willd.
Albizia Durazz.
  adinocephala (J. D. Sm.) Britton & Rose ex Record
  carbonaria Britton
Inga P. Mill.
  callicarpa Zamora
Section I. *Bourgonia* Benth.
  coruscans Willd.
  fagifolia (L.) Willd. ex Benth.
  marginata Willd.
  pezizifera Benth.
Section II. *Inga*
  acuminata Benth.
  cocleensis Pittier
  densiflora Benth.
  oerstediana Benth. ex Seemann
  punctata Willd.
  ruiziana G. Don
  sapindoides Willd.
  squamigera J. León
  thibaudiana DC.
  tonduzii J. D. Sm. [cultivated]
Section III. *Leptinga* Benth.
  heterophylla Willd.
  mortoniana J. León
  paterno Harms
  quaternata Poeppig
  umbellifera (Vahl) Steudel
Mimosa L.
  albida Humb. & Bonpl. ex Willd.
  myriadenia (Benth.) Benth.
  pigra L.
  pudica L.
Pentaclethra Benth.
  macroloba (Willd.) Kuntze
Pithecellobium Martius
  catenatum J. D. Sm.
    (= *Cojoba catenata* (J. D. Sm.) Britton & Rose)
  elegans Ducke
  englesingii (Standley) Standley
    (= *Inga englesingii* Standley)
  gigantifoliolum (Schery) J. León
    (= *Inga gigantifoliola* Schery)
  longifolium (Humb. & Bonpl.) Standley

(= *Zygia longifolia* (Humb. & Bonpl. ex Willd.)
Britton & Rose)
macradenium Pittier
(= *Samanea macradenia* (Pittier) Britton & Rose)
pedicellare (DC.) Benth.
valerioi (Britton & Rose) Standley
(= *Cojoba valerioi* Britton & Rose)
Stryphnodendron Martius
microstachyum Poeppig & Endl.
(= *S. excelsum* Harms)

LEGUMINOSAE—PAPILIONOIDEAE
Aeschynomene L.
americana L.
scabra G. Don
sensitiva Sw.
Andira Juss.
inermis (W. Wright) HBK.
(= *Geofforaea inermis* (W. Wright) W. Wright)
Calopogonium Desv.
coeruleum (Benth.) Sauvalle
Canavalia DC.
oxyphylla Standley & L. O. Williams
Crotalaria L.
micans Link
pallida Aiton
(= *C. mucronata* Desv.)
sagittalis L.
Dalbergia L. f.
cubilquitzensis (J. D. Sm.) Pittier
ecastophylla (L.) Taubert
tucurrensis J. D. Sm.
Desmodium Desv.
adscendens (Sw.) DC.
axillare (Sw.) DC.
heterocarpon (L.) DC. var. strigosum Van Meeuwen
macrodesmum (S. F. Blake) Standley & Steyerm.
triflorum (L.) DC.
Dioclea HBK.
pulchra Moldenke
Dipteryx Schreber
panamensis (Pittier) Record
(= *Coumarouna panamensis* Pittier, *Oleiocarpon panamense* (Pittier) Dwyer)
Dussia Krug & Urban ex Taubert
macroprophyllata (J. D. Sm.) Harms
mexicana (Standley) Harms
Erythrina L.
cochleata Standley
gibbosa Cuf.
poeppigiana (Walp.) O. F. Cook [introduced]
Gliricidia HBK.
sepium (Jacq.) Kunth ex Walp.
Hymenolobium Benth. ex Martius
mesoamericanum H. Lima
Indigofera L.
mucronata Sprengel ex DC.
Lonchocarpus HBK.
oliganthus F. J. Hermann

pentaphyllus (Poiret) DC.
velutinus Benth. ex Seemann
Machaerium Pers.
floribundum Benth.
seemannii Benth.
sp. indet.
Mucuna Adans.
holtonii (Kuntze) Moldenke
(= *M. andreana* M. Micheli)
Ormosia G. Jackson
grandiflora (Tul.) Rudd
macrocalyx Ducke
velutina Rudd
Pachyrhizus L. C. Rich. ex DC.
erosus (L.) Urban
Phaseolus L.
[adenanthus G. F. W. Meyer—see Vigna]
[peduncularis HBK.—see Vigna]
Platymiscium Vogel
pinnatum (Jacq.) Dugand
Pterocarpus L.
officinalis Jacq.
rohrii Vahl
(= *P. hayesii* Hemsley)
sp. indet.
Swartzia Schreber
cubensis (Britton & Wilson) Standley
simplex (Sw.) Sprengel var. continentalis Urban
(= var. *ochnacea* (DC.) Cowan)
Teramnus R. Br.
volubilis (L.) Sw.
Vigna Savi
adenantha (G. F. W. Meyer) Maréchal, Mascherpa & Stainier
(= *Phaseolus adenanthus* G. F. W. Meyer)
?caracalla (L.) Verdcourt
peduncularis (HBK.) Fawc. & Rendle
(= *Phaseolus peduncularis* HBK.)
vexillata (L.) A. Rich.

LOASACEAE
Klaprothia HBK.
fasciculata (K. Presl) Poston
(= *Sclerothrix fasciculata* K. Presl)

LOGANIACEAE
Potalia Aublet
amara Aublet
Spigelia L.
humboldtiana Cham. & Schldl.
Strychnos L.
tabascana Sprague & Sandw.

LORANTHACEAE (and see Viscaceae)
Oryctanthus Eichler
alveolatus (HBK.) Kuijt
occidentalis (HBK.) Blume
Phthirusa Martius
pyrifolia (HBK.) Eichler

Struthanthus Martius
  cansjeraefolius (Oliver) Eichler
  leptostachyus (HBK.) G. Don
  orbicularis (Kunth) Blume
  woodsonii Cuf.

LYTHRACEAE
Cuphea P. Br.
  carthagenensis (Jacq.) Macbr.
  epilobifolia Koehne
    (incl. var. *costaricensis* Koehne)
  hyssopifolia HBK.
  utriculosa Koehne
Lagerstroemia L.
  speciosa Pers. [cultivated]

MAGNOLIACEAE
Talauma Juss.
  gloriensis Pittier
  sambuensis Pittier (or sp. nov.)

MALPHIGIACEAE
Banisteriopsis C. B. Robinson ex Small
  muricata (Cav.) Cuatr.
Bunchosia L. C. Rich. ex HBK.
  macrophylla Rose ex J. D. Sm.
  ocellata Lundell
Byrsonima L. C. Rich. ex HBK.
  crassifolia (L.) HBK. [cultivated]
  crispa Adr. Juss.
Heteropteris HBK.
  laurifolia (L.) Adr. Juss.
  macrostachya Adr. Juss.
Hiraea Jacq.
  fagifolia (DC.) Adr. Juss.
  smilacina Standley
Jubelina Adr. Juss.
  wilburii W. R. Anderson
Lophanthera Adr. Juss.
  hammelii W. R. Anderson
Spachea Adr. Juss.
  correae Cuatr. & Croat
Stigmaphyllon Juss.
  lindenianum Adr. Juss.
  puberum (L. C. Rich.) Adr. Juss.
Tetrapterys Cav.
  donnell-smithii Small
    (= *T. macrocarpa* I. M. Johnston)

MALVACEAE
Abelmoschus Medikus
  moschatus Medikus
Hampea Schldl.
  appendiculata (J. D. Sm.) Standley
Hibiscus L.
  rosa-sinensis L. [cultivated]
Malachra L.
  alceifolia Jacq.
Malvaviscus Fabr.
  arboreus Cav.

Pavonia Cav.
  castaneifolia A. St.-Hil. & Naudin
    (= *P. longipes* Standley)
  rosea Schldl.
Sida L.
  acuta Burm. f.
  rhombifolia L.
Urena L.
  lobata L.
Wissadula Medikus
  exselsior (Cav.) K. Presl

MARCGRAVIACEAE
Marcgravia L.
  membranacea Standley
  nepenthoides Seemann
  serrae de Roon
Norantea Aublet
  costaricensis Gilg
  sessilis L. O. Williams
    (= *Sarcopera sessiliflora* (Triana & Planchon)
    Bedell)
  subsessilis (Benth.) J. D. Sm.
    (= *Marcgraviastrum subsessilis* (Benth.) Bedell)
Souroubea Aublet
  gilgii Al. Richter
  sympetala Gilg

MELASTOMATACEAE
Aciotis D. Don
  levyana Cogn.
  rostellata (Naudin) Triana
Adelobotrys DC.
  adscendens (Sw.) Triana
Arthrostemma Pavón ex D. Don
  ciliatum Ruiz & Pavón
Bellucia Necker ex Raf.
  pentamera Naudin
Blakea P. Br.
  dimorphophylla Almeda *ined.*
  scarlatina Almeda
Centradenia G. Don
  ? inaequilateralis (Schldl. & Cham.) G. Don [probably
    restricted to higher elevations]
Clidemia D. Don
  capitellata (Bonpl.) D. Don
  crenulata Gleason
  densiflora (Standley) Gleason
  dentata D. Don ex DC.
  discolor (Triana) Cogn.
    (= *C. purpureo-violacea* Cogn.)
  epiphytica (Triana) Cogn. var. trichocalyx (S. F. Blake)
    Wurdack
  hammelii Almeda
  hirta (L.) D. Don
  japurensis DC. var. heterobasis (DC.) Wurdack
  ombrophila Gleason
  pubescens Gleason *vel aff.*
  reitziana Cogn. & Gleason ex Gleason

septuplinervia Cogn.
setosa (Triana) Gleason
Conostegia D. Don
bracteata Triana
? icosandra (Sw.) Urban
lasiopoda Benth.
micrantha Standley
montana (Sw.) D. Don
rufescens Naudin
    (= *C. formosa* Macfad., *C. puberula* Cogn.)
setifera Standley
setosa Triana
subcrustulata (Beurl.) Triana
Graffenrieda DC.
galeottii (Naudin) L. O. Williams
Henriettea DC.
tuberculosa (J. D. Sm.) L. O. Williams
Leandra Raddi
[consimilis Gleason—see *Miconia ligulata*
    Almeda]
dichotoma (D. Don) Cogn.
granatensis Gleason
    (= *L. strigosa* Gleason)
grandifolia Cogn.
longicoma Cogn.
mexicana (Naudin) Cogn.
Miconia Ruiz & Pavón
affinis DC.
    (= *M. microcarpa* DC.)
ampla Triana
    (= *M. involucrata* J. D. Sm.)
appendiculata Triana
barbinervis (Benth.) Triana
brenesii Standley
calocoma Almeda *ined.*
centrodesma Naudin
calvescens DC.
dorsiloba Gleason
elata (Sw.) DC.
gracilis Triana
grayumii Almeda
hammelii Almeda *ined.*
impetiolaris (Sw.) D. Don ex DC.
lacera (Bonpl.) Naudin
lateriflora Cogn.
ligulata Almeda
    (= *Leandra consimilis* Gleason)
longifolia (Aublet) DC.
multispicata Naudin
nervosa (J. E. Sm.) Triana
prasina (Sw.) DC.
punctata (Desr.) D. Don ex DC.
serrulata (DC.) Naudin
simplex Triana
smaragdina Naudin *vel aff.*
splendens (Sw.) Griseb.
stevensiana Almeda
trinervia (Sw.) D. Don ex Loud.
Mouriri Aublet
gleasoniana Standley ex Standley & Steyerm.

Nepsera Naudin
aquatica (Aublet) Naudin
Ossaea DC.
brenesii Standley
laxivenula Wurdack
macrophylla (Benth.) Cogn.
micrantha (Sw.) Macf. ex Cogn.
robusta (Triana) Cogn.
Tibouchina Aublet
longifolia (Vahl) Baillon ex Cogn.
Tococa Aublet
guianensis Aublet
Topobea Aublet
dimorphophylla Almeda
maurofernandeziana Cogn.
Triolena Naudin
hirsuta (Benth.) Triana

MELIACEAE
Carapa Aublet
nicaraguensis C. DC.
Cedrela P. Br.
odorata L.
    (= *C. mexicana* Roemer)
Guarea Allamand ex L.
brevianthera C. DC.
bullata Radlk.
grandiflora DC.
    (= *G. chichon* C. DC., *G. pittieri* C. DC.)
guidonia (L.) Sleumer
    (= *G. trichilioides* L.)
macropetala Pennington
microcarpa Radlk.
rhopalocarpa Radlk.
    (= *G. tuisana* C. DC.)
tonduzii C. DC. *vel aff.*
Trichilia P. Br.
pallida Sw.
    (= *T. montana* HBK.)
septentrionalis C. DC.
    (= *T. moritzii* C. DC., *T. polyneura* C. DC.)

[MELIOSMACEAE—see Sabiaceae]

MENISPERMACEAE
Abuta Aublet
panamensis (Standley) Krukoff & Barneby
    (= *Hyperbaena panamensis* Standley)
Anomospermum Miers
reticulatum (Martius) Eichler
Cissampelos L.
grandifolia Triana & Planchon
pareira L.
tropaeolifolia DC.
Disciphania Eichler
calocarpa Standley
Odontocarya Miers
truncata Standley

MOLLUGINACEAE
Mollugo L.
verticillata L.

MONIMIACEAE
Mollinedia Ruiz & Pavón
costaricensis J. D. Sm.
Siparuna Aublet
guianensis Aublet
macra Standley
pauciflora (Beurl.) A. DC.
tonduziana Perkins

MORACEAE
Artocarpus J. R. & J. G. Forst.
altilis (Parkinson) Fosberg [cultivated]
(= *A. communis* J. R. & J. G. Forst., *A. incisus*
(Thunb.) L.f.)
Brosimum Sw.
alicastrum Sw.
guianensis (Aublet) J. E. Huber
lactescens (Moore) C. C. Berg
Castilla Sessé
elastica Sessé ssp. costaricana (Liebm.) C. C. Berg
[Cecropia—see Cecropiaceae]
Clarisia Ruiz & Pavón
biflora Ruiz & Pavón
[Coussapoa—see Cecropiaceae]
Dorstenia L.
choconiana S. Watson
(incl. var. *integrifolia* J. E. Sm.)
contrajerva L.
Ficus L.
cahuitensis C. C. Berg
citrifolia P. Mill.
colubrinae Standley
costaricana (Liebm.) Miq.
crassivenosa W. Burger
donnell-smithii Standley
dugandii Standley
glaucescens (Liebm.) Miq.
insipida Willd.
(= *F. glabrata* HBK.)
maxima P. Mill.
(= *P. radula* Humb. & Bonpl. ex Willd.)
nymphaeifolia P. Mill.
pertusa L.f.
popenoei Standley
schippii Standley
tonduzii Standley
velutina Humb. & Bonpl. ex Willd.
Maquira Aublet
costaricana (Standley) C. C. Berg
(= *Perebea costaricana* Standley)
Naucleopsis Miq.
naga Pittier
(= *Ogcodeia naga* (Pittier) Mildbr.)
Perebea Aublet
angustifolia (Poeppig & Endl.) C. C. Berg
[Pourouma—see Cecropiaceae]

Pseudolmedia Trécul
spuria (Sw.) Griseb.
Sorocea St.-Hil.
pubivena Hemsley
Trophis P. Br.
involucrata W. Burger
racemosa (L.) Urban
ssp. racemosa (reportedly known only from the
West Indies except for a few individuals at La
Selva)?
ssp. ramon (Schldl. & Cham.) W. Burger

MYRISTICACEAE
Compsoneura (A. DC.) Warb.
sprucei (A. DC.) Warb.
Otoba (A. DC.) Karsten
novogranatensis Moldenke
(= *Dialyanthera otoba* (Humb. & Bonpl.) Warb.)
Virola Aublet
koschnyi Warb.
sebifera Aublet

MYRSINACEAE
[*Amatlania* Lundell—see Ardisia Sw.]
Ardisia Sw.
[alstonii Lundell
(= *Icacorea alstonii* (Lundell) Lundell) One
specimen so identified by Lundell is treated here as
Ardisia deminuta Lundell, a species common in the
area.]
auriculata Donn. Sm.
(= *Auriculardisia auricularia* (J. D. Sm.) Lundell)
deminuta Lundell
(= *Icacorea deminuta* (Lundell) Lundell)
fimbrillifera Lundell
(= *Auriculardisia fimbrillifera* (Lundell) Lundell)
nigropunctata Oersted
(= *Auriculardisia nigropunctata* (Oersted)
Lundell)
opegrapha Oersted
(= *Graphardisia opegrapha* (Oersted) Lundell)
pellucida Oersted
(= *A. pectinata* J. D. Sm., *A. myriodonta* Standley,
A. pellucida var. *pectinata* (J. D. Sm.) Lundell,
Amatlania pellucida (Oersted) Lundell, *Amatlania
pellucida* var. *myriodonta* (Standley) Lundell)
proctori Lundell
(= *Icacoria proctori* (Lundell) Lundell)
sarapiquiensis (Lundell) Lundell
(= *Auriculardisia sarapiquiensis* Lundell,
Auriculardisia wilburiana Lundell, *Ardisia
wilburiana* (Lundell) Lundell)
squamata (Lundell) Lundell
(= *Auriculardisia squamata* Lundell)
[subsessilifolia Lundell
(= *Icacorea subsessilifolia* (Lundell) Lundell) A
specimen ± questionably so named by Lundell
(the type is from 7,000 ft in Chiriquí is, in my
opinion, a specimen of *Ardisia proctori* Lundell,
a reasonably common plant at La
Selva)]

wedelii Lundell
   (= *Auriculardisia wedelii* (Lundell) Lundell)
[wilburiana (Lundell) Lundell—see *A.*
   sarapiquiensis Lundell]
[Auriculardisia Lundell—see *Ardisia* Sw.]
Cybianthus Martius
   schlimii (Hook. f.) Agostini
     (= *Weigeltia spectabilis* (Standley) Lundell,
     Ardisia spectabilis Standley, *Correlliana spectabilis*
     (Standley) D'Arcy)
[Graphardisia (Mez) Lundell—see *Ardisia* Sw.]
[*Icacorea* Aublet—see *Ardisia* Sw.]
Parathesis (A. DC.) Hook. f.
   ? chrysophylla Lundell
   longipetiolata Lundell
   microcalyx J. D. Sm.

MYRTACEAE
Eugenia L.
   sarapiquensis P. Sánchez
   uniflora L. [cultivated]
   + about 5 spp. indet.
Myrcia DC. ex Guillemin
   splendens (Sw.) DC.
     (incl. *M. costaricensis* Berg)
Myrciaria Berg
   floribunda (West ex Willd.) Berg
Psidium L.
   guajava L.
Syzygium Gaertner
   malaccensis (L.) Merr. & Perry
     (= *Eugenia malaccensis* L.)

NYCTAGINACEAE
Mirabilis L.
   jalapa L.
Neea Ruiz & Pavón
   amplifolia J. D. Sm.
     (= *N. urophylla* Standley, = *N. elegans* P. H.
     Allen)
   delicatula Standley *vel aff.*
   elegans P. Allen *vel aff.*
   popenoei P. Allen *vel aff.*
   psychotrioides J. D. Sm.
     (= *N. laetevirens* Standley)
   urophylla Standley *vel aff.*
Pisonia L.
   aculeata L.

NYMPHAEACEAE
Nymphaea L.
   ampla (Salisb.) DC.

OCHNACEAE
Cespedesia Goudot
   macrophylla Seemann
Ouratea Aublet
   curvata (St.-Hil.) Engler
     (= *O. costaricensis* Standley, *O. crassinervia*
     Engler)

insulae Riley *vel aff.*
lucens (HBK.) Engler
Sauvagesia L.
   erecta L.

OLACACEAE
Heisteria Jacq.
   concinna Standley
   macrophylla Oersted
   scandens Ducke
Minquartia Aublet
   guianensis Aublet

ONAGRACEAE
Ludwigia L.
   decurrens Walter
     (= *Jussiaea decurrens* (Walter) DC.)
   erecta (L.) Hara
     (= *Jussiaea erecta* L.)
   latifolia (Benth.) Hara
     (= *Jussiaea latifolia* Benth.)
   leptocarpa (Nutt.) Hara
     (= *Jussiaea leptocarpa* Nutt.)
   octovalvis (Jacq.) Raven
     (= *Jussiaea suffruticosa* L.)
   peruviana (L.) Hara
     (= *Jussiaea peruviana* L.)

OXALIDACEAE
Averrhoa L.
   carambola L. [cultivated]
   bilimbi L. [cultivated]
Oxalis L.
   barrelieri L.
   filiformis HBK.

PAPAVERACEAE
Bocconia L.
   ? frutescens L.

PASSIFLORACEAE
Passiflora L.
   ambigua Hemsley
   arbelaezii Uribe
   auriculata HBK.
   biflora Lam.
   coriacea Juss.
   costaricensis Killip
   edulis Sims f. flavicarpa Degener
   foetida L.
   lancearia Masters
   lobata (Killip) Hutchinson ex J. M. MacDougal
     (= *Tetrastylis lobata* Killip)
   menispermifolia HBK.
   nitida HBK.
   oerstedii Masters
   pittieri Masters
   quadrangularis L.
   vitifolia HBK.

PHYTOLACCACEAE
Microtea Sw.
  debilis Sw.
Petiveria L.
  alliacea L.
Phytolacca L.
  rivinoides Kunth & Bouché
Rivina L.
  humilis L.
Trichostigma A. Rich.
    octandrum (L.) H. Watt
    polyandrum (Loes.) H. Watt

PIPERACEAE
Peperomia Ruiz & Pavón
  distachya (L.) A. Dietr.
  ebingeri Yuncker
  emarginella (Sw.) C. DC.
  glabella (Sw.) A. Dietr.
  hernandiifolia (Vahl) A. Dietr.
  macrostachya (Vahl) A. Dietr.
  montecristana Trel.
  obtusifolia (L.) A. Dietr.
  oerstedii C. DC.
  panamensis C. DC. ex Schroeder
  pellucida (L.) HBK.
  pernambucensis Miq.
  rotundifolia (L.) HBK.
  seemanniana Mig.
  serpens (Sw.) Loud.
  urocarpa Fischer & C. A. Meyer
Piper L.
  aduncum L.
  aequale Vahl
  arboreum Aublet
  augustum Rudge
  auritum HBK.
  biolleyi C. DC.
  biseriatum C. DC.
  cenocladum C. DC.
  colonense C. DC.
  concepcionis Trel.—see *P. subsessilifolium*
  darienense C. DC.
  dolichotrichum Yuncker
  dryadum C. DC.
  friedrichsthallii C. DC.
  garagaranum C. DC.
  glabratum Kunth
  glabrescens (Miq.) C. DC.
  hispidum Sw.
  holdridgeianum W. Burger
  imperiale (Miq.) C. DC.
  longispicum C. DC.
      (= *P. euryphyllum* C. DC.)
  melanocladum C. DC.
  multiplinervium C. DC.
  nudifolium C. DC.
  otophorum C. DC.
  peracuminatum C. DC.
  perbrevicaule Yuncker

  phytolaccaefolium Opiz
  pseudoumbratum C. DC.
  reticulatum L.
  sancti-felicis Trel.
  schiedeanum Steudel
      (= *P. carrilloanum* C. DC.)
  silvivagum C. DC.
  sinugaudens C. DC.
  subsessilifolium C. DC.
      (= *P. concepcionis* Trel.)
  terrabanum C. DC.
  tonduzii C. DC.
  trigonum C. DC.
      (= *P. arieianum* C. DC.)
  urophyllum C. DC.
  urostachyum Hemsley
  xanthostachyum C. DC.
Pothomorphe Miq.
  peltata (L.) Miq.
      (= *Piper peltatum* L., *Lepianthes peltata* (L.) Raf.)
  umbellata (L.) Miq.
      (= *Piper umbellatum* L., *Lepianthes umbellata* (L.) Raf.)
Sarcorhachis Trel.
  naranjoana (C. DC.) Trel.

PODOSTEMACEAE
Marathrum Humb. & Bonpl.
  allenii Woodson
Tristicha Thouars
  trifaria (Bory ex Willd.) Tul.

POLYGALACEAE
Moutabea Aublet
  aculeata Poeppig & Endl.
Polygala L.
  paniculata L.
Securidaca L.
  diversifolia (L.) S. F. Blake

POLYGONACEAE
Coccoloba P. Br.
  tuerckheimii J. D. Sm.
  3 spp. indet.
Polygonum L.
  hydropiperoides Michaux
  punctatum Elliott

PORTULACACEAE
Portulaca L.
  oleracea L.

QUIINACEAE
Lacunaria Ducke
  panamensis (Standley) Standley
Quiina Aublet
  schippii Standley

RHAMNACEAE
Colubrina L. C. Rich. ex Brongn.
  spinosa J. D. Sm.

Gouania Jacq.
  lupuloides (L.) Urban
    (= *G. polygama* (Jacq.) Urban)

RHIZOPHORACEAE
  Cassipourea Aublet
    guianensis Aublet
      (= *C. elliptica* (Sw.) Poiret)

[ROSACEAE—all known La Selva representatives are now
  referred to the segregated Chrysobalanaceae]

RUBIACEAE
  Amphidaysa Standley
    ambigua (Standley) Standley
  Bertiera Aublet
    guianensis Aublet
  Borojoa Cuatr.
    panamensis Dwyer
  [Borreria G. F. W. Meyer—see Spermacoce]
  [Cephaelis Sw.—see Psychotria]
  Chimarrhis Jacq.
    parviflora Standley
  Chione DC.
    sylvicola (Standley) W. Burger
      (= *Anisomeris sylvicola* (Standley) Standley,
      Chomelia sylvicola Standley), *Chione costaricensis*
      Standley)
  Coccocypselum P. Br.
    herbaceum P. Br.
  Coffea L.
    liberica Hiern. [cultivated]
  Cosmibuena Ruiz & Pavón
    macrocarpa (Benth.) Klotzsch ex Walp.
  Coussarea Aublet
    hondensis (Standley) C. M. Taylor & W. Burger
      (= *Psychotria hondensis* Standley, *P. ostaurea*
      Dwyer & Hayden)
    impetiolaris J. D. Sm.
    nigrescens C. M. Taylor & Hammel
    psychotrioides C. M. Taylor & Hammel
    talamancana Standley
  Coutarea Aublet
    hexandra (Jacq.) Schumann
  Faramea Aublet
    multiflora A. Rich.
      (= *F. talamancarum* Standley)
    parvibractea Steyerm.
    stenura Standley
    suerrensis J. D. Sm.
  Ferdinandusa Pohl
    panamensis Standley & L. O. Williams
  Genipa L.
    americana L.
      (= *G. caruto* HBK., *G. americana* var. *caruto*
      (HBK.) Schumann)
  Geophila D. Don
    cordifolia Miq.
    macropoda (Ruiz & Pavón) DC.
    repens (L.) I. M. Johnston
      (= *G. herbacea* (Jacq.) Schumann)

Gonzalagunia Ruiz & Pavón
  bracteosa (Donn. Sm.) B. L. Robinson
Hamelia Jacq.
  patens Jacq.
  xerocarpa Kuntze
Hemidiodia Schumann
  ocimifolia (Willd. ex Roemer & Schultes) Schumann
Hillia Jacq.
  grayumii C. M. Taylor
Hippotis Ruiz & Pavón
  albiflora Karsten
Hoffmannia Sw.
  liesneriana L. O. Williams
  valerii Standley
Ixora L.
  nicaraguensis Standley
Ladenbergia Klotzsch
  sericophylla Standley
Lasianthus W. Jack
  panamensis (Dwyer) Robbrecht
    (= *Dressleriopsis panamensis* Dwyer)
Manettia Mutis ex L.
  reclinata L.
Mitracarpus Zucc. ex Schultes & Schultes
  hirtus (L.) DC.
    (= *M. villosus* (Sw.) DC.)
[Montamans Dwyer = *Psychotria aggregata* Standley]
Oldenlandia L.
  corymbosa L.
  lancifolia (Schumacher) DC.
Palicourea Aublet
  crocea (Sw.) Roemer & Schultes
  guianensis Aublet
  triphylla DC.
Pentagonia Benth.
  donnell-smithii (Standley) Standley
Posoqueria Aublet
  coriacea M. Martens & Galeotti
  grandiflora Standley
  latifolia (Rudge) Roemer & Schultes
Psychotria L.
  acuminata Benth.
  aggregata Standley
    (= *P. tonduzii* Standley; *Montamans panamensis*
    Dwyer)
  alfaroana Standley
  angustifolia K. Krause
    (= *Psychotria mima* Standley)
  brachiata Sw.
  brachybotrya Muell. Arg.
  camponutans (Dwyer & Hayden) Hammel
    (= *Cephaelis camponutuns* Dwyer & Hayden)
  capacifolia Dwyer
  chagrensis Standley
  chiapensis Standley
  cooperi Standley
  elata (Sw.) Hammel
    (= *Cephaelis elata* Sw.)
  emetica L.f.
  erecta (Aublet) Standley & Steyerm.
  eurycarpa Standley

glomerulata (J. D. Sm.) Steyerm.
   (= *Cephaelis glomerulata* J. D. Sm.)
graciliflora Benth.
grandis Sw.
guapilensis (Standley) Hammel
   (= *Cephaelis discolor* Polakowsky)
haematocarpa Standley
hebeclada DC.
ipecacuanha (Broteri) Stokes
   (= *Cephaelis ipecacuanha* (Broteri) A. Rich.)
laselvensis C. Hamilton
luxurians Rusby
macrophylla Ruiz & Pavón
marginata Sw.
microbotrys Ruiz ex Standley
officinalis (Aublet) Raeuschel ex Sandwith
orchidearum Standley
panamensis Standley var. compressicaulis (K. Krause)
   C. Hamilton
   (= *P. grandistipula* Standley)
pilosa Ruiz & Pavón
   (= *Psychotria costaricensis* Polakowsky)
pittieri Standley
poeppigiana Muell.Arg.
   (= *Cephaelis tomentosa* (Aublet) Vahl)
polyphlebia J. D. Sm.
psychotrifolia (Seemann) Standley
racemosa (Aublet) Raeuschel
siggersiana Standley
suerrensis J. D. Sm.
[tonduzii Standley—see *P. aggregata* Standley]
uliginosa Sw.
Randia L.
   grandifolia (J. D. Sm.) Standley
     (= *R. diversiloba* Standley)
   mira Dwyer
   pepoformis Dwyer
[Ravnia Oersted—see Hillia]
Richardia L.
   scabra L.
Rudgea Salisb.
   cornifolia (HBK.) Standley
Sabicea Aublet
   panamensis Wernham
     (= *S. costaricensis* Wernham)
   villosa Roemer & Schultes
     (= *S. hirsuta* HBK.)
Simira Aublet
   maxonii (Standley) Steyerm.
     (= *Sickingia maxonii* Standley)
Sommera Schldl.
   donnell-smithii Standley
Spermacoce L.
   assurgens Ruiz & Pavón
     (= *Borreria laevis* auct.)
   gracilis Ruiz & Pavón
     (= *Borreria ocymoides* auct., in part)
   latifolia Aublet
     (= *Borreria latifolia* (Aublet) Schumann)
   prostrata Aublet
     (= *Borreria ocymoides* auct., in part)

Uncaria Schreber
   tomentosa (Willd.) DC.
Warszewiczia Klotzsch
   coccinea (Vahl) Klotzsch

RUTACEAE
Citrus L.
   aurantifolia (Christm.) Swingle [cultivated]
   aurantium L. [cultivated]
   grandis (L.) Osbeck [cultivated]
   limon (L.) Burm. f. [cultivated]
   sinensis (L.) Osbeck [cultivated]
Toxosiphon Baillon
   (= *Erythrochiton lindenii* (Baillon) Hemsley)
Zanthoxylum L.
   mayanum Standley
   panamense P. Wilson

SABIACEAE
Meliosma Blume
   donnell-smithii Urban
   vernicosa (Liebm.) Griseb.

SAPINDACEAE
Allophylus L.
   psilospermus Radlk.
Cupania L.
   guatemalensis (Turcz.) Radlk.
   livida (Radlk.) Croat
Nephelium L.
   ramboutan-ake (Labill.) Lenhouts [cultivated]
     (= *N. mutabile* Blume)
Paullinia L.
   baileyi Standley
   "echinocapsula" *sp. nov. ined.*
   fasciculata Radlk.
   fibrigera Radlk.
   fuscescens HBK.
   grandifolia Benth. ex Radlk.
   pinnata L.
   "pseudostipitata" *sp. nov. ined.*
   pterocarpa Triana & Planchon
   rugosa Benth. ex Radlk.
   serjaniaefolia Triana & Planchon
   venusta Radlk.
Serjania P. Mill.
   atrolineata C. Wright
   decapleuria Croat
   mexicana (L.) Willd.
   rhombea Radlk.
Talisia Aublet
   nervosa Radlk.
Vouarana Aublet
   guianensis Aublet

SAPOTACEAE
Chrysophyllum L.
   cainito L. [widely cultivated and apparently native
     only to the Greater Antilles]
   colombianum (Aubrév.) Pennington
   hirsutum Cronq.

venezuelanense (Pierre) Pennington
  (= *Pouteria lucentifolia* (Standley) Baehni)
Pouteria Aublet
  calistophylla (Standley) Baehni
  campechiana (HBK.) Baehni
  durlandii (Standley) Baehni
  leptopedicellata Pilz
  reticulata (Engler) Eyma
    (= *P. unilocularis* (J. D. Smith) Baehni)
  silvestris Pennington
  torta (Martius) Radlk. ssp. tuberculata (Sleumer)
    Pennington
    (= *P. neglecta* Cronq.)
Pradosia Liais.
  atroviolacea Ducke

SCROPHULARIACEAE
Bacopa Aublet
  salzmannii (Benth.) Wettst. ex Edwall
Lindernia All.
  crustacea (L.) F. Muell.
  diffusa (L.) Wettst.
Mecardonia Ruiz & Pavón
  procumbens (P. Mill.) Small
    (= *Bacopa procumbens* (P. Mill.) Greenman)
[Schlegelia—see Bignoniaceae]
Scoparia L.
  dulcis L.
Stemodia L.
  jorullensis HBK.
  verticillata (P. Mill.) Hassler
Tetranema Benth. ex Lindley
  bicolor L. O. Williams
Torenia L.
  thouarsii (Cham. & Schldl.) Kuntze
    (= *Lindernia thouarsii* (Cham. & Schldl.) G.
    Edwin)

SIMAROUBACEAE
Picramnia Sw.
  cooperi D. M. Porter
Simarouba Aublet
  amara Aublet

SOLANACEAE
Browallia L.
  americana L.
Capsicum L.
  frutescens L.
Cestrum L.
  megalophyllum Dunal
    (= *C. baenitzii* Lingelsh.)
  racemosum Ruiz & Pavón
    (= *C. panamense* Standley)
Cyphomandra Martius ex Sendtner
  hartwegii (Miers) Dunal
Lycianthes (Dunal) Hassler
  multiflora Bitter
  sanctaeclarae (Greenman) D'Arcy
  synanthera (Sendtner) Bitter

Lycopersicon P. Mill.
  esculentum P. Mill.
    (= *L. lycopersicum* (L.) Karsten)
Markea L. C. Rich.
  neurantha Hemsley
Physalis L.
  angulata L.
  pubescens L.
Solanum L.
  americanum P. Mill.
  arboreum Humb. & Bonpl. ex Dunal
    (= *S. enchylozum* Bitter)
  argenteum Dunal ex Poiret
  jamaicense P. Mill.
  lancaeifolium Jacq.
  quitoense Lam.
  rovirosanum J. D. Sm.
  rudepannum Dunal
    (= *S. ochraceo-ferrugineum* (Dunal) Fern.)
  rugosum Dunal
  schlechtendalianum Walp.
  sessiliflorum Dunal
  siparunoides Ewan
  trizygum Bitter
Witheringia L'Her.
  asterotricha (Standley) Hunz.
  cuneata (Standley) Hunz.
  exiguaflora D'Arcy
  solanacea L'Hér.

STERCULIACEAE
Byttneria Loefl.
  aculeata Jacq.
Herrania Goudot
  purpurea (Pittier) R. E. Schultes
Melochia L.
  nodiflora Sw.
Sterculia L.
  recordiana Standley var. papyracea E. Taylor
Theobroma L.
  cacao L.
  mammosum Cuatr. & J. León
  simiarum J. D. Sm.

SYMPLOCACEAE
Symplocos Jacq.
  sp. indet.

THEOPHRASTACEAE
Clavija Ruiz & Pavón
  costaricana Pittier

TILIACEAE
Apeiba Aublet
  membranacea Spruce ex Benth.
Goethalsia Pittier
  meiantha (J. D. Sm.) Burret
Heliocarpus L.
  appendiculatus Turcz.
Luehea Willd.
  seemannii Triana & Planchon

Mortoniodendron Standley & Steyerm.
  anisophyllum (Standley) Standley & Steyerm.
  guatemalense Standley & Steyerm.
    (= *M. costaricense* Standley & L. O. Williams)
Trichospermum Blume
  grewiifolium (A. Rich.) Kosterm.
Triumfetta L.
  lappula L.

ULMACEAE
  Ampelocera Klotzsch
    macrocarpa A. Gentry & Forero
  Celtis L.
    iguanaea (Jacq.) Sarg.
    schippii Trel. ex Standley
  Trema Lour.
    integerrima (Beurling) Standley
    micrantha (L.) Blume

UMBELLIFERAE
  Eryngium L.
    foetidum L.
  Hydrocotyle L.
    mexicana Schldl. & Cham.
    umbellata L.
    verticillata Thunb. var triradiata (A. Rich.) Fern.
      (= *H. prolifera* Kellogg)
  Spananthe Jacq.
    paniculata Jacq.

URTICACEAE
  Boehmeria Jacq.
    aspera Wedd.
    cylindrica (L.) Sw.
  Laportea Gaudich.
    aestuans (L.) Chew
      (= *Fleurya aestuans* (L.) Miq.)
  Myriocarpa Benth.
    longipes Liebm.
      (= *M. yzabalensis* (J. D. Sm.) Killip)
  Phenax Wedd.
    sonneratii (Poiret) Wedd.
  Pilea Lindl.
    diversissima Killip
    hyalina Fenzl
    imparifolia Wedd.
    microphylla (L.) Liebm.
    nummularifolia (Sw.) Wedd.
    pittieri Killip
    ptericlada J. D. Sm.
    quichensis J. D. Sm.
  Urera Gaudich.
    baccifera (L.) Gaudich.
    caracasana (Jacq.) Griseb.

  elata (Sw.) Griseb.
  laciniata (Goudot) Wedd.

VERBENACEAE
  Aegiphila Jacq.
    cephalophora Standley
    elata Sw.
    falcata J. D. Sm.
      (= *Ae. martinicensis* f. *falcata* (J. D. Sm.)
      D. Gibson)
    panamensis Moldenke
  Callicarpa L.
    acuminata HBK.
  Clerodendrum
    ligustrinum (Jacq.) R. Br. var. nicaraguense Moldenke
  Lantana L.
    trifolia L.
  Petrea L.
    rugosa HBK.
    volubilis L.
  Stachytarpheta Vahl
    cayennensis (L. C. Rich.) Vahl
      (= *S. guatemalensis* (L. C. Rich.) Moldenke)
    jamaicensis (L.) Vahl
    mutabilis (Jacq.) Vahl [cultivated]
  Vitex L.
    cooperi Standley

VIOLACEAE
  Gloeospermum Triana & Planchon
    diversipetalum Standley & L. O. Williams
  Rinorea Aublet
    deflexiflora Bartlett

VISCACEAE
  Phoradendron Nutt.
    acinacifolium Eichler
    flavens (Sw.) Griseb.
    piperoides (HBK.) Trel.
    quadrangulare (HBK.) Krug & Urban

VITACEAE
  Cissus L.
    microcarpa Vahl
    pseudosicyoides Croat
    verticillata (L.) D. Nicolson & Jarvis
      (= *C. sicyoides* L.)
  Vitis L.
    tiliifolia Humb. & Bonpl. ex Roemer & Schultes

VOCHYSIACEAE
  Qualea Aublet
    sp. nov. [or an undetermined South American species]
  Vochysia Aublet
    ferruginea Martius
    hondurensis Sprague

APPENDIX 4
Fishes

William A. Bussing

CARCHARHINIDAE
  Carcharhinus leucas "tiburón, shark"*

ELOPIDAE
  Tarpon atlanticus "sábalo real, tarpon"

CHARACIDAE "sardinas, tetras"
  Astyanax fasciatus
  Bramocharax bransfordi
  Brycon guatemalensis "machaca"
  Bryconamericus scleroparius
  Carlana eigenmanni
  Hyphessobrycon tortuguerae
  Roeboides guatemalensis

GYMNOTIDAE "knifefishes"
  Gymnotus cylindricus

PIMELODIDAE "barbudos, catfishes"
  Rhamdia guatemalensis
  Rhamdia nicaraguensis
  Rhamdia rogersi

GOBIESOCIDAE "chupapiedras, clingfishes"
  Gobiesox nudus

RIVULIDAE "olominas, killifishes"
  Rivulus isthmensis

POECILIIDAE "olominas, livebearers"
  Alfaro cultratus
  Brachyrhaphis holdridgei
  Brachyrhaphis parismina
  Neoheterandria umbratilis
  Phallichthys amates
  Phallichthys tico
  Poecilia gillii
  Priapichthys annectens

ATHERINIDAE "sardinas, silversides"
  Atherinella hubbsi

SYNGNATHIDAE "pipefishes"
  Pseudophallus mindii

SYNBRANCHIDAE "anguila, swamp eel"
  Synbranchus marmoratus

CICHLIDAE "mojarras, cichlids"
  Cichlasoma alfari
  Cichlasoma dovii "guapote"
  Cichlasoma loisellei "guapotillo"
  Cichlasoma nicaraguense "moga"
  Cichlasoma nigrofasciatum
  Cichlasoma rostratum
  Cichlasoma septemfasciatum
  Cichlasoma tuba "vieja"
  Herotilapia multispinosa
  Neetroplus nematopus "moga"

MUGILIDAE "lisas, mountain mullets"
  Agonostomus monticola "tepemechín"
  Joturus pichardi "bobo"

HAEMULIDAE
  Pomadasys crocro "roncador, croaker"

CENTROPOMIDAE
  Centropomus undecimalis "róbalo, snook"

ELEOTRIDAE
  Gobiomorus dormitor "guavina, sleeper"

GOBIIDAE "gobies"
  Awaous tajasica
  Sicydium altum "chupapiedra"

*Common names are given in Spanish, followed by English when available.

APPENDIX 5
Amphibians

Maureen A. Donnelly

Scientific Name	Common Name	Abundance[a]	Distribution[b]	Mode[c]
GYMNOPHIONA	Caecilians			
Caeciliidae	Caecilians			
Gymnopis multiplicata		U	M	V[e]
CAUDATA	Salamanders			
Plethodontidae	Lungless salamanders			
Bolitoglossa colonnea		U	M	IID[c]
Oedipina uniformis		U	M	IID[c]
Oedipina pseudouniformis		U	M	IID[c]
ANURA	Frogs and toads			
Bufonidae	True toads			
Bufo coniferus		U	M	1
Bufo haematiticus		C	M	1
Bufo marinus		C	T	1
Centrolenidae	Glass frogs			
Centrolenella albomaculata		U	M	18C
Centrolenella granulosa		R	M	18C
Centrolenella prosoblepon		U	M	18C
Centrolenella pulverata		R	M	18C
Centrolenella spinosa		U	M	18C
Centrolenella valerioi		U	M	18C
Dendrobatidae	Poison-dart frogs			
Dendrobates pumilio		A	M	14
Phyllobates lugubris		U	M	14
Hylidae	Tree frogs			
Agalychnis calcarifer		U	M	18B
Agalychnis callidryas		C	N	18A
Agalychnis saltator		C	M	18A
Hyla ebraccata		C	N	18A
Hyla loquax		C	N	1
Hyla phlebodes		C	M	1
Hyla rufitela		U	M	1
Scinax boulengeri		C	M	1
Scinax elaeochroa		C[d]	M	1
Smilisca baudinii		C	N	1
Smilisca phaeota		U	M	1
Smilisca puma		C	M	1
Smilisca sordida		R	M	2
Leptodactylidae	Tropical frogs			
Eleutherodactylus altae		U	M	17
Eleutherodactylus biporcatus		C	M	17
Eleutherodactylus bransfordii		A	M	17
Eleutherodactylus caryophyllaceus		C	M	17
Eleutherodactylus cerasinus		U	M	17

Appendix 5 (*continued*)

Scientific Name	Common Name	Abundance[a]	Distribution[b]	Mode[c]
Eleutherodactylus crassidigitus		C	M	17
Eleutherodactylus cruentus		U	M	17
Eleutherodactylus diastema		A	M	17
Eleutherodactylus fitzingeri		C	M	17
Eleutherodactylus mimus		C	M	17
Eleutherodactylus noblei		U	M	17
Eleutherodactylus ridens		U	M	17
Eleutherodactylus rugulosus		C	N	17
Eleutherodactylus talamancae		C	M	17
Leptodactylus melanonotus		C	T	8
Leptodactylus pentadactylus		C	T	8
Microhylidae	Narrow-mouthed toads			
Gastrophryne pictiventris		C	M	1
Ranidae	True frogs			
Rana taylori		R	M	1
Rana vaillanti		U	M	1
Rana warschewitschii		C	M	2

[a]Abundance at La Selva (A = abundant, often observed in the appropriate habitat; C = common, frequently observed; U = uncommon, occasionally observed; R = rare, few records for La Selva

[b]Distribution (T = Tropical, Mexico through Central to South America; M = Mesoamerican, Central America to northern Ecuador, N = northern Neotropics, Mexico to southern Central America)

[c]Reproductive mode follows Duellman and Trueb 1986.

[d]Can be abundant at ponds early in the wet season.

[e]Viviparous.

APPENDIX 6
Reptiles

Craig Guyer

Taxon	Abundance[a]	Habitat[b]	Taxon	Abundance[a]	Habitat[b]
Crocodilia			Tropidophiidae		
Crocodylidae			Ungaliophis panamensis[c]	R	Arb
Caiman crocodilus	C	Aqu	Colubridae		
Crocodylus acutus[c]	R	Aqu	Amastridium veliferum	C	Ter
Testudinata			Chironius grandisquamis	C	SA
Kinosternidae			Clelia clelia	U	Ter
Kinosternon angustipons	C	AqM	Coniophanes fissidens	C	Ter
K. leucostomum[d]	R	AqM	Dendrophidion percarinatum	U	SA
Emydidae			D. vinitor	C	SA
Rhinoclemmys annulata	R	Ter	Drymarchon corais[e]	R	Ter
R. funerea	C	Aqu	Drymobius margaritiferus	C	Ter
Sauria			D. melanotropis	R	Ter
Gekkonidae			D. rhombifer[d]	R	Ter
Lepidoblepharis xanthostigma	A	Ter	Enulius sclateri	R	Ter
Sphaerodactylus millepunctatus[c]	U	Ter	Erythrolamprus mimus	C	Ter
S. homolepis	U	Ter	Geophis hoffmani[c]	R	Ter
Thecadactylus rapicaudus	C	Arb	Hydromorphus concolor[c]	R	Aqu
Iguanidae			Imantodes cenchoa	C	Arb
Basiliscus plumifrons	C	AqM	I. inornatus	U	Arb
B. vittatus[d]	E?	AqM	Lampropeltis triangulum[c]	R	Ter
Corytophanes cristatus	U	Arb	Leimadophis epinephalus	R	Ter
Iguana iguana	C	AqM	Leptodeira annulata[c]	R	Arb
Norops biporcatus	U	Arb	L. septentrionalis	A	Arb
N. capito	C	Arb	Leptophis ahaetulla	U	Arb
N. carpenteri	U	Arb	L. depressirostris	C	Arb
N. humilis	A	Ter	L. nebulosus	U	Arb
N. lemurinus	C	Arb	Mastigodryas melanolomus	C	SA
N. limifrons	A	Arb	Ninia maculata	C	Ter
N. lionotus	C	AqM	N. sebae[c]	R	Ter
N. pentaprion	U	Arb	Nothopsis rugosus	C	Ter
Polychrus guttorosus	U	Arb	Oxybelis aeneus	U	Arb
Teiidae			O. brevirostris	U	Arb
Ameiva festiva	A	Ter	O. fulgidus[c]	R	Arb
A. quadrilineata	R	Ter	Oxyrhopus petola	U	Ter
Xantusiidae			Pseustes poecilonotus	C	Arb
Lepidophyma flavimaculata	U	Ter	R. decorata	A	Ter
Scincidae			Scaphiodontophis venustissimus	C	Ter
Mabuya unimarginata	U	Arb?	Sibon annulata	R	Arb
Sphenomorphus cherrei	A	Ter	S. longifrenis[c]	U	Arb
Anguidae			S. nebulata	C	Arb
Celestus hylaius	U	Ter	Spilotes pullatus	C	SA
Diploglossus bilobatus	R	Ter	Tantilla melanocephala	U	Ter
D. monotropis[c]	R	Ter	T. reticulata[e]	U	Ter
Serpentes			Tretanorhinus nigroluteus[d]	R	Aqu
Boidae			Trimetopon pliolepis	U	Ter
Boa constrictor	U	SA	Urotheca decipiens	R	Ter
Corallus annulatus[e]	R	Arb	U. euryzonus	U	Ter

Taxon	Abundance[a]	Habitat[b]
U. guentheri	R	Ter
Xenodon rabdocephalus	C	Ter
Micruridae		
Micrurus alleni	R	Ter
M. mipartitus	R	Ter
M. nigrocinctus	C	Ter
Viperidae		
Bothrops asper	A	Ter
Bothriechis schlegelii	C	Arb
Lachesis muta	U	Ter
Porthidium nasutus	C	Ter

Source: From Scott et al. (1983) except as noted.

Note: This compilation excludes one species (*Leptophis mexicanus*) apparently incorrectly listed for La Selva by Scott et al. (1983). Abundance and habitat categories based on relative encounter rates, not on systematic samples.

[a]Abundance categories: abundant (A); common (C); uncommon (U); rare (R); and extirpated (E).

[b]Habitat categories are aquatic (Aqu), aquatic margin (AqM), arboreal (Arb), semiarboreal (SA), and terrestrial (Ter).

[c]Added to fauna since 1982.

[d]Not seen since 1982.

[e]Not listed by Scott et al. (1983) but known from voucher material.

APPENDIX 7
Birds of La Selva and Vicinity

F. Gary Stiles and Douglas J. Levey

The classification of abundance is empirical, based on encounter frequency, not on actual counts. Thus, the abundances of some very secretive or shy species will be underestimated, and those of some noisy or conspicuous species will be overestimated, relative to their actual numbers.

Scientific name[a]	Common Name	Abundance[b]	Seasonal Status[c]	Habitat and Distribution[d]	Recent Changes[e]
Tinamiformes					
Tinamidae					
Tinamus major	Great Tinamou	C	P	F	—
Crypturellus soui	Little Tinamou	C	P	S	—
C. boucardi	Slaty-breasted Tinamou	U	P	F	—
Podicipediformes					
Podicipedidae					
Tachybaptus dominicus	Least Grebe	X	—	R	—
Pelecaniformres					
Phalacrocoracidae					
Phalacrocorax olivaceus	Olivaceous Cormorant	U	P?	R	—
Anhingidae					
Anhinga anhinga	Anhinga	R	P?	R	—
Fregatidae					
Fregata magnificens	Magnificent Frigatebird	X	—	+	—
Ciconiiformes					
Ardeidae					
Ixobrychus exilis	Least Bittern	X	—	R	—
Tigrisoma lineatum	Rufescent Tiger-heron	R	P	FR	—
Ardea herodias	Great Blue Heron	R	NR	R	—
Casmerodius albus	Great Egret	R	NR	R	—
Egretta thula	Snowy Egret	R	NR	R	—
Egretta caerulea	Little Blue Heron	C	NR	R	—
Egretta tricolor	Tricolored Heron	O	NR	R	—
Bubulcus ibis	Cattle Egret	A	P	O	2
Butorides striatus	Green-backed Heron	U	P	R	—
Agamia agami	Chestnut-bellied Heron	R	P	FR	—
Nycticorax nycticorax	Black-crowned Night Heron	O	V	R	—
Nycticorax violaceus	Yellow-crowned Night Heron	O	V	R	—
Cochlearius cochlearius	Boat-billed Heron	O	V?	FR	3
Threskiornithidae					
Mesembrinibis cayennensis	Green Ibis	U	P	F,R,E	—
Ajaia ajaja	Roseate Spoonbill	X	V	R	—
Ciconiidae					
Mycteria americana	Wood Stork	X	—	+	—

Appendix 7 *(continued)*

Scientific name[a]	Common Name	Abundance[b]	Seasonal Status[c]	Habitat and Distribution[d]	Recent Changes[e]
Anseriformes					
Anatidae					
Dendrocygna autumnalis	Black-bellied Whistling Duck	O	V	R,O	—
Cairina moschata	Muscovy Duck	O	(P)	FR	4
Anas discors	Blue-winged Teal	R	N	R	—
Falconiformes					
Cathartidae					
Coragyps atratus	Black Vulture	A	P	A	—
Cathartes aura	Turkey Vulture	C–A	P,NR,NF,NS	A	—
Sarcorhamphus papa	King Vulture	R	P	A,F	3
Accipitridae					
Pandion haliaetus	Osprey	U	NR	R,A	—
Leptodon cayanensis	Gray-headed Kite	O	P?	F?	3?
Chrondrohierax uncinatus	Hook-billed Kite	X	—	E	—
Elanoides forficatus	American Swallow-tailed Kite	U	S	A,F	—
Elanus caeruleus	Black-shouldered Kite	C	P	O	1,2
Harpagus bidentatus	Double-toothed Kite	U	P	F,E	—
Ictinia mississippiensis	Mississippi Kite	R–O	NF	A	—
Ictinia plumbea	*Plumbeous Kite*	U–R	S	A	—
Circus cyaneus	Northern Harrier	O	NF	O	—
Accipiter superciliosus	Tiny Hawk	R	P	F,S,E	—
Accipiter bicolor	Bicolored Hawk	R	P	F	—
Geranospiza caerulescens	Crane Hawk	R	P	F	—
Leucopternis princeps	**Barred Hawk**	X	V	A	—
L. semiplumbea	Semiplumbeous Hawk	U	P	F,C	—
L. albicollis	White Hawk	U–R	P	F,A	—
Buteogallus anthracinus	Common Black-hawk	U–R	P	E,S,A	3
B. urubitinga	Great Black-hawk	X	V?	A	—
Harpyhaliaetus solitarius	Solitary Eagle	X	V	A	—
Buteo nitidus	Gray Hawk	X	—	E	—
B. magnirostris	Roadside Hawk	O	P?	E	—
B. platypterus	Broad-winged Hawk	A	NR,NF,NS	A,O	—
B. brachyurus	Short-tailed Hawk	O	NF,NS	A	—
B. swainsoni	Swainson's Hawk	O	NF,NS	A	—
B. jamaicensis	Red-tailed Hawk	X	NF	A	—
Harpia harpyja	Harpy Eagle	X	(P)	A	4
Morphnus guianensis	Crested Eagle	X	V?	A	—
Spizastur melanoleucus	Black-and-white Hawk-eagle	O	P?	F,A	4?
Spizaetus tyrannus	Black Hawk-eagle	U	P	F,A	—
S. ornatus	Ornate Hawk-eagle	R–O	P	F,A	3
Falconidae					
Daptrius americanus	Red-throated Caracara	O	P?	F,E	4?
Herpetotheres cachinnans	Laughing Falcon	C–U	P	F,E,C	—
Micrastur ruficollis	Barred Forest-falcon	U	P	F,S	—
M. mirandollei	Slaty-backed Forest-falcon	U–R	P	F,E	2?
M. semitorquatus	Collared Forest-falcon	O	P?V?	F,S	3?
Falco sparverius	American Kestrel	R	NR,NF	O	—
F. rufigularis	Bat Falcon	R	P	R,E	3
F. peregrinus	Peregrine Falcon	O	NF	A	—
Galliformes					
Cracidae					
Ortalis cinereiceps	Gray-headed Chachalaca	U	P	S,E	—
Penelope purpurascens	Crested Guan	U	P	F,E,C	—
Crax rubra	Great Curassow	R	P	F	3
Phasianidae					

Appendix 7 (*continued*)

Scientific name[a]	Common Name	Abundance[b]	Seasonal Status[c]	Habitat and Distribution[d]	Recent Changes[e]
Odontophorus erythrops	Rufous-fronted Wood-quail	R	P	F	3
O. guttatus	Spotted Wood-quail	O	V?	F	—
Rhynchortyx cinctus	Tawny-faced Quail	R?	P?	F	—
Gruiformes					
Rallidae					
Laterallus albigularis	White-throated Crake	C–A	P	S,R	—
L. exilis	Gray-breasted Crake	U–C	P	O	1(?),2
Aramides cajanea	Gray-necked Wood-rail	U	P	R,E,F	—
Amaurolimnas concolor	Uniform Crake	R	P	FR	—
Porphyrula martinica	Purple Gallinule	R	P	R	—
Gallinula chloropus	Common Moorhen	X	—	R	—
Heliornithidae					
Heliornis fulica	Sungrebe	U	P	FR	3?
Eurypygidae					
Eurypyga helias	Sunbittern	R	P?	FR	
Aramidae					
Aramus guarauna	Limpkin	X	V	R	—
Charadriiformes					
Charadriidae					
Charadrius vociferus	Killdeer	U	NF,NR	O,R	—
Jacanidae					
Jacana spinosa	Northern Jacana	C	P	R	—
Scolopacidae					
Tringa melanoleuca	Greater Yellowlegs	X	NR?	R	—
T. solitaria	Solitary Sandpiper	R	NR	R	—
Actitis macularia	Spotted Sandpiper	A	NR	R	—
Laridae					
Stercorarius pomarinus	Pomarine Jaeger	X	V	+	—
Sterna fuscata	Sooty Tern	X	V	+	—
Columbiformes					
Columbidae					
Columba cayennensis	Pale-vented Pigeon	U	P?	O	3?
C. speciosa	Scaled Pigeon	U	P	F,E	2?
C. flavirostris	Red-billed Pigeon	U	P	O,E,S	2
C. nigrirostris	Short-billed Pigeon	A	P	F,E,C,S	—
Zenaida macroura	Mourning Dove	X	NF?	O	—
Columbina talpacoti	Ruddy Ground-dove	A	P	S,O	—
Claravis pretiosa	Blue Ground-dove	C	P	E,S	—
Leptotila verreauxi	White-tipped Dove	X	V?	S	—
L. cassinii	Gray-chested Dove	C–A	P	E,S,C	—
Geotrygon veraguensis	Olive-backed Quail-dive	U	P	F	—
G. montana	Ruddy Quail-dove	O	V?	F	—
Psittaciformes					
Psittacidae					
Aratinga finschi	Crimson-fronted Parakeet	A	P	E,S,C	2
A. nana	Olive-throated Parakeet	C–A	P	F,E,S,C	—
Ara ambigua	Great Green Macaw	U	P?V?	F	3
Brotogeris jugularis	Orange-chinned Parakeet	U	P?	E,S,C	1–2
Pionopsitta haematotis	Brown-hooded Parrot	C	P	F,E,C	—
Pionus senilis	White-crowned Parrot	A	P	F,E,S,C	—
Amazona autumnalis	Red-lored Parrot	C–A	P	F,E,C	—
A. farinosa	Mealy Parrot	C	P	F,E,C	—
Cuculiformes					
Cuculidae					
Coccyzus erythropthalmus	Black-billed Cuckoo	R	NF	E,S	—
C. americanus	Yellow-billed Cuckoo	R	NF,NS	E,S	—
Piaya cayana	Squirrel Cuckoo	C	P	F,E,S,C	—
Tapera naevia	Striped Cuckoo	R	P?	S	1

Appendix 7 (*continued*)

Scientific name[a]	Common Name	Abundance[b]	Seasonal Status[c]	Habitat and Distribution[d]	Recent Changes[e]
Neomorphus geoffroyi	Rufous-vented Ground-cuckoo	O	P	F	3
Crotophaga sulcirostris	Groove-billed Ani	A	P	O	—
Strigiformes					
Tytonidae					
Tyto alba	Common Barn-owl	U	P	O,S	—
Strigidae					
Otus guatemalae	Vermiculated Screech-owl	U	P	F	—
Lophostrix cristata	Crested Owl	U C	P	F,E	—
Pulsatrix perspicillata	Spectacled Owl	U	P	F,C,E	—
Glaucidium minutissimum	Least Pygmy-owl	U	P	F,E,C	—
Ciccaba virgata	Mottled Owl	U–C	P	F,E,S,C	—
C. nigrolineata	Black-and-white Owl	R	P	F	—
Caprimulgiformes					
Caprimulgidae					
Lurocalis semitorquatus	Short-tailed Nighthawk	U	P	E	2
Chordeiles minor	Common Nighthawk	U–R	NF,NS	A	—
Nyctidromus albicollis	Common Pauraque	A	P	S,O,E	—
Caprimulgus carolinensis	Chuck-will's-widow	R	NR?	F,E	—
Nyctibiidae					
Nyctibius grandis	Great Potoo	U–C	P	F	—
N. griseus	Common Potoo	R–O	P?	O,E	—
Apodiformes					
Apodidae					
Cypseloides niger	Black Swift	R	V?	A	—
Streptoprocne zonaris	White-collared Swift	U–C	P	A	—
Chaetura pelagica	Chimney Swift	O	NF	A	—
C. cinereiventris	Gray-rumped Swift	A	P	A	—
Panyptila cayennensis	Lesser Swallow-tailed Swift	U	P	A	—
P. sanctihieronymi	Great Swallow-tailed Swift	X	—	A	—
Trochilidae					
Glaucis aenea	Bronzy Hermit	C	P	S,E	—
Threnetes ruckeri	Band-tailed Barbthroat	C	P	F,E,S	—
Phaethornis guy	Green Hermit	O	V	F,S	—
P. superciliosus	Long-tailed Hermit	A	P	F,E,S,C	—
P. longuemareus	Little Hermit	A	P	F,E,S,C	—
Eutoxeres aquila	White-tipped Sicklebill	U	P	F,E,S	—
Phaeochroa cuvierii	Scaly-breasted Hummingbird	O	V?	E	—
Florisuga mellivora	White-necked Jacobin	U–C	P–D	F,E,C	—
Colibri delphinae	Brown Violet-ear	X	—	E	—
Anthracothorax prevostii	Green-breasted Mango	X	V	O	—
Klais guimeti	Violet-headed Hummingbird	U–C	P	F,E,S,C	—
Lophornis helenae	Black-crested Coquette	R	P?V?	F,E,C	—
Discosura conversii	Green Thorntail	O	V	E,C	—
Thalurania colombica	Crowned Woodnymph	C–A	P	F,E,S,C	—
Hylocharis eliciae	Blue-throated Goldentail	R	V?D	S	—
Amazilia amabilis	Blue-chested Hummingbird	C	P	S,E,F,C	—
A. cyanura	Blue-tailed Hummingbird	X	—	S	—
A. tzacatl	Rufous-tailed Hummingbird	A	P	S,C,E	—
A. rutila	Cinnamon Hummingbird	X	V	S	—
Microchera albocoronata	Snowcap	U–C	V–W	F,E,C	—
Chalybura urochrysia	Bronze-tailed Plumeleteer	C	P	F,E,S,C	—
Heliodoxa jacula	Green-crowned Brilliant	X	V	F	—
Heliothryx barroti	Purple-crowned Fairy	U	P	F,E,S,C	—
Heliomaster longirostris	Long-billed Starthroat	R	P	F,E,C	—
Archilochus colubris	Ruby-throated Hummingbird	X	NF	S	—

Appendix 7 (*continued*)

Scientific name[a]	Common Name	Abundance[b]	Seasonal Status[c]	Habitat and Distribution[d]	Recent Changes[e]
Trogoniformes					
Trogonidae					
Trogon violaceus	Violaceous Trogon	C	P	E,C,S	—
T. collaris	Collared Trogon	X	V	E?	—
T. rufus	Black-throated Trogon	C	P	F,E	—
T. massena	Slaty-tailed Trogon	C–A	P	F,E,C	—
T. clathratus	Lattice-tailed Trogon	R–U	P	F	—
Coraciiformes					
Momotidae					
Momotus momota	Blue-crowned Motmot	X	V	F	—
Baryphthengus ruficapillus	Rufous Motmot	A	P	F,FR,E	—
Electron platyrhynchum	Broad-billed Motmot	A	P	F,FR,E	—
Alcedinidae					
Ceryle torquata	Ringed Kingfisher	U	P	R	—
C. alcyon	Belted Kingfisher	R	NF,NR	R	—
Chloroceryle amazona	Amazon Kingfisher	U	P	R	—
C. americana	Green Kingfisher	C	P	R	—
C. inda	Green-and-rufous Kingfisher	R–O	V?P?	FR	4?
C. aenea	American Pygmy Kingfisher	U	P	FR	
Piciformes					
Bucconidae					
Bucco macrorhynchos	White-necked Puffbird	R	P	F,E,C	—
B. tectus	Pied Puffbird	U	P	F,E,C	—
Malacoptila panamensis	White-whiskered Puffbird	C	P	F,E,C,S	—
Monasa morphoeus	White-fronted Nunbird	C–A	P	F,E,C	—
Galbulidae					
Galbula ruficauda	Rufous-tailed Jacamar	C	P	R,E	—
Jacamerops aurea	Great Jacamar	R	P?	F,E	3?
Ramphastidae					
Aulacorhynchus prasinus	Emerald Toucanet	X	V	E	—
Pteroglossus torquatus	Collared Aracari	A	P	F,E,C,S	—
Selenidera spectabilis	Yellow-eared Toucanet	R	P?V?	F,E	3
Ramphastos sulfuratus	Keel-billed Toucan	A	P	F,E,S,C	—
R. swainsonii	Chestnut-mandibled Toucan	A	P	F,E,S,C	—
Picidae					
Melanerpes pucherani	Black-cheeked Woodpecker	A	P	F,E,S,C	—
Veniliornis fumigatus	Smoky-brown Woodpecker	R	P	E,S	—
Piculus simplex	Rufous-winged Woodpecker	U–C	P	F,E	—
Celeus loricatus	Cinnamon Woodpecker	U–C	P	F,E,C	—
C. castaneus	Chestnut-colored Woodpecker	U–C	P	F,E,C	—
Dryocopus lineatus	Lineated Woodpecker	U–C	P	E,C,S	—
Campephilus guatemalensis	Pale-billed Woodpecker	C	P	F,E,C,S	—
Passeriformes					
Furnariidae					
Synallaxis brachyura	Slaty Spinetail	C	P	S	—
Hyloctistes subulatus	Striped Woodhaunter	U–R	P?	F	—
Automolus ochrolaemus	Buff-throated Foliage-gleaner	C	P	F,E,S,C	—
Xenops minutus	Plain Xenops	R	P?	F,E,C	—
Sclerurus guatemalensis	Scaly-throated Leaftosser	U	P	F	—
Dendrocolaptidae					
Dendrocincla fuliginosa	Plain-brown Woodcreeper	U	P	F,C,E	—
D. homochroa	Ruddy Woodcreeper	X	V	F	—
Deconychura longicauda	Long-tailed Woodcreeper	X	—	F	—

Appendix 7 (*continued*)

Scientific name[a]	Common Name	Abundance[b]	Seasonal Status[c]	Habitat and Distribution[d]	Recent Changes[e]
Glyphorhynchus spirurus	Wedge-billed Woodcreeper	A	P	F,E,C	—
Dendrocolaptes certhia	Barred Woodcreeper	C	P	F,E,C	—
Xiphorhynchus guttatus	Buff-throated Woodcreeper	C	P	E,F,C,S	—
X. lachrymosus	Black-striped Woodcreeper	C	P	F,E,C,S	—
X. erythropygius	Spotted Woodcreeper	U	P	F,E	—
Lepidocolaptes souleyetii	Streak-headed Woodcreeper	C–A	P	E,S,C	—
Formicariidae					
Cymbilaimus lineatus	Fasciated Antshrike	U	P	E,S,C	—
Taraba major	Great Antshrike	U	P	S,E	—
Thamnophilus doliatus	Barred Antshrike	O	P?	S	—
T. punctatus	Slaty Antshrike	C–A	P	F,E,C,S	—
Thamnistes anabatinus	Russet Antshrike	R	P?	F,E	3?
Dysithamnus mentalis	Plain Antvireo	X	V	F	—
D. striaticeps	Streak-crowned Antvireo	C	P	F,E	—
Myrmotherula fulviventris	Checker-throated Antwren	C–A	P	F,E	3?
M. axillaris	White-flanked Antwren	U–C	P	F	3?
Microrhopias quixensis	Dot-winged Antwren	A	P	E,C,F,S	—
Cercomacra tyrannina	Dusky Antbird	C	P	E,S	—
Gymnocichla nudiceps	Bare-crowned Antbird	R	P?	E,S	3?
Myrmeciza exsul	Chestnut-backed Antbird	A	P	F,E,S,C	—
M. immaculata	Immaculate Antbird	O	V	F	—
Hylophylax naevioides	Spotted Antbird	U	P	F,E,C	—
Gymnopithys leucaspis	Bicolored Antbird	U	P	F,E,C	—
Phaenostictus mcleannani	Ocellated Antbird	U	P	F,E	3
Formicarius analis	Black-faced Antthrush	C–A	P	F,E,C	—
Hylopezus perspicillatus	Spectacled Antpitta	C	P	F	—
H. fulviventris	Fulvous-bellied Antpitta	C	P	S,E	—
Tyrannidae					
Zimmerius vilissimus	Mistletoe Tyrannulet	A	P	F,C,S,E	—
Ornithion brunneicapillum	Brown-capped Tyrannulet	C	P	F,E,C	—
Elaenia flavogaster	Yellow-bellied Elaenia	U–C	P	S,E,O	—
Mionectes olivaceus	Olive-striped Flycatcher	U–R	V	F,E,C	—
M. oleagineus	Ochre-bellied Flycatcher	A	P	F,E,S,C	—
Leptopogon amaurocephalus	Sepia-capped Flycatcher	X	V?	C,E	—
Capsiempis flaveola	Yellow Tyrannulet	C–A	P	S,O	—
Myiornis atricapillus	Black-capped Pygmy-tyrant	A	P	F,E,C	—
Oncostoma cinereigulare	Northern Bentbill	C	P	S,E,C	—
Todirostrum sylvia	Slate-headed Tody-flycatcher	C	P	S	—
T. cinereum	Common Tody-flycatcher	A	P	S,E	—
T. nigriceps	Black-headed Tody-flycatcher	A	P	F,E,C	—
Rhynchocyclus brevirostris	Eye-ringed Flatbill	U	P	F,E,C	—
Tolmomyias sulphurescens	Yellow-olive Flycatcher	U–C	P	S,C	—
T. assimilis	Yellow-margined Flycatcher	C	P	F,E,C	—
Platyrinchus coronatus	Golden-crowned Spadebill	C	P	F	—
Onychorhynchus coronatus	Royal Flycatcher	R	P?	F,E	—
Terenotriccus erythrurus	Ruddy-tailed Flycatcher	C	P	F,E,S,C	—
Myiobius	Sulphur-rumped Flycatcher	U	P	F,E	3

Appendix 7 (*continued*)

Scientific name[a]	Common Name	Abundance[b]	Seasonal Status[c]	Habitat and Distribution[d]	Recent Changes[e]
sulphureipygius					
Aphanotriccus capitalis	Tawny-chested Flycatcher	R	P	C,E	—
Contopus borealis	Olive-sided Flycatcher	R	NF,NR?	S,E,C	—
C. sordidulus	Western Wood-pewee	R?	NF?	S,E,C	—
C. virens	Eastern Wood-pewee	A	NF,NR	S,E,C	—
C. cinereus	Tropical Pewee	A	P	S,O	—
Empidonax flaviventris	Yellow-bellied Flycatcher	C	NR	C,E,F,S	—
E. virescens	Acadian Flycatcher	R	NF	C,S	—
E. traillii	Willow Flycatcher	U	NF,NS	C,S,E	—
E. albigularis	White-throated Flycatcher	X	V (NR?)	S	—
E. minimus	Least Flycatcher	R–O	NF	C,S	—
Colonia colonus	Long-tailed Tyrant	A	P	E,C,S	—
Attila spadiceus	Bright-rumped Attila	A	P	F,E,C,S	—
Laniocera rufescens	Speckled Mourner	R	V?	F,E	—
Rhytipterna holerythra	Rufous Mourner	C	P	F,E,C	—
Myiarchus tuberculifer	Dusky-capped Flycatcher	U	P	S,E,C	—
M. crinitus	Great Crested Flycatcher	C	NR	S,E,F	—
Pitangus sulphuratus	Great Kiskadee	C	P	S,E,O	—
Megarhynchus pitangua	Boat-billed Flycatcher	A	P	E,S,C,F	—
Myiozetetes similis	Social Flycatcher	C–A	P	S,E,O	—
M. granadensis	Gray-capped Flycatcher	A	P	S,E	—
Coryphotriccus albovittatus	White-ringed Flycatcher	C–A	P	E,F,C	—
Myiodynastes maculatus	Streaked Flycatcher	X	—	—	—
M. luteiventris	Sulphur-bellied Flycatcher	U	S	E,S	—
Legatus leucophaius	Piratic Flycatcher	U	S	E,S	—
Tyrannus melancholicus	Tropical Kingbird	C–A	P	S,O	—
T. tyrannus	Eastern Kingbird	U	NF,NS	A	—
Pachyramphus cinnamomeus	Cinnamon Becard	A	P	E,C,S	—
P. polychopterus	White-winged Becard	C	P	E,C,S	—
P. aglaiae	Rose-throated Becard	X	—	E?	—
Tityra semifasciata	Masked Tityra	C–A	P	E,F,C,S	—
T. inquisitor	Black-crowned Tityra	U	P	E,F,C	—
Cotingidae					
Lipaugus unirufus	Rufous Piha	C	P	F,E	—
Cotinga amabilis	Lovely Cotinga	X	—	E,F	—
Carpodectes nitidus	Snowy Cotinga	U–O	P?V?	F,E,C	—
Querula purpurata	Purple-throated Fruitcrow	C	P	F,E,C	—
Cephalopterus glabricollis	Bare-necked Umbrellabird	U–R	V	F	—
Procnias tricarunculata	Three-wattled Bellbird	U–R	V	F,E,C	—
Pipridae					
Schiffornis turdinus	Thrushlike Manakin	R	P?	F	—
Piprites griseiceps	Gray-headed Manakin	R	P	F,E	3?
Manacus candei	White-collared Manakin	A	P	E,S,C,F	—
Corapipo leucorrhoa	White-ruffed Manakin	U	V	F,e,S,C	—
Pipra pipra	White-crowned Manakin	X	V	F	—
P. mentalis	Red-capped Manakin	C–A	P	F,E,S,C	—
Hirundinidae					
Progne chalybea	Gray-breasted Martin	U	P	A,O	
Tachycineta albilinea	Mangrove Swallow	A	P	R	—
Stelgidopteryx serripennis	Northern Rough-winged Swallow	U–C	NR	S,R	—
S. ruficollis	Southern Rough-winged Swallow	C–A	P	S,R	—
Riparia riparia	Bank Swallow	R	NF,NS	A	—
Hirundo pyrrhonota	Cliff Swallow	U	NF,NS	A	—

Appendix 7 (*continued*)

Scientific name[a]	Common Name	Abundance[b]	Seasonal Status[c]	Habitat and Distribution[d]	Recent Changes[e]
H. rustica	Barn Swallow	A	NR	A,O	—
Corvidae					
Cyanocorax morio	Brown Jay	U	P	S,E	—
Troglodytidae					
Campylorhynchus zonatus	Band-backed Wren	C–A	P	E,C,S	—
Thryothorus atrogularis	Black-throated Wren	U	P	S,E	—
T. nigricapillus	Bay Wren	A	P	E	—
T. thoracicus	Stripe-breasted Wren	A	P	C,S,E	—
T. modestus	Plain Wren	C	P?	S	1,2
Troglodytes aedon	House Wren	C	P	O,S	1
Henicorhina leucosticta	White-breasted Wood-wren	A	P	F,E,C	—
Microcerculus philomela	Nightingale Wren	U	P	F	—
Cyphorhinus phaeocephalus	Song Wren	U–C	P	F,E	—
Muscicapidae					
Microbates cinereiventris	Tawny-faced Gnatwren	C	P	F,E	—
Ramphocaenus melanurus	Long-billed Gnatwren	C	P	S,E	—
Polioptila plumbea	Tropical Gnatcatcher	A	P	F,E,C	—
Myadestes melanops	Black-faced Solitaire	X	V	E,F	—
Catharus fuscescens	Veery	O	NF	S,C	—
C. minimus	Gray-cheeked Thrush	U	NF,NS	S,C	—
C. ustulatus	Swainson's Thrush	C–A	NF,NR	E,S,C	—
Hylocichla mustelina	Wood Thrush	C	NR	F,E,C,S	—
Turdus obsoletus	Pale-vented Thrush	R–U	V	F,E	—
T. grayi	Clay-colored Robin	C	P	S,E	—
T. assimilis	White-throated Robin	O	V?	F	—
Mimidae					
Dumetella carolinensis	Gray Catbird	U	NR	S	—
Vireonidae					
Vireo flavifrons	Yellow-throated Vireo	C	NR	E,C,F	—
V. griseus	White-eyed Vireo	X	N	E,S	—
V. gilvus	Warbling Vireo	X	NS?	E?	—
V. philadelphicus	Philadelphia Vireo	U	NF,NS	E,C,F	—
V. olivaceus	Red-eyed Vireo	C	NF,NS	E,S,C,F	—
V. flavoviridis	Yellow-green Vireo	U	S	E,C,F	—
Hylophilus ochraceiceps	Tawny-crowned Greenlet	C–A	P	F,E	—
H. decurtatus	Lesser Greenlet	A	P	E,C,F,S	—
Vireolanius pulchellus	Green Shrike-vireo	A	P	F,E,C	—
Emberizidae					
Vermivora pinus	Blue-winged Warbler	O	NF	S	—
V. chrysoptera	Golden-winged Warbler	U	NF,NR	E,S	—
V. peregrina	Tennessee Warbler	A	NR	E,S,C,F	—
Dendroica petechia	Yellow Warbler	A	NR	S,O,E	—
D. pensylvanica	Chestnut-sided Warbler	A	NR	E,S,C,F	—
D. magnolia	Magnolia Warbler	O	NF,NR	E,S	—
D. tigrina	Cape May Warbler	O	NF?	E,S	—
D. coronata	Yellow-rumped Warbler	O	NR?	E,S	—
D. virens	Black-throated Green Warbler	R	NF,NS	E	—
D. fusca	Blackburnian Warbler	C	NF	E,S,C,F	—
D. castanea	Bay-breasted Warbler	U–A	NF,NR	E,S,F,C	—
D. cerulea	Cerulean Warbler	U–C	NF	E,S,C	—
Mniotilta varia	Black-and-white Warbler	U	NF,NS	E,S,C,F	—
Setophaga ruticilla	American Redstart	U	NF,NS	E,F,C	—
Protonotaria citrea	Prothonotary Warbler	U	NF	E,S	—

Appendix 7 (*continued*)

Scientific name[a]	Common Name	Abundance[b]	Seasonal Status[c]	Habitat and Distribution[d]	Recent Changes[e]
Helmitheros vermivorus	Worm-eating Warbler	R	NF	E,F,S	—
Seiurus aurocapillus	Ovenbird	C	NR	F,C,S	—
S. noveboracensis	Northern Waterthrush	C	NR	FR,R,E,C	—
S. motacilla	Louisiana Waterthrush	U	NF,NR	FR	—
Oporornis formosus	Kentucky Warbler	A	NR	E,S,F	—
O. philadelphia	Mourning Warbler	C	NR	S	—
O. tolmiei	MacGillivray's Warbler	O	NR?	S	—
Geothlypis trichas	Common Yellowthroat	X	NF	S	—
G. semiflava	Olive-crowned Yellowthroat	C	P	S,O	—
G. poliocephala	Gray-crowned Yellowthroat	C	P	O	2
Wilsonia citrina	Hooded Warbler	R	NF,NR	E	—
W. pusilla	Wilson's Warbler	C	NF	E,S	—
W. canadensis	Canada Warbler	C–A	NF,NS	E,C,S,F	—
Phaeothlypis fulvicauda	Buff-rumped Warbler	C–A	P	FR,R,E	—
Icteria virens	Yellow-breasted Chat	U–R	NR	S	—
Coereba flaveola	Bananaquit	A	P	S,O,C	—
Tangara inornata	Plain-colored Tanager	U	P	E,S,C	—
T. icterocephala	Silver-throated Tanager	U	V	E,S,C	—
T. gyrola	Bay-headed Tanager	R	V	E,C	—
T. larvata	Golden-masked Tanager	A	P	S,E,C	—
Dacnis venusta	Scarlet-thighed Dacnis	R	V	E,C	—
D. cayana	Blue Dacnis	U–R	P?	e,S,C	—
Chlorophanes spiza	Green Honeycreeper	U	P	E,S,C	—
Cyanerpes lucidus	Shining Honeycreeper	C	P	E,S,C	—
Euphonia luteicapilla	Yellow-crowned Euphonia	A	P	O,S,E	—
E. gouldi	Olive-backed Euphonia	A	P	F,E,S,C	—
E. minuta	White-vented Euphonia	U	P	E,F,C	—
E. anneae	Tawny-capped Euphonia	X	V	F,E	—
Thraupis episcopus	Blue-gray Tanager	C–A	P	S,E,O	2
T. palmarum	Palm Tanager	C	P	S,E,O,C	—
Chlorothraupis carmioli	Olive Tanager	C	P	F,E	3
Lanio leucothorax	White-throated Shrike-tanager	R	P?	F,E	3
Tachyphonus luctuosus	White-shouldered Tanager	U	P	F,E	—
T. delattrii	Tawny-crested Tanager	C	P	F,E	—
T. rufus	White-lined Tanager	U	P	S,E	—
Habia fuscicauda	Red-throated Ant-tanager	C–A	P	E,S,C	—
Piranga rubra	Summer Tanager	C	NR	E,C,S	—
P. olivacea	Scarlet Tanager	U	NF	E,C	—
Ramphocelus sanguinolentus	Crimson-collared Tanager	U	P	S,C	—
R. passerinii	Scarlet-rumped Tanager	A	P	S,E,C	—
Mitrospingus cassinii	Dusky-faced Tanager	A	P	E,S,C	—
Saltator albicollis	Streaked Saltator	X	V	S	—
S. coerulescens	Grayish Saltator	R	P?	S	1
S. maximus	Buff-throated Saltator	A	P	S,E,C,O	—
S. atriceps	Black-headed Saltator	A	P	E,S,C	—
Pitylus grossus	Slate-colored Grosbeak	C	P	F,E,C	—
Caryothraustes poliogaster	Black-faced Grosbeak	C–A	P	E,F,C	—
Pheucticus tibialis	Black-thighed Grosbeak	X	V?	F?	—
P. ludovicianus	Rose-breasted Grosbeak	U–C	NR	E,S,C	—
Cyanocompsa cyanoides	Blue-black Grosbeak	A	P	E,S,F,C	—
Guiraca caerulea	Blue Grosbeak	O	NF,NS	S	—
Passerina cyanea	Indigo Bunting	U–R	NF	S	—
Arremon aurantiirostris	Orange-billed Sparrow	A	P	E,F,C,S	—
Arremonops conirostris	Black-striped Sparrow	A	P	S,O	—

Appendix 7 (*continued*)

Scientific name[a]	Common Name	Abundance[b]	Seasonal Status[c]	Habitat and Distribution[d]	Recent Changes[e]
Volatinia jacarina	Blue-black Grassquit	U	P	O,S	—
Sporophila aurita	Variable Seedeater	A	P	S,O,C	—
S. torqueola	White-collared Seedeater	U–R	P	O,S	2
Oryzoborus nuttingi	Nicaraguan Seed-finch	U–R	P	S	1
O. funereus	Thick-billed Seed-finch	C–U	P	S,O,C	—
Tiaris olivacea	Yellow-faced Grassquit	U	P	S,O	—
Sturnella magna	Eastern Meadowlark	C–A	P	O	—
Quiscalus mexicanus	Great-tailed Grackle	X	–	O	1
Molothrus aeneus	Bronzed Cowbird	U	P?	O,S	2
Scaphidura oryzivora	*Giant Cowbird*	*U–R*	*P*	*F,E,C,O*	*3?*
Icterus dominicensis	Black-cowled Oriole	U	P	S,C	—
I. spurius	Orchard Oriole	U	NF,NR	E,S	—
I. mesomelas	Yellow-tailed Oriole	(R)	(P)	S	4?
I. galbula	Northern Oriole	C–A	NR	E,S,C	—
Amblycercus holosericeus	Yellow-billed Cacique	C	P	S,E	—
Cacicus uropygialis	Scarlet-rumped Cacique	A	P	F,E,C	—
Psarocolius wagleri	Chestnut-headed Oropendola	U	P	E,F,C	3
P. montezuma	Montezuma Oropendola	A	P	E,F,C	—
Passer domesticus	House Sparrow	A	P	E,F,C	—

Source: Based on a 1977 checklist by F. G. Stiles (Organization for Tropical Studies).

[a]Names generally follow the *Checklist of North American Birds* (AOU 1983) and Stiles et al. (1989).

[b]Abundance: A = abundant, many individuals recorded on most days (in appropriate habitat); C = common, a few individuals recorded on most days or many individuals at frequent intervals; U = uncommon; one or several recorded at frequent intervals but usually not daily; R = rare, one or a few recorded at long intervals; O = occasional, one or a few recorded in most years; X = accidental, less than five records.

[c]Seasonal status: P = permanent resident, present year-round; N = northern migrant; NF = fall migrant (September–November); NS = spring migrant (March–May); NR = present all winter as well as during migrations. If both migration and winter resident status are noted, most birds winter farther south and the species is, thus, much more common in migration than winter; S = southern migrant, present February–March to August–September only; V = visitant, usually a water bird (can occur in any month) or a highland bird (most occur between about June and December or January).

[d]Habitat and distribution: O = open country, cultivated areas, pastures, and so forth; A = aerial-strong fliers that pass over most or all habitats present; S = second growth; E = forest-edge habitats—thickets, light gaps, and so forth; C = cacao plantations and similar areas—"semiopen"; F = forest interior or canopy; R = riparian = streams, rivers, ponds, and so forth; FR = forest streams and swamps; + = Species without natural habitat at La Selva.

[e]Changes in status over the 1980s (where known): 1 = species not present before about 1968–70, that is, a recent arrival that may become established in the area; 2 = species that has increased in numbers since 1968 (1 and 2 indicate a species not known before 1968 that has since become established in good numbers); 3 = species that has decreased in numbers since 1968; 4 = species that probably has disappeared since 1968 (no records for at least two years).

APPENDIX 8
Mammals

Robert M. Timm

Scientific Name	Common Name[a]	Abundance[b]	Distribution[c]
Marsupialia	Marsupials		
Didelphidae	American Opossums		
Caluromys derbianus	Wooly opossum Zorro de balsa	uncommon	2
Chironectes minimus	Water opossum Zorro de agua	uncommon	1
Didelphis marsupialis	Southern opossum Zorro pelón or Zarigüeya	abundant	1
Marmosa mexicana	Mexican mouse-opossum Zorra or Zorricí	uncommon	3
Philander opossum	Gray four-eyed opossum Zorro de cuatro ojos	uncommon	1
Chiroptera	Bats		
Emballonuridae	Sac-winged Bats		
Centronycteris maximiliani	Thomas' bat	rare	1
Cormura brevirostris	Wagner's sac-winged bat	rare	1
Cyttarops alecto	Short-eared bat	rare	2
Diclidurus albus	Ghost bat	rare	1
Peropteryx kappleri	Greater doglike bat	uncommon	1
Rhynchonycteris naso	Brazilian long-nosed bat	abundant	1
Saccopteryx bilineata	Greater white-lined bat	abundant	1
Saccopteryx leptura	Lesser white-lined bat	common	1
Noctilionidae	Fishing and Bulldog Bats		
Noctilio albiventris	Lesser bulldog bat	common	1
Noctilio leporinus	Greater bulldog bat	common	1
Mormoopidae	Mustached Bats		
Pteronotus davyi	Davy's naked-backed bat	rare	2
Pteronotus parnellii	Parnell's mustached bat	common	1
Phyllostomidae	Leaf-nosed Bats		
Phyllostominae	Carnivorous Bats		
Chrotopterus auritus	Peter's false vampire bat	rare	1
Macrophyllum macrophyllum	Long-legged bat	rare	1
Micronycteris brachyotis	Dobson's large-eared bat	rare	2
Micronycteris daviesi	Davies' large-eared bat	rare	2
Micronycteris hirsuta	Hairy large-eared bat	rare	2
Micronycteris megalotis	Brazilian large-eared bat	common	1
Micronycteris minuta	Gervais' large-eared bat	rare	1
Micronycteris nicefori	Niceforo's large-eared bat	uncommon	2
Micronycteris schmidtorum	Schmidt's large-eared bat	rare	2
Mimon cozumelae	Cozumel spear-nosed bat	uncommon	1
Mimon crenulatum	Striped spear-nosed bat	uncommon	1

Appendix 8 (*continued*)

Scientific Name	Common Name[a]	Abundance[b]	Distribution[c]
Phylloderma stenops	Northern spear-nosed bat	rare	1
Phyllostomus discolor	Pale spear-nosed bat	rare	1
Phyllostomus hastatus	Spear-nosed bat	rare	1
Tonatia bidens	Spix's round-eared bat	uncommon	1
Tonatia brasiliense	Pygmy round-eared bat	uncommon	1
Tonatia silvicola	D'Orbigny's round-eared bat	rare	1
Trachops cirrhosus	Fringe-lipped bat	common	1
Vampyrum spectrum	False vampire bat	rare	2
Glossophaginae	Nectar-feeding Bats		
Choeroniscus godmani	Godman's bat	rare	2
Glossophaga commissarisi	Commissaris' long-tongued bat	common	3
Glossophaga soricina	Pallas' long-tongued bat	common	1
Hylonycteris underwoodi	Underwood's long-tongued bat	common	3
Lichonycteris obscura	Brown long-nosed bat	rare	2
Lonchophylla robusta	Panama long-tongued bat	uncommon	2
Carollinae	Short-tailed Bats		
Carollia brevicauda	Silky short-tailed bat	abundant	1
Carollia castanea	Allen's short-tailed bat	abundant	2
Carollia perspicillata	Seba's short-tailed bat	abundant	1
Stenoderminae	Fruit-eating Bats		
Artibeus jamaicensis	Jamaican fruit-eating bat	abundant	2
Artibeus lituratus	Big fruit-eating bat	common	1
Artibeus phaeotis	Pygmy fruit-eating bat	common	2
Artibeus watsoni	Thomas' fruit-eating bat	abundant	3
Chiroderma villosum	Shaggy-haired bat	rare	1
Ectophylla alba	Caribbean white bat	common	3
Sturnira lilium	Yellow-shouldered bat	rare	1
Sturnira ludovici	Anthony's bat	rare	2
Uroderma bilobatum	Tent-making bat	common	1
Vampyressa nymphaea	Big yellow-eared bat	uncommon	2
Vampyressa pusilla	Little yellow-eared bat	common	1
Vampyrodes caraccioli	San Pablo bat	rare	1
Vampyrops helleri	Heller's broad-nosed bat	common	1
Desmodontinae	Vampire Bats		
Desmodus rotundus	Vampire bat	common	1
	Vampiro		
Furipteridae	Smoky Bats		
Furipterus horrens	Smoky bat	uncommon	1
Thyropteridae	Disk-winged Bats		
Thyroptera tricolor	Spix's disk-winged bat	rare	1
Vespertilionidae	Vespertilionid Bats		
Eptesicus brasiliensis	Brasilian brown bat	rare	1
Eptesicus furinalis	Argentine brown bat	rare	1
Myotis albescens	Silver-tipped myotis	uncommon	1
Myotis elegans	Elegant myotis	uncommon	3
Myotis nigricans	Black myotis	uncommon	1
Myotis riparius	Riparian myotis	common	1
Rhogeessa tumida	Central American yellow bat	rare	1
Molossidae	Free-tailed Bats		
Molossus bondae	Bond's mastiff bat	uncommon	2
Molossus sinaloae	Allen's mastiff bat	common	2
Primates	Primates		
Cebidae	New World Monkeys		
Alouatta palliata	Mantled howler monkey	abundant	2
	Mono congo or Mono aullador		
Aotus lemurinus	Night monkey	rare?	2
	Mono nocturno		
Ateles geoffroyi	Geoffroy's spider monkey	common	2
	Mono colorado or Mono araña		
Cebus capucinus	White-faced capuchin	abundant	2
	Mono cara blanca		

Appendix 8 (*continued*)

Scientific Name	Common Name[a]	Abundance[b]	Distribution[c]
Edentata	Edentates		
Bradypodidae	Three-toed Sloths		
Bradypus variegatus	Three-toed sloth	common	1
	Perezoso de tres dedos		
Choloepidae	Two-toed Sloths		
Choloepus hoffmanni	Two-toed sloth	uncommon	2
	Perezoso		
Dasypodidae	Armadillos		
Cabassous centralis	Five-toed armadillo	rare	2
	Armadillo zopilote		
Dasypus novemcinctus	Nine-banded armadillo	uncommon	1
	Cusuco		
Myrmecophagidae	Anteaters		
Cyclopes didactylus	Silky anteater	uncommon	1
	Serafín de platanar or Tapacara		
Myrmecophaga tridactyla	Giant anteater	extirpated	1
	Oso caballo or Hormiguero		
Tamandua mexicana	Northern tamandua	common	2
	Oso hormiguero		
Lagomorpha	Rabbits, Hares, and Pikas		
Leporidae	Rabbits		
Sylvilagus brasiliensis	Forest rabbit	uncommon	1
	Conejo		
Rodentia	Rodents		
Sciuridae	Squirrels		
Microsciurus alfari	Alfaro's pygmy squirrel	uncommon	3
	Ardilla or Chiza		
Sciurus granatensis	Red-tailed squirrel	common	2
	Ardilla or Chiza		
Sciurus variegatoides	Variegated squirrel	uncommon	3
	Ardilla or Chiza		
Geomyidae	Pocket Gophers		
Orthogeomys cherriei	Cherrie's pocket gopher	uncommon	3
	Taltusa		
Heteromyidae	Pocket Mice		
Heteromys desmarestianus	Desmarest's spiny pocket mouse	abundant	2
	Ratón semiespinosa		
Muridae	Long-tailed Rats and Mice		
Nyctomys sumichrasti	Sumichrast's vesper rat	rare	3
	Ratón		
Oryzomys alfari	Alfaro's rice rat	rare	3
	Ratón arrocera		
Oryzomys bombycinus	Long-wiskered rice rat	rare	2
	Ratón		
Oryzomys caliginosus	Dusky rice rat	common	2
	Ratón arrocera		
Oryzomys fulvescens	Pygmy rice rat	rare	2
	Ratón		
Tylomys watsoni	Watson's climbing rat	rare	3
	Rata azul		
Erethizontidae	Porcupines		
Coendou mexicanus	Prehensile-tailed porcupine	rare	3
	Puercoespín		
Agoutidae	Pacas		
Agouti paca	Paca	uncommon	1
	Tepezcuintle		
Dasyproctidae	Agoutis		
Dasyprocta punctata	Agouti	abundant	1
	Guatusa		
Echimyidae	Spiny Rats		

Appendix 8 (*continued*)

Scientific Name	Common Name[a]	Abundance[b]	Distribution[c]
Hoplomys gymnurus	Armored rat	uncommon	2
	Ratón		
Proechimys semispinosus	Tomes' spiny rat	common	2
	Ratón		
Carnivora	Carnivores		
Mustelidae	Skunks, Weasels, and Otters		
Conepatus semistriatus	Striped hog-nosed skunk	common	3
	Zorro hediondo		
Eira barbara	Tayra	common	1
	Gato de monte		
Galictis vittata	Grison	rare	1
	Grisón or Tejón		
Lutra longicaudis	Southern river otter	common	1
	Perro de agua or Nutria		
Mustela frenata	Long-tailed weasel	rare	4
	Comadreja		
Procyonidae	Raccoons		
Bassaricyon gabbii	Olingo	rare	2
	Martilla		
Nasua narica	White-nosed coati	common	3
	Pizote		
Potos flavus	Kinkajou	common	1
	Mico de noche		
Procyon lotor	Raccoon	rare	5
	Mapachín		
Fclidac	Cats		
Felis concolor	Puma	rare	4
	Puma or León de montaña		
Felis onca	Jaguar	uncommon	4
	Tigre		
Felis pardalis	Ocelot	rare	4
	Manigordo or Ocelote		
Felis wiedii	Margay	rare	4
	Caucél		
Felis yagouaroundi	Jaguarundi	uncommon	4
	León breñero or Gatillo de monte		
Artiodactyla	Deer and Peccaries		
Tayassuidae	Peccaries		
Tayassu pecari	White-lipped peccary	extirpated	1
	Chancho de monte		
Tayassu tajacu	Collared peccary	common	1
	Saíno		
Cervidae	Deer		
Mazama americana	Red brocket deer	uncommon	1
	Cabro de monte		
Odocoileus virginianus	White-tailed deer	uncommon	4
	Venado cola blanca		
Perissodactyla	Tapirs and Horses		
Tapiridae	Tapirs		
Tapirus bairdii	Baird's tapir	uncommon	2
	Danta		

Note: Night monkeys are added to this list as likely to occur there (see Timm 1988). The smaller species of *Artibeus* and *Vampyressa* are difficult to identify in the field and are often misidentified. I caution investigators to review prepared study specimens carefully before making identifications in the field and to save vouchers whenever possible.

[a]The common name(s) for each species is listed in English and in Spanish immediately below. Spanish names listed herein are those used in this region of Costa Rica. Because nonmammalogists do not distinguish most of the many species of bats, there are few local common names in Spanish for bat species other than direct translations of the published common names.

[b]Abundance: abundant = often observed and/or captured in the appropriate habitats; common = frequently observed in the appropriate habitats; uncommon = only occasionally observed in the appropriate habitats; rare = very few records for La Selva; extirpated = previously known from the area but no longer occurs in the region owing to overhunting and habitat destruction.

Appendix 8 (*continued*)

---

[c]Distribution: 1 = tropical, found throughout the New World tropics often being found as far north as Mexico, through Central America and much of tropical South America; 2 = northern Neotropics, generally found from tropical Mexico through Central America and northern South America; 3 = mesoamerica, found in the Central American countries but not as far south as Colombia; 4 = wide-ranging, found from the United States (and, in some cases, southern Canada) through Mexico, Central America, and most of South America to Argentina; 5 = North American, widely distributed in North America from southern Canada across most of the United States, Mexico, and Central America to Panama.

Acronyms

AAAS   American Association for the Advancement of Science

AOU   American Ornithologists' Union

ASBANA   (Costa Rican) Association for Banana Production

BCCR   Banco Central de Costa Rica

BCI   Barro Colorado Island

BCNP   Braulio Carrillo National Park

CAB   Commonwealth Agricultural Bureaux

CAMCORE   Central American and Mexican Coniferous Resources Cooperative

CATIE   Centro Agronómico Tropical de Investigación y Enseñanza

CITES   Convention on International Trade in Endangered Species

CODESA   Corporación Costarricense de Desarrollo

CNP   Consejo Nacional de Producción

CRNS   Centre National de la Recherche Scientifique

DGEC   Dirección General de Estadísticas y Censos

DGF   Dirección General Forestal

EARTH   Escuela Agropecuaria para la Región Tropical Húmeda

FAO   Food and Agriculture Organization

FDF   Fondo de Desarrollo Forestal

FORESTA   Forest Resources for a Stable Environment Project

FUNDECOR   Fundación de la Cordillera Central

GIS   Geographic Information System

GOES   Geostationary Operational Environmental Satellite

ICBP   International Council for Bird Preservation

ICRAF   International Council for Research in Agroforestry

IDA   Instituto de Desarrollo Agrario

IICA   Instituto Interamericano de Ciencias Agrícolas

INBio   Instituto Nacional de Biodiversidad

INDACO   Instituto de Desarrollo Agropecuario

INDERENA   Instituto Nacional de Recursos Naturales (Colombia)

ITCO   Instituto de Tierras y Colonización

IUFRO   International Union of Forest Research Organizations

JAPDEVA   Junta de Administración Portuaria y de Desarrollo Económico de la Vertiente Atlántico

LSAC   La Selva Advisory Committee

MAB   Man and the Biosphere Program (United Nations)

MAG   Ministerio de Agricultura

MIDEPLAN   Ministerio de Planificación Nacional y Política Económica

MIRENEM   Ministerio de Recursos Naturales, Energía, y Minas

NASA   National Aeronautics and Space Administration

NRC   National Research Council

NSF   National Science Foundation

OET   Organización para Estudios Tropicales (=OTS)

ORSTOM   Institut Français de Recherche Scientifique pour le Développement en Coopération

OTS   Organization for Tropical Studies (=OET)

RPSC   Río Palenque Science Center

RUTA   Unidad Regional de Asistencia Técnica

OTS   Organization for Tropical Studies

SEPSA   Secretaria Ejecutiva de Planificación Sectorial Agropecuaria y de Recursos Naturales Renovables

UCR   University of Costa Rica

UNESCO   United Nations Educational, Scientific, and Cultural Organization

USAID   United States Agency for International Development

USDA   United States Department of Agriculture

BIBLIOGRAPHY

Ackerman, J. D. 1986. Coping with the epiphytic existence: Pollination strategies. *Selbyana* 9:52–60.

Ackery, P. R. 1988. Hostplants and classification: A review of nymphalid butterflies. *Biol. J. Linn. Soc.* 33:95–203.

Ackery, P. R., and R. I. Vane-Wright. 1984. *Milkweed butterflies: Their cladistics and biology,* British Museum (Natural History). Ithaca, N.Y.: Cornell Univ. Press, Comstock Publishing Associates.

Acosta-Solms, M. 1979. The alkaloid content of the genus *Cinchona* in relation to altitude. Pp. 410–13 in *Tropical botany,* ed. K. Larsen and L. B. Holm-Nielsen. New York: Academic Press.

Aichinger, M. 1987. Annual activity patterns of anurans in a seasonal Neotropical environment. *Oecologia* 71:583–92.

Adis, J., K. Furch, and U. Irmler. 1979. Litter production of a central-Amazonian black water inundation forest. *J. Trop. Ecol.* 20:236–45.

Aide, T. M. 1987. Limbfalls: A major cause of sapling mortality for tropical forest plants. *Biotropica* 19:284–85.

———. 1988. Herbivory as a selective agent on the timing of leaf production in a tropical understory community. *Nature* 336:574–75.

Aide, T. M., and E. C. Londoño. 1989. The effects of rapid leaf expansion on the growth and survivorship of a lepidopteran herbivore. *Oikos* 55:66–70.

Akunda, E., and P. A. Huxley. 1990. The application of phenology to agroforestry research. ICRAF Working Paper no. 63:1–50.

Alexandre, D. Y. 1978. La role disséminateur des éléphants en forêt de Tai, Côte-d'Ivoire. *Terre Vie* 32:4772.

Ali, S. A. 1931. The role of the sunbirds and the flowerpeckers in the propagation and distribution of the treeparasite, *Loranthus longiflorus,* in the Konkan. *J. Bombay Nat. Hist. Soc.* 35:144–49.

Alpizar, E., B. Cornejo P., R. Reyes C., and A. Matamorros D. 1985. Evaluación-CORENA. Report, Ministerio de Agricultura y Ganadería, Dirección General Forestal, San José, Costa Rica.

Alston, E. R. 1879–1882. *Biologia Centrali-Americana. Mammalia.* London: Taylor and Francis.

Alvarado, A., and S. W. Buol. 1975. Toposequence relationships of Dystrandepts in Costa Rica. *Soil Sci. Soc. Am. Proc.* 39:932–37.

Alvarado, G. E. 1982. Notas geológicas de La Selva, Puerto Viejo, Heredia. Report, Organization for Tropical Studies, San José, Costa Rica.

Alvarado, G. 1985. Geología de la Estación Biológica La Selva. Report, Organization for Tropical Studies, San José, Costa Rica.

Alvarez-Buylla, E. R., and R. García-Barrios. 1991. Seed and forest dynamics: A theoretical framework and an example from the Neotropics. *Am. Nat.* 137:133–54.

Alverson, W. S. 1985. *Quararibea pumila* (Bombacaceae), a new endemic from Costa Rica. *Brittonia* 36:252–56.

Alvim, P. de T. 1960. Moisture stress as a requirement for flowering of coffee. *Science* 132:354.

Alvim, P. de T., and R. Alvim. 1978. Relation of climate to growth periodicity in tropical trees. Pp. 445–64 in *Tropical trees as living systems,* ed. P. B. Tomlinson and M. H. Zimmerman. Cambridge: Cambridge Univ. Press.

American Ornithologists' Union. 1983. *Check-list of North American birds.* 6th ed. Washington, D.C.: AOU.

Anderson, J. M., and J. S. I. Ingram. 1989. *TSBF: A handbook of methods.* Wallingford, England: CAB International.

Anderson, J. M., and M. J. Swift. 1983. Decomposition in tropical forests. Pp. 287–309 in *The tropical rain forest: Ecology and management,* ed. S. L. Sutton, T. C. Whitmore, and A. D. Chadwick. Oxford: Blackwell Scientific Publications.

Anderson, W. R. 1983. *Lophanthera,* a genus of Malpighiaceae new to Central America. *Brittonia* 35:37–41.

Andreae, M. O., and T. W. Andreae. 1988. The cycle of biogenic sulfur compounds over the Amazon Basin, Part 1. Dry Season. *J. Geophys. Res.* 93:1487–97.

Andrews, C. C. 1983. *Melinaea lilis imitata.* Pp. 736–38 in *Costa Rican Natural History,* ed. D. H. Janzen. Chicago: Univ. of Chicago Press.

Andrews, R. M. 1979. Evolution of life histories: A comparison of *Anolis* lizards from matched island and mainland habitats. *Breviora* 454:1–51.

———. 1988. Demographic correlates of variable egg survival for a tropical lizard. *Oecologia* 76:376–82.

Andrews, R. M., and A. S. Rand. 1974. Reproductive effort in anoline lizards. *Ecology* 55:1317–27.

———. 1982. Seasonal breeding and long-term population fluctuations in the lizard *Anolis limifrons.* Pp. 405–12 in *The ecology of a tropical forest: Seasonal rhythms and long-term changes,* ed. E. G. Leigh, Jr., A. S. Rand, and D. M. Windsor. Washington, D.C.: Smithsonian Institution Press.

———. 1989. Adición: Nuevas percepciones derivadas de la continuación de un estudio a largo plazo de la lagartija *Anolis limifrons.* Pp. 477–79 in *Ecología de un bosque tropical,* ed E. G. Leigh, Jr., A. S. Rand, and D. M. Windsor. Balboa, Panama: Smithsonian Tropical Research Institute.

Andrews, R. M., A. S. Rand, and S. Guerrero. 1983. Seasonal and spatial variation in the annual cycle of a tropical lizard. Pp. 441–54 in *Advances in herpetology and evolutionary biology: Essays in honor of E. E. Williams,* ed. G. J. Rhodin and K. Miyata. Cambridge: Harvard Univ. Press.

Annis, S. 1990. Debt and wrong-way resource flows in Costa Rica. *Ethics Int. Affairs* 4:107–21.

Anonymous. 1973. The functional role of leaf-cutter ants in a tropical ecosystem. OTS 73–2:18g–18j.

Appanah, S. 1981. Pollination in Malaysian primary forests. *Malay. For.* 44:37–42.

———. 1982. Pollination of androdioecious *Xerospermum intermedium* Radlk. (Sapindaceae) in a rain forest. *Biol. J. Linn. Soc.* 18:11–34.

———. 1985. General flowering in the climax rain forests of South-east Asia. *J. Trop. Ecol.* 1:225–40.

———. 1990. Plant-pollinator interactions in Malayasian rain forests. Pp. 85–101 in *Reproductive biology of tropical forest plants,* ed. K. S. Bawa and M. Hadley. Paris: UNESCO; Carnforth, England: Parthenon.

Appanah, S., and H. T. Chan. 1981. Thrips: The pollinators of some dipterocarps. *Malay. For.* 44:234–52.

Archer, S., and J. K. Detling. 1984. The effects of defoliation and competition on regrowth of tillers of two North American mixed-grass prairie graminoids. *Oikos* 43:351–57.

Armbruster, W. S. 1984. Two new species of *Dalechampia* (Euphorbiaceae) from Mesoamerica. *Syst. Bot.* 9:272–78.

———. 1986. Reproductive interactions between sympatric *Dalechampia* species: Are natural assemblages "random" or "organized"? *Ecology* 67:522–33.

Arnold, S. J. 1972. Species densities of predators and their prey. *Am. Nat.* 106:220–36.

Arroyo, J. 1990. Spatial variation of flowering phenology in the Mediterranean shrublands of southern Spain. *Isr. J. Bot.* 39:249–62.

Arroyo, M. T. K. 1979. Comments on breeding systems in Neotropical forests. Pp. 371–80 in *Tropical botany,* ed. K. Larsen and L. B. Holm-Nielson. New York: Academic Press.

———. 1981. Breeding systems and pollination biology in Leguminosae. Pp. 723–69, Part 2, in *Advances in legume systematics,* ed. R. M. Polhill and P. H. Raven. Kew, U.K.: Royal Botanic Gardens.

Artaxo-Neto P., C. Q. Orsini, L. C. Bouhres, and A. Leslie. 1982. Aspectos estruturais do aerossol atmosférico da Bacia amazonica. *Acta Amazonica* (suppl.) 12:39–46.

Ashton, P. S. 1964. Ecological studies on the mixed dipterocarp forests of Brunei state. Oxford Forestry Memoirs, 25. Oxford: Clarendon Press.

———. 1969. Speciation among tropical forest trees: Some deductions in light of recent evidence. *Biol. J. Linn. Soc.* 1:155–96.

———. 1976a. An approach to the study of breeding systems, population structure and taxonomy of tropical trees. Pp. 35–42 in *Tropical trees: Variation, breeding and conservation,* ed. J. Burley and B. T. Styles. London: Academic Press.

———. 1976b. Mixed dipterocarp forest and its variation with habitat in the Malayan lowlands: A re-evaluation at Pasoh. *Malay. For.* 39:56–72.

———. 1984. Biosystematics of tropical woody plants: A problem of rare species. Pp. 497–518 in *Plant biosystematics,* ed. W. F. Grant. New York: Academic Press.

———. 1988. Dipterocarp biology as a window to the understanding of tropical forest structure. *Annu. Rev. Ecol. Syst.* 19:347–70.

Ashton, P. S., Y.-Y. Gan, and F. W. Robertson. 1984. Electrophoretic and morphological comparisons in ten rain forest species of *Shorea* (Dipterocarpaceae). *Bot. J. Linn. Soc.* 89:293–304.

Ashton, P. S., T. J. Givnish, and S. Appanah. 1988. Staggered flowering in the Dipterocarpaceae: New insights into floral induction and the evolution of mast fruiting in the aseasonal tropics. *Am. Nat.* 132:44–66.

Ashton, P. S., E. Soepadmo, and S. K. Yap. 1977. Current research into the breeding systems of rain forest trees and its implications. Pp. 187–92 in Transactions of the International MAB-IUFRO Workshop on Tropical Rainforest Ecosystems and Resources. Hamburg-Reinbek.

Atlantic Zone Programme. 1987. Programa de investigación agropecuaria en la zona Atlántica de Costa Rica. Programme Document no. 3. Turrialba, Costa Rica: Centro Agronómico Tropical de Investigación y Enseñanza and Universidad Agrícola de Wageningen.

Atsatt, P. R. 1981. Lycaenid butterflies and ants: Selection for enemy-free space. *Am. Nat.* 118:638–54.

Atsatt, P. R., and D. J. O'Dowd. 1976. Plant defense guilds. *Science* 193:24–29.

Atwood, J. T. 1987. The vascular flora of La Selva Biological Station, Costa Rica; Orchidaceae. *Selbyana* 10:76–145.

Auerbach, M. J., and S. D. Hendrix. 1980. Insect-fern interactions: Macrolepidopteran utilization and species-area association. *Ecol. Entomol.* 5:99–104.

Auerbach, M. J., and D. R. Strong. 1981. Nutritional ecology of Heliconia herbivores: Experiments with plant fertilization and alternative hosts. *Ecol. Monogr.* 51:63–83.

Augspurger, C. K. 1979. Irregular rain cues and the germination and seedling survival of a Panamanian shrub *(Hybanthus prunifolius). Oecologia* 44:53–59.

———. 1980. Mass-flowering of a tropical shrub *(Hybanthus prunifolius):* Influence on pollinator attraction and movement. *Evolution* 34:475–88.

———. 1981. Reproductive synchrony of a tropical shrub: Experimental studies on effects of pollinators and seed predators on *Hybanthus prunifolius* (Violaceae). *Ecology* 62:775–88.

———. 1982. A cue for synchronous flowering. Pp. 133–49 in *The ecology of a tropical forest: Seasonal rhythms and long-term changes,* ed. E. G. Leigh, Jr., A. S. Rand, and D. M. Windsor. Washington, D.C.: Smithsonian Institution Press.

———. 1983a. Offspring recruitment around tropical trees: Changes in cohort distance with time. *Oikos* 40:189–96.

———. 1983b. Phenology, flowering synchrony, and fruit set of six Neotropical shrubs. *Biotropica* 15:257–67.

———. 1983c. Seed dispersal of the tropical tree, *Platy-*

podium elegans, and the escape of its seedlings from fungal pathogens. *J. Ecol.* 71:759–71.

———. 1984a. Light requirements of Neotropical tree seedlings: A comparative study of growth and survival. *J. Ecol.* 72:777–96.

———. 1984b. Seedling survival among tropical tree species: Interactions of dispersal distance, light-gaps, and pathogens. *Ecology* 65:1705–12.

Augspurger, C. K., and S. E. Franson. 1988. Input of wind-dispersed seeds into light-gaps and forest sites in a Neotropical forest. *J. Trop. Ecol.* 4:239–52.

Augspurger, C. K., and C. K. Kelly. 1984. Pathogen mortality of tropical tree seedlings: Experimental studies of the effects of dispersal distance, seedling density, and light conditions. *Oecologia* 61:211–17.

Austin, M. P., P. S. Ashton, and P. Greig-Smith. 1972. The application of quantitative methods to vegetation survey. Part 3. A re-examination of rain forest data from Brunei. *J. Ecol.* 60:305–24.

Baillie, I. C., P. S. Ashton, M. N. Court, J. A. R. Anderson, E. A. Pitzpatrick, and J. Tinsley. 1987. Site characteristics and the distribution of tree species in mixed dipterocarp forest on Tertiary sediments in central Sarawak, Malaysia. *J. Trop. Ecol.* 3:201–20.

Baker, H. G. 1965. The evolution of the cultivated kapok tree: A probable West African product. Pp. 185–216 in *Ecology and economic development in tropical Africa,* ed. D. Brokesha. Institute for International Studies, Research Series no. 9. Berkeley: Univ. of California Press.

———. 1970. Evolution in the tropics. *Biotropica* 2:101–11.

———. 1973. Evolutionary relationships between flowering plants and animals in American and African tropical forests. Pp. 145–60 in *Tropical forest ecosystems in Africa and South America: A comparative review,* ed. B. J. Meggers, E. S. Ayensu, and W. D. Duckworth. Washington, D.C.: Smithsonian Institution Press.

Baker, H. G., K. S. Bawa, G. W. Frankie, and P. A. Opler. 1983. Reproductive biology of plants in tropical forests. Pp. 183–215 in *Ecosystems of the world.* Vol. 14A, *Tropical rain forest ecosystems: Structure and function,* ed. F. B. Golley. New York: Elsevier Scientific.

Bakus, G. J. 1969. Energetics and feeding in shallow marine waters. *Int. Rev. Gen. Exp. Zool.* 4:275–369.

———. 1974. Toxicity in holothurians: A geographical pattern. *Biotropica* 6:229–36.

Bakus, G. J., and G. Green. 1974. Toxicity in sponges and holothurians: A geographic pattern. *Science* 185:951–53.

Baldwin, I. T., and J. C. Schultz. 1988. Phylogeny and the patterns of leaf phenolics in gap- and forest-adapted *Piper* and *Miconia* understory shrubs. *Oecologia* 75:105–9.

Balick, M. J., D. G. Furth, and G. Cooper-Driver. 1978. Biochemical and evolutionary aspects of arthropod predation on ferns. *Oecologia* 35:55–89.

Barahona R., F. 1980. *Reforma agraria y poder político.* San José: Editorial Universidad de Costa Rica.

Barinaga, M. 1990. Where have all the froggies gone? *Science* 247:1033–34.

Barringer, K. 1983. Notes on Central American Aristolochiaceae. *Brittonia* 35:171–74.

Barry, R. G., and R. J. Chorley. 1987. *Atmosphere, weather and climate.* New York: Methuen.

Barthélémy, D. 1986. Relation entre la position des complexes réitérés sur un arbre et l'expression de leur florasion: L'exemple de trois espèces tropicales, pp. 71–100, Naturalia Monpeliensia, Colloque International sur L'arbre.

Barton, A. M. 1984. Neotropical pioneer and shade-tolerant tree species: Do they partition tree-fall gaps? *J. Trop. Ecol.* 25:196–202.

Barton, A. M., N. Fetcher, and S. Redhead. 1989. The relationship between treefall gap size and light flux in a Neotropical rain forest in Costa Rica. *J. Trop. Ecol.* 5:437–39.

Bates, H. W. 1859. Notes on South American butterflies. *Trans. Entomol. Soc. London* 5:1–11.

———. 1862. Contributions to an insect fauna of the Amazon Valley, Lepidoptera: Heliconidae. *Trans. Entomol. Soc. London* 5:1–11.

Bawa, K. S. 1974. Breeding systems of tree species of a lowland tropical community. *Evolution* 28:85–92.

———. 1979. Breeding systems of trees in a tropical wet forest. *N. Z. J. Bot.* 17:521–24.

———. 1980a. Evolution of dioecy in flowering plants. *Annu. Rev. Ecol. Syst.* 11:15–39.

———. 1980b. Mimicry of male by female flowers and intrasexual competition for pollinators in *Jacaratia dolichaula* (D. Smith) Woodson (Caricaceae). *Evolution* 34:467–74.

———. 1983. Patterns of flowering in tropical plants. Pp. 394–410 in *Handbook of experimental pollination biology,* ed. C. E. Jones and R. J. Little. New York: Van Nostrand Reinhold.

———. 1990. Plant-pollinator interactions in tropical lowland rain forest. *Annu. Rev. Ecol. Syst.* 21.

Bawa, K. S., and P. S. Ashton. 1991. Conservation of rare plant species in tropical rain forests: A genetic perspective. Pp. 62–71 in *Genetics and conservation of rare plants,* ed. D. Falk and K. Holsinger. New York: Oxford Univ. Press.

Bawa, K. S., P. S. Ashton, and S. M. Nor. 1990. Reproductive ecology of tropical forest plants: Management issues. Pp. 3–13 in *Reproductive ecology of tropical forest plants,* ed. K. S. Bawa and M. Hadley. Paris: UNESCO; Carnforth, U.K.: Parthenon.

Bawa, K. S., and J. H. Beach. 1981. Evolution of sexual systems in flowering plants. *Ann. MO Bot. Gard.* 68:254–74.

———. 1983. Self-incompatibility systems in the Rubiaceae of a tropical lowland wet forest. *Am. J. Bot.* 70:1281–88.

Bawa, K. S., S. H. Bullock, D. R. Perry, R. E. Coville, and M. H. Grayum. 1985. Reproductive biology of tropical lowland rain forest trees. Part 2. Pollination systems. *Am. J. Bot.* 72:346–56.

Bawa, K. S., and J. E. Crisp. 1980. Wind-pollination in the understory of a rain forest in Costa Rica. *J. Ecol.* 68:871–76.

Bawa, K. S., and M. Hadley, eds. 1990. *Reproductive ecology of tropical forest plants.* Man and the Biosphere series, vol. 7. Paris: UNESCO; Carnforth, U.K.: Parthenon.

Bawa, K. S., and D. M. O'Malley. 1987. Estudios genéticos de sistemas de cruzamiento en algunas especies arboreas de bosques tropicales. *Rev. Biol. Trop.* 35 (Suppl. 1): 177–88.

Bawa, K. S., and P. A. Opler. 1975. Dioecism in tropical forest trees. *Evolution* 29:167–79.

Bawa, K. S., D. R. Perry, and J. H. Beach. 1985. Reproductive biology of tropical lowland rain forest trees. Part 1. Sexual systems and incompatibility mechanisms. *Am. J. Bot.* 72:331–45.

Baynton, H. W., H. L. Hamilton, Jr., P. E. Sheer, and J. J. B. Worth. 1965. Temperature structure in and above a tropical forest. *J. R. Meteorol. Soc.* 91:225–32.

Bazzaz, F. A. 1979. The physiological ecology of plant succession. *Annu. Rev. Ecol. Syst.* 10:351–71.

———. 1983. Characteristics of populations in relation to disturbance in natural and man-modified ecosystems. Pp. 259–75 in *Disturbance and ecosystems: Components of response,* ed. H. A. Mooney and M. Godron. Berlin: Springer-Verlag.

———. 1984. Dynamics of wet tropical forests and their species strategies. Pp. 233–43 in *Physiological ecology of plants of the wet tropics,* ed. E. Medina, H. A. Mooney, and C. Vázquez-Yánes. The Hague: W. Junk.

Bazzaz, F. A., and S. T. A. Pickett. 1980. Physiological ecology of tropical succession: A comparative review. *Annu. Rev. Ecol. Syst.* 11:287–310.

Beach, J. H. 1981. Pollinator foraging and the evolution of dioecy. *Am. Nat.* 118:572–77.

———. 1982. Beetle pollination of *Cyclanthus bipartitus* (Cyclanthaceae). *Am. J. Bot.* 69:1074–81.

———. 1983. *Posoqueria latifolia.* Pp. 307–8 in *Costa Rican natural history,* ed. D. H. Janzen. Chicago: Univ. of Chicago Press.

———. 1984. The reproductive biology of the peach or "pejibaye" palm *(Bactris gasipaes)* and a wild congener *(B. porschiana)* in the Atlantic lowlands of Costa Rica. *Principes* 28:107–19.

Beach, J. H., and K. S. Bawa. 1980. Role of pollinators in the evolution of dioecy from distyly. *Evolution* 34:1138–42.

Beaman, R. S., P. J. Decker, and J. H. Beaman. 1988. Pollination of *Rafflesia* (Rafflesiaceae). *Am. J. Bot.* 75:1148–62.

Beard, J. S. 1955. The classification of tropical American vegetation-types. *Ecology* 36:89–100.

Beattie, A. J. 1985. The evolutionary ecology of antplant mutualisms. Cambridge: Cambridge Univ. Press.

———. 1989. Myrmecotrophy: Plants fed by ants. *Trends in Ecol. Evol.* 4:172–76.

Beaver, R. A. 1979. Host specificity of temperate and tropical animals. *Nature* 281:139–41.

Becker, P. 1983. Effects of insect herbivory and artificial defoliation on survival of *Shorea* seedlings. Pp. 241–52 in *Tropical rain forest: Ecology and management,* ed. S. L. Sutton, T. C. Whitmore, and A. C. Chadwick. Oxford: Blackwell Scientific.

Becker, P., P. E. Rabenold, J. R. Idol, and A. P. Smith. 1988. Water potential gradients for gaps and slopes in a Panamanian tropical moist forest's dry season. *J. Trop. Ecol.* 4:173–84.

Becker, P., and M. Wong. 1985. Seed dispersal, seed predation, and juvenile mortality of *Aglaia* sp. (Meliaceae) in lowland Dipterocarp rainforest. *Biotropica* 17:230–37.

Becker, R. A., J. M. Chambers, and A. R. Wilks. 1988. *The new S language.* Pacific Grove, Calif.: Wadsworth and Brooks/Cole.

Beehler, B. 1981. Ecological structuring of forest bird communities in New Guinea. *Monogr. Biol.* 42:837–61.

Belsky, A. J. 1986. Does herbivory benefit plants? A review of the evidence. *Am. Nat.* 127:870–92.

Benson, W. W. 1978. Resource partitioning in passion flower butterflies. *Evolution* 32:493–518.

———. 1985. Amazon ant-plants. Pp. 239–66 in *Key environments: Amazonia,* ed. G. H. Prance and T. E. Lovejoy. New York: Pergamon.

Benson, W. W., K. S. Brown, Jr., and L. E. Gilbert. 1975. Coevolution of plants and herbivores: Passion flower butterflies. *Evolution* 29:659–80.

Bentley, B. L. 1976. Plants bearing extrafloral nectaries and the associated ant community: Interhabitat differences in the reduction of herbivore damage. *Ecology* 57:815–20.

———. 1977. Extraflora nectaries and protection by pugnacious bodyguards. *Annu. Rev. Ecol. Syst.* 8:407–27.

Bentley, B. L., and E. J. Carpenter. 1980. The effects of desiccation and rehydration on nitrogen fixation by epiphylls in a tropical rainforest. *Microb. Ecol.* 6:109–13.

———. 1984. Direct transfer of newly fixed nitrogen from free-living epiphyllous microorganisms to their host plant. *Oecologia* 63:52–56.

Bentley, S., and J. B. Whittaker. 1979. Effects of grazing by a chrysomelid beetle *Gastrophysa viridula,* on competition between *Rumex obtusifolius* and *Rumex crispus. J. Ecol.* 67:79–90.

Berenbaum, M. R. 1984. Effects of tannins on growth and digestion in two species of papilionids. *Entomol. Exp. Appl.* 34:245–50.

Berg, C. C. 1989. Classification and distribution of *Ficus. Experientia* 45:605–11.

Bergmark, C. 1985. The river as a potential food source in the Sarapiquí region. Report, OTS 85-4:68.

Bernard-Reversat, F., G. Huttel, and G. Lemee. 1978. Structure and functioning of evergreen rainforest ecosystems of the Ivory Coast. Pp. 557–74 in *Tropical forest ecosystems, A state of knowledge report.* Vendôme, France: Presses Universitaires de France.

Bernier, G. 1988. The control of floral evocation and morphogenesis. *Annu. Rev. Plant Physiol. Plant Mol. Biol.* 39:175–219.

Bernier, G., J.-M. Kinet, and R. M. Sachs. 1981a. *The physiology of flowering.* Vol. 1, *The initiation of flowers.* Boca Raton, Fla.: CRC Press.

———. 1981b. *The physiology of flowering.* Vol. 2, *Transition to reproductive growth.* Boca Raton, Fla.: CRC Press.

Bertsch, F. H. 1986. Manual para interpretar la fertilidad de los suelos en Costa Rica. Report, Escuela de Fitotecnia, Universidad de Costa Rica, San José.

Bertsch, F., and A. Cordero. 1984. Fertilidad de typic dystrandepts de Costa Rica. Part 2. Aniones (N, P, B, S, Mo), materia orgánica y textura. *Turrialba* 34:199–205.

Bertsch, F., A. Cordero, and A. Alvarado. 1984. Fertilidad de typic dystrandepts de Costa Rica. Part 1. Metodología, acidez y cationes (Ca, Mg, K, Fe, Mn, Zn, Cu). *Turrialba* 34:187–97.

Bertsch, F. H., and V. Vega. 1990. Dinámica de nutrimentos en un sistema de producción con bajos insumos en un typic dystropept del trópico muy húmedo, Río Frío, Heredia, Costa Rica. Report, 2d Taller Latinoamericano de Manejo de Suelos Tropicales, Facultad de Agronomía, Universidad de Costa Rica, San José.

Bethel, J. S. 1976. Forests in Central America and Panama: Which kind, how large, and where? *Rev. Biol. Trop.* (suppl. 1): 143–75.

Bien, A. R. 1982. Substrate specificity of leafy liverworts (Hepaticae: Lejeuneaceae) in a Costa Rican rainforest. Master's thesis, State University of New York, Stony Brook.

Bierregaard, R. O., Jr. 1990. Species composition and trophic organization of the understory bird community in a central Amazonian terra firme forest. Pp. 217–36 in *Four Neotropical rainforests,* ed. A. H. Gentry. New Haven, Conn.: Yale Univ. Press.

Binkley, D., and D. Richter. 1987. Nutrient cycles and H+ budgets of forest ecosystems. *Adv. Ecol. Res.* 16:1–51.

Birch, H. F. 1958. The effect of soil drying on humus decomposition and nitrogen availability. *Plant Soil* 10:9–13.

Bishop, J. E. 1973. *Limnology of a small Malayan river Sungai Gombak.* The Hague: W. Junk.

Bjvrkman, O., and M. M. Ludlow. 1972. Characterization of the light climate on the floor of a Queensland rainforest. *Carnegie Inst. Washington Year Book* 71:85–94.

Bjvrkman, O., M. M. Ludlow, and P. S. Morrow. 1972. Photosynthetic performance of rainforest species in their native habitat: An analysis of gas exchange. *Carnegie Inst. Washington Year Book* 71:94–102.

Black, R. W. 1987. The biology of leaf nesting ants in a tropical wet forest. *Biotropica* 19:319–25.

Blake, J. G., and W. G. Hoppes. 1986. Influence of resource abundance on use of treefall gaps by birds in an isolated woodlot. *Auk* 103:328–40.

Blake, J. G., and B. A. Loiselle. 1991. Variation in resource abundance affects capture rates of birds in three lowland habitats in Costa Rica. *Auk* 108:114–30.

Blake, J. G., F. G. Stiles, and B. A. Loiselle. 1990. Birds of La Selva Biological Station: Habitat use, trophic composition, and migrants. Pp. 161–82 in *Four Neotropical rainforests,* ed. A. Gentry. New Haven, Conn.: Yale Univ. Press.

Blakely, N. R., and H. Dingle. 1978. Competition: Butterflies eliminate milkweed bugs from a Caribbean island. *Oecologia* 37:133–36.

Blance, C. A., J. D. Hodges, A. E. Gómez, and E. González. 1991. Seed chemistry of the tropical tree *Vochysia hondurensis* Sprague. *For. Sci.* 37:949–52.

Boardman, N. K. 1977. Comparative photosynthesis of sun and shade plants. *Annu. Rev. Plant Physiol.* 28:355–77.

Boggs, C. L. 1979. Resource allocation and reproductive strategies in heliconiine butterflies. Ph.D. diss., University of Texas, Austin.

———. 1981a. Nutritional and life-history determinants of resource allocation in holometabolous insects. *Am. Nat.* 117:692–709.

———. 1981b. Selection pressures affecting male nutrient investment at mating in heliconiine butterflies. *Evolution* 35:931–40.

Boggs, C. L., and L. E. Gilbert. 1979. Male contribution to egg production in butterflies: Evidence for transfer of nutrients at mating. *Science* 206:83–84.

Boggs, C. L., J. Smiley, and L. E. Gilbert. 1981. Patterns of pollen exploitation by *Heliconius* butterflies. *Oecologia* 48:284–89.

Bonaccorso, F. J. 1979. Foraging and reproductive ecology in a Panamanian bat community. *Bull. Fla. State Mus., Biol. Sci.* 24:359–408.

Bonaccorso, F. J., and T. J. Gush. 1987. Feeding behaviour and foraging strategies of captive phyllostomid fruit bats: An experimental study. *J. Anim. Ecol.* 56:907–20.

Bonaccorso, F. J., and S. R. Humphrey. 1984. Fruit bat niche dynamics: Their role in maintaining tropical forest diversity. Pp. 169–83 in *Tropical rainforest: The Leeds symposium,* ed. A. C. Chadwick and S. L. Sutton. Leeds, U.K.: Leeds Philosophical and Literary Society.

Boppré, M. 1984. Chemically mediated interactions between butterflies. *Symp. R. Entomol. Soc. Lond.* 11:259–75.

———. 1990. Lepidoptera and pyrrolizidine alkaloids. *J. Chem. Ecol.* 16:165–85.

Borchelt, R. 1990. Vanishing species: Frogs, toads and other amphibians in distress. *News Report National Research Council* 40:2–5.

Borchert, R. 1973. Simulation of rhythmic tree growth under constant conditions. *Physiol. Plant.* 29:173–80.

———. 1978. Feedback control and age-related changes of shoot growth in seasonal and nonseasonal climates. Pp. 497–515 in *Tropical trees as living systems,* ed. P. B. Tomlinson and M. H. Zimmerman. Cambridge: Cambridge Univ. Press.

———. 1980. Phenology and ecophysiology of tropical trees: *Erythrina poeppigiana* O. F. Cook. *Ecology* 61:1065–74.

———. 1983. Phenology and control of flowering in tropical trees. *Biotropica* 15:81–89.

———. 1991. Growth periodicity and dormancy. Pp. 219–43 in *Physiology of trees,* ed. A. S. Raghavendra. New York: J. Wiley.

Borges, R., and C. Propper. 1983. Diet selectivity and insectivory in *Heteromys desmarestianus.* OTS 83-3:380–83.

Boring, L. 1979. Seed predation in *Guarea rhopalocarpa.* OTS 79-1:3–5.

Bormann, F. H., and G. Berlyn, eds. 1981. *Age and growth rate of tropical trees: New directions for research.* Yale Univ. Sch. For. Environ. Stud. Bull. no. 94, New Haven, Conn.

Borowicz, V. A. 1988. Do vertebrates reject decaying fruit? An experimental test with *Cornus amomum* fruits. *Oikos* 53:74–78.

Borror, D. J., D. M. DeLong, and C. A. Triplehorn. 1981. *An introduction to the study of insects.* 5th ed. New York: Holt, Rinehart, and Winston.

Boucher, D. H. 1981. The "real" disperser of *Swartzia cubensis. Biotropica* (suppl.) 12:77–78.

Boucher, D. H., J. H. Vandermeer, K. Yih, and N. Zamora.

1990. Contrasting hurricane damage in tropical rain forest and pine forest. *Ecology* 71:2022–24.

Bourgeois, W. W., D. W. Cole, H. Riekerk, and S. P. Gessel. 1972. Geology and soils of comparative ecosystem study areas, Costa Rica. Contribution 11, Tropical Forestry Series, Institute of Forest Products, University of Washington, Seattle.

Bourlière, F. 1989. Mammalian species richness in tropical rainforests. Pp. 153–68 in *Ecological studies*. Vol. 69, *Vertebrates in complex tropical systems*, ed. M. L. Harmelin-Vivien and F. Bourlière. New York: Springer-Verlag.

Bowers, M. D., and Z. Larin. 1989. Acquired chemical defense in the lycaenid butterfly *Eumaeus atala*. *J. Chem. Ecol.* 15:1133–46.

Boyce, M. S. 1984. Restitution of r- and K-selection as a model of density-dependent natural selection. *Annu. Rev. Ecol. Syst.* 15:427–47.

Bradbury, J. W., and S. L. Vehrencamp. 1976a. Social organization and foraging in emballonurid bats. Part 1. Field studies. *Behav. Ecol. Sociobiol.* 1:337–82.

———. 1976b. Social organization and foraging in emballonurid bats. Part 2. A model for the determination of group size. *Behav. Ecol. Sociobiol.* 1:383–404.

———. 1977a. Social organization and foraging in emballonurid bats. Part 3. Mating systems. *Behav. Ecol. Sociobiol.* 2:1–17.

———. 1977b. Social organization and foraging in emballonurid bats. Part 4. Parental investment patterns. *Behav. Ecol. Sociobiol.* 2:19–29.

Bradshaw, A. D. 1974. Environment and phenotypic plasticity. Pp. 75–94 In *Basic mechanisms in plant morphogenesis*, Brookhaven Symp. Biol. 25. Upton, N.Y.: Brookhaven National Laboratory.

Brady, N. C. 1984. *The nature and properties of soils*. 9th ed. New York: Macmillan.

Braithwaite, R. W. 1987. Effects of fire regimes on lizards in the wet-dry tropics of Australia. *J. Trop. Ecol.* 4:77–88.

Braithwaite, R. W., L. Miller, and J. T. Wood. 1988. The structure of termite communities in the Australian tropics. *Aust. J. Ecol.* 13:375–91.

Braker, E. 1991. Natural history of a Neotropical gap-inhabiting grasshopper. *Biotropica* 23:41–50.

Braker, E., and R. L. Chazdon. 1993. Ecological, behavioural, and nutritional factors influencing use of palms as host plants by a Neotropical forest grasshopper. *J. Trop. Ecol.* 9(2):181–95.

Braker, H. E. 1986. Host plant relationships of the Neotropical grasshopper *Microtylopteryx hebardi* Rehn (Acrididae: Ommatolampinae). Ph.D. diss., University of California, Berkeley.

———. 1989a. Evolution and ecology of oviposition on host plants by acridoid grasshoppers. *Biol. J. Linn. Soc.* 38:389–406.

———. 1989b. Oviposition on host plants by a tropical forest grasshopper (*Microtylopteryx herbardi*: Acrididae). *Ecol. Entomol.* 14:141–48.

Brame, A. H., Jr. 1968. Systematics and evolution of the Mesoamerican salamander genus *Oedipina*. *J. Herpetol.* 2:1–64.

Brandani, A., G. S., Hartshorn, and G. H. Orians. 1988. Inter-

nal heterogeneity of gaps and species richness in Costa Rican tropical wet forest. *J. Trop. Ecol.* 4:99–119.

Bray, J. R. 1961. Measurement of leaf utilization as an index of minimum level of primary consumption. *Oikos* 12:70–74.

———. 1964. Primary consumption in three forest canopies. *Ecology* 45:165–67.

Breitsprecher, A., and J. S. Bethel. 1990. Stem-growth periodicity of trees in a tropical wet forest of Costa Rica. *Ecology* 71:1156–64.

Breitswisch, R. 1982. Lekking activity of the white-collared manakin (Pipridae: *Manacus candei*). Report, OTS 823:114–17.

Brinson, M. M., H. D. Bradshaw, and R. N. Holmes. 1983. Significance of floodplain sediments in nutrient exchange between a stream and its floodplain. Pp. 199–221 in *Dynamics of lotic ecosystems*, ed. T. D. Fontaine III and S. M. Bartell. Ann Arbor, Mich.: Ann Arbor Science Publishers.

Briskie, J. V., and S. G. Sealy. 1987. Polygyny and double-brooding in the least flycatcher. *Wilson Bull.* 99:492–94.

Broadhead, E. 1983. The assessment of faunal diversity and guild size in tropical forests with particular reference to the Psocoptera. Pp. 107–19 in *Tropical rain forest: Ecology and management*, ed. S. L. Sutton, T. C. Whitmore, and A. C. Chadwick. Oxford: Blackwell Scientific Publications.

Brockman, C. F. 1968. *Trees of North America*. New York: Golden Press.

Brokaw, N. V. L. 1982. Treefalls: Frequency, timing, and consequences. Pp. 101–8 in *The ecology of a Neotropical forest: Seasonal rhythms and longer-term changes*, ed. E. G. Leigh, Jr., A. S. Rand, and D. M. Windsor. Washington, D.C.: Smithsonian Institution Press.

———. 1983. Groundlayer dominance and apparent inhibition of tree regeneration by *Aechmea magdalenae* (Bromeliaceae) in a tropical forest. *Trop. Ecol.* 24:194–200.

———. 1985a. Gap-phase regeneration in a tropical forest. *Ecology* 66:682–87.

———. 1985b. Treefalls, regrowth, and community structure in tropical forests. Pp. 53–69 in *The ecology of natural disturbance and patch dynamics*, ed. S. T. A. Pickett and P. S. White. Orlando, Fla.: Academic Press.

———. 1986. Seed dispersal, gap colonization, and the case of *Cecropia insignis*. Pp. 323–31 in *Frugivores and seed dispersal*, ed. A. Estrada and T. H. Fleming. Dordrecht: W. Junk.

———. 1987. Gap-phase regeneration of three pioneer tree species in a tropical forest. *J. Ecol.* 75:9–20.

Bronchart, R. 1963. Recherches sur le développement de *Geophila renaris* de Wild. et Th. Dur. dans les conditions écologiques d'un sous-bois forestier équatorial. Influence sur la mise a fleurs d'une perte en eau disponible du sol. *Mèm. Soc. R. Sci. Liège*, 5th ser. 8:1–179.

Bronstein, J. L. 1989. A mutualism at the edge of its range. *Experientia* 45:622–37.

Bronstein, J. L., P. H. Gouyon, C. Gliddon, F. Kjellberg, G. Michaloud. 1990. The ecological consequences of flowering asynchrony in monoecious figs: A simulation study. *Ecology* 71:2145–56.

Brooijmans, W. J. A. M. 1988. La colonización espontánea de

Cocorí, Zona Atlántica de Costa Rica: Un estudio con en-foque socio-histórico. Field Report no. 21, Atlantic Zone Programme CATIE-Agricultural University Wageningen, Ministerio de Agricultura y Ganadería, Costa Rica.

Brooke, A. P. 1987a. The natural history of the Honduran white bat, *Ectophylla alba,* in Costa Rica. Master's thesis, Boston University, Boston.

———. 1987b. Tent construction and social organization in *Vampyressa nymphaea* (Chiroptera: Phyllostomatidae) in Costa Rica. *J. Trop. Ecol.* 3:171–75.

———. 1990. Tent selection, roosting ecology and social organization of the tent-making bat, *Ectophylla alba,* in Costa Rica. *J. Zool. (Lond.)* 221:11–19.

Brooks, D. R., and D. A. McLennan. 1991. *Phylogeny, ecology, and behavior: A research program in comparative biology.* Chicago: Univ. of Chicago Press.

Brosset, A., and C. Erard. 1986. Les oiseaux des regions forestieres du nordest du Gabon. Vol. 1, Ecologie et comparte-ment des espèces. *Rev. Ecol. Suppl.* 3:129.

Brower, L. P. 1984. Chemical defenses in butterflies. *Symp. R. Entomol. Soc. Lond.* 11:110–34.

Brown, J. H. 1981. Two decades of homage to Santa Rosalia: Toward a general theory of diversity. *Am. Zool.* 21:877–88.

Brown, J. H., and D. W. Davidson. 1977. Competition be-tween seed-eating rodents and ants in desert ecosystems. *Science* 196:880–82.

Brown, J. H., D. W. Davidson, and O. J. Reichman. 1979. An experimental study of competition between seed-eating desert rodents and ants. *Am. Zool.* 19:1129–43.

Brown, J. L. 1987. *Helping and communal breeding in birds: Ecology and evolution.* Princeton, N.J.: Princeton Univ. Press.

Brown, K. S., Jr. 1984. Adult-obtained pyrrolizidine alkaloids defend ithomiine butterflies against a predatory spider. *Nature* 309:707–9.

Brown, S., and A. E. Lugo. 1982. The storage and production of organic matter in tropical forests and their role in the global carbon cycle. *Biotropica* 14:161–87.

Browne, F. G. 1933. A note on the defoliation of sendok-sendok *(Endospermum malaccense).* Malay. For. 6:267–69.

———. 1968. *Pests and diseases of forest plantation trees.* Oxford: Oxford Univ. Press.

Brues, C. T. 1920. The selection of food plants by insects, with special reference to lepidopterous larvae. *Am. Nat.* 54:313–32.

———. 1946. *Insect dietary.* Cambridge: Harvard Univ. Press.

Bryant, J. P., F. S. Chapin III, and D. R. Klein. 1983. Carbon/nutrient balance of boreal plants in relation to vertebrate herbivory. *Oikos* 40:357–68.

Bryant, J. P., F. S. Chapin III, P. B. Reichardt, and T. P. Clausen. 1987. Response of winter chemical defense in Alaska paper birch and green alder to manipulation of plant carbon/nutrient balance. *Oecologia* 72:510–14.

Bryant, J. P., T. P. Clausen, P. B. Reichardt, M. C. McCarthy, and R. A. Werner. 1987. Effect of nitrogen fertilization upon the secondary chemistry and nutritional value of quaking aspen (*Populus tremuloides* Michx.) leaves for the large aspen tortrix (*Choristoneura conflictana* [Walker]). *Oecologia* 73:513–17.

Bryant, J. P., P. J. Kuropat, and S. M. Cooper. 1989. Resource availability hypothesis of plant antiherbivore defence tested in South African savanna ecosystem. *Nature* 340:227–29.

Bucher, T. H., M. J. Ryan, and G. A. Bartholomew. 1982. Oxygen consumption during resting, calling, and nest build-ing in the frog *Physalaemus pustulosus. Physiol. Zool.* 55:10–22.

Buchholz, R., and D. J. Levey. 1990. The evolutionary triad of microbes, fruits, and seed dispersers: An experiment in fruit choice by cedar waxwings, *Bombycilla cedrorum. Oikos* 59:200–204.

Buchmann, S. L. 1983. Buzz pollination in angiosperms. Pp. 73–113 in *Handbook of experimental pollination biology,* ed. C. E. Jones and R. J. Little, New York: Van Nostrand Reinhold.

Buckley, D. P., D. M. O'Malley, V. Apsit, G. T. Prance, and K. S. Bawa. 1988. Genetics of Brazil nut (*Bertholletia excelsa* Humb. and Bonpl.: Lecythidaceae). Part 1. Genetic variation in natural populations. *Theor. Appl. Genet.* 76:923–28.

Buckley, R. C. 1983. Interactions involving plants, Homoptera, and ants. *Oecologia* 58:132–36.

Budowski, G. 1961. Studies on forest succession in Costa Rica and Panama. Ph.D. diss., Yale University, New Haven, Conn.

———. 1965. Distribution of tropical American rain forest species in the light of successional processes. *Turrialba* 15:40–42.

Bullock, S. H. 1980. Demography of an undergrowth palm in littoral Cameroon. *Biotropica* 12:247–55.

———. 1981. Notes on the phenology of infloresences and pollination of some rain forest palms in Costa Rica. *Principes* 25:101–5.

———. 1982. Population structure and reproduction in the Neotropical dioecious tree *Compsoneura sprucei.* Oecologia 55:238–42.

———. 1985. Breeding systems in the flora of a tropical de-ciduous forest in Mexico. *Biotropica* 17:287–301.

———. 1986. Allometrics and biomass allocation in *Socratea durissima* (Arecaceae). *Brenesia* 25/26:297–303.

Bullock, S. H., and K. S. Bawa. 1981. Sexual dimorphism and the annual flowering pattern in *Jacaratia dolichaula* (D. Smith) Woodson (Caricaceae) in a Costa Rican rain forest. *Ecology* 62:1494–1504.

Bullock, S. H., J. H. Beach, and K. S. Bawa. 1983. Episodic flowering and sexual dimorphism in *Guarea rhopalocarpa* in a Costa Rican rain forest. *Ecology* 64:851–61.

Bullock, S. H., and J. A. Solís-Magallanes. 1990. Phenology of canopy trees of a tropical deciduous forest in Mexico. *Biotropica* 22:22–35.

Bunnell, P. 1973. Vocalizations in the territorial behavior of the frog *Dendrobates pumilio. Copeia* 1973:277–84.

Buol, S. W., F. D. Hole, and R. J. McCracken. 1980. *Soil genesis and classification.* 2d ed. Ames: Iowa State Univ. Press.

Burcham, J. 1985. Fish communities and environmental char-acteristics of two lowland streams in Costa Rica. Master's thesis, Texas A & M University, College Station.

———. 1988. Fish communities and environmental characteristics of two lowland streams in Costa Rica. *Rev. Biol. Trop.* 36:273–85.

Burger, W. 1971. Piperaceae. In *Flora Costaricensis,* ed. W. Burger. *Fieldiana Bot.* 35:5–227.

———. 1977a. Chloranthaceae. In *Flora Costaricensis,* ed. W. Burger. *Fieldiana Bot.* 40:1–10.

———. 1977b. Moraceae. In *Flora Costaricensis,* ed. W. Burger. *Fieldiana Bot.* 40:94–215.

———. 1977c. Urticaceae. In *Flora Costaricensis,* ed. W. Burger. *Fieldiana Bot.* 40:218–83.

———. 1980. Why are there so many kinds of flowering plants in Costa Rica? *Brenesia* 17:371–88.

Busby, B., J. Beach, R. Chazdon, J. Howard, A. Kurta, and M. Raveret. 1980. Limitations on plant height in *Asterogyne martiana.* Report, OTS 80-3:28–31.

Buschbacher, R. J. 1990. Natural forest management in the humid tropics: Ecological, social, and economic considerations. *Ambio* 19:253–58.

Buskirk, R. E., and W. H. Buskirk. 1976. Changes in arthropod abundance in a highland Costa Rican forest. *Am. Midl. Nat.* 95:288–98.

Buskirk, W. H. 1976. Social systems in a tropical forest avifauna. *Am. Nat.* 110:293–310.

Buskirk, W. H., and M. Lechner. 1978. Frugivory by swallowtailed kites in Costa Rica. *Auk* 95:767–68.

Buskirk, W. H., G. V. N. Powell, J. F. Wittenberger, and T. U. Powell. 1972. Interspecific bird flocks in tropical highland Panama. *Auk* 89:612–24.

Bussing, W. A. 1963. A new poeciliid fish, *Phallichthys tico,* from Costa Rica. *Contrib. Sci. (Los Angel.)* 77:1–13.

———. 1967. New species and new records of Costa Rican freshwater fishes with a tentative list of species. *Rev. Biol. Trop.* 14:205–49.

———. 1976a. Geographic distribution of the San Juan ichthyofauna of Central America with remarks on its origin and ecology. In *Investigations of the ichthyofauna of Nicaraguan lakes,* ed. T. B. Thorson. Lincoln: Univ. of Nebraska Press.

———. 1976b. Taxonomy and biological aspects of the Central American cichlid fishes *Cichlasoma sieboldii* and *C. tuba. Rev. Biol. Trop.* 23:189–211.

———. 1979. Taxonomic status of the atherinid fish genus *Melaniris* in lower Central America, with the description of three new species. *Rev. Biol. Trop.* 26:391–413.

———. 1980. Status of the cyprinodontid fish genus *Rivulus* in Costa Rica, with descriptions of new endemic species. *Brenesia* 17:327–63.

———. 1987. *Peces de las aguas continentales de Costa Rica.* San José: Editorial Universidad de Costa Rica.

———. In press. Fish communities and environmental characteristics in a tropical rainforest river in Costa Rica. *Rev. Biol. Trop.*

Bussing, W. A., and M. Martin. 1975. Systematic status, variation and distribution of four Middle American cichlid fishes belonging to the *Amphilophus* species group, genus *Cichlasoma. Contrib. Sci. (Los Angel.)* 269:1–41.

Butterfield, R. 1990. Native species for reforestation and land restoration: A case study from Costa Rica. Proceedings of the Nineteenth IUFRO World Congress, Montreal, Canada, August 1990. Division 1, 2:3–14.

Byrne, M. M., and D. J. Levey. In press. Removal of seeds from frugivore defecations by ants in a Costa Rican rain forest. *Vegetatio.*

Cachan, P. 1963. Signification écologique des variations microclimatiques verticales dans la forêt sempervirente de basse Côte d'Ivoire. *Ann. Fac. Sci. Univ. Dakar* 8:89–155.

Cachan, P., and J. Duval. 1963. Variations microlimatiques verticales et saisonnières dans la forêt sempervirente de basse Côte d'Ivoire. *Ann. Fac. Sci. Univ. Dakar* 8:5–87.

Cadle, J. E. 1982. Evolutionary relationships among advanced snakes. Ph.D. diss., University of California, Berkeley.

Calder, I. R., I. R. Wright, and D. Murdiyarso. 1986. A study of evapotranspiration from tropical forest—West Java. *J. Hydrol.* 89:13–31.

Camacho, P. 1981. Ensayos de adaptabilidad y rendimiento de especies forestales en Costa Rica. Report, Instituto Tecnológico de Costa Rica and Dirección General Forestal, Cartago, Costa Rica.

Canham, C. D. 1989. Different responses to gaps among shade-tolerant tree species. *Ecology* 70:548–50.

Canham, C. D., J. S. Denslow, W. J. Platt, J. R. Runkle, T. A. Spies, and P. S. White. 1990. Light regimes beneath closed canopies and tree-fall gaps in temperate and tropical forests. *Can. J. For. Res.* 20:620–31.

Caprio, J. M. 1967. Phenological patterns and their use as climatic indicators. Pp. 17–43 in *Ground level climatology,* ed. R. H. Shaw. AAAS Publication no. 86, Washington, D.C.

Carpenter, G. D. H. 1942. Observations and experiments in Africa by the late C. F. M. Swynnerton on wild birds eating butterflies and the preferences shown. *Proc. Linn. Soc. Lond.* 154:10–46.

Carroll, C. R. 1969. An experimental investigation of the extra-floral nectary strategy. Report, OTS 69-6:51–58.

———. 1983. Azteca. Pp. 691–93 in *Costa Rican Natural History,* ed. D. H. Janzen. Chicago: Univ. of Chicago Press.

Carroll, C. R., and C. A. Hoffman. 1980. Chemical feeding deterrent mobilized in response to insect herbivory and counteradaptation by *Epilachna tredecimnotata. Sci.* 209:414–16.

Casebeer, R. S., R. B. Linsky, and C. E. Nelson. 1963. The phyllostomid bats, *Ectophylla alba* and *Vampyrum spectrum,* in Costa Rica. *J. Mammal.* 44:186–89.

Casey, J. J. 1979. *Limón 1880–1940. Un estudio de la industria bananera en Costa Rica.* San José: Editorial Universidad de Costa Rica.

Castillo-Muñoz, R. 1983. Geology. Pp. 47–62 in *Costa Rican Natural History,* ed. D. H. Janzen. Chicago: Univ. of Chicago Press.

Castro, Y., D. Fernández, and N. Fetcher. 1991. Chronic photoinhibition in attached leaves of tropical trees. *Plant Physiol.* 96 (suppl.): 116.

Centro Agronómico Tropical de Investigación y Enseñanza (CATIE). 1990. Descripción de ensayo, experimento #223. Report, Proyecto Mejoramiento Genético Forestal. Turrialba, Costa Rica.

Chabot, B. F., and J. F. Chabot. 1977. Effects of light and temperature on leaf anatomy and photosynthesis in *Fragaria vesca. Oecologia* 26:363–77.

Chai, P. 1986. Field observations and feeding experiments on the response of rufous tailed jacamars *(Galbula ruficauda)* to free-flying butterflies in a tropical rainforest. *Biol. J. Linn. Soc.* 29:166–89.

———. 1987. Patterns of prey selection by an insectivorous bird on butterflies in a tropical rainforest. Ph.D. diss., University of Texas, Austin.

———. 1990. Relationships between visual characteristics of rainforest butterflies and responses of a specialized insectivorous bird. Pp. 31–60 in *Adaptive coloration in invertebrates,* ed. M. Wicksten. College Station: Texas A & M Univ. Press.

Chai, P., and R. Srygley. 1990. Predation and the flight, morphology, and temperature of Neotropical rain-forest butterflies. *Am. Nat.* 135:748–65.

Chan, H. T. 1981. Reproductive biology of some Malaysian dipterocarps. Part 3. Breeding systems. *Malay For.* 44:28–36.

Chan, H. T., and S. Appanah. 1980. Reproductive biology of some Malaysian dipterocarps. Part 1. Flowering biology. *Malay. For.* 43:132–43.

Chapman, J. A. 1983. *Sylvilagus floridanus.* Pp. 492–94 in *Costa Rican natural history,* ed. D. H. Janzen. Chicago: Univ. of Chicago Press.

Charles-Dominique, P. 1971. Eco-ethologie des Prosimiens du Gabon. *Biol. Gabonica* 7:121–228.

———. 1977. *Ecology and behaviour of nocturnal primates.* New York: Columbia Univ. Press.

———. 1978. Ecologie et vie sociale de *Nandinia binotata* (Carnivores, Viverrides): Comparaison avec les prosimiens sympatriques du Gabon. *Terre Vie* 32:477–528.

———. 1986. Interrelations between frugivorous vertebrates and pioneer plants: *Cecropia,* birds and bats in French Guyana. Pp. 119–35 in *Frugivores and seed dispersal,* ed. A. Estrada and T. H. Fleming. Boston: W. Junk.

Charles-Dominique, P., M. Atramentowicz, M. Charles-Dominique, H. Gérard, A. Hladik, C. M. Hladik, and M. F. Prévost. 1981. Les mammifères frugivores arboricoles nocturnes d'une forêt guyanaise: Interrelations plantes-animaux. *Rev. Ecol. Terre Vie* 35:341–435.

Charnov, E. L. 1982. *The theory of sex allocation.* Princeton, N.J.: Princeton Univ. Press.

Chazdon, R. L. 1984. Ecophysiology and architecture of three rain forest understory palm species. Ph.D. diss. Cornell University, Ithaca, N.Y.

———. 1985. Leaf display, canopy structure, and light interception of two understory palm species. *Am. J. Bot.* 72:1493–1502.

———. 1986a. The costs of leaf support in understory palms: Economy versus safety. *Am. Nat.* 127:9–30.

———. 1986b. Light variation and carbon gain in rain forest understory palms. *J. Ecol.* 74:995–1012.

———. 1986c. Physiological and morphological basis of shade tolerance in rain forest understory palms. *Principes* 30:92–99.

———. 1988. Sunflecks and their importance to forest understory plants. *Adv. Ecol. Res.* 18:1–63.

———. 1991a. Effects of leaf and ramet removal on growth and reproduction of *Geonoma congesta,* a clonal understory palm. *J. Ecol.* 79:1137–46.

———. 1991b. Plant size and form in the understory palm genus *Geonoma:* Are species variations on a theme? *Am. J. Bot.* 78:680–94.

———. 1992. Patterns of growth and reproduction of *Geonoma congesta,* a clustered understory palm. *Biotropica* 24:43–51.

Chazdon, R. L., and N. Fetcher. 1984a. Light environments of tropical forests. Pp. 27–36 in *Physiological ecology of plants of the wet tropics,* ed. E. Medina, H. A. Mooney, and C. Vásquez-Yánes. The Hague: W. Junk.

———. 1984b. Photosynthetic light environments in a lowland tropical rain forest in Costa Rica. *J. Ecol.* 72:553–64.

Chazdon, R. L., and C. B. Field. 1987. Determinants of photosynthetic capacity in six rainforest *Piper* species. *Oecologia* 73:222–30.

Chazdon, R. L., and R. W. Pearcy. 1986a. Photosynthetic responses to light variation in rainforest species. Part 1. Induction under constant and fluctuating light conditions. *Oecologia* 69:517–23.

———. 1986b. Photosynthetic responses to light variation in rainforest species. Part 2. Carbon gain and photosynthetic efficiency during lightflecks. *Oecologia* 69:524–31.

———. 1991. The importance of sunflecks for forest understory plants. *Bioscience* 41:760–66.

Chesson, P. L., and R. R. Warner. 1981. Environmental variability promotes coexistence in lottery competitive systems. *Am. Nat.* 117:923–43.

Chew, F. S., and R. K. Robbins. 1984. Egg laying in butterflies. Pp. 65–79 in *The biology of butterflies,* ed. H. Vane-Wright and P. J. Ackery. *Symp. R. Entomol. Soc. Lond.* 11:65–79.

Chiariello, N. 1984. Leaf energy balance in the wet lowland tropics. Pp. 85–98 in *Physiological ecology of plants of the wet tropics,* ed. E. Medina, H. A. Mooney, and C. Vásquez-Yánes. The Hague: W. Junk.

Chiariello, N. R., C. B. Field, and H. A. Mooney. 1987. Midday wilting in a tropical pioneer tree. *Funct. Ecol.* 1:3–11.

Chivers, D. J., ed. 1980. *Malayan forest primates.* New York: Plenum Press.

Christensen, N. L. 1977. Changes in structure, pattern, and diversity associated with climax forest maturation in Piedmont, North Carolina. *Am. Midl. Nat.* 97:176–88.

Chu, K. 1977. Herbivore damage and altitude. Report, OTS 77-3:337–48.

Cifuentes, M. 1983. Reservas de biosfera: Clarificación de su marco conceptual y diseño y amplicación de una metodología para la planificación estratégico de un subsistema nacional. Master's thesis, CATIE, Turrialba, Costa Rica.

Cipollini, M. L., and D. J. Levey. 1991. Why some fruits are green when they are ripe: Carbon balance in fleshy fruits. *Oecologia* 88:371–77.

Clark, D. A. 1988. Research on tropical plant biology at the La Selva Biological Station, Costa Rica. *Evol. Trends Plants* 2:75–78.

Clark, D. A., and D. B. Clark. 1981. Effects of seed dispersal by animals on the regeneration of *Bursera graveolens* (Burseraceae) on Santa Fe Island, Galapagos. *Oecologia* 49:73–75.

———. 1984. Spacing dynamics of a tropical rain forest

tree: Evaluation of the Janzen-Connell model. *Am. Nat.* 124:769–88.

———. 1987a. Análisis de la regeneración de árboles del dosel en bosque muy húmedo tropical: Aspectos teóricos y prácticos. *Rev. Biol. Trop.* 35 (suppl.): 41–54.

———. 1987b. Temporal and environmental patterns of reproduction in *Zamia skinneri,* a tropical rain forest cycad. *J. Ecol.* 75:135–49.

———. 1992. Life history diversity of canopy and emergent trees in a Neotropical rain forest. *Ecol. Monogr.* 62: 315–44.

———. In review. Climate-related variation in tree growth in a Central American rain forest. *Nature.*

———. In review. Response of canopy tree growth in a Neotropical wet forest to year-to-year climatic variation. *Oecologia.*

Clark, D. B. 1988. The search for solutions: Research and education at the La Selva Biological Station and their relation to ecodevelopment. Pp. 209–24 in *Tropical rainforests: Diversity and conservation,* ed. F. Almeda and C. Pringle. San Francisco: California Academy of Science.

———. 1990. La Selva Biological Station: A blueprint for stimulating tropical research. Pp. 9–27 in *Four Neotropical rainforests,* ed. A. H. Gentry. New Haven, Conn.: Yale Univ. Press.

Clark, D. B., and D. A. Clark. 1984. Regeneration of canopy trees in a lowland tropical rain forest: A tolerance continuum. *Bull. Ecol. Soc. America* 65:148 (abstract).

———. 1985. Seedling dynamics of a tropical tree: Impacts of herbivory and meristem damage. *Ecology* 66:1884–92.

———. 1987. Population ecology and microhabitat distribution of *Dipteryx panamensis,* a Neotropical rain forest emergent tree. *Biotropica* 19:236–44.

———. 1988. Leaf production and the cost of reproduction in a tropical rain forest cycad, *Zamia skinneri. J. Ecol.* 76:1153–63.

———. 1989. The role of physical damage in the seedling mortality regime of a Neotropical rain forest. *Oikos* 55:225–230.

———. 1990. Distribution and effects on tree growth of lianas and woody hemiepiphytes in a Costa Rican tropical wet forest. *J. Trop. Ecol.* 6:321–31.

———. 1991a. Herbivores, herbivory and plant phenology: Patterns and consequences in a tropical forest cycad. Pp. 209–26 in *Plant-animal interactions: Evolutionary ecology in tropical and temperate regions,* ed. P. W. Price, T. M. Lewinsohn, G. W. Fernandes, and W. W. Benson. New York: J. Wiley.

———. 1991b. The impact of physical damage on canopy tree regeneration in tropical rain forest. *J. Ecol.* 79:447–57.

Clark, D. B., D. A. Clark, and M. H. Grayum. 1992. Leaf demography of a Neotropical rain forest cycad, *Zamia skinneri* (Zamiaceae). *Am. J. Bot.* 79:28–33.

Clark, G. C., and C. G. C. Dickson. 1971. Life histories of southern African lycaenid butterflies. Cape Town, South Africa: Purnell.

Clark, P. J., and F. C. Evans. 1954. Distance to nearest neighbor as a measure of spatial relationships in populations. *Ecology* 35:445–53.

Cleveland, W. S. 1985. *The elements of graphing data.* Monterey, Calif.: Wadsworth Advanced Books.

Cleveland, W. S., S. J. Devlin, and I. J. Terpenning. 1982. The SABL seasonal and calendar adjustment procedures. Pp. 539–64 in *Time series analysis: Theory and practice 1,* ed. O. D. Anderson. Amsterdam: North Holland.

Cody, M. L. 1966. A general theory of clutch size. *Evolution* 20:174–84.

———. 1971. Ecological aspects of reproduction. Pp. 461–512 in *Avian Biology.* Vol. 1, ed. D. S. Farner and J. R. King. New York: Academic Press.

———. 1975. Towards a theory of continental species diversities. Pp. 214–57 in *Ecology and evolution of communities,* ed. M. L. Cody, and J. M. Diamond. Cambridge: Harvard Univ. Press, Belknap Press.

———. 1986. Diversity, rarity, and conservation in Mediterranean-climate regions. Pp. 122–52 in *Conservation biology: The science of scarcity and diversity,* ed. M. E. Soulé. Sunderland, Mass.: Sinauer Associates.

Cody, M. L., and J. M. Diamond, eds. 1975. *Ecology and evolution of communities.* Cambridge: Harvard Univ. Press, Belknap Press.

Coen, E. 1983. Climate. Pp. 35–46 in *Costa Rican natural history,* ed. D. H. Janzen. Chicago: Univ. of Chicago Press.

Cole, B. J. 1981. Overlap, regularity, and flowering phenologies. *Am. Nat.* 117:993–97.

Cole, D. W., and D. W. Johnson. 1979. Mineral cycling in tropical forests. Publication no. 1269, Environmental Sciences Division, Oak Ridge National Laboratory, Oak Ridge, TN.

Cole, D. W., and M. Rapp. 1981. Elemental cycling in forests. Pp. 341–409 in *Dynamic properties of forest ecosystems,* ed. D. E. Reichle. London: Cambridge Univ. Press.

Coleman, D. C. 1970. Nematodes in the litter and soil of an El Verde rain forest. Pp. 103–4 in *A tropical rain forest: A study of irradiation and ecology at El Verde, Puerto Rico,* ed. H. T. Odum. Washington, D.C.: U.S. Atomic Energy Commission.

Coley, P. D. 1982. Rates of herbivory on different tropical trees. Pp. 123–32 in *The ecology of a tropical forest: Seasonal rhythms and long-term changes,* ed. E. G. Leigh, Jr., A. S. Rand, and D. M. Windsor. Washington, D.C.: Smithsonian Institution Press.

———. 1983a. Herbivory and defense characteristics of tree species in lowland tropical forest. *Ecol. Monogr.* 53:209–33.

———. 1983b. Intraspecific variation in herbivory on two tropical tree species. *Ecology* 64:426–33.

———. 1987. Interspecific variation in plant anti-herbivore properties: The role of habitat quality and rate of disturbance. *New Phytol.* 106 (suppl.): 251–63.

Coley, P. D., and T. M. Aide. 1991. Comparison of herbivory and plant defenses in temperate and tropical broad-leaved forests. Pp. 25–50 in *Plant-animal interactions: Evolutionary ecology in tropical and temperate regions,* ed. P. W. Price, T. M. Lewinsohn, G. W. Fernandes, and W. W. Benson. New York: J. Wiley.

Coley, P. D., J. P. Bryant, and F. S. Chapin III. 1985. Resource availability and plant antiherbivore defense. *Science* 230:895–99.

Colwell, R. K. 1974. Predictability, constancy and contingency of periodic phenomena. *Ecology* 55:1148–53.

———. 1986a. Community biology and sexual selection: Lessons from hummingbird flower mites. Pp. 406–24 in *Community ecology,* ed. J. Diamond and T. J. Case. New York: Harper and Row.

———. 1986b. Population structure and sexual selection for host fidelity in the speciation of hummingbird flower mites. Pp. 475–95 in *Evolutionary processes and theory,* ed. S. Karlin and E. Nevo. New York: Academic Press.

Combe, J., and N. J. Gewald, eds. 1979. *Guia de campo de los ensayos forestales del CATIE en Turrialba, Costa Rica.* Turrialba, Costa Rica: CATIE.

Condon, M. 1974. Interactions between *Cassia fruticosa* and its herbivores. Report, OTS 74-1:129–32.

Connell, J. H. 1971. On the role of natural enemies in preventing competitive exclusion in some marine animals and rain forest trees. Pp. 298–312 in *Dynamics of populations,* ed. P. J. den Boer and G. R. Gradwell. Proceedings of the Advanced Study Institute on Dynamics of Numbers in Populations, Oosterbeek, 1970. Wageningen, Netherlands: Centre for Agricultural Publishing and Documentation.

———. 1978. Diversity in tropical rain forests and coral reefs. *Science* 199:1302–10.

———. 1979. Tropical rain forests and coral reefs as open nonequilibrium systems. Pp. 141–63 in *Population dynamics,* ed. R. M. Anderson, B. D. Turner, and L. R. Taylor. Symposium of the British Ecological Society, London. Oxford: Blackwell Scientific.

Connell, J. H., and M. D. Lowman. 1989. Low-diversity tropical rain forests: Some possible mechanisms for their existence. *Am. Nat.* 134:88–119.

Connell, J. H., J. G. Tracey, and L. J. Webb. 1984. Compensatory recruitment, growth, and mortality as factors maintaining rain forest tree diversity. *Ecol. Monogr.* 54:141–64.

Coombe, D. E. 1960. An analysis of the growth of *Trema guineensis. J. Ecol.* 48:219–31.

Coombe, D. E., and W. Hadfield. 1962. An analysis of the growth of *Musanga cecropioides. J. Ecol.* 50:221–34.

Cooper, S. M., and N. Owen-Smith. 1985. Condensed tannins deter feeding by browsing ruminants in a South African savanna. *Oecologia* 67:142–46.

Cooper-Driver, G. A. 1990. Defense strategies in bracken, *Pteridium aquilinum* (L.) Kuhn. *Ann. MO Bot. Gard.* 77:281–86.

Corbet, S. A., and P. G. Willmer. 1981. The nectar of *Justicia* and *Columnea:* Composition and concentration in a humid tropical climate. *Oecologia* 51:412–18.

Cordero S., R. A. 1988. Effects of light and water stress on the growth and water relations of *Simarouba amara.* P. 4 in OTS International Symposium on Tropical Studies. University of Miami, Coral Gables, Florida (abstract).

Córdova Casillas, B. 1985. Demografía de árboles tropicales. Pp. 103–28 in *Investigaciones sobre la regeneración de selvas altas en Veracruz, México* ed. A. Gómez-Pompa and S. Del Amo. Vol. 2. Mexico City: Editorial Alhambra Mexicana.

Corlett, R. T., and P. W. Lucas. 1990. Alternative seed-handling strategies in primates: Seedspitting by longtailed macaques *(Macaca fascicularis). Oecologia* 82:166–71.

Corn, M. J. 1981. Ecological separation of *Anolis* lizards in a Costa Rican rain forest. Ph.D. diss., University of Florida, Gainesville.

Cornell, H. V. 1985. Local and regional richness of cynipine gall wasps on California oaks. *Ecology* 66:1247–60.

Corner, E. J. H. 1964. *The life of plants.* London: Weidenfeld and Nicolson.

Corporación Costarricense de Desarrollo. 1983. Report, Diagnóstico socio-económico del cantón de Sarapiquí, Provincia de Heredia. Puerto Viejo de Sarapiquí, Costa Rica.

Cort, R., M. Fisher, R. Inouye, L. McIntosh, A. Sauer, and R. Voss. 1977. Report, Ants on Topobea. OTS 77-4:457–60.

Cottrell, C. B. 1984. Aphytophagy in butterflies: Its relationship to myrmecophily. *Zool. J. Linn. Soc.* 79:1–57.

Coville, R. E., and C. Griswold. 1983. Nesting biology of *Trypoxylon xanthandrum* in Costa Rica with observations on its spider prey (Hymenoptera: Sphecidae; Araneae: Senoculidae). *J. Kans. Entomol. Soc.* 56:205–16.

———. 1984. Biology of *Trypoxylon* (Trypargilum) *superbum* (Hymenoptera: Sphecidae), a spider-hunting wasp with extended guarding of the brood by males. *J. Kans. Entomol. Soc.* 57:365–76.

Cox, P. A. 1981. Niche partitioning between sexes of dioecious plants. *Am. Nat.* 117:295–307.

Coyne, J., and S. Schneider. 1974. The foraging behavior of leaf-cutter ants. Report, OTS 74-3:108–120.

Crawley, M. J. 1983. Herbivory: The dynamics of animal-plant interactions. In *Studies in ecology,* vol. 10. Berkeley: Univ. of California Press.

Crease, T., and S. Trott. 1982. Feeding preferences of five species of weevils (Curculionidae) on six species of *Piper* (Piperaceae). Report, OTS 82-1:359–71.

Cressa, C., and C. T. Senior. 1987. Aspects of the chemistry and hydrology of the Orinoco River, Venezuela. *Acta Cient. Venez.* 38:99–105.

Croat, T. B. 1969. Seasonal flowering behavior in central Panama. *Ann. MO Bot. Gard.* 56:295–307.

———. 1974. A case for selection for delayed fruit maturation in *Spondias* (Anacardiaceae). *Biotropica* 6:135–37.

———. 1975. Phenological behavior of habit and habitat classes on Barro Colorado Island (Panama Canal Zone). *Biotropica* 7:270–77.

———. 1978. Flora of Barro Colorado Island. Stanford, Calif.: Stanford Univ. Press.

———. 1979. The sexuality of the Barro Colorado Island flora (Panama). *Phytologia* 42:319–48.

Cromartie, W. J., Jr. 1975. Influence of habitat on colonization of collard plants by *Pieris rapae. Environ. Entomol.* 4:783–84.

Crome, F. H. J. 1975. The ecology of fruit pigeons in tropical northern Queensland. *Aust. Wild. Res.* 2:155–85.

Crowson, R. A. 1981. *The biology of the Coleoptera.* London: Academic Press.

Crozat, G. 1979. Sur l'emission d'un aérosol riche en potassium par la forêt tropicale. *Tellus* 31:52–57.

Crump, M. L. 1974. Reproductive strategies in a tropical anuran community. *Misc. Publ. Mus. Nat. Hist. Univ. Kans.* 61:1–68.

———. 1983. *Dendrobates granuliferus* and *Dendrobates*

pumilio. Pp. 396–98 in *Costa Rican natural history,* ed. D. H. Janzen. Chicago: Univ. of Chicago Press.

Crutzen, P. J. 1987. Role of the tropics in atmospheric chemistry. Pp. 107–30 in *The geophysiology of Amazonia: Vegetation and climate interactions,* ed. R. E. Dickenson. New York: J. Wiley.

da Cruz Alencar, J., R. Aniceta de Almeida, and N. P. Fernandes. 1979. Fenologia de espécies florestais em floresta tropical úmida de terra firme na Amazonia Central. *Acta Amazonica* 9:163–98.

de la Cruz, M., and R. Dirzo. 1987. A survey of the standing levels of herbivory on seedlings from a Mexican rain forest. *Biotropica* 19:98–106.

Culver, D. C., and A. J. Beattie. 1978. Mymecochory in *Viola:* Dynamics of seedant interactions in some West Virginia species. *J. Ecol.* 66:53–72.

Cushing, C. E., C. D. McIntire, J. R. Sedell, K. W. Cummins, G. W. Minshall, R. C. Petersen, and R. L. Vannote. 1980. Comparative study of physical-chemical variables of streams using multivariate analyses. *Arch. Hydrobiol.* 89:343–52.

Dagg, M., and J. R. Blackie. 1970. Estimates of evaporation in East Africa in relation to climatological classification. *Geograph. J.* 36:228–34.

Danilevskii, A. S. 1965. *Photoperiodism and seasonal development of insects.* London: Oliver and Boyd.

D'Arcy, W. G. 1973. *Correlliana* (Myrsinaceae), a new palmoid genus of the tropical rain forest. *Ann. MO Bot. Gard.* 60:442–48.

Darnell, R. M. 1953. An ecological study of the Río Salinas and related waters in southern Tamaulipas, México, with special reference to the fishes. Ph.D. diss., University of Minnesota, Minneapolis.

———. 1962. Fishes of the Río Tamesí and related coastal lagoons in east-central México. *Inst. Mar. Sci.* 8:299–365.

Daubenmire, R. 1972. Phenology and other characteristics of tropical semi-deciduous forest in north-western Costa Rica. *J. Ecol.* 60:147–70.

Davidar, P. 1983. Birds and Neotropical mistletoes: Effects on seedling recruitment. *Oecologia* 60:271–73.

Davies, A. G., E. L. Bennett, and P. G. Waterman. 1988. Food selection by two South-east Asian colobine monkeys (*Presbytis rubicunda* and *Presbytis melalophos*) in relation to plant chemistry. *Biol. J. Linn. Soc.* 34:33–56.

Davis, D. R., D. H. Clayton, D. H. Janzen, and A. P. Brooke. 1986. Neotropical Tineidae, II: Biological notes and descriptions of two new moths phoretic on spiny pocket mice in Costa Rica (Lepidoptera: Tineoidea). *Proc. Entomol. Soc. Wash.* 88:98–109.

DeLapp, J. 1979. Leaf herbivore damage in *Neea* sp. in early successional and mature rain forest. Report, OTS 79-3:155–56.

Delmas, R., J. Baudet, and J. Servant. 1978. Mise en evidence des sources naturelles de sulfate en milieu tropical humide. *Tellus* 30:158–68.

Demmig, B., K. Winter, A. Krüger, and F.-C. Czygan. 1988. Zeaxanthin and the heat dissipation of excess light energy in *Nerium oleander* exposed to a combination of high light and water stress. *Plant Physiol.* 87:17–24.

Denno, R. F., and M. L. Donnelly. 1981. Patterns of herbivory on *Passiflora* leaf tissues and species by generalized and specialized feeding insects. *Ecol. Entomol.* 6:11–16.

Denslow, J. S. 1980a. Gap partitioning among tropical rainforest trees. *Biotropica* 12 (suppl.): 47–55.

———. 1980b. Patterns of plant species diversity during succession under different disturbance regimes. *Oecologia* 46:18–21.

———. 1985. Disturbance-mediated coexistence of species. Pp. 307–23 in *The ecology of natural disturbance and patch dynamics,* S. T. A. Pickett and P. S. White. Orlando, Fla.: Academic Press.

———. 1987a. Fruit removal rates from aggregated and isolated bushes of the red elderberry, *Sambucus pubens. Can. J. Bot.* 65:1229–35.

———. 1987b. Tropical rain forest gaps and tree species diversity. *Ann. Rev. Ecol. Syst.* 18:431–51.

Denslow, J. S., and A. E. Gómez D., 1990. Seed rain to treefall gaps in a Neotropical rain forest. *Can. J. For. Res.* 20:642–48.

Denslow, J. S., and T. C. Moermond. 1982. The effect of accessibility on rates of fruit removal from tropical shrubs: An experimental study. *Oecologia* 54:170–76.

Denslow, J. S., T. C. Moermond, and D. J. Levey. 1986. Spatial components of fruit display in understory trees and shrubs. Pp. 37–44 in *Frugivores and seed dispersal,* ed. A. Estrada and T. H. Fleming. Dordrecht: W. Junk.

Denslow, J. S., E. Newell, and A. M. Ellison. 1991. The effect of understory palms and cyclanths on the growth and survival of *Inga* seedlings. *Biotropica* 23:225–34.

Denslow, J. S., J. C. Schultz, P. M. Vitousek, and B. R. Strain. 1990. Growth responses of tropical shrubs to treefall gap environments. *Ecology* 71:165–79.

Denslow, J. S., P. M. Vitousek, and J. Schultz. 1987. Bioassays of nutrient limitation in a tropical rain forest soil. *Oecologia* 74:370–76.

DeSteven, D. 1989. Genet and ramet demography of *Oenocarpus mapora* ssp. *mapora,* a clonal palm of Panamanian tropical moist forest. *J. Ecol.* 77:579–96.

DeSteven, D., and F. E. Putz. 1984. Impact of mammals on early recruitment of a tropical canopy tree, *Dipteryx panamensis,* in Panama. *Oikos* 43:207–16.

Deuth, D. A. 1980. The protection of *Cassia fruticosa* (Leguminosae) from herbivores by ants foraging at the foliar nectaries. Master's thesis, University of Colorado, Boulder.

DeVries, P. J. 1977. *Eumaeus minyas:* An aposematic lycaenid butterfly. *Brenesia* 12:269–70.

———. 1983. Checklist of butterflies. Pp. 654–78 in *Costa Rican natural history,* ed. D. H. Janzen. Chicago: Univ. of Chicago Press.

———. 1985. Hostplant records and natural history notes on Costa Rican butterflies (Papilionidae, Pieridae and Nymphalidae). *J. Res. Lepid.* 24:290–333.

———. 1987. *The butterflies of Costa Rica and their natural history: Papilionidae, Pieridae, Nymphalidae.* Princeton, N.J.: Princeton Univ. Press.

———. 1988a. The larval ant-organs of *Thisbe irenea* (Riodinidae) and their effects upon attending ants. *Zool. J. Linn. Soc.* 94:379–93.

———. 1988b. Stratification of fruit-feeding nymphalid butterflies in a Costa Rican rainforest. *J. Res. Lepid.* 26:98–108.

———. 1990. Enhancement of symbioses between butterfly caterpillars and ants by vibrational communication. *Science* 248:1104–6.

———. 1991a. Call production by myrmecophilous riodinid and lycaenid butterfly caterpillars (Lepidoptera): Morphological, acoustic, functional, and evolutionary patterns. *Am. Mus. Novit.* 3025:1–23.

———. 1991b. Ecological and evolutionary patterns in riodinid butterflies. Pp. 143–56 in *Interaction between ants and plants,* ed. D. F. Cutler and C. Huxley. Oxford: Oxford Univ. Press.

———. 1991c. The mutualism between *Thisbe irenea* and ants, and the role of ant ecology in the evolution of larval-ant associations. *Biol. J. Linn. Soc.* 43:179–95.

DeVries, P. J., and I. Baker. 1989. Butterfly exploitation of a plant-ant mutualism: Adding insult to herbivory. *J. NY Entomol. Soc.* 97:332–40.

DeVries, P. J., and T. R. Dudley. 1990. Flight physiology of migrating *Urania fulgens* (Uranidae): Flight speeds, body-size, thermoregulation, and lipid reserves in natural free flight. *Physiolog. Zool.* 63:235–51.

DeVries, P. J., I. Kitching, and R. I. Vane-Wright. 1985. The systematic position of *Antirrhea* and *Caerois,* with comments on the higher classification of the Nymphalidae. *Syst. Entomol.* 10:11–32.

Diamond, A. W., and A. C. Hamilton. 1980. The distribution of forest passerine birds and quaternary climatic events in tropical Africa. *J. Zool. (Lond.)* 191:379–402.

Diamond, J. 1986. Biology of birds of paradise and bowerbirds. *Ann. Rev. Ecol. Syst.* 17:17–37.

Dieterlen, F. 1978. *Zur Phänologie des Äquatorialen Regenwaldes im Ost-Zaire (Kivu). Diss. Bot.* 47:5–120.

Dietrich, W. E., D. M. Windsor, and T. Dunne. 1982. Geology, climate, and hydrology of Barro Colorado Island. Pp. 21–46 in *The ecology of a tropical forest, seasonal rhythms and long-term changes,* ed. E. G. Leigh, Jr., A. S. Rand, and D. M. Windsor. Washington, D.C.: Smithsonian Institution Press.

Diggle, P. J. 1983. *Statistical analysis of spatial point patterns.* London: Academic Press.

Dinnerstein, E. 1987. Reproductive ecology of fruit bats and the seasonality of fruit production in a Costa Rican cloud forest. *Biotropica* 18:307–18.

Dirección General de Estadísticas y Censos. 1965. Censo agropecuario 1963. San José, Costa Rica.

———. 1966. Censo de población 1963. San José, Costa Rica.

———. 1974. Censo agropecuario 1973. San José, Cost Rica.

———. 1984a. Censo agropecuario 1983. San José, Costa Rica.

———. 1984b. Censo de población 1983. San José, Costa Rica.

———. 1986. Censo nacional de población. San José. Costa Rica.

———. 1987. Censo agropecuario 1984. San José. Costa Rica.

Dirección General Forestal. 1988a. Censo de la industria forestal 1986–1987. Departamento de Desarrollo Forestal, Ministerio de Recursos Naturales, Energía, y Minas (MIRENEM). San José, Costa Rica.

———. 1988b. Ensayos de técnicas de explotación, regeneración natural y de enriquecimiento del bosque muy húmedo tropical en Sarapiquí. Departamento de Investigaciones Forestales, Ministerio de Recursos Naturales, Energía, y Minas (MIRENEM). San José, Cost Rica.

———. 1989. Boletín estadístico no. 3. Ministerio de Recursos Naturales, Energía, y Minas (MIRENEM). San José, Costa Rica.

———. 1990. Boletín Estadístico Forestal no. 4. Secretaria Técnica de Recursos Naturales, Ministerio de Recursos Naturales, Energía, y Minas (MIRENEM). San José, Costa Rica.

Dirzo, R. 1984. Insect-plant interactions: Some ecophysiological consequences of herbivory. Pp. 209–24 in *Physiological ecology of plants of the wet tropics,* ed. E. Medina, H. A. Mooney, and C. Vásquez-Yánes. The Hague: W. Junk.

———. 1987. Estudios sobre interacciones planta-herbívoro en "Los Tuxtlas," Veracruz. *Rev. Biol. Trop.* 35 (suppl. 1): 119–31.

Dirzo, R., and C. A. Domínguez. 1986. Seed shadows, seed predation and the advantages of dispersal. Pp. 237–49 in *Frugivores and seed dispersal,* ed. A. Estrada and T. H. Fleming. Dordrecht: W. Junk.

Dixon, J. R., and P. Soini. 1975. The reptiles of the upper Amazon Basin, Iquitos Region, Peru. Part 1. Lizards and Amphisbaenians. *Milw. Public Mus. Contrib. Biol. Geol.* 4:1–58.

———. 1977. The reptiles of the upper Amazon Basin, Iquitos Region, Peru. Part 2. Crocodilians, turtles and snakes. *Milw. Public Mus. Contrib. Biol. Geol.* 12:1–91.

Dobat, K. 1985. *Blüten und Fledermäuse.* Frankfurt on the Main: Waldemar Kramer.

Dobkin, D. S. 1984. Flowering patterns of long-lived *Heliconia* inflorescences: Implications for visiting and resident nectarivores. *Oecologia* 64:245–54.

———. 1987. Synchronous flower abscission in plants pollinated by hermit hummingbirds and the evolution of one-day flowers. *Biotropica* 19:90–93.

Dobzhansky, T. 1950. Evolution in the tropics. *Am. Sci.* 38:209–21.

Doctors van Leeuwen, W. M. 1954. On the biology of some Javanese Loranthaceae and the role birds play in their life-history. *Beaufortia* 4:105–205.

Domínguez, C. A., R. Dirzo, and S. H. Bullock. 1989. On the function of floral nectar in *Croton suberosus* (Euphorbiaceae). *Oikos* 56:109–14.

Donnelly, M. A. 1987. Territoriality in the strawberry poison-dart frog *Dendrobates pumilio* (Anura: Dendrobatidae). Ph.D. diss., University of Miami, Coral Gables, Fla.

———. 1989a. Demographic effects of reproductive resource supplementation in a territorial frog, *Dendrobates pumilio. Ecol. Monogr.* 59:207–21.

———. 1989b. Effects of reproductive resource supplementation on space-use patterns in *Dendrobates pumilio. Oecologia* 81:212–18.

———. 1989c. Reproductive phenology and age structure of *Dendrobates pumilio* in northeastern Costa Rica. *J. Herpetol.* 23:362–67.

———. 1991. Feeding patterns of the strawberry poison frog, *Dendrobates pumilio* (Anura: Dendrobatidae). *Copeia* 1991:723–30.

Donnelly, M. A., R. O. de Sa, and C. Guyer. 1990. Description of the tadpoles of *Gastrophyrne pictiventris* and *Nelsonophryne aterrima* (Anura: Microhylidae), with comments on morphological variation in free-swimming microhylid larvae. *Am. Mus. Novit.* 2976:1–19.

Donnelly, M. A., C. Guyer, and R. O. de Sa. 1990. The tadpole of a dart-poison frog *Phyllobates lugubris* (Anura: Dendrobatidae). *Proc. Biol. Soc. Wash.* 103:427–31.

Donnelly, M. A., C. Guyer, D. M. Krempels, and H. E. Braker. 1987. The tadpole of *Agalychnis calcarifer* (Anura: Hylidae). *Copeia* 1987:247–50.

Douglas, M. M. 1986. *The lives of butterflies.* Ann Arbor: Univ. of Michigan Press.

Doutt, R. L. 1960. Natural enemies and insect speciation. *Pan-Pac. Entomol.* 36:1–14.

Downum, K. R., L. S. Swain, and L. J. Faleiro. 1991. Influence of light on plant allelochemicals: A synergistic defense in higher plants. *Arch. Insect. Biochem. Physiol.* 17:201–11.

Dressler, R. L. 1981. Humus collecting shrubs in wet tropical forests. *Proc. Int. Soc. Trop. Ecol.* (Bhopal, India).

Drew, R. A. I. 1988. Amino acid increases in fruit infested by fruit flies of the family Tephritidae. *Zool. J. Linn. Soc.* 93:107–12.

Drewry, G. E. 1970. The role of amphibians in the ecology of Puerto Rican rain forest. Pp. 16–85. In *Puerto Rico nuclear center rain forest project: Annual report.* San Juan.

Dubost, G. 1984. Comparison of the diets of frugivorous forest ruminants of Gabon. *J. Mammal.* 65:298–316.

Dudley, R., and P. J. DeVries. 1990. Flight physiology of migrating *Urania fulgens* (Uraniidae) moths: Kinematics and aerodynamics of natural free flight. *J. Comp. Physiol.* A.167:145–54.

Duellman, W. E. 1970. The hylid frogs of Middle America. *Monogr. Mus. Nat. Hist. Univ. Kans.* 1:1–753.

———. 1978. The biology of an equatorial herpetofauna in Amazonian Ecuador. *Misc. Publ. Mus. Nat. Hist. Univ. Kans.* 65:1–352.

———. 1988. Patterns of species diversity in anuran amphibians in the American tropics. *Ann. MO Bot. Gard.* 75:79–104.

———. 1990. Herpetofaunas in Neotropical rainforests: Comparative composition, history, and resource use. Pp. 455–505 in *Four Neotropical rainforests,* ed. A. H. Gentry. New Haven, Conn.: Yale Univ. Press.

Duellman, W. E., and L. Trueb. 1986. *Biology of amphibians.* New York: McGraw-Hill.

Dunham, A. E., D. B. Miles, and D. N. Reznick. 1988. Life history patterns of squamate reptiles. Pp. 441–522 in *Biology of the Reptilia.* Vol. 16, *Ecology, defense and life histories,* ed. G. Gans and R. B. Huey. London: Academic Press.

Dunn, G. E., and B. I. Miller. 1964. *Atlantic hurricanes.* Baton Rouge: Louisiana State Univ. Press.

Edwards, J. 1983. Diet shifts in moose due to predator avoidance. *Oecologia* 60:185–89.

Edwards, P. J. 1977. Studies of mineral cycling in a montane rain forest in New Guinea. Part 2. The production and disappearance of litter. *J. Ecol.* 65:971–92.

Ehrlich, P. R., and P. H. Raven. 1964. Butterflies and plants: A study in coevolution. *Evolution* 18:586–608.

Eisenberg, J. F. 1983. *Ateles geoffroyi.* Pp. 451–53 in *Costa Rican natural history,* ed. D. H. Janzen. Chicago: Univ. of Chicago Press.

———. 1989. *Mammals of the Neotropics. The northern Neotropics.* Vol. 1, *Panama, Colombia, Venezuela, Guyana, Suriname, French Guiana.* Chicago: Univ. of Chicago Press.

Eisenberg, J. F., and R. W. Thorington, Jr. 1973. A preliminary analysis of a Neotropical mammal fauna. *Biotropica* 5:150–61.

Ellington, C. P. 1984. The aerodynamics of hovering insect flight. Part 3. Kinematics. *Phil. Trans. R. Soc. Lond. B* 305:41–78.

Ellison, A. M., J. S. Denslow, B. A. Loiselle, and D. Brenes M. In press. Seed and seedling ecology of Neotropical Melastomataceae. *Ecology.*

Elton, C. S. 1958. *The ecology of invasions by animals and plants.* London: Methuen.

———. 1973a. *The ecology of animals.* London: Chapman and Hall.

———. 1973b. The structure of invertebrate populations inside Neotropical rain forest. *J. Anim. Ecol.* 42:55–104.

Elwood, J. W., J. D. Newbold, R. V. O'Neill, and W. Van Winkle. 1983. Resource spiraling: An operational paradigm for analyzing lotic ecosystems. Pp. 3–27 in *Dynamics of lotic ecosystems,* ed. T. D. Fontaine III and S. M. Bartell. Ann Arbor, Mich.: Ann Arbor Science Publishers.

Emlen, S. T., and L. W. Oring. 1977. Ecology, sexual selection, and the evolution of mating systems. *Science* 197:215–22.

Emmel, T. C., and J. F. Emmel. 1973. The butterflies of southern California. *Nat. Hist. Mus. Los. Angel. C. Sci. Bull.* 26:1–148.

Emmons, L. 1981. Ecology and resource partitioning among nine species of African rain forest squirrels. *Ecol. Monogr.* 50:31–54.

———. 1984. Geographic variation in densities and diversities of non-flying mammals in Amazonia. *Biotropica* 16:210–22.

———. 1987. Comparative feeding ecology of felids in a Neotropical forest. *Behav. Ecol. Sociobiol.* 20:271–83.

Emmons, L. H., and F. Feer. 1990. *Neotropical rainforest mammals: A field guide.* Chicago: Univ. of Chicago Press.

Enders, R. K. 1935. Mammalian life histories from Barro Colorado Island. *Bull. Mus. Comp. Zool.* 78:384–502.

Endler, J. A. 1986a. Defense against predators. Pp. 109–34 in *Predator-prey relationships,* ed. M. E. Feder and G. Lauder. Chicago: Univ. of Chicago Press.

———. 1986b. *Natural selection in the wild.* Princeton, N.J.: Princeton Univ. Press.

Enright, N. J. 1982. The ecology of *Araucaria* species in New Guinea. Part 3. Population dynamics of sample stands. *Aust. J. Ecol.* 7:227–37.

Ernest, K. A. 1989. Insect herbivory on a tropical understory tree: Effects of leaf age and habitat. *Biotropica* 21:194–99.

Ernst, C. H. 1983. *Rhinoclemmys funerea.* Pp. 417–18 in *Costa Rican natural history,* ed. D. H. Janzen. Chicago: Univ. of Chicago Press.

Erwin, T. L. 1982. Tropical forests: Their richness in Coleoptera and other arthropod species. *Coleopt. Bull.* 36:74–75.

———. 1983a. Beetles and other insects of tropical forest canopies at Manaus, Brazil, sampled by insecticidal fogging. Pp. 59–75 in *Tropical rain forest: Ecology and management,* ed. S. L. Sutton, T. C. Whitmore, and A. C. Chadwick. British Ecological Society Special Publication no. 2. Oxford: Blackwell Scientific Publications.

———. 1983b. Tropical forest canopies: The last biological frontier. *Bull. Entomol. Soc. Am.* 29:14–20.

———. 1985. Tambopata Reserved Zone, Madre de Dios, Peru: History and description of the reserve. *Rev. Peru. Entomol.* 27 (1984): 1–8.

———. 1990. Canopy arthropod diversity: A chronology of sampling techniques and results. *Rev. Peru. Entomol.* 32:71–77.

Erwin, T. L., and J. C. Scott. 1980. Seasonal and size patterns, trophic structure, and richness of Coleoptera in the tropical arboreal ecosystem: The fauna of the tree *Luehea seemannii* Triana and Planch in the Canal Zone of Panama. *Coleopt. Bull.* 34:305–22.

Espinoza C., M., and R. Butterfield. 1990. Adaptabilidad de 13 especies nativas maderables bajo condiciones de plantación en las tierras bajas húmedas del Atlántico, Costa Rica. Pp. 149–72 in *Manejo y aprovechamiento de plantaciones forestales con especies de uso multiple,* ed. R. Salazar. Actas Reunión IUFRO, Guatemala, April 1989. Turrialba, Costa Rica: CATIE.

Espinoza C., M., and E. González J. n.d. Ensayo con *Gmelina arborea* Roxb. y *Pseudobombax septenatum* (Jacq.) O. Report, Organization for Tropical Studies and Dirección General Forestal, San José, Costa Rica.

Estrada, A., and R. Coates-Estrada. 1984. Fruit eating and seed dispersal by howling monkeys (*Alouatta palliata*) in the tropical rain forest of Los Tuxtlas, Mexico. *Am. J. Primatol.* 6:77–91.

Estrada, A., C. Coates-Estrada, and C. Vázquez-Yánes. 1984. Observations on fruiting and dispersers of *Cecropia obtusifolia* at Los Tuxtlas, Mexico. *Biotropica* 16:315–18.

Evans, G. C. 1939. Ecological studies on the rain forest of southern Nigeria. Part 2. The atmospheric environmental conditions. *J. Ecol.* 27:436–82.

Evans, G. C. 1972. *The quantitative analysis of plant growth.* Oxford: Blackwell Scientific.

Evans, J. 1982. *Plantation forestry in the tropics.* New York: Oxford Univ. Press.

Ewel, J. J. 1970. Biomass changes in early tropical succession. *Turrialba* 21:110–12.

———. 1976. Litter fall and leaf decomposition in a tropical forest succession in eastern Guatemala. *J. Ecol.* 64:293–408.

Ewel, J., F. Benedict, C. Berish, and B. Brown. 1982. Leaf area, light transmission, roots, and leaf area damage in nine tropical plant communities. *Agro-Ecosystems* 7:305–26.

Faegri, K., and L. van der Pijl. 1979. *The principles of pollination ecology.* 3d ed. Oxford: Pergamon.

Faeth, S. H. 1986. Indirect interactions between temporally separated herbivores mediated by the host plant. *Ecology* 67:479–94.

Falconer, D. S. 1981. *Introduction to quantitative genetics.* 2d ed. London: Longman.

Farquarson, C. O. 1922. Five years' observations (1914–1918) on the bionomics of southern Nigerian insects, chiefly directed to the investigation of lycaenid life-histories and the relation of Lycaenidae, Diptera, and other insects to ants. *Trans. R. Entomol. Soc. Lond.* 1921:319–448.

Fasehun, F. E., and M. Audu. 1980. Comparative seedling growth and respiration of four tropical hardwood species. *Photosynthetica* (Prague) 14:193–97.

Fauth, J. E., B. I. Crother, and J. B. Slowinski. 1989. Elevational patterns of species richness, evenness, and abundance of the Costa Rican leaf-litter herpetofauna. *Biotropica* 21:178–85.

Fedorov, A. A. 1966. The structure of the tropical rain forest and speciation in the humid tropics. *J. Ecol.* 54:1–11.

Feeny, P. 1976. Plant apparency and chemical defense. *Recent Adv. Phytochem.* 10:1–40.

Feinsinger, P. 1976. Organization of a tropical guild of nectarivorous birds. *Ecol. Monogr.* 46:257–91.

———. 1978. Ecological interactions between plants and hummingbirds in a successional tropical community. *Ecol. Monogr.* 48:269–87.

———. 1980. Asynchronous migration and the coexistence of tropical hummingbirds. Pp. 411–19 in *Migrant birds in the Neotropics: Ecology, behavior, distribution and conservation,* cd. A. Keast and E. S. Morton. Washington, D.C.: Smithsonian Institution Press.

———. 1983. Coevolution and pollination. Pp. 283–310 in *Coevolution,* ed. D. J. Futuyma and M. Slatkin. Sunderland, Mass.: Sinauer Associates.

Feinsinger, P., J. H. Beach, Y. B. Linhart, W. H. Busby, and K. G. Murray. 1987. Disturbance, pollinator predictability, and pollination success among Costa Rican cloud forest plants. *Ecology* 68:1294–1305.

Feinsinger, P., and R. K. Colwell. 1978. Community organization among Neotropical nectar-feeding birds. *Am. Zool.* 18:779–95.

Feinsinger, P., K. G. Murray, S. Kinsman, and W. H. Busby. 1986. Floral neighborhood and pollination success in four hummingbird-pollinated cloud forest plants. *Ecology* 67:449–65.

Feinsinger, P., J. A. Wolfe, and L. A. Swarm. 1982. Island ecology: Reduced hummingbird diversity and the pollination biology of plants on Trinidad and Tobago, West Indies. *Ecology* 63:494–506.

Fernández, D., and N. Fetcher. 1991. Chlorophyll fluorescence and quantum yield in seedlings and saplings of tropical forest. P. 13 in Primer congreso Venezolano de Ecología, Resumenes (abstract). San José, Costa Rica: Editorial Guttenberg.

Fernández G., R. 1985. *Costa Rica en el Siglo XIX: Antología de Viajeros.* San José: Editorial Universidad de Costa Rica.

Fernández-Yépez, A. 1969. Contribución al conocimiento de los cíclidos. *Evencias* 22:1–17.

Fetcher, N. 1979. Water relations of five tree species on Barro Colorado Island, Panama. *Oecologia* 40:229–33.

———. 1986. Integrating photosynthetic data as a measure of acclimation to light availability. Fourth International Congress of Ecology, August 10–16, 1986, Syracuse University, Syracuse, NY. Abstract.

Fetcher, N., S. F. Oberbauer, G. Rojas, and B. R. Strain. 1987. Efectos del regimen de luz sobre la fotosíntesis y el crecimiento en plántulas de árboles de un bosque lluvioso tropical de Costa Rica. *Rev. Biol. Trop.* 35 (suppl.): 97–110.

Fetcher, N., S. F. Oberbauer, and B. R. Strain. 1985. Vegetation effects on microclimate in lowland tropical forest in Costa Rica. *Int. J. Biometeorol.* 29:145–55.

———. n.d. Growth and plasticity of seedlings of five species of tropical trees in response to changing light regimes. Typescript.

Fetcher, N., B. R. Strain, and S. F. Oberbauer. 1983. Effects of light regime on the growth, leaf morphology, and water relations of seedlings of two species of tropical trees. *Oecologia* 58:314–19.

Findley, J. S. 1976. The structure of bat communities. *Am. Nat.* 110:129–39.

Findley, J. S., and D. E. Wilson. 1983. Are bats rare in tropical Africa? *Biotropica* 15:299–303.

Finnegan, B., and C. Sabogal. 1988. El desarrollo de sistemas de producción sostenible en bosques tropicales húmedos de bajura: Un estudio de caso de Costa Rica. *El Chasquí* 17:3–24.

Fisher, E. M. 1985. A preliminary list of the robber flies (Diptera: Asilidae) of the Tambopata Reserved Zone, Madre de Dios, Peru. *Rev. Peru. Entomol.* 27 (1984):25–35.

Fisher, E. M., and H. A. Hespenheide. 1991. Taxonomy and biology of Central American robber flies (Diptera: Asilidae) with an illustrated key to genera. In *Arthropods of Panama and Middle America,* ed. A. Aiello and D. Quintero A. Oxford: Oxford Univ. Press.

Fisher, R. A. [1930] 1978. *The genetical theory of natural selection.* Reprint. New York: Dover.

Fishkind, A. S., and R. W. Sussman. 1987. Preliminary survey of the primates of the Zona Protectora and La Selva Biological Station, northeast Costa Rica. *Primate Conserv.* 8:63–66.

Fitch, H. S. 1973. Population structure and survivorship in some Costa Rican lizards. *Occasional Papers, Mus. Nat. Hist. Univ. Kans.* 18:1–41.

Fleming, T. H. 1970. Notes on the rodent faunas of two Panamanian forests. *J. Mammal.* 51:473–90.

———. 1973. The number of rodent species in two Costa Rican forests. *J. Mammal.* 54:518–21.

———. 1974a. The population ecology of two species of Costa Rican heteromyid rodents. *Ecology* 55:493–510.

———. 1974b. Social organization in two species of Costa Rican heteromyid rodents. *J. Mammal.* 55:543–61.

———. 1977a. Growth and development of two species of tropical heteromyid rodents. *Am. Midl. Nat.* 98:109–23.

———. 1977b. Response of two species of tropical heteromyid rodents to reduced food and water availability. *J. Mammal.* 58:102–6.

———. 1979. Do tropical frugivores compete for food? *Am. Zool.* 19:1157–72.

———. 1981. Fecundity, fruiting pattern and seed dispersal in *Piper amalago* (Piperaceae), a bat-dispersed tropical shrub. *Oecologia* 51:42–46.

———. 1983. *Heteromys desmarestianus.* Pp. 474–75 in

Costa Rican natural history, ed. D. H. Janzen. Chicago: Univ. of Chicago Press.

———. 1986. Opportunism versus specialization: The evolution of feeding strategies in frugivorous bats. Pp. 105–18 in *Frugivores and seed dispersal,* ed. A. Estrada and T. H. Fleming. Dordrecht: W. Junk.

———. 1988. *The short-tailed fruit bat: A study in plant-animal interactions.* Chicago: Univ. of Chicago Press.

———. 1991. The relationship between body size, diet, and habitat use in frugivorous bats, genus *Carollia* (Phyllostomidae). *J. Mammal.* 72:493–501.

Fleming, T. H., R. Breitwisch, and G. H. Whitesides. 1987. Patterns of tropical vertebrate frugivore diversity. *Ann. Rev. Ecol. Syst.* 18:91–109.

Fleming, T. H., and G. J. Brown. 1975. An experimental analysis of seed hoarding and burrowing behavior of two species of Costa Rican heteromyid rodents. *J. Mammal.* 56:301–15.

Fleming, T. H., and E. R. Heithaus. 1981. Frugivorous bats, seed shadows, and the structure of tropical forests. *Biotropica* 13 (suppl.): 45–53.

Fleming, T. H., E. T. Hooper, and D. E. Wilson. 1972. Three Central American bat communities: Structure, reproductive cycles, and movement patterns. *Ecology* 53:555–69.

Fleming, T. H., and J. Maguire. 1988. Patterns of diversity, density, and distribution in a wet tropical forest *Piper* flora. *Bull. Ecol. Soc. Am.* 69:133 (abstract).

Fleming, T. H., and B. L. Partridge. 1984. On the analysis of phenological overlap. *Oecologia* 62:344–50.

Fleming, T. H., C. F. Williams, F. J. Bonaccorso, and L. H. Herbst. 1985. Phenology, seed dispersal, and colonization in *Muntingia calabura,* a Neotropical tree. *Am. J. Bot.* 72:383–91.

Flores R., J. G. 1985. Diagnóstico del sector industrial forestal. San José, Costa Rica: Editorial Universidad Estatal a Distancia.

Flores Silva, E. 1987. *Geografía de Costa Rica.* 2d ed. San José, Costa Rica: Editorial Universidad Estatal a Distancia.

Fogden, M. P. L. 1972. The seasonality and population dynamics of equatorial forest birds in Sarawak. *Ibis* 114:307–43.

Food and Agriculture Organization. 1968. Guía para la descripción de perfiles de suelos. Rome, Italy: FAO.

Forbes, W. T. M. 1960. *Lepidoptera of New York and neighboring states.* Part 4. *Agaristidae through Nymphalidae including butterflies.* Cornell Agricultural Experiment Station Memoirs no. 371.

Foresta, H. de, P. Charles-Dominique, C. Erard, and M. F. Prévost. 1984. Zoochorie et premiers stades de la régénération naturelle après coupe en forêt guyanaise. *Rev. Ecol. Terre Vie* 39:369–400.

Forman, R. T. T. 1975. Canopy lichens with blue-green algae: A nitrogen source in a Columbian rainforest. *Ecology* 56:1176–84.

Forman, R. T. T., and C. D. Hahn. 1980. Spatial patterns of trees in a Caribbean semi-evergreen forest. *Ecology* 61:1267–74.

Foster, M. S. 1975. The overlap of molting and breeding in some tropical birds. *Condor* 77:304–14.

———. 1977. Ecological and nutritional effects of food scarcity on a tropical frugivorous bird and its fruit source. *Ecology* 58:73–85.

———. 1978. Total frugivory in tropical passerines: A reappraisal. *Trop. Ecol.* 2:131–54.

———. 1985. Pre-nesting cooperation in birds: Another form of helping behavior. Pp. 817–28 in *Neotropical ornithology,* ed. P. A. Buckley, M. S. Foster, E. S. Morton, R. S. Ridgely, and F. G. Buckley. Ornithological Monographs vol. 36. Lawrence: AOU.

———. 1990. Factors influencing bird foraging preferences among conspecific fruit trees. *Condor* 92:844–54.

Foster, M. S., and R. M. Timm. 1976. Tent-making by *Artibeus jamaicensis* (Chiroptera: Phyllostomatidae) with comments on plants used by bats for tents. *Biotropica* 8:265–69.

Foster, R. B. 1977. *Tachigalia versicolor* is a suicidal Neotropical tree. *Nature* 268:624–26.

———. 1982a. Famine on Barro Colorado Island. Pp. 201–12 in *The ecology of a tropical forest: Seasonal rhythms and long-term changes,* ed. E. G. Leigh, Jr., A. S. Rand, and D. M. Windsor. Washington, D.C.: Smithsonian Institution Press.

———. 1982b. The seasonal rhythm of fruitfall on Barro Colorado Island. Pp. 151–72 in *The ecology of a tropical forest: Seasonal rhythms and long-term changes,* ed. E. G. Leigh, Jr., A. S. Rand, and D. M. Windsor. Washington, D.C.: Smithsonian Institution Press.

———. 1990. The floristic composition of the Río Manu floodplain forest. Pp. 99–111 in *Four Neotropical rainforests,* ed. A. H. Gentry. New Haven, Conn.: Yale Univ. Press.

Foster, R. B., and S. P. Hubbell. 1990. The floristic composition of the Barro Colorado Island forest. Pp. 85–98 in *Four Neotropical rainforests,* ed. A. Gentry. New Haven, Conn.: Yale Univ. Press.

Fowler, H. P. 1965. *A dictionary of modern English usage.* Oxford: Oxford Univ. Press.

Fowler, S. V., and J. H. Lawton. 1985. Rapidly induced defenses and talking trees: The devil's advocate position. *Am. Nat.* 126:181–95.

Fox, L. R. 1980. Defense and dynamics in plant-herbivore systems. *Am. Zool.* 21:853–64.

Fox, L. R., and P. A. Morrow. 1983. Estimates of damage by herbivorous insects on *Eucalyptus* trees. *Aust. J. Ecol.* 8:139–47.

Frank, S. A. 1989. Ecological and evolutionary dynamics of fig communities. *Experientia* 45:674–80.

Frankel, R., and E. Galun. 1977. *Pollination mechanisms, reproduction and plant breeding.* Berlin: Springer-Verlag.

Frankie, G. W. 1975. Tropical forest phenology and pollinator plant coevolution. Pp. 192–209 in *Coevolution of plants and animals,* ed. L. E. Gilbert and P. H. Raven. Austin: Univ. of Texas Press.

———. 1976. Pollination of widely dispersed forest trees by animals in Central America, with an emphasis on wild bee pollinated systems. *Linn. Soc. Symp. Ser.* 2:151–59.

Frankie, G. W., and H. G. Baker. 1974. The importance of pollinator behavior in the reproductive biology of tropical trees. *An. Inst. Biol. Univ. Nac. Auton. Mex., Ser. Bot.,* no. 45, 1:1–10.

Frankie, G. W., H. G. Baker, and P. A. Opler. 1974a. Comparative phenological studies of trees in tropical wet and dry forest in the lowlands of Costa Rica. *J. Ecol.* 62:881–919.

———. 1974b. Tropical plant phenology: Applications for studies in community ecology. Pp. 287–96 in *Ecological Studies.* Vol. 18, *Phenology and seasonality modeling,* ed. H. Lieth. New York: Springer-Verlag.

Frankie, G. W., and W. A. Haber. 1983. Why bees move among mass-flowering Neotropical trees. Pp. 360–72 in *Handbook of experimental pollination biology,* ed. C. E. Jones and R. J. Little. New York: Van Nostrand Reinhold.

Frankie, G. W., W. A. Haber, P. A. Opler, and K. S. Bawa. 1983. Characteristics and organization of the large bee pollination system in the Costa Rican dry forest. Pp. 411–44 in *Handbook of experimental pollination biology,* ed. C. E. Jones and R. J. Little. New York: Van Nostrand Reinhold.

Frankie, G. W., P. A. Opler, and K. S. Bawa. 1976. Foraging behavior of solitary bees: Implications for outcrossing of a Neotropical forest tree species. *J. Ecol.* 64:1049–57.

Franks, N. G. 1982. Social insects in the aftermath of the swarm raids of the army ant, *Eciton burchelli.* Pp. 275–79 in *The biology of social insects,* ed. M. D. Breed, C. D. Michener, and H. E. Evans. Proceedings of the Ninth International Congress of the International Union for the Study of Social Insects. Boulder, Colo.: Westview Press.

Franks, N. G., and W. H. Bossert. 1983. The influence of swarm raiding army ants on the patchiness and diversity of a tropical leaf litter ant community. Pp. 151–63 in *Tropical rain forest: Ecology and management,* ed. S. L. Sutton, T. C. Whitmore, and A. C. Chadwick. Oxford: Blackwell Scientific Publications.

Freckman, D. C., and E. C. Caswell. 1985. Ecology of nematodes in agroecosystems. *Ann. Rev. Phytopathology* 23:275–98.

Freese, C. H. 1977. Food habits of white-faced capuchins *(Cebus capucinus)* in Santa Rosa National Park, Costa Rica. *Brenesia* 10/11:43–56.

Freese, C. H. 1983. *Cebus capucinus.* Pp. 458–60 in *Costa Rican natural history,* ed. D. H. Janzen. Chicago: Univ. of Chicago Press.

Frith, C. B., and D. W. Frith. 1985. Seasonality of insect abundance in an Australian upland tropical rainforest. *Aust. J. Ecol.* 10:237–48.

Fritz, G. 1979. Foraging patterns of ants on vines. Report, OTS 79-3:101–7.

Fritz, G. N. 1983. Biology and ecology of bat flies (Diptera: Streblidae) on bats in the genus *Carollia. J. Med. Entomol.* 20:1–10.

Fritz, G., A. S. Rand, and C. W. dePamphilis. 1981. The aposematically colored frog, *Dendrobates pumilio,* is distasteful to the large, predatory ant, *Paraponera clavata. Biotropica* 13:158–59.

Frost, S. W. 1965. *Insect life and insect natural history.* New York: Dover.

Fundación Neotrópica. 1988. *Desarrollo socioeconómico y el ambiente natural de Costa Rica,* ed. A. Ramiréz Solera and T. Maldonado Ulloa. San José, Costa Rica: Editorial Heliconia.

Furch, K. 1984. Water chemistry of the Amazonian basin: The distribution of chemical elements among freshwaters. Pp.

167–200 in *The Amazon, limnology and landscape ecology of a mighty tropical river and its basin,* ed. H. Sioli. The Hague: W. Junk.

Futuyma, D. 1976. Food plant specialization and environmental predictability in Lepidoptera. *Am. Nat.* 110:285–92.

———. 1991. Evolution of host specificity in herbivorous insects: Genetic, ecological, and phylogenetic aspects. Pp. 431–54 in *Plant-animal interactions, evolutionary ecology in tropical and temperate regions,* ed. P. W. Price, T. M. Lewinsohn, G. W. Fernandes, and W. W. Benson. New York: J. Wiley.

Futuyma, D. S., and M. Slatkin, eds. 1983. *Coevolution.* Sunderland, Mass.: Sinauer Associates.

Gadgil, M., and W. H. Bossert. 1970. Life historical consequences of natural selection. *Am. Nat.* 104:1–24.

Gage, D. A., and D. R. Strong, Jr. 1981. The chemistry of *Heliconia imbricata* and *H. latispatha* and the slow growth of a hispine beetle herbivore. *Biochem. Syst. Ecol.* 9:79–82.

Gaines, S. D., and J. Lubchenko. 1982. A unified approach to marine plant-herbivore interactions. Part 2. Biogeography. *Ann. Rev. Ecol. Syst.* 13:111–38.

Gallina, S., E. Maury, and V. Serrano. 1978. Hábitos alimenticios del venado colablanco (*Odocoileus virginianus* Rafinesque) en la reserva La Michilia, estado de Durango. Pp. 57–108 in *Reservas de la biosfera en el estado de Durango,* ed. G. Halffler. Publication no. 4, Instituto de Ecología, A. C., Mexico, D.F.

Galloway, J. N., G. E. Likens, and M. E. Hawley. 1984. Acid precipitation: Natural versus anthropogenic components. *Science* 216:829–31.

Galloway, J. N., and G. G. Parker. 1980. Difficulties in measuring wet and dry deposition on forest canopies and soil surfaces. Pp. 57–68 in *Effects of acid precipitation on terrestrial ecosystems,* ed. T. C. Hutchinson and M. Havas. New York: Plenum.

Gan, Y.-Y., F. W. Robertson, P. S. Ashton, E. Soepadmo, and D. W. Lee. 1977. Genetic variation in wild populations of rain-forest trees. *Nature* 269:323–25.

Gan, Y.-Y., F. W. Robertson, and E. Soepadmo. 1981. Isozyme variation in some rain forest trees. *Biotropica* 13:20–28.

Ganzhorn, J. U. 1988. Food partitioning among Malagasy primates. *Oecologia* 75:436–50.

Garber, P. A. 1986. The ecology of seed dispersal in two species of callitrichid primates (*Saguinus mystax* and *Saguinus fuscicollis*). *Am. J. Primatol.* 10:155–70.

Gardella, D. 1981. Current level of settlement and types of agricultural/silvicultural activities in Sarapiquí. Report, USAID Rural Development Office. San José, Costa Rica.

Gardner, A. L. 1983. *Oryzomys caliginosus.* Pp. 483–85 in *Costa Rican natural history,* ed. D. H. Janzen. Chicago: Univ. of Chicago Press.

Gardner, A. L., R. K. LaVal, and D. E. Wilson. 1970. The distributional status of some Costa Rican bats. *J. Mammal.* 51:712–29.

Garita C., D. 1989. Mapa de cobertura boscosa. Report, Unidad de Cartografía y Topografía, DGF, MIRENEM, San José: Costa Rica.

Gartlan, J. S., D. B. McKey, P. G. Waterman, C. N. Mbi, and T. T. Struhsaker. 1980. A comparative study of the phytochemistry of two African rain forests. *Biochem. Syst. Ecol.* 8:401–22.

Gartner, B. L. 1989. Breakage and regrowth of *Piper* species in rain forest understory. *Biotropica* 21:303–7.

Garwood, N. C. 1983. Seed germination in a seasonal tropical forest in Panama: A community study. *Ecol. Monogr.* 53:159–81.

Gascon, C. 1991. Population- and community-level analyses of species occurrences of central Amazonian rainforest tadpoles. *Ecology* 72:1731–46.

Gaston, K. J. 1991. The magnitude of global insect species richness. *Conserv. Biol.* 5:283–96.

Gauld, I. D. 1986. Latitudinal gradients in ichneumonid species-richness in Australia. *Ecol. Entomol.* 11:155–62.

Gautier-Hion, A. 1983. Leaf consumption by monkeys in western and eastern Africa: A comparison. *African J. Ecol.* 21:107–13.

———. 1984. La dissémination des graines par les cercopithèques forestiers africains. *Rev. Ecol. Terre Vie* 39:159–65.

Gautier-Hion, A., J. M. Duplantier, R. Quris, F. Feer, C. Sourd, J. P. Decoux, G. Dubost, L. Emmons, C. Erard, P. Hecketsweiler, A. Moungazi, C. Roussilhon, and J. M. Thiollay. 1985. Fruit characters as a basis of fruit choice and seed dispersal in a tropical forest vertebrate community. *Oecologia* 65:324–37.

Gautier-Hion, A., L. H. Emmons, and G. Dubost. 1980. A comparison of the diets of three major groups of primary consumers of Gabon (primates, squirrels, and ruminants). *Oecologia* 45:182–89.

Gautier-Hion, A., and G. Michaloud. 1989. Are figs always keystone resources for tropical frugivorous vertebrates? A test in Gabon. *Ecology* 70:1826–33.

Gentry, A. H. 1974. Flowering phenology and diversity in tropical Bignoniaceae. *Biotropica* 6:64–68.

———. 1978a. Anti-pollinators for mass-flowering plants? *Biotropica* 10:68–69.

———. 1978b. Floristic knowledge and needs in Pacific tropical America. *Brittonia* 30:134–53.

———. 1982a. Patterns of Neotropical plant species diversity. *Evol. Biol.* 15:1–84.

———. 1982b. Phytogeographic patterns as evidence for a Choco refuge. In *Biological diversification in the tropics,* ed. G. Prance. New York: Columbia Univ. Press.

———. 1986. Endemism in tropical versus temperate plant communities. Pp. 153–81 in *Conservation biology: The science of scarcity and diversity,* ed. M. Soulé. Sunderland, Mass.: Sinauer Associates.

———. 1988. Tree species richness of upper Amazonian forests. *Proc. Nat. Acad. Sci.* 85:156–59.

———, ed. 1990a. *Four Neotropical rainforests.* New Haven, Conn.: Yale Univ. Press.

———. 1990b. Floristic similarities and differences between southern Central America and upper and central Amazonia. Pp. 141–57 in *Four Neotropical rainforests,* ed. A. H. Gentry. New Haven, Conn.: Yale Univ. Press.

Gentry, A. H., and C. Dodson. 1987a. Contribution of nontrees to species richness of a tropical rain forest. *Biotropica* 19:149–56.

———. 1987b. Diversity and biogeography of Neotropical epiphytes. *Ann. MO Bot. Gard.* 74:205–33.

Gentry, A. H., and L. H. Emmons. 1987. Geographic variation in fertility, phenology, and composition of the understory of Neotropical forests. *Biotropica* 19:216–27.

Gershenzon, J. 1984. Changes in the levels of plant secondary metabolites under water and nutrient stress. In *Phytochemical adaptations to stress,* eds. R. Timmerman and C. Steeline. *Rec. Adv. Phytochem.* 10:273–320.

Gessel, S. P., D. W. Cole, D. Johnson, and J. Turner. 1979. The nutrient cycles of two Costa Rican forests. Pp. 623–43 in *Actas del IV Symposium Internacional de Ecología Tropical* Vol. 2. Panama: Univ. of Panama.

———. 1980. The nutrient cycles of two Costa Rican forests. *Prog. Ecol.* 3:23–44.

Gibbs, P. E., J. Semir, and N. Cruz. 1977. Floral biology of *Talauma ovata* St. Hil. (Magnoliaceae). *Cienc. Cult.* (São Paulo) 29:1436–41.

Gilbert, L. E. 1969. Some aspects of the ecology and community structure of ithomiid butterflies in Costa Rica. Report, OTS 69-6:70–92.

———. 1975. Ecological consequences of coevolved mutualism between butterflies and plants. Pp. 210–40 in *Coevolution of animals and plants,* ed. L. E. Gilbert and P. H. Raven. Austin: Univ. of Texas Press.

———. 1977. The role of insect-plant coevolution in the organization of ecosystems. Pp. 339–413 in *Comportement des insectes et milieu trophique.* Colloques Internationaux du CRNS. Paris: Editions de CRNS.

———. 1980. Food web organization and the conservation of Neotropical diversity. Pp. 11–33 in *Conservation biology,* ed. M. Soulé and H. Wilcox. Sunderland, Mass.: Sinauer Associates.

———. 1983. Coevolution and mimicry. Pp. 263–81 in *Coevolution,* ed. D. Futuyma and M. Slatkin. Sunderland, Mass.: Sinauer Associates.

———. 1984. The biology of butterfly communities. *Symp. R. Entomol. Soc. Lond.* 11:41–54.

——— 1991. Biodiversity of a Central American *Heliconius* community: Pattern, process, and problems. Pp. 403–27 in *Plant-animal interactions: Evolutionary ecology in tropical and temperate regions,* ed. P. W. Price, T. M. Lewinsohn, G. W. Fernandes, and W. W. Benson. New York: J. Wiley.

Gilbert, L. E., and M. C. Singer. 1973. Dispersal and gene flow in a butterfly species. *Am. Nat.* 107:58–72.

———. 1975. Butterfly ecology. *Annu. Rev. Ecol. Syst.* 6:365–97.

Gilbert, L. E., and J. T. Smiley. 1978. Determinants of local diversity in phytophagous insects: Host specialists in tropical environments. In *Diversity of insect faunas,* ed. L. A. Mound and N. Waloff. *Symp. R. Entomol. Soc. Lond.* 9:89–105.

Gill, D. E. 1979. The evolution of leks and courtship displays in the manakins: The case of the yellow-thighed manakin, *Pipra mentalis.* Report, OTS 79-3:135–37.

Givnish T. J., and G. J. Vermeij. 1976. Sizes and shapes of liane leaves. *Am. Nat.* 110:743–78.

Glander, K. E. 1981. Feeding patterns in mantled howling monkeys. Pp. 231–57 in *Foraging behavior: Ecological, ethological, and psychological approaches,* ed. A. C. Kamil and T. D. Sargent. New York: Garland Press.

———. 1983. *Aloutta palliata.* Pp. 448–49 in *Costa Rican natural history,* ed. D. H. Janzen. Chicago: Univ. of Chicago Press.

Glanz, W. E. 1982. The terrestrial mammal fauna of Barro Colorado Island: Censuses and long-term changes. Pp. 455–68 in *The ecology of a tropical forest: Seasonal rhythms and long-term changes,* ed. E. G. Leigh, Jr., A. S. Rand, and D. M. Windsor. Washington, D.C.: Smithsonian Institution Press.

———. 1990. Neotropical mammal densities: How unusual is the community on Barro Colorado Island, Panama? Pp. 287–313 in *Four Neotropical rainforests,* ed. A. H. Gentry. New Haven, Conn.: Yale Univ. Press.

Glanz, W. E., R. W. Thorington, J. Giacalone-Madden, and L. R. Heaney. 1982. Seasonal food use and demographic trends in *Sciurus granatensis.* Pp. 239–52 in *The ecology of a tropical forest: Seasonal rhythms and long-term changes,* ed. E. G. Leigh, Jr., A. S. Rand, and D. M. Windsor. Washington, D.C.: Smithsonian Institution Press.

Golley, F. B. 1983a. The abundance of energy and chemical elements. Pp. 101–15 in *Ecosystems of the world.* Vol. 14A, *Tropical rain forest ecosystems,* ed. F. B. Golley. Amsterdam: Elsevier Scientific.

Golley, F. B. 1983b. Nutrient cycling and nutrient conservation. Pp. 137–56 in *Ecosystems of the world.* Vol. 14A, *Tropical rain forest ecosystems,* ed. F. B. Golley. Amsterdam: Elsevier Scientific.

Golley, F. B., ed. 1983c. *Ecosystems of the World.* Vol. 14A, *Tropical rain forest ecosystems.* Amsterdam: Elsevier Scientific.

Golley, F. B., J. T. McGinnis, R. G. Clements, G. I. Child, and M. J. Duever. 1975. *Mineral cycling in tropical forest ecosystems.* Athens: Univ. of Georgia Press.

Golterman, H. L. 1975. Chemistry. In *River ecology,* ed. B. A. Whitton. Berkeley: Univ. of California.

Gómez P., L. D. 1982. The origin of the pteridopyte flora of Central America. *Ann. MO Bot. Gard.* 69:548–56.

Gómez, L. D., and J. Savage. 1983. Searchers on that rich coast: Costa Rican field biology, 1400–1980. Pp. 1–11 in *Costa Rican natural history,* ed. D. H. Janzen. Chicago: Univ. of Chicago Press.

Gómez-Pompa, A. 1971. Posible papel de la vegetación secundaria en la evolución de la flora tropical. *Biotropica* 3:125–35.

Gómez-Pompa, A., and F. W. Burley. 1991. The management of natural tropical forests. Pp. 3–18 in *Rain forest regeneration and management,* ed. A. Gómez-Pompa, T. C. Whitmore, and M. Hadley. Man and Biosphere series, vol. 6. Paris: UNESCO; Carnforth, UK: Parthenon.

Gómez-Pompa, and S. del Amo R., eds. 1985. *Investigaciones sobre la regeneración de selvas altas en Veracruz, Mexico.* Vol. 2. Mexico: Instituto Nacional de Investigaciones sobre Recursos Bióticos.

Gómez-Pompa, A., and C. Vásquez-Yánes. 1974. Studies on the secondary succession of tropical lowlands: The life cycle of secondary species. Pp. 336–42 in Proceedings of the First International Congress on Ecology.

Gómez-Pompa, A., C. Vázquez-Yánes, S. del Amo R., and A. Butanda C., eds. 1976. *Investigaciones sobre la regeneración de selvas altas en Veracruz, México.* Mexico: Companía Editorial Continental.

Gómez-Pompa, A., T. C. Whitmore, and M. Hadley, eds. 1991. *Rain forest regeneration and management.* Man and the Biosphere series, vol. 6. Paris: UNESCO; Carnforth, UK: Parthenon.

González, L. 1988. Tenencia de la tierra. Annex E in report to USAID, Forestry and Wildlands Management Project Identification Document. Tropical Science Center, San José, Costa Rica.

González, M. J., C. Alarcón, and G. Barrantes. 1985. Análisis de herbivoría tomando en cuenta pubescencias, textura de la hoja y el hábito de la planta. Report, OTS 85-2:172–85.

González J., E. 1990. Informe sobre las actividades del programa piloto de la Fase II del Proyecto TRIALS, Financiado por World Wildlife Fund. Report, Organización para Estudios Tropicales, San José, Costa Rica.

———. 1991. Recolección y germinación de semillas de 26 especies arboreas del bosque húmedo tropical. *Rev. Biol. Trop.* 39:1–6.

González Vega, C. 1970. The marketing of agricultural products in Costa Rica. Report, Associated Colleges of the Midwest y Universidad de Costa Rica, San José, Costa Rica.

Goodwin, D. 1970. *Pigeons and doves of the world.* London: British Museum of Natural History.

Goodwin, G. G. 1946. Mammals of Costa Rica. *Bull. Am. Mus. Nat. Hist.* 87:271–474.

Gordon, B. L. 1982. *A Panama forest and shore.* Pacific Grove, Calif.: Boxwood Press.

Gosse, J. P. 1963. Le milieu aquatique et l'écologie des poissons dans la région de Yangambí. *Ann. Mus. R. Afr. Cent. Zool., Ser. Quart. Sci.* 8v, 116:113–270.

Gosz, F. R., G. E. Likens, and F. H. Bormann. 1972. Nutrient content of litter fall on the Hubbard Brook experimental forest, New Hampshire. *Ecology* 53:769–84.

Gottsberger, G. 1978. Seed dispersal by fish in the inundated regions of Humaita, Amazonia. *Biotropica* 10:170–83.

Goulding, M. 1980. *The fishes and the forest. Explorations in Amazonian natural history.* Berkeley: Univ. of California Press.

Gower, S. T. 1987. Relations between mineral nutrient availability and fine root biomass in two Costa Rican tropical wet forests: A hypothesis. *Biotropica* 19:171–75.

Gradwohl, J., and R. Greenberg. 1980. The formation of antwren flocks on Barro Colorado Island, Panama. *Auk* 97:385–95.

———. 1982. The effect of a single species of avian predator on the arthropods of aerial leaf litter. *Ecology* 63:581–83.

Graf, N. R. de. 1986. A silvicultural system for natural regeneration of tropical rain forest in Suriname. *Agric. Univ. Wageningen,* the Netherlands.

Gramzow, R. H., and W. K. Henry. 1972. The rainy pentads of Central America. *J. App. Meteorol.* 11:637–742.

Gray, B. 1972. Economic tropical forest entomology. *Annu. Rev. Entomol.* 17:313–54.

Grayum, M. H., and H. W. Churchill. 1987. An introduction to the pteridophyte flora of Finca La Selva, Costa Rica. *Am. Fern J.* 77:73–89.

———. 1989a. The vascular flora of La Selva Biological Station, Costa Rica: Lycopodiophyta. *Selbyana* 11:61–65.

———. 1989b. The vascular flora of La Selva Biological Station, Costa Rica: Polypodiophyta. *Selbyana* 11:66–118.

Grayum, M. H., and B. E. Hammel. 1982. Three new species of Cyclanthaceae from the Caribbean lowlands of Costa Rica. *Syst. Bot.* 7:221–29.

Greding, E. J., Jr. 1972. Call specificity and hybrid compatibility between *Rana pipiens* and three other *Rana* species in Central America. *Copeia* 1972:383–85.

———. 1976. Call of the tropical American frog *Rana palmipes* Spix (Amphibia, Anura, Ranidae). *J. Herpetol.* 10:263–64.

Green, N. B., and T. K. Pauley. 1987. *Amphibians and reptiles in West Virginia.* Pittsburgh, Pa.: Univ. of Pittsburgh Press.

Greenbaum, I. F., R. J. Baker, and D. E. Wilson. 1975. Evolutionary implications of the karyotypes of the stenodermine genera *Ardops, Ariteus, Phyllops,* and *Ectophylla. Bull. South. Calif. Acad. Sci.* 74:156–59.

Greenberg, R. 1979. Body size, breeding habitat, and winter exploitation systems in *Dendroica. Auk* 96:756–66.

———. 1980. Demographic aspects of long distance migration in birds. Pp. 493–504 in *Migrant birds in the Neotropics: Ecology, behavior, distribution, and conservation,* ed. A. Keast and E. S. Morton. Washington, D.C.: Smithsonian Institution Press.

———. 1981. The abundance and seasonality of forest canopy birds on Barro Colorado Island, Panama. *Biotropica* 13:241–51.

———. 1984. The winter exploitation systems of baybreasted and chestnut-sided warblers in Panama. *Univ. Calif. Pub. Zool.* 116:1–107.

———. 1986. Competition in migrant birds in the nonbreeding season. *Curr. Ornithol.* 3:281–307.

———. 1987. Seasonal foraging specialization in the wormeating warbler. *Condor* 89:158–68.

Greenberg, R., and J. Gradwohl. 1986. Constant density and stable territoriality in some tropical insectivorous birds. *Oecologia* 69:618–25.

Greene, H. W. 1986. Diet and arboreality in the emerald monitor, *Varanus prasinus,* with comments on the study of adaptation. *Fieldiana Zool.* n.s. 31:1–12.

———. 1988. Species richness in tropical predators. Pp. 259–80 in *Diversity and conservation of tropical rainforests,* ed. F. Almeda and C. M. Pringle. San Francisco: California Academy of Science.

———. 1989. Agonistic behavior by three-toed sloths, *Bradypus variegatus. Biotropica* 21:369–72.

Greene, H. W., and F. M. Jaksic. 1983. Food-niche relationships among sympatric predators: Effects of level of prey identification. *Oikos* 40:151–54.

Greene, H. W., and M. Santana. 1983. Field studies of hunting behavior by bushmaster. *Am. Zool.* 23:897.

Gregory, P. T. 1983. Habitat structure affects diel activity pattern in the Neotropical frog *Leptodactylus melanonotus. J. Herpetol.* 17:181–84.

Greig, N. 1991. Ecology of co-occurring species of Neotropical *Piper:* Distribution, reproductive biology, and seed predation. Ph.D. diss., University of Texas, Austin.

———. 1993. Predispersal seed predation on five *Piper* species in tropical rainforest. *Oecologia.*

Greig-Smith, P. 1969. Application of numerical methods to tropical forests. Pp. 195–206 in *Statistical ecology.* Vol. 3, *Populations, ecosystems, and systems analysis,* ed. G. P. Patil, E. C. Pielou, and W. E. Waters. University Park: Pennsylvania State Univ. Press.

———. 1979. Pattern in vegetation. *J. Ecol.* 67:775–79.

Grimm, U., and H. W. Fassbender. 1981. Ciclos biogeoquímicos en un ecosistema forestal de los Andes occidentales de Venezuela III. Ciclo hidrológico y translocación de elementos químicos con el aqua. *Turrialba* 31:89–99.

Griswold, G. 1977. Ants on *Costus* II. Report, OTS 77-3:1–3.

Grove, K. F. 1985. Reproductive biology of Neotropical wet forest understory plants. Ph.D. diss., University of Iowa, Iowa City.

———. 1987. Patterns of nectar production among hermit hummingbird-pollinated plants in a Neotropical wet forest. Typescript.

Grove, K., and A. Sauer. 1977. Predation by parrots on seeds of *Jacaratia costaricensis* (Caricaceae). Report, OTS 77-4:469–72.

Grubb, P. J. 1977a. Control of forest growth and distribution on wet tropical mountains: With special reference to mineral nutrition. *Annu. Rev. Ecol. Syst.* 8:83–107.

———. 1977b. The maintenance of species richness in plant communities: The importance of the regeneration niche. *Biol. Rev.* 52:107–145.

Guevara S., S. A. 1986. Plant species availability and regeneration in a Mexican tropical rain forest. Ph.D. diss., Uppsala University, Uppsala, Sweden.

Guilford, T. 1986. How do "warning colours" work?—Conspicuousness may reduce recognition errors in experienced predators. *Anim. Behav.* 34:286–88.

Gutschick, V. P. 1981. Evolved strategies in nitrogen acquisition by plants. *Am. Nat.* 118:607–37.

Guttman, S. I., T. K. Wood, and A. A. Karlin. 1981. Genetic differentiation along host plant lines in the sympatric *Enchenopa binotata* Say complex (Homoptera: Membracidae). *Evolution* 35:205–17.

Guyer, C. 1986. Seasonal patterns of reproduction of *Norops humilis* (Sauria: Iguanidae) in Costa Rica. *Rev. Biol. Trop.* 34:247–51.

———. 1988a. Food supplementation in a tropical mainland anole, *Norops humilis:* Demographic effects. *Ecology* 69:350–61.

———. 1988b. Food supplementation in a tropical mainland anole, *Norops humilis:* Effects on individuals. *Ecology* 69:362–96.

———. 1990. The herpetofauna of La Selva, Costa Rica. Pp. 371–85 in *Four Neotropical rainforests,* ed. A. H. Gentry. New Haven, Conn.: Yale Univ. Press.

Guyer, C., and J. M. Savage. 1986. Cladistic relationships among anoles (Sauria: Iguanidae). *Syst. Zool.* 35:509–31.

———. 1988a. Reproductive patterns of selected understory trees in the Malaysian rain forest: The apomictic species. *Bot. J. Linn. Soc.* 97:317–31.

Ha, C. O., V. E. Sands, E. Soepadmo, and K. Jong. 1988b.

Reproductive patterns of selected understory trees in the Malaysian rain forest: The sexual species. *Bot. J. Linn. Soc.* 97:295–316.

Haber, W. A. 1978. Evolutionary ecology of tropical mimetic butterflies. Ph.D. diss., University of Minnesota, Minneapolis.

———. 1983. *Solanum siparunoides.* Pp. 326–28 in *Costa Rican natural history,* ed. D. H. Janzen. Chicago: Univ. of Chicago Press.

Haber, W. A., and G. W. Frankie. 1989. A tropical hawkmoth community: Costa Rican dry forest Sphingidae. *Biotropica* 21:155–72.

Haffer, J. 1985. Avian zoogeography of the Neotropical lowlands. Pp. 113–45 in *Neotropical ornithology,* ed. P. A. Buckley, M. S. Foster, E. S. Morton, R. S. Ridgely, and N. G. Buckley. Ornithological Union Monograph, vol. 36. Washington, D.C.: AOU.

Hafner, M. S., and D. J. Hafner. 1987. Geographic distribution of two Costa Rican species of *Orthogeomys,* with comments on dorsal pelage markings in the Geomyidae. *Southwest. Nat.* 32:5–11.

Haines, B. L. 1975. Impact of leaf-cutting ants on vegetation development on Barro Colorado Island. In *Tropical ecological systems,* ed. F. B. Golley and E. Medina. Berlin: Springer-Verlag.

Haines, B. L., and L. McHargue. 1986. Sulfur gas emissions from roots of the rainforest tree *Stryphnodendron excelsum* Harms in Costa Rica. *Ecol. Soc. Am.* 1986:168 (abstract).

Haines, B. L., M. Black, J. Fail, Jr., L. McHargue, and G. Howell. 1987. Potential sulphur gas emissions from a tropical rainforest and a Southern Appalachian deciduous forest. Pp. 599–610 in *Effects of atmospheric pollutants on forests, wetlands, and agricultural ecosystems,* ed. T. C. Hutchinson and K. M. Meema. Berlin: Springer-Verlag.

Hall, C. 1984. Costa Rica, una interpretación geográfica con perspectiva histórica. San José: Editorial Universidad de Costa Rica.

Hall, J. B., and M. D. Swaine. 1981. *Distribution and ecology of vascular plants in a tropical rain forest.* The Hague: W. Junk.

Hallé, F., R. A. A. Oldeman, and P. B. Tomlinson. 1978. *Tropical trees and forest: An architectural analysis.* Berlin: Springer-Verlag.

Hallwachs, W. 1986. Agoutis *(Dasyprocta punctata):* The inheritors of guapinol *(Hymenaea courbaril:* Leguminosae). Pp. 285–304 in *Frugivores and seed dispersal,* ed. A. Estrada and T. H. Fleming. The Hague: W. Junk.

Hamilton, C. 1985. An ecosystematic study and revision of *Psychotria* subgenus *Psychotria* (Rubiaceae) in Mexico and Central America. Ph.D. diss., Washington University, St. Louis, Mo.

Hammel, B. E. 1984. Systematic treatment of the Cyclanthaceae, Marantaceae, Cecropiaceae, Clusiaceae, Lauraceae, and Moraceae for the flora of a wet lowland tropical forest, Finca La Selva, Costa Rica. Ph.D. diss., Duke University, Durham, N.C.

———. 1986a. Characteristics and phytogeographical analysis of a subset of the flora of La Selva (Costa Rica). *Selbyana* 9:149–55.

———. 1986b. New species of Clusiaceae from Central

America with notes on *Clusia* and synonymy in the Clusieae. *Selbyana* 9:112–20.

———. 1986c. New species and notes on Lauraceae from the Caribbean lowlands of Costa Rica. *J. Arnold Arbor. Harv. Univ.* 67:123–36.

———. 1986d. Notes on the Cyclanthaceae of southern Central America including three new species. *Phytologia* 60:5–15.

———. 1986e. The vascular flora of La Selva Biological Station, Costa Rica—Cecropiaceae. *Selbyana* 9:192–95.

———. 1986f. The vascular flora of La Selva Biological Station, Costa Rica—Cyclanthaceae. *Selbyana* 9:196–202.

———. 1986g. The vascular flora of La Selva Biological Station, Costa Rica—Guttiferae. *Selbyana* 9:203–17.

———. 1986h. The vascular flora of La Selva Biological Station, Costa Rica—Lauraceae. *Selbyana* 9:218–33.

———. 1986i. The vascular flora of La Selva Biological Station, Costa Rica—Marantaceae. *Selbyana* 9:234–42.

———. 1986j. The vascular flora of La Selva Biological Station, Costa Rica—Moraceae. *Selbyana* 9:243–59.

Hammel, B. E., and M. Grayum. 1982. Preliminary report on the flora project of La Selva Field Station, Costa Rica. *Ann. MO Bot. Gard.* 69:420–25.

Hamrick, J. L. 1987. Algunos aspectos importantes en el estudio de la genética de poblaciones de plantas en bosques tropicales. *Rev. Biol. Trop.* 35 (suppl. 1): 213–14.

Hamrick, J. L., and M. D. Loveless. 1986. Isozyme variation in tropical trees: Procedures and preliminary results. *Biotropica* 18:201–7.

———. 1989. The genetic structure of tropical tree populations: Associations with reproductive biology. Pp. 129–46 in *The evolutionary ecology of plants,* ed. J. H. Bock and Y. B. Linhart. Boulder, Colo.: Westview Press.

Hamrick, J. L., M. J. W. Godt, D. A. Murawski, and M. D. Loveless. 1991. Correlation between species traits and allozyme diversity: Implications for conservation biology. Pp. 75–86 in *Genetics and conservation of rare plants,* ed. D. Falk and K. Holsinger. New York: Oxford Univ. Press.

Handley, C. O., Jr., D. E. Wilson, and A. L. Gardner, eds. 1991. Demography and natural history of the common fruit bat, *Artibeus jamaicensis,* on Barro Colorado Island, Panama. *Smithson. Contrib. Zool.* 511:1–173.

Harcombe, P. A. 1977. The influence of fertilization on some aspects of succession in a humid tropical forest. *Ecology* 58:1375–83.

Harmon, M. E., J. F. Franklin, F. J. Swanson, P. Sollins, S. V. Gregory, J. D. Lattin, N. H. Anderson, S. P. Cline, N. G. Aumen, J. R. Sedell, G. W. Lienkaemper, K. Cromack, and K. W. Cummins. 1986. Ecology of coarse woody debris in temperate ecosystems. *Adv. Ecol. Res.* 15:303–78.

Harper, J. C., P. H. Lovell, and K. G. Moore. 1970. The shapes and sizes of seeds. *Annu. Rev. Ecol. Syst.* 1:327–56.

Harper, J. L. 1977. *Population biology of plants.* New York: Academic Press.

Harris, L., Jr. 1972. *The butterflies of Georgia.* Norman: Univ. of Oklahoma.

Harrison, A. D., and J. J. Raukin. 1976. Hydrobiological studies of eastern Lesser Antillean Islands. Part 1. St. Vincent: Freshwater habitat and water chemistry. *Arch. Hydrobiol. Suppl.* 50:96–144.

Harrison, S. 1987. Treefall gaps versus forest understory as environments for a defoliating moth on a tropical forest shrub. *Oecologia* 72:65–68.

Harriss, R. C. 1987. Influence of a tropical forest on air chemistry. Pp. 163–73 in *The geophysiology of Amazonia: Vegetation and climate interactions,* ed. R. E. Dickenson. New York: J. Wiley.

Hartshorn, G. S. 1972. The ecological life history and population dynamics of *Pentaclethra macroloba,* a tropical wet forest dominant and *Stryphnodendron excelsum,* an occasional associate. Ph.d. diss., University of Washington, Seattle.

———. 1973. Plant-herbivore relationships. Report, OTS 73–2:18a–f.

———. 1975. A matrix model of tree population dynamics. Pp. 41–51 in *Tropical ecological systems: Trends in terrestrial and aquatic research,* ed. F. B. Golley and E. Medina. New York: Springer-Verlag.

———. 1978. Tree falls and tropical forest dynamics. Pp. 617–38 in *Tropical trees as living systems,* ed. P. B. Tomlinson and M. H. Zimmerman. London: Cambridge Univ. Press.

———. 1980. Neotropical forest dynamics. *Biotropica* 12 (suppl.): 23–30.

———. 1983a. *Pentaclethra macroloba.* Pp. 301–3 in *Costa Rican natural history,* ed. D. H. Janzen. Chicago: Univ. of Chicago Press.

———. 1983b. Plants: Introduction. Pp. 118–83 in *Costa Rican natural history,* ed. D. H. Janzen. Chicago: Univ. of Chicago Press.

———. 1988. Tropical and subtropical vegetation of Meso-America. Pp. 365–90 in *North American terrestrial vegetation,* ed. M. G. Barbour and W. D. Billings. New York: Cambridge Univ. Press.

———. 1989a. Application of gap theory to tropical forest management: Natural regeneration on strip clear cuts in the Peruvian Amazon. *Ecology* 70:567–69.

———. 1989b. Gap-phase dynamics and tropical tree species richness. Pp. 65–73 in *Tropical forests: Botanical dynamics, speciation, and diversity,* ed. L. B. Holm-Nielsen, I. C. Nielsen, and H. Balsev. London: Academic Press.

———. 1990. An overview of Neotropical forest dynamics. Pp. 585–99 in *Four Neotropical rainforests,* ed. A. H. Gentry. New Haven, Conn.: Yale Univ. Press.

Hartshorn, G. S., L. Hartshorn, A. Atmella, L. D. Gómez, A. Mata, R. Morales, R. Ocampo, D. Pool, C. Quesada, C. Solera, R. Solórzano, G. Stiles, J. Tosi, Jr., A. Umaña, C. Villalobos, and R. Wells. 1982. *Costa Rica country environmental profile: A field study.* San José, Costa Rica: Tropical Science Center.

Hartshorn, G. S., and R. Peralta. 1988. Preliminary description of primary forests along the La Selva-Volcán Barva altitudinal transect, Costa Rica. Pp. 281–95 in *Tropical rainforests: Diversity and conservation,* ed. F. Almeda and C. M. Pringle. San Francisco: California Academy of Sciences.

Harvey, D. J. 1987. The higher classification of the Riodinidae (Lepidoptera). Ph.D. diss., University of Texas, Austin.

————. 1991. Appendix B, Higher classification of the Nymphalidae. Pp. 255–73 in *The development and evolution of butterfly wing patterns,* by H. F. Nijhout. Washington, D.C.: Smithsonian Institution Press.

Hayes, M. P. 1985. Nest structure and attendance in the stream-dwelling frog, *Eleutherodactylus angelicus. J. Herpetol.* 19:168–69.

Hazlett, D. L. 1987. Seasonal cambial activity for *Pentaclethra, Goethalsia,* and *Carapa* trees in a Costa Rican lowland forest. *Biotropica* 19:357–60.

Heaney, L. R. 1983. *Sciurus granatensis.* Pp. 489–90 in *Costa Rican natural history,* ed. D. H. Janzen. Chicago: Univ. of Chicago Press.

Heaney, L. R., and R. W. Thorington, Jr. 1978. Ecology of Neotropical red-tailed squirrels, *Sciurus granatensis* in the Panama Canal Zone. *J. Mammal.* 59:846–51.

Heath, J., E. Pollard, and J. Thomas. 1984. *Atlas of butterflies in Britain and Ireland.* London: Viking Press.

Hedley, M. J., J. W. B. Stewart, and B. S. Chauhan. 1982. Changes in inorganic and organic soil phosphorus fractions induced by cultivation practices and laboratory incubation. *Soil Sci. Soc. Am. J.* 46:970–76.

Heinen, J. T. 1992. Comparisons of community characteristics of the herpetofauna of leaf litter in abandoned cacao plantations and primary forest in a lowland tropical rainforest: Some implications for faunal restoration. *Biotropica* 24:420–30.

Heinrich, B., and P. H. Raven. 1972. Energetics and pollination ecology. *Science* 176:597–602.

Heinselman, M. L. 1973. Fire in the virgin forests of the Boundary Waters Canoe Area, Minnesota. *Quat. Res.* (New York) 3:329–82.

Heithaus, E. R. 1974. The role of plant-pollinator interactions in determining community structure. *Ann. MO Bot. Gard.* 61:675–91.

Heithaus, E. R., and T. H. Fleming. 1978. Foraging movements of a frugivorous bat, *Carollia perspicillata* (Phyllostomatidae). *Ecology* 48:127–43.

Heithaus, E. R., T. H. Fleming, and P. A. Opler. 1975. Foraging patterns and resource utilization in seven species of bats in a seasonal tropical forest. *Ecology* 56:841–54.

Henderson, A. 1984. Observations on pollination of *Cryosophila albida. Principes* 28:120–26.

————. 1986. A review of pollination studies in the Palmae. *Bot. Rev.* 52:221–59.

Henderson, R. W., and L. G. Hoevers. 1977. The seasonal incidence of snakes at a locality in northern Belize. *Copeia* 1977:349–55.

Henderson, R. W., J. R. Dixon, and P. Soini. 1978a. On the seasonal incidence of tropical snakes. *Milw. Public Mus. Contrib. Biol. Geol.* 17:1–15.

————. 1978b. Resource partitioning in Amazonian snake communities. *Milw. Public Mus. Contrib. Biol. Geol.* 22:1–11.

Hendrickson, D. A. 1986. Congruence of bolitoglossine biogeography and phylogeny with geologic history: Paleotransport on displaced suspect terranes? *Cladistics* 2:113–29.

Hendrix, S. D. 1980. An evolutionary and ecological perspective of the insect fauna of ferns. *Am. Nat.* 115:171–96.

————. 1988. Herbivory and its impact on plant reproduction. Pp. 246–63 in *Plant reproductive ecology, patterns and strategies,* ed. J. Lovett Doust and L. Lovett Doust. Oxford: Oxford Univ. Press.

Hendrix, S. D., and R. J. Marquis. 1983. Herbivore damage to three tropical ferns. *Biotropica* 15:108–11.

Hendry, C. D., W. Berish, and E. S. Edgerton. 1984. Precipitation chemistry at Turrialba, Costa Rica. *Water Resour. Res.* 20:1677–84.

Herrera, C. M. 1981. Fruit variation and competition for dispersers in natural populations of *Smilax aspera. Oikos* 36:51–58.

————. 1982. Defense of ripe fruit from pests: Its significance in relation to plant-disperser interactions. *Am. Nat.* 120:218–41.

————. 1984. Avian interference of insect frugivory: An exploration into the plant-bird-fruit evolutionary triad. *Oikos* 42:203–10.

————. 1985a. Determinants of plant-animal coevolution: The case of mutualistic dispersal of seeds by vertebrates. *Oikos* 44:132–41.

————. 1985b. Habitat-consumer interactions in frugivorous birds. Pp. 341–65 in *Habitat selection in birds,* ed. M. L. Cody. Orlando, Fla.: Academic Press.

————. 1986. Vertebrate-dispersed plants: Why they don't behave the way they should. Pp. 5–20 in *Frugivores and seed dispersal,* ed. A. Estrada and T. H. Fleming. The Hague: W. Junk.

————. 1989. Seed dispersal by animals: A role in angiosperm diversification? *Am. Nat.* 133:309–22.

Herrera, R., T. Merida, N. Stark, and C. Jordan. 1978. Direct phosphorus transfer from litter to roots. *Naturwissenschaften* 65:208–9.

Hershkovitz, P. 1972. The recent mammals of the Neotropical region: A zoogeographic and ecological review. Pp. 311–431 in *Evolution, mammals, and southern continents,* ed. A. Keast, F. C. Erk, and B. Glass. Albany: State Univ. of New York Press.

Hespenheide, H. A. 1979. Are there fewer parasitoids in the tropics? *Am. Nat.* 113:766–69.

————. 1983a. *Agrilus xanthonotus.* Pp. 681–82 in *Costa Rican natural history,* ed. D. H. Janzen. Chicago: Univ. of Chicago Press.

————. 1983b. *Euchroma gigantea.* P. 719 in *Costa Rican natural history,* ed. D. H. Janzen. Chicago: Univ. of Chicago Press.

————. 1984. New Neotropical species of putative ant-mimicking weevils (Curculionidae: Zygopinae). *Coleopt. Bull.* 38:313–21.

————. 1985. Insect visitors to extrafloral nectaries of *Byttneria aculeata* (Sterculiaceae): Relative importance and roles. *Ecol. Entomol.* 10:191–204.

————. 1986. Mimicry of ants of the genus *Zacryptocerus* (Hymenoptera: Formicidae). *J. NY Entomol. Soc.* 94:394–408.

————. 1987. A revision of *Lissoderes* Champion (Coleoptera: Curculionidae: Zygopinae). *Coleopt. Bull.* 41:41–55.

————. 1991. Bionomics of leaf-mining insects. *Annu. Rev. Entomol.* 36:535–60.

———. In press. Mimicry in the Zygopinae (Coleoptera, Curculionidae). *Mem. Entomol. Soc. Wash.*

———. n.d. Plants with extrafloral nectaries at a second growth site in Costa Rica: Nectary characteristics, relative frequencies, and comparative visitor faunas. Typescript.

Heyer, W. R., R. W. McDiarmid, and D. L. Weigmann. 1975. Tadpoles, predation and pond habitats in the tropics. *Biotropica* 7:100–11.

Heywood, J. S., and T. H. Fleming. 1986. Patterns of allozyme variation in three Costa Rican species of *Piper. Biotropica* 18:208–13.

Hillel, D. 1980. *Fundamentals of soil physics.* New York: Academic Press.

Hilty, S. L. 1980. Flowering and fruiting periodicity in a premontane rain forest in Pacific Colombia. *Biotropica* 12:292–306.

Hladik, A., and C. M. Hladik. 1969. Rapports trophiques entre vegetation et primates dans la forêt de Barro Colorado (Panama). *Terre Vie* 23:25–117.

Holdridge, L. R. 1947. Determination of world plant formations from simple climatic data. *Science* 105(2727): 367–68.

———. 1967. *Life zone ecology.* San José, Costa Rica: Tropical Science Center.

Holdridge, L. R., and L. J. Poveda. 1976. *Arboles de Costa Rica.* Vol. 1. San José, Costa Rica: Tropical Science Center.

Holmes, R. T., J. C. Shultz, and P. Nothnagle. 1979. Bird predation on forest insects: An exclosure experiment. *Science* 206:462–63.

Hoogmoed, M. S., and J. E. Cadle. 1991. Natural history and distribution of *Agalychnis craspedopus* (Funkhouser, 1957) (Amphibia: Anura: Hylidae). *Zool. Meded. (Leiden)* 65:129–42.

Hopkins, M. S., and A. W. Graham. 1983. The species composition of soil seed banks beneath lowland tropical rainforests in north Queensland. *Biotropica* 15:90–99.

Hoppes, W. G. 1988. Seedfall pattern of several species of bird-dispersed plants in an Illinois woodland. *Ecology* 69:320–29.

Horn, S. P., and R. L. Sanford, Jr. 1993. Holocene fires in Costa Rica. *Biotropica* 24:354–61.

Horvitz, C. C. In press. Light environments, stage-structure and dispersal syndromes of Costa Rican Marantaceae: A preliminary survey of ant- and bird-dispersed species. In Oxford Symposium on the Interactions between Ants and Plants.

Horvitz, C. C., and D. W. Schemske. 1984. Effects of ants and an ant-tended herbivore on seed production of a Neotropical herb. *Ecology* 65:1369–78.

———. 1986. Seed dispersal and environmental heterogeneity in a Neotropical herb: A model of population and patch dynamics. Pp. 169–86 in *Frugivores and seed dispersal,* ed. A. Estrada and T. H. Fleming. Dordrecht: W. Junk.

Howard, J. J. 1987. Leafcutting ant diet selection: The role of nutrients, water, and secondary chemistry. *Ecology* 68:503–15.

———. 1988. Leafcutting ant diet selection: Relative influence of leaf chemistry and physical features. *Ecology* 69:250–60.

Howard, J. J., and D. F. Wiemer. 1986. Chemical ecology of host plant selection by the leaf-cutting ant, *Atta cephalotes.* Pp. 260–73 in *Fire ants and leafcutting ants: Biology and management,* ed. C. S. Lofgren and R. K. Vandermeer. Boulder, Colo.: Westview Press.

Howe, H. F. 1977. Bird activity and seed dispersal of a tropical wet forest tree. *Ecology* 58:539–50.

———. 1980. Monkey dispersal and waste of a Neotropical fruit. *Ecology* 61:944–59.

———. 1982. Fruit production and animal activity in two tropical trees. Pp. 189–99 in *The ecology of a tropical forest: Seasonal rhythms and long-term changes,* ed. E. G. Leigh, Jr., A. S. Rand, and D. M. Windsor. Washington, D.C.: Smithsonian Institution Press.

———. 1983. *Ramphastos swainsonii.* Pp. 603–4 in *Costa Rican natural history,* ed. D. H. Janzen. Chicago: Univ. of Chicago Press.

———. 1984a. Constraints on the evolution of mutualism. *Am. Nat.* 123:764–77.

———. 1984b. Implications of seed dispersal by animals for tropical reserve management. *Biol. Conserv.* 30:261–81.

———. 1985. Gomphothere fruits: A critique. *Am. Nat.* 125:853–65.

———. 1986. Seed dispersal by fruit-eating birds and mammals. Pp. 123–90 in *Seed dispersal,* ed. D. R. Murray. New York: Academic Press.

———. 1989. Scatter- and clump-dispersal and seedling demography: Hypothesis and implications. *Oecologia* 79:417–26.

———. 1990. Survival and growth of juvenile *Virola surinamensis* in Panama: Effects of herbivory and canopy closure. *J. Trop. Ecol.* 6:259–80.

Howe, H. F., and D. De Steven. 1979. Fruit production, migrant bird visitation, and seed dispersal of *Guarea glabra* in Panama. *Oecologia* 39:185–96.

Howe, H. F., and G. F. Estabrook. 1977. On intraspecific competition for avian dispersers in tropical trees. *Am. Nat.* 111:817–32.

Howe, H. F., and R. B. Primack. 1975. Differential seed dispersal by birds of the tree *Casearia nitida* (Flacourtiaceae). *Biotropica* 7:278–83.

Howe, H. F., and W. Richter. 1982. Effects of seed size on seedling size in *Virola surinamensis:* A within and between tree analysis. *Oecologia* 53:347–51.

Howe, H. F., E. W. Schupp, and L. C. Westley. 1985. Early consequences of seed dispersal for a Neotropical tree *(Virola surinamensis). Ecology* 66:781–91.

Howe, H. A., and J. Smallwood. 1982. Ecology of seed dispersal. *Annu. Rev. Ecol. Syst.* 13:201–28.

Howe, H. F., and G. A. Vande Kerckhove. 1979. Fecundity and seed dispersal of a tropical tree. *Ecology* 60:180–89.

———. 1980. Nutmeg dispersal by tropical birds. *Science* 210:925–27.

Howe, H. F., and L. C. Westley. 1986. Ecology of pollination and seed dispersal. Pp. 185–215 in *Plant ecology,* ed. M. J. Crawley. Oxford, U.K.: Blackwell Scientific.

Hubbell, S. P. 1979. Tree dispersion, abundance and diversity in a tropical dry forest. *Science* 203:1299–1309.

———. 1980. Seed predation and the coexistence of tree species in tropical forests. *Oikos* 35:214–29.

Hubbell, S. P., and R. B. Foster. 1986a. Biology, chance, and history and the structure of tropical rain forest communities. Pp. 314–29 in *Community ecology,* ed. J. Diamond and T. J. Case. New York: Harper and Row.

———. 1986b. Canopy gaps and the dynamics of a Neotropical forest. Pp. 77–96 in *Plant ecology,* ed. M. J. Crawley. Oxford, U.K.: Blackwell Scientific.

———. 1986c. Commonness and rarity in a Neotropical forest: Implications for tropical tree conservation. Pp. 205–31 in *Conservation biology: The science of scarcity and diversity,* ed. M. E. Soulé. Sunderland, Mass.: Sinauer Associates.

———. 1987a. La estructura espacial en gran escala de un bosque Neotropical. *Rev. Biol. Trop.* 35 (suppl.): 7–22.

———. 1987b. The spatial context of regeneration in a Neotropical forest. Pp. 395–412 in *Colonization, succession, and stability,* ed. A. J. Gray and M. J. Crawley. Oxford, U.K.: Blackwell Scientific.

———. 1990a. The fate of juvenile trees in a Neotropical forest: Implications for the natural maintenance of tropical tree diversity. Pp. 317–41 in *Reproductive biology of tropical forest plants,* ed. K. S. Bawa and M. Hadley. Paris: UNESCO; Carnforth, UK: Parthenon.

———. 1990b. Structure, dynamics, and equilibrium status of old-growth forest on Barro Colorado Island. Pp. 522–41 in *Four Neotropical rainforests,* ed. A. H. Gentry. New Haven, Conn.: Yale Univ. Press.

Hubbell, S. P., J. J. Howard, and D. F. Wiemer. 1984. Chemical leaf repellancy to an attine ant: Seasonal distribution among potential host plant species. *Ecology* 65:1067–76.

Huffaker, C. B., and C. E. Kennett. 1959. A ten-year study of vegetational changes associated with biological control of Klamath weed. *J. Range Manage.* 12:69–82.

Huntington, E. I. 1932. A list of the Rhopalocera of Barro Colorado Island, Canal Zone, Panama. *Bull. Am. Mus. Nat. Hist.* 113:190–230.

Huston, M. A. 1982. The effect of soil nutrients and light on tree growth and interactions during tropical forest succession: Experiments in Costa Rica. Ph.D. diss., Univ. of Michigan, Ann Arbor.

Huxley, C. R., and D. F. Cutler, eds. 1991. *Ant-plant interactions.* Oxford: Oxford Univ. Press.

Huxley, P. A. 1983. Phenology of tropical woody perennials and seasonal crop plants with reference to their management in agroforestry systems. Pp. 503–25 in *Plant research and agroforestry,* ed. P. A. Huxley. Nairobi: ICRAF.

Huxley, P. A., and W. A. Van Eck. 1974. Seasonal changes in growth and development of some woody perennials near Kampala, Uganda. *J. Ecol.* 62:579–92.

Inger, R. F. 1980. Densities of floor-dwelling frogs and lizards in lowland forests of Southeast Asia and Central America. *Am. Nat.* 115:761–70.

Innis, G. 1989. Feeding ecology of fruit pigeons in subtropical rainforests of south-eastern Queensland. *Aust. Wild. Res.* 16:365–94.

Instituto de Desarollo Agrario. 1987. Report, Informe anual, Memoria IDA. 1962–1987. San José, Costa Rica.

Instituto de Tierras y Colonización. 1980. Situación del empleo y los recursos humanos en Río Frío. Serie Estudios no. 40. San José: PREALC-ITCO.

Izor, R. J. 1985. Sloths and other mammalian prey of the Harpy Eagle. Pp. 343–46 in *The evolution and ecology of Armadillos, Sloths, and Vermilinguas,* ed. G. G. Montgomery. Washington, D.C.: Smithsonian Institution Press.

Jacobs, M. 1988. *The tropical rain forest: A first encounter.* Berlin: Springer-Verlag.

Jameson, D. L. 1955. Evolutionary trends in the courtship and mating behavior of Salientia. *Syst. Zool.* 4:105–19.

Janos, D. P. 1977. Vesicular-arbuscular mycorrhizae affect the growth of *Bactris gasipaes. Principes* 21:12.

———. 1980a. Mycorrhizae influence tropical succession. *Biotropica* 12 (suppl.): 56–64.

———. 1980b. Vesicular-arbuscular mycorrhizae affect lowland tropical rainforest plant growth. *Ecology* 61:151–62.

———. 1983. Tropical mycorrhizae, nutrient cycles, and plant growth. Pp. 327–45 in *Tropical rain forest: Ecology and management,* ed. S. L. Sutton, T. C. Whitmore, and A. C. Chadwick. Oxford, U.K.: Blackwell Scientific.

Janson, C. H. 1983. Adaptation of fruit morphology to dispersal agents in a Neotropical forest. *Science* 218:187–89.

Janson, C. H., and L. H. Emmons. 1990. Ecological structure of the nonflying mammal community at Cocha Cashu Biological Station, Manu National Park, Peru. Pp. 314–38 in *Four Neotropical rainforests,* ed. A. H. Gentry. New Haven, Conn.: Yale Univ. Press.

Janzen, D. H. 1966. Coevolution of mutualism between ants and acacias in Central America. *Evolution* 20:249–75.

———. 1967a. Synchronization of sexual reproduction of trees within the dry season in Central America. *Evolution* 21:620–37.

———. 1967b. Why mountain passes are higher in the tropics. *Am. Nat.* 101:233–49.

———. 1969a. Allelopathy by myrmecophytes: The ant *Azteca* as an allelopathic agent of *Cecropia. Ecology* 50:147–53.

———. 1969b. Seed-eaters versus seed size, number, toxicity and dispersal. *Evolution* 23:1–27.

———. 1970. Herbivores and the number of tree species in tropical forests. *Am. Nat.* 104:501–28.

———. 1971a. Euglossine bees as long distance pollinators of tropical plants. *Science* 171:203–5.

———. 1971b. Seed predation by animals. *Annu. Rev. Ecol. Syst.* 2:465–92.

———. 1972. Protection of *Barteria* (Passifloraceae) by *Pachysima* ants (Pseudomyrmecinae) in a Nigerian rainforest. *Ecology* 53:885–92.

———. 1973a. Comments on host-specificity of tropical herbivores and its relevance to species richness. Pp. 201–11 in *Taxonomy and ecology,* ed. V. H. Heywood. New York: Academic Press.

———. 1973b. Host plants as islands. Part 2. Competition in evolutionary and contemporary time. *Am. Nat.* 107:786–90.

———. 1973c. Sweep samples of tropical foliage insects: Effects of seasons, vegetation types, elevation, time of day, and insularity. *Ecology* 54:687–708.

———. 1974. Tropical blackwater rivers, animals, and mast fruiting by the Dipterocarpaceae. *Biotropica* 6:69–103.

———. 1976a. Reduction of *Mucuna andreana* (Leguminosae) seedling fitness by artificial seed damage. *Ecology* 57:826–28.

———. 1976b. Why do bamboos wait so long to flower? *Annu. Rev. Ecol. Syst.* 7:347–91.

———. 1977a. Promising directions of study in tropical animal-plant interactions. *Ann. MO Bot. Gard.* 64:706–36.

———. 1977b. Why fruits rot, seeds mold, and meat spoils. *Am. Nat.* 111:691–713.

———. 1978a. Description of a *Pterocarpus officinalis* (Leguminosae) monoculture in Corcovado National Park, Costa Rica. *Brenesia* 14/15:305–9.

———. 1978b. Seeding patterns of tropical trees. Pp. 83–128 in *Tropical trees as living systems,* ed. P. B. Tomlinson and M. H. Zimmerman. Cambridge: Cambridge Univ. Press.

———. 1979. How to be a fig. *Annu. Rev. Ecol. Syst.* 10:13–51.

———. 1980a. Specificity of seed-attacking beetles in a Costa Rican deciduous forest. *J. Ecol.* 68:929–52.

———. 1980b. When is it coevolution? *Evolution* 34:611–12.

———. 1981a. *Ficus ovalis* seed predation by an orange-chinned parakeet (*Brotogeris jugularis*) in Costa Rica. *Auk* 98:841–44.

———. 1981b. Pattern of herbivory in a tropical deciduous forest. *Biotropica* 13:271–82.

———. 1981c. The peak in North American ichneumonid species richness lies between 38° and 42°N. *Ecology* 62:532–37.

———. 1983a. *Coendou mexicanum.* Pp. 460–61 in *Costa Rican natural history,* ed. D. H. Janzen. Chicago: Univ. of Chicago Press.

———, ed. 1983b. *Costa Rican natural history.* Chicago: Univ. of Chicago Press.

———. 1983c. *Eira barbara.* Pp. 469–70 in *Costa Rican natural history,* ed. D. H. Janzen. Chicago: Univ. of Chicago Press.

———. 1983d. Food webs: Who eats what, why, how, and with what effects in a tropical forest? Pp. 167–82 in *Tropical rain forest ecosystems,* ed. F. B. Golley. New York: Elsevier Scientific.

———. 1983e. Insects: Introduction. Pp. 619–45 in *Costa Rican natural history,* ed. D. H. Janzen. Chicago: Univ. of Chicago Press.

———. 1983f. No park is an island: Increase in interference from outside as park size decreases. *Oikos* 41:402–10.

———. 1983g. *Odocoileus virginianus.* Pp. 481–83 in *Costa Rican natural history,* ed. D. H. Janzen. Chicago: Univ. of Chicago Press.

———. 1983h. *Tapirus bairdii.* Pp. 496–97 in *Costa Rican natural history,* ed. D. H. Janzen. Chicago: Univ. of Chicago Press.

———. 1984. Two ways to be a tropical big moth: Santa Rosa saturniids and sphingids. *Ox. Surv. Evol. Biol.* 1:85–140.

———. 1985. A host plant is more than its chemistry. *Ill. Nat. Hist. Surv. Bull.* 33:141–74.

———. 1986a. *Guanacaste National Park: Tropical ecological and cultural restoration.* San José, Costa Rica: Editorial Universidad Estadal a Distancia.

———. 1986b. Mice, big mammals, and seeds: It matters who defecates what where. Pp. 251–71 in *Frugivores and seed dispersal,* ed. A. Estrada and T. H. Fleming. Dordrecht: W. Junk.

———. 1987a. How moths pass the dry season in a Costa Rican dry forest. *Insect Sci. Appl.* 8:489–500.

———. 1987b. Insect diversity of a Costa Rican dry forest: Why keep it, and how? *Biol. J. Linn. Soc.* 30:343–56.

———. 1988. Ecological characterization of a Costa Rican dry forest caterpillar fauna. *Biotropica* 20:120–35.

Janzen, D. H., and R. M. Adams. 1985. Degradation of tropical forests: A dialogue. *Bull. Entomol. Soc. Am.* 31:10–13.

Janzen, D. H., and P. S. Martin. 1982. Neotropical anachronisms: The fruits the Gomphotheres ate. *Science* 215:19–27.

Janzen, D. H., and C. M. Pond. 1981. A comparison, by sweep sampling, of the arthropod fauna of secondary vegetation in Michigan, England and Costa Rica. *Trans. R. Entomol. Soc. Lond.* 127:33–50.

Janzen, D. H., and T. W. Schoener. 1968. Differences in insect abundance between wetter and drier sites during a tropical dry season. *Ecology* 49:96–110.

Janzen, D. H., and P. G. Waterman. 1984. A seasonal census of phenolics, fibre and alkaloids in foliage of forest trees in Costa Rica: Some factors influencing their distribution and relation to host selection by Sphingidae and Saturniidae. *Biol. J. Linn. Soc.* 21:439–54.

Janzen, D. H., and D. E. Wilson. 1977. Natural history of seed predation by *Rosella sickingiae* Whitehead (Curculionidae) on *Sickingia maxonii* (Rubiaceae) in Costa Rican rainforest. *Coleopt. Bull.* 31:19–23.

———. 1983. Mammals. Pp. 426–42 in *Costa Rican natural history,* ed. D. H. Janzen. Chicago: Univ. of Chicago Press.

Jenny, H. 1941. *The factors of soil formation.* New York: McGraw-Hill.

Jermy, H. 1984. Evolution of insect/host plant relationships. *Am. Nat.* 124:609–30.

Joern, A. 1979. Feeding patterns in grasshoppers (Orthoptera: Acrididae): Factors influencing diet specialization. *Oecologia* 38:325–47.

Johnson, D. W., and D. W. Cole. 1980. Ion mobility in soils: Relevance to nutrient transport from forest ecosystems. *Environ. Int.* 3:79–90.

Johnson, D. W., D. W. Cole, and S. P. Gessel. 1975. Processes of nutrient transfer in a tropical rainforest. *Biotropica* 7:208–15.

———. 1979. Acid precipitation and soil sulfate adsorption properties in a tropical and in a temperate forest soil. *Biotropica* 11:38–42.

Johnson, D. W., D. W. Cole, S. P. Gessel, M. J. Singer, and R. V. Minden. 1977. Carbonic acid leaching in a tropical, temperate, subalpine and northern forest soil. *Arc. Alp. Res.* 9:329–43.

Johnson, D. W., D. D. Richter, H. Van Miegrot, and D. W. Cole. 1983. Contributions of acid deposition and natural processes to cation leaching from forests: A review. *J. Air Pollu. Control Assoc.* 33:1036–41.

Johnson, N., and R. A. Parnell, Jr. 1986. Composition, distribution and neutralization of "acid rain" derived from Masaya volcano, Nicaragua. *Tellus* 36B:106–17.

Johnson, R. A., M. F. Willson, J. N. Thompson, and R. I. Bertin. 1985. Nutritional values of wild fruits and consumption by migrant birds. *Ecology* 66:819–27.

Johnstone, I. M. 1981. Consumption of leaves by herbivores in mixed mangrove stands. *Biotropica* 13:252–59.

Jones, C. 1985. Ants and extrafloral nectaries of *Passiflora pittieri.* Report, OTS 85-1:220–22.

Jones, C. F., and P. C. Morrison. 1952. Evolution of the banana industry in Costa Rica. *Econ. Geography* 28:1–9.

Jones, E. W. 1955. Ecological studies on the rain forest of southern Nigeria. Part 4. The plateau forest of the Okomu forest reserve. *J. Ecol.* 43:564–94.

Jones, T. 1988. The social systems of heteromyid rodents. In *The biology of the rodent family Heteromyidae,* ed. H. H. Genoways and J. H. Brown. American Society of Mammalogists, Special Publication no. 10, Shippensburg, Pa.

Jordan, C. F. 1985. *Nutrient cycling in tropical forest ecosystems.* New York: J. Wiley.

———, ed. 1989. *An Amazonian rain forest: The structure and function of a nutrient stressed ecosystem and the impact of slash-and-burn agriculture.* Man and the Biosphere Series, vol. 2. Paris: UNESCO; Carnforth, U.K.: Parthenon.

Jordan, C. F., F. Golley, J. D. Hall, and J. Hall. 1979. Nutrient scavenging of rainfall by the canopy of an Amazonian rain forest. *Biotropica* 12:61–66.

Jordan, C. F., and R. Herrera. 1981. Tropical rainforests: Are nutrients really critical? *Am. Nat.* 117:167–80.

Jordan, C. F., and J. Heuveldop. 1981. The water budget of an Amazonian rain forest. *Acta Amazonica* 11:87–92.

Jordan, C. F., and C. Uhl. 1978. Biomass of a "tierra firma" forest of the Amazon Basin calculated by a refined allometric relationship. *Oecol. Plant.* 13:387–400.

Jordano, P. 1983. Fig-seed predation and dispersal by birds. *Biotropica* 15:38–41.

———. 1987. Patterns of mutualistic interactions in pollination and seed dispersal: Connectance, dependence asymmetries, and coevolution. *Am. Nat.* 129:657–77.

———. 1988. Diet, fruit choice, and variation in body condition of frugivorous warblers in Mediterranean scrubland. *Ardea* 76:193–209.

———. 1989. Pre-dispersal biology of *Pistacia lentiscus* (Anacardiaceae): Cumulative effects on seed removal by birds. *Oikos* 55:375–86.

———. 1992. Fruits and frugivory. Pp. 105–51 in *Seeds: The ecology of regeneration in natural plant communities,* ed. M. Fenner. London: CAB International.

Judziewicz, E. J., and R. W. Pohl. 1984. Grasses of La Selva, Costa Rica. *Contrib. Univ. Wis. Herb.* 1:1–86.

Kaplan, W. A., S. C. Wofsy, M. Keller, and J. M. de Costa. 1988. Emission of NO and deposition of O_3 in a tropical forest system. *J. Geophys. Res.* 93:1389–95.

Kapp, G. B. 1989. La agroforestería como alternativa de reforestación en la zona Atlántica de Costa Rica. *El Chasquí* 21:6–17.

Kapur, A. P. 1958. A report reviewing entomological problems in the humid tropical regions of South Asia. Pp. 63–85 in *Problems of humid tropical regions.* UNESCO. Paris: Firmin-Didot.

Karban, R., R. Adamchak, and W. C. Schnathorst. 1987. Induced resistance and interspecific competition between spider mites and a vascular wilt fungus. *Science* 235:678–80.

Kareiva, P. 1983. Influence of vegetation texture on herbivore populations: Resource concentration and herbivore movement. Pp. 259–89 in *Variable plants and herbivores in natural and managed systems,* ed. R. F. Denno and M. S. McClure. New York: Academic Press.

Karr, J. R. 1971. Structure of avian communities in selected Panama and Illinois habitats. *Ecol. Monogr.* 41:207–33.

———. 1976a. Seasonality, resource availability, and community diversity in tropical bird communities. *Am. Nat.* 110:973–94.

———. 1976b. Within- and between-habitat avian diversity in African and Neotropical lowland habitats. *Ecol. Monogr.* 46:457–81.

———. 1980. Geographical variation in the avifaunas of tropical forest undergrowth. *Auk* 97:283–98.

———. 1982. Avian extinction on Barro Colorado Island, Panama: A reassessment. *Am. Nat.* 119:220–39.

Karr, J. R., and K. E. Freemark. 1983. Habitat selection and environmental gradients: Dynamics in the "stable" tropics. *Ecology* 64:1481–94.

Karr, J. R., J. D. Nichols, M. K. Klimkiewicz, and J. D. Brawn. 1990. Survival rates of birds of tropical and temperate forests: Will the dogma survive? *Am. Nat.* 136:277–91.

Karr, J. R., S. K. Robinson, J. G. Blake, and R. O. Bierregaard, Jr. 1990. Birds of four Neotropical forest. Pp. 237–69 in *Four Neotropical rainforests,* ed. A. H. Gentry. New Haven, Conn.: Yale Univ. Press.

Karr, J. R., and R. R. Roth. 1971. Vegetation structure and avian diversity in several New World areas. *Am. Nat.* 105:423–35.

Kauck, D., and J. Tosi, Jr. 1989. Public policy decisions and private sector resource use: Changing patterns of deforestation and land use in the Lake Arenal basin, Costa Rica. San José, Costa Rica: Tropical Science Center.

Kaufmann, J. H. 1962. Ecology and social behavior of the coati, *Nasua narica* on Barro Colorado Island Panama. *Univ. Calif. Publ. Zool.* 60:95–203.

———. 1983. *Nasua narica.* Pp. 478–80 in *Costa Rican natural history,* ed. D. H. Janzen. Chicago: Univ. of Chicago Press.

Kaur, A., C. D. Ha, K. Jong, V. E. Sands, H. T. Chan, E. Soepadmo, and P. S. Ashton. 1978. Apomixis may be widespread among trees of the climax rain forest. *Nature* 271:440–41.

Kaur, A., K. Jong, V. E. Sands, and E. Soepadmo. 1986. Cytoembryology of some Malaysian dipterocarps, with some evidence of apomixis. *Bot. J. Linn. Soc.* 92:75–88.

Keast, A., and E. S. Morton, eds. 1980. *Migrant birds in the Neotropics: Ecology, behavior, distribution and conservation.* Washington, D.C.: Smithsonian Institution Press.

Keeler, K. H. 1979. Distribution of plants with extrafloral nectaries and ants at two elevations in Jamaica. *Biotropica* 11:152–54.

———. 1980. Distribution of plants with extrafloral nectaries in temperate communities. *Am. Midl. Nat.* 194:274–80.

Keeler-Wolf, T. 1986. The barred antshrike (*Thamnophilus doliatus*) on Trinidad and Tobago: Habitat niche expansion of a generalist forager. *Oecologia* 70:309–17.

Kelly, C., B. Marschner, W. Reid, N. Staub, and B. Weaver. 1982. Herbivory on *Piper* species (Sperry's) in an upland versus stream habitat. Report, OTS 82-3:60–63.

Kemp, A. C. 1979. A review of the hornbills: Biology and radiation. *Living Bird* 17:105–36.

Kennedy, H. 1978. Systematics and pollination of the "closed-flowered" species of *Calathea* (Marantaceae). *Univ. Calif. Publ. Bot.* 71:1–90.

Kenyi, J. M. 1980. Seasonal defoliation of thickets by larvae of *Belenois aurota.* *African J. Ecol.* 18:123–26.

Keogh, R. M. 1984. Changes in the forest cover of Costa Rica through history. *Turrialba* 34:325–31.

Kiew, R. 1986. Phenological studies of some rain forest herbs in Peninsular Malaysia. *Kew Bull.* 41:733–46.

Kiltie, R. A. 1981. Stomach content of rain forest peccaries (*Tayassu tajacu* and *T. pecari*). *Biotropica* 13:234–36.

Kinet, J-M., R. M. Sachs, and G. Bernier. 1985. *The physiology of flowering.* Vol. 3, *The development of flowers.* Boca Raton, Fla.: CRC Press.

King, D. A. 1987. Load bearing capacity of understory treelets of a tropical wet forest. *Bull. Torrey Bot. Club* 114:419–28.

———. 1991. Correlations between biomass allocation, relative growth rate and light environment in tropical forest saplings. *Funct. Ecol.* 5:485–92.

Kitasako, J. T. 1967. Observations on the biology of *Dendrobates pumilio* Schmidt and *Dendrobates auratus* Girard. Master's thesis, University of Southern California, Los Angeles.

Kitching, I. J. 1985. Early stages and the classification of milkweed butterflies (Lepidoptera: Danainae). *Zool. J. Linn. Soc.* 85:1–97.

Kjellberg, F., P. H. Gouyon, M. Ibrahim, M. Raymond, and G. Valdeyron. 1987. The stability of the symbiosis between dioecious figs and their pollinators: A study of *Ficus carica* L. and *Blastophaga psenes* L. *Evolution* 41:693–704.

Kjellberg, F., and S. Maurice. 1989. Seasonality in the reproductive phenology of *Ficus:* Its evolution and consequences. *Experientia* 45:653–60.

Kleinfeldt, S. E. 1978. Ant-gardens: The interaction of *Codonanthe crassifolia* (Gesneriaceae) and *Crematogaster longispina* (Formicidae). *Ecology* 59:449–56.

Klopfer, P. H., and R. H. MacArthur. 1961. On the causes of tropical species diversity: Niche overlap. *Am. Nat.* 95:223–26.

Knight, D. H. 1975. A phytosociological analysis of species-rich tropical forest on Barro Colorado Island, Panama. *Ecol. Monogr.* 45:259–84.

Knöppel, H-A. 1970. Food of central Amazonian fishes. *Amazoniana* 2:257–352.

Koelmeyer, K. O. 1959. The periodicity of leaf change and flowering in the principal forest communities of Ceylon. *Ceylon For.* 4:157–89; 308–64.

Koford, C. B. 1983. *Felis wiedii.* Pp. 471–72 in *Costa Rican natural history,* ed. D. H. Janzen. Chicago: Univ. of Chicago Press.

Koptur, S. 1985. Alternative defenses against herbivores in *Inga* (Fabaceae: Mimosoideae) over an elevational gradient. *Ecology* 66:1639–50.

Koptur, S., W. A. Haber, G. W. Frankie, and H. G. Baker. 1988. Phenological studies of shrub and treelet species in tropical cloud forests of Costa Rica. *J. Trop. Ecol.* 4:323–46.

Kramer, D. L. 1978. Terrestrial group spawning of *Brycon petrosus* (Pisces: Characidae) in Panama. *Copeia* 1978:536–37.

Kress, W. J. 1983a. Crossability barriers in Neotropical *Heliconia. Ann. Bot.* 52:131–48.

———. 1983b. Self-incompatibility systems in Central American *Heliconia. Evolution* 37:735–44.

———. 1985. Bat pollination of an Old World *Heliconia. Biotropica* 17:302–8.

———. 1986. The systematic distribution of vascular epiphytes: An update. *Selbyana* 9:2–22.

Kress, W. J., and C. S. Roesel. 1991. The influence of floral biology on genetic population structure in Neotropical *Heliconia,* a rain forest understory herb. *Amer. J. Bot.* 78 (Suppl.): 240 (abstract).

Kunkel-Westphal, I., and P. Kunkel. 1979. Litter fall in a Guatemalan primary forest, with details of leaf-shedding by some common tree species. *J. Ecol.* 67:665–86.

Lack, A. J. 1982. The ecology of flowers of chalk grassland and their insect pollinators. *J. Ecol.* 70:773–90.

Lack, D. 1968. *Ecological adaptations for breeding in birds.* London: Methuen.

Laessle, A. M. 1965. Spacing and competition in natural stands of sand pine. *Ecology* 46:65–72.

La Frankie, J. V., and H. T. Chan. 1991. Confirmation of sequential flowering in *Shorea* (Dipterocarpaceae). *Biotropica* 23:200–203.

Lamas, G. 1981. La fauna de mariposas de la Reserva de Tambopata, Madre de Dios, Peru (Lepidoptera, Papilionoidea y Hesperoidea). *Revista Sociedad Mexicana Lepidoptera* 6:23–40.

———. 1985a. The Castniidae and Sphingidae (Lepidoptera) of the Tambopata Reserved Zone, Madre de Dios, Peru: A preliminary list. *Rev. Peru. Entomol.* 27 (1984): 55–58.

——— 1985b. Los Papilionoidea (Lepidoptera) de la Zona Reservada de Tambopata, Madre de Dios, Peru. Part 1. Papilionidae, Pieridae y Nymphalidae (en parte). *Rev. Peru. Entomol.* 27:59–73.

Lambert, J. 1969. *Latin America. Social structures and political institutions.* Berkeley: Univ. California Press.

Lambert, J. D. H., J. T. Arnason, and J. L. Gale. 1980. Leaf-litter and changing nutrient levels in a seasonally dry tropical hardwood forest, Belize. *Plant Soil* 55:429–43.

Land, H. C. 1963. A tropical feeding tree. *Wilson Bull.* 75:199–200.

Lande, R. 1987. Extinction thresholds in demographic models of territorial populations. *Am. Nat.* 130:624–35.

Landsberg, J., and C. Ohmart. 1989. Levels of insect defoliation in forests: Patterns and concepts. *Trends Ecol. Evol.* 4:96–100.

Lang, G. E., and D. H. Knight. 1983. Tree growth, mortality and recruitment, and canopy gap formation during a 10-year period in a tropical moist forest. *Ecology* 64:1075–80.

Lang, G. E., D. H. Knight, and D. A. Anderson. 1971. Sampling the density of tree species with quadrats in a species-rich tropical forest. *For. Sci.* 17:395–400.

Langenheim, J. H. 1984. The roles of plant secondary chemicals in wet tropical ecosystems. Pp. 189–208 in *Physiological ecology of plants of the wet tropics,* ed. E. Medina, H. A. Mooney, and C. Vázquez-Yánes. The Hague: W. Junk.

Langenheim, J. H., S. P. Arrhenius, and J. C. Nascimento. 1981. Relationship of light intensity to leaf resin composition and yield in the tropical genera *Hymenaea* and *Copaifera. Biochem. Syst. Ecol.* 9:27–37.

Langenheim, J. H., C. B. Osmond, A. Brooks, and P. J. Ferrar. 1984. Photosynthetic responses to light in seedlings of selected Amazonian and Australian rainforest tree species. *Oecologia* 63:215–24.

LaVal, R. K. 1977. Notes on some Costa Rican bats. *Brenesia* 10/11:77–83.

LaVal, R. K., and H. S. Fitch. 1977. Structure, movements and reproduction in three Costa Rican bat communities. *Occas. Pap. Mus. Nat. Hist. Univ. Kans.* 69:1–28.

LaVal, R. K., and M. L. LaVal. 1980. Prey selection by a Neotropical foliage-gleaning bat, *Micronycteris megalotis. J. Mammal.* 61:327–30.

Lawson, D. R., and J. W. Winchester. 1979. Sulfur, potassium, and phosphorus associations in aerosols from South American tropical rain forests. *J. Geophys. Res.* 84:3723–27.

Lawton, J. 1982. Vacant niches and unsaturated communities: A comparison of bracken herbivores at sites on two continents. *J. Anim. Ecol.* 51:573–95.

Lawton, J. H. 1978. Host-plant influences on insect diversity: The effects of space and time. *Symp. R. Entomol. Soc. Lond.* 9:105–25.

Lawton, J. H., and D. Schroder. 1977. Effects of plant type, size of geographical range and taxonomic isolation on number of insect species associated with British plants. *Nature* 265:137–40.

Lawton, M. F. 1983. *Cyanocorax morio.* Pp. 573–75 in *Costa Rican natural history,* ed. D. H. Janzen. Chicago: Univ. of Chicago Press.

Lawton, M. F., and C. A. Guindon. 1981. Flock composition, breeding success and learning among brown jays *(Cyanocorax [Psylorhinus] morio). Condor* 83:27–33.

Lawton, R. O., and F. E. Putz. 1988. Natural disturbance and gap-phase regeneration in a wind-exposed tropical cloud forest. *Ecology* 69:764–77.

Lebron, M. L. 1979. An autecological study of *Palicourea riparia* Bentham as related to rain forest disturbance in Puerto Rico. *Oecologia* 42:31–46.

Leck, C. F. 1969. Observations of birds exploiting a Central American fruit tree. *Wilson Bull.* 81:264–69.

———. 1971. Overlap in the diet of some Neotropical birds. *Living Bird* 10:89–106.

———. 1972. Seasonal changes in feeding pressures of fruit and nectar eating birds in the Neotropics. *Condor* 74:54–60.

Leck, C. F., F. I. Ortiz C., and R. Webster. 1980. Las Aves del Centro Científico Río Palenque. *Bol. Inf. Cient. Nac. Ecuador* 20:75–90.

Le Corff, J. 1992. The significance of a mixed reproductive strategy in an understory tropical herb, *Calathea micans* (Marantaceae). Ph.D. diss., University of Miami, Florida.

Lee, D. W. 1986. Unusual strategies of light absorption in rainforest herbs. Pp. 105–31 in *The economics of plant form and function,* ed. T. Givnish. New York: Cambridge Univ. Press.

———. 1987. The spectral distribution of radiation in two Neotropical rainforests. *Biotropica* 19:161–66.

Lee, D. W., R. A. Bone, S. L. Tarsis, and D. Storch. 1990. Correlates of leaf optical properties in tropical forest sun and extreme-shade plants. *Am. J. Bot.* 77:370–80.

Lee, D. W., J. B. Lowry, and B. C. Stone. 1979. Abaxial anthocyanin layer in leaves of tropical rain forest plants: Enhancer of light capture in deep shade. *Biotropica* 11:70–77.

Leigh, C. H. 1978. Slope hydrology and denudation in the Pasoh Forest Reserve. Part 1. Surface wash: Experimental techniques and some preliminary results. *Malay. Nat. J.* 30:179–97.

Leigh, E. G., Jr. 1982. The significance of population fluctuations. Pp. 435–40 in *The ecology of a tropical forest: Seasonal rhythms and long-term changes,* ed. E. G. Leigh, Jr., A. S. Rand, and D. M. Windsor. Washington, D.C.: Smithsonian Institution Press.

Leigh, E. G., Jr., A. S. Rand, and D. M. Windsor, eds. 1982. *The ecology of a tropical forest: Seasonal rhythms and long-term changes.* Washington, D.C.: Smithsonian Institution Press.

Leigh, E. G., Jr., and N. Smythe. 1978. Leaf production, leaf consumption and the regulation of folivory on Barro Colorado Island. Pp. 33–50 in *The ecology of arboreal folivores,* ed. G. G. Montgomery. Washington, D.C.: Smithsonian Institution Press.

Leighton, M., and D. R. Leighton. 1983. Vertebrate responses to fruiting seasonality within a Bornean rain forest. Pp. 181–96 in *Tropical rain forest: Ecology and management,* ed. S. L. Sutton, T. C. Whitmore, and A. C. Chadwick. London: Blackwell Scientific.

León, J. 1943. La agricultura y colonización en Sarapiquí. *Rev. Instit. Def. Café* 13:445–52.

Leonard, H. J. 1987. *Natural resources and economic development in Central America.* International Institute for Environment and Development. Rutgers, N.J.: Transaction Books.

Leopold, A. S. 1977. *Fauna silvestre de México: Aves y mamíferos caza* 2d ed. Mexico: Instituto de Recursos Naturales Renovables.

Lesack, L. F. W., R. E. Hecky, and J. M. Melack. 1984. Transport of carbon, nitrogen, phosphorus and major solutes in the Gambia River, West Africa. *Limnol. Oceanogr.* 29:816–30.

Letourneau, D. L. 1983. Passive aggression: An alternative hypothesis for the *Piper-Pheidole* association. *Oecologia* 60:122–26.

Levey, D. J. 1986. Methods of seed processing by birds and seed deposition patterns. Pp. 147–58 in *Frugivores and seed dispersal,* ed. A. Estrada and T. H. Fleming. Dordrecht: W. Junk.

———. 1987a. Facultative ripening in *Hamelia patens* (Rubiaceae): Effects of removal and rotting. *Oecologia* 74:203–8.

———. 1987b. Seed size and fruit-handling techniques of avian frugivores. *Am. Nat.* 129:471–85.

———. 1987c. Sugar-tasting ability and fruit selection in tropical fruit-eating birds. *Auk* 104:173–79.

———. 1987d. Frugivorous birds as seed dispersers: Detecting past coevolutionary interactions from present community patterns. Pp. 29–32 in *Proceedings of the Third Neotropical Ornithological Congress,* Cali, Colombia.

———. 1988a. Spatial and temporal variation in Costa Rican fruit and fruit-eating bird abundance. *Ecol. Monogr.* 58:251–69.

———. 1988b. Tropical wet forest treefall gaps and distributions of understory birds and plants. *Ecology* 69:1076–89.

———. 1990. Habitat-dependent fruiting behaviour of an understory tree, *Miconia centrodesma,* and tropical treefall gaps as keystone habitats for frugivores in Costa Rica. *J. Trop. Ecol.* 6:409–20.

Levey, D. J., and M. M. Byrne. In press. Complex ant-plant interactions: *Pheidole* as secondary dispersers and post-dispersal seed predators of rain forest plants. *Ecology.*

Levey, D. J., and A. Grajal. 1991. Evolutionary implications of fruit-processing limitations in cedar waxwings. *Am. Nat.* 138:171–89.

Levey, D. J., T. C. Moermond, and J. S. Denslow. 1984. Fruit choice in Neotropical birds: The effect of distance between fruits on preference patterns. *Ecology* 65:844–50.

Levey, D. J., and F. G. Stiles. In press. Evolutionary precursors of long-distance migration: Resource availability and movement patterns in Neotropical landbirds. *Am. Nat.*

Levin, D. A. 1976. Alkaloid-bearing plants: An ecogeographic perspective. *Am. Nat.* 110:261–84.

———. 1978. Alkaloids and geography. *Am. Nat.* 112:1133–34.

Levin, D. A., and H. W. Kerster. 1974. Gene flow in seed plants. Pp. 139–220 in *Evolutionary Biology.* Vol. 7, ed. T. Dobzansky, M. K. Hecht, and W. C. Steere. New York: Plenum Press.

Levin, D. A., and B. M. York. 1978. The toxicity of plant alkaloids: An ecogeographic perspective. *Biochem. Syst. Ecol.* 6:61–76.

Levings, S. C., and D. M. Windsor. 1982. Seasonal and annual variation in litter arthropod populations. Pp. 355–87 in *The ecology of a tropical forest: Seasonal rhythms and long-term changes,* ed. E. G. Leigh, Jr., A. S. Rand, and D. M. Windsor. Washington, D.C.: Smithsonian Institution Press.

Levins, R. 1962. Theory of fitness in a heterogeneous environment. Part 1. The fitness set and adaptive function. *Am. Nat.* 96:361–73.

Lewinsohn, T. M. 1991. Insects in flower heads of Asteraceae in Southeast Brazil: A case study on tropical species richness. Pp. 525–59 in *Plant-animal interactions, evolutionary ecology in tropical and temperate regions,* ed. P. W. Price, T. M. Lewinsohn, G. W. Fernandes, and W. W. Benson. New York: J. Wiley.

Lewis, W. M., Jr. 1986. N and P runoff losses from a nutrient-poor tropical moist forest. *Ecology* 65:1275–82.

Lewis, W. M., Jr., and F. H. Weibezahn. 1976. Chemistry, energy flow, and community structure in some Venezuelan freshwaters. *Arch. Hydrobiol. suppl.* 50:145–207.

Lieb, C. S. 1981. Biochemical and karyological systematics of the Mexican lizards of the *Anolis gadovi* and *A. nebulosus* species groups (Reptilia: Sauria: Iguanidae). Ph.d. diss., University of California, Los Angeles.

Lieberman, D. 1979. Dynamics of forest and thicket vegetation on the Accra plains, Ghana. Ph.D. diss., University of Ghana, Legon.

———. 1982. Seasonality and phenology in a dry tropical forest in Ghana. *J. Ecol.* 70:791–806.

Lieberman, D., G. S. Hartshorn, M. Lieberman, and R. Peralta. 1990. Forest dynamics at La Selva Biological Station, 1969–1985. Pp. 509–21 in *Four Neotropical rainforests,* ed. A. H. Gentry. New Haven, Conn.: Yale Univ. Press.

Lieberman, D., and M. Lieberman. 1987. Forest tree growth and dynamics at La Selva, Costa Rica (1969–1982). *J. Trop. Ecol.* 3:347–58.

Lieberman, D., M. Lieberman, G. Hartshorn, and R. Peralta. 1985a. Growth rates and age-size relationships of tropical wet forest trees in Costa Rica. *J. Trop. Ecol.* 1:97–109.

Lieberman, D., M. Lieberman, R. Peralta, and G. S. Hartshorn. 1985b. Mortality patterns and stand turnover rates in a wet tropical forest in Costa Rica. *J. Ecol.* 73:915–24.

Lieberman, M. 1977. A stochastic model based upon computer simulation using pre-emption to predict size distribution and species equitability. *Bull. Math. Biol.* 39:59–72.

Lieberman, M., and D. Lieberman. 1985. Simulation of growth curves from periodic increment data. *Ecology* 66:632–35.

———. 1991. No matter how you slice it: Reply to Publicover and Vogt. *Ecology* 72:1900–1902.

Lieberman, M., D. Lieberman, G. S. Hartshorn, and R. Peralta. 1985. Small-scale altitudinal variation in lowland wet tropical forest vegetation. *J. Ecol.* 73:505–16.

Lieberman, M., D. Lieberman, and R. Peralta. 1989. Forests are not just Swiss cheese: the canopy stereogeometry of non-gaps in tropical forests. *Ecology* 70:550–52.

Lieberman, M., D. Lieberman, and J. H. Vandermeer. 1988. Age-size relationships and growth behavior of the palm *Welfia georgii. Biotropica* 20:270–73.

Lieberman, S. S. 1986. Ecology of the leaf litter herpetofauna of a Neotropical rainforest: La Selva, Costa Rica. *Acta Zool. Mex.* 15:1–72.

Lieberman, S. S., and C. F. Dock. 1982. Analysis of the leaf litter arthropod fauna of a lowland tropical evergreen forest site (La Selva, Costa Rica). *Rev. Biol. Trop.* 30:27–34.

Lieth, H. 1974. Introduction to phenology and the modeling of seasonality. Pp. 3–19 in *Phenology and seasonality modeling,* ed. H. Lieth. New York: Springer-Verlag.

Limerick, S. 1976. Dietary differences of two sympatric Costa Rican frogs. Master's thesis. University of Southern California, Los Angeles.

———. 1980. Courtship behavior and ovipositoin of the poison-arrow frog *Dendrobates pumilio. Herpetologica* 36:69–71.

Lincoln, D. E., T. S. Newton, P. R. Ehrlich, and K. S. Williams. 1982. Coevolution of the checkerspot butterfly *Euphydryas chalcedon* and its larval food plant *Diplacus aurantiacus:* Larval response to protein and leaf resin. *Oecologia* 52:216–23.

Lindroth, R. L., and G. O. Batzli. 1984. Plant phenolics as chemical defenses: Effects of natural phenolics on survival and growth of prairie voles *(Microtus ochrogaster). J. Chem. Ecol.* 10:229–44.

Linhart, Y. B. 1973. Ecological and behavioral determinants of pollen dispersal in hummingbird pollinated *Heliconia. Am. Nat.* 107:511–23.

Linhart, Y. B., W. H. Busby, J. H. Beach, and P. Feinsinger. 1987. Forager behavior, pollen dispersal, and inbreeding in two species of hummingbird-pollinated plants. *Evolution* 41:679–82.

Lloyd, D. G., and C. J. Webb. 1977. Secondary sex characters in seed plants. *Bot. Rev.* 43:177–216.

———. 1986. The avoidance of interference between the presentation of pollen and stigmas in angiosperms. Part 1. Dichogamy. *N. Z. J. Bot.* 24:135–62.

Loiselle, B. A. 1987a. Birds and plants in a Neotropical rainforest: Seasonality and interactions. Ph.D. diss., University of Wisconsin, Madison.

———. 1987b. Migrant abundance in a Costa Rican lowland forest canopy. *J. Trop. Ecol.* 3:163–68.

———. 1988. Bird abundance and seasonality in a Costa Rican lowland forest canopy. *Condor* 90:761–72.

———. 1990. Seeds in droppings of tropical fruit-eating birds: Importance of considering seed composition. *Oecologia* 82:494–500.

Loiselle, B. A., and J. G. Blake. 1990. Diets of understory fruit-eating birds in Costa Rica: Seasonality and resource abundance. Pp. 91–103 in *Food exploitation by terrestrial birds,* ed. M. L. Morrison, C. J. Ralph, and J. Verner. Studies in Avian Biology no. 13, Cooper Ornithological Society. Los Angeles, Calif.: Cooper Ornithological Society.

———. 1991. Temporal variation in birds and fruits along an elevational gradient in Costa Rica. *Ecology* 72:180–93.

Loiselle, P. V. 1985. *The cichlid aquarium.* Germany: Tetra-Press.

Longman, K. A. 1985. Tropical forest trees. Pp. 23–39 in *CRC handbook of flowering.* Vol. 1., ed. A. H. Halevy. Boca Raton, Fla.: CRC Press.

Longman, K. A., and J. Jenik. 1987. *Tropical forest and its environment.* Singapore: Longman Scientific and Technical.

Lopes, A. S. 1984. Solos sob "cerrado": Características, propriedades e manejo. 2d ed. Piracicaba, Brasil: Associaçao Brasileira para Pesquisa da Potassa e do Fosfato.

Louda, S. M. 1982a. Distribution ecology: Variation in plant dynamics over a gradient in relation to insect seed predation. *Ecol. Monogr.* 52:25–41.

———. 1982b. Limitation of the recruitment of the shrub *Haplopappus squarrosus* (Asteraceae) by flower- and seed-feeding insects. *J. Ecol.* 70:43–54.

Lovejoy, T. E., and R. O. Bierregaard, Jr. 1990. Central Amazonian forests and the minimum critical size of ecosystems project. Pp. 60–71 in *Four Neotropical rainforests,* ed. A. H. Gentry. New Haven, Conn.: Yale Univ. Press.

Lovejoy, T. E., R. O. Bierregaard, Jr., A. B. Rylands, J. R. Malcolm, C. E. Quintela, L. H. Harper, K. S. Brown, Jr., A. H. Powell, G. V. N. Powell, H. O. R. Schubart, and M. B. Hays. 1986. Edge and other effects of isolation on Amazonian forest fragments. Pp. 257–85 in *Conservation biology: The science of scarcity and diversity,* ed. M. E. Soulé. Sunderland, Mass.: Sinauer Associates.

Lowe-McConnell, R. H. 1975. *Fish communities in tropical freshwaters, their distribution, ecology, and evolution.* New York: Longman.

Lowman, M. D. 1984. An assessment of techniques for measuring herbivory: Is rainforest defoliation more intense than we thought? *Biotropica* 16:264–68.

———. 1986. Insect herbivory in Australian rain forests—Is it higher than in the Neotropics? *Proc. Ecol. Soc. Aust.* 14:109–19.

Lowman, M. D., and J. D. Box. 1983. Variation in leaf toughness and phenolic content among five species of Australian rainforest trees. *Aust. J. Ecol.* 8:17–25.

Lugo, A. E., and C. T. Rivera Batlle. 1987. Leaf production, growth rate, and age of the palm *Prestoea montana* in the

Luquillo Experimental Forest, Puerto Rico. *J. Trop. Ecol.* 3:151–61.

Lumer, C. 1980. Rodent pollination of *Blakea* (Melastomataceae) in a Costa Rican cloud forest. *Brittonia* 32:512–17.

———. 1983. *Blakea.* Pp. 194–95 in *Costa Rican natural history,* ed. D. H. Janzen. Chicago: Univ. of Chicago Press.

Lundell, C. L. 1984. Neotropical Myrsinaceae—XII. *Phytologia* 56:19–27.

Luvall, J. C. 1984. Tropical deforestation and recovery: The effects on the evapotranspiration process. Ph.D. diss., University of Georgia, Athens.

Luvall, J. C., and C. F. Jordan. 1984. Weekly stream chemistry in adjacent watersheds on the Vargas Annex. Report, Organization for Tropical Studies, San José, Costa Rica.

Lydolph, P. E. 1985. *The climate of the earth.* Totowa, N.J.: Rowman and Allanheld.

Maas, P. J. M. 1977. Costoideae (additions) (Zingiberaceae). *Flora Neotrop. Monogr.* 18:162–218.

MacArthur, R. 1969. Patterns of communities in the tropics. *Biol. J. Linn. Soc. Lond.* 1:19–30.

———. 1971. Patterns of terrestrial bird communities. Pp. 189–221 in *Avian Biology.* Vol. 1, ed. D. S. Farner and J. R. King. New York: Academic Press.

———. 1972. *Geographic ecology.* New York: Harper and Row.

MacArthur, R. H., H. Recher, and M. Cody. 1966. On the relation between habitat selection and species diversity. *Am. Nat.* 100:319–22.

MacArthur, R. H., and E. O. Wilson. 1967. *The theory of island biogeography.* Princeton, N.J.: Princeton Univ. Press.

Macdonald, D. 1984. *The encyclopedia of mammals.* New York: Facts on File Publications.

MacKinnon, K. S. 1978. Stratification and feeding differences among Malayan squirrels. *Malay. Nat. J.* 30:593–608.

Madison, M. 1977. Vascular epiphytes: Their systematic occurrence and salient features. *Selbyana* 2:1–13.

Madrigal, R. 1970. *Geología del mapa básico Barranca, Costa Rica.* Informes Técnicos y Notas Geológicas, no. 37. San José, Costa Rica: Ministerio de Industria y Comercio.

Madrigal, R., and E. Rojas. 1980. *Manual descriptivo del mapa geomorfológico de Costa Rica.* San José, Costa Rica: Imprenta Nacional.

Maiorana, V. C. 1976. Predation, submergent behavior, and tropical diversity. *Evol. Theory* 1:157–77.

Malcolm, J. R. 1990. Estimation of mammalian densities in continuous forest north of Manaus. Pp. 339–57 in *Four Neotropical rainforests,* ed. A. H. Gentry. New Haven, Conn.: Yale Univ. Press.

Maleret, L., F. Breden, L. Garling, H. Rowell, D. Kennedy, and C. Uhl. 1978. Is there any difference in the insect fauna of mature, building, and/or gap areas of the tropical forest of La Selva? Report, OTS 78-1:85–87.

Mallet, J., and M. C. Singer. 1987. Individual selection, kin selection, and the shifting balance in the evolution of warning colours: Evidence from butterflies. *Biol. J. Linn. Soc.* 32:337–50.

Malmborg, P. K., and M. F. Willson. 1988. Foraging ecology of avian frugivores and some consequences for seed dispersal in an Illinois woodlot. *Condor* 90:173–86.

Manasse, R. S., and H. F. Howe. 1983. Competition for dispersal agents among tropical trees: Influences of neighbors. *Oecologia* 59:185–90.

Manning, S. 1983. Observations on a symbiotic association between ants of the genus *Pheidole* and *Conostegia setosa* (Melastomataceae). Report, OTS 83-3:397–401.

Manokaran, N., and K. M. Kochummen. 1987. Recruitment, growth and mortality of tree species in a lowland dipterocarp forest in Peninsular Malaysia. *J. Trop. Ecol.* 3:315–30.

Mares, M. L., and D. E. Wilson. 1971. Bat reproduction during the Costa Rican dry season. *Bioscience* 21:471–72.

Margalef, R. 1968. *Perspectives in ecological theory.* Chicago: Univ. of Chicago Press.

Marques, J., J. Marden Dossantos, N. A. Villa Nova, and E. Salati. 1977. Precipitable water and water vapor flux between Belem and Manaus. *Acta Amazonica* 7:355–62.

Marquis, R. J. 1984a. Herbivory as a selective force in *Piper arieianum* C. DC. (Piperaceae). Ph.D. diss., University of Iowa, Iowa City.

———. 1984b. Leaf herbivores decrease fitness of a tropical plant. *Science* 226:537–39.

———. 1984c. Natural history of a tropical day-flying Saturniid *Automeris phrynon* Druce (Lepidoptera: Saturniidae: Hemileucinae). *J. Kans. Entomol. Soc.* 57:529–31.

———. 1987. Variación en la herbivoría foliar y su importancia selectiva en *Piper arieianum* (Piperaceae). *Rev. Biol. Trop.* 35 (suppl.): 133–49.

———. 1988. Phenological variation in the Neotropical understory shrub *Piper arieianum:* Causes and consequences. *Ecology* 69:1552–65.

———. 1990. Genotypic variation in leaf damage in *Piper arieianum* (Piperaceae) by a multi-species assemblage of herbivores. *Evolution* 44:104–20.

———. 1991a. Evolution of resistance in plants to herbivores. *Evol. Trends Plants* 5:23–29.

———. 1991b. Herbivore fauna of *Piper* (Piperaceae) in a Costa Rican wet forest: Diversity, specificity and impact. Pp. 179–208 in *Plant-animal interactions: Evolutionary ecology in tropical and temperate regions,* ed. P. W. Price, T. M. Lewinsohn, G. W. Fernandes, and W. W. Benson. New York: J. Wiley.

———. 1992a. A bite is a bite is a bite? Constraints on response to folivory in *Piper arieianum* (Piperaceae). *Ecology* 73:143–52.

———. 1992b. Selective impact of herbivores. Pp. 301–25 in *Ecology and evolution of plant resistance,* ed. R. S. Fritz and E. L. Simms. Chicago: Univ. of Chicago Press.

Marquis, R. J., and G. O. Batzli. 1989. Influence of chemical factors on palatability of forage to voles. *J. Mammol.* 70:503–11.

Marquis, R. J., and H. E. Braker. 1987. Influence of method of presentation on results of plant-host preference tests with two species of grasshopper. *Entomol. Exp. Appl.* 44:59–63.

Marquis, R. J., and D. B. Clark. 1989. Habitat and fertilization effects on leaf herbivory in *Hampea appendiculata* (Malvaceae): Implications for tropical firewood systems. *Agric. Ecosyst. Environ.* 25:165–74.

Marquis, R. J., M. A. Donnelly, and C. Guyer. 1986. Aggregations of calling males of *Agalychnis calcarifer* Boulenger (Anura: Hylidae) in a Costa Rican lowland wet forest. *Biotropica* 18:173–75.

Marquis, R. J., and C. J. Whelan. In press. Insectivorous birds influence growth in oaks through their impact on oak herbivores. *Science.*

Marquis, R. J., H. J. Young, and H. E. Braker. 1986. The influence of understory vegetation cover on germination and seedling establishment in a tropical lowland wet forest. *Biotropica* 4:273–78.

Marrs, R. H., J. Proctor, A. Heaney, and M. D. Mountford. 1988. Changes in soil nitrogen-mineralization and nitrification along an altitudinal transect in tropical rain forest in Costa Rica. *J. Ecol.* 76:466–82.

Marshall, A. G. 1983. Bats, flowers and fruit: Evolutionary relationships in the Old World. *Biol. J. Linn. Soc.* 20:115–35.

Marshall, G. A. K., and E. B. Poulton. 1902. Five years observations and experiments (1896–1901) on the bionomics of south African insects, chiefly mimicry and warning colours. *Trans. R. Entomol. Soc. Lond.* 1902:285–594.

Marshall, L. G., S. D. Webb, J. J. Sepkoski, Jr., and D. M. Raup. 1982. Mammalian evolution and the great American interchange. *Science* 251:1351–57.

Martin, T. E. 1985a. Resource selection by tropical frugivorous birds: integrating multiple interactions. *Oecologia* 66:563–73.

———. 1985b. Selection of second-growth woodlands by frugivorous migrating birds in Panama: An effect of fruit size and plant density? *J. Trop. Ecol.* 1:157–70.

Martin, T. E., and J. R. Karr. 1986a. Patch utilization by migrating birds: Resource oriented? *Ornis Scand.* 17:165–74.

———. 1986b. Temporal dynamics of Neotropical birds with special reference to frugivores in second-growth woods. *Wilson Bull.* 98:38–60.

Martínez-Ramos, M. 1985. Claros, ciclos vitales de los árboles tropicales y regeneración natural de las selvas altas perennifolias. Pp. 191–239 in *Investigaciones sobre la regeneracion de selvas altas en Veracruz, Mexico. Vol. 2,* ed. A. Gómez-Pompa and S. del Amo R. México: Instituto Nacional de Invetigaciones sobre Recursos Bióticos.

Martínez-Ramos, M., J. Sarukhán, and D. Piñero. 1988. The demography of tropical trees in the context of forest gap dynamics. In *Plant population ecology: 28th Symposium of the British Ecological Society,* ed. A. J. Davy, M. J. Hutchings, and A. R. Watkinson. Oxford: Blackwell Scientific Publications.

Masner, L. 1988. Convergent chromatic mimicry among some Neotropical Hymenoptera: A search for the model. P. 12 in Proceedings of the Eighteenth International Congress of Entomology, Vancouver, B.C., Canada (abstract).

Matamoros, A. 1987. *Los recursos forestales.* Report, Borrador de trabajo, estudio del estado del ambiente. San José, Costa Rica: Fundación Neotrópica.

Matson, P. A., P. M. Vitousek, J. J. Ewel, M. J. Mazzarino, and G. P. Robertson. 1987. Nitrogen transformations following tropical forest felling and burning on a volcanic soil. *Ecology* 68:491–502.

Matthes, H. 1964. Les poissons du lac Tumba et de la région d'Ikela. Etude systematique et écologique. *Ann. Mus. R. Afr. Cent. Sci. Quarto Zool.* 126:1–204.

Matthews, R. W. 1983. *Microstigmus comes.* Pp. 739–41 in *Costa Rican natural history,* ed. D. H. Janzen. Chicago: Univ. of Chicago Press.

May, R. M. 1978. The dynamics and diversity of insect faunas. *Symp. R. Entomol. Soc. Lond.* 9:188–204.

———. 1981. *Theoretical ecology.* 2d ed. Oxford: Blackwell Scientific.

———. 1988. How many species are there on earth? *Science* 241:1441–49.

May, R. M., and J. Seger. 1986. Ideas in ecology. *Am. Scientist* 74:256–67.

McCleery, R. H. 1978. Optimal behaviour sequences and decision making. Pp. 377–410 in *Behavioural ecology: An evolutionary approach,* ed. J. R. Krebs and N. B. Davies. Sunderland, Mass.: Sinauer Associates.

McColl, J. G. 1970. Properties of some natural waters in a tropical wet forest in Costa Rica. *Bioscience* 20:1096–1100.

McCoy, E. D. 1984. Colonization by herbivores of *Heliconia* spp. plants (Zingiberales: Heliconiaceae). *Biotropica* 16:10–13.

———. 1985. Interactions among leaf-top herbivores of *Heliconia imbricata* (Zingiberales: Heliconiaceae). *Biotropica* 17:326–29.

McCoy, J. W. 1978. Comments on the geographic distribution of alkaloids in angiosperms. *Am. Nat.* 112:1126–33.

McCoy, M., C. Vaughan, M. Rodríguez, and V. Villalobos. 1983. An interesting feeding habit for the collared peccary (*Tayassu tajacu* Bangs.) in Costa Rica. *Brenesia* 21:456–57.

McCulloch, R., M. Dagg, and J. R. Blackie. 1964. Some forest meteorological observations. Pp. 48–64 in *East African Agriculture and Forestry Research Organization Record of Research Annual Report,* Kikuya, Kenya.

McDade, L. A. 1982. New species of *Justicia* and *Razisea* (Acanthaceae) from Costa Rica, with taxonomic notes. *Syst. Bot.* 7:489–97.

———. 1983. Long-tailed hermit hummingbird visits to inflorescence color morphs of *Heliconia irrasa. Condor* 85:360–64.

———. 1984. Systematics and reproductive biology of the Central American members of the *Aphelandra pulcherrima* complex (Acanthaceae). *Ann. MO Bot. Gard.* 72:104–65.

———. 1985. Breeding systems of Central American *Aphelandra* (Acanthaceae). *Am. J. Bot.* 72:1515–21.

———. 1986. Protandry, synchronized flowering and sequential phenotypic unisexuality in Neotropical *Pentagonia macrophylla* (Rubiaceae). *Oecologia* 68:218–23.

McDade, L. A., and P. Davidar. 1984. Determinants of fruit and seed set in *Pavonia dasypetala* (Malvaceae). *Oecologia* 64:61–67.

McDade, L. A., and S. Kinsman. 1980. The impact of floral parasitism in two Neotropical hummingbird-pollinated plant species. *Evolution* 34:944–58.

McDiarmid, R. W. 1978. Evolution of parental care in frogs. Pp. 127–47 in *The development of behavior: Comparative and evolutionary aspects,* ed. G. M. Burghardt and M. Bekof. New York: New Garland STPM Press.

———. 1983. *Centrolenella fleischmanni* (Ranita de vidrio, glass frog). Pp. 389–90 in *Costa Rican natural history,* ed. D. H. Janzen. Chicago: Univ. of Chicago Press.

McDiarmid, R. W., R. E. Ricklefs, and M. S. Foster. 1977. Dispersal of *Stemmadenia donnell-smithii* (Apocynaceae) by birds. *Biotropica* 9:9–25.

McDonald, E. P., and B. R. Strain. 1991. Photosynthetic responses of four tropical species of *Miconia* to sustained increases in irradiance. *Bull. Ecol. Soc. Am.* 72 (suppl.): 186–87 (abstract).

McDonald, E. P., B. R. Strain, and E. Newell. 1991. Physiological responses of tree and shrub species to canopy opening. Association for Tropical Biology Annual Meeting, San Antonio, Texas (abstract).

McElravy, E. P., V. S. Resh, H. Wolda, and O. S. Flint, Jr. 1981. Diversity of adult trichoptera in a "non-seasonal" tropical environment. Proceedings of the Third International Symposium on Trichoptera, ed. G. P. Moretti. *Ser. Entomol.* (The Hague) 20:149–56.

McGraw, J. B., and R. D. Wulff. 1983. The study of plant growth: A link between the physiological ecology and population biology of plants. *J. Theor. Biol.* 103:21–28.

McHargue, L. A. 1981. Mycorrhizae affect growth and nodulation of two tropical leguminous trees. P. 52 in Program and Abstracts, Fifth North American Conference on Mycorrhizae. Université Laval, Quebec, Canada (abstract).

———. 1987. Análisis comparativo del crecimiento de plántulas de *Pithecellobium catenatum* sin y con nódulos. *Rev. Biol. Trop.* 35 (suppl. 1): 226.

McHargue, L. A., and G. S. Hartshorn. 1983. Seed and seedling ecology of *Carapa guianensis. Turrialba* 33:399–404.

McKey, D. 1974. Ant-plants: Selective eating of an unoccupied *Barteria* by a Colobus monkey. *Biotropica* 6:269–70.

———. 1975. The ecology of coevolved seed dispersal systems. Pp. 159–91 in *Coevolution of animals and plants,* ed. L. E. Gilbert and P. H. Raven. Austin: Univ. of Texas Press.

———. 1979. The distribution of secondary compounds within plants. Pp. 55–133 in *Herbivores: Their interactions with secondary plant metabolites,* ed. G. A. Rosenthal and D. H. Janzen. New York: Academic Press.

———. 1984. Interaction of the ant-plant *Leonardoxa africana* (Caesalpiniaceae) with its obligate inhabitants in a rainforest in Cameroon. *Biotropica* 16:81–99.

McKey, D. B., J. S. Gartlan, P. G. Waterman, and G. M. Choo. 1981. Food selection by black colobus monkeys (*Colobus satanas*) in relation to plant chemistry. *Biol. J. Linn. Soc. Lond.* 16:115–46.

McKey, D., P. G. Waterman, C. N. Mbi, J. S. Gartlan, and T. T. Struhsaker. 1978. Phenolic content of vegetation in two African rain forests: Ecological implications. *Science* 202:61–64.

McMahan, E. A. 1982. Bait-and-capture strategy of a termite-eating assassin bug. *Insectes Soc.* 29:346–51.

———. 1983. Adaptations, feeding preferences, and biometrics of a termite-baiting assassin bug (Hemiptera: Reduviidae). *Ann. Entomol. Soc. Am.* 76:483–86.

McNaughton, S. J. 1986. On plants and herbivores. *Am. Nat.* 128:765–70.

McNab, B. K. 1971. The structure of tropical bat faunas. *Ecology* 52:352–58.

McPherson, A. B. 1985. A biogeographical analysis of factors influencing the distribution of Costa Rican rodents. *Brenesia* 23:97–273.

———. 1986. The biogeography of Costa Rican rodents: An ecological, geological, and evolutionary approach. *Brenesia* 25/26:229–44.

McVaugh, R. 1984. *Compositae.* Vol. 10, pp. 1–1157, *Flora Novo-Galiciana,* ed. W. R. Anderson.

McVey, M. E., R. G. Zahary, D. Perry, and J. MacDougal. 1981. Territoriality and homing behavior in the poison dart frog *(Dendrobates pumilio). Copeia* 1981:1–8.

McVoy, C. W. 1985. Water and solute movement in an aggregated tropical soil: Use of iodide and dyes for a morphological characterization. Master's thesis, University of Florida, Gainesville.

Meagher, T., and J. Damery Parrish. 1975. *Atta* activity cycles at La Selva. Report, OTS 75-2:277.

Medina, E. 1983. Adaptations of tropical trees to moisture stress. Pp. 225–37 in *Tropical rain forest ecosystems: Structure and function,* ed. F. B. Golley. New York: Elsevier Scientific.

Medway, L. 1972. Phenology of a tropical rain forest in Malaya. *Biol. J. Linn. Soc. Lond.* 4:117–46.

———. 1978. *The wild mammals of Malaya (peninsular Malaysia) and Singapore,* 2d ed. Kuala Lumpur: Oxford Univ. Press.

Melcher, J., R. Clayton, J. Stout, C. Allen, C. Blount, and J. Fry. 1983. Changes in stream morphology. Report, OTS 83-1:72–76.

Mendoza, A., D. Piñero, and J. Sarukhán. 1987. Effects of experimental defoliation on growth, reproduction and survival of *Astrocaryum mexicanum. J. Ecol.* 75:545–54.

Menke, A. S. 1988. *Pison* in the New World: A revision (Hymenoptera: Sphecidae: Trypoxylini). *Contrib. Am. Entomol. Inst.* (Ann Arbor) 24(3): i–iii, 1–171.

Michaloud, G. 1988. Aspects de la reproduction des figuiers monoiques en forêt equatoriale africaine. Ph.D. diss., Université des Sciences et Techniques du Languedoc, Montpellier, France.

Miller, H., T. Pratt, M. Turner, W. Copeland, and S. Jones. 1977. An attempt to study active versus passive dispersal along water courses. Report, OTS 77-4:465–68.

Miller, J. S. 1987. Host-plant relationships in the Papilionidae (Lepidoptera): Parallel cladogenesis or colonization? *Cladistics* 3:105–20.

———. 1988. Phylogenetic studies in the Papilioninae (Lepidoptera: Papilionidae). *Bull. Am. Mus. Nat. Hist.* 186:365–512.

Miller, R. R. 1976. Geographical distribution of Central American freshwater fishes. In *Investigations of the ichthyofauna of Nicaraguan lakes,* ed. T. B. Thorson. Lincoln: Univ. of Nebraska Press.

Milton, K. 1979. Factors influencing leaf choice by howler monkeys: A test of some hypotheses of food selection by generalist herbivores. *Am. Nat.* 114:362–78.

———. 1980. *The foraging strategy of howler monkeys: Study in primate economics.* New York: Columbia Univ. Press.

———. 1982. Dietary quality and demographic regulation in a howler monkey population. Pp. 273–89 in *The ecology of a tropical forest: Seasonal rhythms and long-term changes,* ed. E. G. Leigh, Jr., A. S. Rand, and D. M. Windsor. Washington, D.C.: Smithsonian Institution Press.

Milton, K., D. M. Windsor, D. W. Morrison, and M. A. Estribi. 1982. Fruiting phenologies of two Neotropical *Ficus* species. *Ecology* 63:752–62.

Ministerio de Planificación Nacional y Política Económica. 1984. *Plan maestro para el desarollo regional. Región Huetar Atlántica.* San José, Costa Rica.

Misra, R. 1968. Energy transfer along terrestrial food chains. *Trop. Ecol.* 9:105–18.

Mitter, C., B. Farrell, and D. J. Futuyma. 1991. Phylogenetic studies of insect-plant interactions: Insights into the genesis of diversity. *Trends Ecol. Evol.* 6:290–93.

Mittermeier, R. A. 1971. Notes on the behavior and ecology of *Rhinoclemys annulata* Gray. *Herpetologica* 27:485–88.

Mittermeier, R. A., and M. G. M. van Roosmalen. 1981. Preliminary observations on habitat utilization and diet in eight Surinam monkeys. *Folia Primatol.* 36:1–39.

Miyamoto, M. M. 1982. Vertical habitat use by *Eleutherodactylus* frogs (Leptodactylidae) at two Costa Rican localities. *Biotropica* 14:141–44.

Moermond, T. C. 1979. Habitat constraints on the behavior, morphology, and community structure of *Anolis* lizards. *Ecology* 60:152–64.

———. 1981. Cooperative feeding, defense of young, and flocking in the black-faced grosbeak. *Condor* 83:82–83.

———. 1983. Suction-drinking in tanagers and its relation to fruit handling. *Ibis* 125:545–49.

Moermond, T. C., and J. S. Denslow. 1983. Fruit choice in Neotropical birds: Effects of fruit type and accessibility on selectivity. *J. An. Ecol.* 52:407–20.

———. 1985. Neotropical frugivores: Patterns of behavior, morphology and nutrition with consequences for fruit selection. Pp. 865–97 in *Neotropical ornithology,* ed. P. A. Buckley, M. S. Foster, E. S. Morton, R. S. Ridgely, and N. G. Buckley. Ornithological Monographs vol. 36. Lawrence: AOU.

Moermond, T. C., J. S. Denslow, D. J. Levey, and E. Santana-C. 1986. The influence of morphology on fruit choice in Neotropical birds. Pp. 137–46 in *Frugivores and seed dispersal,* ed. A. Estrada and T. H. Fleming. Dordrecht: W. Junk.

———. 1987. The influence of context on choice behavior: Fruit selection by tropical birds. Pp. 229–54 in *Quantitative analysis of behavior: Foraging,* ed. M. L. Commons, A. Kacelnik, and S. Shettleworth. Hillsdale, N.J.: Lawrence Erlbaum Associates.

Moll, E. O., and J. M. Legler. 1971. The life history of a Neotropical slider turtle, *Pseudomys scripta* (Schoepff) in Panama. *Nat. Hist. Mus. Los Angel. C. Sci. Bull.* 11:1–102.

Monge-Nájera, J., and B. Morera Brenes. 1987. Why is the coyote *(Canis latrans)* expanding its range? A critique of the deforestation hypothesis. *Rev. Biol. Trop.* 35:169–71.

Monselise, S. P., and E. E. Goldschmidt. 1982. Alternate bearing in fruit trees. *Hortic. Rev.* 4:128–73.

Montagnini, F. 1990a. Ecology applied to agroforestry in the human tropics. Pp. 49–58 in *Race to save the tropics: Ecol-*

ogy and economics for a sustainable future, ed. R. A. Goodland. Washington, D.C.: Island Press.

———. 1990b. Increasing collaboration in ecological research with academic institutions in Latin America. Proposal to Mellon Foundation, Yale University, New Haven, Conn.

Montagnini, F., F. Sancho, K. Ramstad, and E. Stijfhoorn. 1990. Multipurpose trees for soil restoration in the humid lowlands of Costa Rica. Report, Multipurpose Tree Research in Asia Workshop, 19–23 November 1990, Los Baños, Philippines.

Montagnini, F., E. Stijfhoorn, and F. Sancho. 1989. Soil chemical properties and root biomass under plantations of native tree species, grass cover and secondary forest vegetation in the Atlantic lowlands of Costa Rica. *Below Ground Ecol.* 1:6–8.

Montalvo, A. M., and J. D. Ackerman. 1986. Relative pollinator effectiveness and evolution of floral traits in *Spathiphyllum friedrichsthalii* (Araceae). *Am. J. Bot.* 73:1665–76.

Montgomery, G. G., and M. E. Sunquist. 1975. Impact of sloths in Neotropical energy flow and nutrient cycling. In *Ecological studies.* Pp. 69–98 in vol. 11, *Tropical ecological systems: Trends in terrestrial and acquatic research,* ed. F. B. Golley and E. Medina. New York: Springer-Verlag.

———. 1978. Habitat selection and use by two-toed and three-toed sloths. Pp. 329–59 in *The ecology of arboreal folivores,* ed. G. G. Montgomery. Washington, D.C.: Smithsonian Institution Press.

Monteith, J. L. 1973. *Principles of environmental physics.* London: Edward Arnold.

Moody, S. 1978. Latitude, continental drift, and the percentage of alkaloid-bearing plants in floras. *Am. Nat.* 112:965–68.

Mooney, H. A. 1972. The carbon balance of plants. *Ann. Rev. Ecol. Syst.* 3:315–46.

———. 1977. *Convergent evolution in Chile and California.* Stroudsburg, Pa.: Dowden, Hutchinson and Ross.

Mooney, H. A., O. Björkman, A. E. Hall, E. Medina, and P. B. Tomlinson. 1980. The study of the physiological ecology of tropical plants—current status and needs. *Bioscience* 30:22–26.

Mooney, H. A., and E. L. Dunn. 1970. Convergent evolution of Mediterranean-climate evergreen sclerophyllous shrubs. *Evolution* 25:292–303.

Mooney, H. A., and S. L. Gulmon. 1982. Constraints on leaf structure and function in reference to herbivory. *Bioscience* 32:198–206.

Moore, C. J. 1976. Eddy flux measurements above a pine forest. *Q. J. R. Meteorol. Soc.* 102:913–18.

Moore, L. V., J. H. Myers, and R. Eng. 1988. Western tent caterpillars prefer the sunny side of the tree, but why? *Oikos* 51:321–26.

Moran, V. C. 1980. Interactions among phytophagous insects and their *Opuntia* hosts. *Ecol. Entomol.* 5:153–64.

Morden-Moore, A. L., and M. F. Willson. 1982. On the ecological significance of fruit color in *Prunus* and *Rubus:* Field experiments. *Can. J. Bot.* 60:1554–60.

Morrison, D. W. 1978. Foraging ecology and energetics of the frugivorous bat *Artibeus jamaicensis. Ecology* 59:716–23.

Morrison, G., and D. R. Strong, Jr. 1981. Spatial variations in

egg density and the intensity of parasitism in a Neotropical chrysomelid (*Cephaloleia consanguinea). Ecol. Entomol.* 6:55–61.

Morrow, P. A., and L. R. Fox. 1989. Estimates of presettlement insect damage in Australian and North American forests. *Ecology* 70:1055–60.

Morton, E. S. 1973. On the evolutionary advantage and disadvantages of fruit eating in tropical birds. *Am. Nat.* 107:8–22.

———. 1977. Intratropical migration in the Yellow-green Vireo and Piratic Flycatcher. *Auk* 94:97–106.

———. 1978. Avian arboreal folivores: Why not? Pp. 123–30 in *The ecology of arboreal folivores,* ed. G. G. Montgomery. Washington, D.C.: Smithsonian Institution Press.

Morton, S. R., and C. D. James. 1988. The diversity and abundance of lizards in arid Australia: A new hypothesis. *Am. Nat.* 132:237–56.

Mould, E. D., and C. T. Robbins. 1981. Nitrogen metabolism in elk. *J. Wildlife Management* 45:323–34.

Mound, L. A., and N. Waloff, eds. 1978. Diversity of insect faunas. *Symp. R. Entomol. Soc. Lond.* 9:1–204.

Moynihan, M. 1962. The organization and possible evolution of some mixed-species flocks of Neotropical birds. *Smithson. Misc. Collect.* 143:1–140.

———. 1971. Successes and failures of tropical mammals and birds. *Am. Nat.* 105:371–83.

Mueller, U. G., and B. Wolf-Mueller. 1991. Ephiphyll deterrence to the leafcutter ant *Atta cephalotes. Oecologia* 86:36–39.

Mueller-Dombois, D., and H. Ellenberg. 1974. *Aims and methods of vegetation ecology.* New York: J. Wiley.

Mulcahy, D. L., and G. B. Mulcahy. 1983. Gametophytic self-incompatibility revisited. *Science* 220:1247–51.

Müller, F. 1879. *Ituna* and *Thyridia:* A remarkable case of mimicry in butterflies, trans. R. Mendola. *Proc. R. Entomol. Soc. Lond.* 1879:20–29.

Munn, C. A. 1985. Permanent canopy and understory flocks in Amazonia: Species composition and population density. Pp. 683–712 in *Neotropical ornithology,* ed. P. A. Buckley, M. S. Foster, E. S. Morton, R. S. Ridgely, and N. G. Buckley. Ornithological Monographs vol. 36. Lawrence: AOU.

Munn, C. A., and J. W. Terborgh. 1979. Multi-species territoriality in Neotropical foraging flocks. *Condor* 81:338–47.

Murawski, D. A., and J. L. Hamrick. 1990. Local genetic and clonal structure in the tropical terrestrial bromeliad, *Aechmea magdalenae. Am. J. Bot.* 77:1201–08.

Murawski, D. A., J. L. Hamrick, S. P. Hubbell, and R. B. Foster. 1990. Mating systems of two Bombacaceous trees of a Neotropical moist forest. *Oecologia* 82:501–6.

Mure, V. 1986. Comportement phenologique d'arbres plantes hors de leurs aires d'origine, et notamment de ceux changes d'hemisphere. *Rev. Ecol. Terre Vie* 41:129–71.

Murphy, C. E., Jr., and K. R. Knoerr. 1975. The evaporation of intercepted rainfall from a forest stand: An analysis by simulation. *Water Resour. Res.* 11:273–80.

Murray, K. G. 1987. Selection for optimal fruit-crop size in bird-dispersed plants. *Am. Nat.* 129:18–31.

———. 1988. Avian seed dispersal of three Neotropical gap-dependent plants. *Ecol. Monogr.* 58:271–98.

Myers, C. W., and A. S. Rand. 1969. Checklist of amphibians

and reptiles of Barro Colorado Island, Panama, with comments on faunal change and sampling. *Smithson. Contrib. Zool.* 10:1–11.

Myers, G. S. 1947. The Amazon and its fishes. Part 4. The fish in its environment. *Aquarium J.* 18:8–19, 34.

Myers, N. 1980. *Conversion of tropical moist forests.* Washington, D.C.: National Academy of Sciences, National Research Council.

Naeem, S. 1988. Predator-prey interactions and community structure: Chironomids, mosquitoes and copepods in *Heliconia imbricata. Oecologia* 77:202–9.

Nahrstedt, A., and R. H. Davis. 1981. The occurrence of cyanoglucocides, linamarin and lotaustralin, in *Acraea* and *Heliconius* butterflies. *Comp. Biochem. Physiol.* 68B:575–77.

———. 1983. Occurrence, variation and biosynthesis of the cyanogenic glucosides linamarin and lotaustralin in species of the Heliconiini. *Comp. Biochem. Physiol.* 75B:65–73.

National Research Council, Committee on Research Priorities in Tropical Biology. 1980. *Research priorities in tropical biology.* Washington, D.C.: National Academy of Sciences.

Naveh, Z., and R. H. Whittaker. 1979. Structure and floristic diversity of shrublands and woodlands in northern Israel and other Mediterranean areas. *Vegetatio* 41:171–90.

Needham, J. G., S. W. Frost, and B. H. Tothill. 1928. *Leaf-mining insects.* Baltimore, Md.: Williams and Wilkins.

Neill, D. A. 1987. Trapliners in the trees: Hummingbird pollination of *Erythrina* sect. *Erythrina* (Leguminosae: Papilionoideae). *Ann. MO Bot. Gard.* 74:27–41.

de Nettancourt, D. 1977. *Incompatibility in angiosperms.* New York: Springer-Verlag.

Newbery, D. M., and H. de Foresta. 1985. Herbivory and defense in pioneer gap and understory trees in tropical rain forest in French Guiana. *Biotropica* 17:238–44.

Newbery, D. M., E. Renshaw, and E. F. Brunig. 1986. Spatial pattern of trees in kerangas forest, Sarawak. *Vegetatio* 65:77–89.

Newstrom, L. E., G. W. Frankie, and H. G. Baker. 1991. Survey of long-term flowering patterns in tropical rain forest trees at La Selva, Costa Rica. Pp. 345–66 in *L'arbre. Biologie et développement,* ed. C. Edelin. Naturalia Montpeliensia, n.s. A7, Montpellier, France.

Ng, F. S. P. 1977. Gregarious flowering of dipterocarps in Kepong, 1976. *Malay. For.* 40:126–37.

Nichols-Orians, C. 1991a. Condensed tannins, attine ants, and the performance of a symbiotic fungus. *J. Chem. Ecol.* 17:1177–95.

———. 1991b. The effects of light on foliar chemistry, growth and susceptibility of seedlings of a canopy tree to an attine ant. *Oecologia* 86:552–60.

———. 1991c. Environmentally induced differences in plant traits: Consequences for susceptibility to a leaf-cutter ant. *Ecology* 72:1609–23.

Nichols-Orians, C., and J. C. Schultz. 1990. Interactions among leaf toughness, chemistry, and harvesting by attine ants. *Ecol. Entomol.* 15:311–20.

Nielson, B. O. 1978. Above ground food resources and herbivory in beech forest ecosystem. *Oikos* 31:273–79.

Nobel, P. S., L. J. Zaragoza, and W. K. Smith. 1975. Relation between mesophyll surface area, photosynthetic rate, and illumination level during development for leaves *Plectranthus parviflorus* Henckel. *Plant Physiol.* 55:1067–70.

Núñez-Farfan, J., and R. Dirzo. 1988. Within-gap spatial heterogeneity and seedling performance in a Mexican tropical forest. *Oikos* 51:274–84.

Nye, P. H., and P. J. Greenland. 1960. *The soil under shifting cultivation.* Commonwealth Bureau of Soils Technical Communication no. 51, Commonwealth Agricultural Bureaux, Farnham Royal, Bucks, England.

Oates, J. F., P. G. Waterman, and G. M. Choo. 1980. Food selection by the South Indian leaf monkey, *Presbytis johnii,* in relation to leaf chemistry. *Oecologia* 45:45–56.

Oates, J. F., G. H. Whitesides, A. G. Davies, P. G. Waterman, S. M. Green, G. L. Dasilva, and S. Mole. 1990. Determinants of variation in tropical forest primate biomass: A new evidence from West Africa. *Ecology* 71:328–43.

Oberbauer, S. F. 1983. The ecophysiology of *Pentaclethra macroloba,* a canopy tree species in the rainforests of Costa Rica. Ph.D. diss. Duke University, Durham, N.C.

———. 1985. Plant water relations of selected species in wet and dry tropical lowland forests in Costa Rica. *Rev. Biol. Trop.* 33:137–42.

———. 1987. Diferencias fisiológicas en dos variantes de color de *Triolena hirsuta. Rev. Biol. Trop.* 35 (suppl. 1): 229.

———. 1988. Responses of tropical rain forest tree seedlings to drought. P. 17 in OTS International Symposium on Tropical Studies. University of Miami, Coral Gables, Florida (abstract).

———. 1990. Seed weight and rooting depth of seedlings of Costa Rican wet forest trees. *Rev. Biol. Trop.* 38:473–76.

Oberbauer, S. F., D. A. Clark, D. B. Clark, and M. Quesada. 1989. Comparative analysis of photosynthetic light environments within the crowns of juvenile rain forest trees. *Tree Physiol.* 5:13–23.

Oberbauer, S. F., D. B. Clark, D. A. Clark, and M. Quesada. 1988. Crown light environments of saplings of two species of rain forest emergent trees. *Oecologia* 75:207–12.

Oberbauer, S. F., and M. A. Donnelly. 1986. Growth analysis and successional status of Costa Rican rain forest trees. *New Phytol.* 104:517–21.

Oberbauer, S. F., N. Fetcher, and B. R. Strain. 1983. Effects of light regime on leaf morphology of early and late successional tropical trees. *Bull. Ecol. Soc. Am.* 64:58–59 (abstract).

Oberbauer, S. F., and B. R. Strain. 1984. Photosynthesis and successional status of Costa Rican rain forest trees. *Photosynth. Res.* 5:227–32.

———. 1985. Effects of light regime on the growth and physiology of *Pentaclethra macroloba* (Mimosaceae) in Costa Rica. *J. Trop. Ecol.* 1:303–20.

———. 1986. Effects of canopy position and irradiance on the leaf physiology and morphology of *Pentaclethra macroloba* (Willd.) Kuntz. *Am. J. Bot.* 73:409–16.

Oberbauer, S. F., B. R. Strain, and G. H. Riechers. 1987. Field water relations of a wet-tropical forest tree species, *Pentaclethra macroloba* (Mimosaceae). *Oecologia* 71:369–74.

O'Connell, M. A. 1979. Ecology of didelphid marsupials from northern Venezuela. Pp. 73–96 in *Vertebrate ecology in the northern Neotropics,* ed. J. F. Eisenberg. Washington, D.C.: Smithsonian Institution Press.

O'Donnell, S. 1989. A comparison of fruit removal by bats and birds from *Piper hispidum* Sw. (Piperaceae), a tropical second growth shrub. *Brenesia* 31:25–32.

O'Dowd, D. J. 1979. Foliar nectar production and ant activity on a Neotropical tree, *Ochroma pyramidale:* Ecological implications. *Oecologia* 43:233–48.

———. 1980. Pearl bodies of a Neotropical tree, *Ochroma pyramidale:* Ecological implications. *Am. J. Bot.* 67:543–49.

———. 1982. Pearl bodies as ant food: An ecological role for some leaf emergences of tropical plants. *Biotropica* 14:40–49.

O'Dowd, D. J., and A. M. Gill. 1984. Predator satiation and site alteration following fire: Mass reproduction of alpine ash (*Eucalyptus delegatensis*) in southeastern Australia. *Ecology* 65:1052–66.

Odum, H. T., ed. 1970. *A tropical rain forest: A study of irradiation and ecology at El Verde, Puerto Rico.* Oak Ridge, Tenn.: Division of Technical Information Extension, U.S. Atomic Energy Commission.

Odum, H. T., with J. Ruíz-Reyes. 1970. Holes in leaves and the grazing control mechanism. Pp. I-69–I-80 in *A tropical rain forest: A study of irradiation and ecology at El Verde, Puerto Rico,* ed. H. T. Odum. Oak Ridge, Tenn.: Division of Technical Information Extension, U.S. Atomic Energy Commission.

Ohmart, C. P. 1984. Is insect defoliation in eucalypt forests greater than that in other temperate forests? *Aust. J. Ecol.* 9:413–18.

Ohmart, C. P., L. G. Stewart, and J. R. Thomas. 1983. Leaf consumption by insects in three *Eucalyptus* forest types in Southeastern Australia and their role in short-term nutrient cycling. *Oecologia* 59:322–30.

Oldeman, R. A. A. 1983. Tropical rain forest architecture, silvigenisis and diversity. Pp. 139–50 in *Tropical rain forest: Ecology and management,* ed. S. L. Sutton, T. C. Whitmore, and A. C. Chadwick. Oxford: Blackwell Scientific.

Oliveira, P. S., A. F. de Silva, and A. B. Martins. 1987. Ant foraging on extrafloral nectaries of *Qualea grandiflora* (Vochysiaceae) in cerrado vegetation: Ants as potential antiherbivore agents. *Oecologia* 74:228–30.

O'Malley, D. M., and K. S. Bawa. 1987. Mating system of a tropical rain forest tree species. *Am. J. Bot.* 74:1143–49.

O'Malley, D. M., D. P. Buckley, G. T. Prance, and K. S. Bawa. 1988. Genetics of brazil nut (*Bertholletia excelsa* Humb. & Bonpl.: Lecythidaceae). Part 2. Mating system. *Theor. Appl. Genet.* 76:929–32.

Opler, P. A. 1974. Oaks as evolutionary islands for leaf-mining insects. *Am. Sci.* 62:67–73.

Opler, P. A., H. G. Baker, and G. W. Frankie. 1975. Reproductive biology of some Costa Rican *Cordia* species (Boraginaceae). *Biotropica* 7:234–47.

———. 1980. Plant reproductive characteristics during secondary succession in Neotropical lowland forest ecosystems. *Biotropica* 12 (suppl.): 40–46.

Opler, P. A., and K. S. Bawa. 1978. Sex ratios in tropical forest trees. *Evolution* 32:812–21.

Opler, P. A., G. W. Frankie, and H. G. Baker. 1976. Rainfall

as a factor in the release, timing, and synchronization of anthesis by tropical trees and shrubs. *J. Biogeogr.* 3:231–36.

———. 1980. Comparative phenological studies of treelet and shrub species in tropical wet and dry forests in the lowlands of Costa Rica. *J. Ecol.* 68:167–88.

Opler, P. A., and A. Krizek. 1984. *Butterflies east of the Great Plains.* Baltimore, Md.: Johns Hopkins Univ. Press.

Oppenheimer, J. R. 1982. *Cebus capucinus:* Home range, population dynamics, and interspecific relationships. Pp. 253–72 in *The ecology of a tropical forest: Seasonal rhythms and long-term changes,* ed. E. G. Leigh, Jr., A. S. Rand, and D. M. Windsor. Washington, D.C.: Smithsonian Institution Press.

Organización para Estudios Tropicales and Dirección General Forestal. 1990. *Encuentro regional sobre especies forestales nativas de la zona norte y atlántica, Memoria,* ed. E. González, R. Butterfield, J. Segleau, and M. Espinoza. Cartago: Instituto Tecnológico de Costa Rica.

Organization for Tropical Studies. 1987. Report to the Executive Director of OTS by the Forestry Advisory Committee. Meeting at La Selva Biological Station, January 29–31, 1987. San José, Costa Rica.

Orians, G. H. 1969a. The number of birds species in some tropical forests. *Ecology* 50:783–801.

———. 1969b. On the evolution of mating systems in birds and mammals. *Am. Nat.* 103:589–603.

———. 1982. The influence of tree-falls in tropical forests on tree species richness. *Trop. Ecol.* 23:255–79.

Orians, G. H., and R. T. Paine. 1983. Convergent evolution at the community level. Pp. 431–64 in *Coevolution,* ed. D. J. Futuyma and M. Slatkin. Sunderland, Mass.: Sinauer Associates.

Orians, G. H., and O. T. Solbrig, eds. 1977. *Convergent evolution in warm deserts.* Stroudsburg, Pa.: Dowden, Hutchinson and Ross.

Osonubi, O., and W. J. Davies. 1980. The influence of water stress on the photosynthetic performance and stomatal behaviour of tree seedlings subjected to variation in temperature and irradiance. *Oecologia* 45:3–10.

Otis, G. W., E. Santana C., D. L. Crawford, and M. L. Higgins. 1986. The effect of foraging army ants on leaf-litter arthropods. *Biotropica* 18:56–61.

Owen, D. F. 1971. *Tropical Butterflies.* Oxford: Clarendon Press.

Owen, D. F., and J. Owen. 1974. Species diversity in temperate and tropical Ichneumonidae. *Nature* 249:583–84.

Owen, D. F., and R. G. Wiegert. 1976. Do consumers maximize plant fitness? *Oikos* 27:488–92.

Oyama, K. 1990. Variation in growth and reproduction in the Neotropical dioecious palm *Chamaedorea tepejilote. J. Ecol.* 78:648–63.

Oyama, K., and A. Mendoza. 1990. Effects of defoliation on growth, reproduction, and survival of a Neotropical dioecious palm, *Chamaedorea tepejilote. Biotropica* 22:119–23.

Paaby, P. 1988. Light and nutrient limitation of attached algae in a Costa Rican lowland stream. Ph.D. diss. University of California, Davis.

Paaby, P., D. B. Clark, and H. González. 1991. Training rural residents as naturalist guides: Evaluation of a pilot project in Costa Rica. *Conserv. Biol.* 5:542–46.

Pacala, S., and J. Roughgarden. 1984. Control of arthropod abundance by *Anolis* lizards on St. Eustatius (Netherlands Antilles). *Oecologia* 64:160–62.

Paine, R. T. 1966. Food web complexity and species diversity. *Am. Nat.* 100:65–75.

Palmeirim, J., and K. Etheridge. 1985. The influence of man-made trails on foraging by tropical frugivorous bats. *Biotropica* 17:82–83.

Palmeirim, J. M., D. L. Gorchov, and S. Stoleson. 1989. Trophic structure of a Neotropical frugivore community: Is there competition between birds and bats? *Oecologia* 79:403–11.

Palmer, A. R. 1979. Fish predation and the evolution of gastropod shell sculpture: Experimental and geographic evidence. *Evolution* 33:697–713.

Palmer, B., R. A. Bray, T. M. Ibraham, and M. G. Fulloon. 1989. The effect of the luecaena psyllid on the yield of *Leucaena leucocephala* CV. Cunningham at four sites in the tropics. *Trop. Grassl.* 23:105–7.

Papageorgis, C. 1975. Mimicry in Neotropical butterflies. *Am. Sci.* 63:522–32.

Parfitt, R. L. 1980. Chemical properties of variable charge soils. Pp. 167–94 in *Soils with variable charge,* ed. B. K. G. Theng. Lower Hutt, New Zealand: New Zealand Society of Soil Science.

Parker, G. G. 1983. Throughfall and stemflow in the forest nutrient cycle. *Adv. Ecol. Res.* 13:57–133.

———. 1985. The effect of disturbance on water and solute budgets of hillslope tropical rainforest in northeastern Costa Rica. Ph.D. diss., University of Georgia, Athens.

Parker, M. A., and A. G. Salzman. 1985. Herbivore exclosure and competitor removal: Effects on juvenile survivorship and growth in the shrub *Gutierrezia microcephala. J. Ecol.* 73:903–13.

Parker, T. A. 1981. Distribution and biology of the white-cheeked cotinga, *Zaratornis stresemanni,* a high Andean frugivore. *Bull. Br. Ornithol. Club* 101:256–65.

Parnell, R. A., Jr. 1986. Processes of soil acidification in tropical durandepts, Nicaragua. *Soil Sci. Soc. Am. Proc.* 142:43–55.

Parsons, J. J. 1976. Forest to pasture: Development or destruction? *Rev. Biol. Trop.* 24 (suppl. 1): 121–38.

———. 1983. Beef cattle (ganado). Pp. 77–79 in *Costa Rican natural history,* ed. D. H. Janzen. Chicago: Univ. of Chicago Press.

Partridge, L., and J. Endler. 1987. Life history constraints in sexual selection. In *Sexual selection: Testing the alternatives,* ed. J. W. Bradbury and M. B. Andersson. New York: Wiley Interscience.

Patrick, R. 1964. A discussion of the results of the Catherwood Expedition to the Peruvian headwaters of the Amazon. *Int. Assoc. Theor. Appl. Limnol. Proc.* 15:1084–90.

———. 1966. The Catherwood Foundation Peruvian Amazon Expedition: Limnological and systematic studies. *Monogr. Acad. Nat. Sci. Phila.* 14:1–495.

Patterson, D. T., S. O. Duke, and R. E. Hoagland. 1978. Effects of irradiance during growth on adaptive photo-synthetic characteristics of velvetleaf and cotton. *Plant Physiol.* 61:402–5.

Paulson, D. R. 1985. Odonata of the Tambopata Reserved Zone, Madre de Dios, Peru. *Rev. Peru. Entomol.* 27 (1984): 9–14.

Payne, C. D. 1986. *The generalized linear interactive modelling system. Release 3.77. The GLIM system manual.* Oxford: Royal Statistical Society.

Payne, J., C. M. Francis, and K. Phillips. 1985. *A field guide to the mammals of Borneo.* Malaysia: Sabah Society.

Pearcy, R. W. 1983. The light environment and growth of C3 and C4 tree species in the understory of a Hawaiian forest. *Oecologia* 58:19–25.

———. 1987. Photosynthetic gas exchange responses of Australian tropical forest trees in canopy, gap and understory microenvironments. *Funct. Ecol.* 1:169–78.

Pearcy, R. W., and H. C. Calkin. 1983. Carbon dioxide exchange of C3 and C4 tree species in the understory of a Hawaiian forest. *Oecologia* 58:26–32.

Pearcy, R. W., K. Osteryoung, and H. W. Calkin. 1985. Photosynthetic responses to dynamic light environments by Hawaiian trees: Time course of CO_2 uptake and carbon gain during sunflecks. *Plant Physiol.* 79:896–902.

Pearcy, R. W., and R. H. Robichaux. 1985. Tropical and subtropical forests. Pp. 278–95 in *Physiological ecology of North American plant communities,* ed. B. F. Chabot and H. A. Mooney. New York: Chapman and Hall.

Pearson, D. L. 1972. Un estudio de las aves de Limoncocha, Prov. de Napo, Ecuador. *Bol. Inf. Cient. Nac. Ecuador* 13:3–14.

———. 1977. A pantropical comparison of bird community structure on six lowland forest sites. *Condor* 79:232–44.

———. 1985. The tiger beetles (Coleoptera: Cicindelidae) of the Tambopata Reserved Zone, Madre de Dios, Peru. *Rev. Peru. Entomol.* 27 (1984): 15–24.

Pearson, D. L., D. Tallman, and E. Tallman. 1977. The birds of Limonchocha, Napo Province, Ecuador. Quito, Ecuador: Inst. Lingrustico de Verano.

Pechmann, J. H. K., D. E. Scott, R. D. Semlitsch, J. P. Caldwell, L. J. Vitt, and J. W. Gibbons. 1991. Declining amphibian populations: The problem of separating human impacts from natural fluctuations. *Science* 253:892–95.

Pemadasa, M. A., and C. V. S. Gunatilleke. 1981. Pattern in a rain forest in Sri Lanka. *J. Ecol.* 69:117–24.

Peralta, R., A. I. Barquero, D. Lieberman, M. Lieberman, and G. S. Hartshorn. 1987a. Demografía de plántulas de árboles en parcelas permanentes en La Selva, Sarapiquí, Costa Rica. *Rev. Biol. Trop.* 35 (suppl. 1): 230 (abstract).

———. 1987b. Seguimiento de plántulas de árboles en las parcelas permanentes de La Selva. *Rev. Biol. Trop.* 35 (suppl. 1): 230 (abstract).

Peralta, R., G. Hartshorn, D. Lieberman, and M. Lieberman. 1987. Reseña de estudios a largo plazo sobre composición florística y dinámica del bosque tropical en La Selva, Costa Rica. *Rev. Biol. Trop.* 35 (suppl. 1): 23–39.

Pereira, H. C. 1967. Effects of land use on the water and energy budgets of tropical watersheds. Pp. 435–50 in *International symposium on forest hydrology,* ed. W. E. Sopper and H. W. Lull. Oxford: Pergamon Press.

Pérez, C. 1989. Reforestation in Costa Rica: An introduction for investors. Report, Organization for Tropical Studies. San José, Costa Rica.

Perry, D. R., and A. Starrett. 1980. The pollination ecology and blooming strategy of a Neotropical emergent tree, *Dipteryx panamensis. Biotropica* 12:307–13.

Petr, T., ed. 1983. The Purari-Tropical environment of a high rainfall river basin. The Hague: W. Junk.

Petriceks, J. 1956. Plan de ordenación del bosque de la Finca "La Selva." Master's thesis, Instituto Interamericano de Ciéncias Agrícolas, Turrialba, Costa Rica.

Phillips, E. 1987. Colonización de diatomeas perifíticas sobre sustrato artificial en tres quebradas de la Estación Biológica La Selva, Puerto Viejo de Sarapiquí, Costa Rica. Licenciatura thesis, Universidad de Costa Rica, Costa Rica.

Phillips, K. 1990. Where have all the frogs and toads gone? *Bioscience* 40:422–24.

Pianka, E. R. 1966. Latitudinal gradients in species diversity: A review of concepts. *Am. Nat.* 100:33–46.

———. 1970. On *r* and *K* selection. *Am. Nat.* 104:592–97.

———. 1976. Natural selection of optimal reproductive tactics. *Am. Zool.* 16:775–84.

Pierce, N. E. 1984. Amplified species diversity: A case study of an Australian lycaenid butterfly and its attendant ants. *Symp. R. Entomol. Soc. Lond.* 11:196–200.

———. 1985. Lycaenid butterflies and ants: Selection for nitrogen-fixing and other protein rich food plants. *Am. Nat.* 125:888–95.

———. 1987. The evolution and biogeography of associations between lycaenid butterflies and ants. *Oxf. Surv. Evol. Biol.* 4:89–116.

Pierce, N. E., R. L. Kitching, R. C. Buckley, M. F. J. Taylor, and K. F. Benbow. 1987. The costs and benefits of cooperation between the Australian lycaenid butterfly, *Jalmenus evagoras,* and its attendant ants. *Behav. Ecol. Sociobiol.* 21:237–48.

Pierce, S. M. 1992. Environmental History of La Selva Biological Station: Colonization and deforestation of Sarapiquí Canton, Costa Rica. Pp. 40–57 in *Changing tropical forests: Historical perspectives on today's challenges in Central and South America,* ed. H. K. Steen and R. P. Tucker. Durham, N.C.: Forest History Society.

Pimentel, R. A., and J. D. Smith. 1985. *Biostat II. A multivariate statistical toolbox.* Placentia, Calif.: Sigma Soft.

Piñero, D., M. Martínez-Ramos, and J. Sarukhán. 1984. A population model of *Astrocaryum mexicanum* and a sensitivity analysis of its finite rate of increase. *J. Ecol.* 72:977–91.

Piñero, D., and J. Sarukhán. 1982a. Reproductive behavior and its individual variability in a tropical palm, *Astrocaryum mexicanum. J. Ecol.* 70:461–72.

Piñero, D., J. Sarukhán, and P. Alberdi. 1982. The costs of reproduction in a tropical palm, *Astrocaryum mexicanum. J. Ecol.* 70:473–81.

Piper, J. K. 1986. Effects of habitat and size of fruit display on removal of *Smilacina stellata* (Liliaceae) fruits. *Can. J. Bot.* 64:1050–54.

Pires, J. M., T. Dobzhansky, and G. A. Black. 1953. An esti-

mate of the number of species of trees in an Amazonian forest community. *Bot. Gaz.* 114:467–77.

Pittendrigh, C. S. 1950. The ecotopic specialization of *Anopheles bellator* and *A. momunculus. Ecology* 4:43–63.

Place, S. 1981. Ecological and social consequences of export beef production in Guanacaste Province, Costa Rica. Ph.D. diss. University of California, Los Angeles.

Platt, J., and K. Denman. 1975. Spectral analysis in ecology. *Annu. Rev. Ecol. Syst.* 6:189–210.

Pleasants, J. M. 1980. Competition for bumblebee pollinators in Rocky Mountain plant communities. *Ecology* 61:1446–59.

———. 1990. Null-model tests for competitive displacement: The fallacy of not focusing on the whole community. *Ecology* 71:1078–84.

Pohl, R. W. 1980. Gramineae. In *Flora Costaricensis,* ed. W. Burger. *Fieldiana Bot.,* n.s., 4:1–608.

Poole, R. W., and B. J. Rathcke. 1979. Regularity, randomness, and aggregation in flowering phenologies. *Science* 203:470–71.

Poore, M. E. D. 1968. Studies in Malaysian rain forest. Part 1. The forest on Triassic sediments in Jengka Forest Reserve. *J. Ecol.* 56:143–96.

Popma, J., and F. Bongers. 1988. The effect of canopy gaps on growth and morphology of seedlings of rain forest species. *Oecologia* 75:625–32.

———. 1991. Acclimation of seedlings of three Mexican tropical rain forest tree species to a change in light availability. *J. Trop. Ecol.* 7:85–97.

Popma, J., F. Bongers, M. Martínez-Ramos, and E. Veneklaas. 1988. Pioneer species distributions in treefall gaps in Neotropical rain forest: A gap definition and its consequences. *J. Trop. Ecol.* 4:77–88.

Population Reference Bureau. 1990. 1990 World Population Data Sheet, Washington, D.C.

Porras Z., A. and B. Villarreal. 1985. Deforestación en Costa Rica: Implicaciones sociales, económicas y legales. San José: Editorial Costa Rica.

Portig, W. H. 1976. The climate of Central America. In *World Survey of Climatology.* Pp. 405–78 in vol. 12, *Climates of Central and South America,* ed. W. Schwerdtfeger. New York: Elsevier Scientific.

Poulton, E. B. 1887. Experimental proof of the protective value of colour and markings in insects in reference to their vertebrate enemies. *Proc. Zool. Soc. Lond.* 1887:191–274.

———. 1908. *Essays on evolution.* Oxford: Oxford Univ. Press.

Pounds, J. A. 1988. Ecomorphology, locomotion, and microhabitat structure: Patterns in a tropical mainland *Anolis* community. *Ecol. Monogr.* 58:299–320.

Powell, G. V. N. 1979. Structure and dynamics of interspecific flocks in a mid-elevation Neotropical forest. *Auk* 96:375–90.

———. 1985. Sociobiology and adaptive significance of interspecific foraging flocks in the Neotropics. Pp. 713–32 in *Neotropical ornithology,* ed. P. A. Buckley, M. S. Foster, E. S. Morton, R. S. Ridgely, and N. G. Buckley. Ornithological Monographs vol. 36. Lawrence: AOU.

———. 1989. On the possible contribution of mixed species

flocks to species richness in Neotropical avifaunas. *Behav. Ecol. Sociobiol.* 24:387–93.

Powles, S. B. 1984. Photoinhibition of photosynthesis induced by visible light. *Annu. Rev. Plant Physiol.* 35:15–44.

Powlesland, M. H., M. Philipp, and D. G. Lloyd. 1985. Flowering and fruiting patterns of three species of *Melicytus* (Violaceae) in New Zealand. *N. Z. J. Bot.* 23:581–96.

Prance, G. T. 1972a. Chrysobalanaceae. *Flora Neotrop. Monogr.* 9:1–410.

———. 1972b. Dichapetalaceae. *Flora Neotrop. Monogr.* 10:3–84.

———, ed. 1982. *Biological diversification in the tropics.* New York: Columbia Univ. Press.

Prance, G. T., and T. E. Lovejoy, eds. 1985. *Key environment: Amazonia.* New York: Pergamon.

Prance, G. T., W. A. Rodrigues, and M. F. de Silva. 1976. Inventario florestal de um hectare de mata de terra firme, km 30 de Estrada Manaus-Itacoatiara. *Acta Amazonica* 6:9–35.

Pratt, T. 1977. Fishes and figs. Report, OTS 77-4:417–20.

Pratt, T. K. 1983. Pleistocene seed dispersal. *Science* 216:6.

———. 1984. Examples of tropical frugivores defending fruit-bearing plants. *Condor* 86:123–29.

Pratt, T. K., and E. W. Stiles. 1985. The influence of fruit size and structure on composition of frugivore assemblages in New Guinea. *Biotropica* 17:314–21.

Presch, W. F. 1970. The evolution of macroteiid lizards: An oestological interpretation. Ph.D. diss., University of Southern California, Los Angeles.

Prestwich, G. D., and B. L. Bentley. 1981. Nitrogen fixation by intact colonies of the termite *Nasutitermes corniger. Oecologia* 49:249–51.

Prestwich, G. D., B. L. Bentley, and E. J. Carpenter. 1980. Nitrogen sources for Neotropical nasute termites: Fixation and selective foraging. *Oecologia* 46:397–401.

Price, P. W. 1980. *Evolutionary biology of parasites.* Princeton, N.J.: Princeton Univ. Press.

———. 1991. Patterns in communities along latitudinal gradients. Pp. 51–70 in *Plant-animal interactions: Evolutionary ecology in tropical and temperate regions,* ed. P. W. Price, T. M. Lewinsohn, G. W. Fernandes, and W. W. Benson. New York: J. Wiley.

Price, P. W., C. E. Bouton, P. Gross, B. A. McPheron, J. N. Thompson, and A. E. Weis. 1980. Interactions among three trophic levels: Influence of plants on interactions between insect herbivores and natural enemies. *Annu. Rev. Ecol. Syst.* 11:41–65.

Price, P. W., G. L. Waring, R. Julkunen-Tiito, J. Tahvanainen, H. A. Mooney, and T. P. Craig. 1989. Carbon-nutrient balance hypothesis in within-species phytochemical variation of *Salix lasiolepis. J. Chem. Ecol.* 15:1117–31.

Primack, R. B. 1980. Phenological variation within natural populations: Flowering in New Zealand montane shrubs. *J. Ecol.* 68:849–62.

———. 1985a. Longevity of individual flowers. *Annu. Rev. Ecol. Syst.* 16:15–37.

———. 1985b. Patterns of flowering phenology in communities, populations, individuals, and single flowers. Pp. 571–93 in *The population structure of vegetation,* ed. J. White. Dordrecht: W. Junk.

———. 1987. Relationships among flowers, fruits, and seeds. *Annu. Rev. Ecol. Syst.* 18:409–30.

Primack, R. B., P. Ashton, P. S. Chai, and H. S. Lee. 1985. Growth rates and population structure of Moraceae trees in Sarawak. *Ecology* 66:577–88.

Pringle, C. M. 1988. History of conservation efforts and initial exploration of the lower extension of Parque Nacional Braulio Carrillo, Costa Rica. Pp. 225–41 in *Tropical rainforests: Diversity and conservation,* ed. F. Almeda and C. M. Pringle. San Francisco: California Academy of Sciences.

———. 1990. Nutrient spatial heterogeneity: Effects on community structure, physiognomy, and diversity of stream algae. *Ecology* 71:905–20.

———. 1991. Geothermally modified waters surface at La Selva Biological Station, Costa Rica: Volcanic processes introduce chemical discontinuities into lowland tropical streams. *Biotropica* 23:523–29.

Pringle, C., I. Chacón, M. Grayum, H. Greene, G. Hartshorn, G. Schatz, G. Stiles, C. Gómez, and M. Rodríguez. 1984. Natural history observations and ecological evaluation of the La Selva Protection Zone, Costa Rica. *Brenesia* 22:189–206.

Pringle, C. M., P. Paaby-Hansen, P. D. Vaux, and C. R. Goldman. 1986. In situ nutrient assays of periphyton growth in a lowland Costa Rican stream. *Hydrobiologia* 134:207–13.

Pringle, C., and F. J. Triska. 1991a. Effects of geothermal groundwater on nutrient dynamics of a lowland stream in Costa Rica. *Ecology* 72:951–65.

———. 1991b. Variation in phosphate concentrations of small-order streams draining volcanic landscapes in Costa Rica: sources and implications for nutrient cycling. Pp. 70–83 in *Phosphorous cycles in terrestrial and aquatic ecosystems in Latin America with emphasis on the Amazon basin,* ed. H. Tiessen, D. López-Hernández and I. H. Salcedo. SCOPE Symposium Proceedings. Saskatoon, Canada: Turner-Warwick Communications.

Proctor, J., J. M. Andrews, S. C. L. Fogden, and H. W. Vallack. 1983. Ecological studies in four contrasting lowland rain forests in Gunung Mulu National Park, Sarawak. Part 2. Litterfall, litter standing crop and preliminary observations on herbivory. *J. Ecol.* 71:261–83.

Pullen, A. S. 1987. Adult feeding time, lipid accumulation, and overwintering in *Agalis urticae* and *Inachis io* (Lepidoptera: Nymphalidae). *J. Zool. (Lond.)* 211:631–41.

Pulliam, H. R. 1988. Sources, sinks, and population regulation. *Am. Nat.* 132:652–61.

Pulliam, H. R., and B. J. Danielson. 1991. Sources, sinks, and habitat selection: A landscape perspective on population dynamics. *Am. Nat.* 137 (suppl.): S50–S66.

Putz, F. E. 1979. Aseasonality in Malaysian tree phenology. *Malay. For.* 42:1–24.

———. 1983. Treefall pits and mounds, buried seeds, and the importance of soil disturbance to pioneer trees on Barro Colorado Island, Panama. *Ecology* 64:1069–74.

———. 1984a. How trees avoid and shed lianas. *Biotropica* 16:19–23.

———. 1984b. The natural history of lianas on Barro Colorado Island, Panama. *Ecology* 65:1713–24.

Putz, F. E., P. D. Coley, K. Lu, A. Montalvo, and A. Aiello.

1983. Uprooting and snapping of trees: Structural determinants and ecological consequences. *Can. J. For. Res.* 13:1011–20.

Putz, F. E., and N. M. Holbrook. 1988. Further observations on the dissolution of mutualism between *Cecropia* and its ants: The Malaysian case. *Oikos* 53:121–25.

Putz, F. E., and K. Milton. 1982. Tree mortality rates on Barro Colorado Island. Pp. 95–108 in *The ecology of a tropical forest: Seasonal rhythms and long-term changes,* ed. E. G. Leigh, Jr., A. S. Rand, and D. M. Windsor. Washington, D.C.: Smithsonian Institution Press.

Pyburn, W. F. 1970. Breeding behavior of the leaf-frogs *Phyllomedusa callidryas* and *Phyllomedusa dacnicolor* in Mexico. *Copeia* 1970:209–18.

Quintanilla, I. 1990. Ocupaciones precolombinas en el bosque tropical lluvioso: Evaluación arqueológico de la estación biológica La Selva. Report, Organization for Tropical Studies, San José, Costa Rica.

Rabinowitz, A., and B. G. Nottingham, Jr. 1986. Ecology and behavior of the jaguar *(Panthera onca)* in Belize. *J. Zool.* 210:149–59.

Rabinowitz, D. 1981. Seven forms of rarity. Pp. 205–17 in *The biological aspects of rare plant conservation,* ed. H. Synge. New York: J. Wiley.

Radulovich, R., and P. Sollins. 1985. Compactación de un suelo aluvial de origen volcánico por tráfico de personas. *Agron. Costarric.* 9:143–48.

———. 1987. Improved performance of zero-tension lysimeters. *Soil Sci. Soc. Am. J.* 51:1386–88.

Radulovich, R., E., Solórzano, and P. Sollins. 1989. Soil macropore size distribution from water breakthrough curves. *Soil Sci. Soc. Am. J.* 53:556–59.

Raich, J. W. 1980a. Carbon budget of a tropical soil under mature and young vegetation. Master's thesis, University of Florida, Gainesville.

———. 1980b. Fine roots regrow rapidly after forest felling. *Biotropica* 12:231–32.

———. 1983a. Effects of forest conversion on the carbon budget of a tropical soil. *Biotropica* 15:177–84.

———. 1983b. Understory palms as nutrient traps: A hypothesis. *Brenesia* 21:119–29.

Ramos, M. A. 1983. Seasonal movements of bird populations at a Neotropical study site in southern Veracruz, Mexico. Ph.D. diss., Univ. of Minnesota, Minneapolis.

Rand, A. S. 1978. Reptilian arboreal folivores. Pp. 115–22 in *The ecology of arboreal folivores,* ed. G. G. Montgomery. Washington, D.C.: Smithsonian Institution Press.

Rand, A. S., and C. W. Myers. 1990. The herpetofauna of Barro Colorado Island, Panama: An ecological summary. Pp. 386–409 in *Four Neotropical rainforests,* ed. A. H. Gentry. New Haven, Conn.: Yale Univ. Press.

Rand, A. S., and W. M. Rand. 1982. Variation in rainfall on Barro Colorado Island. Pp. 47–59 in *The ecology of a tropical forest: Seasonal rhythms and long-term changes,* ed. E. G. Leigh, Jr., A. S. Rand, and D. M. Windsor. Washington, D.C.: Smithsonian Institution Press.

Rand, A. S., M. J. Ryan, and K. Troyer. 1983. A population explosion in a tropical tree frog: *Hyla rufitela* on Barro Colorado, Panama. *Biotropica* 15:72–73.

Rappole, J. H., E. S. Morton, T. E. Lovejoy, and J. L. Ruos. 1983. *Nearctic avian migrants in the Neotropics.* U.S. Department of the Interior, Fish and Wildlife Service Publication. Washington, D.C.: USGPO.

Rathbun, S. L. 1980. Latitudinal variation in the spatial patterns of tree species. Master's thesis, Florida State University, Tallahassee.

Rathcke, B. J. 1976. Competition and coexistence within a guild of herbivorous insects. *Ecology* 57:76–87.

———. 1983. Competition and facilitation among plants for pollination. Pp. 305–29 in *Pollination biology,* ed. L. Real. New York: Academic Press.

Rathcke, B. J., and E. P. Lacey. 1985. Phenological patterns of terrestrial plants. *Annu. Rev. Ecol. Syst.* 16:179–214.

Raveh, A., and C. S. Tapiero. 1980. Periodicity, constancy, heterogeneity and the categories of qualitative time series. *Ecology* 61:715–19.

Ray, R. T. 1982. Dynamics of pollen dispersal by hummingbirds in a population of *Heliconia imbricata* (Heliconiaceae). Master's thesis, Univ. of Pennsylvania, Philadelphia.

Ray, T. S., and C. C. Andrews. 1980. Antbutterflies: Butterflies that follow army ants to feed on antbird droppings. *Science* 210:1147–48.

Redfield, A. C. 1958. The biological control of chemical factors in the environment. *Am. Sci.* 46:205–21.

Redford, K. H., G. A. Bouchardet da Fonseca, and T. E. Lacher, Jr. 1984. The relationship between frugivory and insectivory in primates. *Primates* 25:433–40.

Rees, C. J. C. 1983. Microclimate and flying Hemiptera fauna of a primary lowland forest in Sulawesi. Pp. 121–36 in *Tropical rain forest: Ecology and management,* ed. S. L. Sutton, T. C. Whitmore, and A. C. Chadwick. British Ecological Society Special Publication no. 2. Oxford: Blackwell Scientific Publications.

Reese, J. C., B. G. Chan, and A. C. Waiss, Jr. 1982. Effects of cotton condensed tannin, maysin (corn) and pinitol (soybeans) on *Heliothus zea* growth and development. *J. Chem. Ecol.* 8:1429–36.

Regal, P. J. 1977. Ecology and evolution of flowering plant dominance. *Science* 196:622–29.

Regan, C. T. 1906–1908. Pisces. *Biol. Cen.-Am.* 8:1–203.

Rehr, S. S., P. P. Feeny, and D. H. Janzen. 1973. Chemical defence in Central American non-ant-acacias. *J. Anim. Ecol.* 42:405–16.

Reich, P. B., and R. Borchert. 1982. Phenology and ecophysiology of the tropical tree, *Tabebuia neochrysantha* (Bignoniaceae). *Ecology* 63:294–99.

———. 1984. Water stress and tree phenology in a tropical dry forest in the lowlands of Costa Rica. *J. Ecol.* 72:61–74.

Reichle, D. E., and D. A. Crossley, Jr. 1967. Investigation of heterotrophic productivity in forest insect communities. Pp. 563–87 in *Secondary productivity of terrestrial ecosystems,* ed. K. Petrusewicz. Warsaw: Panstwowe Wydawnietwo Naukowe.

Reid, N. 1986. Pollination and seed dispersal of mistletoes (Loranthaceae) by birds in southern Australia. Pp. 64–84 in *The dynamic partnership: Birds and plants in southern Australia,* ed. H. A. Ford and D. C. Paton. South Australia: Woolman, Government Printer.

Remsen, J. V. 1985. Community organization and ecology of

birds of high elevation humid forest in the Bolivian Andes. Pp. 733–56 in *Neotropical ornithology,* ed. P. A. Buckley, M. S. Foster, E. S. Morton, R. S. Ridgely, and F. G. Buckley. Ornithological Monographs vol. 36. Lawrence: AOU.

Renner, S. S. 1986. The Neotropical epiphytic Melastomataceae: Phytogeographic pattern, fruit types, and floral biology. *Selbyana* 9:104–11.

———. 1989. A survey of reproductive biology in Neotropical Melastomataceae and Memecylaceae. *Ann. MO Bot. Gard.* 76:496–518.

Rentz, D. C. 1975. Two new katydids of the genus *Melanonotus* from Costa Rica with comments on their life history strategies (Tettigoniidae: Pseudophyllinae). *Ent. News* 86:129–39.

Restrepo, C. 1987. Aspectos ecológicos de la diseminación de cinco especies de muérdagos por aves. *Humboldtia* 1:65–116.

Reznick, D. 1983. The structure of guppy life histories: The tradeoff between growth and reproduction. *Ecology* 64:862–73.

Reznick, D., and J. A. Endler. 1982. The impact of predation on life history evolution in Trinidadian guppies, genetic basis of observed life history patterns. *Evolution* 36:1236–50.

Rhoades, D. F. 1979. Evolution of plant chemical defense against herbivores. Pp. 1–54 in *Herbivores: Their interactions with secondary plant metabolites,* ed. G. A. Rosenthal and D. H. Janzen. New York: Academic Press.

Rhoades, D. F., and R. G. Cates. 1976. Toward a general theory of plant antiherbivore chemistry. *Recent Adv. Phytochem.* 10:168–213.

Rice, B., and M. Westoby. 1983. Plant species richness at the 0.1 hectare scale in Australian vegetation compared to other continents. *Vegetatio* 52:129–40.

Rich, P. M. 1986. Mechanical architecture of arborescent rain forest palms. *Principes* 30:117–31.

———. 1990. Characterizing plant canopies with hemispherical photographs. *Remote Sens. Rev.* 5:13–29.

Rich, P. M., K. Helenurm, D. Kearns, S. R. Morse, M. W. Palmer, and L. Short. 1986. Height and stem diameter relationships for dicotyledonous trees and arborescent palms of Costa Rican tropical wet forest. *Bull. Torrey Bot. Club* 113:241–46.

Rich, P. V., and T. H. Rich. 1983. The Central American dispersal route: Biotic history and paleogeography. Pp. 12–34 in *Costa Rican natural history,* ed. D. H. Janzen. Chicago: Univ. of Chicago Press.

Richards, P. W. 1952. *The tropical rain forest.* Cambridge: Cambridge Univ. Press.

Richards, P., and G. B. Williamson. 1975. Treefalls and patterns of understory species in wet lowland tropical forest. *Ecology* 56:1226–29.

Ricklefs, R. E. 1966. The temporal component of diversity among species of birds. *Evolution* 20:235–42.

———. 1969. An analysis of nesting morality in birds. *Smithson. Contrib. Zool.* 9:1–48.

———. 1987. Community diversity: Relative roles of local and regional processes. *Science* 235:167–71.

Rickson, F. R., and S. J. Risch. 1984. Anatomical and ultrastructural aspects of the ant-food cell of *Piper cenocladum* C. DC. (Piperaceae). *Am. J. Bot.* 71:1268–74.

Ridley, H. N. 1930. *The dispersal of plants throughout the world.* Ashford, Kent, England: L. Reeve.

Riede, K. 1987. A comparative study of mating behaviour in some Neotropical grasshoppers (Acridoidea). *Ethology* 76:265–96.

Riedl, H. 1983. Analysis of coddling moth phenology in relation to latitude, climate, and food availability. Pp. 233–52 in *Diapause and life cycle strategies in insects,* ed. V. K. Brown and I. Hodek. The Hague: W. Junk.

Riehl, H. 1979. *Climate and weather in the tropics.* New York: Academic Press.

Ripley, B. 1981. *Spatial statistics.* New York: J. Wiley.

Risch, S. J., M. McClure, J. Vandermeer, and S. Waltz. 1977. Mutualism between three species of tropical *Piper* (Piperaceae) and their ant inhabitants. *Am. Midl. Nat.* 98:433–44.

Risch, S. J., and F. R. Rickson. 1981. Mutualism in which ants must be present before plants product food bodies. *Nature* 291:149–50.

Ritland, D. B., and L. P. Brower. 1991. The viceroy butterfly is not a batesian mimic. *Nature* 350:497–98.

Ritter, D. F. 1978. *Process geomorphology.* Dubuque, Iowa: William Brown.

Robbins, C. T., T. A. Hanley, A. E. Hagerman, O. Hjeljord, D. L. Baker, C. C. Schwartz, and W. W. Mautz. 1987. Role of tannins in defending plants against ruminants: Reduction in protein availability. *Ecology* 68:98–107.

Robbins, R. K. 1981. The false head hypothesis: Predation and wing pattern variation in lycaenid butterflies. *Am. Nat.* 118:770–75.

———. 1982. How many butterfly species? *News of the Lepidopt. Soc.* 1982:40–41.

Roberts, T. R. 1972. Ecology of fishes in the Amazon and Congo basins. *Bull. Mus. Comp. Anat., Zool.* 143:117–47.

Robertson, G. P. 1982. Nitrification in forested ecosystems. *Philos. Trans. R. Soc. Lond.* B296:445–57.

———. 1984. Nitrification and nitrogen mineralization in a lowland rainforest succession in Costa Rica, Central America. *Oecologia* 61:99–104.

———. 1989. Nitrification and denitrification in humid tropical ecosystems: Potential controls on nitrogen retention. In *Mineral nutrients in tropical forest and savanna ecosystems,* ed. J. Proctor. Oxford: Blackwell Scientific.

Robertson, G. P., and J. M. Tiedje. 1988. Denitrification in a humid tropical rainforest. *Nature* 336:756–59.

Robichaux, R. H. 1984. Variation in the tissue water relations of two sympatric Hawaiian Dubautia species and their natural hybrid. *Oecologia* 65:75–81.

Robichaux, R. H., P. W. Rundel, L. Stemmermann, J. E. Canfield, S. R. Morse, and E. Friedman. 1984. Tissue water deficits and plant growth in wet tropical environments. Pp. 99–112 in *Physiological ecology of plants of the wet tropics,* ed. E. Medina, H. A. Mooney, and C. Vásquez-Yánes. The Hague: W. Junk.

Robinson, J. G., and K. H. Redford. 1986. Body size, diet, and population density of Neotropical forest mammals. *Am. Nat.* 128:665–80.

Robinson, S. K., and J. Terborgh. 1990. Bird communities of the Cocha Cashu Biological Station in Amazonian Peru. Pp. 199–216 in *Four Neotropical rainforests,* ed. A. H. Gentry. New Haven, Conn.: Yale Univ. Press.

Robyns, A. 1967. Vochysiaceae [Flora of Panama]. *Ann. MO Bot. Gard.* 54:1–7.

Rockwood, L. L. 1973. The effect of defoliation on seed production of six Costa Rican tree species. *Ecology* 54:1363–69.

Rodin, L. E., and N. I. Bazilevich. 1967. Production and mineral cycling in terrestrial vegetation. Edinburgh: Oliver and Boyd.

Rodríguez C., R., and E. Vargas M. 1988. *El recurso forestal en Costa Rica: Políticas públicas y sociedad.* Heredia, Costa Rica: Editorial de la Universidad Nacional.

Rodríguez-M., J. A. 1980. Distribución de la ictiofauna en un sector del Río Puerto Viejo. Report, OTS 80-2:348–53.

Rodríguez, J. V. 1982. *Aves del Parque Nacional "Los Katios."* Bogotá, Colombia: Instituto Nacional de Recursos Naturales (INDERENA).

Rojas, J. M. 1978. Estudio de la actividad cocotera y de la variedad Atlántico alto del cocotero (*Cocos nucifera* L.) en la Región Atlántica (Costa Rica). San José, Costa Rica. Report.

Rollet, B. 1983a. La régénération naturelle dans les trouées. *Rev. Bois For. Trop.* 201:3–34.

———. 1983b. La régénération naturelle dans les trouées. Part 2. *Rev. Bois For. Trop.* 202:18–34.

Romanoff, A. L., and A. J. Romonoff. 1949. *The avian egg.* New York: J. Wiley.

Romero, A. 1981. Studies on feeding behavior on a population of *Cichlasoma tuba* (Pisces: Cichlidae) in shallow waters of the River Puerto Viejo. Report, OTS 81-3:97–99.

Root, R. B. 1967. The niche exploitation pattern of the blue-gray gnatcatcher. *Ecol. Monogr.* 37:317–50.

———. 1973. Organization of a plant-arthropod association in simple and diverse habitats: The fauna of collards (*Brassica oleracea*). *Ecol. Monogr.* 43:95–124.

Rosselli, L. 1989. El ciclo anual de un ave frugívora migratoria altitudinal, *Corapipo leucorrhoa* (Pipridae) y los frutos que consume. Master's thesis, Universidad de Costa Rica, San José.

Rosselli, L., and F. G. Stiles. In press. Consumption of fruits of the Melastomataceae by birds: How diffuse is coevolution? *Vegetatio.*

Rothschild, M., R. J. Nash, and E. A. Bell. 1986. Cycasin in the endangered butterfly *Eumaeus atala florida*. *Phytochemistry* (Oxford) 25:1853–54.

Roughgarden, J. D., D. Haekel, and E. R. Fuentes. 1983. Coevolutionary theory and the biogeography and community structure of *Anolis*. Pp. 371–410 in *Lizard Ecology: Studies of a model organism,* ed. R. B. Huey, E. R. Pianka, and T. W. Schoener. Cambridge: Harvard Univ. Press.

Rowell, H. F. 1978. Food plant specificity in Neotropical rainforest acridids. *Entomol. Exp. Appl.* 24:451–62.

———. 1983a. Checklist of acridoid grasshoppers. Pp. 651–53 in *Costa Rican natural history,* ed. D. H. Janzen. Chicago: Univ. of Chicago Press.

———. 1983b. *Drymophilacris bimaculata.* Pp. 714–16 in *Costa Rican natural history,* ed. D. H. Janzen. Chicago: Univ. of Chicago Press.

———. 1983c. *Osmilia flavolineata.* Pp. 750–51 in *Costa Rican natural history,* ed. D. H. Janzen. Chicago: Univ. of Chicago Press.

———. 1983d. *Tropidacris cristata.* Pp. 772–73 in *Costa Rican natural history,* ed. D. H. Janzen. Chicago: Univ. of Chicago Press.

———. 1985a. The feeding biology of a species-rich genus of rainforest grasshoppers (*Rhachicreagra,* Orthoptera, Acrididae). Part 1. Foodplant use and foodplant acceptance. *Oecologia* 68:87–98.

———. 1985b. The feeding biology of a species-rich genus of rainforest grasshoppers (*Rhachicreagra,* Orthoptera, Acrididae). Part 2. Foodplant preference and its relation to speciation. *Oecologia* 68:99–104.

———. 1987. The biogeography of Costa Rican acridid grasshoppers in relation to their putative phylogenetic origins and ecology. Pp. 470–82 in *Evolutionary biology of orthopteroid insects,* ed. G. Baccetti. Chichester, England: E. Horwood.

Rudolph, S. G., and C. Loudon. 1986. Load size selection by foraging leaf-cutter ants *(Atta cephalotes). Ecol. Entomol.* 11:401–10.

Rundel, P. W., and P. F. Becker. 1987. Cambios estacionales en las relaciones hídricas y en la fenología vegetativa de plantas del estrato bajo del bosque tropical de la Isla de Barro Colorado, Panamá. *Rev. Biol. Trop.* 35 (suppl. 1): 71–84.

Runkle, J. R. 1982. Patterns of disturbance in some old-growth mesic forests of eastern North America. *Ecology* 63:1533–46.

———. 1985. Disturbance regimes in temperate forests. Pp. 17–34 in *The ecology of natural disturbance and patch dynamics,* ed. S. T. A. Pickett and P. S. White. Orlando, Fla.: Academic Press.

Runkle, J. R., and T. C. Yetter. 1987. Treefalls revisited: Gap dynamics in the southern Appalachians. *Ecology* 68:417–24.

Russel, A. E., and J. J. Ewel. 1985. Leaching from a tropical andept during big storms: A comparison of three methods. *Soil Sci.* 139:181–89.

Russell, J. K. 1982. Timing of reproduction by Coatis (*Nasua narica*) in relation to fluctuations in food resources. Pp. 413–31 in *The ecology of a tropical forest: Seasonal rhythms and long-term changes,* ed. E. G. Leigh, Jr., A. S. Rand, and D. M. Windsor. Washington, D.C.: Smithsonian Institution Press.

Rutilio Q., J., J. R. Alegría C., and J. D. Velasco G. 1973. Efecto de los insecticidas en el equilibrio natural de poblaciones de *Rothschildia aroma* Schaus (Lepidoptera: Saturniidae) en El Salvador. *Rev. Biol. Trop.* 21:111–25.

Rutter, A. J. 1963. Studies in the water relations of *Pinus sylvestris* in plantation conditions. Part 1. Measurements of rainfall and interception. *J. Ecol.* 51:191–203.

———. 1967. An analysis of evaporation from a stand of Scots pine. Pp. 403–17 in *International symposium on forest hydrology,* ed. W. E. Sopper and W. H. Lull. Oxford: Pergamon.

Ryan, M. J. 1985. *The Túngara frog. A study in sexual selection and communication.* Chicago: Univ. of Chicago Press.

Rylands, A. B. 1984. Marmosets and tamarins. Pp. 342–45 in *The encyclopedia of mammals,* ed. D. Macdonald. New York: Facts on File Publishers.

Sachs, R. M. 1960. Control of flowering in *Fuchsia. J. R. Hortic. Soc.* 85:491–93.

————. 1977. Nutrient diversion: An hypothesis to explain the chemical control of flowering. *Hortscience* 12:220–22.

Sachs, R. M., and W. P. Hackett. 1983. Source-sink relationships and flowering. Pp. 263–72 in *Strategies of plant reproduction,* ed. W. J. Meudt. *Beltsville Symp. Agric. Res.* no. 6. Totowa, N.J.: Allenheld, Osmum.

Sader, S. A., and A. T. Joyce. 1988. Deforestation rates and trends in Costa Rica, 1940 to 1983. *Biotropica* 20:11–19.

Saldarriaga, J. G., D. C. West, M. L. Tharp, and C. Uhl. 1988. Long-term chronosequence of forest succession in the upper Río Negro of Colombia and Venezuela. *J. Ecol.* 76:938–58.

Sallabanks, R., and S. P. Courtney. 1992. Avoiding the early bird: Frugivory, seed predation, and insect-vertebrate interactions. *Annu. Rev. Entomol.* 37:377–400.

Salthe, S. N., and W. E. Duellman. 1973. Quantitative constraints associated with reproductive mode in anurans. Pp. 229–49 in *Evolutionary biology of the Anurans: Contemporary research on major problems,* ed. J. L. Vial. Columbia: Univ. of Missouri Press.

Salthe, S. N., and J. S. Mecham. 1974. Reproductive and courtship patterns. Pp. 309–521 in *Physiology of the Amphibia,* ed. B. Lofts. Vol. 2. New York: Academic Press.

Sánchez, P. A. 1976. *Properties and management of soils in the tropics.* New York: J. Wiley.

Sancho, F., and R. Mata. 1987. Estudio detallado de suelos. Estación Biológica "La Selva." Report, Organization for Tropical Studies, San José, Costa Rica.

Sandner, G. 1959. La colonización interna, o expansión agrícola en Costa Rica. Progreso o retroceso? Biannual report, *Inst. Geogr. Costa Rica* 2:29–41.

————. 1961. *Aspectos geográficos de la colonización agrícola en el valle del General.* San José: Instituto Geográfico de Costa Rica.

Sanford, R. L., Jr. 1987. Apogeotropic roots in an Amazon rainforest. *Science* 235:1062–64.

————. 1989. Fine root biomass under a tropical forest light gap in Costa Rica. *J. Trop. Ecol.* 5:251–56.

Sanford, R. L., Jr., H. E. Braker, and G. S. Hartshorn. 1986. Canopy openings in a primary Neotropical lowland forest. *J. Trop. Ecol.* 2:277–282.

Sargent, S. 1990. Neighborhood effects on fruit removal by birds: A field experiment with *Viburnum dentatum* (Caprifoliaceae). *Ecology* 71:1289–98.

Sarmiento, G., and M. Monasterio. 1983. Life forms and phenology. In *Ecosystems of the world.* Pp. 79–108 in vol. 13, *Tropical savannas,* ed. F. Boulière. Amsterdam: Elsevier Scientific.

Sarukhán, J. 1978. Studies on the demography of tropical trees. Pp. 163–84 in *Tropical trees as living systems,* ed. P. B. Tomlinson and M. H. Zimmermann. Cambridge: Cambridge Univ. Press.

————. 1980. Demographic problems in tropical systems. Pp. 161–88 in *Demography and evolution in plant populations,* ed. O. T. Solbrig. Botanical Monographs no. 15. Berkeley: Univ. of California Press.

Sarukhán, J., M. Martínez-Ramos, and D. Piñero. 1984. The analysis of demographic variability at the individual level and its population consequences. Pp. 83–106 in *Perspectives on plant population ecology,* ed. R. Dirzo and J. Sarukhán. Sunderland, Mass.: Sinauer Associates.

Sauer, A., and K. Grove. 1977. Spatial distribution of bark beetle (Scolytidae) on *Protium glabrum* in the La Selva arboretum. Report, OTS 77-4:479–85.

Saul, W. G. 1975. An ecological study of fishes at a site in upper Amazonian Ecuador. *Proc. Acad. Nat. Sci. Phila.* 127:93–134.

Savage, J. M. 1966. The origins and history of the Central American herpetofauna. *Copeia* 1966:719–66.

————. 1974. The isthmian link and the evolution of Neotropical mammals. *Los Angel. Cty. Mus. Contrib. Sci.* 260:1–51.

————. 1975. Systematics and distribution of the Mexican and Central American stream frogs related to *Eleutherodactylus rugulosus. Copeia* 1975:254–306.

————. 1980. *A preliminary handlist of the herpetofauna of Costa Rica.* Los Angeles, Calif.: Allan Hancock Foundation.

————. 1982. The enigma of the Central American herpetofauna: Dispersals or vicariance? *Ann. MO Bot. Gard.* 69:464–547.

Savage, J. M., and S. B. Emerson. 1970. Central American frogs allied to *Eleutherodactylus bransfordii* (Cope): A problem of polymorphism. *Copeia* 1970:623–44.

Savage, J. M., and J. Villa. 1986. Introduction to the herpetofauna of Costa Rica. Contributions to Herpetology, no. 3. Oxford, Ohio: Society for the Study of Amphibians and Reptiles.

Sazima, M. 1981. Polinizaçao de duas especies de *Pavonia* (Malvaceae) for beija-flores, na Serra do Cipó, Minas Gerais. *Rev. Bras. Biol.* 41:733–37.

Schal, C. 1982. Intraspecific vertical stratification as a mate-finding mechanism in tropical cockroaches. *Science* 215:1405–7.

Schal, C., and W. J. Bell. 1986. Vertical community structure and resource utilization in Neotropical forest cockroaches. *Ecol. Entomol.* 11:411–23.

Schaller, G. B., and J. M. C. Vasconcelos. 1978. Jaguar predation on capybara. *Z. Saeugetierk.* 43:296–301.

Schatz, G. E. 1985. A new *Cymbopetalum* (Annonaceae) from Costa Rica and Panama with observations on natural hybridization. *Ann. MO Bot. Gard.* 72:535–38.

————. 1987. Systematics and ecological studies of Central American Annonaceae. Ph.D. diss., University of Wisconsin, Madison.

Schatz, G. E., G. B. Williamson, C. M. Cogswell, and A. C. Stam. 1985. Stilt roots and growth of aboreal palms. *Biotropica* 17:206–9.

Schelhas, J. 1991. Socio-economic and biological aspects of land use adjacent to Braulio Carrillo National Park, Costa Rica. Ph.D. diss., University of Arizona, Tucson.

Schemske, D. W. 1980. The evolutionary significance of extrafloral nectar production by *Costus woodsonii* (Zingiberaceae): An experimental analysis of ant protection. *J. Ecol.* 68:959–67.

————. 1981. Floral convergence and pollinator sharing in two bee pollinated tropical herbs. *Ecology* 62:946–54.

————. 1983. Breeding system and habitat effect of fitness

components in three Neotropical *Costus* (Zingiberaceae). *Evolution* 37:523–39.

Schemske, D. W., and N. V. L. Brokaw. 1981. Treefalls and the distribution of understory birds in a tropical forest. *Ecology* 62:938–45.

Schemske, D. W., and C. C. Horvitz. 1984. Variation among floral visitors in pollination ability: A precondition for mutualism specialization. *Science* 225:519–21.

Schlesinger, W. H. 1991. Biogeochemistry: An analysis of global change. San Diego, Calif.: Academic Press.

Schmid, R. 1970. Notes on the reproductive biology of *Asterogyne martiana* (Palmae). Part 2. Pollination by syrphid flies. *Principes* 14:39–49.

Schoener, T. W. 1977. Competition and the niche. Pp. 35–136 in *Biology of the Reptilia*, vol. 7, ed. C. Gans and D. W. Tinkle. New York: Academic Press.

———. 1983. Population and community ecology: Overview. Pp. 233–39 in *Lizard ecology: Studies of a model organism*, ed. R. B. Huey, E. R. Pianka, and T. W. Schoener. Cambridge: Harvard Univ. Press.

Schoener, T. W., and D. A. Spiller. 1987. Effect of lizards on spider populations: Manipulative reconstruction of a natural experiment. *Science* 236:949–52.

Schowalter, T. D., J. W. Webb, and D. A. Crossley, Jr. 1981. Community structure and nutrient content of canopy arthropods in clearcut and uncut forest ecosystems. *Ecology* 62:1010–19.

Schremmer, F. 1982. Flowering and flower visitors in *Carludovica palmata* (Cyclanthaceae)—an ecological paradox. *Plant Syst. Evol.* 140:95–107.

Schultz, J. C., and I. T. Baldwin. 1982. Oak leaf quality declines in response to defoliation by gypsy moth larvae. *Science* 217:149–51.

Schultz, J. C., D. Otte, and F. Enders. 1977. *Larrea* as a habitat component for desert arthropods. Pp. 176–208 in *Creosote bush: Biology and chemistry of Larrea in New World deserts*, ed. T. J. Mabry, J. H. Hunzicker, and D. R. Difeo. Stroudsburg, Pa.: Dowden, Hutchinson and Ross.

Schulz, J. P. 1960. *Ecological studies on rain forest in northern Suriname*. Amsterdam: North Holland.

Schupp, E. W. 1986. *Azteca* protection of *Cecropia*: Ant occupation benefits juvenile trees. *Oecologia* 70:379–85.

———. 1988a. Factors affecting post-dispersal seed survival in a tropical forest. *Oecologia* 76:525–30.

———. 1988b. Seed and early seedling predation in the forest understory and in treefall gaps. *Oikos* 51:71–78.

———. 1990. Annual variation in seedfall, postdispersal predation, and recruitment of a Neotropical tree. *Ecology* 71:504–15.

Schupp, E. W., and D. H. Feener, Jr. 1991. Phylogeny, lifeform, and habitat dependence of ant-defended plants in a Panamanian forest. Pp. 175–97 in *Ant-plant interactions*, ed. C. H. Huxley and D. F. Cutler. Oxford: Oxford Univ. Press.

Schupp, E. W., and E. J. Frost. 1989. Differential predation of *Welfia georgii* seeds in treefall gaps and the forest understory. *Biotropica* 21:200–203.

Schupp, E. W., H. F. Howe, C. K. Augspurger, and D. J. Levey. 1989. Arrival and survival in tropical treefall gaps. *Ecology* 70:562–65.

Schwartz, J. J., and K. D. Wells. 1983. An experimental study of acoustic interference between two species of Neotropical treefrogs. *Anim. Behav.* 31:181–90.

———. 1984. Vocal behavior of the Neotropical treefrog *Hyla phlebodes*. *Herpetologica* 40:452–63.

Scoble, M. J. 1987. The structure and affinities of the Hedyloidea: A new concept of the butterflies. *Bull. Br. Mus. (Nat. Hist.) Entomol.* 53:251–86.

Scott, N. J., Jr. 1969. A zoogeographic analysis of the snakes of Costa Rica. Ph.D. diss., University of Southern California, Los Angeles.

———. 1976. The abundance and diversity of the herpetofauna of tropical forest litter. *Biotropica* 8:41–58.

———. 1983a. *Bufo haematiticus* (Sapo, toad). P. 385 in *Costa Rican natural history*, ed. D. H. Janzen. Chicago: Univ. of Chicago Press.

———. 1983b. *Leptodactylus pentadactylus*, (*Rana ternero*, smoky frog). Pp. 405–6 in *Costa Rican natural history*, ed. D. H. Janzen. Chicago: Univ. of Chicago Press.

Scott, N. J., and S. Limerick. 1983. Reptiles and Amphibians—Introduction. Pp. 351–67 in *Costa Rican natural history*, ed. D. H. Janzen. Chicago: Univ. of Chicago Press.

Scott, N. J., J. M. Savage, and D. C. Robinson. 1983. Checklist of reptiles and amphibians. Pp. 367–74 in *Costa Rican natural history*, ed. D. H. Janzen. Chicago: Univ. of Chicago Press.

Scriber, J. M. 1973. Latitudinal gradients in larval feeding specialization of the world Papilionidae (Lepidoptera). *Psyche* (Cambridge) 80:355–73.

Scriber, J. M., and P. Feeny. 1979. Growth of herbivorous caterpillars in relation to feeding specialization and to the growth form of their food plants. *Ecology* 60:829–50.

Seavey, S. R., and K. S. Bawa. 1986. Late-acting self-incompatibility in angiosperms. *Bot. Rev.* 52:195–219.

Secretaria Ejecutiva de Planificación Sectorial Agropecuaria y de Recursos Naturales Renovables. 1982. Información básica del sector agropecuario de Costa Rica, no. 2. San José, Costa Rica: Ministerio de Agricultura y Ganadería.

Segleau, J. 1985. *La deforestación*. San José, Costa Rica: Dirección General Forestal, Ministerio de Agricultura y Ganadería.

Seifert, R. P., and F. H. Seifert. 1976. A community matrix analysis of *Heliconia* insect communities. *Am. Nat.* 110:461–83.

———. 1979. Utilization of *Heliconia* (Musaceae) by the beetle *Xenarescus monoceros* (Oliver) (Chrysomelidae: Hispinae) in a Venezuelan forest. *Biotropica* 11:51–59.

Seitz, A. E. 1916–20. *Macrolepidoptera of the world*. Vol. 5. Stuttgart: Alfred Kernan.

Semple, K. S. 1974. Pollination in Piperaceae. *Ann. MO Bot. Gard.* 61:868–71.

Sexton, O. J., J. Heatwole, and E. Meseth. 1963. Seasonal population changes in the lizard, *Anolis limifrons*, in Panama. *Am. Midl. Nat.* 69:482–91.

Sexton, O. J., E. P. Ortleb, L. M. Hathaway, R. E. Ballinger, and P. Licht. 1971. Reproductive cycles of three species of anoline lizards from the isthmus of Panama. *Ecology* 52:201–15.

Seyfried, M. S., and P. S. C. Rao. 1987. Solute transport in undisturbed columns of an aggregated tropical soil: Preferential flow effects. *Soil Sci. Soc. Am. J.* 51:1434–44.

Shapiro, A. M. 1975. The temporal component of butterfly species diversity. Pp. 181–95 in *Ecology and evolution of communities,* ed. M. Cody and J. M. Diamond. Cambridge: Cambridge Univ. Press.

Sharpe, J. M. 1988. Growth, demography, tropic responses and apical dominance in the Neotropical fern *Danaea wendlandii* Reichenb. (Marattiaceae). Ph.D. diss., University of Georgia, Athens.

Sharpe, J. M., and J. A. Jernstedt. 1990. Leaf growth and phenology of the dimorphic herbaceous layer fern *Danaea wendlandii* (Marattiaceae). *Am. J. Bot.* 77:1040–49.

Sherry, T. W. 1982. Ecological and evolutionary inferences from morphology, foraging behavior, and diet of sympatric insectivorous Neotropical flycatchers (Tyrannidae). Ph.D. diss., University of California, Los Angeles.

———. 1983a. *Mionectes oleaginea.* Pp. 586–87 in *Costa Rican natural history,* ed. D. H. Janzen. Chicago: Univ. of Chicago Press.

———. 1983b. *Monasa morpheoeus.* Pp. 587–90 in *Costa Rican natural history,* ed. D. H. Janzen. Chicago: Univ. of Chicago Press.

———. 1984. Comparative dietary ecology of sympatric insectivorous Neotropical flycatchers (Tyrannidae). *Ecol. Monogr.* 54:313–38.

Sherry, T. W., and L. A. McDade. 1982. Prey selection and handling in two Neotropical hover-gleaning birds. *Ecology* 63:1016–28.

Short, L. 1984. Yet another ant-plant mutualism explored. Report, OTS 84-1:70–72.

Shull, E. M. 1987. *The butterflies of Indiana.* Bloomington: Indiana Academy of Science.

Shuttleworth, J. et al. 1984. Eddy correlation measurements of energy partition for Amazonian forest. *Q. J. R. Meteorol. Soc.* 110:1143–62.

Silberglied, R. E. 1984. Visual communication and sexual selection among butterflies. *Symp. R. Entomol. Soc. Lond.* 11:208–23.

Silva, J. L., J. Valdez, and J. Ojasti. 1985. Algunos aspectos de una comunidad de ofidios del norte de Venezuela. *Biotropica* 17:112–25.

Silvertown, J. W. 1980. The evolutionary ecology of mast seeding in trees. *Biol. J. Linn. Soc.* 14:235–50.

Simberloff, D. S., and E. F. Connor. 1979. Q-mode and R-mode analyses of biogeographic distributions: Null hypotheses based on random colonization. Pp. 123–38 in *Contemporary ecology and related econometrics,* ed. G. P. Patil and M. Rosenzweig. International Co-operative Publishing House.

Simms, E. L., and R. S. Fritz. 1990. The ecology and evolution of host-plant resistance to insects. *Trends Ecol. Evol.* 5:356–60.

Simpson, B. B., and J. Haffer. 1978. Speciation patterns in the Amazonian forest birds. *Annu. Rev. Ecol. Syst.* 9:497–518.

Singer, M. C., and J. Mandrachia. 1982. On the failure of two butterfly species to respond to the presence of conspecific eggs prior to oviposition. *Ecol. Entomol.* 7:327–30.

Singh, B., and G. Szeicz. 1979. The effect of intercepted rainfall on the water balance of a hardwood forest. *Water Resour. Res.* 15:131–38.

Sioli, H. 1975. Tropical rivers as expressions of their terrestrial environments. Pp. 275–88 in *Tropical ecological systems: Trends in terrestrial and agronomic research,* ed. F. D. Golley and E. Medina. New York: Springer-Verlag.

Sisk, T., and C. Vaughan. 1984. Notes on some aspects of the natural history of the giant pocket gopher. *Orthogeomys Merriam,* in Costa Rica. *Brenesia* 22:233–47.

Skutch, A. F. 1933. The aquatic flowers of a terrestrial plant, *Heliconia bihai* L. *Am. J. Bot.* 20:535–44.

———. 1935a. Helpers at the nest. *Auk* 52:257–73.

———. 1935b. The white-throated magpie-jay. *Wilson Bull.* 65:68–74.

———. 1949. Do tropical birds rear as many young as they can nourish? *Ibis* 91:430–55.

———. 1950. The nesting seasons of Central American birds in relation to climate and food supply. *Ibis* 92:185–222.

———. 1954. *Pacific Coast Avifauna.* Vol 31, *Life histories of Central American birds.* Berkeley, Calif.: Cooper Ornithological Society.

———. 1959. Life history of the groove-billed ani. *Auk* 76:281–317.

———. 1960a. *Pacific Coast Avifauna.* Vol. 34, *Life histories of Central American birds II.* Berkeley, Calif.: Cooper Ornithological Society.

———. 1960b. Roosting and nesting of aracari toucans. *Condor* 60:201–19.

———. 1961. Helpers among birds. *Condor* 63:198–226.

———. 1967a. Adaptive limitation of the reproductive rate of birds. *Ibis* 109:579–99.

———. 1967b. Life histories of Central American highland birds. Publications of the Nuttal Ornithological Club no. 7. Cambridge, Mass.: Nuttal Ornithological Club.

———. 1969. *Pacific Coast Avifauna.* Vol. 35, *Life histories of Central American birds III.* Berkeley, Calif.: Cooper Ornithological Society.

———. 1971. Life history of the keel-billed toucan. *Auk* 88:381–424.

———. 1972. Studies of tropical American birds. Publications of the Nuttal Ornithological Club no. 10. Cambridge, Mass.: Nuttal Ornithological Club.

———. 1976. Parent birds and their young. Austin: Univ. of Texas Press.

———. 1985. Clutch size, nesting success, and predation on nests of Neotropical birds, reviewed. Pp. 575–94 in *Neotropical ornithology,* ed. P. A. Buckley, M. S. Foster, E. S. Morton, R. S. Ridgely, and F. G. Buckley. Ornithological Monographs vol. 36. Lawrence: AOU.

———. 1987. *Helpers at birds' nests: A worldwide survey of cooperative breeding and related behavior.* Iowa City: Univ. Iowa Press.

Sleumer, H. O. 1980. Flacourtiaceae. *Flora Neotrop. Monogr.* 22:1–499.

Slobodkin, L. B., and H. L. Sanders. 1969. On the contribution of environmental predictability to species diversity. Pp. 82–93 in *Diversity and stability in ecological systems,* Brookhaven Symposium in Biology no. 22.

Slowinski, J. B., B. I. Crother, and J. E. Fauth. 1987. Diel differences in leaf-litter abundances of several species of reptiles and amphibians in an abandoned cacao grove in Costa Rica. *Rev. Biol. Trop.* 35:349–50.

Slud, P. 1960. The birds of Finca "La Selva," a tropical wet forest locality. *Bull. Am. Mus. Nat. Hist.* 121:49–148.

———. 1964. The birds of Costa Rica: Distribution and ecology. *Bull. Am. Mus. Nat. Hist.* 128:1–430.

———. 1976. Geographical and climatic relationships of avifaunas, with special reference to avian distribution in the Neotropics. *Smithson. Contrib. Zool.* 212:1–149.

Smiley, J. T. 1978a. The host plant ecology of *Heliconius* butterflies in northeastern Costa Rica. Ph.D. diss., Univ. of Texas, Austin.

———. 1978b. Plant chemistry and the evolution of host specificity: New evidence from *Heliconius* and *Passiflora*. *Science* 201: 745–47.

———. 1982. The herbivores of *Passiflora*: Comparison of monophyletic and polyphyletic feeding guilds. Pp. 325–30 in Proceedings of the Fifth International Symposium on Insect-Plant Relationships, ed. J. H. Visser and A. K. Minks. Wageningen: Centre for Agricultural Publishing and Documentation.

———. 1985a. Are chemical barriers necessary for evolution of butterfly-plant associations? *Oecologia* 65:580–83.

———. 1985b. *Heliconius* caterpillar mortality during establishment on plants with and without attending ants. *Ecology* 66:845–49.

Smith, A., and I. Hume, eds. 1984. *Possums and gliders*. Chipping Norton, NSW, Australia: Surrey Beatty and Sons.

Smith, A. P. 1973. Stratification of temperate and tropical forests. *Am. Nat.* 107:671–83.

———. 1987. Respuestas de hierbas del sotobosque tropical a claros ocasionados por la caída de árboles. *Rev. Biol. Trop.* 35 (suppl.): 111–18.

Smith, H. 1982. Light quality, photoreception, and plant strategy. *Annu. Rev. Plant Physiol.* 33:481–518.

Smith, N. G. 1972. Migration of the day-flying moth *Urania fulgens* in Central and South America. *Caribb. J. Sci.* 12:45–51.

———. 1983. *Urania fulgens*. Pp. 775–77 in *Costa Rican natural history*, ed. D. H. Janzen. Chicago: Univ. of Chicago Press.

Smithe, F. B. 1966. *The birds of Tikal*. Garden City, N.Y.: Natural History Press.

Smythe, N. 1970. Relationships between fruiting seasons and seed dispersal methods in a Neotropical forest. *Am. Nat.* 104:25–35.

———. 1978. The natural history of the Central American agouti *(Dasyprocta punctuatus)*. *Smithson. Contrib. Zool.* 257:1–57.

———. 1983. *Dasyprocta punctata* and *Agouti paca*. Pp. 463–65 in *Costa Rican natural history*, ed. D. H. Janzen. Chicago: Univ. of Chicago Press.

———. 1986. Competition and resource partitioning in the guild of Neotropical terrestrial frugivorous mammals. *Annu. Rev. Ecol. Syst.* 17:169–188.

Smythe, N., W. E. Glanz, and E. G. Leigh, Jr. 1982. Population regulation in some terrestrial frugivores. Pp. 227–338 in *The ecology of a tropical forest: Seasonal rhythms and long-term changes*, ed. E. G. Leigh, Jr., A. S. Rand, and D. M. Windsor. Washington, D.C.: Smithsonian Institution Press.

Snarskis, M. J. 1975. Excavaciones estratigráficas en la vertiente Atlántica de Costa Rica. *Vínculos* 1:2–17.

———. 1976. La vertiente Atlántica de Costa Rica. *Vínculos* 2:101–4.

Snow, B. K., and D. W. Snow. 1971. The feeding ecology of tanagers and honeycreepers in Trinidad. *Auk* 88:291–322.

Snow, D. W. 1962a. A field study of the black and white manakin, *Manacus manacus*, in Trinidad. *Zoologica* 47:65–104.

———. 1962b. A field study of the golden-headed manakin, *Pipra erythrocephala*, in Trinidad. *Zoologica* 47:183–98.

———. 1962c. The natural history of the oilbird, *Steatornis caripensis*, in Trinidad, W.I. Part 2. Population, breeding ecology, food. *Zoologica* 47:199–221.

———. 1965. A possible selective factor in the evolution of fruiting seasons in tropical forest. *Oikos* 15:274–81.

———. 1971. Evolutionary aspects of fruit-eating by birds. *Ibis* 113:194–202.

———. 1976. *The web of adaptation*. New York: Quadrangle, New York Times Books.

———. 1980. Regional differences between tropical floras and the evolution of frugivory. *Acta 17th Congr. Int. Ornithol.* 2:1192–98.

———. 1981. Tropical frugivorous birds and their food plants: A world survey. *Biotropica* 13:1–14.

———. 1982. *The cotingas*. Ithaca, N.Y.: Cornell Univ. Press.

Sobrevila, C., and M. T. K. Arroyo. 1982. Breeding systems in a montane tropical cloud forest in Venezuela. *Plant Syst. Evol.* 140:19–38.

Soderstrom, T. R., and C. E. Calderon. 1971. Insect pollination in tropical rain forest grasses. *Biotropica* 3:1–16.

Sokal, R. R., and F. J. Rohlf. 1981. Biometry. 2d ed. New York: W. H. Freeman.

Sollins, P. 1989. Factors affecting nutrient cycling in tropical soils. Pp. 85–95 in *Mineral nutrients in tropical forests and savanna ecosystems*, ed. J. Proctor. British Ecological Society Special Publication no. 9. Oxford: Blackwell Scientific Publications.

Sollins, P., and R. Radulovich. 1988. Effects of soil physical structure on solute transport in a weathered tropical soil. *Soil Sci. Soc. Am. J.* 52:1168–73.

Sollins, P., G. P. Robertson, and J. Uehara. 1988. Nutrient mobility in variable- and permanent-charge soils. *Biogeochemistry* (Dordrecht) 6:181–89.

Sollins, P., G. Spycher, and C. A. Glassman. 1984. Net nitrogen mineralization from light- and heavy-fraction forest soil organic matter. *Soil Biol. Biochem.* 16:31–37.

Sollins, P., G. Spycher, and C. Topik. 1983. Processes of soil organic-matter accretion at a mudflow chronosequence, Mt. Shasta, California. *Ecology* 64:1273–82.

Sorensen, A. E. 1983. Taste aversion and frugivore preference. *Oecologia* 56:117–20.

———. 1984. Nutrition, energy and passage time: Experiments with fruit preference in European blackbirds *(Turdus merula)*. *J. Anim. Ecol.* 53:545–57.

Sork, V. L. 1987. Effects of predation and light on seedling establishment in *Gustavia superba*. *Ecology* 68:1341–50.

Soulé, M. E. 1986. Section 2. Patterns of diversity and rarity: Their implications for conservation. Pp. 117–21 in *Conservation biology, the science of scarcity and diversity*, ed. M. E. Soulé. Sunderland, Mass.: Sinauer Associates.

Soulé, M. E., and K. A. Kohm. 1989. *Research priorities for conservation biology.* Covelo, Calif.: Island Press.

Southwood, T. R. E. 1977. The stability of the trophic milieu, its influence on the evolution of behaviour and of responsiveness to trophic signals. Pp. 471–93 in *Comportement des insectes et milieu trophique.* Colloques Internationaux du CRNS no. 265. Paris: Edition de CRNS.

Southwood, T. R. E., V. K. Brown, and P. M. Reader. 1986. Leaf palatability, life expectancy and herbivore damage. *Oecologia* 70:544–48.

Southwood, T. R. E., V. C. Moran, and C. E. J. Kennedy. 1982. The assessment of arboreal insect fauna: Comparisons of knockdown sampling and faunal lists. *Ecol. Entomol.* 7:331–40.

Sowls, L. K. 1983. *Tayassu tajacu.* Pp. 497–98 in *Costa Rican natural history,* ed. D. H. Janzen. Chicago: Univ. of Chicago Press.

———. 1984. *The peccaries.* Tucson: Univ. of Arizona Press.

Sprugel, D. G. 1984. Density, biomass, productivity, and nutrient cycling changes during stand development in wave-generated balsam fir forests. *Ecol. Monogr.* 54:165–86.

Spycher, G., P. Sollins, and S. Rose. 1983. Carbon and nitrogen in the light fraction of a forest soil: Vertical distribution and seasonal patterns. *Soil Sci.* 135:79–87.

Srygley, R. B., and P. Chai. 1990a. Flight morphology of Neotropical butterflies: Palatability and distribution of mass to the thorax and abdomen. *Oecologia* 84:491–99.

———. 1990b. Predation and the elevation of thoracic temperature in Neotropical butterflies. *Am. Nat.* 135:766–87.

Stallings, J. R. 1984. Notes on feeding habits of *Mazama gouazoubira* in the Chaco boreal of Paraguay. *Biotropica* 16:155–57.

Stamp, N. E. 1980. Egg deposition patterns in butterflies: Why do some species cluster their eggs rather than deposit them singly? *Am. Nat.* 115:367–80.

Stamps, J., and S. Tanaka. 1981. The influence of food and water on growth rates in a tropical lizard *(Anolis aeneus).* *Ecology* 62:33–40.

Standley, P. C. 1933. The flora of Barro Colorado Island. *Contrib. Arnold Arbor.* 5:1–178.

———. 1937. Flora of Costa Rica. *Field Mus. Nat. Hist. Publ. Bot. Ser.* 18:5–63.

Stanton, M. L. 1983. Spatial patterns in the plant community and their effects upon insect search. Pp. 125–57 in *Herbivorous insects, host-seeking behavior, and mechanisms,* ed. S. Ahmad. New York: Academic Press.

Stapanian, M. A. 1982. A model for fruiting display: Seed dispersal by birds of mulberry seeds. *Ecology* 63:1432–43.

Stark, N. M., and C. F. Jordan. 1978. Nutrient retention by the root mat of an Amazonian rain forest. *Ecology* 59:434–37.

Starmuhlner, F. 1984. Checklist and longitudinal distribution of the meso- and macrofauna of mountain streams of Sri Lanka (Ceylon). *Arch. Hydrobiol.* 101 (1/2):303–25.

Starmuhlner, F., and I. Therezien. 1982. Résultats de la mission hydrobiologique austro-française de 1979 aux îles de la Guadalupe, de la Dominique et de la Martinique (Petite Antilles). Part 1. Etude générale de la Guadalupe. *Rev. Hydrobiol. Trop.* 15:131–50.

Starrett, A., and R. S. Casebeer. 1968. Records of bats from Costa Rica. *Los Angel. Cty. Mus. Contrib. Sci.* 148:1–21.

Stearns, F. W. 1974. Phenology and environmental education. Pp. 425–29 in *Phenology and seasonality modeling,* ed. H. Lieth. New York: Springer-Verlag.

Stearns, S. C. 1976. Life history tactics: A review of the ideas. *Q. Rev. Biol.* 51:3–49.

———. 1981. On measuring fluctuating environments: Predictability, constancy and contingency. *Ecology* 62:185–99.

———. 1989. Trade-offs in life-history evolution. *Funct. Ecol.* 3:259–68.

Steiner, K. E. 1981. Nectarivory and potential pollination by a Neotropical marsupial. *Ann. MO Bot. Gard.* 68:505–13.

———. 1983. Pollination of *Mabea occidentalis* (Euphorbiaceae) in Panama. *Syst. Bot.* 8:105–17.

Stephenson, A. G. 1981. Flower and fruit abortion: Proximate causes and ultimate functions. *Annu. Rev. Ecol. Syst.* 12:253–79.

Sterner, R. W., C. A. Ribic, and G. E. Schatz. 1986. Testing for life historical changes in spatial patterns of four tropical tree species. *J. Ecol.* 74:621–33.

Stewart, J. B. 1977. Evaporation from the wet canopy of a pine forest. *Water Resour. Res.* 13:915–21.

Stiles, F. G. 1975. Ecology, flowering phenology, and hummingbird pollination of some Costa Rican *Heliconia* species. *Ecology* 56:285–301.

———. 1977. Coadapted competitors: The flowering seasons of hummingbird-pollinated plants in a tropical forest. *Science* 196:1177–78.

———. 1978a. Ecological and evolutionary implications of bird pollination. *Am. Zool.* 18:715–27.

———. 1978b. Possible specialization for hummingbird-hunting in the tiny hawk. *Auk* 95:550–53.

———. 1978c. Temporal organization of flowering among the hummingbird foodplants of a tropical wet forest. *Biotropica* 10:194–210.

———. 1979a. El ciclo anual en una comunidad coadaptada de colibríes y flores en el bosque tropical húmedo de Costa Rica. *Rev. Biol. Trop.* 27:75–101.

———. 1979b. Notes on the natural history of *Heliconia* (Musaceae) in Costa Rica. *Brenesia* 15 (suppl.): 151–80.

———. 1979c. Reply to Poole and Rathcke. *Science* 203:471.

———. 1980a. The annual cycle in a tropical wet forest hummingbird community. *Ibis* 122:322–43.

———. 1980b. Evolutionary implications of habitat relations between permanent and winter resident land-birds in Costa Rica. Pp. 421–36 in *Migrant birds in the Neotropics: Ecology, behavior, distribution and conservation,* ed. A. Keast and E. S. Morton. Washington, D.C.: Smithsonian Institution Press.

———. 1983a. Birds: Introduction. Pp. 502–29 in *Costa Rican natural history,* ed. D. H. Janzen. Chicago: Univ. of Chicago Press.

———. 1983b. Checklist of birds. Pp. 530–44 in *Costa Rican natural history,* ed. D. H. Janzen. Chicago: Univ. of Chicago Press.

———. 1985a. Conservation of forest birds in Costa Rica: Problems and perspectives. Pp. 141–68 in *Conservation of tropical forest birds,* ed. A. W. Diamond and T. Lovejoy. ICBP Technical Paper no. 4. Cambridge: ICBP.

———. 1985b. On the role of birds in the dynamics of Neotropical forests. Pp. 49–59 in *Conservation of tropical for-*

est birds, ed. A. W. Diamond and T. E. Lovejoy. ICBP Technical paper no. 4. Cambridge: ICBP.

———. 1985c. Seasonal patterns in the hummingbird-flower community of a Costa Rican subtropical forest. Pp. 757–87 in *Neotropical ornithology,* ed. P. A. Buckley, M. S. Foster, E. S. Morton, R. S. Ridgely, and F. G. Buckley. Ornithological Monographs vol. 36. Washington, D.C.: AOU.

———. 1988. Altitudinal movements of birds on the Caribbean slope of Costa Rica: Implications for conservation. Pp. 243–58 in *Tropical rain forest: Diversity and conservation,* ed. F. Almeda and C. M. Pringle. San Francisco: California Academy of Sciences.

Stiles, F. G., and D. A. Clark. 1989. Conservation of tropical rain forest: A case study from Costa Rica. *Am. Birds* 43:420–28.

Stiles, F. G., and D. J. Levey. 1988. The gray-breasted crake *(Laterallus exilis)* in Costa Rica: Vocalizations, distribution, and interactions with white-throated crakes *(L. albigularis). Condor* 90:607–12.

Stiles, F. G., A. F. Skutch, and D. Gardner. 1989. *A guide to the birds of Costa Rica.* Ithaca, N.Y.: Cornell Univ. Press.

Stiles, F. G., and L. L. Wolf. 1970. Hummingbird territoriality at a tropical flowering tree. *Auk* 87:467–91.

———. 1974. A possible circannual molt rhythm in a tropical hummingbird. *Am. Nat.* 108:341–54.

———. 1979. *The ecology and evolution of lek mating behavior in the long-tailed hermit hummingbird.* Ornithological Monographs vol. 27. Washington, D.C.: AOU.

Stone, D. 1977. Pre-Columbian man in Costa Rica. Cambridge, Mass.: Peabody Museum Press.

Stone, D. E. 1988. The Organization for Tropical Studies (OTS): A success story in graduate training and research. Pp. 143–87 in *Tropical rainforest diversity and conservation,* ed. F. Almeda and C. M. Pringle. San Francisco: California Academy of Sciences and Pacific Division, AAAS.

Stork, N. E. 1988. Insect diversity: Facts, fiction and speculation. *Biol. J. Linn. Soc.* 35:321–37.

Stout, J. 1978. Migration of the aquatic hemipteran *Limnocoris insularis* (Naucoridae) in a tropical lowland stream (Costa Rica, Central America). *Brenesia* 14/15:1–11.

———. 1979a. An association of an ant, a mealy bug, and an understory tree from a Costa Rican rain forest. *Biotropica* 11:309–11.

———. 1979b. The influence of biotic and abiotic factors on two species of stream inhabiting hemiptera (Family: Naucoridae). Ph.D. diss., Univ. of Michigan, Ann Arbor.

———. 1980. Leaf decomposition rates in some lowland tropical rainforest streams. *Biotropica* 12:264–72.

———. 1981a. How abiotic factors affect the distribution of two species of tropical predaceous aquatic bugs (Family: Naucoridae). *Ecology* 63:1170–78.

———. 1981b. Photometric determination of leaf input into tropical streams. *J. Freshwater Ecol.* 1:287–93.

Stout, R. J. 1982. Effects of a harsh environment on the life history patterns of two species of tropical aquatic Hemiptera (Family: Naucoridae). *Ecology* 63:75–83.

———. 1983. *Limnocoris insularis.* Pp. 733–34 in *Costa Rican natural history,* ed. D. H. Janzen. Chicago: Univ. of Chicago Press.

Stout, J., and J. Vandermeer. 1975. Comparison of species richness for stream inhabiting insects in tropical and mid-latitude streams. *Am. Nat.* 109:263–80.

Strahan, R. 1983. *The Australian Museum complete book of Australian mammals.* Sydney, Australia: Angus and Robertson.

Strickland, T. C., and P. Sollins. 1987. Improved method for separating light- and heavy-fraction organic material from soil. *Soil Sci. Soc. Am. J.* 51:1390–93.

Strickland, T. C., P. Sollins, D. S. Schimel, and E. A. Kerle. 1988. Aggregation and aggregate stability in forest and range soils. *Soil Sci. Soc. Am. J.* 52:829–33.

Strong, D. R., Jr. 1974. Rapid asymptotic species accumulation in phytophagous insect communities: The pests of cacao. *Science* 185:1064–66.

———. 1977a. Insect species richness: Hispine beetles of *Heliconia latispatha. Ecology* 58:573–82.

———. 1977b. Rolled-leaf hispine beetles (Chrysomelidae) and their Zingiberales host plants in Middle America. *Biotropica* 9:156–69.

———. 1981. The possibility of insect communities without competition: Hispine beetles on *Heliconia.* Pp. 183–94 in *Insect life history patterns: Habitat and geographic variation,* ed. R. F. Denno and H. Dingle. Berlin: Springer-Verlag.

———. 1982a. Harmonious coexistence of hispine beetles on *Heliconia* in experimental and natural communities. *Ecology* 63:1039–49.

———. 1982b. Potential interspecific competition and host specificity: Hispine beetles on *Heliconia. Ecol. Entomol.* 7:217–20.

———. 1983. Exorcising the ghost of competition past: Phytophagous insects. Pp. 28–41 in *Ecological communities: Conceptual issues and the evidence,* ed. D. R. Strong, D. S. Simberloff, L. G. Abele, and A. B. Thistle. Princeton, N.J.: Princeton Univ. Press.

Strong, D. R., Jr., J. H. Lawton, and T. R. E. Southwood. 1984. *Insects on plants: Community patterns and mechanisms.* Cambridge: Harvard Univ. Press.

Strong, D. R., Jr., and D. A. Levin. 1979. Species richness of plant parasites and growth form of their hosts. *Am. Nat.* 114:1–22.

Strong, D. R., Jr., E. D. McCoy, and J. R. Rey. 1977. Time and the number of herbivore species: The pests of sugarcane. *Ecology* 58:167–75.

Strong, D. R., Jr., D. Simberloff, L. G. Abele, and A. B. Thistle, eds. 1983. *Ecological communities: Conceptual issues and the evidence.* Princeton, N.J.: Princeton Univ. Press.

Strong, D. R., Jr., and M. D. Wang. 1977. Evolution of insect life histories and host plant chemistry: Hispine beetles on *Heliconia. Evolution* 31:854–62.

Sutton, G. M. 1951. Mistletoe dispersal by birds. *Wilson Bull.* 63:235–37.

Sutton, S. L. 1983. The spatial distribution of flying insects in tropical rain forest. Pp. 77–91 in *Tropical rain forest: Ecology and management,* ed. S. L. Sutton, T. C. Whitmore, and A. C. Chadwick. British Ecological Society Special Publication no. 2. Oxford: Blackwell Scientific Publications.

Swaine, M. D., and J. B. Hall. 1986. Forest structure and

dynamics. Pp. 47–93 in *Plant ecology in West Africa,* ed. G. W. Lawson. New York: J. Wiley.

Swaine, M. D., J. B. Hall, and I. J. Alexander. 1987. Tree population dynamics at Kade, Ghana (1968–1982). *J. Trop. Ecol.* 3:331–45.

Swaine, M. D., D. Lieberman, and F. E. Putz. 1987. The dynamics of tree populations in tropical forest: A review. *J. Trop. Ecol.* 3:359–66.

Swaine, M. D., and T. C. Whitmore. 1988. On the definition of ecological species groups in tropical rain forests. *Vegetatio* 75:81–86.

Sweeney, B. M. 1987. Rhythmic phenomena in plants. 2d ed. San Diego, Calif.: Academic Press.

Swynnerton, C. F. M. 1919. Experiments and observations bearing on the explanation of form and colouring, 1908–1913. *Zool. J. Linn. Soc.* 33:203–385.

Sytsma, K. J., and B. A. Schaal. 1985. Genetic variation, differentiation and evolution in a species complex of tropical shrubs based on isozymic data. *Evolution* 39:582–93.

Szelistowski, W. A. 1985. Unpalatability of the poison arrow frog *Dendrobates pumilio* to the ctenid spider *Cupiennius coccineus. Biotropica* 17:345–46.

Tahvanainen, J. O., and R. B. Root. 1972. The influence of vegetational diversity on the population ecology of a specialized herbivore, *Phyllotreta cruciferae* (Coleoptera: Chrysomelidae). *Oecologia* 10:321–46.

Taigen, T. L., and F. H. Pough. 1985. Metabolic correlates of anuran behavior. *Am. Zool.* 25:987–97.

Talbot, J. J. 1979. Time budget, niche overlap, inter- and intra-specific aggression in *Anolis humilis* and *Anolis limifrons* from Costa Rica. *Copeia* 1979:472–81.

Talbot, R. W., M. O. Andeae, T. W. Andreae, and R. C. Harriss. 1988. Regional aerosol chemistry of the Amazon Basin during the dry season. *J. Geophys. Res.* 93:1499–1508.

Tanner, E. V. J. 1982. Species diversity and reproductive mechanisms in Jamaican trees. *Biol. J. Linn. Soc.* 18:263–78.

Tanner, E. V. J., and V. Kapos. 1982. Leaf structure of Jamaican upper Montane rain-forest trees. *Biotropica* 14:16–24.

Tasaico, H. 1959. *La fisionomía de las hojas de árboles en algunas formaciones tropicales.* Master's thesis, Instituto Interamericano de Ciencias Agrícolas, Turrialba, Costa Rica.

Tattersall, I. 1982. *The primates of Madagascar.* New York: Columbia Univ. Press.

Tauber, M. J., C. A. Tauber, and S. Masaki. 1986. *Seasonal adaptations of insects.* Oxford: Oxford Univ. Press.

Te Boekhorst, I. J. A., C. L. Schurmann, and J. Sugardjito. 1990. Residential status and seasonal movements of wild orang-utans in the Gunung Leuser Reserve (Sumatra, Indonesia). *Anim. Behav.* 39:1098–1109.

Temple, S. A. 1977. Plant-animal mutualism: Coevolution with dodo leads to near extinction of plant. *Science* 197:885–86.

Terborgh, J. 1977. Bird species diversity along an Andean elevational gradient. *Ecology* 56:562–76.

———. 1980. Causes of tropical species diversity. Acta 17th Congress International Ornithologica Berlin 955–61.

———. 1983. *Five New World primates: A study in comparative ecology.* Princeton, N.J.: Princeton Univ. Press.

———. 1985a. Habitat selection in Amazonian birds. Pp. 311–38 in *Habitat selection in birds,* ed. M. L. Cody. Orlando, Fla.: Academic Press.

———. 1985b. The vertical component of plant species diversity in temperate and tropical forests. *Am. Nat.* 126:760–76.

———. 1986a. Community aspects of frugivory in tropical forests. Pp. 371–84 in *Frugivores and seed dispersal,* ed. A. Estrada and T. H. Fleming. Amsterdam: W. Junk.

———. 1986b. Keystone plant resources in the tropical forest. Pp. 330–44 in *Conservation biology,* ed. M. E. Soulé. Sunderland, Mass.: Sinauer Associates.

———. 1988. The big things that run the world—A sequel to E. O. Wilson. *Conserv. Biol.* 2:402–3.

———. 1990a. An overview of research at Cocha Cashu Biological Station. Pp. 48–59 in *Four Neotropical rainforests,* ed. A. H. Gentry. New Haven, Conn.: Yale Univ. Press.

———. 1990b. The role of felid predators in Neotropical forests. *Vida Silvestre Neotrop.* 2:3–5.

Terborgh, J., S. K. Robinson, T. A. Parker III, C. A. Munn, and N. Pierpont. 1990. Structure and organization of an Amazonian forest bird community. *Ecol. Monogr.* 60:213–38.

Terwilliger, V. 1978. Natural history of Baird's tapir on Barro Colorado Island, Panama Canal Zone. *Biotropica* 10:211–20.

———. 1981. A comparison of *Atta* ant leaf damage on understory vegetation in mature forest and a recently slashed successional plot. Report, OTS 81-3:121–27.

Thiollay, J. M. 1988. Comparative foraging success of insectivorous birds in tropical and temperate forests: Ecological implications. *Oikos* 53:17–30.

Thomas, C. D. 1990. Herbivore diets, herbivore colonization, and the escape hypothesis. *Ecology* 71:610–15.

Thomas, J. A., G. W. Elmes, J. C. Wardlaw, and M. Woyciechowski. 1989. Host specificity among *Maculinea* butterflies in *Myrmica* ant nests. *Oecologia* 79:452–57.

Thompson, J. N. 1982. *Interaction and coevolution.* New York: J. Wiley.

Thompson, J. N., and M. F. Willson. 1978. Disturbance and the dispersal of fleshy fruits. *Science* 200:1161–63.

Thorington, R. W., Jr., B. Tannenbaum, A. Tarak, and R. Rudran. 1982. Distribution of trees on Barro Colorado Island: A five hectare sample. Pp. 83–94 in *The ecology of a tropical forest: Seasonal rhythms and long-term changes,* ed. E. G. Leigh, Jr., A. S. Rand, and D. M. Windsor. Washington, D.C.: Smithsonian Institution Press.

Tilman, D. 1982. *Resource competition and community structure.* Princeton, N.J.: Princeton Univ. Press.

———. 1988. *Plant strategies and the dynamics and structure of plant communities.* Princeton, N.J.: Princeton Univ. Press.

Timm, R. M. 1982. *Ectophylla alba. Mamm. Species* 166:1–4.

———. 1984. Tent construction by *Vampyressa* in Costa Rica. *J. Mammal.* 65:166–67.

———. 1985. *Artibeus phaeotis. Mamm. Species* 235:1–6.

———. 1987. Tent construction by bats of the genera *Artibeus* and *Uroderma.* In *Studies in Neotropical mammalogy: Essays in honor of Philip Hershkovitz,* ed. B. D. Patterson and R. M. Timm. *Fieldiana: Zool.,* n.s. 39:187–212.

————. 1988 [1989]. A review and reappraisal of the night monkey, *Aotus lemurinus* (Primates: Cebidae), in Costa Rica. *Rev. Biol. Trop.* 36(2B): 537–40.

Timm, R. M., and B. L. Clauson. 1990. A roof over their feet: Tent-making bats of the New World tropics turn leaves into living quarters. *Nat. Hist.* 3/90:54–59.

Timm, R. M., and L. H. Kermott. 1982. Subcutaneous and cutaneous melanins in *Rhabdomys:* Complementary ultraviolet radiation shields. *J. Mammal.* 63:16–22.

Timm, R. M., and J. Mortimer. 1976. Selection of roost sites by Honduran white bats, *Ectophylla alba* (Chiroptera: Phyllostomatidae). *Ecology* 57:385–89.

Timm, R. M., D. E. Wilson, B. L. Clauson, R. K. LaVal, and C. S. Vaughan. 1989. Mammals of the La Selva-Braulio Carrillo complex, Costa Rica. *N. Am. Fauna* 75:1–162.

Ting, I. P., E. M. Lord, L. da S., L. Sternberg, and M. J. DeNiro. 1985. Crassulacean acid metabolism in the strangler *Clusia rosea* Jacq. *Science* 229:969–71.

Tinkle, D. W. 1969. The concept of reproductive effort and its relation to the evolution of life histories in lizards. *Am. Nat.* 103:501–16.

Tinkle, D. W., and A. E. Dunham. 1986. Comparative life histories of two syntopic sceloporine lizards. *Copeia* 1986:1–18.

Tisdall, J. M., and J. M. Oades. 1982. Organic matter and water-stable aggregates in soil. *J. Soil Sci.* 33:141–63.

Toft, C. A. 1980a. Feeding ecology of thirteen syntopic species of anurans in a seasonal tropical environment. *Oecologia* 45:131–41.

————. 1980b. Seasonal variations in populations of Panamanian litter frogs and their prey: A comparison of wetter and drier sites. *Oecologia* 47:34–38.

————. 1981. Feeding ecology of Panamanian litter anurans: Patterns in diet and foraging mode. *J. Herpetol.* 15:139–44.

————. 1985. Resource partitioning in amphibians and reptiles. *Copeia* 1985:1–20.

Toft, C. A., and S. C. Levings. 1977. Tendencias estacionales relacionadas con las poblaciones de arthrópodos en la hojarasca. In *Acta del IV Symposium Internacional de Ecología Tropical,* ed. H. Wolda, March 7–11, 1977.

Toft, C. A., A. S. Rand, and M. Clark. 1982. Population dynamics and seasonal recruitment in *Bufo typhonius* and *Colostethus nubicola* (Anura). Pp. 397–403 in *The ecology of a tropical forest: Seasonal rhythms and long-term changes,* ed. E. G. Leigh, Jr., A. S. Rand, and D. M. Windsor. Washington, D.C.: Smithsonian Institution Press.

Tomlinson, P. B. 1980. *The biology of trees native to tropical Florida.* Allston, Mass.: Harvard Univ. Printing Office.

Torquebiau, E. F. 1987. Mosaic patterns in dipterocarp rain forest in Indonesia, and their implications for practical forestry. *J. Trop. Ecol.* 2:301–25.

Torres, S. de K. 1985. Pruebas de espaciamientos en especies forestales de Costa Rica. Pp. 255–64 in *Técnica de producción de leña en fincas pequeñas y recuperación de sitios degradados por medio de la silvicultura intensiva,* ed. R. Salazar. Turrialba: CATIE.

Tosi, J. A., Jr. 1969. *Mapa ecológico, República de Costa Rica: Según la clasificación de zonas de vida del mundo de L. R. Holdridge.* San José, Costa Rica: Tropical Science Center.

————. 1971. *El recurso forestal como base potencial para el desarrollo industrial de Costa Rica.* San José, Costa Rica: Tropical Science Center.

————. 1972. *Una clasificación y metodología para la determinación y levantamiento de mapas para la capacidad de uso de la tierra.* San José, Costa Rica: Tropical Science Center.

————. 1974. *Los recursos forestales de Costa Rica.* San José, Costa Rica: Tropical Science Center.

Townsend, D. S., M. M. Stewart, F. H. Pough, and P. F. Brussard. 1981. Internal fertilization in an oviparous frog. *Science* 212:469–71.

Townsend, D. S., M. M. Stewart, and F. H. Pough. 1984. Male parental care and its adaptive significance in a Neotropical frog. *Anim. Behav.* 32:421–31.

Traveset, A. 1991. Pre-dispersal seed predation in Central American *Acacia farnesiana:* Factors affecting the abundance of co-occurring bruchid beetles. *Oecologia* 87:570–76.

Traylor, M. A., Jr., and J. W. Fitzpatrick. 1982. A survey of the tyrant flycatchers. *Living Bird* 19:7–50.

Trudgill, S. T. 1977. *Soil and vegetation systems.* Oxford: Clarendon Press.

Tufte, E. R. 1983. *The visual display of quantitative information.* Cheshire, Conn.: Graphics Press.

Turner, F. B. 1960. Population structure and dynamics of the western spotted frog, *Rana p. pretiosa* Baird & Girand, in Yellowstone Park, Wyoming. *Ecol. Monogr.* 30:251–78.

Turner, I. M. 1989. The seedling survivorship and growth of three *Shorea* species in a Malaysian tropical rain forest. *J. Trop. Ecol.* 6:469–78.

Turner, J. R. G. 1984. Mimicry: The palatability spectrum and its consequences. *Symp. R. Entomol. Soc. Lond.* 11:41–161.

Udvardy, M. D. F. 1975. A classification of the biogeographical provinces of the world. International Union for the Conservation of Nature and Natural Resources, Occasional Paper no. 18. Morges, Switzerland.

Ueckert, D. N., and R. M. Hansen. 1971. Dietary overlap of grasshoppers on sandhill rangeland in northeastern Colorado. *Oecologia* 8:276–95.

Uehara, G., and G. Gillman. 1981. *The mineralogy, chemistry, and physics of tropical soils with variable charge clays.* Boulder, Colo.: Westview Press.

Ugalde, M. A. 1981. Estudio detallado de suelos y clasificación por capacidad de uso de las tierras de la Estación Experimental de Río Frío. Master's thesis, Facultad de Agronomía, Universidad de Costa Rica, San José.

Uhl, C. 1982. Tree dynamics in a species rich Tierra Firme forest in Amazonia, Venezuela. *Acta Cient. Venez.* 33:72–77.

Uhl, C., K. Clark, N. Dezzeo, and P. Maquirino. 1988. Vegetation dynamics in Amazonian treefall gaps. *Ecology* 69:751–63.

Uhl, C., and P. G. Murphy. 1981. Composition, structure, and regeneration of tierra firme forest in the Amazon Basin of Venezuela. *Trop. Ecol.* 22:219–37.

Uhl, N. W., and H. E. Moore, Jr. 1977. Correlations of inflorescence, flower structure, and floral anatomy with pollination in some palms. *Biotropica* 9:170–90.

United Nations Educational, Scientific, and Cultural Organization. 1978a. Pests and diseases in forests and plantations.

Pp. 286–314 in *Tropical forest ecosystems: A state of knowledge report,* UNESCO, UNEP, and FAO. Natural Resources Research no. 14. Paris: UNESCO.

———. 1978b. *Tropical forest ecosystems: A state of knowledge report,* UNESCO, UNEP, and FAO. Natural Resources Research no. 14. Paris: UNESCO.

———. 1984. Action plan for biosphere reserves. *Nat. Resour.* 20:1–12.

United States Department of Agriculture, Soil Conservation Service. 1972. *Soil survey laboratory methods and procedures for collecting soil samples.* USDA SSIR 1. Rev. ed. Washington, D.C.: USGPO.

United States Department of Agriculture, Soil Survey Staff. 1975. *Soil taxonomy.* Washington, D.C.: USGPO.

———. 1987. *Keys to soil taxonomy.* USDA Soil Management Support Service Technical Monograph no. 6. Ithaca, N.Y.: Cornell University.

———. 1990. *Keys to soil taxonomy.* USDA Soil Management Support Service Technical Monograph no. 19. Blacksburg: Virginia Polytechnic Institute.

Universidad de Costa Rica. 1984. *Caracterización ambiental y de los sistemas de cultivo en fincas de la región de Río Frío, Horquetas, Sarapiquí, Heredia, Costa Rica.* San José, Costa Rica: Facultad de Agronomía.

Ureña H., A. I. 1983. Un estudio de caso para diagnosticar algunas variables socio-económicas en la región de Río Frío. Master's thesis, Universidad de Costa Rica, Facultad de Agronomía, San José.

Valdeyron, G., and D. G. Lloyd. 1979. Sex differences and flowering phenology in the common fig, *Ficus carica* L. *Evolution* 33:673–85.

Valerio, C. E. 1983. Insect visitors to the inflorescence of the aroid *Dieffenbachia oerstedii* (Araceae) in Costa Rica. *Brenesia* 22:139–46.

Van Berkum, F. H. 1986. Evolutionary patterns of the thermal sensitivity of sprint speed in *Anolis* lizards. *Evolution* 40:594–604.

———. 1988. Latitudinal patterns of the thermal sensitivity of sprint speed in lizards. *Am. Nat.* 132:327–43.

Van Breemen, N., J. Mulder, and C. T. Driscoll. 1983. Acidification and alkalinization of soils. *Plant Soil* 75:283–308.

Vandermeer, J. H. 1972. Seed predation in weedy plants and its relation to the spatial distribution of *Sida acuta* (Malvaceae). Report, OTS 72-3:147–53.

———. 1977. Notes on density dependence in *Welfia georgii* Wendl. ex Burret (Palmae), a lowland rainforest species in Costa Rica. *Brenesia* 10/11:9–15.

———. 1979. Hoarding behavior of captive *Heteromys desmarestianus* (Rodentia) on the fruits of *Welfia georgii,* a rainforest dominant palm in Costa Rica. *Brenesia* 16:107–16.

———. 1983. Pejibaye palm. Pp. 98–101 in *Costa Rican natural history,* ed. D. H. Janzen. Chicago: Univ. of Chicago Press.

Vandermeer, J. H., J. Stout, and S. Risch. 1979. Seed dispersal of a common Costa Rican rain forest palm *(Welfia georgii). Trop. Ecol.* 20:17–26.

Van der Pijl, L. 1982. *Principles of dispersal in higher plants.* Berlin: Springer-Verlag.

Van Devender, R. W. 1983. *Basiliscus basiliscus.* Pp. 379–80 in *Costa Rican natural history,* ed. D. H. Janzen. Chicago: Univ. of Chicago Press.

Vane-Wright, R. I. 1978. Ecological and behavioural origins of diversity in butterflies. *Symp. R. Entomol. Soc. Lond.* 9:56–70.

Vane-Wright, R. I., and P. R. Ackery, eds. 1989. *The biology of butterflies. Symp. R. Entomol. Soc. Lond.* 11:1–429.

Van Schaik, C. P. V. 1986. Phenological changes in a Sumatran rain forest. *J. Trop. Ecol.* 2:237–247.

Van Sluys, F. R., H. Waaijenberg, W. G. Wielemaker, and J. F. Wienk. 1989. *Agricultura en la Zona Atlántica de Costa Rica.* Serie Técnica, Informe Técnico no. 141, Programme Paper no. 4. Turrialba, Costa Rica: Centro Agronómico Tropical de Investigación y Enseñanza and Universidad Agrícola de Wageningen.

Van Steenis, C. G. G. J. 1958. Rejuvenation as a factor for judging the status of vegetation types: The biological nomad theory. Pp. 212–15 in *Study of tropical vegetation.* Proceedings of the Kandy Symposium. Paris: UNESCO.

Van Wambeke, A., and R. Dudal. 1978. Macrovariability of soils of the tropics. Pp. 13–28 in *Diversity of soils in the tropics,* ed. M. Stelly. Madison, Wis.: American Society of Agronomy.

Vasconselos, H. L. 1991. Mutualism between *Maieta guianensis* Aubl., a myrmecophytic melastome and one of its ant inhabitants: Ant protection against insect herbivores. *Oecologia* 87:295–98.

Vaughan, C. 1983. Coyote range expansion in Costa Rica and Panama. *Brenesia* 21:27–32.

Vaux, P., C. Pringle, and P. Paaby. 1984. A study of the feeding ecology of fish in La Selva streams. Report, Organization for Tropical Studies, San José, Costa Rica.

Vázquez-Yánes, C., and A. Orozco-Segovia. 1984. Ecophysiology of seed germination in the tropical humid forests of the world: A review. Pp. 37–50 in *Physiological ecology of plants of the wet tropics,* ed. E. Medina, H. A. Mooney, and C. Vázquez-Yánes. The Hague: W. Junk.

———. 1985. Posibles efectos del microclima de los claros de la selva sobre la germinación de tres especies de árboles pioneros: *Cecropia obtusifolia, Heliocarpus donnellsmithii,* y *Piper auritum.* Pp. 241–53 in *Frugivores and seed dispersal,* ed. A. Estrada and T. H. Fleming. Dordrecht: W. Junk.

———. 1987. Fisiología ecológica de semillas en la Estación de Biología Tropical "Los Tuxtlas," Veracruz, Mexico. *Rev. Biol. Trop.* 35 (suppl.): 85–96.

———. 1990. Ecological significance of light controlled seed germination in two contrasting tropical habitats. *Oecologia* 83:171–75.

Vehrencamp, S. L. 1977. Relative fecundity and parental effort in communally nesting anis, *Crotophaga sulcirostris. Science* 197:403–5.

———. 1978. The adaptive significance of communal nesting in groove-billed anis, *Crotophaga sulcirostris. Behav. Ecol. Sociobiol.* 4:1–33.

———. 1983. *Crotophaga sulcirostris.* Pp. 570–71 in *Costa Rican natural history,* ed. D. H. Janzen. Chicago: Univ. of Chicago Press.

Vermeij, G. J. 1978. *Biogeography and adaptation, patterns of marine life.* Cambridge: Harvard Univ. Press.

Vermeij, G. J., and J. A. Vail. 1978. A latitudinal pattern in bivalve shell gaping. *Malacologia* 17:57–61.

Verner, J., and M. F. Willson. 1966. The influence of habitats on mating systems of North American passerine birds. *Ecology* 47:143–47.

Vial, J. L. 1968. The ecology of the tropical salamander, *Bolitoglossa subpalmata,* in Costa Rica. *Rev. Biol. Trop.* 15:13–115.

Villa, J. 1977. A symbiotic relationship between frog (Amphibian, Anura, Centrolenidae) and fly larvae (Drosophilidae). *J. Herpetol.* 11:317–22.

———. 1979. Two fungi lethal to frog eggs in Central America. *Copeia* 1979:650–55.

———. 1980. "Frogflies" from Central and South America with notes on other organisms of the amphibian egg microhabitat. *Brenesia* 17:49–68.

———. 1984. Biology of a Neotropical glass frog *Centrolenella fleischmanni* (Boettger), with special reference to its frogfly associates. *Milw. Public Mus. Contrib. Biol. Geol.* 55:1–60.

Villa, J., and D. S. Townsend. 1983. Viable frog eggs eaten by phorid fly larvae. *J. Herpetol.* 17:278–81.

Villa Nova, N. A., E. Salati, and E. Matusi. 1976. Estimativa da evapotranspiraçao na Bacia Amazônica. *Acta Amazônica* 6:215–28.

Vince-Prue, D. 1984. *Light and the flowering process—Setting the scene.* Pp. 3–15 in *Light and the flowering process,* ed. D. Vince-Prue, B. Thomas, and K. E. Cockshull. London: Academic Press.

Vitousek, P. M. 1984. Litterfall, nutrient cycling and nutrient limitation in tropical forests. *Ecology* 65:285–98.

Vitousek, P. M., and J. S. Denslow. 1986. Nitrogen and phosphorus availability in treefall gaps of a lowland tropical forest. *J. Ecol.* 74:1167–78.

———. 1987. Differences in extractable phosphorus among soils of the La Selva Biological Station, Costa Rica. *Biotropica* 19:167–70.

Vitousek, P. M., and P. A. Matson. 1988. Nitrogen transformations in a range of tropical forest soils. *Soil Biol. Biochem.* 20:361–67.

Vitousek, P. M., and R. L. Sanford, Jr. 1987. Nutrient cycling in moist tropical forest. *Annu. Rev. Ecol. Syst.* 17:137–67.

Vitt, L. J. 1986. Reproductive tactics of sympatric gekkonid lizards with a comment on the evolutionary and ecological consequences of invariant clutch size. *Copeia* 1986:773–86.

Vitt, L. J., J. P. Caldwell, H. M. Wilbur, and D. C. Smith. 1990. Amphibians as harbingers of decay. *Bioscience* 40:418.

Vitt, L. J., and T. E. Lacher, Jr. 1981. Behavior, habitat, diet, and reproduction of the iguanid lizard *Polychrus acustirostris* in the caatinga of northeastern Brazil. *Herpetologica* 37:53–63.

Vitt, L. J., and L. D. Vangilder. 1983. Ecology of a snake community in northeastern Brazil. *Amphibia-Reptilia* 4:273–96.

Vogt, K. A., C. C. Grier, and D. J. Vogt. 1986. Production, turnover, and nutrient dynamics of above- and belowground detritus of world forests. *Adv. Ecol. Res.* 15:303–78.

Voss, R., M. Turner, R. Inouye, M. Fisher, and R. Cort. 1980. Floral biology of *Markea neurantha* Hemsley (Solanaceae), a bat-pollinated epiphyte. *Am. Midl. Nat.* 103:262–68.

Waage, J. K., and R. C. Best. 1985. Arthropod associates of sloths. Pp. 297–311 in *The evolution and ecology of armadillos, sloths, and vermilinguas,* ed. G. G. Montgomery. Washington, D.C.: Smithsonian Institution Press.

Wagner, M., and C. Scherzer. 1856. *La República de Costa Rica en Centro America.* Vol. 1 (1974 trans.). San José: Ministerio de Cultura, Juventud y Deportes.

Wake, D. B. 1991. Declining amphibian populations. *Science* 253:860.

Wake, D. B., and P. Elias. 1983. New genera and a new species of Central American salamanders, with a review of the tropical genera (Amphibia, Caudata, Plethodontidae). *Los Angel. Cty. Mus. Nat. Hist. Contrib. Sci.* 345:1–19.

Wake, D. B., and J. L. Lynch. 1976. The distribution, ecology, and evolutionary history of plethodontid salamanders in tropical America. *Bull. Los Angel. Cty. Mus. Nat. Hist. Sci.* 25:1–65.

Wake, M. H. 1977. The reproductive biology of caecilians: An evolutionary perspective. Pp. 73–101 in *The reproductive biology of amphibians,* ed. D. H. Taylor and S. I. Guttman. New York: Plenum Press.

———. 1983. *Gymnopis multiplicata, Dermophis mexicanus,* and *Dermophis parviceps (Soldas, Suelda con suelda, Dos cabezas, caecilians).* Pp. 400–401 in *Costa Rican natural history,* ed. D. H. Janzen, Chicago: Univ. of Chicago Press.

Walker, E. P. 1975. *Mammals of the world.* 3d ed. Baltimore, Md.: Johns Hopkins Univ. Press.

Wallace, A. R. 1878. *Tropical nature and other essays.* London: Macmillan.

Walter, H. 1973. *Vegetation of the earth in relation to climate and the ecophysiological conditions.* New York: Springer-Verlag.

Walters, M. B., and C. B. Field. 1987. Photosynthetic light acclimation in two rainforest *Piper* species with different ecological amplitudes. *Oecologia* 72:449–56.

Waltz, S. A. 1984. *Comparative study of predictability, value, and defenses of leaves of tropical wet forest trees (Costa Rica).* Ph.d. diss., University of Washington, Seattle.

Waring, R. H., and W. H. Schlesinger. 1985. *Forest ecosystems—Concepts and management.* New York: Academic Press.

Warkentin, B. P., and T. Maeda. 1980. Physical and mechanical characteristics of Andisols. Pp. 281–301 in *Soils with variable charge,* ed. B. K. G. Theng. Lower Hutt, New Zealand: New Zealand Society of Soil Science.

Waser, N. M. 1983. Competition for pollination and floral character differences among sympatric plant species: A review of evidence. Pp. 277–93 In *Handbook of experimental pollination biology,* ed. C. E. Jones and R. J. Little. New York: Scientific and Academic Editions.

Waterman, P. G. 1983. Distribution of secondary metabolites in rain forest plants: Toward an understanding of cause and effect. Pp. 167–79 in *Tropical rain forest: Ecology and management,* ed. S. L. Sutton, T. C. Whitmore, and A. C. Chadwick. Oxford: Blackwell Scientific.

———. 1986. A phytochemist in the African rain forest. *Phytochemistry* 25:3–17.

Waterman, P. G., and G. M. Choo. 1981. The effects of digestibility-reducing compounds in leaves on food selection by some Colobinae. *Malays. Appl. Biol.* 10:147–62.

Waterman, P. G., J. A. M. Ross, E. L. Bennett, and A. G. Davies. 1988. A comparison of the floristics and leaf chemistry of the tree flora in two Malaysian rain forests and the influence of leaf chemistry on populations of colobine monkeys in the Old World. *Biol. J. Linn. Soc.* 34:1–32.

Way, M. J. 1963. Mutualism between ants and honeydew-producing Homoptera. *Annu. Rev. Entomol.* 8:307–44.

Weaver, P. L. 1989. Forest changes after hurricanes in Puerto Rico's Luquillo Mountains. *Interciencia* 14:181–92.

Weaver, P. L., R. A. Birdsey, and A. E. Lugo. 1987. Soil organic matter in secondary forests of Puerto Rico. *Biotropica* 19:17–23.

Webb, L. J. 1959. A physiognomic classification of Australian rain forests. *J. Ecol.* 47:551–70.

———. 1969. Edaphic differentiation of some forest types in eastern Australia. Part 2. Soil chemical factors. *J. Ecol.* 57:817–30.

Webb, S. L., and M. F. Willson. 1985. Spatial heterogeneity in post-dispersal predation on *Prunus* and *Uvularia* seeds. *Oecologia* 67:150–53.

Weekes, A. A. 1992. *La Flaminea at La Selva Biological Station, the Organization for Tropical Studies.* Master's thesis, University of Florida, Gainesville.

Wells, K. D. 1977a. The courtship of frogs. Pp. 233–62 in *The reproductive biology of amphibians,* ed. D. H. Taylor and S. I. Guttman. New York: Plenum Press.

———. 1977b. The social behaviour of anuran amphibians. *Anim. Behav.* 25:666–93.

———. 1988. The effect of social interactions on anuran vocal behavior. Pp. 433–54 in *The evolution of the amphibian auditory system,* ed. B. Fritzsch, M. J. Ryan, W. Wilczynski, T. E. Hetherington, and W. Walkowiak. New York: J. Wiley.

Wells, K. D., and K. M. Bard. 1987. Vocal communication in a Neotropical treefrog, *Hyla ebraccata:* responses of females to advertisement and aggressive calls. *Behaviour* 101:200–210.

Went, F. W., and N. Stark. 1968a. The biological and mechanical role of soil fungi. *Proc. Nat. Acad. Sci.* 60:497–504.

———. 1968b. Mycorrhiza. *Bioscience* 18:1035–39.

Werman, S. D. 1986. Phylogenetic systematics and biogeography of Neotropical pit vipers: A cladistic analysis of biochemical and anatomical characters. Ph.D. diss., University of Miami, Coral Gables, Fla.

Werner, P. 1984. Changes in soil properties during tropical wet forest succession in Costa Rica. *Biotropica* 16:43–50.

———. 1985. La Reconstitution de la Forêt Tropicale Humide au Costa Rica: Analyse de Croissance et Dynamique de la Végétation. Ph.D. diss., Université de Lausanne, Switzerland.

Westoby, M., and B. Rice. 1982. Evolution of the seed plants and inclusive fitness of plant tissues. *Evolution* 36:713–24.

Wetmore, A. 1914. The development of the stomach in the Euphonias. *Auk* 31:458–61.

———. 1926. *The migration of birds.* Cambridge: Harvard Univ. Press.

Wetzel, R. M. 1983. *Dasypus novemcinctus.* Pp. 465–67 in *Costa Rican natural history,* ed. D. H. Janzen. Chicago: Univ. of Chicago Press.

Weygoldt, P. 1980. Complex brood care and reproductive behavior in captive poison-arrow frogs, *Dendrobates pumilio* O. Schmidt. *Behav. Ecol. Sociobiol.* 7:329–32.

Wheelwright, N. T. 1983. Fruits and the ecology of resplendent quetzals. *Auk* 100:286–301.

———. 1985a. Competition for dispersers, and the timing of flowering and fruiting in a guild of tropical trees. *Oikos* 44:465–77.

———. 1985b. Fruit size, gape width, and the diets of fruit-eating birds. *Ecology* 66:808–18.

———. 1988. Fruit-eating birds and bird-dispersed plants in the tropics and temperate zone. *Trends Ecol. Evol.* 10:270–74.

Wheelwright, N. T., W. A. Haber, K. G. Murray, and C. Guindon. 1984. Tropical fruit-eating birds and their food plants: A survey of a Costa Rican lower montane forest. *Biotropica* 16:173–92.

Wheelwright, N. T., and C. H. Janson. 1985. Colors of fruit displays of bird-dispersed plants in two tropical forests. *Am. Nat.* 126:777–99.

Wheelwright, N. T., and G. H. Orians. 1982. Seed dispersal by animals: Contrasts with pollen dispersal, problems of terminology, and constraints on coevolution. *Am. Nat.* 119:402–13.

White, M. J. D. 1978. *Modes of speciation.* San Francisco, Calif.: W. H. Freeman.

White, P. S., M. D. Mackenzie, and R. T. Busing. 1985. A critique on overstory/understory comparisons based on transition probability analysis of an old growth spruce-fir stand in the Appalachians. *Vegetatio* 64:37–45.

White, R. E. 1985. The influence of macropores on the transport of dissolved and suspended matter through soil. *Adv. Soil Sci.* 3:95–120.

White, S. C. 1974. Ecological aspects of growth and nutrition in tropical fruit-eating birds. Ph.D. diss., University of Pennsylvania, College Station.

Whiteside, T. J. 1985. *Central American climatology.* USAFETAC/TN-85/002.

Whitham, T. G., J. Maschinski, K. C. Larson, and K. N. Paige. 1991. Plant responses to herbivory: The continuum from negative to positive and underlying physiological mechanisms. Pp. 227–56 in *Plant-animal interactions: Evolutionary ecology in tropical and temperate regions,* ed. P. W. Price, T. M. Lewinsohn, G. W. Fernandes, and W. W. Benson. New York: J. Wiley.

Whitmore, T. C. 1978a. Gaps in the forest canopy. Pp. 639–55 in *Tropical trees as living systems,* ed. P. B. Tomlinson and M. H. Zimmerman. Cambridge: Cambridge Univ. Press.

———. 1978b. *Tropical rain forests of the far east.* Oxford: Clarendon Press.

———. 1984. *Tropical rain forests of the far east.* 2d ed. Oxford: Clarendon Press.

———. 1989a. Canopy gaps and the two major groups of forest trees. *Ecology* 70:536–38.

———. 1989b. Changes over twenty-one years in the Kolombangara rainforests. *J. Ecol.* 77:469–83.

———. 1991. Tropical rain forest dynamics and its implications for management. Pp. 67–89 in *Tropical rain forest regeneration and management,* ed. A. Gómez-Pompa, T. C. Whitmore and M. Hadley. Paris: UNESCO; Carnforth, U.K.: Plenum.

Whitmore, T. C., and W. K. Gong. 1983. Growth analysis of the seedlings of balsa, *Ochroma lagopus. New Phytol.* 95:305–11.

Whittaker, R. H. 1977. Evolution of species diversity in land communities. *Evol. Biol.* 10:1–67.

Whittaker, R. H., and G. M. Woodwell. 1969. Structure, production and diversity of the oak-pine forest at Brookhaven, New York. *J. Ecol.* 57:155–74.

Wibmer, G. J. 1989. Revision of the weevil genus *Tyloderma* Say (Col.: Curculionidae) in Mexico, Central America, South America, and the West Indies. *Evol. Monogr.* 11:1–118.

Wiebes, J. T. 1979. Co-evolution of figs and their insect pollinators. *Annu. Rev. Ecol. Syst.* 10:1–12.

Wiehler, H. 1983. A synopsis of the Neotropical Gesneriaceae. *Selbyana* 6:1–219.

Wilbur, R. L. 1986. The vascular flora of La Selva Biological Station, Costa Rica—Introduction. *Selbyana* 9:191.

Wilkerson, R. C., and G. B. Fairchild. 1985. A checklist and generic key to the Tabanidae (Diptera) of Peru with special reference to the Tambopata Reserved Zone, Madre de Dios. *Rev. Peru. Entomol.* 27 (1984): 37–53.

Williams, C. F. 1986. Social organization of the bat, *Carollia perspicillata* (Chiroptera: Phyllostomidae). *Ethology* 71:265–82.

Williams. G. C. 1966. *Adaptation and natural selection.* Princeton, N.J.: Princeton Univ. Press.

Williams, N. H., and R. R. Dressler. 1976. Euglossine pollination of *Spathiphyllum (Araceae). Selbyana* 1:349–56.

Willis, E. O. 1966. Competitive exclusion and birds at fruiting trees in Western Colombia. *Auk* 83:479–80.

———. 1974. Populations and local extinctions of birds on Barro Colorado Island, Panama. *Ecol. Monogr.* 44:153–69.

———. 1976. Seasonal changes in the invertebrate litter fauna on Barro Colorado Island, Panama. *Rev. Bras. Biol.* 36:643–57.

———. 1980. Ecological roles of migratory and resident birds on Barro Colorado Island, Panama. Pp. 205–25 in *Migrant birds in the Neotropics,* ed. A. Keast and E. S. Morton. Washington, D.C.: Smithsonian Institution Press.

Willis, E. O., and E. Eisenmann. 1979. A revised list of the birds of Barro Colorado Island, Panama. *Smithson. Contrib. Zool.* 291:1–31.

Willis, E. O., and Y. Oniki. 1978. Birds and army ants. *Annu. Rev. Ecol. Syst.* 9:243–63.

Willson, M. F. 1983. *Plant reproductive ecology.* New York: J. Wiley.

———. 1986. Avian frugivory and seed dispersal in eastern North America. *Curr. Ornithol.* 3:223–79.

———. 1988. Spatial heterogeneity of post-dispersal survivorship of Queensland rainforest seeds. *Aust. J. Ecol.* 13:137–45.

———. 1991. Birds and fruits: How does this mutualism matter? *Acta 20th Congr. Int. Ornithol.* 3:1630–35.

Willson, M. F., and N. Burley. 1983. *Mate choice in plants.* Princeton, N.J.: Princeton Univ. Press.

Willson, M. F., and F. H. J. Crome. 1989. Patterns of seed rain at the edge of a tropical Queensland rain forest. *J. Trop. Ecol.* 5:301–8.

Willson, M. F., and W. G. Hoppes. 1986. Foliar "flags" for avian frugivores: Signal or serendipity? Pp. 55–70 in *Frugivores and seed dispersal,* ed. A. Estrada and T. H. Fleming. Amsterdam: W. Junk.

Willson, M. F., A. K. Irvine, and N. G. Walsh. 1989. Vertebrate dispersal syndromes in some Australian and New Zealand plant communities, with geographic comparisons. *Biotropica* 21:133–47.

Willson, M. F., E. A. Porter, and R. S. Condit. 1982. Avian frugivore activity in relation to forest light gaps. *Caribb. J. Sci.* 18:1–6.

Willson, M. F., and J. N. Thompson. 1982. Phenology and ecology of color in bird-dispersed fruits, or why some fruits are red when they are "green." *Can. J. Bot.* 60:701–13.

Willson, M. F., and C. J. Whelan. 1990. The evolution of fruit color in fleshy-fruited plants. *Am. Nat.* 136:790–809.

Wilson, D. E. 1979. Reproductive patterns. Pp. 317–18 in *Biology of bats of the New World family Phyllostomatidae,* ed. R. J. Baker, J. K. Jones, Jr., and D. C. Carter. Part 3. Special Publications of the Museum no. 16., Texas Tech University, Lubbock.

———. 1983. Checklist of mammals. Pp. 443–47 in *Costa Rican natural history,* ed. D. H. Janzen. Chicago: Univ. of Chicago Press.

———. 1990. Mammals of La Selva, Costa Rica. Pp. 273–86 in *Four Neotropical rainforests,* ed. A. H. Gentry. New Haven, Conn.: Yale Univ. Press.

Wilson, E. O. 1971. *The insect societies.* Cambridge: Harvard Univ. Press, Belknap Press.

Wilson, J. W. III. 1974. Analytical zoogeography of North American mammals. *Evolution* 28:124–40.

Windsor, D. M. 1978. The feeding activities of tropical insect herbivores on some deciduous forest legumes. Pp. 101–13 in *The ecology of arboreal folivores,* ed. G. G. Montgomery. Washington, D.C.: Smithsonian Institution Press.

Windsor, D. M., D. W. Morrison, M. A. Estribi, and B. de Leon. 1989. Phenology of fruit and leaf production by "strangler" figs on Barro Colorado Island, Panama. *Experientia* (Basel) 45:647–53.

Winnett-Murray, K. 1981. Body length in leafcutter ants foraging at different times of day. Report, OTS 81-1:82–83.

Wint, G. R. W. 1983. Leaf damage in tropical rain forest canopies. Pp. 229–39 in *Tropical rain forest: Ecology and management,* ed. S. L. Sutton, T. C. Whitmore, and A. C. Chadwick. Oxford: Blackwell Scientific.

Wolda, H. 1978a. Fluctuations in abundance of tropical insects. *Am. Nat.* 112:1017–45.

———. 1978b. Seasonal fluctuations in rainfall, food, and abundance of tropical insects. *J. Anim. Ecol.* 36:643–57.

———. 1979. Abundance and diversity of Homoptera in the canopy of a tropical forest. *Ecol. Entomol.* 4:181–90.

———. 1982. Seasonality of Homoptera on Barro Colorado Island. Pp. 319–30 in *The ecology of a tropical rain forest: Seasonal rhythms and long-term changes,* ed. E. G. Leigh, Jr., A. S. Rand, and D. M. Windsor. Washington, D.C.: Smithsonian Institution Press.

———. 1983. Spatial and temporal variation in abundance in tropical animals. Pp. 93–105 in *Tropical rain forest: Ecology and management,* ed. S. L. Sutton, T. C. Whitmore, and A. C. Chadwick. Oxford: Blackwell Scientific.

———. 1988. Insect seasonality: Why? *Annu. Rev. Ecol. Syst.* 19:1–18.

Wolda, H., F. W. Fisk, and M. Estribi. 1983. Faunistics of Panamanian cockroaches (Blattaria). *Uttar Pradesh J. Zool.* 3:1–9.

Wolda, H., and R. Foster. 1978. *Zunacetha annulata* (Lepidoptera: Dioptidae), an outbreak insect in a Neotropical forest. *Geo-Eco-Tropo* 2:443–54.

Wolf, L. L., F. R. Hainsworth, and F. G. Stiles. 1972. Energetics of foraging: Rate and efficiency of nectar extraction by hummingbirds. *Science* 186:1351–52.

Wong, M. 1983. Understory phenology of the virgin and regenerating habitats in Pasoh Forest Reserve, Negeri Sembilan, West Malaysia. *Malay. For.* 46:198–223.

———. 1984. Understory foliage arthropods in the virgin and regenerating habitats of Pasoh Forest Reserve, West Malaysia. *Malay. For.* 47:43–69.

Wong, M., S. J. Wright, S. P. Hubbell, and R. B. Foster. 1990. The spatial pattern and reproductive consequences of outbreak defoliation in *Quararibea asterolepis,* a tropical tree. *J. Ecol.* 78:579–88.

Wood, T. K. 1974. Aggregating behavior of *Umbronia crassicornis* (Homoptera: Membracidae). *Can. Entomol.* 106:169–73.

———. 1976. Alarm behavior of brooding female *Umbronia crassicornis* (Homoptera: Membracidae). *Ann. Entomol. Soc. Am.* 69:340–44.

———. 1977. Role of parent females and attendant ants in the maturation of the treehopper, *Entylia bactriana* (Homoptera: Membracidae). *Sociobiology* 2:257–72.

———. 1978. Parental care in *Guayaquila compressa* Walker (Homoptera: Membracidae). *Psyche* 85:135–45.

———. 1980. Divergence in the *Enchenopa binotata* Say complex (Homoptera: Membracidae) effected by host plant adaptation. *Evolution* 34:147–60.

———. 1982. Selective factors associated with the evolution of membracid sociality. Pp. 175–79 in *The biology of social insects,* ed. M. D. Breed, C. D. Michener, and H. E. Evans. Boulder, Colo.: Westview Press.

———. 1983. *Umbonia crassicornis.* Pp. 773–75 in *Costa Rican natural history,* ed. D. H. Janzen. Chicago: Univ. of Chicago Press.

———. 1984. Life history patterns of tropical membracids (Homoptera: Membracidae). *Sociobiology* 8:299–344.

Wood, T. K., and S. I. Guttman. 1981. The role of host plants in the speciation of treehoppers: An example from the *Enchenopa binotata* complex. Pp. 39–54 in *Insect life history patterns,* ed. R. F. Denno and H. Dingle. New York: Springer-Verlag.

———. 1983. *Enchenopa binotata* complex: Sympatric speciation? *Science* 220:310–12.

Wood, T. K., and M. C. Keese. 1990. Host-induced assortative mating in *Echenopa* treehoppers. *Evolution* 44:619–28.

Wood, T. K., and K. L. Olmstead. 1984. Latitudinal effects on treehopper species richness (Homoptera: Membracidae). *Ecol. Entomol.* 9:109–15.

Wood, T. K., K. L. Olmstead, and S. I. Guttman. 1990. Insect phenology mediated by host-plant water relations. *Evolution* 44:629–36.

Woodman, N. 1992. Biogeographical and evolutionary relationships among Central American small-eared shrews of the genus *Cryptotis* (Mammalia: Insectivora: Soricidae). Ph.D. diss., University of Kansas, Lawrence.

Woolfenden, G. E., and J. W. Fitzpatrick. 1984. *The Florida scrub jay: Demography of a cooperative-breeding bird.* Princeton, N.J.: Princeton Univ. Press.

Wooten, J. T., and I. Sun. 1990. Bract liquid as a herbivore defense mechanism for *Heliconia wagneriana* inflorescences. *Biotropica* 22:155–59.

Worthington, A. 1982. Population sizes and breeding rhythms of two species of manakins in relation to food supply. Pp. 213–26 in *The ecology of a tropical forest: Seasonal rhythms and long-term changes,* ed. E. G. Leigh, Jr., A. S. Rand, and D. M. Windsor. Washington, D.C.: Smithsonian Institution Press.

———. 1989. Adaptations for avian frugivory: Assimilation efficiency and gut transit time of *Manacus vitellinus* and *Pipra mentalis. Oecologia* 80:381–89.

Worthington, A. H., and R. M. Olberg. 1990. Random encounter model predicts fruit selected by Neotropical understory birds. *Bull. Ecol. Soc. Am.* 71:373–74 (abstract).

Wright, S. 1943. Isolation by distance. *Genetics* 28:114–38.

———. 1969. *Evolution and the genetics of populations.* Vol. 2, *The theory of gene frequencies.* Chicago: Univ. of Chicago Press.

Wright, S. J. 1979. Competition between insectivorous lizards and birds in central Panama. *Am. Zool.* 19:1145–56.

———. 1983. The dispersion of eggs by a bruchid beetle among *Scheelea* palm seeds and the effect of distance to the parent plant. *Ecology* 64:1016–21.

———. 1991. Seasonal drought and the phenology of understory shrubs in a tropical moist forest. *Ecology* 72:1643–57.

Wright, S. J., and F. H. Cornejo. 1990a. Seasonal drought and leaf fall in a tropical forest. *Ecology* 71:1165–75.

———. 1990b. Seasonal drought and the timing of flowering and leaf fall in a Neotropical forest. Pp. 49–61 in *Reproductive ecology of tropical forest plants,* ed. K. S. Bawa and M. Hadley. Man and the Biosphere series. Paris: UNESCO.

Wrobel, D. 1979. A comparison of insect community structure between early successional and primary rainforest communities. Report, OTS 79-3:94–96.

Wyatt, R. 1974. *Miconia* sp.: A new ant plant for La Selva. Report, OTS 74-1:148–53.

Yap, S. K., and H. T. Chan. 1990. Phenological behaviour of some *Shorea* species in peninsular Malaysia. Pp. 21–35 in *Reproductive ecology of tropical forest plants,* ed. K. S. Bawa and M. Hadley. Man and the Biosphere series. Paris: UNESCO.

Yarbro, L. A. 1983. The influence of hydrologic variations on phosphorus cycling and retention in a swamp stream ecosystem. Pp. 223–45 in *Dynamics of lotic ecosystems,* ed. T. D. Fontaine III and S. M. Bartell. Ann Arbor, Mich.: Ann Arbor Science Publishers.

Yatskievych, G., and K. M. Yatskievych. 1987. A floristic survey of the Yellow Birch Ravine Nature Preserve, Crawford County, Indiana. *Indiana Acad. Sci.* 96:435–45.

Yeaton, R. I. 1978. Competition and spacing in plant communities: Differential mortality of white pine (*Pinus strobus* L.) in a New England woodlot. *Am. Midl. Nat.* 100: 285–93.

Yih, K., D. H. Boucher, J. H. Vandermeer, and N. Zamora. 1991. Recovery of the rain forest of southeastern Nicaragua after destruction by Hurricane Joan. *Biotropica* 23:106–13.

Young, A. M. 1971a. Foraging of vampire bats *(Desmodus rotundus)* in Atlantic wet lowland Costa Rica. *Rev. Biol. Trop.* 18:73–88.

———. 1971b. Mimetic associations in a natural population of tropical papilionid butterflies (Lepidoptera: Papilionidae). *J. N.Y. Entomol. Soc.* 79:210–24.

———. 1971c. Notes on the gregarious roosting in tropical butterflies in the genus *Morpho. J. Lepid. Soc.* 25:223–34.

———. 1971d. Wing coloration and reflectance in *Morpho* butterflies as related to reproductive behavior and escape from avian predators. *Oecologia* 7:209–22.

———. 1972a. Cicada ecology in a Costa Rican tropical rain forest. *Biotropica* 4:152–59.

———. 1972b. Community ecology of some tropical rain forest butterflies. *Am. Midl. Nat.* 87:146–57.

———. 1972c. An experimental study on the relation of mortality of young to adult members in colonies of the lesser sac-winged bat, *Saccopteryx bilineata perspicillifer. Am. Midl. Nat.* 87:158–64.

———. 1973a. Cicada populations on palms in a tropical rain forest. *Principes* 17:3–9.

———. 1973b. Notes on the life cycle and natural history of *Parides arcas mylotes* (Papilionidae) in a Costa Rican premontane wet forest. *Psyche* 80:1–21.

———. 1973c. Studies on comparative ecology and ethology in adult populations of several species of *Morpho* butterflies in Costa Rica (Lepidoptera: Morphidae). *Studies Neotrop. Fauna* 8:17–50.

———. 1974. Further observations on the natural history of *Philaethria dido dido* (Lepidoptera: Nymphalidae: Heliconiinae). *J. N.Y. Entomol. Soc.* 82:30–41.

———. 1975. Correction on mortality in colonies of lesser sac-winged bat, *Saccopteryx bilineata perspicillifer. Am. Midl. Nat.* 93:512.

———. 1976. Notes on the faunistic complexity of cicadas (Homoptera; Cicadidae) in northern Costa Rica. *Rev. Biol. Trop.* 24:267–79.

———. 1979a. Notes on a migration of *Urania fulgens* (Lepidoptera: Uraniidae) in Costa Rica. *J. N.Y. Entomol. Soc.* 78:60–70.

———. 1979b. Weather and regulation of *Hypothyris euclea* (Lepidoptera: Nymphalidae: Ithomiinae) populations in northeastern Costa Rica. *J. Lepid. Soc.* 34:36–47.

———. 1980. Environmental partitioning in lowland tropical rain forest cicadas. *J. N.Y. Entomol. Soc.* 88:86–101.

———. 1982a. Notes on the natural history of *Morpho granadensis polybaptus* Butler (Lepidoptera: Nymphalidae: Morphinae), and its relation to that of *Morpho peleides limpida* Butler. *J. N.Y. Entomol. Soc.* 90:35–54.

———. 1982b. *Population biology of tropical insects.* New York: Plenum Press.

———. 1983. *Fidicina mannifera.* Pp. 724–26 in *Costa Rican natural history,* ed. D. H. Janzen. Chicago: Univ. of Chicago Press.

Young, H. J. 1986a. Beetle pollination of *Dieffenbachia longispatha* (Araceae). *Am. J. Bot.* 73:931–44.

———. 1986b. Pollination of *Dieffenbachia longispatha* (Araceae): Effects of beetles on reproductive success, gene flow, and gender. Ph.D. diss., State University of New York, Stony Brook.

———. 1988a. Differential importance of beetle species pollinating *Dieffenbachia longispatha* (Araceae). *Ecology* 69:832–44.

———. 1988b. Neighborhood size in a beetle pollinated tropical aroid: Effects of low density and asynchronous flowering. *Oecologia* 76:461–66.

———. 1990. Pollination and reproductive biology of an understory Neotropical aroid. Pp. 151–64 in *Reproductive biology of tropical forest plants,* ed. K. S. Bawa and M. Hadley. Paris: UNESCO.

Young, T. P., and S. P. Hubbell. 1991. Crown asymmetry, treefalls, and repeat disturbance of broad-leaved forest gaps. *Ecology* 72:1464–1471.

Zamora, N. 1989. *Flora Arborescente de Costa Rica.* Cartago: Editorial Tecnológico de Costa Rica.

Zapata, J. R., and M. T. K. Arroyo. 1978. Plant reproductive ecology of a secondary deciduous tropical forest in Venezuela. *Biotropica* 10:221–30.

Zimmerman, B. L., and M. T. Rodrigues. 1990. Frogs, snakes, and lizards of the INPA-WWF Reserves near Manaus, Brazil. Pp. 426–54 in *Four Neotropical rainforests,* ed. A. H. Gentry. New Haven, Conn.: Yale Univ. Press.

Zimmerman, P. R., J. P. Greenburg, and C. E. Westberg. 1988. Measurements of atmospheric hydrocarbons and biogenic emission fluxes in the Amazon boundary layer. *J. Geophys. Res.* 93:1407–16.

Zug, G. 1983. *Bufo marinus (sapo grande, sapo, giant toad, marine toad).* Pp. 386–87 in *Costa Rican natural history,* ed. D. H. Janzen. Chicago: Univ. of Chicago Press.

Contributors

K. S. Bawa
Department of Biology
University of Massachusetts
Boston, Massachusetts 02125

James H. Beach
Harvard University Herbaria
Museum of Comparative Zoology
22 Divinity Avenue
Cambridge, Massachusetts 02138

H. Elizabeth Braker
Department of Biology
Occidental College
Los Angeles, California 90041

William A. Bussing
Escuela de Biología y Centro de
    Investigación en Ciencias del Mar y
    Limnología (CIMAR)
Universidad de Costa Rica
Costa Rica

Rebecca P. Butterfield
Organización para Estudios Tropicales
Apartado 676-2050
San Pedro
Costa Rica

Robin L. Chazdon
Department of Ecology
    and Evolutionary Biology
University of Connecticut
Storrs, Connecticut 06268

Deborah A. Clark
Organización para Estudios Tropicales
Apartado 676-2050
San Pedro
Costa Rica

R. K. Colwell
Department of Ecology
    and Evolutionary Biology
University of Connecticut
Storrs, Connecticut 06269-3042

Julie Sloan Denslow
Department of Ecology, Evolution
    and Organismal Biology
Tulane University
New Orleans, Louisiana 70118

Philip J. DeVries
Museum of Comparative Zoology
Harvard University
Cambridge, Massachusetts 02138

Maureen A. Donnelly
Department of Biology
University of Miami
Coral Gables, Florida 33124

Ned Fetcher
Department of Biology
University of Puerto Rico
P. O. Box 23360
San Juan, PR 00931-3360

G. W. Frankie
Department of Integrative Biology
University of California
Berkeley, California 94720

Rodrigo Gámez
Centro de Investigación
    en Biología Celular y Molecular
Universidad de Costa Rica
Ciudad Universitaria 2060
Costa Rica

Harry W. Greene
Museum of Vertebrate Zoology and
    Department of Integrative Biology
University of California
Berkeley, California 94720

Craig Guyer
Department of Zoology and Wildlife
    Science
Auburn University
Auburn, Alabama 36849-5414

Barry E. Hammel
Missouri Botanical Garden
St. Louis, Missouri 63166-0299

Gary S. Hartshorn
World Wildlife Fund
1250 24th Street, NW
Washington, D.C. 20057

Henry A. Hespenheide
Department of Biology
University of California
Los Angeles, California 90024-1606

W. John Kress
Department of Botany, NHB-166
Smithsonian Institution
Washington, D.C. 20560

Douglas J. Levey
Department of Zoology
223 Bartram Hall
University of Florida
Gainesville, Florida 32611

Milton Lieberman
Department of Biology
University of North Dakota
Grand Forks, North Dakota 58202-
8238

Diana Lieberman
Department of Biology
University of North Dakota
Grand Forks, North Dakota 58202-
8238

Jeffrey C. Luvall
NASA
John C. Stennis Space Center
Stennis Space Center, Mississippi
39529

Rafael Mata Ch.
Facultad de Agronomía
Universidad de Costa Rica
San José
Costa Rica

Lucinda A. McDade
Department of Ecology and
Evolutionary Biology
University of Arizona
Tucson, Arizona 85721

Robert J. Marquis
Department of Biology
University of Missouri, St. Louis
8001 Natural Bridge Road
St. Louis, Missouri 63121-4499

Timothy C. Moermond
Department of Zoology
University of Wisconsin
Madison, Wisconsin 53706

Florencia Montagnini
Yale University
School of Forestry
and Environmental Studies
370 Prospect Street
New Haven, Connecticut 06511

L. E. Newstrom
Department of Entomology and
Parasitology
University of California
Berkeley, California 94720

Steven F. Oberbauer
Department of Biological Sciences
Florida International University
Miami, Florida 3319

Gordon H. Orians
Department of Zoology and
Institute for Environmental Studies
University of Washington
Seattle, Washington 98195

Pia Paaby
Organizació para Estudios Tropicales
Apartado 676-2050
San Pedro
Costa Rica

Geoffrey G. Parker
Smithsonian Environmental
Research Center
P.O. Box 28
Edgewater, Maryland 21037

Eugenie Phillips
Instituto Nacional de
Biodiversidad
Apartado 22-3100
Santo Domingo de Heredia
Costa Rica

Freddy Sancho M.
Cornell University
Ithaca, New York 14853

Robert L. Sanford, Jr.
Department of Biological Sciences
University of Denver
Denver, Colorado 80208

Phillip Sollins
Department of Forest Science
Oregon State University
Corvallis, Oregon 97331

F. Gary Stiles
Instituto de Ciencias Naturales
Universidad Nacional de Colombia
Bogotá
Colombia

Robert M. Timm
Museum of Natural History and
    Department of Systematics & Ecology
University of Kansas
Lawrence, Kansas 66045-2454

Robert L. Wilbur
Department of Botany
Duke University
Durham NC 27706

PART-TITLE PHOTOGRAPH LEGENDS

Special thanks to all photographers who generously offered their work, to those who bravely took rolls of black and white film into the jungle, and to those who helped with the legends. You all know who you are!

All legends are clockwise from upper right.

INTRODUCTION

Upper right. Duplex cabins at La Selva provide housing for long-term researchers. Each building has two private rooms upstairs and shared bath below. Windows on all sides ensure good ventilation and (reasonably) comfortable temperatures. Five of these cabinas are located in the research clearing (i.e., the pejibaye grove planted by Holdridge) on the West Bank.—D. E. Stone

Lower Right. Researchers and other visitors now arrive at La Selva via the East Bank where the dining hall (shown here) and housing for short-term visitors are located. These buildings were completed in the mid-1980s. The young tree behind the welcome sign is *Ochroma lagopus, balsa,* a very fast growing member of the Bombacaceae. The wood is very light and useful for a variety of purposes including model airplanes.—D. E. Stone

Lower left. The forest at La Selva is accessible via a network of trails. The most heavily used of these have been planked since the mid-1970s in an effort to protect them from erosion and widening (which occurs when people go around wet spots). The planks are covered with hardware cloth for traction. In the early 1990s, the planking on the most heavily used trails was replaced with cement sidewalks, which are longer lived and will permit access by disabled persons.—R. J. Marquis

Upper left. In 1982, a 100 m long suspension footbridge was completed to link the newer East Bank facilities to *old* La Selva on the west bank of the Puerto Viejo river. The walkway is some 14 m above the river (except during floods) and incidentally serves as a fine place from which to view the riparian habitat: *Brycon* fish waiting for fruits beneath fig trees, turtles sunning on logs, the occasional river otter, kingfishers moving up and downstream.—D. E. Stone

PART I

Upper right. La Selva is drained by several small stream systems that are tributaries of the Puerto Viejo, Sarapiquí and Peje rivers (see chapter 3).—J. S. Denslow

Lower center. Researchers C. Pringle (right) and F. Triska (left) at work on stream insects in a tributary of the Río Santo Domingo at ca. 2000 m elevation in Braulio Carrillo National Park behind La Selva. In the tropics, roots that grow into the flowing water of streams can provide important habitats for many benthic organisms.—G. Dimijian

Center left. View of the Río Puerto Viejo from the suspension footbridge that connects the east bank to the buildings and reserve on the west bank of the river.—R. J. Marquis

Upper left. Surface of a leaf cutter ant next, *Atta cephalotes.* The nests are mounds that can be several meters in diameter and often have several entry ways. The ants remove most vegetation from the nest area.—R. J. Marquis

PART II

Upper right. Piper sancti-felicis (Piperaceae) is a common understory shrub at La Selva. In fact, with nearly 50 species, *Piper* is the most species-rich genus at La Selva (see chapter 6). This and other species of *Piper* have been the subject of extensive research on population biology (see chapter 7) and herbivory (see chapter 21). Note the erect, pencil-like inflorescences. The shape is characteristic of the family, but species vary in the orientation of these structures, which may be erect or pendulous, and may also change orientation as flowers and then fruits develop.—R. J. Marquis

Center right. The monkey ladder vine, *Bauhinia guianensis* (Fabaceae) is a large woody liana at La Selva. The biology of this plant is described in chapter 6 (p. 87).—R. J. Marquis

Lower right. The understory of primary forests at La Selva is characterized by an abundance of palms (Arecaceae) and cyclanths (Cyclanthaceae), which look superficially similar at least vegetatively. Their flowers and fruits are quite different, however, with palms looking like typical monocots and cyclanths looking like a cross between a plant and a pre-Columbian artifact.—J. S. Denslow

Lower left. Strangler figs (*Ficus,* Moraceae) usually begin life as epiphytes, developing woody roots and stems that eventually surround and strangle the host.—R. J. Marquis

Upper left. Socratea exorrhiza is a "fish tail palm" (i.e., the segments of the leaves have uneven, scalloped edges like the margins of fish fins). It is also notable for its stilt roots, which may raise the trunk of the tree 2 m or more above the ground. These palms are common at La Selva (see chapters 6, 7, 8).—J. S. Denslow

PART III

Upper right. The original Holdridge farmhouse at Finca La Selva was built in 1956. In the early 1960s, the first story of the building was walled in to provide for a kitchen and dining hall sufficient for groups. A minimalist laboratory with researcher housing above was built to the west of and perpendicular to the original house in 1975. Both buildings have since been extensively remodeled and rechristened as the River Station to serve mostly as researcher housing.—L. R. Holdridge

Center right. Brown leaf katydid (Orthoptera: Tettigoniidae) with forewings that mimic dead leaves. Note the "veins" in the "leaf" and even a "hole" in the "leaf" to mimic herbivore damage.—C. W. Rettenmeyer

Lower right. The Tamandua, *Tamandua mexicana,* is a small anteater that is reasonably common at La Selva. Two other anteaters are native to La Selva. The Giant anteater, *Myrmecophaga tridactyla,* originally occurred at La Selva but has not been seen since the mid-70s and is presumed extirpated. *Cyclopes didactylus,* the Silky anteater, is nocturnal, arboreal and seen only rarely.—C. W. Rettenmeyer

Lower left. Army ants (*Eciton burchelli* and several other species) are common at La Selva (sometimes even invading the buildings where their effect on pest populations makes them almost welcome). In their foraging columns, they are fully capable of spanning gaps with their own bodies, and their bivouacs, which protect the queen and brood, have "walls" of solid ants.—C. W. Rettenmeyer

Upper left. Caterpillars of species of *Automeris* (Lepidoptera: Saturniidae) have urticating hairs, presumably as an antipredator defense.—R. J. Marquis

PART IV

Upper right. At the base of the petioles of *Cecropia* leaves, there are pads of tissue that produce glycogen containing food bodies (Mullerian bodies) for *Azteca* ants that inhabit the plant. These ants have a protective function: they have a nasty bite (coupled with a squirt of formic acid), and plants that lack them suffer demonstrably more herbivory than those occupied by ants.—C. W. Rettenmeyer

Center right. Leaf cutter ants, *Atta cephalotes,* at work. The ants cut pieces of leaves and often flowers from selected trees that may be some distance from their nests. These pieces are then carried to their nests and "fed" to the fungal colony main-

tained by the ants. Pieces of the fungus are in turn used as food by the ants.—S. J. Krasseman

Lower right. Papilionid caterpillar mimicking bird dropping, presumably as an antipredator defense mechanism.—C. W. Rettenmeyer

Lower left. Eulaema meriana (Apidae: Euglossini) scraping the scent chemical cineole from a paper bait. Bees in this group are known for their foraging habits, which involve "trap-lining" among distantly spaced flowers, often of the orchid family. Their habit of scent collecting has been described for some time but it is still uncertain how the scents are used by the bees.—C. W. Rettenmeyer

Center left. The green iguana, *Iguana iguana,* is a common largely arboreal lizard at La Selva. Egg-laying occurs during the dry season, when females must come to the ground. While on the ground, females are vulnerable to predators and, in fact, constitute a significant portion of the diet of jaguars during the dry season (see chapter 20).—R. J. Marquis

Upper left. The giant black ant, *Paraponera clavata* (Formicidae: Ponerinae), is common at La Selva. Its sting is so painful that its common name in Costa Rica (*bala*) means bullet. Despite their ferocious sting, these ants seem to be largely nectarivorous.—C. W. Rettenmeyer

PART V

Upper right. Settlement developed by the Costa Rican Instituto de Desarollo Agrícola (IDA), El Búfalo. A number of farmers in this settlement are cooperating with the OTS Trials project and are establishing experimental plantations of promising tree species on their land.

Lower right. This plantation of *Vochysia guatemalensis* was established on a local farm by the Trials project (see chapter 25) in 1989. The trees are three years old in the photo. —R. Butterfield

Lower left. This footbridge over the Rio Sarapiquí provides access to the IDA settlement of El Búfalo. The Trials project has several experimental plots on the land of cooperating farmers in this settlement.—R. Butterfield

Center left. A remnant forest patch on an IDA settlement in Sarapiquí was replaced by this banana plantation a few months before this photo was taken.—R. Butterfield

Upper left. Plantation of *Vochysia guatemalensis* at 6.5 years since planting.—R. Butterfield

INDEX

Abracris, 265, 272
Abrams, S., x, 236
Acacia, 77, 276, 323, 325, 327, 368
Acalypha, 76, 177, 365
Acanthaceae, ix, 77, 79, 82–85, 162, 163, 164, 166, 168, 169, 172, 174–175, 350, 360
Acanthosceloides, 269
Acarina, 207, 213, 239, 261, 278
Acción Nacional de Trabajo, 301
Accipiter, 220, 385
Accipiters. *See* Accipitridae
Accipitridae, 220, 223, 226, 235, 285, 385
Achiote. See *Bixa*
Aciotis, 178, 370
Ackerman, J. D., 172, 173, 182
Ackery, P. R., 187
Acosta-Solís, M., 261
Acraeinae (subfamily of Nymphalidae), 188, 189
Acrididae, 246, 249, 264–266, 272, 278
Acridoidea (superfamily of Orthoptera), 185, 259, 264–266, 278
Actinidiaceae, 360
Adams, R. M., 238
Adelia, 75, 112, 177, 365
Adiantaceae, 79, 267
Adiantum, 79, 267, 351
Adis, J., 267
Administration, La Selva, 345–346
Adzuko bean. See *Vigna*
Aechmea, 173, 354–355
Aegiphila, 182, 378
Africa, 21, 25, 30, 32, 61, 81, 82, 96, 97, 102, 117, 135, 137, 146, 158, 196, 198, 215, 232, 241, 242, 272, 277, 287, 288, 321, 323, 330, 333, 335, 336. *See also specific regions and countries*
African oil palm (*Elaeis*), 310
Agalychnis, 200, 204, 246, 380
Agaonidae, 155, 281
Agavaceae, 311, 312
Agency for International Development (AID), U.S., 12, 239, 306, 321
Agonostomus, 197, 198, 380
Agouti. See *Dasyprocta*
Agouti (pacas), 122, 229, 230, 234, 235, 262, 269, 396
Agoutidae, 396
Agriculture, vii, viii, ix, 4, 217, 263, 297–300, 301, 303, 306–318, 324, 337, 338, 341
Agrilinae (subfamily of Buprestidae), 239, 241–242
Agromyzidae, 280
Aichinger, M., 208

AID. *See* Agency for International Development
Aide, T. M., 104, 153, 261, 267, 268
Akunda, E., 144, 158
Alajuela Province, Costa Rica, 86, 314
Albizzia, 326, 328, 368
Alcedinidae, 223, 388
Alchornea, 77, 85, 365
Alexandre, D. Y., 287
Aleyrodidea (superfamily of Homoptera), 279
Alfaro, 196, 197, 379
Ali, S. A., 292
Almeda, F., 85, 350
Alouatta, 231, 233, 234, 235, 255, 262, 272, 286, 395
Alpizar, E., 321
Alston, E. R., 235
Alternanthera, 175, 360
Altos de Pipe, Venezuela, 171
Alvarado, G., 19, 25, 26, 27, 37, 39, 45, 55, 59, 62
Alvarado, I., 44, 92, 299, 306
Alvarez-Buylla, E. R., 127
Alverson, W. S., 57, 85, 172, 175, 176, 182, 350
Alvim, P. de T., 158
Alvim, R., 158
Alydidae, 279
Amaranthaceae, 175, 265, 360
Amaryllidaceae, 77, 173, 353
Amastridium, 382
Amazilia, 223, 387
Amazon River, 7, 32, 164, 196, 197, 198, 219, 286, 308
Amazon Region, 56, 60, 61, 79, 81, 82, 100, 164, 210–212, 233–234, 239, 241, 243, 330, 331
Ameiva, 210, 214, 382
American Museum of Natural History, 209, 294
American Ornithologists' Union (AOU), 393
Amphibians, 199–209, 220, 239, 240, 241, 243, 245, 252, 254, 285, 343, 344, 380–381. *See also entries for orders, families, and genera*
Amphidasya, 86, 375
Amphisbaenidae, 211
Amydria, 232
Anacardiaceae, 142, 175, 286, 361
Ananas (pineapple), 312, 355
Anartia, 248, 254
Anatidae, 385
Anaxogorea, 74, 137, 175, 361
Ancyluris, 193

Andean region, South America, 82, 217, 230. *See also entries for specific countries*
Anderson, A., 331
Anderson, J. A. R., 116
Anderson, J. M., 52, 60, 123
Anderson, S. D., x, 237
Anderson, W. R., 85, 350
Andira, 146, 151, 153, 156, 158, 177, 369
Andreae, M. O., 56
Andreae, T. W., 56
Andrews, C. C., 280
Andrews, R. M., 213–214, 215, 246, 248, 334
Angiosperms, 282, 353–378. *See also entries for families and genera*
Anguidae, 212, 382
Anhingidae, 384
Aniba, 81
Aniliidae, 212
Annelida, 58, 63
Annex A (La Selva), 9, 60, 76
Annex B (La Selva), 10
Annis, S., 318
Annona, 85, 86, 175, 361
Annonaceae, 74, 77, 80, 82, 84, 85, 86, 88, 137, 146, 147, 149, 162, 163, 164, 166, 167, 172, 175, 326, 328, 350, 361
Anoline lizards, 255. See also *Norops*
Anolis, 213, 334. See also *Norops*
Anomalepidae, 212
Anomospermum, 135, 371
Anotheca, 200
Anseriformes, 385
Antbirds, 219, 221, 223, 226, 228, 243, 255, 284. *See also* Formicariidae
Anteaters. *See* Myrmecophagidae
Antelopes, 287
Anthomyiidae, 281
Anthomyzidae, 280
Anthophoridae, 164, 281
Anthurium, 79, 80, 81, 85, 173, 220, 284, 292, 353–354
Antilles. *See* West Indies
Antonio, T., 172, 179, 182
Ants. *See* Formicidae
Antvireos, 228. *See also* Formicariidae
Antwrens, 226, 228, 334. *See also* Formicariidae
Anura (frogs and toads), 199–202, 204, 207, 238, 240, 241, 250, 252, 253, 331, 380–381. *See also entries for families and genera*
AOU. *See* American Ornithologists' Union
Aotus, 395, 397
Apatelodidae, 280

Apaturinae (subfamily of Nymphalidae), 188, 189
Apeiba, 76, 181, 377
Aphanotriccus, 218, 390
Aphelandra, 162, 174, 360
Aphidoidea (superfamily of Homoptera), 279
Apiaceae, 378
Apocynaceae, 82, 175, 269, 270, 328, 361
Apodidae, 223, 228, 244, 387
Apodiformes, 387
Apoidea (superfamily of Hymenoptera, bees), 87, 88, 155, 156, 161–170, 220, 255, 281. *See also entries for families and genera*
Appanah, S., 144, 151, 152, 153, 155, 170
Apples, 307
Apterygota, 278
Aquatic ecology 17, 27–32
Aquifoliaceae, 79, 175, 327, 361
Ara, 92, 225, 226, 269, 386
Araceae, 73, 74, 76–81, 83, 84, 85, 88, 92, 93, 122, 126, 161–167, 172, 173, 220, 265, 284, 292, 308, 310, 312, 313, 315, 350, 353–354
Arachnida, 88, 207, 208, 220, 251, 252
Araliaceae, 74, 76, 175, 283, 284, 286, 361
Aramidae, 285, 386
Araneida, 88, 207, 208, 220, 252
Araucaria, 327
Araucariaceae, 327
Arawacus, 192
Arboretum, Holdridge (La Selva), ix, 6, 75, 77, 100, 347
Archaeoprepona, 268
Archer, S., 270
Arctictis, 287
Arctiidae, 192, 193, 280
Arctogalidia, 287
Ardeidae, 226, 384
Ardisia, 77, 88, 179, 286, 372–373
Arecaceae, 9, 11, 57, 58, 74–80, 82, 83, 86, 87, 95, 100, 102, 108, 112, 113, 115, 118, 122, 125, 132, 133, 134, 139, 140, 154, 161–167, 172, 173, 176, 189, 215, 217, 220, 231, 253, 267, 269, 270, 273, 276, 283, 286, 291, 300, 307, 308, 310–315, 334, 359–360
Arenal, Costa Rica, 309
Argentina, 335, 398
Argidae, 281
Arias, O. (president of Costa Rica), 12
Aristolochia, 85, 361–362
Aristolochiaceae, 82, 85, 361–362
Armadillos. *See* Dasypodidae
Armbruster, W. S., 85, 154, 155
Armesto, J., 127
Army ants, 221, 241, 251, 255
Aroids. *See* Araceae
Arnold, S. J., 215
Arremon, 221, 224, 228, 284, 292, 392
Arroyo, J., 155
Arroyo, M. T. K., 163, 165, 166, 169, 170, 171
Artaxo Neto P., C. Q., 56
Arthropods, 208, 212, 214, 220, 225, 239, 240, 251, 252, 266, 331, 332. *See also entries for classes, orders, families, and genera*
Artibeus, 234, 286, 395, 397
Artiodactyla, 233, 262, 269, 272, 397

Artocarpus, 310, 315, 372
ASBANA (Association for Banana Production), Costa Rica, 311
Asclepiadaceae, 362
Ashton, P. S., x, 34, 52, 67, 70, 72, 106–116, 118, 127, 142, 144, 151, 152, 153, 155, 158, 161, 168, 170, 172, 330, 337
Asia, 73, 81, 170, 171, 287, 288, 321, 335, 336. *See also entries for specific regions and countries*
Asilidae, 241
Aspleniaceae, 79, 267. *See also* Polypodiaceae
Asplundia, 74, 80, 85, 87, 122, 174, 355
Assasin bugs. *See* Reduviidae
Asteraceae, 78, 79, 80, 82, 83, 84, 136, 175, 265, 364
Asterogyne, 57, 58, 74, 87, 95, 122, 132, 134, 139, 140, 173, 273, 359
Astrocaryum, 74, 75, 76, 95, 102, 139, 165, 173, 286, 359
Astyanax, 196, 197, 198, 379
Ateles, 231, 234, 235, 262, 286, 395
Atherinella, 197, 198, 379
Atherinidae, 197, 198, 379
Atlantic lowlands region: Costa Rica, 7, 8, 84, 86, 87, 195, 218, 230, 231, 242, 247, 297, 299, 301, 307–316, 322, 333; México and Central America, 84, 86, 217, 218, 230. *See also entries for specific countries*
Atlantic Zone Programme, CATIE and Wageningen Agricultural University, 316
Atractosteus, 195, 196, 304
Atsatt, P. R., 192, 273
Atta, 58, 63, 122, 255, 259, 267, 273, 274, 277, 281, 325
Attelabinae (subfamily of Curculionidae), 268
Attila, 228, 390
Attine ants. *See Atta*
Atwood, J. T., 83, 84, 350
Auburn University, 216
Audu, M., 138
Auer, B., 316
Auerbach, M. T., 119, 252, 263, 266, 267, 279, 280
Augspurger, C. K., 104, 114, 118, 120, 125, 135, 138, 142, 144, 150, 151, 153, 156, 157, 158, 160, 270, 271, 276, 289, 290, 294, 330
Auriculardisia, 85, 179, 372, 373
Austin, M. P., 330
Australia, 242, 267, 272, 287, 330, 331, 333, 336
Auer, B., 316
Automeris, 248
Automolus, 219, 228, 388
Averrhoa, 312, 373
Aves, 87, 162, 163, 169, 170, 215, 238–241, 243–246, 249, 250, 252, 253, 254, 255, 259, 262, 268, 269, 277, 282–286, 288–294, 330, 332, 333, 336, 337, 343, 344, 384–393. *See also entries for orders, families, genera, and common names*
Awaous, 197, 379
Azteca, 193

Bactris, 9, 11, 74, 75, 78, 173, 308, 312–315, 359
Bactris gasipaes, 9, 11, 74, 78, 308, 312–315, 359

Bactroid palms, 265. *See also entries for genera*
Baham, J., 46
Baillie, I. C., 34, 330, 331
Baker, H. G., 142, 143, 153, 154, 156, 158, 161, 169, 261, 266, 270, 284
Baker, I., 153, 159, 192
Bakus, G. J., 240
Baldwin, I. T., 273, 276, 277
Balick, M. J., 267
Balsaminaceae, 78, 362
Bamboo, 160, 312
Bambusa, 312, 356
Bananas. *See Musa*
Banco Nacional de Costa Rica (BNCR), 307
Banco National Park, Ivory Coast, Africa, 21, 25, 61
Barahona R., F., 301
Barbee, A., x
Barbets, 219
Bard, K. M., 209
Baridinae (subfamily of Curculionidae), 263
Barinaga, M., 209
Barquero, A. I., 96, 103, 108
Barra del Colorado, Costa Rica, 299
Barra del Colorado Wildlife Refuge, Costa Rica, 304
Barranca, Costa Rica, 300
Barro Colorado Island, Panama, 8, 21, 52, 68, 70, 78, 79, 83, 88, 91, 92, 95, 96, 97, 100, 102, 103, 104, 117, 120, 121, 122, 124, 125, 130, 135, 136, 137, 150, 170, 171, 185, 188, 189, 194, 200–202, 204, 205, 211–215, 219, 220, 227, 233, 234, 236, 239–242, 247, 248, 259, 261, 262, 263, 267, 268, 277, 286, 290, 291, 330, 331, 333–337, 345
Barry, R. G., 20, 21
Barthélémy, D., 158
Barthell, J., 159
Barton, A. M., 100, 128, 248
Barva (Volcano), Costa Rica, 6, 11, 79, 322, 333
Basiliscus, 210, 262, 286, 382
Basilisk lizard. See *Basiliscus*
Bassaricyon, 262, 287
Batáan-Matina, Costa Rica, 312
Bates, H. W., 3
Bat flies. *See* Streblidae
Bats. *See* Chiroptera
Battus, 188
Batzli, G. O., 272
Bauhinia, 87, 96, 368
Bawa, K. S., ix, 3, 71, 72, 91, 97, 103, 105, 115, 119, 142, 143, 144, 151, 152, 153, 156, 157, 159, 161, 162, 163, 165–182, 277, 332, 344
Bazilevich, N. I., 63
Bazzaz, F. A., 60, 125, 129, 136, 139
BCI. *See* Barro Colorado Island
BCNP. *See* Braulio Carrillo National Park
BDFFP. *See* Biological Diversity of Forest Fragments Project, Manaus, Brazil
Beach, J. H., ix, 92, 161–167, 169, 172–182, 332
Beaman, R. S., 170
Beans, 299, 302, 309, 311, 312, 313, 315
Beard, J. S., 120
Bears (Ursidae), 230, 287

Beattie, A. J., 276, 282, 288
Beaver, R. A., 332, 264
Becker, P., 124, 270, 271, 276, 287
Becker, P. F., 135, 136
Becker, R. A., 143
Bedell, H. G., 88
Beef. *See* Cattle; Pasture
Bees. *See* Apoidea; *and entries for families and genera*
Beehler, B., 287
Beetles. *See* Coleoptera
Begonia, 175, 362
Begoniaceae, 175, 362
Bejuco (El) Station, Costa Rica, 200, 304
Bejuco Stream (La Selva), 56, 59
Belize, 192, 235, 242, 253
Bell, W. J., 249
Belsky, A. J., 270
Belt, T., 3
Benson, W. W., 188, 189, 264, 274, 276
Bentley, B. L., 56, 136, 137, 273, 274, 275
Bentley, S., 270
Berenbaum, M. R., 272, 278
Berg, C. C., 147, 155
Bergmark, C., 195, 196
Berlyn, G., 91, 101
Bernard-Reversat, F., 25
Bernier, G., 142, 157, 158
Bertiera, 180, 375
Bertsch, F. H., 59, 308, 316
Berytidae, 279
Besleria, 74, 177, 366
Best, R. C., 232
Bethel, J. S., 121, 135, 142
Bibionidae, 280
Bien, A. R., 56, 119, 194
Bierregaard, R. O., Jr., 331, 336
Bignoniaceae, 79, 80, 82, 83, 150, 175, 320, 325, 326, 328, 362
Binkley, D., 50
Biological Diversity of Forest Fragments Project, Manaus, Brazil, 8
Biotropica, 253–254
Birch, H. F., 62
Birds. *See* Aves
Birds of paradise (Paradisaeidae), 287
Birds of prey. *See* Raptors
Bishop, J. E., 32, 198
Bixa, 312
Bixaceae, 312
Bjorkman, O., 122, 133
Black, R. W., 249
Black-faced Grosbeak, 246
Blackie, J. R., 27
Black pepper. See *Piper nigrum*
Blake, J. R., 217, 220, 221, 222, 225, 250, 255, 284, 287–293, 344
Blakea, 85, 370
Blakely, N. R., 266
Blakey, A., 194
Blance, C. A., 321
Blattoidea, 249
Blue Mountains, Jamaica, 169, 171
BNCR. *See* Banco Nacional de Costa Rica
Boa, 210, 382
Boardman, N. K., 133, 141
Bocconia, 78, 373
Boggs, C. L., 187, 247, 250, 251, 254
Boidae, 211, 382

Bolaños, A. L., x
Bolitoglossa, 200, 206, 380
Bolitoglossini, 200
Bomarea, 77, 353
Bombacaceae, 57, 75, 78, 82, 85, 100, 120, 121, 126, 131, 132, 134, 135, 138, 139, 140, 151, 158, 162, 163, 171, 175, 274, 276, 283, 286, 323, 326, 328, 350, 362
Bombacopsis, 328
Bombak River, Malaysia 30
Bombycidae, 280
Bonaccorso, F. J., 285, 286, 291, 293
Bongers, F., 125, 138
Boppré, M., 191
Boraginaceae, 9, 74, 75, 77, 82, 131, 134, 135, 137, 138, 176, 276, 286, 320, 323, 325, 326, 327, 362–363
Borchelt, R., 209
Borchers, J., 46
Borchert, R., 135, 142, 144, 146, 152, 157, 158, 159
Borden Company, 313
Borges, R., 270
Boring, L., 269
Bormann, F. H., 91, 101
Bornemisza, E., 53, 308, 316
Borneo, 170, 206
Borowicz, A., 293
Borror, D. J., 281
Bossert, W. H., 247, 251
Bostrichidae, 279
Bothriechis, 253, 382
Bothrops, 216, 382
Boucher, D. H., 120, 286
Bourgeois, W. W., 35, 36, 54, 108, 124
Bourlière, F., 233
Bowers, M. D., 191
Box, J. D., 272
Boyce, M. S., 244
Bracher, G., 53
Brachiaria, 314, 356
Brachyglenis, 191
Brachyrhaphis, 196, 197, 379
Brachys, 242
Bradbury, J. W., 231, 246, 249, 250, 253
Brady, N. C., 35
Bradypodidae, 262, 396
Bradypus (Three-toed sloth), 231, 232, 235, 255, 262, 296
Braithwaite, R. W., 331
Braker, (H.) E., 185, 211, 246, 249, 250, 252, 253, 254, 259, 262–268, 273, 278
Brame, A. H., Jr., 200, 210
Bramocharax, 196, 197, 379
Brandani, A., 100, 121, 126, 248, 330
Brassicaceae, 78, 187, 365
Brassolinae (subfamily of Nymphalidae), 188, 189, 264
Braulio Carrillo National Park (BCNP), Costa Rica, vii, 5, 6, 7, 11–14, 27, 62, 85, 87, 88, 92, 127, 187, 188, 190, 210, 225, 226, 230, 231, 235, 243, 250, 255, 276, 301, 303–306, 321, 322, 333, 343
Bravaisia, 77, 175, 360
Bray, J. R., 267
Brazil, 79, 117, 208, 209, 232, 234, 253, 262, 263, 320, 330, 331, 334, 336
Bread fruit. See *Artocarpus*
Breitsprecher, A., 135, 142

Breitswich, R., 246
Brenes, D., 119
Brentidae, 279
Brinson, M. M., 30
Briskie, J. V., 246
Broadhead, E., 238
Brockman, C. F., 322
Brokaw, N. V. L., x, 100, 116, 120–128, 220, 293, 330
Bromeliaceae, 79, 80, 81, 83, 162, 173, 203, 205, 211, 245, 248, 252, 253, 312, 354–355
Bronchart, R., 158
Bronstein, J. L., 155, 156
Broojimans, W. J. A. M., 302
Brooke, A. P., x, 231, 232, 236, 246
Brooks, D. R., 254
Brosimum, 77, 138, 179, 286, 327, 328, 372
Brosset, A., 286, 287
Brower, L. P., 190, 191
Brown, C., 194
Brown, G. J., 231
Brown, J. H., 243, 336
Brown, J. L., 245, 333
Brown, K. S., Jr., 191, 194
Brown, S., 57
Brown, W. H., 99
Browne, F. G., 277
Bruchidae, 242, 269, 270, 279
Brues, C. T., 267, 281
Brunei, 170
Bryant, J. P., 274
Brycon, 196, 197, 198, 286, 379
Bryconamericus, 196, 379
Bucconidae, 217, 223, 228, 247, 388
Bucerotidae (hornbills), 287
Bucher, T. H., 209
Buchholz, R., 285, 293
Buchmann, S. L., 174, 182
Buckley, D. P., 72
Budowski, G., 9, 73, 120, 322
Bufo, 200, 202, 203, 206, 207, 209, 245, 380
Bufonidae, 201, 202, 203, 207, 380
Bull, J. J., 194
Bullock, S. H., x, 90, 91, 95, 97, 102, 103, 105, 115, 135, 142, 147–148, 151, 152, 156–163, 165, 166, 168, 170–182, 269, 283, 290, 333
Bunchosia, 178, 370
Bunnell, P., 204
Buol, S. W., 34, 62
Buprestidae, 155, 239, 241–242, 279, 281
Burcham, J., 28, 29, 195, 196, 197
Burger, W., 81, 87, 93, 166, 172, 176, 179, 180, 182
Burley, F. W., 321, 323
Burley, N., 332
Bursera, 58, 77, 131, 134, 135, 176, 315, 363
Burseraceae, 58, 74, 75, 76, 77, 81, 82, 108, 131, 134, 135, 176, 283, 286, 315, 363
Busby, B., 140
Buschbacher, R. J., 321
Bushdog (*Speothos*), 230
Bushmaster. See *Lachesis*
Buskirk, R. E., 213
Buskirk, W. H., 213, 225, 284
Bussing, W. A., 29, 185, 195, 196, 197
Buteo, 221

Butterfield, R., 53, 237, 269, 297, 298, 320, 323, 325, 328
Butterflies, 156, 163, 164, 167–170, 185, 187–194, 238, 239, 241, 247–250, 252, 259, 263, 264, 265, 268, 277, 278, 280, 331, 332, 346. *See also entries for families, subfamilies, and genera*
Byrne, M. M., 289, 293
Byrsonima, 86, 178, 370
Byttneria, 259, 264, 275, 377
Byturidae, 279

Cabassous, 396
Cacao. See *Theobroma cacao*
Cachan, P., 128
Cacho Negro Volcano, Costa Rica, 11
Cacicus, 220, 221, 225, 228, 393
Caciques, 228. *See also entries for genera*
Cactaceae, 83, 176, 363
Cadle, J. E., 204, 212
Caecilians. *See* Gymnophiona *and entries for families and genera*
Caeciliidae, 201, 380
Caesalpinaceae, Caesalpinoideae. *See* Fabaceae
CAF. *See* Certificado de Abonos Forestales
Caiman, 210, 215, 380
Calathea, 74, 76, 80, 92, 121, 122, 174, 220, 231, 269, 357
Calder, I. R., 27
Calderón, C. E., 86, 174, 182
California, U.S., 264 , 265
Calkin, H. C., 125
Callitrichidae, 288
Calophyllum, 323, 325, 326, 327, 367
Caluromys, 262, 287, 394
Calyptrogyne, 173, 359
CAMCORE. *See* Central American and Mexican Coniferous Resources Cooperative
Camacho, P., 320
Campanulaceae, 363
Campbell, A. F., 231, 285
Campelia, 173, 355
Campephilus, 218, 388
Camponotus, 193
Campylorhynchus, 247, 391
Canada, 319, 327, 398
Canadian Embassy (in Costa Rica), 327
Cañas, Costa Rica, 211–214, 239
Canet, G., 319, 327
Canham, C. D., 121, 127
Canidae, 230
Canis, 230
Canna, 173, 355
Cannaceae, 166, 173, 355
Cantharidae, 279
Capparaceae, 75, 176, 363
Capparis, 75, 176, 363
Caprifoliaceae, 78, 363
Caprimulgidae, 223, 387
Caprimulgiformes, 387
Caprio, J. M., 142
Capsicum, 312, 314, 377
Capybaras, 253
Carambola (*Averrhoa*), 312, 373
Carapa, 75, 76, 91, 97, 100, 102, 103, 108, 127, 131, 133, 138, 269, 298, 321, 327, 371
Carapa Creek, La Selva, 28

Carazo, R. (president of Costa Rica), 11
Carcharhinidae, 197, 379
Carcharhinus, 197, 379
Cardamom. See *Elletaria*
Cardinalinae (subfamily of Emberizidae), 285
Cariari, Costa Rica, 312
Caribbean lowlands region, Central America and México. *See* Atlantic lowlands region; *and entries for specific countries*
Caribbean Sea, 55, 215
Caribbean white bats. See *Ectophylla*
Cariblanco, Costa Rica, 299
Carica, 312, 363
Caricaceae, 97, 103, 146, 149, 150, 151, 157, 162, 165, 176, 269, 312, 363
Carlana, 197, 198, 379
Carludovica, 77, 87, 122, 174, 355
Carnivora, 230, 231, 233, 234, 236, 262, 284–287, 334, 397
Carollia, 231, 232, 234, 286, 395
Carolliinae (subfamily of Phyllostomidae), 285–286, 395
Carpenter, C. J., 56, 136, 137, 273
Carpenter, G. D. H., 192
Carpodectes, 221, 390
Carposinidae, 280
Carpotroche, 74, 177, 366
Carr (Archie) Postdoctoral Fellowship, 294
Carroll, C. R., 273, 275, 276
Carter A., x
Caryophyllaceae, 78, 363
Caryothraustes, 221, 228, 247, 284, 392
Casearia, 76, 77, 81, 104, 112, 116, 177, 269, 271, 289, 291, 292, 366
Casebeer, R. S., 230, 231
Casey, J. J., 309, 310
Cassava. See *Manihot*
Cassia, 149, 151, 177, 269, 275, 368
Cassipourea, 74, 180, 375
Castertine, G., 159
Castilla, 77, 100, 299, 327, 372
Castro, M., 301, 302, 306
Castro, Y., 132
Caswell, E. C., 263
Cates, R. G., 264, 271, 276
Cathartes, 221
Cathartidae, 223, 284, 285, 385
Catharus, 221, 223, 284, 385
CATIE. *See* Centro Agronómico Tropical de Investigación y Enseñanza
Cats. *See* Felidae
Cattle, 297, 301, 302, 307, 309, 311, 313, 314, 318, 321, 324
Caudata (salamanders), 199, 200, 201, 202, 208, 380
Caura River, Venezuela, 32
Caviomorph rodents, 287
Cayaponia, 85, 365
Cayenne, French Guyana, 120, 287, 293
Cebidae, 235, 262, 285, 286, 288, 395
Cebus, 231, 233, 234, 235, 262, 286, 395
Cecidomyiidae, 280
Cecropia, 76, 77, 78, 97, 100, 101, 111, 120, 121, 126, 131, 132, 283, 286, 363, 372
Cecropiaceae, 75–80, 82, 84, 97, 100, 101, 111, 120, 121, 126, 131, 132, 140, 276, 283, 286, 350, 363
Cedrela, 135, 179, 299, 322, 327, 371
Ceiba, 134, 151, 158, 175, 276, 283, 362

Celastraceae, 363
Celestus, 382
Censo Agropecuario, Costa Rica, 301
Central America, 82–85, 87, 88, 190, 195, 196, 199, 200, 202, 211, 212, 213, 216–220, 222, 223, 229, 230, 235, 241, 244, 284, 285, 291, 301, 316, 317, 321, 322, 330, 331, 344, 381, 398. *See also entries for specific regions and countries*
Central American and Mexican Coniferous Resources Cooperative (CAMCORE), 323
Central American Fuelwood Project, CATIE, 323
Central Valley, Costa Rica, 9, 13, 309, 318, 319
Centro Agronómico Tropical de Investigación y Enseñanza (CATIE), Costa Rica, 308, 314, 316, 320, 322, 323, 326
Centrolenella, 200, 206, 380
Centrolenidae, 200, 201, 204, 208, 380
Centropomidae, 197, 379
Centropomus, 197, 379
Cephaelis, 80, 180, 375. See also *Psychotria*
Cephaloleia, 251
Cephidae, 281
Cerambycidae, 239, 279
Cercopithidae, 288
Cerro Coronel, Costa Rica, 88
Cerro de la Muerte, Costa Rica, 211–214
Certificado de Abonos Forestales (CAF), Costa Rican Forestry Bonds, 319
Cervidae, 229, 262, 285, 287, 397
Cespedezia, 180, 373
Cestrum, 181, 377
Chabot, B. F., 131
Chabot, J. F., 131
Chachalacas (*Ortalis*), 285, 385
Chacon, I. A., 194
Chaetura, 228, 387
Chagres River, Panama, 233
Chai, P., 187, 191, 192, 194
Chalcidae, 242, 251, 281
Chamela Biological Station, México, 170, 171, 240
Chamaedorea, 77, 164, 173, 359
Chan, H. T., 144, 151, 152, 155, 158, 170, 171
Chapman Fund, 294
Chapman, J. A., 262, 263
Characidae, 196, 197, 198, 286, 379
Charadriidae, 386
Charadriiformes, 386
Charaxinae (subfamily of Nymphalidae), 188, 189
Charles-Dominique, P., 120, 286, 287
Charnov, E. L., 332
Chavarría, M. M., 7
Chavarría, R., ix
Chavarría, V., 44, 300, 306
Chazdon, R. L., 68, 69, 95, 120–123, 125, 127, 129–134, 139, 140, 141, 158, 265, 267, 273, 278
Checklist of North American Birds, 393
Chelobasis, 251
Chelonethidae, 251
Chelonia. *See* Testudinata
Chesson, P. L., 126
Chevrotains, 287
Chew, F. S., 188, 247
Chiarello, N., 128, 135, 136

Chickens. *See* Poultry
Chilamate, Costa Rica, 86, 200, 302, 304
Chilis (*Capsicum*), 312, 314
Chione, 75, 375
Chironimidae, 251
Chironius, 382
Chiroptera, 87, 88, 156, 162, 163, 164, 167, 168, 170, 229–234, 236, 239, 240, 241, 249, 250, 252, 253, 282–285, 287–288, 291, 292, 293, 394–395
Chivers, D. J., 288
Chloranthaceae, 79, 176, 363
Chloroceryle, 225
Chlorophanes, 88, 228, 392
Chloropidae, 280
Chlorothraupis, 226, 228, 392
Chocó region, Colombia (and Panama), 81, 219
Chocolate. See *Theobroma cacao*
Choloepidae, 262, 396
Choloepus (two-toed sloth), 235, 255, 262, 396
Choo, G. M., 272
Chorley, R. J., 20, 21
Christensen, N. L., 116
Chrotopterus, 229, 394
Chrysobalanaceae, 80, 81, 82, 85, 86, 176, 363
Chrysomelidae, 263, 266, 279
Chu, K., 275
Churchill, H. W., 78, 83, 84, 350
Cicadoidea (superfamily of Homoptera), 279
Cicadas, 249, 250. *See also* Cicadoidea
Cichlasoma, 195–198, 379
Cicindelidae, 241
Cichlidae, 195–198, 379
Ciconiidae, 384
Ciconiiformes, 384
Cifuentes, M., 304
Cimbicidae, 281
Cionosicyos, 85, 365
Cipollini, M. L., 284
Cissia, 188
CITES. *See* Convention on International Trade in Endangered Species
Citrus, 273, 307, 310, 312, 376
Clarisia, 87, 372
Clark, D. A., x, 6, 26, 37, 53, 68, 69, 70, 90, 91, 94–104, 114, 116, 118, 125, 137, 138–139, 141, 142, 151, 157, 158, 172, 214, 216, 237, 240, 268, 270, 271, 276, 277, 279, 280, 282, 289, 290, 331, 333, 344, 347
Clark, D. B., x, 6, 8, 14, 53, 70, 90, 91, 94–105, 114, 116, 118, 125, 137, 138–139, 141, 142, 151, 157, 158, 172, 214, 216, 237, 240, 268–271, 273, 276, 277, 279, 280, 282, 289, 290, 331, 333, 344, 347
Clark, G. C., 192
Clark, P. J., 110, 112
Clauson, B. L., 231, 237
Clelia, 210, 382
Clethra, 363
Clethraceae, 79, 363
Cleveland, W. S., 143
Clibadium, 175, 364
Clidemia, 77, 80, 85, 165, 171, 178, 276, 370–371
Climate: general, 19–20, 329, 334, 337; La

Selva and vicinity, 6, 17, 18, 20–27, 32–33, 62, 73–74, 129–131, 134–135, 143, 147, 148, 266, 308
Clusia, 77, 80, 85, 87, 88, 176, 220, 286, 367
Clusiaceae, 74, 77, 79, 80, 82–85, 87, 88, 171, 176, 220, 286, 323, 325, 326, 327, 350, 367
Clusiella, 88, 367
Cnemidophorus, 214
Coates-Estrada, C., 286
Coati(mundi). See *Nasua*
Coccidae, 275–276
Coccoidea (superfamily of Homoptera), 279
Cocha Cashú, Perú, 8, 234, 330, 331, 334, 336, 337
Cochlearius, 226
Cochlospermaceae, 151
Cochlospermum, 151
Cochylidae, 280
Cockroaches, 249
Cocoa. See *Theobroma cacao*
Coconuts. See *Cocos*
Cocos, 299, 307, 310, 311, 312, 315, 359
CODESA (Corporación Costarricense de Desarrollo), Costa Rica, 301
Codonanthe, 275, 276, 366
Cody, M. L., 242, 244, 251, 266, 332, 334, 335
Coen, E., 19, 202, 261
Coendou, 262, 396
Coffea, 9, 299, 307, 309, 311, 312, 315, 375
Coffee. See *Coffea*
Cole, B. J., 155
Cole, D. W., 56, 57, 58, 62
Coleman, D. C., 263
Coleophoridae, 280
Coleoptera, 87, 153, 154, 155, 161, 163, 164, 165, 167–170, 172, 207, 238, 239, 241–242, 251, 252, 259, 263, 266–270, 277, 279, 281
Coley, P. D., 125, 261, 267, 268, 272, 273, 274, 276, 277, 278
Coliadinae (subfamily of Pieridae), 188, 189
Collared peccaries. See *Tayassu tajacu*
Collembola, 278
Colobine monkeys, 272
Colocasia, 312, 313, 354
Colombia, 56, 80, 220, 230, 308, 398
Colonia, 220, 390
Colorado, U.S., 265
Colostethus, 205
Colubridae, 211, 212, 382
Colubrina, 76, 180, 375
Columba, 219, 223, 224, 284, 287, 386
Columbidae, 218, 219, 245, 262, 284, 285, 287
Columbiformes, 262, 386
Columbina, 223, 386
Columnea, 177, 366
Colwell, R. K., 144, 146, 147, 153, 156, 160, 239
Combe, J., 320
Combretaceae, 63, 77, 132, 138, 176, 323, 325–328, 363
Commelinaceae, 86, 88, 162, 163, 164, 172, 173, 355
Compositae. *See* Asteraceae
Compsoneura, 97, 103, 146, 148, 149, 150, 156, 157, 158, 165, 170, 179, 372

Condon, M., 269
Congo River, Africa, 196, 198
Coniophanes, 382
Connaraceae, 104, 364
Connarus, 104
Connell, J. H., 70, 104, 106, 114, 118, 119, 252, 261, 264, 266, 270, 271, 282, 331
Connor, E. F., 199
Conservation, La Selva and vicinity, 9–12, 303–306, 317
Conostegia, 178, 371
Consejo Nacional de Producción (CNP), Costa Rica, 312, 313
Contopus, 221, 252, 390
Convention on International Trade in Endangered Species (CITES), 235
Convolvulaceae, 364–365
Coombe, D. E., 138
Cooper, S. M., 272
Cooperación en los Sectores Forestal y Maderero (COSEFORMA), Costa Rica, 326
Cooper-Driver, G. A., 276
Copepoda, 251
Copiocerinae (subfamily of Acrididae), 265
Copra (*Cocos*), 299
Coraciiformes, 388
Corallus, 382
Corapipo, 250, 284, 390
Corbet, S. A., 175, 182
Corcovado National Park, Costa Rica, 73, 188, 189, 191, 200–202, 209, 241, 267
Cordero, A., 59
Cordero S., R. A., 136
Cordia, 9, 74, 75, 77, 131, 134, 135, 137, 138, 176, 276, 286, 320, 323, 325, 326, 327, 362
Cordillera Central, Costa Rica, 73, 217, 230, 308
Cordillera de Talamanca, Costa Rica, 200, 308
Cordillera de Tilarán, Costa Rica, 200, 230
Cordillera Volcánica Central Biosphere Reserve, 5, 7, 14, 297, 299, 304, 305, 306
Cordillera Volcánica Central Forest Reserve, 301, 305
Córdova Cassillas, B., 103, 104
Coreidae, 279
CORENA Project, 321
Corimelaenidae, 279
Corinto River, Costa Rica, 322
Corlett, R. T., 288
Corn. See *Zea*
Corn, M. J., 215
Cornejo, F. H., 135, 158
Cornell, H. V., 263
Corner, E. J. H., 147, 282
Correlliana, 88
Cort, R., 276
Corvidae, 223, 247, 285, 390
Coryphotriccus, 228, 390
Corytophanes, 382
COSEFORMA. *See* Cooperación en los Sectores Forestal y Maderero
Cosmopterigidae, 280
Cossidae, 280
Costa Rican Natural History, 230, 238, 341, 342
Costaceae, 162, 163, 164, 166, 168, 173–174, 275, 355
Costus, 173–174, 275, 355

Cotinga, 390
Cotingas. *See* Cotingidae
Cotingidae, 219, 223, 247, 284, 285, 287, 289, 390
Cottrell, C. B., 192
Couepia, 81, 363
Courtney, S. P., 293
Coussarea, 85, 165, 375
Coville, R. E., 245, 252
Cowpeas. See *Vigna*
Cox, P. A., 63
Coyne, J., 277
Coyotes (Canidae), 230
Cracidae, 285, 385
Crane hawks, 253
Crawley, M. J., 270
Crax, 92, 284, 285, 385
Crease, T., 279
Crematogaster, 275
Cressa, C., 32
Crinum, 173, 353
Crisp, J. E., 153, 165, 169, 179
Croat, T. B., 78, 83, 85, 88, 91, 282, 283, 284, 290, 291, 332, 350
Crocodilia, 210, 211, 213, 214, 215, 253, 382
Crocodilians. *See* Crocodilia
Crocodylidae, 382
Crocodylus, 210, 382
Cromartie, W. J., Jr., 273
Crome, F. H. J., 287, 290, 291
Crops, 302, 304, 307–316. *See also entries for specific crops*
Crossley, D. A., Jr., 267
Crother, B. I., 209
Croton, 177, 192, 365–366
Crotophaga, 246, 247
Crowson, R. A., 281
Crozat, G., 56
Cruciferae. *See* Brassicaceae
Crump, M. L., x, 202, 205, 208, 209, 246
Crutzen, P. J., 56
Cruz Alençar, J. da, 142
Cryosophila, 140, 164, 165, 173, 359
Cryphocricos, 251
Cryptochloa, 174, 356
Ctenidae, 208
Cuckoos. *See* Cuculidae
Cuculidae, 223, 224, 228, 247, 386–387
Cuculiformes, 386–387
Cucurbitaceae, 82, 84, 85, 365
Culver, D. C., 282
Cupania, 181, 376
Cuphea 29, 77, 370
Cupiennius, 208
Curassows. See *Crax*
Curculionidae, 239, 263, 268, 269, 279
Cushing, C. E., 30
Cutler, D. F., 192
Cyanerpes, 228, 392
Cyanobacteria, 137
Cyanocompsa, 228, 392
Cyanocorax, 247, 390
Cyatheaceae, 78, 351
Cybianthus, 88, 373
Cycadaceae, 353. *See also* Zamiaceae
Cyclanthus, 92, 122, 162, 174, 273, 355
Cyclanthaceae, 29, 73, 74, 77, 80, 83, 84, 85, 87, 92, 122, 135, 161–167, 174, 273, 276, 350, 355

Cydnidae, 279
Cymbopetalum, 85, 175, 361
Cynipidae, 281
Cynodon, 314, 356
Cyperaceae, 81, 83, 84, 355–356
Cyperus, 81, 355–356
Cyphomandra, 181, 377
Cyphorhinus, 228, 391
Cypseloides, 228, 387

Dacnis, 88, 228, 392
Dagg, M., 27
Dairy farming (milk production), 302, 309, 313–314
Dalbergia, 77, 177, 322, 325–328, 369
Dalceridae, 280
Dalechampia, 85, 366
Damery Parrish, J., 277
Danaea, 76, 79, 92, 93
Danainae (subfamily of Danaidae), 188, 189, 191, 264
Danaus, 189
Danielson, B. J., 333
Danilevskii, A. S., 189
Daptrius, 225, 385
D'Arcy, W. G., 57, 88
Darién Province, Panamá, 60, 61, 86, 204
Darnell, R. M., 197
Darwin, C., 3
Dasypodidae, 122, 234, 253, 286, 396
Dasyprocta (agoutis), 122, 229, 230, 233, 234, 235, 255, 262, 269, 286, 287, 396
Dasyproctidae, 262, 285, 396
Dasypus, 122, 234, 253, 286, 396
Daubenmire, R., 134, 158, 212
Davidar, P., 178, 182, 292
Davidson, D. W., 336
Davies, A. G., 272, 277
Davis, D. R., 232
Davis, M., 194
Davis, R. H., 191
de Abate, J., 210
Dee Lite, T., 194
Deer. *See* Cervidae
de la Cruz, M., 267
del Amo R., S., 120
DeLapp, J., 267
Delmas, R., 56
Delonix, 143
de los Angeles Molina, M., 119
Demmig, B., 141
Dendrobates, 200, 203–209, 240, 245, 246, 248–252, 254, 255, 380
Dendrobatidae, 201, 202, 205, 207, 208, 246, 380
Dendrocincla, 228, 388
Dendrocolaptes, 228, 389
Dendrocolaptidae (woodcreepers), 218, 219, 221, 223, 226, 228, 284, 388–389
Dendropanax, 74, 76, 175, 283, 284, 286, 361
Dendrophidion, 382
Denman, K., 159
Denno, R. F., 187, 189, 278
Denslow, J. S., x, 35, 36, 38, 44, 45, 51, 52, 53, 56, 59, 60, 62, 63, 68, 69, 70, 91, 94, 99, 105, 116, 120–127, 129, 132, 135, 136, 138, 139, 141, 158, 172, 217, 220, 225, 232, 248, 269, 270, 273, 274, 276,

277, 283, 284, 287–292, 327, 330, 336, 344
Departamento de Ingeniería Forestal, Costa Rica, 109
Department of Entomology, University of California, Berkeley, 159
Depranidae, 280
de Sa, R. O., 203
Desmarest's spiny pocket mouse. See *Heteromys*
Desmodontinae (subfamily of Phyllostomidae), 285, 395
Desmodus, 231, 250, 395
Desmoncus, 80, 359
Desmopsis, 175, 361
DeSteven, D., x, 92, 104, 125, 270, 288, 293
Detling, J. K., 270
Deuth, D. A., 275
DeVries, P. J., 185, 187–193, 238–242, 249, 250, 264, 277, 280
DGEC. *See* Dirección General de Estadísticas y Censos
DGF. See *Dirección General Forestal*
Diaethria, 191
Dialyanthera. See *Otoba*
Diamond, A. W., 334
Diamond, J., 266, 287
Dichapetalaceae, 87, 176, 268, 365
Dichapetalum, 87, 176, 268, 365
Dichorisandra, 86, 88, 173, 355
Dickson, C. G. C., 192
Dicots, 265, 294, 360–378. *See also entries for families and genera*
Dicotyles (Tayassu tajacu), 262
Dicranopygium, 29, 85, 174, 355
Didelphidae, 234, 235, 253, 285, 286, 287, 394
Didelphis, 234, 262, 394
Dieffenbachia, 77, 85, 92, 122, 162, 173, 354
Dieterlen, F., 142, 144, 151, 152
Dietrich, W. E., 25, 27, 261
Diggle, P. J., 106
Dilleniaceae, 82, 96, 365
Dilodendron, 328
Dinerstein, E., 283, 286, 293
Dingle, H., 266
Dioptidae, 280
Dioptinae, 191
Dioscoreaceae, 356
Diphysa, 315
Diploglossus, 382
Diptera, 156, 163, 164, 165, 167, 168, 169, 170, 232, 239, 241, 251, 274, 275, 277, 280–281
Dipterocarpaceae, 99, 116, 117, 142, 150, 152, 154, 155, 158, 170, 171, 172, 206, 271, 336, 337
Dipterocarpus, 170, 171
Dipterodendron. See *Dilodendron*
Dipteryx, 63, 75, 76, 91, 97–104, 118, 131, 132, 134, 135, 137, 138, 146, 149, 150, 157, 162, 163, 164, 166, 177, 268, 269, 271, 286, 323, 325, 326, 327, 331, 369
Dirección General de Estadísticas y Censos (DGEC), Costa Rica, 301, 307, 311
Dirección General Forestal (DGF), Costa Rica, 307, 312, 316–324, 326, 327
Dirzo, R., 125, 267, 270, 271, 290
Disk-winged bats, 230

Dismorphiinae (subfamily of Pieridae), 188, 189
Dixon, J. R., 211
Dobat, K., 163
Dobkin, D. S., 167
Dobzhansky, T., 244, 250, 266
Dock, C. F., 213, 214
Doctors van Leeuwen, W. M., 292
Dodson, C., 81, 82, 86, 172
Dole Fresh Fruit Company, 310
Doliocarpus, 96, 365
Domínguez, C. A., 270, 290
Donnelly, M. A., 137, 138, 187, 189, 200, 203, 204, 205, 207, 208, 216, 239, 240, 245, 246, 248–252, 255, 278, 344
Dorosoma, 195
Dos Pinos Company, 313
Double-toothed kites, 253
Douglas, M. M., 187
Douglasiidae, 280
Doutt, R. L., 261
Doves, 284
Downum, K. R., 277
Dracaena, 311
Dracontium, 173, 354
Dressler, R. L., 88, 173, 182
Drew, R. A. I., 293
Drewry, G. E., 204
Driprodontidae, 287
Drosophilidae, 281
Dry forest habitats (tropical), 245, 248, 253, 261, 266–267, 275, 333. *See also entries for specific sites and regions*
Dryas, 188, 247
Drymarchon, 382
Drymobius, 382
Drymonia, 80, 366
Dubautia, 136
Dubost, G., 287
Ducula, 287
Dudal, R., 35
Dudley, T. R., 191, 194
Duellman, W. E., 199, 200, 202, 203, 204, 206, 208, 210, 211, 331, 381
Dulcedo, 189
Dunham, A. E., 215, 254
Dunn, E. L., 336
Dunn, G. E., 19, 20
Durango State, México, 262, 263
Dussia, 177, 369
Duval, J., 128
Dwyer, J. D., 85
Dysithamnus, 218, 225, 228, 389
Dysmicoccus, 275
Dystovomita, 74, 367

Eagles, 235
EARTH (Escuela Agropecuaria para la Región Tropical Húmeda), Costa Rica, 308, 326
Earth Resources Laboratory, NASA, U.S., 108, 119
Earthworms, 58, 63
East Bank (La Selva), 6, 10, 13, 21, 78
Eastern region, U.S., 121, 127, 188, 189, 242, 243
Echenopa, 264
Echeveria, R., 159
Echimyidae, 229, 262, 285, 396–397

Ecological Reserve (La Selva). *See* Reserva Ecológica Rafael Chavarría
Ecology, 253–254
Ecology of a Tropical Forest, The, 241
Ectophylla, 230, 231, 240, 395
Ecuador, 79, 81, 86, 87, 196, 200–202, 204, 205, 208, 211–214, 219, 220, 227, 381
Edentata, 229, 233, 236, 241, 262, 285, 396
Edentates. *See* Edentata
Edwards, J., 261
Edwards, P. J., 267
Ehrlich, P. R., 192, 267
Eira, 235, 292, 286, 287, 397
Eisenberg, J. F., 230, 233, 286
Eisenmann, E., 217, 219, 227, 239, 241, 243
Elachistidae, 280
Elaeocarpaceae, 75, 82, 131, 286, 350, 365
Elaeis, 310, 359
Elanoides, 284
Elateridae, 279
Electron, 228, 388
Eleotridae, 197, 379
Elephantidae, 287
Elephants. *See* Elephantidae
Eleutherodactylus, 200, 201, 203–204, 206–208, 240, 249, 255, 380–381
Ellenberg, H., 73
Elletaria, 312, 313, 314
Ellington, C. P., 191
Ellison, A. M., 125
El Niño Southern Oscillation (ENSO), 333, 335, 336
Elopidae, 304, 379
El Salvador, 242, 266
Elton, C. S., 240, 249, 251
El Verde, Puerto Rico, 4, 61
Elwood, J. W., 30
Elytrostachys, 80, 356
El Yunque, Puerto Rico, 337
Emballonuridae, 231, 246, 394
Emberizidae, 245, 247, 285, 287, 289, 391–393
Emberizinae, 284, 285, 289
Emerson, S. B., 207
Emlen, S. T., 244
Emmel, T. C., x, 264
Emmel, J. F., 264
Emmons, L. H., 230, 233, 253, 282, 287, 288, 290
Emydidae, 262, 285, 382
Empidonax, 217, 221, 390
Enders, R. K., 262, 263
Endler, J. A., 191, 244, 254
Enright, N. J., 102
Enterolobium, 328
Entylia, 245
Enulius, 382
Environmental Education and Community Relations Program (OTS) 14, 315
Epermeniidae, 280
Epidendrum, 80, 81, 358
Epipedobates, 205
Epiphyllum, 176, 363
Episcia, 177, 367
Erard, C., 286, 287
Erbessa, 191
Erechtites, 78, 364
Erethizontidae, 229, 230, 262, 285, 396
Ericaceae, 79, 365

Eriocraniidae, 279
Eriophyidae, 278
Ernest, K. A., 268, 279
Ernst, C. H., 262, 263, 286
Erwin, T. L., 190, 239, 241, 249
Erythrina, 77, 79, 163, 169, 177, 327, 369
Erythrolamprus, 382
Erythroxylaceae, 365
Eschweilera, 81, 178, 368
Escuela Agropecuaria para la Región Tropical Húmeda (EARTH), Costa Rica, 308, 326
Espinoza C., M., 323, 325, 328
Esquina Stream (La Selva), 44, 60, 86
Esquivel, C., 159
Estabrook, G. F., 291, 292
Esthemopsis, 191
Estrada, A., 286
Estrella, Costa Rica, 310
Etheridge, K., 231
Eucalyptus, 267, 315, 320, 326, 327, 331
Eugenia, 179, 312, 373
Euglossinae, 88, 156, 163, 164, 168, 255
Eulechriops, 269
Eumaeus, 191, 268, 277
Eumasticidae, 278
Euphonia, 220, 228, 284, 292, 392
Euphorbiaceae, 74–80, 82–86, 97, 98, 100, 101, 112, 138, 162, 163, 165, 177, 192, 286, 302, 308–313, 315, 323, 325, 326, 327, 350, 365–366
Eurema, 188
Europe, 299, 308, 309, 310, 329, 335, 336, 343. *See also entries for specific countries*
Eurypygidae, 386
Eurytomidae, 281
Euterpe, 74, 75, 140, 164, 165, 173, 231, 359
Evans, F. C., 110, 112
Evans, G. C., 128, 138
Evans, G. C., 137
Evans, J., 320
Ewel, J. J., 11, 46, 54, 60, 120, 212
Eye-lash pitvipers (*Bothriechis*), 253

Fabaceae, 52, 56, 58, 63, 68, 73–83, 86–91, 96–104, 108, 112, 113, 114, 116, 117, 118, 121, 125, 130, 131, 132, 134–141, 143, 146, 149, 150, 151, 153, 156, 157, 158, 160, 162, 163, 164, 166, 169, 177, 188, 268–271, 274, 275, 276, 283, 286, 312, 315, 316, 320, 322, 323, 325–328, 331, 368–369
Faden, R. B., 86
Faeth, S. H., 266, 273
Fagaceae, 271
Fairchild, G. B., 239, 241
Falconer, D. S., 128, 245
Falconidae, 223, 226, 385
Falconiformes, 385
Falcons. *See* Falconidae
FAO. *See* Food and Agriculture Organization (U.N.)
Faramea, 85, 140, 180, 375
Far East, Asia, 73
Farquarson, C. O., 192
Fasehun, F. E., 138
Fassbender, H. W., 27
Fauth, J. E., 206
FDF. *See* Fondo de Desarrollo Forestal
Federov, A. A., 161, 168

Feener, D. H., Jr., 194, 274
Feeny, P., 264, 271, 272, 273, 276
Feer, F., 230
Feinsinger, P., x, 153, 154, 156, 168, 169, 172, 222
Felidae, 229, 233, 235, 253, 284, 285, 286, 397
Felis: concolor (puma, mountain lion), 233, 284, 397; *onca* (jaguar), 229, 235, 253 (as *Panthera*); *pardalis* (ocelot), 235, 397; *wiedii* (margay), 286, 397; *yagouroundi* (jaguarundi), 397
Fernández, G. R., 299
Fernández-Yépez, A., 195
Ferns. *See* Pteridophytes
Fetcher, N., 68, 69, 70, 91, 101, 120–123, 125, 127, 128, 129, 131–139, 248, 322
Ficus, 76, 77, 79, 80, 81, 143, 146–147, 155–156, 179, 197, 259, 283, 284, 286, 291, 372
Fidicina (Cicada), 249, 250
Field, C. B., x, 125, 139, 141
Figs. See *Ficus*
Fiji, 320
Finca Commelco. *See* Palo Verde Biological Station
Finches, 223, 284
Findley, J. S., 232, 243
Finnegan, B., 323
Fisher, E. M., 239, 241
Fisher, R. A., 332
Fishes, 77, 195–198, 208, 238, 240, 241, 254, 286, 299, 307, 379
Fishkind, A. S., 234
Fitch, H. S., 214, 231, 232, 235
Fitzpatrick, J. W., 245
Fitzpatrick, J. W., 284
Flacourtiaceae, 74, 76, 77, 81, 82, 83, 104, 112, 116, 177, 269, 271, 289, 291, 292, 326, 328, 366
Flea beetles. *See* Chrysomelidae
Fleming, T. H., 72, 94, 106, 155, 171, 231, 246, 250, 252, 253, 254, 262, 263, 270, 271, 282–286, 288, 291–294
Flies. *See* Diptera
Flores R., J. G., 319, 320
Flores Silva, E., 307
Florida, U.S., 312
Floristics (La Selva), 78–84
Floscopa, 173, 355
Flycatchers, 217, 223, 228, 284. *See also* Tyrannidae
Fogden, M. P. L., 216, 224, 225, 282
Fondo de Desarrollo Forestal (FDF), Costa Rican Forestry Fund, 319
Food and Agriculture Organization (U.N., FAO), 34, 35, 320
Forbes, W. T. M., 264
FORESTA (Forest Resources for a Stable Environment Project), 12, 316, 321
Foresta, H. de, 120, 273, 293
Forest falcons, 226
Forest industry, Costa Rica, 318–319
Forest structure, 68, 106–119, 220–221
Forestry, viii, ix, 11, 277–278, 297, 298, 303, 306, 307, 309, 314–328, 338, 341, 342
Forman, R. T. T., 56, 106, 116, 118
Formicariidae, 225, 389
Formicarius, 224, 389

Formicidae (ants), 192, 193, 207, 245, 249, 251, 255, 259, 274–277, 281, 331, 336; relationships with butterflies, 192–193; relationships with plants, 274–276
Foster, M. S., x, 224, 232, 246, 284, 288, 290, 291
Foster, R. B., 68, 70, 78, 90, 96, 102, 104, 115, 119, 120, 122, 124, 126, 127, 155, 160,, 261, 267, 293, 330, 331, 333, 335, 336
Fowler, H. P., 147
Fowler, S. V., 273
Fox, L. R., 267, 273
Foxes. *See* Canidae
Frank, S. A., 155, 156
Frankel, R., 164
Frankie, G. W., 90, 134, 142, 143, 144, 146, 147, 150–153, 155, 156, 158, 159, 160, 161, 167, 168, 169, 212, 213, 214, 220, 225, 226, 248, 249, 261, 282, 283, 290, 291, 344
Frankie, J., 159
Franks, N. G., 251
Franson, S. E., 330
Freckman, D. C., 263
Freemark, K. E., 220, 222, 293
Freese, C. H., 286
Fregatidae, 384
French Guyana, 120, 287, 293
Fringillidae, 218, 219, 262
Frith, C. B., 266
Frith, D. W., 266
Fritz, G., 207, 275
Fritz, G. N., 232
Fritz, R. S., 277
Frogs. *See* Anura
Frost, E. J., 269, 270, 290, 293
Frost, S. W., 281
Frugivory, 125, 282–294
Fruit pigeons, 287
Fulbright Hayes Fellowships, 194
Fulgoridae, 252
Fulgoroidea (superfamily of Homoptera), 279
Fundación para el Desarrollo de la Cordillera Volcánica Central (FUNDECOR), 321, 326
Fundación Neotrópica, 318
FUNDECOR. *See* Fundación para el Desarrollo de la Cordillera Volcánica Central
Fungi, 125, 281, 309–313. *See also* Mycorrhizae
Furch, K., 32
Furipteridae, 395
Furnariidae, 219, 245, 388
Fusarium, 309, 313
Futuyma, D. S., 264, 282
Fynbos, South Africa, 330

Gabon, Africa, 146, 288
Gaceta Oficial, Government of Costa Rica, 310
Gadgil, M., 247
Gage, D. A., 266, 277
Gaines, S. D., 240
Galago, 288
Galbulidae, 191, 223, 226, 388
Gall-forming insects, 242. *See also entries for families*
Galliformes, 223, 262, 385–386

Gallina, S., 262, 263
Galloway, J. N., 55, 56
Galun, E., 164
Gambia River, Africa, 30, 32
Gamboa, J., 194
Gámez, R., 338
Gan, Y.-Y., 172
Ganzhorn, J. V., 277
Gaps, treefall, 57, 62, 68, 69, 74, 91, 99–101, 104, 116, 120–129, 221, 249, 252, 259, 263, 266, 268, 270, 272–273, 275, 277, 290, 293, 322, 330, 343
Gar. *See Atractosteus*
Garber, P. A., 288
García-Barrios, R., 127
Garcinia, 171, 367
Gardella, D., 301
Gardner, A. L., 230, 262, 263
Garita, C. D., 317, 319, 321
Gartlan, J. S., 277
Gartner, B. L., 94
Garwood, N. C., 126
Gascon, C., 208–209
Gaston, K. J., 190, 194
Gastrophryne, 200, 203, 206, 207, 381
Gauld, I. D., 242
Gautier-Hion, A., 155, 286, 287, 288, 291, 292, 293
Gekkonidae, 212, 382
Gelechiidae, 280
Genipa, 180, 326, 327, 375
Gentianaceae, 366
Gentry, A. H., x, 3, 8, 67, 79, 82, 83, 86, 87, 144, 147, 150, 153, 156, 167, 172, 185, 241, 253, 282, 283, 284, 286, 330, 334
Geographic Information System (GIS), 12, 17, 18, 26, 29, 30, 345
Geometridae, 263, 280
Geomorphology (La Selva), 19, 25–28
Geomyidae, 396
Geonoma, 57, 74, 86, 95, 122, 132, 134, 139, 140, 173, 231, 273, 359–360
Geonomoid palms, 265, 267, 273. *See also entries for genera*
Geophila, 158, 180, 375
Geophis, 382
Georgia, U.S., 264, 265
Geotrygon, 221, 386
Geranospiza, 253
Gershenzon, J., 273
Gessel, S. P., 54, 56, 57, 58
Gewald, N. J., 320
Ghana, Africa, 97, 117, 135, 137
Gibbs, P. E., 178, 182
Giant Anteater. *See Myrmecophaga*
Gilbert, L. E., 156, 188, 189, 190, 194, 240, 243, 263, 264, 266, 270, 279, 282
Gill, A. M., 336
Gill, D. E., 246
Gillman, G., 34, 35, 46, 49, 51, 52, 53, 59, 124
Ginger (spice, *Zingiber*), 310
Gingers. *See* Zingiberaceae
GIS. *See* Geographic Information System
Givnish, T. J., 136
Glander, K. E., 262, 263

Glanz, W. E., 233, 241, 243, 286, 287, 291
Gleicheniaceae, 78, 351
Gliricidia, 315, 326, 328, 369
Gloeospermum, 182, 378
Glossophaga, 234, 395
Glossophaginae (subfamily of Phyllostomi-
dae), 395
Glyphipterigidae, 280
Glyphorhynchus, 224, 228, 389
Gmelina, 315, 320, 323, 325, 327
Gnatcatchers. See *Polioptila*
Gnetaceae, 80, 85, 131, 135, 353
Gnetum, 80, 85, 131, 135, 353
Gobiesocidae, 197, 379
Gobiesox, 197, 379
Gobiidae, 197, 198, 379
Gobiomorus, 197, 379
Godman, F. D. C., 185
Goethalsia, 76, 77, 96, 99–102, 112, 116, 131,
132, 181, 323, 327, 377
Golden Toad, 209
Goldschmidt, E. E., 142, 157, 158
Golfo Dulce Region, Costa Rica, 199. *See
also* Osa Peninsula; Rincón de Osa
Golley, F. B., 3, 60, 63
Golterman, H. L., 32
Gombak River, Malaysia, 32
Gómez D., A. E., 126
Gómez P., L. D., 73, 83, 185, 241, 242
Gómez-Pompa, A., 3, 70, 120, 169, 321, 323
Gong, W. K., 138
Gonzalagunia, 180, 375
González, G., 267
González, L., 301
González J., E., 53, 324, 326
González Vega, C., 311
Goodwin, D., 287
Goodwin, G. G., 230, 231
Gordon, B. L., 91
Gosse, J. P., 196
Gottsberger, G., 286
Goulding, M., 197, 198, 286
Governance, La Selva, 346–349
Gower, S. T., 30, 54
Gracillariidae, 280
Gradwohl, J., 225, 255, 277, 334
Graf, N. R. de, 322
Graham, A. W., 126
Grajal, A., 290, 292
Gramineae. *See* Poaceae
Gramzow, R. H., 21
Grapes, 307
Grasses. *See* Poaceae
Grasshoppers, 185, 249, 250, 252, 259, 266,
278, 332. *See also entries for families and
genera*
Grasslands, 330
Gray, G., 267, 278
Grayum, M. H., x, 73, 78, 79, 81, 83, 84, 85,
88, 103, 162, 172, 276, 350
Great Britain, 264, 265
Great Smoky Mountains National Park, U.S.,
122, 124
Greding, E. J., Jr., 203
Green, G., 240
Green, N. B., 199
Greenbaum, I. F., 231
Greenberg, R., x, 220, 222, 225, 226, 250,
255, 277, 334

Greene, H. W., x, 185, 194, 208, 210, 211,
215, 216, 231, 234, 252–255, 262, 263,
344, 346
Green iguanas. See *Iguana*
Greenland, P. J., 62
Greenlets. See *Hylophilus*
Green Revolution, 320
Gregory, P. T., 203
Greig, N., 94, 194, 269, 270, 274, 278, 279
Greig-Smith, P., 106, 116
Grias, 75, 368
Grimaldi, D., 194
Grimm, U., 27
Griswold, C., 245, 252
Griswold, G., 275
Groove-billed anis. See *Crotophaga*
Grosbeaks, 228, 246
Ground-doves, 229. *See also* Columbidae
Grove, K. F., 164–166, 172, 174, 175, 177,
178, 182, 269, 279
Grubb, P. J., 116, 131
Gruiformes, 386
Gryllacrididae, 278
Guácimo, Costa Rica, 312
Guácimo River, Costa Rica, 321
Guadaloupe Island, Antilles, 32
Guaiacum, 189
Guanacaste Province, Costa Rica, 54, 86, 117,
134, 143, 146, 151, 156, 158, 170, 171,
172, 185, 188, 189, 223, 232, 239, 241,
242, 248, 253, 261, 262, 263, 266–267,
277, 301, 309, 318, 319, 333, 334
Guans. See *Penelope*
Guápiles, Costa Rica, 35, 300, 314, 322, 326
Guapote, 196
Guarea, 74, 77, 97, 103, 146–149, 150, 156,
157, 158, 165, 179, 269, 276, 283, 284,
371
Guatemala, 196, 199, 219, 220, 227, 235, 242,
308, 323
Guatteria, 175, 361
Guatuso, Costa Rica, 307
Guazuma, 323, 328
Guerrant, E. O., 159
Guevara S., S. A., 126
Guilford, T., 192
Guindon, C. A., 247
Gulmon, S. L., 272, 274
Gunatilleke, C. V. S., 106
Gush, T. J., 293
Gutiérrez, T., 194
Gutschick, V. P., 136
Guttiferae. *See* Clusiaceae
Guttman, S. I., 264
Guyana, 235
Guyanas, 230. *See also entries for specific
countries*
Guyer, C., x, 199, 200, 201, 203, 208, 209,
210, 213, 214, 215, 239, 240, 246, 248–
252, 255, 344
Guzmania, 173, 355
Gymnocichla, 228, 389
Gymnophiona (caecilians), 199–203, 208,
255, 381
Gymnophthalmidae, 211, 212
Gymnopis, 199, 200, 202, 380
Gymnopithys, 228, 389
Gymnosperms, 78, 80, 85, 94, 95, 131, 133,
135, 138–139, 151, 157, 158, 240, 268,

277, 315, 320, 325, 326, 327, 353. *See
also entries for families and genera*
Gymnostinops, 221, 225
Gymnotidae, 197, 379
Gymnotus, 197, 379

Ha, C. O., 171
Haber, W. A., 147, 153, 155, 156, 159, 167,
181, 182, 188, 189
Habia, 225, 288, 392
Hackett, W. P., 157
Hades, 191
Hadfield, W., 138
Hadley, M., 3
Haemodoraceae, 174, 357
Haemulidae, 197, 379
Haffer, J., 219, 241, 334
Hafner, D. J., 230
Hafner, M. S., 230
Hahn, C. D., 106, 116, 118
Haines, B. L., 53, 59, 63
Hairstreak butterflies. *See* Lycaenidae
Halictidae, 164
Hall, C., 308, 309
Hall, J. B., 97, 119, 137
Hall, P., 159
Hallé, F., 142, 157
Hallwachs, W., 287
Hamelia, 144, 146, 147, 149, 150, 156, 158,
180, 268, 291, 375
Hamilton, A. C., 334
Hammel, B. E., 67, 73, 78, 81, 83, 84, 85, 88,
90, 91, 95, 162, 166, 172, 174–179, 181,
182, 332, 350
Hampea, 100, 131, 132, 138, 178, 277, 327,
362, 370
Hamrick, J. L., 72, 105, 165, 168, 171, 172
Hands, M., 53
Hansen, R. M., 265
Haplomys, 103
Harcombe, P. A., 54, 63
Harmon, M. E., 58
Harpagus, 228
Harper, J. C., 283
Harper, J. L., 270
Harpia (Harpy eagles), 225, 233, 235, 243,
244
Harris, L., 264
Harrison, A. D., 32,
Harrison, S., 125
Harriss, R. C., 56
Hartshorn, G. S., ix, 6, 9, 11, 12, 32, 53, 57,
63, 67–70, 73, 75, 77, 78, 80, 90, 91, 95–
104, 108, 110, 112–116, 119, 120, 122,
124–128, 134, 136, 158, 159, 232, 248,
267, 269, 270, 272, 275, 279, 280, 283,
301, 302, 306, 317, 320, 322, 330, 333,
343, 346
Harvey, D. J., 187, 194
Hasseltia, 177, 366
Hawaii, 46, 134, 136
Hawk moths. *See* Sphingidae
Hawks. *See* Accipitridae
Hayes, M. P., 205, 211, 216
Hazlett, D. L., 97, 99, 135
Heaney, L. R., 262, 263, 286, 287
Heath, J., 264
Hedley, M. J., 59
Hedyosmum, 176, 363

Heinen, J. T., 206, 209
Heinrich, B., 156
Heinselman, M. L., 124
Heinz (H. John III) Charitable Trust, 327
Heithaus, E. R., 106, 153, 154, 271, 282, 285, 286, 293
Heliconia, 92, 121, 122, 161, 167, 168, 174, 230, 231, 249, 250, 251, 252, 259, 266, 267, 273, 277, 357
Heliconiaceae, 78, 84, 92, 121, 122, 161–164, 166–169, 172, 174, 230, 231, 249–252, 259, 265, 266, 267, 273, 277, 283, 350, 357
Heliconiinae (subfamily of Nymphalidae), 188–191, 247, 248, 259, 263, 264
Heliconius, 156, 189, 190, 191, 247–252, 254, 255, 259, 263
Heliocarpus, 78, 100, 126, 131, 132, 135, 137, 138, 140, 181, 377
Heliodinidae, 280
Heliornis, 226, 386
Heliornithidae, 386
Heliozelidae, 279
Hemileucinae (subfamily of Saturniidae), 248
Hemiptera, 29, 250, 251, 269, 270, 278, 279
Henderson, A., 161, 164, 165, 166, 173, 182
Henderson, R. W., 216
Hendrickson, D. A., 200
Hendrix, S. D., 267, 270
Hendry, C. D., 55
Henicorhina, 228, 391
Henriettea, 178, 371
Henry, W. K., 21
Hepialidae, 279
Heredia Province, Costa Rica, 307. *See also* Puerto Viejo de Sarapiquí; Sarapiquí Region
Hermann, S. M., 216, 287
Hernandia, 77, 116, 177, 327, 367
Hernandiaceae, 77, 116, 177, 327, 367
Herpetacanthus, 175, 360
Herpetofauna. *See entries for orders, families and genera*
Herrania, 74, 181, 377
Herrera, C. M., 282, 288, 290, 292, 293
Herrera, R., 54, 55, 58
Hershey Chocolate Co., 10, 11
Hershkovitz, P., 230
Hespenheide, H. A., ix, 32, 105, 216, 220, 238–242, 264, 269, 274, 275, 276, 279
Hesperiidae, 187, 280
Heteromyidae, 229, 246, 248, 262, 270, 285, 286, 396
Heteromys, 231–232, 240, 246, 248, 250, 251, 254, 255, 261, 262, 269, 270, 286, 396
Heteropsis, 80, 354
Heteroptera, 278
Heterothripidae, 278
Heterotilapia, 197, 379
Heuveldop, J., 23, 25, 27
Hevea, 309, 315, 326, 327, 366
Heyer, W. R., 203
Heywood, J. S., 72
High elevation tropical habitats, 211–214, 245, 275. *See also entries for specific sites*
Hillia, 85, 375
Hillis, D. M., 209
Hilty, S. L., 282
Hippocrateaceae, 367

Hirtella, 86, 176, 363
Hirundinidae, 218, 219, 223, 225, 245, 285, 390
Hispine beetles, 92, 251, 252, 259, 263, 266, 277
History of La Selva: general, 8–13; pre-Columbian, 8, 308
Hladik, A., 262, 263, 283, 286, 288
Hladik, C. M., 262, 263, 283, 286, 288
Hoevers, L. G., 216
Hoffman, C. A., 273
Hoffmania, 85, 181, 375
Holbrook, N. M., 267
Holdridge, L. R., ix, 9, 11, 67, 73, 74, 80, 113, 218, 229, 299, 303, 322, 343
Holland, 312, 316
Holmes, R. T., 277
Homalomena, 85, 88, 354
Homoptera, 220, 242, 245, 249, 250, 252, 264, 275–276, 278, 279, 332
Honduras, 219, 230, 231, 242, 308, 323
Honeycreepers, 88, 228. *See also entries for genera*
Hoogmoed, M. S., 204
Hopea, 170, 171
Hopkins, M. S., 126
Hoplomys, 262, 397
Hoppes, W. G., 283, 290, 293
Horn, S. P., 8, 91
Hornbills (Bucerotidae), 287
Horquetas (Las), Costa Rica, 6, 10, 11, 13, 14, 300, 301, 302, 304, 306, 307, 314, 321, 349
Horvitz, C. C., x, 93, 127, 164, 174, 182, 192, 209, 270
Hovore, F. T., 129
Howard, J. J., 272, 277
Howe, H. F., 104, 106, 126, 246, 269, 270, 271, 276, 282, 284, 286–294
Howler monkeys. See *Alouatta*
Hubbell, S. P., 68, 70, 78, 90, 96, 102, 104, 106, 115, 116, 118, 119, 120, 122, 124, 126, 127, 271, 277, 293, 330, 331, 333
Huffaker, C. B., 273
Huft, M. J., 350
Hume, I., 287
Humiriaceae, 367
Hummingbirds. *See* Trochilidae
Humphrey, S. R., 293
Hunter, R., 11
Hunting, 234–235, 306
Huntington, E. I., 188, 189
Hura, 77, 366
Huston, M. A., 35, 99, 103, 138, 141, 322
Huxley, C. R., 192
Huxley, P. A., 144, 146, 157, 158
Hybanthus, 150, 270
Hydromorphus, 382
Hyeronima, 86, 97, 98, 100, 101, 177, 323, 325, 326, 327, 366
Hyla, 200, 202–206, 380
Hylaeogena, 242
Hylidae, 200–203, 208, 380
Hylocichla, 221, 391
Hyloctistes, 228, 388
Hylopezus, 221, 389
Hylophilus, 226, 228, 391
Hylophylax, 228, 389

Hymenolobium, 75, 97, 98, 100, 101, 103, 134, 177, 327, 369
Hymenophyllaceae, 78, 79, 351
Hymenoptera, 87, 88, 155–156, 161–170, 192, 193, 207, 220, 238, 242, 245, 249, 251, 252, 255, 259, 268, 274–277, 281, 331, 336
Hyphessobrycon, 379
Hypsipyla, 103, 269

Icacinaceae, 367
Ichneumonidae, 242
Icteridae, 223. *See also* Emberizidae
Icterinae, 225, 284, 285. *See also* Emberizidae
Icterus, 225, 393
IDA. *See* Instituto de Desarrollo Agrario
Iguana, 253, 262, 286, 382
Iguanidae, 246, 253, 262, 285, 286, 382
Ilex, 175, 327, 361
Imantodes, 204, 382
INBio. *See* Instituto Nacional de Biodiversidad
Incurvariidae, 279
INDACO. *See* Industrias de Desarrollo Agropecuario
India, 277
Indiana, U.S., 80, 264, 265
Indo-Malayan Region, 206, 287. *See also specific countries and sites*
Indonesia, 81. *See also specific islands*
Industrias de Desarrollo Agropecuario (INDACO), Costa Rica, 312
Inga, 77, 80, 81, 88, 102, 116, 125, 132, 274, 276, 325–328, 368
Inger, R. F., 206
Ingram, J. S. I., 52
Innis, G., 291
Insectivores, 233
Insects, 162, 163, 164, 167–170, 187–194, 207, 232, 238–242, 245, 246, 252, 254, 259, 261, 268, 270, 274, 275, 278, 284, 293, 310, 312, 332, 333, 334. *See also entries for orders, families, and genera*
Institut Français de Recherche Scientifique pour le Développement en Coopération (ORSTOM), 120
Instituto de Desarrollo Agrario (IDA), Costa Rica, 299, 301–304, 306–309, 311–314, 317, 320
Instituto de Tierras y Colonización (ITCO), Costa Rica, 301, 304, 320
Instituto Interamericano para Ciencias Agrícolas (Inter-American Institute for Agricultural Sciences [IICA]), 73
Instituto Nacional de Biodiversidad (INBio), Costa Rica, 12, 239
Instituto Tecnológico de Costa Rica (ITCR), 108, 326
Inter-American Institute for Agricultural Sciences (IICA), 73
Invertebrates, 185, 213, 238, 241, 254, 261, 263, 278, 284, 293, 331. *See also entries for phyla, classes, orders, families, and genera*
Iquitos, Perú, 211–214
Irazú Volcano National Park, 11, 305, 321
Iriartea, 75, 76, 100, 112, 113, 122, 140, 164, 165, 173, 231, 360
Ischnosiphon, 174, 357

Isopods, 251
Isoptera, 56, 58, 63, 252, 275, 278, 323, 331
Isthmus of Tehuántepec, México, 201
Itabo (*Yucca*), 312
ITCO. *See* Instituto de Tierras y Colonización
Ithomiinae (subfamily of Nymphalidae), 188, 189, 190, 191, 264
Ivory Coast, Africa, 21, 25
Izor, R. J., 235

Jacamars. *See* Galbulidae
Jacamerops, 226
Jacana, 223, 386
Jacanidae, 386
Jacaranda, 320, 326, 328, 362
Jacaratia, 97, 146, 149, 150, 151, 157, 165, 176, 269, 363
Jacobs, M., 3
Jacobson, S., 159
Jaguar. See *Felis onca*
Jaguarundi (*Felis yagouroundi*), 397
Jaksic, F. M., 255
Jalisco State, México, 170, 171, 241
Jamaica, 169, 170, 171, 310
Jamaican fruit bat. See *Artibeus*
James, C. D., 331
Jamesbondia, 361
Jameson, D. L., 202
Janos, D. P., 56, 58, 63, 119, 124, 136
Janson, C. H., 282, 283, 284, 288, 290, 292
Janzen, D. H., 7, 70, 84, 92, 103, 104, 106, 114, 142, 144, 151, 153, 155, 156, 157, 160, 164, 168, 169, 175, 182, 185, 191, 194, 211, 213, 215, 225, 230, 238, 242, 250, 252, 261, 262, 263, 264, 266, 267, 269, 270, 271, 276, 277, 279, 282, 284–287, 289, 292, 293, 322, 331, 333, 336, 341, 342
JAPDEVA (Junta de Administratión Portuaria y de Desarrollo Económico de la Vertiente Atlántica), 307
Java, 81
Jays. *See* Corvidae
Jedlovec, G., 32
Jengka, Malaysia, 117
Jenik, J., 3, 151, 157, 158
Jenny, H., 35, 62
Jepson, D., 32
Jermy, H., 270
Jernstedt, J. A., 92
Joern, A., 265
Johnson, D. W., 51, 54, 58
Johnson, N., 56
Johnson, R. A., 288, 290
Johnstone, I. M., 267
Jones, C., 275
Jones, C. F., 310
Jones, E. W., 102, 120
Jones, T., 246
Jordan, C. F., 4, 23, 25, 27, 54, 55, 57–63, 123, 124, 267
Jordano, P., 282, 284, 286, 288, 292, 293
Josia, 191
Joturus, 197, 198, 299, 379
Joyce, A. T., 108, 119, 299, 317
Judziewicz, E. J., 78 86, 91, 350
Junonia, 188
Junta de Administratión Portuaria y de Desar-

rollo Económico de la Vertiente Atlántica (JAPDEVA), Costa Rica, 307
Justicia, 77, 85, 175, 360

Kakachi forest, India, 277
Kaplan, W. A., 56
Kapos, V., 336
Kapp, G. B., 315
Kapur, A. P., 278
Karban, R., 266
Kareiva, P., 273
Karr, J. R., 219–222, 226, 243, 249, 282, 287, 288, 290, 293, 331, 332, 334
Kauck, D., 306, 317, 318
Kaufmann, J. H., 262, 263, 286, 291
Kaur, A., 171
Keast, A., 222
Keeler, K. H., 275
Keeler-Wolf, T., 284
Keese, M. C., 264
Keller, M., 53
Kellison, R., 327
Kelly, C. K., 114, 125, 270, 271, 273, 278, 290
Kemp, A. C., 287
Kennedy, H., 164, 174, 182
Kennett, C. E., 273
Kenyi, J. M., 267
Keogh, R. M., 318
Kermott, L. H., 231
Kerster, H. W., 282
Kevan, P. G., 154
Kibale, West Africa, 272
Kiew, R., 92
Kiltie, R. A., 286
Kiltie, G., 294
Kinet, J.-M., 157, 158
King, D. A., 69, 138, 140
Kingfishers. *See* Alcedinidae
Kinkajous. See *Potos*
Kinosternidae, 382
Kinosternon, 382
Kinsman, S., 175, 182
Kirkpatrick, S., 209
Kitasako, J. T., 204
Kitching, I. J., 187, 194
Kjellberg, F., 155, 156
Kleinfeldt, S. E., 275
Klopfer, P. H., 251, 252
Kluge, A. G., 210
Knight, D. H., 97, 102, 124, 128, 137
Knoerr, K. R., 24
Knoppel, H.-A., 198
Kochummen, K. M., 97
Koelmeyer, K. O., 142, 143, 144, 151
Koford, C. B., 286
Kohm, K. A., 255
Koptur, S., 142, 159, 270, 274, 276
Kramer, D. L., 196
Kramer, E., 53
Kress, W. J., 81, 161, 164–168, 172–182, 350
Kricogonia, 189
Krizek, A., 188, 189
Kunkel, P., 212
Kunkel-Westphal, I., 212

Labiatae. *See* Lamiaceae
Lacey, E. P., 142, 143, 151, 153–157
Lacher, T. E., Jr., 262, 263

Lachesis, 215, 235, 253, 382
Lacistemataceae, 177, 366, 367
Lack, A. J., 160
Lack, D., 333
Lacmellea, 175, 328, 361
Ladenbergia, 79, 85, 375
Laessle, A. M., 116
Laetia, 76, 77, 177, 326, 328, 366
La Flaminea (La Selva), 6, 11, 14, 316, 349
La Frankie, J. V., 155
Lagerstroemia, 143, 370
Lagomorpha, 229, 233, 234, 262, 296, 396
La Guaria Annex (La Selva), 6, 10, 13, 26, 37, 38, 45, 46, 60, 307, 316, 323, 326
Lahanas, P. N., 201, 209, 211, 216
Lamas, G., 187, 239, 241
Lambert, J., 309
Lambert, J. D. H., 212
Lamīaceae, 177, 367
Lampropeltis, 210, 382
Land, H. C., 283, 291
Land Use: La Selva, 6, 347; Sarapiquí Cantón, Costa Rica, 297, 299–306
Lande, R., 194, 334
LANDSAT, 299, 319
Landsberg, J., 267
Lang, G. E., 97, 106, 116, 118, 124
Langenheim, J. H., 132, 261, 273
Languriidae, 279
Lanio, 226, 228, 392
La Pacífica, Costa Rica, 232
Larentiinae (subfamily of Geometridae), 263
Largidae, 279
Laridae, 218, 219, 386
Larin, Z., 191
Las Alturas Biological Station, Costa Rica, 190
Las Cruces Biological Station, Costa Rica, 211–214, 304
La Selva Advisory Committiee (LSAC), 346, 347
Lasianthaea, 175, 364
Lasiocampidae, 192, 280
Las Vegas Annex (La Selva), 11, 42, 207, 217
Latin America. *See specific regions and countries*
La Tirimbina, Costa Rica, 323
Lauraceae, 74, 77, 79–82, 84, 85, 87, 103, 104, 116, 162, 163, 166, 177–178, 268, 275, 276, 328, 334, 350, 367–368
LaVal, D. L., 231, 252, 255
LaVal, R. K., 230, 231, 232, 235, 237, 252, 255
Lavenberg, R. J., 210
La Virgen, Costa Rica, 35, 300, 301, 302, 304, 307, 323
Lawson, D. R., 56
Lawton, J., 331, 337
Lawton, J. H., 188, 263, 273
Lawton, M. F., 247
Lawton, R. O., 121, 122, 124
Leaf-cutting ants. See *Atta*
Leaf-miners, 240, 242, 279, 280. *See also entries for families of leaf-miners*
Leandra, 165, 178, 371
Lebron, M. L., 138
Lechner, M., 284
Lechriopini (Tribe of Curculionidae), 239
Leck, C. F., 219, 220, 227, 289, 291

Le Corff, J., 93
Lecythidaceae, 75, 80, 81, 97, 98, 100, 101,
    131, 132, 134, 178, 268, 283, 328, 368
Lecythis, 97, 98, 100, 101, 131, 132, 134, 178,
    268, 283, 328, 368
Lee, D. W., 130, 133
Legatus, 222, 390
Legler, J. M., 262, 263
Leguminosae. *See* Fabaceae
Leigh, C. H., 337
Leigh, E. G., Jr., x, 4, 32, 105, 194, 233, 241,
    267, 334
Leighton, D. R., 283, 287, 291
Leighton, M., 283, 287, 291
Leimadophis, 382
Leiopleura, 242
Lemurs, 288
León, J., 309
Leonard, H. J., 301, 317
Leonel Stream (La Selva), 77
Leopold, A. S., 262, 263
Lepidoblepharis, 215, 382
Lepidophyma, 382
Lepidoptera, 103, 153, 156, 162, 163, 164,
    167–170, 185, 187–194, 232, 238, 239,
    241, 247–250, 252, 259, 263–266, 268,
    269, 277–280, 331, 332, 346; sloth, 232
Leporidae, 262, 396
Leptodactylidae, 200, 201, 202, 208, 380–381
Leptodactylus, 200, 202, 203, 206, 381
Leptodeira, 204, 210, 216, 382
Leptomerinthoprora, 272
Leptophis, 210, 216, 382, 383
Leptotylophidae, 212
Lesack, L. F., 32
Letourneau, D. L., 270, 275, 276
Leucaena, 320
Levey, D. J., 94, 125, 142, 147, 217, 218,
    220–223, 243, 248, 249, 252, 259, 283–
    286, 288–293, 344
Levin, D. A., 261, 263, 270, 282, 331
Levings, S. C., 212, 213
Levins, R., 334
Lewinsohn, T. M., 267
Lewis, W. M., Jr., 32
Liberia, Africa, 241
Libytheana, 189
Libytheinae (subfamily of Nymphalidae),
    188, 189
Licania, 81, 85, 363
Lieb, C. S., 210
Lieberman, D., 58, 68, 69, 70, 95–99, 101,
    104, 106, 108, 110, 113–116, 118–121,
    124, 125, 128, 135, 137, 162, 331, 344
Lieberman, M., 44, 63, 68, 69, 70, 73, 75, 91,
    96, 98, 99, 108, 112, 113, 115, 118, 119,
    121, 137, 162, 331, 344
Lieberman, S. S., 200, 206, 207, 208, 210,
    213, 214, 225, 252, 255
Liesner, R., 87
Lieth, H., 142
Limacodidae, 280
Limerick, S., 203, 205, 207, 245, 262, 263
Limnocorid bugs, 250
Limnocoris, 251
Limoncocha, Ecuador, 219, 220, 227
Limón Province and city, Costa Rica, 88, 215,
    307, 309, 310, 312, 314
Limón River, Venezuela, 32

Lincoln, D. E., 272
Lindroth, R. L., 272
Linhart, Y. B., 167, 168, 172, 217
Liomys, 231, 248
Lipaugus, 228, 390
Litchi, 312
Litter, Leaf, 57–58, 61, 212–214, 240, 324
Lius, 242
Liverworts, 73
Livestock. *See* Cattle; Poultry; Suidae (pigs)
Lizards. *See* Sauria
Lloyd, D. G., 155, 157, 159
Loasaceae, 78, 369
Lobotrachelini (Tribe of Curculionidae), 239
Loganiaceae, 85, 166, 178, 369
Loiselle, B. A., 217, 220, 221, 222, 225, 226,
    249, 250, 255, 284, 287–293, 344
Loiselle, P. V., 195
Lonchocarpus, 77, 112, 166, 177, 325, 369
Londoño, E. C., 268
Longino, J., x, 194, 239
Longman, K. A., 3, 142, 151, 157, 158
Long-tailed hermit hummingbird. See *Phaeth-
    ornis*
Lopes, A. S., 50
López, E., 92
Lophanthera, 85, 86, 370
Loranthaceae, 78, 80, 83, 369–370
Lorisidae, 288
Los Arrepentidos Hills, Costa Rica, 42, 46
Los Chiles, Costa Rica, 307
Los Diamantes Experiment Station, Costa
    Rica, 314
Los Katíos, Ecuador, 219, 220, 227
Los Tuxtlas, México, 52, 102, 103, 104, 120,
    121, 200–202, 204, 205, 209, 241, 267
Louda, S. M., x, 276, 278
Loudon, C., 277
Lovejoy, T. E., 3, 243, 331
Loveless, M. D., 72, 165, 168, 171
Lowe-McConnell, R. H., 198
Lowman, M. D., 267, 270, 272
Lozania, 177, 366, 367
LSAC (La Selva Advisory Committee), 346,
    347
Lubchenko, J., 240
Lucas, P. W., 288
Ludlow, M. M., 122
Luehea, 75, 77, 181, 377
Lugo, A. E., 57, 95
Lumer, C., 164
Lundberg, J. G., x
Lutra, 235, 284, 286, 396
Luvall, J. C., x, 19, 21, 23, 24, 27, 33, 38, 59,
    62, 73
Lycaenidae, 187, 188, 190–193, 268, 277, 280
Lycianthes, 181, 377
Lycopodiaceae, 78, 350
Lycopodiophyta, 78, 350–351
Lydolph, P. E., 20
Lygaeidae, 279
Lymantriidae, 280
Lynch, J. L., 200, 201, 202
Lyonetiidae, 280
Lyropteryx, 191
Lythraceae, 29, 77, 143, 370

Maas, P. J. M., 174, 182
Mabea, 74, 177, 366

Mabuya, 382
Macadamia, 307, 312, 313, 314
MacArthur Foundation (John D. and Cather-
    ine T.), 11, 194, 266, 327
MacArthur, R. H, 32, 241, 244, 249, 251, 252,
    261, 332
Macaws. See *Ara*
Macdonald, D., 285
MacDougal, J. M., 350
MacKinnon, K. S., 287
Macropodidae, 287
Madagascar, 288
Madison, M., 82
Madrigal, F., 299, 300, 302, 306
Madrigal, R., 25, 308
Maeda, T., 46
MAG. *See* Ministerio de Agricultura
Magnoliaceae, 79, 178, 370
Magsasay, Costa Rica, 301
Maguire, J., 94
Mahogany (*Swietenia*), 320, 322
Maiorana, V. C., 200, 240
Malanga. See *Colocasia*
Maleret, L., 273
Malaysia, 30, 32, 61, 97, 116, 117, 137, 142,
    151, 152, 155, 158, 159, 170, 171, 172,
    198, 206, 241, 271, 277, 288, 331, 337
Malcolm, J. R., 234
Malmborg, P. K., 290, 293
Malpighiaceae, 81, 82, 85, 86, 178, 350, 370
Malvaceae, 76, 100, 131, 132, 138, 178, 269,
    277, 327, 370
Malvaviscus, 76, 178, 370
Mammalia, 92, 103–104, 122, 215, 229–237,
    238–241, 243, 245, 246, 248, 252, 254,
    259, 261, 262, 268, 277, 282, 283, 285–
    288, 331, 334, 336, 337, 394–398. *See
    also entries for families, genera, and com-
    mon names*
Mammals. *See* Mammalia
Mammals of Costa Rica, 230, 231
Mamón Chino (*Litchi*), 312
Manacus, 221, 224, 225, 246, 284, 390
Manakins. *See* Pipridae; *and entries for
    genera*
Man and the Biosphere program (MAB, UN-
    ESCO), 11, 63, 297, 304, 305, 306, 338
Manasse, R. S., 288, 290
Manatee, 304
Manaus, Brazil, 79, 234, 330, 331, 334, 336
Mandrachia, J., 187
Manettia, 85, 375
Manihot, 308, 310–313, 315, 366
Manioc. See *Manihot*
Manning, S., 276
Manokaran, N., 97
Manu, Perú, 79, 215, 234, 236. *See also* Co-
    cha Cashu
Maquira, 179, 372
Marantaceae, 74, 76, 79, 80, 83, 84, 92, 93,
    121, 122, 162, 163, 164, 166, 168, 220,
    231, 265, 269, 350, 357
Maranthes, 85, 363
Marattiaceae, 74, 76, 78, 79, 92, 93
Marcgravia, 88, 370
Marcgraviaceae, 83, 88, 370
Mares, M. L., 231, 250
Margalef, R., 251
Margays (*Felis wiedii*), 286, 397

Markea, 181, 286, 377
Marmosets, 288
Marpesia, 188
Marques, J., 23, 27
Marquis, R. J., 94, 96, 98, 122, 125, 147, 157, 194, 204, 243, 246, 259, 263–274, 276–280, 283, 290, 298
Marrs, R. H., 62
Marshall, A. G., 287, 288, 291, 293
Marshall, G. A. K., 192
Marshall, L. G., 230
Marsupialia, 229, 233, 234, 236, 241, 252, 262, 287, 394
Marsupials. *See* Marsupialia
Martin, C., x
Martin, M., 195
Martin, P. S., 282, 287, 293, 336
Martin, T. E., 282, 288, 289, 290
Martínez-Ramos, M., 120, 121, 125, 139
Mary Flagler Cary Charitable Trust, 73
Masner, L., 240
Master Plan for Development and Administration, La Selva, 346–348
Mastigodryas, 382
Mata, R., ix, 18, 35–38, 42–45, 51, 53, 55, 57, 60, 62, 91
Matamoros D., A., 317
Matisia, 175, 362
Matson, P. A., 51, 54, 56
Matthes, H., 198
Matthews, R. W., 245
Maurice, S., 155, 156
May, J., 190, 194
May, R. M., 187, 255, 266
McCleery, R. H., 288
McColl, J. G., 54, 58
McCoy, E. D., 261, 266, 273
McCulloch, R., 27
McDade, L. A., 32, 53, 69, 85, 104, 141, 159, 162, 164, 166, 167, 172, 174, 175, 178, 181, 182, 216, 217, 220, 237, 306, 316, 327, 350
McDiarmid, R. W., x, 209, 211, 216, 245, 284
McDonald, E. P., 132
McElravy, E. P., 238
McElwain, K., 32
McGraw, J. B., 137
McHargue, L. A., 56, 59, 90, 91, 97, 100, 103, 136, 137, 232, 269, 270
McKey, D., 220, 270, 272, 276, 277, 283, 291, 292
McLennan, D. A., 254
McMahan, E. A., 252
McNab, B., 287, 294
McNaughton, S. J., 270
McPherson, A. B., 230
McVaugh, R., 175, 182
McVey, M. E., 205
McVoy, C. W., 49
Meagher, T., 277
Mecham, J. S., 203
Medicinal plants, 312. *See also entries for genera*
Medina, E., 134
Mediterranean habitat, 335, 336. *See also specific countries*
Medway, L., 142, 144, 151, 287, 288
Megachilidae, 164, 281
Megachiroptera, 287–288

Megalopygidae, 192, 280
Megarhynchus, 228, 284, 390
Melastomataceae, 59, 74, 77–85, 88, 94, 124, 125, 126, 131, 132, 133, 138, 162–166, 171, 172, 178, 273, 274, 276, 283, 286, 292, 328, 350, 370–371
Melcher, J., 195
Meliaceae, 74–77, 82, 91, 97, 100, 102, 103, 108, 127, 131, 133, 135, 138, 146–150, 156, 157, 158, 162, 163, 165, 178, 179, 269, 276, 283, 284, 286, 298, 299, 320, 321, 322, 327, 371
Melicytus, 151
Meliosma, 181, 231, 376
Meliponinini, 164
Melitaeinae (subfamily of Nymphalidae), 188, 189
Mellon (Andrew J.) Foundation, 327
Meloidae, 279
Melson, W., 53
Membracidae, 242, 245, 264, 278, 332
Mendoza, A., 139, 140, 270
Menispermaceae, 135, 371
Menke, A. S., 238
Mephitis, 287
Mesene, 191
Mesoamerica. *See* Central America; *and entries for specific countries and regions*
Metaxya, 86, 351
Metaxyaceae, 86, 351
México, 52, 87, 102, 103, 104, 120, 121, 135, 138, 139, 170, 171, 196, 197, 199, 200–202, 204, 205, 209, 212, 217, 218, 219, 229, 240, 241, 242, 262, 263, 267, 308, 322, 381, 398
Mice. *See* Rodentia; *and entries for families and genera*
Michaloud, G., 142, 146, 155, 291, 293
Miconia, 59, 74, 79–82, 85, 94, 124, 125, 126, 131, 132, 133, 138, 165, 171, 178, 273, 274, 276, 286, 328, 371
Micosphaerella, 309, 312
Micrastur, 228, 385
Microbates, 228, 391
Microcerculus, 228, 391
Microchiroptera, 232, 287–288
Microhylidae, 201, 203, 381
Micronycteris, 231, 252, 394
Micropterigidae, 279
Microrhopias, 228, 389
Microstigmus, 245
Microtine rodents, 272
Microtylopteryx, 246, 250, 252, 254, 265
Micruridae, 212, 382
Middle America. *See* Central America; *and entries for specific countries and regions*
MIDEPLAN (Ministerio de Planificación Nacional y Política Económica), 312
Migration: birds, 221–222, 250, 333; fish, 198; insects, 360, 333
Milk production. *See* Dairy farming
Miller, B. I., 19, 20
Miller, J., 172, 176, 182
Miller, J. S., 187, 189, 191
Miller, R. R., 195, 196, 197
Milton, K., 96, 120, 142, 155, 156, 234, 262, 263, 272, 286
Mimicry, 240; lepidoptera, 190–192
Mimidae, 285, 391

Mimosaceae, Mimosoideae. *See* Fabaceae
Ministerio de Agriculture (MAG), Costa Rica, 307, 310, 312, 314, 316, 320
Ministerio de Planificación Nacional y Política Económica (MIDEPLAN), 312
Minnick, G., 327
Minquartia, 97, 98, 100, 101, 133, 328, 373
Mionectes, 220, 221, 224, 225, 246, 284, 292
Miridae, 278
Misra, R., 267
Missouri, U. S., 80
Mites, 207, 213, 239, 261, 278
Mitter, C., 264
Mittermeier, R. A., 288
Mitrospingus, 221, 225, 247, 392
Miyamoto, M. M., 200, 208, 249
Moermond, T. C., 125, 215, 217, 220, 232, 246, 269, 270, 283, 284, 285, 287–292, 336
Mokukua, Gabon, Africa, 146
Molden, S., 32
Moll, E. O., 262, 263
Molluginaceae, 78, 372
Molossidae, 395
Momotidae, 223, 285, 388
Momphidae, 279, 280
Monarch butterfly. See *Danaus*
Monasa, 228, 247, 388
Monasterio, M., 142, 144, 147
Monge, L. A. (president of Costa Rica), 11
Monge-Nájera, J., 230
Monilia, 310–313
Monimiaceae, 83, 179, 220, 286, 372
Monk, T. W., 194
Monkeys. *See* Primates
Monocots, 265, 353–360. *See also entries for families and genera*
Monselise, S. P., 142, 157, 158
Monstera, 80, 85, 284, 354
Montagnini, F., 53, 63, 297, 315, 323, 326
Montalvo, A. M., 173, 182
Monte (xeric habitat, Argentina), 335
Monteith, J. L., 23, 33
Monteverde, Costa Rica, 121, 122, 124, 164, 172, 208, 211–216, 232, 241, 247, 292
Montgomery, G. G., 262, 263
Moody, S., 261
Mooney, H. A., 128, 133, 272, 274, 335, 336
Moore, C. J., 24
Moore, H. E., Jr., 164, 173, 182
Moore, L. V., 273
Moraceae, 63, 76–84, 86, 87, 88, 100, 120, 138, 143, 146–147, 149, 151, 153, 155, 157, 162–166, 169, 179, 197, 259, 276, 283, 284, 286, 291, 292, 299, 310, 315, 327, 328, 350, 372
Moran, V. C., 331
Mordellidae, 279
Morden-Moore, A. L., 288
Morera Brenes, B., 230
Mormopidae, 394
Morphinae (subfamily of Nymphalidae), 188, 189, 264
Morphnus, 217, 385
Morrison, D. W., 285, 286, 293
Morrison, G., 251, 266
Morrison, P. C., 310
Morrow, P. A., 267
Mortimer, J., 231

Morton, E. S., x, 222, 225, 284, 284, 293
Morton, S. R., 331
Mosquitoes, 234, 251
Mosses, 73
Moths, 153, 156, 162, 163, 164, 167–170, 185, 191, 192, 193, 232, 241, 247, 248, 250, 263, 266, 268, 269, 277, 279, 280, 331; sloth, 232
Motmots. See Momotidae
Mould, E. D., 271
Mound, L. A., 187
Mountain lion. See Felis concolor
Mouse-deer, 287
Moynihan, M., 225, 240
Mucuna, 163, 316 (as Styzolobium), 369
Mueller, U. G., 273
Mueller-Dombois, D., 73
Mugilidae, 197, 198, 299, 379
Mulcahy, D. L., 166
Mulcahy, G. B., 166
Mules, 299
Muller, F., 191
Munn, C. A., 225
Murawski, D. A., 165
Mure, V., 146
Muridae, 229, 262, 285, 396
Murphy, C. E., Jr., 24
Murphy, P. G., 120, 125
Murray, K. G., 283, 288, 290
Musa, 74, 266, 297, 298, 299, 301, 302, 307, 309, 310–315, 358
Musaceae, 74, 78, 266, 297, 298, 299, 301, 302, 307, 309–315, 358
Muscicapidae, 245, 285, 287, 391
Museo Nacional de Costa Rica, 194, 239
Museum of Vertebrate Zoology, University of California, Berkeley, 210
Mustelidae, 229, 262, 285, 287, 397
Mustelids. See Mustelidae
Mycorrhizae, 58, 63, 124, 136–137
Myers, C. W., 202, 211, 239, 241, 243
Myers, G. S., 196
Myers, N., 73
Myiobius, 226, 228, 389
Myiodynastes, 222, 284, 390
Myiornis, 252, 389
Myiozetetes, 219, 221, 228, 284, 390
Myrcia, 77, 179, 373
Myriocarpa, 164, 182, 378
Myristicaceae, 75, 97, 103, 108, 116, 131, 137, 146, 148, 149, 150, 156, 157, 158, 165, 170, 179, 270, 283, 286, 325, 326, 328, 372
Myrmeciza, 221, 389
Myrmecophaga, 229, 235, 243, 396
Myrmecophagidae, 229, 235, 243, 396
Myrmelachista, 275
Myrmotherula, 219, 220, 225, 228, 389
Myroxylon, 328
Myrsinaceae, 77, 83, 85, 88, 162–166, 179, 286, 372–373
Myrtaceae, 58, 77, 82, 179, 267, 312, 315, 320, 326, 327, 331, 373

Naeem, S., 144, 147, 159, 251
Nahrstedt, A., 191
Nandinia, 287
NASA, 108, 110, 111, 119
Nasua, 231, 233, 234, 235, 262, 286, 287, 396

National Geographic Society, 119, 172
National Parks Foundation (Costa Rica), 11
National Research Council (NRC), U.S. National Academy of Science, 13, 119
National Science Foundation (NSF), U.S., ix, 12, 13, 105, 119, 127, 141, 159, 172, 239, 278, 294, 342, 343
Naturalist Guides (at La Selva), 13, 348
Nature Conservancy, 11
Naucleopsis, 86, 372
Naucoridae, 29, 251
Naveh, Z., 330
Nebraska, U. S., 275
Nectandra, 77, 85, 103, 104, 328, 367–368
Neea, 58, 179, 267, 277, 286, 373
Needham, J. G., 281
Neetroplus, 196, 197, 379
Neill, D. A., 163, 169, 176, 182
Nelder design, 323–324
Nelson, M., 32
Nemathelminthes. See Nematodes
Nematodes, 239, 263, 298, 310
Neoheterandria, 196, 197, 198, 379
Neomorphus, 228, 387
Neotrachys, 242
Neotropical Lowlands Research Program, Smithsonian Institution, 172
Neotropics. See entries for specific regions and countries
Nepticulidae, 279
Nettancourt, D. de, 165, 166, 167
Newbery, D. McC., 106, 116, 118, 119, 273
Newell, E., 53
New Guinea, 102, 287
Newstrom, L. E., 103, 142, 143, 148, 152, 154, 159
New York, U.S., 264, 265
New Zealand, 151
Nicaragua, 84, 86, 87, 219, 230, 231, 242, 299, 304, 319
Nichols, J. D., 334
Nichols-Orians, C., 270, 274, 276, 277, 281
Nielson, B. O., 267
Nigeria, 102
Nightjars, 223, 387
Night monkeys. See Aotus
Ninia, 382
Nitao, J., 278
Nitrogen-fixing Tree Project, CATIE, 316
Nitulidae, 87
Nobel, P. S., 131
Noctilionidae, 394
Noctuidae, 192, 280
Norops, 213, 214, 215, 240, 246, 248–252, 254, 255, 334, 382. See also Anolis
North America, 213, 216, 218, 219, 221, 241, 278, 322, 335, 336, 345, 398. See also entries for specific countries, regions, and states
North Carolina State University, 316
Nothopsis, 382
Notodontidae, 191, 280
Nottingham, B. G., Jr., 253
Noyes Foundation (Jesse Smith), 141, 209, 216
NSF. See National Science Foundation
Nunbirds. See Monasa
Nuñez, J., 53
Nuñez-Farfán, J., 125

Nutrients, 49–52, 54–63, 68–69, 123–124, 272–274, 283
Nyctaginaceae, 58, 83, 179, 265, 267, 277, 286, 373
Nyctibiidae, 387
Nycticebus, 288
Nyctomys, 262, 396
Nye, P. H., 62
Nymphaeaceae, 373
Nymphalidae, 187–191, 247, 248, 264, 268, 280
Nymphalinae (subfamily of Nymphalidae), 188, 189, 248
Nymphidium, 193

Oades, J. M., 46
Oates, J. F., 272, 273, 277
Oberbauer, S. F., 101, 129–133, 135, 137, 139, 140, 141
Occophoridae, 280
Ocelots (Felis pardalis), 235, 297
Ochnaceae, 180, 373
Ochroma, 78, 100, 120, 121, 126, 131, 135, 139, 140, 175, 274, 326, 328, 362
Ocotea, 80, 81, 85, 116, 177–178, 268, 275, 276, 328, 368
O'Connell, M. A., 262, 263
O'Donnell, S., 232
O'Dowd, D. J., 273, 274, 336
Odocoileus, 262, 397
Odonata, 239, 241
Odontonema, 175, 360
Odum, H. T., 4, 267, 337
Oedipina, 200, 206, 380
OET. See Organization for Tropical Studies
Ohmart, C. P., 267
Oilbirds (Steatornis), 292
Olacaceae, 97, 98, 100, 101, 133, 328, 373
Olberg, R. M., 289
Oldeman, R. A. A., 120, 158
Old World, 83, 85, 171, 172, 206, 241, 335. See also entries for specific regions and countries
Oliveira, P. S., 270
Olmstead, K. L., 242, 264, 279
Ololygon. See Scinax
O'Malley, D. M., 161, 166, 168, 171, 172
Ommatolampinae (subfamily of Acrididae), 265
Onagraceae, 373
Onychophora, 255
Ophioglossaceae, 78, 351
Opler, P. A., 90, 115, 134, 143, 146, 156, 158, 160, 161, 167, 170, 171, 188, 189, 220, 225, 261, 263, 282, 283, 284, 286, 290, 291, 332
Opomyzidae, 280
Oporornis, 221, 392
Opossums. See Didelphidae
Opostegidae, 279
Oppenheimer, J. R., 286, 291
Orchidaceae, 78–81, 83, 84, 162, 172, 350, 358–359
Orchid bees. See Euglossinae
Organización para Estudios Tropicales. See Organization for Tropical Studies
Organization for Tropical Studies (OTS) vii, ix, 3, 6, 9, 10, 11, 13, 18, 24, 67, 73, 77, 95–99, 101, 102, 105, 119, 120, 121, 127,

141, 159, 195, 206, 209, 217, 229, 237, 278, 294, 297, 298, 299, 304, 306, 308, 315, 316, 319, 320, 323, 325, 326, 327, 341, 342, 343, 345–348
Orians, G. H., x, 5, 100, 101, 105, 126, 159, 216, 220, 244, 249, 292, 330, 335, 336
Oring, L. W., 244
Orinoco River, Venezuela, 7, 308
Orituco River, Venezuela 32
Ormosia, 78, 328, 369
Ornamental plants (as cash crop for export), 307, 311–315
Ornithion, 221, 228, 284, 389
Oropendolas. See *Psaricolius*
Orophus, 272
Orozco Segovia, A., 130
ORSTOM (Institut Français de Recherche Scientifiques pour le Développement en Coopération), France, 120
Ortalis, 285, 385
Orthoclada, 174, 357
Orthogeomys, 229, 230, 323, 396
Orthoptera, 185, 207, 246, 249, 252, 253, 255, 259, 264–266, 272, 278
Oryza, 299, 300, 302, 307, 309, 311, 312, 313, 314, 315, 316, 357
Oryzoborus, 221, 228, 393
Oryzomys, 236, 262, 396
Ortiz, R., 53
Osa Peninsula, Costa Rica, 85, 86, 189, 200–202, 204, 205, 206, 209, 211–214, 216, 217, 219, 220, 227, 262, 263, 274. *See also entries for specific sites*
Oscine passerines, 218, 219, 221, 223, 227
Osmilia. See *Abracris*
Osmond, B., x
Ossaea, 178, 371
Othnacidae, 279
Otis, G. W., 251, 255
Otitidae, 280
Otoba, 75, 116, 328, 372
OTS. *See* Organization for Tropical Studies
Otter. *See Lutra*
Ouratea, 180, 373
Ovenbirds, 221, 223
Owen, D. F., 106, 242, 270
Owen, J., 242
Owen-Smith, N., 272
Owls. *See* Strigidae
Oxalidaceae, 312, 373
Oxybelis, 210, 382
Oxyrhopus, 382
Oyama, K., 139, 140, 270

Paaby, P., 14, 19, 28, 29, 31, 32, 348
Paca. See *Agouti*
Pacala, S., 213, 215
Pachira, 75, 175, 362
Pachyschelus, 242
Pacific Slope, Central America, 218, 219, 242, 310. *See also entries for specific countries and sites*
Paguma, 287
Paine, R. T., 244, 252, 336
Palacazú Valley, Perú, 127
Paleotropics, 82, 170, 171, 172, 287, 335. *See also entries for specific continents, regions and countries*
Palmae. *See* Arecaceae

Palm, C. A., x, 53
Palm civets (Viverridae), 287
Palmeirim, J. M., 231, 283, 286
Palmer, B., 240, 278
Palms. *See* Arecaceae
Palo Verde Biological Station and National Park, 54, 239, 322
Panama, 8, 21, 52, 60, 61, 68, 78, 84–88, 91, 104, 117, 120, 124, 126, 130, 135, 142, 150, 170, 171, 188, 191, 192, 196, 199, 200, 204, 207, 209, 212, 219, 220, 225, 227, 229, 230, 232, 233, 238, 239, 241, 242, 247, 248, 261, 262, 263, 267, 268, 272, 277, 278, 285, 286, 288, 290, 293, 316, 331, 334, 398
Panama Canal, 233
Panama disease (*Fusarium*), 309, 313
Pantano Creek, La Selva, 31
Panthera (jaguar). See *Felis onca*
Pantophthalmidae, 281
Panyptila, 228
Papageorgis, C., 190, 249
Papaveraceae, 78, 373
Papaya (*Carica*), 312
Papilionaceae. *See* Fabaceae
Papilionidae, 187–192, 248, 264, 280
Papilioninae (subfamily of Papilionidae), 248
Papilio, 188
Papilionoideae. *See* Fabaceae
Papua New Guinea, 30, 32
Paradoxurinae, 287
Paradoxurus, 287
Paradisaeidae, 287
Paraponera, 207
Parashorea, 99
Pará State, Brazil, 117
Parfitt, R. L., 59
Pariana, 86, 174, 357
Parides, 248
Parker, G. G., x, 19, 21, 24, 35, 38, 53–60, 62, 123, 124
Parker, M. A., 270, 271
Parker, T. A., 292
Parnell, R. A., Jr., 56
Parrita, Costa Rica, 309
Parrots. *See* Psittacidae
Parsons, J. J., 309, 314
Partridge, L., 155, 244
Parulidae, Parulinae. *See* Emberizidae
Pasoh (Forest Reserve), Malaysia, 61, 170, 337
Passeriformes, 218, 219, 225, 262, 333, 388–393. *See also entries for families and genera*
Passerines. *See* Passeriformes
Passerina, 228, 392
Passiflora, 189, 190, 251, 259, 263, 264, 275, 276, 312, 314, 373
Passifloraceae, 84, 189, 190, 259, 263, 264, 275, 276, 312, 314, 350, 373
Passion fruit (*Passiflora*), 312, 314
Passoa, S., 278
Pasture, 301, 302, 304, 306, 307, 309, 311, 314, 315, 318. *See also* Cattle
Patrick, R., 32
Patterson, D. T., 131
Pauley, T. K., 199
Paulson, D. R., 239, 241
Pavonia, 178, 370

Payne, C. D., 152
Payne, J., 287, 288
Pearcy, R. W., 122, 125, 129, 132, 134
Pearson, D. L., 219, 227, 241
Peccaries. *See* Tayassuidae
Peje Annex (La Selva), 11, 316, 324
Peje River (La Selva). *See* Rio Peje
Pejibaye palm. See *Bactris gasipaes*
Pelecaniformes, 384
Pemadasa, M. A., 106
Penelope, 92, 285, 385
Pennisetum, 314, 357
Pentaclethra, 52, 56, 58, 63, 68, 73–76, 80, 96–104, 108, 112, 114, 116, 117, 118, 121, 130, 131, 132, 134–137, 139, 140, 141, 162, 163, 166, 177, 269, 270, 275, 322, 325, 328, 368
Pentagonia, 181, 268
Pentaneura, 251
Pentatomidae, 279
Peperomia, 79, 80, 81, 374
Peralta, R., 6, 12, 73, 75, 96, 103, 108, 110, 112, 114, 119, 321, 322, 327, 344
Perebea, 179, 372
Pereira, H. C., 27
Pérez, C., 319
Pergidae, 281
Pericopidae, 191
Perisasama, 191
Perissodactyla, 233, 262, 397
Perodicticus, 288
Peropteryx, 246, 249, 394
Perrhybris, 248
Perry, D. R., 115, 144, 150, 163, 165, 166, 173–177, 179–182, 194
Perú, 79, 127, 187, 190, 204, 207, 208, 211–215, 225, 234, 235, 236, 239, 241, 262, 263, 288, 290, 334
Pests and pesticides, 310, 313
Petr, T., 32
Petriceks, J., 9, 26, 35, 73, 322
Phaenostictus, 226, 228, 389
Phaeochlaena, 191
Phaethornis, 220, 221, 224, 226, 240, 246, 249, 255, 387
Phalacridae, 279
Phalacrocoracidae, 384
Phalanderidae, 287
Phalangers. *See* Phalanderidae
Phallichthys, 196, 379
Pharomachrus, 287
Pharus, 174, 357
Phaseolus, 299, 302, 309, 311, 312, 313, 315, 369
Phasianidae, 262, 386
Phasmastidae, 278
Phasmotodea (suborder of Orthoptera), 278
Pheidole, 274–275
Philander, 262, 394
Philippines, 206
Phillips, E., 19, 28, 31
Phillips, K., 209
Philodendron, 79, 80, 81, 85, 88, 173, 354
Phoebe, 85, 368
Phoebis, 188
Phryganodes, 268
Phyllobates, 200, 203, 206, 380
Phyllostomidae, 232, 246, 252, 284, 285–288, 394–395

Phyllostominae (subfamily of Phyllostomi-
dae), 235, 394–395
Phytolacca, 59, 78, 124, 136, 269, 374
Phytolaccaceae, 59, 78, 124, 136, 265,
269, 374
Pianka, E. R., 244, 245, 247, 251, 329
Piasus, 269
Piazurini (Tribe of Curculionidae), 239
Picidae, 218, 219, 223, 226, 227, 228, 245,
284, 285, 388
Piciformes, 388
Pickett, S. T. A., x, 125, 127, 136
Pierce, N. E., 192
Pierce, S. M., 303
Pieridae, 187, 188, 189, 191, 241, 248,
264, 280
Pierinae (subfamily of Pieridae), 188, 189,
248
Pigeons. See *Columba*
Pigs. *See* Suidae
Pilea, 182, 378
Pimelodidae, 197, 379
Pimentel, R. A., 201
Pinaceae, 315, 320, 325, 326, 327
Pineapple (*Ananas*), 312
Piñero, D., 95, 102, 103, 120, 125, 139, 142
Pinkwae, Ghana, 117
Pinus, 315, 320, 325, 326, 327
Pionus, 269, 386
Piper, 59, 77, 79–82, 88, 94, 124, 125, 126,
131, 132, 133, 136, 140, 147, 157, 171,
180, 232, 263–264, 266, 267, 269, 270–
271, 273–277, 283, 286, 310, 374
Piper nigrum, 310
Piper Stream (La Selva), 39
Piperaceae, 59, 77–83, 88, 94, 124, 125, 126,
131, 132, 133, 136, 140, 147, 157, 162–
165, 169, 171, 172, 180, 232, 263–264,
266, 267, 269, 270–271, 273–277, 283,
286, 310, 374
Piper, J. K., 290, 293
Pipidae, 201
Pipra, 221, 225, 246, 250, 284, 390
Pipridae, 219, 221, 223, 224, 225, 245, 246,
250, 284, 285, 287, 289, 292, 390
Pires, J. M., 106, 118
Pitangus, 284, 390
Pithecellobium, 75, 76, 77, 134, 136, 163,
166, 177, 269, 276, 323, 325, 326, 328,
368–369
Pittendrigh, C. S., 249
Place, S., 318
Plantains. See *Musa*
Plástico (El), Costa Rica, 301
Plataspididae, 279
Platt, J., 159
Platypodidae, 279
Platypodium, 104
Platyrinchus, 221, 389
Platystomidae, 280
Pleasants, J. M., 154, 155
Pleiostachya, 174, 357
Plethodontidae, 200, 201, 213, 208, 380
Pleuranthodendron, 77, 366
Pleurothallis, 80, 359
Ploceidae, 218, 219
Plutellidae, 280
Poaceae, 78, 79, 80, 83, 84, 86, 91, 160, 162,
163, 164, 172, 174, 265, 299, 300, 302,
307, 308, 309, 311–316, 350, 356–357
Poás Volcano National Park, 9, 321
Pocket gopher. See *Orthogeomys*
Podicipedidae, 384
Podicipediformes, 384
Podocarpus, 85
Podostemaceae, 374
Poecilia, 196, 197, 198, 379
Poeciliidae, 196, 197, 198, 379
Pohl, R. W., 78, 86, 91, 174, 182, 350
Polar habitats, 254
Polioptila, 223, 226, 228, 391
Polybotrya, 79, 267, 352
Polychrus, 262, 382
Polygalaceae, 374
Polygonaceae, 374
Polypodiaceae, 78, 80, 81, 351–353
Polypodiophyta, 350, 351–353
Polypodium, 80, 81, 352–353
Pomadasys, 197, 379
Pontederiaceae, 360
Poole, R. W., 155
Poore, M. E. D., 90, 106, 116, 118
Popma, J., 125, 138, 331
Population Reference Bureau, 318
Porcupines. *See* Erethizontidae
Pork, 299. *See also* Suidae
Porras Z., A., 317
Porthidium, 382
Portico, S.A., 127, 298, 321, 327
Portig, W. H., 20, 21
Portulaceae, 374
Posoqueria, 77, 181, 375
Possums (Pseudocheiridae), 287
Potalia, 85, 369
Pothomorphe, 79, 374
Potos, 262, 286, 287, 397
Pough, F. H., 209
Poulton, E. B., 191, 192
Poultry, 309, 313
Pounds, J. A., 211, 215, 216
Pouteria, 132, 328, 377
Pourouma, 75, 363, 372
Poveda, L. J., 80, 113, 322
Powell, B., 194
Powell, G. V. N., 225, 245
Power, M. E., x
Powles, S. B., 132, 141
Powlesland, M. H., 151
Prance, G. T., 3, 85, 176, 182, 330, 334
Pratt, T. K., 195, 197, 287, 292
Presch, W. F., 210, 212
Prestoea, 164, 173, 360
Prestwich, G. D., 56
Priapichthys, 196, 379
Price, P. W., 240, 242, 264, 274, 332
Primack, R. B., x, 104, 105, 137, 142,
143, 144, 153, 154, 156, 160, 269, 271,
289, 290
Primates, 229, 231, 233–236, 241, 262, 283,
286, 287, 288, 291, 292, 395
Pringle, C. M., 6, 11, 17, 19, 28, 29, 31, 32,
54, 59, 303, 304
Proboscideae, 287
Proctolabinae (subfamily of Acrididae), 265
Proctor, J., 267
Procyon, 396
Procyonidae, 122, 229, 233, 262, 285, 287,
396, 397
Proechimys, 215, 253, 262, 397
Progne, 228, 390
Propper, C., 270
Proteaceae, 307, 312, 313, 314
Protium, 74, 75, 76, 81, 108, 176, 283, 363
Psaricolius, 78, 87, 393
Pseudechinolaena, 78, 357
Pseudobombax, 323, 328
Pseudocheiridae (possums), 287
Pseudophallus, 379
Pseudoscorpionida, 251
Pseustes, 382
Psidium, 77, 373
Psilidae, 280
Psittacidae, 103, 219, 223, 244, 249, 262, 269,
284, 285, 386
Psittaciformes, 262, 386
Psocoptera, 238
Psychotria, 75, 77–82, 85, 88, 94, 181, 269,
276, 350, 375–376
Psyloidea, 279
Pteridophytes, 76, 78–81, 83, 84, 86, 92, 93,
108, 267, 312, 337, 350–353
Pterocarpus, 75, 76, 112, 134, 135, 177, 283,
325, 328, 369
Pteroglossus, 247, 388
Pteronotus, 229, 394
Pteropodidae (Pteropidae), 232, 288
Ptilinopus, 287
Puerto Rico, 4, 46, 52, 61, 337
Puerto Viejo de Sarapiquí, Costa Rica, vii, 9,
10, 13, 43, 44, 53, 86, 87, 298, 321, 349
Puerto Viejo River. *See* Rio Puerto Viejo
Puffbirds. *See* Bucconidae
Pullen, A. S., 190
Pulliam, H. R., 333
Pumas. See *Felis concolor*
Puntarenas, Costa Rica, 300
Puriscal, Costa Rica, 309
Putz, F. E., x, 91, 92, 95, 96, 104, 105, 120,
121, 122, 124, 126, 142, 151, 159, 160,
267, 270, 293
Pyburn, W. F., 246
Pycnonotidae, 287
Pyralidae, 268, 269, 280
Pyrochroidae, 279
Pyrrhocoridae, 279
Pyschidae, 280

Qualea, 86, 378
Quararibea, 57, 85, 175, 286, 362
Quebradas (streams) at La Selva. *See entries
for proper names*
Quercus, 271
Querula, 225, 247, 390
Quiinaceae, 374
Quiscalus, 393
Quintanilla, I., 8

Rabbits. *See* Lagomorpha
Rabinowitz, A., 253
Rabinowitz, D., 118
Radulovich, R., 49
Rafael Chavarría Ecological Reserve (La
Selva). *See* Reserva Ecológica Rafael Cha-
varría

Rafflesia, 170
Rafflesiaceae, 170
Raich, J. W., 54, 55, 57, 58, 59, 60, 124
Rails, 284, 386
Rallidae, 284, 386
Ramos, M. A., 222
Ramphastidae, 221, 223, 225, 226, 246, 247,
   284, 285, 287, 292, 388
Ramphastos, 221, 225, 246, 388
Ramphocelus, 221, 223, 224, 225, 228,
   291, 392
Rana, 200, 202, 203, 206, 301
Rand, A. S., 202, 211, 213, 215, 239, 241,
   243, 246, 248, 262, 263, 286, 334, 335
Rand, W. M., 213, 335
Randia, 85, 181, 376
Ranidae, 201, 202, 203, 301
Rao, P. S. C., 49
Raphia, 217
Rapp, M., 57
Rappole, J. H., 222
Raptors, 223, 226, 227, 233, 253, 334. *See
   also entries for families and genera*
Rara Avis Reserve, Costa Rica, 301, 304
Rathbun, S. L., 106
Rathcke, B. J., 266
Rats. *See* Rodentia; *and entries for families
   and genera*
Raukin, J. J., 32
Rauvolfia (Rauwolfia), 270, 361
Raveh, A., 160
Raven, P. H., 156, 192, 267
Razisea, 85, 175, 360
Ray, R. T., 168, 280
Redfield, A. C., 28
Redford, K. H., 252, 293
Reduviidae, 252
Rees, C. J. C., 249
Reese, J. C., 272
Regal, P. J., 267, 282
Rehr, S. S., 276
Reich, P. B., 135, 144, 157, 158
Reichle, D. E., 267
Reid, N., 292
Reiners, W. A., x, 32, 53, 63
Remsen, J. V., 282
Renealmia, 174, 360
Renner, S. S., 164, 171, 172, 178, 182
Rentz, D. C., 278
Reproductive biology: animals (general), 244–
   248; amphibians, 202–206, 208, 209;
   birds, 222–225, 333; fish, 196, 197, 198;
   plants, 71–72, 102–104, 115, 126, 142–
   182, 298
Reptiles. *See* Reptilia
Reptilia, 207, 210–216, 220, 239, 240, 241,
   243, 246, 252, 253, 254, 262, 285, 286,
   343, 344, 382–383. *See also entries for or-
   ders, families, and genera*
Research Swamp, La Selva, 203
Reserva Ecológica Rafael Chavarría (La
   Selva), 8, 44, 347
Restrepo, C., 292
Reznick, D., 247, 254
Rhacicreagra, 265
Rhadinaea, 216, 382
Rhamdia, 197, 379
Rhamnaceae, 76, 180, 374–375

Rheingans, R., 316
Rhinoclemmys, 215, 262, 286, 382
Rhinophrynidae, 201
Rhinotermitidae, 278
Rhizobium, 56, 136–137
Rhizophagidae, 279
Rhizophoraceae, 74, 180, 375
Rhoades, D. F., 264, 271, 272, 273, 276
Rhopalidae, 279
Rhynchocyclus, 228, 389
Rhynchonycteris, 249, 394
Rhychophorinae (subfamily of Curculioni-
   dae), 263
Rhytipterna, 228, 390
Rice, B., 330, 332, 336
Rice. See *Oryza*
Rich, P. M., 95, 113, 122, 140, 141
Rich, P. V., 230
Rich, T. H., 230
Richards, P., 106
Richards, P. W., 3, 58, 73, 74, 99, 106, 120,
   142, 143, 144, 147, 151, 152, 157
Richter, D., Jr., x, 50, 53
Richter, W., 294
Ricklefs, R. E., 251, 252, 331, 334
Rickson, F. R., 274
Ridley, H. N., 292
Riede, K., 246
Ried!, H., 189
Riehl, H., 19, 20, 21
Rincón de Osa, Costa Rica, 189, 200–202,
   209, 211–214, 216, 219, 227
Rinorea, 74, 112, 182, 378
Río Caura, Venezuela, 32
Río Chagres, Panama, 233
Río Corinto, Costa Rica, 322
Río Cuarto, Costa Rica, 307
Riodinidae, 187–193
Río Frío, Costa Rica, 35, 44, 215, 300, 301,
   304, 308, 310–314, 316
Río Guácimo, Costa Rica, 321
Río Orinoco, Venezuela, 7, 308
Río Orituco, Venezuela 32
Río Palenque Science Center, Ecuador, 79, 81,
   86, 87, 219, 220, 227
Río Peje (La Selva), 6, 11, 36, 37, 39, 85
Río Puerto Viejo (La Selva), ix, 6, 9, 10, 13,
   17, 27, 35, 36, 42, 44, 46, 52, 57, 74, 75,
   77, 195, 196, 198, 321
Río San Carlos, Costa Rica, 308
Río San Juan, Costa Rica and Nicaragua, 195,
   196, 299, 308
Río Sarapiquí (La Selva), 6, 9, 10, 27, 35, 36,
   38, 39, 43–46, 52, 57, 77, 195, 217, 299,
   300, 301, 307, 308, 309
Río Sucio, Costa Rica, 301, 310
Río Usumacinta, México and Guatemala, 195
Ripley, B., 106
Risch, S. J., 274
Ritland, D. B., 190
Rivera Batlle, C. T., 95
Rivulidae, 197, 379
Rivulus, 197, 379
Robakiewicz, P., 205
Robbins, R. K., x, 187, 188, 192, 194, 247
Robbins, C. T., 271
Roberts, T. R., 198
Roberts, W., 204

Robertson, G. P., 44, 49–53, 56, 58, 59, 60
Robichaux, R. H., 122, 125, 136
Robinson, D. C., 210
Robinson, J. G., 331
Robinson, S. K., 252, 331
Robyns, A., 86
Rockwood, L. L., 270
Rodentia, 164, 191, 229–236, 241, 248, 253,
   261, 262, 269, 286, 287, 289, 292, 336,
   396–397. *See also entries for families and
   genera*
Rodin, L. E., 63
Rodríguez C, R., 317, 318, 320
Rodríguez-M., J. A., 195, 196
Rodríguez, J. V., 219, 227
Roeboides, 197, 379
Rojas, E., 308
Rojas, J. M., 312
Rollinia, 77, 326, 328, 361
Romaleidae, 264–266, 278
Romanoff, A. J., 225
Romanoff, A. L., 225
Romero, A., 195, 196
Roosmalen, M. G. M. van., 288
Root, R. B., 220, 273, 336
Rosaceae, 375
Rosella, 269
Rosenthal, J., 159
Rosewood. See *Dalbergia*
Rosselli, M., 222, 291, 293
Roth, R. R., 249, 332
Rothschild, M., 191, 266
Roughgarden, J. D., 213, 215
Rowell, H. F., 249, 264, 265, 266, 278
RPSC. *See* Río Palenque Science Center
Rubiaceae, 9, 75–86, 88, 94, 112, 140, 144,
   146, 147, 149, 150, 156, 158, 162–167,
   169, 172, 180–181, 268, 269, 276, 283,
   286, 291, 299, 307, 309, 311, 312, 315,
   326, 327, 350, 375–376
Rubber: cultivated (see *Hevea*); wild (see *Cas-
   tilla*)
Rudd, N., 53
Rudgea, 181, 376
Rudolph, S. G., 277
Ruellia, 175, 360
Rufous-tailed Jacamar, 191
Ruíz-Reyes, J., 267
Ruminants, 272
Rundel, P. W., 135, 136
Runkle, J. R., 121, 122, 124, 293
Russel, A. E., 46, 54, 60
Russell, J. K., 291
RUTA (Unidad Regional de Asistencia Téc-
   nica, IDA), Costa Rica, 311
Rutaceae, 77, 181, 273, 307, 310, 312, 326,
   328, 376
Rutilio Q., J., 266
Rutter, A. J., 24
Ryan, M. J., x, 204, 208
Rylands, A. B., 288

Sabal Forest Reserve, Sarawak, Malaysia,
   117
Sábalo-Esquina Stream (La Selva), 6, 9, 27,
   29, 32, 44, 195, 196, 197, 251
Sabiaceae, 181, 231, 376
Sabogal, C., 323

Saccharum (sugar cane), 307, 309, 311, 332, 357
Saccopteryx, 231, 246, 249, 394
Sachs, R. M., 147, 157
Sader, S. A., 299, 317
Saguinus (Tamarinds), 233, 288
St. John, Virgin Islands, U.S., 117
St. Vincent, Antilles 32
Salamanders. *See* Caudata
Saldarriaga, J. G., 124
Sallabanks, R., 293
Saltators. See *Saltator*
Saltator, 221, 228, 392
Salthe, S. N., 202, 203
Salto Stream (La Selva) 27, 28, 29, 31, 38, 39, 44, 59, 75
Salvin, O., 185
Salyavata, 252
Salzman, A. G., 270, 271
San Carlos, Costa Rica, 86, 301, 307, 308, 313
San Carlos, Venezuela, 4, 21, 61, 120
San José, Costa Rica. *See* Central Valley, Costa Rica
San Juan River. *See* Río San Juan
Sánchez, P. A., 34, 35, 51, 59
Sancho, F., ix, 18, 35–39, 43, 44, 45, 51, 53, 55, 57, 60, 62, 91, 316
Sanders, H. L., 266
Sandner, G., 309
Sanford, R. L., Jr., 8, 30, 35, 52–55, 57, 58, 59, 63, 91, 120, 121, 122, 126, 135, 330
San Isidro de Peñas Blancas, Costa Rica, 307
San Miguel, Costa Rica, 35, 86
Santa Cecilia, Ecuador, 200–202, 204, 205, 208, 211–214
Santana, M., 215, 216, 253
Santa Rita Ridge, Panama, 84
Santa Rosa National Park, Costa Rica, 92, 172, 188, 189, 215, 239, 241, 242
Sapindaceae, 81, 82, 171, 172, 181, 312, 328, 376
Sapium, 177, 366
Sapotaceae, 79, 80, 82, 132, 162, 163, 166, 328, 376–377
Sapranthus, 175
Sarapiquí Annex (La Selva), 10, 26, 38, 43, 51, 56, 59, 77, 78, 86, 277, 326
Sarapiquí Region, Costa Rica, 73, 229, 230, 234, 235, 297, 299–315, 317, 321, 323, 324, 326. *See also entries for specific towns and geographical locations*
Sarapiquí River. *See* Río Sarapiquí
Sarawak, Malaysia, 116, 117, 137, 170, 331
Sarcorhachis, 88, 374
Sargent, S., 288, 290
Sarmiento, G., 142, 144, 147
Sarukhán, J., 90, 95, 120, 125, 139, 142, 294
Saturniidae, 248, 266, 280
Satyrinae (subfamily of Nymphalidae), 188, 189
Sauer, A., 269, 279
Saul, W. G., 196
Sauria, 210–216, 249, 253, 262, 331, 346, 382
Savage, J. M., 73, 185, 199–203, 207, 209–213, 216, 230, 241, 242, 344
Savannahs, 333, 334
Sawfly, 268
Sazima, M., 178, 182

Scaphiodontophis, 382
Scarabaeidae, 279
Schaal, B. A., 72
Schal, C., 249
Schaller, G. B., 253
Schatz, G. E., 85, 119, 140, 162, 164, 166, 172, 175, 182, 350
Schelhas, J., 302
Schemske, D. W., 127, 164, 173, 174, 182, 192, 220, 270, 293
Scherzer, C., 9
Schizolobium, 325
Schlegelia, 175, 362, 377
Schlesinger, W. H., x, 35, 57
Schmid, R., 166, 173, 182
Schnell, C. E., x, 53, 236, 306, 316, 327
Schoener, T. W., 213, 215
Schowalter, T. D., 267
Schremmer, F., 87
Schröder, D., 263, 331
Schultz, J. C., x, 273, 274, 276, 277, 278, 281, 332
Schupp, E. J., 270, 290, 293
Schupp, E. W., 263, 269, 271, 274, 276, 294
Schwartz, J. J., 209
Scinax, 200, 203, 380
Scincidae, 212, 380
Sciuridae, 262, 285, 286, 287, 396
Sciurus, 103, 229, 230, 233, 234, 235, 262, 286, 396
Sclerolobium, 328, 368
Scoble, M. J., 187
Scolopacidae, 386
Scolytidae, 279
Scott, J. C., 239
Scott, N. J., Jr., x, 200, 203, 206, 210, 211, 216, 239, 245, 252, 262, 263, 383
Scriber, J. M., 189, 264, 272
Scrophulariaceae, 175, 377
Scutellaria, 177, 367
Scutelleridae, 279
Sealy, S. G., 246
Seavey, S. R., 167
Sección de Inventario de Tierras, IDA, Costa Rica, 303, 304, 306
Sedges. *See* Cyperaceae
Seger, J., 255
Segleau, J., 318
Seifert, R. H., 266
Seifert, R. P., 266
Seitz, A. E., 191
Selaginella, 79, 350–351
Selaginellaceae, 78, 79, 350–351
Selenidera, 226, 388
Selva Tica Reserve, Costa Rica, 304
Selva Verde Reserve, Costa Rica, 304
Semple, K. S., 165, 180, 182
Senior, C. T., 32
SEPSA (Secretaria Ejecutiva de Planificación Sectorial Agropecuaria y de Recursos Naturales Renovables), Costa Rica, 307
Serpentes, 210–215, 219, 240, 241, 252, 253, 255, 331, 346, 382–383
Sesiidae, 269, 280
Sexton, O. J., 213
Seyfried, M. S., 49
Sharpe, J. M., 92, 93
Sherry, T. W., ix, x, 217, 220, 221, 246, 252, 255

Shorea, 116, 150, 154, 155, 158, 170, 171, 271
Short, L., 276
Shrews (Soricidae), 230
Shull, E. M., 264
Shuttleworth, J., 23, 24, 27, 33
Sibaria, 269
Sibon, 216, 382
Sicydium (fish), 197, 198, 379
Sicydium (plant), 365
Sida, 269, 370
Sierra Madre Oriental, México, 201
Sigatoka disease (*Micosphaerella*), 309, 312
Silberglied, R. E., 187
Silva, J. L., 216
Silvertown, J. W., 144, 156
Simarouba, 63, 97, 101, 136, 181, 328, 377
Simaroubaceae, 63, 97, 101, 136, 181, 328, 377
Simberloff, D. S., 199
Simira, 77, 181, 269, 376
Simms, E. L., 277
Simpson, B. B., x, 334
Singer, M. C., 187, 188, 191, 194
Singh, B., 24
Sioli, H., 32
Siparuna, 179, 220, 286, 372
Siprocta, 248
Siquirres, Costa Rica, 312
Siricidae, 281
Sisk, T., 323
Skippers. *See* Hesperiidae
Skutch, A. F., 213, 217, 222, 223, 244–247, 251, 252, 284
Slatkin, M., 282
Sleumer, H. O., 177, 182
Sloanea, 75, 131, 286, 365
Slobodkin, L. B., 266
Sloths, 231, 232, 233, 235, 255, 396. See also *Bradypus; Choloepus*
Slowinski, J. B., 200
Slud, P., 9, 185, 217, 218, 222, 225, 226, 230, 235, 343
Smallwood, J., 106, 126, 282, 288, 292
Smilacaceae, 360
Smiley, J. T., 187, 188, 189, 191, 240, 249–252, 263, 264, 279
Smilisca, 200, 203, 380
Sminthuridae, 278
Smith, A., 287
Smith, A. P., 91, 92, 120, 130
Smith, H., 162
Smith, J. D., 201
Smith, N. G., 250
Smith, D. A., 350
Smithe, F. B., 219, 227
Smithsonian Environmental Research Program, 63
Smithsonian Institution, 172, 194, 241
Smithsonian Tropical Research Institute (STRI), 120, 345. *See also* Barro Colorado Island
Smythe, N., 153, 213, 262, 263, 267, 283, 286, 287, 291
Snakes. *See* Serpentes
Snarskis, M. J., 308
Snow, B. K., 220, 282
Snow, D. W., 153, 220, 245, 247, 282, 283, 284, 286, 289, 291, 292, 294

Sobrevila, C., 165, 169–172
Socratea, 74, 75, 76, 95, 100, 112, 113, 122, 140, 173, 231, 270, 360
Soderstrom, T. R., 86, 174, 182
Soils, vii, ix, 6, 17, 18, 34–53, 54–63, 91, 123, 124, 277, 308, 313, 323–326, 329, 330, 331, 337, 343; consociations described, 37–45
Soini, P., 211
Solanaceae, 79, 82, 83, 162–165, 172, 181, 265, 267, 276, 286, 312, 314, 377
Solanum, 181, 267, 276, 286, 377
Solbrig, O. T., 335
Solimoes River, Brazil, 32
Sollins, P., x, 26, 34, 35, 43, 44, 46, 49–55, 59, 60, 62, 63, 316
Sonoran Desert, North America, 335, 336
Sorensen, A. E., 288, 290
Soricidae, 230
Sork, V. L., 92, 294
Sorocea, 164, 179, 372
Soulé, M. E., 242, 255
Souroubea, 88, 370
South Africa, 272, 330
South America, 82–88, 196, 199, 202, 204, 213, 216, 217, 219, 229, 230, 235, 241, 242, 244, 291, 308, 321, 322, 330, 331, 336, 381, 398. *See also entries for specific regions and countries*
Southeast Asia, 96, 170, 288, 336. *See also entries for specific countries and islands*
Southwestern United States, 242
Southwood, T. R. E., 239, 261, 267, 276
Sowls, L. K., 286
Spachea, 86, 370
Spain, 155
Spanish Cedar. See *Cedrela*
Spathiphyllum, 74, 76, 126, 173, 354
Species diversity/richness (factors generating), 70–71, 241–243, 329–332, 335. *See also entries for particular taxa*
Spectacled bear (*Tremarctos*), 230
Spector, P., 159
Speothos, 230
Sphaeradenia, 85, 355
Sphaerodactylus, 382
Sphecidae, 245, 252
Sphenomorphus, 215, 382
Sphingidae, 153, 167, 185, 241, 280
Spider monkeys. See *Ateles*
Spiders. *See* Araneida
Spigelia, 178, 369
Spiller, D. A., 213, 215
Spilotes, 382
Spondias, 142, 175, 286, 361
Sporophila, 221, 224, 228, 392
Sprugel, D. G., 58
Spycher, G., 46
Squamata. *See* Sauria
Squirrels. *See* Sciuridae
Sri Lanka, 32
Srygley, R. B., 191
Stamp, N. E., 247
Stamps, J., 213–214
Standard Fruit Company, 300, 310, 311
Standley, P. C., 78, 83
Stanford University, 194
Stanton, M. L., 273
Stapanian, M. A., 288

Stark, N., 58
Stark, N. M., 58
Starmuhlner, F., 32
Starrett, A., 144, 150, 163, 166, 177, 182, 230
Stearns, F. W., 142
Stearns, S. C., 160, 244, 247
Steiner, K. E., 177, 182
Stemmadenia, 175, 269, 361
Stenoderminae (Subfamily of Phyllostomidae), 285–286, 395
Stephenson, A. G., 142
Sterculiaceae, 9, 11, 74, 181, 259, 264, 275, 297, 299, 302, 307–313, 315, 323, 328, 332, 377
Sterna, 217, 386
Sterner, R. W., 106, 118, 271
Stewart, J. B., 24
Stiles, E. W., 287
Stiles, F. G., 6, 9, 92, 147, 153–156, 159, 161–164, 167, 168, 169, 172–177, 180, 181, 182, 185, 194, 217–227, 239, 242, 243, 245, 246, 248, 249, 250, 252, 255, 262, 263, 282, 284, 286, 287, 293, 344, 393
Stone, B. L., x, 14
Stone, D. E., x, 6, 9, 10, 11, 13, 14, 236, 299, 342
Stone, D., 308
Stoner, K., 231, 234, 237
Stork, N. E., 239
Stout, (R.) J., x, 19, 28, 29, 32, 250, 251, 275, 279
Strahan, R., 287
Strain, B. R., x, 130, 131, 136, 137, 139, 140
Streblidae, 232
Streptoprocne, 228, 387
Strickland, T. C., 46, 49, 52
Strigidae, 223, 387
Strigiformes, 223, 387
Strong, D. R., x, 251, 252, 254, 263, 266, 267, 277–281, 331, 332
Sturnella, 218, 393
Stryphnodendron, 96, 98, 102, 103, 104, 116, 137, 177, 269, 270, 320, 323, 325, 326, 328, 369
Styzolobium, 316
Suboscine passerines, 219, 223, 227. *See also entries for relevant orders, families, and genera*
Successional Strips (La Selva), 6, 9, 77–78, 96, 347
Sucio River. *See* Rio Sucio
Sugar (cane), 307, 309, 311, 332
Suidae, 287, 309, 313
Sun, I., 277
Sunbirds, 170
Sunquist, M. E., 262, 263
Surá Stream (La Selva), 27, 28, 29, 32, 38, 39, 75, 195, 196, 197, 251
Surinam, 288, 322
Sussman, R. W., 231, 234
Sutton, G. M., 292
Sutton, S. L., 249
Swaine, M. D., x, 96–99, 119, 128, 137
Swallows (Hirundinidae), 223, 225
Swallow-tailed kite (*Elanoides*), 284
Swartzia, 81, 86, 89, 166, 177, 286, 369
Sweeney, B. M., 159
Sweet potatoes, 310

Swietenia, 320, 322
Swift, M. J., 60, 122
Swifts (Apodidae), 223, 228
Swynerton, C. F. M., 191, 192
Sylvilagus, 234, 262
Symmachia, 191
Synallaxis, 223, 388
Synargis, 193
Sylviidae, 223
Symphonia, 74, 87, 176, 367
Symplocaceae, 79, 377
Synechanthus, 173, 360
Syngonium, 85, 284
Syrphidae, 280
Sytsma, K. J., 72
Szeicz, G., 24
Szelitowski, W. A., 208

Tabanidae, 241
Tabebuia, 150, 175, 325, 328, 362
Tabernaemontana, 175, 361
Tachigalia, 160
Tachygoninae (subfamily of Curculionidae), 239
Tachyphonus, 228, 392
Taconazo Stream (La Selva), 39
Taigen, T. L., 209
Talauma, 178, 370
Talbot, I. J., 56, 246, 249
Tamarins (*Saguinus*), 233, 288
Tamaulipas State, México, 199, 212
Tambopata Reserve, Perú, 187, 239, 241
Tanagers, 219, 221, 223, 228, 246, 284, 287, 289, 292. *See also* Emberizidae
Tanaka, S., 213–214
Tangara, 219, 223, 228, 247, 392
Tanner, E. V. J., 169, 170, 171, 336
Tantilla, 382
Taphrocerus, 242
Tapiero, C. S., 160
Tapir. See *Tapirus*
Tapiridae, 262, 285, 287, 397
Tapirus, 92, 229, 235, 261, 262, 287, 397
Tarpon, 304
Tasaico, H., 73
Tauber, M. J., 189
Tahvanainen, J. O., 273
Tayassu: pecari (white-lipped peccary), 92, 122, 229, 233, 235, 244, 261, 286, 397; *tajacu* (collared peccary), 92, 235, 262, 286, 397
Tayassuidae, 92, 122, 229, 233, 235, 244, 261, 262, 285, 286, 287, 397
Taylor, C. M., 85, 350
Taylor, E. H., 210, 343
Tayras. See *Eira*
Te Boekhurst, I. J., 291
Tectaria, 79, 353
Tectona, 320, 321, 327
Teiidae, 212, 382
Telchau, B., 32
Temperate zone, 245, 254, 261, 267, 271, 276, 278, 282, 327, 329, 332, 334, 338. *See also entries for specific continents, regions, and countries*
Temple, S. A., 282
Tenebrionidae, 279
Tenthredinidae, 281
Tephritidae, 280

Terborgh, J., x, 155, 220, 221, 225, 226, 234, 240, 243, 252, 253, 282, 283, 284, 287, 288, 291, 293, 334, 336, 337
Terenotriccus, 220, 252, 389
Terminalia, 63, 77, 132, 138, 176, 323, 325–328, 363
Termitidae, 278
Termites. *See* Isoptera
Terwilliger, V., 277
Testudinata, 210, 211, 213, 214, 215, 253, 262, 382
Tetragastris, 108, 176, 286, 363
Tetrigidae, 278
Tettigoniidae, 255, 272, 278
Texas, 265
Thailand, 206
Thalurania, 222, 224, 246, 250, 255, 387
Thamnistes, 226, 389
Thecadactylus, 382
Thelypteridaceae, 79, 80, 81, 267. *See also* Polypodiaceae
Thelypteris, 79, 80, 81, 267, 353
Theobroma, 377
Theobroma cacao, 9, 11, 74, 77, 297, 299, 302, 307–313, 315, 332, 377
Theophrastaceae, 377
Therezien, I., 32
Thiollay, J. M., 255
Thisbe, 189, 192
Thomas, C. D., 263, 264, 270
Thomas, J. A., 192
Thompson, J. N., 282, 288, 290
Thorington, R. W., Jr., 106, 118, 233, 262, 263, 286, 331
Thraupidae. *See* Emberizidae
Thraupinae (subfamily of Emberizidae), 285, 287, 289. *See also* Emberizidae
Three-toed sloth. See *Bradypus*
Threnetes, 220, 387
Threskiornithidae, 384
Thripidae, 278
Thrips. *See* Thysanoptera
Thryothorus, 221, 391
Thyrididae, 280
Thryopteridae, 395
Thysanoptera, 154, 155, 156, 164, 169, 170, 171, 278
Tiaris, 228, 393
Tibouchina, 79, 371
Ticks, 213
Tico-Fruiti Company, 312
Tiedje, J. M., 56, 59, 60
Tikal, Guatemala, 219, 220, 227
Tiliaceae, 75–78, 96, 99–102, 112, 116, 126, 131, 132, 135, 137, 138, 140, 181, 323, 327, 377–378
Tilley, J., 53
Tilman, D., 330
Timm, R. M., 229–232, 235, 238, 239, 397
Timmerman, W., 216
Tinamidae, 218, 219, 223, 224, 262, 284, 285
Tinamiformes, 262, 384
Tinamous. *See* Tinamidae
Tineidae, 232, 280
Ting, I. P., 134
Tingidae, 278
Tinkle, D. W., 244, 247
Tiquisque. See *Xanthosoma*

Tischeriidae, 279
Tisdall, J. M., 46
Tityra, 223, 228, 291
Tityras. See *Tityra*
Toads. *See* Bufonidae
Toft, C. A., 154, 155, 156, 169, 170, 171, 282
Tolmomyias, 228, 389
Tomlinson, P. B., 142, 144
Topobea, 276, 371
Topography, 12, 18, 26, 29, 30, 37
Torres, S. de K., 324
Torquebiao, E. F., 120
Tortricidae, 280
Tortuguero National Park, Costa Rica, 87, 200–202, 204, 205, 209, 211–212, 213, 214, 216, 230, 304, 321
Tosi, J. A., Jr., 63, 308, 309, 317, 318
Toucans, 223, 292. *See also* Ramphastidae
Tovomitopsis, 85, 176, 367
Townsend, D. S., 203, 204
Trachyinae (Subfamily of Buprestidae), 239, 241–242
Tragulidae, 287
Traveset, A., 267
Travis, J., x
Traylor, M. A., Jr., 284
Tree ferns, 108
Treehoppers. *See* Membracidae
Trema, 78, 182, 378
Tremarctos, 230
Treron, 287
Trichechus, 304
Trials Project (La Selva, OTS), 38, 43, 323–327, 338, 342
Trichilia, 179, 286
Trichomanes, 79
Trichoptera, 238
Trigona, 220
Trigonid bees, 87
Trimetopon, 382
Trinidad, West Indies, 146, 247
Triolena, 133, 178, 371
Tripogandra, 173, 355
Triska, F. J., 17, 19, 28
Trochilidae, 88, 92, 153–156, 161, 163, 164, 166–169, 172, 217, 220–224, 226, 240, 246, 249, 250, 255, 387
Troglodytes, 218, 223, 391
Troglodytidae, 221, 223, 226, 247, 284, 391
Trogon, 228, 388
Trogonidae, 219, 223, 228, 284, 285, 287, 289, 292, 388
Trogoniformes, 388
Trogons. *See* Trogonidae
Trophis, 63, 146, 149, 151, 153, 157, 162–165, 169, 179, 372
Tropidophiidae, 212, 382
Trott, S., 279
Trudgill, S. T., 55
Trueb, L., 200, 202, 203, 204, 381
Trypoxylon, 245, 252
Tufte, E. R., 143
Turdinae (subfamily of Emberizidae), 284. *See also* Emberizidae
Turdus, 221, 391
Turner, F. B., 209
Turner, I. M., 276
Turner, J. R. G., 190, 191
Turrialba, Costa Rica, 54, 55, 312

Turrialba Volcano, Costa Rica, 312
Turrúbares, Costa Rica, 309
Turtles. *See* Testudinata
Two-toed sloth. See *Choloepus*
Tylophidae, 212
Tyrannidae, 218, 219, 245, 246, 252, 285, 287, 289, 389–390
Tyrant Flycatchers, 287. *See also* Tyrannidae; *and entries for genera*
Tytonidae, 387

UCR. *See* University of Costa Rica
Ueckert, D. N., 265
Uehara, G., 34, 35, 46, 49, 51, 53, 59, 124
Uhl, C., 100, 120, 121, 124, 125, 126, 267
Uhl, N. W., 164, 173, 182
Ulmaceae, 78, 100, 120, 126, 182, 378
Umbelliferae. *See* Apiaceae
Umbonia, 245
Ungalophis, 283
Unidad Regional de Asistencia Técnica (RUTA), IDA, Costa Rica, 311
United Brands, 309
United Fruit Company, 309, 310
United Nations Development Program (UNDP), 312
United Nations Educational, Scientific, and Cultural Organization (UNESCO), 11, 73, 278, 297, 304, 305, 306, 338
United States, 80, 121, 127, 188, 189, 199, 217, 219, 229, 242, 243, 264, 265, 275, 310, 312, 319, 322, 338, 343, 344, 345, 398. *See also entries for specific regions and states*
University of Costa Rica (UCR), 210, 301, 302, 308, 311, 312, 314, 316
University of Florida, 294
University of Kansas, 210
University of Miami, 209, 210
University of Texas, 194
University of Washington, 322
University of Wisconsin, 294
Unonopsis, 74, 85, 146, 147, 149, 361
Upala, Costa Rica, 310, 326
Uranesis, 191
Urania, 250
Uraniid moths, 250
Ureña H., A. I., 304
Urera, 87, 89, 182, 378
Urotheca, 382
Ursidae, 230, 287
Urticaceae, 84, 87, 89, 163, 164, 166, 182, 265, 378
USAID. *See* Agency for International Development
USDA Soil Survey Staff, 34, 36, 42, 43, 46, 48, 49
USDA Competitive Grants Program, 141
Usumacinta River, México and Guatemala, 195

Vail, J. A., 240
Valdeyron, G., 155
Valerio, C. E., 173, 182
Valle del General, Costa Rica, 309
Vampire bats. See *Desmodus*
Vampyressa, 231, 246, 395, 397
Vampyrum, 236, 395
Van Berkum, F. H., 214, 215

Van Breemen, N., 50
Vande Kerckhove, G. A., 288, 292
Vandermeer, J. H., 32, 95, 100, 113, 125, 215, 232, 253, 262, 263, 269, 270, 286, 291
Van del Pijl, L., 283, 291, 292
Van Devander, R. W., 262, 263, 286
Van Eck, W. A., 157
Vanessa, 189
Vane-Wright, R. I., 187, 192, 194
Vanglider, L. D., 216
Van Schaik, C. P. V., 142, 151, 158
Van Sluys, F. R., 311–314
Van Steenis, C. G. G. J., 125
Van Wambeke, A., 35
Vara Blanca, Costa Rica, 299, 306
Vargas M., E., 317, 318, 320
Vasconcelos, J. M. C., 253
Vasconcelos, H. L., 270
Vatica, 170
Vaughan, C., 230, 323
Vaux, P., 195, 196
Vázquez-Yánes, C., 126, 130, 169
Vega, G., 119
Vega, V., 316
Vegetation types (La Selva), 73–78
Vehrencamp, S. L., 231, 246, 247, 249, 250, 253
Venezuela, 4, 21, 32, 60, 61, 100, 120, 121, 169, 170, 171, 230, 262, 263, 308
Veracruz, México, 139, 200, 241
Verbenaceae, 82, 182, 315, 320, 321, 323, 325, 327, 378
Vermeij, G. J., 136, 240
Verner, J., 333
Vertebrates, 185, 208, 213, 215, 220, 234, 238–242, 252, 253, 259, 261, 270, 278, 282, 284–294, 331, 334, 346. *See also entries for classes, orders, families, and genera*
Vespertilionidae, 395
Vial, J. L., 203
Vigna, 312, 316, 369
Villa, J., 199, 200, 201, 204, 210
Villa Nova, N. A., 27
Villarreal, B., 317
Vince-Prue, D., 158
Violaceae, 74, 112, 150, 151, 182, 270, 378
Viperidae, 212, 382
Víquez, M., 119
Viral disease (of plants), 312
Vireo, 222, 391
Vireolanius, 228, 391
Vireonidae, 222, 223, 228, 285, 391
Vireos. *See* Vireonidae
Virgin Islands, 117
Virola, 75, 108–109, 131, 137, 179, 270, 283, 286, 325, 326, 328, 372
Viscaceae, 78, 80, 83, 378
Vismia, 176, 367
Vitaceae, 307, 378
Vitex, 182, 325, 378
Vitousek, P. M, 30, 35, 36, 38, 44, 45, 51–57, 59–62, 91, 120, 122, 124, 135, 136, 330
Vitt, L. J., x, 185, 209, 216, 240, 262, 263
Viverridae, 287
Vochysia, 109, 182, 323, 325, 326, 328, 378
Vochysiaceae, 83, 86, 108, 182, 323, 325, 326, 328, 378
Vogt, K. A., 57, 209

Volatinia, 228, 392
von Humboldt, A., 3
Voss, R., 181, 182, 232
Vultures. *See* Cathartidae

Waage, J. K., 232
Wageningen Agricultural University, Holland, 316
Wagner, M., 9
Wake, D. B., 200–203, 209
Wake, M. H., 199, 200, 202, 203
Walker, E. P., 262, 263
Wallabies (Macropodidae), 287
Wallace, A. R., 3, 106
Waloff, N., 187
Walter, H., 73
Walters, M. B., 125
Waltz, S. A., 265, 267, 268, 272, 273, 276, 277, 278
Wang, M. D., 266, 279
Warblers, 221, 223, 228. *See also* Emberizidae; *and entries for genera*
Waring, R. H., 57
Warkentin, B. P., 46
Warner, R. R., 126
Warrea, 79, 359
Warscewiczia, 75, 112, 167, 181, 376
Waser, N. M., 155
Wasps, 88, 155–156, 163, 164, 167, 168, 170, 238, 242, 245, 251, 252, 255, 275, 281. *See also entries for families and genera*
Water apple (*Eugenia*), 179, 312
Waterbirds, 223, 227. *See also entries for families and genera*
Water bugs. *See* Naucoridae
Waterman, P. G., 272, 273, 277
Way, M. J., 275
Weaver, P. L., 52, 124
Webb, C. J., 157
Webb, L. J., 34, 120
Webb, S. L., 290
Webster, G. L., 350
Weekes, A. A., 14
Weevils, 238, 241, 263, 268, 269. *See also entries for families and genera*
Weibsahn, F. H., 32
Weigeltia, 88, 179, 373
Welfia, 74, 75, 76, 95, 100, 112, 113, 118, 122, 164, 165, 176, 189, 215, 231, 253, 269, 270, 283, 286, 291, 360
Wells, K. D., 203, 204, 209
Went, F. W., 58
Werman, S. D., 210, 212
Werner, P., 35, 44, 46, 60, 91, 95, 96, 97, 99, 100, 101
West Africa, 81, 323
West Indies, 4, 30, 32, 46, 52, 61, 84, 87, 117, 146, 169, 170, 171, 216, 247, 310, 337. *See also entries for specific islands and countries*
Westoby, M., 330, 332, 336
West Virginia, 199, 322
Wetmore, A., 222, 292
Wetzel, R. M., 286
Weyerhaeuser Company Foundation, 327
Weygoldt, P., 203, 245
Wheat, 307
Wheelwright, N. T., 154, 282, 283, 284, 287, 288, 289, 292

Whelan, C. J., 277, 284, 293
White, M. J. D., 172
White, P., 127
White, R. E., 49
White, S. C., 283
White-faced capuchin monkeys. See *Cebus*
White-lipped peccaries. See *Tayassu pecari*
White-ruffed manakin. See *Corapipo*
Whiteside, T. J., 20, 21
White-tailed manakin. See *Pipra*
Whitham, T. G., 270
Whitmore, T. C., 3, 70, 91, 99, 102, 116, 120, 124, 125, 126, 128, 138, 151, 157, 273, 286
Whittaker, J. B., 270
Whittaker, R. H., 267, 330
Wibmer, G. H., 238
Wiebes, J. T., 165
Wiegert, R. G., 270
Wiehler, H., 164, 177, 182
Wiemer, D. F., 277
Wilbur, H. M., x
Wilbur, R. L., 78, 94
Wilkerson, R. C., 239, 241
Williams, C. F., 246
Williams, G. C., 244
Williams, N. H., 173, 182
Williamson, G. B., 106
Willis, E. O., 213, 217, 219, 221, 226, 227, 239, 241, 243, 282, 283, 284, 291
Willmer, P. G., 175, 182
Willson, M. F., 282, 283, 284, 286, 288, 290, 292, 293, 332, 333
Wilson Botanical Garden. See Las Cruces Biological Station
Wilson, D. E., 229–232, 236, 239, 250, 252, 263, 269, 279
Wilson, E. O., 244, 245, 251
Wilson, J. W., III, 241
Winchester, J. W., 56
Windsor, D. M., 155, 156, 212, 213, 266
Winnett-Murray, K., 277
Wint, G. W., 267
Witheringia, 181, 377
Wolda, H., 71, 213, 225, 238, 249, 266, 267, 334
Wolf, L. L., 169, 217, 220, 224, 226, 246, 248
Wolfe, L., 277
Wolf-Mueller, B., 273
Wong, M., 151, 266, 267, 287
Wood, T. K., 242, 245, 264, 279
Woodcreepers. See Dendrocolaptidae
Woodman, N., 230
Woodpeckers. See Picidae
Wood production and products. See Forestry
Woodwell, G. M., 267
Woolfenden, G. E., 245
Wooten, J. T., 277
World Wildlife Fund, 11, 327
Worthington, A. H., x, 225, 289, 290, 294
Wrens. See Troglodytidae
Wright, S. J., x, 135, 158, 160, 213, 270
Wright, S., 119
Wrobel, D., 273
Wulff, R. D., 137
Wyatt, R., 276

Xanthosoma, 308, 310, 312, 313, 315, 354
Xantusiidae, 212, 382

Xenodon, 382
Xenops, 228, 228, 388
Xiphidium, 174, 357
Xyelidae, 281
Xylopia, 175, 361

Yale University, 326
Yams, 310
Yap, S. K., 152, 155, 158
Yarbro, L. A., 30
Yatskievych, G., 80
Yatskievych, K. M., 80
Yeaton, R. I., 116
Yellow fever, 234
Yetter, T. C., 293
Yih, K., 120
York, B. M., 261

Young, A. M., 187, 189, 231, 249, 250, 264, 279, 280
Young, H. J., 90, 92, 153, 161, 162, 164, 166, 172, 173, 182
Young, T. P., 293
Yponomeutidae, 280
Yucatán Peninsula, 242
Yucca (Itabo), 312
Yucca (root crop). See *Manihot*

Zamia, 85, 94, 95, 133, 138–139, 151, 157, 158, 240, 268, 277, 353
Zamiaceae, 85, 94, 95, 133, 138–139, 151, 157, 158, 240, 268, 277
Zamora, N., 322
Zanthoxylum, 77, 181, 326, 328, 376
Zea, 299, 308, 309, 311–315

Zebrina, 173, 355
Zimmerius, 284, 389
Zimmerman, P. R., 56
Zingiber, 310
Zingiberaceae, 162, 163, 164, 166, 168, 172, 174, 275, 310, 312, 313, 314, 360
Zingiberales, 263. *See also entries for families and genera*
Zoning. *See* Land use
Zug, G., 203
Zurquí hills, Costa Rica, 26
Zygainidae, 280
Zygophyllaceae, 189
Zypoginae (subfamily of Curculionidae), 239, 241
Zygopini (tribe of Curculionidae), 239